Treatment and Utilization of Combustion and Incineration Residues

Treatment and Utilization of Combustion and Incineration Residues

Edited by

Lei Wang
State Key Laboratory of Clean Energy Utilization, Zhejiang University, Hangzhou, China

Dan Tsang
State Key Laboratory of Clean Energy Utilization, Zhejiang University, Hangzhou, China; Department of Civil and Environmental Engineering, The Hong Kong University of Science and Technology, Hong Kong, China

Jianhua Yan
State Key Laboratory of Clean Energy Utilization, Zhejiang University, Hangzhou, China

ELSEVIER

Elsevier
Radarweg 29, PO Box 211, 1000 AE Amsterdam, Netherlands
125 London Wall, London EC2Y 5AS, United Kingdom
50 Hampshire Street, 5th Floor, Cambridge, MA 02139, United States

ISBN: 978-0-443-21536-0

For information on all Elsevier publications visit our website at https://www.elsevier.com/books-and-journals

Publisher: Candice Janco
Acquisitions Editor: Jessica Mack
Editorial Project Manager: Joshua Mearns
Production Project Manager: Rashmi Manoharan
Cover Designer: Greg Harris

Typeset by TNQ Technologies

Contents

4. Alkali-activated materials for the stabilization/solidification of heavy metals and radioactive substances in incineration residues

Masaki Takaoka

5. Sintering and melting of combustion/incineration residues

Jingqi Sun, Yike Zhang and Zengyi Ma

6. Hydrothermal treatment of combustion/incineration residues

Yaqian Shi, Qiang Zeng, Jianhua Yan and Lei Wang

7. Chemical agent–based immobilization of combustion/incineration residues

Lizhi Tong

8. Mechanochemical treatment of combustion/incineration residues

Yaqi Peng

Part IV
Recycling of combustion/incineration residues into SCMs and aggregates

15. Recycling of pulverized fuel ash as supplementary cementitious materials (SCMs) and aggregates in concrete production

Muhammed Bayram, Ömer Faruk Kuranlı, Anıl Niş and Togay Ozbakkaloglu

16. Recycling of biomass combustion ash into SCMs and aggregates

Huanyu Li, Jian Yang, Lei Wang, Ning Zhang, Qingyuan Wang and Viktor Mechtcherine

17. Recycling of incineration sewage sludge ash as SCM and aggregate

Miao Lu, Zhenhao Song, Yan Xia, Guoqing Geng and Lei Wang

Contributors

Quan An, State Key Laboratory of Technologies for Efficient Utilization of Coal Waste Resources, Shanxi University, Taiyuan, China

Muhammed Bayram, Ingram School of Engineering, Texas State University, San Marcos, TX, United States

Zhiliang Chen, Department of Environmental Science, College of Environment and Ecology, Chongqing University, Chongqing, China

Fangqin Cheng, State Key Laboratory of Technologies for Efficient Utilization of Coal Waste Resources, Shanxi University, Taiyuan, China; Centre for Energy (M473), The University of Western Australia, Crawley, WA, Australia

Francesco Colangelo, Department of Engineering, University Parthenope of Naples, Centro Direzionale, Naples, Italy

Hongguang Cui, The First Affiliated Hospital, Zhejiang University School of Medicine, Hangzhou, China

Jinglei Cui, State Key Laboratory of Technologies for Efficient Utilization of Coal Waste Resources, Shanxi University, Taiyuan, China

Sabino De Gisi, Department of Civil, Environmental, Land, Building Engineering and Chemistry (DICATECh), Polytechnic University of Bari, Bari, Italy

Bin Du, Key Laboratory of Special Equipment Safety Testing Technology of Zhejiang Province, Zhejiang Academy of Special Equipment Science, Hangzhou, China

Weiwei Duan, Sustainable Infrastructure and Resource Management (SIRM), UniSA STEM, University of South Australia, Adelaide, SA, Australia

Yuan Fan, State Key Laboratory of Technologies for Efficient Utilization of Coal Waste Resources, Shanxi University, Taiyuan, China

Le Fang, RDC for Watershed Environmental Eco-Engineering, Advanced Institute of Natural Sciences, Beijing Normal University at Zhuhai, Zhuhai, China

Ilenia Farina, Department of Engineering, University Parthenope of Naples, Centro Direzionale, Naples, Italy

Valerio Funari, Institute of Marine Sciences, National Research Council of Italy, Venice, Italy; Department of Ecosustainable Marine Biotechnologies, Stazione Zoologica Anton Dohrn, Naples, Italy

Jian-ming Gao, State Key Laboratory of Technologies for Efficient Utilization of Coal Waste Resources, Shanxi University, Taiyuan, China

Peng Gao, School of Civil Engineering, Hefei University of Technology, Hefei, China; Engineering Research Center of Low-carbon Technology and Equipment for Cement-based Materials, Ministry of Education, Hefei University of Technology, Hefei, China; Anhui Low-carbon Technologies Engineering Center for Cement-based Materials, Hefei, China

Hongyu Gao, State Key Laboratory of Technologies for Efficient Utilization of Coal Waste Resources, Shanxi University, Taiyuan, China

Guoqing Geng, Department of Civil and Environmental Engineering, National University of Singapore, Singapore, Singapore

Junaid Ghani, Institute of Marine Sciences, National Research Council of Italy, Venice, Italy; Department of Biological Geological and Environmental Science, University of Bologna, Bologna, Italy

Binglin Guo, School of Civil Engineering, Hefei University of Technology, Hefei, China; Engineering Research Center of Low-carbon Technology and Equipment for Cement-based Materials, Ministry of Education, Hefei University of Technology, Hefei, China; Anhui Low-carbon Technologies Engineering Center for Cement-based Materials, Hefei, China

Qianqian Guo, Institute of Energy & Power Engineering, Zhejiang University of Technology, Hangzhou, China

Li Hong, School of Civil Engineering, Hefei University of Technology, Hefei, China; Engineering Research Center of Low-carbon Technology and Equipment for Cement-based Materials, Ministry of Education, Hefei University of Technology, Hefei, China

Yanjun Hu, Institute of Energy & Power Engineering, Zhejiang University of Technology, Hangzhou, China

Gang Huang, State Key Laboratory of Clean Energy Utilization, Zhejiang University, Hangzhou, China

Xinyu Jiang, Department of Civil and Environmental Engineering, The Hong Kong Polytechnic University, Hong Kong, China

Ömer Faruk Kuranlı, Civil Engineering Department, Yıldız Technical University, Istanbul, Turkey

Claudia Labianca, Department of Civil and Environmental Engineering, The Hong Kong Polytechnic University, Hong Kong, China

Huanyu Li, School of Ocean and Civil Engineering, Shanghai Jiao Tong University, Shanghai, China; Institute of Construction Materials, Technische Universität Dresden, Dresden, Germany

Lufan Li, Department of Civil Engineering, Hangzhou City University, Hangzhou, China

Jining Li, Institute of Eco-environmental and Soil Sciences, Guangdong Academy of Sciences, GuangZhou, China

Huiyan Li, The First Affiliated Hospital, Zhejiang University School of Medicine, Hangzhou, China

Jianbo Li, Key Laboratory of Low-grade Energy Utilization Technologies and Systems, Ministry of Education of PRC, Chongqing University, Chongqing, China

Songsong Lian, School of Civil Engineering, NingboTech University, Ningbo, China

Tung-Chai Ling, College of Civil Engineering, Hunan University, Changsha, China

Kang Liu, Department of Civil and Environmental Engineering, The Hong Kong Polytechnic University, Hong Kong, China

Yue Liu, Sustainable Infrastructure and Resource Management (SIRM), UniSA STEM, University of South Australia, Adelaide, SA, Australia

Lei Liu, Institute of industrial Science, The University of Tokyo, Tokyo, Japan

Miao Lu, State Key Laboratory of Clean Energy Utilization, Zhejiang University, Hangzhou, China

Zhibin Ma, State Key Laboratory of Technologies for Efficient Utilization of Coal Waste Resources, Shanxi University, Taiyuan, China; Centre for Energy (M473), The University of Western Australia, Crawley, WA, Australia

Zengyi Ma, State Key Laboratory of Clean Energy Utilization, Zhejiang University, Hangzhou, China; Ningbo Innovation Center, Zhejiang University, Ningbo, China; Institute for Carbon Neutrality, Ningbo Innovation Center, Zhejiang University, Ningbo, China

Luciana Mantovani, Department of Chemistry, Life Sciences and Environmental Sustainability, University of Parma, Parma, Italy

Viktor Mechtcherine, Institute of Construction Materials, Technische Universität Dresden, Dresden, Germany

Hengjun Mei, The First Affiliated Hospital, Zhejiang University School of Medicine, Hangzhou, China

Yuan Meng, Department of Civil and Environmental Engineering, The Hong Kong Polytechnic University, Hong Kong, China

Anıl Niş, Civil Engineering Department, Istanbul Gelisim University, Istanbul, Turkey

Xinyi Niu, School of Human Settlements and Civil Engineering, Xi'an Jiaotong University, Xi'an, China

Michele Notarnicola, Department of Civil, Environmental, Land, Building Engineering and Chemistry (DICATECh), Polytechnic University of Bari, Bari, Italy

Togay Ozbakkaloglu, Ingram School of Engineering, Texas State University, San Marcos, TX, United States

Yaqi Peng, State Key Laboratory of Clean Energy Utilization, Institute for Thermal Power Engineering, Zhejiang University, Hangzhou, China

Zezhi Peng, Department of Environmental Science and Engineering, Xi'an Jiaotong University, Xi'an, China

Huyong Qin, School of Civil Engineering, Hefei University of Technology, Hefei, China

Qili Qiu, School of Environmental Engineering, Nanjing Institute of Technology, Nanjing, China

Pengfei Ren, College of Civil Engineering, Hunan University, Changsha, China

Shaoqin Ruan, College of Civil Engineering and Architecture, Zhejiang University, Hangzhou, China

Keiko Sasaki, Department of Earth Resources Engineering, Kyushu University, Fukuoka, Japan

Yaqian Shi, State Key Laboratory of Clean Energy Utilization, Zhejiang University, Hangzhou, China

Narinder Singh, Department of Engineering, University Parthenope of Naples, Centro Direzionale, Naples, Italy

Zhenhao Song, State Key Laboratory of Clean Energy Utilization, Zhejiang University, Hangzhou, China

Huiping Song, State Key Laboratory of Technologies for Efficient Utilization of Coal Waste Resources, Shanxi University, Taiyuan, China

Jian Sun, Department of Environmental Science and Engineering, Xi'an Jiaotong University, Xi'an, China

Jingqi Sun, State Key Laboratory of Clean Energy Utilization, Zhejiang University, Hangzhou, China

Xinlei Sun, School of Civil and Environmental Engineering, Nanyang Technological University, Singapore, Singapore

Hanyang Sun, State Key Laboratory of Clean Energy Utilization, Zhejiang University, Hangzhou, China

Chen Sun, State Key Laboratory of Clean Energy Utilization, Zhejiang University, Hangzhou, China; Institute of Zhejiang University-Quzhou, Quzhou, China

Masaki Takaoka, Department of Environmental Engineering, Graduate School of Engineering, Kyoto University, Kyoto, Japan

Pei Tang, State Key Laboratory of Silicate Materials for Architectures, Wuhan University of Technology, Wuhan, China

Quanzhi Tian, National Engineering Research Center of Coal Preparation and Purification, China University of Mining and Technology, Xuzhou, China

Francesco Todaro, Department of Civil, Environmental, Land, Building Engineering and Chemistry (DICATECh), Polytechnic University of Bari, Bari, Italy

Lizhi Tong, State Environmental Protection Key Laboratory of Environmental Pollution Health Risk Assessment, South China Institute of Environmental Sciences, Ministry of Ecology and Environment (MEE), Guangzhou, China

Daniel C.W. Tsang, State Key Laboratory of Clean Energy Utilization, Zhejiang University, Hangzhou, China; Department of Civil and Environmental Engineering, The Hong Kong University of Science and Technology, Hong Kong, China

Mucteba Uysal, Civil Engineering Department, Yıldız Technical University, Istanbul, Turkey

Xujiang Wang, National Engineering Laboratory for Reducing Emissions from Coal Combustion, Shandong University, Jinan, China; Shenzhen Research Institute of Shandong University, Shenzhen, Guangdong, China; Shandong-Nottingham Joint Research Centre for Green Energy & Materials, Shandong University, Jinan, China; School of Energy and Power Engineering, Shandong University, Jinan, China

Lei Wang, State Key Laboratory of Clean Energy Utilization, Zhejiang University, Hangzhou, China

Yanan Wang, Institute of Energy & Power Engineering, Zhejiang University of Technology, Hangzhou, China

Qingyuan Wang, School of Ocean and Civil Engineering, Shanghai Jiao Tong University, Shanghai, China

Jinpeng Wu, State Key Laboratory of Clean Energy Utilization, Zhejiang University, Hangzhou, China

Kai Wu, Key Laboratory of Advanced Civil Engineering Materials of Ministry of Education, School of Materials Science and Engineering, Tongji University, Shanghai, China

Yan Xia, State Key Laboratory of Clean Energy Utilization, Zhejiang University, Hangzhou, China

Yingying Xiong, Institute for Carbon Neutrality, Ningbo Innovation Center, Zhejiang University, Ningbo, China; State Key Laboratory of Clean Energy Utilization, Zhejiang University, Hangzhou, China

Xinni Xiong, School of Environmental Science and Engineering, Guangzhou University, Guangzhou, China

Caicai Xu, Institute of Zhejiang University-Quzhou, Quzhou, China

Jianhua Yan, State Key Laboratory of Clean Energy Utilization, Zhejiang University, Hangzhou, China

Jian Yang, School of Ocean and Civil Engineering, Shanghai Jiao Tong University, Shanghai, China

Ken Yang, Key Laboratory of Advanced Civil Engineering Materials of Ministry of Education, School of Materials Science and Engineering, Tongji University, Shanghai, China

Yonggan Yang, School of Civil Engineering, Hefei University of Technology, Hefei, China; Anhui Low-carbon Technologies Engineering Center for Cement-based Materials, Hefei, China

Bin Yang, Institute of Zhejiang University-Quzhou, Quzhou, China

Yaolin Yi, School of Civil and Environmental Engineering, Nanyang Technological University, Singapore, Singapore

Fan Yu, Institute of Energy & Power Engineering, Zhejiang University of Technology, Hangzhou, China

Qijun Yu, School of Civil Engineering, Hefei University of Technology, Hefei, China; Engineering Research Center of Low-carbon Technology and Equipment for Cement-based Materials, Ministry of Education, Hefei University of Technology, Hefei, China; Anhui Low-carbon Technologies Engineering Center for Cement-based Materials, Hefei, China

Qiang Zeng, State Key Laboratory of Clean Energy Utilization, Zhejiang University, Hangzhou, China; Department of Civil Engineering and Architecture, Zhejiang University, Hangzhou, China

Binggen Zhan, School of Civil Engineering, Hefei University of Technology, Hefei, China; Engineering Research Center of Low-carbon Technology and Equipment for Cement-based Materials, Ministry of Education, Hefei University of Technology, Hefei, China; Anhui Low-carbon Technologies Engineering Center for Cement-based Materials, Hefei, China

Dongke Zhang, State Key Laboratory of Technologies for Efficient Utilization of Coal Waste Resources, Shanxi University, Taiyuan, China; Centre for Energy (M473), The University of Western Australia, Crawley, WA, Australia

Yike Zhang, State Key Laboratory of Clean Energy Utilization, Zhejiang University, Hangzhou, China; Ningbo Innovation Center, Zhejiang University, Ningbo, China

Yuying Zhang, Department of Civil and Environmental Engineering, The Hong Kong University of Science and Technology, Hong Kong, China

Ning Zhang, Leibniz Institute of Ecological Urban and Regional Development (IOER), Dresden, Germany

Bojun Zhao, Key Laboratory of Special Equipment Safety Testing Technology of Zhejiang Province, Zhejiang Academy of Special Equipment Science, Hangzhou, China

Ruolin Zhao, State Key Laboratory of Clean Energy Utilization, Zhejiang University, Hangzhou, China; Institute of Eco-environmental and Soil Sciences, Guangdong Academy of Sciences, GuangZhou, China

Lingqin Zhao, Institute of Energy & Power Engineering, Zhejiang University of Technology, Hangzhou, China

Yunan Zhou, Aviation Key Laboratory of Science and Technology on Aero Electromechanical System Integration, Nanjing Engineering Institute of Aircraft System, Nanjing, China

Yan Zhuge, Sustainable Infrastructure and Resource Management (SIRM), UniSA STEM, University of South Australia, Adelaide, SA, Australia

Part I

Overview of combustion/ incineration residues

Chapter 1

Characteristics of combustion residues, waste incineration residues, various slags

Yuying Zhang[2], Lei Wang[1] and Daniel C.W. Tsang[1,2]
[1]State Key Laboratory of Clean Energy Utilization, Zhejiang University, Hangzhou, China; [2]Department of Civil and Environmental Engineering, The Hong Kong University of Science and Technology, Hong Kong, China

1. Introduction

Combustion and incineration processes have emerged as predominant waste disposal and energy generation methods worldwide (Krishnamoorthi et al., 2019; Saikia et al., 2021; Zhang et al., 2021). Combustion, a fundamental chemical process, involves the exothermic reaction between fuel and oxygen, resulting in the release of heat and the transformation of the fuel into combustion products (Adanez et al., 2012; Akram et al., 2021; Kohse-Höinghaus, 2023; Novitskaya et al., 2021). This combustion process occurs in various forms, such as the combustion of solid fuels (e.g., coal, biomass), liquid fuels (e.g., oil), or gaseous fuels (e.g., natural gas) (Li et al., 2023a; Tillman et al., 2009; Zhang et al., 2010). It relies on the interplay of factors, including encompassing fuel, oxygen, and heat (Kohse-Höinghaus, 2023). Incineration is a controlled combustion of waste materials, encompassing a range of systems, including mass burn incineration, fluidized bed incineration, rotary kiln incineration, and pyrolysis/gasification (Jiang et al., 2019a; Liang et al., 2021; Schnell et al., 2020). These processes generate significant amounts of residues, including combustion residues (coal fly ash, coal bottom ash, and flue gas desulphurization gypsum), waste incineration residues (incineration fly ash and bottom ash), and various slags (ferrous slags and nonferrous slags), which are the primary focus of this chapter. As these residues can pose potential environmental risks during the valorization process (de Titto and Savino, 2019; Dong et al., 2019; Huang et al., 2023; Liang et al., 2021), it is of paramount importance to understand their characteristics, manage their disposal, and expand the potential utilization opportunities.

The treatment and utilization of combustion residues, waste incineration residues, and various slags hold significant scientific and environmental importance. The primary goal of residue treatment is to mitigate the hazardous properties of the residues and facilitate their safe valorization or disposal (Chen et al., 2020; Wang et al., 2019). Physical, chemical, and biological processes, such as solidification/stabilization and vitrification, have been widely applied to treat these residues (Xiong et al., 2022; Zhang et al., 2021). Additionally, valorizing these residues in various applications, such as construction materials, soil amendments, and catalysts, offers an avenue for conserving natural resources and reducing landfill waste (He et al., 2022; Jiang et al., 2018; Xue and Liu, 2021). Mismanagement of combustion, incineration residues, and various slags can lead to adverse environmental impacts (Dong et al., 2019; Luo et al., 2019; Zhu et al., 2021). Effective treatment and utilization of these residues can minimize environmental risks and contamination potential (Luo et al., 2019; Silva et al., 2019). Furthermore, by extracting useful materials from residues, the overall waste stream can be reduced, leading to reduced landfill requirements and associated environmental burdens (Bakalár et al., 2021; Zhai et al., 2021; Zhang et al., 2021). Moreover, diverting residues from landfills contributes to sustainable waste management practices and helps alleviate the strain on limited landfill capacities (Mancini et al., 2021; Sapkota and Pariatamby, 2023).

Understanding the physical, chemical, and mineralogical properties of combustion residues is crucial for their subsequent valorization and disposal. This chapter will provide a comprehensive overview of the characteristics of combustion residues, waste incineration residues, and various slags. The physical and chemical properties and mineralogy of combustion residues, waste incineration residues, and various slags will be critically discussed. Additionally, state-of-the-art methods and technologies for the treatment and utilization of these residues will also be reviewed. By understanding the characteristics of these residues and the current practices for their management, readers will gain insights into the

Treatment and Utilization of Combustion and Incineration Residues. https://doi.org/10.1016/B978-0-443-21536-0.00037-X

challenges and opportunities in the recycling of combustion residues, waste incineration residues, and various slags, ultimately contributing to more sustainable practices in the combustion and incineration industries.

2. Combustion residues

This section provides a comprehensive review of the characteristics of combustion residues, focusing on two main types: coal fly ash (CFA) and bottom ash (CBA). The discussion encompasses their definition, formation, chemical composition, mineralogy, particle size, classifications, and morphology properties.

2.1 Coal fly ash

Coal serves as a major energy source worldwide, with its share of the energy supply expected to remain around 24% by 2035 despite the increasing use of renewable energy sources (Bhatt et al., 2019; Panda and Dash, 2020; Yao et al., 2015). China is the world's largest coal consumer, accounting for 50% of the global coal consumption in 2017 (Luo et al., 2021). Nearly 80% of coal in China is used for thermal power generation annually, resulting in 686 million tons of CFA in 2017, 715 million tons in 2018, and 748 million tons in 2019 (Wang et al., 2021a). CFA, a fine particulate waste, is generated during coal combustion in thermal power plants and collected using electrostatic precipitators or other particle filtration equipment (Bhatt et al., 2019; Yao et al., 2015). Despite the significant amount of CFA generation, its utilization remains relatively low in some countries (Gollakota et al., 2019). Fig. 1.1 illustrates the simplified schematic of an atmospheric circulating fluidized-bed (CFB) boiler power plant (a) and a pressurized fluidized-bed combustion (PFBC) power plant (b) for coal-supplied energy generation. In the CFB configuration (a), pulverized coal is introduced into the combustion chamber, which mixes with limestone or other desulphurizing agents. The combustion process occurs at relatively lower temperatures, promoting efficient fuel burnout and minimizing nitrogen oxide (NOx) emissions. CFA, consisting of fine particulate matter, along with other combustion residues, is entrained in the flue gas. It is then collected using electrostatic precipitators or other particulate control technologies. In the PFBC setup (b), coal is again introduced into a fluidized-bed reactor, but in this case, at elevated pressure. This facilitates more thorough combustion and higher overall efficiency. Similar to the CFB process, CFA is entrained in the flue gas and subsequently separated for collection. Fig. 1.2 offers an overview of global CFA production and utilization. India leads in the CFA production with approximately 110 Mt/yr, followed closely by China with around 100 Mt/yr, America with nearly 75 Mt/yr, and Germany with almost 40 Mt/yr (see Fig. 1.1A). Regarding the CFA utilization rate, European countries, including France, Germany, Denmark, Italy, and Netherlands, exhibit notably high CFA utilization rates ranging from 80% to 100%, while America follows with 62%. Intriguingly, India and China, as the top CFA generators, only have less than 50% utilization rates, highlighting the substantial potential to increase the utilization rate.

A trend of highly efficient valorization of CFA in different sectors has been increasing dramatically, with significant economic and environmental benefits. Previous studies have successfully demonstrated the adoption of CFA in soil amendments, ceramic manufacturing, catalysis, adsorbents, zeolite synthesis, and valuable metal recovery (Awoyemi et al., 2019; Ding et al., 2017; Li et al., 2021a; Makgabutlane et al., 2020; Verrecchia et al., 2020; Zhang et al., 2019a, 2022). In particular, using CFA as supplementary cementitious materials (SCMs) in cement and concrete areas accounts for more than one-third of all valorization methods (Al-Shmaisani et al., 2022; Fan et al., 2021; Sigvardsen and Ottosen, 2019). Using CFA as SCMs can reduce water usage, enhance workability, reduce the cost of concrete manufacturing, and reduce CO_2 emissions during cement manufacturing (Li et al., 2021a; Wang et al., 2020a; Yin et al., 2018). A sufficient understanding of the physicochemical properties and mineral composition of CFA will help to further advance its efficient recycling and scientific management (Grabias-Blicharz and Franus, 2023; Luo et al., 2023; Panda and Dash, 2020).

2.1.1 Classification

The classification of CFA is based on the type of coal burnt and the contents of its major oxides. According to the American Society for Testing and Materials (ASTM C 16), CFA can be categorized into class C and class F types (ASTM, 2010). Class F CFA type results from the burning of the harder, older anthracite and bituminous coal, having a low calcium (Ca) content (less than 20%) and low pozzolanic and cementitious nature (Bhatt et al., 2019; Liu et al., 2020a). The total content of $SiO_2 + Al_2O_3 + Fe_2O_3$ in Class F type CFA is greater than 70% (Liu et al., 2020a). Class C CFA results from burning younger lignite or subbituminous coal in thermal power plants, characterized by high calcium and magnesium (Mg) contents (more than 20%) (Gollakota et al., 2019). These fly ashes have high pozzolanic properties and some self-cementing characteristics, with a total content of $SiO_2 + Al_2O_3 + Fe_2O_3$ less than 70% (Li et al., 2022a; Yoon et al., 2022).

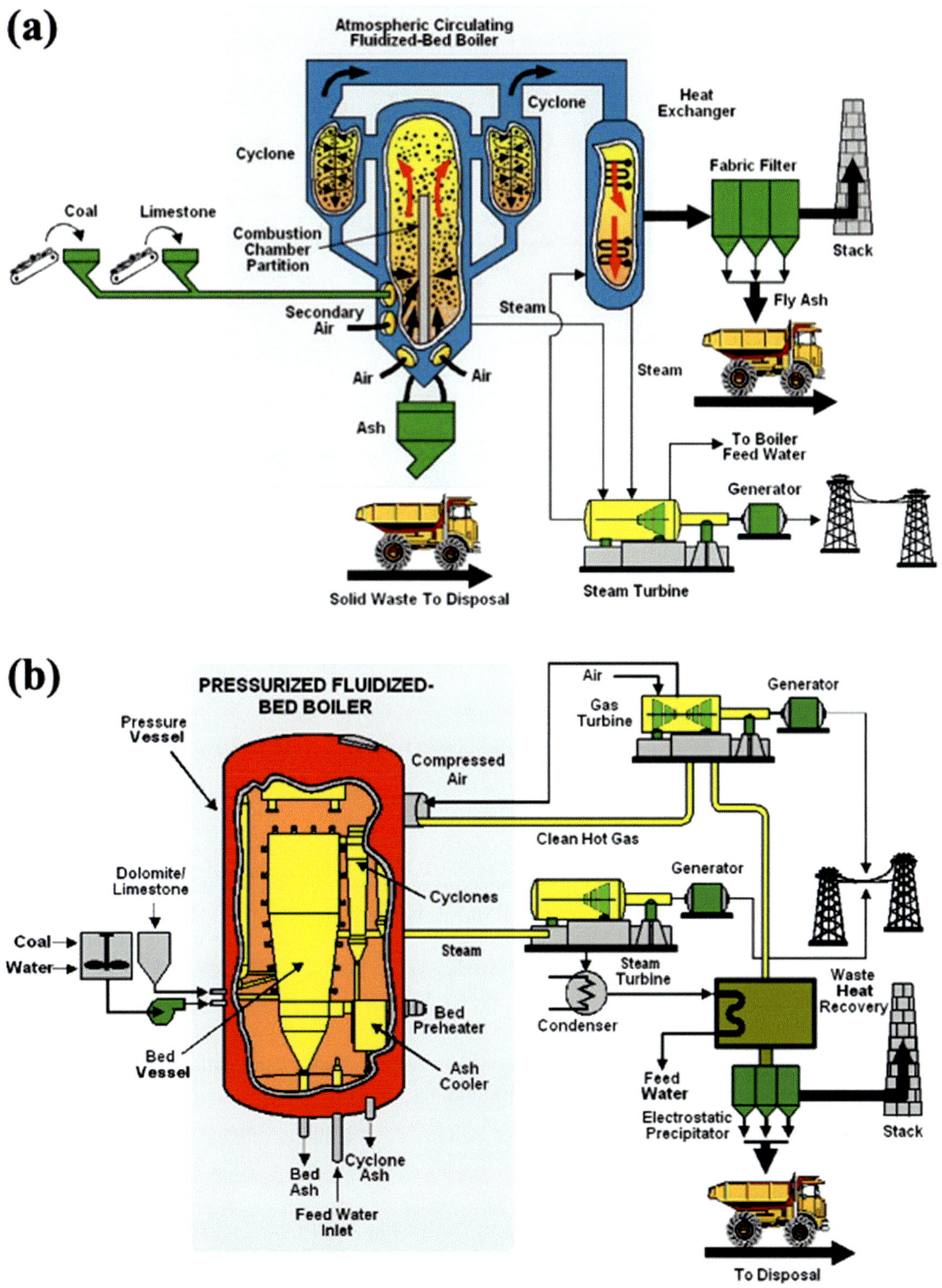

FIGURE 1.1 Simplified schematic of an atmospheric circulating fluidized-bed (CFB) boiler power plant (A) and a pressurized fluidized-bed combustion (PFBC) power plant (B). *Source: NETL, 2010c, CCPI/Clean Coal Demonstrations Tidd PFBC Demonstration Project, Project Fact Sheet.*

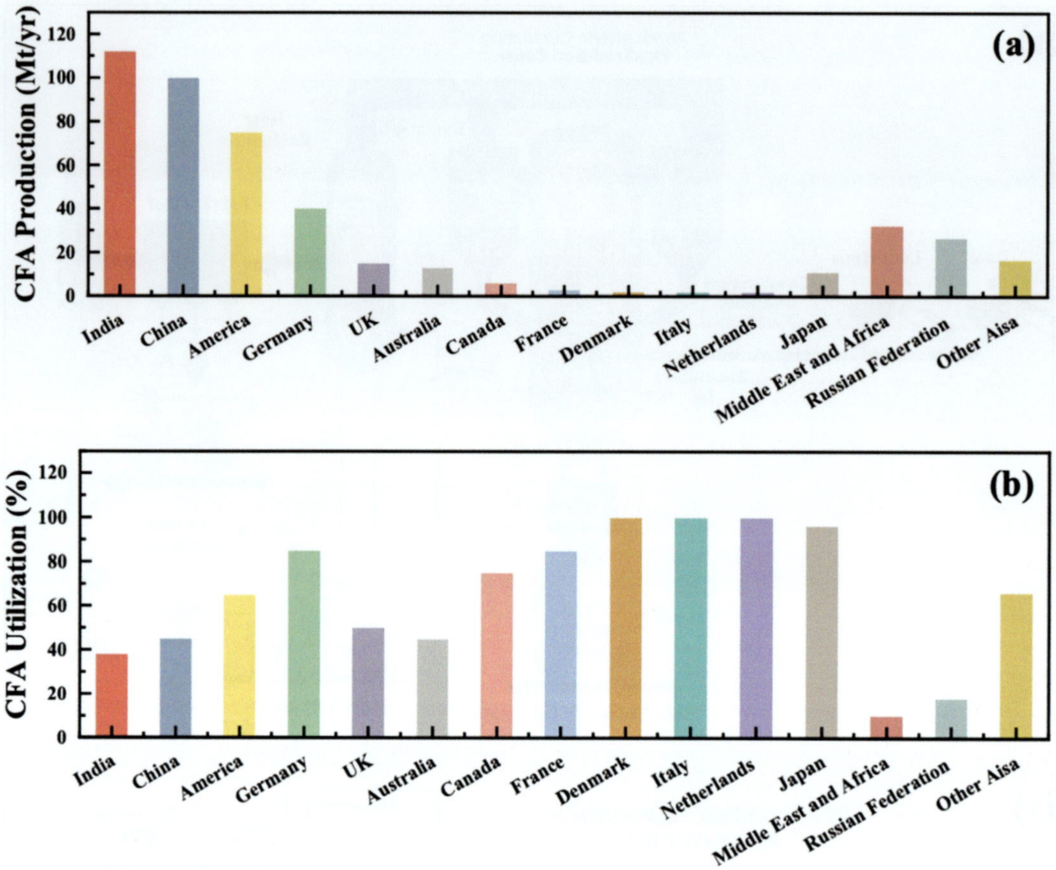

FIGURE 1.2 Global CFA production and utilization. *Data from Gollakota, A.R.K., Volli, V., Shu, C.M., 2019. Progressive utilisation prospects of coal fly ash: a review. Sci. Total Environ. 672, 951–989.*

2.1.2 Chemical compositions

Summarizes the major chemical compositions from CFA reported in previous studies. The major chemical composition of CFA includes SiO_2 (18.1–60.1 wt.%), Al_2O_3 (7.6–34.0 wt.%), Fe_2O_3 (1.7–17.3 wt.%), MgO (1.3–3.3 wt.%), Na_2O (0.1–9.12 wt.%), TiO_2 (0.6–1.7 wt.%), and P_2O_5 (n.a.–2.3 wt.%). Si, Al, and Fe are the main elements in CFA, while some potentially toxic elements (PTEs) are also found in CFA, such as Pb (3.1–5000 mg/kg), Cd (0.7–130 mg/kg), Cr (10–1000 mg/kg), Cu (14–2800 mg/kg), Zn (10–3500 mg/kg), and even radioactive elements (^{238}U, ^{232}Th, ^{226}Ra, etc.)

TABLE 1.1 Major chemical compositions and PTEs concentrations reported in the literature.

SiO_2	Al_2O_3	Fe_2O_3	CaO	Na_2O	MgO	TiO_2	P_2O_5	Mn_2O_3
18.1–60.1	7.6–34.0	1.7–17.3	1.8–18.2	0.1–9.2	1.3–3.3	0.6–1.7	n.a.–2.3	n.a.–0.6

Data sources: Bhatt, A., Priyadarshini, S., Acharath Mohanakrishnan, A., Abri, A., Sattler, M., Techapaphawit, S., 2019. Physical, chemical, and geotechnical properties of coal fly ash: a global review. Case Stud. Constr. Mater. 11, e00263; Blissett, R.S., Rowson, N.A., 2012. A review of the multi-component utilisation of coal fly ash. Fuel 97, 1–23; Fan, C., Wang, B., Ai, H., Liu, Z., 2022. A comparative study on characteristics and leaching toxicity of fluidized bed and grate furnace MSWI fly ash. J. Environ. Manage. 305, 114345; Gollakota, A.R.K., Volli, V., Shu, C.M., 2019. Progressive utilisation prospects of coal fly ash: a review. Sci. Total Environ. 672, 951–989; Hoyos-Montilla, A.A., Tobón, J.I., Puertas, F., 2023. Role of calcium hydroxide in the alkaline activation of coal fly ash. Cem. Concr. Compos. 137, 104925; Li, M.Y., Peng, Z.X., Zhang, B.W., Wang, S.K., Wang, X.S., 2023b. Morphology, mineralogical composition, and heavy metal enrichment characteristics of magnetic fractions in coal fly ash. Environ. Earth Sci. 82, 227. https://doi.org/10.1007/s12665-023-10908-0; Panda, L., Dash, S., 2020. Characterization and utilization of coal fly ash: a review. Emerg. Mater. Res. 9, 921–934; Rosita, W., Bendiyasa, I.M., Perdana, I., Anggara, F., 2020. Sequential particle-size and magnetic separation for enrichment of rare-earth elements and yttrium in Indonesia coal fly ash. J. Environ. Chem. Eng. 8, 103575; Wang, N., Sun, X., Zhao, Q., Yang, Y., Wang, P., 2020b. Leachability and adverse effects of coal fly ash: a review. J. Hazard Mater. 396; Yao, Z.T., Ji, X.S., Sarker, P.K., Tang, J.H., Ge, L.Q., Xia, M.S., Xi, Y.Q., 2015. A comprehensive review on the applications of coal fly ash. Earth-Science Rev.; Zhang, J.B., Li, S.P., Li, H.Q., He, M.M., 2016. Acid activation for pre-desilicated high-alumina fly ash. Fuel Process. Technol. 151, 64–71.

(Wang et al., 2020b). Factors influencing CFA's chemical composition are complex and include many aspects such as region, the type/treatment/combustion process, CFA collection and disposal methods, and CFA aging time, resulting in variations depending on the source of the coal and combustion conditions (Bhatt et al., 2019; Wang et al., 2020b; Yao et al., 2015).

2.1.3 Specific surface area

Specific surface area (SSA) of CFA significantly impacts its reactivity and potentials for valorization, typically ranging from 0.16 to 6.10 m^2/g (Bhatt et al., 2019; Li et al., 2021a), revealing the significant variations between the different CFAs. The SSA of raw CFA can be increased significantly by various modifications or pretreatment methods (Grabias-Blicharz and Franus, 2023). Acid-relevant modification methods dissolve the SiO_2 and Al_2O_3 to the maximum possible extent, reducing the internal impurities and volatile substances in the CFA, resulting in increased SSA (Zhang et al., 2016). Alkali-relevant modification methods, using alkaline activators, break Si—O and Al—O bonds, leading to a significant SSA increase (Huang et al., 2020; Zhao et al., 2020). Ball mill grinding is another method that can also be used to physically modify CFA, boosting the increased SSA (Yuan et al., 2021). The particle size distribution (or fineness) of CFA also reflects its reactivity, resulting in varying reactivity of different particle size fractions (Li et al., 2021a; Panda and Dash, 2020; Sigvardsen and Ottosen, 2019).

2.1.4 Mineralogy

Major mineral phases in CFA include mullite and quartz (Tennakoon et al., 2015). The typical peaks of mullite are observed at $2\theta = 16.4$ degrees, 25.9 degrees, 26.2 degrees, 30.9 degrees, 33.1 degrees, 35.2 degrees, 39.2 degrees, 40.8 degrees, 42.5 degrees, 53.9 degrees, and 60.6 degrees etc, while those of quartz are at $2\theta = 20.9$ degrees, 26.7 degrees, 36.6 degrees, 39.4 degrees, 50.2 degrees, 60.1 degree, etc. (Rawat and Yadav, 2019; Wang et al., 2020b). Other crystals, such as magnetite (30.2 degrees, 35.7 degrees, 43.1 degrees, 57.0 degrees, 62.8 degrees, etc.), hematite, anhydrite, calcite, anorthite, corundum, rutile, thenardite, and srebrodolskite, are also sometimes observed but less common (Li et al., 2023b; Rawat and Yadav, 2019). Some factors, including types of coal (subbituminous, lignite), the combustion process of coal, collection methods, the aging period, and the treatment process used (chemical and physical modification), might affect the mineralogy of CFA (Gopinathan et al., 2022).

2.1.5 Particle size and morphology

The particle sizes of CFA generally range from less than 1—150 μm (Li et al., 2023c; Pan et al., 2020), and the particle size distribution of fluidized bed CFA and the pulverized CFA exhibited similar characteristics (Rosita et al., 2020). Previous studies also found that the particle size is related to the chemical composition of CFA, where SiO_2, Al_2O_3, and C contents in the CFA are closely linked to particle size distribution, and finer particles contain more SiO_2 (Bhatt et al., 2019). Furthermore, the particle size distribution would also contribute to the various SSA of CFA, necessitating investigations on the simultaneous modification mechanisms of CFA related to the particle size and the SSA (Fernández-Jiménez et al., 2019; Zhang et al., 2020).

CFA particles have three typical morphologies, namely spherical particles, elliptical particles, and irregular particles, and possess a glassy appearance, constituting both fine and coarse fractions of amorphous and crystalline particles (Sow et al., 2015; Wang et al., 2020b). Fig. 1.3A exhibited the typical and simplest morphology of CFA, demonstrating a regular spherical shape often observed in the CFA particles. Hollow CFA particles with a large amount of small or fine particles encapsulated inside (Fig. 1.3C) and joint CFA particles with multiparticles melting together are also sometimes observed in the previous study. Meanwhile, CFA particles with cracked, mullite frameworks and aggregated forms of many small clusters have also been reported before (Fig. 1.3D), but this morphology of CFA is uncommon. Despite the fact that the morphologies of irregular CFA particles are diverse, they have in common a porous surface (Rawat and Yadav, 2019). Also, a significant portion of the SSA of CFA is due to irregularly shaped CFA particles.

Fig. 1.4 depicts the glass structure in CFA, classified into three regions based on the level of alkali cations: (1) high level, (2) medium level, and (3) low level (Jin et al., 2020). The Si-to-Al ratio in the glass phase of CFA is related to the alkali metal content, playing a vital role in CFA's subsequent adoption as pozzolanic additives and precursors for geopolymer fabrication (Lin et al., 2020; Luo et al., 2023; Yang et al., 2022a). Previous studies found that the reactive Si-to-Al ratio varies from 1.4 to 2.9, depending on the CFA types, reaction conditions, and characterization methods (Ambikakumari Sanalkumar et al., 2019). Amorphous aluminosilicates form a molecular skeleton consisting of interwoven silica—oxygen tetrahedra and aluminum—oxygen polyhedra, with distributed alkali cations, including Ca^{2+}, Fe^{3+}, Na^+,

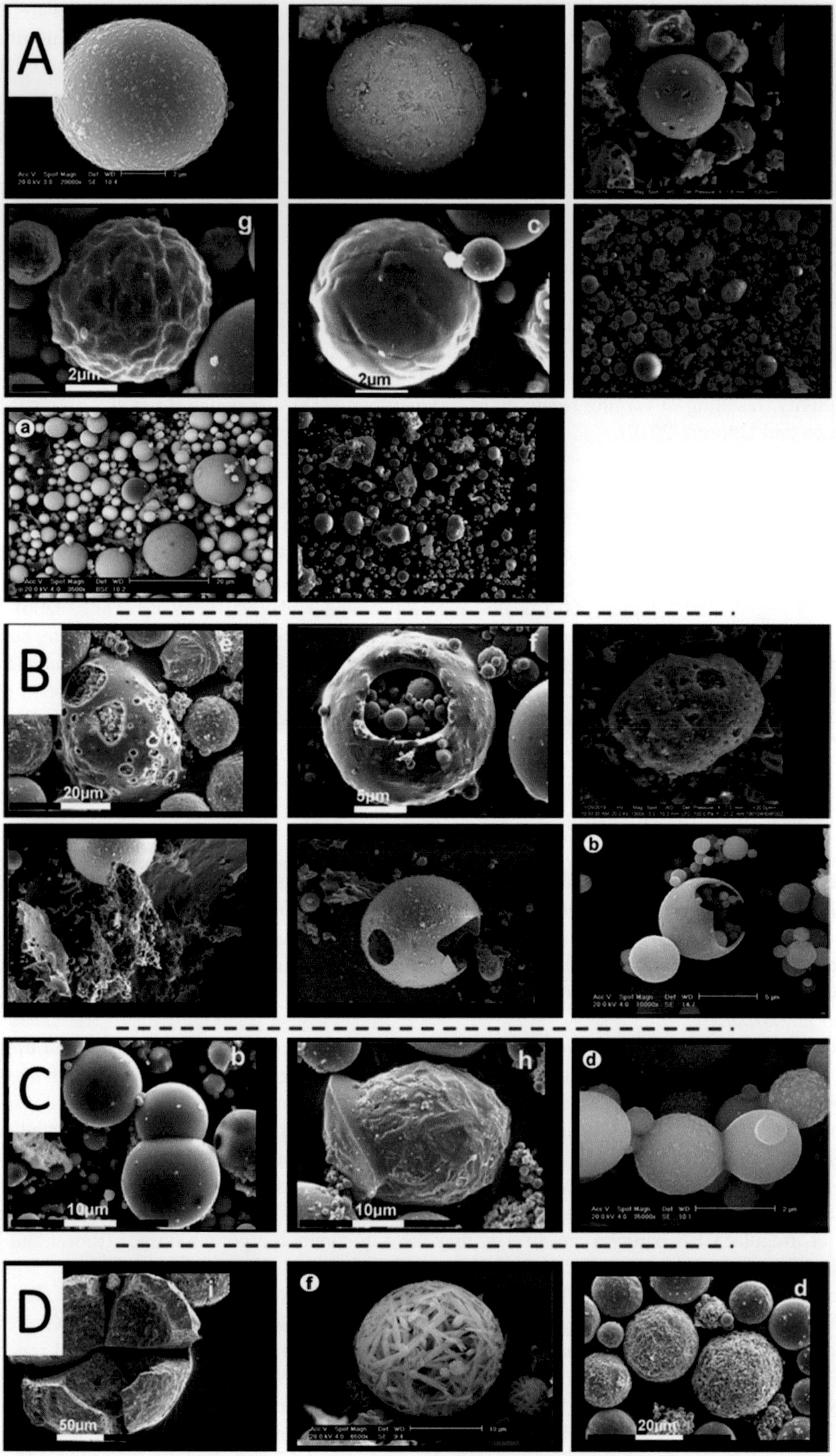

FIGURE 1.3 Morphology of coal fly ash. *Adapted from Wang, N., Sun, X., Zhao, Q., Yang, Y., Wang, P., 2020b. Leachability and adverse effects of coal fly ash: a review. J. Hazard Mater. 396.*

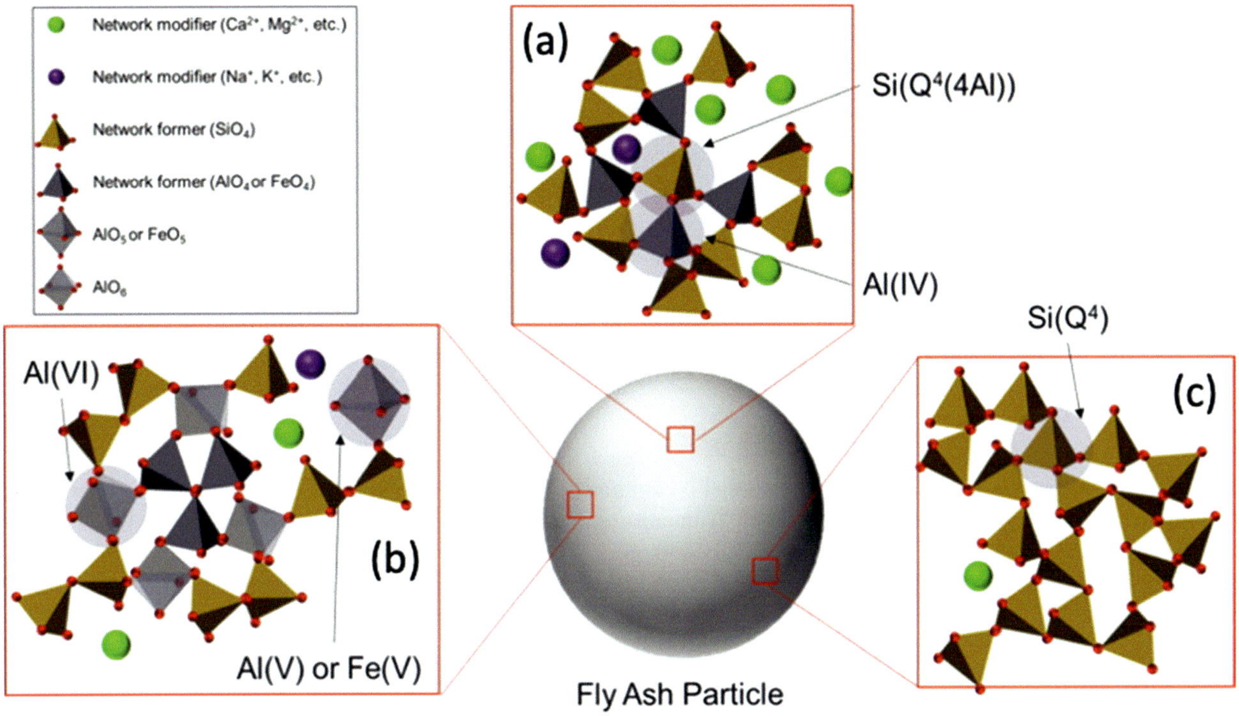

FIGURE 1.4 Schematic illustration of the glass structure in CFA. *Adapted from Jin, Y., Feng, W., Zheng, D., Dong, Z., Cui, H., 2020. Structure refinement of fly ash in connection with its reactivity in geopolymerization. Waste Manag. 118, 350–359.*

K^+, and Mg^{2+}, acting as charge balancers (Li et al., 2022b; Wei et al., 2023). Si is most abundant in the molecular skeleton, while Al is relatively low but crucial for activation reactions (Hoyos-Montilla et al., 2023; Wei et al., 2023).

2.1.6 Other properties

CFA color varies from tan to gray to black, determined by the amount of unburnt carbon remaining in the ash (Panda and Dash, 2020; Rosita et al., 2020). Light-colored CFA exhibits higher lime percentages due to lower unburned carbon levels. CFA's specific gravity generally varies from 2.0 to 3.0, and its porosity ranges from 30% to 65% (Blissett and Rowson, 2012; Panda and Dash, 2020). Table 1.2 summarizes the different physical characteristics of CFA reported in the literature.

TABLE 1.2 Different physical characteristics of CFA.

Parameter	Range	Refs.
Bulk density (g/cm^3)	1.9–2.9	Bhatt et al. (2019), Blissett and Rowson (2012), Fan et al. (2022), Gollakota et al. (2019), Hoyos-Montilla et al. (2023), Li et al. (2023b), Panda and Dash (2020), Rosita et al. (2020), Wang et al. (2020b), Yao et al. (2015), Zhang et al. (2016, 2020)
Particle size (μm)	0.01–200	
Specific gravity	2.0–3.0	
Specific surface area (m^2/g)	0.16–6.10	
Porosity (%)	30–65	
Water-holding capacity (%)	40–60	
Permeability (cm/s)	10^{-4}–10^{-6}	

2.2 Bottom ash

Combustion of coal for electricity production inevitably yields a number of by-products. The fine and light particles undergo flue gas treatment system, resulting in CFA. Larger and heavier particles generated from the bottom of coal furnaces are known as coal bottom ash (CBA). Historically, CBA was primarily treated through landfill disposal, leading to resource wastage and land occupation (Zhou et al., 2022a).

With the increasing landfill cost and the potential environmental risk associated with dumping and pollution, more efforts have been devoted to exploring various high-efficient valorization methods. Recycling CBA into SCMs and aggregate for cement and concrete manufacturing is deemed the most effective approach (Gooi et al., 2020; Muthusamy et al., 2020; Singh et al., 2020). Coarse CBA particles, resembling fine aggregates and sand, can be used as a substitution for natural fine aggregates and sand in concrete (Khaw et al., 2022; Muthusamy et al., 2020). Furthermore, pulverizing CBA particles to a particle size similar to Portland cement (PC) allows efficient utilization as SCMs, exhibiting desirable pozzolanic reactivity for low-carbon cement manufacturing (Khaw et al., 2022; Oruji et al., 2017; Tamanna et al., 2023).

2.2.1 Chemical compositions and mineralogy

CBA is a coarse, granular, and porous waste material predominantly composed of unburnt carbon, inorganic minerals, and agglomerated particles (Rafieizonooz et al., 2022; Tamanna et al., 2023). Its chemical composition varies based on coal source, types, and combustion process (Singh and Siddique, 2013). For instance, anthracite coal and bituminous coal contribute to different contents in SiO_2, Al_2O_3, Fe_2O_3, and loss on ignition (LOI) level in CBA (Ankur and Singh, 2021; Hamada et al., 2022; Rathee and Singh, 2022). Table 1.3 presents the typical chemical compositions of CBA, where Si, Al, and Fe are major elements in CBA, accounting for 70–90 wt.%. The Si and Al contents in CBA vary from 45 wt.% to 75 wt.% and 10 wt.% to 35 wt.%, respectively, with silicon phosphate and mullite as the predominant mineral phases (Ankur and Singh, 2021; Singh et al., 2020). Consequently, the pulverized CBA also exhibits pozzolanic reactivity similar to CFA due to its high available Si and Al contents. Fe in CBA mainly exists as magnetite (Fe_3O_4) and hematite (Fe_2O_3) (Guan et al., 2023). The high LOI or carbon contents in the CBA are attributed to insufficient combustion in boilers (Singh et al., 2020). Though PTEs from CBA are typically at low levels, it is still necessary to test PTEs' leachability of CBA used in construction materials to ensure long-term safety (Hashemi et al., 2019; Rafieizonooz et al., 2022).

2.2.2 Physical properties

CBA particles are larger and have an asymmetric, irregular, and angular shape, with a dark gray color and spongy texture (Ankur and Singh, 2021; Ge et al., 2018; Muthusamy et al., 2020). Fig. 1.5 presents the different sizes (Fig. 1.5A–C) and the SEM image (Fig. 1.5D–E) of CBA. Various-sized pits and air holes can be found on the coarser CBA surface, while only a few pits may be present on the finer CBA surface (Rafieizonooz et al., 2016). Table 1.4 lists the physical properties of CBA reported in the literature, including specific gravity, specific surface area, fineness modulus, moisture absorption, etc. The specific gravity of CBA lies in the range of 1.3–3.0 due to excessive voids (Ankur and Singh, 2021; Zhou et al., 2022a). The SSA of CBA is a good indicator of its reactivity, which varies from 310 m^2/kg to 14,180 m^2/kg (Gooi et al.,

TABLE 1.3 Chemical compositions of CBA (wt.%).

SiO_2	Al_2O_3	Fe_2O_3	CaO	LOI	MgO	SO_3	K_2O	Mn_2O_3
42–65	15–32	3.2–11	0.5–18	n.a.–13	0.8–6	0.1–3	0.6–4	n.a.–1

n.a. means not available.
Data source: Ankur, N., Singh, N., 2021. Performance of cement mortars and concretes containing coal bottom ash: a comprehensive review. Renew. Sustain. Energy Rev. 149, 111361; Gooi, S., Mousa, A.A., Kong, D., 2020. A critical review and gap analysis on the use of coal bottom ash as a substitute constituent in concrete. J. Clean. Prod. 268, 121752; Guan, X., Wang, L., Mo, L., 2023. Effects of ground coal bottom ash on the properties of cement-based materials under various curing temperatures. J. Build. Eng. 69, 106196; Hashemi, S.S.G., Mahmud, H. B., Ghuan, T.C., Chin, A.B., Kuenzel, C., Ranjbar, N., 2019. Safe disposal of coal bottom ash by solidification and stabilization techniques. Constr. Build. Mater. 197, 705–715; Kasaniya, M., Thomas, M.D.A., Moffatt, E.G., 2021. Efficiency of natural pozzolans, ground glasses and coal bottom ashes in mitigating sulfate attack and alkali-silica reaction. Cem. Concr. Res. 149, 106551; Muthusamy, K., Rasid, M.H., Jokhio, G.A., Mokhtar Albshir Budiea, A., Hussin, M.W., Mirza, J., 2020. Coal bottom ash as sand replacement in concrete: a review. Construct. Build. Mater. 236, 117507; Oruji, S., Brake, N.A., Nalluri, L., Guduru, R.K., 2017. Strength activity and microstructure of blended ultra-fine coal bottom ash-cement mortar. Constr. Build. Mater. 153, 317–326; Singh, N., Shehnazdeep, Bhardwaj, A., 2020. Reviewing the role of coal bottom ash as an alternative of cement. Constr. Build. Mater. 233, 117276; Tamanna, K., Raman, S.N., Jamil, M., Hamid, R., 2023. Coal bottom ash as supplementary material for sustainable construction: a comprehensive review. Construct. Build. Mater. 389, 131679; Zhou, H., Bhattarai, R., Li, Y., Si, B., Dong, X., Wang, T., Yao, Z., 2022a. Towards sustainable coal industry: Turning coal bottom ash into wealth. Sci. Total Environ. 804, 149985.

FIGURE 1.5 Different sizes (A–C) and SEM images (D–E) of CBA. *Adapted from Rafieizonooz, M., Mirza, J., Salim, M.R., Hussin, M.W., Khankhaje, E., 2016. Investigation of coal bottom ash and fly ash in concrete as replacement for sand and cement. Constr. Build. Mater. 116, 15–24.*

TABLE 1.4 Physical characteristics of CBA.

Parameter	Range	Ref.
Specific gravity	1.3–3.0	Gooi et al. (2020), Hashemi et al. (2019), Rafieizonooz et al. (2016), Rathee and Singh (2022), Reijnders (2005), Saffarzadeh et al. (2009), Singh and Siddique (2013), Singh et al. (2019), Tamanna et al. (2023)
Specific surface area (m^2/kg)	310–14180	
Dry bulk density (kg/m^3)	700–1600	
Moisture absorption (wt.%)	0.8–32	
Fineness modulus	1.4–3.5	
Porosity (vol.%)	5–13	
Uncompacted void content (vol.%)	30–50	
Permeability (cm/s)	10^{-4}–10^{-7}	

2020). The wide range of SSA is attributed to differences in the combustion process, collection, and treatment methods. Fineness modulus of CBA ranges from 1.4 to 3.5, mostly within the typical range of 1.6–2.2 for fine sand (Hamada et al., 2022; Zhang and Poon, 2015). The identical particle size distribution of CBA makes it a suitable candidate for natural fine aggregate replacement. Moisture absorption values of CBA vary from 0.8 wt.% to 37 wt.% (Gooi et al., 2020; Singh et al., 2016), influenced by the different porous structures of CBA and the water-trapping carbon and impurities in CBA.

2.3 Flue gas desulphurization gypsum

FGD gypsum is a by-product generated during the flue gas desulfurization process in the coal-fired power plant (Aakriti et al., 2023; Koralegedara et al., 2019). This process aims to remove sulfur dioxide (SO_2) emissions from the flue gases by reacting SO_2 gas with lime slurry under strong oxidation conditions (Córdoba, 2015, 2017). As environmental sustainability concerns grow, recycling FGD gypsum becomes increasingly important to mitigate its environmental impact. A comprehensive understanding of the characteristics of FGD gypsum would boost the efficiency of upcycling FGD gypsum.

2.3.1 Chemical compositions and morphology

FGD gypsum shares similar properties and textures with natural gypsum. Both contain major chemical compositions such as calcium sulfate dihydrate ($CaSO_4 \cdot 2H_2O$) and calcium sulfate hemihydrate ($CaSO_4 \cdot 0.5H_2O$) (Liu et al., 2020b; Wang and Yang, 2018). Additionally, FGD gypsum might contain unreacted calcium hydroxide, calcium carbonate, and quartz (Fig. 1.6A). However, the solubility of calcium sulfate dihydrate crystals in the FGD gypsum makes it susceptible to damage in high-humidity environments, necessitating modification methods to improve its water resistance for specific applications (Aakriti et al., 2023; Li et al., 2021c; Wu et al., 2019). Besides Ca and S, FGD gypsum also contains various elements, including Mg, Al, Ti, Fe, Si, P, Na, Cl, and trace amounts of PTEs, such as Zn, Pb, As, Hg, and Cd (Fu et al., 2019; Liu et al., 2020b). Normally, the FGD gypsum contains 35%–56% of SO_3, 29%–47% of CaO, and 0.01%–3.8% of MgO and SiO_2 with a small content of other oxides due to contamination with CFA during desulphurization process (Koralegedara et al., 2019; Liu et al., 2021). Table 1.5 presents the chemical compositions of FGD gypsum in the literature. Particles of FGD gypsum appeared in the form of rectangular bars, with the length of the bars ranging from 10 to 200 μm, as shown in Fig. 1.6B. Notably, the chemical compositions and morphologies of FGD gypsum may vary depending on the source and types of coal along with the FGD process (Aakriti et al., 2023).

2.3.2 Physical characteristics

The typical particle size of FGD gypsum lies in the range of 0.1–400 μm (Liu et al., 2021; Wang and Yang, 2018), making it generally larger than natural gypsum particles, which fall between 10 and 60 μm (Córdoba, 2015). FGD gypsum exhibits loose particles with relatively high moisture content (10%–20%), similar to natural gypsum. Normally, the color of FGD gypsum is primarily white or slightly yellow, although the presence of some impurities, such as CFA, during desulfurization can cause it to turn gray (Kun and Xiaoping, 2019; Liu et al., 2021). The specific surface area of dried FGD gypsum is approximately 320–400 m^2/kg, while the specific gravity of FGD gypsum lies at 2.3–3.5 (Bakshi et al., 2022). The moisture content in FGD gypsum is comparable with that of natural gypsum, approximately 8.0%–13% (Jiang et al., 2019b). LOI of FGD gypsum varies from 5% to 20%, and the value of LOI reflects lower volatile compound concentration and total moisture content (Kim and Lee, 2015). Differences in LOI values also indicate the presence of various hydrate

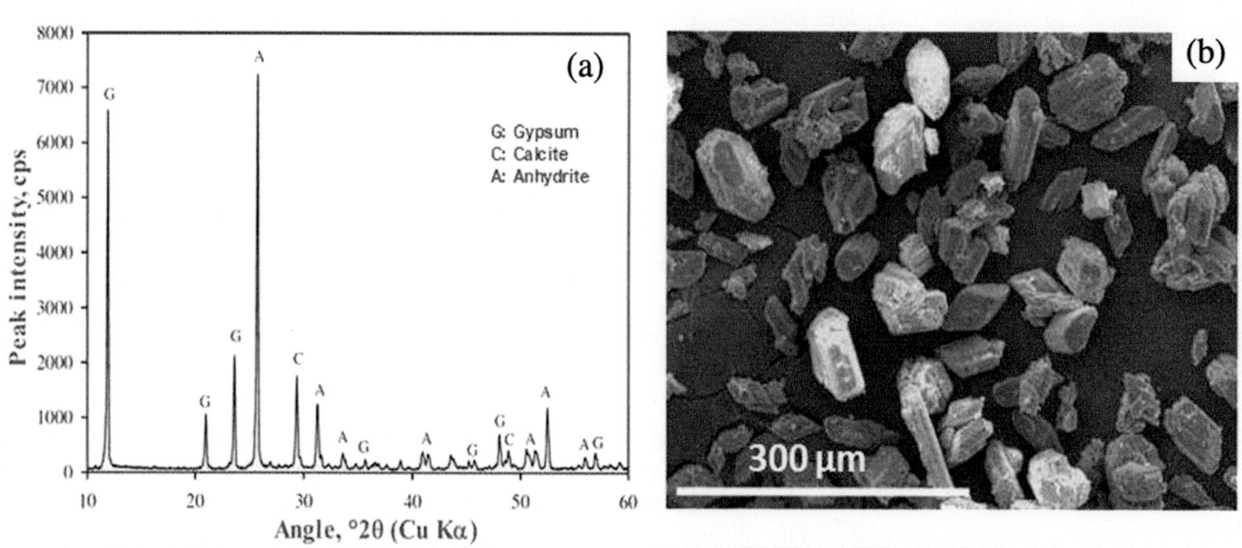

FIGURE 1.6 XRD pattern (A) and SEM image (B) of FGD gypsum. *Adapted from Aakriti, Maiti, S., Jain, N., Malik, J., 2023. A comprehensive review of flue gas desulphurized gypsum: production, properties, and applications. Construct. Build. Mater. 393, 131918.*

TABLE 1.5 Chemical composition of FGD gypsum (wt.%).

SO_3	CaO	MgO	SiO_2	Fe_2O_3	Al_2O_3	P_2O_5	K_2O	Na_2O	LOI
35–56	29–47	0.01–3.8	0.2–4.4	0.1–1.0	0.1–2.3	n.a–0.2	n.a.–0.6	n.a.–0.9	1.9–23

Data source: Aakriti, Maiti, S., Jain, N., Malik, J., 2023. A comprehensive review of flue gas desulphurized gypsum: production, properties, and applications. Construct. Build. Mater. 393, 131918; Córdoba, P., 2017. Partitioning and speciation of selenium in wet limestone flue gas desulphurisation systems: a review. Fuel 202, 184–195; Koralegedara, N.H., Al-Abed, S.R., Arambewela, M.K.J., Dionysiou, D.D., 2017. Impact of leaching conditions on constituents release from Flue Gas Desulfurization Gypsum (FGDG) and FGDG-soil mixture. J. Hazard Mater. 324, 83–93; Koralegedara, N.H., Pinto, P.X., Dionysiou, D.D., Al-Abed, S.R., 2019. Recent advances in flue gas desulfurization gypsum processes and applications – a review. J. Environ. Manag. 251, 109572; Liu, Z., Hao, Y., Zhang, J., Wu, S., Pan, Y., Zhou, J., Qian, G., 2020b. The characteristics of arsenic in Chinese coal-fired power plant flue gas desulphurisation gypsum. Fuel 271, 117515; Liu, S., Liu, W., Jiao, F., Qin, W., Yang, C., 2021. Production and resource utilization of flue gas desulfurized gypsum in China - a review. Environ. Pollut. 288, 117799; Wang, J., Yang, P., 2018. Potential flue gas desulfurization gypsum utilization in agriculture: a comprehensive review. Renew. Sustain. Energy Rev. 82, 1969–1978; Xing, G., Wang, W., Zhao, S., Qi, L., 2023. Application of Ca-based adsorbents in fixed-bed dry flue gas desulfurization (FGD): a critical review. Environ. Sci. Pollut. Res. 30, 76471–76490; Yan, Y., Li, Q., Sun, X., Ren, Z., He, F., Wang, Y., Wang, L., 2015. Recycling flue gas desulphurization (FGD) gypsum for removal of Pb(II) and Cd(II) from wastewater. J. Colloid Interface Sci. 457, 86–95.

calcium sulfate types in FGD gypsum. The bulk density of FGD gypsum generally varies from 0.90 to 1.40 g/cm^3, influenced by its geographical origin, coal type, and desulfurization process (Bakshi et al., 2022; Koralegedara et al., 2019). Compared with natural gypsum, FGD gypsum particles have a relatively lower average bulk density due to their larger size.

FGD gypsum exhibits rapid setting times due to its natural crystallization process, with initial and final setting times of unmodified FGD gypsum without additives measured at 9 and 13 min, respectively (Wu et al., 2023a). The higher solubility of FGDG and the presence of a small quantity of $CaSO_3 \cdot 0.5H_2O$ contribute to prolonged setting times (Caillahua and Moura, 2018). The water adsorption capacity of FGD gypsum can vary depending on particle size, purity, and specific manufacturing processes (Yang et al., 2020). Higher water absorption can lead to reduced mechanical strength and exert a dead load on structures, posing challenges to its widespread application (Wu et al., 2023a). For example, FGD gypsum blocks may lose 75% of their strength after water absorption and are prone to warping (Wu et al., 2019). Therefore, various studies have explored methods to reduce FGD gypsum's water absorption capacity and enhance its waterproofing to expand its use as a construction material.

3. Waste incineration residues

Waste incineration technology is a waste treatment method that involves combusting waste materials at high temperatures in controlled conditions. This technology is designed to reduce the volume of waste, destroy harmful substances, and recover energy through the process of incineration. It is commonly used to treat various types of waste, including municipal solid waste (MSW), industrial waste, and hazardous waste. In this section, a comprehensive overview of the physical and chemical characteristics of residues from MSW incineration and sewage sludge incineration will be presented.

3.1 Municipal solid waste incineration residues

The generation of MSW is increasing rapidly due to urbanization and improved living standards (Fan et al., 2022; Sustainable Org, 2019; Zhang et al., 2021). Globally, the annual MSW generation currently stands at 2.01 billion tons and is projected to reach 3.40 billion tons by 2050 (Kaza et al., 2018). In China, the MSW generation reached 249 million tons in 2021, a 5.7% increase compared with 2020 (China NBS, 2022). Incineration and waste-to-energy facilities have been adopted as means to manage this growing waste issue, significantly reducing the volume of MSW by approximately 85%–90%, mass by 60%–90%, and organic matter by almost 100% (Huang et al., 2022; Leckner, 2015). Fig. 1.7A shows the trend of MSW generation, sanitary landfill, and incineration treatment in China, indicating a rising proportion of MSWI with increasing MSW generation. Between 2011 and 2021, the amount of MSW treated by incineration rapidly increased from 25 to 180 million tons. MSWI is also widely used for MSW treatment in European countries, as shown in Fig. 1.7B.

MSWI produces two main by-products: bottom ash (BA) and fly ash (FA) (Phua et al., 2019; Verbinnen et al., 2017). The MSWI process can be illustrated using fluidized bed (Fig. 1.8A) and grate furnace (Fig. 1.8B) configurations, both generating BA and FA in the air pollution control (APC) process. In this section, we provide a comprehensive overview of the physical and chemical characteristics of FA and BA.

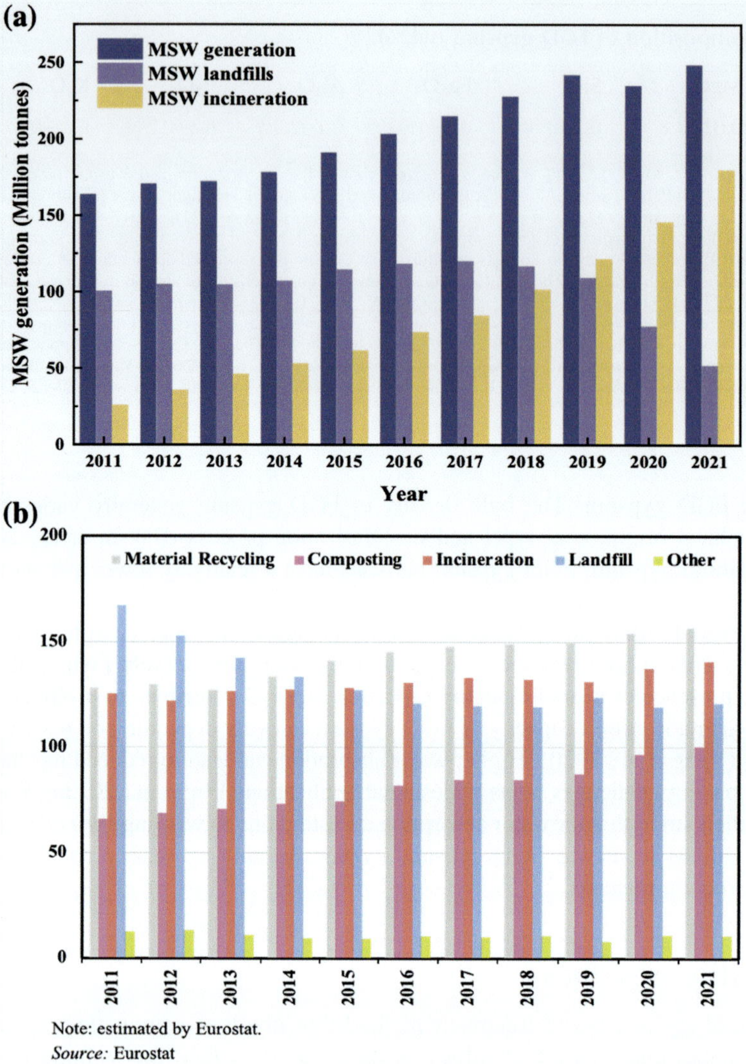

FIGURE 1.7 MSW generation and incineration treatment in China and MSW treatment in Europe ranging from 2011 to 2021 (unit: kg per capita, B). *(A) Data from NBS China, 2022.*

3.1.1 Fly ash

MSWI FA is the residue collected from APC devices, such as bag filters, dry/semidry/wet scrubbers, electrostatic precipitators, fabric filters, and cyclones, in MSWI plants (Ghouleh and Shao, 2018; Vavva et al., 2017), It constitutes about 3%−15% of the total MSW, depending on the type of MSWI furnace, such as moving grate furnace or fluidized bed furnace (Fan et al., 2022; Funari et al., 2015).

3.1.1.1 Chemical compositions and morphology

The chemical characteristics of MSWI FA highly depend on the nature of waste, combustion unit type, and APC devices (Phua et al., 2019). MSWI FA may exhibit high concentrations of Na, K, and Cl when unsorted kitchen waste, wood, and polyvinyl chloride are incinerated (Zhang et al., 2021). Table 1.6 summarizes the varied compositions of MSWI FA reported in the literature, with major components including CaO, SiO_2, Al_2O_3, Fe_2O_3, MgO, Na_2O, and K_2O. CaO, SiO_2, and Al_2O_3 dominate, collectively accounting for over 50% of the weight. The major phases in MSWI FA encompass portlandite, calcium carbonate, calcium chloride hydroxide, halite, sylvite, and quartz (Tang et al., 2020a), mostly formed as reaction products from acidic flue gas and injected lime (Phua et al., 2019).

MSWI FA frequently contains a considerable amount of PTEs, including Zn, Pb, Cd, Cr, Cu, Hg, Ni, As, and Sb (Cui et al., 2023; Zhang et al., 2021), with Zn and Pb being particularly abundant (Lederer et al., 2017). Fig. 1.9 displays the

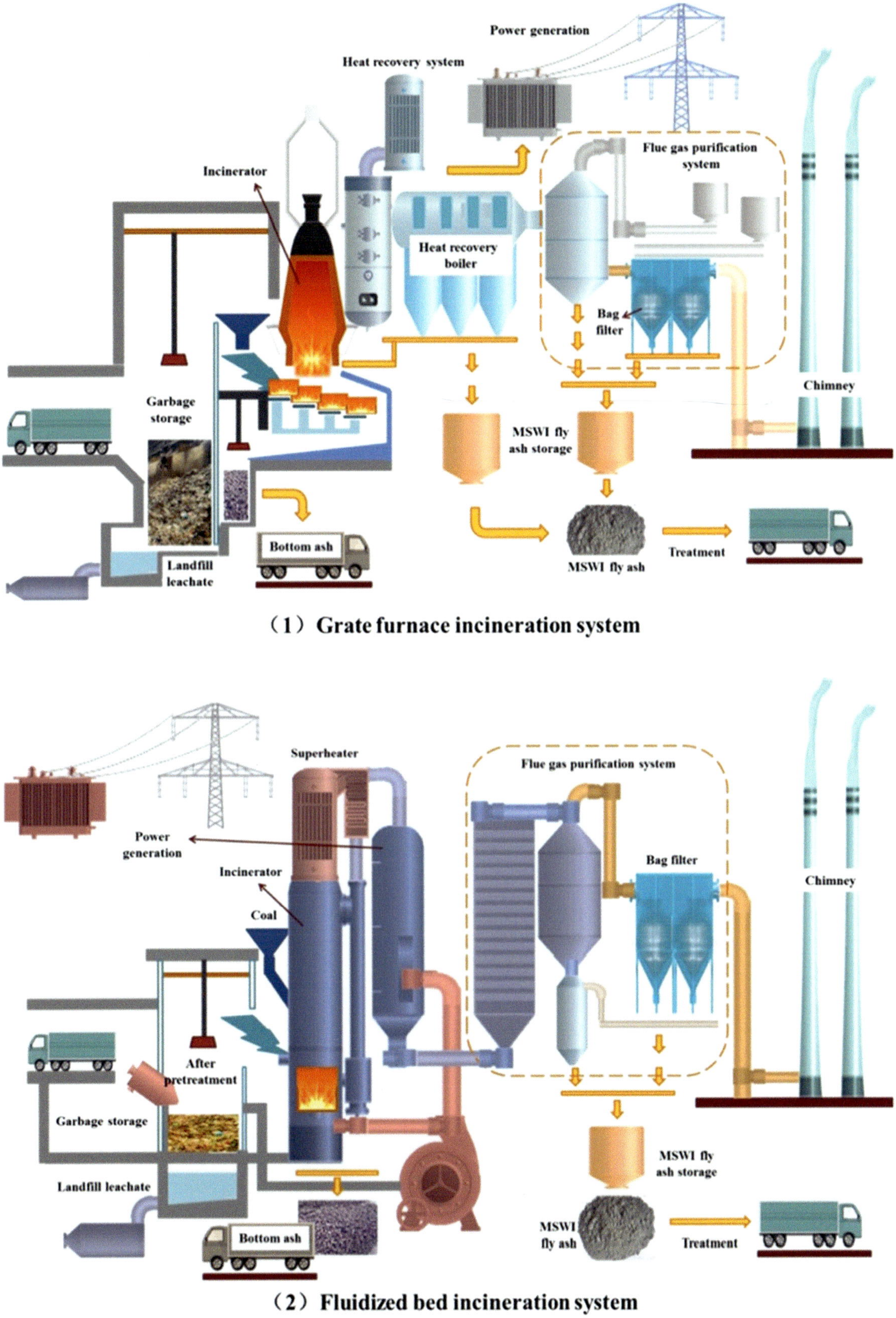

FIGURE 1.8 Municipal solid waste incineration process: (A) fluidized bed; (B) grate furnace. *Adapted from Fan, C., Wang, B., Ai, H., Liu, Z., 2022. A comparative study on characteristics and leaching toxicity of fluidized bed and grate furnace MSWI fly ash. J. Environ. Manage. 305, 114345.*

TABLE 1.6 Chemical composition in FA/APC MSWI residues based on literature.

Major components/%	CaO	SiO_2	Al_2O_3	Fe_2O_3	MgO	Na_2O	K_2O	SO_3	Cl	LOI	References
FA	18–33	8.3–18	4.1–14	2.2–3.7	1.6–6.7	8.5–13	2.2–7	2.0–16	3.5–12	3.2–20	Ashraf et al. (2019), Ferraro et al. (2019), Phua et al. (2019), Saqib and Bäckström (2016), Ghouleh and Shao (2018)
APC	34–56	1.8–7.8	0.17–8.4	0.15–3.2	0.4–4.9	0–5	0.2–6.8	0.31–10.7	0.6–21	5.4–35	Ashraf et al. (2019), Bogush et al. (2020), Cristelo et al. (2020), Phua et al. (2019), Tang et al. (2020a), Ghouleh and Shao (2018)

Trace elements/ (mg/kg)	Zn	Cu	Pb	Cr	Cd	Ni	As	Sb	Hg	Mn	References
FA	320–19000	108–3200	108–16800	26–450	20–350	9–260	6.2–60	26–1100	0.7–110	400–1500	Guo et al. (2014), Loginova et al. (2019), Phua et al. (2019)
APC	282–13000	43–6200	65–2300	25–190	0.80–180	17–160	3.6–210	200–682	0.7–110	270–710	Bogush et al. (2020), Chen et al. (2020), Huber et al. (2018), Lane et al. (2020)

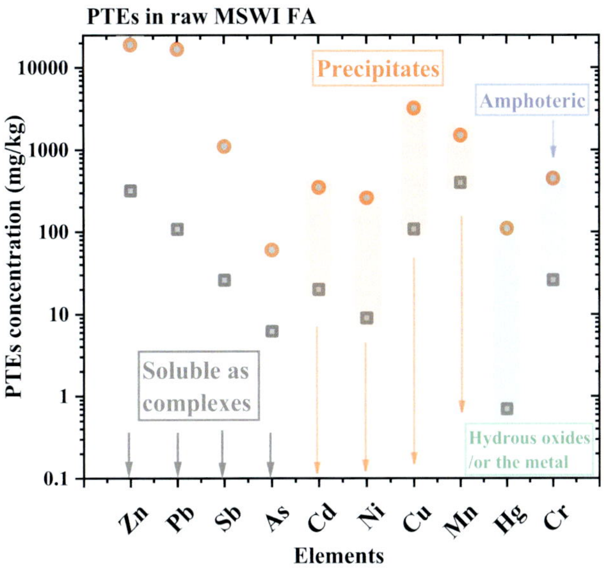

FIGURE 1.9 PTEs in raw MSWI FA and their solubility under the oxidation states (pH 12–13). *Data source from Cui, S., Peng, Y., Yang, X., Gao, X., Guan, C.Y., Fan, B., Zhou, X., Chen, Q., 2023. Comprehensive understanding of guest compound intercalated layered double hydroxides: design and applications in removal of potentially toxic elements. Crit. Rev. Environ. Sci. Technol. 53, 457–482.*

PTEs contents in raw MSWI FA and their solubility under the oxidation states. LOI of MSWI FA ranges from 11% to 35%, indicating a significant unburned organic carbon content. The alkaline nature of MSWI FA, with a pH range of 10.5–13.5, is attributed to the use of hydrated lime or sodium carbonate for acid gas neutralization (Cristelo et al., 2020; Quina et al., 2018).

MSWI FA exhibits an irregular flocculent distribution with a loose structure and rough surface, as observed in SEM images (Fig. 1.10). Specifically, the grate furnace MSWI FA typically exhibits a spherical and elliptical vitreous microbead shape (Ji et al., 2019a). Lots of NaCl crystals can be found in the microstructure morphology of grate furnace MSWI fly ash. Meanwhile, the fluidized bed MSWI fly ash appears irregular, with no apparent crystal structure.

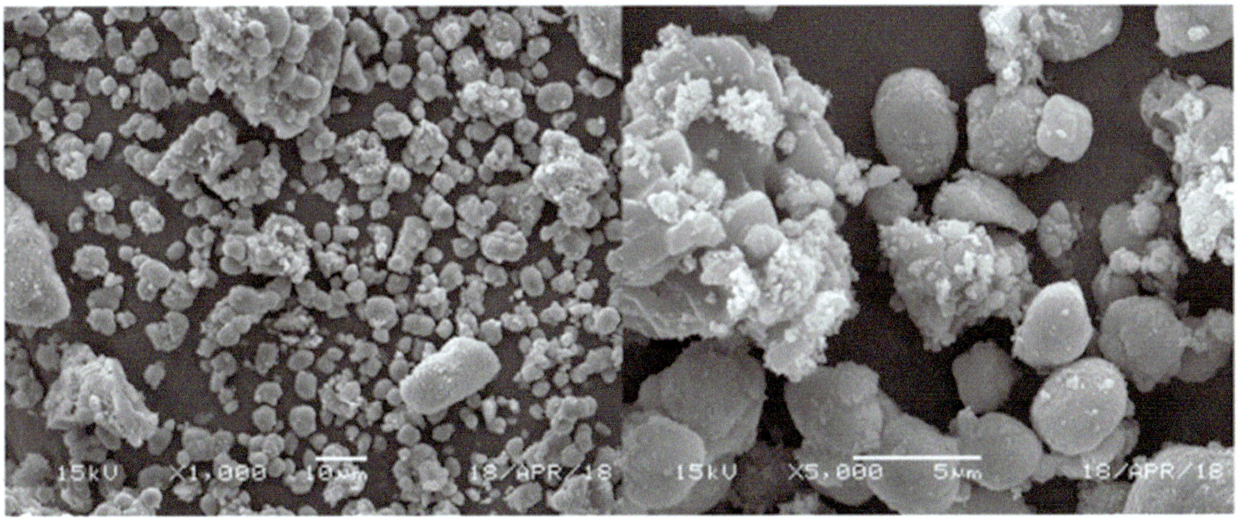

FIGURE 1.10 Morphology of the MSWI FA. *Adapted from Ji, H., Huang, W., Xing, Z., Zuo, J., Wang, Z., Yang, K., 2019a. Experimental study on removing heavy metals from the municipal solid waste incineration fly ash with the modified electrokinetic remediation device. Sci. Rep. 1–9.*

3.1.1.2 Physical characteristics

MSWI FA consists of micron-scale fine particles, with over 90% being smaller than 300 μm (Phua et al., 2019). The particle size is influenced by the MSWI operational conditions of the MSWI process and the collection methods. Particle sizes can range from 0.5 to 166.41 μm (D50), with smaller particles showing higher concentrations of potentially toxic elements (PTEs) due to reduced mixing efficiency (Bernasconi et al., 2022). The appearance of MSWI FA differs depending on the type of incinerator used. MSWI FA collected from grate furnaces appears as a fine powder, largely due to the addition of white lime powder for acid gas neutralization, while fluidized bed MSWI FA particles present a sandy yellow appearance, attributed to the burning of municipal waste with quartz sand or powdered coal (Bai et al., 2023; Chuai et al., 2022; Fan et al., 2022; Yue et al., 2019). Consequently, fluidized bed MSWI FA contains higher concentrations of Si and Al, giving rise to its sand-granular nature (Fan et al., 2022). Table 1.7 compares the major physical characteristics of grate furnace MSWI FA and fluidized bed MSWI FA. In comparison, grate furnace MSWI FA exhibits higher alkalinity, lower density, and a higher specific surface area (SSA) than fluidized bed MSWI FA. The typical MSWI FA is characterized by its corrosive nature and low moisture content.

3.1.2 Municipal solid waste incineration bottom ash

MSWI BA, a composite material comprised of stone, brick, ceramics, glass, and residual organic matter (wood, plastic, fibers, etc.), is usually collected from the lower section of the incineration furnace (Fig. 1.11) (Chen et al., 2023; Lindberg et al., 2015). It can constitute approximately 80−90 wt% of the overall MSWI residue (Brunner and Rechberger, 2015; Chen et al., 2023). Since the amount of MSW being incinerated is rising, the pressure to dispose of waste incineration residues will inevitably increase. With the escalating incineration of MSW, the necessity to manage waste incineration by-products is becoming more pronounced. Given the widespread availability of MSWI BA, delving into its intrinsic characteristics becomes imperative to explore the technical viability of recycling this industrial by-product.

3.1.2.1 Chemical compositions and morphology

The elemental constituents of primary significance in MSWI BA encompass SiO_2, CaO, Al_2O_3, and Fe_2O_3 (Table 1.8), collectively constituting over 60% of the total mass of MSWI BA (Clavier et al., 2020; Šyc et al., 2020; Tang et al., 2015). The mineralogical composition of MSWI BA is largely governed by factors such as the original waste composition, incineration methodologies, and subsequent weathering mechanisms (Dou et al., 2017). The mineral diversity in MSWI BA spans 11 categories, encompassing silicon dioxide (SiO_2), iron oxides (FeO_x), silicates, carbonates, sulfates, chloride salts, phosphates, nonferrous metal oxides, hydroxides, sulfides, and other minerals (Chen et al., 2023). Quartz content within MSWI BA fluctuates between 4.7 and 21 wt.%, surpassing the weight proportion of other discerned crystalline phases (Alam et al., 2019; Le et al., 2018; Yan et al., 2020). Notably, calcite, a crystalline phase, can range from 0.9 to 22.7 wt.% in MSWI BA, exhibiting regional variations (Alam et al., 2019). Among the silicates prevalent in MSWI BA, gehlenite, melilite group's akermanite, feldspar group's albite and anorthite, and pyroxene group's diopside and wollastonite are commonly identified. The total silicate content in MSWI BA typically remains below 15 wt.% (del Valle-Zermeño et al., 2017). Iron oxide content within MSWI BA, usually less than 5 wt.%, predominantly comprises magnetite (del Valle-Zermeño et al., 2017; Flesoura et al., 2019). The amorphous phases constitute 30.8−81.3 wt.% of the MSWI BA, aligning with its presence in CFA (50−95 wt.%) (Wei et al., 2014; Zhu et al., 2021). Contrarily, compared with blast furnace slag (BFS) with more than 90 wt% amorphous phases, MSWI BA has a lower amorphous phase content.

TABLE 1.7 Physical characteristics of MSWI FA collected from grate furnace and fluidized bed processing (Fan et al., 2022).

	Grate furnace	Fluidized bed
Corrosivity	Yes	Yes
pH	11−12.8	10.5−12.3
Moisture content (%)	0.2−2.0	0.20.5
Density (g/cm^3)	2.2−2.5	2.4−2.6
SSA (m^2/kg)	720−790	230−340

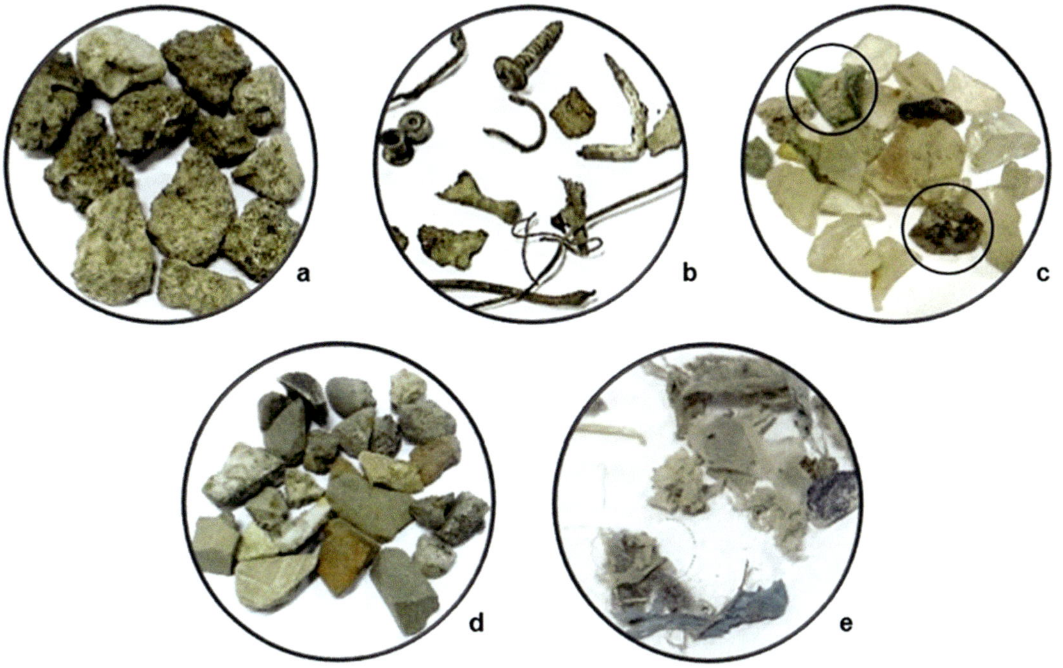

FIGURE 1.11 Common fractions of MSWI BA: (A) slag, (B) metals, (C) glass, (D) refractory, and (E) unburned size fractions 4.0–31.5 mm. *Adapted from van de Wouw, P.M.F., Loginova, E., Florea, M.V.A., Brouwers, H.J.H., 2020. Compositional modelling and crushing behaviour of MSWI bottom ash material classes. Waste Manag. 101, 268–282.*

TABLE 1.8 Chemical composition of MSWI BA (wt.%).

SiO_2	CaO	Al_2O_3	Fe_2O_3	MgO	Na_2O	K_2O	P_2O_5	SO_3	Cl	LOI
5.4 –59.3	13.6–50.4	1.2–18.0	1.2–20.2	1.5–3.3	0–17.2	0.4–7.5	0.3–6.9	0.3 –12.7	0.2–5	2.7–30

Data source: Antoun, M., Becquart, F., Gerges, N., Aouad, G., 2020. The use of calcium sulfo-aluminate cement as an alternative to Portland Cement for the recycling of municipal solid waste incineration bottom ash in mortar. Waste Manag. Res. 38, 868–875; Clavier, K.A., Paris, J.M., Ferraro, C.C., Townsend, T.G., 2020. Opportunities and challenges associated with using municipal waste incineration ash as a raw ingredient in cement production – a review. Resour. Conserv. Recycl. 160, 104888; Cristelo, N., Segadães, L., Coelho, J., Chaves, B., Sousa, N.R., de Lurdes Lopes, M., 2020. Recycling municipal solid waste incineration slag and fly ash as precursors in low-range alkaline cements. Waste Manag.. 104, 60–73; Dou, X., Ren, F., Nguyen, M.Q., Ahamed, A., Yin, K., Chan, W.P., Chang, V.W.C., 2017. Review of MSWI bottom ash utilization from perspectives of collective characterization, treatment and existing application. Renew. Sustain. Energy Rev. 79, 24–38; Flesoura, G., Garcia-Banos, B., Catala-Civera, J.M., Vleugels, J., Pontikes, Y., 2019. In-situ measurements of high-temperature dielectric properties of municipal solid waste incinerator bottom ash. Ceram. Int. 45, 18751–18759; Le, N.H., Razakamanantsoa, A., Nguyen, M.L., Phan, V.T., Dao, P.L., Nguyen, D.H., 2018. Evaluation of physicochemical and hydromechanical properties of MSWI bottom ash for road construction. Waste Manag.. 80, 168–174; Tang, P., Florea, M.V.A., Spiesz, P., Brouwers, H.J.H., 2015. Characteristics and application potential of municipal solid waste incineration (MSWI) bottom ashes from two waste-to-energy plants. Constr. Build. Mater. 83, 77–94; Verbinnen, B., Billen, P., Van Caneghem, J., Vandecasteele, C., 2017. Recycling of MSWI bottom ash: a review of chemical barriers, engineering applications and treatment technologies. Waste and Biomass Valorization 8, 1453–1466; Wei, Y., Saffarzadeh, A., Shimaoka, T., Zhao, C., Peng, X., Gao, J., 2014. Geoenvironmental weathering/deterioration of landfilled MSWI-BA glass. J. Hazard Mater. 278, 610–619; Xuan, D., Tang, P., Poon, C.S., 2018. Limitations and quality upgrading techniques for utilization of MSW incineration bottom ash in engineering applications – a review. Construct. Build. Mater. 190, 1091–1102; Yan, K., Sun, H., Gao, F., Ge, D., You, L., 2020. Assessment and mechanism analysis of municipal solid waste incineration bottom ash as aggregate in cement stabilized macadam. J. Clean. Prod. 244, 118750; Yang, Z., Ji, R., Liu, L., Wang, X., Zhang, Z., 2018. Recycling of municipal solid waste incineration by-product for cement composites preparation. Constr. Build. Mater. 162, 794–801.

The amorphous phase emerges as the pivotal reactive component in MSWI BA, with its reactivity potentially linked to its chemical composition (Chen et al., 2023; Clavier et al., 2020). Notably, previous studies primarily quantified the amorphous phase, paying scant attention to its chemical attributes.

The morphology of MSWI BA is depicted in Fig. 1.12. Fine particle fractions of MSWI BA (Fig. 1.12A) primarily exhibit diminutive dimensions along with substantial surface area. Larger fractions (Fig. 1.12B–F) also feature well-defined surfaces, with particles measuring at least 500 μm displaying distinct attributes identified as quench products (Clavier et al., 2020; Loginova et al., 2019). These quench products play a role in the leaching of PTEs when MSWI BA is

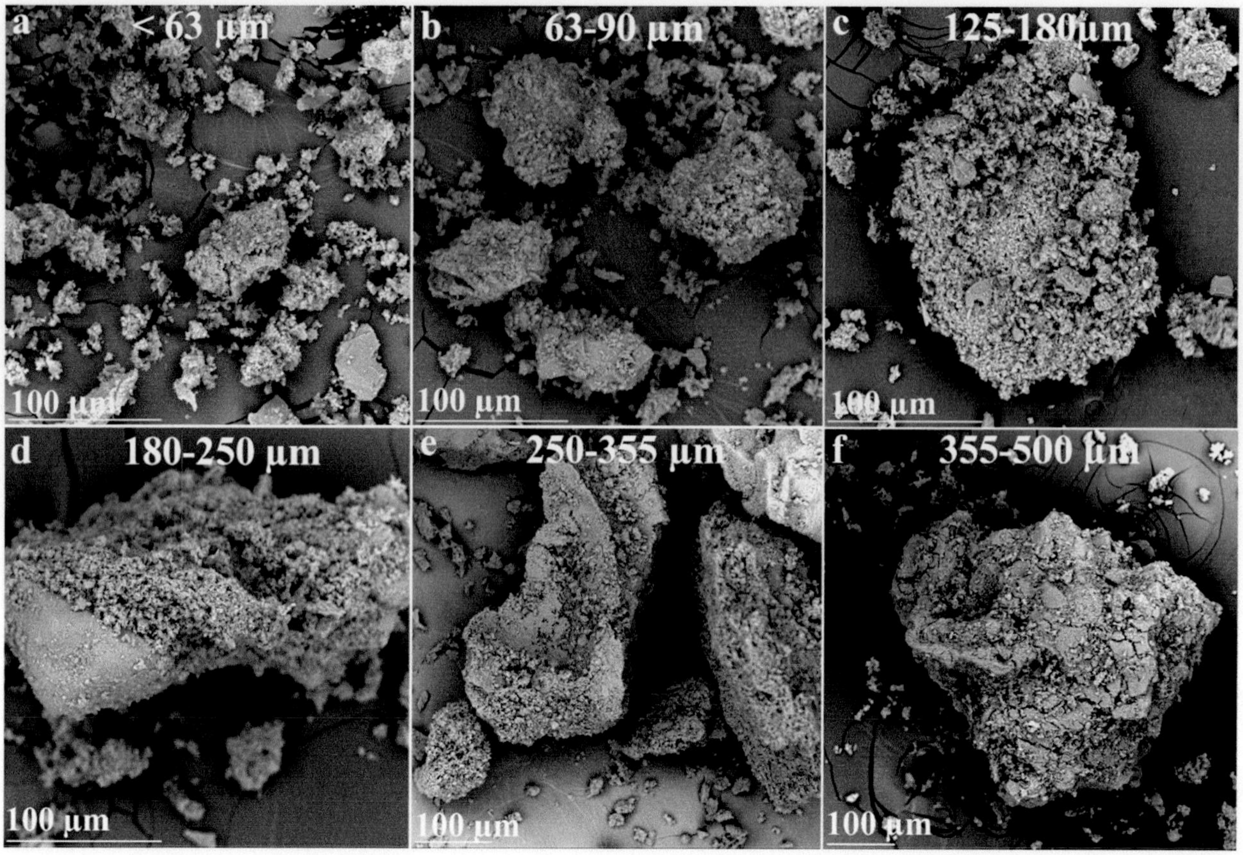

FIGURE 1.12 SEM images of MSWI BA fine particle size fractions. *Adapted from Loginova, E., Volkov, D.S., van de Wouw, P.M.F., Florea, M.V.A., Brouwers, H.J.H., 2019. Detailed characterization of particle size fractions of municipal solid waste incineration bottom ash. J. Clean. Prod. 207, 866–874.*

employed in construction materials. Notably, the morphology of smaller particles remains largely consistent throughout the fine fraction.

3.1.2.2 Physical characteristics

MSWI BA assumes a heterogenous gray or dark gray appearance, with a bulk density typically ranging between 1200 kg/m^3 and 1800 kg/m^3 (Loginova et al., 2019; Tang et al., 2015). Fine particles exhibit specific gravities within 1.5–2.0, while coarse particles range from 1.8 to 2.4 (Alam et al., 2021; Singh and Kumar, 2020). Water absorption capacity spans from 2.4% to 18%, averaging at 10.2% (Singh and Kumar, 2020). The pH value of MSWI BA ranges between 10.5 and 12.0 (Alam et al., 2021; Chen et al., 2023; Sun and Yi, 2020). Moisture content is contingent on discharge type, distinguishing between wet and dry discharges. Wet-discharged MSWI BA holds moisture levels of approximately 18%–25%, predominantly within the fine fraction (Pienkoß et al., 2022; Šyc et al., 2020). Conversely, dry-discharged MSWI BA retains moisture content below 1% (Back and Sakanakura, 2022; Mehr et al., 2021). The particle size distribution in MSWI BA encompasses micrometers to centimeters, with around 30%–40% of particles smaller than 2 mm and approximately 20% larger than 2 cm (Kumar and Singh, 2023; Loginova et al., 2021).

3.2 Sewage sludge incineration ash

Sewage sludge incineration ash (ISSA) emerges as a by-product from the incineration of sewage sludge, sourced from wastewater treatment plants (Chang et al., 2020; Fang et al., 2021). The gradual increase in ISSA production is attributed to expanding populations and heightened hygiene standards. An illustrative depiction of the sewage sludge incineration process is presented in Fig. 1.13, highlighting the origin of sewage sludge incineration fly ash from the electrostatic precipitator and the bottom ash from the fluidized bed incinerator (Donatello and Cheeseman, 2013). The utilization of incineration for sewage sludge treatment has become predominant, significantly curbing the volume of sewage sludge. In

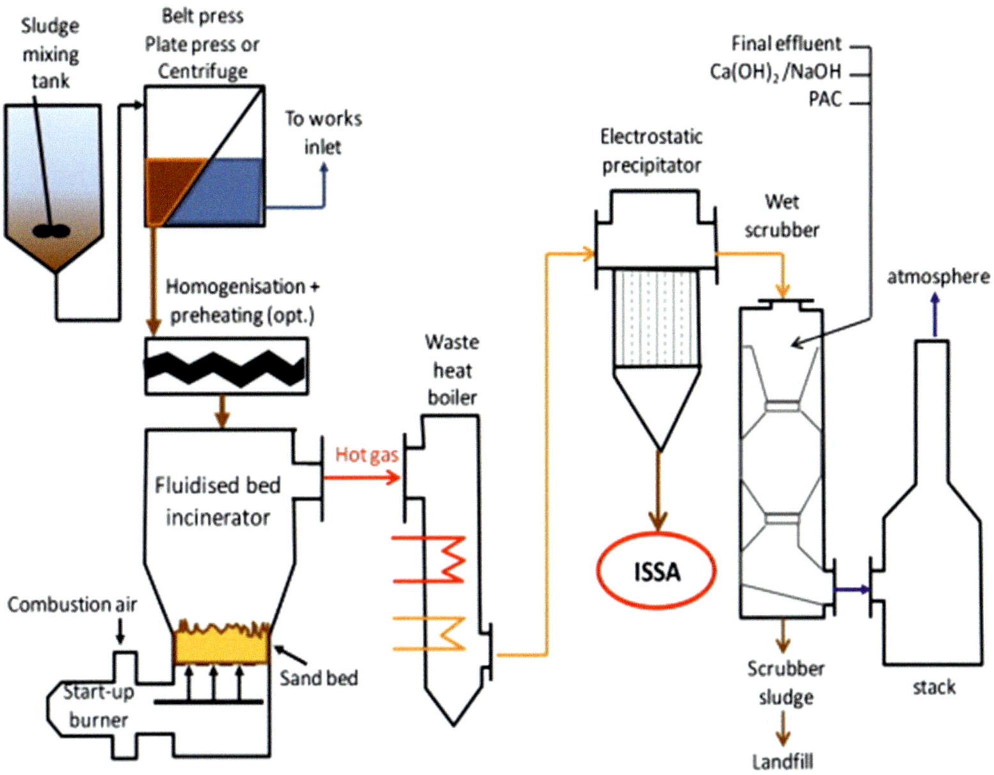

FIGURE 1.13 Overview of the sewage sludge incineration process. *Adapted from Donatello, S., Cheeseman, C.R., 2013. Recycling and recovery routes for incinerated sewage sludge ash (ISSA): a review. Waste Manag.*

line with the emphasis on sustainable waste management practices and the inclination to transform waste incineration residues into value-added commodities, a comprehensive understanding of the fundamental attributes of sewage sludge incineration residues becomes essential.

3.2.1 Chemical compositions and morphology

The intrinsic attributes of ISSA are influenced by the original sewage sludge composition and the incineration process, encompassing variables such as burner technology, combustion temperature, and residence time. Divergent incineration techniques induce varying degrees of sewage sludge burnout, thereby influencing carbonization of resultant ashes (Bogush et al., 2020; Clavier et al., 2019, 2020). The principal oxides identified in ISSA include SiO_2, Al_2O_3, CaO, Fe_2O_3, NaO, MgO, and P_2O_5 (Table 1.9). Among the crystalline minerals present in ISSA, one can identify silicon dioxide minerals (e.g., quartz), whitlockite, mullite, calcium sulfate minerals (e.g., anhydride), magnetite, feldspars, and micas (Bogush et al., 2020; Clavier et al., 2020; Sarmiento et al., 2019). Notably, approximately 45% of ISSA mass constitutes the amorphous phase (Świerczek et al., 2021; Zhou et al., 2020, 2022b). ISSA also comprises various inorganic PTEs, including Cd, Cu, Ni, Pb, Cr, Mn, Fe, As, Sb, and Hg, presenting varying concentrations (Benassi et al., 2019; Hušek et al., 2022). Incineration-induced reduction in sewage sludge volume and mass enhances the concentration of PTEs in resultant solid residues.

ISSA particles exhibit irregular shapes with rough surfaces (Fig. 1.14), distinguishing them from the more spherical CFA particles and the more angular limestone, PC, and clay particles. Additionally, ISSA features a porous microstructure due to the formation of gas bubbles during the heating and cooling phases of the incineration process (Chang et al., 2022a). The combination of irregular particle shape and porous microstructure yields a higher specific surface area, implying augmented absorption properties.

3.2.2 Physical characteristics

ISSA is characterized by a free-flowing, fine-grained nature, showcasing a color spectrum ranging from yellow to red, brown, and gray(Chang et al., 2022b; Strandberg et al., 2021). Color variations correspond to the particular sewage sludge

TABLE 1.9 Chemical composition of ISSA (wt.%).

SiO_2	CaO	Al_2O_3	Fe_2O_3	MgO	Na_2O	K_2O	P_2O_5	SO_3	TiO_2	LOI
12.4–29.0	18.5–38.0	2.3–17.9	1.7–7.0	2.3–4.5	0.6–4.4	0–4.1	5.3–13.2	2.0–6.6	0.7–3.8	2.7–25

Data source: Benassi, L., Zanoletti, A., Depero, L.E., Bontempi, E., 2019. Sewage sludge ash recovery as valuable raw material for chemical stabilization of leachable heavy metals. J. Environ. Manage. 245, 464–470; Bogush, A.A., Stegemann, J.A., Zhou, Q., Wang, Z., Zhang, B., Zhang, T., Zhang, W., Wei, J., 2020. Co-processing of raw and washed air pollution control residues from energy-from-waste facilities in the cement kiln. J. Clean. Prod. 254; Chang, Z., Long, G., Zhou, J.L., Ma, C., 2020. Valorization of sewage sludge in the fabrication of construction and building materials: a review. Resour. Conserv. Recycl. 154, 104606; Clavier, K.A., Paris, J.M., Ferraro, C.C., Townsend, T.G., 2020. Opportunities and challenges associated with using municipal waste incineration ash as a raw ingredient in cement production – a review. Resour. Conserv. Recycl. 160, 104888; Fang, L., Wang, Q., Li, J. S., Poon, C.S., Cheeseman, C.R., Donatello, S., Tsang, D.C.W., 2021. Feasibility of wet-extraction of phosphorus from incinerated sewage sludge ash (ISSA) for phosphate fertilizer production: a critical review. Crit. Rev. Environ. Sci. Technol. 51, 939–971; Hušek, M., Moško, J., Pohořelý, M., 2022. Sewage sludge treatment methods and P-recovery possibilities: current state-of-the-art. J. Environ. Manag. 315, 115090; Sarmiento, L.M., Clavier, K.A., Paris, J.M., Ferraro, C.C., Townsend, T.G., 2019. Critical examination of recycled municipal solid waste incineration ash as a mineral source for portland cement manufacture – a case study. Resour. Conserv. Recycl. 148, 1–10; Świerczek, L., Cieślik, B.M., Konieczka, P., 2021. Challenges and opportunities related to the use of sewage sludge ash in cement-based building materials – a review. J. Clean. Prod. 287, 125054; Zhou, Y., Cai, G., Cheeseman, C., Li, J., Poon, C.S., 2022b. Sewage sludge ash-incorporated stabilisation/solidification for recycling and remediation of marine sediments. J. Environ. Manage. 301, 113877; Zhou, Y. F., Li, J. S., Lu, J. X., Cheeseman, C., Poon, C.S., 2020. Sewage sludge ash: a comparative evaluation with fly ash for potential use as lime-pozzolan binders. Construct. Build. Mater. 242, 118160.

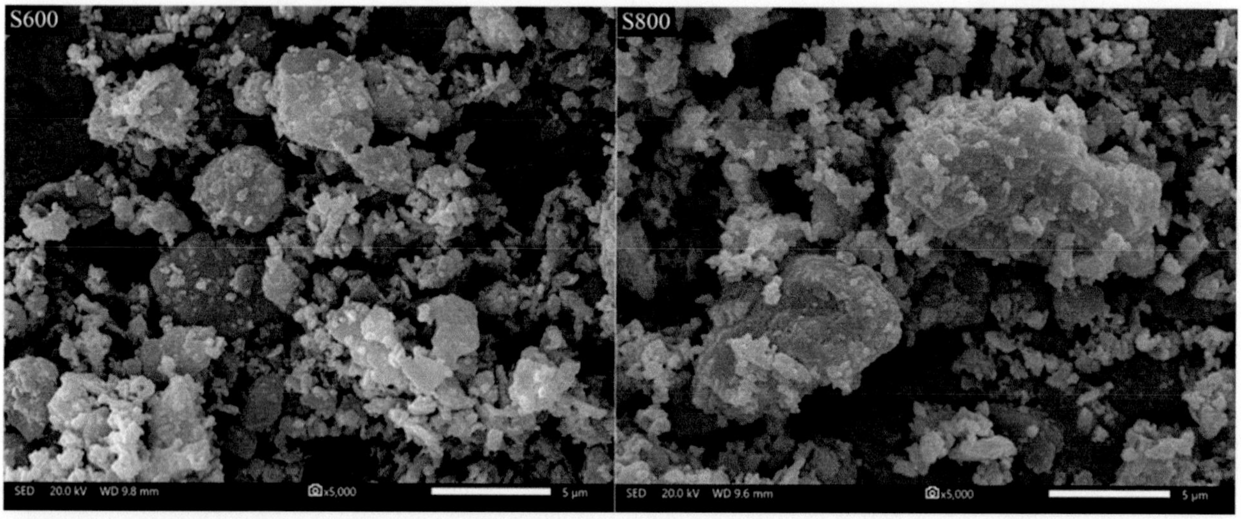

FIGURE 1.14 The SEM images of ISSA.

processing, where the use of ferric chloride or ferrous sulfate in enhancing raw sludge thickening and dewatering properties imparts a reddish hue due to heightened Fe_2O_3 content (Ottosen et al., 2020). Particle fineness and grading in ISSA significantly influence its performance in construction applications. ISSA predominantly comprises silt (2.5–62.5 μm) and fine sand (62.5–250 μm) size fractions, rendering it suitable as a filler or fine aggregate in concrete (Świerczek et al., 2021; Tang et al., 2020b; Zhou et al., 2022c). The pozzolanic reactivity of ISSA aligns closely with particle fineness, with finer ISSA displaying greater reactivity, rendering it favorable for construction applications (Chang et al., 2022b). Mechanical activation proves effective in enhancing the pozzolanic reactivity of ISSA, given the increased reactive surface area (Chang et al., 2020; Zhou et al., 2022b). The Blaine fineness of ISSA approached 1000 m^2/kg after 60 min of grinding, demonstrating minimal increase with longer grinding durations (de Azevedo Basto et al., 2019; Kumar et al., 2021). The incineration process parameters, particularly temperature and duration, influence ISSA characteristics, particularly porosity (Chang et al., 2022a). ISSA exhibits a specific gravity spanning 1.8–3.1, averaging around 2.5 (de Azevedo Basto et al., 2019). The ISSA density lies in 2.5–3.1 g/cm^3 (Kumar et al., 2021; Lynn et al., 2015). Variations in density may be attributed to differences in chemical composition, particularly the presence and influence of denser metals such as Fe, Zn, Cu, Pb, Cr, and Ni (densities ranging from 7.9 to 11.4 g/cm^3) in determining overall material density.

4. Various slags

Slag is a residual product originating from the pyrometallurgical treatment of diverse ores. These slags are typically comprised of metal oxides, silicates, and other impurities that are separated from the metal during the production process. They are categorized into several types, including blast furnace slag, steel slag, and nonferrous slag, based on the specific metal production methods. This section presents an extensive analysis of the attributes of different slags, encompassing their definition, formation, production processes, chemical composition, mineralogy, as well as particle size and morphology.

4.1 Blast furnace slag

Blast furnace slag (BFS) is a by-product generated during pig iron production within the blast furnace, primarily containing silicates and aluminosilicates of calcium and other bases (Oge et al., 2019; Özbay et al., 2016). Fig. 1.15 shows the scheme of the iron-making process and the BFS generation. The formation of BFS occurs when fluxes such as limestone or dolomite react with impurities within iron ore, coke, and scrap metal. This interaction produces molten slag that rests atop molten iron (Oge et al., 2019). Three prominent types of BFS—granulated, air-cooled, and expanded slag—can be manufactured, contingent upon the cooling and solidification methods employed for the molten slag (Tripathy et al., 2020). Granulated blast furnace slag (GBFS) is swiftly quenched using high-pressure water jets, yielding glassy granular particles, usually smaller than 5 mm (Özbay et al., 2016). GBFS is often further processed by drying and grinding to create ground granulated blast furnace slag (GGBS), a finely powdered material utilized as a direct substitute for ordinary cement, with replacement ratios varying between 30% and 85% (Cheah et al., 2021; Zhu et al., 2020). Air-cooled blast furnace slag (ABFS) forms when molten slag is solidified in open conditions, resulting in a crystalline, rock-like structure (Tripathy et al., 2020; Verian and Behnood, 2018). ABFS is a robust, dense material deployed for railroad ballast, roadbed stabilization, concrete aggregate, and other applications requiring a solid base (de Matos et al., 2020; Gan et al., 2012). Expanded slag is produced using mechanical devices and minimal cooling water, yielding a lightweight, dry material. It finds use in brick production, insulation composites, and as an aggregate in lightweight concrete.

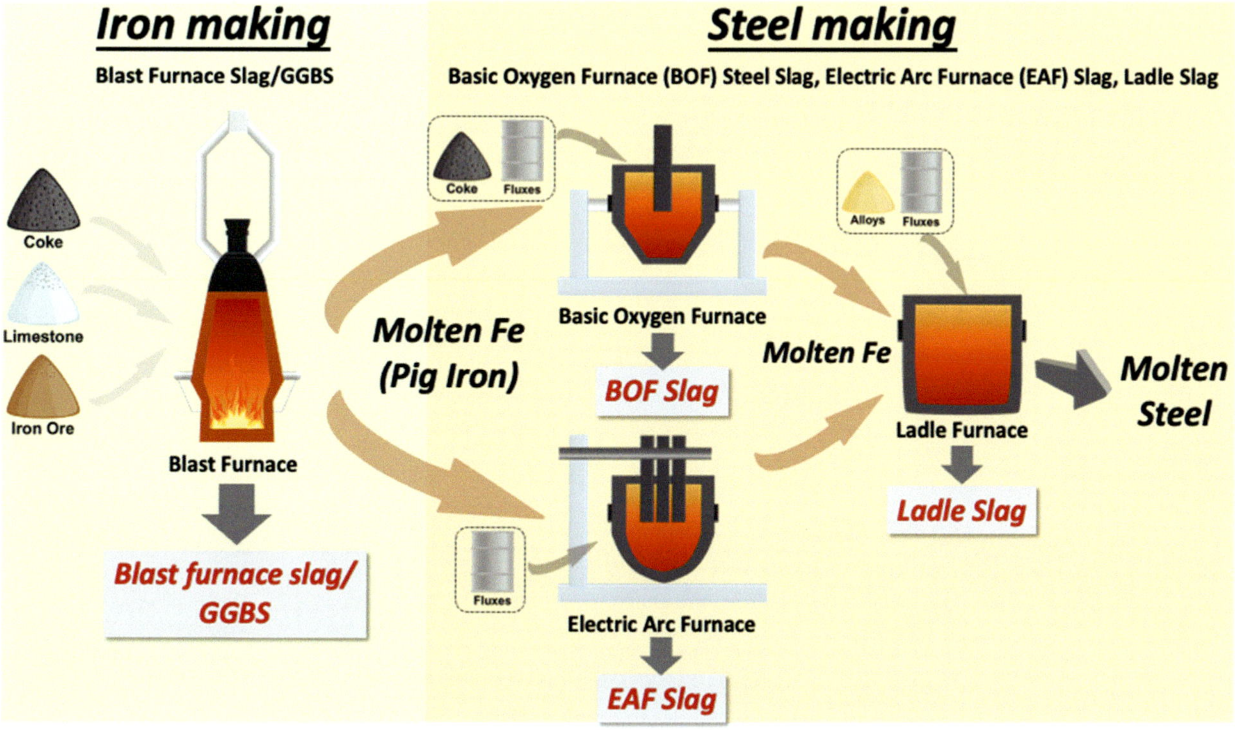

FIGURE 1.15 Scheme of blast furnace slag (GGBS) and steel slags production.

4.1.1 Chemical compositions

The composition of BFS varies based on ores, fluxing agents, and impurities in the coke fed into the blast furnace. Typically, silica, calcium, aluminum, magnesium, and oxygen constitute more than 95% of BFS composition (Haha et al., 2011, 2012; Li et al., 2023c). Table 1.10 shows the chemical composition of GGBS used in the literature. Reactivity of slag hinges on factors such as components composition, chemical composition, geometrical properties, temperature, glass structure, and alkali environment. CaO/SiO_2 ratio is the simplest basicity indices, and it was mentioned that the calcium over siliceous oxide ratio should be greater than 1 for better reactivity (Cheah et al., 2021; Zhu et al., 2020). Water-cooled BFS and air-cooled BFS display differing reactivity levels, with the hydraulic activity of air-cooled BFS notably lower (Mostafa et al., 2001; Oge et al., 2019). In spite of a large number of scientific and practical experiments relating to integrating the chemical and physical characteristics of the GBFS with its hydraulic activity, this issue continues to be the subject of various studies and discussions (Ballari et al., 2023). The influence of magnesia oxide on BFS reactivity is moderated by the basicity of GGBS and the magnesia oxide concentration. Variations in MgO concentration of approximately 8–10 wt.% likely have little effect on strength development (Ahmad et al., 2022; Haha et al., 2011). However, magnesium oxide above 10 wt.% may adversely affect strength development. As indicated in Table 1.10, most researchers observed that MgO concentrations in GGBS are less than 10%. Pozzolanic materials can form when chemicals such as silica, calcium, alumina, magnesia, and iron accumulate to concentrations exceeding 70% (Ganesh and Murthy, 2019; Jamil et al., 2020; Zhu et al., 2022). This renders GGBS a credible pozzolanic material suitable for replacing PC with high CO_2 emissions in concrete, making the low-carbon concrete feasible. The anhydrous GGBSs are mainly composed of glassy phases, which show a broad hump at around $2\theta = 30$ degrees in the XRD pattern (He et al., 2023; Shi et al., 2022; Zhu et al., 2020). In GGBS, a small proportion of crystalline phases, such as akermanite ($Ca_2MgSi_2O_7$), gehlenite ($Ca_2Al_2SiO_7$), merwinite ($Ca_3MgSi_2O_8$), and calcite ($CaCO_3$), were also found (Poornima et al., 2022; Xu et al., 2023a; Yadav et al., 2023). With a high MgO content, akermanite and merwinite may be found in GGBS.

4.1.2 Physical characteristics and morphology

BFS appears gray, while GGBS can be gray or white depending on blast furnace feedstocks and processing methods. GGBS exhibits Blaine fineness ranging from 300 m^2/kg to 550 m^2/kg, with a specific gravity of 2.4–3.0 (Cheah et al.,

TABLE 1.10 Chemical compositions of GGBS, BOF steel slag, EAF slag, and LS (wt.%) used in the literature.

	GGBS	*BOF steel slag*	*EAF slag*	*LS*
CaO	30.2–43.9	37.99–56.4	15.0–50.0	43.3–57.0
MgO	5.5–17.3	3.81–9.3	1.0–21.4	3.2–11.5
SiO_2	19.0–34.6	10.55–18.9	5.038.6	2.3–30.5
Al_2O_3	11.3–16.6	1.4–5.5	5.0–51.0	1.1–30.3
Fe/FeO/ Fe_2O_3	0.3–1.5	8.9–31.54	n.a.–50.0	n.a.–4.3
Na_2O	0.1–0.8	0.11–0.16	n.a.	n.a.
K_2O	0.3–1.4	0.04–0.05	n.a.	n.a.
TiO_2	0.5–0.7	0.88–1.47	n.a.–2.2	n.a.–0.3
SO_3	1.9–2.5	0.1–0.9	n.a.	n.a.–1.7
References	Ahmad et al. (2022), Cheah et al. (2021), Haha et al. (2011, 2012), Jamil et al. (2020), O'Shea et al. (2019), Poornima et al. (2022), Shi et al. (2022), Wan et al. (2004), Xu et al. (2023a), Zhu et al. (2020, 2022)	Alex et al. (2021), Jiang et al. (2018), Kovtun et al. (2021), Martins et al. (2021), Nunes and Borges (2021), O'Connor et al. (2021), Yan et al. (2022)	Adegoloye et al. (2016), Cristelo et al. (2022), Jiang et al. (2018), Meshram et al. (2023), Muhmood et al. (2009), Sukmak et al. (2023), Shi (2004), Yildirim and Prezzi (2022)	Adesanya et al. (2017, 2018), Dash et al. (2022), Gollapalli et al. (2020), Kriskova et al. (2012), Najm et al. (2021), Singh et al. (2021), Wu et al. (2023b)

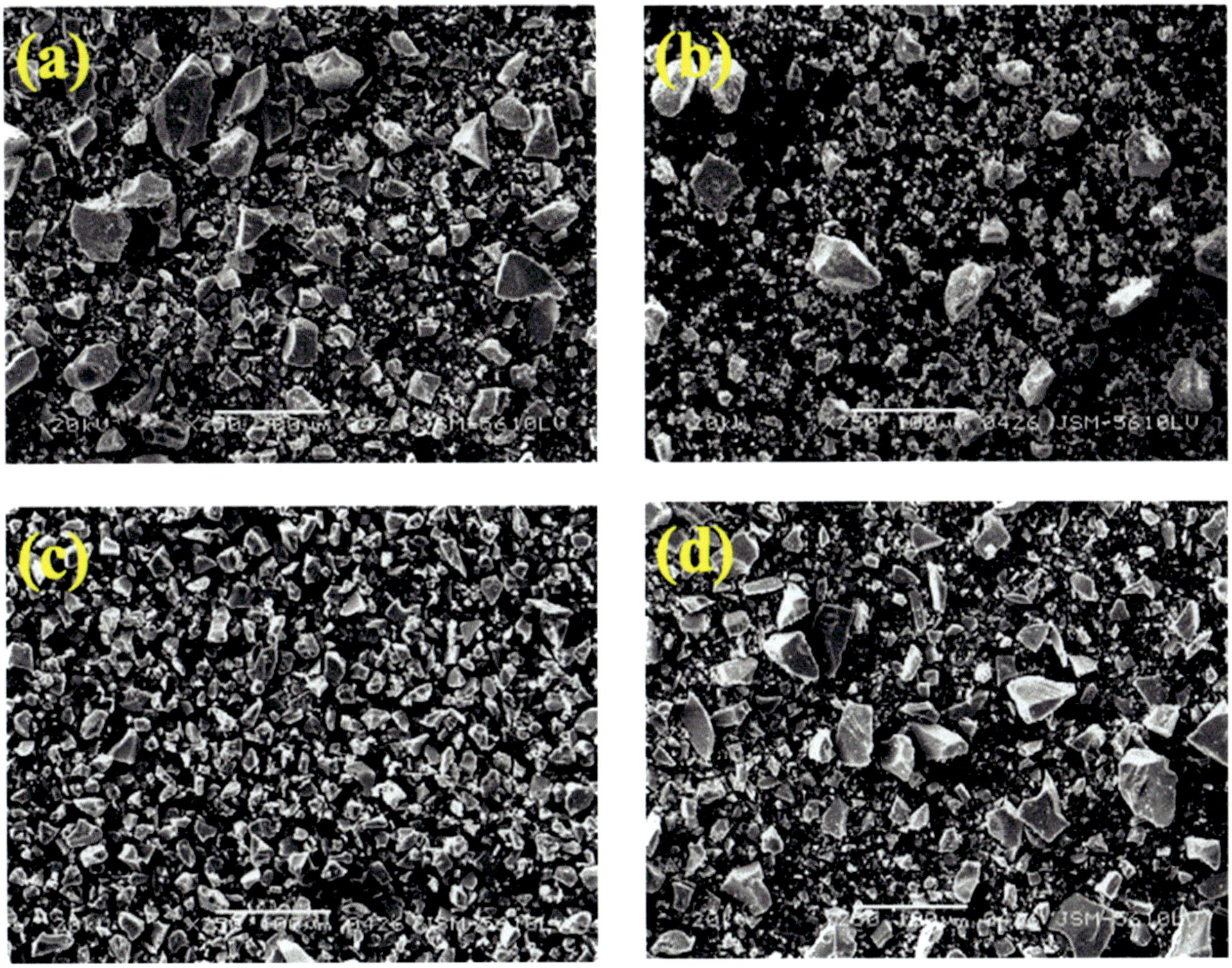

FIGURE 1.16 Morphologies of GGBSs after (A) industrial ball mill; (B) laboratory vibromill; (C) industrial airflow mill; (D) identical ball mill. *Adapted from Wan, H., Shui, Z., Lin, Z., 2004. Analysis of geometric characteristics of GGBS particles and their influences on cement properties. Cem. Concr. Res. 34, 133–137.*

2021). Its bulk density varies between 1200 kg/m^3 and 1670 kg/m^3, roughly equal to that of cement at 1440 kg/m^3 (Ahmad et al., 2022). BFS manifests a broad range of particle sizes, from fine powders to coarse aggregates, with morphologies varying from irregular, angular particles to more rounded shapes (Tripathy et al., 2020; Yang et al., 2022b). The morphology of GGBS, as depicted in Fig. 1.16, varies based on grinding methods, with irregular and angular particles commonly observed (He et al., 2021a). Particle size directly impacts GGBS reactivity and dissolution rate, with fineness enhancement generally leading to increased reactivity (Liu et al., 2019). Particle size and morphology play pivotal roles in the physical, mechanical, and environmental characteristics of slags, thereby influencing their potential applications.

4.2 Steel slags

Steel slags, a by-product of the steel manufacturing process, represent a substantial annual output and can be considered valuable resources. Modern steel production encompasses various categories, including carbon, alloy, stainless, and tool steels. Carbon steel is produced using either a basic oxygen furnace (BOF) or an electric arc furnace (EAF), followed by refining in a ladle furnace (LF) for improved quality (Martins et al., 2021; O'Connor et al., 2021; Song et al., 2021). Fig. 1.15 illustrates the iron manufacturing process and the resultant steel slag generation. In the manufacturing of carbon and stainless steels, a significant quantity of steel slag is produced, constituting around 15%–20% by weight of the total steel output (Gencel et al., 2021; Jiang et al., 2018; Nunes and Borges, 2021). The generated steel slags can be categorized into BOF slag, EAF slag, and LF slag based on their composition and production process.

4.2.1 BOF slag

4.2.1.1 Generation process

In China, BOF steel slag constitutes approximately 70% of the yearly steel slag production (Jiang et al., 2018). In the BOF process (depicted in Fig. 1.15), molten iron from ironmaking, minor steel scrap, and fluxes (such as lime or dolomite) are introduced into the furnace (Alex et al., 2021; Yan et al., 2022). Intense oxidation reactions are initiated at temperatures of 1600−1650°C by injecting 99% pure oxygen at supersonic speeds using a lance (Babenko et al., 2020). Once the desired chemical composition is attained, oxygen supply ceases, and the resulting slag, comprising impurities combined with burned lime or dolomite, emerges as a floating layer atop the molten steel (Alex et al., 2021; Yan et al., 2022). Various cooling methods, including natural air cooling, water spraying, air quenching, and shallow box chilling, are available for steel slag cooling (Naidu et al., 2020). These methods vary in cooling medium and rate, leading to diverse compositions, morphologies, hydration characteristics, and leaching properties of the produced steel slag. Before utilization, steel slags may undergo metal recovery processes (e.g., crushing, screening, magnetic separation) to retrieve valuable components such as iron (de Matos et al., 2020).

4.2.1.2 Physicochemical characteristics

The chemical compositions of BOF slag are highly variable due to the diversity of iron ores, additives, steel-making techniques, and cooling processes. Table 1.10 summarizes the chemical properties of BOF slag as documented in the literature. BOF slag predominantly comprises 40%−60% CaO, 10%−20% SiO_2, 20%−30% Fe_2O_3 (FeO/Fe), 1%−6% Al_2O_3, and 2%−10% MgO. Minor oxides present include MnO, P_2O_5, Na_2O, SO_3, etc. (Alex et al., 2021; Jiang et al., 2018; Martins et al., 2021; Piatak et al., 2015; Yan et al., 2022). The high CaO and MgO contents in BOF slag result from using substantial fluxes to minimize impurities, while the iron oxides originate from unconverted iron during the conversion of molten iron into steel (Jiang et al., 2018b; Kovtun et al., 2021).

BOF slag is generally characterized by high basicity (alkaline oxides to acidic oxides ratio) and comprises various mineral phases, including tricalcium silicate (C3S), dicalcium silicate (C2S), dicalcium ferrite (C2F), MgO, CaO, and a solid solution of CaO, MgO, MnO, and FeO (RO phase) (Kovtun et al., 2021; Martins et al., 2021). The abundance of CaO-based alkaline oxides in BOF steel slag contributes to its potential applications in wastewater treatment, soil pH neutralization, and carbon capture (Nunes and Borges, 2021; O'Connor et al., 2021). The lime content (f-CaO) can reach up to 10%, higher than other steel slags (Ji et al., 2019b; Jiang et al., 2018). Iron phases such as wustite and magnetite, with limited cementitious properties are also present (Balaguera and Botero, 2022; Jawad Ahmed et al., 2023). Additionally, BOF steel slag contains considerable levels of beneficial elements such as Ca, Si, P, and various trace elements that could aid in crop growth, especially after high-temperature calcination (Masindi et al., 2021; O'Connor et al., 2021).

The microstructure of BOF steel slag exhibits rough textures and irregular fine particles attributed to unbalanced thermodynamic conditions during slag formation (Alex et al., 2021; Zepper et al., 2023). The density of basic oxygen furnace (BOF) slag typically ranges between 3.1 and 3.6 g/cm^3, characterized by elevated iron content and notable wear resistance (Martins et al., 2021; Tozsin et al., 2023). These attributes render it a favorable candidate for road construction materials, serving effectively as foundational and backfill components for roadways. The mean specific gravity of this slag variant approximates 3.4, signifying an approximately 30% increase relative to standard aggregates (Cárdenas Balaguera and Gómez Botero, 2020; Martins et al., 2021). Notably, it exhibits a low crushing value, indicative of heightened hardness, and encompasses a markedly porous structure. Upon scrutinizing the microstructure of BOF steel slag, discernible features such as rough textures and irregularly shaped fine particles manifest on both BOF and electric arc furnace (EAF) steel slag surfaces (Jiang et al., 2018).

4.2.2 Electric arc furnace slag

4.2.2.1 Generation processes

EAF slag is produced during steel-making using an EAF, where high-power electric arcs replace gaseous fuels (as depicted in Fig. 1.15). Steel scrap forms the primary feed material in EAF steel-making, along with limited iron scrap, pig iron, and direct reduced iron, reducing the amount of molten iron used (Sukmak et al., 2023). EAF processes generate two types of steel slag: EAF-C slag from carbon steel production and EAF-S slag from stainless steel production (Piatak et al., 2015; Piemonti et al., 2021).

4.2.2.2 Physicochemical characteristics

The EAF process, reliant on steel scrap recycling, yields EAF slag with a wider compositional range compared with BOF slag. As shown in Table 1.10, EAF slag's major constituents include iron oxides (FeO, Fe_2O_3), lime (CaO), silica (SiO_2), magnesia (MgO), and alumina (Al_2O_3), with minor components such as chromium, manganese, and phosphorus oxides, along with small free lime amounts (Motevalizadeh et al., 2020; Zhang et al., 2019b). EAF-C slag shares similarities with BOF slag, including primary oxides, mineral phases, and visual aspects. In contrast, EAF-S slag from stainless steel production boasts a lower FeO content and higher Cr content (Shi, 2004; Yildirim and Prezzi, 2022). Mineral phases identified in EAF slag encompass merwinite ($3CaO \cdot MgO \cdot 2SiO_2$), wustite (solid solution of FeO), olivine, C2S, and C3S (Adegoloye et al., 2016; Cristelo et al., 2022; Muhmood et al., 2009).

EAF slag's mechanical properties, including soundness, abrasion resistance, and stability, are conducive to its use as construction aggregates (Jiang et al., 2018; Meshram et al., 2023; Sukmak et al., 2023). Its bulk density and specific gravity (around 3.5 g/cm^3), often higher than conventional aggregates (2.4–3.0 g/cm^3), contribute to its suitability for road materials, base materials, and backfilling (Cárdenas Balaguera and Gómez Botero, 2020). The water absorption of EAF slag is less than 4%, exhibiting better water resistance than limestone (<10%) while worse than quartzite and granite (<2%) (Jiang et al., 2018). EAF slag's microstructure exhibits rough, porous, and angular particles, making it an ideal candidate for wastewater treatment as an adsorbent.

4.2.3 Ladle slag

4.2.3.1 Generation processes

Ladle slag (LS) emerges from refining procedures carried out in a ladle furnace after initial steelmaking. The ladle furnace (LF) process encompasses deoxidation, alloying, temperature and composition homogenization, desulphurization, enhancement of steel cleanliness, inclusion flotation, and regulation of sulfide and oxide configurations (Najm et al., 2021). The incorporation of fluxes such as calcium aluminate or CaF_2 during LF processing introduces differences in composition and properties compared with BOF and EAF slags (Shi, 2002, 2004; Varanasi et al., 2019).

4.2.3.2 Physicochemical characteristics

Table 1.10 outlines LS's chemical properties, characterized by CaO, SiO_2, MgO, and Al_2O_3. Notably, the CaO content within LS spans 43.3-57.0 wt.% (Adesanya et al., 2017; Wu et al., 2023b). This elevated CaO concentration distinguishes LS from BOF and EAF slags, while iron-bearing components exhibit lower levels. Minor quantities of TiO_2 and Cr_2O_3 can be observed in LS, likely due to the inclusion of alloys to achieve desired compositions (Jiang et al., 2018; Kriskova et al., 2012). The distinctive secondary refining process of LF, along with applied cooling techniques, imparts a high crystalline structure to LS, featuring calcium aluminates (CAs) as the principal mineral phases (Najm et al., 2021; Sáez-De-Guinoa Vilaplana et al., 2015; Shi, 2002). The prevalence of calcium aluminates such as mayenite (C12A7) and tricalcium aluminate (C3A) in LS varies across different plants due to disparities in deoxidation methods and oxide types used (Adesanya et al., 2017, 2018). The deoxidation process during secondary steel production can result in significant quantities of belite (Ca_2SiO_4) or gehlenite ($2CaO \cdot Al_2O_3 \cdot SiO_2$) phases, influenced by silicon or aluminum levels (Isteri et al., 2022; Najm et al., 2021). Cooling methods employed subsequent to slag separation from molten steel, specifically water and air cooling, impact the mineral composition. Water cooling reduces γ-C2S content and prompts the precipitation of secondary mineral phases, while air cooling yields mayenite, larnite, and periclase (Dash et al., 2022; Shi, 2002, 2004).

LS exhibits hydraulic properties, implying a reaction with water to generate cement-like hydration products. While γ-C2S demonstrates limited hydraulic activity under normal hydration conditions, C12A7 displays high hydraulic behavior, reacting rapidly upon contact with water (Dash et al., 2022; Yildirim and Prezzi, 2011). The significant calcium and aluminum content in LS suggests the probable formation of calcium aluminate hydrates as primary phases during hydration. Typically, LS manifests as a white powder, resulting from self-pulverization (or dusting) during the cooling process, where β-C2S transforms into γ-C2S, expanding the volume by approximately 10% (Dash et al., 2022; Shi, 2002, 2004). This self-pulverization of LS introduces potential challenges related to handling and storage. To address this issue, stabilizers such as borates and phosphates can be utilized (Gollapalli et al., 2020; Li and Dai, 2018). Enhancing the fineness of LS has a pronounced impact on its reaction and hardened properties. Preceding research frequently subjects the received fine-grained LS to grinding and milling procedures to decrease nominal particle size and achieve heightened activation levels (Adesanya et al., 2017; Singh et al., 2021).

4.3 Copper slag

Copper slag is a form of solid waste produced during the copper metal production process. Around 2–3 tons of copper slag are generated for every 1 ton of copper metal produced (Klaffenbach et al., 2023; Tian et al., 2021; Xu et al., 2023b). Neglected copper slag can exacerbate PTEs leaching, contaminating nearby soil and groundwater (He et al., 2021b). Therefore, the efficient management and utilization of copper slag demand immediate attention. A comprehensive characterization of copper slag is essential to comprehend its potential hazards.

4.3.1 Physical characteristics

Copper slag usually appears black and glassy, exhibiting increased vesicular properties when granulated. The features of copper slag vary based on the cooling process utilized (Wang et al., 2020c). Two primary slag forms are observed: air-cooled and granulated slag. Slow cooling of molten slag at ambient temperature results in the formation of a coarse-grained crystalline substance known as air-cooled slag (Nazer et al., 2016). Air-cooled copper slag has a glassy appearance, with specific gravity ranging from 2.8 to 3.8, contingent on iron content. Granulated slag is rapidly quenched in water, yielding angular particles with regular shapes due to swift solidification (Kundu et al., 2023; Wang et al., 2021b). Granulated copper slag possesses higher absorption capacity and lower specific gravity compared with air-cooled copper slag (Feng et al., 2019). Both air-cooled and granulated copper slags exhibit favorable mechanical properties suitable for construction aggregates, including excellent soundness, abrasion resistance, and stability (Klaffenbach et al., 2023; Tian et al., 2021; Wang et al., 2021b). Table 1.11 summarizes key physical and mechanical attributes of copper slag documented in the literature.

4.3.2 Chemical characteristics

Copper slag primarily originates from stages of matte smelting and subsequent copper matte blowing. Consequently, the physicochemical composition of copper slag samples varies significantly (Tables 1.12 and 1.13), influenced by the smelting process, smelted ores, fluxes, and additives introduced during pyrometallurgical processes. Predominant elements in the waste slag are Fe and Si. Less prevalent components encompass Cu, Zn, Ca, Mg, Al, and possibly Ni, Co, and Cr. Copper slags may also contain As and Pb. Notably, copper slag features considerable amounts of Cu (0.5 wt.%–2.7 wt.%) and Fe (31.7 wt.%–40.3 wt.%), potentially exceeding or matching content levels found in general Cu and Fe ore mines.

The phase composition of copper slags displays considerable variation due to factors tied to the pyrometallurgical process, such as characteristics of the processed copper mine, types of furnaces used, additives or fluxes employed, and cooling methodologies adopted (cooling rate, final cooling temperature, cooling medium), among others (Phiri et al., 2022; Wang et al., 2021b). Furthermore, weathering can influence phase composition, especially if copper slag

TABLE 1.11 Physical properties of BOF slags and natural aggregates.

	BOF steel slag	Crushed granite	Gravel	Sand
Specific gravity	3.35–3.42	2.69	2.73	2.55–2.72
Water absorption (%)	2.0–3.31	0.5	0.75	0.4–3.99
Crushing value (%)	21	24	–	–
Impact value (%)	16	21	–	–
Los Angeles test (%)	11–18	20	18	–
References	Alex et al. (2021), Jiang et al. (2018), Martins et al. (2021), Masindi et al. (2021), Nunes and Borges (2021), O'Connor et al. (2021), Piatak et al. (2015), Yan et al. (2022)	Palankar et al. (2016)	Pellegrino et al. (2013)	Jiang et al. (2018), Palankar et al. (2016), Pellegrino et al. (2013)

TABLE 1.12 Physical and mechanical properties of copper slag.

Appearance	Black, glassy, more vesicular when granulated
Bulk density	2.3–2.6 g/cm^3
Water absorption	0.13%
Specific gravity	2.8–3.8
Hardness	6–7 Moh
Moisture	<5%
Water soluble chloride	<50 ppm
Abrasion loss	24.1%

Data source: He, R., Zhang, S., Zhang, X., Zhang, Z., Zhao, Y., Ding, H., 2021b. Copper slag: the leaching behavior of heavy metals and its applicability as a supplementary cementitious material. J. Environ. Chem. Eng. 9, 105132; Kundu, T., Senapati, S., Das, S.K., Angadi, S.I., Rath, S.S., 2023. A comprehensive review on the recovery of copper values from copper slag. Powder Technol.. 426, 118693; Marsh, A.T.M., Yang, T., Adu-Amankwah, S., Bernal, S.A., 2021. Utilization of metallurgical wastes as raw materials for manufacturing alkali-activated cements. In: de Brito, J., Thomas, C., Medina, C., Agrela, F. (Eds.), Waste and Byproducts in Cement-Based Materials. Woodhead Publishing, pp. 335–383; Siddique, R., Singh, M., Jain, M., 2020. Recycling copper slag in steel fibre concrete for sustainable construction. J. Clean. Prod. 271, 122559; Tian, H., Guo, Z., Pan, J., Zhu, D., Yang, C., Xue, Y., Li, S., Wang, D., 2021. Comprehensive review on metallurgical recycling and cleaning of copper slag. Resour. Conserv. Recycl. 168, 105366; Wang, R., Shi, Q., Li, Y., Cao, Z., Si, Z., 2021b. A critical review on the use of copper slag (CS) as a substitute constituent in concrete. Constr. Build. Mater. 292, 123371.

TABLE 1.13 Chemical composition of copper slag (wt.%).

SiO_2	CaO	Al_2O_3	MgO	Fe	Cu	Na_2O	K_2O	SO_3
19.0–40.0	1.0–12.0	2.5–11.0	n.a.–6.0	7.0–52.0	1.6–2.5	0–2.5	0–5.0	0–1.4

Data source: He, R., Zhang, S., Zhang, X., Zhang, Z., Zhao, Y., Ding, H., 2021b. Copper slag: the leaching behavior of heavy metals and its applicability as a supplementary cementitious material. J. Environ. Chem. Eng. 9, 105132; Kundu, T., Senapati, S., Das, S.K., Angadi, S.I., Rath, S.S., 2023. A comprehensive review on the recovery of copper values from copper slag. Powder Technol.. 426, 118693; Marsh, A.T.M., Yang, T., Adu-Amankwah, S., Bernal, S.A., 2021. Utilization of metallurgical wastes as raw materials for manufacturing alkali-activated cements. In: de Brito, J., Thomas, C., Medina, C., Agrela, F. (Eds.), Waste and Byproducts in Cement-Based Materials. Woodhead Publishing, pp. 335–383; Siddique, R., Singh, M., Jain, M., 2020. Recycling copper slag in steel fibre concrete for sustainable construction. J. Clean. Prod. 271, 122559; Tian, H., Guo, Z., Pan, J., Zhu, D., Yang, C., Xue, Y., Li, S., Wang, D., 2021. Comprehensive review on metallurgical recycling and cleaning of copper slag. Resour. Conserv. Recycl. 168, 105366; Wang, R., Shi, Q., Li, Y., Cao, Z., Si, Z., 2021b. A critical review on the use of copper slag (CS) as a substitute constituent in concrete. Constr. Build. Mater. 292, 123371.

remains deposited for extended periods in complex environments (Phiri et al., 2022; Piatak et al., 2015). Fayalite (Fe_2SiO_4) constitutes the principal phase, representing the major compounds within copper slag. Olivine-group phases, such as forsterite (Mg_2SiO_4) and kirschsteinite ($CaFeSiO_4$), are also commonly observed (Klaffenbach et al., 2023). Other phases belonging to silicate-group minerals, such as hedenbergite ($CaFe(SiO_3)_2$), diopside ($CaMg(SiO_3)_2$), essonite ($Ca_3Al_2(SiO_4)_3$), and willemite (Zn_2SiO_4) are also prevalent in copper slag (Tian et al., 2021). Moreover, elements such as Cu, Fe, Zn, Pb, As, and Ni in copper slag are often associated with copper sulfides such as chalcocite (Cu_2S), chalcopyrite ($CuFeS_2$), and bornite (Cu_5FeS_4), as well as other sulfide phases such as pyrrhotite ($Fe_{(1-x)}S$), sphalerite ((Zn, Fe)S), galena (PbS), arsenones (As_2S_3), and pentlandite ($(Fe,Ni)_9S_8$) (Ettler et al., 2022; He et al., 2021b; Kundu et al., 2023). These associations are commonly linked to copper mines. Moreover, weathering effects in certain copper slags give rise to various secondary phases, including carbonates such as cerussite ($PbCO_3$), conichalcite ($CaCu(AsO_4)(OH)$), hydrozincite ($Zn_5(CO_3)_2(OH)_2$), hydrocerussite ($Pb_3(CO_3)_2(OH)_2$), and bayldonite ($PbCu_3(AsO_4)_2(OH)_2$) (Men et al., 2023; Tian et al., 2021). The formation of secondary phases is influenced by external factors such as redox properties, pH levels, and the heightened reactivity of specific compounds and ions, accelerating the generation of these secondary phases (Ettler et al., 2023) (Table 1.13).

4.4 Nickel slag

Nickel slag is a type of granulated slag that forms through the natural cooling or water quenching of a melt produced during the smelting process of nickel metal (Li et al., 2021b). In China, approximately 6–16 tons of nickel slag are generated and discharged for every ton of nickel produced (Wu et al., 2018). Nickel slag is highly siliceous and can contain

TABLE 1.14 Chemical composition of nickel slag (wt.%).

SiO_2	CaO	Al_2O_3	MgO	Fe	Ni	Na_2O	K_2O	SO_3
29.0–50.5	1.8–4.0	<0.1	1.5–26.7	20.7–53.1	0.1–0.2	0–0.03	0–0.03	0–0.4

Data source: Choi, Y.C., Choi, S., 2015. Alkali-silica reactivity of cementitious materials using ferro-nickel slag fine aggregates produced in different cooling conditions. Constr. Build. Mater. 99, 279–287; Han, F., Li, Y., Jiao, D., 2023. Understanding the rheology and hydration behavior of cement paste with nickel slag. J. Build. Eng. 73, 106724; Marsh, A.T.M., Yang, T., Adu-Amankwah, S., Bernal, S.A., 2021. Utilization of metallurgical wastes as raw materials for manufacturing alkali-activated cements. In: de Brito, J., Thomas, C., Medina, C., Agrela, F. (Eds.), Waste and Byproducts in Cement-Based Materials. Woodhead Publishing, pp. 335–383; Saha, A.K., Majhi, S., Sarker, P.K., Mukherjee, A., Siddika, A., Aslani, F., Zhuge, Y., 2021. Non-destructive prediction of strength of concrete made by lightweight recycled aggregates and nickel slag. J. Build. Eng. 33, 101614; Wang, Z.J., Ni, W., Jia, Y., Zhu, L.P., Huang, X.Y., 2010. Crystallization behavior of glass ceramics prepared from the mixture of nickel slag, blast furnace slag and quartz sand. J. Non-Cryst. Solids 356, 1554–1558; Wu, Q., Wu, Y., Tong, W., Ma, H., 2018. Utilization of nickel slag as raw material in the production of Portland cement for road construction. Construct. Build. Mater. 193, 426–434; Wu, Q., Chen, Q., Huang, Z., Gu, B., Zhu, H., Tian, L., 2020. Preparation and characterization of porous ceramics from nickel smelting slag and metakaolin. Ceram. Int. 46, 4581–4586.

significant amounts of silica, with levels reaching up to 50%–55% (Wang et al., 2010). It also possesses a notable magnesium content (Li et al., 2022c; Yang et al., 2017). In comparison with copper slag, nickel slag contains lower concentrations of iron, lime, and alumina. The chemical composition of nickel slag is outlined in Table 1.14. Notably, due to its limited lime and alumina content, the hydration of nickel slag does not yield the formation of silicate and hydrated lime alumina, which are the primary elements responsible for the hydraulic properties of ground granulated blast furnace slag (GGBS) (Han et al., 2023; Sun et al., 2023; Wu et al., 2018).

Nickel slag presents a reddish-brown to black color with a massive, amorphous texture. Its unit weight is around 3500 kg/m^3(Li et al., 2022b; Saha et al., 2021). Air-cooled nickel slag has a brownish-black color and tends to crush into angular particles, retaining a smooth and glassy texture (Wang, 2016). The specific gravity of air-cooled nickel slag can be as high as 3.5, while absorption is quite low at approximately 0.37% (Marsh et al., 2021). The unit weight of nickel slag is somewhat greater than that of conventional aggregate. Nickel slag primarily comprises a significant vitreous phase concentrated in smaller particles and a less abundant crystalline phase, which is more prevalent in larger particles (Wang et al., 2022). This distinction is visible to the naked eye, where one can observe torn particles resembling broken glass alongside larger, softer, friable particles displaying honeycomb-like surfaces (Elahi et al., 2020; Wang et al., 2022). Free oxides are absent, and the reduced crystalline phase mainly consists of enstatite and forsterite (Choi and Choi, 2015; Wu et al., 2020). The vitreous nature of nickel slag makes it abrasive and relatively resistant to friability. This vitreous phase can potentially consist of enstatites (chain silicates-$MgSiO_2$), marking a substantial difference from the potential mineralogical composition of conventional iron slag.

5. Conclusion

In the pursuit of a more sustainable industrial landscape, the valorization of combustion residues, waste incineration residues, and diverse slags emerges as a pivotal strategy within the framework of a circular economy. These materials, often seen as liabilities or discarded waste, can transform into valuable resources, contingent on their intricate physical attributes, chemical compositions, mineralogical properties, and responses to varying environmental conditions. A comprehensive evaluation of combustion residues, waste incineration residues, and slags becomes imperative, necessitating a nuanced understanding of their characteristics and potential environmental implications facilitated through meticulous laboratory analyses and real-world observations. This assessment is notably intricate due to the multifaceted nature of these materials, shaped by evolving feedstocks and ore-processing methodologies, thereby demanding a refined approach to their characterization.

The journey forward necessitates dedicated research efforts aimed at bridging the gap between controlled laboratory assessments and real-world behavior, enabling us to reliably extrapolate the potential implications of these materials. The pursuit of methods for extracting residual metals from combustion residues, waste incineration residues, and slags, coupled with their expanded use in raw materials or treatment applications, stands as a promising avenue toward mitigating the depletion of natural resources and generating environmentally conscious by-products. This dual benefit not only curtails the volume of waste requiring disposal but also curbs the reliance on conventional resource extraction through mining, paving the way for a more sustainable future.

Among the diverse avenues of valorization, integration into cement and concrete production emerges as particularly promising. However, it is equally essential to comprehend the long-term weathering behavior of combustion residues, waste incineration residues, and slags, whether they are components of end products or resting in disposal sites. The refinement of our ability to assess the environmental risks and resource potential associated with these materials undoubtedly paves the road toward a reduced ecological footprint, ushering in an era of greater sustainability, resource efficiency, and responsible waste management for the benefit of present and future generations.

References

Aakriti, Maiti, S., Jain, N., Malik, J., 2023. A comprehensive review of flue gas desulphurized gypsum: production, properties, and applications. Construct. Build. Mater. 393, 131918.

Adanez, J., Abad, A., Garcia-Labiano, F., Gayan, P., De Diego, L.F., 2012. Progress in chemical-looping combustion and reforming technologies. Prog. Energy Combust. Sci. 38, 215–282.

Adegoloye, G., Beaucour, A.L., Ortola, S., Noumowe, A., 2016. Mineralogical composition of EAF slag and stabilised AOD slag aggregates and dimensional stability of slag aggregate concretes. Construct. Build. Mater. 115, 171–178.

Adesanya, E., Ohenoja, K., Kinnunen, P., Illikainen, M., 2017. Alkali activation of ladle slag from steel-making process. J. Sustain. Metall. 3, 300–310.

Adesanya, E., Sreenivasan, H., Kantola, A.M., Telkki, V.V., Ohenoja, K., Kinnunen, P., Illikainen, M., 2018. Ladle slag cement – characterization of hydration and conversion. Construct. Build. Mater. 193, 128–134.

Ahmad, J., Kontoleon, K.J., Majdi, A., Naqash, M.T., Deifalla, A.F., Ben Kahla, N., Isleem, H.F., Qaidi, S.M.A., 2022. A comprehensive review on the ground granulated blast furnace slag (GGBS) in concrete production. Sustain. Times.

Akram, W., Sanjay, Hassan, M.A., 2021. Chemical looping combustion with nanosize oxygen carrier: a review. Int. J. Environ. Sci. Technol. 18, 787–798.

Al-Shmaisani, S., Kalina, R.D., Douglas Ferron, R., Juenger, M.C.G., 2022. Assessment of blended coal source fly ashes and blended fly ashes. Construct. Build. Mater. 342, 127918.

Alam, P., Singh, D., Kumar, S., 2021. Incinerated municipal solid waste bottom ash bricks: a sustainable and cost-efficient building material. Mater. Today Proc. 49, 1566–1572.

Alam, Q., Schollbach, K., van Hoek, C., van der Laan, S., de Wolf, T., Brouwers, H.J.H., 2019. In-depth mineralogical quantification of MSWI bottom ash phases and their association with potentially toxic elements. Waste Manag. 87, 1–12.

Alex, T.C., Mucsi, G., Venugopalan, T., Kumar, S., 2021. BOF steel slag: critical assessment and integrated approach for utilization. J. Sustain. Metall. 7, 1407–1424.

Ambikakumari Sanalkumar, K.U., Lahoti, M., Yang, E.H., 2019. Investigating the potential reactivity of fly ash for geopolymerization. Construct. Build. Mater. 225, 283–291. https://doi.org/10.1016/j.conbuildmat.2019.07.140.

Ankur, N., Singh, N., 2021. Performance of cement mortars and concretes containing coal bottom ash: a comprehensive review. Renew. Sustain. Energy Rev. 149, 111361.

Antoun, M., Becquart, F., Gerges, N., Aouad, G., 2020. The use of calcium sulfo-aluminate cement as an alternative to Portland Cement for the recycling of municipal solid waste incineration bottom ash in mortar. Waste Manag. Res. 38, 868–875.

Ashraf, M.S., Ghouleh, Z., Shao, Y., 2019. Production of eco-cement exclusively from municipal solid waste incineration residues. Resour. Conserv. Recycl. 149, 332–342.

ASTM, 2010. ASTM C 618 -01 standard specification for coal fly ash and raw or calcined natural pozzolan for use as a mineral admixture in concrete. Annu. B. ASTM Stand. 4, 3–6.

Awoyemi, O.M., Adeleke, E.O., Dzantor, E.K., 2019. Arbuscular mycorrhizal fungi and exogenous glutathione mitigate coal fly ash (CFA)-induced phytotoxicity in CFA-contaminated soil. J. Environ. Manag. 237, 449–456.

Babenko, A.A., Smirnov, L.A., Protopopov, E.V., Mikhailova, L.Y., 2020. Theoretical basics and technology of smelting steel semiproduct in basic oxygen furnaces and electric arc furnaces under magnesian slags. Steel Transl. 50, 427–433.

Back, S., Sakanakura, H., 2022. Comparison of the efficiency of metal recovery from wet- and dry-discharged municipal solid waste incineration bottom ash by air table sorting and milling. Waste Manag. 154, 113–125.

Bai, M., Zhang, L., Zhao, Y., Sun, S., Du, S., Qiu, P., Zhang, W., Feng, D., 2023. Numerical simulation on the deposition characteristics of MSWI fly ash particles in a cyclone furnace. Waste Manag. 161, 203–212.

Bakalár, T., Pavolová, H., Hajduová, Z., Lacko, R., Kyšeľa, K., 2021. Metal recovery from municipal solid waste incineration fly ash as a tool of circular economy. J. Clean. Prod. 302, 126977.

Bakshi, P., Pappu, A., Kumar Bharti, D., 2022. Transformation of flue gas desulfurization (FGD) gypsum to β-$CaSO4 \cdot 0.5H2O$ whiskers using facile water treatment. Mater. Lett. 308, 131177.

Balaguera, C.A.C., Botero, M.A.G., 2022. Multiphase phosphate cements from steel slags. J. Sustain. Cem. Mater. 11, 21–40.

Ballari, S.O., Raffikbasha, M., Shirgire, A., Thakur, L.S., Thenmozhi, S., Sarath Chandra Kumar, B., 2023. Replacement of coarse aggregates by industrial slag. Mater. Today Proc.

Benassi, L., Zanoletti, A., Depero, L.E., Bontempi, E., 2019. Sewage sludge ash recovery as valuable raw material for chemical stabilization of leachable heavy metals. J. Environ. Manag. 245, 464–470.

Bernasconi, D., Caviglia, C., Destefanis, E., Agostino, A., Boero, R., Marinoni, N., Bonadiman, C., Pavese, A., 2022. Influence of speciation distribution and particle size on heavy metal leaching from MSWI fly ash. Waste Manag. 138, 318–327.

Bhatt, A., Priyadarshini, S., Acharath Mohanakrishnan, A., Abri, A., Sattler, M., Techapaphawit, S., 2019. Physical, chemical, and geotechnical properties of coal fly ash: a global review. Case Stud. Constr. Mater. 11, e00263.

Blissett, R.S., Rowson, N.A., 2012. A review of the multi-component utilisation of coal fly ash. Fuel 97, 1–23.

Bogush, A.A., Stegemann, J.A., Zhou, Q., Wang, Z., Zhang, B., Zhang, T., Zhang, W., Wei, J., 2020. Co-processing of raw and washed air pollution control residues from energy-from-waste facilities in the cement kiln. J. Clean. Prod. 254, 119924.

Brunner, P.H., Rechberger, H., 2015. Waste to energy - key element for sustainable waste management. Waste Manag. 37, 3–12.

Caillahua, M.C., Moura, F.J., 2018. Technical feasibility for use of FGD gypsum as an additive setting time retarder for Portland cement. J. Mater. Res. Technol. 7, 190–197.

Cárdenas Balaguera, C.A., Gómez Botero, M.A., 2020. Characterization of steel slag for the production of chemically bonded phosphate ceramics (CBPC). Construct. Build. Mater. 241, 118138.

Chang, Z., Long, G., Xie, Y., Zhou, J.L., 2022a. Chemical effect of sewage sludge ash on early-age hydration of cement used as supplementary cementitious material. Construct. Build. Mater. 322, 126116.

Chang, Z., Long, G., Xie, Y., Zhou, J.L., 2022b. Pozzolanic reactivity of aluminum-rich sewage sludge ash: influence of calcination process and effect of calcination products on cement hydration. Construct. Build. Mater. 318, 126096.

Chang, Z., Long, G., Zhou, J.L., Ma, C., 2020. Valorization of sewage sludge in the fabrication of construction and building materials: a review. Resour. Conserv. Recycl. 154, 104606.

Cheah, C.B., Tan, L.E., Ramli, M., 2021. Recent advances in slag-based binder and chemical activators derived from industrial by-products – a review. Construct. Build. Mater. 272, 121657.

Chen, B., Perumal, P., Illikainen, M., Ye, G., 2023. A review on the utilization of municipal solid waste incineration (MSWI) bottom ash as a mineral resource for construction materials. J. Build. Eng. 71, 106386.

Chen, L., Liao, Y., Ma, X., Niu, Y., 2020. Effect of co-combusted sludge in waste incinerator on heavy metals chemical speciation and environmental risk of horizontal flue ash. Waste Manag. 102, 645–654.

China NBS, 2022. China Statistical Yearbook of 2022, Chinese Statistical Bureau. China Statistical Press, Beijing.

Choi, Y.C., Choi, S., 2015. Alkali-silica reactivity of cementitious materials using ferro-nickel slag fine aggregates produced in different cooling conditions. Construct. Build. Mater. 99, 279–287.

Chuai, X., Yang, Q., Zhang, T., Zhao, Y., Wang, J., Zhao, G., Cui, X., Zhang, Y., Zhang, T., Xiong, Z., Zhang, J., 2022. Speciation and leaching characteristics of heavy metals from municipal solid waste incineration fly ash. Fuel 328, 125338.

Clavier, K.A., Paris, J.M., Ferraro, C.C., Townsend, T.G., 2020. Opportunities and challenges associated with using municipal waste incineration ash as a raw ingredient in cement production – a review. Resour. Conserv. Recycl. 160, 104888.

Clavier, K.A., Watts, B., Liu, Y., Ferraro, C.C., Townsend, T.G., 2019. Risk and performance assessment of cement made using municipal solid waste incinerator bottom ash as a cement kiln feed. Resour. Conserv. Recycl. 146, 270–279.

Córdoba, P., 2017. Partitioning and speciation of selenium in wet limestone flue gas desulphurisation systems: a review. Fuel 202, 184–195.

Córdoba, P., 2015. Status of Flue Gas Desulphurisation (FGD) systems from coal-fired power plants: overview of the physic-chemical control processes of wet limestone FGDs. Fuel 144, 274–286.

Cristelo, N., Coelho, J., Rivera, J., Garcia-Lodeiro, I., Miranda, T., Fernández-Jiménez, A., 2022. Application of electric arc furnace slag as an alternative precursor to blast furnace slag in alkaline cements. J. Sustain. Cem. Mater. 1–13.

Cristelo, N., Segadães, L., Coelho, J., Chaves, B., Sousa, N.R., de Lurdes Lopes, M., 2020. Recycling municipal solid waste incineration slag and fly ash as precursors in low-range alkaline cements. Waste Manag. 104, 60–73.

Cui, S., Peng, Y., Yang, X., Gao, X., Guan, C.Y., Fan, B., Zhou, X., Chen, Q., 2023. Comprehensive understanding of guest compound intercalated layered double hydroxides: design and applications in removal of potentially toxic elements. Crit. Rev. Environ. Sci. Technol. 53, 457–482.

Dash, A., Chanda, P., Tripathy, P.K., Kumar, N., 2022. A review on stabilization of ladle furnace slag-powdering issue. J. Sustain. Metall. 8, 1435–1449.

de Azevedo Basto, P., Savastano Junior, H., de Melo Neto, A.A., 2019. Characterization and pozzolanic properties of sewage sludge ashes (SSA) by electrical conductivity. Cem. Concr. Compos. 104, 103410.

de Matos, P.R., Oliveira, J.C.P., Medina, T.M., Magalhães, D.C., Gleize, P.J.P., Schankoski, R.A., Pilar, R., 2020. Use of air-cooled blast furnace slag as supplementary cementitious material for self-compacting concrete production. Construct. Build. Mater. 262, 120102.

de Titto, E., Savino, A., 2019. Environmental and health risks related to waste incineration. Waste Manag. Res. 37, 976–986.

del Valle-Zermeño, R., Gómez-Manrique, J., Giro-Paloma, J., Formosa, J., Chimenos, J.M., 2017. Material characterization of the MSWI bottom ash as a function of particle size. Effects of glass recycling over time. Sci. Total Environ. 581–582, 897–905.

Ding, J., Ma, S., Shen, S., Xie, Z., Zheng, S., Zhang, Y., 2017. Research and industrialization progress of recovering alumina from fly ash: a concise review. Waste Manag. 60, 375–387.

Donatello, S., Cheeseman, C.R., 2013. Recycling and recovery routes for incinerated sewage sludge ash (ISSA): a review. Waste Manag.

Dong, J., Tang, Y., Nzihou, A., Chi, Y., 2019. Key factors influencing the environmental performance of pyrolysis, gasification and incineration Waste-to-Energy technologies. Energy Convers. Manag. 196, 497–512.

Dou, X., Ren, F., Nguyen, M.Q., Ahamed, A., Yin, K., Chan, W.P., Chang, V.W.C., 2017. Review of MSWI bottom ash utilization from perspectives of collective characterization, treatment and existing application. Renew. Sustain. Energy Rev. 79, 24–38.

Elahi, M.M.A., Hossain, M.M., Karim, M.R., Zain, M.F.M., Shearer, C., 2020. A review on alkali-activated binders: materials composition and fresh properties of concrete. Construct. Build. Mater. 260, 119788.

Ettler, V., Mihaljevič, M., Culka, A., 2023. Contaminant release from massive copper metallurgical slags: insights from long-term monolithic leaching tests. Chemosphere 335, 139079.

Ettler, V., Mihaljevič, M., Drahota, P., Kříbek, B., Nyambe, I., Vaněk, A., Penížek, V., Sracek, O., Natherová, V., 2022. Cobalt-bearing copper slags from Luanshya (Zambian Copperbelt): mineralogy, geochemistry, and potential recovery of critical metals. J. Geochem. Explor. 237, 106987.

Fan, C., Wang, B., Ai, H., Liu, Z., 2022. A comparative study on characteristics and leaching toxicity of fluidized bed and grate furnace MSWI fly ash. J. Environ. Manag. 305, 114345.

Fan, C., Wang, B., Ai, H., Qi, Y., Liu, Z., 2021. A comparative study on solidification/stabilization characteristics of coal fly ash-based geopolymer and Portland cement on heavy metals in MSWI fly ash. J. Clean. Prod. 319, 128790.

Fang, L., Wang, Q., Li, J.S., Poon, C.S., Cheeseman, C.R., Donatello, S., Tsang, D.C.W., 2021. Feasibility of wet-extraction of phosphorus from incinerated sewage sludge ash (ISSA) for phosphate fertilizer production: a critical review. Crit. Rev. Environ. Sci. Technol. 51, 939–971.

Feng, Y., Yang, Q., Chen, Q., Kero, J., Andersson, A., Ahmed, H., Engström, F., Samuelsson, C., 2019. Characterization and evaluation of the pozzolanic activity of granulated copper slag modified with CaO. J. Clean. Prod. 232, 1112–1120.

Fernández-Jiménez, A., Garcia-Lodeiro, I., Maltseva, O., Palomo, A., 2019. Mechanical-chemical activation of coal fly ashes: an effective way for recycling and make cementitious materials. Front. Mater. 6, 51.

Ferraro, A., Farina, I., Race, M., Colangelo, F., Cioffi, R., Fabbricino, M., 2019. Pre-treatments of MSWI fly-ashes: a comprehensive review to determine optimal conditions for their reuse and/or environmentally sustainable disposal. Rev. Environ. Sci. Bio/Technol. 18, 453–471.

Flesoura, G., Garcia-Banos, B., Catala-Civera, J.M., Vleugels, J., Pontikes, Y., 2019. In-situ measurements of high-temperature dielectric properties of municipal solid waste incinerator bottom ash. Ceram. Int. 45, 18751–18759.

Fu, B., Liu, G., Mian, M.M., Sun, M., Wu, D., 2019. Characteristics and speciation of heavy metals in fly ash and FGD gypsum from Chinese coal-fired power plants. Fuel 251, 593–602.

Funari, V., Braga, R., Bokhari, S.N.H., Dinelli, E., Meisel, T., 2015. Solid residues from Italian municipal solid waste incinerators: a source for "critical" raw materials. Waste Manag. 45, 206–216.

Gan, L., Zhang, C., Shangguan, F., Li, X., 2012. A differential scanning calorimetry method for construction of continuous cooling transformation diagram of blast furnace slag. Metall. Mater. Trans. B Process Metall. Mater. Process. Sci. 43, 460–467.

Ganesh, P., Murthy, A.R., 2019. Tensile behaviour and durability aspects of sustainable ultra-high performance concrete incorporated with GGBS as cementitious material. Construct. Build. Mater. 197, 667–680.

Ge, X., Zhou, M., Wang, H., Liu, Z., Wu, H., Chen, X., 2018. Preparation and characterization of ceramic foams from chromium slag and coal bottom ash. Ceram. Int. 44, 11888–11891.

Gencel, O., Karadag, O., Oren, O.H., Bilir, T., 2021. Steel slag and its applications in cement and concrete technology: a review. Construct. Build. Mater. 283, 122783.

Ghouleh, Z., Shao, Y., 2018. Turning municipal solid waste incineration into a cleaner cement production. J. Clean. Prod. 195, 268–279.

Gollakota, A.R.K., Volli, V., Shu, C.M., 2019. Progressive utilisation prospects of coal fly ash: a review. Sci. Total Environ. 672, 951–989.

Gollapalli, V., Tadivaka, S.R., Borra, C.R., Varanasi, S.S., Karamched, P.S., Venkata Rao, M.B., 2020. Investigation on stabilization of ladle furnace slag with different additives. J. Sustain. Metall. 6, 121–131.

Guo, X., Hu, W., Shi, H., 2014. Microstructure and self-solidification/stabilization (S/S) of heavy metals of nano-modified CFA-MSWIFA composite geopolymers. Construct. Build. Mater. 56, 81–86.

Gooi, S., Mousa, A.A., Kong, D., 2020. A critical review and gap analysis on the use of coal bottom ash as a substitute constituent in concrete. J. Clean. Prod. 268, 121752.

Gopinathan, P., Santosh, M.S., Dileepkumar, V.G., Subramani, T., Reddy, R., Masto, R.E., Maity, S., 2022. Geochemical, mineralogical and toxicological characteristics of coal fly ash and its environmental impacts. Chemosphere 307, 135710.

Grabias-Blicharz, E., Franus, W., 2023. A critical review on mechanochemical processing of fly ash and fly ash-derived materials. Sci. Total Environ. 860, 160529.

Guan, X., Wang, L., Mo, L., 2023. Effects of ground coal bottom ash on the properties of cement-based materials under various curing temperatures. J. Build. Eng. 69, 106196.

Haha, M.B., Lothenbach, B., Le Saout, G., Winnefeld, F., 2012. Influence of slag chemistry on the hydration of alkali-activated blast-furnace slag - Part II: effect of Al2O3. Cement Concr. Res. 42, 74–83.

Haha, M.B., Lothenbach, B., Le Saout, G., Winnefeld, F., 2011. Influence of slag chemistry on the hydration of alkali-activated blast-furnace slag - Part I: effect of MgO. Cement Concr. Res. 41, 955–963.

Hamada, H., Alattar, A., Tayeh, B., Yahaya, F., Adesina, A., 2022. Sustainable application of coal bottom ash as fine aggregates in concrete: a comprehensive review. Case Stud. Constr. Mater. 16, e01109.

Han, F., Li, Y., Jiao, D., 2023. Understanding the rheology and hydration behavior of cement paste with nickel slag. J. Build. Eng. 73, 106724.

Hashemi, S.S.G., Mahmud, H.B., Ghuan, T.C., Chin, A.B., Kuenzel, C., Ranjbar, N., 2019. Safe disposal of coal bottom ash by solidification and stabilization techniques. Construct. Build. Mater. 197, 705–715.

He, M., Xu, Z., Hou, D., Gao, B., Cao, X., Ok, Y.S., Rinklebe, J., Bolan, N.S., Tsang, D.C.W., 2022. Waste-derived biochar for water pollution control and sustainable development. Nat. Rev. Earth Environ. 3, 444–460.

He, P., Drissi, S., Hu, X., Liu, J., Shi, C., 2023. Investigation on the influential mechanism of FA and GGBS on the properties of CO2-cured cement paste. Cem. Concr. Compos. 142, 105186.

He, R., Zhang, S., Zhang, X., Zhang, Z., Zhao, Y., Ding, H., 2021b. Copper slag: the leaching behavior of heavy metals and its applicability as a supplementary cementitious material. J. Environ. Chem. Eng. 9, 105132.

He, T., Li, Z., Zhao, S., Zhao, X., Qu, X., 2021a. Study on the particle morphology, powder characteristics and hydration activity of blast furnace slag prepared by different grinding methods. Construct. Build. Mater. 270, 121445.

Hoyos-Montilla, A.A., Tobón, J.I., Puertas, F., 2023. Role of calcium hydroxide in the alkaline activation of coal fly ash. Cem. Concr. Compos. 137, 104925.

Huang, B., Gan, M., Ji, Z., Fan, X., Zhang, D., Chen, X., Sun, Z., Huang, X., Fan, Y., 2022. Recent progress on the thermal treatment and resource utilization technologies of municipal waste incineration fly ash: a review. Process Saf. Environ. Protect.

Huang, X., Yang, Z., Liu, M., He, J., Li, L., Cui, C., Huang, Z., Wang, S., Yan, D., 2023. Release characteristics of organic pollutants during co-firing coal liquefaction residue in a circulating fluidized bed boiler. Fuel 332, 125958.

Huang, X., Zhao, H., Hu, X., Liu, F., Wang, L., Zhao, X., Gao, P., Ji, P., 2020. Optimization of preparation technology for modified coal fly ash and its adsorption properties for Cd2+. J. Hazard Mater. 392, 122461.

Huber, F., Laner, D., Fellner, J., 2018. Comparative life cycle assessment of MSWI fly ash treatment and disposal. Waste Manag. 73, 392–403.

Hušek, M., Moško, J., Pohořelý, M., 2022. Sewage sludge treatment methods and P-recovery possibilities: current state-of-the-art. J. Environ. Manag. 315, 115090.

Isteri, V., Ohenoja, K., Hanein, T., Kinoshita, H., Kletti, H., Rößler, C., Tanskanen, P., Illikainen, M., Fabritius, T., 2022. Ferritic calcium sulfoaluminate belite cement from metallurgical industry residues and phosphogypsum: clinker production, scale-up, and microstructural characterisation. Cement Concr. Res. 154, 106715.

Jamil, N.H., Abdullah, M.M.A.B., Che Pa, F., Mohamad, H., Ibrahim, W.M.A., Chaiprapa, J., 2020. Influences of SiO2, Al2O3, CaO and MgO in phase transformation of sintered kaolin-ground granulated blast furnace slag geopolymer. J. Mater. Res. Technol. 9, 14922–14932.

Jawad Ahmed, M., Franco Santos, W., Brouwers, H.J.H., 2023. Air granulated basic oxygen furnace (BOF) slag application as a binder: effect on strength, volumetric stability, hydration study, and environmental risk. Construct. Build. Mater. 367, 130342.

Ji, H., Huang, W., Xing, Z., Zuo, J., Wang, Z., Yang, K., 2019a. Experimental study on removing heavy metals from the municipal solid waste incineration fly ash with the modified electrokinetic remediation device. Sci. Rep. 1–9.

Ji, X., Hou, J., Liu, Y., Liu, J., 2019b. Effect of CaO-FeO-MnO system solid solution on the hydration activity of tri-component f-CaO in steel slag. Construct. Build. Mater. 225, 476–484.

Jiang, B., Xie, Y., Xia, D., Liu, X., 2019b. A potential source for PM2.5: analysis of fine particle generation mechanism in Wet Flue Gas Desulfurization System by modeling drying and breakage of slurry droplet. Environ. Pollut. 246, 249–256.

Jiang, X., Li, Y., Yan, J., 2019a. Hazardous waste incineration in a rotary kiln: a review. Waste Dispos. Sustain. Energy 1, 3–37.

Jiang, Y., Ling, T.C., Shi, C., Pan, S.Y., 2018. Characteristics of steel slags and their use in cement and concrete—a review. Resour. Conserv. Recycl. 136, 187–197.

Jin, Y., Feng, W., Zheng, D., Dong, Z., Cui, H., 2020. Structure refinement of fly ash in connection with its reactivity in geopolymerization. Waste Manag. 118, 350–359.

Kasaniya, M., Thomas, M.D.A., Moffatt, E.G., 2021. Efficiency of natural pozzolans, ground glasses and coal bottom ashes in mitigating sulfate attack and alkali-silica reaction. Cement Concr. Res. 149, 106551.

Kaza, S., Yao, L., Bhada-Tata, P., Van Woerden, F., 2018. What a Waste 2.0: A Global Snapshot of Solid Waste Management to 2050, what a Waste 2.0: A Global Snapshot of Solid Waste Management to 2050. World Bank, Washington, DC.

Khaw, L.P.K., Cheah, C.B., Liew, J.J., Siddique, R., Tangchirapat, W., Megat, J.,M.A.B., 2022. Coal bottom ash as constituent binder and aggregate replacement in cementitious and geopolymer composites: a review. J. Build. Eng. 52, 104369.

Kim, H.K., Lee, H.K., 2015. Coal bottom ash in field of civil engineering: a review of advanced applications and environmental considerations. KSCE J. Civ. Eng. 19, 1802–1818.

Klaffenbach, E., Montenegro, V., Guo, M., Blanpain, B., 2023. Sustainable and comprehensive utilization of copper slag: a review and critical analysis. J. Sustain. Metall.

Kohse-Höinghaus, K., 2023. Combustion, chemistry, and carbon neutrality. Chem. Rev. 123, 5139–5219.

Koralegedara, N.H., Al-Abed, S.R., Arambewela, M.K.J., Dionysiou, D.D., 2017. Impact of leaching conditions on constituents release from Flue Gas Desulfurization Gypsum (FGDG) and FGDG-soil mixture. J. Hazard Mater. 324, 83–93.

Koralegedara, N.H., Pinto, P.X., Dionysiou, D.D., Al-Abed, S.R., 2019. Recent advances in flue gas desulfurization gypsum processes and applications – a review. J. Environ. Manag. 251, 109572.

Kovtun, O., Karbayev, M., Korobeinikov, I., Srishilan, C., Shukla, A.K., Volkova, O., 2021. Phosphorus partition between liquid crude steel and high-basicity basic oxygen furnace slags. Steel Res. Int. 92, 2000607.

Krishnamoorthi, M., Malayalamurthi, R., He, Z., Kandasamy, S., 2019. A review on low temperature combustion engines: performance, combustion and emission characteristics. Renew. Sustain. Energy Rev. 116, 109404.

Kriskova, L., Pontikes, Y., Cizer, Ö., Mertens, G., Veulemans, W., Geysen, D., Jones, P.T., Vandewalle, L., Van Balen, K., Blanpain, B., 2012. Effect of mechanical activation on the hydraulic properties of stainless steel slags. Cement Concr. Res. 42, 778–788.

Kumar, M., Shreelaxmi, P., Kamath, M., 2021. Review on characteristics of sewage sludge ash and its partial replacement as binder material in concrete. In: Das, B.B., Nanukuttan, S.V., Patnaik, A.K., Panandikar, N.S. (Eds.), Lecture Notes in Civil Engineering. Springer, Singapore, pp. 65–78.

Kumar, S., Singh, D., 2023. From waste to resource: evaluating the possibility of incinerator bottom ash composites for geotechnical applications. Int. J. Environ. Sci. Technol.

Kun, H., Xiaoping, L., 2019. Inhibiting effects of flue gas desulfurization gypsum on soil phosphorus loss in Chongming Dongtan, southeastern China. Environ. Sci. Pollut. Res. 26, 17195–17203.

Kundu, T., Senapati, S., Das, S.K., Angadi, S.I., Rath, S.S., 2023. A comprehensive review on the recovery of copper values from copper slag. Powder Technol. 426, 118693.

Lane, D.J., Sippula, O., Koponen, H., Heimonen, M., Peraniemi, S., Lahde, A., Kinnunen, N.M., Nivajarvi, T., Shurpali, N.J., Jokiniemi, J., 2020. Volatilisation of major, minor, and trace elements during thermal processing of fly ashes from waste- and wood-fired power plants in oxidising and reducing gas atmospheres. Waste Manag. 102, 698–709.

Le, N.H., Razakamanantsoa, A., Nguyen, M.L., Phan, V.T., Dao, P.L., Nguyen, D.H., 2018. Evaluation of physicochemical and hydromechanical properties of MSWI bottom ash for road construction. Waste Manag. 80, 168–174.

Leckner, B., 2015. Process aspects in combustion and gasification Waste-to-Energy (WtE) units. Waste Manag. 37, 13–25.

Lederer, J., Trinkel, V., Fellner, J., 2017. Wide-scale utilization of MSWI fly ashes in cement production and its impact on average heavy metal contents in cements: the case of Austria. Waste Manag. 60, 247–258.

Li, B., Rong, T., Du, X., Shen, Y., Shen, Y., 2021b. Preparation of Fe3O4 particles with unique structures from nickel slag for enhancing microwave absorption properties. Ceram. Int. 47, 18848–18857.

Li, C., Zhou, C., Li, W., Zhu, W., Shi, J., Liu, G., 2023c. Enrichment of critical elements from coal fly ash by the combination of physical separations. Fuel 336, 127156.

Li, J., Cao, J., Ren, Q., Ding, Y., Zhu, H., Xiong, C., Chen, R., 2021c. Effect of nano-silica and silicone oil paraffin emulsion composite waterproofing agent on the water resistance of flue gas desulfurization gypsum. Construct. Build. Mater. 287, 123055.

Li, J., Ma, Z., Gao, J., Guo, Y., Cheng, F., 2022b. Synthesis and characterization of geopolymer prepared from circulating fluidized bed-derived fly ash. Ceram. Int. 48, 11820–11829.

Li, J., Sun, Z., Wang, L., Yang, X., Zhang, D., Zhang, X., Wang, M., 2022c. Properties and mechanism of high-magnesium nickel slag-fly ash based geopolymer activated by phosphoric acid. Construct. Build. Mater. 345, 128256.

Li, L., Xie, J., Zhang, B., Feng, Y., Yang, J., 2023c. A state-of-the-art review on the setting behaviours of ground granulated blast furnace slag- and metakaolin-based alkali-activated materials. Construct. Build. Mater. 368, 130389. https://doi.org/10.1016/j.conbuildmat.2023.130389.

Li, M.Y., Peng, Z.X., Zhang, B.W., Wang, S.K., Wang, X.S., 2023b. Morphology, mineralogical composition, and heavy metal enrichment characteristics of magnetic fractions in coal fly ash. Environ. Earth Sci. 82, 227. https://doi.org/10.1007/s12665-023-10908-0.

Li, P., Cheng, P., Wang, G., Wang, F., Cheong, K.P., Liu, Z., Mi, J., 2023a. Mild combustion of solid fuels: its definition, establishment, characteristics, and emissions. Energy Fuel. 37, 9998–10022.

Li, Y., Dai, W.B., 2018. Modifying hot slag and converting it into value-added materials: a review. J. Clean. Prod. 175, 176–189.

Li, Y., Wu, B., Wang, R., 2022a. Critical review and gap analysis on the use of high-volume fly ash as a substitute constituent in concrete. Construct. Build. Mater. 341, 127889.

Li, Z., Xu, G., Shi, X., 2021a. Reactivity of coal fly ash used in cementitious binder systems: a state-of-the-art overview. Fuel 301, 121031.

Liang, Y., Xu, D., Feng, P., Hao, B., Guo, Y., Wang, S., 2021. Municipal sewage sludge incineration and its air pollution control. J. Clean. Prod. 295, 126456.

Lin, W.Y., Prabhakar, A.K., Mohan, B.C., Wang, C.H., 2020. A factorial experimental analysis of using wood fly ash as an alkaline activator along with coal fly ash for production of geopolymer-cementitious hybrids. Sci. Total Environ. 718, 135289.

Lindberg, D., Molin, C., Hupa, M., 2015. Thermal treatment of solid residues from WtE units: a review. Waste Manag. 37, 82–94.

Liu, J., Yu, Q., Zuo, Z., Yang, F., Han, Z., Qin, Q., 2019. Reactivity and performance of dry granulation blast furnace slag cement. Cem. Concr. Compos. 95, 19–24.

Liu, P., Wang, Q., Jung, H., Tang, Y., 2020a. Speciation, distribution, and mobility of hazardous trace elements in coal fly ash: insights from Cr, Ni, and Cu. Energy Fuel. 34, 14333–14343.

Liu, S., Liu, W., Jiao, F., Qin, W., Yang, C., 2021. Production and resource utilization of flue gas desulfurized gypsum in China - a review. Environ. Pollut. 288, 117799.

Liu, Z., Hao, Y., Zhang, J., Wu, S., Pan, Y., Zhou, J., Qian, G., 2020b. The characteristics of arsenic in Chinese coal-fired power plant flue gas desulphurisation gypsum. Fuel 271, 117515.

Loginova, E., Schollbach, K., Proskurnin, M., Brouwers, H.J.H., 2021. Municipal solid waste incineration bottom ash fines: transformation into a minor additional constituent for cements. Resour. Conserv. Recycl. 166, 105354.

Loginova, E., Volkov, D.S., van de Wouw, P.M.F., Florea, M.V.A., Brouwers, H.J.H., 2019. Detailed characterization of particle size fractions of municipal solid waste incineration bottom ash. J. Clean. Prod. 207, 866–874.

Luo, H., Cheng, Y., He, D., Yang, E.H., 2019. Review of leaching behavior of municipal solid waste incineration (MSWI) ash. Sci. Total Environ. 668, 90–103.

Luo, Y., Brouwers, H.J.H., Yu, Q., 2023. Understanding the gel compatibility and thermal behavior of alkali activated Class F fly ash/ladle slag: the underlying role of Ca availability. Cement Concr. Res. 170, 107198.

Luo, Y., Wu, Y., Ma, S., Zheng, S., Zhang, Y., Chu, P.K., 2021. Utilization of coal fly ash in China: a mini-review on challenges and future directions. Environ. Sci. Pollut. Res. 28, 18727–18740.

Lynn, C.J., Dhir, R.K., Ghataora, G.S., West, R.P., 2015. Sewage sludge ash characteristics and potential for use in concrete. Construct. Build. Mater. 98, 767–779.

Makgabutlane, B., Nthunya, L.N., Nxumalo, E.N., Musyoka, N.M., Mhlanga, S.D., 2020. Microwave irradiation-assisted synthesis of zeolites from coal fly ash: an optimization study for a sustainable and efficient production process. ACS Omega 5, 25000–25008.

Mancini, G., Luciano, A., Bolzonella, D., Fatone, F., Viotti, P., Fino, D., 2021. A water-waste-energy nexus approach to bridge the sustainability gap in landfill-based waste management regions. Renew. Sustain. Energy Rev. 137, 110441.

Marsh, A.T.M., Yang, T., Adu-Amankwah, S., Bernal, S.A., 2021. Utilization of metallurgical wastes as raw materials for manufacturing alkali-activated cements. In: de Brito, J., Thomas, C., Medina, C., Agrela, F. (Eds.), Waste and Byproducts in Cement-Based Materials. Woodhead Publishing, pp. 335−383.

Martins, A.C.P., Franco de Carvalho, J.M., Costa, L.C.B., Andrade, H.D., de Melo, T.V., Ribeiro, J.C.L., Pedroti, L.G., Peixoto, R.A.F., 2021. Steel slags in cement-based composites: an ultimate review on characterization, applications and performance. Construct. Build. Mater. 291, 123265.

Masindi, V., Ramakokovhu, M.M., Osman, M.S., Tekere, M., 2021. Advanced application of BOF and SAF slags for the treatment of acid mine drainage (AMD): a comparative study. Mater. Today Proc. 38, 934−941.

Mehr, J., Haupt, M., Skutan, S., Morf, L., Raka Adrianto, L., Weibel, G., Hellweg, S., 2021. The environmental performance of enhanced metal recovery from dry municipal solid waste incineration bottom ash. Waste Manag. 119, 330−341.

Men, D., Yao, J., Li, H., Jordan, G., Zhao, B., Cao, Y., Ma, B., Liu, B., Sun, Y., Ban, J., 2023. The potential environmental risk implications of two typical non-ferrous metal smelting slags: contrasting toxic metal(loid)s leaching behavior and geochemical characteristics. J. Soils Sediments 23, 1944−1959.

Meshram, S., Raut, S.P., Ansari, K., Madurwar, M., Daniyal, M., Khan, M.A., Katare, V., Khan, A.H., Khan, N.A., Hasan, M.A., 2023. Waste slags as sustainable construction materials: a compressive review on physico mechanical properties. J. Mater. Res. Technol. 23, 5821−5845.

Mostafa, N.Y., El-Hemaly, S.A.S., Al-Wakeel, E.I., El-Korashy, S.A., Brown, P.W., 2001. Characterization and evaluation of the hydraulic activity of water-cooled slag and air-cooled slag. Cement Concr. Res. 31, 899−904.

Motevalizadeh, S.M., Sedghi, R., Rooholamini, H., 2020. Fracture properties of asphalt mixtures containing electric arc furnace slag at low and intermediate temperatures. Construct. Build. Mater. 240, 117965.

Muhmood, L., Vitta, S., Venkateswaran, D., 2009. Cementitious and pozzolanic behavior of electric arc furnace steel slags. Cement Concr. Res. 39, 102−109.

Muthusamy, K., Rasid, M.H., Jokhio, G.A., Mokhtar Albshir Budiea, A., Hussin, M.W., Mirza, J., 2020. Coal bottom ash as sand replacement in concrete: a review. Construct. Build. Mater. 236, 117507.

Naidu, T.S., Sheridan, C.M., van Dyk, L.D., 2020. Basic oxygen furnace slag: review of current and potential uses. Miner. Eng. 149, 106234.

Najm, O., El-Hassan, H., El-Dieb, A., 2021. Ladle slag characteristics and use in mortar and concrete: a comprehensive review. J. Clean. Prod. 288, 125584.

Nazer, A., Payá, J., Borrachero, M.V., Monzó, J., 2016. Use of ancient copper slags in Portland cement and alkali activated cement matrices. J. Environ. Manag. 167, 115−123.

Novitskaya, E., Kelly, J.P., Bhaduri, S., Graeve, O.A., 2021. A review of solution combustion synthesis: an analysis of parameters controlling powder characteristics. Int. Mater. Rev. 66, 188−214.

Nunes, V.A., Borges, P.H.R., 2021. Recent advances in the reuse of steel slags and future perspectives as binder and aggregate for alkali-activated materials. Construct. Build. Mater. 281, 122605.

O'Connor, J., Nguyen, T.B.T., Honeyands, T., Monaghan, B., O'Dea, D., Rinklebe, J., Vinu, A., Hoang, S.A., Singh, G., Kirkham, M.B., Bolan, N., 2021. Production, characterisation, utilisation, and beneficial soil application of steel slag: a review. J. Hazard Mater. 419, 126478.

O'Shea, M., Gray, K., Murphy, J., 2019. An investigation into the suitability of GGBS and OPC as low percentage single-component binders for the stabilisation and solidification of harbour dredge material mildly contaminated with metals. J. Mar. Sci. Eng. 7.

Oge, M., Ozkan, D., Celik, M.B., Sabri Gok, M., Cahit Karaoglanli, A., 2019. An overview of utilization of blast furnace and steelmaking slag in various applications. Mater. Today Proc. 11, 516−525.

Oruji, S., Brake, N.A., Nalluri, L., Guduru, R.K., 2017. Strength activity and microstructure of blended ultra-fine coal bottom ash-cement mortar. Construct. Build. Mater. 153, 317−326.

Ottosen, L.M., Bertelsen, I.M.G., Jensen, P.E., Kirkelund, G.M., 2020. Sewage sludge ash as resource for phosphorous and material for clay brick manufacturing. Construct. Build. Mater. 249, 118684.

Özbay, E., Erdemir, M., Durmuş, H.I., 2016. Utilization and efficiency of ground granulated blast furnace slag on concrete properties - a review. Construct. Build. Mater. 105, 423−434.

Palankar, N., Ravi Shankar, A.U., Mithun, B.M., 2016. Durability studies on eco-friendly concrete mixes incorporating steel slag as coarse aggregates. J. Clean. Prod. 129, 437−448.

Pan, J., Nie, T., Vaziri Hassas, B., Rezaee, M., Wen, Z., Zhou, C., 2020. Recovery of rare earth elements from coal fly ash by integrated physical separation and acid leaching. Chemosphere 248, 126112.

Panda, L., Dash, S., 2020. Characterization and utilization of coal fly ash: a review. Emerg. Mater. Res. 9, 921−934.

Pellegrino, C., Cavagnis, P., Faleschini, F., Brunelli, K., 2013. Properties of concretes with black/oxidizing electric arc furnace slag aggregate. Cem. Concr. Compos. 37, 232−240.

Phiri, T.C., Singh, P., Nikoloski, A.N., 2022. The potential for copper slag waste as a resource for a circular economy: a review − Part I. Miner. Eng. 180, 107474.

Phua, Z., Giannis, A., Dong, Z.L., Lisak, G., Ng, W.J., 2019. Characteristics of incineration ash for sustainable treatment and reutilization. Environ. Sci. Pollut. Res. 26, 16974−16997.

Piatak, N.M., Parsons, M.B., Seal, R.R., 2015. Characteristics and environmental aspects of slag: a review. Appl. Geochem. 57, 236−266.

Piemonti, A., Conforti, A., Cominoli, L., Sorlini, S., Luciano, A., Plizzari, G., 2021. Use of iron and steel slags in concrete: state of the art and future perspectives. Sustain. Times.

Pienkoß, F., Abis, M., Bruno, M., Grönholm, R., Hoppe, M., Kuchta, K., Fiore, S., Simon, F.G., 2022. Heavy metal recovery from the fine fraction of solid waste incineration bottom ash by wet density separation. J. Mater. Cycles Waste Manag. 24, 364–377.

Poornima, N., Sivasakthi, M., Jeyalakshmi, J., 2022. Microstructure investigation of the Na/Ca aluminosilicate hydrate gels and its thermal compatibility in fly ash–GGBS cementitious binder. J. Build. Eng. 50, 104168.

Quina, M.J., Bontempi, E., Bogush, A., Schlumberger, S., Weibel, G., Braga, R., Funari, V., Hyks, J., Rasmussen, E., Lederer, J., 2018. Technologies for the management of MSW incineration ashes from gas cleaning: new perspectives on recovery of secondary raw materials and circular economy. Sci. Total Environ. 635, 526–542.

Rafieizonooz, M., Khankhaje, E., Rezania, S., 2022. Assessment of environmental and chemical properties of coal ashes including fly ash and bottom ash, and coal ash concrete. J. Build. Eng. 49, 104040.

Rafieizonooz, M., Mirza, J., Salim, M.R., Hussin, M.W., Khankhaje, E., 2016. Investigation of coal bottom ash and fly ash in concrete as replacement for sand and cement. Construct. Build. Mater. 116, 15–24.

Rathee, M., Singh, N., 2022. Durability properties of copper slag and coal bottom ash based I-shaped geopolymer paver blocks. Construct. Build. Mater. 347, 128461.

Rawat, K., Yadav, A.K., 2019. Characterization of coal and fly ash (generated) at coal based thermal power plant. Mater. Today Proc. 26, 1406–1411.

Reijnders, L., 2005. Disposal, uses and treatments of combustion ashes: a review. Resour. Conserv. Recycl. 43, 313–336.

Rosita, W., Bendiyasa, I.M., Perdana, I., Anggara, F., 2020. Sequential particle-size and magnetic separation for enrichment of rare-earth elements and yttrium in Indonesia coal fly ash. J. Environ. Chem. Eng. 8, 103575.

Sáez-De-Guinoa Vilaplana, A., Ferreira, V.J., López-Sabirón, A.M., Aranda-Usón, A., Lausín-González, C., Berganza-Conde, C., Ferreira, G., 2015. Utilization of Ladle Furnace slag from a steelwork for laboratory scale production of Portland cement. Construct. Build. Mater. 94, 837–843.

Saffarzadeh, A., Shimaoka, T., Motomura, Y., Watanabe, K., 2009. Characterization study of heavy metal-bearing phases in MSW slag. J. Hazard Mater. 164, 829–834.

Saha, A.K., Majhi, S., Sarker, P.K., Mukherjee, A., Siddika, A., Aslani, F., Zhuge, Y., 2021. Non-destructive prediction of strength of concrete made by lightweight recycled aggregates and nickel slag. J. Build. Eng. 33, 101614.

Saikia, B.K., Hower, J.C., Islam, N., Sharma, A., Das, P., 2021. Geochemistry and petrology of coal and coal fly ash from a thermal power plant in India. Fuel 291, 120122.

Sapkota, B., Pariatamby, A., 2023. Pharmaceutical waste management system – are the current techniques sustainable, eco-friendly and circular? A review. Waste Manag. 168, 83–97.

Saqib, N., Bäckström, M., 2016. Chemical association and mobility of trace elements in 13 different fuel incineration fly ashes. Fuel 165, 193–204.

Sarmiento, L.M., Clavier, K.A., Paris, J.M., Ferraro, C.C., Townsend, T.G., 2019. Critical examination of recycled municipal solid waste incineration ash as a mineral source for portland cement manufacture – a case study. Resour. Conserv. Recycl. 148, 1–10.

Schnell, M., Horst, T., Quicker, P., 2020. Thermal treatment of sewage sludge in Germany: a review. J. Environ. Manag. 263, 110367.

Shi, C., 2004. Steel slag—its production, processing, characteristics, and cementitious properties. J. Mater. Civ. Eng. 16, 230–236.

Shi, C., 2002. Characteristics and cementitious properties of ladle slag fines from steel production. Cement Concr. Res. 32, 459–462.

Shi, X., Zhang, C., Wang, X., Zhang, T., Wang, Q., 2022. Response surface methodology for multi-objective optimization of fly ash-GGBS based geopolymer mortar. Construct. Build. Mater. 315, 125644.

Siddique, R., Singh, M., Jain, M., 2020. Recycling copper slag in steel fibre concrete for sustainable construction. J. Clean. Prod. 271, 122559.

Sigvardsen, N.M., Ottosen, L.M., 2019. Characterization of coal bio ash from wood pellets and low-alkali coal fly ash and use as partial cement replacement in mortar. Cem. Concr. Compos. 95, 25–32.

Silva, R.V., de Brito, J., Lynn, C.J., Dhir, R.K., 2019. Environmental impacts of the use of bottom ashes from municipal solid waste incineration: a review. Resour. Conserv. Recycl. 140, 23–35.

Singh, D., Kumar, A., 2020. Factors affecting properties of MSWI bottom ash employing cement and fiber for geotechnical applications. Environ. Dev. Sustain. 22, 6891–6905.

Singh, M., Siddique, R., 2013. Effect of coal bottom ash as partial replacement of sand on properties of concrete. Resour. Conserv. Recycl. 72, 20–32.

Singh, M., Siddique, R., Ait-Mokhtar, K., Belarbi, R., 2016. Durability properties of concrete made with high volumes of low-calcium coal bottom ash as a replacement of two types of sand. J. Mater. Civ. Eng. 28, 4015175.

Singh, N., Mithulraj, M., Arya, S., 2019. Utilization of coal bottom ash in recycled concrete aggregates based self compacting concrete blended with metakaolin. Resour. Conserv. Recycl. 144, 240–251.

Singh, N., Shehnazdeep, Bhardwaj, A., 2020. Reviewing the role of coal bottom ash as an alternative of cement. Construct. Build. Mater. 233, 117276.

Singh, S.K., Jyoti, Vashistha, P., 2021. Development of newer composite cement through mechano-chemical activation of steel slag. Construct. Build. Mater. 268, 121147.

Song, Q., Guo, M.Z., Wang, L., Ling, T.C., 2021. Use of steel slag as sustainable construction materials: a review of accelerated carbonation treatment. Resour. Conserv. Recycl. 173, 105740.

Sow, M., Hot, J., Tribout, C., Cyr, M., 2015. Characterization of spreader stoker coal fly ashes (SSCFA) for their use in cement-based applications. Fuel 162, 224–233.

Strandberg, A., Skoglund, N., Thyrel, M., 2021. Morphological characterisation of ash particles from co-combustion of sewage sludge and wheat straw with X-ray microtomography. Waste Manag. 135, 30–39.

Sukmak, P., Sukmak, G., De Silva, P., Horpibulsuk, S., Kassawat, S., Suddeepong, A., 2023. The potential of industrial waste: electric arc furnace slag (EAF) as recycled road construction materials. Construct. Build. Mater. 368, 130393.

Sun, X., Yi, Y., 2020. pH evolution during water washing of incineration bottom ash and its effect on removal of heavy metals. Waste Manag. 104, 213–219.

Sun, Z., Li, J., Yang, X., Wang, M., 2023. Investigation on high-magnesium nickel slag treated by phase-separated activation as cementitious material. J. Build. Eng. 69, 106265.

Sustainable Org, 2019. Waste to Energy: Incineration, Gasification and Pyrolysis.

Świerczek, L., Cieślik, B.M., Konieczka, P., 2021. Challenges and opportunities related to the use of sewage sludge ash in cement-based building materials – a review. J. Clean. Prod. 287, 125054.

Šyc, M., Simon, F.G., Hykš, J., Braga, R., Biganzoli, L., Costa, G., Funari, V., Grosso, M., 2020. Metal recovery from incineration bottom ash: state-of-the-art and recent developments. J. Hazard Mater. 393.

Tamanna, K., Raman, S.N., Jamil, M., Hamid, R., 2023. Coal bottom ash as supplementary material for sustainable construction: a comprehensive review. Construct. Build. Mater. 389, 131679.

Tang, P., Florea, M.V.A., Spiesz, P., Brouwers, H.J.H., 2015. Characteristics and application potential of municipal solid waste incineration (MSWI) bottom ashes from two waste-to-energy plants. Construct. Build. Mater. 83, 77–94.

Tang, P., Xuan, D., Li, J., Cheng, H.W., Poon, C.S., Tsang, D.C.W., 2020b. Investigation of cold bonded lightweight aggregates produced with incineration sewage sludge ash (ISSA) and cementitious waste. J. Clean. Prod. 251, 119709.

Tang, P., Chen, W., Xuan, D., Cheng, H., Poon, C.S., Tsang, D.C.W., 2020a. Immobilization of hazardous municipal solid waste incineration fly ash by novel alternative binders derived from cementitious waste. J. Hazard Mater. 393, 122386.

Tennakoon, C., Sagoe-Crentsil, K., San Nicolas, R., Sanjayan, J.G., 2015. Characteristics of Australian brown coal fly ash blended geopolymers. Construct. Build. Mater. 101, 396–409.

Tian, H., Guo, Z., Pan, J., Zhu, D., Yang, C., Xue, Y., Li, S., Wang, D., 2021. Comprehensive review on metallurgical recycling and cleaning of copper slag. Resour. Conserv. Recycl. 168, 105366.

Tillman, D.A., Duong, D., Miller, B., 2009. Chlorine in solid fuels fired in pulverized fuel boilerss-sources, forms, reactions, and consequences: a literature review. Energy Fuel. 23, 3379–3391.

Tozsin, G., Yonar, F., Yucel, O., Dikbas, A., 2023. Utilization possibilities of steel slag as backfill material in coastal structures. Sci. Rep. 13, 4318.

Tripathy, S.K., Dasu, J., Murthy, Y.R., Kapure, G., Pal, A.R., Filippov, L.O., 2020. Utilisation perspective on water quenched and air-cooled blast furnace slags. J. Clean. Prod. 262, 121354.

van de Wouw, P.M.F., Loginova, E., Florea, M.V.A., Brouwers, H.J.H., 2020. Compositional modelling and crushing behaviour of MSWI bottom ash material classes. Waste Manag. 101, 268–282.

Varanasi, S.S., More, V.M.R., Rao, M.B.V., Alli, S.R., Tangudu, A.K., Santanu, D., 2019. Recycling ladle furnace slag as flux in steelmaking: a review. J. Sustain. Metall. 5, 449–462.

Vavva, C., Voutsas, E., Magoulas, K., 2017. Process development for chemical stabilization of fly ash from municipal solid waste incineration. Chem. Eng. Res. Des. 125, 57–71.

Verbinnen, B., Billen, P., Van Caneghem, J., Vandecasteele, C., 2017. Recycling of MSWI bottom ash: a review of chemical barriers, engineering applications and treatment technologies. Waste and Biomass Valorization 8, 1453–1466.

Verian, K.P., Behnood, A., 2018. Effects of deicers on the performance of concrete pavements containing air-cooled blast furnace slag and supplementary cementitious materials. Cem. Concr. Compos. 90, 27–41.

Verrecchia, G., Cafiero, L., de Caprariis, B., Dell'Era, A., Pettiti, I., Tuffi, R., Scarsella, M., 2020. Study of the parameters of zeolites synthesis from coal fly ash in order to optimize their CO2 adsorption. Fuel 276, 118041.

Wan, H., Shui, Z., Lin, Z., 2004. Analysis of geometric characteristics of GGBS particles and their influences on cement properties. Cement Concr. Res. 34, 133–137.

Wang, D., Wang, Q., Huang, Z., 2020c. Reuse of copper slag as a supplementary cementitious material: reactivity and safety. Resour. Conserv. Recycl. 162, 105037.

Wang, G.C., 2016. Usability criteria for slag use in rigid matrices. In: Wang, G.C. (Ed.), The Utilization of Slag in Civil Infrastructure Construction. Woodhead Publishing, pp. 275–304.

Wang, J., Yang, P., 2018. Potential flue gas desulfurization gypsum utilization in agriculture: a comprehensive review. Renew. Sustain. Energy Rev. 82, 1969–1978.

Wang, M., Liu, Y., Feng, C., Zhang, D., Zhang, X., Jiao, G., 2022. Pozzolanic activity enhancement of magnesium-rich nickel slag and geopolymer preparation. J. Mater. Cycles Waste Manag. 24, 2598–2607.

Wang, N., Sun, X., Zhao, Q., Yang, Y., Wang, P., 2020b. Leachability and adverse effects of coal fly ash: a review. J. Hazard Mater. 396.

Wang, P., Hu, Y., Cheng, H., 2019. Municipal solid waste (MSW) incineration fly ash as an important source of heavy metal pollution in China. Environ. Pollut. 252, 461–475.

Wang, C., Xu, G., Gu, X., Gao, Y., Zhao, P., 2021a. High value-added applications of coal fly ash in the form of porous materials: a review. Ceram. Int. 47, 22302–22315.

Wang, R., Shi, Q., Li, Y., Cao, Z., Si, Z., 2021b. A critical review on the use of copper slag (CS) as a substitute constituent in concrete. Construct. Build. Mater. 292, 123371.

Wang, Y., Liu, C., Tan, Y., Wang, Y., Li, Q., 2020a. Chloride binding capacity of green concrete mixed with fly ash or coal gangue in the marine environment. Construct. Build. Mater. 242, 118006.

Wang, Z.J., Ni, W., Jia, Y., Zhu, L.P., Huang, X.Y., 2010. Crystallization behavior of glass ceramics prepared from the mixture of nickel slag, blast furnace slag and quartz sand. J. Non-Cryst. Solids 356, 1554−1558.

Wei, G., Dong, B., Hong, S., Fang, G., Xing, F., Wang, Y., 2023. Deep insight into reactive chemical structures of Class F fly ash. Cem. Concr. Compos. 142, 105195.

Wei, Y., Saffarzadeh, A., Shimaoka, T., Zhao, C., Peng, X., Gao, J., 2014. Geoenvironmental weathering/deterioration of landfilled MSWI-BA glass. J. Hazard Mater. 278, 610−619.

Wu, C., He, J., Wang, K., Yang, L., Wang, F., 2023a. Enhance the mechanical and water resistance performance of flue gas desulfurization gypsum by quaternary phase. Construct. Build. Mater. 387, 131565.

Wu, L., Li, H., Mei, H., Rao, L., Wang, H., Lv, N., 2023b. Generation, utilization, and environmental impact of ladle furnace slag: a minor review. Sci. Total Environ. 895, 165070.

Wu, Q., Chen, Q., Huang, Z., Gu, B., Zhu, H., Tian, L., 2020. Preparation and characterization of porous ceramics from nickel smelting slag and metakaolin. Ceram. Int. 46, 4581−4586.

Wu, Q., Ma, H., Chen, Q., Gu, B., Li, S., Zhu, H., 2019. Effect of silane modified styrene-acrylic emulsion on the waterproof properties of flue gas desulfurization gypsum. Construct. Build. Mater. 197, 506−512.

Wu, Q., Wu, Y., Tong, W., Ma, H., 2018. Utilization of nickel slag as raw material in the production of Portland cement for road construction. Construct. Build. Mater. 193, 426−434.

Xing, G., Wang, W., Zhao, S., Qi, L., 2023. Application of Ca-based adsorbents in fixed-bed dry flue gas desulfurization (FGD): a critical review. Environ. Sci. Pollut. Res. 30, 76471−76490.

Xiong, X., Zhang, Y., Wang, L., Tsang, D.C.W., 2022. In: Tsang, D.C.W., Wang, L. (Eds.), Chapter 1 - Overview of Hazardous Waste Treatment and Stabilization/Solidification Technology. Elsevier, pp. 1−14.

Xu, L., Liu, Y., Chen, M., Wang, N., Chen, H., Liu, L., 2023b. Production of green, low-cost and high-performance anorthite-based ceramics from reduced copper slag. Construct. Build. Mater. 375, 130982.

Xu, L., Wang, J., Li, K., Hao, T., Li, Z., Li, L., Ran, B., Du, H., 2023a. New insights on dehydration at elevated temperature and rehydration of GGBS blended cement. Cem. Concr. Compos. 139, 105068.

Xuan, D., Tang, P., Poon, C.S., 2018. Limitations and quality upgrading techniques for utilization of MSW incineration bottom ash in engineering applications − a review. Construct. Build. Mater. 190, 1091−1102.

Xue, Y., Liu, X., 2021. Detoxification, solidification and recycling of municipal solid waste incineration fly ash: a review. Chem. Eng. J. 420, 130349.

Yadav, P., Raju, M.K., Samudrala, R.K., Gangadhar, M., Pani, J., Borkar, H., Azeem, P.A., 2023. Cost-effective akermanite derived from industrial waste for working electrodes in supercapacitor applications. New J. Chem. 47, 3255−3265.

Yan, K., Sun, H., Gao, F., Ge, D., You, L., 2020. Assessment and mechanism analysis of municipal solid waste incineration bottom ash as aggregate in cement stabilized macadam. J. Clean. Prod. 244, 118750.

Yan, Y., Li, Q., Sun, X., Ren, Z., He, F., Wang, Y., Wang, L., 2015. Recycling flue gas desulphurization (FGD) gypsum for removal of Pb(II) and Cd(II) from wastewater. J. Colloid Interface Sci. 457, 86−95.

Yan, Z., Li, Z., Manocha, S., Ponchon, F., Nafornita, S., 2022. Value in use of lime in BOF steelmaking process. Ironmak. Steelmak. 49, 42−48.

Yang, L., Jing, M., Lu, L., Zhu, X., Zhao, P., Chen, M., Li, L., Liu, J., 2020. Effects of modified materials prepared from wastes on the performance of flue gas desulfurization gypsum-based composite wall materials. Construct. Build. Mater. 257, 119519.

Yang, T., Wu, Q., Zhu, H., Zhang, Z., 2017. Geopolymer with improved thermal stability by incorporating high-magnesium nickel slag. Construct. Build. Mater. 155, 475−484.

Yang, J., Zhang, Q., He, X., Su, Y., Zeng, J., Xiong, L., Zeng, L., Yu, X., Tan, H., 2022a. Low-carbon wet-ground fly ash geopolymer activated by single calcium carbide slag. Construct. Build. Mater. 353, 129084.

Yang, X., Zhou, H., Li, Q., Tan, Z., Zhang, Y., 2022b. Characterization of blast furnace slag particles generated by nitrogen jet granulation. Can. J. Chem. Eng. 100, 3600−3607.

Yang, Z., Ji, R., Liu, L., Wang, X., Zhang, Z., 2018. Recycling of municipal solid waste incineration by-product for cement composites preparation. Construct. Build. Mater. 162, 794−801.

Yao, Z.T., Ji, X.S., Sarker, P.K., Tang, J.H., Ge, L.Q., Xia, M.S., Xi, Y.Q., 2015. A comprehensive review on the applications of coal fly ash. Earth-Science Rev.

Yildirim, I.Z., Prezzi, M., 2022. Subgrade stabilisation mixtures with EAF steel slag: an experimental study followed by field implementation. Int. J. Pavement Eng. 23, 1754−1767.

Yildirim, I.Z., Prezzi, M., 2011. Chemical, mineralogical, and morphological properties of steel slag. Adv. Civ. Eng. 2011.

Yin, K., Ahamed, A., Lisak, G., 2018. Environmental perspectives of recycling various combustion ashes in cement production − a review. Waste Manag. 78, 401−416.

Yoon, J., Jafari, K., Tokpatayeva, R., Peethamparan, S., Olek, J., Rajabipour, F., 2022. Characterization and quantification of the pozzolanic reactivity of natural and non-conventional pozzolans. Cem. Concr. Compos. 133, 104708.

Yuan, Q., Zhang, Y., Wang, T., Wang, J., Romero, C.E., 2021. Mechanochemical stabilization of heavy metals in fly ash from coal-fired power plants via dry milling and wet milling. Waste Manag. 135, 428−436.

Yue, Y., Zhang, J., Sun, F., Wu, S., Pan, Y., Zhou, J., Qian, G., 2019. Heavy metal leaching and distribution in glass products from the co-melting treatment of electroplating sludge and MSWI fly ash. J. Environ. Manag. 232, 226−235.

Zepper, J.C.O., van der Laan, S.R., Schollbach, K., Brouwers, H.J.H., 2023. Reactivity of BOF slag under autoclaving conditions. Construct. Build. Mater. 364, 129957.

Zhai, J., Burke, I.T., Stewart, D.I., 2021. Beneficial management of biomass combustion ashes. Renew. Sustain. Energy Rev. 151, 111555.

Zhang, B., Poon, C.S., 2015. Use of Furnace Bottom Ash for producing lightweight aggregate concrete with thermal insulation properties. J. Clean. Prod. 99, 94–100.

Zhang, J.B., Li, S.P., Li, H.Q., He, M.M., 2016. Acid activation for pre-desilicated high-alumina fly ash. Fuel Process. Technol. 151, 64–71.

Zhang, L., Xu, C., Champagne, P., 2010. Overview of recent advances in thermo-chemical conversion of biomass. Energy Convers. Manag. 51, 969–982.

Zhang, L., Zhang, W., Li, M., Li, P., Zheng, X., Chang, C., Zou, W., 2022. Coal fly ash reinforcement for the property enhancement of crude glycerol-based polyurethane foam composites. Waste Dispos. Sustain. Energy 4, 271–282.

Zhang, N., Wu, L., Liu, X., Zhang, Y., 2019b. Structural characteristics and cementitious behavior of basic oxygen furnace slag mud and electric arc furnace slag. Construct. Build. Mater. 219, 11–18.

Zhang, P., Liao, W., Kumar, A., Zhang, Q., Ma, H., 2020. Characterization of sugarcane bagasse ash as a potential supplementary cementitious material: comparison with coal combustion fly ash. J. Clean. Prod. 277, 123834.

Zhang, Y., Wang, L., Chen, L., Ma, B., Zhang, Y., Ni, W., Tsang, D.C.W., 2021. Treatment of municipal solid waste incineration fly ash: state-of-the-art technologies and future perspectives. J. Hazard Mater. 411, 125132.

Zhang, Z., Wang, J., Liu, L., Shen, B., 2019a. Preparation and characterization of glass-ceramics via co-sintering of coal fly ash and oil shale ash-derived amorphous slag. Ceram. Int. 45, 20058–20065.

Zhao, H., Huang, X., Zhang, G., Li, J., He, Z., Ji, P., Zhao, J., 2020. Possibility of removing cadmium pollution from the environment using a newly synthesized material coal fly ash. Environ. Sci. Pollut. Res. 27, 4997–5008.

Zhou, H., Bhattarai, R., Li, Y., Si, B., Dong, X., Wang, T., Yao, Z., 2022a. Towards sustainable coal industry: turning coal bottom ash into wealth. Sci. Total Environ. 804, 149985.

Zhou, X., Chen, Y., Liu, C., Wu, F., 2022c. Preparation of artificial lightweight aggregate using alkali-activated incinerator bottom ash from urban sewage sludge. Construct. Build. Mater. 341, 127844.

Zhou, Y., Cai, G., Cheeseman, C., Li, J., Poon, C.S., 2022b. Sewage sludge ash-incorporated stabilisation/solidification for recycling and remediation of marine sediments. J. Environ. Manag. 301, 113877.

Zhou, Y.F., Li, J.S., Lu, J.X., Cheeseman, C., Poon, C.S., 2020. Sewage sludge ash: a comparative evaluation with fly ash for potential use as lime-pozzolan binders. Construct. Build. Mater. 242, 118160.

Zhu, H., Chen, W., Cheng, S., Yang, L., Wang, S., Xiong, J., 2022. Low carbon and high efficiency limestone-calcined clay as supplementary cementitious materials (SCMs): multi-indicator comparison with conventional SCMs. Construct. Build. Mater. 341, 127748.

Zhu, J., Wei, Z., Luo, Z., Yu, L., Yin, K., 2021. Phase changes during various treatment processes for incineration bottom ash from municipal solid wastes: a review in the application-environment nexus. Environ. Pollut. 287, 117618.

Zhu, X., Zhang, M., Yang, K., Yu, L., Yang, C., 2020. Setting behaviours and early-age microstructures of alkali-activated ground granulated blast furnace slag (GGBS) from different regions in China. Cem. Concr. Compos. 114, 103782.

Chapter 2

Regulations and policies for combustion/incineration residues treatment and utilization

Ruolin Zhao[1,2], Xinyu Jiang[3], Jining Li[2] and Lei Wang[1]

[1]*State Key Laboratory of Clean Energy Utilization, Zhejiang University, Hangzhou, China;* [2]*Institute of Eco-environmental and Soil Sciences, Guangdong Academy of Sciences, GuangZhou, China;* [3]*Department of Civil and Environmental Engineering, The Hong Kong Polytechnic University, Hong Kong, China*

1. Introduction

Incineration, as an integral part of solid waste treatment, has been increasingly popular due to the urbanization, industrialization, and consequently the limited landfill storage (Kallio et al., 2023). The attendant problem of substantial combustion/incineration residues management has aroused great concern. The residues generated by coal combustion and solid waste incineration are normally include fly ash (e.g., coal fly ash [CFA], municipal solid waste incineration fly ash [MSWI FA], and incineration sewage sludge ash [ISSA]), bottom ash (e.g., municipal solid waste incineration bottom ash), and furnace slags (e.g., ground granulated blast furnace slag [GGBS] and steel furnace slag [SFS]) (Zhang et al., 2021). According to statistics, there are 6.0−8.0 billion tons of fly ash and 1.5 to 2.0 billion tons of bottom ash generated globally (Wang et al., 2020). Among the total ash produced, CFA accounts for 65%−90% of the total volume, with a production of over 1 billion metric tons annually. With the dramatically high production of residues worldwide every year, the gap between the generation and utilization is fairly high.

Furthermore, potential toxic elements such as lead (Pb), cadmium (Cd), nickel (Ni), etc. and other contaminants such as dioxins, polycyclic aromatic hydrocarbons (PAHs), sulfate, chloride, etc. are commonly found in most combustion/incineration residues (Jayaranjan et al., 2014), thus causing significant environmental concern. According to the concentrations of contaminants and toxicity detection specifications, combustion/incineration residues could be classified into general waste and hazardous waste, of which the latter should be strictly treated before disposal (Zhang et al., 2021).

With regard to the utilization, some combustion/incineration residues, such as CFA and furnace slags, have become valuable building-material products, which are commonly used in mortar and concrete. There are standards and regulations regarding their applications; thus those residues have to object to corresponding requirements. However, substantial part of residues is not utilized widely, and corresponding regulations are in lack as well. As for different countries and regions, there are regulations regarding utilization and management for the residues, which are in huge differences on the strength and legal foundation. Not only economic and technology development but also resources and national circumstances will influence the establishment of national laws and regulations. In general, developed countries, with better economic and technical conditions, have set up more comprehensive and stricter standards for waste utilization compared with those of developing countries.

Under sustainable development strategy, the management of combustion/incineration residues has gained critical concern. Consequently, improving the utilization rate of residues, and exploring environmentally friendly, cost-effective treatment methods are desperately in need. Accordingly, relevant standards/regulations on utilization play an essential role as to facilitate and regulate the reuse and recycle. Taking into account the differences in the utilization and management of combustion/incineration residues in various countries and regions, there is a huge difference on the local test methods, standard system, standard limits, etc. On the whole, there are still some common deficiencies of development in

Treatment and Utilization of Combustion and Incineration Residues. https://doi.org/10.1016/B978-0-443-21536-0.00001-0

standard systems from all over the world. Therefore, it is of great significance to summarize and compare different standards and regulations, which can give inspirations and references to countries and regions.

In this chapter, relevant regulations and standards mainly focus on seven types of combustion/incineration ashes, including CFA, MSWI FA, ISSA, coal bottom ash (CBA), municipal solid waste incineration bottom ash (MSWI BA), ground granulated blast furnace slag (GGBS), and steel furnace slag (SFS). The comparisons are basically among the United States (US), the European Union (EU), China, and Japan. With a comprehensive view of regulations regarding the utilization of combustion/incineration residues outside of landfills, this chapter summarizes the requirements that have to be met in different aspects and discusses the similarities and differences of those requirements among countries. By comparison, it is possible to provide guidance and suggestions for improving the standard system and facilitating the utilization of combustion/incineration residues.

2. Fly ashes

2.1 Coal fly ash

CFA has obvious resource attributes and can be widely used in various products and materials. The comprehensive utilization level of CFA in various countries and regions is affected by local economic conditions, energy policies, solid waste emissions, environmental awareness, and other factors, and there is a certain gap between the comprehensive utilization rate and utilization technology.

The classification standards of CFA are distinct among countries and regions. For China and the United States, the classification is mainly based on the content of calcium oxide (CaO). To be specific, CFA is classified into three classes to meet the applicable requirements in the United States, including Class N, Class F, and Class C. And in China, the classification includes Class F and Class C. In addition to pozzolanic properties, Class C fly ash, with relatively high calcium, has considerable cementitious properties, in comparison with Class F fly ash. In Europe, the classification criteria are primarily based on loss on ignition (LOI) and fineness, with five classes: Type A (LOI $\leq 5\%$), Type B (LOI $\leq 7\%$), Type C (LOI $\leq 9\%$), Type N (amount retained on 45 μm sieve $\leq 40\%$), and Type S (amount retained on 45 μm sieve $\leq 12\%$). The distinct classification criteria and various limited value influence the final utilization.

Regarding the utilization, CFA is basically reused in civil engineering, including cement and concrete, road construction, etc. CFA can also be used in other fields, such as high value−added product (e.g., glass-ceramics production), ecological restoration (e.g., soil stabilization, wastewater treatment), agriculture (e.g., fertilizer), etc. Presently, however, relevant standards are mainly focused on utilization in cement and concrete, and those on other aspects need to be issued and regulated for facilitating further use. As for CFA use in cement and concrete, Table 2.1 and Table 2.2 summarize the chemical−physical requirements among countries. Specifically, the American Society for Testing and Materials (ASTM) specifies the standard for CFA used in concrete in ASTM C618 (ASTM, 2022); the European Committee for Standardization (CEN) issues European Norm (EN) (EN 450) for using CFA as partial replacement of cement in concrete in Europe; in China, the national standard of fly ash used for cement and concrete (GB/T 1596, 2017) specifies specific requirements of CFA; in Japan, the Japanese Industrial Standards Committee (JIS) specifies standards of CFA for use in concrete (JIS A6201, 2015).

Taking account of the characteristics of CFA and the domestic situation, the emphasis on the selected factors is different among countries and regions. To be specific, the monitoring of environmental indicators is stricter in Europe, which consequently specifies more composition indicators of CFA. Additionally, the use of fly ash produced from cocombustion is allowed, which may cause differences from conventional CFA. As such, some indicators (e.g., phosphate (P_2O_5) and free CaO), which are known to cause significant set retardation and expansion, respectively, in supplementary cementitious materials (SCMs), are limited particularly (Suraneni et al., 2021). With the consideration of these, the strength activity index (SAI) is limited to $\geq 75\%$ at 28 days, and to 85% at 91 days (EN 450-1, 2012). Other items such as soluble P_2O_5, initial setting time, reactive silicon dioxide (SiO_2), reactive and free CaO, and magnesium oxide content (MgO) are also specified. As for China, it is worth mentioning that the newest standard, which is still on trial specifies values, varied in use for whether mix mortar and concrete or cement active mix material. Due to the special location in the seismic zone, Japan focuses more on the strength parameters of CFA. Items for chemical compositions such as sulfur trioxide (SO_3) are not limited instead.

On the field of ecological restoration, most countries haven't issued relevant standards or limit the utilization of CFA, which is related to national legislation to some extent. Take the European Union as an example firstly, the legislation for application of fertilizers, liming agents, soil improvers, etc. has come into force (EU 1009, 2019). This should/will include biomass ashes for application in top soils. The base of EU legislation should be either protection of top soil quality or allow

TABLE 2.1 Chemical requirements of CFA for use in concrete.

Item	US	EU	China: For mix mortar and concrete	China: For cement active mix material	Japan
SiO_2+Al_2O_3+Fe_2O_3, min, %	70 50 50	70	70 50	70 50	–
Free CaO, max, %	–	1.5	1.0 4.0	1.0 4.0	–
SO_3, max, %	4.0 5.0 5.0	3.0	3.0	3.5	–
Moisture content, max, %	3.0 3.0 3.0	–	–	1.0	1.0
LOI, max, %	10.0 6.0 6.0	5 7 9	5.0 8.0 10.0	8.0	3.0 5.0 8.0 5.0
Reactive CaO, max, %	–	10	–	–	–
Chloride, max, %	–	0.1	–	–	–
MgO, max, %	–	4.0	–	–	–
P_2O_5, max, %	–	5.0	–	–	

TABLE 2.2 Physical requirements of CFA for use in concrete.

Item		US	EU	China: For mix mortar and concrete	China: For cement active mix material	Japan
Fineness: amount retained when wet-sieved on 45 μm sieve, max, %		34 34 34	40 12	12 30 45	–	10 40 70
Strength activity index: With Portland cement, min, percent of control	7 days	75	–	–	–	–
	28 days	75	75	70	70	90 80 60
	91 days	–	85	–	–	100 90 70
Water requirement, max, percent of control		115 105 105	95	95 105 115	–	105 95 85 75
Density, max, g/cm^3		–	–	2.6	2.6	1.95
Soundness, max, mm		–	10	5	5	–

for the recycling of biomass ash from pure biomass back to top soils. In general, biomass ashes will not meet the limits of the new EU Fertilizer Regulation 2019. In China, there was a national standard of CFA for use in agriculture (GB 8173, 1987), which has been abolished yet.

2.2 Municipal solid waste incineration fly ash

MSWI FA is classified as a hazardous waste on account of its substantial leachable heavy metals, salts, chlorides, and toxic organic pollutants such as dioxins and furans (Liu et al., 2022). For heavy metals, chromium (Cr), copper (Cu), hydrargyrum (Hg), Ni, Cd, zinc (Zn), and Pb are the most commonly found in MSWI ash, among which Zn and Pb usually cover the largest amounts. Additionally, the dioxin values in fly ash are higher than 1 ng I-toxic equivalent quantity (TEQ)/g in most countries (Bie et al., 2007), while the international emission standard limit for dioxin concentration in flue gas is 0.1 ng I-TEQ/m^3 (Lam et al., 2010). Moreover, the high concentration of chlorides would cause the corrosion in production equipment, and the deterioration in the performance of MSWI FA cementitious materials. With regard to these, it is common to conduct pretreatments to reduce the environmental impact before utilizing, which includes processes such as decreasing the total concentration of the contaminants by washing, and reducing the leachability/leaching rate of contaminants by stabilization/solidification (Lam et al., 2010). In practice, strategies may vary based on different applications.

As for MSWI FA utilization, using MSWI FA to produce cementitious materials is one of the most widely applications (Zhang et al., 2021). Table 2.3 summarizes the regulations for products with MSWI FA. In particular, China issued the latest regulations for MSWI FA disposal (HJ 1134, 2020), in which the treatment products should meet the accordingly pollution control requirements for use in cement clinker production or other ways. Specifically, the concentration of heavy metals in MSWI FA for cement clinker production into kiln should meet the requirements conforming to GB/T 30760 (GB/T 30760, 2014). Moreover, MSWI FA is allowed for further applications if it meets the requirements of toxic equivalents of dioxins (<50 ng TEQ/kg), leachable heavy metals content (GB 8978), and soluble chlorine content. It is noteworthy that MSWI FA is not allowed to be used in the brick production in China. In Europe, the European Union regulates the leaching concentration of heavy metals and the biological toxicity for MSWI FA production. In the United States, the additive amount of MSWI FA in the cement production is regulated by American Concrete Institute (ACI) and ASTM as well. Overall, it pays large attention to the concentrations of soluble heavy metals, alkalis, and chlorides in most countries, while the dioxin content is only regulated in the latest standard of China.

2.3 Incineration sewage sludge ash

The safe treatment of sewage sludge has become an increasingly significant issue because of its nonuniform composition. Sludge contains a wide range of household and industrial pollutants, including heavy metals, organic pollutants and nutrients, pathogens, or microplastics (Hušek et al., 2022). Some conventional methods such as agriculture use, composting, or landfilling are being given up or under strict limitation due to environmental concerns and the strategy of circular economy. Incineration, as one of the most common treatment processes for handling sewage sludge, is increasingly popular due to its excellent effect on large volume reduction, odor minimization, and thermal destruction of toxic organic compounds (Zhou et al., 2022). As a result, incinerated sewage sludge ash (ISSA) is generated instead.

TABLE 2.3 Regulations regarding MSWI FA for cement production.

Region	Requirements	Regulations
China	Leaching of heavy metals The toxic equivalence of dioxins (<50 ng TEQ/kg) for utilization in other aspects excluding cement preparation	HJ 1134 HJ 662 (HJ 662, 2013) GB/T 30760 GB/T 30810 (GB/T 30810, 2014)
US	Alkali Chloride Mix up to 3.7% or 0.3% of MSWI FA	ACI318-05/318R-05 (ACI, 2014) ASTM, C150/C150M (ASTM, 2015a) ASTM, C114 (ASTM, 2015b)
EU	The biological toxicity of products Leaching of heavy metals.	EN 12457-2 EN ISO 11348-2

According to statistics and literature, the annual sludge production of 2019 was 39.04 million tons in China, and 26.7% of which was incinerated (Wei et al., 2020). The United States produces ISSA about 0.5−1.0 million tons annually (Zhou et al., 2022). It was supposed that about 265 kilo tons of ISSA was generated annually in the European countries. Supposedly, the output of ISSA is increasing over the next few years. Consequently, large quantities of ISSA are still disposed in landfill without further resourceful utilization in most countries and regions.

The various components of sewage sludge cause the complicated composition in ISSA, which makes the treatment and utilization complicated. To be specific, it is common to find substances include SiO_2, CaO, Al_2O_3, Fe_2O_3, MgO, and P_2O_5 in ISSA, but the precise composition may vary greatly with the source of the wastewater, and the additives introduced into the sludge conditioning operation (Smol et al., 2015). In addition, ISSA may contain leachable heavy metals, and toxic organic substances such as dioxins, furans, and polycyclic aromatic hydrocarbons; thus, it requires pretreatment before utilization.

To achieve the sustainable management of waste materials, appropriate recycling and reusing methods of ISSA have been developed. For its similar particle size as sand, ISSA has been used as a direct replacement for sand or clay (Zhou et al., 2022). Additionally, because of its pozzolanic properties and high Si-content, ISSA can be used as a partial replacement for cement in concrete applications as an alternative to SCMs. Specifically, the major elements in Portland cement are Ca, Si, Al, and Fe, which compare reasonably well to the major elements in ISSA, with the notable exception of P (Shane and Christopher, 2013). As such, ISSA has been employed for cement and concrete, and for producing some high value−added products such as bricks and glass ceramics (Lynn et al., 2016). Fig. 2.1 summarizes utilization of ISSA in concrete-related applications.

Apart from used in concrete-related applications, it was proven that ISSA worked successfully for the removal of Pb, Cr, Cu, and Zn as a heavy metal's adsorbent from polluted water and soil (Wang et al., 2019a, 2019b). Additionally, phosphate recovery from ISSA is also an attractive alternative as fertilizers (Fang et al., 2018). However, the absence of defined regulations or standards related to other application fields, such as resource recovery, and manufactured products, limits the development of treatment and utilization to a great extent.

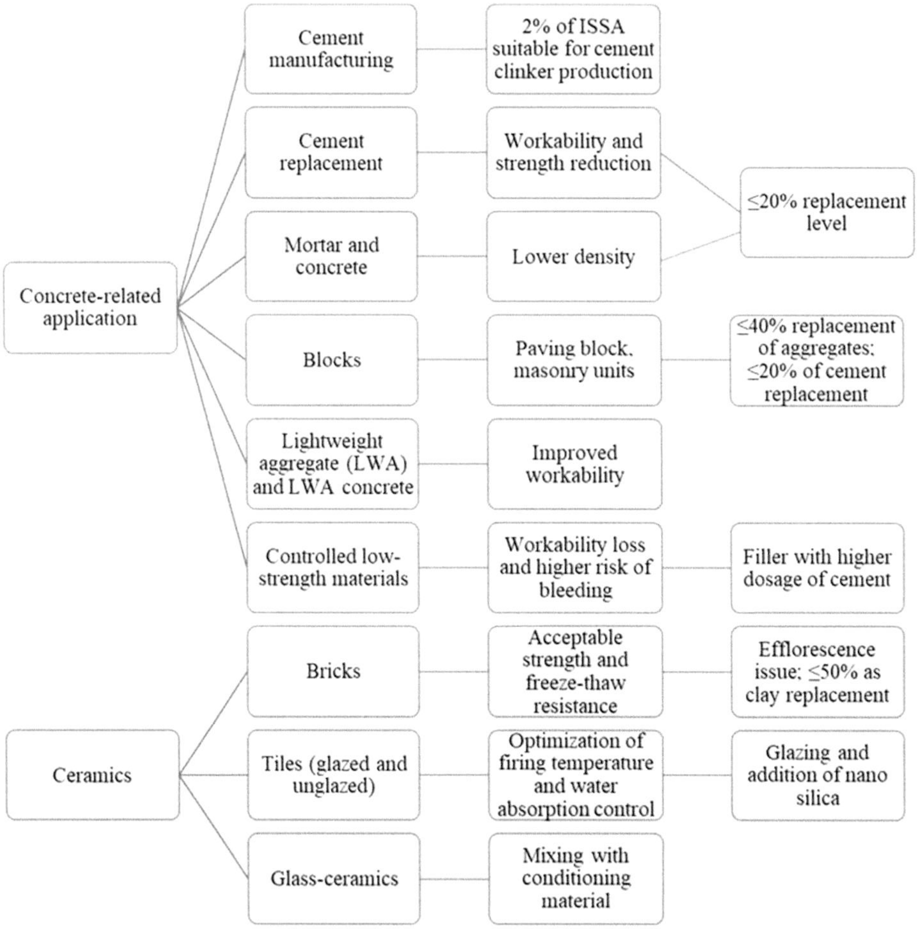

FIGURE 2.1 Illustrative diagram of the use of incinerated sewage sludge ash in concrete-related applications (Zhou et al., 2022).

TABLE 2.4 Requirements for application of ISSA in China (draft for comment).

Items		Limit values	
		Used for ready-mixed mortar	Used for ready-mixed mortar
Chemical requirements	SiO_2, min, %	–	40.0
	SO_3, max, %	3.0	3.0
	Moisture content, max, %	1.0	–
	LOI, max, %	10.0	10.0
	Chloride, max, %	–	0.06
	P_2O_5, max, %	10	–
Physical requirements	Fineness, max, %	30 (amount retained when wet-sieved on 45 μm sieve)	25 (amount retained when wet-sieved on 80 μm sieve)
	Strength activity index at 28 days, min, %	60	–
	Water requirement, max, percent of control	115	–
	Density, max, g/cm^3	2.6	–

Nowadays, most regulations of ISSA recycling focus on cement construction, and most countries have not even issued specific standards for ISSA utilization. As for Europe, the use of ISSA is related to the restrictions contained in the applicable regulations implementing Directive EU/2010/75 (Rutkowska, 2023). Up to now, no guidelines or standards have been issued regarding the application of ISSA in the cement and concrete production. In China, China Association for Engineering Construction Standardization (CECS) has completed the draft of "Technical specification for application of sewage sludge ash." The draft specifies the requirement of ISSA, which is used either for ready-mixed mortar or autoclaved concrete as shown in Table 2.4. The detection of heavy metal toxicity of ISSA should comply with the current national standard (GB 5085.3, 2007).

3. Bottom ashes and furnace slags

3.1 Municipal solid waste incineration bottom ash

Bottom ashes are mainly composed of the inorganic components of the solid fuel remain in the combustor (Li and Poon, 2017), including coal bottom ashes (CBA) and municipal solid waste incineration bottom ashes (MSWI BA). CBA, classified as nonhazardous, solid or inert wastes in many countries, is commonly used in construction despite the lack of legal certainty. As to municipal solid wastes, bottom ash covers the most of residues (about 80%–90% by volume and 30% by mass) generated from the process of incineration (Yin et al., 2018). Generally, MSWI BA has a more modest environmental impact than MSWI FA basically due to its lower heavy metal content, and much lower chloride content. The utilization of MSWI BA for alternative cement material was a prospective method, because of the similar component between them (Li et al., 2016). Currently, MSWI BA has been in practical applications including road materials, bricks, and cement concrete aggregates (Kumar et al., 2023).

As for normative use, the EU has the most comprehensive regulations and standards in the utilization of MSWI BA, the application of which is especially extensive in European countries, e.g., the United Kingdom (UK) (up to 45% for blended cement), France (up to 50% for aggregates in concrete) and Portugal (up to 50% for cement clinker) (Lam et al., 2010; Yin et al., 2018). In Europe, there is no harmonized test method and related limit values at EU level for the utilization of MIBA as secondary raw material in construction. The majority of the countries have developed their own rules to regulate the utilization of MSWI BA outside of landfills, which are on the basis of legislation (decrees, regulations, and ordinances) and include limitations on total and leaching content and other requirements that have to be considered in the permitted fields of application. In those countries where MSWI BA utilization is practiced, the utilization rate for MSWI BA varies between

20 wt% and 100 wt%. In China, MSWI BA has limited utilization rate and is not as much widely used as in Europe, thus lacking in comprehensive specification regarding utilization in various fields. Presently, the issued standards only focus on MSWI BA aggregate used for road subgrade, cushion, subbase, base, and nonreinforcement coagulation (GB/T 25032-2010).

Table 2.5 summarizes limited parameters and requirements of MSWI BA in different utilization fields. In Europe, MSWI BA is most commonly used in road construction, followed by different forms of earth works, cement process, and for foundations of structures. The necessary requirements are mainly physical—chemical parameters, and other terms such as classification as nonhazardous waste, maximum content of ferrous and nonferrous metals, etc. If MSWI BA is allowed to use in a more general application in bound or unbound form, which is not just applied for road construction, requirements for the total content of contaminants become especially important. Notably, specific limit values for the utilization of MSWI BA in the cement process are defined in Austria and Switzerland, to limit the dispersion of pollutants (heavy metals, persistent organic pollutants (POPs), etc.) (Blasenbauer et al., 2020). Additionally, leaching behavior can be neglected when MSWI BA serves as replacement of primary raw material in the cement kiln, and the total content of volatile components such as Hg comes under the spotlight instead. For instance, the test of leaching behavior of MSWI BA in Italy is not mandatory in the cement manufacturing process.

3.2 Furnace slag

3.2.1 Ground granulated blast furnace slag

In general, slag is a waste product from the pyrometallurgical processing of various ores in the blast furnace (Piatak et al., 2015). GGBS is a principal by-product produced by steel and iron productions. According to research, GGBS mostly consists of silicate and aluminosilicate of molten calcium and has a large quantity of amorphous calcium, silica, and alumina, making it an excellent binder for cement concrete manufacture (Aydõn and Baradan, 2014). As it improves the

TABLE 2.5 Requirements for application of MSWI BA.

Region	Requirements	Application
EU	(1) Total content Heavy metals LOI Total organic carbon Benzene, toluene, xylene Benzene, toluene, ethylbenzene, xylene Extractable halogens inorganic bonding Polycyclic aromatic hydrocarbon Polychlorinated biphenyl Polychlorinated dibenzodioxins/furans (2) Leaching content Leaching of heavy metals Chloride Sulfate Dissolved organic carbon Chemical oxygen demand Total dissolved solids pH (3) Other requirements Nonhazardous waste Content of ferrous and nonferrous metals Content of floating and nonfloating contaminants Particle size	Road construction Earth works Cement process Foundations of structures.
China	Radioactivity Leaching of heavy metals Particle size Impurity content Water content Cylinder compressive strength	Aggregate for road subgrade, cushion, subbase, base, and nonreinforcement coagulation

strength and reduces penetrability by increasing the boundary with the aggregate, GGBS, with both cementitious and pozzolanic characteristics, is commonly used as a cementitious material, providing significant cost savings and environmental advantages (Wang and Lee, 2010). For more than a century, GGBS was the primary supplemental cementing material extensively employed in various civil engineering tasks, including concrete manufacture. Various research studies have been performed on the impact of GGBS on the performance of various kinds of concrete and mortars (Dinakar et al., 2013; Abbass et al., 2021). It is proven that the GGBS substitution of OPC decreases the discharge of harmful gases and the use of superfluous electricity (Ahmad et al., 2022). In addition to its cost-effectiveness and being eco-friendly, its strength and durability characteristics are equivalent to those of cement.

As statistics, the worldwide output of GGBS is about 530 million tons, with 65% of it being used by the building sector (Sharma and Sivapullaiah, 2016). GGBS has been used widely in cement and concrete, and relevant standards have been issued. In the United Kingdom, BS EN 15167-1:2006 specifies requirements for the chemical−physical properties as well as quality control procedures for GGBS for use as a type II addition in the production of concrete (particular cast-in situ or prefabricated structural concrete conforming to EN 206-1), mortars, and grouts. Worth mentioning, GGBS containing any added materials other than grinding aids is beyond the scope of this standard. Provisions governing the practical application of GGBS in the production of concrete, mortar, or grout are not included in this standard, such as requirements concerning composition, mixing, placing, curing, and so on. About these terms, it can be referred to other European or national standards, such as EN 206-1. As for the United States, the specification of ASTM C989/C989M-22 covers slag cement for use as a cementitious material in concrete and mortar. In addition, ISO 22904-2020 includes the requirement of GGBS as addition for production of concrete. And in China, GB/T 18046-2017 specifies requirements of GGBS used for cement, mortar, and concrete (Table 2.6).

TABLE 2.6 Requirements of the GGBS for use in concrete, mortar, and grout.

	Property		US	UK	ISO	China
Chemical properties	MgO, max, %		–	18	10	–
	Sulfide, max, %		2.5	2.0	2.0	4.0
	Sulfate, max, %			2.5	4.0	
	LOI, corrected for oxidation of sulfide, max, %		–	3.0	3.0	1.0
	Chloride, max, %		–	0.10	0.1	0.06
	Moisture content, max, %		–	1.0	1.0	–
	Mass fraction of undissolved substance, %		–	–	–	3.0
	Vitreous content, %		–	–	–	85
Physical properties	Fineness		20 (amount retained when wet screened on a 45-μm sieve)	≥275 m^2/kg	≥275 m^2/kg	–
	Specific surface area, m^2/kg		–	–	–	500, 400, 300
	Density		–	–	–	2.8 g/cm^3
	Mobility ratio, %		–	–	–	95
	Water content, %		–	–	–	1.0
	Air content of slag mortar, max %		12	–	–	–
	Initial setting time ratio, %		–	200	200	200
	Activity index, min, %	7 days	–	45	45	95, 70, 55
		28 days	70, 90, 110	70	70	105, 95, 75

3.2.2 Steel furnace slag

As reported, there will be about one ton of SFS produced per three tons of steel products, accounting for a large part of solid wastes. Based on data from the World Steel Association in 2022, there was a total of around 1885 million tons of crude steel production (World Steel in Figures, 2023). Nevertheless, the current utilization rate of SFS is heterogenous among different countries. In the developed countries, a half of slag has been applied for road construction while the other half for sintering and recycling of iron (Gao et al., 2011).

SFS is largely used in road construction, cement production, internal recycling, civil engineering, and agriculture. According to the data in the United States (United States Geological Survey, 2021), SFS output reached 8.1 million tons, about 44.8% of which was used for road construction and 12.4% was used for asphalt concrete, only 12.9% of which was final placed in a landfill (Guo et al., 2018). Thus, SFS has basically reached a balance of discharge in the United States. The standard of SFS aggregates for bituminous paving mixtures is specified in ASTM D5106–22.

Information about SFS utilization in Europe is published by the Euroslag Association every 2 years. In 2018, SFS utilization was about 11.8 million tons, and the utilization rate was 72.4% (EUROSLAG, 2018a). In the European Union, SFS shall be considered as by-products in most cases with regard to the Waste Framework Directive. In other words, SFS shall no longer be regarded as waste if it is initially generated as waste following processed in such treatments that it can be used for specific purposes. Treatments such as granulation, foaming, controlled solidification, separation, crushing, sieving, and milling are allowed by the EU commission to be applicable industrial treating processes helping SFS to lose its waste status. According to statistics in 2018 (EUROSLAG, 2018b), SFS were mainly applied in the production of aggregates for road construction (70.6%) and for metallurgical purposes (13.1%). However, the utilization rate for other purposes such as fertilizer, and cement and concrete construction, summed up to just 1.3%–5.4%. When it comes to standards for utilization of SFS, issued standards focus on three main aspects, including cement and concrete production, road construction, and agriculture, as shown in Table 2.7. These standards meet high levels of technical issues, which also apply to primary raw materials and secondary materials. With regard to the health and environmental problems, national regulations are still to

TABLE 2.7 The relevant standards for the utilization of steel slag.

Region	Utilization	Standard	Standard Title
US	For roads	ASTM D5106	Steel slag aggregates for asphalt paving mixtures
EU	For cement concrete	EN 197-1	Cement—part 1: Composition, specifications and conformity criteria for common cements
		EN 206	Concrete
		EN 12620	Aggregates for concrete
		EN 13139	Aggregates for mortar
	For roads	EN 13043	Aggregates for bituminous mixtures and surface treatments for roads, airfields and other trafficked areas
		EN 13242	Aggregates for unbound and hydraulically bound materials for use in civil engineering work and road construction
		EN 13282	Hydraulic road binders—composition, specifications, and conformity criteria
		EN 13450	Aggregates for railway ballast
	For soil	EN 12945	Fertilizer
China	For cement concrete	GB/T 20491	Steel slag powder used for cement and concrete
	For roads	GB/T 24765	Steel slag for wearing asphalt pave
	For engineering backfill	YB/T 801	Steel slag for engineering backfill
	For concrete products	GB/T 24763	Steel slag for foamed concrete blocks
Japan	For roads	JIS A 5015	Iron and steel slag for road construction

be applied owing to no common European standards or regulations about the harmful substances assessment. Moreover, most of the European countries have developed their own directives and regulations as a basis for marketing by-products or secondary raw materials. Actually, the given chemical, physical, and ecological requirements are consistently achieved by the generated slag that is intended to be placed on the market. Some slag-derived products even exceed the technical properties of competitive natural products by far. All European standards about SFS for specific purposes contain conformity criteria mainly based on a factory production control (FPC) system, made by the producer for continuous production controls. For instance, the FPC is mentioned in the published Construction Product Regulation 305/2011 as a means of ensuring the fulfillment of the relevant technical specifications. In most EU countries, the FPC system is certified by a third party, and most slag products are labeled and traded with a CE-mark.

As in China, the utilization of SFS has achieved remarkable progress, and a relatively comprehensive standard system has been established, especially in the product field. Corresponding standardization of the application of SFS includes cement and concrete, engineering backfills, rod construction, etc. (Li et al., 2012). To be exactly, the SFS applications for cement concrete, roads, engineering backfill, and concrete products are specified in GB/T 20491, GB/T 24765, YB/T 801, and GB/T 24763, respectively.

In Japan, according to the official data (Nippon Slag Association, 2017), SFS production reached about 14.0 million tons in 2017, and the utilization rate was 98.5%, basically achieving output utilization balance. Of this, SFS was mostly employed in road construction and civil applications and covered 67.2% of the total applications, including backfills, earth coverings, road subgrades, and embankments. In addition, SFS has been suggested to improve marine environment, such as reducing the hydrogen sulfide content in seawater (Hayashi et al., 2014), and creating marine forests (Matsuura et al., 2022). In general, most SFS in Japan is recycled by magnetic separation after crushing, and the remaining tailings are almost all used in cement, road engineering, concrete aggregate, and civil materials. However, there is only one standard issued for the utilization of SFS in Japan (JIS A 5015-2018).

4. Conclusions and perspectives

Relevant regulations of the treatment and utilization of combustion/incineration residues are reviewed in this chapter. The degree of utilization varies greatly in different countries and regions, which is related to their level of economy and technology. The areas of application are also various, relating to local conditions and environmental laws. Among various residues, some mature and commercialized residues, such as ground granulated blast furnace slag and steel slag, have clear product standards; while for other residues, such as ISSA, relevant utilization standards are in lack, and most of which are disposed as general industrial solid waste. Currently, the main regulations and standards are focused on civil engineering applications, which include road construction, embankments, construction, geopolymer applications, cement, etc. With support of the incorporation of different types of additions to cement production in regulations (e.g., CFA and SFS), blended cement has been commonly used in the European Union, the United States, China, and many other countries/regions. These regulations cover different aspects with different specifications, including raw material qualification, industrial processes and production, production standards, and environmental supervision. Establishing systematic regulations and standards for utilization of combustion/incineration residues is crucial to ensure quality control and standardized assessments, including aspects such as chemical constituents, reactivity of amorphous substances, particle sizes, specific surface area, etc. The regulatory items and limits are various based on local circumstances and application aspects. For instance, biological assessment and chemical parameters of the incineration residues will play a significant role when they are used as a fertilizer for growing crops. Besides, the physical characteristics may place a strong emphasis on when residues are used for soils improvement with specific needs. In any case, as the qualified key limiting chemical factors, the limit values of heavy metals have been thoroughly regulated, which is conducive to their benign application.

Nevertheless, the utilization of combustion/incineration residues is a dynamically developing issue without sufficiently understanding; thus the policy support is in crying need. The management of residues should be based mainly on the chemical composition, in particular their content of contaminants. In the future, the classification system of residues may be improved to achieve better graded utilization. More attention should be paid on treating techniques to those residues for the sake of both economic and environment. In addition, the relevant standards and regulations on the utilization in aspects of agriculture and recovery of valuable compounds and rare elements apart from construction industry may be enhanced so as to improve the comprehensive applications. Moreover, the relevant policy and local standards, which is the most important motivation, for facilitating the utilization of residues and encouraging the industrial application may be issued and promoted.

References

Abbass, M., Singh, D., Singh, G., 2021. Properties of hybrid geopolymer concrete prepared using rice husk ash, fly ash and GGBS with coconut fiber. Mater. Today: Proc. 45, 4964–4970.

ACI, 2014. Building Code Requirements for Structural Concrete (ACI 318-05) and Commentary (ACI 318R-05) (USA).

Ahmad, J., Kontoleon, K.J., Majdi, A., et al., 2022. A comprehensive review on the ground granulated blast furnace slag (GGBS) in concrete production. Sustainability 14 (14), 8783.

ASTM, 2015a. Standard Specification for Portland Cement (ASTM C150/C150M) (USA).

ASTM, 2015b. Standard Test Methods for Chemical Analysis of Hydraulic Cement (ASTM C114) (USA).

ASTM, 2022. Standard Specification for Coal Fly Ash and Raw or Calcined Natural Pozzolan for Use in Concrete (ASTM C618) (USA).

Aydõn, S., Baradan, B., 2014. Effect of activator type and content on properties of alkali-activated slag mortars. Compos. Part B-Eng. 57, 166–172.

Bie, R.S., Li, S.Y., Wang, H., 2007. Characterization of PCDD/Fs and heavy metals from MSW incineration plant in Harbin. Waste Manage. (Tucson, Ariz.) 27 (12), 1860–1869.

Blasenbauer, D., Huber, F., Lederer, J., et al., 2020. Legal situation and current practice of waste incineration bottom ash utilisation in Europe. Waste Manage. (Tucson, Ariz.) 102, 868–883.

Dinakar, P., Sethy, K.P., Sahoo, U.C., 2013. Design of self-compacting concrete with ground granulated blast furnace slag. Mater. Des. 43, 161–169.

EN 450-1, 2012. Fly Ash for Concrete - Part 1: Definition, Specifications and Conformity Criteria.

EU 1009, 2019. Fertiliser Regulation.

EUROSLAG, 2018a. The European Slag Association. Position Paper on the Status of Ferrous Slag. https://www.euroslag.com/products/statistics/statistics-2018/.

EUROSLAG, 2018b. The European Association Representing Metallurgical Slag Producers and Processors. https://www.euroslag.com/products/statistics/statistics-2018/.

Fang, L., Li, J.S., Donatello, S., et al., 2018. Recovery of phosphorus from incinerated sewage sludge ash by combined two-step extraction and selective precipitation. Chem. Eng. J. 348, 74–83.

Gao, J.T., Li, S.Q., Zhang, Y.T., et al., 2011. Process of Re-resourcing of converter slag. J. Iron Steel Res. Int. 18, 32–39.

GB 8173, 1987. Control Standards of Pollutants in Fly Ash for Agricultural Use (China).

GB/T 1596, 2017. Fly Ash Used for Cement and Concrete (China).

GB/T 30760, 2014. Technical Specification for Co-processing of Solid Waste in Cement Kiln (China).

GB/T 30810, 2014. Test Methods for Leachable Ions of Heavy Metals in Cement Mortar (China).

Guo, J.L., Bao, Y.P., Wang, M., 2018. Steel slag in China: treatment, recycling, and management. Waste Manage. (Tucson, Ariz.) 78, 318–330.

Hayashi, A., Asaoka, S., Watanabe, T., et al., 2014. Mechanism of suppression of sulfide ion in seawater using steelmaking slag. ISIJ Int. 54 (7), 1741–1748.

HJ 1134, 2020. Technical Specification for Pollution Control of Fly-Ash from Municipal Solid Waste Incineration (On Trial) (China).

HJ 662, 2013. Environmental Protection Technical Specification for Co-processing of Solid Wastes in Cement Kiln (China).

Hušek, M., Moško, J., Pohořelý, M., 2022. Sewage sludge treatment methods and P-recovery possibilities: current state-of the-art. J. Environ. Manag. 315.

Jayaranjan, M.L.D., van Hullebusch, E.D., Annachhatre, A.P., 2014. Reuse options for coal fired power plant bottom ash and fly ash. Rev. Environ. Sci. Biotechnol. 13, 467–486.

JIS A6201, 2015. Fly Ash for Use in Concrete (Japan).

Kallio, A., Virtanen, S., Leikoski, N., et al., 2023. Radioactivity of residues from waste incineration facilities in Finland. J. Radiol. Prot. 43 (2), 021502.

Kumar, A., Abbas, S., Saluja, S., 2023. Utilization of incineration ash as a construction material: a review. Mater. Today: Proc. 2214.

Lam, C.H.K., Ip, A.W.M., Barford, J.P., et al., 2010. Use of incineration MSW ash: a review. Sustainability 2 (7), 1943–1968.

Li, J.S., Poon, C.S., 2017. Innovative solidification/stabilization of lead contaminated soil using incineration sewage sludge ash. Chemosphere 173, 143–152.

Li, Z.B., Zhao, S.Y., Zhao, X.G., et al., 2012. Leaching characteristics of steel slag components and their application in cementitious property prediction. J. Hazard Mater. 448–452.

Li, Y., Hao, L., Chen, X., 2016. Analysis of MSWI bottom ash reused as alternative material for cement production. Procedia Environ. Sci. 31, 549–553.

Liu, J., Wang, Z.D., Xie, G.M., et al., 2022. Resource utilization of municipal solid waste incineration fly ash cement and alkali-activated cementitious materials: a review. Sci. Total Environ. 852, 158254.

Lynn, C.J., Dhir, R.K., Ghataora, G.S., 2016. Sewage sludge ash characteristics and potential for use in bricks, tiles and glass ceramics. Water Sci. Technol. 74, 17–29.

Matsuura, H., Yang, X., Li, G.Q., et al., 2022. Recycling of ironmaking and steelmaking slags in Japan and China. Inter. J. Min. Met. Mater. 29 (4), 739–749.

Nippon Slag Association, 2017. Production and Uses of Steel Slag in Japan.

Piatak, N.M., Parsons, M.B., Seal Ii, R.R., 2015. Characteristics and environmental aspects of slag: a review. Appl. Geochem. 57, 236–266.

Rutkowska, G., 2023. Assessment of fly ash from thermal treatment of sewage sludge according to the applicable standards. J. Ecol. Eng. 24 (3), 20–34.

Sharma, A.K., Sivapullaiah, P.V., 2016. Ground granulated blast furnace slag amended fly ash as an expansive soil stabilizer. Soils Found. 56, 205–212.

Shane, D., Christopher, R.C., 2013. Recycling and recovery routes for incinerated sewage sludge ash (ISSA): a review. Waste Manage. (Tucson, Ariz.) 33 (11), 2328–2340.

Smol, M., Joanna, K., Anna, H., et al., 2015. The possible use of sewage sludge ash (SSA) in the construction industry as a way towards a circular economy. J. Clean. Prod. 95, 45–54.

Suraneni, P., Burris, L., Shearer, C.R., et al., 2021. ASTM C618 fly ash specification: comparison with other specifications, shortcomings, and solutions. ACI Mater. J. 118, 157–167.

United States Geological Survey, 2021. Minerals Yearbook-Slag, Iron and Steel.

Wang, X.Y., Lee, H.S., 2010. Modeling the hydration of concrete incorporating fly ash or slag. Cement Concr. Res. 40, 984–996.

Wang, Q., Li, J.S., Poon, C.S., 2019a. Recycling of incinerated sewage sludge ash as an adsorbent for heavy metals removal from aqueous solutions. J. Environ. Manag. 247, 509–517.

Wang, Q., Li, J.S., Poon, C.S., 2019b. Using incinerated sewage sludge ash as a high-performance adsorbent for lead removal from aqueous solutions: performances and mechanisms. Chemosphere 226, 587–596.

Wang, N., Sun, X., Zhao, Q., et al., 2020. Leachability and adverse effects of coal fly ash: a review. J. Hazard Mater. 396, 122725.

Wei, L., Zhu, F., Li, Q., et al., 2020. Development, current state and future trends of sludge management in China: based on exploratory data and CO_2-equivaient emissions analysis. Environ. Int. 144, 106093.

World Steel in Figures, 2023. World Steel Association.

Yin, K., Ahamed, A., Lisak, G., 2018. Environmental perspectives of recycling various combustion ashes in cement production – a review. Waste Manage. (Tucson, Ariz.) 78, 401–416.

Zhang, Y., Labianca, C., Chen, L., et al., 2021. Sustainable ex-situ remediation of contaminated sediment: a review. Environ. Pollut. 287, 117333.

Zhou, Y.F., Li, J.S., Poon, C.S., 2022. Sustainable utilization of incinerated sewage sludge ash. In: Low Carbon Stabilization and Solidification of Hazardous Wastes, pp. 211–225.

Part II

Purification and detoxification of combustion/incineration residues

Chapter 3

Cement-based immobilization of combustion/incineration residues

Binglin Guo[1,2,3], Huyong Qin[1], Quanzhi Tian[4], Peng Gao[1,2,3], Yonggan Yang[1,3], Li Hong[1,2], Binggen Zhan[1,2,3], Qijun Yu[1,2,3], Lei Liu[5] and Keiko Sasaki[6]

[1]*School of Civil Engineering, Hefei University of Technology, Hefei, China;* [2]*Engineering Research Center of Low-carbon Technology and Equipment for Cement-based Materials, Ministry of Education, Hefei University of Technology, Hefei, China;* [3]*Anhui Low-carbon Technologies Engineering Center for Cement-based Materials, Hefei, China;* [4]*National Engineering Research Center of Coal Preparation and Purification, China University of Mining and Technology, Xuzhou, China;* [5]*Institute of industrial Science, The University of Tokyo, Tokyo, Japan;* [6]*Department of Earth Resources Engineering, Kyushu University, Fukuoka, Japan*

1. Introduction

On December 14, 2022, the 77th session of the United Nations General Assembly passed a resolution claiming March 30 of each year as International Zero Waste Day. With the acceleration of urbanization, the amount of municipal solid waste (MSW) is increasing. More than two billion tons of MSW globally produced annually, and it is estimated that by 2050, the MSW produced by humans will rise to nearly four billion tons. The Pacific region and East Asia produce most of the world's waste, ca. 23%. Currently, more than half of the waste is openly dumped, and landfilling is the second choice. The generation of waste would have significant impacts on the environment, public health, and economic growth, thus requiring necessary and urgent action (World Bank Group, 2018; Li and Xu, 2022) (Fig. 3.1).

The incineration is currently one of commonly applied methods for the treatment of MSW. This is because this technology can not only effectively reduce the volume of waste but also generate energy via the form of electricity and heat. The process involves burning the waste at high temperatures, which can destroy harmful substances, such as pathogens and toxic organic compounds. In addition, incineration can be a more cost-effective waste management option

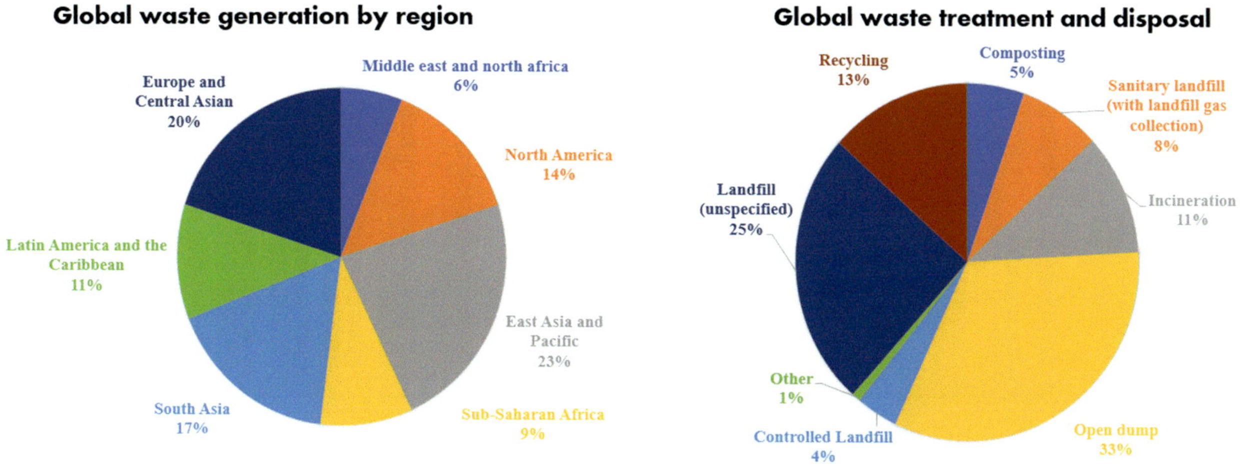

FIGURE 3.1 The global waste generation, treatment, and disposal. *Modified from World Bank Group, 2018. What a Waste 2.0: A Global Snapshot of Solid Waste Management to 2050-The Urban Development Series. Int. Bank Reconstr. Dev./World Bank 1818 H Str. NW, Washington, DC. Available at: https://openknowledge.worldbank.org/handle/10986/30317. (Accessed June 2023).*

Treatment and Utilization of Combustion and Incineration Residues. https://doi.org/10.1016/B978-0-443-21536-0.00021-6

than other methods, such as landfilling, particularly in areas with limited space for landfills. However, it is worth noting that the incineration process should proceed as a last resort for waste management, as it can have negative environmental and health impacts if not properly managed.

The Ministry of Housing and Urban-Rural Development of China released the "2021 Urban and Rural Construction Statistical Yearbook" and "2021 Urban Construction Statistical Yearbook" on October 12, 2022. According to statistics, there were 583 domestic waste incineration plants in 2021, with a daily processing capacity of 719,533 tons. Compared with the two most widely used types of domestic waste incinerators, the fly ash (FA) production rate (3%–5%) of the grate furnace incineration process is much lower than that of the fluidized bed incineration process, with an FA production rate of 15%–20%. However, even if the median FA production rate of the grate furnace incineration process is 4%, according to the statistics of incineration capacity in 2021, the national daily output of FA is 28,781 tons, and the national annual output of FA is about 10 million tons. The trend of urban waste generation and treatment methods in China is shown in Fig. 3.2 (National Bureau of Statistics, 2021).

Fly ash (FA) and bottom ash (BA) are two types of residues that are generated during the combustion or incineration of solid waste. Both FA and BA could be treated and managed through various methods, such as stabilization/solidification with cement or other binders, landfilling, or reuse in construction materials. However, it is important to properly manage these residues to prevent environmental contamination and protect public health. Under incineration, heavy metals or chlorides in domestic waste are generally evaporated in a high-temperature atmosphere and finally accumulated in the structure of municipal solid waste incineration fly ash (MSWI FA), which makes MSWI BA less toxic than MSWI FA, and thus the treatment of MSWI BA is relatively simple, even realizing the application in engineering (Sormunen et al., 2017; Abdulmatin et al., 2018; Ke et al., 2018; Maldonado Alameda et al., 2020; Phutthimethakul and Supakata, 2022). Treatment of MSWI FA has gradually become a pressing problem to be solved in waste treatment since this waste contains high concentrations of hazardous trace elements (HTEs), such as Pb, Zn, As, Pb, Cr, soluble salts, and some organic substances including dioxins and furans (Jiang et al., 2022; Li et al., 2022; Liu et al., 2022). When the National Catalog of Hazardous Wastes was revised, it was clarified that "domestic waste incineration fly ash" is classified as hazardous waste, and the category is HW18. In addition, the environmental risks and implications of MSWI FA treatment are also a growing concern.

The content of the literature collected in this chapter is mainly about MSWI FA. For the disposal and treatment of MSWI FA, common methods include cement solidification, landfill, resource recovery, reuse, etc. Cementation is a common and economical treatment to cure MSWI FA into a stable material by mixing with cement and other additives to reduce its potential harm to the environment (Maldonado-Alameda et al., 2020). Cementation can improve the physical and

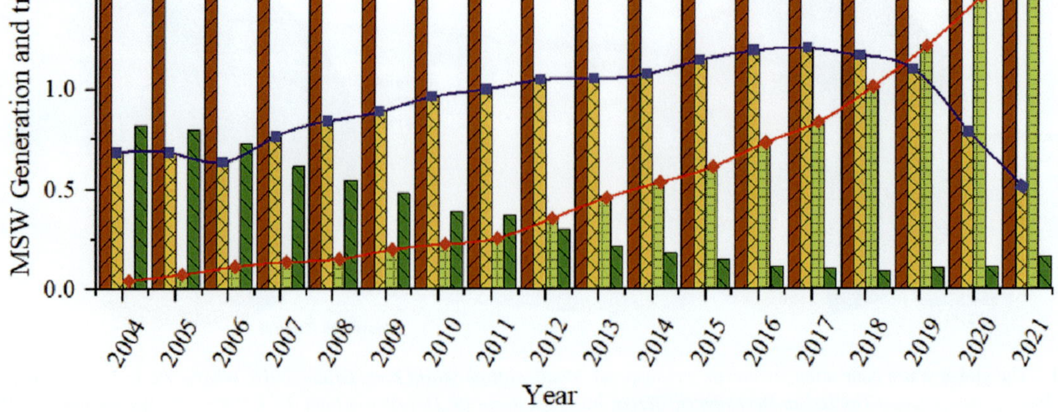

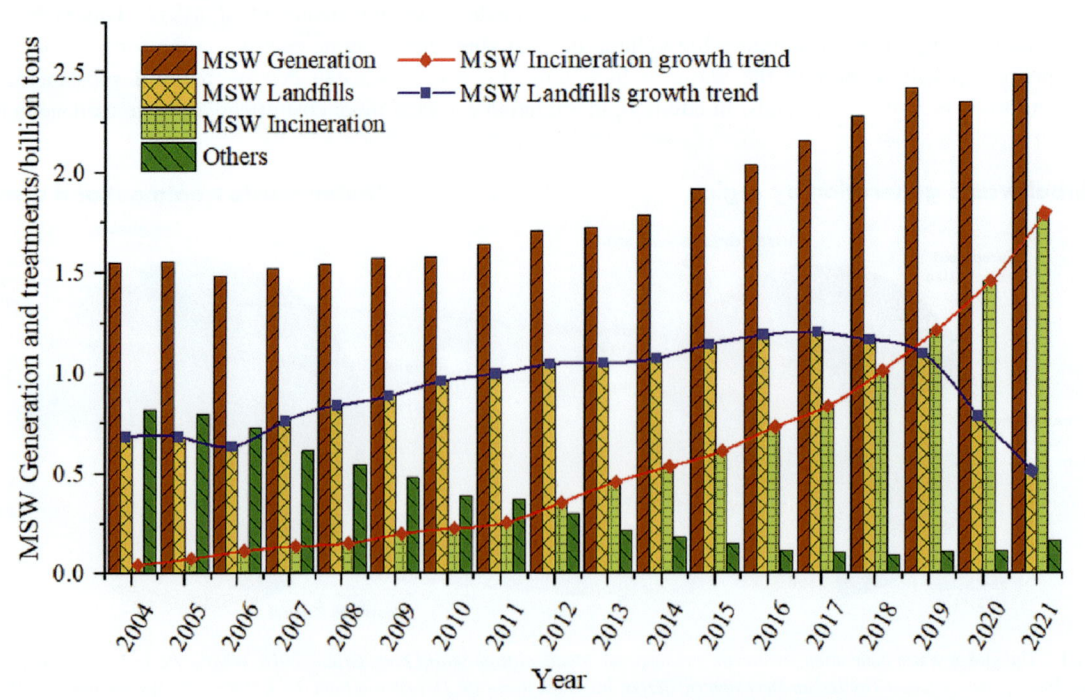

FIGURE 3.2 Trends in the generation and treatment of urban household waste in China from 2004 to 2021 (National Bureau of statistics, 2021).

chemical stability of MSWI FA and make it a controllable material. However, most of the solidified MSWI FA is transported to a landfill to be landfilled. Under the various geochemical factors, HTEs would leach from solidified MSWI FA in landfill, causing potential harm to the environment and ecosystem. Therefore, it is necessary to review the existing research on MSWI FA to comprehensively understand its properties, key factors influencing the leaching characteristics of HTEs from MSWI FA after immobilization, stability analysis of various cement solidification technologies, potential environmental risk assessment, and sustainability research.

2. Generation and characteristics of municipal solid waste incineration residues

2.1 Generation of municipal solid waste incineration bottom ash and fly ash

MSW incineration is a common waste disposal method for solid waste generated in cities and urban areas. As mentioned in Section 1, when MSW is incinerated, two main solid wastes are produced after high-temperature combustion: BA and FA. The general process of MSWI is shown in Fig. 3.3.

The process can be classified as below.

(1) Garbage collection and pretreatment: This step ensures that the waste is properly handled, sorted, and prepared before it undergoes incineration. Firstly, MSW is collected from households, businesses, and other sources within a city or community and then transported to a waste processing facility. Before incineration, MSW must undergo pre-treatment to improve the efficiency of the incineration process and reduce potential environmental impacts. The pre-treatment process typically involves the sorting into different categories (e.g., recyclables, organic waste, and non-recyclable waste), size reduction of the waste items, drying process (improving combustion efficiency), and removal of hazardous materials (e.g., batteries, chemicals, and medical waste). After pre-treatment, the MSW is ready for incineration. Proper garbage collection and pre-treatment are essential for ensuring the safe, efficient, and environmentally friendly disposal of MSW through incineration.

(2) Furnace feeding and combustion: Furnace feeding and combustion are the second steps in the MSWI process. This step involves introducing the pre-treated waste into the incinerator and completing combustion to minimize environmental impacts. The pre-treated MSW is fed into the incinerator, typically using a conveyor or a crane with a grabber. The waste is introduced into the combustion chamber, where it is evenly distributed to ensure proper combustion and to

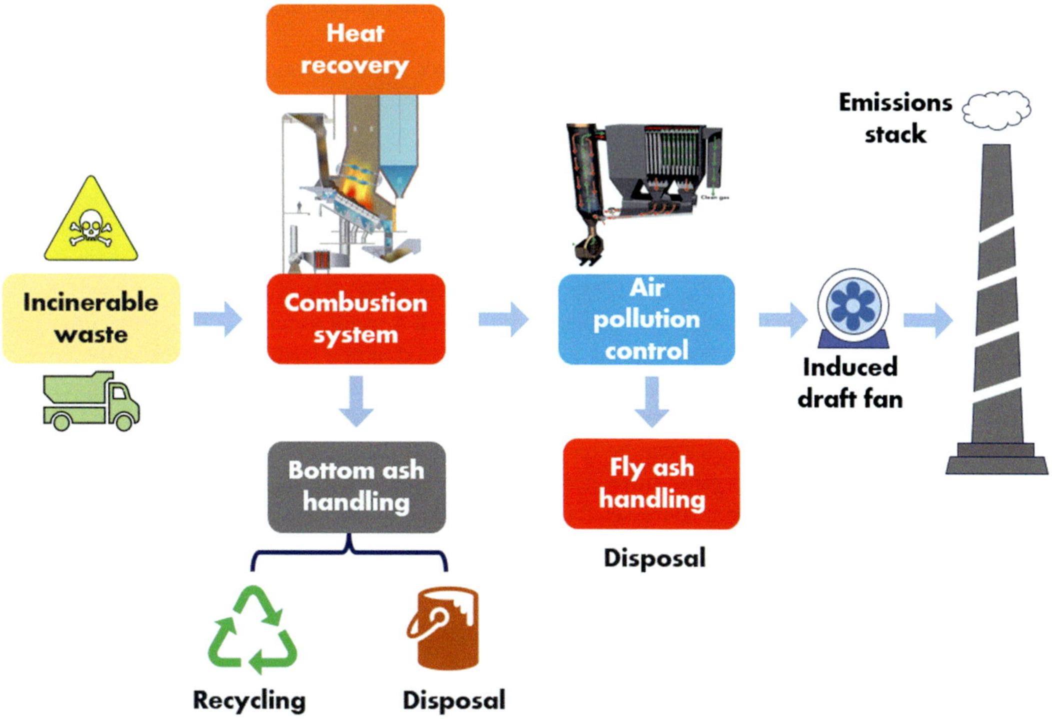

FIGURE 3.3 General process of MSW incineration.

prevent the formation of cold spots that could hinder the process. The combustion process involves burning the waste at high temperatures to break down the organic materials, reduce the waste volume, and generate heat. Proper control of the combustion process is important for ensuring efficient waste conversion and minimizing the generation of harmful emissions. Factors such as temperature, air supply, and waste feed rate must be closely monitored and controlled to optimize the combustion process. Following the combustion process, the resulting ash, flue gas, and heat are managed through various post-combustion treatments, such as ash handling, flue gas cleaning, and energy recovery systems. Proper furnace feeding and combustion control are essential for ensuring that MSWI is an efficient and environmentally friendly waste disposal method.

(3) Energy recovery: Energy recovery is a valuable aspect of MSWI, as it allows for the conversion of waste into valuable energy in the form of heat or electricity. This process not only helps to reduce the environmental impact of waste disposal but also contributes to energy sustainability and reduces the dependence on traditional energy sources. During the combustion process, a significant amount of heat is generated. This heat can be captured and utilized for various purposes, such as steam generation for various industrial processes or to generate electricity, district providing heating and hot water to nearby residential and commercial buildings. In addition to generating electricity, waste heat from the incineration process can also be used for other applications, such as preheating combustion air or drying incoming waste. This further improves the whole energy efficiency of the waste incineration system.

(4) Flue gas treatment: Flue gas treatment helps to minimize the environmental impact of the incineration process by reducing air pollution and ensuring compliance with regulatory standards. The primary goal of flue gas treatment is to remove harmful pollutants, such as particulate matter, heavy metals, acidic gases, and dioxins, from the flue gas before it is released into the atmosphere. The first stage of flue gas treatment involves removing particulate matter from the flue gas. This is typically achieved using one or more of the following techniques: electrostatic precipitators and fabric filters (baghouses). Then, some harmful substances including acidic gases, heavy metals, and NOx considered to be removed. Acidic gases, such as hydrogen chloride (HCl) and sulfur dioxide (SO_2), could be removed from the flue gas using wet or dry scrubbing techniques. The acidic gases react with the alkaline solution or sorbent to form neutral salts, which are then collected along with the particulate matter in the dust removal stage. Heavy metals (e.g., mercury, lead) and dioxins are harmful pollutants that need to be removed from the flue gas. This can be achieved by using activated carbon injection or other specialized adsorbents. NOx emissions can be reduced using selective catalytic reduction or selective non-catalytic reduction techniques (Masera and Hossain, 2021). These processes involve injecting reducing agents, such as urea or ammonia, which react with NOx to form nitrogen gas and water without harm. Finally, the treated flue gas may need to be cooled before being released into the atmosphere. After undergoing these treatment stages, the processed flue gas is released into the atmosphere through a stack, ensuring minimal environmental impact from the MSWI process.

(5) Bottom ash treatment: The residue at the bottom of the incinerator is called bottom ash (BA), which contains incompletely burned substances during the incineration process. BA treatment is an important aspect of MSW incineration because it deals with the management and disposal of the solid residues produced during the combustion process. BA is the non-combustible material that accumulates at the bottom of the incinerator and typically consists of glass, ceramics, metals, and other inert materials. Proper treatment of bottom ash is crucial to minimize its environmental impact and to recover valuable materials for recycling or reuse. After the combustion process, the hot BA is removed from the incinerator and cooled, either by air or water quenching. Then the BA is passed through a magnetic separator to recover ferrous metals, such as iron and steel. The recovered metals can be recycled and reused in various industries. metals, such as aluminum, copper, and brass, could also be recovered from BA using techniques like eddy current separation or sensor-based sorting. Once the BA has been treated, it can be disposed of in a landfill, used as a daily cover material for landfills, or repurposed for other applications, such as road construction, concrete production, or aggregates in various civil engineering projects. Proper bottom ash treatment is essential for minimizing the environmental impact of MSWI and for recovering valuable materials that can be recycled or reused.

(6) Fly ash treatment: FA is rich in inorganic substances and heavy metals and needs to be treated. FA treatment is the most important part of MSWI because it addresses the fine particulate matter generated during the combustion process. Fly ash exhibits small particles that are carried along with some flue gas and contain hazardous substances, such as heavy metals, dioxins, and furans, which makes BA less toxic than FA. Proper treatment of FA is crucial to minimize its environmental impact and to ensure compliance with regulatory standards. To immobilize any potentially hazardous substances present in the FA, such as heavy metals, it may be necessary to stabilize and solidify the material. This process typically involves mixing the FA with binders, such as cement, lime, or other chemicals, to create a stable, solid matrix that reduces the potential for leaching of contaminants into the environment. After stabilization and solidification, the treated FA is typically stored in a designated area or container before being transported to a suitable disposal facility. Disposal options for FA include secure landfills specifically designed for hazardous waste or encapsulation within other waste materials (e.g., as a component in concrete or asphalt). In some cases, FA can be recycled or reused for various applications, depending on its chemical composition and physical properties. For example, FA with high calcium content can be used as a substitute for cement in concrete production. However, recycling and reuse of FA are generally limited due to the potential presence of hazardous substances. It is important to note that FA treatment methods and disposal options may vary depending on local regulations and the specific characteristics of the ash produced by a particular waste incineration facility. Proper management of FA is essential for minimizing the environmental impact of MSWI and ensuring compliance with air quality standards and waste disposal regulations.

2.2 Characteristics of municipal solid waste incineration residues

2.2.1 Bottom ash

Account for ca. 80% of MSWI residues are considered as BA. BA generally exhibits heterogeneous granulates with different particle sizes (Fig. 3.4A). This waste is generated in the largest amount ca. 25% of the total MSW mass while ca. 70% consists of waste flue gas and ca. 5% is from other systems. BA is mainly consisting of stone, metals, ceramics, and glasses. BA is the components consist of SiO_2, CaO, Al_2O_3, glass and ceramics, and other amorphous phases (Tang et al., 2020) (Fig. 3.5A). After grounded finely of BA, it could exhibit pozzolanic activity and thus applied in the civil engineering (Ayobami, 2021). In addition, the size fraction influenced the metal concentrations in the BA (Huber et al., 2020). However, the compositions of BA vary from different countries and regions due to numerous factors covering waste incineration technology, geography, etc. As mentioned previously, due to its specific properties, BA is regarded as a non-hazardous residue from MSWI residues that is also directly used in various civil engineering (Poulikakos et al., 2017; Mohammed et al., 2021).

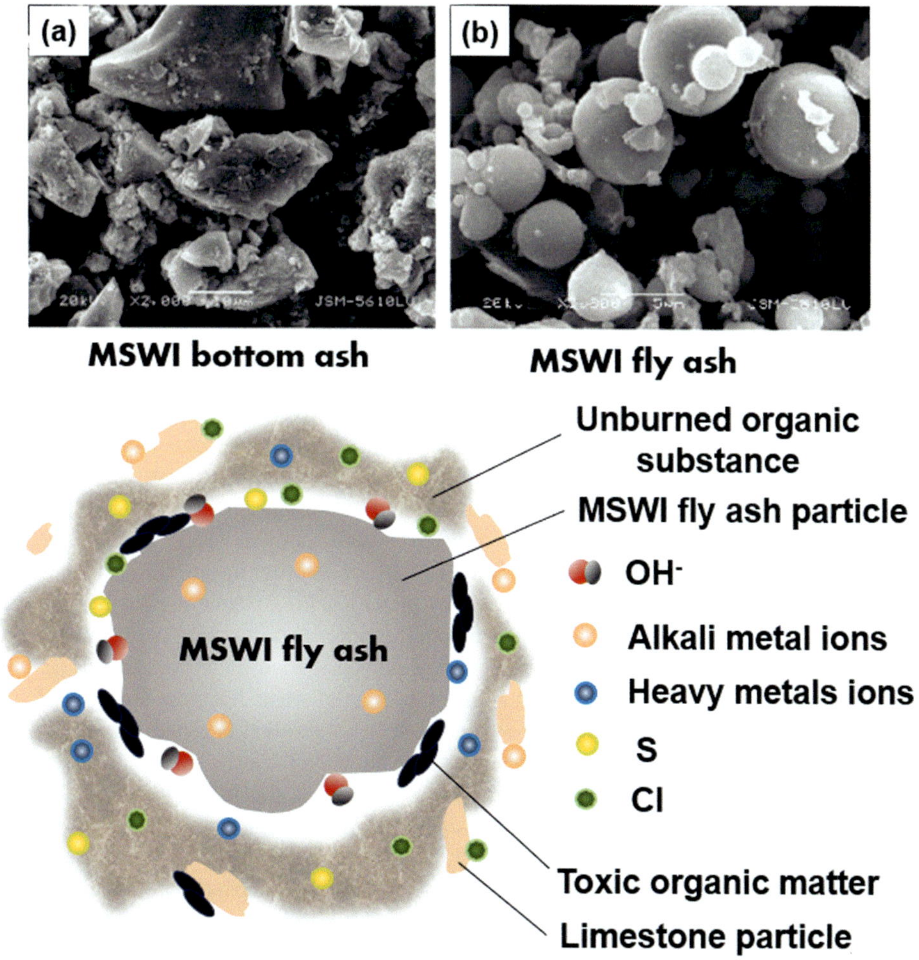

FIGURE 3.4 SEM images of MSWI bottom ash, fly ash, and the illustration of MSWI fly ash. *SEM images modified from Li, X.G., Lv, Y., Ma, B.G., Chen, Q.B., Yin, X.B., Jian, S.W., 2012. Utilization of municipal solid waste incineration bottom ash in blended cement. J. Clean. Prod. 32, 96–100.*

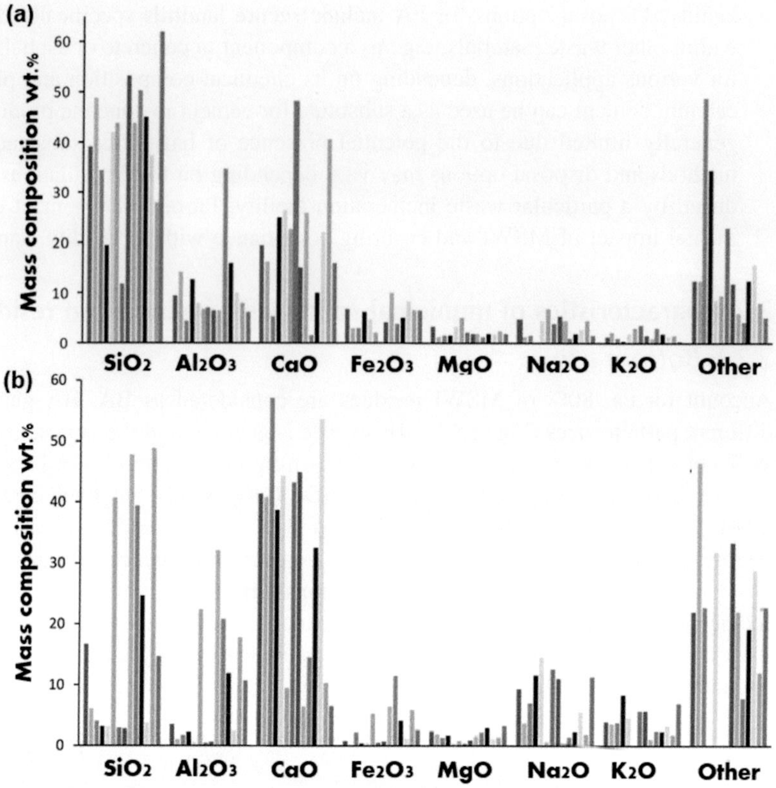

FIGURE 3.5 Compilation of various studies showing elemental composition in (A) MSWI-BA, (B) MSWI-FA. Note: The bars in the figure indicate different research data collected from 2010 to 2020. *Modified from Al-Ghouti, M.A., Khan, M., Nasser, M.S., Al-Saad, K., Heng, O.E., 2021. Recent advances and applications of municipal solid wastes bottom and fly ashes: insights into sustainable management and conservation of resources. Environ. Technol. Innov. 21, 101267.*

2.2.2 Fly ash

Different from the MSWI BA, MSWI FA is also a byproduct of the combustion of mixed waste materials and is classified as a hazardous material in some regions. The pollution properties of MSWI FA can vary depending on the composition of the waste, combustion process, and efficiency of pollution control equipment. MSWI FA generally exhibits as off-white or dark gray fine powder, which has large specific surface area, low moisture content, irregular particle size, and high porosity (Fig. 3.4B).

The MSWI FA has an extremely complex chemical composition, commonly composed of inorganic substances, including silicates, sulfates, oxides, chlorides, calcium hydroxide, and other alkaline compounds, etc. (Figs. 3.4 and 3.5B). These components mainly come from the inorganic substances and additives in the garbage. Other ingredients include HTEs such as Pb, Hg, As, Cr, and others (Geng et al., 2020). In terms of heavy metals, they generally exist as oxides, sulfates, and chlorides and are accumulated on the surface of MSWI FA particles (Huang et al., 2022). These heavy metals mainly come from metal substances and additives in the garbage and exist in an amorphous form (Kirkelund et al., 2020). The concentration of HTEs in MSWI FA depends on the compositions and sources of the wastes as well as the combustion conditions. Due to the danger and mobility of HTEs, there is a lot of interest in handling them in an environmentally compliant manner. Therefore, all countries have relevant heavy metal leaching regulations as the evaluation standards for HTEs treatment in waste incineration plants. The United States introduced the toxicity characteristic leaching procedure (TCLP) (EPA, 1990), the European Union applied the EU hazardous waste identification standard (EN 12457), and China adopted the acetic acid buffer solution as modified TCLP method for solid waste leaching (Jiang et al., 2022). In addition, MSWI FA may also contain a certain amount of organic matter, such as residual organic matter, traces of dioxin, furans, polychlorinated biphenyls (PCBs), polycyclic aromatic hydrocarbons (PAHs), etc. These substances have various toxic effects, which have the potential to be accumulated in the human body via the food chain, and have irreversible carcinogenic and mutagenic properties (Rajan et al., 2021). The content of these organic substances is low, and these organic pollutants can be removed by chemical or thermal treatment. When deacidifying the flue gas produced in the process of domestic waste incineration, it is necessary to inject a large number of alkaline substances such as slaked lime, resulting in the composition of MSWI FA containing more than half of calcium compounds at most, and MSWI FA exhibits high the pH value between 8 and 12, and has shown strong acid neutralization ability (Chen et al., 2022). However, the specific pH value also depends on many factors, including the temperature during the incineration process, the composition of the waste components, and the combustion conditions. The differences in the components in MSWI FA are also sources by the composition of incinerated waste, regions, and incineration atmospheres.

2.3 Leaching properties of the municipal solid waste incineration fly ash

As described before, HTEs such as mercury (Hg), arsenic (As), and manganese (Mn) pose significant environmental and health risks when released into the environment during combustion. These elements contribute to pollution in air, water, and soil and have been identified by several global bodies, including the United States Clean Air Act Amendments (CAAA) in 1990 (EPA, 1990), the European Union, and the Canadian Environmental Protection Agency, as significant environmental pollutants. China's "12th Five-Year Plan" (MEP, 2011) also highlights these concerns and addresses emission control. Although landfilling is a common disposal method for MSWI FA, increasing production has gradually saturated landfill capacity, making landfill expansion both spatially demanding and costly. To address this, the 2020 Technical Specification for Pollution Control of Fly Ash from Domestic Waste Incineration proposed utilizing fly ash resources. This policy emphasizes that heavy metal concentration in fly ash during comprehensive use should be less than standard limits, encouraging research on how to enhance fly ash's resource utilization capacity, diversify its treatment and disposal methods, and alleviate landfill pressure.

One way to evaluate fly ash's environmental risk is by assessing the leaching toxicity of its heavy metals to determine its suitability for resource utilization. Chai tested four types of MSWI FA, examining their physicochemical properties, heavy metal enrichment and leaching patterns, and leaching toxicity. The results proved that the main mineral components of FA were chloride and calcium salts (Chuai et al., 2022). Heavy metal content initially decreased and then increased with the decrease in MSWI FA particle size. Furthermore, increasing the calcium oxide content can render fly ash alkaline, promoting the formation of compounds that reduce the leaching rate of arsenic and lead. The study suggested optimizing calcium oxide addition in the acid gas cleaning process in the future to minimize the environmental risks associated with MSW fly ash emissions. Landfilling, despite being the primary disposal method for hazardous solid waste such as fly ash in China, poses potential environmental risks due to the long-term leaching of hazardous substances from fly ash into the soil. To study the possible release of certain heavy metals from fly ash, researchers conducted a continuous leaching test using a sodium acetate solution as an extractant. Their findings showed that the leaching amount of each heavy metal, except zinc, decreased with an increasing liquid-to-solid ratio. Heavy metal concentration significantly decreased with increasing the initial pH of the leachate, staying low when the pH was above 6. Long-term leaching behaviors indicated that the toxicity of heavy metals could generally be controlled in natural environments; however, the risk increased with the presence of reducing agents and pH changes in the surrounding environment. Consequently, treating fly ash to solidify its heavy metals is crucial to reduce environmental risks (Liu et al., 2005).

Sintering and melting processes are generally efficient in immobilizing most heavy metals in waste fly ash. During these high-temperature treatments, heavy metals in the fly ash are primarily confined within a glassy substance that forms as the melt cools (Gu et al., 2022). However, some heavy metals may enhance their leaching toxicity post these high-temperature treatments. For instance, it has been reported that chromium (Cr) in solid waste can be oxidized at high temperatures into Cr (VI), a highly toxic and easily leachable form. Simultaneously, under high-temperature conditions, Cr may react with soluble salts to create soluble chromium salts, such as K_4CrO_4 and Na_2CrO_4. To mitigate the risk of Cr leaching, additives are frequently incorporated during the heat treatment to control Cr's conversion. Compounds such as silicates, phosphates, and iron oxides are considered capable of enhancing Cr's stability during high-temperature treatment, thereby reducing its leaching toxicity (Colangelo et al., 2012). Moreover, geological polymers, prepared using FA as a raw material, can immobilize the harmful components in the ash. In geological polymers made by Zhou et al. (2020), using a mixture of municipal solid waste fly ash and red mud, the chemical forms of Pb, Zn, Cr, and Cu were transformed into stable states such as residual and oxidized forms. This suggests that most heavy metals in the geological polymer solids are successfully fixed within the hydration phase and the structure of the geological polymers, transforming from effective components to stable ones. Therefore, the long-term safety of heavy metals in the RM-MSWIFA binary system geological polymers is deemed secure.

3. Pretreatment of the MSWI FA

3.1 Function of the pretreatment for the MSWI FA

Since MSWI FA contains high contents of sulfates, chlorides, and alkali metal, which can cause the formation of porous microstructure and decrease of mechanical strength of cement, the compressive strength of final products will sharply decrease with increasing the MSWI FA content. The content of alkali metals and chloride determine the allowable amount of MSWI FA in cement (Sarmiento et al., 2019). The presence of chloride not only corrodes production equipment but also causes adverse effects on the performance of cementitious materials after adding MSWI FA. Therefore, it is necessary to carry out pretreatment to minimize the impact of alkali metal, sulfates, and chloride content for utilization of MSWI FA

(Clavier et al., 2021). Pretreatment processes can remove, reduce, and stabilize the concentration of hazardous substances in the MSWI FA. This reduces the leaching potential of these contaminants and mitigates the environmental risks when the fly ash is disposed of or used in other applications. In addition, pretreatment of MSWI FA can enable the recovery of valuable resources, such as metals, which can be recycled or used in other industrial processes.

3.2 Washing pretreatment of MSWI FA

Nowadays, washing is the most common pretreatment method for the pretreatment of MSWI FA (Caviglia et al., 2022). Washing pretreatment is the use of various kinds of solution, covering water, acid, and alkali, as a detergent to remove the content of hazardous substances in MSWI FA. This process can reduce the leaching ability of various hazardous substances. However, MSWI FA washing pretreatment can only wash away soluble substances, but cannot reduce the content of dioxins in MSWI FA (Hu et al., 2012), but washing away chlorides can prevent MSWI FA from producing dioxins in the process of cement production (Cieplik et al., 2006). The cement produced by MSWI FA pretreated by water washing performed better than untreated cement (Mangialardi et al., 2004).

3.3 Heat treatment of MSWI FA

As described before, MSWI FA may also contain a certain amount of heavy metals and organic matter, such as dioxin, furans, PAHs, and PCBs. The organic pollutants are very stable lipophilic solid organic substance, which is extremely difficult to dissolve in water and difficult to remove by only washing with water. Thus, heating at high temperatures to destroy organic pollutants could be the ideal solution to this problem, these will be removed in a high-temperature atmosphere (Xia et al., 2017). Heat treatment methods for MSWI FA cover thermal desorption (general heating temperature range of 300−600°C under an inert or oxidative atmosphere), vitrification (hearting temperatures above 1000°C), sintering (heating temperatures between 800 and 1200°C under controlled conditions), plasma treatment (high-temperature plasma ~10,000°C generated by an electric arc or radio frequency energy), and gasification. Heat treatment of MSWI fly ash can significantly reduce the environmental risks associated with its disposal or reuse. However, it is essential to consider the energy consumption, greenhouse gas emissions, and potential formation of toxic by-products during the heat treatment process.

3.4 Other pretreatment methods

Based on the properties of MSWI FA and target, the treatment method could be classified as follows: (1) separations, (2) stabilization and solidification, and (3) heat treatment (Fig. 3.6). These pretreatment methods all focus on the removal or immobilization of hazardous substances in MSWI FA. Each pretreatment method has its advantages and limitations, and the choice of an appropriate method depends on the characteristics of MSWI FA, as well as the intended end-use or disposal requirements.

4. Cement-based stabilization and solidification of combustion/incineration residues

As mentioned in Section 3.3, stabilization and solidification (S/S) is a widely accepted technique for the management and disposal of combustion and incineration residues, including FA from MSWI residues, and other hazardous wastes. This method involves mixing hazardous wastes with various binder materials to immobilize contaminants, particularly HTEs, and reduce their leaching potential. Finally, hazardous wastes can be converted to low-toxicity materials for disposal or further application in civil engineering (Kirkelund et al., 2020; Marieta et al., 2021). In terms of binder materials, ordinary Portland cement (OPC) is a commonly available and relatively economical material with unique properties and characteristics, making it an economically viable option for waste immobilization (Guo et al., 2017a,b,c, 2020; Vinter et al., 2016; Chen et al., 2023). Apart from OPC, various specific cementitious materials, such as magnesia cement, geopolymers, etc., are also developed to immobilize MSWI FA.

4.1 Ordinary Portland cement

OPC is the most traditional type of cement used in various civil engineering applications. It is a hydraulic cement made by grinding clinker, a mixture of calcium silicates and aluminates, primarily consisting of calcium silicates (tricalcium silicate [C3S] and dicalcium silicate [C2S]), tricalcium aluminate (C3A), and tetracalcium aluminoferrite (C4AF), with a small

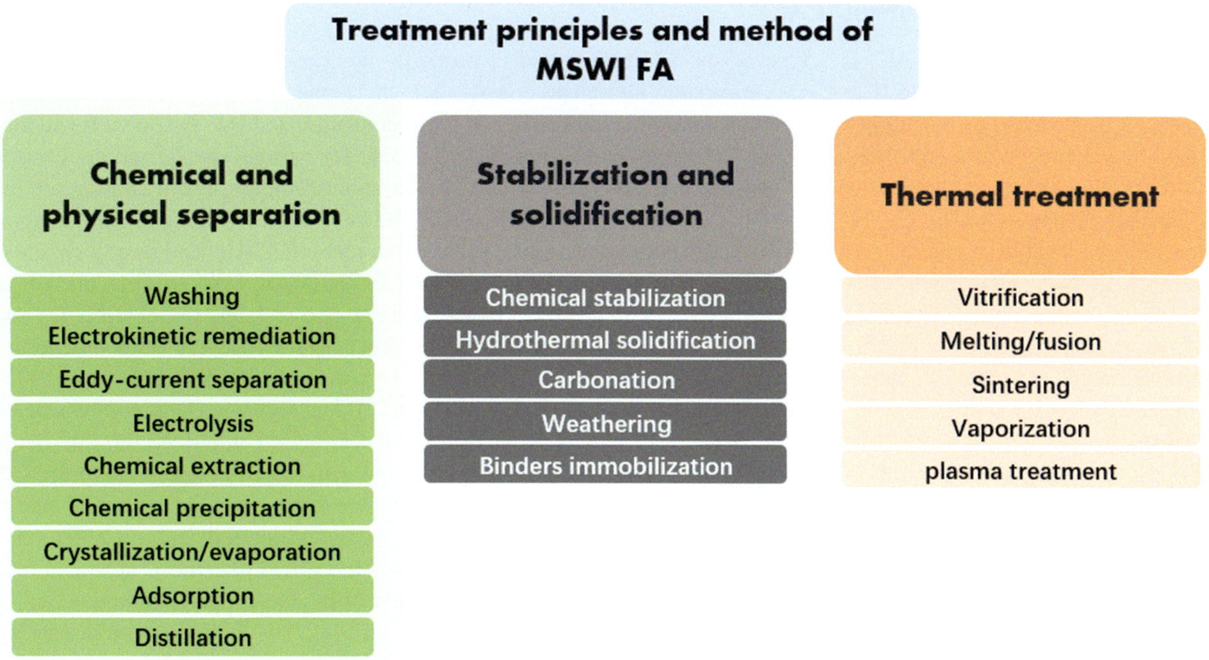

FIGURE 3.6 Summary of treatment principles and methods for MSW ashes. *Modified from Luo, H., Cheng, Y., He, D., Yang, E., 2019. Review of leaching behavior of municipal solid waste incineration (MSWI) ash. Sci. Total Environ. 668, 90–103 and Al-Ghouti, M.A., Khan, M., Nasser, M.S., Al-Saad, K., Heng, O.E., 2021. Recent advances and applications of municipal solid wastes bottom and fly ashes: insights into sustainable management and conservation of resources. Environ. Technol. Innov. 21, 101267.*

amount of gypsum to control the setting time. The clinker is produced by heating a mixture of limestone and clay in a rotary kiln at high temperatures. When OPC is mixed with water, a chemical reaction is conducted named hydration. The calcium silicates hydrate (C-S-H) to form C-S-H gel, which contributes strength to the cement matrix, and $Ca(OH)_2$, which provides the high alkalinity of the cement paste, and other hydration products (monosulfate-type hydrated calcium aluminate sulfate [AFm], trisulfate-type hydrated calcium aluminate sulfate [Aft], hydrated calcium aluminosilicate [CASH], etc.). CSH gel, AFt, AFm, and CASH are insoluble hydration products that are proven to be an important fraction in the immobilization HTEs. As OPC is commonly applied for the treatment and disposal of hazardous wastes, the immobilization of MSWI FA is well investigated. However, the mechanical properties of the OPC block could be influenced after adding MSWI FA. The compressive strength of OPC will be decreased by increasing the added MSWI FA amount (Li et al., 2018a). Since the HTEs, chlorides, and sulfates co-exist in MSWI FA, the hydration behavior of OPC could be significantly altered. It is also reported that the maximum allowable content for the MSWI FA in OPC blocks to achieve appreciable mechanical strength was ca. 20 wt%. Specifically, the speciation of HTEs in MSWI FA could modify the working ability and mechanical strength of OPC blocks after immobilization of the MSWI FA (Wei et al., 2011; Wan et al., 2018). MSWI FA can be utilized in the manufacturing of cement composites with improved durability properties by frost durability tests (Girskas et al., 2021). It is observed that some cations, such as Cr, could enhance the C3S hydration, while some cations, such as Zn, could delay the hydration process of C3S. When using OPC to immobilize MSWI FA, the interior of OPC is a highly alkaline environment. Since most heavy metals, such as Pb, Cr, Cu, etc., could be precipitated as relatively low solubility products (e.g., K_{sp} of $Pb(OH)_2$ is 1.4×10^{-20}) in high pH environments, the leaching of HTEs in MSWI FA can be reduced. The leachabilities of HTEs from the blocks were highly dependent on the pH (Arickx et al., 2010; Weibel et al., 2021). Moreover, the solubility of the formed mineral was also one of the crucial factors to influence HTEs leaching (Polettini et al., 2001; Du et al., 2019; Chen et al., 2019, Dontriros et al., 2020). Considering most of the heavy metal ions could form low solubility products under highly alkaline conditions, the OPC matrix exhibits a hyperalkaline condition, which could precipitate heavy metal ions from MSWI FA and thus decrease the leachability of the HTEs (Lan et al., 2022). Notability, some anions, such as Se and I, exhibit high mobility under alkaline conditions (Guo et al., 2017b, 2019). According to the study by Du et al., under natural conditions, the fixation of the HMs in MSWI FA by OPC was highly influenced by carbonates, hydrous Fe oxides, and dissolved organic carbon. The long-term security of OPC-immobilized MSWI FA in a natural environment should be concerned (Du et al., 2019).

4.2 Calcium aluminate cement (CAC)

Calcium aluminate cement (CAC) is a type of special cement that is primarily composed of monocalcium aluminate ($CaAl_2O_4$), dicalcium aluminate ($Ca_2Al_2O_5$), tricalcium aluminate ($Ca_3Al_2O_6$), and tetracalcium aluminoferrite ($Ca_4Al_2Fe_2O_{10}$). It is produced by calcining a mixture of limestone and bauxite at high temperatures. CAC is known for its rapid strength development, and high fire, acid, and sulfate resistance, making it suitable for various specialized applications. Different from OPC, the hydration reaction and the reaction condition of CAC are more complex. The hydration products of CAC are CAH_{10}, C_2AH_8, C_3AH_6, and AH_3 (Taylor, 1997; Midgley and Midgley, 1975). Among them, hexagonal metastable phases (CAH_{10} and C_2AH_8), which will finally convert into stable cubic phase (C_3AH_6) and poorly crystalline phase AH_3-gibbsite, cause the porosity increasing and strength reduction (Shirani et al., 2020). CAC exhibits outstanding performance on the rapid setting and hardening, resistance to chemical attack, and low porosity, and these properties make CAC as the ideal alternative binder materials for hazardous wastes. CAC has been applied for the immobilization of hazardous waste, covering hexavalent chromium, strontium, and chloride ions, etc. (Jin et al., 2016; Irisawa et al., 2022). Some studies compared the immobilization performance between CAC and OPC. CAC exhibits excellent performance for stabilizing heavily contaminated soils containing various HTEs (i.e., Ba, As, Be, Cd, Co, Cr, Cu, Hg, Ni, Pb, Sb, Se, Sn, Tl, V, and Zn) (Calgaro et al., 2021). The CAC system has a moderate pH compared with the OPC system, favoring CAC to immobilize some amphoteric metals, such as Zn and Al (Kinoshita et al., 2013). Due to the ettringite formation and low porosity of the CAS, crystallization and physical encapsulation could result in a successful immobilization (Calgaro et al., 2021). Application of CAC for the stabilization/solidification of MSWI FA is a viable method to achieve; it is also found that MSWI FA incorporated into CAC had better workability due to the formation of monocarbonate aluminate, and the material strength increased by 48% compared with the control CAC (López-Zaldívar et al., 2017). Cr(VI) could react with Ca^{2+} to form low solubility products on the surface of the clinker, hindering the hydration reaction, and thus reducing the strength of CAC cement by prolonging the setting time (Pang et al., 2022). Considering the high toxicity of the MSWI FA, some researchers also immobilized MIFA by phosphate-modified CAC (Chen et al., 2021). It is reported that the immobilization efficiency of Pb from MSWI FA by using phosphate-modified CAC is significantly increased, from 89.4% to 98.5% compared with OPC.

4.3 Magnesia cement

Magnesia cement is a kind of cement whose main component is light-burned magnesia. It is a gelatinous material produced by burning magnesite, dolomite, etc. into light-burned magnesia, and then mixing it with an aqueous solution of magnesium chloride or magnesium sulfate.

Among them, magnesium phosphate cement (MPC) is originally developed as an immobilization material for hazardous waste immobilization among magnesia cement (Haque and Chen, 2019). Unlike OPC, which relies on the hydration of calcium silicates, MPC relies on the formation of magnesium phosphate compounds to harden. MPC is a type of rapid-setting cement that is produced by the reaction between magnesium oxide (MgO) and ammonium dihydrogen phosphate ($NH_4H_2PO_4$) in water. Based on the different phosphate sources, MPC hydration products mainly include struvite ($NH_4MgPO_4 \cdot 6H_2O$), flint $(NH_4)_2Mg(HPO_4)_2 \cdot 4H_2O$, limonite ($NH_4MgPO_4 \cdot H_2O$), and hard rock ($NaNH_4HPO_4 \cdot 4H_2O$). MPC has the advantages of high early strength development, rapid setting and hardening, high chemical resistance, and low shrinkage (Hou et al., 2016), and MPC is very effective in immobilizing HTEs. However, MPC has some limitations, such as low long-term durability and susceptibility to freeze-thaw damage. Due to these obvious advantages, MPC has been developed and utilized for the immobilization of various hazardous wastes. It is found that less than 40% of MSWI FA can be stabilized by MPC and detoxified, and less than 20% of MSWI FA can be used as additives of MPC for civil engineering (Su et al., 2016). However, when a high content of HTEs (such as Zn) exists, although the leached concentrations of HTEs are still lower than the regulatory, it will prolong the setting time of MPC and reduce the compressive strength (Wang et al., 2017). A flocculant may need to be added to speed up its coagulation. In the follow-up study of its immobilization mechanism, it was found that some HTEs, such as Pb, immobilized via the coprecipitation of pyroblende, resulting in a low leaching ability of MPC (Wang et al., 2020).

Magnesium silicate hydrate (M-S-H) cement is developed as the novel magnesia cement, which has high early strength development, low shrinkage, high durability, low carbon footprint, and excellent adsorption capacity (Nied et al., 2016). Because M-S-H cement immobilizes PTEs with a pH range of 9.5–10.5, M-S-H cement immobilizes hazardous wastes longer than OPC, and its immobilization efficiency for Zn and Pb was reported to be over 95% (Wang et al., 2020). MgO-based binder is developed for solidifying MSWI FA. In the binder, $Mg(OH)_2$ played a major role in the immobilization of some HTEs through alkaline precipitation, while M-C-S-H gels improve the stability of the blocks (Duan et al., 2023). MgO also can be applied for activating ground granulated blast furnace slag (GGBS) to form M-S-H gel for immobilization of MSWI FA (Sun et al., 2023).

Magnesium oxysulfate cement (MOSC) is also a new type of magnesia cement, which is formed by the reaction of MgO and $MgSO_4$ solutions (Walling and Provis, 2016), and its main composition is magnesium hydroxide sulfate. MOSC has several advantages over traditional Portland cement, including high early strength development, rapid setting and hardening, and high fire resistance. Compared with calcium-based cement, MOSC exhibits better passivation ability with HTEs, which can be applied to immobilize high-concentration HTEs in hazardous waste. Therefore, MOSCs exhibited excellent immobilization effects in the immobilization of zinc-rich hazardous wastes (Guo et al., 2021). In addition, in terms of doping amount, a small amount of active silicon in MSWI FA interfered with the reaction and adversely affected the immobilization efficiency of Pb, As, and other elements. By using 33 wt% of a specially prepared MOSC to treat MSWI fly ash, a relatively excellent fixation performance can be obtained (Wang et al., 2022).

4.4 Alkali-activated cement

Alkali-activated cement (AAC), also named geopolymer, is a special type of cementitious material, which is different from the traditional OPC. AAC is a green cementitious material produced by using alkaline activators to activate active powders (such as metallurgical slag, fly ash, fly ash, etc.). Compared with OPC, AAC has very excellent durability in harsh environments (Shi et al., 2019) and AAC can reduce CO_2 emissions by 9%−97% during the production process (Provis and Bernal, 2014). AAC is a promising method for immobilizing MSWI FA due to its ability to form a stable, solid matrix that can effectively encapsulate contaminants and reduce their potential for leaching into the environment. The use of AAC for MSWI FA immobilization has been investigated in several studies, which have demonstrated its potential for improving the environmental performance of fly ash disposal and reuse. Shi and Fernández-Jiménez provide an overview of the use of AAC for the stabilization/solidification of hazardous and even radioactive wastes, including MSWI FA (Shi and Fernández-Jiménez, 2006). The mechanisms of immobilization and the factors influencing the performance of AAC-based materials also attracted many concerns (Zhu et al., 2022).

The effects of washing on the immobilization of MSWI FA using AAC are also discussed, showing that water washing can improve mechanical properties and reduce the leachability of HTEs in the geopolymer matrix (Liu et al., 2017). The stabilization and solidification of MSWI fly ash through AAC by using a calcium-looping process−derived CaO-based activator is possible, and this approach can effectively immobilize heavy metals and reduce their leaching potential (Li et al., 2018b). In the process of AAC immobilization, alumina and silicate are dissolved under the hyperalkaline condition to generate a 3D network structure, which is also considered as the precursors of the zeolite (Kiventerä et al., 2018), and HTEs are bound in the 3D network skeleton (Guo et al., 2017a,b,c). Meanwhile, negatively charged ions would form chemical bonds or be adsorbed in the pore structures of amorphous aluminum silicate member rings (Winkler et al., 2004; Wang et al., 2019). It is noting that the immobilization of AAC on MSWI FA is significantly affected by the pH value, and the long-term stability of its immobilization needs to be systematically studied.

5. Immobilization mechanisms

Cement solidification FA technology refers to the process of mixing FA and cementitious materials for the formation of a stable matrix. During the cementitious materials hydration process, both chemical immobilization and physical encapsulation of HTEs occurred, such as physical bonding, adsorption, isomorphic substitution, composite material precipitation, and passivation. The HTEs in the MSWI FA are immobilized in the form of hydroxides or complexes in the hydrated silicate generated after the hydration reaction, forming a low heavy metal leaching toxicity and long-term stable hydration product such as block C-S-H, C-A-S-H, and ettringite, etc. (Guo et al., 2019; Shao et al., 2020).

C-S-H has a strong cation exchange capacity. Compared with alkali metal ions, C-S-H often has stronger adsorption for heavy metal cations, such as Cr^{3+}, Cd^{2+}, Pb^{2+}, and Co^{2+} (Bernard et al., 2021). Therefore, heavy metal cations easily undergo ion exchange and then stabilize in the crystal structure of the C-S-H. In addition, cement hydration products are mainly composited by Ca^{2+}, Al^{3+}, and Si^{4+}, which have the potential to be substituted by other cations; SO_4^{2-} and OH^-in the hydration products could be substituted by pollutant anions (Żak and Deja, 2015). Apart from C-S-H, AFt is a secondary hydration product that forms during the early stages of Portland cement hydration. However, when there is a lack of chloride or sulfate, the AFt phase easily transforms into AFm. Both AFt and AFm exhibit as scavengers in the immobilization of HTEs through ions substitution, formation of chemical bonding, and chemical adsorption. Numerous studies have proved that both AFt and AFm are ion exchangers and have abilities to immobilize HTEs. Some divalent cations, such as Zn^{2+}, Cd^{2+}, Pb^{2+}, and Ni^{2+}, can replace Ca^{2+} sites, while some trivalent or higher valence cations, including, Ti^{3+}, Fe^{3+}, Cr^{3+}, and even Si^{4+}, can substitute Al^{3+} sites in AFt and AFm. Moreover, AFt and AFm also have anionic sites, which are occupied by Cl^- or SO_4^{2-}. The anions, such as SeO_4^{2-}, CO_3^{2-}, CrO_4^{2-}, and $AsO_4^{3}-$, can substitute SO_4^{2-} or

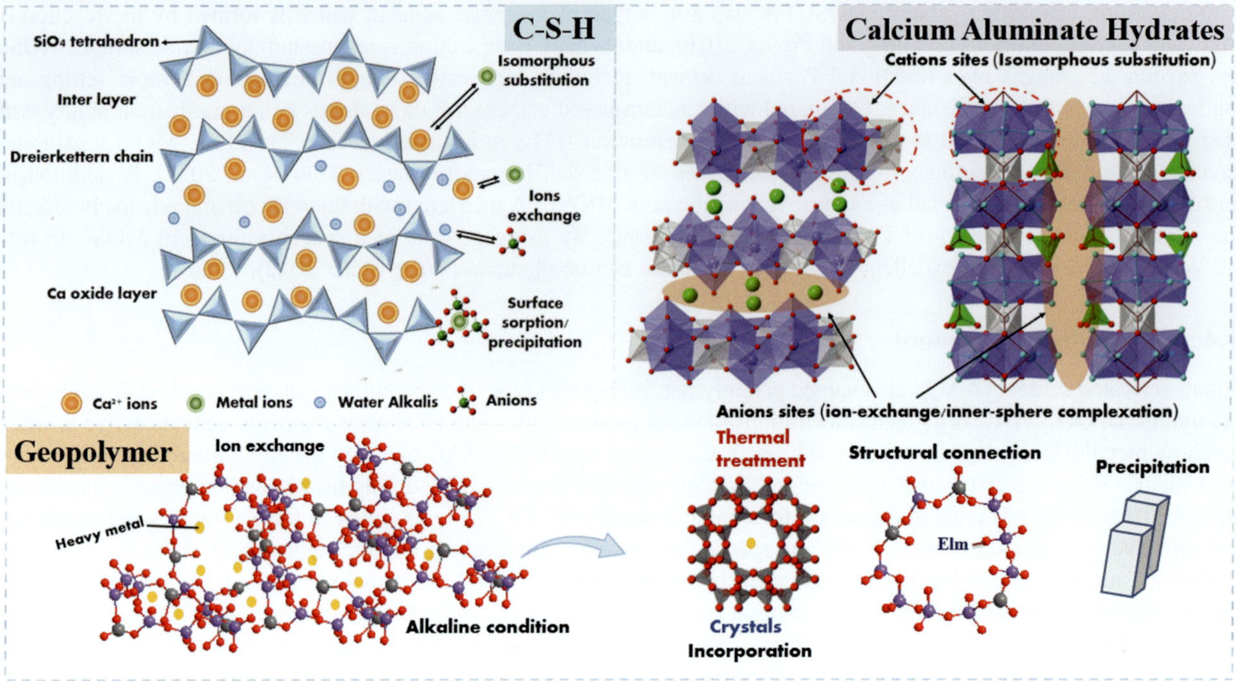

FIGURE 3.7 Summary of immobilization mechanism of various cementitious materials to MSWI FA and HTEs. *Modified from Guo, B., Sasaki, K., Hirajima, T., 2017a. Characterization of the intermediate in formation of selenate-substituted ettringite. Cem. Concr. Res. 99, 30–37; Du, B., Li, J., Fang, W., Liu, J., 2019. Comparison of long-term stability under natural ageing between cement solidified and chelator-stabilised MSWI fly ash. Environ. Pollut. 250, 68–78; and Baldermann, A., Landler, A., Mittermayr, F., Letofsky-Papst, I., Steindl, F., Galan, I., Dietzel, M., 2019. Removal of heavy metals (Co, Cr, and Zn) during calcium–aluminium–silicate–hydrate and trioctahedral smectite formation. J. Mater. Sci. 54, 9331–9351.*

Cl^- in anionic sites (De Weerdt et al., 2014; Saeed et al., 2015). In terms of geopolymer immobilization, due to its compact structure and high strength, physical encapsulation acts as the basic mechanism for waste S/S, avoiding the immobilized HTEs leached out. Moreover, the structure of geopolymer consists of the Al tetrahedrons, which exhibit negative charges and allow cations exchange with Na^+ or K^+. It is worth noting that some cations have the potential to be substituted with Al and directly link with Si during the geopolymerization process. On the other hand, the precipitation of cations has also another important role for the geopolymer-based S/S (Baldermann et al., 2019; Tian et al., 2022). Moreover, the structure of geopolymer can be converted from amorphous to crystals and thus promotes the immobilization of HTEs (Tian et al., 2020). The interactions between C-S-H, AFt, AFM, and geopolymers with HTEs are shown in Fig. 3.7.

Please note that the aforementioned content only mentions some possible mechanisms of cement fixation of MSWI FA. The fixation mechanisms of different HTEs vary significantly. For example, Pb can be effectively fixed through the chemical precipitation of hydroxides and the physical encapsulation (Pan et al., 2019), while AsO_4^{3-} can be fixed by forming Ca-As precipitate and adsorbing it on C-S-H gel (Guo et al., 2019). In addition, anionic species (such as 129I) can be stabilized by replacing IO_3^- in the AFt structure, while I^- is difficult to be immobilized in AFt (Guo et al., 2020).

6. Summary and future trends

In recent years, the treatment of MSWI FA using various cement materials has gained significant attention. This is because MSWI FA is considered hazardous waste and poses a potential threat to the environment and human health if not properly managed. Cement-based immobilization has been identified as an effective method for stabilizing MSWI FA, thereby reducing its mobility and potential for leaching.

This chapter summarizes some methods of effectively immobilization of MSWI residues, especially for MSWI FA. Several types of cement materials have been summarized for their potential to immobilize MSWI FA, including OPC, CAC, and geopolymer cement. Each of these materials has unique properties that make them suitable for immobilizing MSWI FA under different conditions. OPC is the most used cement material for MSWI FA immobilization. It has been shown to effectively immobilize heavy metals and reduce the leaching of both metals and organic compounds. However, the long-term stability of OPC-based immobilization remains a concern due to the potential for carbonation and the formation of secondary minerals that may release immobilized contaminants. CAC has been identified as a promising alternative to OPC for MSWI FA immobilization. This material has a higher resistance to acid attack and can form stable mineral phases that are less prone to leaching. In addition, CAC can be used in combination with other materials such as fly ash and slag to improve its performance. AAC is another promising material for MSWI FA immobilization. It is produced from industrial by-products such as fly ash, slag, and metakaolin, which makes it an environmentally friendly alternative to traditional cement materials. AAC has been shown to effectively immobilize heavy metals and reduce the leaching of both metals and organic compounds. However, further research is needed to optimize the performance of geopolymer cement for MSWI FA immobilization.

In addition to the development of new cement materials, future trends in MSWI FA immobilization will also focus on improving the performance of existing materials. This includes optimizing the composition of cement-based immobilization matrices, improving mixing techniques, and developing more effective curing methods. In addition, the use of additives such as carbon materials may be explored to enhance the performance of cement-based immobilization. The durability of cementitious after immobilization of MSWI FA in harsh environments should also be considered. In the case of actual landfill filling, the long-term leaching of HTEs in the final product remains to be verified under the acidic, various salts, and more complex environmental conditions (atmospheric carbon dioxide, acid rain, etc.). Therefore, future research also needs to consider the leaching of real landfills and cement into the working environment to better understanding the leaching behavior and environmental risks of HTEs in MSWI FA cementitious materials system under complex conditions.

Overall, the immobilization of MSWI FA using various cement materials is a promising approach for reducing the potential environmental and health risks associated with this hazardous waste. While OPC remains the most used material for MSWI FA immobilization, alternative materials such as CAC and AAC show great potential for improving the long-term stability and effectiveness of immobilization. As research in this field continues, it is likely that new and innovative approaches to MSWI FA immobilization will emerge, further improving our ability to manage this hazardous waste.

References

Al-Ghouti, M.A., Khan, M., Nasser, M.S., Al-Saad, K., Heng, O.E., 2021. Recent advances and applications of municipal solid wastes bottom and fly ashes: insights into sustainable management and conservation of resources. Environ. Technol. Innov. 21, 101267.

Ayobami, A.B., 2021. Performance of wood bottom ash in cement-based applications and comparison with other selected ashes: overview. Resour. Conserv. Recycl. 166, 105351.

Abdulmatin, A., Khongpermgoson, P., Jaturapitakkul, C., Tangchirapat, W., 2018. Use of eco-friendly cementing material in concrete made from bottom ash and calcium carbide residue Arab. J. Sci. Eng. 43, 1617−1626.

Arickx, S., De Borger, V., Van Gerven, T., Vandecasteele, C., 2010. Effect of carbonation on the leaching of organic carbon and of copper from MSWI bottom ash. Waste Manage. (Tucson, Ariz.) 30, 1296−1302.

Bernard, E., Yan, Y., Lothenbach, B., 2021. Effective cation exchange capacity of calcium silicate hydrates (CSH). Cem. Concr. Res. 143, 106393.

Baldermann, A., Landler, A., Mittermayr, F., Letofsky-Papst, I., Steindl, F., Galan, I., Dietzel, M., 2019. Removal of heavy metals (Co, Cr, and Zn) during calcium−aluminium−silicate−hydrate and trioctahedral smectite formation. J. Mater. Sci. 54, 9331−9351.

Chen, L., Wang, Y.S., Wang, L., Zhang, Y., Li, J., Tong, L., Hu, Q., Dai, J.G., Tsang, D.C.W., 2021. Stabilisation/solidification of municipal solid waste incineration fly ash by phosphate-enhanced calcium aluminate cement. J. Hazard Mater. 408, 124404.

Clavier, K.A., Paris, J.M., Ferraro, C.C., Bueno, E.T., Tibbetts, C.M., Townsend, T.G., 2021. Washed waste incineration bottom ash as a raw ingredient in cement production: implications for lab-scale clinker behavior. Resour. Conserv. Recycl. 169, 105513.

Caviglia, C., Destefanis, E., Pastero, L., Bernasconi, D., Bonadiman, C., Pavese, A., 2022. MSWI fly ash multiple washing: kinetics of dissolution in water, as function of time, temperature and dilution. Minerals 12, 742.

Calgaro, L., Contessi, S., Bonetto, A., Badetti, E., Ferrari, G., Artioli, G., Marcomini, A., 2021. Calcium aluminate cement as an alternative to ordinary Portland cement for the remediation of heavy metals contaminated soil: mechanisms and performance. J. Soils Sediments 21, 1755−1768.

Chen, J., Lin, X., Li, M., Mao, T., Li, X., Yan, J., 2022. Heavy metal solidification and CO_2 sequestration from MSWI fly ash by calcium carbonate oligomer regulation. J. Clean. Prod. 359, 132044.

Colangelo, F., Cioffi, R., Montagnaro, F., Santoro, L., 2012. Soluble salt removal from MSWI fly ash and its stabilization for safer disposal and recovery as road basement material. Waste Manage. (Tucson, Ariz.) 32, 1179−1185.

Chuai, X., Yang, Q., Zhang, T., Zhao, Y., Wang, J., Zhao, G., Zhang, J., 2022. Speciation and leaching characteristics of heavy metals from municipal solid waste incineration fly ash. Fuel 328, 125338.

Cieplik, M.K., De Jong, V., Bozovič, J., Liljelind, P., Marklund, S., Louw, R., 2006. Formation of dioxins from combustion micropollutants over MSWI fly ash. Environ. Sci. Technol. 40, 1263−1269.

Chen, Z., Lu, S., Tang, M., Ding, J., Buekens, A., Yang, J., Yan, J., 2019. Mechanical activation of fly ash from MSWI for utilization in cementitious materials. Waste Manage. (Tucson, Ariz.) 88, 182−190.

Chen, L., Nakamura, K., Hama, T., 2023. Review on stabilization/solidification methods and mechanism of heavy metals based on OPC-based binders. J. Environ. Manage. 332, 117362.

Duan, Y., Liu, X., Zheng, L., Khalid, Z., Long, L., Jiang, X., 2023. MgO-based binders with different formulations for solidifying Pb and Cd in MSWI fly ash: solidification effect and related mechanisms. Process Saf. Environ. Prot 175, 160–167.

Du, B., Li, J., Fang, W., Liu, J., 2019. Comparison of long-term stability under natural ageing between cement solidified and chelator-stabilised MSWI fly ash. Environ. Pollut. 250, 68–78.

Dontriros, S., Likitlersuang, S., Janjaroen, D., 2020. Mechanisms of chloride and sulfate removal from municipal-solid-waste-incineration fly ash (MSWI FA): effect of acid-base solutions. Waste Manage. (Tucson, Ariz.) 101, 44–53.

De Weerdt, K., Orsáková, D., Geiker, M.R., 2014. The impact of sulphate and magnesium on chloride binding in Portland cement paste. Cem. Concr. Res. 65, 30–40.

Geng, C., Liu, J., Wu, S., Jia, Y., Du, B., Yu, S., 2020. Novel method for comprehensive utilization of MSWI fly ash through co-reduction with red mud to prepare crude alloy and cleaned slag. J. Hazard Mater. 384, 121315.

Girskas, G., Kizinievič, O., Kizinievič, V., 2021. Analysis of durability (frost resistance) of MSWI fly ash modified cement composites. Arch. Civ. Mech. Eng. 21, 1–12.

Gu, F., Zhang, Y., Tu, Y., Wu, X., Zhu, Y., Long, Y., Shen, D., 2022. Assessing magnesia effect on preparing refractory materials from ferrochromium slag. Ceram. Int. 48, 13100–13107.

Guo, B., Sasaki, K., Hirajima, T., 2017a. Characterization of the intermediate in formation of selenate-substituted ettringite. Cem. Concr. Res. 99, 30–37.

Guo, B., Sasaki, K., Hirajima, T., 2017b. Selenite and selenate uptaken in ettringite: immobilization mechanisms, coordination chemistry, and insights from structure. Cem. Concr. Res. 100, 166–175.

Guo, B., Xiong, Y., Chen, W., Saslow, S.A., Kozai, N., Ohnuki, T., Sasaki, K., 2020. Spectroscopic and first-principles investigations of iodine species incorporation into ettringite: implications for iodine migration in cement waste forms. J. Hazard Mater. 389, 121880.

Guo, B., Nakama, S., Tian, Q., Pahlevi, N.D., Hu, Z., Sasaki, K., 2019. Suppression processes of anionic pollutants released from fly ash by various Ca additives. J. Hazard Mater. 371, 474–483.

Guo, B., Liu, B., Yang, J., Zhang, S., 2017c. The mechanisms of heavy metal immobilization by cementitious material treatments and thermal treatments: a review. J. Environ. Manage. 193, 410–422.

Guo, B., Tan, Y., Wang, L., Chen, L., Wu, Z., Sasaki, K., Tsang, D.C., 2021. High-efficiency and low-carbon remediation of zinc contaminated sludge by magnesium oxysulfate cement. J. Hazard Mater. 408, 124486.

Hu, Y., Zhang, P., Chen, D., Zhou, B., Li, J., Li, X.W., 2012. Hydrothermal treatment of municipal solid waste incineration fly ash for dioxin decomposition. J. Hazard Mater. 207, 79–85.

Huang, B., Gan, M., Ji, Z., Fan, X., Zhang, D., Chen, X., Fan, Y., 2022. Recent progress on the thermal treatment and resource utilization technologies of municipal waste incineration fly ash: a review. Process Saf. Environ. Prot. 159, 547–565.

Huber, F., Blasenbauer, D., Aschenbrenner, P., Fellner, J., 2020. Complete determination of the material composition of municipal solid waste incineration bottom ash. Waste Manage. (Tucson, Ariz.) 102, 677–685.

Haque, M.A., Chen, B., 2019. Research progresses on magnesium phosphate cement: a review. Constr. Build. Mater. 211, 885–898.

Hou, D., Yan, H., Zhang, J., Wang, P., Li, Z., 2016. Experimental and computational investigation of magnesium phosphate cement mortar. Constr. Build. Mater. 112, 331–342.

Irisawa, K., Namiki, M., Taniguchi, T., Garcia-Lodeiro, I., Kinoshita, H., 2022. Solidification and stabilization of strontium and chloride ions in thermally treated calcium aluminate cement modified with or without sodium polyphosphate. Cem. Concr. Res. 156, 106758.

Jiang, X., Zhao, Y., Yan, J., 2022. Disposal technology and new progress for dioxins and heavy metals in fly ash from municipal solid waste incineration: a critical review. Environ. Pollut. 119878.

Jin, S.H., Yang, H.J., Hwang, J.P., Ann, K.Y., 2016. Corrosion behaviour of steel in CAC-mixed concrete containing different concentrations of chloride. Constr. Build. Mater. 110, 227–234.

Ke, Y., Ashiq, A., Grzegorz, L., 2018. Environmental perspectives of recycling various combustion ashes in cement production – a review. Waste Manag. 78, 401–416.

Kiventerä, J., Lancellotti, I., Catauro, M., Dal Poggetto, F., Leonelli, C., Illikainen, M., 2018. Alkali activation as new option for gold mine tailings inertization. J. Clean. Prod. 187, 76–84.

Kinoshita, H., Swift, P., Utton, C., Carro-Mateo, B., Marchand, G., Collier, N., Milestone, N., 2013. Corrosion of aluminium metal in OPC- and CAC-based cement matrices. Cem. Concr. Res. 50, 11–18.

Kirkelund, G.M., Skevi, L., Ottosen, L.M., 2020. Electrodialytically treated MSWI fly ash use in clay bricks. Constr. Build. Mater. 254, 119286.

Liu, Y., Shi, C., Zhang, Z., Ou, Z., 2017. Immobilization of MSWI fly ash through geopolymerization: effects of water-wash. Waste Manage. (Tucson, Ariz.) 61, 157–164.

Li, X., Liu, Y., Zhang, Z., Gao, X., Wang, H., 2018a. Stabilization/solidification of municipal solid waste incineration fly ash using calcium-looping process-derived CaO-based activator. J. Clean. Prod. 185, 692–701.

Li, J., Zeng, M., Ji, W., 2018b. Characteristics of the cement-solidified municipal solid waste incineration fly ash. Environ. Sci. Pollut. Res. 25, 36736–36744.

Li, W., Yu, Q., Gu, K., Sun, Y., Wang, Y., Zhang, P., Bian, R., 2022. Stability evaluation of potentially toxic elements in MSWI fly ash during carbonation in view of two leaching scenarios. Sci. Total Environ. 803, 150135.

Liu, Q., Huang, Q., Zhao, Y., Liu, Y., Wang, Q., Khan, M.A., Wang, J., 2022. Dissolved organic matter (DOM) was detected in MSWI plant: an investigation of DOM and potential toxic elements variation in the bottom ash and fly ash. Sci. Total Environ. 828, 154339.

López-Zaldívar, O., Lozano-Díez, R.V., Verdú-Vázquez, A., Llauradó-Pérez, N., 2017. Effects of the addition of inertized MSW fly ash on calcium aluminate cement mortars. Constr. Build. Mater. 157, 1106–1116.

Liu, F., Liu, J., Yu, Q., Jin, Y., Nie, Y., 2005. Leaching characteristics of heavy metals in municipal solid waste incinerator fly ash. J. Environ. Sci. Health 40, 1975–1985.

Li, X.G., Lv, Y., Ma, B.G., Chen, Q.B., Yin, X.B., Jian, S.W., 2012. Utilization of municipal solid waste incineration bottom ash in blended cement. J. Clean. Prod. 32, 96–100.

Li, J., Xu, G., 2022. Circular economy towards zero waste and decarbonization. Circ. Economy 1, 100002.

Luo, H., Cheng, Y., He, D., Yang, E., 2019. Review of leaching behavior of municipal solid waste incineration (MSWI) ash. Sci. Total Environ. 668, 90–103.

Lan, T., Meng, Y., Ju, T., Chen, Z., Du, Y., Deng, Y., Jiang, J., 2022. Synthesis and application of geopolymers from municipal waste incineration fly ash (MSWI FA) as raw ingredient-a review. Resour. Conserv. Recycl. 182, 106308.

Midgley, H.G., Midgley, A., 1975. The conversion of high alumina cement. Mag. Concr. Res. 27, 59–77.

Maldonado-Alameda, A., Giro-Paloma, J., Svobodova-Sedlackova, A., Formosa, J., Chimenos, J.M., 2020. Municipal solid waste incineration bottom ash as alkali-activated cement precursor depending on particle size. J. Clean. Prod. 242, 118443.

Mohammed, S.A., Koting, S., Katman, H.Y.B., Babalghaith, A.M., Abdul Patah, M.F., Ibrahim, M.R., Karim, M.R., 2021. A review of the utilization of coal bottom ash (CBA) in the construction industry. Sustainability 13, 8031.

Masera, K., Hossain, A.K., 2021. Modified selective non-catalytic reduction system to reduce NOx gas emission in biodiesel powered engines. Fuel 298, 120826.

Ministry of Environmental Protection of the People's Republic of China (MEP), 2011. The 12th Five-Year Plan for Comprehensive Prevention and Control of Heavy Metals Pollution. Beijing, China (in Chinese).

Mangialardi, T., 2004. Effects of a washing pre-treatment of municipal solid waste incineration fly ash on the hydration behaviour and properties of ash—Portland cement mixtures. Adv. Cem. Res. 16, 45–54.

Marieta, C., Guerrero, A., Leon, I., 2021. Municipal solid waste incineration fly ash to produce eco-friendly binders for sustainable building construction. Waste Manage. (Tucson, Ariz.) 120, 114–124.

National Bureau of Statistics C, 2021. China Statistical Yearbook of 2021. NBS. http://www.stats.gov.cn/.

Nied, D., Enemark-Rasmussen, K., L'Hopital, E., Skibsted, J., Lothenbach, B., 2016. Properties of magnesium silicate hydrates (MSH). Cem. Concr. Res. 79, 323–332.

Pang, F., Wei, C., Zhang, Z., Wang, W., Wang, Z., 2022. The migration and immobilization for heavy metal chromium ions in the hydration products of calcium sulfoaluminate cement and their leaching behavior. J. Clean. Prod. 365, 132778.

Phutthimethakul, L., Supakata, N., 2022. Partial replacement of municipal incinerated bottom ash and PET pellets as fine aggregate in cement mortars. Polymers 14 (13), 2597.

Poulikakos, L.D., Papadaskalopoulou, C., Hofko, B., Gschösser, F., Falchetto, A.C., Bueno, M., Partl, M.N., 2017. Harvesting the unexplored potential of European waste materials for road construction. Resour. Conserv. Recycl. 116, 32–44.

Provis, J.L., Bernal, S.A., 2014. Geopolymers and related alkali-activated materials. Annu. Rev. Mater. Res. 44, 299–327.

Pan, Y., Rossabi, J., Pan, C., Xie, X., 2019. Stabilization/solidification characteristics of organic clay contaminated by lead when using cement. J. Hazard Mater. 362, 132–139.

Polettini, A., Pomi, R., Sirini, P., Testa, F., 2001. Properties of Portland cement—stabilised MSWI fly ashes. J. Hazard Mater. 88, 123–138.

Rajan, S., Rex, K.R., Pasupuleti, M., Muñoz-Arnanz, J., Jiménez, B., Chakraborty, P., 2021. Soil concentrations, compositional profiles, sources and bioavailability of polychlorinated dibenzo dioxins/furans, polychlorinated biphenyls and polycyclic aromatic hydrocarbons in open municipal dumpsites of Chennai city, India. Waste Manage. (Tucson, Ariz.) 131, 331–340.

Shi, C., Fernández-Jiménez, A., 2006. Stabilization/solidification of hazardous and radioactive wastes with alkali-activated cements. J. Hazard Mater. 137, 1656–1663.

Su, Y., Yang, J., Liu, D., Zhen, S., Lin, N., Zhou, Y., 2016. Effects of municipal solid waste incineration fly ash on solidification/stabilization of Cd and Pb by magnesium potassium phosphate cement. J. Environ. Chem. Eng. 4, 259–265.

Saeed, K.A., Kassim, K.A., Nur, H., Yunus, N.Z.M., 2015. Strength of lime-cement stabilized tropical lateritic clay contaminated by heavy metals. KSCE J. Civ. Eng. 19, 887–892.

Sarmiento, L.M., Clavier, K.A., Paris, J.M., Ferraro, C.C., Townsend, T.G., 2019. Critical examination of recycled municipal solid waste incineration ash as a mineral source for Portland cement manufacture—a case study. Resour. Conserv. Recycl. 148, 1–10.

Shi, C., Qu, B., Provis, J.L., 2019. Recent progress in low-carbon binders. Cem. Concr. Res. 122, 227–250.

Shao, N., Li, S., Yan, F., Su, Y., Liu, F., Zhang, Z., 2020. An all-in-one strategy for the adsorption of heavy metal ions and photodegradation of organic pollutants using steel slag-derived calcium silicate hydrate. J. Hazard Mater. 382, 121120.

Shirani, S., Cuesta, A., De la Torre, A.G., Diaz, A., Trtik, P., Holler, M., Aranda, M.A.G., 2020. Calcium aluminate cement conversion analysed by ptychographic nanotomography. Cem. Concr. Res. 137, 106201.

Sormunen, L.A., Kalliainen, A., Kolisoja, P., Rantsi, R., 2017. Combining mineral fractions of recovered MSWI bottom ash: improvement for utilization in civil engineering structures. Waste Bio. Valori. 8, 1467–1478.

Sun, C., Ge, W., Zhang, Y., Wang, L., Xia, Y., Lin, X., Yan, J., 2023. Designing low-carbon cement-free binders for stabilization/solidification of MSWI fly ash. J. Environ. Manage. 339, 117938.

Taylor, H.F.W., 1997. Calcium Aluminate, Expansive and Other Cements, Cement Chemistry, second ed. Thomas Telford, London, UK, pp. 295–322.

Tang, P., Chen, W., Xuan, D., Zuo, Y., Poon, C.S., 2020. Investigation of cementitious properties of different constituents in municipal solid waste incineration bottom ash as supplementary cementitious materials. J. Clean. Prod. 258, 120675.

Tian, Q., Bai, Y., Pan, Y., Chen, C., Yao, S., Sasaki, K., Zhang, H., 2022. Application of geopolymer in stabilization/solidification of hazardous pollutants: a review. Molecules 27, 4570.

Tian, Q., Guo, B., Sasaki, K., 2020. Immobilization mechanism of Se oxyanions in geopolymer: effects of alkaline activators and calcined hydrotalcite additive. J. Hazard Mater. 387, 121994.

U.S. Environmental Protection Agency (EPA), 1990. Clean Air Act Amendments of 1990; 1st Congress (1989−1990). U.S. EPA, Washington DC.

Vinter, S., Montañés, M.T., Bednarik, V., Hrivnova, P., 2016. Stabilization/solidification of hot dip galvanizing ash using different binders. J. Hazard Mater. 320, 105−113.

Wei, G.X., Liu, H.Q., Zhang, S.G., 2011. Using of different type cement in solidification/stabilization of MSWI fly ash. In: Advanced Materials Research, vol 291. Trans Tech Publications Ltd, pp. 1870−1874.

Wan, S., Zhou, X., Zhou, M., Han, Y., Chen, Y., Geng, J., Hou, H., 2018. Hydration characteristics and modeling of ternary system of municipal solid wastes incineration fly ash-blast furnace slag-cement. Constr. Build. Mater. 180, 154−166.

Weibel, G., Zappatini, A., Wolffers, M., Ringmann, S., 2021. Optimization of metal recovery from MSWI fly ash by acid leaching: findings from laboratory-and industrial-scale experiments. Processes 9, 352.

Wang, P., Xue, Q., Yang, Z., Li, J., Zhang, T., Huang, Q., 2017. Factors affecting the leaching behaviors of magnesium phosphate cement-stabilized/solidified Pb-contaminated soil, part II: dosage and curing age. Environ. Prog. Sustain. Energy 36, 1351−1357.

Wang, L., Chen, L., Guo, B., Tsang, D.C., Huang, L., Ok, Y.S., Mechtcherine, V., 2020. Red mud-enhanced magnesium phosphate cement for remediation of Pb and As contaminated soil. J. Hazard Mater. 400, 123317.

Walling, S.A., Provis, J.L., 2016. Magnesia-based cements: a journey of 150 years, and cements for the future? Chem. Rev. 116, 4170−4204.

Winkler, A., Horbach, J., Kob, W., Binder, K., 2004. Structure and diffusion in amorphous aluminum silicate: a molecular dynamics computer simulation. J. Chem. Phys. 12, 384−393.

Wang, L., Cho, D.W., Tsang, D.C., Cao, X., Hou, D., Shen, Z., Poon, C.S., 2019. Green remediation of As and Pb contaminated soil using cement-free clay-based stabilization/solidification. Enviro. Int. 126, 336−345.

Wang, L., Zhang, Y., Chen, L., Guo, B., Tan, Y., Sasaki, K., Tsang, D.C., 2022. Designing novel magnesium oxysulfate cement for stabilization/solidification of municipal solid waste incineration fly ash. J. Hazard Mater. 423, 127025.

World Bank Group, 2018. What a Waste 2.0: A Global Snapshot of Solid Waste Management to 2050-The Urban Development Series. Int. Bank Reconstr. Dev./World Bank 1818 H Str. NW, Washington, DC. Available at: https://openknowledge.worldbank.org/handle/10986/30317. (Accessed June 2023).

Xia, Y., He, P., Shao, L., Zhang, H., 2017. Metal distribution characteristic of MSWI bottom ash in view of metal recovery. J. Environ. Sci 52, 178−189.

Żak, R., Deja, J., 2015. Spectroscopy study of Zn, Cd, Pb and Cr ions immobilization on C−S−H phase. Spectrochim. Acta: Mol. Biomol. Spectrosc. 134, 614−620.

Zhou, X., Zhang, T., Wan, S., Hu, B., Tong, J., Sun, H., Hou, H., 2020. Immobilizatiaon of heavy metals in municipal solid waste incineration fly ash with red mud-coal gangue. J. Mater. Cycles Waste Manag. 22, 1953−1964.

Zhu, Y., Zheng, Z., Deng, Y., Shi, C., Zhang, Z., 2022. Advances in immobilization of radionuclide wastes by alkali activated cement and related materials. Cem. Concr. Compos. 126, 104377.

Chapter 4

Alkali-activated materials for the stabilization/solidification of heavy metals and radioactive substances in incineration residues

Masaki Takaoka
Department of Environmental Engineering, Graduate School of Engineering, Kyoto University, Kyoto, Japan

1. Introduction

Solid waste incineration, a popular waste management technology worldwide, has several advantages. The high temperatures generated during incineration prevent the rotting of organic matter and spread of diseases and pests (Takeda et al., 2014). Incineration also reduces waste volume and weight, thereby reducing the landfilling area to 1/10th to 1/20th of the area needed for nonincinerated waste. Additionally, methane emissions from landfills of organic wastes are reduced. Finally, the energy recovered during waste incineration can be used for power generation and heating. However, incineration has a notable disadvantage: the generation and discharge of hazardous substances (Takeda and Takaoka, 2013).

Incineration generates several types of solid waste, including ash and slag, boiler ash, filter dust, flue gas cleaning residues (e.g., $CaCl_2$ and NaCl), and wastewater treatment sludge. The types and quantities of incineration residues considerably vary according to incinerator design, operation, and waste input.

Incineration residues are collected at the bottom of the combustion chamber (i.e., bottom ash: 15%–25% by mass of the original waste and approximately 80%–90% of the total residues) and from the components through which gases pass before atmospheric emission. The residues collected after chemical cleaning of flue gases are known as air pollution control (APC) residues. Ash entrained by gases outside the furnace, known as fly ash, is collected before the gases are chemically cleaned. APC residues are sometimes distinguished from fly ash; however, the term "incineration fly ash" typically also includes APC residues (ISWA Report, 2008).

Incineration residues contain various groups of contaminants, including metal ions, amphoteric metals, oxyanionic species, and persistent organic pollutants. Fly ash is regarded as hazardous waste, and its potential release of contaminants into the environment is a concern (Liu et al., 2022). In contrast, bottom ash contains relatively few contaminants and has various uses in European countries (ISWA Report, 2016). Fly ash and APC residues are usually treated before landfilling. The treatment techniques can be grouped into four categories: extraction and separation, thermal treatment, chemical stabilization, and solidification (ISWA Report, 2008). The extraction and separation technique is utilized to remove or recover heavy metals and salts from residues, typically using water or acids (Hong et al., 2000). Although this technique is relatively simple, it generates process water with high metal and salt contents. Thermal treatments such as vitrification, melting, and sintering destroy dioxins and produce very stable products with good leaching properties; however, such treatments are expensive.

Both chemical stabilization and solidification are commonly used to treat incineration residues. Stabilization processes alter the constituents of hazardous waste, transforming it into nonhazardous waste. Solidification processes use additives to change the physical state of waste (e.g., from liquid to solid) without altering its chemical properties (EU Commission Decision, 2015). Various chemical stabilization techniques have been developed that significantly reduce leaching from residues; most of these techniques use inorganic or organic chemical reagents in simple, inexpensive processes (Lundtorp

Treatment and Utilization of Combustion and Incineration Residues. https://doi.org/10.1016/B978-0-443-21536-0.00028-9

et al., 2002; Ecke et al., 2003; Hjelmar et al., 2001; Eighmy et al., 1997; Youcai et al., 2002; Mizutani et al., 2000). However, several concerns have recently been identified. First, fly ash samples treated with chelating agents may releach Pb after treatment if appropriate conditions are not maintained (Sakanakura, 2007). Additionally, nitrification inhibition in leachate treatment has become apparent because of the generation of thiourea-like substances by organic chelating agents (Uchida et al., 2012). Solidification processes use cement, asphalt, and gypsum to solidify waste (Polettini et al., 2001), improving the mechanical properties of the waste and decreasing leaching. However, solidification treatments increase the mass and volume of waste; the physical integrity of the final product may deteriorate over time, resulting in increased metal leaching.

Alkali-activated materials (AAMs) show potential as a new chemical stabilization/solidification method. When aluminosilicates (i.e., filler) react with an alkaline solution (i.e., alkaline activator), Si and Al dissolve into the alkaline solution, which solidifies via gelation, reorganization, and polymerization. During this solidification process, metals are stabilized through fixation within the AAM structure. Since Davidovits proposed geopolymers in 1979, research concerning the immobilization of toxic substances using AAMs has been limited and mostly focused on construction material applications (e.g., industrial by-product applications). A few studies regarding the fixation of radioactive substances and toxic metals were published beginning around 1988 (Comrie and Davidovits, 1988), but the literature remained sparse until the mid-2000s; thereafter, the number of such studies rapidly increased through the 2010s, and particularly increased within the past 5 years. Among the various factors that may have influenced this research trend, conventional cement and heavy metal stabilizers (e.g., chelating agents) are not sufficiently immobilizing (e.g., alkali metals such as Cs in cement solidification) or have other environmental impacts (e.g., emission of CO_2 or organic chelate degradation products). Notably, industrial by-products containing aluminosilicates also often include toxic elements; their use as AAMs fillers may confer the dual benefits of fixing these toxic elements while reusing construction materials. In Japan, since the Great East Japan Earthquake, there has been increasing interest in the application of geopolymers to the disposal of radioactive waste, including residues from the thermal treatment of waste contaminated with radioactive substances (Cantarel et al., 2017).

In this chapter, the immobilization mechanism of hazardous substances is summarized, and its application of AAMs to radioactive substances as an example is discussed.

2. Immobilization of toxic elements by AAMs

2.1 Immobilization mechanism

Fig. 4.1 summarizes the previously proposed immobilization mechanisms of AAMs (Van Jaarsverld and Van Deventer, 1999; Provis and Van Deventer, 2009; El-eswed et al., 2017; Ji and Pei, 2019; El-eswed, 2020; Japan Concrete Institute, 2020; Sato, 2021), which can immobilize various heavy metals and radioactive elements; here, Pb is used as a representative metallic element. Five main mechanisms for immobilization of toxic elements are identified. These mechanisms are described separately for convenience; in reality, they are often complex and difficult to distinguish. In addition to the mechanisms described here, impurities in the target waste, aluminosilicate material, or alkaline activator may affect the adsorption of hazardous substances.

(1) Formation of insoluble compounds

AAMs form a three-dimensional (3D) network of structures; separate from that network, heavy metals generally form hydroxides in the presence of concentrated alkali. Some metals form insoluble hydroxide precipitates, which may then react with atmospheric CO_2 to form insoluble carbonates (Bouzar and Mamindy-Pajany, 2023). In Fig. 4.1, Pb is shown as $PbCO_3$; note that the solubilities of $Pb(OH)_2$ and $PbCO_3$ differ by a factor of at least 1000. Pb may also be fixed in insoluble silicate compounds (Shiota et al., 2017).

(2) Chemical fixation to the network structure

This mechanism is summarized under the reaction process in Fig. 4.1. Aluminosilicates such as metakaolin are first dissolved, followed by elution of Si and Al ions in the highly alkaline solution. Under such conditions, gelation, reorganization, and dehydration–condensation polymerization occur, resulting in the connection of SiO_4 and AlO_4 that share oxygen in a four-coordinated form, forming a 3D structure (Bouzar and Mamindy-Pajany, 2023). During this process, Pb may be fixed as a replacement for Al in the bridging structure, or it may be fixed at the end of the matrix via covalent or coordination bonds (Shiota et al., 2017; Luo et al., 2022). In addition, zeolites or zeolite-like materials often form, immobilizing the target toxic elements in the zeolite structure.

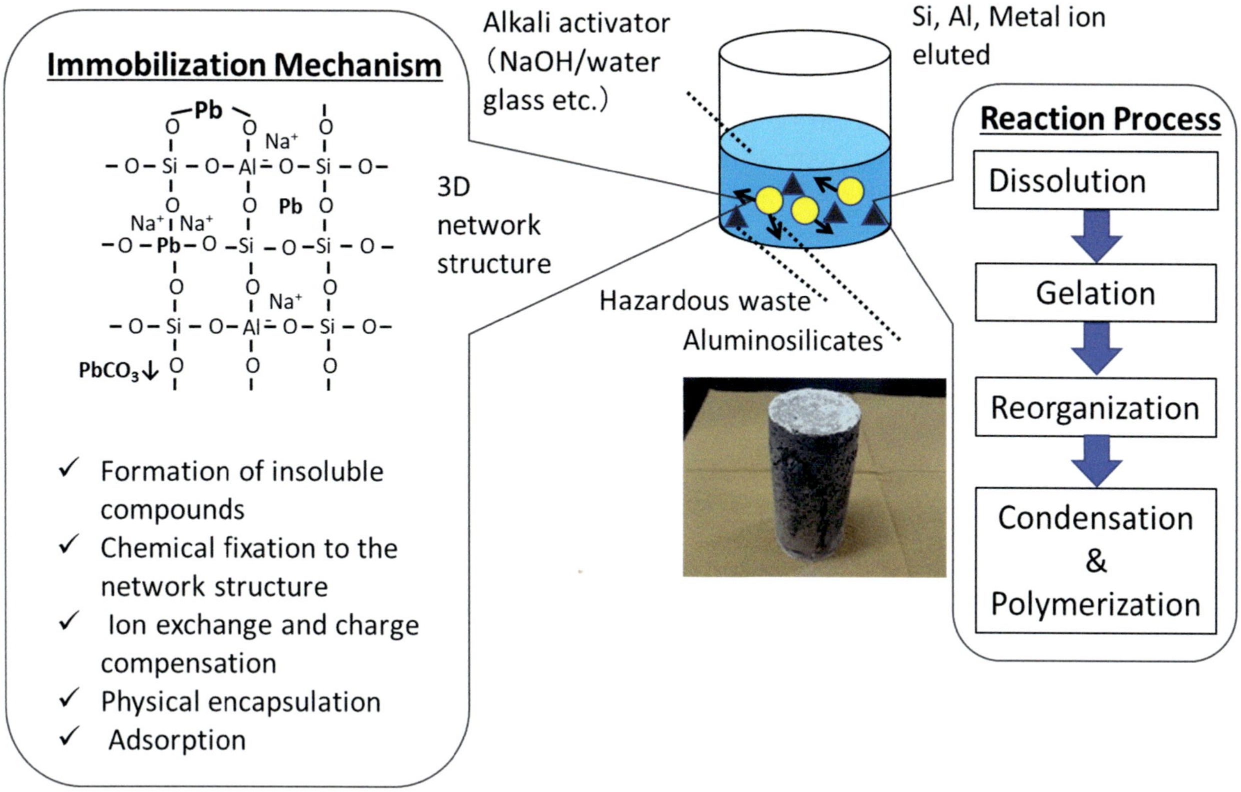

FIGURE 4.1 Mechanism of toxic element fixation by AAMs.

(3) Ion exchange and charge compensation

Monovalent ions (e.g., Na^+) are presumed to compensate for the lack of charge at Al sites in the Si—O—Al 3D network structure. Therefore, the number of ion exchange sites in AAMs varies according to the Si/Al ratio and amount of alkaline metals present. Furthermore, solidified AAMs may form zeolites and other structures in addition to the amorphous 3D network structure. Thus, the target elements are fixed by ion exchange (El-eswed et al., 2017; Sato et al., 2021). Note that this description refers only to the mechanism of uptake during AAM formation; it differs from adsorption, which is described in the following.

(4) Physical encapsulation

During monomer polymerization, hydroxide ions and insoluble salts of target elements may be physically encapsulated within the developing 3D mesh structure. Because water cannot readily infiltrate dense AAMs, the leaching of target elements is limited. Physical encapsulation is one of the main immobilization mechanisms; AAMs may be stronger than cement under optimal formation conditions (e.g., aluminosilicate material, formulation, and curing conditions) (Long et al., 2021; Luo et al., 2022). For instance, minimal leaching of elements from AAMs has been reported in cases where hydroxides or oxyanions did not exhibit low solubility.

(5) Adsorption

In contrast to the previously described four mechanisms, through which elements are immobilized during AAM formation, the adsorption of elements occurs after AAMs have been formed. For instance, heavy metals may be removed from wastewater through adsorption and fixation on AAMs, whereby zeolites and zeolite-like substances within the amorphous 3D network structure have ion exchange capacity and can fix the target elements (He et al., 2021; Yang et al., 2022). However, this process does not necessarily rely on ion exchange alone; it is more accurately described as adsorption. After the disposal of AAM-treated hazardous waste, leaching tests are conducted, and it is assumed that adsorbed target element ions will elute from the AAMs into the leaching test solvent; however, it is also assumed that they will resorb onto the AAM surface.

2.2 Target elements and reaction conditions

Fig. 4.2 shows the elements in the periodic table that have been studied for immobilization using AAMs; the data were primarily compiled by Provis and Van Deventer (2009), and supplemented with other works (Mattigod et al., 2011; Chen et al., 2019), in the cases of I, Tc, Re, and Hg. Some studied elements (gray and black shading, Fig. 4.2) are not readily immobilized. For instance, Sb displays increased mobility in AAMs (Provis and Van Deventer, 2009). Moreover, when highly alkaline activators are used during the production of AAMs, oxyanions (e.g., SeO_4 and SbO_4) are present in the solution (Álvarez-Ayuso, 2008). Elements present as oxyanions are difficult to fix, even by ion exchange or charge compensation; therefore, additional fixation mechanisms are necessary.

AAMs are widely applicable; they can be used to immobilize incinerator residue, slag, sludge, slurry, wastewater, contaminated soil, organic oil, zeolite, ion-exchange resin, and process residues (i.e., salts). AAMs are prepared by mixing an aluminosilicate filler with an alkaline activator. The curing conditions after mixing, especially curing time and temperature, substantially influence the immobilization of elements. Notably, an increase in the curing temperature decreases the necessary curing duration; however, the immobilization rate does not uniformly increase with increasing temperature. The elemental composition ratio of AAMs also influences the properties of the produced AAMs. For instance, the Si/Al ratio of AAMs significantly influences ion exchange, charge compensation, and zeolite formation. Many studies have investigated the immobilization-specific effects of changes in the Si/Al ratio of aluminosilicates and waste materials. Considering the AAM reaction process, the Si/Al ratio of the test materials is important, but the extent to which Si and Al leach from the materials during the initial dissolution process affects the subsequent condensation polymerization reaction and other processes. Na and K alkaline solutions are used as alkali activators. Depending on the alkali concentration, the Si and Al elution rates from aluminosilicates vary and zeolite formation is altered. Alkali metal ions play an important role in AAM formation by first aligning water molecules and then dissolving silicates to form the nuclei of AAMs; they also influence the ratio of hydroxide ions of the target elements. The effect of each of these factors is discussed in Section 3, using Cs immobilization as an example.

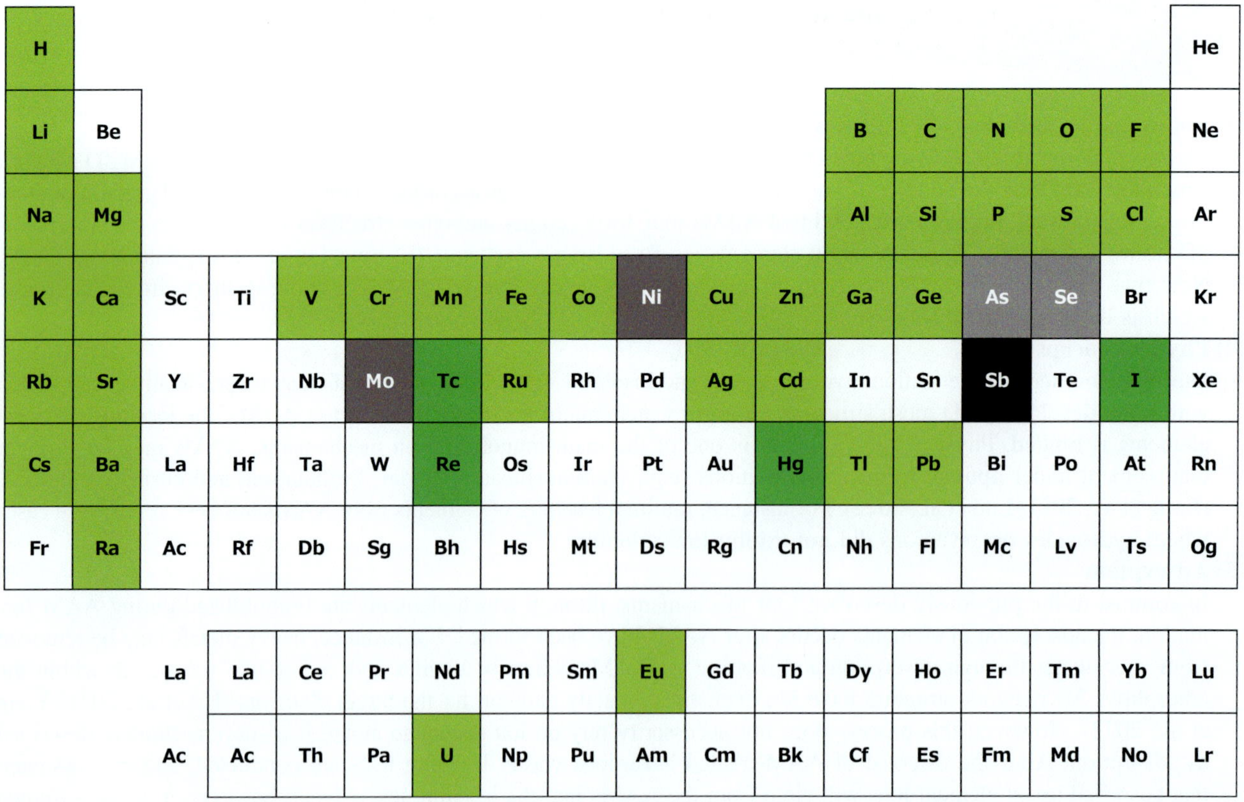

FIGURE 4.2 Periodic table showing elements that have been immobilized in geopolymer-solidified products (Provis and Van Deventer, 2009; Mattigod et al., 2011; Chen et al., 2019). *Light green* (Provis and Van Deventer, 2009) and *dark green* (Mattigod et al., 2011; Chen et al., 2019) shading indicates elements that have been fixed by AAMs. *Gray* and *black* shading indicates elements that are difficult to fix, although some studies have shown success. Unshaded elements have not been studied.

3. Application of AAMs to radioactive waste

3.1 Significance of AAM use

In Japan, cement (i.e., Portland cement and blast furnace cement), asphalt, and plastic materials are specified as materials that can be used for radioactive waste solidification (Meguro and Sato, 2016). Safety assessments have ensured that the use of cement materials for radioactive waste solidification does not affect the surrounding environment, even if all radioactive substances were to leak from the solidified product into the environment after underground disposal (Meguro and Sato, 2016). Regardless, it is important for solidification, as the first among multiple protective measures, to provide maximal prevention of radioactive substance leaching. AAMs are presumed to have better radioactive substance containment performance than cement materials; they have been used in other countries. In addition, whereas cement solidification occurs through a hydration reaction, AAMs undergo a dehydration–condensation–polymerization reaction, yielding a final solidified product with lower water content. This is expected to be an advantage because a lower water content would prevent hydrogen generation via radiolysis of water, thereby preventing decomposition of the solidified product (Cantarel et al., 2019). The SIAL matrix, created by Amec Foster Wheeler (now WOOD), is the most highly regarded geopolymer material for radioactive waste immobilization (Sato, 2021; Meguro and Sato, 2016; Onozaki, 2017; Lichvar et al., 2013). This geopolymer is an alkali AAM with high Al content and low Ca content. It was developed for the on-site solidification of sludge contaminated with radioactive Sr, Cs, and trans-U from the 1976 and 1977 fuel meltdowns at the Bohunice Nuclear Power Plant (NPP) in former Czechoslovakia (present-day Slovakia). It was subsequently tested on waste from the Dukovany and Temelín NPPs in the Czech Republic and then licensed as a solidification material in Slovakia in 2003 and in the Czech Republic in 2006. Additionally, in a pilot-scale plant of a water treatment facility for U mines in Germany, the geopolymer material "Géopolymère" was used for the treatment of U-containing sludge (Sato, 2021; Meguro and Sato, 2016; Hermann et al., 1999).

In Japan, radioactive substances diffused into the environment during and after the nuclear accident at the Fukushima Daiichi Nuclear Power Station (FDNPS). This required the treatment of radioactive contaminated waste in the flow of normal municipal and industrial waste, rather than the conventional radioactive waste treatment and disposal flow. In Fukushima Prefecture, 1.45 million tons of contaminated waste has been treated using temporal waste incinerators (Ministry of the Environment Japan, 2023); during this process, the radioactive substances become concentrated in the incineration residues and require further immobilization.

Because many studies have been conducted regarding the fixation of Cs as a radioactive element, the following section focuses on studies of the influencing factors of Cs fixation and the immobilization mechanism.

3.2 Solidification of radioactive materials

(1) Effect of the Si/Al ratio

Although a lower Si/Al ratio generally supports greater ion exchange and charge compensation, this relationship does not show a constant trend because unique zeolite minerals may form and incorporate Cs. For instance, Aly et al. (2008) prepared geopolymers by mixing silica fume, NaOH solution, and metakaolin, adding small amounts of CsOH and $SrOH_2$, varying the Si/Al ratio from 1.5 to 4.0 (where the H_2O/Na_2O ratio was 7–10), curing at 60°C for 24 h, and removing free water at 400°C for 2 h. Leaching tests of Si, Al, Na, Cs, and Sr showed that the amount of leached Cs reached a minimum at a Si/Al ratio of 2.0; as the Si/Al ratio increased, the leaching of Si, Na, and Sr increased, whereas Al decreased, until the Si/Al ratio exceeded 3.0. Ichikawa et al. (2018) added CsCl to geopolymer adjusted with metakaolin, waterglass (i.e., Na silicate), and NaOH, while varying the Si/Al ratio from 1.34 to 2.04 to investigate the effect on the Cs/Na selectivity coefficient; the zeolite produced at different Si/Al ratios influenced the Cs adsorption rate. Tsujii et al. (2021) prepared geopolymer-solidified materials by varying the Si/Al ratio using several aluminosilicate reagents to investigate its effect on Cs immobilization; they found that the use of Si and Al dissolved in alkali, rather than the Si/Al ratio, described the change in the balance of Si and Al that directly contributed to Cs immobilization. Koide et al. (2021) prepared AAMs by mixing CsCl with Na aluminate and waterglass at room temperature to identify the optimal conditions that could support Cs immobilization. The leaching rate was lower at smaller Si/Al ratios. Using a 15% Cs solution, a Si/Al ratio of 1.5, and 12 days of curing, the Cs leaching rate was 5.1%. Cs-containing zeolites formed even at room temperature; it was assumed that the Cs leaching rate decreased because of zeolite formation, in addition to fixation within the amorphous network.

(2) Effect of curing conditions

Although room-temperature curing is often utilized during the stabilization of other metals and the production of construction materials using AAMs, temperature substantially alters Cs fixation, and many studies have been conducted at relatively high temperatures. For instance, Jiménez et al. (2005) investigated the effect of Cs immobilization in AAMs consisting of coal ash, 8 M NaOH solution, and CsOH or $CsNO_3$ to achieve a Cs concentration of 1% in the material. They reported higher Cs leaching rates at 120°C than at 85°C; Cs leaching rates were lower at 7 days of curing time than at 5 h of curing time. Li et al. (2013) compared the immobilization of Cs with AAMs using coal ash and cement; they prepared solidified products by adding waterglass and stable Cs to coal ash and ion-exchanged water and stable Cs to Portland cement, respectively, and then cured the mixture at 60°C for 28 days. The amount of Cs leached from AAMs was much lower than the amount leached from cement-solidified products; AAMs were superior in terms of acid resistance and compressive strength. Nakamura et al. (2017) prepared AAMs using model fly ash that contained Cs along with metakaolin, waterglass, and NaOH; they investigated the effects of curing temperatures of 80°C, 90°C, and 105°C. Cs leaching from the solidified product at 105°C was reduced to 20% of the amount leaching at 80°C. Using X-ray absorption fine structure (XAFS) analysis, Nakamura et al. also determined that the state of Cs in AAMs at 105°C was closer to the state of pollucite ($Cs(Si_2Al)O_6 \cdot nH_2O$), suggesting that curing temperature was an important factor in suppressing leaching. Takaoka et al. (2018) prepared AAMs from Cs-containing melting fly ash, using metakaolin, waterglass, and NaOH as materials; they utilized a curing temperature of 105°C and varied the curing time. A curing time of 6 h resulted in a Cs leaching rate of approximately 20%, whereas a curing time exceeding 12 h resulted in a leaching rate below 5%, indicating that longer curing times improve Cs immobilization. Finally, Koide et al. (2021) prepared AAMs by mixing CsCl with Na aluminate and waterglass at room temperature and then investigated the effect of curing time on the Cs leaching rate; the elution rate decreased from approximately 30% after 3 h of curing to 1.7% after 60 days of curing.

(3) Effect of the alkaline activator

Higher alkalinity is considered advantageous because it improves aluminosilicate dissolution; however, higher alkalinity does not necessarily result in less leaching of the target elements, which ultimately depends on the final AAM-solidified products. Takaoka et al. (2018) prepared solidified products of Cs-containing melting fly ash, using metakaolin, waterglass, and NaOH; they utilized a curing temperature of 105°C and a curing time of 24 h, while varying the NaOH concentration from 1 to 14 M, and then conducted leaching tests on the solidified products. The lowest Cs leaching rate (<5%) was achieved when 10 M NaOH was used; at all other NaOH concentrations, the Cs leaching rate exceeded 20%. Tsujii et al. (2020) investigated the effect of NaOH concentration in greater detail; the Cs leaching rate rapidly decreased to below 10% at NaOH concentrations of 7−10 M, and it was 2.6% at the lowest tested NaOH concentration of 8 M. X-ray diffraction (XRD) analysis confirmed the formation of chabazite-Na under conditions in which the Cs leaching rate was below 10%. Similarly, nuclear magnetic resonance (NMR) analysis revealed the formation of zeolite with a crystalline structure in the 8 and 10 M NaOH solidified products. They concluded that NaOH concentration had a significant effect on Cs immobilization and AAM structure. Tian et al. (2022) used NaOH and sodium silicate as alkali activators and investigate differences in the leaching of Cs from geopolymers, which suggested the leaching of Cs from geopolymers with sodium silicate was lower. Studies examining the effect of different types of alkaline activators on Cs fixation are limited, but in general, other alkaline activators are also widely used in the preparation of AAMs (Ishwarya et al., 2019) and should be explored further.

(4) Mechanism of Cs immobilization

The results of many studies have suggested that the primary Cs fixation mechanism involves formation of the insoluble Cs compound pollucite. However, researchers have expressed different views concerning the formation temperature necessary to yield pollucite. He et al. (2013) performed thermogravimetry/differential scanning calorimetry analysis of the solidified products of an alkali silicate solution prepared by mixing SiO_2 sol and aqueous CsOH solution with synthetic metakaolin, cured at 50°C for 48 h; the results revealed an endothermic peak at 97.7°C, caused by water evaporation, and an exothermic peak at 1216.3°C, which originated from pollucite crystallization. Moreover, XRD analysis showed an amorphous phase up to 1000°C and the formation of pollucite above 1100°C.

Chlique et al. (2015) prepared an alkali silicate solution using CsOH or NaOH in deionized water with amorphous silica and then mixed it with metakaolin to produce a solidified product that was subsequently cured at room temperature for 5 days; finally, it was heated at 1100°C. XRD analysis of the final solidified product revealed an amorphous phase. In the solidified product produced using only CsOH, primarily pollucite was formed; only small amounts of Cs aluminosilicate and amorphous silica phase were detected. Furthermore, in solidified products containing Na, both nepheline and pollucite were detected, the pollucite-derived peaks became smaller with increasing Na content, and the formation of Cs aluminosilicate was inhibited.

Blackford et al. (2007) performed NMR, transmission electron microscopy/energy dispersive X-ray (TEM/EDX), and XRD analyses of solidified products prepared at 40°C using metakaolin, waterglass, and deionized water at a Si/Al ratio of 2.0 and Na/Al ratio of 1.0; stable Cs was added at 5 wt% in the form of nitrate or hydroxide. XRD and TEM/EDX results indicated that the produced geopolymer was amorphous and that Cs was incorporated into the amorphous phase; no crystalline material was observed.

Shiota et al. (2017) prepared AAMs using municipal solid waste incineration fly ash, calcined pyrophyllite, waterglass, and NaOH at a curing temperature of 105°C; they conducted leaching tests that confirmed a low leaching rate. XRD, SEM, and XAFS analyses revealed that Cs was insoluble because of pollucite formation. Additionally, samples were prepared under various conditions; pollucite content was assessed by XAFS analysis to investigate its relationship with the Cs leaching rate. A negative correlation was observed between the Cs leaching rate and the pollucite content, suggesting that the incorporation of Cs into pollucite was the main cause of immobilization (Shiota et al., 2018).

Thus, the formation of insoluble Cs compounds during the network formation reaction reduces the leachability of Cs. On the other hand, it has been suggested that Cs is immobilized in the amorphous network structure and suppresses leaching, which suggests that the amorphous network also contributes to the suppression of leaching. In the future, the contribution ratio of these immobilization mechanisms, as well as the AAM preparation conditions, should be clarified.

(5) Sr and Cs fixation

In addition to Cs, Sr was detected near the FDNPS after the nuclear accident (Ministry of the Environment, Japan, 2023), although a much smaller amount of Sr was released. Thus, there is a need to consider the fixation of both Cs and Sr when treating highly concentrated radioactive waste. For this purpose, He et al. (2019) performed long-term leaching tests on AAMs created from metakaolin cured at 25 and 60°C, using silica sol and NaOH or KOH solutions as activators to immobilize simulated radioactive Cs^+ and Sr^{2+}. Compared with K-based AAMs, Na-based AAMs yielded much lower Cs^+ leaching rates under otherwise equivalent conditions. Furthermore, when NaCl solution was used in the leaching test (rather than deionized water), Cs^+ and Sr^{2+} leaching rates were enhanced. The researchers speculated that this was related to AAM corrosion by NaCl, based on the appearance of cracks and pores during the leaching test.

Nakamura et al. (2016) also studied the simultaneous immobilization of Cs and Sr. They prepared AAMs using municipal solid waste incineration fly ash, calcined pyrophyllite, NaOH, and waterglass. The leaching rate of Sr remained below 0.7% under all conditions. However, the leaching rate of Cs considerably varied; when the proportion of municipal solid waste incineration fly ash in the total mixed material was 25% and 49%, the Cs leaching rate was 0.83% and 7.2%, respectively. These findings indicate that the optimization of immobilization conditions depends more on Cs than on Sr.

4. Conclusions and perspectives

AAMs show great potential for the stabilization/solidification of heavy metals and radioactive substances in incineration residues. They offer several advantages over traditional cement-based binders, including improved durability, enhanced chemical resistance, and reduced CO_2 emissions. AAMs can effectively immobilize heavy metals and radioactive substances, preventing leaching and minimizing the risk of pollution.

A key advantage of AAMs is their high alkalinity, which promotes the formation of stable mineral phases that encapsulate contaminants. The alkaline environment helps to precipitate heavy metals and radioactive substances, reducing their mobility and leachability. AAMs also possess a greater binding capacity compared with conventional binders, facilitating the incorporation of a larger quantity of contaminants within the matrix.

However, there are several challenges that must be addressed when using AAMs for the stabilization/solidification of heavy metals and radioactive substances in incineration residues. First, the composition and characteristics of the incineration residues can substantially vary, hindering the development of a standardized approach for AAM formulation. The presence of different types and concentrations of contaminants requires careful consideration to optimize the AAM mixture design and ensure effective immobilization. Second, the long-term durability and stability of AAMs should be thoroughly assessed. Although AAMs have shown promising performance in short-term laboratory tests, their resistance to leaching and degradation over extended periods in field conditions must be evaluated. Comprehensive long-term studies are essential to validate the effectiveness and reliability of AAMs as a sustainable solution for the containment of heavy metals and radioactive substances.

Another challenge is the cost of raw materials required for AAM production. Compared with cement, AAMs are expensive to produce. Various industrial by-products can be used as fillers, but few substitutes (e.g., waste alkali) are available as alkali activators. If these limitations can be solved, the use of by-products could reduce the overall cost and environmental footprint of AAM production.

Finally, there is a need to establish regulatory frameworks and ensure that AAMs are accepted as a viable technology for the treatment of incineration residues. Guidelines and standards should be developed to ensure the safety and compliance of AAM-based stabilization/solidification methods. Moreover, public acceptance and stakeholder engagement are crucial in efforts to gain trust and support for the implementation of AAMs in waste management practices.

Acknowledgments

This manuscript is based on research conducted by the Concrete Committee of the Japan Society of Civil Engineers (JSCE) on infrastructure construction in a low-carbon society using new alkali-active materials. In addition, many of the studies were promoted by the Environment Research and Technology Development Fund (3K122106, JPMEERF20143K09, and JPMEERF20171002) of the Environmental Restoration and Conservation Agency provided by the Ministry of Environment of Japan and the Grant-in-Aid for Scientific Research(A)24246092 and (B) 22H03772.

References

Álvarez-Ayuso, E., Querol, X., Plana, F., Alastuey, A., Moreno, N., Izquierdo, M., Font, O., Moreno, T., Diez, S., Vázquez, E., Barra, M., 2008. Environmental, physical and structural characterisation of geopolymer matrixes synthesised from coal (co-)combustion fly ashes. J. Hazard Mater. 54, 175−183.

Aly, Z., Vance, E.R., Perera, D.S., Hanna, J.V., Griffith, C.S., Davis, J., Durce, D., 2008. Aqueous leachability of metakaolin-based geopolymers with molar ratios of Si/Al=1.5−4. J. Nucl. Mater. 378, 172−179.

Blackford, M.G., Hanna, J.V., Pike, K.J., Vance, E.R., Perera, D.S., 2007. Transmission electron microscopy and nuclear magnetic resonance studies of geopolymers for radioactive waste immobilization. J. Am. Ceram. Soc. 90, 1193−1199.

Bouzar, B., Mamindy-Pajany, Y., 2023. Immobilization study of As, Cr, Mo, Pb, Sb, Se and Zn in geopolymer matrix: application to shooting range soil and biomass fly ash. Int. J. Environ. Sci. Technol. 1−22.

Cantarel, V., Arisaka, M., Yamagishi, I., 2019. On the hydrogen production of geopolymer waste forms under irradiation. J. Am. Ceram. Soc. 102, 7553−7563.

Cantarel, V., Motooka, T., Yamagishi, I., 2017. Geopolymers and their potential applications in the nuclear waste management field - a bibliographical study. JAEA-Rev. 14.

Chen, S., Qi, Y., Cossa, J.J., Salomao Dos, S.I.D., 2019. Efficient removal of radioactive iodide anions from simulated wastewater by HDTMA-geopolymer. Prog. Nucl. Energy 117, 102112.

Chlique, C., Lambertin, D., Antonucci, P., Frizon, F., Deniard, P., 2015. XRD analysis of the role of cesium in sodium-based geopolymer. J. Am. Ceram. Soc. 98, 1308−1313.

Comrie, D.C., Davidovits, J., 1988. Long term durability of hazardous toxic and nuclear waste disposals. Geopolymer '88 First Eur. Conf. Soft Mineralurgy. 1, 125−134.

Ecke, H., Menad, N., Lagerkvist, A., 2003. Carbonation of municipal solid waste incineration fly ash and the impact on metal mobility. J. Environ. Eng. 435−440.

Eighmy, T.T., Crannell, B.S., Butler, L.G., Cartledge, F.K., Emery, E.F., Oblas, D., Krzanowski, J.E., Eusden, J.D., Shaw, E.L., Francis, C.A., 1997. Heavy metal stabilization in municipal solid waste combustion dry scrubber residue using soluble phosphate. Environ. Sci. Technol. 31, 3330−3338.

El-eswed, B.I., 2020. Chemical evaluation of immobilization of wastes containing Pb, Cd, Cu and Zn in alkali-activated materials: a critical review. J. Environ. Chem. Eng. 8, 104194.

El-eswed, B.I., Aldagag, O.M., Khalili, F.I., 2017. Efficiency and mechanism of stabilization/solidification of Pb(II), Cd(II), Cu(II), Th(IV) and U(VI) in metakaolin based geopolymers. Appl. Clay Sci. 140, 148−156.

EU Commission Decision, 2015. Commission Decision of 3 May 2000 Replacing Decision 94/3/EC Establishing a List of Wastes Pursuant to Article 1(a) of Council Directive 75/442/EEC on Waste and Council Decision 94/904/EC Establishing a List of Hazardous Waste Pursuant to Article 1(4) of Council Directive 91/689/EEC on Hazardous Waste.

He, P., Cui, J., Wang, M., Fu, S., Yang, H., Sun, C., Duan, X., Yang, Z., Jia, D., Zhou, Y., 2019. Interplay between storage temperature, medium and leaching kinetics of hazardous wastes in Metakaolin-based geopolymer. J. Hazard Mater. 384, 121377.

He, P., Jia, D., 2013. Low-temperature sintered pollucite ceramic from geopolymer precursor using synthetic metakaolin. J. Mater. Sci. 48, 1812−1818.

He, P., Zhang, Y., Zhang, X., Chen, H., 2021. Diverse zeolites derived from a circulating fluidized bed fly ash based geopolymer for the adsorption of lead ions from wastewater. J. Clean. Prod. 312, 127769.

Hermann, E., Kunze, C., Gatzweiler, R., Kiessig, G., Davidovits, J., 1999. Solidification of various radioactive residues by géopolymère with special emphasis on long-term-stability. Géopolymère '99 Proc. 211−228.

Hjelmar, O., Birch, H., Hansen, J.B., 2001. Treatment of APC residues from MSW incineration: development and optimisation of a treatment process in pilot scale. In: Christensen, T.H., Cossu, R., Stegmann, R. (Eds.), 8th Inter. Waste Manage. Landfill Symposium, pp. 667−675.

Hong, K.J., Tokunaga, S., Ishigami, Y., Kajiuchi, T., 2000. Extraction of heavy metals from MSW incinerator fly ash using saponins. Chemosphere 41, 345−352.

Ichikawa, T., Yamada, K., Watanabe, S., Haga, K., Osako, M., 2018. Cesium holding capacity of geopolymers made from metakaolin and water glass. In: Proc. Of the 7th Annual Conf. on Remediation of Radioactive Contamination in the Environ, vol. 15 (in Japanese).

Ishwarya, G., Singh, B., Deshwal, S., Bhattacharyya, S.K., 2019. Effect of sodium carbonate/sodium silicate activator on the rheology, geopolymerization and strength of fly ash/slag geopolymer pastes. Cement Concr. Compos. 97, 226–238.

ISWA Report (International Solid Waste Association Working Group on Thermal Treatment of Waste, Subgroup on APC Residues From W-t-E plants), 2008. Management of APC Residues From W-T-E Plants - an Overview of Management Options and Treatment Methods, second ed.

ISWA Report (International Solid Waste Association Working Group on energy recovery), 2016. Bottom Ash from WTE Plants: Metal Recovery and Utilization.

Japan Concrete Institute, 2020. Report on Application of Cement and Concrete Technology to Hazardous and Radioactive Waste Disposal (in Japanese).

Ji, Z., Pei, Y., 2019. Bibliographic and visualized analysis of geopolymer research application in heavy metal immobilization: a review. J. Environ. Manag. 231, 256–267.

Jiménez, A.F., Macphee, D.E., Lachowski, E.E., Palomo, A., 2005. Immobilization of cesium in alkaline activated fly ash matrix. J. Nucl. Mater. 346, 185–193.

Koide, M., Shiota, K., Fujimori, T., Oshita, K., Takaoka, M., 2021. Immobilization of liquid waste containing cesium using alkali activated aluminosilicate materials. J. Soc. Remediat. Radioac. Contaminat. Environ. 9, 173–182 (in Japanese).

Li, Q., Sun, Z., Tao, D., Xu, Y., Li, P., Cui, H., Zhai, J., 2013. Immobilization of simulated radionuclide $133Cs^{+}$ by fly ash-based geopolymer. J. Hazard Mater. 262, 325–331.

Lichvar, P., Rozloznik, M., Sekely, S., 2013. Behavior of Aluminosilicate Inorganic Matrix SIAL® During and After Solidification of Radioactive Sludge and Radioactive Spent Resins and Their Mixtures. IAEA/INIS. *RN-44122421*.

Liu, J., Wang, Z., Xie, G., Li, Z., Fan, X., Zhang, W., Xing, F., Tang, L., Ren, J., 2022. Resource utilization of municipal solid waste incineration fly ash -cement and alkali-activated cementitious materials: a review. Sci. Total Environ. 852, 158254.

Long, W., Lin, C., Ye, T., Dong, B., Xing, F., 2021. Stabilization/solidification of hazardous lead glass by geopolymers. Construct. Build. Mater. 294, 123574.

Lundtorp, K., Jensen, D.L., Christensen, T.H., 2002. Stabilization of APC residues from waste incineration with ferrous sulfate on a semi-industrial scale. J. Air Waste Manage. Assoc. 52, 722–731.

Luo, S., Zhao, S., Zhang, P., Li, J., Huang, X., Jiao, B., Li, D., 2022. Co-disposal of MSWI fly ash and lead–zinc smelting slag through alkali-activation technology. Construct. Build. Mater. 327, 127006.

Mattigod, S.V., Westsik, J.H., Chung, C.W., Lindberg, M.J., Parker, K.E., 2011. Waste Aacceptance Testing of Secondary Waste Forms: Cast Stone, Ceramicrete and Duralith. Pacific North Laboratory. *PNNL-20632*.

Meguro, Y., Sato, J., 2016. Solidification of radioactive wastes using alkali-activated materials "Geopolymer". J. RANDEC. 54, 48–55 (in Japanese).

Ministry of the Environment, Japan, 2023a. Environment Monitoring of Deposition of Other Radioactive Compounds. https://www.env.go.jp/chemi/rhm/h29kisoshiryo/h29kiso-07-09-01.html. (Accessed 25 June 2023) (in Japanese).

Ministry of the Environment, Japan, 2023b. Progress of Disposal of Disaster Waste Directly Governed by the National Government in Designated Areas in Fukushima Prefecture. http://shiteihaiki.env.go.jp/initiatives_fukushima/waste_disposal/pdf/progress_230630.pdf. (Accessed 30 June 2023) (in Japanese).

Mizutani, S., Van der Sloot, H., Sakai, S., 2000. Evaluation of treatment of gas cleaning residues from MSWI with chemical agents. Waste Manage. (Tucson, Ariz.) 20, 233–240.

Nakamura, Y., Takaoka, M., Suzuki, Y., Kikuchi, T., Ishida, Y., Ichimura, T., Suzuki, T., 2017. Geopolymer solidification of fly ash with easily soluble cesium compounds (2) relationship between chemical form of Cs and its immobilization. In: Proc. Of Annual Spring Conf. on Atomic Energy Soc. of Japan, p. 2L21 (in Japanese).

Nakamura, Y., Takaoka, M., Shiota, K., Oshita, K., Fujimori, T., 2016. Immobilization of strontium in fly ash from MSW incineration by geopolymer. Proc. of the 27th Annual Conf. of Japan Soc. of Mater. Cycles Waste Manage. 479–480 (in Japanese).

Onozaki, K., 2017. Characteristic test of "SIAL®" of alkali-activated materials. J. RANDEC. 55, 28–35 (in Japanese).

Polettini, A., Pomi, R., Sirini, P., Testa, F., 2001. Properties of portland cement stabilised MSWI fly ashes. J. Hazard Mater. B88, 123–138.

Provis, J., Van Deventer, J.S.J., 2009. Geopolymers Structure, Processing, Properties and Industrial Applications. Woodhead publishing limited.

Sakanakura, H., 2007. Formation and durability of dithiocarbamic metals in stabilized air pollution control residue from municipal solid waste incineration and melting processes. Environ. Sci. Technol. 41, 1717–1722.

Sato, J., 2021. Study on Solidification/stabilization of Heavy Metal and Radioactive Nuclide in Alumino-Silicate Solid. Kyoto University Doctor Dissertation (in Japanese).

Shiota, K., Nakamura, T., Oshita, K., Fujimori, T., Takaoka, M., 2018. Quantitative cesium speciation and leaching properties in alkali-activated municipal solid waste incineration fly ash and pyrophyllite-based systems. Chemosphere 213, 578–586.

Shiota, K., Nakamura, T., Takaoka, M., Aminuddin, S.F., Oshita, K., Fujimori, T., 2017. Stabilization of lead in an alkali-activated municipal solid waste incineration fly ash - pyrophyllite-based system. J. Environ. Manag. 201, 327–334.

Shiota, K., Nakamura, T., Takaoka, M., Aminuddin, S.F., Oshita, K., Fujimori, T., 2017. Stabilization of cesium in alkali-activated municipal solid waste incineration fly ash and a pyrophyllite-based system. Chemosphere 187, 188–195.

Takaoka, M., Nakamura, Y., Shiota, K., Kusakabe, T., Fujimori, T., Oshita, K., 2018. Stabilization of Cs in melting fly ash by geopolymerization method. In: Proc. Of the 24th Symposium on Soil and Groundwater Contamination on Remediation (in Japanese).

Takeda, N., Takaoka, M., 2013. An assessment of dioxin contamination from the intermittent operation of a municipal waste incinerator in Japan and associated remediation. Environ. Sci. Pollut. Res. 20, 2070–2080.

Takeda, N., Wang, W., Takaoka, M., 2014. Solid Waste Management. Kyoto University Press.

Tian, Q., Wang, H., Pan, Y., Bai, Y., Chen, C., Yao, S., Guo, B., Zhang, H., 2022. Immobilization mechanism of cesium in geopolymer: effects of alkaline activators and calcination temperature. Environ. Res. 215, 114333.

Tsuji, H., Nakamura, Y., Shiota, K., Fujimori, T., Oshita, K., Takaoka, M., 2021. The influence that dissolution properties of aluminosilicates to alkali solutions have on the immobilization of cesium in fly ash by geopolymer solidification. J. Japan Soc. Mater. Cycles Waste Manage. 32, 136−146 (in Japanese).

Tsujii, H., Shiota, K., Kusakabe, T., Fujimori, T., Oshita, K., Takaoka, M., 2020. Effect of NaOH concentration on immobilization of cesium in melting fly ash by geopolymerization. Proc. 31st Annual Conf. Japan Soc. Mate. Cycles Waste Manage. 427−428 (in Japanese).

Uchida, M., Tameda, K., Higuchi, S., Nishio, S., Hamasaki, Y., Yuda, K., Kusuda, T., 2012. Impact of chelating agent used for treatment of municipal solid waste incineration fly ash on landfill management (Part 3). Proc. 23rd Annual Conf. Japan Soc. Mater. Cycles Waste Manage. 491−492 (in Japanese).

Van Jaarsverld, J.G.S., Van Deventer, J.S.J., 1999. The effect of metal contaminants on the formation and properties of waste-based geopolymer. Cement Concr. Res. 29, 1189−1200.

Yang, S., Yang, L., Gao, M., Bai, H., Nagasaka, T., 2022. Synthesis of zeolite-geopolymer composites with high zeolite content for Pb(II) removal by a simple two-step method using fly ash and metakaolin. J. Clean. Prod. 378, 134528.

Youcai, Z., Lijie, S., Guojian, L., 2002. Chemical stabilization of MSW incinerator fly ashes. J. Hazard Mater. B95, 47−63.

Chapter 5

Sintering and melting of combustion/incineration residues

Jingqi Sun[1], Yike Zhang[1,2] and Zengyi Ma[1,2]
[1]*State Key Laboratory of Clean Energy Utilization, Zhejiang University, Hangzhou, China;* [2]*Ningbo Innovation Center, Zhejiang University, Ningbo, China*

1. Introduction

The increasing population, along with the progress of urbanization and industrialization in modern society, has resulted in the generation of large amounts of solid waste. This situation poses a serious problem to the sustainable development of the economy and society (Das et al., 2019; Pujara et al., 2019). Municipal solid waste (MSW) is a significant component of this waste stream, and its quantity continues to rise. According to the World Bank's publication "*What a Waste 2.0: A Global Snapshot of Solid Waste Management to 2050*," global waste generation is projected to increase by 70% from 2016 to 2050, driven by factors such as increased prosperity and urbanization. The report also estimated that waste generation would grow from 2.24 billion tons in 2020 to 3.88 billion tons by 2050.

In light of these challenges, incineration emerges as a promising alternative to landfilling for the treatment of this vast quantity of MSW. Incineration not only reduces the volume and weight of solid waste by 90% and 70%, respectively, but also generates energy and heat that can be utilized for various purposes such as power generation and resource recovery. However, municipal solid waste incineration (MSWI) generates residues, including fly ash (FA), air pollution control (APC) residues, and bottom ash (BA) (Phua et al., 2019). The treatment of FA and APC residues has been reviewed by Phua et al., while Dou et al. provided a review of the utilization, treatment, and application of MSWIBA (Dou et al., 2017). These residues often contain heavy metals, soluble salts, and organic pollutants such as dioxins. Hence, the detoxification and harmless disposal of MSWI residues have become pressing issues.

Among various disposal and degradation technologies, thermal treatment methods have shown promise in efficiently degrading dioxins and immobilizing heavy metals, with reported degradation rates of dioxins and removal rates of volatile heavy metals reaching up to 99% (Katou et al., 2001; Lindberg et al., 2015).

Thermal treatment involves the utilization of high temperatures to process incineration residues, transforming them into environmentally stable substances. This process entails the decomposition, combustion, and vaporization of organic matter present in the residues, while the inorganic matter undergoes a transformation into more stable glass slags. Through thermal treatment, organic pollutants such as dioxins in incineration residues are thermally decomposed and destroyed, while low-boiling-point heavy metal salts are vaporized, with the majority being transferred to glass slags. As a result, the leaching potential of heavy metals is significantly reduced.

The objective of thermal treatment methods is to reduce the volume of residues and produce more stable products suitable for a wide range of applications, such as construction materials. High-temperature thermal treatment encompasses two approaches: sintering and melting (Li et al., 2023). Sintering involves heating the residues to temperatures ranging from 700 to 1200°C (below the melting temperature) to facilitate bonding and recrystallization of small particles into dense polycrystalline materials. Sintering yields products characterized by reduced porosity, increased strength and density, and lower leaching rates of heavy metals such as cadmium (Cd), lead (Pb), and mercury (Hg). In contrast, melting treatment typically operates at higher temperatures exceeding 1200°C. The specific operating temperature of the melting process surpasses the melting point of the incineration residues, resulting in the destruction of crystalline structures and the formation of a homogenous liquid phase without additives. Melting treatment significantly enhances the density of

Treatment and Utilization of Combustion and Incineration Residues. https://doi.org/10.1016/B978-0-443-21536-0.00018-6

incineration residues, leading to a reduction in ash volume by over 50%. Additionally, metal recovery from ash slag becomes feasible, and the stable slag can be utilized as roadbed material, thereby achieving effective utilization goals.

In this chapter, we will examine the sintering and melting methods for combustion/incineration residues. This will encompass defining sintering and melting technologies, comparing their capacities for the removal or fixation of heavy metals and dioxins, analyzing their advantages and disadvantages, exploring current application progress, and discussing future prospects. The authors have done a lot of research on melting, so there are more details on melting than sintering in this chapter.

2. Sintering technology

2.1 The principle of sintering

The sintering process involves heating combustion/incineration residues to temperatures typically ranging from 700 to 1200°C (below the melting point of the major components) to facilitate the reconfiguration of solid materials. High-temperature sintering treatment enables the transformation of residues into various stable materials, such as glass, concrete (Clavier et al., 2019), and ceramics (Zhan et al., 2021, 2022). These products exhibit reduced porosity, increased strength, and higher density compared with the raw residues. Consequently, the decrease in porosity in the sintered material leads to a decrease in the leachability of harmful components from the end products. Moreover, high-temperature sintering facilitates the decomposition of PCDD/Fs (polychlorinated dibenzo-p-dioxins and furans). To prevent the resynthesis of dioxins, rapid cooling of the sintering fuel gas is necessary.

Before the sintering process, pretreatment steps such as washing, drying, and sieving are often required (De Casa et al., 2007). It has been confirmed that pretreatment steps enhance the quality of sintered products, improve their chemical and mechanical characteristics, reduce the sintering temperature and time, and lower energy costs (Wey et al., 2006). Additionally, additives can be introduced during the sintering process to enhance the performance of the sintered products.

2.2 Effect of sintering on heavy metal immobilization

Different operational conditions, such as temperature and sintering time, have varying levels of influence on the solidification of heavy metals. To assess the immobilization effect of heavy metals, the residual and leaching rates of heavy metals are commonly evaluated. Li et al. (2015) developed an overall pollution toxicity index (OPTI) and an integrated method of heavy metal pollution control based on this index. The results showed that the residual rate of six heavy metals in the sintering experiments initially increased and then decreased with the increasing amount of NaCl additive. The use of NaCl solution exhibited a better inhibition effect on heavy metal residues compared with dry NaCl powder. However, the addition of NaCl solution to Ni promoted residual formation. With the increase in sintering process time, the residual rate of most heavy metals gradually decreased, except for Pb, which showed slight fluctuations. This could be attributed to heavy metals that had become volatile and then deposited on the sintered body again during the sintering process. Notably, the residual rates of Cr and Ni exhibited opposite trends with increasing sintering temperature.

Furthermore, the prewashing of MSWI residues can impact the leaching ability of heavy metals. Liu et al. (2009) compared the sintering results of raw fly ash (RFA) and washed fly ash (WFA) and discovered that washing pretreatment reduces the leaching rate of Cd, Pb, and Ni, but increases the leaching rate of Cr. The leaching concentrations of heavy metals in the sintering products from both RFA and WFA met the limit levels.

Sintering method has a good effect on heavy metal immobilization. The powdered nature of FA allows it to be combined with other raw materials (e.g., electrolytic manganese residue) for sintering, making it suitable for use as a lightweight aggregate in building materials.

2.3 Effect of sintering on dioxin destruction

The sintering method effectively degrades dioxins in MSWIFA, achieving a stable degradation rate of over 95% (Chen et al., 2020). To enhance the value of treated FA, it is typically mixed with other raw materials. Dioxins decompose into smaller molecules during the sintering process and oxidize to H_2O and CO_2 in the presence of oxygen (Lindberg et al., 2015). Washing before sintering can reduce the chlorine content in FA. At a sintering temperature of 500°C, chlorine transforms into complex and stable crystal structures, and continued heating promotes the dechlorination of dioxins (Wang et al., 2021). He et al. (2022a,b) observed an 88.9% degradation of dioxins in unwashed FA and a 99.1% degradation in washed FA under the same sintering conditions. Moreover, Peng et al. (2020) found that the dioxin toxicity level in secondary fly ash (SFA), the solid ash captured after flue gas condensation, was much higher (reaching 14.3 μg I-TEQ/kg) than that in flue gas (0.019−0.025 ng I-TEQ/Nm^3) and sintered FA (0.002−0.008 μg I-TEQ/kg). Chen et al. (2020) investigated the decomposition and reformation pathways of PCDD/Fs under different conditions, including three heating temperatures (500°C, 800°C, and 1100°C) and 2 atmospheres (oxidative and inert). The results revealed that over 95 wt.% of PCDD/Fs in ashes decompose primarily through cyclic-skeleton destruction along with dechlorination. Increasing the treatment temperature or changing the atmosphere from air to N_2 had little effect on the decomposition process. He et al. (2022a,b) inhibited the formation of PCDD/Fs during iron ore cosintering with MSWI FA through adjusting the sinter raw mix (CaO and $CaCO_3$).

Moreover, it is crucial to prevent the resynthesis of dioxins. Studies have demonstrated that rapid cooling of the sintering flue gas can reduce the risk of low-temperature resynthesis of PCDD/Fs by reducing the residence time for resynthesis in the low-temperature range (Peng et al., 2020). Chen et al. found a positive correlation between the amount of PCDD/Fs generated and the oxygen concentration (Chen et al., 2020). Therefore, the production of PCDD/Fs is lower under reducing and inert atmospheres compared with an oxidizing atmosphere.

3. Melting technology

3.1 The principle of melting

Compared with the sintering process, the melting of MSWI residues typically requires a high operating temperature, usually above 1200°C (Wang et al., 2008). The objective of melting MSWI residues is to encapsulate harmful elements within a glass matrix, preventing their leaching.

During the melting treatment of MSWI residues, a small portion of heavy metal salts with low boiling points volatilizes into the flue gas, while the majority is transferred to the glassy slag. The chemical composition of MSWI residues greatly influences their melting treatment, particularly the presence of Cl and S (Huang et al., 2022). Chlorides begin to volatilize at temperatures ranging from 500 to 550°C and almost completely volatilize at 700−800°C. The volatilization temperature range of sulfates is more extensive, with sulfur dioxide being produced gradually during the degradation of sulfates from 600 to 1300°C. Chlorides in MSWI residues can promote the volatilization of metal elements, especially Pb, into the SFA (Okada and Tomikawa, 2012).

Studies on FA melting (Li et al., 2007) have demonstrated that the melting process involves three main endothermic reactions: dehydration, polymorphic transition, and fusion, occurring within temperature ranges of 100−200°C, 480−670°C, and 1101−1244°C, respectively (as shown in Fig. 5.1). Fig. 5.2 illustrates the DTA and DSC curves of an FA sample under an oxygen atmosphere at a heating rate of 10°C/min. Three absorption peaks are observed in Fig. 5.2, representing the processes of dehydration, polymorphic transition, and fusion, respectively.

The mechanism of melting MSWI FA can be described as follows (Li et al., 2007): When multiple crystal states of the same substance exist, classic thermodynamic considerations indicate that the polymorph with the lowest free enthalpy is the most stable. As the temperature increases, the crystal becomes mechanically unstable or unattainable, leading to a polymorphic transformation and equilibrium change from one phase to another.

During the melting treatment, the heat and temperature required for melting primarily depend on the composition of the FA, and the interaction between its internal components also affects the melting temperature. Research by Li et al. (2007) demonstrates that the addition of SiO_2 increases the viscosity of the liquid slag, thereby raising the melting temperature. In cases where the FA has relatively low alkalinity, CaO can help reduce the viscosity of the liquid slag, lower the melting temperature, and accelerate the melting and crystallization process. In a study on the influence of CaO and Fe_2O_3 on the melting temperature of MSWIFA (Shi et al., 2018), it was observed that as the proportion of CaO/Fe_2O_3 increases, the melting temperature of the FA initially increases and then gradually stabilizes. Regarding other elements, Wang et al. (2019) demonstrated that both V and Ni can increase the melting temperature. Additionally, Zhang et al. (2013) found that P can lower the melting temperature only in FA containing high Al_2O_3. In conclusion (Liu et al., 2018), for MSWIFA, acidic oxides (such as SiO_2, P_2O_5, TiO_2) generally increase the melting temperature, while basic oxides (such as Na_2O,

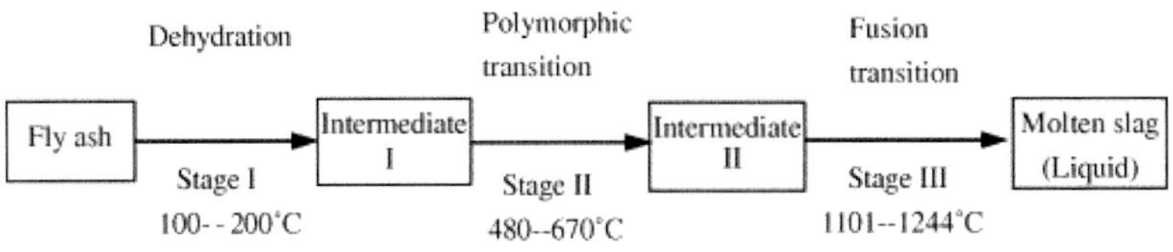

FIGURE 5.1 Schematic illustration of melting mechanism of MSWI FA (Li et al., 2007).

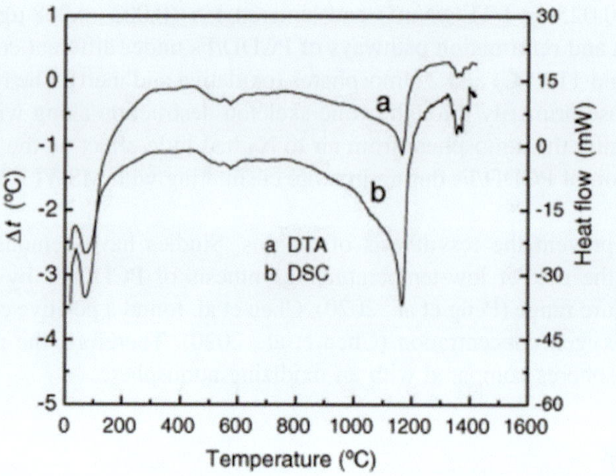

FIGURE 5.2 Typical plots of DTA-DSC of FA melting at 10°C/min (Li et al., 2007).

K_2O, MgO, CaO, MnO_2) generally lower the melting temperature. As for the amphoteric oxides Al_2O_3 and Fe_2O_3, Al_2O_3 exhibits characteristics of an acidic oxide that increases the melting temperature, while Fe_2O_3 exhibits characteristics of a basic oxide that lowers the melting temperature.

The weight loss curve of MSWIFA during the melting process, ranging from room temperature to 1400°C at a heating rate of 10°C/min, is presented in Fig. 5.3. It can be observed that the melting of FA is characterized by a continuous weight loss process at varying rates. From room temperature to 1400°C, the total weight loss amounts to approximately 40%. Specifically, the first stage occurs from room temperature to 470°C, with a weight loss of approximately 3%. Throughout this stage, the weight loss rate remains relatively stable, fluctuating around 0.1%/min. The weight loss peak at 100°C corresponds to water evaporation. In the second stage, spanning 470–740°C, the weight loss reaches 9%, accompanied by a rapid weight loss phase. The maximum weight loss rate is observed at 640°C, reaching 0.90%/min. The third stage, occurring between 740 and 970°C, results in a weight loss of 10% at an average rate of approximately 0.5%/min. The fourth stage, ranging from 970 to 1100°C, leads to an 8% weight loss, with a maintained rate of approximately 0.7%/min. The fifth stage, from 1100 to 1255°C, represents another rapid weight loss phase, resulting in a 10% weight loss and peaking at a weight loss rate of 1.00%/min at 1196°C. Finally, the sixth stage, spanning 1255–1400°C, does not involve any weight loss, indicating the completion of the melting stage.

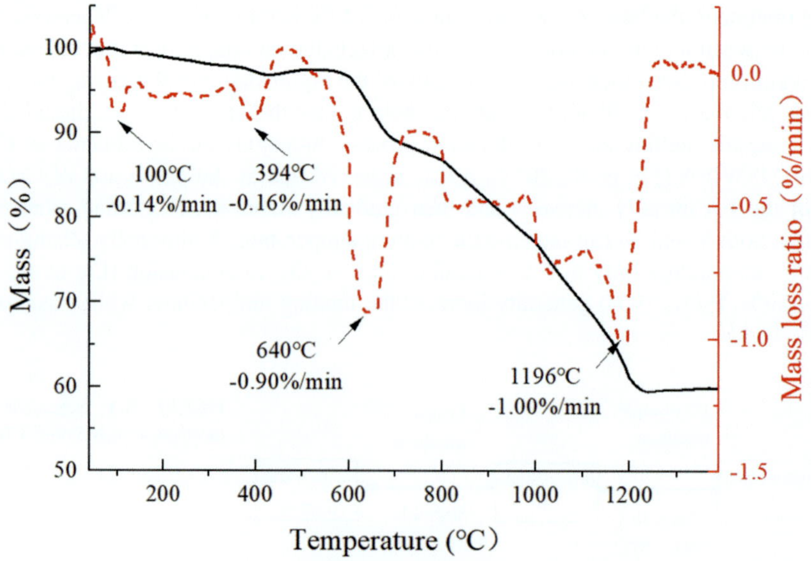

FIGURE 5.3 Melting weight loss curve of MSWIFA.

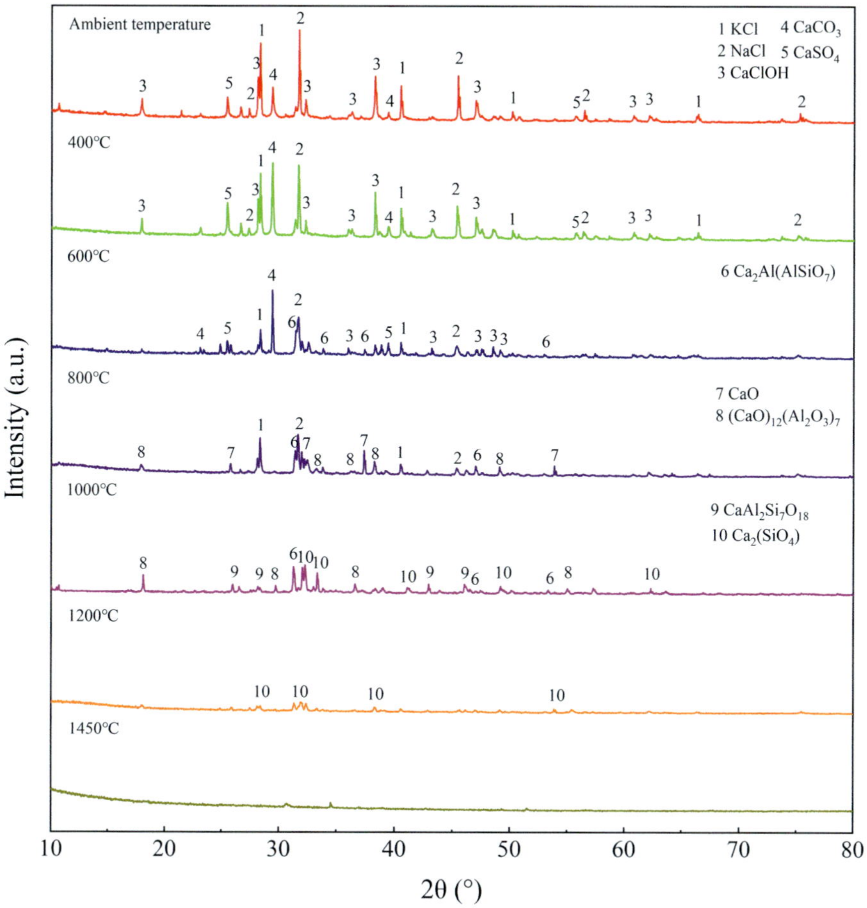

FIGURE 5.4 XRD patterns of slag at different temperatures of FA melting.

To determine the crystal composition and relative proportions at different temperatures during the FA melting process, sampling at various temperature intervals, X-ray diffraction (XRD) analysis, and Rietveld refinement techniques can be employed. The obtained results are presented in Figs. 5.4 and 5.5.

The main components of FA at room temperature primarily consist of chlorine-containing and calcium-containing salts, including NaCl, KCl, CaClOH, $CaCO_3$, and $CaSO_4$. Calcium-containing salts account for 60% of the main components, with CaClOH being the most prevalent at 31%. This is primarily attributed to the excessive $Ca(OH)_2$ released from deacidification systems such as dry and semidry deacidification, reacting with HCl, SO_2, and CO_2 in the flue gas to form these salts. The reaction equations are as follows:

$$Ca(OH)_2+HCl \rightarrow CaClOH + H_2O$$

$$Ca(OH)_2+SO_2 \rightarrow CaSO_4+H_2O$$

$$Ca(OH)_2+CO_2 \rightarrow CaCO_3+H_2O$$

In contrast, most of the chlorinated salts in FA originate from the waste source, resulting from the incineration of kitchen waste or the incineration of concentrated leachate water.

As the melting temperature increases to 400°C, the first stage of weight loss (room temperature to 470°C), the content of CaClOH in the FA significantly decreases from 31% to 17%. This indicates that the weight loss peak observed around 394°C in the first stage is caused by the decomposition of CaClOH. Some of the newly generated CaO combines with CO_2 and SO_2 released from the air or FA melt to form $CaCO_3$ and $CaSO_4$, resulting in a significant increase in their content. The reactions involved are as follows:

$$CaClOH + SO_2 \rightarrow CaSO_4+HCl$$

$$CaClOH + CO_2 \rightarrow CaCO_3+HCl$$

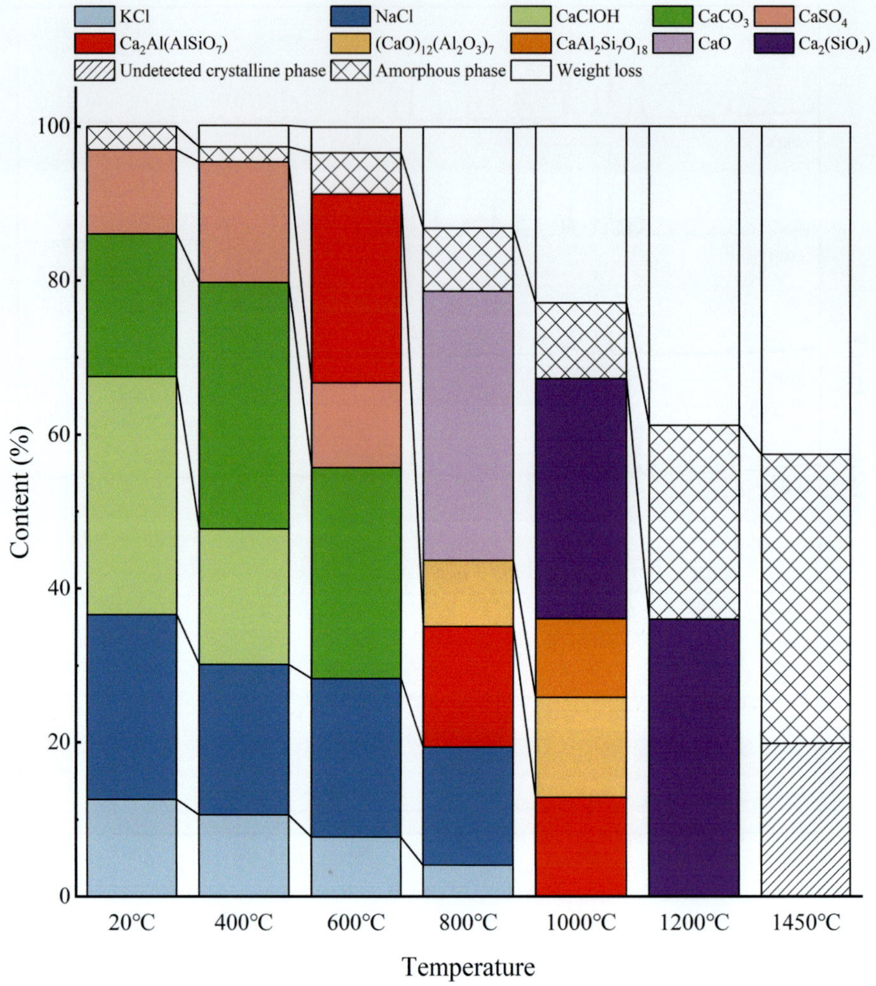

FIGURE 5.5 Composition distribution of slag components at different temperatures of FA melting.

When the melting temperature reaches 600°C, the decomposition of CaClOH is complete. $CaCO_3$ and $CaSO_4$ also begin to destabilize, leading to the second stage of weight loss. The slag produced contains calcium yellow feldspar $Ca_2Al(AlSiO_7)$, a component of the Ca—Si—Al system. The amorphous phase material in the FA exceeds 5%, indicating the transformation of inorganic minerals into an amorphous state. At 800°C, $CaCO_3$ and $CaSO_4$ in the FA are completely decomposed, generating CaO at high temperatures. The increased amount of CaO in the system contributes to the formation of calcium aluminate $(CaO)_{12}(Al_2O_3)_7$. The release of gaseous components CO_2 and SO_2 initiates the third stage of weight loss.

When the melting temperature reaches 1000°C, chloride salts such as NaCl and KCl in the FA are completely volatilized, marking the fourth stage of weight loss. Simultaneously, all CaO in the FA transforms into Ca—Si—Al compounds, including calcium silicate $Ca_2Al(AlSiO_7)$ and calcium silicate products $CaAl_2Si_7O_{18}$ and $Ca_2(SiO_4)$. The content of amorphous phase components in the FA increases to 13%. Beyond 1000°C, all Ca—Si—Al compounds in the FA transform into amorphous phase material, with only calcium silicate $Ca_2(SiO_4)$ remaining as a detectable crystalline component. At 1200°C, the amorphous phase component in the FA reaches 41% in the slag. By 1450°C, the FA completely melts, and the amorphous phase component in the residue after melting amounts to 65%.

Based on the aforementioned analysis, the FA melting process primarily involves three reactions: (1) volatilization of chloride salt, (2) decomposition of calcium salt, and (3) generation of glassy slag of the Ca—Si—Al system. As shown in Fig. 5.6, the volatilization of chloride salt occurs from the beginning of the melting process, with the volatilization rate peaking at approximately 800°C until the volatilization of all chloride salts is complete at 1000°C. Calcium salt, including CaClOH, $CaCO_3$, and $CaSO_4$, undergoes decomposition, with CaClOH decomposing before 600°C, while $CaCO_3$ and $CaSO_4$ are completely converted to CaO at 800°C. This excess CaO acts as a source for the formation of Ca—Si—Al

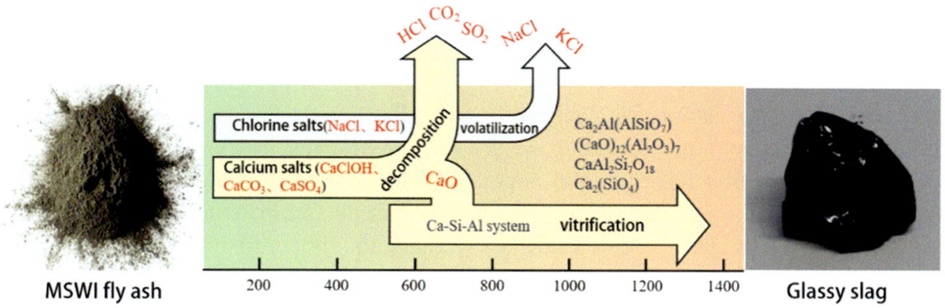

FIGURE 5.6 Reaction mechanism of FA high-temperature melting process.

compounds, which begins around 600°C. The Ca—Si—Al system compounds, such as calcium xanthite $Ca_2Al(AlSiO_7)$, calcium aluminate $(CaO)_{12}(Al_2O_3)_7$, calcium—silica—alumina product $CaAl_2Si_7O_{18}$, and calcium silicate product $Ca_2(SiO_4)$, are progressively generated as the melting temperature increases. When the temperature exceeds 1450°C, the FA is completely melted, and no distinct crystal composition can be detected. The resulting slag has a vitreous content of approximately 65%. To further increase the vitreous content, materials rich in silica—alumina components need to be added to the molten raw material.

3.2 Effect of melting on heavy metal immobilization

Li et al. (2022) conducted a study on the melting characteristics and leaching behavior of heavy metals. They observed a significant reduction in the leaching concentration of Cu, Zn, Cd, Pb, Cr, and Ni after the molten vitrification treatment, meeting the leaching standards. This indicates that the glass structure effectively inhibits the leaching of heavy metals. The favorable effect can be attributed to the composition of SiO_4, which consists of O^{2-} at the four vertices and Si^{4+} at the center. The O^{2-} forms tetrahedral silicate groups by sharing O^{2-} with other Si^{4+}. During the molten vitrification process, heavy metals such as Pb, Cd, Zn, Cu, Cr, and Ni react with Al^{3+} and Ca^{2+} to form a chain of SiO_4, which becomes fixed in the silicate network structure, thereby inhibiting the leaching of heavy metals. Gao et al. (2020) found that the addition of B_2O_3 as a cosolvent can lower the melting temperature of MSWIFA and promote the formation of the glass phase, leading to increased mass transfer resistance of heavy metals and inhibition of the volatilization of Pb, Cd, and Zn. Additionally, Lin et al. (2021) studied a $CaO—SiO_2—Al_2O_3$ ternary system molten glass obtained by coheat treatment of MSWIFA and sludge. They found that heavy metals can enter calcium-containing minerals or silicon through processes such as ion exchange, migration, and filling. The aluminate glass bodies achieved high-efficiency solidification, meeting the national standard for leaching rates of Cd, Cr, Cu, Ni, Pd, and Zn. Particularly, the solidification rates of Zn and Cu reached over 98%.

Melting method operates at high temperature, forms more stable glassy slags, and has higher integrated control efficiency (Li et al., 2015) than that of sintering process.

3.3 Effect of melting on dioxin destruction

High-temperature melting technology is implemented using plasma furnaces and natural gas furnaces. Plasma furnaces generate an electric arc in the plasma, which mixes with gas to form a high-temperature and high-activity plasma (Huang et al., 2022). Plasma technology has been shown to effectively decompose PCDD/Fs, enabling a harmless, reduced, and resourceful treatment of MSWIFA. Commercially used melting/vitrification technologies include those developed by HITACHI, JFE Corporation, EER, Plasco, and others (Gao et al., 2019). Furthermore, besides the stand-alone treatment of MSWI residues, the high-temperature heat treatment of MSWI residues in combination with other solid wastes such as sewage sludge, electrolytic manganese residue, and red mud has gained attention. Chen et al. (2021) codisposed sewage sludge and FA, achieving the simultaneous reuse of various solid wastes and reducing the content of PCDD/Fs in the product to a safe level below 50 ng/g. The decomposition of PCDD/Fs during the coprocessing process was primarily attributed to the cracking of oxygen bridges. Further, the thermal decomposition of PCDF and PCDD molecules is likely to primarily happen at the oxygen bridge contained in furan and 1,4-dioxin rings, which are the most unstable rings (Chen et al., 2021).

4. Case study for melting technology

4.1 Fuel burning melting and electric melting

Fuel-burning melting typically involves the use of natural gas, fuel oil, or coke. Sakai and Hiraoka (2000) summarized various fuel-burning melting/vitrification furnace systems, including surface melting systems, swirling-flow melting systems, coke-bed melting furnaces, rotary kiln melting furnaces, and internal melting furnaces. Wang et al. (2008) constructed a pilot-scale melting furnace that utilized a spraying tower to rapidly reduce the temperature of the flue gas to below 200°C, preventing the resynthesis of dioxins. The testing results showed that the dioxin toxicity equivalent in the flue gas generated by the FA melting process was 0.053 ng-TEQ/m^3, meeting the requirements of the national standards. Electric melting/vitrification treatments of FA, as summarized by Sakai and Hiraoka (2000), include electric-arc melting furnaces, electric resistance melting furnaces, plasma melting furnaces, and induction melting furnaces. Compared with fuel-burning methods, electric melting/vitrification of FA requires electricity and is particularly suitable for implementation in coal-fired power plants, where large amounts of electricity are generated.

4.2 Plasma melting

Plasma technology is currently a hot research topic in the field of FA electric melting treatment. Plasma is the fourth state of matter, alongside solid, liquid, and gas, characterized by the coexistence of electrons, negatively charged free electrons, and positively charged ions when matter is free from atomic nucleus attraction. Plasma can be observed in phenomena such as lightning, fluorescent lights, and even the sun, as they all contain matter in the plasma state. Plasma exhibits high temperature, high enthalpy, and high energy density characteristics. The use of a plasma torch for FA melting significantly increases the melting rate and reduces the residence time of FA, enabling rapid and continuous melting. Researchers from Zhejiang University conducted a pilot test study on the plasma melting of FA from domestic waste incineration, based on the basic theory of FA melting characteristics in MSWI. They verified the feasibility of using plasma melting for the harmless treatment of FA at a large scale.

Fig. 5.7 depicts the setup used for the pilot test of plasma melting. The system consists of two 150-kW DC nontransfer arc plasma torches, which generate a high-temperature plasma torch for the purpose of melting the FA. Air is utilized as the carrier gas in the plasma torch, while nitrogen serves as the protective gas. The FA is fed into the melting furnace through the feeder, where it is heated and melted by the plasma torch. The resulting molten slurry overflows from the discharge port and enters the slag chiller, completing the harmless disposal process of the FA. The flue gas and SFA generated during the melting process are discharged in compliance with standards after undergoing treatment in the tail gas purification device.

Fig. 5.8 shows the appearance of the plasma melting test samples, including FA, SFA and FA slag. The test was conducted using fluidized-bed FA obtained from the Zhuji Bafang waste power plant, which exhibits a yellowish-brown appearance. The quantity of SFA produced during the melting process accounts for 15% of the mass of the initial FA. The SFA appears as a white powder that readily dissolves in water, forming a blue solution. The molten glass slag constitutes 72% of the mass of the original FA and appears as black, glossy particles that are easily breakable.

The utilization of plasma technology enables the rapid melting of FA, resulting in a glassy slag material of the Ca—Si—Al system, which accounts for 72% of the total FA mass. After plasma melting, the leaching of heavy metals from the

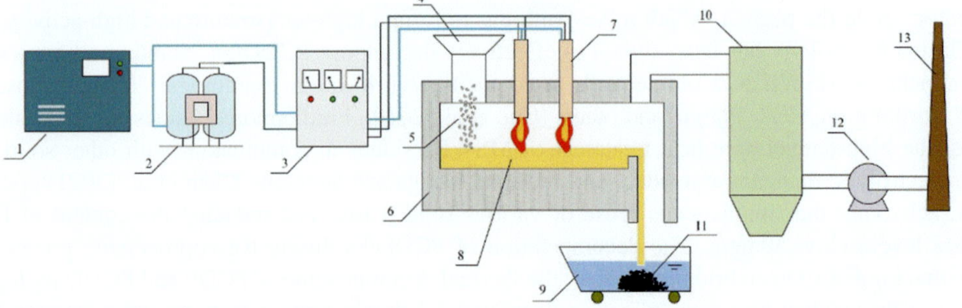

FIGURE 5.7 Plasma melting device flowchart (1. air compressor, 2. nitrogen generator, 3. plasma torch control device, 4. feeder, 5. fly ash, 6. plasma melting furnace, 7. plasma torch, 8. fly ash molten slurry, 9. slag cooler, 10. tail gas purification device, 11. fly ash slag, 12. induced draft fan, 13. chimney).

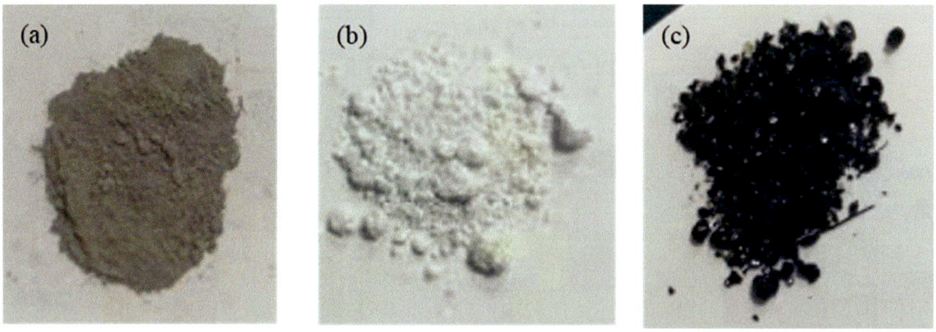

FIGURE 5.8 Plasma melting test samples: (A) FA, (B) SFA, (C) FA slag.

TABLE 5.1 Heavy metal leaching of plasma melting test samples.

Heavy metal leaching(mg/L)	FA	SFA	Slag	Leaching limits (GB/T 14,848)
Cr	0.283	9.057	0.011	0.050
Ni	0.003	1.405	0.006	0.020
Cu	0.002	315.900	0.010	1.000
Zn	0.109	67.020	0.010	1.000
As	0.021	0.327	0.005	0.010
Cd	0.001	19.490	0.001	0.005
Pb	1.898	1.279	0.001	0.010
Mn	0.002	3.395	0.004	0.100

slag conforms to the national standard of GB/T 14848−2017 and allows for disposal in landfills as general industrial solid waste. This indicates that plasma melting achieves the harmless treatment of FA. The SFA generated during plasma melting presents potential for subsequent resource utilization, as it contains salt substances and heavy metal components that were volatilized during the melting process. Determining the appropriate approach for the resource utilization of SFA is a crucial challenge that needs to be addressed in future plasma melting treatments of FA (Table 5.1).

4.3 Oxygen-enriched melting

The use of fuel combustion for the harmless treatment of MSWIFA necessitates the consumption of significant amounts of fossil fuels due to the high melting point of FA. Reducing the energy consumption during the melting process is a key issue in FA fuel melting. Oxygen-enriched melting involves employing air with oxygen content greater than the normal atmospheric level (>21%) as the combustion air. Oxygen-enriched combustion offers several advantages. Firstly, it enhances the fuel combustion process, increasing the theoretical combustion temperature of the flame, improving heat transfer in the melting furnace, and reducing energy consumption during melting. Secondly, oxygen-enriched combustion reduces the nitrogen content in the flue gas, thereby decreasing the volume of flue gas generated during the melting process, resulting in reduced fan power and operational energy consumption. Moreover, the reduced amount of flue gas carries away less heat, significantly minimizing heat loss during the melting process. Therefore, the implementation of oxygen-enriched melting technology effectively achieves energy savings and consumption reduction in the FA melting process.

Fig. 5.9 depicts the flowchart of the oxygen-enriched melting plant, encompassing the oxygen production system, the oxygen-enriched melting system, the flue gas cooling system, and the tail flue gas cleaning system. The oxygen generation system employs a variable pressure adsorption oxygen generator to produce oxygen-enriched air containing 90% oxygen. Oxygen-enriched air and natural gas are directed into the oxygen-enriched lances separately through distinct channels to prevent premixing and enhance the safety of the ignition and combustion process. The five oxygen-enriched lances are evenly arranged diagonally downward along the perimeter of the furnace, generating a tangentially circulating flame to

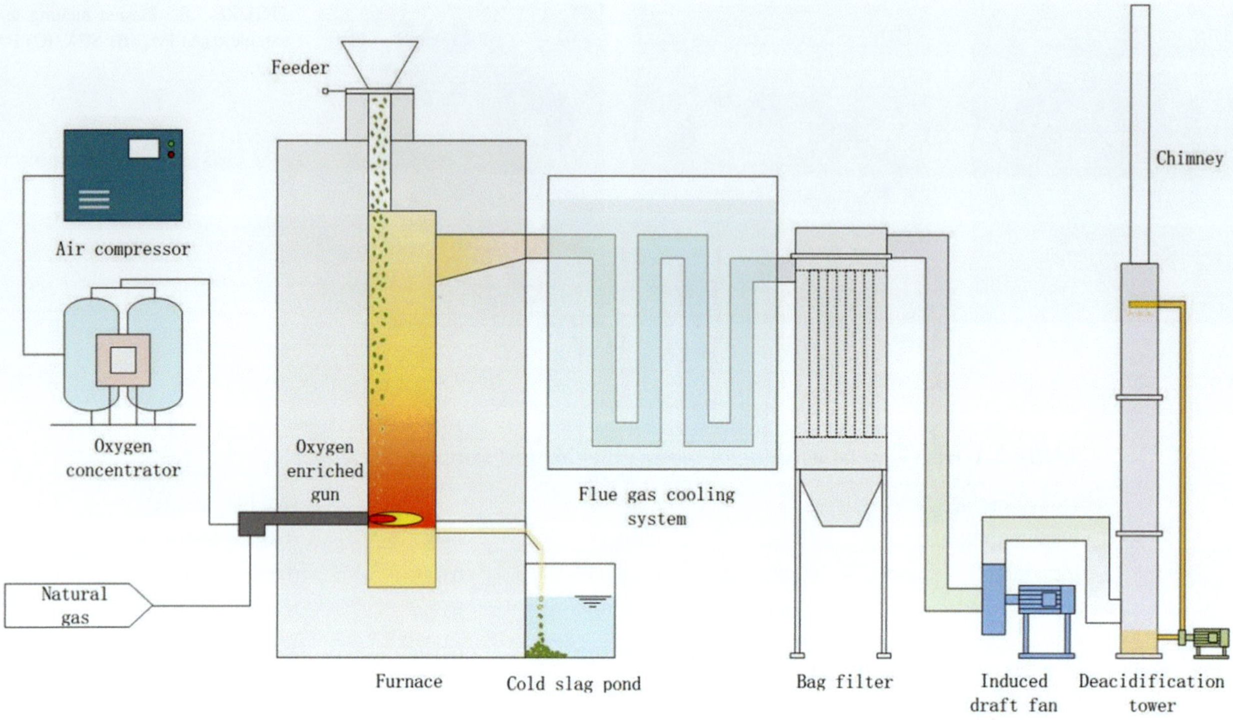

FIGURE 5.9 Flowchart of oxygen-rich melting device.

improve flame distribution within the furnace. Simultaneously, the flames directly brush against and stir the molten slurry, thereby enhancing heat transfer. The resulting high-temperature molten slurry is then directed to the cold slag pond for cooling. The flue gas is cooled via the flue gas cooling system, subsequently passing through a bag filter to eliminate the SFA and through a deacidification tower to remove acidic gases such as HCl and SO_2, thus complying with emission standards.

For comparative analysis, as seen in Table 5.2, three operating conditions were selected: air + natural gas melting, oxygen-enriched + natural gas melting, and oxygen-enriched + coal melting. The test was conducted with an FA feed rate of 75 kg/h. When air was employed as the combustion air, the natural gas consumption ranged between 48 and 51 m^3/h. Conversely, with the use of oxygen enrichment, the natural gas consumption decreased to 16–17 m^3/h. This indicates a significant reduction in fuel usage through the application of oxygen enrichment. Under the same conditions of oxygen-

TABLE 5.2 Operating conditions for the different experiments.

Experimental condition	Air + natural gas	Oxygen + natural gas	Oxygen + coal
Heating method	Five natural gas–air combustors	Five natural gas–oxygen-enriched air combustors	Five oxygen-enriched air gun jets
Energy source	Natural gas	Natural gas	Coal
Energy consumption	48–51 m^3/h (total consumption of five combustors)	16–17 m^3/h (total consumption of five combustors)	25 kg/h
Feeding rate	75 kg/h	75 kg/h	75 kg/h
Combustion air	Air, 570–600 Nm^3/h	85%–90% oxygen-enriched air, 37–40 Nm^3/h	85%–90% oxygen-enriched air, 37–40 Nm^3/h
Operating temperature	1473°C	1520°C	1495°C

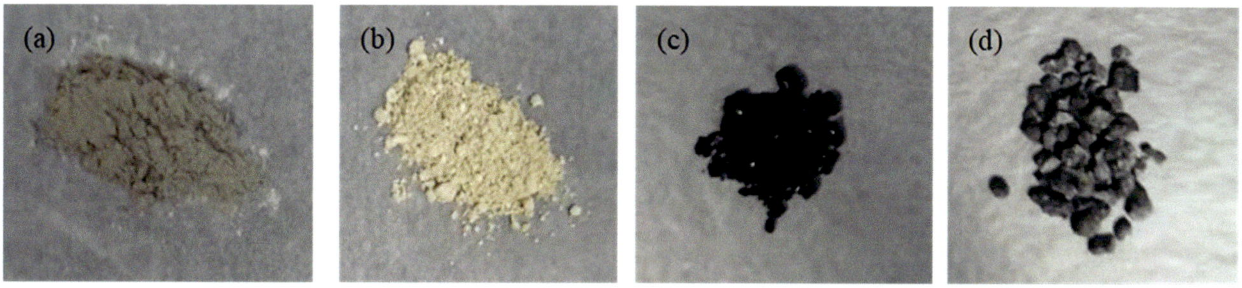

FIGURE 5.10 Oxygen-enriched melting test samples: (A) FA, (B) SFA, (C) FA water-cooled slag, (D) FA air-cooled slag.

enriched combustion air (37–40 Nm^3/h), the test also successfully achieved melting using coal as fuel, with a fuel consumption of 21 kg/h. By comparing the operating temperatures of the melting process, it is evident that the oxygen-enriched + natural gas method attains a higher melting temperature, reaching 1520°C, compared with 1473°C when air is used as the combustion air. Thus, it is apparent that oxygen-enriched melting enables higher melting temperatures with reduced fuel consumption.

The test sample of molten FA was obtained from the Zhongke Cixi waste incineration plant, and the resulting product is presented in Fig. 5.10. The FA exhibits a gray-brown appearance and possesses a fine powdery granular texture. The SFA obtained after melting is yellowish white, easily soluble in water, and accounts for 25% of the total FA mass. The molten residue accounts for 62% of the total mass of FA. The water-cooled slag and air-cooled slag were studied separately, with the former appearing as black, glossy, and friable, while the latter appeared as dark gray and dense.

According to the national standard GB/T 18046–2017, the glass phase content of the water-cooled slag was calculated under three operating conditions based on the XRD pattern (Fig. 5.11). For the water-cooled slag, the glass phase content of the molten slag under the air + natural gas condition and the oxygen + natural gas condition is 65.42% and 77.67%, respectively. These values do not meet the requirement of 85% glass phase content stipulated in national standard GB/T 41015–2021. However, the glass phase content of the molten slag obtained using oxygen + coal for melting is 95.55%, significantly higher than the previous two conditions. This is due to the high Ca content in the FA and the insufficient Si and Al components. Direct melting results in a tendency for crystallization, leading to a decrease in the glass phase content of the molten slag. FA resulting from coal combustion is rich in Si and Al components, which compensate for the shortage of Si and Al during the melting process of FA. It also inhibits the cooling crystallization phenomenon after melting, resulting in a higher glass phase content of the molten slag. Furthermore, a higher cooling rate favors the formation of the

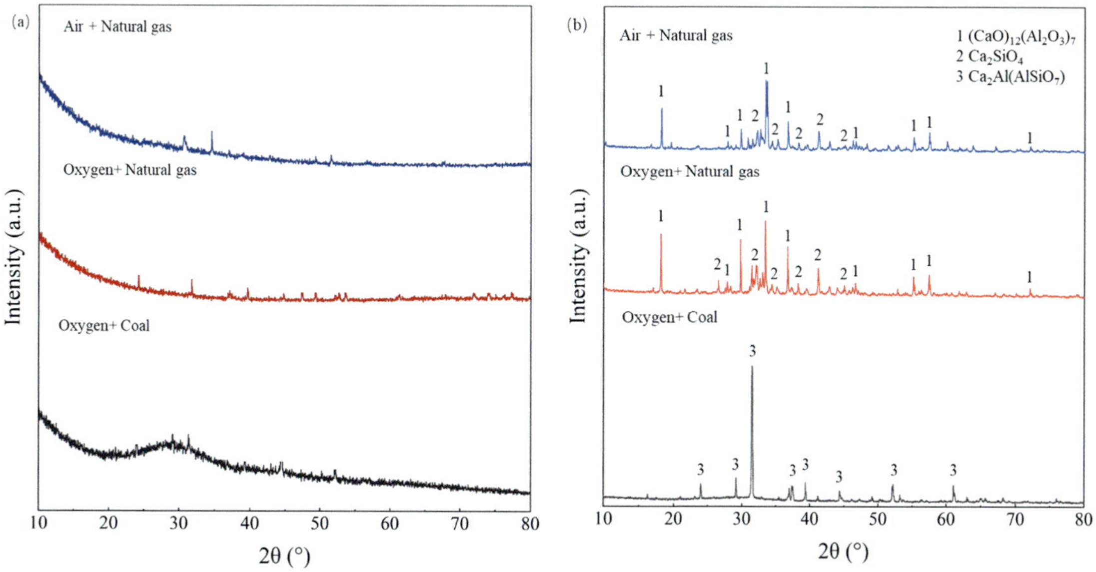

FIGURE 5.11 XRD pattern of slag in different experiment conditions: (A) water-cooled slag and (B) air-cooled slag.

TABLE 5.3 Heavy metal leaching of FA, SFA, and slag in three experiment conditions.

		Air + natural gas		Oxygen + natural gas		Oxygen + coal		Limits
	FA	SFA	Slag	SFA	Slag	SFA	Slag	GB/T 14,848
Element	Heavy metal leaching (mg/L)							
Cr	0.078	0.014	0.014	0.029	0.035	1.3943	0.005	0.050
Ni	0.000	0.555	0.000	2.305	0.000	0.0021	0.000	0.020
Cu	0.466	1.808	0.001	19.717	0.001	0.0228	0.000	1.000
Zn	2.048	1142.056	0.001	1562.717	0.000	70.590	0.000	1.000
As	0.013	0.01	0.001	0.005	0.001	0.007	0.000	0.010
Cd	0.001	40.011	0.000	23.515	0.000	0.125	0.000	0.005
Pb	37.860	26.509	0.000	28.855	0.000	1.643	0.000	0.010
Mn	0.010	8.814	0.005	38.645	0.0037	0.007	0.002	0.100

glass phase. Therefore, the water-cooling method is preferable to the air-cooling method for achieving harmless melt treatment of FA.

However, the melting of FA leads to the volatilization of volatile heavy metals, especially Zn and Pb, which condense in the SFA. A comparison of heavy metal leaching under the three conditions (Table 5.3) reveals that the heavy metal leaching from the SFA obtained using oxygen + natural gas is higher, with particularly high levels of Zn and Pb leaching, reaching 1562 mg/L and 29 mg/L, respectively, surpassing the standard requirements. These results indicate that heavy metal leaching is significantly influenced by the rise in melting temperature. However, the use of coal as fuel leads to a significant decrease in heavy metal leaching, with Zn and Pb levels of only 70 mg/L and 1.64 mg/L, respectively, compared with the oxygen + natural gas condition. This suggests that FA resulting from coal combustion not only promotes the formation of the glass phase but also effectively immobilizes heavy metals in FA slag, thereby reducing heavy metal leaching.

The advantage of oxygen-enriched melting technology is the good energy-saving effect for MSWIFA melting. As illustrated in Table 5.4, when the feeding of MSWI FA remain constantly at 75 kg/h, the natural gas consumption in air + natural gas condition is 49.34 m^3/h, while in oxygen + natural gas condition is 16.82 m^3/h, reduced by 66%. Thus, the total operating cost is dropped from 1880.23 RMB/ton FA to 838.49 RMB/ton FA. Moreover, more money will be saved when using cheaper fuel, such as coal, during enrichment melting. The operating cost will be 464.74 RMB/ton FA.

TABLE 5.4 Operating cost of three melting conditions.

Experiment conditions		N + A	N + O	C + O
Feeding rate (kg/h)		75	75	75
Electricity consumption	Other equipment (kWh)	13.87	7.93	8.42
	Oxygenerator (kWh)	–	32.07	32.07
Fuel consumption	Nature gas (m^3/h)	49.34	16.82	–
	Coal (kg/h)	–	–	21
Electricity price (RMB/kWh)		0.42		
Gas price (RMB/m^3)		2.74		
Coal price (RMB/kg)		0.85		
Total cost (RMB/ton FA)		1880.23	838.49	464.74

5. Conclusion

This chapter provides an overview of high-temperature treatment methods for combustion/incineration residues, including sintering and melting. The principles, benefits, and challenges of sintering and melting technologies have been systematically discussed. Both sintering and melting are effective in reducing the volume of incineration residues and producing more stable products. The advantage of sintering method is that it can be performed at lower temperatures compared by melting method, thus reducing energy consumption and operating cost. Melting of MSWI residues often requires higher operating temperature to transform the residues into stable glass materials. The melting method has a significant impact on dioxin decomposition, but it is less effective in fixing volatile heavy metals such ash Zn and Pb. Advanced melting technology such as plasma melting and oxygen-enriched melting are discussed in this chapter. Both plasma melting and oxygen-enriched melting achieve the harmless disposal of MSWIFA. The leachability of all heavy metals in treated slag meets the national standard. Compared with plasma melting, the advantage of oxygen-enriched melting technology is the low operating cost and the variability of fuel choice. Cheaper fuel such as coal can further reduce the operating cost. Meanwhile, high Si and Al content in coal FA can help the formation of Ca−Si−Al system to immobilize heavy metals. Results proves that oxygen-enriched melting has a wider range of application scenarios and more opportunities for its future industrial application.

Acknowledgments

Authors would like to thank the support by Science and Technology Innovation Cooperation Project between Science and Technology Department of Guangxi Province and Zhejiang University (Grant No. 2020ZD003).

References

Chen, Z., Zhang, S., Lin, X., Li, X., 2020. Decomposition and reformation pathways of PCDD/Fs during thermal treatment of municipal solid waste incineration fly ash. J. Hazard Mater. 394, 122526.

Chen, Z., Lin, X., Zhang, S., Xiangbo, Z., Li, X., Lu, S., Yan, J., 2021. Thermal cotreatment of municipal solid waste incineration fly ash with sewage sludge for PCDD/Fs decomposition and reformation suppression. J. Hazard Mater. 416, 126216.

Clavier, K.A., Watts, B., Liu, Y., Ferraro, C.C., Townsend, T.G., 2019. Risk and performance assessment of cement made using municipal solid waste incinerator bottom ash as a cement kiln feed. Resour. Conserv. Recycl. 146, 270−279.

Das, S., Lee, S.H., Kumar, P., Kim, K.H., Lee, S.S., Bhattacharya, S.S., 2019. Solid waste management: scope and the challenge of sustainability. J. Clean. Prod. 228, 658−678.

De Casa, G., Mangialardi, T., Paolini, A.E., Piga, L., 2007. Physical-mechanical and environmental properties of sintered municipal incinerator fly ash. Waste Manag. 27, 238−247.

Dou, X., Ren, F., Nguyen, M.Q., Ahamed, A., Yin, K., Chan, W.P., Chang, V.W.-C., 2017. Review of MSWI bottom ash utilization from perspectives of collective characterization, treatment and existing application. Renew. Sustain. Energy Rev. 79, 24−38.

Gao, X., Ji, G., Bhatia, S.K., Nicholson, D., 2019. Special issue on “transport of fluids in nanoporous materials”. Processes 7.

Gao, J., Dong, C., Zhao, Y., Hu, X., Qin, W., Wang, X., Zhang, J., Xue, J., Zhang, X., 2020. Vitrification of municipal solid waste incineration fly ash with B(2)O(3) as a fluxing agent. Waste Manag. 102, 932−938.

He, H., Lu, S., Peng, Y., Tang, M., Zhan, M., Lu, S., Xu, L., Zhong, W., Xu, L., 2022a. Emission characteristics of dioxins during iron ore Co-sintering with municipal solid waste incinerator fly ash in a sintering pot. Chemosphere 287, 131884.

He, H., Guo, X., Jin, L., Peng, Y., Tang, M., Lu, S., 2022b. The effect of adjusting sinter raw mix on dioxins from iron ore Co-sintering with municipal solid waste incineration fly ash. Energies 15.

Huang, B., Gan, M., Ji, Z., fly ashn, X., Zhang, D., Chen, X., Sun, Z., Huang, X., fly ashn, Y., 2022. Recent progress on the thermal treatment and resource utilization technologies of municipal waste incineration fly ash: a review. Process Saf. Environ. Protect. 159, 547−565.

Katou, K., Asou, T., Kurauchi, Y., Sameshima, R., 2001. Melting municipal solid waste incineration residue by plasma melting furnace with a graphite electrode. Thin Solid Films 386, 183−188.

Li, R., Wang, L., Yang, T., Raninger, B., 2007. Investigation of MSWI fly ash melting characteristic by DSC-DTA. Waste Manag. 27, 1383−1392.

Li, R., Li, Y., Yang, T., Wang, L., Wang, W., 2015. A new integrated evaluation method of heavy metals pollution control during melting and sintering of MSWI fly ash. J. Hazard Mater. 289, 197−203.

Li, Y., Feng, D., Bai, C., Sun, S., Zhang, Y., Zhao, Y., Li, Y., Zhang, F., Chang, G., Qin, Y., 2022. Thermal synergistic treatment of municipal solid waste incineration (MSWI) fly ash and fluxing agent in specific situation: melting characteristics, leaching characteristics of heavy metals. Fuel Process. Technol. 233.

Li, W., Yan, D., Li, L., Wen, Z., Liu, M., Lu, S., Huang, Q., 2023. Review of thermal treatments for the degradation of dioxins in municipal solid waste incineration fly ash: proposing a suitable method for large-scale processing. Sci. Total Environ. 875, 162565.

Lin, X., Mao, T., Chen, Z., Chen, J., Zhang, S., Li, X., Yan, J., 2021. Thermal cotreatment of municipal solid waste incineration fly ash with sewage sludge: phases transformation, kinetics and fusion characteristics, and heavy metals solidification. J. Clean. Prod. 317.

Lindberg, D., Molin, C., Hupa, M., 2015. Thermal treatment of solid residues from WtE units: a review. Waste Manag. 37, 82–94.

Liu, Y., Zheng, L., Li, X., Xie, S., 2009. SEM/EDS and XRD characterization of raw and washed MSWI fly ash sintered at different temperatures. J. Hazard Mater. 162, 161–173.

Liu, Z., Zhang, T., Zhang, J., Xiang, H., Yang, X., Hu, W., Liang, F., Mi, B., 2018. Ash fusion characteristics of bamboo, wood and coal. Energy 161, 517–522.

Okada, T., Tomikawa, H., 2012. Leaching characteristics of lead from melting furnace fly ash generated by melting of incineration fly ash. J. Environ. Manag. 110, 207–214.

Peng, Z., Weber, R., Ren, Y., Wang, J., Sun, Y., Wang, L., 2020. Characterization of PCDD/Fs and heavy metal distribution from municipal solid waste incinerator fly ash sintering process. Waste Manag. 103, 260–267.

Phua, Z., Giannis, A., Dong, Z.L., Lisak, G., Ng, W.J., 2019. Characteristics of incineration ash for sustainable treatment and reutilization. Environ. Sci. Pollut. Res. Int. 26, 16974–16997.

Pujara, Y., Pathak, P., Sharma, A., Govani, J., 2019. Review on Indian Municipal Solid Waste Management practices for reduction of environmental impacts to achieve sustainable development goals. J. Environ. Manag. 248, 109238.

Sakai, S.-i., Hiraoka, M., 2000. Municipal solid waste incinerator residue recycling by thermal processes. Waste Manag. 20, 249–258.

Shi, W.-J., Kong, L.-X., Bai, J., Xu, J., Li, W.-C., Bai, Z.-Q., Li, W., 2018. Effect of CaO/Fe2O3 on fusion behaviors of coal ash at high temperatures. Fuel Process. Technol. 181, 18–24.

Wang, Q., Tian, S., Wang, Q., Huang, Q., Yang, J., 2008. Melting characteristics during the vitrification of MSWI fly ash with a pilot-scale diesel oil furnace. J. Hazard Mater. 160, 376–381.

Wang, Z.-G., Kong, L.-X., Bai, J., Li, H.-Z., He, C., Yan, T.-G., Guo, Z.-X., Bai, Z.-Q., Li, W., 2019. Effect of vanadium and nickel on iron-rich ash fusion characteristics. Fuel 246, 491–499.

Wang, X., Zhang, L., Zhu, K., Li, C., Zhang, Y., Li, A., 2021. Distribution and chemical species transition behavior of chlorides in municipal solid waste incineration fly ash during the pressure-assisted sintering treatment. Chem. Eng. J. 415.

Wey, M.Y., Liu, K.Y., Tsai, T.H., Chou, J.T., 2006. Thermal treatment of the fly ash from municipal solid waste incinerator with rotary kiln. J. Hazard Mater. 137, 981–989.

Zhan, X., Wang, L., Wang, L., Gong, J., Wang, X., Song, X., Xu, T., 2021. Co-sintering MSWI fly ash with electrolytic manganese residue and coal fly ash for lightweight ceramisite. Chemosphere 263, 127914.

Zhan, X., Wang, L., Wang, J., Yue, Z., Deng, R., Wang, Y., Xu, X., 2022. Roasting mechanism of lightweight low-aluminum-silicon ceramisite derived from municipal solid waste incineration fly ash and electrolytic manganese residue. Waste Manag. 153, 264–274.

Zhang, Q., Liu, H., Qian, Y., Xu, M., Li, W., Xu, J., 2013. The influence of phosphorus on ash fusion temperature of sludge and coal. Fuel Process. Technol. 110, 218–226.

Chapter 6

Hydrothermal treatment of combustion/incineration residues

Yaqian Shi[1], Qiang Zeng[1,2], Jianhua Yan[1] and Lei Wang[1]

[1]*State Key Laboratory of Clean Energy Utilization, Zhejiang University, Hangzhou, China;* [2]*Department of Civil Engineering and Architecture, Zhejiang University, Hangzhou, China*

1. Introduction

Combustion/incineration is commonly used in the treatment of municipal solid waste (MSW), coal, etc. (Pan et al., 2013), since it has advantages in the reduction of mass and volume, as well as the generation of both heat and electricity. During the combustion/incineration process, a large amount of fly ash (FA) and bottom ash (BA) was generated. Therefore, appropriate methods for further treatment of combustion/incineration residues are required to accomplish immobilization/stabilization, harmlessness, and resource utilization. The MSW incineration bottom ash (MSWI BA) is primarily the residue collected on the furnace bed after the incineration of garbage, whereas the municipal solid waste incineration fly ash (MSWI FA) is primarily fine particulate powder material collected in a garbage incinerator's flue gas purification system. MSWI FA and MSWI BA have been shown to include a variety of metals and refractory organic compounds, including polycyclic aromatic hydrocarbons (PAHs) and polychlorinated dibenzodioxins/furans (PCDD/Fs). Landfilling, resource recovery, composting, and gasification are used to treat MSWI FA and MSWI BA. They can be used as aggregates, backfill materials, soil amendments, and cement additives after appropriate pretreatment (Chen and Lin, 2006). Coal fly ash (CFA) and coal bottom ash (CBA), which are produced at the top and bottom of the combustion furnace, respectively, are used worldwide for soil improvement, the production of Portland cement, ceramics, zeolites, and fibers, as well as the production of catalysts, polymer fillers, and wastewater treatment adsorbents (Gollakota et al., 2019).

Hydrothermal treatment technology has many advantages in safely treating and disposing of combustion/incineration residues. In the 1840s, Schafhat conducted a hydrothermal synthesis of minerals (Schafhat, 1845). After 200 years of continuous development, hydrothermal technology had been applied to the solidification of MSWI ash (Jing et al., 2007) and coal fly ash (Jing et al., 2006). For the "compaction first, then hydrothermal" treatment, the strength development of solidified samples mainly depends on the formation of tobermorite, which indicates the ability to stabilize heavy metals in solidified bodies. There are very few tobermorite in nature, but it's easy to be synthesized under hydrothermal reaction conditions. In the hydration process of Portland cement, tobermorite-like calcium silicate hydrates (C—S—H) are considered the principal phase to produce strength. In recent years, hydrothermal solidification, particularly subcritical hydrothermal treatment, has been deeply concerned in the field of solid waste treatment and is regarded as one of the most promising ways of solid waste utilization (Guo and Song, 2018). Hydrothermal synthesis and autoclave curing are also used in the production and manufacturing of aerated concrete. As the main hydrothermal product of autoclaved aerated concrete (AAC), tobermorite also affects its performance, because the synthesis of tobermorite is influenced by the autoclaving temperature, autoclaving time, and the molar ratio of CaO/SiO_2 (C/S) (Chen et al., 2017; Galvánková et al., 2016). For "liquid-phase hydrothermal" treatment, zeolites are often synthesized and widely used as adsorbents, ion exchangers, and catalysts due to their remarkable adsorption and cation exchange capabilities, large specific surface area, abundant micropores, and good thermal stability (Deng et al., 2016).

Treatment and Utilization of Combustion and Incineration Residues. https://doi.org/10.1016/B978-0-443-21536-0.00027-7

2. Hydrothermal treatment

2.1 Equipment and classification

Hydrothermal reaction refers to the chemical reaction that occurs in a closed reactor under high pressure and temperature in fluids such as water, aqueous solution, or steam. The classification of hydrothermal reaction is diverse and can be approached from different perspectives. Firstly, based on the reaction temperature, hydrothermal reactions can be categorized into low (below 100 °C), medium (100–300 °C), and high-temperature reaction (above 300°C). Another way to classify hydrothermal reactions is based on the reaction conditions, which leads to the segmentation into subcritical hydrothermal reaction and supercritical hydrothermal reaction. The reaction temperature of the former is between 100 and 240°C, suitable for industrial or laboratory operations. The reaction temperature of the latter can reach up to 1000°C, which is sufficient to utilize the properties of supercritical water and the characteristics of reaction materials under the condition of high temperature and high pressure for synthesis reactions. Additionally, hydrothermal reactions can also be classified based on the type of equipment used. This leads to the division between the ordinary hydrothermal method and the special hydrothermal method. The former uses electricity, hot steam, and hot oil as the heating medium. Also, a special hydrothermal method has been developed, and it refers to adding other action fields, such as direct current field, magnetic field, and microwave field, to the hydrothermal reaction systems (Yang and Park, 2019).

The existing research on the hydrothermal treatment of combustion/incineration residues can be divided into two categories. One method can be called "compaction first, then hydrothermal" treatment, which refers to first mixing the residues with additives to obtain a mixture for compression treatment, and then performing hydrothermal curing in high-pressure steam, as shown in Fig. 6.1. The other method can be called "liquid-phase hydrothermal" treatment, which is to mix the residues with water or other solutions at a certain solid–liquid ratio and put them into an autoclave for hydrothermal reaction with or without stirring, as shown in Fig. 6.2.

When treating combustion/incineration residues, these hydrothermal methods have the following advantages. Firstly, the hydrothermal reaction medium is usually water or aqueous solution. Due to its green nature, water is almost pollution-free to the environment (Kumar and Gupta, 2009). Secondly, water is not only a reactant but also a reaction medium and a catalyst, which can increase the reaction rate (Zhang and Itoh, 2006). Water has characteristics of dissolving nonpolar organic compounds, low dielectric properties, and high concentration of H^+ and OH^- under high temperature and pressure, which can participate in reactions as chemical components. Therefore, water serves as a mineralizer, pressure transfer medium, and powerful catalyst (Shen et al., 2016). Thirdly, the reactor has reliable airtightness, avoiding secondary pollution caused by raw material reactions.

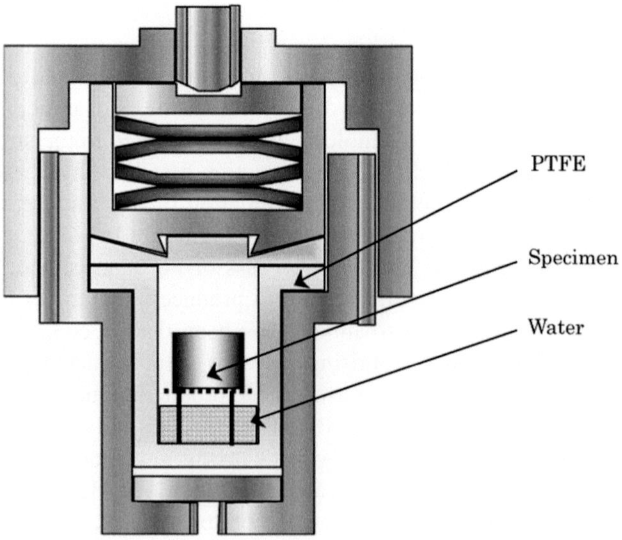

FIGURE 6.1 Hydrothermal apparatus (Jing et al., 2010).

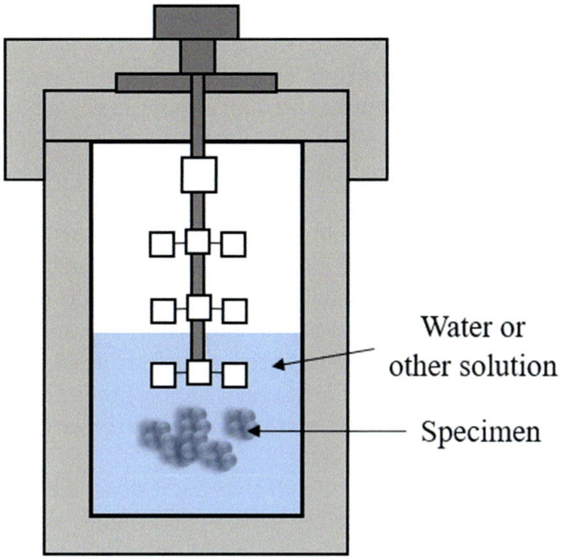

FIGURE 6.2 Schematic of typical hydrothermal method equipment.

2.2 Hydrothermal reaction mechanism

Hydrothermal synthesis refers to the use of high-temperature and high-pressure aqueous solution to dissolve or react to those substances that are difficult to dissolve or are insoluble under atmospheric conditions. This process takes place in a specially made closed reaction vessel, relying on an aqueous solution as the reaction system, with oxides or hydroxides, or gels as the precursor. It is an effective method to produce convection by controlling the temperature difference in the autoclave to make the solution reach the supersaturated state, and then synthesize materials through recrystallization. Hydrothermal treatment is a very useful way of activating reactions, and its mechanism is believed to involve dissolution and/or precipitation processes (Jing et al., 2007). Some reports also show that hydrothermal treatment can significantly increase the pozzolanic reaction of fly ash (Goi et al., 2003; Ma and Brown, 1997). The obtained compounds can be used to immobilize toxic waste. In addition, it is reported that hydrothermal treatment of MSWI FA with water or alkaline solution can decompose dioxins, such as PCDDs and PCDFs, through dechlorination, especially at high temperatures (Hashimoto et al., 2004). The process of hydrothermal synthesis of zeolite is to dissolve the silicon source and aluminum source in an alkaline solution and then mix them to form a hydrogel. After aging for some time, zeolite crystals nucleate and grow under appropriate temperature and pressure in an autoclave, and pure zeolite can be obtained after washing and calcining.

The main principle of hydrothermal curing of combustion/incineration residues is to use the residues as a siliceous raw material to provide SiO_2, and the additive CaO or $Ca(OH)_2$ generates C−S−H under hydrothermal conditions. The main products are tobermorite or tobermorite-like C−S−H. As for AAC, which is usually made of the mixture of fine quartz sand, Portland cement, lime, calcium sulfate, aluminum powder, and water, the initial fluid mixture is cast into the mold, and then the reaction of calcium hydroxide, aluminum, and water raises the temperature and generates hydrogen, which causes foam in the first hour. Subsequently, the hydration of Portland cement and lime causes the hardened “green cake” to harden and finally solidify. At this stage, poorly crystallized C−(A)−S−H (C/S ratio = 1.7−1.8) and ettringite provide initial strength. Next, these blocks will undergo hydrothermal curing at 190−200 °C. The most important reaction for AAC hardening is the partial transformation of C−(A)−S−H into well-crystalline tobermorite (C/S ratio = 0.75) (Mesecke et al., 2023).

2.3 Influencing factors of hydrothermal reaction

For the reaction of hydrothermal synthesis of zeolite, factors such as hydrothermal time, temperature, liquid-solid ratio of raw materials, alkali concentration, and activator can all affect the synthesis process of zeolite products. Research has shown that reaction temperature and time are the most important variables affecting the synthesis and adsorption capacity of zeolites (Qiu et al., 2018). Alkali plays a decisive role in the hydrothermal synthesis of zeolite. The results of hydrothermal synthesis of zeolite from fly ash show that NaOH has the strongest solubility in fly ash, and the conversion rate of fly ash into zeolite is also the highest, followed by KOH, and the worst at Na_2CO_3 (Murayama et al., 2002). The differences in the composition of combustion/incineration residues also have a certain impact on the synthesized zeolite products, which contain various small and trace elements, leading to poor uniformity in the properties of the synthesized products. The addition of other substances can also affect the synthesis of the final zeolite product, such as the addition of rice husk ash (RHA) as a silica source during the hydrothermal treatment of CFA, which changes the crystalline phase of product zeolite (Fukui et al., 2007).

In the study of synthesizing tobermorite fibers using a mixture of CFA and lime as raw materials through the hydrothermal method, the addition or absence of NaOH also affects the synthesized tobermorite fibers (Cao et al., 2020). When NaOH is not added, the dissolution of silicon-containing components in CFA is limited, and only small tobermorite fibers are synthesized. When 20 g/L NaOH is added, the dissolution of silicon and aluminum-containing components is promoted. The obtained lath-like tobermorite fiber can be further transformed into a spherical structure, and the porous aggregation structure formed makes the synthetic material conducive to the production of energy-saving building materials with low thermal conductivity and apparent density. It is worth noting that excessive addition of high concentration NaOH (above 40 g/L) can damage the porous aggregation structure.

When the hydrothermal treatment is applied to solidify combustion residue, the experimental results show that NaOH solution, $Ca(OH)_2$ content, compaction pressure, autoclave curing temperature, and time have significant effects on the strength of the resultant solidified body. The C/S ratio is considered to be the most important factor in the synthesis of tobermorite. When the C/S ratio of a starting material approaches 0.83, the tobermorite becomes more favorable and readily achievable (Jing et al., 2006).

3. Utilization of combustion residues as value-added materials

3.1 Coal fly ash

Coal ash is a by-product of thermal power plants, which can be divided into fly ash and bottom ash. Generally, coal fly ash (CFA) accounts for over 80% of the total coal ash volume. The increasing demand for coal as a global energy source and the significant instability of alternative energy sources have led to a surge in the use of coal-based energy. Therefore, a large amount of CFA has been generated around the world, posing a serious threat to ecology due to its storage and disposal limitations. Unlike most ordinary industrial by-products, CFA is toxic and difficult to handle due to its complex composition. CFA has always been a major contributor to the cement industry because it is used as a raw material or additive for cement manufacturing. Because of its inherent pozzolanic characteristics, CFA is used to partially replace the clinkers in ordinary Portland cement. Approximately 15% of CFA produced worldwide is used as raw materials for cement and concrete production. The application of CFA in mixed cement is an innovation. However, its use must follow strict guidelines. Another main application of CFA is to prepare mesoporous materials or zeolites. However, zeolite synthesis is facing a key issue related to the existence of a large number of potentially toxic coal-derived compounds. Although CFA is also used for various other applications, such as road base materials, most of it is still used for landfill. At present, hydrothermal treatment has been successfully applied to synthesize zeolite and adsorbent from CFA. Hydrothermal technology is considered very attractive as it can produce very tough and durable products while processing and recycling waste on a large scale (Jing et al., 2006). The current research status of CFA hydrothermal treatment in recent years is summarized in Table 6.1.

3.1.1 CFA-based zeolite as adsorbent

The similarity in composition between CFA and zeolite provides the possibility for the conversion of CFA into zeolite. However, their crystallinity is different: CFA is mainly composed of an amorphous structure, while zeolite has a clear crystalline structure (Belviso et al., 2014). The ordered microstructure of zeolites makes them possess higher cation exchange capacity (CEC), larger surface area, and excellent thermal stability, further improving utilization efficiency (Wang et al., 2009). Hydrothermal treatment can increase the adsorption capacity of CFA by 2–25 times (Nascimento et al., 2009). The synthesis of zeolite products using CFA is a good way to utilize and dispose of CFA with great economic and environmental benefits. The research shows that CFA can be transformed into different zeolite structures by using appropriate synthesis conditions, such as analcime (Penilla et al., 2006), chabazite, cancrinite, gmelinite, Na-P1 (Ferrarini et al., 2016; Ma et al., 2022), ZSM-5, ZSM-28, Na-X, Na-Y, phillipsite (Fukui et al., 2007), and sodalite (Wdowin et al., 2014). The hydrothermal reaction of synthesizing zeolite through CFA hydrothermal treatment involves three steps:

TABLE 6.1 Summary of the research status on the hydrothermal treatment of CFA in recent years.

Material	Sample preparation	T/°C	t/h	Product	References
CFA (class F), NaOH	Samples (mixture of 3 g CFA and 7.5 mL NaOH solution) were aged for 2 days at room temperature and then treated hydrothermally.	100	144	NaP-type zeolite	Ma et al. (1998)
CFA, NaOH, Na_2CO_3, KOH	Solid/liquid ratio was 100 g/400 cm^3	120	3, 24	Zeolite P and chabazite	Murayama et al. (2002)
CFA, NaOH	50 mL of NaOH aqueous solution and 2 g of CFA	120	–	Phillipsite	Fukui et al. (2006)
CFA (class F), 1 M NaOH	Solution to CFA ratio was 10:1 by weight	150, 200	12	150°C: gismondine-type Na zeolite P1 200 °C: analcime-C and sodalite, traces of tobermorite11 Å	Penilla et al. (2006)
CFA, 2 M NaOH, rice husk ash	40 g of CFA was dissolved into 1.0 L of aqueous NaOH solution	120	–	Zeolite	Fukui et al. (2007)
CFA, NaOH	Mass ratio of the CFA to NaOH powder was 1:1.3	100	5, 24	Zeolite A and X	Wang et al. (2009)
CFA, 1 M NaOH	Solution to CFA ratio of 10:1 by weight	85, 150	24	Zeolite	Wu et al. (2012)
CFA, La^{3+} solution	–	–	–	Lanthanum-doped fly ash (La-FA) adsorbent	Wang et al. (2013)
CFA, NaOH, NaOCl	CFA:NaOH:NaOCl:H_2O mass ratio of 3.03:1.00:1.14:1.00	–	–	Geopolymer	Nyale et al. (2013)
CFA, NaOH		80	36	Na–P1 zeolite	Wdowin et al. (2014)
CFA, 3 M NaOH	L/S ratio of 6 L/kg	100	24	Zeolite Na–P1 and 4A	Ferrarini et al. (2016)
CFA, slag, and alkaline solution	After mixing fly ash and slag with the alkaline solution, the geopolymer pastes were cast into 25 mm cubic molds	90	24	Mesoporous geopolymers containing zeolite phases	Lee et al. (2016)
Carbide slag and acid residues of fly ash	–	200–260	12–60	Xonotlite fiber	Zou et al. (2016)
CFA, alkaline activation	–	80	24	Sodalite	Król et al. (2017)
CFA, NaOH	–	90–110		Zeolite	Tauanov et al. (2018)
CFA, quicklime, NaOH	C/S ratio of 1.0, L/S ratio of 20 mL/g	180–220	0.2–7	Tobermorite	Cao et al. (2020)
CFA, lime, quartz sand	C/S ratio of 0.73	190	16	Tobermorite	Drochytka and Černý (2020)
CFA, NaOH	S/L ratio of 1:10 g/mL	110	9	Na–P1 zeolite	Ma et al. (2022)
CFA, $Ca(OH)_2$, KOH	C/S ratio of 1, L/S ratio of 20 mL/g	200	1–6	Tobermorite fiber	Hou et al. (2023)

dissolution of Si^{4+} and Al^{3+}, condensation of silicate and aluminate ions in an alkali solution to prepare aluminosilicate gel, and crystallization of aluminosilicate gel to generate zeolite crystals (Murayama et al., 2002).

The hydrothermal treatment of CFA to synthesize zeolite mainly includes an ordinary hydrothermal synthesis method and a microwave assisted synthesis method. The latter also belongs to hydrothermal synthesis, but with microwave-assisted crystallization, the reaction speed can be increased and the synthesis time can be greatly shortened (Fukui et al., 2007). Research has found that the particle size of phillipsite obtained by microwave was smaller than that obtained by conventional methods. The application of zeolite synthesized by CFA is the same as that of natural zeolite and zeolite made from natural materials (such as kaolin, diatomites) or other aluminosilicate industrial wastes (Ferrarini et al., 2016). CFA-based zeolite is also used to remove heavy metals from wastewater (Wu et al., 2012). It has adsorption properties for heavy metal cations and can be used as a resource for separating, immobilizing, and treating radioactive Cs and Sr (Ma et al., 1998). CFA-based zeolite as an adsorbent can remove Pb^{2+} from an aqueous solution (Ma et al., 2022).

For zeolite synthesis, temperature and time are important variables that affect the adsorption of cations. Most zeolites formed at 150°C are gismondine-type P1-Na species, which are converted to analcime-C zeolite at 200 °C. Gismondine-type Na zeolite P has high selectivity for Cs and is the best candidate for fixing radioactive waste. Gismondine and analcime-C zeolites also show high Cd selectivity (Penilla et al., 2006). CFA-based zeolite is also used to adsorb dyes in aqueous solutions. Adsorbents FA-ZA and FA-ZX synthesized from aluminosilicate gel prepared from CFA have the potential to remove MB cationic dyes from aqueous solutions as adsorbents (Wang et al., 2009). The preparation of lanthanum-doped fly ash (La-FA) adsorbent by hydrothermal method can be used to remove reactive red B (WRR-B) dye for wool from an aqueous solution (Wang et al., 2013).

3.1.2 Geopolymer

Geopolymers are considered precursors of zeolites, so zeolites are often detected in geopolymer gel. CFA can be used as a matrix in geological polymerization. Davidovits first used the term "geopolymer" to refer to a three-dimensional aluminosilicate material (Davidovits, 1999). The geological polymerization process is considered to have two steps. First, it involves the dissolution of silicon and aluminum species from the surface of raw material, and the surface hydration of undissolved particles of raw material. Then the active surface groups and soluble species polymerize form gel, and then the hardened geopolymer structure is formed (as shown in Fig. 6.3). Dissolved silicon and aluminum species require alkali metal salts and/or hydroxide components (Swanepoel and Strydom, 2002). Geopolymer materials can be used in various fields because of their fire resistance, durability, and excellent mechanical properties, such as ceramics, cement, hazardous waste stabilized matrices, refractory materials, asbestos-free materials, and high-tech materials (Cioffi et al., 2003). Sodalite hydrothermally synthesized from alkali-activated fly ash can be used for Pb^{2+} cation adsorption (Luo et al., 2016). Foam geopolymer can be synthesized by alkali activation of CFA and subsequent hydrothermal treatment, which may be applicable to various applications including the construction industry (Nyale et al., 2013). Mesoporous geopolymers can be synthesized by hydrothermal treatment using CFA, slag, and alkaline solutions. The results show that the geopolymer mixed with CFA and slag is composed of Na-P1 and sodalite zeolite phases, which have the characteristics of mesoporous materials, and its maximum compressive strength is 16.57 MPa (Luo et al., 2016).

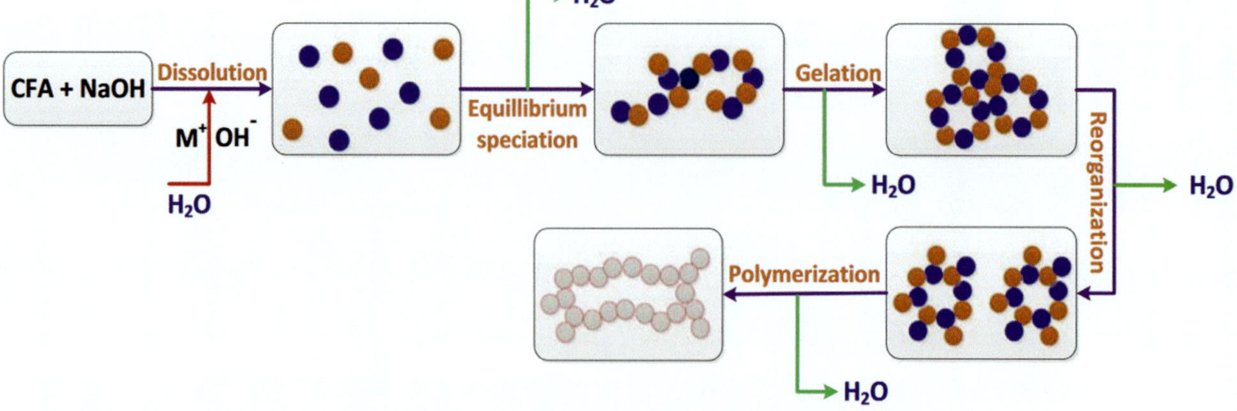

FIGURE 6.3 Schematic representation and mechanism of geopolymerization (Gollakota et al., 2019).

3.1.3 High strength block

The solidification of CFA has been carried out using hydrothermal treatment. CFA is first compacted in a mold at 20–50 MPa and then subjected to hydrothermal curing in an autoclave. Hydrothermal curing is usually carried out at 150–250°C for a duration of 10–72 h. The experimental results show that NaOH solution, $Ca(OH)_2$ content, compaction pressure, autoclave curing temperature, and time have significant effects on the strength of the cured body. In the solidified bodies produced by CFA, the most important component to produce strength is tobermorite, or tobermorite-like C–S–H. When the C/S ratio of the raw material approaches 0.83, tobermorite is easily formed, and the generated tobermorite improves the strength of solidified bodies (Jing et al., 2006). This hydrothermal treatment method may provide a high potential for large-scale recovery of CFA.

3.1.4 High-value fiber materials

Acid extraction residues of CFA and carbide slag has been successfully prepared into well-crystallized xonotlite fibers using the hydrothermal method (Zou et al., 2016). Acid extraction residues of CFA refer to the solid waste obtained from acid leaching of aluminum from CFA, with a silica content of 50%–85%. The chemical formula of xonotlite is $Ca_6Si_6O_{17}(OH)_2$, which is a novel type of ultralight fiber synthesized from calcium and silicon and is widely used in the fields of superinsulation and flame retardancy. This material is usually synthesized through hydrothermal treatment, using different types of high-purity siliceous and calcareous materials, such as quartz, sodium silicate, calcium oxide, and calcium carbonate. If industrial wastes such as combustion/incineration residues with high calcium or silicon content are used as the source of calcium or silicon, the synthesis cost of xonotlite fibers can be significantly reduced. CFA and lime can be synthesized into tobermorite fibers through hydrothermal treatment, and the well-crystallized tobermorite fibers aggregate into porous spherical structures, which makes the synthetic materials beneficial to the production of energy-saving building materials with low thermal conductivity and apparent density (as shown in Fig. 6.4).

3.2 Coal bottom ash

Coal bottom ash (CBA) contains abundant aluminosilicate components, which are the precursor of mesoporous materials (also known as zeolites). Therefore, transforming CBA into zeolite can reduce environmental pollution and improve economic values, which is also a sustainable CBA treatment method.

3.2.1 Autoclaved aerated concrete

The basic raw materials for producing AAC include pulverized fuel ash or sand, cement, gypsum, lime, and water, as well as a small amount of aluminum powder as a pore-forming agent, reinforced with rust-resistant steel bars, and cured by high-temperature, high-pressure, and steam. The green cake of AAC is autoclaved by a hydrothermal process under steam pressure, and well-crystallized C–S–H and 1.1-nm tobermorite are the main combined phases. It is a type of building material with superior performance (Arun Kumar et al., 2022). AAC has many advantages, such as low density, low thermal conductivity, low shrinkage, reduced dead load, faster construction speed, and low transportation costs (Wongkeo and Chaipanich, 2010). In the reaction system, aluminum reacts with hydroxide of calcium or alkali to release hydrogen gas and form bubbles, as shown in Eq. (6.1). The speed of bubble formation is very important to the success of the final aerated concrete product (Holt and Raivio, 2005).

$$2Al + 3Ca(OH)_2 + 6H_2O \rightarrow 3CaO \cdot Al_2O_3 \cdot 6H_2O + 3H_2 \text{ (gas)} \tag{6.1}$$

The use of CBA in AAC formulas results in the reduction of the unit weight of produced AAC for all relevant CBA substitution rates. However, it has been observed that the use of CBA in concrete is beneficial for increasing the strength of AAC. As the CBA replacement rate increases, the thermal conductivity decreases (Kurama et al., 2009). Therefore, CBA can be used as a substitute for Portland cement to produce lightweight concrete (LWC) through the autoclave aerated concrete method. Moreover, as the CBA content increases, the compressive strength, flexural strength, and thermal conductivity increase, which is due to the formation of the tobermorite. From the comparison with Portland cement concrete, it can be seen that using CBA as a cement substitute is beneficial for improving the strength of AAC while achieving relatively low thermal conductivity (Wongkeo et al., 2012).

Compared with air curing, steam pressure curing has many advantages, especially in shortening the curing time. Research has found that the compressive strength of CBA-based LWC after 6 h of steam compression is similar to that after 28 days of air curing, and both have slightly higher thermal conductivity than Portland cement-controlled concrete.

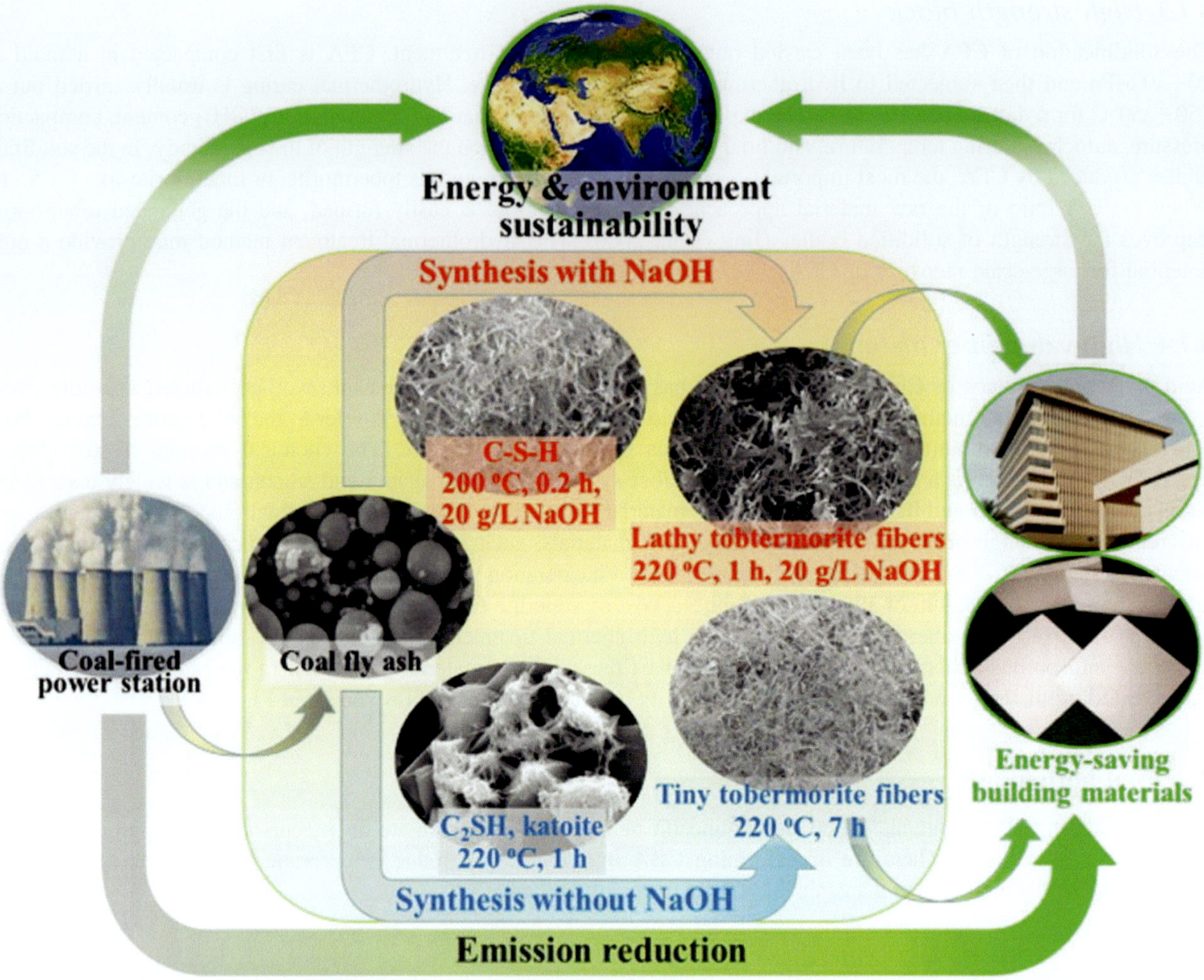

FIGURE 6.4 Coal fly ash recycle process (Cao et al., 2020).

Furthermore, adding silica fume to the raw material has a significant effect on improving compressive strength. Compared with the fiber C–S–H detected in Portland cement-controlled concrete, the tobermorite detected in high-pressure autoclaved bottom ash concrete containing silica fume has a denser microstructure (Wongkeo and Chaipanich, 2010).

3.2.2 Zeolite and mesoporous silica synthesis

CBA can be converted into zeolite due to its high silicon and aluminum content. The commonly used method for converting CBA into zeolite is hydrothermal conversion. CBA is treated with strong alkaline solutions such as NaOH and KOH and then heated (20–200 °C) to form zeolite materials. This way, the synthesis process at relatively low temperatures can avoid excessive energy loss. The hydrothermal process includes three stages: dissolution stage, condensation or gel stage, and crystallization stage (Yao et al., 2015). CBA-based zeolite can be used as a dye adsorbent, such as zeolite prepared from CBA and CFA by hydrothermal activation in NaOH solution, and using it for the adsorption of cationic dye Crystal Violet (Bertolini et al., 2013). Another example is the synthesis of SSZ-13 zeolite using CBA as raw material after hydrothermal treatment to selectively adsorb Alizarin Red S (ARS) dye (Gollakota et al., 2021).

Many mesoporous silicas have been synthesized using CBA due to the abundant presence of silica in CBA. Mesoporous silica has a large and accessible surface area, which can overcome the shortcomings of zeolite materials. The first mesoporous silica material was prepared by hydrothermal treatment of aluminosilicate gel with a crystal liquid template formed by surfactants (Kresge et al., 1992), which started the application of mesoporous silica in catalysis, adsorption, sensor, separation, or gas storage (Al-Shehri et al., 2019). Using CBA as raw material, various MCM-41 were synthesized by alkaline melt hydrothermal method, adjusting the raw material ratio (NaOH/CBA) and calcination temperature. In the potentially toxic element adsorption experiment, the synthesized MCM-41 removed up to 99.4% of Pb^{2+} from the aqueous solution (Vu et al., 2020).

3.3 Municipal solid waste incineration bottom ash

MSWI ash can be used as raw material for cement production. However, due to its presence of chlorine and heavy metals, pretreatment is necessary before being used in the cement industry (Kikuchi, 2001). Slag formed by molten incineration ash has been used as concrete aggregate (Jimbo, 1996). Incineration ash can also be used to produce ceramic materials (Cheeseman et al., 2003) and zeolites. Generally speaking, only a small portion of incineration ash is utilized by other industries, and the majority is still being disposed of in landfills. The disposal of incineration ash has become increasingly expensive because there are fewer landfill sites for waste disposal, and incineration ash contains toxic substances that usually need to be stabilized before final disposal (Jing et al., 2007). Therefore, there is an urgent need for more innovative and effective processing techniques to solve the problem of incineration ash. Hydrothermal treatment MSWI BA has been applied as an innovative technology. The current research status of MSWI BA hydrothermal treatment in recent years is summarized in Table 6.2.

3.3.1 Sorbent material

MSWI BA has the potential to serve as an adsorbent for wastewater treatment and soil/sediment remediation applications through hydrothermal treatment, which is a sustainable pathway for MSWI BA reuse. By subjecting MSWI BA to a hydrothermal process under strong alkaline conditions, it can be converted into adsorption materials with zeolite-like properties. The bottom ash generated from the combustion of urban solid waste in fluidized beds is subjected to alkaline hydrothermal treatment in NaOH solution, and the transformed zeolites and other compounds (tobermorite, bauxite, and $Ca_3AlFe(SiO_4)(OH)_8$) can replace various ionic species, making MSWI BA suitable for the fixation of toxic and radioactive waste. Zeolite and tobermorite can fix ion species through ion exchange mechanisms, opening up new opportunities to be used for fixing other toxic and radioactive waste such as cement and asphalt (Penilla et al., 2003). After hydrothermal treatment, the BET surface area of MSWI BA increased from 4.6 to 22.1 m^2/g. The increase in surface area generates strong adsorption capacity. Research has found that the adsorption performance of zeolite synthesized by MSWI BA on heavy metal solutions and contaminated sediments is generally better than natural zeolite (Peña et al., 2006). The removal rate of Cd, Co, Ni, Pb, and Zn in sediment pore water by MSWI BA at a dosage of 50 mg/g adsorbent exceeds 84%, while natural zeolites are effective only for As and Pb (Chiang et al., 2012). The adsorption capacity of zeolite transformed by MSWI BA hydrothermal treatment is of great significance for the treatment of printing and dyeing wastewater containing organic dyes (Wang et al., 2016).

MSWI BA and NaOH solutions can also be used to synthesize tobermorite-based porous materials through hydrothermal treatment technology. In the absence of additional foaming agents, the NaOH solution in the raw material can be used as both reaction solvent and foaming agent. NaOH solution, as a reaction solvent, first promotes the formation of tobermorite during hydrothermal treatment and forms a large number of small pores. Then, as a foaming agent, it is released from the cured sample in the form of steam and forms large pores after drying. The BET-specific surface area of the obtained product is 60 m^2/g, and the pore size distribution is 0.01−10 mm. The MSWI BA treated with hydrothermal treatment shows no significant leachability, so when it is used for wastewater treatment, the adsorbent is most likely to retain more elements instead of releasing the elements originally existing in the material (Jing et al., 2007).

3.3.2 High-strength block

The hydrothermal treatment technology used for MSWI BA curing is similar to the previous hydrothermal technology used for CFA curing. The raw materials are first compacted in the mold and then hydrothermally cured in an autoclave for 10−72 h under the saturated steam pressure of 150−250°C. In addition to the obvious influence of hydrothermal curing temperature and time on the development of the tensile strength of the cured body, adding NaOH solution and fresh cement to MSWI BA also has a significant influence on the tensile strength. The development of strength is mainly due to the formation of 1.1 nm tobermorite, which densifies the matrix and promotes the development of strength. Mixing MSWI BA with other substances will also result in different hydrothermal curing effects. Compared with mixed quartz, hydrated lime, and water-cooled blast furnace slag (WBFS), the result shows that adding WBFS enhances the strength the most, which is better than the addition of quartz. The reason for this is that during the initial hydrothermal process, WBFS has higher reactivity than quartz, providing more silica for hardening and curing samples. The solidified bricks also exhibit a very low heavy metal leaching rate, and the hydrothermal treatment method has a certain potential for recovering MSWI BA (Jing

TABLE 6.2 Summary of the research status on the hydrothermal treatment of MSWI BA in recent years.

Material	Sample preparation	T/°C	t/h	Product	References
MSWI BA, NaOH	–	50–200	12	Zeolite, aluminum tobermorite, andradite, and $Ca_3AlFe(SiO_4)(OH)_8$	Penilla et al. (2003)
MSWI BA, 2 M NaOH	The mixture (MSWI BA mixed with 40 mass % NaOH solution) was compacted with a pressure of 1–5 MPa	180	12	Tobermorite	Jing et al. (2007)
MSWI BA, NaOH	–	150	24	Sodium aluminum silicate hydrate and tobermorite ($Ca_5Si_6O_{16}(OH)_2\cdot 4H_2O$)	Chiang et al. (2012)
MSWI BA, 1 M NaOH	15 g of MSWI BA was mixed with 150 mL of NaOH aqueous solution	150	1	Aeolite, sodium aluminum silicon phosphorus oxide hydrate, katoite, and potassium aluminum silicate hydrate	Wang et al. (2016)
Cement, quicklime, quartz sand, MSWI BA, gypsum	After mixing, add the Al powder solution slowly into the slurry.	190	10	–	Li et al. (2018)
MSWI BA, CFA, CaO, SiO_2	Water content 30% Compaction pressure 5 MPa	180	10	–	Rożek et al. (2019)

et al., 2007). The solidification in the autoclave resulted in the immobilization of almost all heavy metals in the obtained matrix (more than 99%). In addition, the hydrothermal solidification of MSWI BA seems to have several sustainable benefits: it saves space in landfills, eliminates the need for CO_2 generation and energy-consuming materials such as Portland cement, and minimizes wastewater produced during pretreatment washing of MSWI BA (Rożek et al., 2019).

3.3.3 Autoclaved aerated concrete

The main components of MSWI BA include amorphous silica (usually over 50 wt%), alumina, iron oxide, and calcium oxide. When MSWI BA is used in the production of AAC, it can replace expensive aluminum powder as an aerator, partially replace silica powder or fly ash as the source of silica, and replace some Portland cement in AAC (Wongkeo et al., 2012; Song et al., 2015; Li et al., 2018). Replacing quartz sand with MSWI BA can shorten the gas foaming time and increase the gas-foaming amount. The fineness of the MSWI BA also has a significant impact on the gas foaming performance of AAC. The amount of gas formed decreases as the particle size of MSWI BA decreases. Compared with quartz sand, the density, compressive strength, and thermal conductivity of AAC often decrease when MSWI BA is added. The leaching toxicity result shows that AAC samples containing MSWI BA do not pose any harm to the environment. Therefore, it is suitable to use MSWI BA to manufacture AAC blocks.

3.4 Municipal solid waste incineration fly ash

Hydrothermal treatment of MSWI FA is considered a promising technology showing significant advantages in terms of economy, technology, and environment (Ferreira et al., 2003). As an emerging method, hydrothermal methods have been successfully applied in MSWI FA treatments, especially in detoxification (Xie et al., 2010) and solidification (Jin et al., 2013). NaOH solution containing MSWI FA can be successfully synthesized with aluminum-substituted tobermorite and different types of zeolites through hydrothermal treatment (Yao et al., 1999; Miyake et al., 2004). In the hydrothermal treatment of MSWI FA, researchers usually add additives or pretreat the fly ash. Adding alkali during the hydrothermal process can convert MSWI FA into stable minerals and immobilize heavy metal ions (Bayuseno et al., 2009). Quartz and high silicon−aluminum additives can also be used to synthesize tobermorite and zeolite, respectively, to stabilize heavy metals in MSWI FA (Shan et al., 2011). Synergistic effects occur when pretreated MSWI FA is mixed with urban sewage sludge for hydrothermal treatment. MSWI FA promotes the dehydration of municipal sewage sludge, which in return improves the release of chlorine in MSWI FA (Chen et al., 2019). After hydrothermal treatment, the dioxins in MSWI FA are degraded, and heavy metals are stabilized. Therefore, hydrothermal treatment of MSWI FA has a promising application prospect. The current research status of MSWI FA hydrothermal treatment in recent years is summarized in Table 6.3.

3.4.1 Synthesis of zeolite

Due to the presence of SiO_2 and Al_2O_3 components, MSWI FA is suitable for the preparation of zeolite materials. Although the content of SiO_2 and Al_2O_3 is relatively low (15%−30%), different types of zeolites can also be synthesized from MSWI FA through fusion or hydrothermal processes (Yang and Yang, 1998). The process of synthesizing zeolite using MSWI FA through two methods: ordinary hydrothermal treatment and fusion-hydrothermal treatment is shown in Fig. 6.5. Although the molar ratio of Si/Al in MSWI FA is mostly not suitable for synthesizing zeolite, the yield of zeolite can be improved by adding additional aluminum or silicon-containing materials. Adding solid waste such as waste glass and alumina powder as silicon and aluminum sources to MSWI FA, followed by hydrothermal and fusion processes, can synthesize zeolite-like materials. From the results, it can be seen that the crystallization performance of zeolite products synthesized by the hydrothermal method is similar to that of natural zeolite, which is higher than that of zeolite products synthesized by fusion methods. The thermal stability of hydrothermal products is also higher than that of fusion products (Deng et al., 2016).

The microwave-assisted hydrothermal process has also been applied to treat MSWI FA to obtain zeolite-like materials (Qiu et al., 2018). Quartz and aluminosilicate glass in MSWI FA are the main components in the synthesis of zeolites. To improve the purity of zeolite products, pretreatment can be carried out to remove soluble salts and calcium minerals from MSWI FA.

3.4.2 Synthesis of tobermorite

The calcium component, quartz, and aluminosilicate glass in MSWI FA can be used as components for synthesizing tobermorite (Fan et al., 2023). Tobermorite has a wide range of industrial applications, such as being used as thermal insulation material due to its excellent thermal conductivity (Bai et al., 2015), enhancing the strength of building

TABLE 6.3 Summary of the research status on the hydrothermal treatment of MSWI FA in recent years.

Material	Sample preparation	T/°C	t/h	Product	References
MSWI FA, NaOH	FA/NaOH solution of 1/25	180	20	Tobermorite	Yao et al. (1999)
MSWI FA, acid solution	Prewashing; liquid/solid ratios of 3:1, 5:1, 10:1 and 20:1 (mL:g)	100–200	1–7	–	Zhang and Itoh (2006)
MSWI FA, 0.5 M NaOH	Prewashing	180	48	Tobermorite-11 Å and katoite, small quantities of hydroxycancrinite and analcime	Bayuseno et al. (2009)
MSWI FA, reductant carbohydrazide (CHZ)	–	245, 260		–	Xie et al. (2010)
MSWI FA, MSWI BA, quartz, 2 M NaOH solution	The mixture was compacted by uniaxial pressing in a rectangular-shaped mold with compaction pressure of 30 MPa.	0–250	48	Tobermorite	Shan et al. (2011)
MSWI FA, ferric and/or ferrous additives	L/S ratios were between 2 and 2.5	290	1	–	Hu et al. (2015)
MSWI FA, waste glass powder, Al_2O_3 powder	L/S ratio of 10	60	24	Zeolites (perlialite, zeolite A and Y)	Deng et al. (2016)
MSWI FA, bentonite and kaolin, NaOH	L/S ratio was adjusted to be 10 mL/g	150, 200	48	–	Shi et al. (2017)
MSWI FA, CFA, NaOH	L/S ratio was 25 mL/g	220	10	Tobermorite	Guo and Song (2018)
MSWI FA, Na_2HPO_4	1 mol/L Na_2HPO_4 was added at a liquid–solid ratio of 3 mL/g	200	0.5	Zeolite-like materials	Qiu et al. (2018)
MSWI FA, municipal sewage sludge	–	180	1	–	Chen et al. (2019)
MSWI FA, CFA, tobermorite seed crystals	L/S ratio of 10 mL/g	150, 200	48	–	Shi et al. (2019)
MSWI FA, NaCl, $NaNO_3$, NaOH	A molding pressure of 20 MPa	220	10	–	Guo and Zhang (2020).
MSWI FA, NaOH	Alkali fusion at 400°C for 40 min, and hydrothermal at 105°C for 48 hours	105	48	ZSM-23 zeolite	Lin and Chen (2021)
MSWI FA, Ce–Mn catalyst	Original pH = 8.5, L/S = 4 mL/g	280	2	–	Zhang et al. (2022)
MSWI FA, hydroxyapatite	"Combined hydrothermal method of water-bath with microwave-assisted" water-bath: hydrothermal method (35 °C, 24 h); microwave-assisted hydrothermal method (140°C, 45 min)			–	Tong et al. (2022)
MSWI FA, Na_2S, Na_2SiO_3, $Na_2B_4O_7$, H_3BO_3	Microwave hydrothermal, L/S ratio of 2–8 mL/g	100–175	10–25	Zeolites	Xu et al. (2023)
MSWI FA, NaOH, quartz sand	Microwave hydrothermal, L/S ratio of 10 mL/g, C/S ratio of 1.1	260	3	Tobermorite	Yuan et al. (2023)

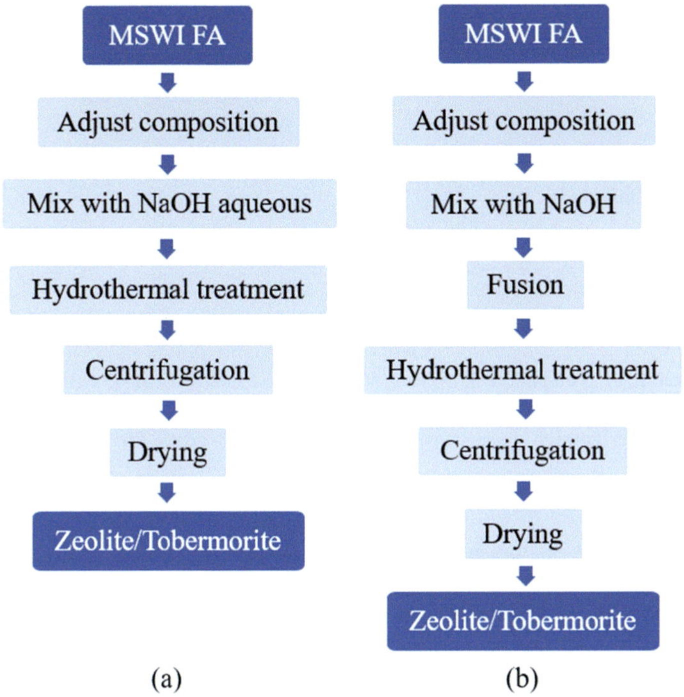

FIGURE 6.5 The process of zeolite/tobermorite preparation: (A) ordinary hydrothermal treatment; (B) fusion-hydrothermal treatment (Fan et al., 2023).

materials through crystallization revulsive (Maruyama et al., 2021), and treating industrial wastewater as an effective and low-cost adsorbent (Wang et al., 2022). Tobermorite can be crystallized by C—S—H through three steps: (1) forming amorphous and semicrystalline C—S—H; (2) semicrystalline tobermorite growth; and (3) recrystallization of tobermorite (Houston et al., 2009). The tobermorite synthesized using MSWI FA can serve as an ion exchanger and exhibit absorption behavior for Cs^+ and NH_4^+ (Tong et al., 2022). During the hydrothermal treatment process, adding tobermorite seeds to the reaction system can accelerate the crystallization rate of tobermorite and increase production (Shi et al., 2016).

3.4.3 Building materials

MSWI FA has the potential to be used as siliceous and calcareous materials to prepare solid wastes—based autoclaved wall blocks (SW-AWBs) (Silva et al., 2017). Recycling and utilization of MSWI by hydrothermal method to prepare SW-AWBs is a green and healthy method (Guo and Zhang, 2020). As a new type of wall material, SW-AWBs not only integrate the advantages of heat preservation and light weight but also have implications for waste management and environmental protection (Pedro et al., 2017).

4. Conclusions and perspectives

Sustainable management of combustion/incineration residues has been a global environmental issue, and hydrothermal methods have many advantages in treating such residues. This chapter introduces the hydrothermal treatment technology of combustion/incineration residues and illustrates the great potential of hydrothermal methods in synthesizing zeolite and tobermorite, and using them as building and adsorption materials. Overall, the preparation of materials using hydrothermal methods is still in the active exploration and development stage, and there is relatively little hydrothermal research on nonaqueous solvent systems. Therefore, studying hydrothermal equipment, the physical and chemical properties of solvents and mineralizers, as well as the mechanism of chemical reactions in hydrothermal treatment, and the mechanism of heat and mass transfer in hydrothermal systems, is of great significance for promoting the preparation of hydrothermal materials. In addition, advanced characterization technology is an important tool for deepening our understanding of hydrothermal treatment mechanisms and promoting the design of new technologies for the treatment and disposal of combustion/incineration residues. With the development of modern materials science and engineering research, the application fields and basic theories of hydrothermal treatment will be further developed. At the same time, with the intersection of disciplines, the combination of hydrothermal methods with other methods will be a development trend, which will lead to a wider application of hydrothermal methods for these residues.

Acknowledgments

The authors gratefully acknowledge the financial support from the Baima Lake Laboratory Joint Funds of the Zhejiang Provincial Natural Science Foundation of China under Grant No. BMHZ24E020004 for this study.

References

Al-Shehri, B.M., Khder, A., Ashour, S.S., Hamdy, M.S., 2019. A review: the utilization of mesoporous materials in wastewater treatment. Mater. Res. Express 6 (12), 122002.

Arun Kumar, M., Prasanna, K., Chinna Raj, C., Parthiban, V., Kulanthaivel, P., Narasimman, S., Naveen, V., 2022. Bond strength of autoclaved aerated concrete manufactured using partial replacement of fly ash with fibers—a review. Mater. Today 65, 581—589.

Bai, J., Li, Y., Ren, L., Mao, M., Zeng, M., Zhao, X., 2015. Thermal insulation monolith of aluminum tobermorite nanosheets prepared from fly ash. ACS Sustain. Chem. Eng. 3 (11), 2866—2873.

Bayuseno, A.P., Schmahl, W.W., Müllejans, T., 2009. Hydrothermal processing of MSWI fly ash-towards new stable minerals and fixation of heavy metals. J. Hazard Mater. 167 (1), 250—259.

Belviso, C., Cavalcante, F., Di Gennaro, S., Lettino, A., Palma, A., Ragone, P., Fiore, S., 2014. Removal of Mn from aqueous solution using fly ash and its hydrothermal synthetic zeolite. J. Environ. Manage. 137, 16—22.

Bertolini, T.C.R., Izidoro, J.C., Magdalena, C.P., Fungaro, D.A., 2013. Adsorption of crystal violet dye from aqueous solution onto zeolites from coal fly and bottom ashes. Electron. J. Chem 3.

Cao, P., Li, G., Luo, J., Rao, M., Jiang, H., Peng, Z., Jiang, T., 2020. Alkali-reinforced hydrothermal synthesis of lathy tobermorite fibers using mixture of coal fly ash and lime. Constr. Build. Mater. 238, 117655.

Cheeseman, C.R., Monteiro da Rocha, S., Sollars, C., Bethanis, S., Boccaccini, A.R., 2003. Ceramic processing of incinerator bottom ash. Waste Manage. (Tucson, Ariz.) 23 (10), 907—916.

Chen, B.Y., Lin, K.L., 2006. Biotoxicity assessment on reusability of municipal solid waste incinerator (MSWI) ash. J. Hazard Mater. 136 (3), 741—746.

Chen, Y.L., Chang, J.E., Lai, Y.C., Chou, M.I.M., 2017. A comprehensive study on the production of autoclaved aerated concrete: effects of silica-lime-cement composition and autoclaving conditions. Constr. Build. Mater. 153, 622—629.

Chen, Z., Yu, G., Wang, Y., Liu, X., Wang, X., 2019. Research on synergistically hydrothermal treatment of municipal solid waste incineration fly ash and sewage sludge. Waste Manage. (Tucson, Ariz.) 100, 182—190.

Chiang, Y.W., Ghyselbrecht, K., Santos, R.M., Meesschaert, B., Martens, J.A., 2012. Synthesis of zeolitic-type adsorbent material from municipal solid waste incinerator bottom ash and its application in heavy metal adsorption. Catal. Today 190 (1), 23—30.

Cioffi, R., Maffucci, L., Santoro, L., 2003. Optimization of geopolymer synthesis by calcination and polycondensation of a kaolinitic residue. Resour. Conserv. Recy. 40 (1), 27—38.

Davidovits, J., 1999. Chemistry of geopolymeric systems, terminology. Geopolymer 9—39.

Deng, L., Xu, Q., Wu, H., 2016. Synthesis of zeolite-like material by hydrothermal and fusion methods using municipal solid waste fly ash. Procedia Environmental Sciences 31, 662—667.

Drochytka, R., Černý, V., 2020. Influence of fluidized bed combustion fly ash admixture on hydrothermal synthesis of tobermorite in the mixture with quartz sand, high temperature fly ash and lime. Constr. Build. Mater. 230, 117033.

Fan, X., Yuan, R., Gan, M., Ji, Z., Sun, Z., 2023. Subcritical hydrothermal treatment of municipal solid waste incineration fly ash: a review. Sci. Total Environ. 865, 160745.

Ferrarini, S.F., Cardoso, A.M., Paprocki, A., Pires, M., 2016. Integrated synthesis of zeolites using coal fly ash: element distribution in the products, washing waters and effluent. J. Brazil. Chem. Soc. 27.

Ferreira, C., Ribeiro, A., Ottosen, L., 2003. Possible applications for municipal solid waste fly ash. J. Hazard Mater. 96 (2), 201—216.

Fukui, K., Arai, K., Kanayama, K., Yoshida, H., 2006. Phillipsite synthesis from fly ash prepared by hydrothermal treatment with microwave heating. Adv. Powder Technol. 17 (4), 369—382.

Fukui, K., Arai, K., Kanayama, K., Yamamoto, T., Yoshida, H., 2007. Effects of microwave irradiation on the crystalline phase of zeolite synthesized from fly ash by hydrothermal treatment. Adv. Powder Technol. 18 (4), 381—393.

Galvánková, L., Másilko, J., Solný, T., Štěpánková, E., 2016. Tobermorite synthesis under hydrothermal conditions. Procedia Eng. 151, 100—107.

Gollakota, A.R.K., Munagapati, V.S., Volli, V., Gautam, S., Wen, J.C., Shu, C.M., 2021. Coal bottom ash derived zeolite (SSZ-13) for the sorption of synthetic anion alizarin red s (ARS) dye. J. Hazard Mater. 416, 125925.

Gollakota, A.R.K., Volli, V., Shu, C.M., 2019. Progressive utilisation prospects of coal fly ash: a review. Sci. Total Environ. 672, 951—989.

Goi, S., Guerrero, A., Luxán, M.P., Macías, A., 2003. Activation of the fly ash pozzolanic reaction by hydrothermal conditions. Cement Concr. Res. 33 (9), 1399—1405.

Guo, X., Song, M., 2018. Micro-nanostructures of tobermorite hydrothermal-synthesized from fly ash and municipal solid waste incineration fly ash. Constr. Build. Mater. 191, 431—439.

Guo, X., Zhang, T., 2020. Utilization of municipal solid waste incineration fly ash to produce autoclaved and modified wall blocks. J. Clean. Prod. 252, 119759.

Hashimoto, S., Watanabe, K., Nose, K., Morita, M., 2004. Remediation of soil contaminated with dioxins by subcritical water extraction. Chemosphere 54 (1), 89–96.

Holt, E., Raivio, P., 2005. Use of gasification residues in aerated autoclaved concrete. Cement Concr. Res. 35 (4), 796–802.

Hou, X., Ma, S., Wang, X., Ou, Y., Liu, R., 2023. Effects of alkali activation and hydrothermal processes on the transformation of fly ash into Al-substituted tobermorite fiber. Constr. Build. Mater. 397, 132372.

Houston, J.R., Maxwell, R.S., Carroll, S.A., 2009. Transformation of meta-stable calcium silicate hydrates to tobermorite: reaction kinetics and molecular structure from XRD and NMR spectroscopy. Geochem. Trans. 10 (10), 1.

Hu, Y., Zhang, P., Li, J., Chen, D., 2015. Stabilization and separation of heavy metals in incineration fly ash during the hydrothermal treatment process. J. Hazard Mater. 299, 149–157.

Jimbo, H., 1996. Plasma melting and useful application of molten slag. Waste Manage. (Tucson, Ariz.) 16 (5–6), 417–422.

Jin, Y., Ma, X., Jiang, X., Liu, H., Li, X., Yan, J., Cen, K., 2013. Effects of hydrothermal treatment on the major heavy metals in fly ash from municipal solid waste incineration. Energy Fuel. 27 (1), 394–400.

Jing, Z., Jin, F., Yamasaki, N., Ishida, E.H., 2007. Hydrothermal synthesis of a novel tobermorite-based porous material from municipal incineration bottom ash. Ind. Eng. Chem. Res. 4.

Jing, Z., Matsuoka, N., Jin, F., Hashida, T., Yamasaki, N., 2007. Municipal incineration bottom ash treatment using hydrothermal solidification. Waste Manage. (Tucson, Ariz.) 27 (2), 287–293.

Jing, Z., Matsuoka, N., Jin, F., Yamasaki, N., Suzuki, K., Hashida, T., 2006. Solidification of coal fly ash using hydrothermal processing method. J. Mater. Sci. 41 (5), 1579–1584.

Jing, Z., Ran, X., Jin, F., Ishida, E.H., 2010. Hydrothermal solidification of municipal solid waste incineration bottom ash with slag addition. Waste Manage. (Tucson, Ariz.) 30 (8–9), 1521–1527.

Kikuchi, R., 2001. Recycling of municipal solid waste for cement production: pilot-scale test for transforming incineration ash of solid waste into cement clinker. Resour. Conserv. Recy. 31 (2), 137–147.

Kresge, C.T., Leonowicz, M.E., Roth, W.J., Vartuli, J.C., Beck, J.S., 1992. Ordered mesoporous molecular sieves synthesized by a liquid-crystal template mechanism. Nature 359.

Król, M., Rożek, P., Mozgawa, W., 2017. Synthesis of the sodalite by geopolymerization process using coal fly ash. Pol. J. Environ. Stud. 26 (6), 2611–2617.

Kumar, S., Gupta, R.B., 2009. Biocrude production from switchgrass using subcritical water. Energy Fuel. 23 (10), 5151–5159.

Kurama, H., Topçu, İ.B., Karakurt, C., 2009. Properties of the autoclaved aerated concrete produced from coal bottom ash. J. Mater. Process. Tech. 209 (2), 767–773.

Lee, N.K., Khalid, H.R., Lee, H.K., 2016. Synthesis of mesoporous geopolymers containing zeolite phases by a hydrothermal treatment. Micropor. Mesopor. Mat. 229, 22–30.

Li, X., Liu, Z., Lv, Y., Cai, L., Jiang, D., Jiang, W., Jian, S., 2018. Utilization of municipal solid waste incineration bottom ash in autoclaved aerated concrete. Constr. Build. Mater. 178, 175–182.

Lin, Y.J., Chen, J.C., 2021. Resourcization and valorization of waste incineration fly ash for the synthesis of zeolite and applications. J. Environ. Chem. Eng. 9 (6), 106549.

Luo, J., Zhang, H., Yang, J., 2016. Hydrothermal synthesis of sodalite on alkali-activated coal fly ash for removal of lead ions. Procedia Environ. Sci. 31, 605–614.

Ma, W., Brown, P.W., 1997. Hydrothermal synthesis of tobermorite from fly ashes. Adv. Cem. Res. 9 (33), 9–16.

Ma, W., Brown, P.W., Komarneni, S., 1998. Characterization and cation exchange properties of zeolite synthesized from fly ashes. J. Mater. Res. 13 (1), 3–7.

Ma, Z., Zhang, X., Lu, G., Guo, Y., Song, H., Cheng, F., 2022. Hydrothermal synthesis of zeolitic material from circulating fluidized bed combustion fly ash for the highly efficient removal of lead from aqueous solution. Chin. J. Chem. Eng. 47, 193–205.

Maruyama, I., Rymeš, J., Aili, A., Sawada, S., Kontani, O., Ueda, S., Shimamoto, R., 2021. Long-term use of modern portland cement concrete: the impact of Al-tobermorite formation. Mater. Design 198, 109297.

Mesecke, K., Malorny, W., Warr, L.N., 2023. Understanding the effect of sulfate ions on the hydrothermal curing of autoclaved aerated concrete. Cement Concr. Res. 164, 107044.

Miyake, M., Tamura, C., Matsuda, M., 2004. Resource recovery of waste incineration fly ash: synthesis of zeolites A and P. J. Am. Ceram. Soc. 85 (7), 1873–1875.

Murayama, N., Yamamoto, H., Shibata, J., 2002. Mechanism of zeolite synthesis from coal fly ash by alkali hydrothermal reaction. Int. J. Miner. Process. 64 (1), 1–17.

Nascimento, M., Soares, P.S.M., Souza, V., 2009. Adsorption of heavy metal cations using coal fly ash modified by hydrothermal method. Fuel 88 (9), 1714–1719.

Nyale, S.M., Babajide, O.O., Birch, G.D., Böke, N., Petrik, L.F., 2013. Synthesis and characterization of coal fly ash-based foamed geopolymer. Procedia Environmental Sciences 18, 722–730.

Pan, Y., Wu, Z., Zhou, J., Zhao, J., Ruan, X., Liu, J., Qian, G., 2013. Chemical characteristics and risk assessment of typical municipal solid waste incineration (MSWI) fly ash in China. J. Hazard Mater. 261, 269–276.

Pedro, R., Tubino, R., Anversa, J., Col, D.D., Lermen, R., Silva, R., 2017. Production of aerated foamed concrete with industrial waste from the gems and jewels sector of rio grande do sul-Brazil. Appl. Sci. 7 (10), 985.

Peña, R., Guerrero, A., Goñi, S., 2006. Hydrothermal treatment of bottom ash from the incineration of municipal solid waste: Retention of Cs(I), Cd(II), Pb(II) and Cr(III). J. Hazard. Mater. 129 (1–3), 151–157.

Penilla, R.P., Bustos, A.G., Elizalde, S.G., 2003. Zeolite synthesized by alkaline hydrothermal treatment of bottom ash from combustion of municipal solid wastes. Chem. Inform. 34 (50).

Penilla, R.P., Bustos, A.G., Elizalde, S.G., 2006. Hydrothermal treatment of bottom ash from the incineration of municipal solid waste: retention of Cs(I), Cd(II), Pb(II) and Cr(III). J. Hazard Mater. 129 (1), 151–157.

Penilla, R.P., Bustos, A.G., Elizalde, S.G., 2006. Immobilization of Cs, Cd, Pb and Cr by synthetic zeolites from Spanish low-calcium coal fly ash. Fuel 85 (5–6), 823–832.

Qiu, Q., Jiang, X., Lv, G., Chen, Z., Lu, S., Ni, M., Yan, J., Deng, X., 2018. Adsorption of heavy metal ions using zeolite materials of municipal solid waste incineration fly ash modified by microwave-assisted hydrothermal treatment. Powder Technol. 335, 156–163.

Rożek, P., Król, M., Mozgawa, W., 2019. Solidification/stabilization of municipal solid waste incineration bottom ash via autoclave treatment: structural and mechanical properties. Constr. Build. Mater. 202, 603–613.

Schafhat, K.F.E., 1845. Galehrte Angeigen Bayer. Acad 20 (557), 3.

Shan, C., Jing, Z., Pan, L., Zhou, L., Pan, X., Lu, L., 2011. Hydrothermal solidification of municipal solid waste incineration fly ash. Res. Chem. Intermediat. 37 (2–5), 551–565.

Shen, Y., Zhao, R., Wang, J., Chen, X., Ge, X., Chen, M., 2016. Waste-to-energy: dehalogenation of plastic-containing wastes. Waste Manag. 49, 287–303.

Shi, D., Zhang, C., Zhang, J., Li, P., 2016. Seed-assisted hydrothermal treatment with composite silicon aluminum additive for solidification of heavy metals in municipal solid waste incineration fly ash. Energy & Fuels 12 (30).

Shi, D., Hu, C., Zhang, J., Li, P., Zhang, C., Wang, X., Ma, H., 2017. Silicon-aluminum additives assisted hydrothermal process for stabilization of heavy metals in fly ash from MSW incineration. Fuel Process. Technol. 165, 44–53.

Shi, D., Ma, J., Wang, H., Wang, P., Hu, C., Zhang, J., Gu, L., 2019. Detoxification of PCBs in fly ash from MSW incineration by hydrothermal treatment with composite silicon-aluminum additives and seed induction. Fuel Process. Technol. 195, 106157.

Silva, R.V., Brito, J.D., Lynn, C.J., Dhir, R.K., 2017. Use of municipal solid waste incineration bottom ashes in alkali-activated materials, ceramics and granular applications: a review. Waste Manage. (Tucson, Ariz.) 68, 207–220.

Song, Y., Li, B., Yang, E.H., Liu, Y., Ding, T., 2015. Feasibility study on utilization of municipal solid waste incineration bottom ash as aerating agent for the production of autoclaved aerated concrete. Cement. Concrete Comp. 56, 51–58.

Swanepoel, J.C., Strydom, C.A., 2002. Utilisation of fly ash in a geopolymeric material. Appl Geochem. 17 (8), 1143–1148.

Tauanov, Z., Shah, D., Inglezakis, V., Jamwal, P.K., 2018. Hydrothermal synthesis of zeolite production from coal fly ash: a heuristic approach and its optimization for system identification of conversion. J. Clean. Prod. 182, 616–623.

Tong, H., Shi, D., Cai, H., Liu, J., Lv, M., Gu, L., Luo, L., Wang, B., 2022. Novel hydroxyapatite (HAP)-assisted hydrothermal solidification of heavy metals in fly ash from MSW incineration: effect of HAP liquid-precursor and HAP seed crystal derived from eggshell waste. Fuel Process. Technol. 236, 107400.

Vu, D.H., Bui, H.B., Bui, X.N., An-Nguyen, D., Le, Q.T., Do, N.H., Nguyen, H., 2020. A novel approach in adsorption of heavy metal ions from aqueous solution using synthesized MCM-41 from coal bottom ash. Int. J. Environ. An. Ch. 100 (11), 1226–1244.

Wang, C., Li, B.G., Mi, J., 2013. Preparation of fly ash/rare earth adsorbent and its adsorption for reactive dye from aqueous solution. Mater. Sci. Forum 743–744, 409–413.

Wang, C., Li, J., Sun, X., Wang, L., Sun, X., 2009. Evaluation of zeolites synthesized from fly ash as potential adsorbents for wastewater containing heavy metals. J. Environ. Sci. 21 (1), 127–136.

Wang, C., Li, J., Wang, L., Sun, X., Huang, J., 2009. Adsorption of dye from wastewater by zeolites synthesized from fly ash: kinetic and equilibrium studies. Chin. J. Chem. Eng. 17 (3), 513–521.

Wang, Y., Huang, L., Lau, R., 2016. Conversion of municipal solid waste incineration bottom ash to sorbent material for pollutants removal from water. J. Taiwan Inst. Chem. E. 60, 275–286.

Wang, Z., Xu, L., Wu, D., Zheng, S., 2022. Hydrothermal synthesis of mesoporous tobermorite from fly ash with enhanced removal performance towards Pb2+ from wastewater. Colloid. Surface. 632, 127775.

Wdowin, M., Franus, M., Panek, R., Badura, L., Franus, W., 2014. The conversion technology of fly ash into zeolites. Clean Technol. Environ. Policy 16 (6), 1217–1223.

Wongkeo, W., Chaipanich, A., 2010. Compressive strength, microstructure and thermal analysis of autoclaved and air cured structural lightweight concrete made with coal bottom ash and silica fume. Mater. Sci. Eng. A 527 (16–17), 3676–3684.

Wongkeo, W., Thongsanitgarn, P., Pimraksa, K., Chaipanich, A., 2012. Compressive strength, flexural strength and thermal conductivity of autoclaved concrete block made using bottom ash as cement replacement materials. Mater. Design 35, 434–439.

Wu, C.N., Tang, Y.C., Tang, L.H., 2012. Removal of heavy metal from wastewater using zeolite from fly ash. Adv. Mater. Res. 518 (523), 2736–2739.

Xie, J., Hu, Y., Chen, D., Zhou, B., 2010. Hydrothermal treatment of MSWI fly ash for simultaneous dioxins decomposition and heavy metal stabilization. Front. Environ. Sci. En. 4, 108–115.

Xu, Z., Liang, Z., Shao, H., Zhao, Q., 2023. Heavy metal stabilization in MSWI fly ash using an additive-assisted microwave hydrothermal method. J. Ind. Eng. Chem. 117, 352–360.

Yang, G.C.C., Yang, T.Y., 1998. Synthesis of zeolites from municipal incinerator fly ash. J. Hazard Mater. 62 (1), 75–89.

Yang, G., Park, S.J., 2019. Conventional and microwave hydrothermal synthesis and application of functional materials: a review. Materials 12 (7), 1177.

Yao, Z.T., Ji, X.S., Sarker, P.K., Tang, J.H., Ge, L.Q., Xia, M.S., Xi, Y.Q., 2015. A comprehensive review on the applications of coal fly ash. Earth Sci. Rev. 141, 105–121.

Yao, Z., Tamura, C., Matsuda, M., Miyake, M., 1999. Resource recovery of waste incineration fly ash: synthesis of tobermorite as ion exchanger. J. Mater. Res. 14 (11), 4437–4442.

Yuan, R., Fan, X., Gan, M., Zhao, Q., Lu, S., Li, S., 2023. Facile treatment of municipal solid waste incineration fly ash by one-step microwave hydrothermal method: hazards detoxification and tobermorite synthesis. J. Environ. Chem. Eng. 11 (3), 109768.

Zhang, F.S., Itoh, H., 2006. Extraction of metals from municipal solid waste incinerator fly ash by hydrothermal process. J. Hazard Mater. 136 (3), 663–670.

Zhang, T., Yang, Y., Zhou, K., Liu, B., Tian, G., Zuo, W., Zhou, H., Bian, B., 2022. Hydrothermal oxidation degradation of dioxins in fly ash with water-washing and added Ce–Mn catalyst. J. Environ. Manage. 317, 115430.

Zou, J., Guo, C., Jiang, Y., Wei, C., Li, F., 2016. Structure, morphology and mechanism research on synthesizing xonotlite fiber from acid-extracting residues of coal fly ash and carbide slag. Mater. Chem. Phys. 172, 121–128.

Chapter 7

Chemical agent—based immobilization of combustion/incineration residues

Lizhi Tong

State Environmental Protection Key Laboratory of Environmental Pollution Health Risk Assessment, South China Institute of Environmental Sciences, Ministry of Ecology and Environment (MEE), Guangzhou, China

1. Introduction

With the growth of social and economic levels and the improvement of human living standards, solid waste production worldwide has increased substantially in recent decades (He et al., 2022). Global municipal solid waste (MSW) production is expected to increase by about 70% to 3.4 billion tons by 2050 (Chen et al., 2020). As a result, MSW has become a significant problem affecting the daily life of residents and the normal development of society. Various treatment technologies, such as combustion/incineration processes, are well established and widely used to treat MSWs, with the merits of energy recovery and rapid waste reduction. However, these processes considerably generate a massive number of residues, including incineration fly ash (IFA), air pollution control (APC) residue, and bottom ash (Istrate et al., 2023).

Many studies have been performed concerning the physicochemical properties and the distribution of toxic metals in combustion/incineration residues (CIRs), such as bottom ash or IFA. Although the bottom ash has a large proportion, the average content of the heavy metal is low. As bottom ash can potentially replace all or a portion of the natural aggregates used in concrete, it holds considerable potential to be used as internal curing of high-performance concretes (Nakararoj et al., 2022). IFA or APC residues, which are collected in flue gas purification and heat recovery system after waste incineration, usually contain the main components such as silica oxides, chloride salts, calcium salts, and trace components such as heavy metals with lower boiling points (e.g., Zn, Pb, Cd, etc.) and organic pollutants (Lou et al., 2023). Given the high leachability of heavy metals and soluble salts, IFA or APC residue is generally considered a hazardous material in many countries, e.g., China, Japan, the European Union, and the United States (Tong et al., 2020; Bolan et al., 2023).

To effectively prevent heavy metals in CIRs from leaching by water and minimize their adverse impacts on human health and the environment, many technologies have been developed to treat CIRs before their final disposal. Depending on how they work, the treatment techniques at home and abroad can be roughly grouped into four categories: (1) extraction and separation, (2) chemical agent—based solidification/stabilization, (3) thermal/hydrothermal treatment, and (4) mechanical treatment (ball milling) (Ajorloo et al., 2022; Ates et al., 2023; Yuan et al., 2023). Meanwhile, many combinations of treatment, disposal, and utilization solutions are also used to manage these residues safely (Labianca et al., 2022). Extraction and separation technology can recover valuable metals, as well as remove heavy metals and salts from CIRs under the action of specific chemical agents (e.g., acid or organic collectors) (Wu et al., 2023). Solidification/stabilization technology can be sorted into cement-based solidification technology and chemical agent—based immobilization technology, which can reduce heavy metals' mobility through physical processes or chemical reactions (Chen et al., 2021). Thermal treatment technology can solidify most heavy metals with low boiling points in hard sinter at high temperatures (around 1000°C), while hydrothermal treatment technology can immobilize heavy metals by synthesizing aluminosilicate minerals under alkaline conditions (Sanito et al., 2023; Zhang et al., 2023a). Mechanical treatment, which has attracted attention in recent years, can achieve the degradation of dioxins and the stabilization of heavy metals via ball milling (Chen et al., 2023).

The comparison between the technologies mentioned before has been developed on a whole or pilot scale. As shown in Table 7.1, chemical agent—based immobilization of heavy metals in CIRs has drawn worldwide attention, given its high

Treatment and Utilization of Combustion and Incineration Residues. https://doi.org/10.1016/B978-0-443-21536-0.00042-3

TABLE 7.1 Comparison of the advantages and disadvantages of typical CIRs treatment technologies.

Technology	Advantages	Disadvantages
Extraction and separation	Valuable metals can be recovered	Complicated operation
		High acid consumption
		Untreated dioxins
Cement-based solidification	Simple operation	Large volume growth
	Low cost	Poor long-term stability and acid resistance
		Untreated dioxins
Chemical agent-based immobilization	Simple operation	Long-term stability varies depending on the chemicals
	Small volume increase	
	Moderate cost	
Thermal treatment	Excellent long-term stability	High cost
	Dioxin destruction	High energy consumption
	Significant volume reduction	
Hydrothermal treatment	Dioxin destruction	Relatively high cost
		Poor long-term stability
Mechanical treatment	Dioxin decomposition	High energy consumption
		Uncertain long-term stability

efficiency, low cost, and easy operation after treatment (Ajorloo et al., 2022). As it displays excellent advantages and development prospects in engineering applications, this technology has been well used in Japan, China, the United States, and other developed countries. Chemical agent–based immobilization technology uses chelating agents, including organic or inorganic agents, to make heavy metal ions react to form various chelating precipitates. The heavy metals (e.g., Pb, Cu, Cd, etc.) in these chelating precipitates have low toxicity and minimal solubility. They are challenging to be leached out to the surrounding waterbody or soil. At present, inorganic chelating agents (e.g., soluble phosphates, sodium sulfide) and organic chelating agents (e.g., dithiocarbamate [DTC] and sodium dimethyl dithiocarbamate [SDD]) are commonly used for toxic heavy metal stabilization of CIRs.

Since most chemicals are generally selective for heavy metals, they cannot stabilize all the toxic heavy metals in CIRs; it is essential to analyze the research and application status of different chemical agents. This chapter reviews the types, reaction mechanisms, and latest research progress of chemical stabilization agents used to immobilize heavy metals in CIRs. In the meantime, the benefits and challenges of different chemical agents used in stabilizing CIRs are discussed. Finally, chemical agent–based immobilization technology research topics are proposed, such as developing novel chelating agents and sustainable management of stabilized CIRs.

2. Inorganic chelating agent–based immobilization processes

2.1 Types and performances of inorganic chelating agents

Common inorganic agents that are selected for fixing heavy metals in CIRs include sulfides, soluble phosphates, amorphous silicon material, ferrous salts, and carbonates (Chen et al., 2022). Fig. 7.1 shows two main reaction mechanisms, precipitation–dissolution and adsorption/encapsulation, reported for these chemical agents. According to the precipitation–dissolution equilibrium theory, using the aforementioned inorganic chemicals to form precipitates with positively low solubility of heavy metals can achieve the goal of minimizing the mobility of heavy metals in CIRs. In addition, some amorphous silicon materials and ferrite agents, due to their large surface area, can also effectively adsorb heavy metals and organic pollutants or bind heavy metals firmly in the lattice (Nazeer et al., 2023).

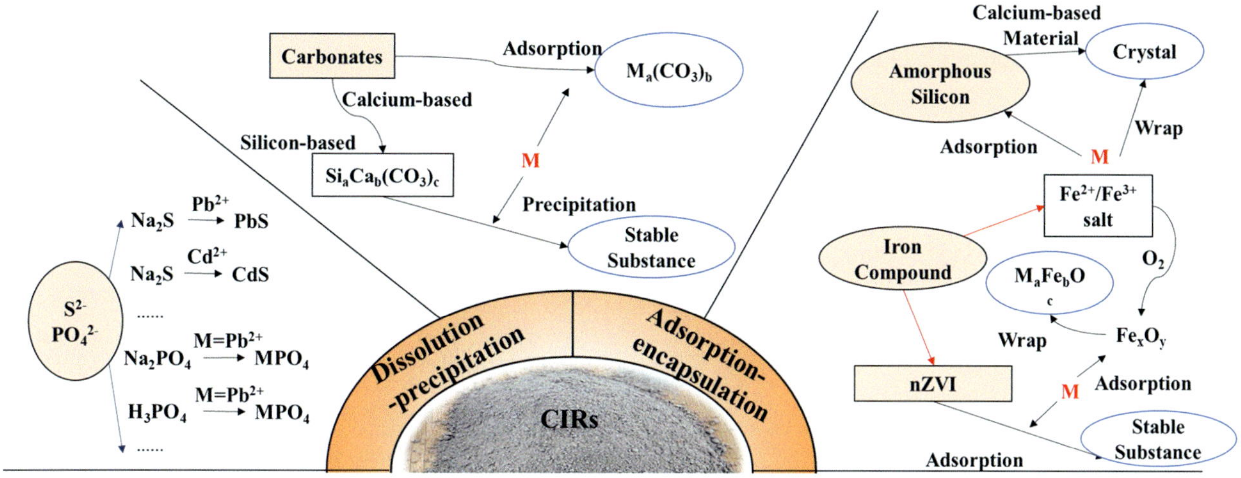

FIGURE 7.1 Types of inorganic chemical agents and their reaction mechanisms to heavy metals.

Sulfide precipitation has better treatment performance than hydroxide precipitation for removing toxic metals from various industrial wastewater or acid mining drainage. The high reactivity of sulfides (S^{2-}, HS^{-}) with heavy metal ions (e.g., Pb^{2+} and Cd^{2+}) and the insolubility of heavy metal sulfides over a broad pH range (3–14) are also attractive features for heavy metal immobilization in solid wastes. With the addition of 2% sodium sulfide, the leaching concentration of Pb in stabilized IFA can be reduced to 0.1 mg/L, far below the limit (0.25 mg/L) regulated in the Chinese national standard for pollution control on the landfill site for domestic waste (GB 16,889–2008) (Ma et al., 2019). With the optimum dosage of sodium, sulfide is 0.125‱, and the optimum time of ball milling reaction is 5 min; the concentrations of Cd, Cr, Cu, Pb, and Zn in the solidified fly ash also can meet the requirements of GB 16,889–2008 (Ye et al., 2023).

Soluble phosphates, including phosphate acid, sodium phosphate, etc., also can minimize the mobility of many heavy metals in CIRs. According to the solubility of metal phosphate, the stabilization efficiency of heavy metal is in the order of Pb > Cu > Cd > Hg. It is reported that the leaching amount of Pb in CIRs is meager after the addition of phosphates, in the pH range of 4~13. The stabilizing effect of heavy metals is closely related to the types of phosphate ions. It is found that using 10% w/w phosphoric acid (acid to ash ratio) followed by water washing can successfully stabilize the metals in medical waste incineration ash (Vavva et al., 2020). Besides, hydroxyapatite has a much better stabilization effect on Pb than phosphate because the former can preferably bind Pb and form stable lead-hydroxyapatite minerals (Nag et al., 2020).

Amorphous silicon materials include silica gel, silica fume, diatomite, rice husk ash, etc. Among them, silica gel, with relatively high porosity and specific surface area, has a strong adsorption effect on heavy metal ions. At the same time, the amorphous silicon material can react with the calcium-based material in fly ash to form a stable silicate crystal phase, which can stabilize the heavy metal in the structure. For example, Tan et al. treated fly ash with 10% rice husk ash at 600°C for 2 hours and found that with the extension of leaching time, the leaching concentration of Pb continuously decreased to 0.075 mg/L, and the Pb chelation rate reached 97.58% (Tan et al., 2020).

Ferrous salts, such as ferrous sulfate ($FeSO_4$), can be oxidized into iron oxides (e.g., Fe_2O_3, FeOOH, etc.) or transformed into hydrated iron oxide crystals under alkaline heating conditions. It is demonstrated that after mixing ferrous salts/iron sulfate with various solid wastes, the heavy metals are firmly fixed in the newly formed crystal lattices, and the metal leaching concentration decreases dramatically (Sun et al., 2023). In addition, carbonate ions also can precipitate with heavy metal ions to reduce the leaching concentration of heavy metals. Meanwhile, the generated calcium carbonate, calcium silicate hydrate, and other materials may also adsorb Pb^{2+}, Cd^{2+}, and other heavy metal ions on the surface to form coprecipitation (Chen et al., 2022). Table 7.2 summarizes the latest reported stabilization effect of common inorganic chemicals on heavy metals in CIRs.

2.2 State-of-the-art of inorganic chelating agents

The inorganic chemical agent–based immobilization technology to stabilize heavy metals in CIRs or other hazardous wastes has been applied relatively early, given that raw materials are readily available and the principle of removing heavy metals is simple. However, it should be noted that the type of leachate (e.g., water or dilute acetic acid), pH, liquid/solid ratio, mixing time and temperature, particle size, chemical agent purity, and physiochemical characteristics of CIRs can

TABLE 7.2 Stabilization effect of inorganic chemicals on heavy metals in CIRs.

Chemical agent	Reaction condition (hr)	Chelating agent dosage (w.t)	Performance	References
Na_2S	Water ratio: 20%, 25°C, 0.5	8%	All heavy metals meet the regulated limit of GB 16889–2008	Zhu et al. (2020)
H_3PO_4	Water ratio: 30%–40%, 25°C, 1	5%	Over 80% of Pb was stabilized	Liu and Chen (2018)
Na_3PO_4	Water ratio: 20%, room temperature, 0.5	4%	Over 50% of Pb was stabilized, but still failing to meet the regulated limit of GB 16889–2008 (0.25 mg/L)	Sun et al. (2019)
NaH_2PO_4	Water ratio: 30%, 25°C, 0.5	5%	The leaching concentration of Pb was 5.36 mg/L, and over 57.14% of Pb was in the residual fraction	Zhu et al. (2021)
Na_2CO_3	Liquid/solid ratios (L/S) in mL/g were 30, 25°C, 2	30 g/L	All heavy metals meet the regulated limit of GB 16889–2008	Chen et al. (2022)
Alkaline silica sol	Heat treatment: 700°C, 2	20%	All heavy metals meet the regulated limit of GB 16889–2008	Yu et al. (2021)

The leaching test was followed by Chinese solid waste-extraction procedure for leaching toxicity-acetic acid buffer solution method (HJ/T 300-2007), and the concentration of heavy metals in the leachate was compared with the Chinese national standard for pollution control on the landfill site for domestic waste (GB 16889–2008).

affect the metal stabilization efficiency. For example, heavy metals in stabilized CIRs are easily leached out in an acidic environment in the long term. The maximum leaching amount of Pb, Cd, and Ni for the scenario of colandfill of phosphate-stabilized fly ash and MSW exceeded the corresponding regulated limit in GB 16,889–2008 (Xin et al., 2022). Therefore, inorganic agents have progressively become supplementary agents in stabilizing heavy metals, and the research on combining these inorganic agents with inorganic binders (e.g., cement) or other organic chelating agents is increasing moderately (Chen et al., 2021). For example, it is shown that 1.2% Na_2S + 1.2% NaH_2PO_4 + 0.8% ADDP (ammonium dibutyl dithiophosphate) had a cost advantage over other stabilization schemes and had a more outstanding market application value (Zhu et al., 2020).

2.3 Benefits and challenges of inorganic agent–based S/S technology

The most significant advantage of inorganic agent-based S/S technology is its low price. However, to further promote the application of inorganic chemicals, the cost and long-term stability are substantial concerns. It has been considered adequate because of metal sulfides' relative stability and the sulfide ions' reductive character. Due to the amount of toxic hydrogen sulfide generated, sodium sulfide is not widely used. Besides, with the popularization of waste incineration/combustion technology, more and more solid wastes with combustible characteristics, such as activated sludges, industrial solid wastes, medical soil wastes, etc., are incinerated or combusted. The physicochemical properties of the residues generated from these processes are diverse. For example, the latest data show that medical soil wastes IFA have at least five times more Zn and Cu than MSWs IFA (Shen et al., 2022). The change in heavy metal content increases the uncertainty of the long-term stability of this technology. Therefore, developing more efficient inorganic chemical agents to meet the long-term stable safety requirements of CIRs is a big challenge.

3. Organic chelating agent–based immobilization processes

3.1 Types and performances of organic chelating agents

Common organic chelating agents mainly include dithiocarbamates (DTCs), thiourea, chitosan, piperazine, 2,4,6-trimercaptotriazine (TMT), and their derivatives (Ma et al., 2019; Zhang et al., 2020). As shown in Fig. 7.2, except for ion exchange, surface-complex precipitation is the primary reaction mechanism widely reported for these organic chemical

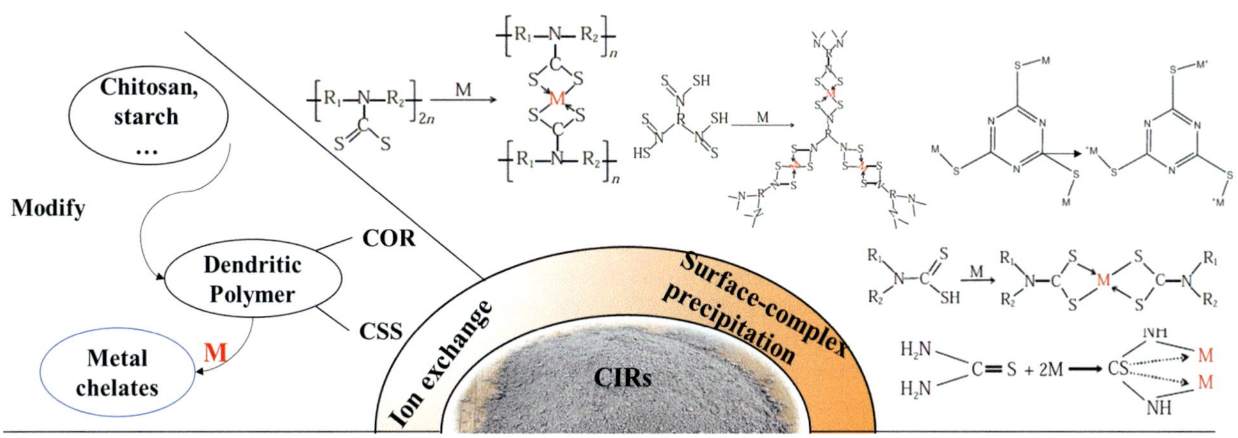

FIGURE 7.2 Types of organic chemical agents and their reaction mechanisms to heavy metals.

agents. The surface-complex precipitation process is based on the complexation mechanism, where specific organic agents, represented by organic chelators with C—S or C—N groups, coordinate with heavy metal ions to form stable coordinate bonds. Unlike inorganic chemical agents, organic chelating agents, especially macromolecular organic polymers, have a more vital binding force with heavy metal ions (e.g., Hg, Cu, etc.), forming more stable insoluble precipitates (Zhang et al., 2020).

Thiourea (CH_4N_2S) is an organosulfur compound similar to urea in which a sulfur atom replaces the oxygen atom. Thiourea and its derivatives contain S and N atoms, which can form stable two-dimensional complexes with heavy metal ions. As a result, with the addition of 2% thiourea, the leaching concentration of heavy metals in CIRs can be minimized to less than 0.1 mg/L, successfully meeting the limits of Pb in GB 16889—2008 (0.25 mg/L) (Ma et al., 2019). However, due to the disadvantage that thiourea is easy to decompose and produce toxic gas when heated, the current research began to turn to the cross-linked chelate resin formed by the reaction of thiourea and other organic substances. For example, the thiourea—formaldehyde chelate resin can be developed through hydroxymethylation and methylene reaction, and it has a complex branched chain structure. This unique structure can form heavy metal chelates with stable three-dimensional structures (Liu et al., 2021).

Piperazine ($C_4H_{10}N_2$) is an organic compound that consists of a six-membered ring containing two nitrogen rings, first used as an anthelmintic in the 1950s. In recent years, piperazine has been applied to stabilize heavy metals in CIRs because it releases very little CS_2 when heated and has an excellent heavy metal stabilization effect. Compared with dimethyl dithiocarbamic acid and diethyl dithiocarbamic acid, piperazine chelates have a better stability effect on heavy metals. When the dosage is 4%, Pb and Cd leaching concentrations are 0.15 and 0.01 mg/L, respectively, much lower than the corresponding regulated limit in GB 16889—2008 (Sun et al., 2019). This result is because piperazine chelators have a stable structure and contain two terminal thiocarboxylic groups, which can form a linear structure with heavy metals, compared with a single-point structure, which is more stable.

2,4,6-Trimercaptotriazine (trithiocyanuric acid, $C_3H_3N_3S_3$, denoted further as H_3TMT) and its trisodium salt (TMT) have three thiol and three nitrogen groups. Owing to the role of three N and S donors, it displays a great versatility of coordination with transition metals. Therefore, it has been practiced for precipitating divalent and univalent heavy metals to immobilize them in soil or hazardous wastes. It is reported that 4.2% TMT could decrease the leaching concentration to meet the limits set in GB 16889—2008 (Zhu et al., 2020).

DTCs are compounds with one or multiple -N(H)-CSS- functional groups. DTCs are synthesized by the nucleophilic reaction of small molecule amines (e.g., ethylenediamine, diethylenetriamine, etc.) or polyamines with primary or secondary amine function groups and carbon disulfide in basic media or alcoholic solution (Tong, 2020). Compared with inorganic chemicals, such as sodium sulfide, DTCs have more significant advantages in the stability and efficiency of seizing heavy metals in CIRs. Based on the number of -N(H)-CSS- functional groups in the molecule, DTCs can be grouped into single-DTC chelating agents and multi-DTC linear chelating agents. The macromolecule DTCs usually have numbers of -N(H)-CSS- functional groups, thus owning more sites to react with heavy metals. The resulting metal chelates are more complex in structure and can firmly embed heavy metals in them, reducing the leaching concentration of toxic metals (Tong, 2020).

Many studies have been focused on exploring the possibilities of increasing the number of key functional groups of DTCs or surface modification with DTCs on natural or synthetic composites to improve metal removal efficiency. These macromolecule chelating agents act as heavy metal collectors that introduce -N(H)-CSS- functional group through cross-linking, etherification, and polyamine substitution. Natural polymers include starch, chitosan, cellulose, lignin, and so on (Sadeghifar and Ragauskas, 2022; Biswas and Biswas, 2023). It is shown that amphoteric starch derivative bearing dual functional groups (amino and carboxyl groups), prepared through etherification of starch with 2-chloro-4,6-diglycino-[1,3,5]-triazine, displayed outstanding stability and reproducibility on Cu and Zn removal (Akinterinwa et al., 2022). Polymer DTC chelating resins, synthesized by grafting -N(H)-CSS- functional groups on various aforementioned polymer matrixes with an amine group (-NH(H)-) or an amine group after modification, are mainly adopted to separate and recover heavy metals (Nakakubo et al., 2022). However, due to the high cost and water-insoluble characteristics, DTC chelating resins are seldom used to dispose of CIRs or other hazardous wastes in practical applications.

Chitosan is prepared from chitin deacetylated and has good biodegradability and nontoxicity. It contains many amino and hydroxyl groups, which are easily complexed with heavy metals and have recently been applied in treating heavy metals (Zhang et al., 2022). Chitosan derivatives include isobutyl chitosan, hydroxypropyl chitosan, carboxymethyl chitosan, etc. Carboxymethyl chitosan was synthesized by chitosan and chloroacetic acid under alkaline conditions. The product introduced carboxylic groups, introducing more reaction sites for heavy metal ions. However, the easy leaching of single chitosan under acidic conditions dramatically limits its application, so the current research focus has gradually shifted to the synthesis of cross-linked compounds of chitosan and other macromolecular organic compounds. Table 7.3 summarizes the stabilization effect of common organic chemicals on heavy metals in CIRs.

3.2 State-of-art of organic chelating agent

Due to their relatively high efficiency and tolerance in various environments, organic chelating agents have attracted worldwide attention in minimizing the ecological and environmental risks of toxic heavy metal pollution. In particular, DTCs have been extensively investigated because of their strong coordination affinity with toxic metals. However, the stability of different organic agents to toxic metals varies significantly due to their mechanism of action and molecular weight. The comparative study between single-DTC chelating agents (SDD) and multi-DTC linear chelating agents with different molecular weights in disposing of IFA has shown that the latter can bind more effectively to heavy metals and display an outstanding overall curing performance due to their multiple hydrosulfide (-SH) groups (Tong, 2020). Also, the decomposition rate of dithiocarbamate is weaker as the alkyl chain length increases (Shen et al., 2019). Therefore, many studies have been focused on exploring the possibilities of increasing the number of key functional groups of DTCs or surface modification with DTCs on natural or synthetic composites, such as cellulose and starch (Akinterinwa et al., 2022; Nakakubo et al., 2022). As these raw organic polymer materials are abundant, cheap, and easy to degrade, their modified products have been used to trap toxic metal ions with high efficiency and low toxicity. Due to the consideration of economic cost and environmental friendliness, natural organic polymer materials may become a promising research hotspot.

TABLE 7.3 Stabilization effect of typical organic chemicals on heavy metals in CIRs.

Chemical agent	Reaction condition (hr)	Chelating agent dosage (w.t) (%)	Performance	References
TMT	Water ratio: 50%, 25°C, 0.5	4.2	All heavy metals meet the regulated limit of GB 16889–2008	Zhu et al. (2018)
Piperazine	Water ratio: 50%, room temperature, 1	4	All heavy metals meet the regulated limit of GB 16889–2008	Sun et al. (2019)
DTC	Water ratio: 30%, 25°C, 0.5	2	Pb meets the regulated limit of GB 16889–2008	Zhu et al. (2021)
Dithiocarboxylate-functionalized polyaminoamide dendrimer	Water ratio: 50%, room temperature, 1	1	All heavy metals meet the regulated limit of GB 16889–2008	Zhang et al. (2020)
Dithiocarboxylate functionalized polymer	Water ratio: 50%, 25°C, 0.5	3	All heavy metals meet the regulated limit of GB 16889–2008	Li et al. (2019)

3.3 Benefits and challenges of organic agent-based S/S technology

Due to cost and long-term stability merits, organic agent—based S/S technology with DTC and TMT has been widely used for CIRs (such as IFA) treatment in many waste incineration power plants or landfills. After being stabilized, the fly ash is pressed into a block, cured, and finally formed into a solidified body, which can then be transported to a sanitary landfill for disposal. The dissolution and migration behavior of the products created by chemical agents and heavy metals based on the precipitation—dissolution or adsorption principle is easily affected by environmental conditions (e.g., rainfall), while the products formed by the chemical agents and heavy metal ions via a surface-complex mechanism usually show stronger environmental resistance.

Based on the aforementioned discussion, two significant challenges exist in using organic chelating agents to stabilize CIRs. On the one hand, since the stabilization effect of heavy metals is related to the dosage, determining the optimal dosage is very important. In general, the stabilization efficiency of heavy metals in CIRs increases with the increase in chemical dosage. When the dosage of chemical additives exceeds a certain level, the leaching amount of heavy metals in the treated CIRs is almost stable (Zhang et al., 2020). There is an optimal additive dosage to achieve this state. Moreover, in developed countries or big cities with high population density, landfill construction and operation space are very scarce, and the proportion of volume growth must be reduced as much as possible. Therefore, the chemical that can achieve the goal of stabilization under the minimum dosage has a more excellent prospect of popularization and application.

On the other hand, macromolecular organic chelators present a better heavy metal stabilization effect due to a large number of coordinating groups; it is essential to explore more organic chelating agents with different molecular weights and figure out the reaction mechanisms between chemical agents and heavy metal ions. Furthermore, some macromolecular organic chemical agents can form two-dimensional and three-dimensional structures (Li et al., 2019). Besides, the functional group types on macromolecular organic chelators will also affect the agent's selectivity to heavy metals. Recent research on chemical stabilization mainly focuses on organic chelators with N, P, and S as coordinating atoms, and the corresponding groups include amine ($-NH_2$), sulfhydryl (-SH), and dithiocarboxylate acid (-RC(S)SH) groups (Ahmad et al., 2020). However, related studies have achieved better results, which has led to the research trend of integrating multiple groups into one chelating agent.

4. Composite chelating agent—based immobilization processes

4.1 Types and performances of composite chelating agents

The total cost of the organic chelating agent is high, and it isn't easy to popularize the application in large areas. Generally, the price of organic chelating agents is double or more than inorganic chelating agents. In addition, organic chelating agents are selective in stabilizing heavy metals, and their stabilization effects vary to some extent. Therefore, it isn't easy to guarantee the long-lasting impact of all the heavy metals in CIRs. The environment affects the inorganic reagent, and the precipitates' long-term stability is relatively poor, but the cost is low. Therefore, the combination of organic and inorganic chelating agents has been paid attention to. Cost analysis showed that a compound agent of 0.9% SDD and 3% sodium dihydrogen phosphate could reduce the cost by 26.72% compared with a single SDD agent (Zhu et al., 2021). Heavy metals' chelation efficiency and stability in CIRs can be improved using an appropriate mixture of inorganic and organic chemicals (Zhang et al., 2023b). Also, the product volume growth is small, and the economic benefit is considerable. Depending on the type of agent mixed, one or more mechanisms of heavy metal stabilization may be involved, including chelation, precipitation, ion exchange, and adsorption.

4.2 State-of-the-art of composite chelating agents

Compared with inorganic chemicals, many studies have shown that organic chelating agents, such as DTCs, can steadily work within a broad pH range (3~11). The chelation effect of DTCs on most heavy metals (e.g., Cu, Cd, etc.) keeps reliable and stable under various environmental conditions (Zheng et al., 2019). A comparative study shows that macromolecular DTC chelating agents can reach a better overall stabilization effect on heavy metals at low doses (3%) and prevent them from leaching in a wide pH range (2—13), which is not achieved by small molecular chelators (Zhang et al., 2020). Zhu et al. designed a series of single and composite chemical agents to treat IFA. It was found that a single reagent could reduce the leaching concentration of heavy metals in fly ash below the prescribed limit value, the dosage of the composite reagent was lower than that of a single reagent, and the overall cost was saved nearly 10% (Zhu et al., 2020).

It should be noted that the residues treated by chemical agent—based stabilization technology have a large surface area and loose structure, which is conducive to releasing heavy metals and other harmful substances (Ma et al., 2019). To minimize the migration of heavy metals, some researchers have conducted experimental studies in recent years by combining the stabilization process of reagents with the solidification technology (Chen et al., 2021). After chemical treatments are used to convert heavy metals into less toxic forms, cement curing or heat treatment can improve the physical strength of the solidified CIRs and reduce the porosity of the solidified CIRs, thereby reducing the mobility of contaminants. The primary mechanism of this process is that adding cement to CIRs produces C−S−H gel, which can coat heavy metals in cement-based cured products. This process, mixed with cement or other binders, reduces the treatment cost, prolongs the life cycle of the stabilized blocks, and provides technical and economic support for the industrial treatment of CIRs.

4.3 Benefits and challenges of composite chemicals−based S/S technology

Most organic chelators with excellent metal stabilization effects are made in laboratories or patented are expensive. The complementarity of different chemicals commonly available on the market can minimize the mobility of various metals in CIRs, thus reducing costs. Combining two or more agents is a new trend that warrants further investigation. Similar to the problems encountered in using and promoting single inorganic and organic chemical agents, the future development of composite chemical agents will also meet some challenges. For example, how to find the optimal dose while maintaining the stabilization effect to reduce the cost. How to reduce the impact of salts in residues on stabilizing heavy metals. And how to evaluate the environmental behavior of heavy metals in fly ash after composite agent stabilization.

5. Summary and prospect

Waste incineration is the general trend of waste reduction treatment in the world. The main environmental concern of CIRs is the potential for residues to leach pollutants (such as heavy metals, salts, and dioxins) during landfill or utilization. Heavy metals in the residues will leach out under various environmental conditions if disposed of incorrectly. Thus, stabilizing the heavy metals in CIRs is necessary before utilization or disposal. Among the current treatment technologies, chemical agent−based stabilization technology is the earliest studied and most widely used. Many studies have shown that the technology is feasible regarding the economy and long-term stability.

This chapter summarizes the types of existing chemical agents that have been studied or used and discusses the performances and mechanisms of these chemical agents in reducing the mobility of heavy metals. Two types of chemical agents, including inorganic and organic, have been studied or used in treating CIRs. The common inorganic agents are sulfides, soluble phosphates, amorphous silicon material, ferrous salts, and carbonates, while the common organic agents include thiourea, chitosan, piperazine, TMT, DTC, and their derivatives. Even though the cost of inorganic chelating is lower, organic chelating agents have the advantages of shorter reaction time, higher chelating efficiency, and relatively small dosage. In addition, combining two or more stabilization agents broadens the general applicability of the agents to heavy metals and reduces the cost.

With increasingly stringent environmental standards, the further study of chemical agent−based immobilization technology of CIRs mainly includes the following three aspects.

Firstly, developing novel environmentally friendly and cost-effective chemical agents to stabilize the heavy metals in CIRs is very important. DTC has the advantages of low dosage and promising effects in treating heavy metals in fly ash, but it is easy to release toxic substances in the reaction process. At the same time, the organic coordination groups acting on heavy metals are easy to oxidize in the air, so the leaching toxicity may exceed the standard in the late curing period. Therefore, future studies should also focus on improving the morphology and structure of DTC substances, such as modification, adding branch chain or ring substance to the original substance to complicate its structure, cross-linking branches with silica gel material with high porosity, and developing new efficient chelating agents with long-term stability.

Secondly, reasonably adding pretreatment steps such as water or dilute acid washing can improve the uniformity of physical and chemical properties of CIRs, which is conducive to the solidification and stabilization of harmful heavy metals in residues. Soluble salts such as sodium chloride in fly ash hinder the stabilization process of heavy metals. The presence of a large number of chloride ions increases the solubility of Cd, Pb, Zn, and other heavy metals in processed products, thus increasing the difficulty in reducing the migration of heavy metals. Therefore, it is essential to investigate the effect and mechanism of heavy metal stabilization after water-washing pretreatment. In addition to chemical agent−based stabilization technology, this pretreatment step has also been applied to thermal treatment and other resource utilization technologies (e.g., cement production).

Finally, it is of great concern to the environmental behavior of heavy metals and soluble salts in chemically stabilized fly ash and to establish relevant standards to improve sustainable management. During the past two decades, landfill mixed with domestic waste has been the primary treatment process of CIRs. This operation is easy to cause the collection and drainage of leachate to be blocked, and the deteriorating water quality dramatically affects the regular operation of the end treatment. With the increase in the proportion of waste incineration and the generated CIRs, it is regulated that the stabilized CIRs must meet the limits before disposal in a landfill. However, due to the lack of notable research and design specifications, there are still some problems in the operation and management of landfills, such as the primary impervious membrane being easily damaged, the pollution diversion design being complicated, and the landfill body being easy to slip.

In general, chemical chelation technology is economical and practical. Therefore, this technology is still the primary treatment process for CIRs until new and alternative technologies mature.

References

Ahmad, S.Z.N., Wan, W.N., Ismail, A.F., Yusof, N., Mohd, M.Z., Aziz, F., 2020. Adsorptive removal of heavy metal ions using graphene-based nanomaterials: toxicity, roles of functional groups and mechanisms. Chemosphere 248, 126008.

Ajorloo, M., Ghodrat, M., Scott, J., Strezov, V., 2022. Heavy metals removal/stabilization from municipal solid waste incineration fly ash: a review and recent trends. J. Mater. Cycles. Waste. 24 (5), 1693–1717.

Akinterinwa, A., Reuben, U., Atiku, J.U., Adamu, M., 2022. Focus on the removal of lead and cadmium ions from aqueous solutions using starch derivatives: a review. Carbohydr. Polym. 290, 119463.

Ates, F., Park, K.T., Kim, K.W., Woo, B.-H., Kim, H.G., 2023. Effects of treated biomass wood fly ash as a partial substitute for fly ash in a geopolymer mortar system. Construct. Build. Mater. 376, 131063.

Biswas, S., Biswas, R., 2023. Chitosan-the miracle biomaterial as detection and diminishing mediating agent for heavy metal ions: a mini review. Chemosphere 312 (Pt 1), 137187.

Bolan, S., Padhye, L.P., Kumar, M., Antoniadis, V., Sridharan, S., Tang, Y., Singh, N., Hewawasam, C., Vithanage, M., Singh, L., Rinklebe, J., Song, H., Siddique, K.H.M., Kirkham, M.B., Wang, H., Bolan, N., 2023. Review on distribution, fate, and management of potentially toxic elements in incinerated medical wastes. Environ. Pollut. 321, 121080.

Chen, D.M.-C., Bodirsky, B.L., Krueger, T., Mishra, A., Popp, A., 2020. The world's growing municipal solid waste: trends and impacts. Environ. Res. Lett. 15 (7), 074021.

Chen, J., Shen, Y., Chen, Z., Fu, C., Li, M., Mao, T., Xu, R., Lin, X., Li, X., Yan, J., 2023. Accelerated carbonation of ball-milling modified MSWI fly ash: migration and stabilization of heavy metals. J. Environ. Chem. Eng. 11 (2), 109396.

Chen, L., Wang, Y.S., Wang, L., Zhang, Y., Li, J., Tong, L., Hu, Q., Dai, J.G., Tsang, D.C.W., 2021. Stabilisation/solidification of municipal solid waste incineration fly ash by phosphate-enhanced calcium aluminate cement. J. Hazard Mater. 408, 124404.

Chen, W., Wang, Y., Sun, Y., Fang, G., Li, Y., 2022. Release of soluble ions and heavy metal during fly ash washing by deionized water and sodium carbonate solution. Chemosphere 307 (Pt 2), 135860.

He, R., Sandoval-Reyes, M., Scott, I., Semeano, R., Ferrão, P., Matthews, S., Small, M.J., 2022. Global knowledge base for municipal solid waste management: framework development and application in waste generation prediction. J. Clean. Prod. 377, 134501.

Istrate, I.-R., Galvez-Martos, J.-L., Vázquez, D., Guillén-Gosálbez, G., Dufour, J., 2023. Prospective analysis of the optimal capacity, economics and carbon footprint of energy recovery from municipal solid waste incineration. Resour. Conserv. Recycl. 193, 106943.

Labianca, C., Ferrara, C., Zhang, Y., Zhu, X., De Feo, G., Hsu, S.-C., You, S., Huang, L., Tsang, D.C.W., 2022. Alkali-activated binders – a sustainable alternative to OPC for stabilization and solidification of fly ash from municipal solid waste incineration. J. Clean. Prod. 380, 134963.

Li, R., Zhang, B., Wang, Y., Zhao, Y., Li, F., 2019. Leaching potential of stabilized fly ash from the incineration of municipal solid waste with a new polymer. J. Environ. Manag. 232, 286–294.

Liu, G., Chen, F., 2018. Chelating effect of heavy metals in MSWI fly ash with several kinds of stabilizing agents. Environ. Eng. 36 (9), 139–143.

Liu, S., Miao, C., Yao, S., Ding, H., Zhang, K., 2021. Soil stabilization/solidification (S/S) agent—water-soluble thiourea formaldehyde (WTF) resin: mechanism and performance with cadmium (II). Environ. Pollut. 272, 116025.

Lou, Y., Jiang, S., Du, B., Dai, X., Wang, T., Wang, J., Zhang, Y., 2023. Leaching morphology characteristics and environmental risk assessment of 13 hazardous trace elements from municipal solid waste incineration fly ash. Fuel 346, 1287.

Ma, W., Chen, D., Pan, M., Gu, T., Zhong, L., Chen, G., Yan, B., Cheng, Z., 2019. Performance of chemical chelating agent stabilization and cement solidification on heavy metals in MSWI fly ash: a comparative study. J. Environ. Manag. 247, 169–177.

Nag, M., Saffarzadeh, A., Nomichi, T., Shimaoka, T., Nakayama, H., 2020. Enhanced Pb and Zn stabilization in municipal solid waste incineration fly ash using waste fishbone hydroxyapatite. Waste Manag. 118, 281–290.

Nakakubo, K., Endo, M., Sakai, Y., Biswas, F.B., Wong, K.H., Mashio, A.S., Taniguchi, T., Nishimura, T., Maeda, K., Hasegawa, H., 2022. Cross-linked dithiocarbamate-modified cellulose with enhanced thermal stability and dispersibility as a sorbent for arsenite removal. Chemosphere 307 (Pt 1), 135671.

Nakararoj, N., Nhat Ho Tran, T., Sukontasukkul, P., Attachaiyawuth, A., Tangchirapat, W., Chee Ban, C., Rattanachu, P., Jaturapitakkul, C., 2022. Effects of High-Volume bottom ash on Strength, Shrinkage, and creep of high-strength recycled concrete aggregate. Construct. Build. Mater. 356, 129233.

Nazeer, M., Kapoor, K., Singh, S.P., 2023. Strength, durability and microstructural investigations on pervious concrete made with fly ash and silica fume as supplementary cementitious materials. J. Build. 69, 106275.

Sadeghifar, H., Ragauskas, A., 2022. Lignin as a bioactive polymer and heavy metal absorber- an overview. Chemosphere 309 (Pt 1), 136564.

Sanito, R.C., Bernuy-Zumaeta, M., Wang, W.-C., Yang, H.-H., You, S.-J., Wang, Y.-F., 2023. Optimization of metals degradation and vitrification from fly ash using Taguchi design combined with plasma pyrolysis and recycling in cement construction. J. Clean. Prod. 387, 135930.

Shen, W., Zhu, N., Xi, Y., Huang, J., Li, F., Wu, P., Dang, Z., 2022. Effects of medical waste incineration fly ash on the promotion of heavy metal chlorination volatilization from incineration residues. J. Hazard Mater. 425, 128037.

Shen, Y., Nagaraj, D.R., Farinato, R., Somasundaran, P., Tong, S., 2019. Decomposition of flotation reagents in solutions containing metal ions. Part III: comparison between xanthates and dithiocarbamates. Miner. Eng. 139, 105898.

Sun, Y., Xu, C., Yang, W., Ma, L., Tian, X., Lin, A., 2019. Evaluation of a mixed chelator as heavy metal stabilizer for municipal solid-waste incineration fly ash: behaviors and mechanisms. J. Chin. Chem. Soc. 66, 188−196.

Sun, Y., Zhang, P., Li, Z., Chen, J., Ke, Y., Hu, J., Liu, B., Yang, J., Liang, S., Su, X., Hou, H., 2023. Iron-calcium reinforced solidification of arsenic alkali residue in geopolymer composite: wide pH stabilization and its mechanism. Chemosphere 312 (Pt 2), 137063.

Tan, J., Wu, X., Li, J., al, e., 2020. Stabilization of heavy metals in MSWI fly ash by thermal treatment at intermediate temperatures with rice husk ash. China Environ. Sci. 40 (7), 3054−3060.

Tong, L., 2020. Optimization of the Chelation Efficiency Evaluation Method and Advanced Synthesis of Novel Macromolecule Chelating Agents for Incineration Fly Ash Treatment. Harbin Institute of Technology.

Tong, L., He, J., Wang, F., Wang, Y., Wang, L., Tsang, D.C.W., Hu, Q., Hu, B., Tang, Y., 2020. Evaluation of the BCR sequential extraction scheme for trace metal fractionation of alkaline municipal solid waste incineration fly ash. Chemosphere 249, 126115.

Vavva, C., Lymperopoulou, T., Magoulas, K., Voutsas, E., 2020. Chemical stabilization of fly ash from medical waste incinerators. Environ. Proc. 7 (2), 421−441.

Wu, M., Qi, C., Chen, Q., Liu, H., 2023. Evaluating the metal recovery potential of coal fly ash based on sequential extraction and machine learning. Environ. Res. 224, 115546.

Xin, M., Sun, Y., Wu, Y., Li, W., Yin, J., Long, Y., Wang, X., Wang, Y.-n., Huang, Y., Wang, H., 2022. Stabilized MSW incineration fly ash co-landfilled with organic waste: leaching pattern of heavy metals and related influencing factors. Process Saf. Environ. Protect. 165, 445−452.

Ye, G., Deng, Y., Wu, H., Li, M., 2023. Solidification of heavy metals in waste incineration fly ash by sodium sulfide ball milling reaction. Non-Met. Mines. (Chinese) 46 (1), 40−44.

Yu, Z., Liu, W., Peng, Z., Zhang, L., Bu, S., Xu, W., Xu, C., Yao, H., Ma, Z., 2021. Experimental study on the stabilization of heavy metals in fly ash from municipal solid waste incineration by N-30 alkaline silica sol. Process Saf. Environ. Protect. 148, 1367−1376.

Yuan, R., Fan, X., Gan, M., Zhao, Q., Lu, S., Li, S., 2023. Facile treatment of municipal solid waste incineration fly ash by one-step microwave hydrothermal method: hazards detoxification and tobermorite synthesis. J. Environ. Chem. Eng. 11 (3), 109768.

Zhang, J., Chen, T., Li, H., Tu, S., Zhang, L., Hao, T., Yan, B., 2023a. Mineral phase transition characteristics and its effects on the stabilization of heavy metals in industrial hazardous wastes incineration (IHWI) fly ash via microwave-assisted hydrothermal treatment. Sci. Total Environ. 877, 162842.

Zhang, J., Zhang, H., Yuan, C., Lu, A., 2023b. Chelating effect of heavy metals in MSWI fly ash with several kinds of stabilizing agents. Environ. Eng. 41, 430−435.

Zhang, M., Guo, M., Zhang, B., Li, F., Wang, H., Zhang, H., 2020. Stabilization of heavy metals in MSWI fly ash with a novel dithiocarboxylate-functionalized polyaminoamide dendrimer. Waste Manag. 105, 289−298.

Zhang, R., Liu, B., Ma, J., Zhu, R., 2022. Preparation and characterization of carboxymethyl cellulose/chitosan/alginic acid hydrogels with adjustable pore structure for adsorption of heavy metal ions. Eur. Polym. J. 179, 111577.

Zheng, L., Wang, W., Li, Z., Zhang, L., Cheng, S., 2019. Immobilization of heavy metal using dithiocarbamate agent. J. Mater. Cycles Waste Manag. 21, 652−658.

Zhu, J., Hao, Q., Chen, J., Hu, M., Tu, T., Jiang, C., 2020. Distribution characteristics and comparison of chemical stabilization ways of heavy metals from MSW incineration fly ashes. Waste Manag. 113, 488−496.

Zhu, J., Li, M., Zheng, D., 2018. Distribution and chemical stabilization of heavy metals in municipal solid waste incineration fly ash of Chongqing. Environ. Chem. 37 (4), 880−888.

Zhu, Z., Guo, Y., Zhao, Y., Chen, W., 2021. Stabilization of Pb and characteristic agents in municipal solid waste incineration fly ash. China Environ. Sci. 41 (6), 2737−2743.

Chapter 8

Mechanochemical treatment of combustion/incineration residues

Yaqi Peng

State Key Laboratory of Clean Energy Utilization, Institute for Thermal Power Engineering, Zhejiang University, Hangzhou, China

1. Introduction

With the increase in population and urbanization, energy consumption is rapidly growing. Coal, as the largest fuel resource, is used to generate electricity and prepare the chemical raw materials (Yuan et al., 2023a). A large amount of coal ash is generated during the combustion process, accounting for 5−20 wt.% of the feed coal by weight. Additionally, the substantial increase in urbanization has put tremendous pressure on the disposal of the municipal solid waste. According to the estimates by the World Bank, the worldwide municipal solid waste generated will reach to 6.1 million tons per day (Lin et al., 2022). In 2021, the amount of municipal solid waste has increased to 248 million tons, with 72.3% of the waste being incinerated in China (NBSC, 2022). Municipal solid waste incineration (MSWI) can effectively reduce the volume by 90% and weight by 60%−90% while saving land and recovering heat or power (Tian et al., 2023). However, it produces bottom ash and fly ash, with the latter accounting for 3−15 wt.% of the incinerated solid waste (Pan et al., 2013; Mao et al., 2020). The growing amount of combustion/incineration residues, often hazardous, has attracted great attention and requires proper management.

Fly ash, as a typical residue, comprises powdery fine particles with a chemical composition of SiO_2, Al_2O_3, and CaO, which is significantly influenced by the fuel components, combustion/incineration conditions and air pollution control system. The ash from combustion/incineration is regarded as general solid waste; however, fly ash from MSWI has been classified as hazardous waste due to the large amount of harmful heavy metal (Pb, Hg, Cu, etc.), persistent organic pollutants (POPs, such as dioxins, polychlorinated biphenyls, etc.), and chlorinated residues (Jin et al., 2023). Currently, fly ash is mainly disposed of through landfilling, which is far from being environmentally friendly as the long-term accumulation can lead to leaching of heavy metals, posing environmental risk to the soils and waters (Ma et al., 2017; Huang et al., 2022). Therefore, the government encourages research and development of new technologies for the detoxification and recycling utilization of fly ash (Lin et al., 2022).

Harmless treatment followed by recycling utilization of the residues is the main trend for achieving environmentally friendly waste management. Highlighting the threats and potential benefits of the residues, it is essential to develop new applications that reduce environmental impacts and increase recycling utilization. Combustion residues recycling aligns with the overall circular economy model and sustainable development (Huang et al., 2017; Kurda et al., 2018; Nie et al., 2022). To date, there have been numerous methods to treat the residues, including thermal treatment (such as sintering, melt vitrification, hydrothermal, etc.) and nonthermal treatment (cement/chemical solidification/stabilization, biological/chemical leaching, etc.), showing different pros and cons. In addition, the treated residues can be used in construction industry, environmental engineering, and agriculture. To date, various research and review papers have been published on combustion/incineration residues, their properties, treatment methods, and applications. Ahmaruzzaman (2010) summarized the properties of coal fly ash and its utilization as adsorbents for cleaning of flue gas and removing heavy metal and inorganic/organic compounds. He also reviewed the conversion of fly ash into zeolites, its application as road subbase, and lightweight aggregate in construction industry. Yao et al. (2015) focused on the application of fly ash on the soil amelioration, catalysis and zeolites synthesis, and construction industry. Muir et al. (2021) reviewed the properties of fly ash and fly ash−based porous materials for volatile organic compounds (VOCs) adsorption. Kurda et al. (2018) used the

Treatment and Utilization of Combustion and Incineration Residues. https://doi.org/10.1016/B978-0-443-21536-0.00023-X

life cycle assessment (LCA) method and evaluated the environmental impacts and economic of incorporating fly ash in concrete production. In recent years with the release of the standard "Technical specification for pollution control of fly-ash from municipal solid waste incineration" in China (HJ1134, 2020), researchers paid attention to the treatment and recycling of MSWI fly ash, and many review papers were published. Dou et al. (2017) reviewed the characterization, treatment, and application of MSWI bottom ash. Huang et al. (2022) summarized the thermal detoxification technology of MSWI fly ash, such as melting/vitrification, hydrothermal treatment, thermal plasma, and the recycling of fly ash such as cement, asphalt mixture, and so on. Xue and Liu (2021) analyzed the detoxification of dioxins, solidification of heavy metals, and removal and utilization of chlorides in the MSWI fly ash. However, traditional methods often require extreme conditions such as high temperature and/or high pressure, high cost, and low recycling efficiency. Mechanochemical treatment with its simplicity, mild reaction conditions, and solvent-freeness offers a green approach to treat the combustion/incineration residues. To maximize the economic, environmental benefits of residues recycling, it is necessary to address the potential application of mechanochemical treatment.

Mechanochemistry (MC) has been rediscovered as a promising approach for solid waste treatment and materials synthesis. It has been defined as "a chemical reaction induced by mechanical energy" by the International Union of Pure and Applied Chemistry (IUPAC). In the milling process, shear and impact forces induce the formation of defects and changes in the crystal structure, considerably modifying and improving the reactivity of the treated solids. Since the pioneering research on the MC destroying chlorinated pollutants conducted by Rowlands et al. (1994), MC has been widely studied for the destruction of persistent organic pollutants, such as chlorobenzenes (Lou et al., 2023), chlorophenols (Aresta et al., 2003), PCDD/Fs (Chen et al., 2019c), and PCBs (Wang et al., 2017). Moreover, MC also shows good stability to heavy metal in soil and MSWI fly ash via wet milling and dry milling (Yuan et al., 2021; Li et al., 2023). The low reactivity of the residues inhibits their recycling application, but MC can realize the particle size reduction and surface activation, which has been recognized as an important method to improve the reactivity of the residues (Grabias-Blicharz and Franus, 2023).

This chapter aims to introduce the MC treatment of the combustion/incineration residues and highlight the destruction of persistent organic pollutants after MC treatment. It also addresses the importance of leaching and stabilization of heavy metals from the residues through MC treatment. Additionally, this chapter provides an overview of improving the physical and chemical properties of milled residues for construction or environmental applications. This chapter endeavors to offer an overview of the advancements achieved in the mechanochemical treatment of combustion/incineration residues. It is expected that this comprehensive examination will not only enhance the understanding and application of mechanochemical methods in this particular field but also inspire further research into mechanochemical methods.

2. General aspects of mechanochemistry

2.1 Principles of mechanochemistry

The field of mechanochemistry, which involves the use of mechanical grinding or milling, is experiencing a tremendous surge. In fact, it has been recognized by IUPAC as one of the top 10 technologies that will revolutionize the world (Gomollón-Bel, 2019). MC is an environmentally friendly and efficient technology that utilizes chemical reactions to modify the internal structure, crystal phase, and surface properties of materials. This is achieved through the collision, extrusion, shear, and friction between grinding media and materials. Under the influence of MC, the solid materials undergo fragmentation, breaking down the surface chemical bonds and creating new, active surfaces with unsaturated bonds. This process introduces lattice defects and enhances the internal energy of the materials, rendering them in an unstable and reactive state. Consequently, MC triggers chemical reactions in the materials (Liu et al., 2022).

The MC technology operates through three primary effects. Firstly, it induces physical alterations by refining particle or grain size and altering their specific surface area (SSA). Additionally, it induces crystallization, resulting in the formation of lattice defects and distortion, which in turn lead to decreases in crystallinity and transformations of crystals. Lastly, it initiates chemical reaction such as the removal of crystal water or hydroxyl groups, the breaking of chemical bonds, and the reduction of activation energy required for reactions (Dong et al., 2022). Unlike the conventional method of mechanical grinding that relies on milling action to break materials, MC utilizes the collision and stirring of grinding balls to induce repeated collisions, compression, cold-welding, and cracking of the materials so that the materials are diffused on the atomic scale (Zhang et al., 2013).

Mechanochemistry has seen a consistent rise since the mid-20th century. The sheer volume of research papers on this subject has grown so much that it is now impractical to mention more than a small selection of them. Gaining insights into the underlying characteristics of mechanochemical reactions remains a crucial goal for researchers from both theoretical and experimental backgrounds. However, accomplishing this is challenging due to the intricate and multifaceted nature of mechanochemical processes, which encompass various scales of length and time. Additionally, these reactions are highly dependent on specific systems and occur under a diverse range of circumstances.

2.2 Milling tools

The most important tool of practical mechanochemistry is the mill equipment, and a number of mechanical tools are available on the market (Lomovskiy et al., 2020). Automated electrical mills are manufactured with a wide range of designs. Those commonly employed for laboratory-level treatment can be classified into several types, as shown in Fig. 8.1 (Dong et al., 2022). The high-speed mechanical extrusion crusher is a type of crusher that utilizes a stator and rotor to crush and modify materials. The stator and rotor move and generate colliding, extruding, and shearing the materials. In contrast, the high-speed impact crusher does not have any moving parts. Instead, it relies on a high-speed flowing gas to carry the materials. As the materials are carried by the gas, they collide and shear at an extremely high speed, effectively changing their size and physicochemical properties.

Planetary ball mill is the most common equipment used in the laboratory. It consists of jars, containing milling tools such as stainless steel or zirconia balls, and the materials to be milled. These jars rotate on their own axis and revolve on the central axis at the same time. This dual motion allows the mill to provide both shear and impact force through the collisions between the milling balls or between the balls and the jar wall. By adjusting the speed of rotation and revolution, the ratio of shear to impact force can be controlled, giving flexibility in the milling process. In addition, the vertical rotation of the main spindle plays a crucial role in preventing aggregation caused by gravity, thereby ensuring a comprehensive and efficient grinding process (Peng et al., 2018). The tumbler ball mill consists of a cylindrical drum with a horizontal axis that is partially filled with grinding media, such as metal or ceramic balls. As the cylinder rotates, the balls are lifted up and then cascade back down, crushing and grinding the material between them. This equipment offers significant benefits such as reduced noise, minimal pollution, enhanced efficiency, and the ability to facilitate large-scale production. The tumbler ball mill has found extensive application in various industries, including coal mining, cement production, and mineral dressing (Sitotaw et al., 2021).

The vibration ball mill is a type of mill that utilizes mechanical vibration to facilitate the grinding process (Bulgakov et al., 2018). Unlike traditional ball mills, the vibration ball mill incorporates additional forces generated through the oscillation or vibration of the mill chamber, resulting in improved grinding efficiency and finer particle distribution. The attrition ball mill is designed with a vertical rotating spindle and horizontal impellers that are securely attached to the spindle. It works by impelling the feed material against a rotating surface within the mill chamber. This impelling action is facilitated by the rotation of the grinding media, typically steel balls or ceramic beads.

Each of these mills exhibits their individual advantages and disadvantages concerning treatment efficiency, cost, and environmental impact. In comparison with other mills, planetary ball mills offer versatility as the shear-to-impact ratio can be controlled, making them extensively employed in laboratory settings.

3. Mechanochemical degradation of organic pollutants in residues

3.1 Persistent organic pollutants in residues

Numerous organic pollutants are formed in the combustion or incineration process, including chlorobenzenes, chlorophenols, polycyclic aromatic hydrocarbons (PAHs), polychlorinated biphenyls (PCBs), and polychlorinated dibenzo-p-dioxins and polychlorinated dibenzo-furans (PCDD/Fs, also refer to dioxins) (Peng et al., 2016). Especially, dioxins is a key assessment indicator for the proper utilization of fly ash, as emphasized in the Chinese standard (Jin et al., 2023). Their levels should be maintained below 50 ng TEQ/kg prior to any usage (HJ1134, 2020). The investigation of dioxins commenced in the 1970s following their discovery in fly ashes produced by waste incineration (Olie et al., 1977). By adhering to the principle of "3T + E" (Temperature above 850°C, residence Time over 2 s, Turbulence, and Excessive air in the furnace), solid waste incineration effectively destroys the majority of pollutants. However, it is important to note that PCDD/Fs may still be generated in the postcombustion zone as the flue gas cools down. The formation of dioxins occurs through three pathways: high-temperature homogeneous synthesis, de novo synthesis, and precursor synthesis (Peng et al., 2016). And over 95% of the PCDD/Fs generated in the combustion/incineration process are found predominantly in fly ash.

The toxic equivalent quantities (TEQ) of the MSWI fly ash were reported in the range of 34~250 ng TEQ/kg (Pan et al., 2022). In another study conducted in southeast China, the TEQ concentration in fly ash ranged from 115 to 645 ng

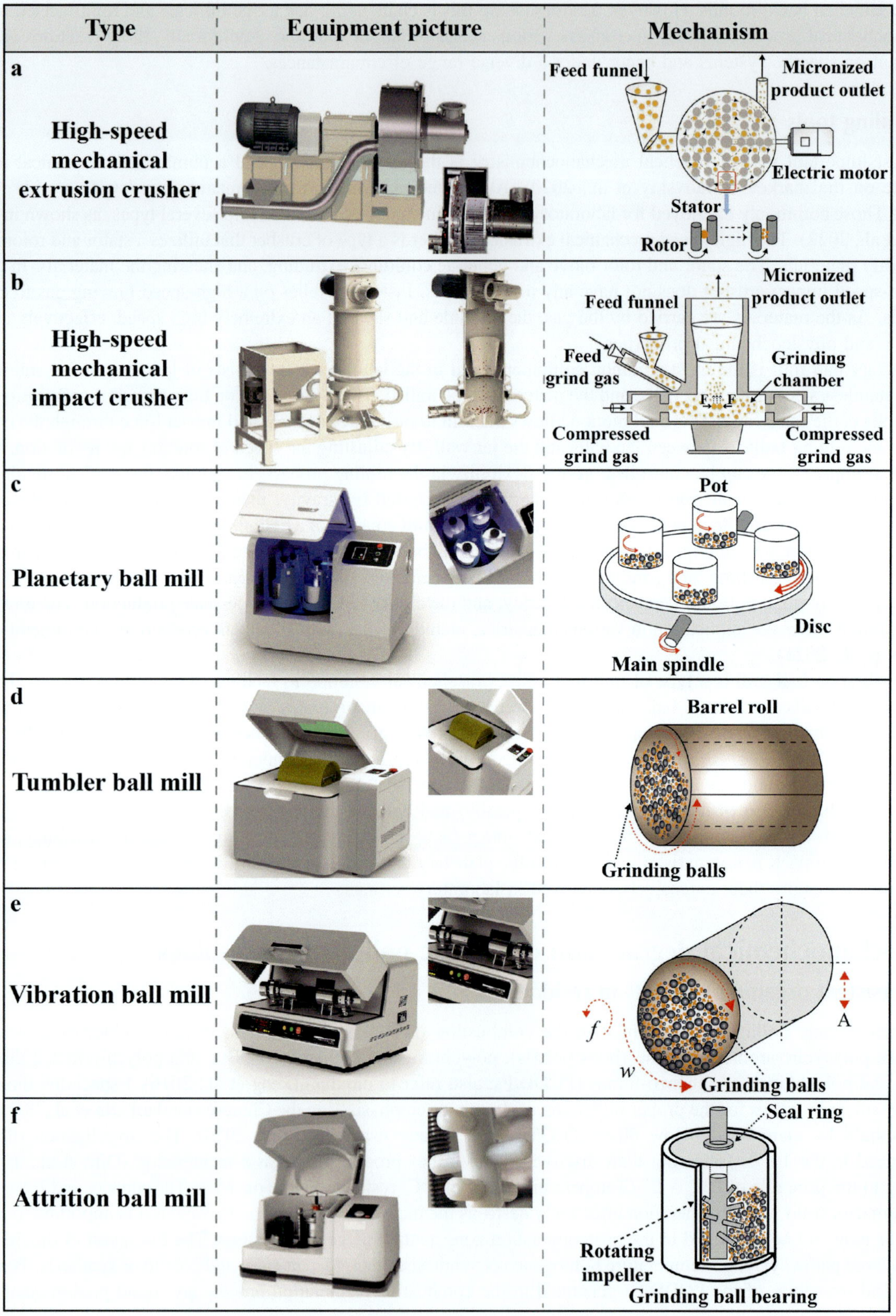

FIGURE 8.1 MC equipment and working principles (Dong et al., 2022).

TEQ/kg (Pan et al., 2013). The PCDD/Fs concentrations are closely associated with the composition of waste, incineration conditions, and air pollution control system.

3.2 Destruction of dioxins by MC treatment

Calcium oxide (CaO) is widely utilized as an additive in the degradation of persistent organic pollutants through mechanochemical processes (Table 8.1). This preference is primarily attributed to its accessibility and low cost. Nomura et al. (2005) studied the MC destruction of 4-chlorobiphenyl (4-CB), octachlorodibenzo-p-dioxin (OCDD), and octachlorodibenzofuran (OCDF) as the model compounds to destruct PCDD/Fs. With the aid of CaO, 95% of the OCDD/F were degraded rapidly in the first 6 min, resulting in the generation and subsequent degradation of tetra-to hepta-CDD/F congeners. The simultaneous peak of these low-chlorinated dioxins in the initial reaction suggests that it is not the result of consecutive dechlorination. Moreover, the cleavage of the OCDD/F structure led to the formation of chlorinated phenols and quinones. Another study discovered that a lower rotation speed resulted in a lower PCDD/F degradation efficiency (Yan et al., 2007). To study the degree of chlorination, the average number of chlorine atoms in the PCDD/F was calculated, which is defined as the sum of the products of the mole fraction and the number of chlorine atoms for each homologue. They observed a decrease in the degree of chlorination from 6.05 to 5.82 through the MC treatment process, confirming dechlorination.

TABLE 8.1 Summary of the contaminants destruction by MC treatment.

Contaminant	Additives	Reagent ratio[a]	Rotation speed, duration time (h)	Destruction efficiency (%)	References
OCDD/F	CaO	PCDD/F: CaO = 1:20 (wt)	700 rpm, 2	100	Nomura et al. (2005)
PCDD/F (fly ash)	CaO	Fly ash: CaO = 6:1 (wt)	350 rpm, 2	60	Peng et al. (2010)
PCDD/F, PCBs (fly ash)	Ca + CaO	Fly ash: Ca + CaO = 100:1 (wt)	400 rpm, 20	100	Mitoma et al. (2011)
2,4,6-trichlorophenol	CaO + SiO_2	Trichlorophenol: CaO = 1:16 (mol)	400 rpm, 6	99.0	Lu et al. (2012)
Hexachlorocyclohexane	CaO	Hexachlorocyclohexane: CaO = 1:10/1:60	700 rpm, 2	100	Nomura et al. (2012)
PCDD/F (fly ash)	CaO + Al	Fly ash: CaO + Al = 6:1 (wt)	600 rpm, 10	93.2	Chen et al. (2017)
PCDD/F (fly ash)	Fe/Ni-SiO_2	Fly ash: Fe/Ni-SiO_2 = 20:0.6 (wt)	400 rpm, 24	93.2	Li et al. (2017)
PCDD/F, PAH (fly ash)	/	Fly ash = 100%	350 rpm, 6	39.6, 41.7	Peng et al. (2018)
Hexachlorobenzene	Fe/Fe_3O_4	Hexachlorobenzene: Fe/Fe_3O_4 = 1:13.5 (wt)	600 rpm, 6	98.7	Hu et al. (2019)
BDE, PCB (fly ash)	nZIVI-CaO-$Ca_3(PO_4)_2$	Fly ash: nZIVI-CaO-$Ca_3(PO_4)_2$ = 12.5:1 (wt)	550 rpm, 6	99.6, >87.0	Yuan et al. (2023b)
Hexachlorobenzene (fly ash)	n-Al/CaO	Fly ash: n-Al/CaO = 9:1 (wt)	600 rpm, 3	82.4	Yu et al. (2020)
Hexachlorobenzene	Na_2SO_3	Hexachlorobenzene: Na_2SO_3 = 1:20 (wt)	600 rpm, 6	97.4	Lou et al. (2023)

[a]*Weight (wt) or molar (mol) ratio expressed as fly ash/contaminants amount over additive.*

Without the additives, the TEQ tends to increase due to the transformation of highly chlorinated CDD/F with lower toxic equivalent factor (TEF) into low chlorinated CDD/F with higher TEF (Mitoma et al., 2011). The addition of CaO and metallic calcium as additives resulted in destruction efficiencies of 44.2% and 82.7%, respectively. However, when a combination of Ca and CaO additives was used for the treatment of MC, a degradation efficiency of 100% for PCDD/F was achieved (Mitoma et al., 2011). In the milling process, free electrons were generated by the tribo-emission phenomenon (Nakayama, 2007). In the absence of additives, the steel balls only generate heat, and limited hydrodechlorination occurs to part of the chlorine atom, yielding low chlorinated CDD/F. However, the presence of Al, Ca, and CaO leads to higher tribo-plasma fields, generating more electrons to break the C-Cl bond. Consequently, a notable decrease in the TEQ is observed (Mitoma et al., 2011).

Dechlorination has been demonstrated as a significant route for the degradation of chlorinated aromatics in the milling process, leading to the removal of chlorine atoms from the molecular structure. While only a portion of the organochlorine transformed to inorganic chloride, the remaining proportion underwent degradation into various other organic chlorinated compounds. In the 2,4,6-trichlorophenol destruction experiment, tetrachlorophenols, along with monochlorophenols and dechlorophenols, were generated (Lu et al., 2012). The crystal structure of CaO undergoes a transformation to an amorphous phase as time is extended. In addition, calcium hydroxychloride (CaOHCl) was formed. The intermediates from chlorinated pollutants destruction were also found (Nomura et al., 2005; Yu et al., 2020). The degradation of γ-hexachlorocyclohexane gave rise to the formation of monochlorobenzene (MCBz), dichlorobenzene (DCBz), trichlorobenzene (TriCBz), and tetrachlorobenzene (TeCBz) (Nomura et al., 2012). Mono- to heptachlorinated naphthalenes (CN) were formed as by-products during the destruction process of octa-CN (Nomura et al., 2013a). The production of soluble chloride ions escalated with longer milling durations, achieving an 85% efficiency after 1 h of milling and ultimately reaching a maximum efficiency of 100% after 3 h. These findings demonstrate that all the organic chlorine derived from polychlorinated naphthalene (PCN) underwent a complete conversion into inorganic chloride, indicating the thorough degradation of PCNs through dechlorination during the MC treatment (Nomura et al., 2013a).

The quantity of CaO also exhibits a significant impact on the efficiency of pollutants removal. The investigation illustrated that in cases where the hexachlorocyclohexane (HCH) to CaO ratio was 1:10, the chlorine removal efficiency merely reached 50% even after an 8-h MC treatment. In contrast, an increased efficiency of chlorine removal exceeding 60% was promptly achieved within a mere 0.25 h when the ratio was adjusted to 1:60. Moreover, complete elimination of chlorine occurred within 2 h at the equivalent ratio (Nomura et al., 2012). Compared with CaO, the addition of SiO_2, Al_2O_3, and metallic aluminum (Al) to CaO demonstrates a substantial improvement in reactivity. CaO utilizes an oxygen atom as a carrier to transfer electrons, facilitating the attack and breakage of the C-Cl bond. Additionally, aluminum serves as an electron donor to CaO, enhancing the release of electrons and thus expediting the degradation process of PCDD/Fs (Chen et al., 2018).

3.3 Reaction mechanism

The destruction of chlorinated pollutants, such as chlorobenzenes, chlorophenols, and dioxins in the residues, was carried out following a specific procedure, as shown in Fig. 8.2 (Cagnetta et al., 2016b). Initially, the pollutants were adsorbed onto the surface of the solid reagent through the milling process. Subsequently, the activated surfaces underwent reduction or oxidation reactions with the pollutants. As the milling time increased, the organic compounds underwent carbonization or fragmentation.

Reagents play a vital role in mechanochemical reactions involving pollutants. There are four distinct categories of reagents commonly used in mechanochemistry: Lewis bases, such as metal oxides such as CaO; neutral reagents such as SiO_2 and alumina; reducing reagents, which include zero valent metals and hydrides; oxidizing agents, such as MnO_2 and $S_2O_8^{2-}$. Cagnetta et al. (2016b) provided a comprehensive summary of the dehalogenation mechanism of CaO in the MC treatment process. When the reagents are subjected to mechanical force, it results in an increase in structural defects on the surface of the solids. Subsequently, upon cleaving of CaO crystals, oxide ions facilitate the transfer of an electron to the carbon atom of the pollutant, leading to the generation of an anion radical. The expulsion of chlorine is facilitated by its electronegativity and the relative fragility of the C-Cl bond, resulting in the formation of an oxide radical and a pollutant-free radical. The chlorine radical can potentially be trapped within the CaO lattice and subsequently converted into a halide through the acquisition of an additional electron. Examples of such halides include CaOHCl and $CaCl_2$. The inorganic halides can subsequently be extracted using water. The organic radical undergoes further reaction by bonding with an oxygen ion or oxygen radical, resulting in the formation of a C—O bond. In the case of oxidizing reagents such as δ-MnO_2 (birnessite), Mn^{4+} undergoes reduction to Mn^{3+} and Mn^{2+}, while pollutant molecules are oxidized to radicals and cations. Neutral reagents such as SiO_2 can lead to the attack of the C-Cl bond by quartz E′ centers ($\equiv Si\cdot$) and nonbridging oxygen

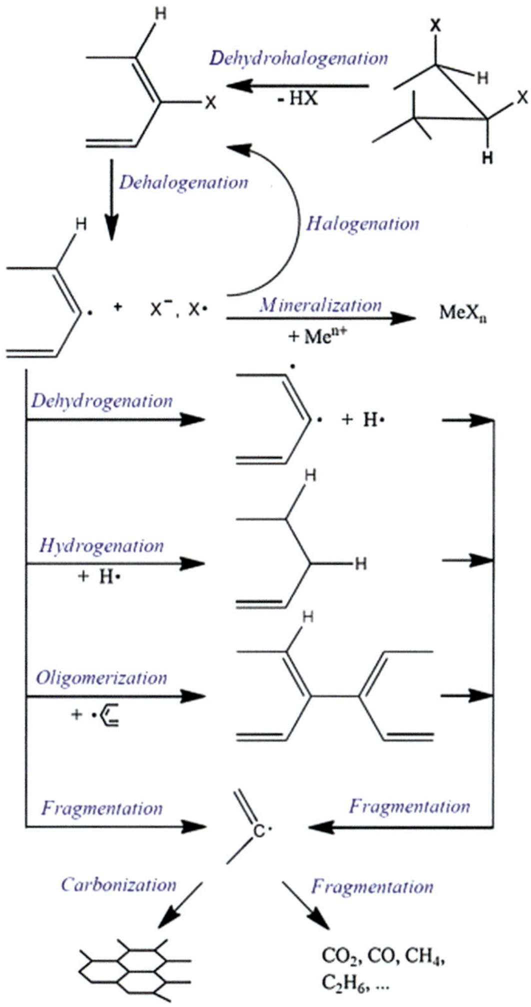

FIGURE 8.2 Scheme of degradation process of halogenated pollutants (Cagnetta et al., 2016b).

hole centers (≡Si-O·). The pollutant carbon skeleton, upon experiencing partial or complete dehalogenation processes resulting in the formation of radicals, can engage in various reactions, including dehydrogenation, hydrogenation, oligomerization, fragmentation, and ultimately graphitization. This transformation leads to a combination of amorphous and graphitic carbon. Additionally, it has been observed that samples subjected to MC treatment exhibit a darkening effect due to carbonization (Yan et al., 2007; Chen et al., 2017).

Cagnetta et al. (2016a) constructed a kinetic model to investigate the destruction mechanism of persistent organic pollutants by MC. The reagent (C) undergoes activation through ball milling (forming C*), prior to reacting with the pollutant (P). Fig. 8.3A illustrates that the initial conversion rate is hindered due to the limited proportion of activated centers (C*). However, as the quantity of C* increases exponentially, the fraction of P undergoes a corresponding decrease. Eventually, the fraction of P becomes small, leading to a deceleration of the reaction rate. The authors examined the impact of reagent ratio and specific dose on the kinetics. It was observed that a higher reagent proportion resulted in greater efficiency in pollutant destruction.

4. Mechanochemical stabilization of the heavy metals

4.1 Heavy metals in residues

Heavy metals are a significant environmental pollutant in combustion/incineration residues, in addition to persistent organic pollutants. These metals can undergo leaching in acidic or water environments, resulting in harm to soils and water. The incineration process results in a redistribution of heavy metal elements due to the varying chemical and physical

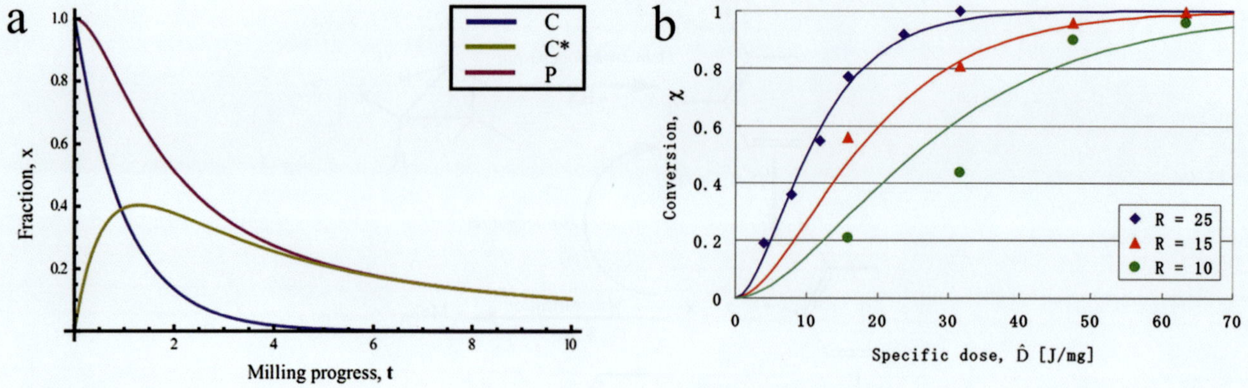

FIGURE 8.3 Numerical solution for the differential equations system (A) model fitting lines of dechlorane plus comilled with CaO, with different reagent ratios (R) (B) (Cagnetta et al., 2016a).

properties of different metals, as well as the impact of active elements such as Cl and S (Lin et al., 2022). Nonvolatile metals such as Co, Cr, Mn, and Ni tend to accumulate in the bottom ash. On the other hand, heavy metals with lower boiling points are enriched in the flue gas and subsequently captured by the filter before being transferred into fly ash. In the past in China, the fly ash was landfilled after chemical/cement stabilization/solidification following the standard for pollution control on the landfill of municipal solid waste (GB16889, 2008). In 2020, the technical specification for pollution control of fly ash from MSWI was launched in China (HJ1134, 2020). The well-treated fly ash could be classified as general solid waste for further recycling if it fulfills the requirement. Table 8.2 shows the heavy metals leaching concentrations in the standards.

Fig. 8.4 displays the concentration and leaching concentration of the main heavy metals found in the fly ash. Zn and Pb exhibit the most significant concentration. Moreover, the leaching concentration of Zn, Pb, Cd, and Cr exceeds the corresponding standard value for leaching toxicity in solid waste (Huang et al., 2022). To mitigate the potential harmful effects on the environment, it is imperative to promptly immobilize the heavy metals.

4.2 Heavy metals immobilization in residues by MC treatment

Research on the mechanochemical immobilization of heavy metals initially focused on soil remediation. Milling can induce and accelerate the reaction between heavy metals and additives, leading to a conversion of species from soluble to insoluble forms, ultimately producing a residue. Nomura et al. (2010) utilized X-ray absorption fine structure spectroscopy to examine the chemical state of Pb before and after undergoing mechanochemical treatment. By introducing CaO as an additive, $PbCl_2$ was successfully converted into the form of Pb_3O_4, which exhibits low solubility in water. They also reported that Pb leaching concentration in fly ash reduced by 93% with the presence of CaO after MC treatment (Nomura et al., 2008, 2010). Furthermore, a crystal transformation of PbO from an orthorhombic to a tetragonal structure was observed, resulting in a decrease in solubility. Additionally, the reduction of $PbCl_2$ and PbO to Pb was verified through analysis of X-ray diffraction patterns. Notably, iron particles abraded from the stainless steel mill jar were found to play a significant role in the reduction process, resulting in low solubility (Nomura et al., 2013b). In the MSWI, Pb was oxidized to Pb_2O_3, Pb_3O_4 and α-PbO as intermediates, leading eventually to the production of β-PbO. Conversely, the milling process realizes the opposite redox reduction reactions, and β-PbO was converted to α-PbO and Pb (Nomura et al., 2013b).

TABLE 8.2 Heavy metals leaching concentration limit (mg/L).

	As	Ba	Be	Cd	Cr	Cr^{6+}	Cu	Hg	Ni	Pb	Se	Zn
HJ 1134[a]	0.3	/	0.005	0.05	1.5	0.5	2	0.005	1	0.5	0.5	5
GB16889[b]	0.3	25	0.02	0.15	4.5	1.5	40	0.05	0.5	0.25	0.1	100

[a]*Extracted by horizontal vibration method according to Chinese Standard HJ 557.*
[b]*Extracted by acetic acid buffer solution method according to Chinese Standard HJ/T 300.*

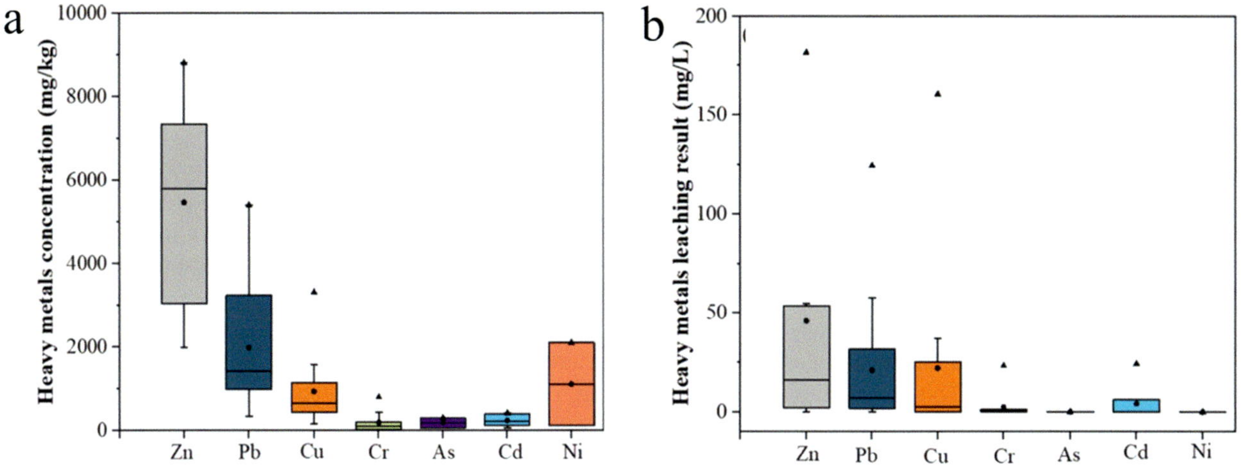

FIGURE 8.4 Boxplots of heavy metal (A) and heavy metals leaching concentration (B) in municipal solid waste incineration fly ash (Lan et al., 2022).

In the milling process, the particle size of fly ash was reduced from tens of microns to several microns. However, after milling for 0.5 h, the rate of particle size reduction gradually decreased, and smaller particles tended to agglomerate, resulting in the formation of larger particles. Subsequently, the milling energy was utilized to disrupt the crystalline structure of the particles. Fragmentation and agglomeration occurred repeatedly during the milling process, leading to the formation of defective surfaces. Thus, aggregates served as a network with a crystalline structure, effectively trapping heavy metals within their structure (Yuan et al., 2021).

The leaching of heavy metals is also highly influenced by the pH level. The MC treatment can elevate the pH of the leachate by causing particle cracking and enhancing the solubility of its components. Consequently, this process leads to an increased leaching of sensitive elements such as Cu, Cr, and Pb, all of which are particularly responsive to pH changes. To further enhance the stabilization efficiency, prewashing is recommended as it effectively eliminates soluble salts and increases the proportion of acid buffering materials in fly ash (Chen et al., 2016). Moreover, the addition of CaO can elevate the pH level, which promotes the formation of metal oxide and carbonate precipitates (Mallampati et al., 2016). This, in turn, reduces the mobility of metals. When CaO reacts with water or atmospheric moisture, it results in the formation of $Ca(OH)_2$, creating a high pH environment that is conducive to the retention of heavy metals. Furthermore, the formation of calcium carbonate also contributes to the immobilization of heavy metals by generating insoluble pozzolanic-like cement.

Wang et al. (2018) studied the stabilization of heavy metals in fly ash from grate furnace incinerator (GFI) and fluidized bed incinerator (FBI) by wet milling. The treatment resulted in a significant decrease of the relative leaching rates of Pb, reducing from 72.64% to 1.12% and 8.92% to 0.02% for the GFI and FBI fly ash, respectively. The European Community Bureau of Reference (BCR) sequential extraction results demonstrated that the heavy metals transferred from water- and acid-soluble and reducible to residual fraction. The water- and acid-soluble heavy metals are easily released under acidic conditions. The reducible form, including amorphous iron-manganese oxide and hydrated oxide, can be released in the reducing conditions. The oxidizing fraction is predominantly bound in an organic form. The residual state, where heavy metals are incorporated into the mineral lattice, is generally resistant to release. The addition of additives also exerts a significant influence on the transformation of heavy metal species. A decrease in the water- and acid-soluble fractions of Cr, Cu, and Pb from 1.39%, 12.3%, and 1.13% to 0.54%, 5.37%, and 0.47% respectively, was observed through ball milling without any additives (Chen et al., 2019b). Subsequently, milling for 10 h with $Ca_3(PO_4)_2$ led to a further reduction in the water- and acid-soluble fractions, which reduced to 0.17%, 0.14%, and 0.12%. The formation of phosphates played a significant role in the increase of the residual fraction of chromium and lead, which rose from 90.1% and 62.2% to 96.6% and 95.8%, respectively.

The additive has a remarkable influence on the stabilization. The reduction of Cr(VI) can potentially be achieved through the utilization of ball milling technique with Na_2S. This method can effectively release Cr(VI) from the residues through high-energy milling and subsequently facilitate its reduction to Cr(III) by Na_2S (Sun et al., 2021). Compared with $CaSO_4$, $Ca_3(PO_4)_2$, and NaH_2PO_4, the mixture of NaH_2PO_4 and CaO exhibited the highest inhibition efficiency for Cr, Cd, Cu, Ni, Pb, and Zn, with percentages of 95.21%, 46.84%, 99.42%, 49.91%, 99.83%, and 99.55%, respectively (Jin et al., 2023). It was found that PO_4^{3+} reacted with Pb to form $Pb_3(PO_4)_2$ and $Pb_xCa_{10-x}(PO_4)_6(OH)_2$, which are precipitates with low solubility and bioavailability (Zhang et al., 2020). In addition, NaH_2PO_4 reacted with calcium present in the fly ash, resulting in the formation of hydroxyapatite (HAP). Through ion exchange, heavy metals replaced calcium ions, leading to the precipitation of stable M-apatite. Simultaneously, CaO provided an alkaline environment that effectively inhibited leaching (Jin et al., 2023).

A comparative analysis of dry milling and wet milling techniques to evaluate their impact on the stabilization of heavy metals in fly ash was conducted (Yuan et al., 2021). Notably, after a period of 10 h of wet milling, the relative leaching rates of Cu, Cr, Pb, Zn, Cd, and Ni decreased by 58.11%, 70.92%, 89.64%, 23.26%, 10.59%, and 30.77%, respectively. Additionally, examining the particle size distribution and scanning electron microscope images, it appears that the presence of water during wet milling may mitigate the agglomeration of fine fly ash particles compared with dry milling. However, no significant difference was observed between the two methods in terms of their effectiveness in stabilizing Cu, Cr, and Pb. Wet milling on the MSWI bottom ash was conducted using citric acid and ethylenediaminetetraacetic acid (EDTA) as additives (Xiang et al., 2022). These acids were found to effectively remove 1%−10% of the heavy metals present in the ash in advance. Following a treatment time of 60 min, the leaching concentrations of Zn, As, and Ni were found to be below the standard values. This wet milling process successfully stabilizes the exchangeable and residual heavy metals, enhancing their reaction with CO_3^{2-} to form stable compounds.

4.3 Stabilization mechanism of heavy metals by MC treatment

The ball milling process subsequently fractures the fly ash, resulting in the production of fine particles and the activation of its surface through the generation of additional defects. This enhances the inherent ability of the fly ash to adsorb heavy metals. In addition to the creation of fresh surfaces, the milling process also facilitates particle aggregation and agglomeration. Consequently, particle breakage yields new surfaces containing dangling bonds that readily participate in chemical reactions. The leachable heavy metals engage with the additives present, leading to the formation of stable precipitates. In essence, the stabilization of heavy metals in fly ash can be achieved through MC treatment with suitable additives via the following mechanisms as shown in Fig. 8.5: (1) entrapment, which involves the encapsulation of heavy metal ions within the fine particles, thus rendering them immobile; (2) adsorption, whereby the milling process significantly enlarges the surface area of the fly ash and induces the generation of defects, subsequently enabling the adsorption and stabilization of heavy metals; (3) diffusion, in which high-energy milling facilitates the migration of heavy metal ions to nonsoluble crystals, thereby forming stable structures; (4) chelation, during which organic compounds act as ligands that react with heavy metals in response to the milling force; and (5) formation of insoluble species, wherein heavy metals react with additives and transition from a mobile to a stable state (Chen et al., 2019b; Dong et al., 2022; Grabias-Blicharz and Franus, 2023).

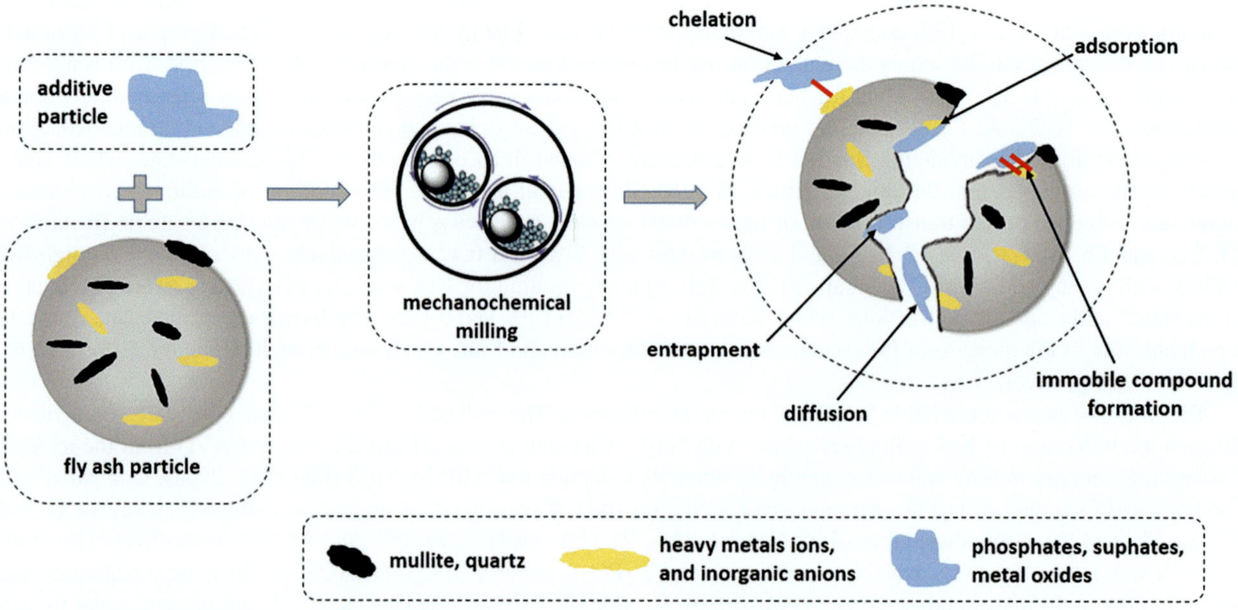

FIGURE 8.5 Mechanisms of immobilization of heavy metals in FA by mechanochemistry (Grabias-Blicharz and Franus, 2023).

5. Mechanical activation of the residues

5.1 Physicochemical properties improvement by MC treatment

Coal fly ash can serve as a supplementary cementitious material in the construction industry. Nevertheless, the extensive recycling of fly ash at a large scale is constrained by its inadequate reactivity. Fly ash is composed of stable crystalline phases, including quartz and mullite, alongside reactive glassy phases. Throughout the reaction process, the reactive phases undertake dissolution, condensation, and polymerization, while the nonreactive fraction acts as an aggregate. The limited reactivity of coal fly ash presents challenges for its widespread application in recycling efforts. Thus, the activation and modification of FAs are necessary to enhance their reactivity. Mechanochemical treatment, being an environmentally friendly approach, has garnered significant attention as a method to activate fly ash.

As discussed in Section 4, the technique of ball milling has the capacity to fracture particles and significantly reduce their size. For example, after 70 min of mechanical treatment, the circulating fluidized combustion fly ash can be reduced to approximately 1 μm in size. Grinding can also raise the pH of the fly ash due to the larger surface area and the presence of free CaO, while simultaneously decreasing the amount of water required for cementitious material preparation (Mukhtar et al., 2023). A similar phenomenon was observed, where just 3 min of grinding resulted in a reduction of particle size from the range of 30 to 200 μm to only a few μm. However, it should be noted that extended milling can cause particle aggregation, leading to an increase in their size (Molnár et al., 2014). The researchers also observed a substantial reduction in the average particle size of a coal fly ash sample (Rajak et al., 2017). They achieved this reduction by subjecting the sample to 2.5 h of grinding using a planetary ball mill. Prior to grinding, the average particle size was measured to be 134.5 μm, but after the grinding process, it decreased to 13.4 μm. The grinding process resulted in an increase in the surface area of the coal fly ash from 7.31 to 28.08 m^2/g. Additionally, the study found that the percentage of crystallinity of the fly ash and the crystallite of quartz decreased by 26.12% and 40.1%, respectively. Moreover, the milling can significantly enhance the amorphous content of fly ash with an increase from 73% to 83% (Molnár et al., 2014).

To mitigate the accumulation of milled particles, the addition of ethyl acetate as a surfactant was incorporated into the milling procedure (Patil et al., 2015). Through the utilization of high-energy ball milling, the mechanochemical activation of fly ash resulted in an average particle size of 329 nm and a specific surface area measuring at 8.73 m^2/g. Furthermore, the milling process caused a reduction in the crystallite size of the quartz phase from 28 to 7.7 nm after 48 h. The Fourier transform infrared spectrometer (FTIR) analysis exhibited an expansion in the Si-O-Si stretching vibrations, along with an increase in the -OH terminal group on the surface of the treated fly ash due to the surfactant. Moreover, the contact angle test demonstrated a diminished wettability. These micelles introduced steric repulsive forces, disentangling the agglomerated milled fly ash.

Mechanochemical treatment offers several advantages in terms of reactivity enhancement, including an increase in surface area, a reduction in average particle size, a decrease in crystallite size, and a modification of the crystallite type. To optimize reactivity, the duration of the milling process, the rotational speed, and the inclusion of surfactants play crucial roles.

5.2 Utilization of activated residues as construction material

Cement, being an essential component in construction, is accountable for the emission of approximately 1.5 billion tons of CO_2 per year (Amran et al., 2020). Introducing partial cement replacement during concrete production has the potential to decrease CO_2 emissions and contribute to the implementation of a circular economy model (Sun et al., 2022). By employing the MC treatment to enhance reactivity, a more substantial proportion of cement can be substituted with fly ash, consequently enhancing the performance of the construction material.

The workability effect of fly ash on cementitious materials has garnered significant attention. The addition of fly ash can substantially reduce the consistency of the cement paste, whereas the incorporation of milled fly ash can alleviate this decline in consistency (Hamzaoui et al., 2016). To assess the workability, the flow diameter of raw fly ash and milled ultrafine fly ash was measured, resulting in an equivalent flow score for both standard mortar and ultrafine fly ash—modified mortars. Furthermore, the particle size influences the variation in workability of masonry mortars. The addition of milled fly ash can enhance the flow characteristics of mortars, with an optimal particle size of 2.5 μm (Li and Wu, 2005). The mechanical performance findings concerning modified cement mortars demonstrate the successful replacement of cements with milled fly ashes up to a maximum of 50%. This substitution leads to significantly high levels of compressive strength, reaching up to 70 MPa, owing to the enhanced pozzolanic reactivity (Hamzaoui et al., 2016).

The fly ash produced from MSWI is considered hazardous waste, which has impeded its usage in various industries. Nonetheless, numerous studies have explored the potential utilization of this fly ash. Utilizing the MC treatment holds significant benefits, as it not only effectively eliminates dioxins and stabilizes heavy metals but also enhances reactivity. Chlorine can be readily eliminated from fly ash through wet milling, resulting in a profound chlorine removal. Furthermore, aluminum present in the fly ash has the potential to react with the alkaline environment in the wet milling and release the hydrogen in advance, effectively resolving the expansion issue. Consequently, the compressive strength of the 28-day mortar incorporating a 20% addition of both raw and milled fly ash showcased strengths of 25.5 and 56.9 MPa, respectively (Chen et al., 2019a). In addition to its pozzolanic effect, the filling effect contributes to the enhancement of compressive strength during the early stage. The particle size of the raw fly ash was reduced to a few micrometers, significantly smaller than that of the cement. Consequently, the fly ash particles are capable of filling the pores within the cement, thereby further enhancing the overall performance of the mortar (Jin et al., 2023).

5.3 Utilization of activated residues for chemical synthesis

Fly ash possesses a multitude of valuable elements and exhibits a diverse structure. Studies have primarily concentrated on its utilization in construction materials, encompassing supplementary cementitious materials, geopolymer, and lightweight aggregates. Additionally, researchers have explored its potential in synthesizing environmentally friendly chemicals, notably adsorbents and catalysts. Moreover, investigations have been conducted to evaluate the effectiveness of mechanochemical treatment in the realm of green chemical synthesis.

Mechanochemical activation can significantly improve the adsorption characteristics of the coal fly ash with a high energy ball milling. The best performing sample for the adsorption of phenol from aqueous solution was obtained from the fly ash with high carbon content treated for 4 h in N_2 atmosphere. This material had more favorable adsorption isotherms, improved specific adsorption capacity, and faster adsorption rates compared with powdered activated carbon (Stellacci et al., 2009). Mechanochemical treatment in combination with bromide modification can be used to enhance the mercury adsorption capabilities of fly ash sorbents (Zhang et al., 2017, 2019; Hu et al., 2020). The process involved grinding, which led to a reduction in particle size and an increase in surface area, thereby exposing more unburned carbon and enhancing the adsorption of mercury. Additionally, the addition of bromide in the fly ash facilitated the oxidation of Hg^0 and improved the sorbent's capacity to capture mercury, even at lower temperatures. Subsequently, the researchers tested the newly developed adsorbents on a 300 and 1000 MW plant to reduce mercury emissions. The findings revealed a significant reduction in the final concentration of mercury to 1.15 $\mu g/m^3$, resulting in an overall Hg removal efficiency of 94.8% (Zhang et al., 2017; Wang et al., 2019).

The process of mechanochemical activation holds promise for the production of catalysts using fly ash. In a study conducted by Czuma et al. (2022), nickel was loaded onto the surface of fly ash–based zeolite through ball milling. Notably, at a lower methanation temperature of 350°C, a significant conversion of approximately 55%–58% of CO_2 into methane was observed. This outcome can be attributed to the increased surface area of micropores and subsequent dispersion of nickel species within the ball-milled samples. The mechanism of ball milling promotes the dispersal of nickel into smaller particles, which enhances CO_2 adsorption capacity.

6. Challenges and perspectives

In the past century, mechanochemical treatment of the combustion/incineration residues as a green and eco-friendly method has been investigated. This approach aims to efficiently eliminate persistent organic pollutants, stabilize heavy metals, and enhance residue reactivity. Extensive research has focused on identifying key milling parameters and understanding the mechanisms for pollutant control and reactivity improvement. Despite these advancements, mechanochemical treatment for residue recycling in industries has not yet been widely utilized due to safety concerns and economic viability.

Firstly, previous studies primarily utilized laboratory planetary ball mills for mechanochemical treatment of residues. As discussed in Section 2, the equipment utilized plays a crucial role in pollutant destruction by influencing impact and shear forces. The operating conditions obtained from planetary ball mill may be not suitable for other mills. Secondly, the operating conditions derived from laboratory studies are not readily applicable in industrial settings. The high quantities of reagent additions or prolonged operating times may pose challenges when operating at a large scale. Thirdly, it should be noted that distinct reagents are required for the destruction of organic pollutants as opposed to the stabilization of heavy metals. In addition, the use of surfactants is essential to mitigate the agglomeration of fine particles. It is challenging to achieve simultaneous control of organic pollutants and heavy metals through the employment of a single reagent in mechanochemical treatment. However, the interactions between these reagents and the potential synergistic effects of their combinations have been infrequently investigated. Fourthly, the precise mechanism of the reaction process in ball milling remains elusive when employing diverse reagents and varying conditions. To address this knowledge gap, it is imperative to incorporate novel characterization techniques or numerical calculations to elucidate the reaction mechanism properly. Fifthly, the economic feasibility and environmental impact of the MC treatment of residues are crucial considerations for its practical implementation. It is vital to conduct a life cycle assessment to thoroughly evaluate the treatment process.

7. Conclusions

This chapter examines the mechanochemical treatment of combustion/incineration residues, highlighting recent advancements in the destruction of persistent organic pollutants, the stabilization of heavy metals, and the improvement of reactivity. The key findings can be summarized as follows:

1. Mechanochemical treatment is an environmentally friendly and sustainable technology that is attracting significant interest for controlling pollutants in combustion/incineration residues.
2. Milling equipment is crucial for facilitating the mechanochemical reaction, and it is important to investigate appropriate tools for both laboratory and large-scale applications.
3. The degradation of chlorinated aromatics through dichlorination has been established as a notable pathway in the milling process. The destruction of chlorinated organic pollutants involves several steps, such as reagent adsorption and activation, dehalogenation/dehydrohalogenation, and carbonation.
4. For heavy metals stabilization, the selection of appropriate reagents is crucial, as the formation of immobilized metal species plays a significant role in this process.
5. Reactivity holds significant importance in the successful recycling of residues. MC treatment not only promotes the reuse for construction materials but also enhances their transformation from waste into valuable products.

In conclusion, despite the multiple uncertainties, the implementation of MC treatment holds promise for recycling combustion/incineration residues. This approach contributes to the attainment of both sustainability objectives and the circular economy model.

Acknowledgment

This work is financially supported by the “Pioneer” and “Leading Goose” R&D Program of Zhejiang (2022C03056) and National Natural Science Foundation of China(52206176).

References

Ahmaruzzaman, M., 2010. A review on the utilization of fly ash. Prog. Energy Combust. Sci. 36, 327−363.

Amran, Y.M., Alyousef, R., Alabduljabbar, H., El-Zeadani, M., 2020. Clean production and properties of geopolymer concrete: a review. J. Clean. Prod. 251, 119679.

Aresta, M., Caramuscio, P., De Stefano, L., Pastore, T., 2003. Solid state dehalogenation of PCBs in contaminated soil using NaBH4. Waste Manag. 23, 315−319.

Bulgakov, V., Pascuzzi, S., Ivanovs, S., Kaletnik, G., Yanovich, V., 2018. Angular oscillation model to predict the performance of a vibratory ball mill for the fine grinding of grain. Biosyst. Eng. 171, 155−164.

Cagnetta, G., Huang, J., Wang, B., Deng, S., Yu, G., 2016a. A comprehensive kinetic model for mechanochemical destruction of persistent organic pollutants. Chem. Eng. J. 291, 30−38.

Cagnetta, G., Robertson, J., Huang, J., Zhang, K., Yu, G., 2016b. Mechanochemical destruction of halogenated organic pollutants: a critical review. J. Hazard Mater. 313, 85−102.

Chen, Z., Lu, S., Mao, Q., Buekens, A., Chang, W., Wang, X., Yan, J., 2016. Suppressing heavy metal leaching through ball milling of fly ash. Energies 9, 524.

Chen, Z., Lu, S., Mao, Q., Buekens, A., Tang, M., Wang, Y., Ding, J., Wu, A., Yan, J., 2018. Accelerating dechlorination using calcium oxide with the assistance of metallic aluminum in mechanochemical reaction. Chem. Lett. 47, 40−43.

Chen, Z., Lu, S., Tang, M., Ding, J., Buekens, A., Yang, J., Qiu, Q., Yan, J., 2019a. Mechanical activation of fly ash from MSWI for utilization in cementitious materials. Waste Manag. 88, 182−190.

Chen, Z., Lu, S., Tang, M., Lin, X., Qiu, Q., He, H., Yan, J., 2019b. Mechanochemical stabilization of heavy metals in fly ash with additives. Sci. Total Environ. 694, 133813.

Chen, Z., Mao, Q., Lu, S., Buekens, A., Xu, S., Wang, X., Yan, J., 2017. Dioxins degradation and reformation during mechanochemical treatment. Chemosphere 180, 130−140.

Chen, Z., Tang, M., Lu, S., Buekens, A., Ding, J., Qiu, Q., Yan, J., 2019c. Mechanochemical degradation of PCDD/Fs in fly ash within different milling systems. Chemosphere 223, 188−195.

Czuma, N., Samojeden, B., Zarebska, K., Motak, M., Da Costa, P., 2022. Modified fly ash, a waste material from the energy industry, as a catalyst for the CO2 reduction to methane. Energy 243, 122718.

Dong, D., Zhang, Y., Shan, M., Yin, T., Wang, T., Wang, J., Gao, W., 2022. Application of mechanochemical technology for removal/solidification pollutant and preparation/recycling energy storage materials. J. Clean. Prod. 348, 131351.

Dou, X., Ren, F., Nguyen, M.Q., Ahamed, A., Yin, K., Chan, W.P., Chang, V.W.-C., 2017. Review of MSWI bottom ash utilization from perspectives of collective characterization, treatment and existing application. Renew. Sustain. Energy Rev. 79, 24–38.

GB16889, 2008. Standard for Pollution Control on the Landfill Site of Municipal Solid Waste. Ministry of Environmental Protection, China.

Gomollón-Bel, F., 2019. Ten chemical innovations that will change our world: IUPAC identifies emerging technologies in chemistry with potential to make our planet more sustainable. Chem. Int. 41, 12–17.

Grabias-Blicharz, E., Franus, W., 2023. A critical review on mechanochemical processing of fly ash and fly ash-derived materials. Sci. Total Environ. 860, 160529.

Hamzaoui, R., Bouchenafa, O., Guessasma, S., Leklou, N., Bouaziz, A., 2016. The sequel of modified fly ashes using high energy ball milling on mechanical performance of substituted past cement. Mater. Des. 90, 29–37.

HJ1134, 2020. Technical Specification for Pollution Control of Fly-Ash from Municipal Solid Waste Incineration. Ministry of Environmental Protection, China.

Hu, J., Geng, X., Duan, Y., Zhao, W., Zhu, M., Ren, S., 2020. Effect of mechanical–chemical modification process on mercury removal of bromine modified fly ash. Energy Fuels 34, 9829–9839.

Hu, J., Huang, Z., Yu, J., 2019. Highly-effective mechanochemical destruction of hexachloroethane and hexachlorobenzene with Fe/Fe3O4 mixture as a novel additive. Sci. Total Environ. 659, 578–586.

Huang, B., Gan, M., Ji, Z., Fan, X., Zhang, D., Chen, X., Sun, Z., Huang, X., Fan, Y., 2022. Recent progress on the thermal treatment and resource utilization technologies of municipal waste incineration fly ash: a review. Process Saf. Environ. Protect. 159, 547–565.

Huang, T.Y., Chiueh, P.T., Lo, S.L., 2017. Life-cycle environmental and cost impacts of reusing fly ash. Resour. Conserv. Recycl. 123, 255–260.

Jin, L., Chen, M., Wang, Y., Peng, Y., Yao, Q., Ding, J., Ma, B., Lu, S., 2023. Utilization of mechanochemically pretreated municipal solid waste incineration fly ash for supplementary cementitious material. J. Environ. Chem. Eng. 11, 109112.

Kurda, R., Silvestre, J.D., de Brito, J., 2018. Toxicity and environmental and economic performance of fly ash and recycled concrete aggregates use in concrete: a review. Heliyon 4, e00611.

Lan, T., Meng, Y., Ju, T., Chen, Z., Du, Y., Deng, Y., Song, M., Han, S., Jiang, J., 2022. Synthesis and application of geopolymers from municipal waste incineration fly ash (MSWI FA) as raw ingredient - a review. Resour. Conserv. Recycl. 182, 106308.

Li, B., Deng, Z., Wang, W., Fang, H., Zhou, H., Deng, F., Huang, L., Li, H., 2017. Degradation characteristics of dioxin in the fly ash by washing and ball-milling treatment. J. Hazard Mater. 339, 191–199.

Li, G., Wu, X., 2005. Influence of fly ash and its mean particle size on certain engineering properties of cement composite mortars. Cement Concr. Res. 35, 1128–1134.

Li, W., Wang, W., Wu, D., Yang, S., Fang, H., Sun, S., 2023. Mechanochemical treatment with red mud added for heavy metals solidification in municipal solid waste incineration fly ash. J. Clean. Prod. 398, 136642.

Lin, S., Jiang, X., Zhao, Y., Yan, J., 2022. Disposal technology and new progress for dioxins and heavy metals in fly ash from municipal solid waste incineration: a critical review. Environ. Pollut. 311, 119878.

Liu, X., Li, Y., Zeng, L., Li, X., Chen, N., Bai, S., He, H., Wang, Q., Zhang, C., 2022. A review on mechanochemistry: approaching advanced energy materials with greener force. Adv. Mater. 34, 2108327.

Lomovskiy, I., Bychkov, A., Lomovsky, O., Skripkina, T., 2020. Mechanochemical and size reduction machines for biorefining. Molecules 25, 5345.

Lou, Z., Song, L., Liu, W., Chen, H., Yan, C., Yu, J., Xu, X., 2023. Sulfite as a green co-milling agent for mechanochemical destruction of polychlorinated aromatics: working mechanism and structural dependence. ACS ES&T Eng. 3, 944–954.

Lu, S., Huang, J., Peng, Z., Li, X., Yan, J., 2012. Ball milling 2,4,6-trichlorophenol with calcium oxide: dechlorination experiment and mechanism considerations. Chem. Eng. J. 195–196, 62–68.

Ma, W., Fang, Y., Chen, D., Chen, G., Xu, Y., Sheng, H., Zhou, Z., 2017. Volatilization and leaching behavior of heavy metals in MSW incineration fly ash in a DC arc plasma furnace. Fuel 210, 145–153.

Mallampati, S.R., Lee, B.H., Mitoma, Y., Simion, C., 2016. Dual mechanochemical immobilization of heavy metals and decomposition of halogenated compounds in automobile shredder residue using a nano-sized metallic calcium reagent. Environ. Sci. Pollut. Res. Int. 23, 22783–22792.

Mao, Y., Wu, H., Wang, W., Jia, M., Che, X., 2020. Pretreatment of municipal solid waste incineration fly ash and preparation of solid waste source sulphoaluminate cementitious material. J. Hazard Mater. 385, 121580.

Mitoma, Y., Miyata, H., Egashira, N., Simion, A.M., Kakeda, M., Simion, C., 2011. Mechanochemical degradation of chlorinated contaminants in fly ash with a calcium-based degradation reagent. Chemosphere 83, 1326–1330.

Molnár, Z., Kristály, F., Mucsi, G., 2014. Mechanical activation of deposited brown coal fly ash in stirred media mill. Acta Phys. Pol., A 126, 988–993.

Muir, B., Sobczyk, M., Bajda, T., 2021. Fundamental features of mesoporous functional materials influencing the efficiency of removal of VOCs from aqueous systems: a review. Sci. Total Environ. 784, 147121.

Mukhtar, H.A., Abbas, I.S., Canakci, H., 2023. Influence of mechanochemical activation on the rheological, fresh and mechanical properties of one-part geopolymer grout. Adv. Cement Res. 35, 96–110.

Nakayama, K., 2007. Microplasma generated in a gap of sliding contact. J. Vac. Soc. Japan. 49, 618–623.

NBSC, 2022. China Statistical Yearbook. China Statistical Press, Beijing.

Nie, Y., Shi, J., He, Z., Zhang, B., Peng, Y., Lu, J., 2022. Evaluation of high-volume fly ash (HVFA) concrete modified by metakaolin: technical, economic and environmental analysis. Powder Technol. 397, 117121.

Nomura, Y., Aono, S., Arino, T., Yamamoto, T., Terada, A., Noma, Y., Hosomi, M., 2013a. Degradation of polychlorinated naphthalene by mechanochemical treatment. Chemosphere 93, 2657–2661.

Nomura, Y., Fujiwara, K., Akihiko, T., Naski, S., Hosomi, M., 2013b. An immobilisation mechanism for lead in fly ash subjected to mechanochemical treatment. Int. J. Environ. Waste Manag. 12, 14.

Nomura, Y., Fujiwara, K., Takada, M., Nakai, S., Hosomi, M., 2008. Lead immobilization in mechanochemical fly ash recycling. J. Mater. Cycles Waste Manag. 10, 14–18.

Nomura, Y., Fujiwara, K., Terada, A., Nakai, S., Hosomi, M., 2010. Prevention of lead leaching from fly ashes by mechanochemical treatment. Waste Manag. 30, 1290–1295.

Nomura, Y., Fujiwara, K., Terada, A., Nakai, S., Hosomi, M., 2012. Mechanochemical degradation of γ-hexachlorocyclohexane by a planetary ball mill in the presence of CaO. Chemosphere 86, 228–234.

Nomura, Y., Nakai, S., Hosomi, M., 2005. Elucidation of degradation mechanism of dioxins during mechanochemical treatment. Environ. Sci. Technol. 39, 3799–3804.

Olie, K., Vermeulen, P., Hutzinger, O., 1977. Chlorodibenzo-p-dioxins and chlorodibenzofurans are trace components of fly ash of some municipal incinerators in The Netherlands. Chemosphere 6, 455–459.

Pan, S., Yao, Q., Cai, W., Peng, Y., Luo, Y., Wang, Z., Jiang, C., Li, X., Lu, S., 2022. Characterization of dioxins and heavy metals in chelated fly ash. Energies 15, 4868.

Pan, Y., Yang, L., Zhou, J., Liu, J., Qian, G., Ohtsuka, N., Motegi, M., Oh, K., Hosono, S., 2013. Characteristics of dioxins content in fly ash from municipal solid waste incinerators in China. Chemosphere 92, 765–771.

Patil, A.G., Shanmugharaj, A.M., Anandhan, S., 2015. Interparticle interactions and lacunarity of mechano-chemically activated fly ash. Powder Technol. 272, 241–249.

Peng, Y., Buekens, A., Tang, M., Lu, S., 2018. Mechanochemical treatment of fly ash and de novo testing of milled fly ash. Environ. Sci. Pollut. Res. Int. 25, 19092–19100.

Peng, Y., Chen, J., Lu, S., Huang, J., Zhang, M., Buekens, A., Li, X., Yan, J., 2016. Chlorophenols in municipal solid waste incineration: a review. Chem. Eng. J. 292, 398–414.

Peng, Z., Ding, Q., Sun, Y., Jiang, C., Gao, X., Yan, J., 2010. Characterization of mechanochemical treated fly ash from a medical waste incinerator. J. Environ. Sci. (China) 22, 1643–1648.

Rajak, D.K., Raj, A., Guria, C., Pathak, A.K., 2017. Grinding of Class-F fly ash using planetary ball mill: a simulation study to determine the breakage kinetics by direct- and back-calculation method. S. Afr. J. Chem. Eng. 24, 135–147.

Rowlands, S., Hall, A., McCormick, P., Street, R., Hart, R., Ebell, G., Donecker, P., 1994. Destruction of toxic materials. Nature 367, 223.

Sitotaw, Y.W., Habtu, N.G., Gebreyohannes, A.Y., Nunes, S.P., Van Gerven, T., 2021. Ball milling as an important pretreatment technique in lignocellulose biorefineries: a review. Biomass Convers. Biorefin. 1–24.

Stellacci, P., Liberti, L., Notarnicola, M., Bishop, P.L., 2009. Valorization of coal fly ash by mechano-chemical activation: Part I. Enhancing adsorption capacity. Chem. Eng. J. 149, 11–18.

Sun, C., Wang, L., Lin, X., Lu, S., Huang, Q., Yan, J., 2022. Low-carbon stabilization/solidification of municipal solid waste incineration fly ash. Waste Dispos. Sustain. Energy 4, 69–74.

Sun, Y., Du, Y., Lan, J., Zhan, W., Zhang, T.C., 2021. A new method (ball milling and sodium sulfide) for mechanochemical treatment of soda ash chromite ore processing residue. J. Hazard Mater. 415, 125601.

Tian, Y., Dai, S., Wang, J., 2023. Environmental standards and beneficial uses of waste-to-energy (WTE) residues in civil engineering applications. In: Waste Disposal & Sustainable Energy, pp. 1–28.

Wang, H., Hwang, J., Huang, J., Xu, Y., Yu, G., Li, W., Zhang, K., Liu, K., Cao, Z., Ma, X., Wei, Z., Wang, Q., 2017. Mechanochemical remediation of PCB contaminated soil. Chemosphere 168, 333–340.

Wang, S., Zhang, Y., Gu, Y., Wang, J., Yu, X., Wang, T., Sun, Z., liu, Z., Romero, C.E., Pan, W.-p., 2019. Coupling of bromide and on-line mechanical modified fly ash for mercury removal at a 1000 MW coal-fired power plant. Fuel 247, 179–186.

Wang, W., Gao, X., Li, T., Cheng, S., Yang, H., Qiao, Y., 2018. Stabilization of heavy metals in fly ashes from municipal solid waste incineration via wet milling. Fuel 216, 153–159.

Xiang, J., Qiu, J., Li, Z., Chen, J., Song, Y., 2022. Eco-friendly treatment for MSWI bottom ash applied to supplementary cementing: mechanical properties and heavy metal leaching concentration evaluation. Construct. Build. Mater. 327, 127012.

Xue, Y., Liu, X., 2021. Detoxification, solidification and recycling of municipal solid waste incineration fly ash: a review. Chem. Eng. J. 420, 130349.

Yan, J.H., Peng, Z., Lu, S.Y., Li, X.D., Ni, M.J., Cen, K.F., Dai, H.F., 2007. Degradation of PCDD/Fs by mechanochemical treatment of fly ash from medical waste incineration. J. Hazard Mater. 147, 652–657.

Yao, Z.T., Ji, X.S., Sarker, P.K., Tang, J.H., Ge, L.Q., Xia, M.S., Xi, Y.Q., 2015. A comprehensive review on the applications of coal fly ash. Earth Sci. Rev. 141, 105–121.

Yu, S., Du, B., Baheiduola, A., Geng, C., Liu, J., 2020. HCB dechlorination combined with heavy metals immobilization in MSWI fly ash by using n-Al/CaO dispersion mixture. J. Hazard Mater. 392, 122510.

Yuan, Q., Zhang, Y., Wang, T., Wang, J., 2023a. Characterization of heavy metals in fly ash stabilized by carbonation with supercritical CO2 coupling mechanical force. J. CO2 Util. 67, 102308.

Yuan, Q., Zhang, Y., Wang, T., Wang, J., Romero, C.E., 2021. Mechanochemical stabilization of heavy metals in fly ash from coal-fired power plants via dry milling and wet milling. Waste Manag. 135, 428–436.

Yuan, W., Xie, J., Wang, X., Huang, Q., Huang, K., 2023b. Mechanochemical remediation of soil contaminated with heavy metals and persistent organic pollutants by ball milling with nZVI-CaO-Ca3(PO4)2 additives. Chem. Eng. J. 466, 143109.

Zhang, T., Huang, J., Zhang, W., Yu, Y., Deng, S., Wang, B., Yu, G., 2013. Coupling the dechlorination of aqueous 4-CP with the mechanochemical destruction of solid PCNB using Fe–Ni–SiO2. J. Hazard Mater. 250, 175–180.

Zhang, Y., Mei, D., Wang, T., Wang, J., Gu, Y., Zhang, Z., Romero, C.E., Pan, W.P., 2019. In-situ capture of mercury in coal-fired power plants using high surface energy fly ash. Environ. Sci. Technol. 53, 7913–7920.

Zhang, Y., Zhang, Z., Liu, Z., Norris, P., Pan, W.-p., 2017. Study on the mercury captured by mechanochemical and bromide surface modification of coal fly ash. Fuel 200, 427–434.

Zhang, Z., Yuan, W., Li, P., Song, Q., Wang, X., Xu, W., Zhu, X., Zhang, Q., Yue, J., Bai, J., Wang, J., 2020. Mechanochemical immobilization of lead contaminated soil by ball milling with the additive of Ca(H(2)PO(4))(2). Chemosphere 247, 125963.

Chapter 9

Washing, electrochemical, and carbonation treatment of combustion and incineration residues

Gang Huang[1], Miao Lu[1], Kang Liu[2], Lei Wang[1] and Jianhua Yan[1]

[1]State Key Laboratory of Clean Energy Utilization, Zhejiang University, Hangzhou, China; [2]Department of Civil and Environmental Engineering, The Hong Kong Polytechnic University, Hong Kong, China

1. Introduction

The accelerating urbanization and improving living standards have been accompanied by a steady increase in the production of municipal solid wastes. Moreover, the negative impact of these wastes on the environment has become a growing concern among local communities (Li et al., 2016). To mitigate this problem, incineration has emerged as an effective solution, capable of reducing waste volume by up to 90% and generating heat for power or steam production (Quina et al., 2008; Arena, 2012). Nevertheless, incineration also leads to the generation of inorganic pollutants, such as incineration bottom ash and incineration fly ash, which require immediate treatment to prevent further pollution (Sabbas et al., 2003; Cappai et al., 2012). The treatment of these by-products has therefore become a crucial issue. In response to the challenges posed by municipal solid waste incineration (MSWI) residues, a range of technologies have been developed to address their hazardous characteristics and promote reuse (Quina et al., 2008; Bogush et al., 2019; Chen et al., 2022a). Among these, water washing, accelerated carbonation, and electrokinetic remediation are particularly effective in reducing leaching toxicity and enhancing the reuse efficiency of these residues.

Washing treatment is a cost-effective and straightforward approach that can reduce the levels of heavy metals in incineration residues by adjusting pH or adding additives, consequently decreasing their leachabilities (Jadhav et al., 2018). Accelerated carbonation treatment, on the other hand, mimics natural mineral weathering processes and is recognized as a solidification/stabilization technique that immobilizes heavy metals through encapsulation, chemical reactions, precipitation, adsorption, and speciation transformations (Yuan et al., 2023). Meanwhile, electrokinetic remediation is carried out through electrodialysis, electromigration, and electrophoresis to purify contaminated media by using low direct current or potential to stimulate pollutant migration toward the electrode chamber (Hu et al., 2015). Despite their different underlying principles, the primary goal of these three methods is to treat the soluble components in incineration residues. In this chapter, we will comprehensively review these three approaches, examining the principles underlying each method, their respective applications and influencing factors, as well as the auxiliary additives and technologies utilized to optimize their effectiveness.

2. Washing treatment of combustion and incineration residues

Washing treatment, including water and acid washing, is a widely used method for treating residues arising from municipal solid wastes, such as incineration bottom ash, incineration fly ash, and air pollution control residues (which come from flue gas cleaning and are usually considered as a type of incineration fly ash) (Jadhav et al., 2018; Phua et al., 2019; Liu et al., 2021; Sun and Yi, 2021). Additionally, MSWI fly ash presents a higher risk and demands specific treatment protocols when compared with other MSWI by-products. So, this section will focus primarily on the washing treatment of MSWI fly ash. Table 9.1 summarizes the washing treatment discussed in the current research.

Treatment and Utilization of Combustion and Incineration Residues. https://doi.org/10.1016/B978-0-443-21536-0.00043-5

TABLE 9.1 Summary of washing treatment for MSWI fly ash.

Liquid-to-solid ratio	Duration	Auxiliary means	Specific performance	References
3, 5, 10, 20, 50	10 min	/	Pb, Zn, Cu, and Cr showed a significant decreasing trend within a liquid-to-solid ratio range of 3–50.	Wang et al. (2010)
2, 3, 5,10	30 min	/	Chlorides and sulfates decreased by 80%–100%.	Colangelo et al. (2012)
8	5 h	Stir in an agitation apparatus at 200 ± 2 rpm	Zn, Pb, and Cu decreased by 33.9%, 22.7%, and 28.7%, respectively.	Wang et al. (2015)
3	5 min	/	More than 70% of chlorides and nearly 25% of sulfates were removed.	Yang et al. (2017)
8	40 min	30 rpm	Pb decreased by 70% and Cr increased by 30%.	Liang et al. (2020)
3	Stir three times for 5, 10, and 10 min		Double and triple washing could remove 88.0% and 95.5% of chlorine.	Wang et al. (2023)

"/" means nothing or not mentioned.

2.1 Washing with water

Wet pretreatment is a cost-effective and widely used method in the treatment of MSWI fly ash. Multiple studies have systematically examined the impact of using the water washing process to remove chlorides, sulfates, and heavy metals, with the ultimate goal of facilitating better reuse and disposal of MSWI fly ash (Chen et al., 2016; Yang et al., 2017; Bogush et al., 2019).

After the postwashing of MSWI fly ash, the leaching concentrations of all heavy metals, except chromium, decreased substantially. The concentration of Pb went down by 70% (exceeding the limit value), whereas the concentration of Cr increased by almost 30% (approaching the limit value) (Liang et al., 2020). This increase can be attributed to the accumulation of insoluble Cr-containing compounds caused by the weight loss of raw MSWI fly ash (Wang et al., 2015). The decrease of the other heavy metals can be attributed to the removal of a part of soluble forms in washing. Except for Pb and Cr, the heavy metals leaching concentrations were well below the limit value. As a result, subsequent chemical treatment of washed MSWI fly ash would be focused on the immobilization of Cr and Pb.

The efficacy of washing treatment in removing targeted compounds was found to be significantly influenced by the liquid-to-solid (L/S) ratio and the number of washing steps, particularly for chloride removal (Wang et al., 2010a,b). In cases where the L/S ratio is not optimal, there is a risk of producing a significant amount of wastewater with high concentrations of chlorides (Chen et al., 2016). The effectiveness of washing conditions on MSWI fly ash processing can vary depending on the specific study. While some studies suggest that increasing the L/S ratio can have a substantial influence on extraction yield (Wang et al., 2010b), others have found that significant effects on extraction yield can be achieved at lower L/S ratios (Colangelo et al., 2012; Yang et al., 2017). Researchers have investigated the effects of double or triple washing on heavy metals in MSWI fly ash (Wang et al., 2023). These results showed that double washing reduced the chlorine content from 29.1 wt% to 3.49 wt%, achieving a removal percentage of 88.0%. Triple washing further decreased the chlorine content to 1.30 wt% with a total removal percentage of 95.5%. Moreover, to investigate the effect of temperature on washing treatment, Wang et al. (2023) conducted a third water washing in a water bath at 80°C, which resulted in a chlorine concentration of 1.19 wt%.

The use of water washing treatment can efficiently remove some of the soluble chlorine from MSWI fly ash, thus allowing the washed ash to be applied as an additive for cement clinker production. Mao et al. (2020) suggested utilizing water-washed MSWI fly ash, flue-gas desulfurization gypsum, and aluminum ash to prepare sulfoaluminate cementitious materials. By including MSWI fly ash as the primary raw material (up to 35 wt%), high-quality sulfoaluminate cement clinker can be produced from a single solid waste source.

In addition to heavy metals and soluble chlorine, important steps for the safe treatment of MSWI fly ash include the degradation of dioxins. However, compared with the raw MSWI fly ash, the concentration of dioxins and PCDD/Fs homologs in water-washed MSWI fly ash increased substantially (Huang et al., 2023a). This is mainly due to the washing away of soluble substances in the MSWI fly ash (chlorine content less than 1.0 wt%), while dioxins and PCDD/Fs

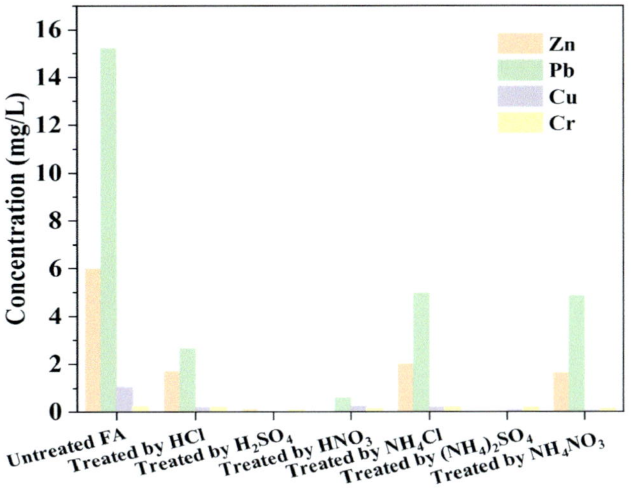

FIGURE 9.1 Heavy metal leaching concentrations of MSWI fly ash after washing with a strong or weak acid (Kang et al., 2021).

congeners are extremely persistent and are not easily dissolved by water. Instead, they tend to combine with organic matter, oils, and other organic substances in MSWI fly ash to form compounds that are not easily decomposed by water, resulting in a relative increase in their residual content. Therefore, proper treatment of MSWI fly ash before or after washing is still necessary.

2.2 Washing with chemical additives

While water-washing treatment can reduce the chlorine content of MSWI fly ash, it has limitations in the removal of heavy metals and other pollutants (as demonstrated in Section 2.1). Thus, it is necessary to consider incorporating chemical substances, such as acid and alkali in the washing process to improve the removal efficiency of heavy metals.

The influence of incorporating acids during the washing treatment process has also been studied. Kang et al. (2021) experimented with strong and weak acid solutions (HCl, H_2SO_4, HNO_3, NH_4Cl, $(NH_4)_2SO_4$, and NH_4NO_3) during washing of MSWI fly ash. The experimental results were shown in Fig. 9.1. After washing treatment, the concentration of Zn, Pb, Cu, and Cr in MSWI fly ash was reduced. Some heavy metals such as Sn were not detected in any solution, possibly due to their strong bond with oxygen atoms. For weak acid washing treatment, the concentration of heavy metals was higher compared with strong acid treatments, especially for Pb. This is because these components were not transformed in the presence of NH_4^+. Moreover, their findings revealed that $(NH_4)_2SO_4$ was the most effective at leaching heavy metals. Additionally, researchers have observed that adding acid to the washing process can effectively remove precipitates that are not quickly soluble during water washing, such as calcium carbonate and calcium hydroxide (Wang et al., 2015).

The change in the overall heavy metal content of MSWI fly ash after washing pretreatment is not solely reliant on the elimination of soluble salts and acid-soluble precipitates but also on the chemical type of heavy metals. Additionally, the stability of heavy metal speciation in water or acid solutions also plays a role in the process. Wang et al. (2015) subjected four MSWI fly ash samples to HNO_3 solutions (0.5 mol/L) in a beaker at an L/S ratio of 20. The outcomes (as displayed in Fig. 9.2) revealed that HNO_3 solutions removed a higher quantity of exchangeable and acid-soluble fractions, as well as the reducible fraction of Zn and Cu compared with other nonmetal soluble salts.

Researchers have treated MSWI fly ash with lactic and citric acid during the triple washing process, which means washing three times for 5, 10, and 10 min, respectively (Wang et al., 2023). The results revealed that citric acid had a similar dechlorination effect to that of lactic acid. After lactic acid and citric treatment, the chlorine content of MSWI fly ash is reduced to 1.10 wt% and 1.07 wt%, respectively. Citric acid is a tricarboxylic acid capable of strong chelation effects, which can lead to form stable water-soluble chelates between heavy metals and ligands. After the citric acid washing treatment, the removal percentages of heavy metals were 43.6%, 39.6%, 87.0%, 11.9%, and 17.2% for Zn, Pb, Cd, Cu, and Cr, respectively.

The coprocessing of wastes can be effectively achieved by washing fly ash with acid produced by different waste disposal methods. For instance, Zhang et al. (2022) presented a novel coprocessing method for organic anaerobic fermentation broth and MSWI fly ash. Acetic acid, the most prevalent organic acid in anaerobic fermentation, was utilized to investigate the dechlorination effects of low concentrations of organic acid lotion. The best washing pretreatment

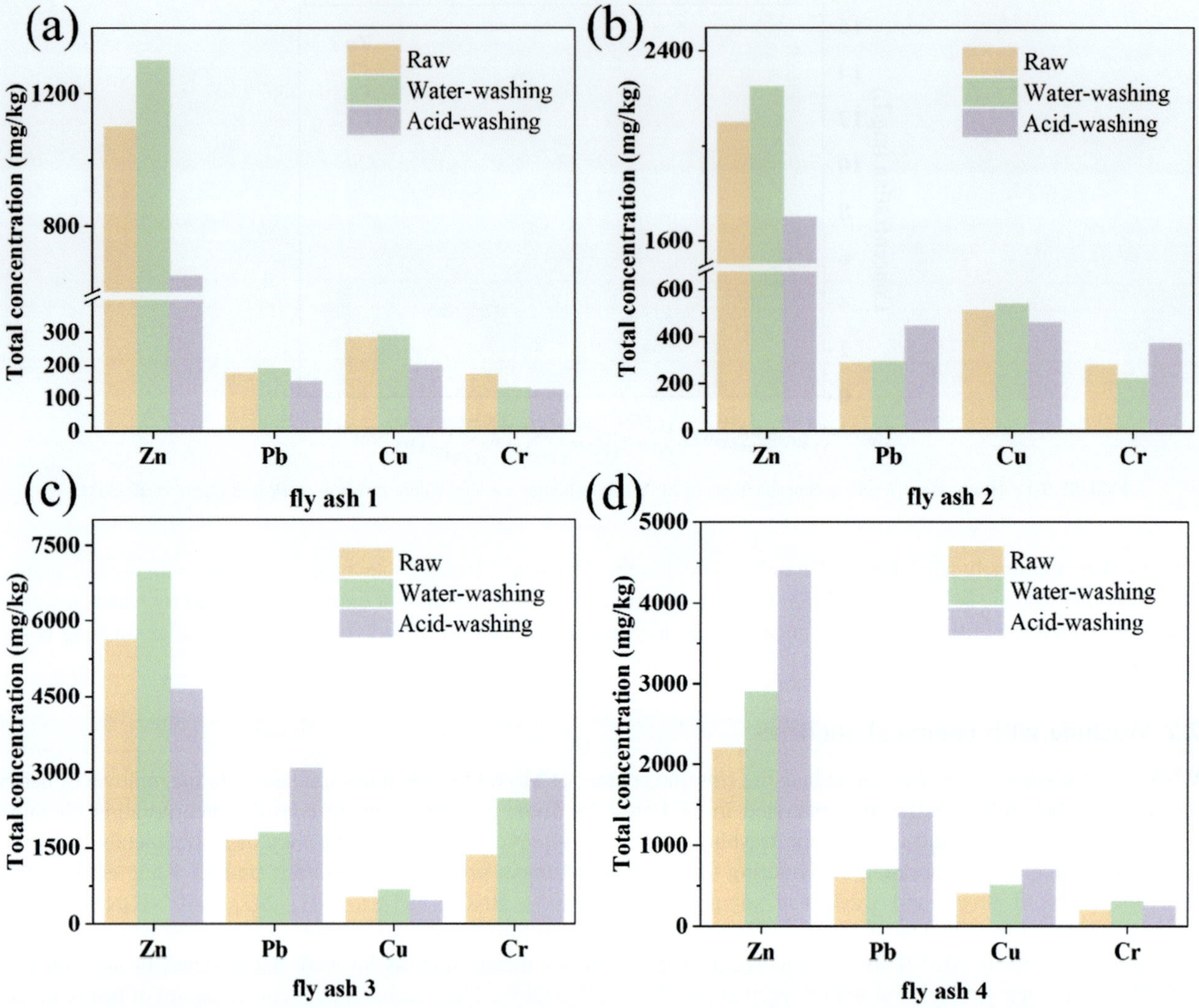

FIGURE 9.2 The total contents of heavy metals in raw and HNO_3-treated MSWI fly ash (Wang et al., 2015).

reduced the chlorine content of MSWI fly ash to 0.82 wt%. The melting point of the washed MSWI fly ash fell by nearly 30°C. After calcination at 1100°C, only 0.06 wt% of chlorine remained, considerably reducing the release of trace heavy metals during thermal treatment, resulting in a suitable feedstock for cement production.

As discussed in Section 2.1, the leachability of Cr and Pb in the washed MSWI fly ash was a crucial concern. To address this issue, Liang et al. (2020) studied the leaching of targeted heavy metals through various solidification techniques employing chemical curing agents and chelating agents such as EDTA-2Na, $NaPO_3$, Na_2S, and $FeSO_4$. Results showed that with the increase in $NaPO_3$ dosage, the leaching concentration of Cr showed a significant upward trend, while with the increase in the dosage of EDTA-2Na, Na_2S, and $FeSO_4$, the leaching concentration of Cr generally exhibited a downward trend. All four chemical agents can reduce the leaching concentration of Pb, but the effect of Na_2S is very limited. After solidification with Na_2S, the leaching concentration of Pb remains higher than the limit. Additionally, Liang et al. (2020) also revealed that the addition of 4 wt% ferrous sulfate to the washing treatment decreased the leaching concentrations of Pb and Cr by 89% and 67%, respectively. As a result, the final leachabilities could be under the limit value of the national standard (GB 5085.3-2007).

Recent literature report on the cosintering of MSWI fly ash and waste glass containing Cr, including the impact of washing treatments using different additives on MSWI fly ash composition and ceramsite characteristics (Huang et al., 2023b). The outcomes indicated that the washing procedure increased the CaO content while decreasing the soluble salt content. The washing pretreatment was found to lower ceramsite's environmental risk, and adding phosphoric acid can further improve heavy metal immobilization.

2.3 Disposal of wastewater after washing treatment

It is widely recognized that the washing leachate generated through various treatments contains high amounts of salts and moderate amounts of metals. For removing and recovering some heavy metals from the washing effluents, conventional methods such as pH adjustment, electrolysis, and solvent extraction can be employed (Tang et al., 2019). Additionally, chloride salts can be recovered from the leachate using techniques such as evaporation and precipitation (Xie et al., 2022). These methods can aid in reducing the environmental impact of wastewater and potentially provide a source of valuable resources to a certain extent.

Adsorption can be used to recover valuable metals, such as Cu and Zn, from leachate. Iron-based adsorbents can then be introduced to remove the remaining heavy metals from the wastewater effectively (Tang et al., 2019; Patil and Kothavale, 2022). Tang et al. (2019) employed acid leaching, extraction, and adsorption to treat Cu, Zn, Pd, and Cd in MSWI fly ash leachate. Cu and Zn could be further recovered through electrolytic extraction, while the remaining aqueous phase could be treated with adsorbents to remove Cd and Pb. Overall, the study suggested that the treated effluent may meet acceptable emission standards, with 95% of the Cu and 82% of the Zn in the leachates being recovered under optimal conditions.

The washing liquid typically contains soluble chlorides such as KCl, NaCl, and $CaCl_2$, which can be utilized as fertilizers and chemical materials. However, separating KCl and NaCl from the mixture via evaporative crystallization poses a challenge due to their similar solubilities (Xie et al., 2022). $CaCl_2$ can be effectively removed using Na_2CO_3 (Mangialardi, 2003). Recent research suggests that evaporative crystallization can effectively separate the KCl and NaCl mixture from $CaCl_2$, achieving theoretical recoveries of approximately 92% for KCl and NaCl after five concentration cycles (Xie et al., 2022). However, considering the energy and time consumption, and the unstable recovery rate, there are still several limitations in using evaporation and precipitation technology to recover chloride salts.

3. Accelerated carbonation treatment of combustion and incineration residues

The concept of CO_2 carbonation has been proposed for the treatment of MSWI fly ash in literature. While the reaction rate of natural carbonation is generally very low, accelerated carbonation techniques can achieve complete carbonation in just a few hours or days (Costa et al., 2007). Industrial alkaline solid wastes, such as incineration fly ash and steel slag, usually contain high calcium oxide and occasional magnesium oxide content. These can react with CO_2 to form dense carbonate and achieve the purpose of carbon sequestration (Pan et al., 2012). And it will accelerate the dissolution of CO_2 to form carbonate in the alkaline substrate (Chang et al., 2017). Additionally, carbonation is an exothermic process that produces thermodynamically stable carbonate phases under ambient conditions with minimal energy consumption (Sorrentino et al., 2023). Accelerated carbonation has been utilized as a pretreatment technique for solid wastes before reuse or final disposal, resulting in chemically stable material formation with low leachability (Costa et al., 2007; Zha et al., 2015; Yuan et al., 2023).

3.1 Mechanism of accelerated carbonation treatment

Carbonation is a three-phase reaction involving gas, solid, and liquid phases. The process and reaction equation of carbonation can be summarized as follows (Peter et al., 2008; Saillio et al., 2021):

(1) CO_2 in the atmosphere diffuses through the pores of materials, dissolves in pore water, and forms carbonate anions;

$$CO_2\ (g) + H_2O\ (l) \rightarrow 2H^+\ (aq) + CO_3^{2-}\ (aq) \tag{9.1}$$

(2) Anions react with carbonate solutes, mostly calcium, and magnesium ions in the pore solution dissolved from the solid matrix, resulting in insoluble carbonate;

$$(CaO, MgO)_x\ (SiO_2)_y + 2H^+\ (aq) \rightarrow\ x\ (Ca, Mg)^{2+}(aq) + y\ SiO_2\ (s) + H_2O\ (l) \tag{9.2}$$

$$(Ca, Mg)^{2+}(aq) + CO_3^{2-}\ (aq) \rightarrow (Ca, Mg)CO_3\ (s) \tag{9.3}$$

(3) The carbonates precipitate quickly onto the surface of the solid matrix.

Carbonation has been proven effective in stabilizing heavy metals, such as Pb, Zn, Cr, Cu, and Mo, reducing their mobility (Cappai et al., 2012). Carbonation impacts the leaching of heavy metals due to mineralogical and chemical transformations during the process (Cappai et al., 2012; Yuan et al., 2023). Four types of mechanisms have been identified

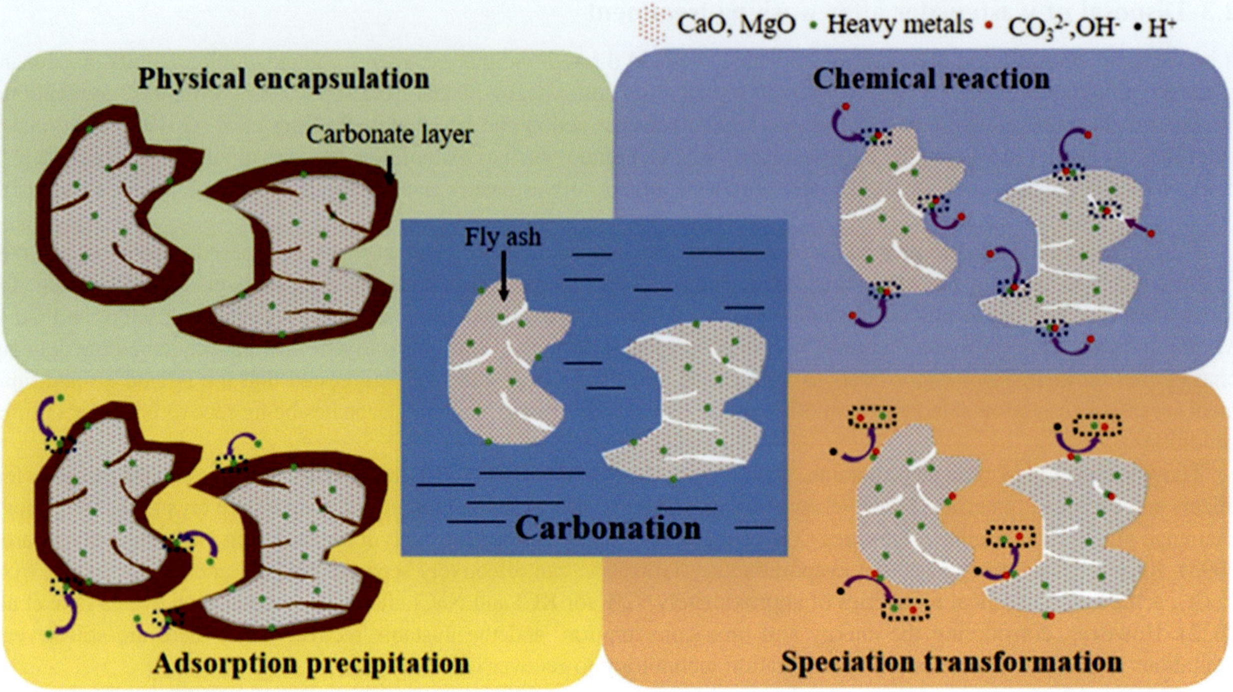

FIGURE 9.3 Four types of mechanisms for stabilizing heavy metals during carbonation progress (Yuan et al., 2023).

(Fig. 9.3). Firstly, physical encapsulation occurs where carbonates are formed on the surface of MSWI fly ash during carbonation, creating a physical barrier that impedes heavy metal leaching from the interior (Gerven et al., 2005). Secondly, chemical reactions take place, where Pb, Zn, and Cu ions react with CO_3^{2-} to produce stable carbonate precipitates (Ni et al., 2017). Thirdly, adsorption precipitation occurs where Pb, Zn, and Cu ions are adsorbed onto the surface of $CaCO_3$ precipitates during carbonation due to their isomorphism with calcium (Ecke et al., 2002). Fourthly, speciation transformation can occur when the heavy metal ions precipitate in the form of hydroxides and carbonate as the pH in the solution reaches the appropriate value. Low or high pH will lead to the leaching of heavy metals, mainly some amphoteric heavy metals such as lead (II). The dissolution of CO_2 in water results in the formation of H^+ ions, which regulate the suitable pH for heavy metals immobilization (Yuan et al., 2023). However, with the increase in the concentration of H^+ ions, heavy metals in the form of carbonates and hydrogen and oxygen compounds can decompose and change from a precipitate to a leaching form (Cappai et al., 2012).

The carbonation process has demonstrated a positive impact on stabilizing heavy metals, including Pb, Zn, and Cu, primarily through adsorption precipitation and chemical reactions (Polettini et al., 2016). Furthermore, physical encapsulation is a nonselective, physical phenomenon that has an excellent stabilization effect on all heavy metals (Yuan et al., 2023). Finally, speciation transformation has a significant effect on stabilizing Pb, Zn, Cd, and Ni (Ecke et al., 2002).

3.2 The application of accelerated carbonation treatment

Carbonation methods are broadly classified as either direct or indirect. A summary of various carbonation treatment methods is listed in Table 9.2. Direct carbonation involves a single process step, while indirect carbonation involves the extraction of calcium and magnesium elements, followed by carbonation (Chen et al., 2021b). Direct carbonation can be further divided into two categories: gas—solid (dry) carbonation and aqueous (wet) carbonation. The former refers to placing the solid waste into a reaction kettle filled with CO_2 gas, while the latter refers to passing of CO_2 gas into a solution containing the solid waste or putting the solid waste into a solution already containing carbonates, such as Na_2CO_3 and $NaHCO_3$ solution (Lim et al., 2010). The gas—solid carbonation process is generally considered more cost-effective than the aqueous carbonation method. However, it has some limitations such as slow kinetics and significant energy input for permeation, which has hindered its large-scale adoption (IPCC, 2005; Zevenhoven et al., 2008).

Dry carbonation is known to have a slow reaction rate and low conversion efficiency; therefore, wet carbonation has gained more attention in recent research (Dananjayan et al., 2016). Researchers have investigated the effectiveness of an

TABLE 9.2 Summary of accelerated carbonation treatment for MSWI fly ash.

Conditions	Leaching concentration reduction of heavy metals	References
MSWI air pollution control residues + CO_2 gas	Pb: 4-order of magnitude reduction; Zn, Cu: 2-order of magnitude reduction; Cr, Mo: 1-order of magnitude reduction	Cappai et al. (2012)
MSWI fly ash + oxy-fuel combustion flue gas	69% of Zn; 25% of Cu; 33% of As; 11% of Cr; 21% of Hg	Ni et al. (2017)
MSWI fly ash + high-gravity rotating packed bed	TCLP results indicated that there was a low leaching concentration of heavy metals	Chen et al. (2021a)
MSWI fly ash + calcium carbonate oligomers	99.1% of Zn; 100% of Pb; 91.3% of Cu; 95.6% of Ni	Chen et al. (2022b)
MSWI fly ash + ultrasound	99.99% of Pb; 99.99% of Zn	Chen et al. (2022a)
Using $NaHCO_3$ and Na_2CO_3 for GFA carbonation	Below National limits for nonhazardous waste identification	Qin et al. (2022)

aqueous-phase accelerated carbonation treatment for air pollution control residues obtained from MSWI (Cappai et al., 2012). The carbonation route process was conducted in an open vessel by passing CO_2 through the continually mixed ash slurry at 20°C and ambient pressure. This carbonation treatment resulted in significant reductions of pollutants with a 4-order of magnitude reduction in Pb release, a 2-order of magnitude reduction in Zn and Cu, and 1-order of magnitude reduction in the mobility of Cr and Mo.

Wet carbonation is a highly effective process for washing MSWI fly ash. Introducing CO_2 during the washing process can lead to a carbonation reaction, which allows the nonwashed heavy metals in MSWI fly ash to be better immobilized (Lu et al., 2024). A recent study (Chen et al., 2022b) proposed a novel MSWI fly ash carbonation treatment that utilizes calcium carbonate oligomers to control the process at ambient temperature and pressure, as illustrated in Fig. 9.4. The method resulted in significant reductions of leaching concentrations of 99.1%, 100.0%, 91.3%, and 95.6%, for Zn, Pb, Cu, and Ni, respectively. The carbonation reaction converted heavy metals into carbonate, which was then adsorbed and enclosed by the calcium carbonate polymer, leading to improve stabilization of heavy metals.

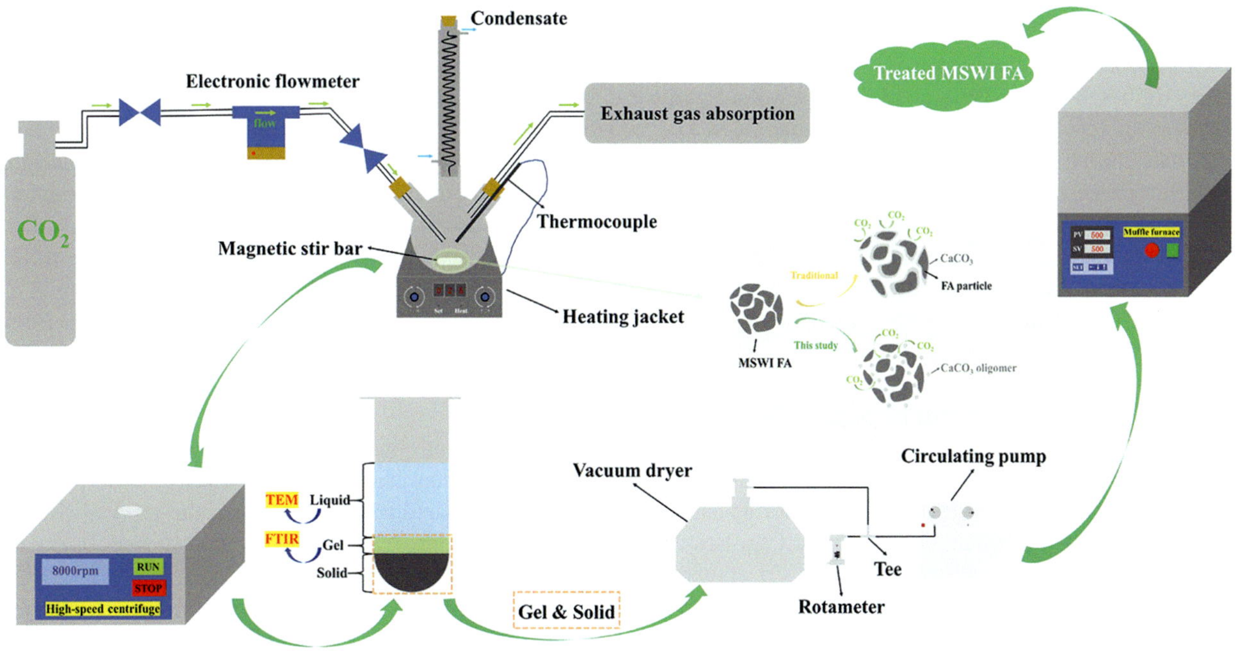

FIGURE 9.4 MSWI fly ash carbonation treatment using calcium carbonate oligomers (Chen et al., 2022b).

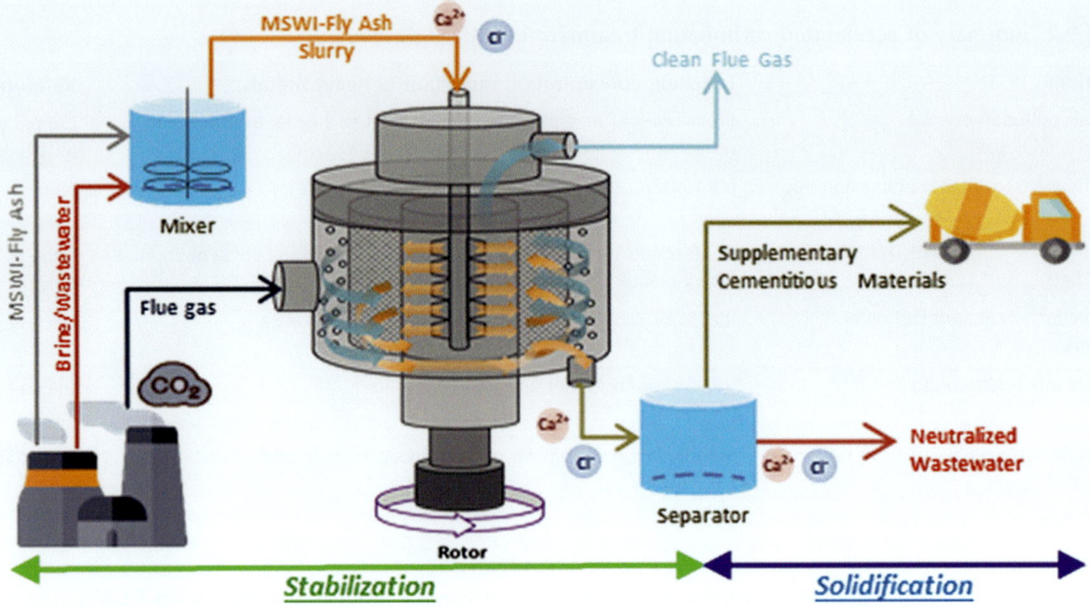

FIGURE 9.5 Treatment process of the combined wet extraction and accelerated carbonation (Chen et al., 2021a).

Chen et al. (2021a) investigated the impact of wet carbonation treatment on the utilization of MSWI fly ash, through its curing stabilization in a high-gravity rotating packed bed. Wet extraction led to an increase in the micropore area, micropore volume, and specific surface area of MSWI fly ash, while carbonation resulted in a decrease in these parameters. Results of the toxicity characteristic leaching procedure (TCLP) revealed low concentrations of heavy metal leaching from the carbonated MSWI fly ash. Combining wet extraction and accelerated carbonation of MSWI fly ash using the high-gravity device could serve as a valuable technique for waste stabilization and utilization in a waste-to-resource supply chain (Fig. 9.5).

Ultrasound has been employed in the disposal of incineration residues to enhance chloride salt removal during pickling and facilitate the extraction of some heavy metals (Huang et al., 2018; Nguyen et al., 2022). Chen et al. (2022a) conducted a study using ultrasound to enhance the immobilization of heavy metals during the wet carbonation process of MSWI fly ash. Results indicated that ultrasonic treatment significantly improved the CO_2 capture capacity of MSWI fly ash, with a carbonation efficiency increase from 14.5% to 17.4%, and a faster reaction percentage compared with conventional methods. Furthermore, the ultrasonic treatment demonstrated a curing efficiency of 99.99% for Pb and 99.99% for Zn in raw MSWI fly ash. This was attributed to the high-efficiency carbonation reaction and generation of nano-$CaCO_3$, which effectively converted more heavy metals into stable carbonate forms. Additionally, it is worth noting that the use of ultrasonic-assisted treatment will increase the cost of the entire process. It is necessary to measure the relationship between the increased benefits of contaminant removal and the increased costs.

Controlling the carbonation ratio of direct carbonation treatment can be challenging as excessive CO_2 can lower the pH value to around 7 and increase the leachability of heavy metals, such as Zn, Cd, and Cu (Ecke et al., 2002; Wang et al., 2010a,b). Additionally, direct CO_2 carbonation is not effective in stabilizing or removing chloride and sulfate (De Boom et al., 2014). An alternative approach involves the use of carbonate ion-containing solutions as a carbonation agent. This method can be carried out in an open environment and can achieve heavy metal immobilization and chloride and sulfate removal simultaneously (Dontriros et al., 2020; Qin et al., 2022).

Researchers have performed carbonation treatment by employing different concentrations of $NaHCO_3$ and Na_2CO_3 solutions for gasification fly ash (GFA) (Qin et al., 2022). Results indicated that Na_2CO_3 or $NaHCO_3$ solution treatment could effectively promote the removal of chloride and sulfate in GFA. Compared to water washing, the residual chlorine content of $NaHCO_3$-treated GFA and Na_2CO_3-treated GFA decreased from 1.22 wt% to 0.16 wt% and 0.26 wt%, respectively. For GFA with high Pb and Zn contents, $NaHCO_3$ was found to be more effective than Na_2CO_3 for immobilizing heavy metals. The results showed that for GFAs treated with $NaHCO_3$ solution having a concentration of 50 g/L or above for 1 h, the leaching concentrations of all selected heavy metals were below the national limits for nonhazardous waste identification.

Carbonation treatment has proven to be effective in reducing heavy metal leachability in incineration residues. However, using pure CO_2 for carbonation treatment can be expensive. To address this issue, researchers explored low-cost carbonation methods. Oxy-fuel combustion technology is a promising clean coal technology that enriches flue gas with CO_2, thereby providing a cost-effective approach to carbonation (Zheng et al., 2015). Ni et al. (2017) conducted a study using simulated oxy-fuel combustion flue gas for carbonation treatment of MSWI fly ash. The study found that after carbonation, leaching concentrations of Pb decreased below the legal limit set in China. Furthermore, the leaching concentrations of Zn, Cu, As, Cr, and Hg decreased by 69%, 25%, 33%, 11%, and 21%, respectively.

3.3 Factors influencing the accelerated carbonation

3.3.1 Pressure of CO_2 gas in the carbonation reactor

For gas–solid carbonation treatment, the carbonation efficiency of the incineration residues increases when the CO_2 pressure in the reactor increases. This is because the increase in pressure enhances the solubility of CO_2 in water, effectively promoting the generation of CO_3^{2-} and the carbonation reaction (Xu and Yi, 2022). Moreover, it increases the occurrence of drainage phenomenon and the likelihood of CO_2 diffusion into particle gaps (Wu et al., 2020). Furthermore, when the pressure of CO_2 increased enough to make it reach the supercritical state (31.26°C, 7.29 MPa), the carbonation efficiency will further increase. Supercritical CO_2 has 10 times greater solubility in water than nonsupercritical CO_2 (Zha et al., 2015) and possesses strong diffusion characteristics and low viscosity stress, allowing it to enter particle gaps and enhance carbonation (Zha et al., 2019).

The supercritical CO_2 can improve carbonation efficiency and then promote the stabilization of heavy metals by enhancing the effects of physical encapsulation, adsorption precipitation, and chemical reaction as described in Section 3.1. Researchers revealed that when CO_2 in the reactor is converted from a nonsupercritical state (1 MPa) to a supercritical state (8 MPa), two trends in heavy metal stability of MSWI fly ash can be observed (Yuan et al., 2022): one trend shows a gradual increase in stability, and the other displays an initial increase and then is followed by a decrease after reaching a peak at the critical point. However, converting CO_2 to the supercritical state can also result in further consumption of OH^- in the solution and lead to an increase in H^+ concentration due to the increased solubility of CO_2, which may increase the leachability of some heavy metals through the speciation transformation mechanism as described in Section 3.1. Additionally, the cost of supercritical CO_2 is high, and it is not economical to use it for carbonation treatment (Yuan et al., 2023).

3.3.2 Temperature in the carbonation reactor

When the temperature increases, the reaction rate of carbonation could be accelerated, and the dissolution of heavy metal ions in the pore solution of incineration residues increases, which is conducive to the immobilization of heavy metals by forming carbonates. Ni et al. (2017) examined leaching concentrations of heavy metals under different reaction temperatures, and their findings are presented in Fig. 9.6. An increase in temperature accelerates the dissolution of Ca^{2+} and heavy metal irons such as Pb^{2+}, Zn^{2+}, and Cu^{2+} and their subsequent reaction with CO_3^{2-}, resulting in the decreased leachabilities of heavy metals. However, excessively high temperatures (up to 100°C) negatively influence the dissolution of CO_2 in water, hindering CO_2 dissolution and leading to unsatisfactory immobilization of heavy metals.

3.3.3 pH of the reaction solution

A pH range of 9.5–10.5 was found to be most effective for metal immobilization during carbonation treatment of incineration residues (Wang et al., 2010a,b). Researchers evaluated the effects of accelerated carbonation treatment of air pollution control residues at various pH levels in the aqueous phase (Cappai et al., 2012). Leaching trends for both untreated and carbonated air pollution control residues were significantly influenced by the pH value of solution. Pb, Zn, and Cu exhibited V-shaped release characteristics, and pH dependence was also noted for Cr (higher mobility at higher pH) and Sb (higher mobility at lower pH). Based on the immobilization effects observed for Pb, Zn, Cu, and Cr, a final pH of around 10.5 was estimated to be the optimal pH for carbonation treatment (Wang et al., 2010a).

4. Electrokinetic remediation treatment of combustion and incineration residues

The electrokinetic remediation (EKR) involves inserting electrodes into a contaminated medium and applying low direct current (10–100 mA/cm^2) or electric field intensity (1–3 V/cm) to drive pollutants such as organic pollutants and heavy metal ions to migrate from the affected area to the electrode chamber, thus removing the pollutants (Hu et al., 2015). EKR

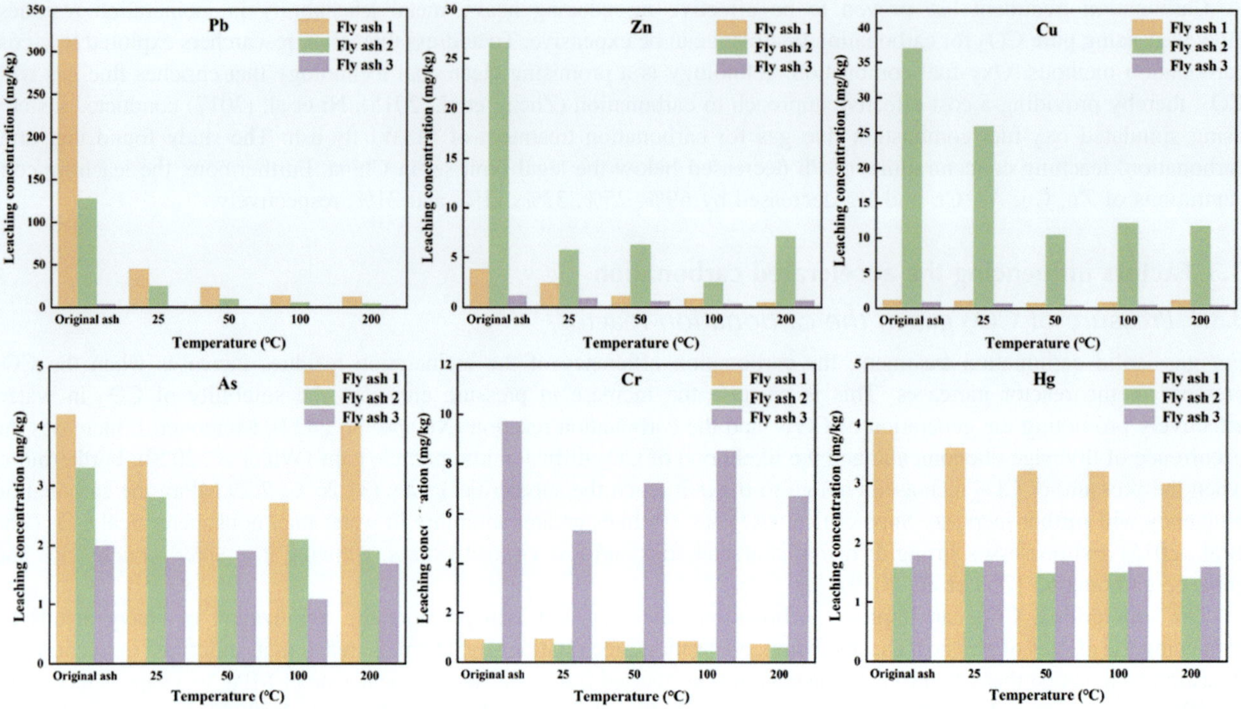

FIGURE 9.6 Leachability of heavy metals at various carbonation temperatures (Ni et al., 2017).

was initially used for restoring polluted soils (Pedersen, 2002). Researchers discovered that it has the advantage of high efficiency and low consumption in treating pollutants with heavy metals (Peng et al., 2011). It is currently being used in the treatment of sludge, sediment, electronic waste, incineration fly ash, and other fields (Yeung, 2011).

4.1 Mechanism of electrokinetic remediation treatment

Fig. 9.7 demonstrates the schematic diagram of the principle of electrokinetic remediation (Pedersen, 2002). An anode and cathode are placed in an electrolyte chamber, and a direct current field is applied. Water is electrolyzed at both electrodes, producing H^+ through oxidation at the anode and OH^- through H^+ reduction at the cathode. The H^+ and OH^- ions move toward electrodes with opposite charges, changing the pH of the sample area and gradually increasing the pH value in the electrolytic cell from the anode to the cathode. During the time, the electrochemical oxidation–reduction reaction is induced between the substrates, driving the transfer of organic pollutants and heavy metals in the sample area through electrodialysis, electromigration, and electrophoresis (Hu et al., 2015). The ions in the sample area (compartment III) eventually migrate to the electrodes in the electric field and enter the electrolyte solution in compartment II or IV, where they may be separated using traditional techniques such as electroplating, ion exchange, or precipitation.

The migration of pollutants is primarily influenced by two mechanisms, electromigration and electrodialysis. When there is an electric field present and suspensions with low solid content, electrophoresis also plays a role in the migration of heavy metals (Hu et al., 2015). Electromigration refers to the quick migration of inorganic salt ions that are dissolved in the

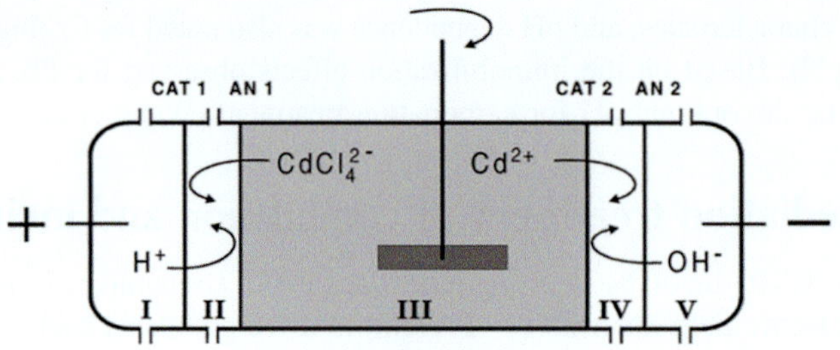

FIGURE 9.7 The principle of electrokinetic remediation (Pedersen., 2002).

pore solution toward the opposite-charged electrode due to the electric field. The migration rate depends mainly on the voltage gradient, ion charge, and diffusion coefficient, and it is less affected by the medium properties (Page and Page, 2002). Electrodialysis refers to the migration of saturated liquid that is charged, primarily pore water and dissolved substances in the medium, toward the electrode chamber. The rate of electrodialysis is affected by the strength of the electric field, the viscosity of the pore liquid, and the zeta potential of the medium interface but is not significantly affected by the pore size of the medium. For systems where pollutants are primarily present in ionic form, electrodialysis contributes much less to pollutant removal than electromigration (Yeung et al., 1997; Vane and Zang, 1997; Hu et al., 2015).

Despite its benefits, electrokinetic remediation treatment has certain limitations, as it can only remove heavy metals that are dissolved in the pore solution with a mobile potential (Pedersen, 2002). In the electrolytic cell, organic pollutants and heavy metals typically exist in four different forms: some are adsorbed on the surface of particles or encapsulated inside particles, some are adsorbed on the surface of suspended colloids, some are dissolved in the electrolyte, and some are enriched on the surface of solid substrates as precipitation. However, only heavy metals that are adsorbed on the surface of suspended colloids and dissolved in the electrolyte solution can be effectively migrated under the action of electroosmotic flow and electromigration (Yang et al., 2014; Xu et al., 2017). Other pollutants that are strongly immobilized or deeply encrusted in the matrix cannot be removed efficiently by electrokinetic remediation treatment alone. Furthermore, the treatment may also face challenges caused by technical difficulties, such as the establishment of proper electrode spacing and controlling the current and voltage levels during operation (Hu et al., 2015; Yang et al., 2014).

4.2 Factors influencing the effect of electrokinetic remediation

The effect of the EKR process is influenced by several factors, including the pH of the medium, current density, duration, liquid-to-solid ratio, and the type of assisting agent (Pazos et al., 2010; Kirkelund and Jensen, 2018). Tao et al. (2014) conducted a study to investigate the impact of liquid-to-solid ratio, initial pH, and EKR treatment duration on heavy metal concentrations in leachate from MSWI fly ash (Fig. 9.8). The heavy metal concentrations in the leaching solution decreased, while the recovery efficiency increased with an increase in the liquid-to-solid ratio from 5:1 to 30:1 after leaching for 18 h (Fig. 9.8a). Typically, a lower initial pH results in a higher metal concentration, and the negative impact on solubility with an increase in pH follows the order of Pb < Zn < Cu, as illustrated in Fig. 9.8b (Sawyer et al., 1994; Tao et al., 2014). And the Zn concentration peaked at 8 h after mixing with the leaching agent and then decreased, while the concentrations of Pb and Cu peaked at 10 h (Fig. 9.8c).

4.3 Use of assisting agents in electrokinetic remediation

The addition of a suitable assisting agent to the electrolytic cell during EKR treatment can enhance the desorption of certain heavy metals from MSWI fly ash, as compared with basic experiments using water only as an assisting agent (Yeung, 2011). When selecting an enhancement agent, it is crucial to consider several characteristics. Firstly, the assisting agents should be able to react with contaminants in the incineration residues to form insoluble substances. Secondly, the formed insoluble substance should be chemically stable across a broad range of pH and can be efficiently transported by a direct-current electric field. Thirdly, the agent should be cost-effective and readily available to ensure that this EKR process can

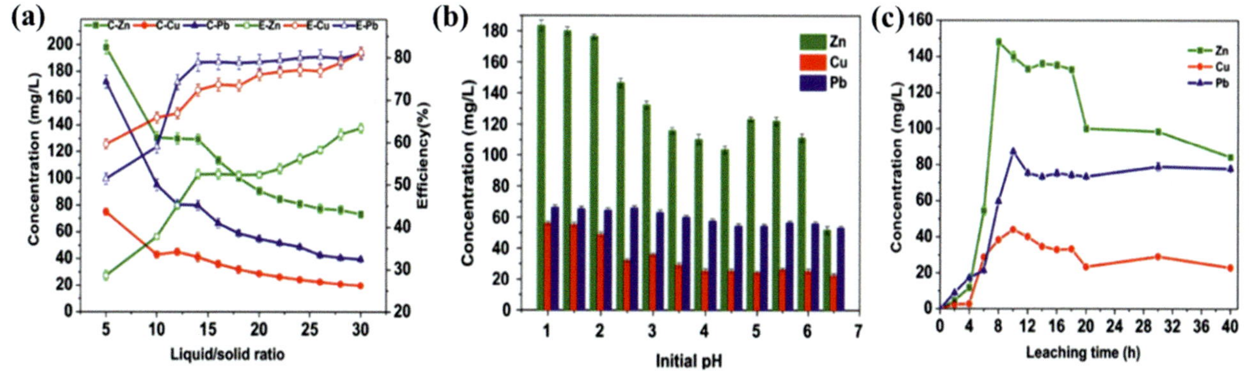

FIGURE 9.8 The leaching concentrations of Zn, Cu, and Pb: (a) versus different liquid-to-solid ratios; (b) with different initial pH; (c) at different leaching times (Tao et al., 2014).

TABLE 9.3 Summary of various conditions and results of EKR treatment.

	Removal percentages of heavy metals (%)					
Auxiliary means	**Zn**	**Pb**	**Cu**	**Cd**	**Cr**	**References**
Water	49.9	21.2	22.7	41.3	/	Li et al. (2016)
0.1 mol/L HNO_3	57.9	45.0	47.5	57.9	/	
0.25 mol/L ammonium citrate	62.0	20.0	81.0	86.0	44.0	Pedersen et al. (2005)
PRB	27.0	45.5	28.1	28.1	/	Huang et al. (2015)
BES	95.4	98.1	98.5	/	/	Rabaey et al. (2009)
Ultrasound	69.8	64.2	67.7	59.9	/	Huang et al. (2018)

"/" means not mentioned.

be widely used. Lastly, if feasible, the agent should be selectively complex with targeted contaminants (Pedersen et al., 2005; Yeung and Gu, 2011).

Table 9.3 summarizes various EKR-related treatments. The simplest EKR technology involves using water as the solvent without adding other additives. Li et al. (2016) conducted a study using the conventional EKR method with water as the assisting agent to extract heavy metals from MSWI fly ash. The remediation was carried out for 15 days in a rectangular electrolytic cell with a voltage gradient of 2 V/cm. Results revealed that compared with the untreated MSWI fly ash, the leaching reduction percentages of Zn, Pb, Cu, and Cd through EKR treatment were 49.9%, 21.2%, 22.7%, and 41.3%, respectively.

During the EKR process, a phenomenon called focusing occurs due to the rapid change in pH value near the electrodes. This effect is typically caused by the accumulation of positively or negatively charged ions near the pH gradients and can limit the extraction or removal of contaminants (Yeung, 2006; Li et al., 2012). To address this issue, Li et al. (2016) investigated the use of pH adjustment as a feasible method to alleviate the focusing phenomenon during the EKR process. They added 0.1 mol/L nitric acid (HNO_3) near the anode in the electrolytic cell. Results showed that compared with the leaching outcomes of the untreated MSWI fly ash, the leaching reduction percentages of Zn, Pb, Cu, and Cd through electrokinetic remediation were 57.9%, 45.0%, 47.5%, and 57.9%, respectively. Importantly, the reduction effect of HNO_3-assisted EKR was better than that achieved by using water as leachate.

Inorganic chelating agents such as iodide ions, chloride ions, ammonia, and hydroxide ions are commonly used to regulate the pH in the electrolytic cell and generate soluble complex ions with metal ions, including $[HgI_4]^{2-}$, $[CuCl_2]^-$, $[CuCl_4]^{2-}$, $[Cu(NH_3)_4]^{2+}$, $[Zn(OH)_4]^{2-}$, $[Cr(OH)_4]^-$, and $[Cr(OH)_3]^{2-}$, among others (Genc et al., 2009; Estabragh et al., 2016). Pedersen et al. (2005) found that using a 0.25 mol/L ammonium citrate solution as an assisting agent in the EKR treatment resulted in removal percentages of 86%, 20%, 62%, 81%, and 44% for Cd, Pb, Zn, Cu, and Cr in MSWI fly ash, respectively. By analyzing the migration direction of different metals in the experiment, it was revealed that Pb is mainly chelated with citrate to form a negatively charged and soluble chelate ion, while Cu, Zn, and Cd were preferentially formed positively charged complexes with ammonium. And Cr was removed as a negatively charged species in all conditions, presumably as chromate ions.

4.4 Coupling enhancement technology of electrokinetic remediation

Coupling the EKR technology with a permeable reactive barrier (PRB) made of activated carbon has been suggested to enhance the removal efficiency of heavy metals from MSWI fly ash (Fig. 9.9). The PRB system involves retaining or breaking down pollutants in fluids through adsorption, precipitation, and redox reactions between pollutants and the barrier (Gibert et al., 2003). Huang et al. (2015) conducted a study demonstrating that the leaching toxicities of Zn, Pb, Cu, and Cd from MSWI fly ash after EKR-PRB treatment decreased by 27.0%, 45.5%, 28.1%, and 28.1%, respectively, compared with samples treated with EKR alone.

The electrodes of the EKR technology have been modified to improve electrochemical efficiency. Tao et al. (2014) developed a two-stage process that combined a bioelectrochemical system (BES) and an electrolytic reactor to extract

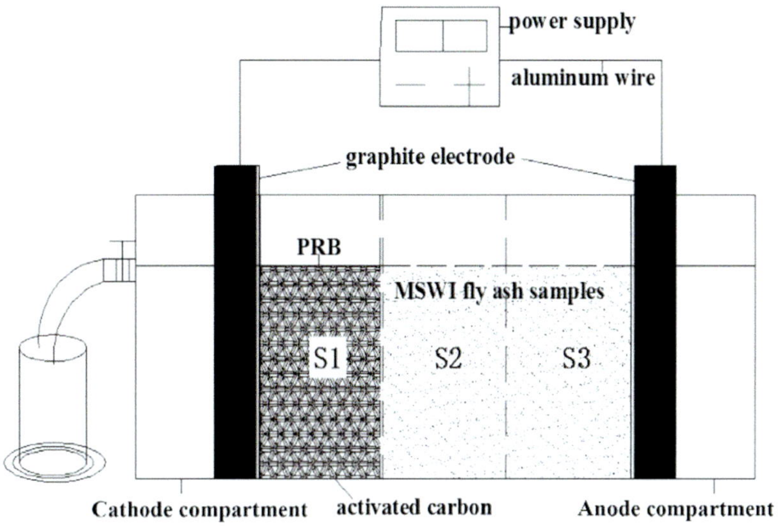

FIGURE 9.9 A device diagram of EKR coupled with PRB treatment (Huang et al., 2015).

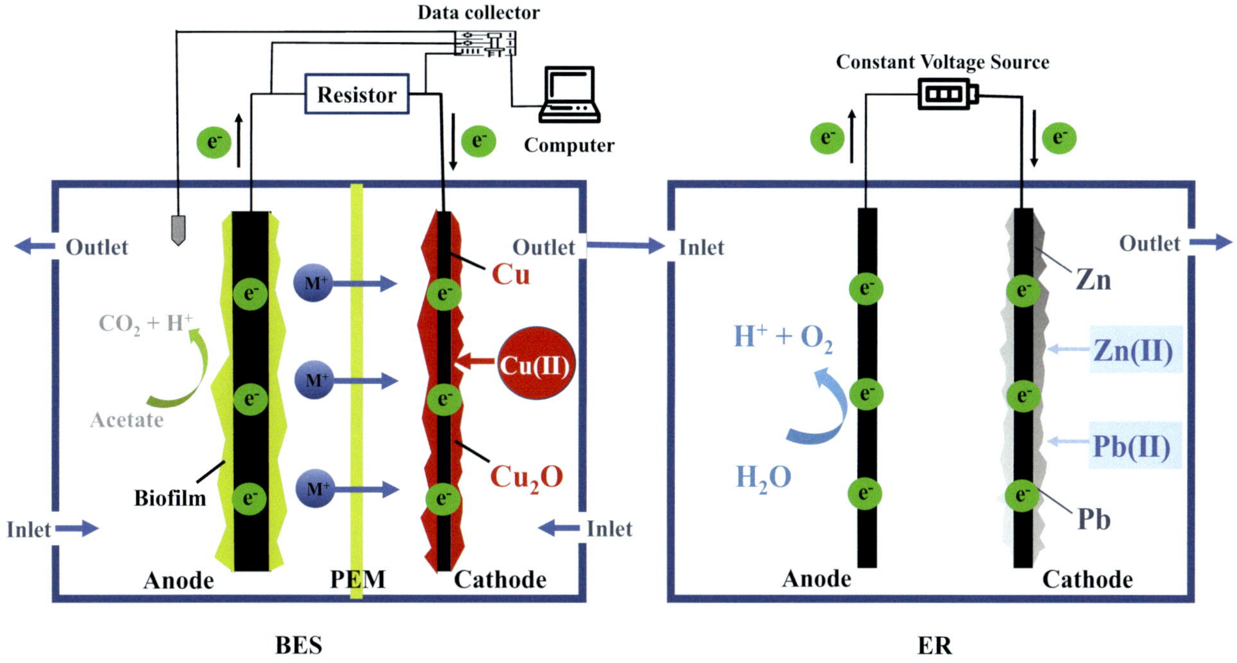

FIGURE 9.10 EKR-coupled BES treatment device diagram (Tao et al., 2014).

heavy metals from MSWI fly ash leachate, as shown in Fig. 9.10. The BES used microorganisms to catalyze an oxidation and reduction reaction on an anodic or cathodic electrode, respectively (Rabaey et al., 2009). Results showed that the combined use of BES and electrolytic reactor could achieve high heavy metal removal efficiency and the contents of Cu, Zn, and Pb decreased by 98.5%, 95.4%, and 98.1%, respectively (Tao et al., 2014).

Low-frequency ultrasound has been utilized by several researchers to activate MSWI fly ash and enhance the removal of heavy metals within the EKR process (Huang et al., 2018). During sonication, cavitation bubbles were generated at the interfaces between the solid particles and the electrolyte. These bubbles violently shocked the solid matrices in the irradiation area through the microjet of liquid and the acoustic streaming, when they unevenly implode. As a result, contaminants adsorbed at the solid grains of the MSWI fly ash were continually desorbed, and the solubility of the inorganic salt significantly increased (Chen and Huang, 2014; Kim et al., 2015). The maximum removal percentages of Zn, Pb, Cu, and Cd by the coupling treatment of ultrasound and EKR were 69.8%, 64.2%, 67.7%, and 59.9%, respectively (Huang et al., 2018). The findings highlight the potential of low-frequency ultrasound as a promising technique to improve the removal of heavy metals in the EKR process.

Additionally, it is worth noting that the use of some assisted technology, such as PRB, BES, and ultrasonic, will increase the cost of the entire process. It is necessary to balance the relationship between the increased benefits of contaminant removal and the increased costs.

5. Conclusions

This chapter provides a comprehensive overview of the principles and applications of washing, accelerated carbonation, and electrokinetic remediation technologies for the treatment of incineration residues. It is important to note that each of these methods has unique advantages and limitations, and their effectiveness in removing various heavy metals and organic pollutants may vary. Water washing treatment can reduce the soluble chlorine content and some heavy metals content in the incineration residue. Accelerated carbonation and electrokinetic remediation treatments exhibit a positive effect on reducing the leaching toxicity of heavy metals, but a poor effect on chloride removal. Additionally, these three methods are mainly suitable for the removal of soluble components from incineration residues. To achieve better removal of heavy metals and organic pollutants in incineration residues, combining multiple methods is a promising strategy.

It is important to consider that a large amount of waste-washing effluents containing chlorine and heavy metals will be produced after washing treatment technologies. The treatment of washing waste liquid should concentrate on removing heavy metals and recovering chloride salts. Cost-effective strategies can be proposed by codisposing incineration residues with various waste liquids and organic acids. Furthermore, the development of more efficient chelating agents for different types of heavy metals is also a potential research direction. The accelerated carbonation treatment reduces the leaching toxicity of certain heavy metals by immobilizing them in carbonated compounds. For future research, it is necessary to find suitable sources of carbon dioxide and optimal carbonation conditions to achieve low-cost and high-efficiency carbonation treatment. The electrokinetic remediation treatment drives pollutants, including organic pollutants and heavy metal ions, to migrate from the sample area to the electrode chamber through electrochemical reactions to remove pollutants. Usually, it can only remove heavy metals that are dissolved in the void liquid with a mobile potential. Currently, electrokinetic remediation technology is mainly studied in the field of soil remediation, and there are limited studies on the disposal of incineration residues. And the recent research of electrokinetic remediation technology primarily focuses on developing auxiliary complexing agents or chelating agents and coupled technologies to enhance the removal efficiencies of heavy metals.

Acknowledgments

The authors gratefully acknowledge the financial support from the National Natural Science Foundation of China (Grant No. 52206174 and 52236008) and the Zhejiang Provincial Natural Science Foundation of China (Grant No. LZ23E060004) for this study.

References

Arena, U., 2012. Process and technological aspects of municipal solid waste gasification: a review. Waste Manage. (Tucson, Ariz.) 32, 625–639.

Bogush, A.A., Stegemann, J.A., Roy, A., 2019. Changes in composition and lead speciation due to water washing of air pollution control residue from municipal waste incineration. J. Hazard Mater. 361, 187–199.

Cappai, G., Cara, S., Muntoni, A., Piredda, M., 2012. Application of accelerated carbonation on MSW combustion APC residues for metal immobilization and CO_2 sequestration. J. Hazard Mater. 207–208, 159–164.

Chang, R., Choi, D., Kim, M.H., Park, Y., 2017. Tuning crystal polymorphisms and structural investigation of precipitated calcium carbonates for CO_2 mineralization. ACS Sustainable Chem. Eng. 5, 1659–1667.

Chen, J., Fu, C., Mao, T., Shen, Y., Li, M., Lin, X., Li, X., Yan, J., 2022a. Study on the accelerated carbonation of MSWI fly ash under ultrasonic excitation: CO_2 capture, heavy metals solidification, mechanism and geochemical modelling. Chem. Eng. J. 450, 138418.

Chen, J., Lin, X., Li, M., Mao, T., Li, X., Yan, J., 2022b. Heavy metal solidification and CO_2 sequestration from MSWI fly ash by calcium carbonate oligomer regulation. J. Clean. Prod. 359, 132044.

Chen, W.S., Huang, C.P., 2014. Decomposition of nitrotoluenes in wastewater by sonoelectrochemical and sonoelectro-Fenton oxidation. Ultrason. Sonochem. 21, 840–845.

Chen, X., Bi, Y., Zhang, H., Wang, J., 2016. Chlorides removal and control through water-washing process on MSWI fly ash. Procedia Environ. Sci. 31, 560–566.

Chen, T.L., Chen, Y.H., Dai, M.Y., Chiang, P.C., 2021a. Stabilization-solidification-utilization of MSWI fly ash coupling CO_2 mineralization using a high-gravity rotating packed bed. Waste Manage. (Tucson, Ariz.) 121, 412–421.

Chen, Z., Cang, Z., Yang, F., Zhang, J., Zhang, L., 2021b. Carbonation of steelmaking slag presents an opportunity for carbon neutral: a review. J. CO2 Util. 54, 101738.

Colangelo, F., Cioffi, R., Montagnaro, F., Santoro, L., 2012. Soluble salt removal from MSWI fly ash and its stabilization for safer disposal and recovery as road basement material. Waste Manage. (Tucson, Ariz.) 32, 1179–1185.
Costa, G., Baciocchi, R., Polettini, A., Pomi, R., Hills, C.D., Carey, P.J., 2007. Current status and perspectives of accelerated carbonation processes on municipal waste combustion residues. Environ. Monit. Assess. 135, 55–75.
Dananjayan, R.R.T., Kandasamy, P., Andimuthu, R., 2016. Direct mineral carbonation of coal fly ash for CO_2 sequestration. J. Clean. Prod. 112, 4173–4182.
De Boom, A., Aubert, J.E., Degrez, M., 2014. Carbonation of municipal solid waste incineration electrostatic precipitator fly ashes in solution. Waste Manag. Res. 32, 406–413.
Dontriros, S., Likitlersuang, S., Janjaroen, D., 2020. Mechanisms of chloride and sulfate removal from municipal-solid-waste-incineration fly ash (MSWI FA): effect of acid-base solutions. Waste Manage. (Tucson, Ariz.) 101, 44–53.
Ecke, H., Menad, N., Lagerkvist, A., 2002. Treatment-oriented characterization of dry scrubber residue from municipal solid waste incineration. J. Mater. Cycles Waste Manag. 4, 117–126.
Estabragh, A.R., Bordbar, A.T., Ghaziani, F., Javadi, A.A., 2016. Removal of MTBE from a clay soil using electrokinetic technique. Environ. Technol. 37 (14), 1745–1756.
Genc, A., Chase, G., Foos, A., 2009. Electrokinetic removal of manganese from river sediment. Water Air Soil Pollut. 197, 131–141.
Gerven, T.V., Keer, E.V., Arickx, S., Jaspers, M., Wauters, G., Vandecasteele, C., 2005. Carbonation of MSWI-bottom ash to decrease heavy metal leaching, in view of recycling. Waste Manage. (Tucson, Ariz.) 25, 291–300.
Gibert, O., de Pablo, J., Cortina, J.L., Ayora, C., 2003. Evaluation of municipal compost/limestone/iron mixtures as filling material for permeable reactive barriers for in-situ acid mine drainage treatment. J. Chem. Technol. Biotechnol. 78, 489–496.
Hu, Y., Xu, Z., Wang, W., Ji, Z., 2015. Research progress in removal of chromium in environment by electrokinetic remediation. Chin. J. Rare Met. 34 (10), 941–947.
Huang, B., Gan, M., Ji, Z., Fan, X., Wang, G., Sun, Z., Zhao, Q., Wu, Y., Lu, S., 2023a. Co-treating MSWI fly ash in iron ore sintering process: influence of water-washing and roll forming pretreatment on dioxins emission. Process. Saf. Environ. 173, 143–153.
Huang, J., Chu, X., Shu, Z., Ma, X., Jin, Y., 2023b. The effects of washing solvents on the properties of ceramsite and heavy metal immobilization via the cosintering of municipal solid waste incineration fly ash and Cr-containing waste glass. J. Environ. Chem. Eng. 11, 109089.
Huang, T., Li, D., Liu, K., Zhang, Y., 2015. Heavy metal removal from MSWI fly ash by electrokinetic remediation coupled with a permeable activated charcoal reactive barrier. Sci. Rep. 5, 15412.
Huang, T., Zhou, L., Liu, L., Xia, M., 2018. Ultrasound-enhanced electrokinetic remediation for removal of Zn, Pb, Cu and Cd in municipal solid waste incineration fly ashes. Waste Manage. (Tucson, Ariz.) 75, 226–235.
IPCC, 2005. IPCC Special Report on Carbon Dioxide Capture and Storage.
Jadhav, U.U., Biswal, B.K., Chen, Z., Yang, E., Hocheng, H., 2018. Leaching of metals from incineration bottom ash using organic acid. J. Sustain. Metall. 4 (1), 115–125.
Kang, D., Yoo, Y., Park, J., 2021. Stabilization of heavy metals in municipal solid waste incineration fly ash via CO2 uptake procedure by using various weak acids. J. Ind. Eng. Chem. 94, 472–481.
Kim, K., Cho, E., Thokchom, B., Cui, M., Jang, M., Khim, J., 2015. Synergistic sonoelectrochemical removal of substituted phenols: implications of ultrasonic parameters and physicochemical properties. Ultrason. Sonochem. 24, 172–177.
Kirkelund, G.M., Jensen, P.E., 2018. Electrodialytic treatment of Greenlandic municipal solid waste incineration fly ash. Waste Manage. (Tucson, Ariz.) 80, 241–251.
Li, D., Huang, T., Liu, K., 2016. Near-anode focusing phenomenon caused by the coupling effect of early precipitation and backward electromigration in electrokinetic remediation of MSWI fly ashes. Environ. Technol. 37, 216–227.
Li, D., Xiong, Z., Nie, Y., Niu, Y.Y., Wang, L., Liu, Y.Y., 2012. Near-anode focusing phenomenon caused by the high anolyte concentration in the electrokinetic remediation of chromium(VI)-contaminated soil. J. Hazard Mater. 229, 282–291.
Liang, S., Chen, J., Guo, M., Feng, D., Liu, L., Qi, T., 2020. Utilization of pretreated municipal solid waste incineration fly ash for cement-stabilized soil. Waste Manage. (Tucson, Ariz.) 105, 425–432.
Lim, M., Han, G.C., Ahn, J.W., You, K.S., 2010. Environmental remediation and conversion of carbon dioxide (CO_2) into useful green products by accelerated carbonation technology. Int. J. Environ. Res. Publ. Health 7, 203–228.
Liu, F., Liu, H., Yang, N., Wang, L., 2021. Comparative study of municipal solid waste incinerator fly ash reutilization in China: environmental and economic performances. Resour. Conserv. Recycl. 169, 105541.
Lu, M., Ge, W., Xia, Y., Sun, C., Lin, X., Tsang, D.C.W., Ling, T.C., Hu, Y., Wang, L., Yan, J., 2024. Upcycling MSWI fly ash into green binders via flue gas-enhanced wet carbonation. J. Clean. Prod. 440, 141013.
Mangialardi, T., 2003. Disposal of mswi fly ash through a combined washing-immobilisation process. J. Hazard Mater. 98 (1–3), 225–324.
Mao, Y., Wu, H., Wang, W., Jia, M., Che, X., 2020. Pretreatment of municipal solid waste incineration fly ash and preparation of solid waste source sulphoaluminate cementitious material. J. Hazard Mater. 385, 121580.
Nguyen, T.T., Tsai, C.K., Horng, J.J., 2022. Sustainable recovery of valuable nanoporous materials from high-chlorine MSWI fly ash by ultrasound with organic acids. Molecules 27, 2289.
Ni, P., Xiong, Z., Tian, C., Li, H., Zhao, Y., Zhang, J., Zheng, C., 2017. Influence of carbonation under oxy-fuel combustion flue gas on the leach ability of heavy metals in MSWI fly ash. Waste Manage. (Tucson, Ariz.) 67, 171–180.
Page, M.M., Page, C.L., 2002. Electroremediation of contaminated soils. J. Environ. Eng. 128, 208.

Pan, S., Chang, E.E., Chiang, P., 2012. CO_2 capture by accelerated carbonation of alkaline wastes: a review on its principles and applications. Aerosol Air Qual. Res. 12, 770–791.

Patil, P.B., Kothavale, V.P., 2022. Functionalized magnetic iron oxide-based composites as adsorbents for the removal of heavy metals from wastewater. In: Advances in Metal Oxides and Their Composites for Emerging Applications, vol 12, pp. 401–424.

Pazos, M., Kirkelund, G.M., Ottosen, L.M., 2010. Electrodialytic treatment for metal removal from sewage sludge ash from fluidized bed combustion. J. Hazard Mater. 176, 1073–1078.

Pedersen, A.J., 2002. Evaluation of assisting agents for electrodialytic removal of Cd, Pb, Zn, Cu and Cr from MSWI fly ash. J. Hazard Mater. 95, 185–198.

Pedersen, A.J., Ottosen, L.M., Villumsen, A., 2005. Electrodialytic removal of heavy metals from municipal solid waste incineration fly ash using ammonium citrate as assisting agent. J. Hazard Mater. 122, 103–109.

Peng, X., Jiao, B., Yu, L., Li, D., Yang, K., 2011. Enhanced electrokinetic removal of heavy metals in MSWI fly ash assisted by cation exchange membranes. Appl. Mech. Mater. 84–85, 626–630.

Peter, M.A., Muntean, A., Meier, S.A., Böhm, M., 2008. Competition of several carbonation reactions in concrete: a parametric study. Cement Concr. Res. 38, 1385–1393.

Phua, Z., Giannis, A., Dong, Z., Lisak, G., Ng, W.J., 2019. Characteristics of incineration ash for sustainable treatment and reutilization. Environ. Sci. Pollut. Res. 26, 16974–16997.

Polettini, A., Pomi, R., Stramazzo, A., 2016. CO_2 sequestration through aqueous accelerated carbonation of BOF slag: a factorial study of parameters effects. J. Environ. Manag. 167, 185–195.

Qin, J., Zhang, Y., Yi, Y., Fang, M., 2022. Carbonation treatment of gasification fly ash from municipal solid waste using sodium carbonate and sodium bicarbonate solutions. Environ. Pollut. 299, 118906.

Quina, M.J., Bordado, J.C., Quinta-Ferreira, R.M., 2008. Treatment and use of air pollution control residues from MSW incineration: an overview. Waste Manage. (Tucson, Ariz.) 28, 2097–2121.

Rabaey, K., Angenent, L., Schröder, U., Keller, J., 2009. Bioelectrochemical Systems: From Extracellular Electron Transfer to Biotechnological Application. IWA Publishing, London, UK.

Sabbas, T., Polettini, A., Pomi, R., Astrup, T., Hjelmar, O., Mostbauer, P., Cappai, G., Magel, G., Salhofer, S., Speiser, C., Heuss-Assbichler, S., Klein, R., Lechner, P., 2003. Management of municipal solid waste incineration residues. Waste Manage. (Tucson, Ariz.) 23, 61–88.

Saillio, M., Baroghel-Bouny, V., Pradelle, S., Bertin, M., Vincent, J., d'Espinose de Lacaillerie, J.-B., 2021. Effect of supplementary cementitious materials on carbonation of cement pastes. Cement Concr. Res. 142, 106358.

Sawyer, C.N., McCarty, P.L., Parkin, G.F., 1994. Chemistry for Environmental Engineering, fourth ed. McGraw-Hill, New York.

Sorrentino, G.P., Zanoletti, A., Ducoli, S., Zacco, A., Lora, P., Invernizzi, C.M., Marcoberardino, G.D., Depero, L.E., Bontempi, E., 2023. Accelerated and natural carbonation of a municipal solid waste incineration (MSWI) fly ash mixture: basic strategies for higher carbon dioxide sequestration and reliable mass quantification. Environ. Res. 217, 114805.

Sun, X., Yi, Y., 2021. Acid washing of incineration bottom ash of municipal solid waste: effects of ph on removal and leaching of heavy metals. Waste Manage. (Tucson, Ariz.) 120, 183–192.

Tang, J., Su, M., Wu, Q., Wei, L., Wang, N.,, Xiao, E., Zhang, H., Wei, L., Wang, N., Xiao, E., Zhang, H., Wei, Y., Liu, Y., Ekberg, C., Steenari, B.M., Xiao, T., 2019. Highly efficient recovery and clean-up of four heavy metals from MSWI fly ash by integrating leaching, selective extraction and adsorption. J. Clean. Prod. 234, 139–149.

Tao, H.C., Lei, T., Shi, G., Sun, X.N., Wei, X.Y., Zhang, L.J., Wu, W.M., 2014. Removal of heavy metals from fly ash leachate using combined bioelectrochemical systems and electrolysis. J. Hazard Mater. 264, 1–7.

Vane, L.M., Zang, G.M., 1997. Effect of aqueous phase properties on clay particle zeta potential and electro-osmotic permeability: implications for electro-kinetic soil remediation processes. J. Hazard Mater. 55, 1–22.

Wang, H., Zhao, B., Zhu, F., Chen, Q., Zhou, T., Wang, Y., 2023. Study on the reduction of chlorine and heavy metals in municipal solid waste incineration fly ash by organic acid and microwave treatment and the variation of environmental risk of heavy metals. Sci. Total Environ. 870, 161929.

Wang, L., Jin, Y., Nie, Y., 2010a. Investigation of accelerated and natural carbonation of MSWI fly ash with a high content of Ca. J. Hazard Mater. 174, 334–343.

Wang, L., Jin, Y., Nie, Y., Li, R., 2010b. Recycling of municipal solid waste incineration fly ash for ordinary Portland cement production: a real-scale test. Resour. Conserv. Recycl. 54, 1428–1435.

Wang, Y., Pan, Y., Zhang, L., Yue, Y., Zhou, J., Xu, Y., Qian, G., 2015. Can washing-pretreatment eliminate the health risk of municipal solid waste incineration fly ash reuse? Ecotoxicol. Environ. Saf. 111, 177–184.

Wu, J., Wu, X., Wang, J., Wang, T., Zhang, Y., Pan, W.P., 2020. Speciation analysis of Hg, As, Pb, Cd, and Cr in fly ash at different ESP's hoppers. Fuel 280, 118688.

Xie, Q., Wang, D., Fu, D., Tao, H., Liu, S., 2022. Recovery of soluble chlorides from municipal solid waste incineration fly ash using evaporative crystallisation and flotation methods. Separ. Sci. Technol. 57 (14), 2276–2286.

Xu, B., Yi, Y., 2022. Treatment of ladle furnace slag by carbonation: carbon dioxide sequestration, heavy metal immobilization, and strength enhancement. Chemosphere 287, 132274.

Xu, H., Ding, M., Shen, K., Cui, J., Chen, W., 2017. Removal of aluminum from drinking water treatment sludge using vacuum electrokinetic technology. Chemosphere 173, 404–410.

Yang, J.S., Kwon, M.J., Choi, J., Baek, K., O'Loughlin, E.J., 2014. The transport behavior of As, Cu, Pb, and Zn during electrokinetic remediation of a contaminated soil using electrolyte conditioning. Chemosphere 117, 79–86.

Yang, Z., Tian, S., Ji, R., Liu, L., Wang, X., Zhang, Z., 2017. Effect of water-washing on the co-removal of chlorine and heavy metals in air pollution control residue from MSW incineration. Waste Manage. (Tucson, Ariz.) 68, 221–231.

Yeung, A.T., 2006. Contaminant extractability by electrokinetics. Environ. Eng. Sci. 23, 202–224.

Yeung, A.T., 2011. Milestone developments, myths, and future directions of electrokinetic remediation. Sep. Purif. Technol. 79, 124–132.

Yeung, A.T., Gu, Y.Y., 2011. A review on techniques to enhance electrochemical remediation of contaminated soils. J. Hazard Mater. 195, 11–29.

Yeung, A.T., Hsu, C., Menon, R.M., 1997. Physicochemical soil-contaminant interactions during electrokinetic extraction. J. Hazard Mater. 55, 221–237.

Yuan, Q., Yang, G., Zhang, Y., Wang, T., Wang, J., Romero, C.E., 2022. Supercritical CO_2 coupled with mechanical force to enhance carbonation of fly ash and heavy metal solidification. Fuel 315, 123154.

Yuan, Q., Zhang, Y., Wang, T., Wang, J., 2023. Characterization of heavy metals in fly ash stabilized by carbonation with supercritical CO_2 coupling mechanical force. J. CO_2 Util. 67, 102308.

Zevenhoven, R., Teir, S., Eloneva, S., 2008. Heat Optimisation of a staged gas-Solid mineral carbonation process for long-term CO_2 storage. Energy 33, 362–370.

Zha, X., Ning, J., Saafi, M., Dong, L., Dassekp, J.B.M., Ye, J.Q., 2019. Effect of supercritical carbonation on the strength and heavy metal retentionof cement-solidified fly ash. Cement Concr. Res. 120, 36–45.

Zha, X., Yu, M., Ye, J., Feng, G., 2015. Numerical modeling of supercritical carbonation process in cement-based materials. Cement Concr. Res. 72, 10–20.

Zhang, J., Mao, Y., Wang, W., Wang, X., Li, J., Jin, Y., Pang, D., 2022. A new co-processing mode of organic anaerobic fermentation liquid and municipal solid waste incineration fly ash. Waste Manage. (Tucson, Ariz.) 151, 70–80.

Zheng, Z., Wang, H., Guo, Y., Yang, L., Guo, S., Wu, S., 2015. Comparisons of fly ash and deposition between air and oxy-fuel combustion in bench-scale fluidized bed with limestone addition. J. Harbin Inst. Technol. 22, 78–84.

Chapter 10

Toxicity evaluation and environmental risk assessment methodology on combustion/incineration residues

Jian Sun[1], Le Fang[2], Zezhi Peng[1], Xinyi Niu[3], Hengjun Mei[4], Huiyan Li[4] and Hongguang Cui[4]
[1]*Department of Environmental Science and Engineering, Xi'an Jiaotong University, Xi'an, China;* [2]*RDC for Watershed Environmental Eco-Engineering, Advanced Institute of Natural Sciences, Beijing Normal University at Zhuhai, Zhuhai, China;* [3]*School of Human Settlements and Civil Engineering, Xi'an Jiaotong University, Xi'an, China;* [4]*The First Affiliated Hospital, Zhejiang University School of Medicine, Hangzhou, China*

1. Introduction

Combustion/incineration residues, generally referring to fly ash (FA, including coal combustion FA, sewage sludge, and municipal solid waste incineration [MSWI] FA) and bottom ash (BA, the same sources as FA), are produced by coal combustion and solid waste incineration (Bilibio et al., 2021; Zhang et al., 2021b). During the combustion process, the coal/sludge/waste forms molten mineral residues at high temperature, which then cool with flue gases and harden to form ashes. The larger and heavier ash particles fall to the bottom of the combustion chamber and thus named BA, while the lighter ones remain in the flue gas and are collected in the air pollution control device and are called FA (Smichowski et al., 2008; Verbinnen et al., 2017). During the ash formation process, potential toxic elements (PTEs) (lead, cadmium, mercury, molybdenum, nickel, selenium, etc.) and other contaminants (dioxins, PAHs, sulfate, chloride and acids, etc.) are easily attached to their surface due to their small size (Gan et al., 2021; Liu et al., 2022a). Additionally, the extremely small size promotes their abilities in harming ecological environment and human health. Hence, combustion/incineration residues are a typical industrial solid waste produced worldwide that causes severe environmental pollution due to its physical and chemical properties.

Combustion/incineration residues are massively produced worldwide. In terms of combustion, the statistics revealed that 150−200 million tons of BA and 600−800 million tons of FA were produced globally (Wang et al., 2020). For sewage sludge and municipal waste incineration, increasing number of BA and SA production has been recorded in recent years (Danish and Ozbakkaloglu, 2022; Wang et al., 2022). According to the internal concentrations of contaminants, combustion/incineration residues could be classified into general waste and hazardous waste, in which the former ones are widely used as secondary raw materials in the construction area while the latter ones should be strictly treated before disposal (Zhang et al., 2021a; Sun et al., 2022). The dramatically high production of these residues results in greatly economic investment, as well as negative impact on the environment and human health. Consequently, more economic-effective treatments on combustion/incineration residues are urgently needed, and the standards/regulations change accordingly. For instance, in August 2020, the first technical specification for pollution control of MSWI FA (HJ 1134, 2020) has been launched in China. The well-treated MSWI FA could be classified as a general waste or even resource for further applications if it fulfills the specific requirements in the national standard (GB 34330, 2017). More and more policies are tending to reuse the residues other than disposal.

The aforementioned forces draw forth lots of new treatment and recycling technologies in the past years. However, the majority of FA and BA ecotoxicity assessments are based on the toxicity characteristic leaching procedure (TCLP), which serves to represent the leachability and mobility of PTEs in hazardous waste (Intrakamhaeng et al., 2020). The TCLP procedures could neither reflect the biological magnification effect of PTEs in the food chain and water circulation system (Garg and Mishra, 2010; Alimba et al., 2016) nor provide the potential impacts to human health (Sun et al., 2022). In this

Treatment and Utilization of Combustion and Incineration Residues. https://doi.org/10.1016/B978-0-443-21536-0.00029-0

chapter, the existing leaching-based ecotoxicity assessments and relevant evaluation protocols are conducted to appraise the environmental safety of combustion/incineration residues.

2. Potential hazardous effects of compositions in combustion/incineration residues

The potential hazardous effects of combustion/incineration residues on the environment, ecology, and human health vary significantly depending on the type of fuel burned, the combustion conditions, and the disposal methods. Once enter into the environment, by either disposal or utilization, the combustion/incineration residues could post adverse impacts on air, water, soil, and even human health, which highly depend on the substance in residues.

2.1 Toxicity and environmental impacts of Pb

Lead (Pb) is a high-density metal with a relatively low melting point that is frequently employed in the fabrication processes of pipes, batteries, projectiles, and a diverse range of other items. Moreover, Pb finds applications in corrosion prevention, coatings, and shielding against radiation. Nevertheless, it is crucial to recognize that lead and its associated compounds possess toxic characteristics, thereby potentially engendering substantial hazards to the ecosystem. The contamination of the environment by Pb can manifest in multiple forms, including atmospheric deposition, aquatic systems, and soil, which, in turn, implicate deleterious consequences for organisms inhabiting these environments.

Pb, owing to its strong negative charge, readily forms covalent bonds with various elements such as iron, aluminum, manganese, oxides, organic matter, and carbonates. Consequently, its absorption by plants can detrimentally impact cell membranes while disrupting physiological and biochemical processes, thereby precipitating metabolic disorders. Additionally, airborne Pb particles have the potential to infiltrate plants via deposition (Zhou et al., 2019). Human exposure to Pb can occur through inhalation, ingestion, and dermal contact. Notably, children are particularly susceptible to atmospheric Pb exposure, which can pose grave harm to all bodily organs, particularly during the critical stages of childhood development. Importantly, it should be acknowledged that the effects of Pb on tissues and organs are enduring and irreversible, even in the presence of environmental pollution mitigation efforts.

Coal deposits often contain minerals that may include lead-bearing minerals. When coal is burned, these lead-bearing minerals are released and form FA, along with coal smoke and combustion by-products. Geological conditions and groundwater contamination can contribute to elevated lead levels in FA. In some areas, high lead concentrations exist in groundwater and soil. When underground water in coal interacts with mineral deposits, lead can dissolve and be released into FA through combustion.

2.2 Toxicity and environmental impacts of Cu

Copper (Cu) is a versatile industrial metal with excellent electrical and thermal conductivity, malleability, and corrosion resistance. It is widely used in various industries, electrical engineering, thermal applications, construction, manufacturing, and the arts. Cu alloys with other metals and is also utilized in craftsmanship and decoration. Furthermore, Cu is an essential trace element necessary for normal physiological functions in organisms.

While Cu is beneficial to the human body in appropriate amounts, exposure to high concentrations or prolonged exposure can be toxic. Its toxicity primarily stems from its impact on redox reactions and interactions with other heavy metals. Elevated Cu levels can induce oxidative stress, generate reactive oxygen species, and lead to cell damage and inflammation. Cu can also disrupt normal cellular functions and gene expression by binding to proteins and DNA. Acute Cu poisoning symptoms, such as nausea, vomiting, diarrhea, and abdominal pain, can occur from short-term ingestion of high doses. Chronic Cu toxicity resulting from long-term exposure to high concentrations can affect the liver, kidneys, and central nervous system.

During coal combustion, if the burned coal contains Cu or its compounds, Cu can potentially enter FA. Cu can exist in FA as an elemental form or in various compounds, such as Cu oxide and Cu sulfide, formed by reactions with other metallic and nonmetallic elements. The sources of Cu in FA primarily include the Cu content in the fuel itself, reactions with gases and particles during combustion, and the presence of Cu vapor in the flue gas. Cu vapor can condense into particulate matter as the flue gas cools and be collected with FA.

2.3 Toxicity and environmental impacts of As

Arsenic (As) is a nonmetallic element. As and its compounds are used in pesticides, herbicides, insecticides, and many kinds of alloys. As content in soil is increased by the use of As containing pesticides and sewage sludge (Tsang and Yip, 2014; Niazi et al., 2018; Shakoor et al., 2018). In addition, As in soil can come from tailings, waste, and sediments from metallurgical processes.

As is highly toxic and widely distributed in the environment. Trivalent arsenic (As(III)) is characterized by greater toxicity when compared with pentavalent arsenic (As (V)), while inorganic As exhibits higher toxicity levels than organic As compounds (Niazi et al., 2018). When As enters plant tissues from the soil, it disrupts metabolic processes. Arsenate interferes with mitochondrial phosphorylation, while arsenate reacts with sulfhydryl proteins, inactivating enzymes and impeding water uptake in plants. The detrimental impacts of As on plants are manifested through its influence on the absorption of water and nutrients, resulting in the accumulation of As in plant tissues and subsequent damage to chlorophyll. Moreover, the presence of As-laden tailings can contaminate soil, groundwater, and surface water, thereby posing hazards to plants, animals, and humans alike (Mandal and Suzuki, 2002). Organisms are susceptible to As exposure through water pollution as well as the consumption of plants, dairy products, and meat that have been contaminated with As. Arsenic poisoning can engender a range of adverse health conditions, including skin-related problems and an increased risk of developing certain types of cancers (Singh et al., 2015). As accumulates primarily in hair, nails, and teeth in the human body, with absorption levels varying based on the species and solubility of As (Singh et al., 2015).

During coal combustion, if the coal contains As or its compounds, As can be released into the flue gas and captured by FA particles. As can exist in different forms in FA, depending on factors such as coal composition, combustion conditions, and pollution control technologies. Concerns arise regarding the potential environmental impacts of As in FA. Improper treatment and disposal of FA may result in leaching of As and other pollutants into the surrounding environment, including soil and water bodies.

2.4 Toxicity and environmental impacts of Cd

Cadmium (Cd) is a bluish-white heavy metal with a relatively low melting point. It is naturally found in mineral form, often associated with zinc ores. Cd's ductility and malleability make it suitable for manufacturing alloys and batteries, and it is used in various industrial applications such as electroplating, coatings, plastics, and glass production. The widespread use of industrial materials containing Cd, including alloys, coatings, batteries, pigments, and plastic stabilizers, contributes to human exposure to Cd and its compounds through consumption.

However, Cd is highly toxic to humans. Exposure to Cd can occur through inhalation of Cd vapor or dust, ingestion of Cd-contaminated food or water, and dermal contact. High levels of Cd exposure have detrimental effects on various bodily systems, with the reproductive system, lungs, DNA, and kidneys being particularly vulnerable. Children exposed to high concentrations of Cd may experience learning difficulties, impaired cognition, behavioral issues, and deficits in neuromotor skills (Rizwan et al., 2018). Cd has a high capacity for accumulation in the human body, and even prolonged exposure to low doses can result in serious health problems. It can cause kidney dysfunction, leading to kidney disease and chronic renal failure. Cd is also associated with skeletal issues such as osteoporosis and bone fractures, cardiovascular diseases, respiratory problems, reproductive abnormalities, and certain types of cancer. Cd pollution in the environment is a significant concern, with industrial activities, waste disposal, coal combustion, and agricultural practices contributing to its release and accumulation. Cd accumulates in soil, enters crops and water sources, and ultimately threatens ecosystems and biodiversity.

During the incineration process, volatile Cd species from municipal solid waste (MSW) sources are primarily transferred to FA. During the combustion of MSW and coal, Cd is released into the flue gas. Some of the Cd condenses or gets adsorbed onto particles present in the flue gas. These particles, along with the Cd, are then retained in the FA (Zhang et al., 2018). Cd makes up $\sim$20% of the metal content found in FA and is primarily present in the nonmineral phase. This characteristic of Cd in FA contributes to its relatively higher mobility and release compared with other metals present in the ash. Due to its nonmineral nature, Cd can be more readily released into the environment under certain conditions (Du et al., 2018). In the flue gas, Cd primarily exists in particulate form and is collected as FA using fabric bags or electrostatic precipitators (Ahmad et al., 2018). Cd exhibits a greater potential for volatilization during incineration and gasification processes, while demonstrating moderate volatilization during pyrolysis and hydrothermal processes.

2.5 Toxicity and environmental impacts of Cr

Chromium (Cr) is a chemical element that exists primarily in trivalent and hexavalent forms. It has extensive applications in industries such as leather production, electroplating, chemical manufacturing, ceramics, alloys, pigments, and more (Tsang and Yip, 2014). Cr can be sourced from electronic waste, which contains a significant concentration of Cr. However, the disposal of electronic waste through landfilling restricts the recycling and reutilization of Cr-containing materials (Ashraf et al., 2017).

The toxicity of Cr is contingent upon its valence state and morphology. Hexavalent chromium (Cr(VI)) exhibits high toxicity and carcinogenic properties, surpassing trivalent chromium (Cr(III)) by a factor of 100 in terms of toxicity levels (Mandal et al., 2017; Xia et al., 2019). Leaching of Cr(VI) compounds can lead to the contamination of soil, groundwater, and rivers, thereby giving rise to ecological issues (Rajapaksha et al., 2013, 2018; Xia et al., 2019). Additionally, Cr(VI) poses significant risks to human health. Adverse health effects can occur as a result of inhalation of contaminated air, ingestion of food or water contaminated with (Cr(VI), or dermal contact with soil or water that is likewise contaminated. It can cause respiratory issues, gastrointestinal problems, and damage to the liver and kidneys.

Cr in FA refers to the presence of Cr elements and compounds in the solid waste generated during the combustion process. It primarily originates from the combustion of Cr-containing substances like coal, Cr ores, and alloy materials. Cr can exist in various forms in FA, including hexavalent and trivalent Cr. The concentration of Cr in FA depends on factors such as fuel composition, combustion conditions, waste gas treatment technologies, and FA handling methods. Coal-fired power plants and industrial boilers are the main sources of Cr in FA. Residues produced in these processes may contain soluble hexavalent Cr, which can leach into the soil and groundwater, posing risks to ecosystems and human health.

2.6 Toxicity and environmental impacts of Hg

Mercury (Hg) is a dense metallic element that exists in various forms, including Hg0 and Hg(II) species. It is widely used in household items, electronics, lighting, and dental applications, contributing to its release into the environment (Bartov et al., 2013; Hu et al., 2018).

Metallic Hg is highly toxic and poses risks to human health. It can cause neurological disorders, kidney damage, lung diseases, cardiovascular issues, liver damage, immune system dysfunction, and reproductive system abnormalities. Children and fetuses are particularly vulnerable to Hg's neurotoxic effects (Hu et al., 2018).

In flue gas, Hg is predominantly present in its oxidized state, with a small fraction as elemental Hg (Hg0) (Hu et al., 2018). Incineration processes release Hg from waste and supplementary fuels, such as coal. The presence of chloride in MSW plays a facilitating role in the oxidation of Hg and the subsequent development of chloromercury compounds within flue gas. Reactive oxidized mercury species, specifically Hg^{2+}, have the ability to adsorb onto carbon particles and can be effectively eliminated through the utilization of particulate matter control devices or wet scrubbers. However, volatile Hg0 is difficult to remove from the flue gas, but it can be oxidized to Hg^{2+} by selective catalytic reduction devices (Hu et al., 2018). Consequently, Hg concentrations in FA and flue gas desulphurization by-products are comparable (O'Connor et al., 2019). Hg in coal combustion residues is mainly associated with sulfur, chlorine, and carbon, with concentrations ranging from 0.001 to 10 mg/kg. The capture of Hg increases with lower flue gas temperatures and increased activated carbon presence (Deonarine et al., 2013; Hood et al., 2017).

2.7 Toxicity and environmental impacts of dioxin

Dioxins, including polychlorinated dibenzo-p-dioxins (PCDDs) and polychlorinated dibenzofurans (PCDFs), are a group of compounds with similar chemical properties, characterized by two benzene rings connected by oxygen atoms. They have high melting points, low vapor pressures, and low solubility in water but a strong affinity for particulate matter surfaces. The water solubility of dioxins decreases with increasing chlorine content, while their solubility in organic solvents and fats increases.

Dioxins have been shown to pose various health risks to humans and vertebrates. These risks include cancer, immune deficiency, nervous system disorders, endocrine disruption, pulmonary function impairment, altered testosterone levels, eye and gum pathology, nausea, skin rashes, liver damage, elevated cholesterol and triglycerides, and enamel hypomineralization in children (Kociba et al., 1976).

Dioxins are formed during combustion processes involving the incomplete burning of chlorine-containing organic compounds such as plastics, chlorinated solvents, and certain chemicals. When these substances are present in the fuel or waste being burned, complex chemical reactions and thermal degradation occur, leading to dioxin formation. The primary pathway for dioxin formation in FA is the adsorption of gas-phase dioxins onto fine ash particles. The concentration of dioxins in FA varies depending on factors such as fuel or waste composition, combustion conditions, and the efficiency of air pollution control devices. Higher levels of dioxin formation in FA can be attributed to the presence of chlorinated compounds and higher combustion temperatures.

2.8 Toxicity and environmental impacts of other toxic substances

Combustion and incineration residues commonly contain a variety of toxic substances, including polychlorinated biphenyls (PCBs), polyaromatic hydrocarbons (PAHs), monomethyl and dimethyl sulfate, and others (Sahu et al., 2004). The presence of high levels of PAHs in FA can be attributed to the abundant organic matter in coal (Sahu et al., 2009). Lower-molecular-weight PAHs have also been reported in FA (Shaheen et al., 2014).

PCBs and PAHs are harmful substances with significant environmental impacts. PCBs are synthetic organic chemicals widely used in industries such as electrical equipment, hydraulic fluids, and flame retardants. They are persistent in the environment, bioaccumulate in the food chain, and are classified as probable human carcinogens. PCB exposure can lead to cancer, particularly affecting the liver and skin. It can also cause immune system dysfunction, reproductive and developmental disorders, and neurological effects, including cognitive impairments and behavioral abnormalities. PAHs, on the other hand, are organic compounds formed during incomplete combustion or pyrolysis of organic materials such as coal and petroleum. They are released into the environment through various sources, including vehicle emissions, industrial processes, and fossil fuel burning. Several PAHs are classified as carcinogens and can enter the human body through inhalation, ingestion, or skin contact. PAH exposure is associated with an increased risk of cancer, particularly lung, skin, bladder, and gastrointestinal cancers. It can also cause respiratory problems and skin irritations and have adverse effects on the immune and reproductive systems. Both PCBs and PAHs are persistent organic pollutants that can have long-lasting effects on the environment.

Radioactive elements are also present in combustion and incineration residues, although their concentrations are generally not high. Uranium and thorium are commonly observed in FA, with average concentrations ranging from 15 to 110 mg/kg for uranium and about 10–25 mg/kg for thorium. These radioactive elements are mainly found in the finer particles of FA. FA may also contain radioactive isotopes of ruthenium, such as ^{222}Ru and ^{220}Ru.

Uranium and thorium are products of radioactive decay, releasing alpha particles, beta particles, and gamma rays. These radioactive particles can penetrate biological tissues and interact with cells, posing potential risks to human health. Prolonged exposure to high concentrations of uranium and thorium can result in genetic mutations, cell damage, and increased cancer risks, particularly lung and bone cancer. Ingestion or inhalation of high concentrations of these radioactive elements can lead to radioactive poisoning, with symptoms such as nausea, vomiting, diarrhea, dehydration, and kidney damage. Long-term exposure to high concentrations can cause radiation-induced skin damage and lung diseases, such as radiation dermatitis, skin ulcers, and pulmonary fibrosis.

3. Ecological environment and human health assessment methodology

Understanding and quantifying the adverse impacts of combustion/incineration residues on ecosystems, the environment, and human health is indeed crucial. In this section, we introduce some of the most prevalent evaluation methods used for this purpose.

3.1 Leaching-based assessment methods

The leaching testing is the widely used to evaluation the environmental risk assessment for solid waste, concrete or soil, etc. (Liu et al., 2022b). In natural environment, the leached PTEs may cause water or soil pollution therefore posing risk to plants, animals, and humans (Kim et al., 2009). Leaching is the vertical profile of PTEs effusing from studied solids triggered by natural water flushing (Li et al., 2022), which is determined by the adsorption–desorption behavior or distribution of PTEs between liquid and solid (Zhai et al., 2019). Therefore, the leaching-based assessment could simulate the rainfall leaching and study the migration patterns and leaching kinetics of PTEs.

TCLP is a standard extraction procedure developed by the United States Environmental Protection Agency (USEPA), which applied two kinds of leaching fluids to determine the contamination levels (Mantis et al., 2005; Sima et al., 2015). To be specific, the experimental subject is crushed into particles with a maximum particle size of 9.5 mm (US EPA TCLP) or 2.4 mm (Australian Standard); then, acetic acid solution ($pH = 2.88 \pm 0.05$) and acetate buffer solution ($pH = 4.92 \pm 0.05$) are successively used to leach the treated samples (Poon and Lio, 1997; Halim et al., 2003). TCLP is extensively used in environmental remediation field to evaluate the leachability of heavy metal or organics in amended contaminated soils or industrial wastes (Halim et al., 2003; Tsang et al., 2013). In codisposal landfills, the TCLP represented the worst-case scenarios. However, the applicability of TCLP in evaluating of some waste disposals is challenged due to the neglecting of some importation factors, such as pH, redox potential, biological activity, retention time, etc. (Sima et al., 2015). Taking an example of assessing heavy metal leachability in P-amended soils by TCLP, fluid chemicals in soil would affect the chemical speciation of heavy metals, therefore causing deviations of results. Hooper et al. found that the concentrations of elements that form oxyanions (Sb, As, Mo, Se, V) would be underestimated due to the complexation with acetate ions in leaching fluids, especially at high pH values (Halim et al., 2003). In assessing the heavy metals leaching from cementitious waste, a high acid neutralizing capacity of cementitious wastes quickly cause heavy metal precipitation. Therefore, the TCLP may not correctly reflect the hazards of wastes, more procedures are necessary to combat the deficiency of TCLP under specific condition, such as leaching environmental assessment framework (LEAF), which can characterize metal leachability in multiple disposal scenarios for a wider range of assessing materials (Kosson et al., 2014). It was identified that the LEAF method can lead to overall more resource efficient, cost-effective secondary material usage, and disposal of wastes (Garrabrants et al., 2021). In addition, the indoor column experiments can be used to study the long-term releasing behaviors of PTEs.

To correctly reflect the leaching assessment of specific conditions, the approach should take consideration of (i) detail requirement level for the assessment; (ii) available supporting information/data; (iii) the uncertainty with characterization of materials; and (iv) the uncertainty related to field conditions under the possible leaching scenarios (Garrabrants et al., 2021). The scenario-based assessment can correct and refine assess the leaching characteristics of studied materials.

3.2 USEPA risk assessment

The United States Environmental Protection Agency (USEPA) developed environmental risk assessment (ERA) as a tool to evaluate the potential health impacts of environmental contaminants (USEPA, 1989). ERA involves identifying, characterizing, and assessing the risks associated with exposure to hazardous substances, including chemicals, radiation, and biological agents. ERA has become a widely used tool for environmental risk assessment, including those exposure to combustion/incineration residues (Hung et al., 2009; Zhou et al., 2015). The ERA model involves a quantitative evaluation of the potential risks associated with exposure to toxic substances. The calculation method for this risk model is as follows.

Step 1: Hazard identification
In this step, the potential hazards or adverse effects associated with exposure to a particular chemical or substance are identified through a review of scientific literature and research, which could be assessed in USEPA guideline as well (https://iris.epa.gov). These hazards are typically expressed in terms of the type and severity of adverse health effects, i.e., cancer and noncancer effects.

Step 2: Dose—response assessment
Quantitative methods are used to determine the relationship between the magnitude of exposure to a chemical or substance and the likelihood and severity of adverse health effects. Dose—response models are developed, which estimate the probability of adverse effects occurring at different levels of exposure.

Step 3: Exposure assessment
The magnitude and duration of exposure to the chemical or substance are estimated, along with the potential routes of exposure (i.e., inhalation, oral, dermal). Exposure levels are determined through measurements in different environmental media (e.g., air, water, soil) and personal monitoring for high-risk groups. The predicted average daily dose (mg element kg^{-1} bodyweight day^{-1}) received through ingestion, inhalation, and dermal absorption was calculated according to the following equations:

$$D_{ing} = \frac{C \times IngR \times EF \times ED}{BW \times AT} \times 10^{-6} \tag{10.1}$$

$$D_{inh} = \frac{C \times InhR \times EF \times ED}{PEF \times BW \times AT} \times 10^{-6} \tag{10.2}$$

$$D_{der} = \frac{C \times SL \times SA \times ABS \times EF \times ED}{BW \times AT} \times 10^{-6} \tag{10.3}$$

where C stands for the metal or toxics concentrations in combustion residues (mg/kg); IngR is ingestion rate (children: 200 mg/d, adults: 100 mg/d); Inh represents inhalation rate (children: 7.63 m^3/d adults: 20 m^3/d; EF is the exposure frequency (depends on exposure scenarios, unit: d/y; ED represents the exposure duration (e.g., 6 years for children and 24 years for adults); BW is the body weight of exposed individual (children: 15 kg; adults: 70 kg); AT is the time period over which the dose was averaged (noncarcinogens: ED × 365 d; carcinogens: 70 × 365 = 25,550 d); PEF is the emission factor (1×10^{-9} m^3/kg), transforming the solid-based concentration into air-based concentration; SL is skin adherence factor (children: 0.2 $mg/cm^2/d$; adults: 0.7 $mg/cm^2/d$); SA represents the exposed skin surface area (children: 2800 cm^2; adults: 5700 cm^2); and finally ABS stands for the dermal absorption factor (0.001) (Zhou et al., 2015; Huang et al., 2016). Further parameters were listed in ERA Guideline by USEPA (https://www.epa.gov/expobox).

Step 4: Risk characterization

The information generated from the previous steps is integrated to assess the potential risks associated with exposure to the chemical or substance. Uncertainties and variability in the data are also considered in the risk characterization. The result of the risk characterization is expressed as a numerical estimate of risk, typically in terms of likelihood or probability of adverse health effects occurring, as follows:

$$HI = \sum HQ_i = \sum \frac{D_i}{RfD_i} \tag{10.4}$$

$$ILCR = \sum CR_i = \sum D_i \times SF_i \tag{10.5}$$

The hazard index (HI) is the sum of hazard quotients (HQs) and estimates the health risk of different exposure pathways. The reference dose (RfD_i) (mg/kg/d) estimates the maximum permissible risk to a human population through daily exposure during a lifetime. The SF and RfD values were obtained from the United States Department of Energy's Risk Assessment Information System (RAIS) compilation (Energy, 2004), and values of HQ and HI > 1 indicate a high probability of the occurrence of adverse health effects (Li and Ji, 2017).

For an estimation of the carcinogenic risk, the dose is multiplied by the corresponding slope factor (SF) to produce an estimate of cancer risk. Similarly, the incremental lifetime carcinogenic risk (ILCR) was calculated by summing the individual cancer risk across different exposure pathways. The acceptable threshold value of the cancer risk is 1.0×10^{-4}, while the tolerable ILCR for regulatory purposes is in the range of 1.0×10^{-6}–1.0×10^{-4} (USEPA, 1996).

Overall, the risk model developed by USEPA is a scientifically rigorous and quantitative method for evaluating the risks associated with chemical exposure. It provides a framework for decision-making aimed at minimizing the potential risks to human health and the environment. Both cancer and noncancer risk assessments could be used in evaluating the risk exposure to combustion/incineration residues combustion/incineration residues. These calculations help in understanding the potential risk posed by the contaminants in the residues ether in direct emission or in treated status.

3.3 Biological assessment

The biological assessment is essential for evaluating the biological activity of given raw materials or semiproduct regarding putrescibility, emission of odor compounds, and leaching of ions/cations (Bayard et al., 2010). This assessment procedure and testing factors are more assisted with their application situations. For organic wastes applying in fermentation or animal feeding, the biological activity assessing methods such as respiration test were normally used. And the respirometric techniques including dynamic respiration index (DRI), static respiration index (SRI), and specific oxygen uptake rate (SOUR) test could determine the biological stability of solid materials (Lasaridi and Stentiford, 1998; Adani et al., 2006; Barrena et al., 2009). In these testing processes, hydrolytic enzymes are essential for the depolymerization of organic wastes, which includes cellulases, hemicellulases, proteases, lipases, phosphatases, and arlysulfatases (Goyal et al., 2005). Cow dung possesses rich microflora and is effective in degradation of organic waste, which is added to biomass to improve the chemical and physical properties with composting process (Randhawa and Kullar, 2011; Mishra and Yadav, 2022). In addition, studies found that the C/N content and particle size are important factor for composting, and the heat generated in time and the consumption of dissolved oxygen are evaluated, studied, and fitted by models. To be specific, the finer composes release more nitrogen and phosphorus that coarse composts. And the fine fraction has higher N mineralization than that in coarse fraction.

Many recyclable waste materials or treated semiproduct, especially for P contained wastes (such as biochar), are suggested for using in soil improvement, landscaping, and carbon sequestration for plant growing or combating soil degradation (Heiskanen et al., 2022). For safety soil application of these materials, their effect on soil structure, fertility, and plant growth is increasingly emphasized (Esse et al., 2001; Goyal et al., 2005). Therefore, methods of germination index testing, composting in detection of microflora and enzyme activities, and pot experiment of plant cultivation testing are applied. To be specific, the germination index is considered as a reliable indirect quantification of compost maturity because the plant germ is highly sensitive to toxic and harmful substances released from wastes. Pot experiment evaluates the transformation of heavy metals from soil to plants and their enrichment in plants, therefore assessing the potential risk of heavy metals with the food chain enrichment. Hence, pot experiment affords more convictive and robust data for evaluating the risk of wastes, compared with directly applying the maximum allowable concentration of regulations, such as (<4 mg/kg), Cd (<1 mg/kg), and Pb (<30 mg/kg), which was regulated for animal feeding.

Besides pot experiment or fermentation experiments, the potential risk and the modified potential ecological risk index (RI) could be evaluated based on the metal fraction experimental results. Prior to calculation, four heavy metal fractions of wastes were tested: acid-soluble (F1), reducible (F2), oxidizable (F3), and residual (F4). Then the potential risk and RI could be calculated as follows (Eqs. 10.1–10.3):

$$C_f = C_i/C_n \tag{10.6}$$

$$E_r = T_r \times C_f \tag{10.7}$$

$$RI = \sum E_r \tag{10.8}$$

where C_f is the contamination factor of heavy metal; C_i is the concentration of active heavy metal fraction (F1 + F2 + F3); C_n is the heavy metal content of biochar in residual fraction (F4); E_r is the potential ecological index for heavy metals; T_r is the toxic factor of metals.

To sum up, the biological assessment is necessary for the reuse of raw materials or semiproduct and the laboratory experiment should take in consideration of their reuse scenario in practical application.

3.4 In vitro and in vivo cytotoxicity assessment

Combustion/incineration residues are released into the environment directly in effluents from combustion/incineration process, or indirectly after various treatment. The toxic chemicals (e.g., heavy metals and toxic organics) in the residues may pose hazards to workers/residents who have opportunity to exposure to the residues (Netkueakul et al., 2022; Gao et al., 2023). For instance, coal combustion FA could pose an inhalation hazard to exposure workers (Borm, 1997; Zierold and Odoh, 2020). As traditional risk model or leaching test could not reflect the toxicity to certain tissues/organs in human, in vitro and in vivo toxicity evaluation methods are thus developed.

Taking in vitro study as an example, the procedures are as follows:

(1) Sample preparation
Generally, two situations are considered for different preparations, one is residue intake or direct into human body and the other is indirect from water, soil for food. For the first situation, solid residues could be dissolved into biological buffer system (e.g., phosphate-buffered saline (PBS) solution) with certain concentrations. For the second situation, residue leaching should be conducted. Because the leaching solutions of TCLP contain additional acids and salts, which are all potentially toxic, a nontoxic blank solution is needed for the investigation of cytotoxicity of dissolution material from residues. Alternatively, the residue samples are extracted by distilled and deionized water (DDW, a resistivity of 18 MΩ cm) with the certain ratio (typically 1 g in 10 mL DDW for 16 h). Then, the filtered leaching solution (i.e., distilled and deionized water leaching [DDWL]) is freeze-dried or nitrogen-blow to remove the solvent. The remaining extracts are dissolved in biological buffer system (e.g., phosphate-buffered saline [PBS] solution), and diluted to different concentrations for dose-dependent experiments. Usually, the PBS diluent alone should be applied as a control (Sun et al., 2020). The concentration of final solution was calculated according to the exposure dose of residues, referring exposure realistic scenarios or published studies (Abbas et al., 2019; Sun et al., 2022).

(2) Cell culture and exposure
Target cell should be selected according to the potential exposure pathway of residues. In inhalation pathway, the A549 cells (human alveolar basal epithelial cells in lung cancer) are widely used to determine the toxicity to human lung, which could be purchased commercially. A549 cells were cultured using F-12 cell culture medium (Thermo Fisher Scientific Inc., Waltham, MA, United States) supplemented with 10% heat-inactivated fetal bovine serum (Biowest, Riverside, MO, United States) and 1% antibiotics penicillin/streptomycin (100 units/mL), where a humidified incubator supplied with 5% CO_2 at 37°C was employed. The A549 cells were inoculated at a density of 1×10^5 cells/mL into 24-well transwells (Thermo Fisher Scientific Inc., Waltham, MA, United States) and incubated for 24 h. The cultured cells were then exposed to DDWL of MIFA for 24 h before bioreactivity measurement. For different cells, the cell culture should strictly follow the manufacturer's specification.

(3) Bioreactivity and cytotoxicity measurement
After exposure for certain duration, the cells are then prepared for bioreactivity and cytotoxicity measurement. Bioreactivities, defined as the reactivities of cells in response to the stimulation of residues or residue leachates, involve cell variability, cell cycle, cell apoptosis, cell morphology, etc. Cell variability is usually measured by 3-(4,5-dimethylthiazol-2-yl)-2,5-diphenyltetrazolium bromide (MTT) method, while cell cycle and cell apoptosis are determined by flow cytometry. Cell growth and morphological changes were observed under the inverted microscope. In addition, cytotoxicity, defined as expression of cytokines, is determined as well. Cytotoxicity could be roughly classified into two groups, namely, oxidative stress and inflammation factors. Oxidative stress could be determined by the expression of cellular reactive oxygen species (ROS), malondialdehyde (MDA), superoxide dismutase (SOD), as well as some oxidative products such as 8-hydroxy-2′-deoxyguanosine (8-OHdg). They reflect the oxidative stress level in cells stimulated by residues or residue leachates. On the other hand, inflammation cytokines include monocyte chemoattractant protein-1 (MCP-1), interleukins (e.g., IL-1β, IL-6, IL-8, etc.), TNF-α, etc. They reflect the inflammation level in cells induced by residues or residue leachates.

Generally, both bioreactivities and cytotoxicity should be determined in in vitro studies; however, the detail parameters and cytokines could be highly case-dependent. The general experimental process of an in vitro study on combustion/incineration residues is shown in Fig. 10.1.

For in vivo studies, procedures are as follows:

(1) Sample preparation
Follow the same procedure of in vitro.

(2) Animal culture and exposure

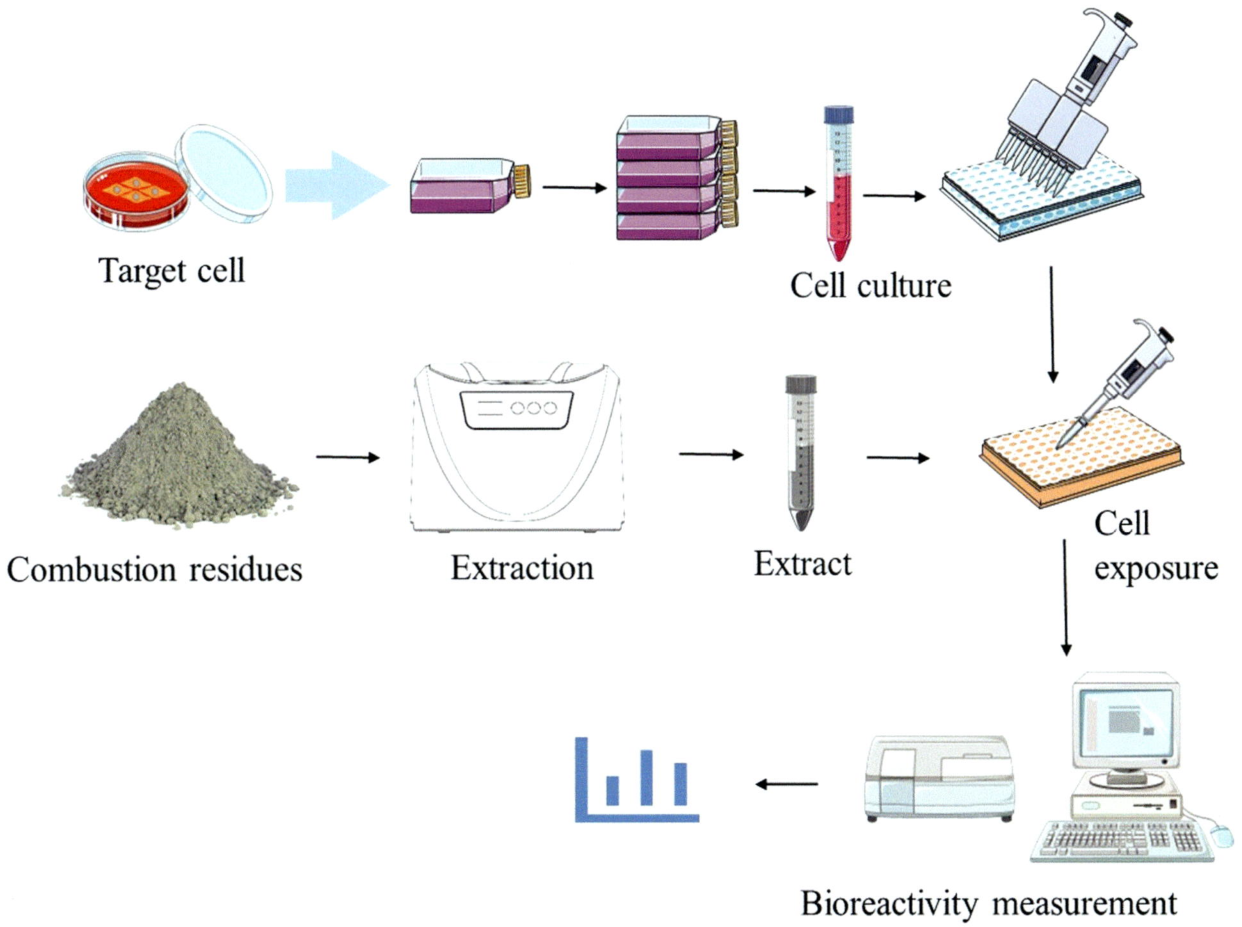

FIGURE 10.1 General flowchart of the in vitro assessment on toxicity of combustion/incineration residues.

Animal model should be selected according to the research purpose and potential exposure pathway of residues. Rat and mouse are the most commonly used animals in vivo (Wang et al., 2020). After purchase, animals should be fed for several days to fit the experimental environment. The well-prepared animals are then classified into different groups randomly; the animal count in each group should be odd number (generally ≥ 5). According to the exposure pathway of residues, the prepared samples could be fed in different methods, such as nasal drip simulating the inhalation pathway and intragastric administration for ingestion pathway. For each group, the residue type and dose should be constant during the experiment and at least one group should be set as control group. After certain duration of exposure, euthanasia would be done to the animals for further pathological and toxicological measurement.

(3) Pathological and toxicological measurement

In vivo toxicological measurements are similar to those in vitro. The pathological measurements are mainly based on the and sectioning techniques (León-Mejía et al., 2018; Saba et al., 2020). In addition, some new techniques such as in vivo imaging system and isotope labeling are getting popular, which help to locate the metabolic and cumulative pathways of certain pollutants (Cieplik et al., 2006; El-Sayed et al., 2013). Detail methods and development of pathological and toxicological measurement are shown in relative studies (Flora et al., 2013; León-Mejía et al., 2018). The general experimental process of an in vivo study on combustion/incineration residues is shown in Fig. 10.2.

3.5 Epidemiological study

Due to ethical limitations, toxicological experiments cannot be conducted on human subjects. However, epidemiological investigations are encouraged and getting popular as they cold directly reflect the toxicity of residues to human and provide useful information of how the toxicity happened in human body. The following are some of the methods that can be used for the epidemiological investigation of combustion/incineration residues.

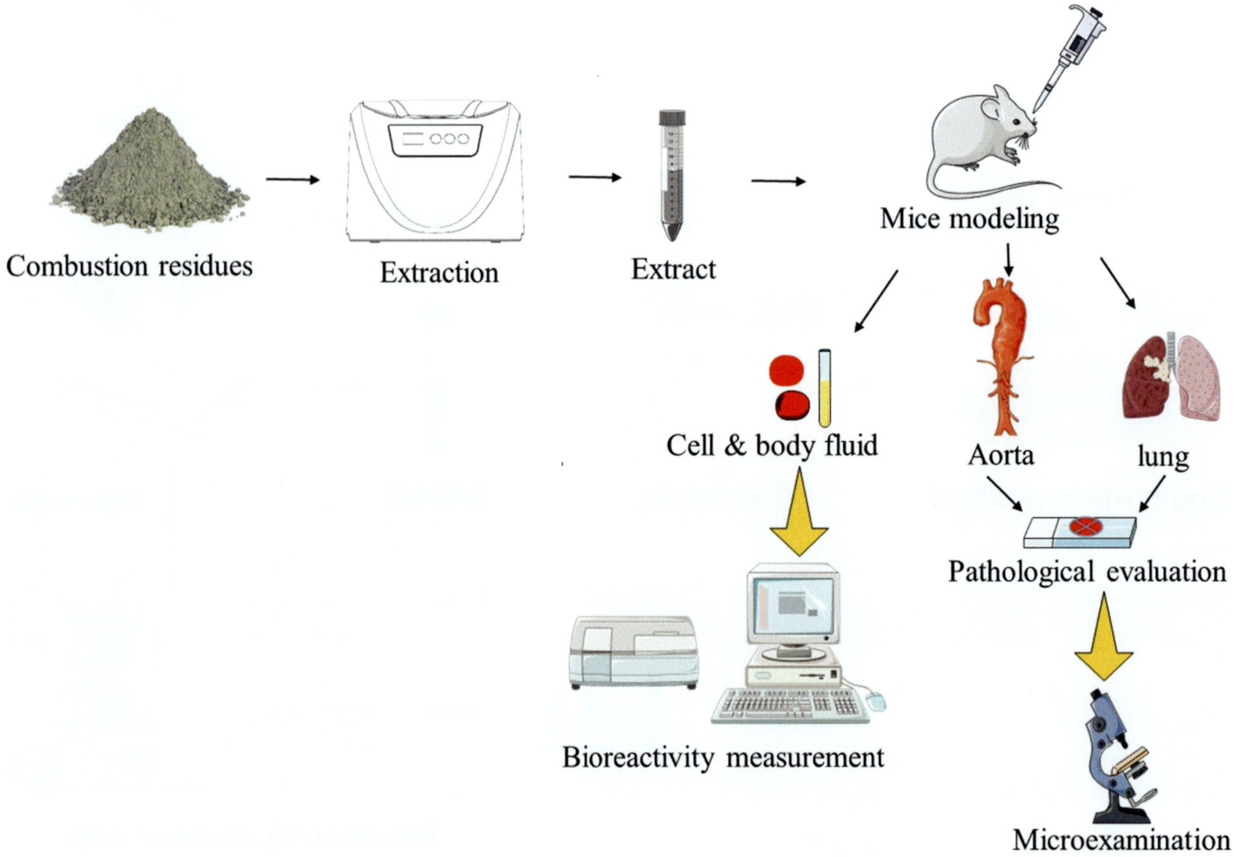

FIGURE 10.2 General flowchart of the in vivo assessment on toxicity of combustion/incineration residues.

(1) Case-control study

This type of study compares individuals who develop a specific health outcome that is potentially related to exposure to combustion/incineration residues, with individuals who have not developed the outcome, to determine whether exposure to combustion residues is a risk factor for that condition. Exposure dose of combustion/incineration residues is required to be measurable, such as environmental monitoring stations or personal monitors. More information could be provided by questionnaires, medical records, or other data sources. By using appropriate statistical models, the association between exposure to combustion/incineration residues and the health outcome could be determined. Meanwhile, the statistical results should be interpreted in light of the study design and the limitations of the study. The strengths and weaknesses of the study should be identified, and their implications for public health policies and interventions should be discussed.

Case–control studies on combustion/incineration residues provide important information on the potential linkage between residue exposure and health effects. The small sample number of participants involved in case–control study makes it easy to conduct; however, it results in some shortcomings at the same time, such as the unreliable statistical results, the effect of accidental error, etc. (Ghosh et al., 2019). Therefore, studies with more completed experimental design, more participants, and longer duration are developed.

(2) Cohort study

Cohort study is an observational research method that divides a certain population into different subgroups according to whether they are exposed to a suspicious factor or the degree of exposure, tracks the occurrence of outcomes (such as diseases) among members of two or more groups, compares the differences in the incidence of outcomes among groups, and determines whether there is a causal correlation between these factors and the outcome and the degree of correlation (Tait et al., 2020). This type of study involves following a group of individuals who have been exposed to combustion/incineration residues over a period of time and monitoring their health outcomes. Briefly, cohort studies on combustion/incineration residues can be broken down into the following steps:

Step 1: Define the study population—The study population should be defined based on the exposure to combustion residues. This can include people living in areas with high levels of residue pollution, workers exposed to combustion residues in industrial settings, or individuals living near waste incineration facilities.

Step 2: Recruitment of participants—Participants should be recruited from the defined study population. Recruitment can be done through community events, advertisements, and medical facilities.

Step 3: Measurement of exposure—Exposure to combustion/incineration residues can be measured using various tools, such as environmental monitoring stations or personal monitors. Biological markers can also be used to assess exposure.

Step 4: Baseline assessment—A baseline assessment is conducted at the beginning of the study to collect information on potential confounders, such as lifestyle factors and preexisting medical conditions.

Step 5: Follow-up assessments—Follow-up assessments are conducted at regular intervals to collect information on health outcomes and changes in exposure to combustion/incineration residues.

Step 6: Data analysis—The results should be analyzed using appropriate statistical methods to determine the association between exposure to combustion/incineration residues and health outcomes.

Step 7: Interpretation of results—The results should be interpreted in light of the study design and the limitations of the study. The strengths and weaknesses of the study should be identified, and their implications for public health policies and interventions should be discussed.

Cohort studies on combustion/incineration residues provide important information on the long-term health effects of exposure to these pollutants. Considering the strong health impacts of incineration residues, cohort studies on incineration residues are frequently reported (Ranzi et al., 2011; Ghosh et al., 2019), while those on combustion residues (FA and BA) are rarely seen. With the wide application of FA/BA-based materials, cohort studies should be done to track their impacts on human health in a relatively long period.

In addition to the aforementioned methods, various biomarkers and environmental monitoring tools can be used to measure exposure to combustion/incineration residues. Overall, the epidemiological investigation can provide insights of combustion/incineration residues induced health hazards and inform public health policies and interventions.

4. Summary and future trends

The large amount of combustion/incineration residues causes huge environmental burden, enlarging their influence on both ecological environment and human health. Recycling combustion/incineration residues into building materials is an attractive technology, which relieves environmental impacts and generates value-added products. However, it also greatly enhances the opportunity of contacting these residues with workers and even public. In this situation, the risk assessment of combustion/incineration residues and their products are of great significance. There are different protocols and models to evaluate the risks of residues. Leaching-based tests are the most widely used due to the fast measurement efficiency and handleability. Noted that different countries/organizations developed different standards for leaching tests, none of which could fully guarantee the safety to human health. USEPA human health risk model focuses on the adverse impacts exposure to toxics in residues, while the statistical model cannot directly quantify the impacts to individuals and ignores the impacts to ecological environment. Biological and in vitro/in vivo methods are specifically designed for target species or tissues, which provide useful toxicological data. To get closer to the actual human reaction in response to residue exposure, epidemiological studies on human beings are employed. Based on large participants and long period, the cohort study exhibits its excellence than any other methods in evaluating the impacts on human health and build the associations between residue exposure to certain outcome. It also brings great difficulties in implement, such as the extremely long period of tracking study and large number of participants. It is strongly suggested to conduct cohort studies on the promotion and utilization of residue-based products considering the unpredictable impacts to environment and human health. With no doubt, human-related assessment methods are advanced; however, studies on combustion/incineration residue are still free to select relative methods due to their realistic needs.

References

Abbas, I., Badran, G., Verdin, A., Ledoux, F., Roumie, M., Lo Guidice, J.-M., Courcot, D., Garçon, G., 2019. In vitro evaluation of organic extractable matter from ambient PM2.5 using human bronchial epithelial BEAS-2B cells: cytotoxicity, oxidative stress, pro-inflammatory response, genotoxicity, and cell cycle deregulation. Environ. Res. 171, 510–522.

Adani, F., Ubbiali, C., Generini, P., 2006. The determination of biological stability of composts using the dynamic respiration index: the results of experience after two years. Waste Manage. 26, 41–48.

Ahmad, T., Park, J., Keel, S., Yun, J., Lee, U., Kim, Y., Lee, S.-S., 2018. Behavior of heavy metals in air pollution control devices of 2,400 kg/h municipal solid waste incinerator. Korean J. Chem. Eng. 35, 1823–1828.

Alimba, C.G., Gandhi, D., Sivanesan, S., Bhanarkar, M.D., Naoghare, P.K., Bakare, A.A., Krishnamurthi, K., 2016. Chemical characterization of simulated landfill soil leachates from Nigeria and India and their cytotoxicity and DNA damage inductions on three human cell lines. Chemosphere 164, 469–479.

Ashraf, A., Bibi, I., Niazi, N.K., Ok, Y.S., Murtaza, G., Shahid, M., Kunhikrishnan, A., Li, D., Mahmood, T., 2017. Chromium(VI) sorption efficiency of acid-activated banana peel over organo-montmorillonite in aqueous solutions. Int. J. Phytoremediat. 19, 605–613.

Barrena, R., d'Imporzano, G., Ponsá, S., Gea, T., Artola, A., Vázquez, F., Sánchez, A., Adani, F., 2009. In search of a reliable technique for the determination of the biological stability of the organic matter in the mechanical–biological treated waste. J. Hazard Mater. 162, 1065–1072.

Bartov, G., Deonarine, A., Johnson, T.M., Ruhl, L., Vengosh, A., Hsu-Kim, H., 2013. Environmental impacts of the Tennessee valley Authority Kingston coal ash Spill. 1. Source apportionment using mercury stable isotopes. Environ. Sci. Technol. 47, 2092–2099.

Bayard, R., de Araújo Morais, J., Ducom, G., Achour, F., Rouez, M., Gourdon, R., 2010. Assessment of the effectiveness of an industrial unit of mechanical–biological treatment of municipal solid waste. J. Hazard Mater. 175, 23–32.

Bilibio, C., Retz, S., Schellert, C., Hensel, O., 2021. Drainage properties of technosols made of municipal solid waste incineration bottom ash and coal combustion residues on potash-tailings piles: a lysimeter study. J. Clean. Prod. 279, 123442.

Borm, P.J.A., 1997. Toxicity and occupational health hazards of coal fly ash (CFA). A review of data and comparison to coal mine dust. Ann. Occup. Hyg. 41, 659–676.

Cieplik, M.K., De Jong, V., Bozovič, J., Liljelind, P., Marklund, S., Louw, R., 2006. Formation of dioxins from combustion micropollutants over MSWI fly ash. Environ. Sci. Technol. 40, 1263–1269.

Danish, A., Ozbakkaloglu, T., 2022. Greener cementitious composites incorporating sewage sludge ash as cement replacement: a review of progress, potentials, and future prospects. J. Clean. Prod. 371, 133364.

Deonarine, A., Bartov, G., Johnson, T.M., Ruhl, L., Vengosh, A., Hsu-Kim, H., 2013. Environmental impacts of the Tennessee valley authority Kingston coal ash Spill. 2. Effect of coal ash on methylmercury in historically contaminated river sediments. Environ. Sci. Technol. 47, 2100–2108.

Du, B., Li, J., Fang, W., Liu, Y., Yu, S., Li, Y., Liu, J., 2018. Characterization of naturally aged cement-solidified MSWI fly ash. Waste Manage. 80, 101–111.

El-Sayed, R., Eita, M., Barrefelt, Å., Ye, F., Jain, H., Fares, M., Lundin, A., Crona, M., Abu-Salah, K., Muhammed, M., Hassan, M., 2013. Thermostable luciferase from Luciola cruciate for imaging of carbon nanotubes and carbon nanotubes carrying doxorubicin using in vivo imaging system. Nano Lett. 13, 1393–1398.

Energy, U.D.o., 2004. RAIS: Risk Assessment Information System. US Department of Energy, Washington, DC, U S A.

Esse, P.C., Buerkert, A., Hiernaux, P., Assa, A., 2001. Decomposition of and nutrient release from ruminant manure on acid sandy soils in the Sahelian zone of Niger, West Africa. Agr. Ecosys. Environ. 83, 55–63.

Flora, G., Gupta, D., Tiwari, A., 2013. Preventive efficacy of bulk and nanocurcumin against lead-induced oxidative stress in mice. Biol. Trace Elem. Res. 152, 31–40.

Gan, M., Wong, G., Fan, X., Ji, Z., Ye, H., Zhou, Z., Wang, Z., 2021. Enhancing the degradation of dioxins during the process of iron ore sintering co-disposing municipal solid waste incineration fly ash. J. Clean. Prod. 291, 125286.

Gao, B., Jiang, H., Chen, H., Peng, M., Zhang, W., Hu, L., Mao, L., 2023. The introduction of sulfates to suppress Cr(III) oxidation during incineration of tannery sludge and reduce leachability toxicity of incineration residue. J. Clean. Prod. 382, 135272.

Garg, A., Mishra, A., 2010. Wet oxidation—an option for enhancing biodegradability of leachate derived from municipal solid waste (MSW) landfill. Indus. Eng. Chem. Res. 49, 5575—5582.

Garrabrants, A.C., Kosson, D.S., Brown, K.G., Fagnant, D.P., Helms, G., Thorneloe, S.A., 2021. Demonstration of the use of test results from the leaching environmental assessment framework (LEAF) to develop screening-level leaching assessments. Waste Manag. 121, 226—236.

Ghosh, R.E., Freni-Sterrantino, A., Douglas, P., Parkes, B., Fecht, D., de Hoogh, K., Fuller, G., Gulliver, J., Font, A., Smith, R.B., Blangiardo, M., Elliott, P., Toledano, M.B., Hansell, A.L., 2019. Fetal growth, stillbirth, infant mortality and other birth outcomes near UK municipal waste incinerators; retrospective population based cohort and case-control study. Environ. Int. 122, 151—158.

Goyal, S., Dhull, S.K., Kapoor, K.K., 2005. Chemical and biological changes during composting of different organic wastes and assessment of compost maturity. Bioresour. Technol. 96, 1584—1591.

Halim, C.E., Amal, R., Beydoun, D., Scott, J.A., Low, G., 2003. Evaluating the applicability of a modified toxicity characteristic leaching procedure (TCLP) for the classification of cementitious wastes containing lead and cadmium. J. Hazard Mater. 103, 125—140.

Heiskanen, J., Ruhanen, H., Hagner, M., 2022. Effects of compost, biochar and ash mixed in till soil cover of mine tailings on plant growth and bio-accumulation of elements: a growing test in a greenhouse. Heliyon 8, e08838.

Hood, M., Taggart, R., Smith, R., Hsu-Kim, H., Henke, K., Graham, U., Groppo, J., Unrine, J., Hower, J., 2017. Rare earth element distribution in fly ash derived from the fire Clay coal, Kentucky. Coal Combus. Gasification Prod. 9, 22—33.

Hu, Y., Cheng, H., Tao, S., 2018. The growing importance of waste-to-energy (WTE) incineration in China's anthropogenic mercury emissions: emission inventories and reduction strategies. Renew. Sust. Energ. Rev. 97, 119—137.

Huang, C.-L., Bao, L.-J., Luo, P., Wang, Z.-Y., Li, S.-M., Zeng, E.Y., 2016. Potential health risk for residents around a typical e-waste recycling zone via inhalation of size-fractionated particle-bound heavy metals. J. Hazard Mater. 317, 449—456.

Hung, M.-L., Wu, S.-Y., Chen, Y.-C., Shih, H.-C., Yu, Y.-H., Ma, H.-w., 2009. The health risk assessment of Pb and Cr leachated from fly ash monolith landfill. J. Hazard Mater. 172, 316—323.

Intrakamhaeng, V., Clavier, K.A., Townsend, T.G., 2020. Hazardous waste characterization implications of updating the toxicity characteristic list. J. Hazard Mater. 383, 121171.

Kim, J., Koo, S.Y., Kim, J.Y., Lee, E.H., Lee, S.D., Ko, K.S., Ko, D.C., Cho, K.S., 2009. Influence of acid mine drainage on microbial communities in stream and groundwater samples at Guryong Mine, South Korea. Environ. Geology 58, 1567—1574.

Kociba, R.J., Keeler, P.A., Park, C.N., Gehring, P.J., 1976. 2,3,7,8-Tetrachlorodibenzo-p-dioxin (TCDD): results of a 13-week oral toxicity study in rats. Toxicol. Appl. Pharmacol. 35, 553—574.

Kosson, D.S., Garrabrants, A.C., DeLapp, R., van der Sloot, H.A., 2014. pH-dependent leaching of constituents of potential concern from concrete materials containing coal combustion fly ash. Chemosphere 103, 140—147.

Lasaridi, K.E., Stentiford, E.I., 1998. A simple respirometric technique for assessing compost stability. Water Res. 32, 3717—3723.

León-Mejía, G., Machado, M.N., Okuro, R.T., Silva, L.F.O., Telles, C., Dias, J., Niekraszewicz, L., Da Silva, J., Henriques, J.A.P., Zin, W.A., 2018. Intratracheal instillation of coal and coal fly ash particles in mice induces DNA damage and translocation of metals to extrapulmonary tissues. Sci. Total Environ. 625, 589—599.

Li, F., Yu, T., Huang, Z., Jiang, T., Wang, L., Hou, Q., Tang, Q., Liu, J., Yang, Z., 2022. Leaching experiments and risk assessment to explore the migration and risk of potentially toxic elements in soil from black shale. Sci. Total Environ. 844, 156922.

Li, H., Ji, H., 2017. Chemical speciation, vertical profile and human health risk assessment of heavy metals in soils from coal-mine brownfield, Beijing, China. J. Geochem. Explor. 183, 22—32.

Liu, Q., Huang, Q., Zhao, Y., Liu, Y., Wang, Q., Khan, M.A., Che, X., Li, X., Bai, Y., Su, X., Lin, L., Zhao, Y., Chen, Y., Wang, J., 2022a. Dissolved organic matter (DOM) was detected in MSWI plant: an investigation of DOM and potential toxic elements variation in the bottom ash and fly ash. Sci. Total Environ. 828, 154339.

Liu, Q., Wang, X., Gao, M., Guan, Y., Wu, C., Wang, Q., Rao, Y., Liu, S., 2022b. Heavy metal leaching behaviour and long-term environmental risk assessment of cement-solidified municipal solid waste incineration fly ash in sanitary landfill. Chemosphere 300, 134571.

Mandal, B.K., Suzuki, K.T., 2002. Arsenic round the world: a review. Talanta 58, 201—235.

Mandal, S., Sarkar, B., Bolan, N., Ok, Y.S., Naidu, R., 2017. Enhancement of chromate reduction in soils by surface modified biochar. J. Environ. Manag. 186, 277—284.

Mantis, I., Voutsa, D., Samara, C., 2005. Assessment of the environmental hazard from municipal and industrial wastewater treatment sludge by employing chemical and biological methods. Ecotoxicol. Environ. Saf. 62, 397—407.

Mishra, S.K., Yadav, K.D., 2022. Assessment of the effect of particle size and selected physico-chemical and biological parameters on the efficiency and quality of composting of garden waste. J. Environ. Chem. Eng. 10, 107925.

Netkueakul, W., Chortarea, S., Kulthong, K., Li, H., Qiu, G., Jovic, M., Gaan, S., Hannig, Y., Buerki-Thurnherr, T., Wick, P., Wang, J., 2022. Airborne emissions from combustion of graphene nanoplatelet/epoxy composites and their cytotoxicity on lung cells via air-liquid interface cell exposure in vitro. Nano. Impact. 27, 100414.

Niazi, N.K., Bibi, I., Shahid, M., Ok, Y.S., Burton, E.D., Wang, H., Shaheen, S.M., Rinklebe, J., Lüttge, A., 2018. Arsenic removal by perilla leaf biochar in aqueous solutions and groundwater: an integrated spectroscopic and microscopic examination. Environ. Pollut. 232, 31—41.

O'Connor, D., Hou, D., Ok, Y.S., Mulder, J., Duan, L., Wu, Q., Wang, S., Tack, F.M.G., Rinklebe, J., 2019. Mercury speciation, transformation, and transportation in soils, atmospheric flux, and implications for risk management: a critical review. Environ. Int. 126, 747–761.

Poon, C.S., Lio, K.W., 1997. The limitation of the toxicity characteristic leaching procedure for evaluating cement-based stabilised/solidified waste forms. Waste Manag. 17, 15–23.

Rajapaksha, A.U., Alam, M.S., Chen, N., Alessi, D.S., Igalavithana, A.D., Tsang, D.C.W., Ok, Y.S., 2018. Removal of hexavalent chromium in aqueous solutions using biochar: chemical and spectroscopic investigations. Sci. Total Environ. 625, 1567–1573.

Rajapaksha, A.U., Vithanage, M., Ok, Y.S., Oze, C., 2013. Cr(VI) formation related to Cr(III)-muscovite and birnessite interactions in ultramafic environments. Environ. Sci. Technol. 47, 9722–9729.

Randhawa, G.K., Kullar, J.S., 2011. Bioremediation of pharmaceuticals, pesticides, and petrochemicals with gomeya/cow dung. ISRN Pharmacol. 2011, 362459.

Ranzi, A., Fano, V., Erspamer, L., Lauriola, P., Perucci, C.A., Forastiere, F., 2011. Mortality and morbidity among people living close to incinerators: a cohort study based on dispersion modeling for exposure assessment. Environ. Health. 10, 22.

Rizwan, M., Ali, S., Zia ur Rehman, M., Rinklebe, J., Tsang, D.C.W., Bashir, A., Maqbool, A., Tack, F.M.G., Ok, Y.S., 2018. Cadmium phytoremediation potential of Brassica crop species: a review. Sci. Total Environ. 631–632, 1175–1191.

Saba, E., Lee, Y.-s., Yang, W.-K., Lee, Y.Y., Kim, M., Woo, S.-M., Kim, K., Kwon, Y.-S., Kim, T.-H., Kwak, D., Park, Y.-C., Shin, H.J., Han, C.K., Oh, J.-W., Lee, Y.C., Kang, H.-S., Rhee, M.H., Kim, S.-H., 2020. Effects of a herbal formulation, KGC3P, and its individual component, nepetin, on coal fly dust-induced airway inflammation. Sci. Rep. 10, 14036.

Sahu, S.K., Bhangare, R.C., Ajmal, P.Y., Sharma, S., Pandit, G.G., Puranik, V.D., 2009. Characterization and quantification of persistent organic pollutants in fly ash from coal fueled thermal power stations in India. J. Microchem. 92, 92–96.

Sahu, S.K., Pandit, G.G., Sadasivan, S., 2004. Precipitation scavenging of polycyclic aromatic hydrocarbons in Mumbai, India. Sci. Total Environ. 318, 245–249.

Shaheen, S.M., Hooda, P.S., Tsadilas, C.D., 2014. Opportunities and challenges in the use of coal fly ash for soil improvements – a review. J. Environ. Manag. 145, 249–267.

Shakoor, M.B., Niazi, N.K., Bibi, I., Shahid, M., Sharif, F., Bashir, S., Shaheen, S.M., Wang, H., Tsang, D.C.W., Ok, Y.S., Rinklebe, J., 2018. Arsenic removal by natural and chemically modified water melon rind in aqueous solutions and groundwater. Sci. Total Environ. 645, 1444–1455.

Sima, J.K., Cao, X.D., Zhao, L., Luo, Q.S., 2015. Toxicity characteristic leaching procedure over- or under-estimates leachability of lead in phosphate-amended contaminated soils. Chemosphere 138, 744–750.

Singh, R., Singh, S., Parihar, P., Singh, V.P., Prasad, S.M., 2015. Arsenic contamination, consequences and remediation techniques: a review. Ecotoxicol. Environ. Saf. 112, 247–270.

Smichowski, P., Polla, G., Gómez, D., Fernández Espinosa, A.J., López, A.C., 2008. A three-step metal fractionation scheme for fly ashes collected in an Argentine thermal power plant. Fuel 87, 1249–1258.

Sun, J., Shen, Z., Niu, X., Zhang, Y., Zhang, B., Zhang, T., He, K., Xu, H., Liu, S., Ho, S.S.H., Li, X., Cao, J., 2020. Cytotoxicity and potential pathway to vascular smooth muscle cells induced by PM2.5 emitted from raw coal chunks and clean coal combustion. Environ. Sci. Technol. 54, 14482–14493.

Sun, J., Wang, L., Yu, J., Guo, B., Chen, L., Zhang, Y., Wang, D., Shen, Z., Tsang, D.C.W., 2022. Cytotoxicity of stabilized/solidified municipal solid waste incineration fly ash. J. Hazard Mater. 424, 127369.

Tait, P.W., Brew, J., Che, A., Costanzo, A., Danyluk, A., Davis, M., Khalaf, A., McMahon, K., Watson, A., Rowcliff, K., Bowles, D., 2020. The health impacts of waste incineration: a systematic review. Aust. Nz. J. Publ. Heal. 44, 40–48.

Tsang, D.C.W., Olds, W.E., Weber, P.A., Yip, A.C.K., 2013. Soil stabilisation using AMD sludge, compost and lignite: TCLP leachability and continuous acid leaching. Chemosphere 93, 2839–2847.

Tsang, D.C.W., Yip, A.C.K., 2014. Comparing chemical-enhanced washing and waste-based stabilisation approach for soil remediation. J. Soil. Sediment. 14, 936–947.

USEPA, 1989. Risk-Assessment Guidance for Superfund. In: Human Health Evaluation Manual. Part A, vol. 1. Interim report (Final).

USEPA, 1996. https://www.epa.gov/air-emissions-factors-and-quantification/ap-42-compilation-air-emission-factors.

Verbinnen, B., Billen, P., Van Caneghem, J., Vandecasteele, C., 2017. Recycling of MSWI bottom ash: a review of chemical barriers, engineering applications and treatment technologies. Waste. Biomass. Valori. 8, 1453–1466.

Wang, L., Zhang, Y., Chen, L., Guo, B., Tan, Y., Sasaki, K., Tsang, D.C.W., 2022. Designing novel magnesium oxysulfate cement for stabilization/solidification of municipal solid waste incineration fly ash. J. Hazard Mater. 423, 127025.

Wang, N., Sun, X., Zhao, Q., Yang, Y., Wang, P., 2020. Leachability and adverse effects of coal fly ash: a review. J. Hazard Mater. 396, 122725.

Xia, S., Song, Z., Jeyakumar, P., Shaheen, S.M., Rinklebe, J., Ok, Y.S., Bolan, N., Wang, H., 2019. A critical review on bioremediation technologies for Cr(VI)-contaminated soils and wastewater. Crit. Rev. Environ. Sci. Technol. 49, 1027–1078.

Zhai, H., Xue, M.Y., Du, Z.K., Wang, D., Zhou, F., Feng, P.Y., Liang, D.L., 2019. Leaching behaviors and chemical fraction distribution of exogenous selenium in three agricultural soils through simulated rainfall. Ecotoxicol. Environ. Saf. 173, 393–400.

Zhang, S., Lin, X., Chen, Z., Li, X., Jiang, X., Yan, J., 2018. Influence on gaseous pollutants emissions and fly ash characteristics from co-combustion of municipal solid waste and coal by a drop tube furnace. Waste Manag. 81, 33–40.

Zhang, Y., Labianca, C., Chen, L., De Gisi, S., Notarnicola, M., Guo, B., Sun, J., Ding, S., Wang, L., 2021a. Sustainable ex-situ remediation of contaminated sediment: a review. Environ. Pollut. 287, 117333.

Zhang, Y., Wang, L., Chen, L., Ma, B., Zhang, Y., Ni, W., Tsang, D.C.W., 2021b. Treatment of municipal solid waste incineration fly ash: state-of-the-art technologies and future perspectives. J. Hazard Mater. 411, 125132.

Zhou, J., Du, B., Wang, Z., Zhang, W., Xu, L., Fan, X., Liu, X., Zhou, J., 2019. Distributions and pools of lead (Pb) in a terrestrial forest ecosystem with highly elevated atmospheric Pb deposition and ecological risks to insects. Sci. Total Environ. 647, 932–941.

Zhou, J., Wu, S., Pan, Y., Su, Y., Yang, L., Zhao, J., Lu, Y., Xu, Y., Oh, K., Qian, G., 2015. Mercury in municipal solids waste incineration (MSWI) fly ash in China: chemical speciation and risk assessment. Fuel 158, 619–624.

Zierold, K.M., Odoh, C., 2020. A review on fly ash from coal-fired power plants: chemical composition, regulations, and health evidence. Rev. Environ. Health 35, 401–418.

Part III

Recycling of combustion/incineration residues into cement clinker

Chapter 11

Recycling of incineration sewage sludge ash into cement clinker

Songsong Lian[1] and Shaoqin Ruan[2]

[1]*School of Civil Engineering, NingboTech University, Ningbo, China;* [2]*College of Civil Engineering and Architecture, Zhejiang University, Hangzhou, China*

1. Introduction

1.1 Sewage sludge and the disposal practices

At present, the effective management of sludge has become one of the most severe challenges in the field of sewage treatment to the economy and the environment (Di Iaconi et al., 2020). Propelled by global urbanization, there's an escalating discharge of sewage, and consequently, an increase in sludge generation. Therefore, handling and disposing of sludge effectively has transformed into a pressing and complex issue within the scope of global environmental protection.

Sewage sludge is essentially the sediment derived from various stages of wastewater treatment undertaken by urban sewage treatment plants. The quantity of sludge produced typically corresponds to around 4%–6% of the volume of the processed sewage, with a moisture content oscillating between 97% and 98%. Sewage sludge manifests a complex composition that amalgamates diverse constituents such as sediments, fibrous materials, residues of animal and plant origin, alongside an array of microorganisms with their affiliated organic matter, in addition to heavy metal elements and salts. The salient characteristic of sludge lies in its high water content, elevated organic matter, propensity to decompose and produce foul odors, in addition to its low specific gravity, minuscule particle size, and colloidal fluidity.

The random dumping and discarding of large volumes of untreated sludge precipitate further contamination of the ecological environment. As such, devising effective methods for sludge disposal have emerged as a subject of global attention. The prevailing consensus advocates for the transformation of sludge into a resource to be used effectively, based on a foundation of harmless treatment, thereby converting waste into a valuable asset while mitigating pollution.

Currently, prevalent sludge treatment strategies encompass composting, landfill, and incineration. Composting necessitates a specific time duration and stringent operating conditions, while landfill demands sufficient land capacity. Moreover, both methods might engender secondary pollution due to the inherent hard-to-degrade organics, pathogens, and heavy metals in the sludge (Bairq et al., 2018). Hence, the composting and landfill disposal methods for sludge harbor certain inherent risks. Contrarily, the incineration of sewage sludge results in effective sterilization and considerable volume reduction, and the residual dry sludge, comprising approximately 30%–88% organic components and providing a calorific value ranging from 11.0 to 25.5 kJ/kg, attests to its potential as a significant biomass energy source (Han et al., 2012; Wang et al., 2016). Therefore, due to its superior performance in volumetric reduction, thorough sterilization, and energy recovery (Zhou et al., 2020), sludge incineration is increasingly being recognized as the leading solution for sludge management worldwide (Li et al., 2018; Samolada and Zabaniotou, 2014; Zhao et al., 2016).

1.2 Sewage sludge incineration and its byproduct ash

Incineration emerges as an effective mechanism for the management of municipal sewage sludge, offering key benefits in terms of waste neutralization and volume reduction (Yang et al., 2015; Mininni et al., 2015). This process capitalizes on the substantial organic component and appreciable calorific value of sludge. An incineration-focused strategy is arguably the most thorough, enabling full carbonization of organic materials, extermination of pathogens, and significant minimization

Treatment and Utilization of Combustion and Incineration Residues. https://doi.org/10.1016/B978-0-443-21536-0.00024-1

of sludge volume and weight. Additionally, it touts a fast treatment pace, negates the need for long-term storage, and circumvents odor issues. However, it also presents issues such as substantial investment in treatment facilities, high processing costs, and the necessity for handling the resultant flue gas. To prevent the release of harmful substances such as dioxins during incineration, a swift heating process and a sustained operating temperature above 850°C are required.

During the incineration process, sludge generates incineration ash, accounting for 3%−10% of the sludge volume (Lynn et al., 2016). This incineration sewage sludge ash (ISSA) represents the solid residue remaining after the total carbonization of organic substances and fibrous lignin present in the sludge, mainly composed of inorganic components. Given the large base quantity of sludge, the resultant ISSA is sizable. As an illustration, the world's most substantial municipal sewage sludge incineration plant (located in Hong Kong, China) produces an estimated 240,000 tons of ISSA per annum. Currently, the principal disposal methods for ISSA involve depositing and landfilling (Zhang et al., 2015). However, these methods exhibit distinct limitations. The incineration process can lead to the concentration of heavy metals within the ash, presenting a potential hazard to groundwater supplies, especially during rainfall events. Landfill practices, albeit prevalent, consume considerable land resources and contribute to soil degradation. Also, the fine granular nature of the ISSA can escalate dust pollution, posing potential health risks. Given these limitations, it is imperative to pursue more resourceful and sustainable methods of treating incinerated sludge ash. Studies indicate that ISSA contains a significant quantity of inorganic elements such as aluminum, silicon, calcium, and iron. With appropriate adjustments, these elements can be employed in the manufacturing of building materials such as bricks and cement (Smol et al., 2015). The exploration of ISSA's application in cement clinker encompasses two major domains: (1) as a raw feed for cement clinker manufacture and (2) as a kind of supplementary cementitious materials. This chapter will systematically delve into both these aspects.

2. Characteristics of incineration sewage sludge ash

2.1 Physical properties

The physical characteristics of sewage sludge incineration ash (SSIA) vary depending on the specific properties of the original sludge and the conditions under which it is incinerated. However, some general characteristics are typically observed.

ISSA is a complex material comprising various particles ranging from silt to fine sand. These particles exhibit curves similar to those found in filler or fine aggregate materials (Hu et al., 2013), which predominantly falls in the silt (2.5−62.5 μm) and fine sand (62.5−250 μm) size fractions (Donatello et al., 2010; Ottosen et al., 2014). ISSA has a mean density comparable with light sand, much lower than that of the original sludge. This is because the incineration process removes much of the water and organic material from the sludge, leaving primarily mineral and metal oxides. The microscopic form of ISSA manifests in diverse irregular particles, predominantly comprising amorphous phases interspersed with various crystalline phase granules. As illustrated in Fig. 11.1, these irregular particles of ISSA demonstrate rough surface textures and embody a porous microstructure, thereby facilitating high water absorption.

These physical characteristics can have significant impacts on how ISSA is handled and disposed of, as well as on its potential for reuse in other applications. For example, its small particle size and porosity can make it challenging to handle, but can also enhance its reactivity in certain applications.

2.2 Chemical composition

The ISSA is complex and highly dependent on the specific wastewater treatment processes and incineration methods employed. It typically contains a mix of mineral and metal oxides, with elements such as silicon, calcium, aluminum, and iron being common. Generally, ISSA primarily consists of SiO_2, Al_2O_3, CaO, P_2O_5, and Fe_2O_3 as the main oxides (Cong et al., 2020). Lynn et al. (2015) conducted a statistical analysis of the SiO_2, Al_2O_3, and CaO content data extracted from the aforementioned 76 publications, as depicted in Fig. 11.2. The calculated mean, standard deviation (St. Dev.), and coefficient of variation (CV) values for SiO_2, Al_2O_3, and CaO are also given. This figure shows that the majority of results fall around the latent hydraulic and pozzolanic regions, suggesting potential for ISSA use as a cementitious component in concrete. Therefore, the unique composition of ISSA imparts latent hydraulic and pozzolanic properties, making it a promising by-product for various applications in the construction sector.

Nevertheless, ISSA also houses potentially hazardous elements such as lead, cadmium, and mercury, along with a residual organic fraction that remains after combustion, contingent on the specific properties of the sludge (Yu et al., 2021). Table 11.1 presents the total contents of the toxic trace elements in ISSA, with a primary focus on heavy metals such as, Cd, Cr, Cu, Ni, Pb, and Zn. The coefficient of variations displayed in Table 11.1 indicates a significant variability in the

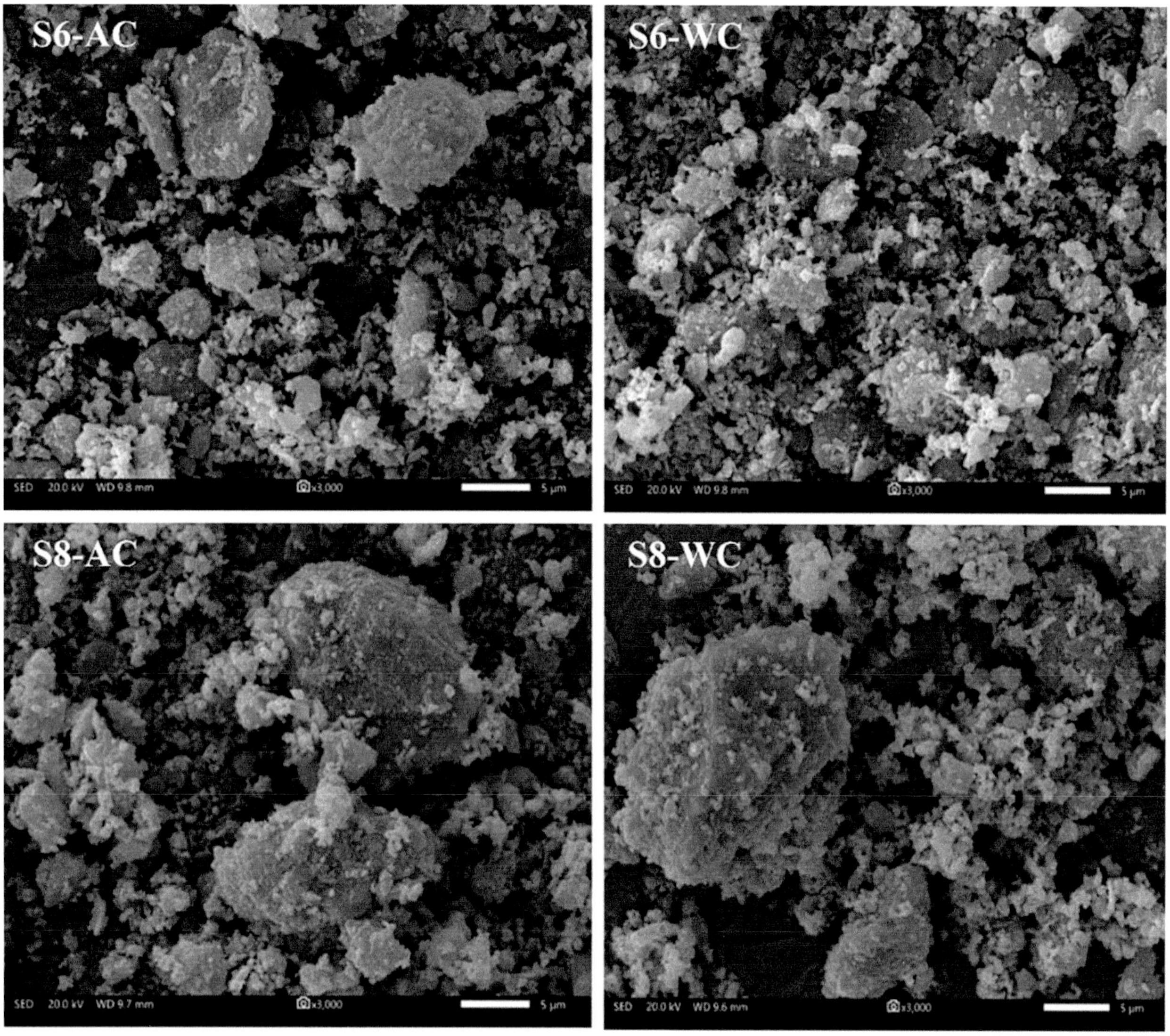

FIGURE 11.1 Microstructure of incineration sewage sludge ash (ISSA) (Chang et al., 2022).

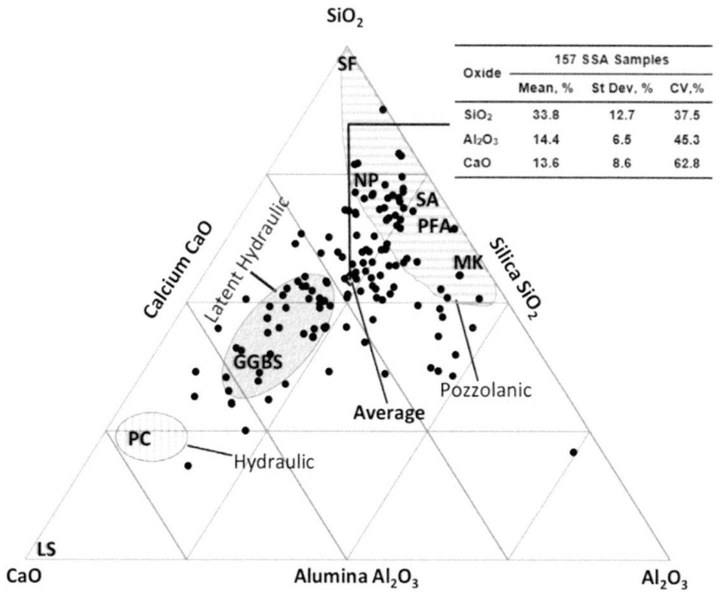

Oxide	157 SSA Samples		
	Mean, %	St Dev, %	CV,%
SiO_2	33.8	12.7	37.5
Al_2O_3	14.4	6.5	45.3
CaO	13.6	8.6	62.8

FIGURE 11.2 Ternary plot of SiO_2, Al_2O_3, and CaO contents for ISSA (Lynn et al., 2015). *GGBS*, ground granulated blast furnace slag; *LS*, limestone; *MK*, metakaolin; *NP*, natural pozzolan; *PC*, Portland cement; *PFA*, pulverized fuel ash; *SA*, shale ash; *SF*, silica fume.

TABLE 11.1 Total contents of the toxic trace elements in ISSA (Lynn et al., 2018).

Element	No. of samples	Mean, mg/kg	STDEV, mg/kg	CV, %	Limits mg/kg
Fe	48	79,578	55,333	70	–
Al	45	48,253	27,668	57	–
Cl	31	1241	3043	245	–
Zn	103	2964	3257	110	10,000
Cu	117	1673	2713	162	7000
Pb	115	321	402	125	6000
Ba	27	1663	1174	71	–
Cr	106	477	928	195	2000
Sr	12	441	173	39	–
Sb	11	33	24	73	–
Ni	96	198	325	164	500
V	30	135	129	96	–
Se	11	57	154	270	–
As	47	30	52	173	–
Co	17	137	172	126	–
Mo	36	29	30	101	–
Cd	84	17	57	328	20
Hg	44	2.2	3.3	148	–

element concentrations among ISSA samples worldwide. Several factors contribute to this variability, including variations in waste and wastewater composition, processing techniques, incineration conditions, and the methodologies employed to analyze the element contents.

2.3 Pollution characteristics

Beyond the evaluation of total trace toxic elements contained within ISSA, as shown in Table 11.1, the analysis of leachable contents is pivotal, providing a genuine insight into the pollution attributes of ISSA. Numerous scholars have conducted extensive testing and research on the leaching concentration of ISSA (Wei, 2015; Li et al., 2017; Chen et al., 2018; Chen and Poon, 2017; Zhang et al., 2015). However, the test results exhibit significant disparities, as shown in Fig. 11.3. These variations primarily stem from substantial inherent differences in ISSA obtained from diverse sources, as well as the influence of various testing methodologies employed. Given that the leaching behavior of ISSA is influenced by the lack of standard harmonization and the challenge of replicating real-life usage conditions in a laboratory setting, the Toxicity Characteristic Leaching Procedure (TCLP) has been widely adopted as the most common method for assessing leaching behavior in ISSA samples. Lynn et al. (2018) summarized previous research findings and identified the following patterns: ISSA demonstrated a low capacity for acid neutralization. Its overall leaching of elements was highest under acidic pH conditions. Leaching of Ni, Pb, Cu, Co, Mo, Se, As, Mg, and Ca increased in increasingly acidic conditions, while Cd and Cr leaching was higher in neutral conditions. Zn exhibited amphoteric properties, with high leaching observed in both strongly acidic and alkaline conditions. The leachability of Na and K appeared to be less dependent on pH. Latosińska and Gawdzik (2012), based on their experiments, found that the leaching of Cd, Cr, Pb, and Ni was not affected by the incineration temperature of sewage sludge. However, the mobility of Cu and Zn significantly increased with higher temperatures, specifically from 850 to 1000°C.

Furthermore, substances of concern such as dioxins, which can be generated during the sewage sludge incineration process, have the potential to permeate into by-products such as fly ash and slag. Dioxins, more formally referred to as polychlorinated dibenzodioxins (PCDD) and polychlorinated dibenzofurans (PCDF), are a subset of chlorinated polycyclic

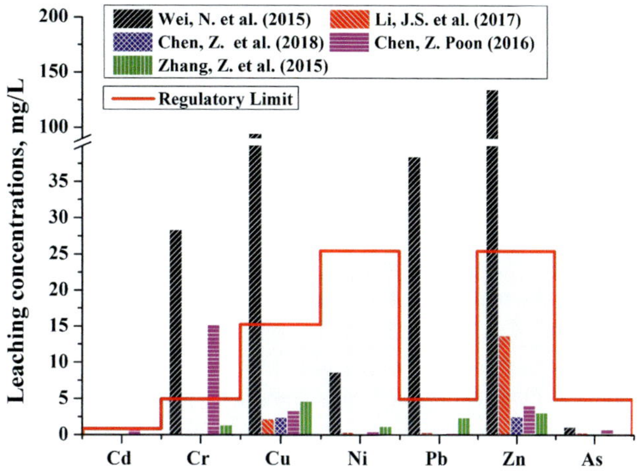

FIGURE 11.3 Leaching concentration of ISSA samples. *Data were adopted from Wei, N., 2015. Leachability of heavy metals from lightweight aggregates made with sewage sludge and municipal solid waste incineration fly ash. Int. J. Environ. Res. Publ. Health 12 (5), 4992–5005; Li, J.S., Xue, Q., Fang, L., Poon, C.S., 2017. Characteristics and metal leachability of incinerated sewage sludge ash and air pollution control residues from Hong Kong evaluated by different methods. Waste Manage. (Tucson, Ariz.) 64, 161–170; Chen, Z., Li, J.S., Poon, C. S., 2018. Combined use of sewage sludge ash and recycled glass cullet for the production of concrete blocks. J. Clean. Prod. 171, 1447–1459; Chen, Z., Poon, C.S., 2017. Comparing the use of sewage sludge ash and glass powder in cement mortars. Environ. Technol. 38 (11), 1390–1398; Zhang, Z., Zhang, L., Yin, Y., Liang, X., Li, A., 2015. The recycling of incinerated sewage sludge ash as a raw material for CaO–Al2O3–SiO2–P2O5 glass-ceramic production. Environ. Technol. 36 (9), 1098–1103.*

aromatic compounds. These substances are typically solid at room temperature, exhibit high melting points, and have low vapor pressure and water solubility, thus showing a propensity for adsorption onto particulates. If the levels of toxic and harmful substances in the fly ash and slag from incineration are too high, or if these by-products are improperly managed, they pose potential detrimental impacts on the atmosphere, soil, surface water, and groundwater.

Overall, the characteristics of ISSA play a pivotal role in shaping its environmental impact and potential application value. Recognizing the duality of ISSA's nature—as a potential pollutant and a valuable resource—is key. Therefore, understanding these characteristics and devising ways to mitigate the environmental risks while maximizing utility is paramount for sustainable ISSA management.

3. Utilization of incineration sewage sludge ash in cement clinker production

The organic constituents within ISSA contribute an additional thermal value to the cement kiln, effectively positioning it as a pragmatic alternative fuel source for cement production. In a more specific analysis, the comprehensive calorific value of ISSA resides at approximately 8.3 MJ/kg, resoundingly meeting and surpassing the stipulated 6.25 MJ/kg benchmark set by the cement industry (Liang et al., 2021). Consequently, in conjunction with its reduced energy consumption and decreased greenhouse gas emissions, this trend has sparked a notable upswing in the popularity of coincinerating SSA alongside raw cement materials within cement kilns (Huang et al., 2018). After deducting combustion losses, its major components closely resemble those of cement raw materials, primarily consisting of CaO, SiO_2, Fe_2O_3, and Al_2O_3. This similarity enables sludge to have the potential for partially substituting raw materials in cement production, while also providing additional heat value. Meanwhile, utilizing cement kilns for sludge treatment offers significant advantages. The firing temperatures in cement kilns exceed 1400°C, effectively eliminating harmful microorganisms such as bacteria and pathogens present in the sludge. Furthermore, a significant portion of heavy metal elements is incorporated into the crystal lattice of cement clinker minerals. Therefore, the utilization of dried or even dewatered sewage sludge in the production of eco-cement has received considerable attention in the literature. This approach allows for the simultaneous utilization of sewage sludge's calorific value to reduce fuel requirements and the reduction of inorganic content in the sludge, contributing to a decrease in cement raw material demands.

A wealth of research indicates that sewage sludge can be utilized as a partial raw material for clinker production (Guo et al., 2022; Rezaee et al., 2019; Chang et al., 2020, 2022b). Experimental results demonstrate that the primary mineral phases in the clinker, including C3S, C2S, C3A, and C4AF, maintain their mineral structures identical to those in ordinary Portland cement clinker. Shih et al. (2005) demonstrated that using heavy metal-laden sludge as a replacement for raw materials in cement production was feasible, as it didn't pose a leaching risk from sintered clinkers, as long as the sludge addition was kept under 15%. Chang et al. (2020) also concurred that the quantity of sewage sludge used as raw materials in cement clinker production needs to be rigorously managed within a certain range ($\leq$15 wt%) to ensure strength requirements and environmental safety. However, although the mineralogical structures of cement produced from sewage sludge resemble those of standard Portland cement clinker, the incinerated sewage sludge ash exerts varying degrees of influence on the proportion of cement components, the hydration rate of the cement, and its mechanical properties.

3.1 Burnability of eco-cement raw

The formation of C_3S plays a crucial role in the burning process of cement clinker minerals. The widely accepted mechanism for C_3S formation involves the melting of C_2S and CaO into the liquid phase, leading to the generation of C_3S. Consequently, the temperature, amount of liquid phase, and properties of the liquid phase significantly impact the formation of C_3S during the clinker burning process, thereby affecting the quality of the final product.

Sludge contains a small amount of alkali and trace heavy metal elements. The introduction of sludge into the molten system of clinker burning introduces certain impurity ions. On one hand, the inclusion of trace elements, such as heavy metals, from the sludge brings impurity ions with significant residual bonding forces. These forces impact the molten system by causing lattice distortion and lowering the system's melting point, thus reducing the formation temperature of C_3S. A study by Xu et al. (2014) indicated that the introduced trace elements in mixing sludge played the role as mineralizers and cosolvents in cement sintering by reducing the eutectic point of the system and increasing the amount of liquid phase, which contributes the formation of C_3S. However, research also indicates that cements incorporating ISSA contain less C_3S and more C_2S in comparison with ordinary Portland cement, particularly when calcination occurs at lower temperatures and when the proportion of ISSA in the cement is higher. This situation is primarily due to a considerable decrease in the content of CaO in eco-cement, while the reduction in SiO_2 content is relatively less substantial. Furthermore, an increase in P_2O_5 can trigger the decomposition of C_3S into α-C_2S.

On the other hand, sludge contains alkali metal elements, which have lower melting points. This can result in the formation of localized regions with lower melting points during the sintering process. Due to the nonuniformity of these regions, liquid phases can form at lower temperatures, thereby reducing the temperature at which the liquid phase appears. Hence, an appropriate amount of alkali is beneficial for clinker sintering. However, excessive alkali content increases the viscosity of the liquid phase during sintering, hindering particle diffusion and negatively affecting the sintering reaction. Additionally, high alkali content leads to the reaction of alkali metals primarily with sulfuric acid to form potassium sulfate, and sometimes calcium aluminate or sodium–potassium nitrate. The excess alkali also reacts with clinker minerals, resulting in the formation of solid solutions containing alkali minerals. This causes K_2O and Na_2O to replace CaO, leading to the formation of alkali-containing compounds and the precipitation of CaO. As a result, the reabsorption of CaO by C_2S to form C_3S becomes challenging, while the increased presence of free lime in the clinker reduces its overall quality. Naamane et al. (2016) conducted tests to analyze the chemical composition of cements produced using varying amounts of ISSA that had been calcined at temperatures ranging from 300 to 800°C. The study findings revealed that compared with ordinary Portland cement, cements derived from sewage sludge demonstrated lower percentages of SiO_2, Al_2O_3, Fe_2O_3, CaO, and K_2O, while exhibiting higher contents of MgO, SO_3, Na_2O, P_2O_5, and LOI (loss on ignition).

Therefore, the inclusion of sewage sludge and incinerated sewage sludge ash in the cement production process presents a dual effect on the clinker's quality. While trace elements from the sludge can serve as mineralizers, potentially enhancing the formation of C_3S, the core component of cement clinker, the substantial reduction in CaO, and slight decrease in SiO_2 content, along with increased P_2O_5, can lead to a higher amount of C_2S and less C_3S. Furthermore, although the alkali content in sludge may benefit clinker sintering by creating localized low-melting regions, an excess can lead to high viscosity during sintering and subsequent challenges in the formation of quality clinker. Thus, the careful control of sludge content is crucial in utilizing it for cement production.

3.2 Hydration of eco-cement

Experimental studies were conducted by researchers to investigate the hydration process of clinker pastes produced with varying amounts of added sewage sludge, and most of the conclusions were drawn that the addition of sewage sludge into raw meal can potentially cause delayed hydration of the clinker pastes. Lin et al. (2012) found that, as the curing period was extended from 1 to 28 days, the hydration levels remained similar for the clinker pastes produced with 0%, 5.0%, and 10.0% sewage sludge. However, a noticeable delay in the hydration process was observed for the clinker paste manufactured with a 15.0% addition of sewage sludge. This delay could potentially be attributed to an increase in C_2S and alterations in the majority of the elements present in the eco-cement clinkers.

Incorporation of soluble salts into C_3S significantly influences the initial stages of cement hydration. However, their precise roles are contingent upon their impact on different stages of the hydration process. Broadly, alkali compounds, including Na_2O and K_2O, expedite the initial hydration rate of silicate cement but inhibit the continuation into later stages of hydration, which leads to enhanced early-stage compressive strength at the expense of reduced long-term strength. Stephan's studies (1999) also discovered that chromium (Cr) accelerates the hydration process and reduces the initial setting time of cement, resulting in diminished strength, alterations in the free lime content, and modifications in the composition of C3S within clinkers (Stephan et al., 1999). However, the elevated phosphate and possibly sulfate contents of ISSA inhibit alite formation (Lam et al., 2010)

Meanwhile, the presence of certain elements, such as Zn, Pb, Cu, and Sn, has been found to delay cement hydration. Specifically, Zn^{2+} can act as a retarder for cement, stabilizing the suspension and slowing the development of the initial structure of the slurry. The addition of Zn^{2+} can create an insoluble film on the surface of cement particles, impeding further hydration. In addition, Shih et al. (2005) pointed out that heavy metals such as Zn, Cr, and Ni only have an impact on cement hydration when present in large amounts. When present in small quantities, they basically have no effect on hydration. Interestingly, Cr can speed up the hydration rate of cement and shorten the setting time, while Ni has basically no effect on cement hydration.

Therefore, the addition of sewage sludge to cement production can delay clinker hydration, particularly when the sludge proportion exceeds 15%. This is linked to an increase in C_2S and changes in the elements within eco-cement clinkers. The hydration process is also impacted by soluble salts and heavy metals. Alkali compounds speed up initial hydration but inhibit later stages. In contrast, elements such as Zn, Pb, Cu, and Sn can delay cement hydration, depending on their concentration levels.

3.3 Properties of the eco-cement pastes

In general, cements produced from sewage sludge calcined at lower temperatures tend to exhibit slower strength development and lower overall strength properties. This is primarily due to the relatively low contents of CaO, SiO_2, and Al_2O_3 in the sludge. Meanwhile, the increase in the P_2O_5 content, particularly with a higher rate of ISSA in cement, has been found to potentially reduce the compressive strength of mortars and prolong the setting time (Lin et al., 2004; Asavapisit et al., 2005).

Rezaee et al. (2019) conducted a study on eco-cement, where they investigated its physical, chemical, and mechanical characteristics. The research focused on the utilization of dry municipal sewage sludge as a partial substitute for traditional raw materials in eco-cement production. Various substitution levels, ranging from 5% to 15%, were examined. The analysis revealed that the major chemical components of eco-cements were similar to those of ordinary Portland cement. However, it was observed that the water demand increased and both the initial and final setting times were prolonged in the eco-cement formulations.

Besides, the setting times for cement pastes containing ISSA increase as the temperature of calcination decreases and the ISSA content in the cement increases. In Lin's study, it was found that the initial setting time for the control mortar was 170 min. However, when the substitution rate of ISSA was 10% and the calcination temperatures were 800 and 300°C, the setting times increased to 256 and 198 min, respectively (Lin et al., 2004). This result can also be explained by the high amounts of P_2O_5, SO_3, and C_2S in cements with ISSA compared with the control paste, which affect the rate of the pozzolanic reactions (Lin et al., 2009).

3.4 Environmental assessment

Environmental assessment of production has become increasingly relevant in many industries, with the cement industry being no exception. Several researchers have investigated the potential use of sewage sludge as an alternative fuel and raw material for cement production, assessing the environmental implications of such an approach.

Valderrama et al. (2013) implemented the life cycle assessment methodology to gauge the environmental impacts of employing sewage sludge as an alternative fuel and raw material in cement production. The findings indicated a decrease in CO_2 emissions as the rate of fuel substitution escalated from 5% to 15%. Notably, alongside CO_2, a reduction in NO_x emissions was observed during the coprocessing of sewage sludge in cement kilns (Lv et al., 2016). Fang et al. (2015) advocated for the use of sewage sludge as a denitrification agent and secondary fuel in cement production, which was beneficial for NO_x reduction. While the usage of sewage sludge as an alternative fuel in cement plants leads to resource conservation, it is essential to pay attention to environmental risk assessment and potential human health risks.

During thermal treatments, heavy metals typically undergo phase transformations. High levels of SiO_2, CaO, Al_2O_3, and Fe_2O_3 in sewage sludge can yield a large production of amorphous and crystalline phases during these thermal processes (Tang and Shih, 2015), which gives heavy metals the chance to chemically incorporate into the amorphous structures and crystalline phases or to evolve into a new crystalline phase (Chang et al., 2022b). In the study by Weng et al. (2003), it was proposed that the minimal contamination leaching level in sludge-enhanced bricks is due to high firing temperatures that immobilize harmful substances within silicate structures. Furthermore, as the sintering temperature increases from 950 to 1050°C, the leaching levels of certain heavy metals such as Cd, Cr, Cu, and Pb diminish as these metals become part of new crystalline phases at higher sintering temperatures (Liu et al., 2018).

Therefore, a plethora of studies (Cusidó and Cremades, 2012; Rusănescu et al., 2022) have substantiated that employing sewage sludge as a combustible material offers environmental enhancements without introducing any supplementary health risks, making it an admissible practice in line with international standards.

4. Utilization of incineration sewage sludge ash as supplementary cementitious materials

Supplementary cementitious materials (SCMs), often used as a partial substitute for clinker in cement, are a common addition to concrete mixtures. Their effectiveness hinges on two central processes: self-cementing and the pozzolanic reaction (Gomes et al., 2019). Both of these processes can be stimulated when ISSA is utilized as an alternative to traditional cementitious materials in concrete. Nonetheless, the inclusion of ISSA into concrete can affect multiple characteristics of the resulting product, such as its workability, mechanical properties, and leaching behavior, among others (Liang et al., 2022). These impacts are controlled by the physiochemical properties of the ISSA, which include its chemical makeup, particle size distribution, and particle morphology.

4.1 Effect of ISSA on the workability of cementitious materials

It is crucial to ensure that ISSA exhibits good workability performance when used in cementitious materials. The presence of ISSA may affect the workability of the concrete mix, as shown in Fig. 11.4. The angular and irregular particles of ISSA can increase the water demand of the mix, decreasing workability (Adamczyk et al., 2023; Xia et al., 2023).

A consensus emerges from most studies, indicating that higher ISSA content decelerates the cement hydration process, resulting in prolonged initial and final setting times (Fig. 11.5). In Haustein 's research (2022), substituting 20% of cement with ash extends the initial setting time by 3.34 h and the final setting time by about 10.88 h in comparison with the control, which was ascribed to the hindrance of C_3S hydration caused by the adsorption of orthophosphate ions at its dissolution sites. Additionally, the setting time of ISSA-cement pastes extends as the fineness of ISSA increases, which can be elucidated by the fact that the highly developed surface of ISSA particles facilitates the absorption of calcium ions, consequently leading to an inhibition of the hydration process (Ottosen et al., 2022).

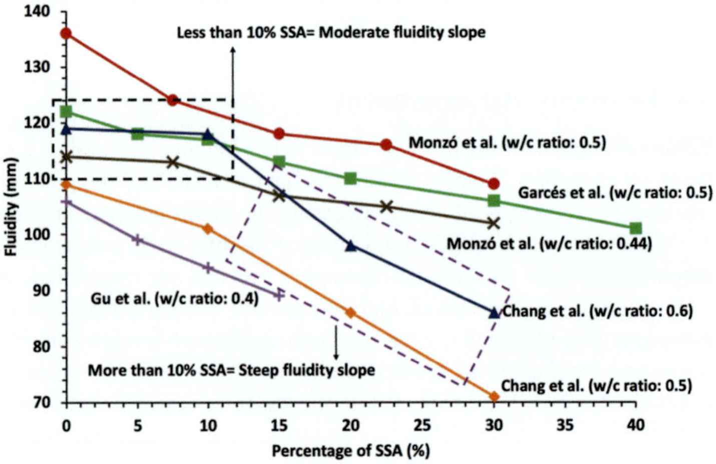

FIGURE 11.4 Fluidity of SSA-modified cementitious composites (Danish and Ozbakkalohlu, 2022).

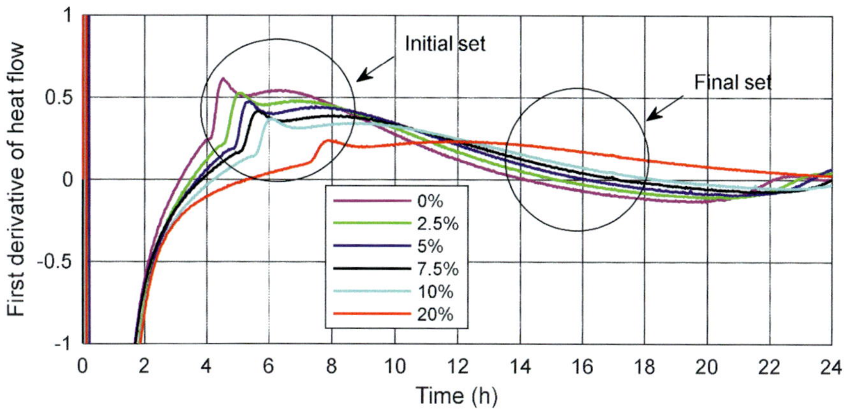

FIGURE 11.5 Initial and final setting time for the cement mortars without and with sewage sludge ash (Haustein et al., 2022).

The different disposal processes of ISSA result in distinct particle characteristics, such as mineral composition, particle fineness, water absorption rate, etc. These variations can have diverse effects on the workability performance of cementitious materials. To ensure good workability performance of ISSA in cementitious materials, some measures could be taken (Chakraborty et al., 2017; Dyer et al., 2011), such as follows:

(1) Control the particle fineness of incinerated sewage sludge ash: Adjust the particle size and distribution of the ash through appropriate grinding processes to ensure uniform dispersion in the cementitious materials, avoiding agglomeration and cohesion.
(2) Consider the cement-to-ash ratio: Determine the appropriate ratio to achieve the desired workability performance. Excessive ash content may decrease the plasticity and flowability of the cementitious materials, while insufficient content may not fully utilize the potential benefits of the ash.
(3) Consider the use of admixtures: Introduce suitable admixtures such as water reducers or fly ash to improve the workability performance of the cementitious materials.

4.2 Effect of ISSA on the mechanical performance of cementitious materials

The assessment of ISSA's effect on the mechanical performance of cementitious materials plays a crucial role in determining its suitability as a replacement material in cement applications. Chen and Poon (2017) conducted a study on the application of ISSA in cement mortar specimens, indicating that ISSA exhibits long-term volcanic ash activity. When replacing 20% of cement with ISSA, there is minimal impact on the mechanical properties of cement mortar specimens. Baeza-Brotons et al. (2014) conducted experiments where ISSA obtained from incineration at 800°C was used to partially replace cement in the production of cement mortar specimens and concrete specimens. The ISSA dosages in the experiments were 5%, 10%, 15%, and 20%, respectively. The research findings indicate that the addition of ISSA to concrete had minimal effect on the density and mechanical properties of concrete specimens. Moreover, Zdeb's research (2022) revealed that a relatively small addition of sewage sludge ash, i.e., not exceeding 10% of cement weight, even has a positive effect on both the mechanical properties of cement mortars and their tightness, which was attributed to the physical sealing of the composite structure (Zdeb et al., 2022). However, once the content of ISSA surpasses a certain threshold, the mechanical properties exhibit a decreasing trend, which could be attributed to two factors: the low pozzolanic reactivity of ISSA and the formation of excessive ettringite, leading to increased porosity, ultimately changing the morphology of hydration products and causing a notable mechanical reduction (Gu et al., 2021). Similar conclusions were drawn in the research by Prabhakar et al. (2022), as shown in Fig. 11.6.

Overall, ISSA exhibits a certain level of volcanic ash activity. However, compared with well-known cementitious materials such as cement and fly ash, its volcanic ash activity is relatively weaker (Rusănescu et al., 2022). The pozzolanic reactivity of sludge ash is intricately associated with its chemical composition, notably the presence of amorphous and crystalline phases. This interdependence is influenced by a range of factors, encompassing the source of the sludge, the additives used during treatment, and the specifics of the incineration process (Chang et al., 2022a). In a general trend, it has been noted that ash with finer particle dimensions typically exhibits an elevated level of pozzolanic reactivity

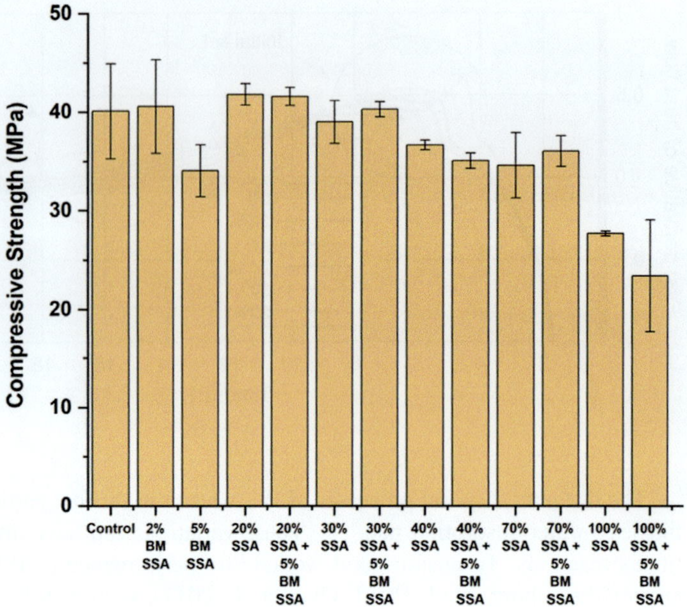

FIGURE 11.6 Compressive strength of the various mortar formulations (Prabhakar et al., 2022). *BM SSA*, ball milled sewage sludge ash (<75 μm); *SSA*, sewage sludge ash.

(de Azevedo Basto et al., 2019). Therefore, enhancing its volcanic ash activity through pretreatment methods has become a recent research focus. It is recommended to extract the phosphorus elements from ISSA before its utilization to mitigate any negative effects caused by the presence of phosphorus in the ash. Additionally, enhancing the volcanic ash activity of ISSA can be achieved by increasing the incineration temperature and duration of sewage sludge, as well as reducing the particle fineness of ISSA through mechanical ball milling. These methods can effectively improve the volcanic ash activity of ISSA, making it more suitable for cementitious applications (Baeza et al., 2014; Dyer et al., 2011).

4.3 Effect of ISSA on the leaching behavior of cementitious materials

The leaching behavior of heavy metals from ISSA is an important consideration for its environmental impact. It is crucial to conduct thorough testing and analysis to evaluate the leaching behavior of heavy metals from ISSA before considering its utilization as a substitute material in cement applications.

Cementitious components within cement have the ability to bind with heavy metals during the process of solidification (Wu et al., 2021). Chen et al. (2013) assessed the environmental impact of ISSA concrete using leaching tests. The results showed that when ISSA was incorporated into concrete, the leaching of heavy metals decreased compared with the original ISSA, and the previously exceeded metal concentrations were below the regulatory limits. The findings reported by Coutand et al. (2006) suggest that when utilizing 25% of ISSA, the leaching concentration of certain elements remained below the limits set for drinking water by the World Health Organization. Chen and Lin (2009) found that the leaching concentration determined by the Toxicity Characteristic Leaching Procedure (TCLP) for unbound material was below the allowable leaching concentrations for those respective elements. These studies indicate that under the tested conditions and concentrations, the leaching of elements from ISSA did not exceed the regulatory limits or pose significant environmental risks.

Thus, solidification through cementitious materials has been recognized as an efficient technique for the immobilization of heavy metal contaminants (Guo et al., 2017; Wang et al., 2023), which can be entrapped and assimilated within the structure of cement hydration products such as the layered calcium–silicate–hydrate (C–S–H) and needle-shaped ettringite (Vespa et al., 2014). Moreover, heavy metals present in sludge can engage in chemical reactions with components of hydrated products, leading to precipitate formation during cement hydration (Lasheras-Zubiate et al., 2012). Various simultaneous interactions such as adsorption, precipitation, complexation, and encapsulation play significant roles in the immobilization of contaminants (Chang et al., 2022a). The compact structure of the hardened cement paste establishes a low-permeability barrier, effectively impeding the leaching behavior of heavy metals (Chang et al., 2020).

Therefore, it can be concluded that using ISSA in cementitious materials is considered safe and feasible. However, it is essential to consider site-specific conditions, leaching test methods, and local regulations to assess the potential leaching behavior and environmental impact of ISSA in specific applications.

In summary, the use of ISSA as supplementary cementitious material not only facilitates the resource utilization of sludge ash but also effectively immobilizes toxic and hazardous heavy metal contaminants present in the ash. This opens up promising research directions for the treatment and disposal of sludge ash, offering great potential for further development.

5. Challenges and potential solutions

Despite the promising use of ISSA as a raw material in cement clinker production and as a supplementary cementitious material, it's not without potential drawbacks. The significant phosphorus content in ISSA poses challenges when it is used excessively in the production process. Overreliance on ISSA may adversely affect the performance characteristics of both cement clinker and cementitious materials. For example, Lam's findings indicate a substantial decrease in the active amount of tricalcium silicate (C_3S)—down to 19.3 wt%—in clinker when the addition of ISSA increases to 8 wt%. The high content of phosphorus and sulfur trioxide in ISSA could be the underlying reason for this outcome, as it may inhibit the formation of the main phase, C_3S (Lam et al., 2010).

As with all innovations, it is important to strike a balance between harnessing the benefits of new processes and materials while managing potential downsides. Future research should, therefore, focus on optimizing the use of ISSA in cement clinkers to maximize its benefits while minimizing any potential adverse effects on the product's performance.

One of the most effective strategies to optimize the use of ISSA in cement clinkers is to pretreat the ash, aiming to reduce its phosphate content (Jama-Rodzeńska et al., 2021). Numerous studies have proposed the recovery of phosphate from ISSA through wet acid leaching process (Kasina, 2023; Liang et al., 2021; Fournie et al., 2022; Li et al., 2021). Luyckx and Van Caneghem (2022) proposed that the addition of aluminum chloride, iron(III) chloride, magnesium chloride, or aluminum sulfate to solid waste prior to incineration provided the most favorable balance between high phosphorus extraction and low coextraction of heavy metals, as shown in Fig. 11.7.

Recently, many researchers have been studying the use of nanomaterials for the modification and improvement of the properties of ISSA. The addition of nanomaterials has proven effective in enhancing the microstructure and density of cement pastes incorporating ISSA, leading to an improvement in their mechanical performance. It is also expected that other materials such as silica fume or metakaolin could be used effectively alongside ISSA in cementitious materials (Luo et al., 2015).

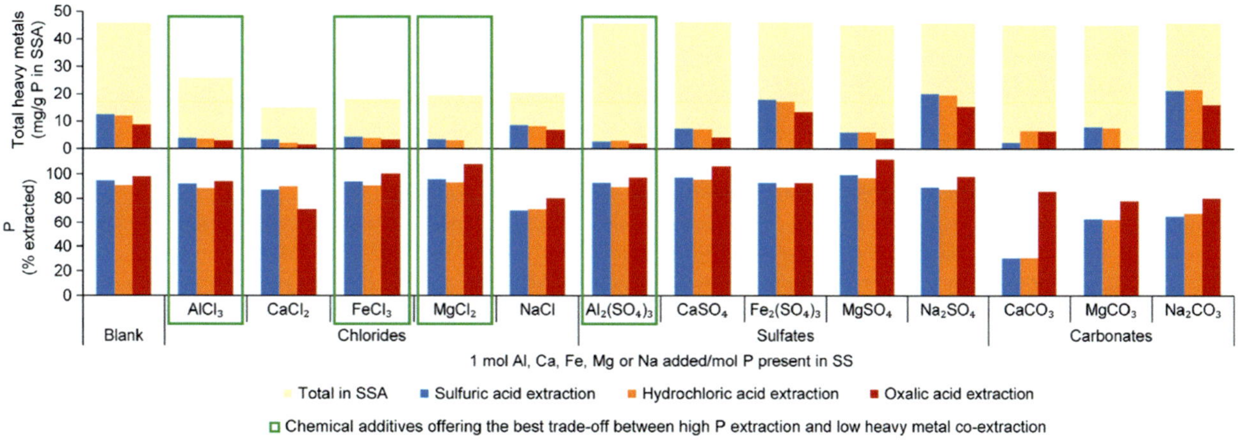

FIGURE 11.7 P extraction and total heavy metal coextraction from the ISSA samples to which different chemicals were added (Luyckx and Caneghem, 2022).

6. Conclusions

Upon conducting an exhaustive and critical review of existing literature on the utilization of sewage sludge and incinerated sewage sludge ash (ISSA) into cement clinker, it has been determined that this type of solid waste can serve two primary functions: (1) as a raw ingredient for cement clinker production and (2) as a supplementary cementitious material. The key conclusions drawn are as follows:

(1) The primary oxides in sewage sludge and sludge ash are SiO_2, Al_2O_3, and CaO, essential components for cementitious materials. Incineration of sewage sludge results in higher oxide content due to the removal of volatile components and the decomposition of organic matter. ISSA displays the pozzolanic activity due to its large specific surface area and amorphous phase.

(2) ISSA can be safely used up to a 15% replacement level in the production of eco-cement, delivering comparable performance to traditional Portland cement. However, using excessive amounts of sludge can negatively affect the compressive strength, increase water demand, and prolong the final setting time of the eco-cement paste.

(3) In cementitious materials, heavy metal contaminants present in ISSA can be incorporated into the structure of hydration products or form stable precipitates with the hydrated products.

Therefore, an extensive analysis of the application of sewage sludge ash (SSA) in concrete and related products demonstrated that ISSA can be used both as a raw feed for cement clinker production and as a supplementary cementitious material. ISSA has potential suitability as a cementitious material due to its oxide composition and amorphous content. While ISSA contains toxic elements, the amounts generally fall below the target limits for construction materials. Derived from the review, several pivotal future research directions can be identified: (1) recovery and utilization of phosphorous from ISSA; (2) strategies to mitigate leaching of heavy metals; and (3) methods to enhance the performance of cement clinker with ISSA incorporation. These investigations could offer a theoretical foundation for effective ISSA recycling into cement clinker under the dual considerations of environmental protection and utilization efficiency.

Acknowledgments

The authors gratefully acknowledge the financial support provided by the "Pioneer" and "Leading Goose" R&D Program of Zhejiang (2023C04033) and Ningbo Public Welfare Science And Technology Plan Project (Grant No. 2023S056).

References

Adamczyk, M., Zdeb, T., Tracz, T., 2023. Physico-chemical properties of sewage sludge ash and its influence on the chemical shrinkage of cement pastes. Mater. Process. 13 (1), 26.

Asavapisit, S., Naksrichum, S., Harnwajanawong, N., 2005. Strength, leachability and microstructure characteristics of cement-based solidified plating sludge. Cement Concr. Res. 35 (6), 1042–1049.

Baeza, F., Payá, J., Galao, O., Saval, J.M., Garcés, P., 2014. Blending of industrial waste from different sources as partial substitution of Portland cement in pastes and mortars. Construct. Build. Mater. 66, 645–653.

Baeza-Brotons, F., Garcés, P., Payá, J., Saval, J.M., 2014. Portland cement systems with addition of sewage sludge ash. Application in concretes for the manufacture of blocks. J. Clean. Prod. 82, 112–124.

Bairq, Z.A.S., Li, R., Li, Y., Gao, H., Sema, T., Teng, W., Liang, Z., 2018. New advancement perspectives of chloride additives on enhanced heavy metals removal and phosphorus fixation during thermal processing of sewage sludge. J. Clean. Prod. 188, 185–194.

Chakraborty, S., Jo, B.W., Jo, J.H., Baloch, Z., 2017. Effectiveness of sewage sludge ash combined with waste pozzolanic minerals in developing sustainable construction material: an alternative approach for waste management. J. Clean. Prod. 153, 253–263.

Chang, Z., Long, G., Zhou, J.L., Ma, C., 2020. Valorization of sewage sludge in the fabrication of construction and building materials: a review. Resour. Conserv. Recycl. 154, 104606.

Chang, Z., Long, G., Xie, Y., Zhou, J.L., 2022a. Pozzolanic reactivity of aluminum-rich sewage sludge ash: influence of calcination process and effect of calcination products on cement hydration. Construct. Build. Mater. 318, 126096.

Chang, Z., Long, G., Xie, Y., Zhou, J.L., 2022b. Recycling sewage sludge ash and limestone for sustainable cementitious material production. J. Build. Eng. 49, 104035.

Chen, M., Blanc, D., Gautier, M., Mehu, J., Gourdon, R., 2013. Environmental and technical assessments of the potential utilization of sewage sludge ashes (SSAs) as secondary raw materials in construction. Waste Manage. (Tucson, Ariz.) 33 (5), 1268–1275.

Chen, Z., Li, J.S., Poon, C.S., 2018. Combined use of sewage sludge ash and recycled glass cullet for the production of concrete blocks. J. Clean. Prod. 171, 1447–1459.

Chen, L., Lin, D.F., 2009. Stabilization treatment of soft subgrade soil by sewage sludge ash and cement. J. Hazard Mater. 162 (1), 321–327.

Chen, Z., Poon, C.S., 2017. Comparing the use of sewage sludge ash and glass powder in cement mortars. Environ. Technol. 38 (11), 1390–1398.

Cong, X., Lu, S., Gao, Y., Yao, Y., Elchalakani, M., Shi, X., 2020. Effects of microwave, thermomechanical and chemical treatments of sewage sludge ash on its early-age behavior as supplementary cementitious material. J. Clean. Prod. 258, 120647.

Coutand, M., Cyr, M., Clastres, P., 2006. Use of sewage sludge ash as mineral admixture in mortars. Proc. Inst. Civ. Eng.-Co. 159 (4), 153–162.

Cusidó, J.A., Cremades, L.V., 2012. Environmental effects of using clay bricks produced with sewage sludge: leachability and toxicity studies. Waste Manage. (Tucson, Ariz.) 32 (6), 1202–1208.

Danish, A., Ozbakkaloglu, T., 2022. Greener cementitious composites incorporating sewage sludge ash as cement replacement: a review of progress, potentials, and future prospects. J. Clean. Prod. 133364.

de Azevedo Basto, P., Junior, H.S., de Melo Neto, A.A., 2019. Characterization and pozzolanic properties of sewage sludge ashes (SSA) by electrical conductivity. Cement Concr. Compos. 104, 103410.

Di Iaconi, C., De Sanctis, M., Altieri, V.G., 2020. Full-scale sludge reduction in the water line of municipal wastewater treatment plant. J. Environ. Manag. 269, 110714.

Donatello, S., Tong, D., Cheeseman, C.R., 2010. Production of technical grade phosphoric acid from incinerator sewage sludge ash (ISSA). Waste Manage. (Tucson, Ariz.) 30 (8–9), 1634–1642.

Dyer, T.D., Halliday, J.E., Dhir, R.K., 2011. Hydration chemistry of sewage sludge ash used as a cement component. J. Mater. Civ. Eng. 23 (5), 648–655.

Fang, P., Tang, Z.J., Huang, J.H., Cen, C.P., Tang, Z.X., Chen, X.B., 2015. Using sewage sludge as a denitration agent and secondary fuel in a cement plant: a case study. Fuel Process. Technol. 137, 1–7.

Fournie, T., Rashwan, T.L., Switzer, C., Gerhard, J.I., 2022. Phosphorus recovery and reuse potential from smouldered sewage sludge ash. Waste Manage. (Tucson, Ariz.) 137, 241–252.

Gomes, S.D.C., Zhou, J.L., Li, W., Long, G., 2019. Progress in manufacture and properties of construction materials incorporating water treatment sludge: a review. Resour. Conserv. Recycl. 145, 148–159.

Gu, C., Ji, Y., Zhang, Y., Yang, Y., Liu, J., Ni, T., 2021. Recycling use of sulfate-rich sewage sludge ash (SR-SSA) in cement-based materials: assessment on the basic properties, volume deformation and microstructure of SR-SSA blended cement pastes. J. Clean. Prod. 282, 124511.

Guo, B., Liu, B., Yang, J., Zhang, S., 2017. The mechanisms of heavy metal immobilization by cementitious material treatments and thermal treatments: a review. J. Environ. Manag. 193, 410–422.

Guo, X., Yuan, S., Xu, Y., Qian, G., 2022. Effects of phosphorus and iron on the composition and property of Portland cement clinker utilized incinerated sewage sludge ash. Construct. Build. Mater. 341, 127754.

Han, R., Liu, J., Zhang, Y., Fan, X., Lu, W., Wang, H., 2012. Dewatering and granulation of sewage sludge by biophysical drying and thermo-degradation performance of prepared sludge particles during succedent fast pyrolysis. Bioresour. Technol. 107, 429–436.

Haustein, E., Kuryłowicz-Cudowska, A., Łuczkiewicz, A., Fudala-Książek, S., Cieślik, B.M., 2022. Influence of cement replacement with sewage sludge ash (SSA) on the heat of hydration of cement mortar. Materials 15 (4), 1547.

Hu, S.H., Hu, S.C., Fu, Y.P., 2013. Recycling technology—artificial lightweight aggregates synthesized from sewage sludge and its ash at lowered comelting temperature. Environ. Prog. Sustain. 32 (3), 740–748.

Huang, Y., Li, H., Jiang, Z., Yang, X., Chen, Q., 2018. Migration and transformation of sulfur in the municipal sewage sludge during disposal in cement kiln. Waste Manage. (Tucson, Ariz.) 77, 537–544.

Jama-Rodzeńska, A., Sowiński, J., Koziel, J.A., Białowiec, A., 2021. Phosphorus recovery from sewage sludge ash based on cradle-to-cradle approach—mini-review. Minerals 11 (9), 985.

Kasina, M., 2023. The assessment of phosphorus recovery potential in sewage sludge incineration ashes—a case study. Environ. Sci. Pollut. Res. 30 (5), 13067–13078.

Lam, C.H., Barford, J.P., McKay, G., 2010. Utilization of incineration waste ash residues in Portland cement clinker. Chem. Eng. 21, 757–762.

Lasheras-Zubiate, M., Navarro-Blasco, I., Fernández, J.M., Alvarez, J.I., 2012. Encapsulation, solid-phases identification and leaching of toxic metals in cement systems modified by natural biodegradable polymers. J. Hazard Mater. 233, 7–17.

Latosińska, J., Gawdzik, J., 2012. The effect of incineration temperatures on mobility of heavy metals in sewage sludge ash. Environ. Protect. Eng. 38 (3), 31–44.

Li, J.S., Xue, Q., Fang, L., Poon, C.S., 2017. Characteristics and metal leachability of incinerated sewage sludge ash and air pollution control residues from Hong Kong evaluated by different methods. Waste Manage. (Tucson, Ariz.) 64, 161–170.

Li, J.S., Chen, Z., Wang, Q.M., Fang, L., Xue, Q., Cheeseman, C.R., Poon, C.S., 2018. Change in re-use value of incinerated sewage sludge ash due to chemical extraction of phosphorus. Waste Manage. (Tucson, Ariz.) 74, 404–412.

Li, S., Zeng, W., Ren, Z., Jia, Z., Peng, X., Peng, Y., 2021. A novel strategy to capture phosphate as high-quality struvite from the sewage sludge ash: process, mechanism and application. J. Clean. Prod. 322, 129162.

Liang, S., Yang, L., Chen, H., Yu, W., Tao, S., Yuan, S., Yang, J., 2021. Phosphorus recovery from incinerated sewage sludge ash (ISSA) and reutilization of residues for sludge pretreated by different conditioners. Resour. Conserv. Recycl. 169, 105524.

Liang, Y., Xu, D., Feng, P., Hao, B., Guo, Y., Wang, S., 2021. Municipal sewage sludge incineration and its air pollution control. J. Clean. Prod. 295, 126456.

Liang, C., Le, X., Fang, W., Zhao, J., Fang, L., Hou, S., 2022. The utilization of recycled sewage sludge ash as a supplementary cementitious material in mortar: a review. Sustainability 14 (8), 4432.

Lin, K.L., Wang, K.S., Tzeng, B.Y., Wang, N.F., Lin, C.Y., 2004. Effects of Al2O3 on the hydration activity of municipal solid waste incinerator fly ash slag. Cement Concr. Res. 34 (4), 587–592.

Lin, K.L., Lin, D.F., Luo, H.L., 2009. Influence of phosphate of the waste sludge on the hydration characteristics of eco-cement. J. Hazard Mater. 168 (2–3), 1105–1110.

Lin, Y., Zhou, S., Li, F., Lin, Y., 2012. Utilization of municipal sewage sludge as additives for the production of eco-cement. J. Hazard Mater. 213, 457–465.

Liu, M., Wang, C., Bai, Y., Xu, G., 2018. Effects of sintering temperature on the characteristics of lightweight aggregate made from sewage sludge and river sediment. J. Alloys Compd. 748, 522–527.

Luo, H.L., Lin, D.F., Shieh, S.I., You, Y.F., 2015. Micro-observations of different types of nano-Al2O3 on the hydration of cement paste with sludge ash replacement. Environ. Technol. 36 (23), 2967–2976.

Luyckx, L., Van Caneghem, J., 2022. Recovery of phosphorus from sewage sludge ash: influence of chemical addition prior to incineration on ash mineralogy and related phosphorus and heavy metal extraction. J. Environ. Chem. Eng. 10 (4), 108117.

Lv, D., Zhu, T., Liu, R., Lv, Q., Sun, Y., Wang, H., Zhang, F., 2016. Effects of co-processing sewage sludge in cement kiln on NOx, NH3 and PAHs emissions. Chemosphere 159, 595–601.

Lynn, C.J., Dhir, R.K., Ghataora, G.S., West, R.P., 2015. Sewage sludge ash characteristics and potential for use in concrete. Construct. Build. Mater. 98, 767–779.

Lynn, C.J., Obe, R.K.D., Ghataora, G.S., 2016. Municipal incinerated bottom ash characteristics and potential for use as aggregate in concrete. Construct. Build. Mater. 127, 504–517.

Lynn, C.J., Dhir, R.K., Ghataora, G.S., 2018. Environmental impacts of sewage sludge ash in construction: leaching assessment. Resour. Conserv. Recycl. 136, 306–314.

Mininni, G., Blanch, A.R., Lucena, F., Berselli, S., 2015. EU policy on sewage sludge utilization and perspectives on new approaches of sludge management. Environ. Sci. Pollut. Res. 22, 7361–7374.

Naamane, S., Rais, Z., Taleb, M., 2016. The effectiveness of the incineration of sewage sludge on the evolution of physicochemical and mechanical properties of Portland cement. Construct. Build. Mater. 112, 783–789.

Ottosen, L.M., Jensen, P.E., Kirkelund, G.M., 2014. Electrodialytic separation of phosphorus and heavy metals from two types of sewage sludge ash. Separ. Sci. Technol. 49 (12), 1910–1920.

Ottosen, L.M., Thornberg, D., Cohen, Y., Stiernström, S., 2022. Utilization of acid-washed sewage sludge ash as sand or cement replacement in concrete. Resour. Conserv. Recycl. 176, 105943.

Prabhakar, A.K., Krishnan, P., Lee, S.S.C., Lim, C.S., Dixit, A., Mohan, B.C., Wang, C.H., 2022. Sewage sludge ash-based mortar as construction material: mechanical studies, macrofouling, and marine toxicity. Sci. Total Environ. 824, 153768.

Rezaee, F., Danesh, S., Tavakkolizadeh, M., Mohammadi-Khatami, M., 2019. Investigating chemical, physical and mechanical properties of eco-cement produced using dry sewage sludge and traditional raw materials. J. Clean. Prod. 214, 749–757.

Rusănescu, C.O., Voicu, G., Paraschiv, G., Begea, M., Purdea, L., Petre, I.C., Stoian, E.V., 2022. Recovery of sewage sludge in the cement industry. Energies 15 (7), 2664.

Samolada, M.C., Zabaniotou, A.A., 2014. Comparative assessment of municipal sewage sludge incineration, gasification and pyrolysis for a sustainable sludge-to-energy management in Greece. Waste Manage. (Tucson, Ariz.) 34 (2), 411–420.

Shih, P.H., Chang, J.E., Lu, H.C., Chiang, L.C., 2005. Reuse of heavy metal-containing sludges in cement production. Cement Concr. Res. 35 (11), 2110–2115.

Smol, M., Kulczycka, J., Henclik, A., Gorazda, K., Wzorek, Z., 2015. The possible use of sewage sludge ash (SSA) in the construction industry as a way towards a circular economy. J. Clean. Prod. 95, 45–54.

Stephan, D., Maleki, H., Knöfel, D., Eber, B., Härdtl, R., 1999. Influence of Cr, Ni, and Zn on the properties of pure clinker phases: Part II. C3A and C4AF. Cement Concr. Res. 29 (5), 651–657.

Tang, Y., Shih, K., 2015. Mechanisms of zinc incorporation in aluminosilicate crystalline structures and the leaching behaviour of product phases. Environ. Technol. 36 (23), 2977–2986.

Valderrama, C., Granados, R., Cortina, J.L., Gasol, C.M., Guillem, M., Josa, A., 2013. Comparative LCA of sewage sludge valorisation as both fuel and raw material substitute in clinker production. J. Clean. Prod. 51, 205–213.

Vespa, M., Dähn, R., Wieland, E., 2014. Competition behaviour of metal uptake in cementitious systems: an XRD and EXAFS investigation of Nd-and Zn-loaded 11 Å tobermorite. Phys. Chem. Earth, Parts A/B/C 70, 32–38.

Wang, X., Deng, S., Tan, H., Adeosun, A., Vujanović, M., Yang, F., Duić, N., 2016. Synergetic effect of sewage sludge and biomass co-pyrolysis: a combined study in thermogravimetric analyzer and a fixed bed reactor. Energy Convers. Manag. 118, 399–405.

Wang, Q., Li, J.S., Poon, C.S., 2023. Production of sorptive granules from incinerated sewage sludge ash and upcycling in cement mortar. Sep. Purif. Technol. 309, 123046.

Wei, N., 2015. Leachability of heavy metals from lightweight aggregates made with sewage sludge and municipal solid waste incineration fly ash. Int. J. Environ. Res. Publ. Health 12 (5), 4992–5005.

Weng, C.H., Lin, D.F., Chiang, P.C., 2003. Utilization of sludge as brick materials. Adv. Environ. Res. 7 (3), 679–685.

Wu, Z., Jiang, Y., Guo, W., Jin, J., Wu, M., Shen, D., Long, Y., 2021. The long-term performance of concrete amended with municipal sewage sludge incineration ash. Environ. Technol. Inno. 23, 101574.

Xia, Y., Liu, M., Zhao, Y., Chi, X., Lu, Z., Tang, K., Guo, J., 2023. Utilization of sewage sludge ash in ultra-high performance concrete (UHPC): microstructure and life-cycle assessment. J. Environ. Manag. 326, 116690.

Xu, W., Xu, J., Liu, J., Li, H., Cao, B., Huang, X., Li, G., 2014. The utilization of lime-dried sludge as resource for producing cement. J. Clean. Prod. 83, 286−293.

Yang, G., Zhang, G., Wang, H., 2015. Current state of sludge production, management, treatment and disposal in China. Water Res. 78, 60−73.

Yu, S., Zhang, H., Lu, F., Shao, L., He, P., 2021. Flow analysis of major and trace elements in residues from large-scale sewage sludge incineration. J. Environ. Sci. 102, 99−109.

Zdeb, T., Tracz, T., Adamczyk, M., 2022. Physical, mechanical properties and durability of cement mortars containing fly ash from the sewage sludge incineration process. J. Clean. Prod. 345, 131055.

Zhang, Z., Zhang, L., Yin, Y., Liang, X., Li, A., 2015. The recycling of incinerated sewage sludge ash as a raw material for CaO−Al2O3−SiO2−P2O5 glass-ceramic production. Environ. Technol. 36 (9), 1098−1103.

Zhao, X.G., Jiang, G.W., Li, A., Li, Y., 2016. Technology, cost, a performance of waste-to-energy incineration industry in China. Renew. Sust. Energ. Rev. 55, 115−130.

Zhou, Y.F., Li, J.S., Lu, J.X., Cheeseman, C., Poon, C.S., 2020. Recycling incinerated sewage sludge ash (ISSA) as a cementitious binder by lime activation. J. Clean. Prod. 244, 118856.

Chapter 12

Recycling of municipal solid waste incineration fly ash into cement clinker

Yingying Xiong[1,2], Yan Xia[2], Yuan Meng[3], Gang Huang[2], Zengyi Ma[1,2], Lei Wang[2] and Jianhua Yan[2]

[1]*Institute for Carbon Neutrality, Ningbo Innovation Center, Zhejiang University, Ningbo, China;* [2]*State Key Laboratory of Clean Energy Utilization, Zhejiang University, Hangzhou, China;* [3]*Department of Civil and Environmental Engineering, The Hong Kong Polytechnic University, Hong Kong, China*

1. Introduction

The production of cement clinker primarily relies on the presence of essential elementals (calcium, silicon, iron, and aluminum) within raw materials, which are typically derived from natural sources, such as limestone, sand, and bauxite (Viczek et al., 2020). These materials usually undergo the calcination process within a rotary kiln operating at temperatures ranging from 1400 to 1500°C, leading to the formation of cement clinker. It is worth noting that the cement production sector represents the major contributor to industrial carbon dioxide emissions and demands a substantial amount of energy input (Benhelal et al., 2021). For every metric ton of Portland cement production, approximately equal mass of CO_2 is generated (Andrew, 2018). To reduce CO_2 emissions in this process is one of the reasons for seeking alternative raw materials in cement production. Additionally, the substantial demand for limestone has prompted the exploration of alternative raw materials for cement clinker manufacturing. Furthermore, changes in the proportion of coal-fired power generation have influenced the availability of frequently used kiln ash, thus causing a scarcity of appropriate raw materials for cement clinker production (Diaz-Loya et al., 2019).

Recycling municipal solid waste incineration fly ash (MSWI FA) producing cement clinker is feasible solution. The increase in municipal solid waste (MSW) generation results in a rise in MSWI FA production (Cucchiella et al., 2017). Currently, the global annual production of MSW amounts to about 2.01 billion tons, with projections indicating an increase to 3.4 billion tons by 2050 (Zhang et al., 2022). Incineration stands as the prevailing approach for MSW management, offering the dual benefits of energy generation through MSW utilization and waste volume reduction of up to 90% (Leckner, 2015). Undoubtedly, incineration remains the favored choice for MSW management. However, the appropriate handling of MSWI ashes poses a crucial challenge. MSWI ashes are categorized into two categories: MSWI bottom ash (BA) and fly ash (FA) (Tang et al., 2015). MSWI FA, as mainly composed of CaO, SiO_2, Al_2O_3, and Fe_2O_3, exhibits the elemental composition required for cement clinker production, which makes it become a potential substitute for traditional cement raw materials. However, MSWI FA is classified as hazardous waste in many countries owing to its content of potentially toxic elements (PTEs) such as lead (Pb), cadmium (Cd), mercury (Hg), molybdenum (Mo), and nickel (Ni), as well as other pollutants including dioxins, sulfates, chlorides, and acids (Marieta et al., 2021). Consequently, the appropriate incorporation of MSWI FA into cement raw materials presents a notable challenge meriting thorough consideration.

With its predominant composition abundant in silica and lime, MSWI FA emerges as a valuable mineral source well-suited for cement production (Tang et al., 2018). It holds the potential for partially substitute limestone or clay as a cement raw material, thereby contributing to environmental preservation. The elevated temperature and prolonged residence time during the cement manufacturing process effectively eliminates toxic and hazardous pollutants such as dioxin in the MSWI FA (Cheng et al., 2007). Simultaneously, the process facilitates the stabilization of heavy metal content within the ash through structural binding in the resulting clinker phase (Ghouleh and Shao, 2018). The advantages of employing MSWI FA as an alternative to conventional raw materials for cement clinker production are manifold. Consequently, numerous studies have focused on investigating the utilization of fly ash as a raw material in laboratory-scale or real-scale cement

Treatment and Utilization of Combustion and Incineration Residues. https://doi.org/10.1016/B978-0-443-21536-0.00039-3

production (Zheng et al., 2022). However, as previously emphasized, the direct inclusion of MSWI FA into raw materials is not feasible due to regional variations in the composition of MSWI FA and the presence of environmentally and human health-hazardous constituents. Some pretreatments are necessary to mitigate the influence of hazardous substances on the quality of cement clinker modified with MSWI FA. And the decomposition of dioxin, stabilization of heavy metals, and removal of chlorides are crucial aspects in the conversion of MSWI FA into a viable raw material for civil engineering applications. Therefore, it is imperative to summarize and discuss these aspects to attain a comprehensive understanding of the related challenges and opportunities.

It is essential to summarize the properties, mineral composition, and strength of clinker products resulting from the incorporation of MSWI FA in the raw materials. And evaluating the feasibility of reusing pretreated MSWI FA as cement raw materials, along with assessing the environmental impacts of these processes, including heavy metal and dioxin analysis, as well as leaching tests of the produced clinker, is necessary. Throughout this endeavor, it is crucial to consider the potential environmental pollution resulting from various treatment processes. Research on MSWI FA as an alternative cement raw material is still evolving, and future developments are expected to bring about expanded applications and standardized treatment regulations.

1.1 Components and pollutants in MSWI FA: Mineral composition

The MSWI FA exhibits consistent main components across various regions, characterized by significant oxides including Al_2O_3, CaO, Fe_2O_3, and SiO_2. These oxides are akin to those found in cement clinker and belong to the $CaO-SiO_2-SO_3-Al_2O_3$ system (Wang et al., 2015). These constituents allow MSWI FA to react with water and lead to the formation of calcium silicate hydrates (C−S−H), which provides compelling evidence for the potential utilization of MSWI FA in cement clinker production. However, diverse lifestyles, waste recycling practices, and regional variations contribute to significant differences in the composition of MSWI FA. The chemical and physical characteristics, as well as the composition of MSWI FA, are commonly influenced by the original waste composition, operating condition, incinerator type, and flue gas cleaning system (Vaitkus et al., 2018). Table 12.1 provides a summary of the predominant oxide contents in various MSWI FAs utilized in cement-related studies. It has revealed that fly ash prominently features SiO_2, Al_2O_3, CaO, and Fe_2O_3, thereby qualifying as viable alternatives to the limited portion of conventional cement raw materials that typically contain these elements.

TABLE 12.1 The composition of major oxides in MSWI FA.

	Unit	Zhang et al. (2022)	Marieta et al. (2021)	Mao et al. (2020)	Ghouleh and Shao (2018)	Guo et al. (2015)	Pan et al. (2008)
CaO	wt%	36.06	36.98	49.80	20.23	39.10	45.42
SiO_2	wt%	24.7	10.8	9.20	18.28	15.9	13.60
Al_2O_3	wt%	8.23	5.28	1.61	14.47	4.35	0.92
Fe_2O_3	wt%	7.24	1.35	3.24	2.30	1.91	3.83
SO_3	wt%	7.49	–	5.89	–	7.24	6.27
MgO	wt%	7.79	1.94	0.98	6.73	1.64	3.16
P_2O_5	wt%	2.59	1.83	–	3.50	1.08	1.72
Na_2O	wt%	2.28	4.21	0.40	12.66	3.43	4.16
K_2O	wt%	1.00	4.04	2.25	2.21	3.68	3.83
MnO	wt%	–	–	–	0.09	–	–
TiO_2	wt%	–	–	0.52	1.76	–	–

TABLE 12.2 The composition of heavy metals in MSWI FA.

	Wu et al. (2023) (mg/L)	Marieta et al. (2021) (mg/kg)	Wang et al. (2021) (mg/kg)	Bogush et al. (2020) (mg/kg)	Guo et al. (2015) (mg/kg)
Cr	6.11	165.6	49.7	863.7	175
Mn	–	–	–	1092.72	–
Zn	1.32	5099.1	4950.0	3764.95	–
Cu	1.32	679.4	702.2	1123.79	–
Ni	–	92.5	–	118.22	–
As	–	–	–	51.92	116
Pb	83.12	1153.5	111.2	765.41	–
Cd	0.08	–	108.1	66.31	86.5

1.2 Inorganic pollutants

Table 12.2 shows the presence of various common heavy metal elements, such as chromium (Cr), cuprum (Cu), mercury (Hg), nickel (Ni), cadmium (Cd), zinc (Zn), lead (Pb), etc. Calcium (Ca) and aluminum (Al) are the most prevalent metals found in MSWI FA, and their presence does not pose environmental risks. However, it is noteworthy that studies conducted on MSWI FA consistently demonstrate similar types and concentrations of heavy metals, indicating that the boiling point of these metals plays a significant role in their composition (He et al., 2023). The concentrations of these trace metals far exceed those found in traditional raw materials fed into the kiln (Quina et al., 2008). Moreover, MSWI FA contains a substantial amount of soluble salts, primarily chlorides, and sulfates such as sodium chloride, calcium chloride, and sodium sulfate (Saikia et al., 2007). Proper treatment of MSWI FA is necessary to eliminate these soluble salts effectively. And incorporating MSWI FA as a partial substitute for cement raw materials offers a viable approach to mitigating its environmental and health hazards. However, it is crucial to ensure the effective immobilization of the aforementioned heavy metals in the solid phase. Otherwise, issues such as the leaching of metals and the potential impact on cement hydration and mechanical properties may emerge (Shao et al., 2021). Notably, Zn and Pb can influence the normal development of cement properties during hydration (Niu et al., 2018). The presence of chlorides can interfere with cement setting and hardening processes, leading to compromised passivation of embedded steel bars and early degradation of concrete integrity (Coumes and Courtois, 2003). And sulfates can contribute to unexpected expansion and cracking, potentially attributed to the delayed formation of ettringite and the gradual deterioration of concrete results (Cyr et al., 2012).

As previously discussed, the utilization of MSWI FA as a raw material in cement production will pose challenges due to the presence of elements, such as Cl and S, which have the potential to cause concrete damage and steel corrosion. To address these issues, a potential approach is to restrict the quantity of MSWI FA added to the raw material mixture (Lederer et al., 2017). In terms of mineral composition, the average permissible mixing ratio of MSWI FA is 67%, leading to the production of suitable clinker. However, the presence of alkali imposes a limitation, decreasing this value to 3.72%, and when alkali and chloride are simultaneously present, the allowable mixing ratio drops further to 0.33% (Sarmiento et al., 2019). Consequently, it is necessary to implement a washing pretreatment process for MSWI FA before using it as cement raw materials, which can increase the proportion of incorporated MSWI FA while ensuring the quality of the resulting cement clinker.

The application of pretreatment methods can decrease the Cl, S, and other element contents such as Na, K, and P in MSWI FA (Lederer et al., 2017). Research indicated that the feasibility of MSWI FA for secondary recovery applications relied on the effectiveness of the pretreatment technology. Among the various pretreatment methods, water washing is the most commonly employed, which can effectively decrease the content of soluble salts, such as chloride and sulfate, as well as the chlorine crystalline phases. And the washing pretreatment exhibits a favorable impact by eliminating elements that might impede the subsequent stabilization process (Ferraro et al., 2023). Numerous other pretreatment methods have also been proposed continuously. For instance, alkali washing and thermal quenching could effectively eliminate aluminum alloy from MSWI FA through chemical reactions with an alkaline washing solution or thermal oxidation during thermal quenching. These methods also effectively removed chloride and sulfate from the ash. Although the leaching concentration of Cd and Pb from MSWI FA after alkali washing still exceeded the permissible limit for nonhazardous waste, the leaching of these heavy metals is significantly reduced. In the case of thermal quenching, the reduction of soluble sulfate leads to the formation of $PbSnS_3$. Furthermore, the combination of alkaline washing and thermal quenching is favorable for MSWI FA treatment (Chen et al., 2023). Moreover, Zhang et al. (2022) proposed a low-concentration anaerobic fermentation organic acid synergistic washing method for MSWI FA. Through optimized washing pretreatment, the chlorine content in MSWI FA can be reduced to 0.82%, facilitating the release of trace elements.

1.3 Organic pollutants

MSWI FA contains significant quantities of persistent organic pollutants (POPs), including polychlorinated dibenzo-dioxins (PCDDs) and polychlorinated dibenzofurans (PCDFs), collectively known as dioxins (Cobo et al., 2009). These compounds are exceedingly toxic, stable, lipophilic, difficult to degrade, and prone to bioaccumulation, which poses a substantial risk to both human health and the environment (Wong et al., 2021). Organic pollutants are typically decomposed at temperatures ranging from 700 to 1200°C. Fortunately, the calcination temperature in cement production can reach the decomposition temperature of dioxin degradation. Moreover, these harmful compounds volatilized at clinker temperatures and were eliminated from the feedstock during cement kiln operations (Liu et al., 2015). Cement shaft kilns can degrade up to 90.1% of dioxins in MSWI FA, with gas phase dioxin accounting for only 2.5% of the total dioxins, which is lower than the national emission standard (Li et al., 2023). When the sintering temperature ranges from 700 to 1200°C, the degradation rate of dioxin can reach 95%. This aspect represents an advantage for utilizing MSWI FA in cement production (Pan et al., 2008).

2. Application of MSWI FA in cement production

2.1 Ordinary Portland cement

2.1.1 Preparation technology

Table 12.3 shows an overview of relevant studies investigating the utilization of MSWI FA for the production of OPC clinker. MSWI FA presents a viable option as a secondary raw material for the production of OPC clinker in cement kilns (Lam et al., 2011). Similar to other essential cement ingredients such as blast furnace slag and coal fly ash, the incorporation of MSWI FA should take place before cement milling to ensure uniform particle size in the resulting cement (Shi et al., 2009). The technological feasibility of using MSWI FA for OPC cement clinker production has been explored. Given the similar chemical composition to marl, MSWI FA was utilized as a complete replacement for marl during the research process. The sintering process was conducted at 1450°C, and the raw material substitution for MSWI FA ranged from 14.5% to 34%. MSWI FA with high chlorine and calcium content exhibited potential for OPC clinker production in both laboratory and industrial testing scales (Diliberto et al., 2020). Researchers have employed MSWI FA combined with pure compounds such as $CaCO_3$, SiO_2, Al_2O_3, and Fe_2O_3 as raw materials. The mixture was calcined at 900°C for 30 min, followed by heating to 1450°C at a rate of 20°C/min and rapid cooling to room temperature. The OPC clinker

TABLE 12.3 Performances of OPC with MSWI FA as raw materials.

Materials	Compressive strength	Calcination conditions	References
MSWI FA, with limestone, marl, clay, bauxite, iron oxides	–	A heating rate of 10°C/min and maintained at 1200°C 1300°C, 1400°C, 1450°C for 20 min, respectively	Diliberto et al. (2020)
MSWI FA, $CaCO_3$, SiO_2, Al_2O_3, and Fe_2O_3.	Similar to strength of the control sample	Calcined at 900°C for 30 min, at 1450°C for 2 h	Bogush et al. (2020)
MSWI FA, SiO_2, Al_2O_3, Fe_2O_3	>32.5 MPa after 28-day curing	20°C/min heating rate to 1450°C and maintained for 1 h	Wang et al. (2010)
MSWI FA, with limestone, iron slag, clay, and gypsum	Greater than the ASTM values of type II Portland cement	10°C/min heating rate to 1450°C; maintained for 30 min	Pan et al. (2008)
MSWI FA, with cement raw mix and magnet-repelled ash	12.4, 19.3, and 27.6 MPa at 3, 7, and 28 days	Heated at 900°C for 1 h, to 1400°C for 3 h	Shih et al. (2003)

incorporating 35% MSWI FA demonstrated its successful incorporation into the cement matrix (Bogush et al., 2020). This utilization method enhances the burnability of the OPC raw mixture, diminishes the leaching of heavy metals in hydration products, and exhibits favorable enforceability characteristics.

2.1.2 Mineral compositions

The production of OPC clinker involves the high-temperature reaction of calcium, aluminum, iron, and silica oxides (Lam et al., 2011). This reaction leads to the formation of four primary mineral phases: alite, which is tricalcium silicate, has the chemical formula Ca_3SiO_5 and is commonly known as C_3S; belite, which is dicalcium silicate, has the chemical formula Ca_2SiO_4 and is commonly known as C_2S; aluminate, which is tricalcium aluminate, has the chemical formula $Ca_3Al_2O_6$ and is commonly known as C_3A; and ferrite, which is tetracalcium aluminoferrite, has the chemical formula $Ca_4Al_2Fe_2O_{10}$ and is commonly known as C_4AF. The essential elements for clinker production are supplied by natural resources such as limestone, sand, and clay, as well as industrial by-products such as coal fly ash, waste glass, and steelmaking residues (Schneider et al., 2011). Given the widespread utilization of waste materials in cement production globally, the substantial content of calcium oxide, aluminum oxide, iron oxide, and silica oxide in MSWI FA presents another potential viable alternative or supplement to conventional cement raw materials (Saikia et al., 2007). The capacity of MSWI FA to form these clinker phases directly influences the physical properties of modified cement products; therefore, it is necessary to understand the impact of MSWI FA on clinker phase formation. Many researches have demonstrated that the mineral composition and hydration characteristics of cement clinker produced with MSWI FA as an alternative raw material are essentially identical to those of OPC cement manufactured without MSWI FA, and the addition of MSWI FA has no discernible effect on the chemical composition of the clinker (Li et al., 2016). Researchers found that the substitution rate of MSWI FA in cement raw materials could reach up to 30%, depending on the contents of harmful elements (Diliberto et al., 2020). Moreover, the substitution of MSWI FA did not impact the properties of cement clinker under a favorable chemical composition balance of the raw materials. Additionally, the produced clinker displayed the typical mineral phases found in OPC clinker, with only a small amount of free lime (f-CaO), while there was an increase in the formation of C_2S and a decrease in the formation of C_3S and C_3A (Bogush et al., 2020). The final product exhibited nearly zero chloride content. The f-CaO accounted for 1.14 wt.%−1.56 wt.% of the cement clinker (Wang et al., 2010). Furthermore, the components of C_3S, C_2S, C_3A, and C_4AF in the cement clinker did not show significant changes when a minor amount of washed MSWI FA was cotreated with ordinary cement clinker. Moreover, the content of some mineral components was significantly influenced by the amount of MSWI FA incorporated. The impact of increasing the fly ash content from 5% to 35% on the formation of minerals was investigated. Results revealed a decrease in the proportion of alite and an increase in the proportion of belite in the cement (Bogush et al., 2020).

2.1.3 Hydration and mechanical characteristics

The properties of cement include some important factors, such as compressive strength, setting time, hydration rate, and other parameters of concrete and mortar. When incorporating MSWI FA as a substitute for material production, it is essential to thoroughly assess the influence of MSWI FA on the properties of the resulting concrete and mortar (Clavier et al., 2020).

As for cement clinker hydration behavior, the presence of alkali metals accelerates hydration, while Zn, Pb, and Cd can hinder the hydration of cement. The MSWI FA contains significant amounts of chloride and sulfate, indicating the formation of ettringite phase during the hydration stage. Consequently, the hydration reaction slows down as the MSWI FA content increases (Saikia et al., 2007). Moreover, through washing pretreatment, the alkali metal content decreases, leading to a lower hydration rate in clinker containing washed FA compared with that containing original FA (Lam et al., 2011). And high content of P_2O_5 (>0.5%) in clinker can result in a decline in cement strength. The presence of f-CaO in clinker serves as an indicator, and its low hydration rate results in an increase in volume due to the formation of $Ca(OH)_2$, ultimately reducing the compressive strength of concrete. Hence, the f-CaO content should be below 1% in cement clinker (Ampadu and Torii, 2001).

Compressive strength serves as a crucial parameter for assessing the mechanical properties of concrete. MSWI FA-incorporated cement exhibits either lower or equal compressive strengths to normal cement. There is a generally decrease in compressive strength with the increase of MSWI FA dosage (Clavier et al., 2020). A comparison was made between the properties such as setting time and compressive strength of cement clinker produced with and without MSWI FA. The setting time of clinker produced with MSWI FA was only slightly shorter than that of clinker without MSWI FA. After 28-day curing, the compressive strength of the concrete exceeded the minimum requirement of 32.5 MPa as specified in the Chinese standard (GB 175-2007) for OPC. The mechanical properties of concrete are mainly affected by the pretreatment of MSWI FA, the dosage, and the composition of the concrete mix. Increasing the content of MSWI FA resulted in a decrease in compressive strength (Wang et al., 2010).

In certain applications, the utilization of cement with lower early strength may be deemed acceptable. The influence related to compressive strength arises from the inadequate formation of aluminite and calcium oxide in clinker amended with MSWI FA. Conversely, there is a partial increase in the late strength of MSWI FA–incorporated specimens, which can be attributed to the higher proportion of belite (Bogush et al., 2020). A particular study reported a higher compressive strength for the MSWI FA incorporated samples; however, this result could potentially be attributed to variations in the amount of calcium sulfate or gypsum added (Guo et al., 2015).

2.2 Calcium sulfoaluminate cement

2.2.1 Preparation technology

Table 12.4 presents a summary of relevant studies investigating the utilization of MSWI FA for the production of CSA cement clinker. Researchers successfully generated CSA clinker by incorporating MSWI FA with limestone, bauxite, desulfurization gypsum, and silicon dioxide (Wu et al., 2011). A comparative study involving nine representative sample groups was conducted, involving the variation in firing temperatures ranging from 1150 to 1300°C, and durations spanning 2–4 h. The results indicated that the finest quality of CSA clinker was obtained by firing the raw mix at 1250°C for 2 h, with an average heating rate of 30°C/min. Similarly, Guo et al. (2014) also accomplished the laboratory production of CSA cement by employing raw materials consisting of 31.14% MSWI FA, 29.71% $CaCO_3$, 14.23% $CaSO_4$, and 24.92% Al_2O_3. The materials were sintered at a firing temperature of 1200°C for 120 min with an average heating rate of 30°C/min. The prepared CSA cement exhibited commendable quality with heavy metal leaching concentrations within acceptable thresholds. Additionally, it demonstrated favorable water permeability, resistance to drying shrinkage, and carbonation resistance. A portion of soluble chloride could be sintered within the cement clinker and subsequently stabilized in the hydration product. By utilizing 35% washed MSWI FA, 37.5% flue gas desulfurization (FGD) gypsum, and 27.5% aluminum ash, solid waste-based CSA cement could be successfully produced through a 30-minute calcination process at 1250°C. It has been demonstrated that CSA cementitious materials can be effectively prepared by using prewashed MSWI FA, flue gas desulfurization gypsum, and aluminum ash to eliminate chloride leaching. The MSWI FA content in raw materials can reach up to 35%. While qualified CSA cement could be achieved by elevating the temperature to 1250°C, 1270°C, or 1300°C at a heating rate of 5°C/min and maintaining for 30 min, the optimal calcination temperature was 1270°C (Mao et al., 2020).

2.2.2 Mineral compositions

The distinction between CSA and OPC clinker primarily lies in their mineralogical composition. CSA lacks high-temperature phases C_3S and C_3A, while it is enriched in $3CaO \cdot 3Al_2O_3 \cdot CaSO_4$ (C_4A_3S) and $2CaO \cdot SiO_2$ (C_2S) (Singh et al., 2008). The addition of MSWI FA to CSA cement does not alter its primary mineral phase, which retains robust hydraulic properties. However, changes in alkalinity can influence the presence and content of $CaSO_4$, C_2AS, $C_{11}A_7 \cdot CaCl_2$, and other minerals. Insufficient alkalinity results in unreacted $CaSO_4$ remaining in the clinker, while

TABLE 12.4 Performances of CSA with MSWI FA as raw materials.

Materials	Compressive strength	Calcination conditions	References
MSWI FA with limestone, bauxite, gypsum, and silicon dioxide	CSA addition decreased the compressive strengths	Heating rate of 30°C/min; firing; temperature at 1150–1300°C for 2 h	Wu et al. (2011)
31.1% MSWI FA, with 29.7% $CaCO_3$, 14.2% $CaSO_4$, 24.9% Al_2O_3	High strength at an early age,	Heating rate of 30°C/min, firing temperature of 1200°C for 120 2 h	Guo et al. (2014)
35% MSWI FA, with 37.5% gypsum and 27.5% aluminum ash	1, 3, and 28 days reach 37.4, 57.2, and 75.5 MPa, respectively	Heating rate of 5°C/min, 1250°C for 30 min	Mao et al. (2020)
35% MSWI FA, with 37.5% gypsum and 27.5% aluminum ash	1, 3, and 28 days reach 44.1, 69.8, and 93.4 MPa, respectively	Heating rate of 5°C/min, 1270°C for 30 min	Mao et al. (2020)
35% MSWI FA, with 37.5% gypsum and 27.5% aluminum ash	1, 3, and 28 days reach 40.6, 63.2, and 83.2 MPa, respectively	Heating rate of 5°C/min, 1300°C for 30 min	Mao et al. (2020)

excessive alkalinity leads to an excess of CaO in the system and the potential existence of a small amount of $C_{11}A_7 \cdot CaCl_2$. The quantity of MSWI FA also impacts the mineral phase of the clinker. An excessive amount of MSWI FA may induce the formation of the $C_{11}A_7 \cdot CaCl_2$ mineral phase and decrease the aluminum-silicon ratio in the raw material composition, which could adversely affect the performance of CSA cement clinker (Wu et al., 2011). The chemical composition of CSA cement shows minimal variation with different calcination temperatures, whereas the mineral composition can exhibit some divergence. The cement fired at 1250°C comprises 54.3% calcium sulfoaluminate, 22.1% dicalcium silicate, and 6.1% calcium sulfate; the cement fired at 1270°C consists of 62.9% calcium sulfoaluminate, 29.1% dicalcium silicate, and 4.8% calcium sulfate; the cement at 1300°C is composed of 56.8% calcium sulfoaluminate, 26.5% dicalcium silicate, and 3.2% calcium sulfate (Mao et al., 2020). Moreover, increasing the calcination temperature results in the release of gaseous SO_2, which leads to a decrease in sulfur content. Therefore, there is a discernible difference in the SO_3 content. It is worth noted that it is crucial to address the chloride present in the MSWI FA used for CSA cement production, as the untreated product exhibits a chloride content of 0.07%.

2.2.3 Hydration and mechanical characteristics

A significant disparity exists in the hydration heat between CSA cement calcined with MSWI FA and conventional CSA cement. The trace heavy metals present in MSWI FA impede hydration and diminish the rate of hydration reaction (Wu et al., 2012). During the hydration process, there are two significant factors. Firstly, besides gypsum's reaction with the calcium sulfoaluminate phase to form AFt, the excess gypsum serves as a sulfate activator for cement admixtures. Secondly, the hydration of C_2S and lime powder in this system produces $Ca(OH)_2$, which serves as an alkali activator. Due to the combined effects of these two factors, $A1_2O_3$ rapidly reacts with $Ca(OH)_2$ and $CaSO_4$ to generate AFt. Simultaneously, the surplus $Ca(OH)_2$ reacts with SiO_2 to form C−S−H. Consequently, the density of the cement mortar increases and its resistance to carbonation improves. The hydration products derived from cement admixtures, such as AFt and C−S−H, fill the voids within the cement mortar, thereby enhancing its overall density (Mao et al., 2020). Furthermore, it was observed that the presence of sulfate solution facilitates the hydration process of CSA cement, leading to the formation of a substantial quantity of gel-like material, which is hypothesized to be alumina gel generated through hydration reactions (Guo et al., 2014). The utilization of MSWI FA in the firing process of CSA cement results in higher early-stage compressive strength, with the 3-day strength accounting for 90% of the 28-day strength, while also exhibiting smooth development in the later stages (Wang et al., 2019). Similarly, cement fired with MSWI FA exhibits rapid development of early age strength, followed by a slower strength increase after 7 days (Guo et al., 2014).

The hardness of clinker increases with the MSWI FA content increases. And the compressive strength of products incorporating MSWI FA is lower compared with those produced with common raw materials. Specifically, CSA with 44.3% MSWI FA achieves a compressive strength of only 30.3 MPa at 1 day, corresponding to 59.6% of that made from traditional raw materials. At 28 days, the strength reaches 73.2 MPa, equivalent to 65.1% of CSA made from traditional raw materials (Ampadu and Torii, 2001). Although a minor amount of chloride ions can enhance early strength, it is essential to remove chloride ions during the pretreatment stage to ensure the safety of production equipment and the quality of cementitious materials (Mao et al., 2020).

3. Environmental impacts

3.1 Heavy metal behavior

MAWI FA accumulated high content of heavy metals (Wang et al., 2019). Special processes, such as utilizing as a substitute for cement raw materials, can be employed to immobilize heavy metals in the final product (Park et al., 2005). However, the retention of heavy metals in the clinker contributes to an environmental risk in the long term. The harmful elements present in cement materials mainly include Cr, Cd, Pb, As, Zn, and Cu. The heavy metal content of clinker produced by MSWI FA is provided in Table 12.5. Heavy metals not only pose a risk to the environment but also have the potential to change valence, thus leading to cycle through various media and cause persistent pollution. Therefore, comprehending the specific chemical forms of heavy metals during the clinker sintering process is crucial for assessing the environmental impact associated with MSWI FA utilization in cement clinker production.

Based on the material balance within the suspension preheater rotary kiln, the German Cement Plant Association categorizes trace elements into four groups: high volatile elements, volatile elements, difficult volatile elements, and nonvolatile elements (Fig. 12.1). Among them, Cd and Pb belong to the nonvolatile elements group. Approximately 90% of these two elements remain in the clinker during the calcination process. Researchers employed proprietary thermochemical software to simulate the thermodynamic behavior and one-step fixation of Cd and Pb in industrial cement kilns.

TABLE 12.5 The heavy metal content of clinker produced by MSWI FA (mg/kg).

	Mao et al. (2020)	Diliberto et al. (2020)	Wu et al. (2012)	Wu et al. (2011)
Cr	195	200	316.8	386.3
Zn	755	500	413.5	616.5
Cu	598	200	976.4	1022
Ni	25	200	54.0	65.63
As	44	50	–	13.31
Pb	82	300	32.5	63.48
Cd	21	3	4.3	10.38

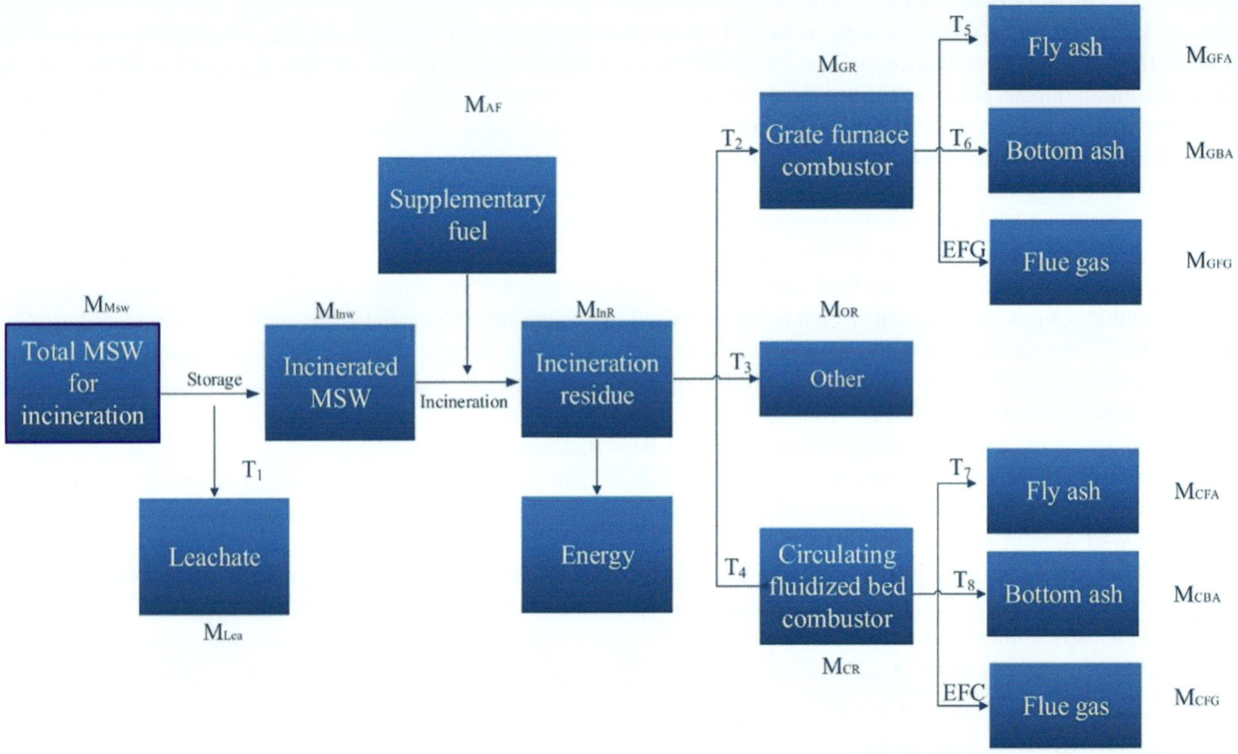

FIGURE 12.1 Schematic diagram of heavy metal mass flow during MSW incineration (Wang et al., 2019).

The impact of Cd, Pb, and Cl loading on the fixation and/or evaporation of Cd and Pb during the sintering process was analyzed using data obtained from an industrial cement kiln. During cement calcination, Cd exhibits a strong tendency to form aluminates and silicates; however, the presence of a substantial amount of f-CaO in the kiln atmosphere hinders this reaction. The behavior of Pb is closely associated with the presence of Cl^-, with nearly 90% of Pb volatilizing in the form of $PbCl_2$ vapor (Wang et al., 2022). The presence of Cl^- in the clinker leads to a decrease in the proportion of Pb and Cd, while an increase in Cl^- results in higher levels of Pb and Cd in kiln dust. The kiln dust recovery model calculations demonstrate a sharp increase in the concentration of Pb and Cd in both kiln dust and clinker after the recovery process. Under unstable conditions, the concentration of Pb and Cd in kiln dust rises, leading to the reentry of heavy metals into the cement kiln. The clinker contains more than 1% of the Cl element, predominantly in the form of $C_{11}A_7 \cdot CaCl_2$. This study found that all other elements except copper were undetectable at 28 days. The leaching concentrations were significantly below the limits specified in GB 5085.3–2007, indicating that the sintering sulfoaluminate effectively immobilizes these elements. This could be attributed to the integration of some elements into the clinker mineral structure during calcination, as well as the strong capability of AFt and gel produced during hydration to immobilize heavy metal ions. The leaching behavior of these heavy metals further confirms the environmental safety of using MSWI FA for sulfoaluminate production (Peysson et al., 2005).

3.2 Destruction and formation of PCDD/Fs

The presence of dioxin in MSWI FA can be attributed to the following sources. Firstly, MSW inherently contains dioxin with a portion persisting in FA after the high-temperature incineration of MSW. Then domestic waste contains various precursors that can contribute to the synthesis of dioxin. Additionally, the decomposition of macromolecular dioxin compounds in MSW at high temperatures leads to the formation of precursor substances for dioxins, subsequently contributing to the synthesis of dioxins (Cieplik et al., 2006). PCDD/Fs in MSWI FA primarily involve high-temperature gas-phase reactions and low-temperature multiphase reactions (Fig. 12.2).

Dioxins pose a serious problem due to their resistance to destruction through chemical means. Fortunately, the high-temperature treatment involved in the manufacturing process of cement clinker using MSWI FA addresses these challenges effectively. During the high-temperature process of cement clinker production, the majority of organic components undergo decomposition and volatilization. The increased combustion temperature and extended residence time within the cement kiln ensure the efficient destruction of organic matter present in the MSWI FA (Pan et al., 2008). Furthermore, the relatively faster cooling rate prevents the "de novo synthesis" of dioxin compounds, leading to their complete eradication, which serves as a notable advantage in utilizing MSWI FA for cement production. The degradation rate of dioxin reaches an impressive 95% when the sintering temperature ranges from 700 to 1200°C (Li et al., 2023). Moreover, at calcination

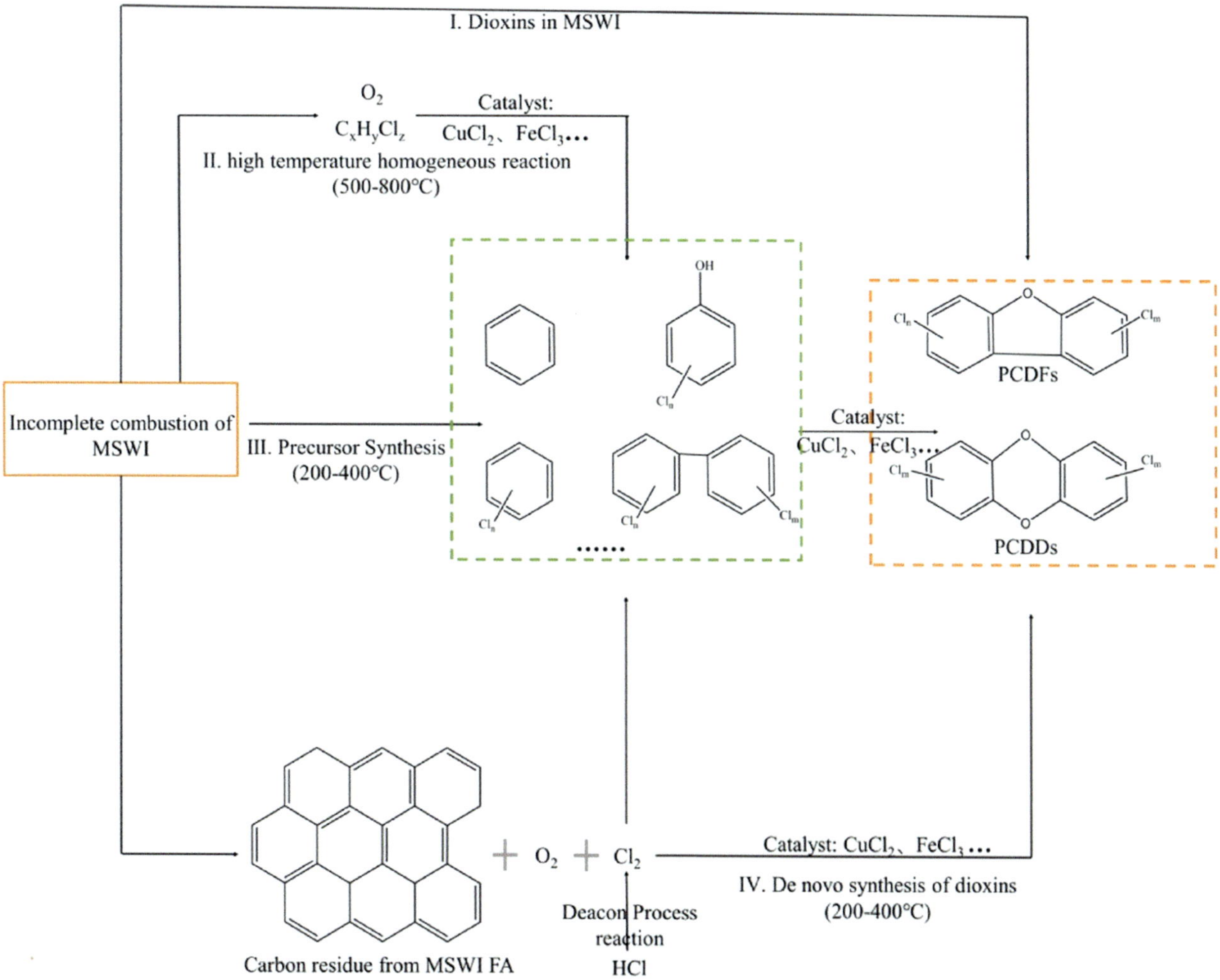

FIGURE 12.2 Main paths of dioxin generation in MSWI FA (Shunda et al., 2022).

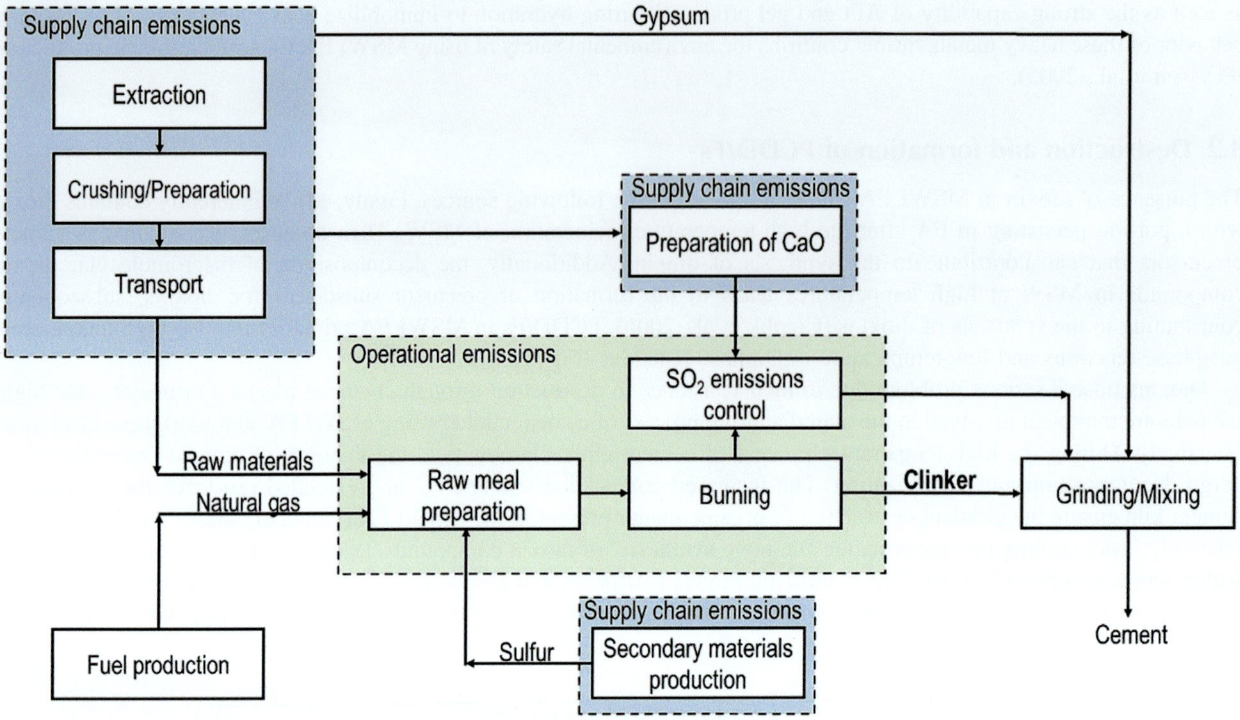

FIGURE 12.3 Clinker system flowchart used for the calculation of life cycle assessment greenhouse gases emissions (Hanein et al., 2018).

temperatures above 1450°C, dioxins undergo complete decomposition. Simultaneously, harmful substances such as heavy metals can be consolidated within the clinker phase (Ghouleh and Shao, 2018).

3.3 Life cycle assessment

Life cycle assessment (LCA) serves as a fundamental tool for quantifying the environmental performance throughout the entire life cycle of a product or service, encompassing both direct and indirect impacts (Selih and Sousa, 2007). It is considered the comprehensive approach to evaluate the overall environmental footprint and validate the sustainability of MSWI FA resource-saving solutions and FA-derived building materials (Zheng et al., 2022). Therefore, employing the LCA methodology is crucial to comprehensively study the resource-energy-environmental impact assessment of utilizing MSWI FA for cement clinker production. A comprehensive LCA analysis of CSA cement has demonstrated its considerable potential as a sustainable and realistic alternative (Hanein et al., 2018). Fig. 12.3 depicts the clinker production system for facilitating the calculation of greenhouse gas (GHG) emissions in LCAs. This offers a structured approach to quantify emissions associated with various stages of clinker production, aiding in informed decision-making for sustainable practices. The LCA findings revealed that the production of OPC resulted in significant environmental impact, primarily attributable to raw material imports and fossil fuel combustion. The utilization of alternative materials such as MSWI FA can effectively alleviate environmental repercussions (Hossain et al., 2017).

The LCA database contains pertinent information regarding the evaluation of reuse treatment, which guides the utilization of pretreatment MSWI FA as a raw material in cement production. The utilization of MSWI FA as a raw material for cement production is associated with various environmental impacts, among which CO_2 emissions are recognized as the most significant factor. Following CO_2 emissions, the pretreatment process of MSWI FA and the electricity consumption during cement generation also contribute to environmental effects (Huang and Chuieh, 2015). In the pursuit of ecological cement production, the reaction of certain heavy metals with CO_2 offers the potential for their removal. Several studies have indicated that incorporating 5%−10% MSWI FA as raw materials in cement production can lead to a reduction of 25−49 kg of CO_2 per ton of clinker (Yan et al., 2018). It demonstrated that the concrete derived from ecological cement can be activated using locally captured carbon dioxide from the emission stack, thereby mitigating the carbon footprint of operations and facilitating the transfer of carbon emissions away from the atmosphere (Ashraf et al., 2019). When considering the utilization of MSWI FA in cement clinker production, stakeholders must undertake a comprehensive assessment that takes into account all costs, including financial and environmental costs, to evaluate the feasibility of the process or identify the most optimal approach (Huang and Chuieh, 2015). It is essential to consider the broader spectrum of costs to make informed judgments and ensure selecting the most rational and sustainable process.

4. Perspectives

According to the current research on using MSWI FA to manufacture cement clinker, the following aspects warrant consideration:

1. For the cement rotary kiln waste treatment, it is imperative to ensure that the waste does not cause any damage to the production equipment and does not disrupt the normal operation of the system. Previous research has indicated that the presence of chloride in MSWI FA can corrode the kiln. Additionally, during the recalcination process, the potential formation of volatile condensation cycles may cause challenges such as crust formation, blockages, clinker agglomeration, and ring formation within the kiln. Consequently, the optimal utilization of MSWI FA for cement clinker production demands careful consideration and further discussion.
2. Investigating the mechanisms of heavy metals transfer from MSWI FA and their bonding state in cement clinker during recalcination using MSWI FA as raw materials is essential. Research in this area holds significant importance as it affects not only the properties of the resulting sintering cement clinker but also the safety considerations for its engineering applications. Therefore, further discussion and analysis are warranted to delve into these aspects.
3. It is essential to further explore the utilization of MSWI FA in the production of cleaner cement. The utilization ratio of MSWI FA to raw material in OPC and CSA ranged from 10% to 40% (Zhang Li, 2012). Recent studies have successfully utilized MSWI FA to produce carbon dioxide–activated eco-cement, which exhibits comparable binding capabilities to OPC when activated by carbon dioxide. With MSWI FA constituting up to 85% of the raw materials and being calcined within the typical incinerator's maximum operating temperature range, this approach can achieve an immediate compressive strength exceeding 50 MPa following carbonation activation (Ghouleh and Shao, 2018). However, concerns have been raised regarding the accidental nature of this research due to its reliance on a single source of residues. Therefore, based on this premise, efforts have been made to optimize the production of ecological cement through a series of steps and procedures, aimed at enhancing reproducibility and composition consistency of products. Novel methods should be developed to harness the energy and residues from incineration to produce ecological cement at a temperature of approximately 1100°C, thereby achieving structural bond strength upon reaction with carbon dioxide (Ashraf et al., 2019). MSWI FA has the potential to produce environmentally friendly ecological cement, facilitating carbon dioxide sequestration, and conserving energy and natural resources. However, further research is necessary to evaluate the impact of variations in MSWI FA composition on the performance of ecological cement and the consistency of ecological cement binders.
4. Novel approaches are required to address the environmental risks associated with elevated levels of Pb and Cl in MSWI FA. Researchers developed a pioneering strategy for the preparation of magnesium potassium phosphate cement (MKPC) using MSWI FA. This approach enables MSWI FA recovery and synergistic immobilization of Pb and Cl without the washing pretreatment. The obtained MKPC, containing 10% MSWI FA, exhibits a commendable compressive strength of 28.4 MPa, while effectively reducing the leaching toxicity of Pb and Cl by over 99.2% (Cao et al., 2023). This resourceful utilization of MSWI FA presents a groundbreaking strategy and innovative concept for future investigations into the synergistic immobilization mechanism of Pb and Cl. Moreover, it provides valuable insights and references for the environmentally benign treatment and resource utilization of cement clinker produced using chlorine-rich MSWI FA.

5. Conclusions

MSWI FA presents a promising practical application prospect as a substitute for conventional raw materials in cement manufacturing. Its utilization holds the potential to conserve energy and resources, leading to waste reduction during clinker manufacturing. This study provides a comprehensive overview of this approach based on the existing literature, with the following relevant findings.

1. Undoubtedly, MSWI FA contains essential elements and components necessary for cement production. However, the presence of other constituents, such as chloride, alkali content, heavy metals, dioxins, etc., imposes certain constraints on its utilization as a substitute raw material for OPC and CSA. It becomes crucial to identify an appropriate mixing ratio that ensures the desired properties while addressing these concerns.

2. The pretreatment of MSWI FA is essential, and water washing is a commonly employed method. The crucial aspect of pretreatment lies in addressing the presence of heavy metals in MSWI FA. On the other hand, the treatment of dioxins through high-temperature processes during cement clinker production allows for their decomposition, eliminating concerns about their subsequent environmental impact. This advantage further underscores the benefits of utilizing MSWI FA as the raw material for cement clinker manufacturing.
3. However, trace elements, which are not present in traditional raw materials in MSWI FA, can affect the properties of the cement products. The addition of trace elements can ensure the acquisition of basic clinker mineralogy substances.
4. Under proper treatment conditions, the heavy metals in MSWI FA can be effectively immobilized in the solid phase, and its presence has an insignificant effect on the mineral composition of the clinker. This integrated waste management approach has important implications for reducing heavy metal pollution and maximizing the use of waste resources.
5. Expanding beyond OPC and CSA, the concept of utilizing MSWI FA for eco-friendly cement production gained widespread attention. Thus, the studies for the incorporation of MSWI FA as a substitute for conventional raw materials increase, showing the immense potential of cost-effective and environmentally friendly materials.
6. Based on LCA results, the substitution of raw materials with MSWI FA proves to be highly effective in mitigating greenhouse gas emissions associated with cement clinker production. Therefore, recycling MSWI FA into eco-friendly cement is a promising route for MSWI FA management.

Acknowledgments

The authors gratefully acknowledge the financial support from the Major Project of Science and Technology Innovation in Ningbo (Grants No. 2023Z148) for this study.

References

Ampadu, K.O., Torii, K., 2001. Characterization of ecocement pastes and mortars produced from incinerated ashes. Cement Concr. Res. 31, 431–436.

Andrew, R.M., 2018. Global CO_2 emissions from cement production, 1928-2017. Earth Syst. Sci. Data 10, 2213–2239.

Ashraf, M.S., Ghouleh, Z., Shao, Y., 2019. Production of eco-cement exclusively from municipal solid waste incineration residues. Resour. Conserv. Recycl. 149, 332–342.

Benhelal, E., Shamsaei, E., Rashid, M.I., 2021. Challenges against CO_2 abatement strategies in cement industry: a review. J. Environ. Sci. (China) 104, 84–101.

Bogush, A.A., Stegemann, J.A., Zhou, Q., Wang, Z., Zhang, B., Zhang, T., Zhang, W., Wei, J., 2020. Co-processing of raw and washed air pollution control residues from energy-from-waste facilities in the cement kiln. J. Clean. Prod. 254.

Cao, X., Zhang, Q., Yang, W., Fang, L., Liu, S., Ma, R., Guo, K., Ma, N., 2023. Lead-chlorine synergistic immobilization mechanism in municipal solid waste incineration fly ash (MSWIFA)-based magnesium potassium phosphate cement. J. Hazard Mater. 442, 130038.

Chen, Z., Li, J.S., Poon, C.S., Jiang, W.H., Ma, Z.H., Chen, X., Lu, J.X., Dong, H.X., 2023. Physicochemical and pozzolanic properties of municipal solid waste incineration fly ash with different pretreatments. Waste Manag. 160, 146–155.

Cheng, T.W., Huang, M.Z., Tzeng, C.C., Cheng, K.B., Ueng, T.H., 2007. Production of coloured glass-ceramics from incinerator ash using thermal plasma technology. Chemosphere 68, 1937–1945.

Cieplik, M.K., De Jong, V., Bozovic, J., Liljelind, P., Marklund, S., Louw, R., 2006. Formation of dioxins from combustion micropollutants over MSWI fly ash. Environ. Sci. Technol. 40, 1263–1269.

Clavier, K.A., Paris, J.M., Ferraro, C.C., Townsend, T.G., 2020. Opportunities and challenges associated with using municipal waste incineration ash as a raw ingredient in cement production - a review. Resour. Conserv. Recycl. 160.

Cobo, M., Galvez, A., Conesa, J.A., Montes de Correa, C., 2009. Characterization of fly ash from a hazardous waste incinerator in Medellin, Colombia. J. Hazard Mater. 168, 1223–1232.

Coumes, C.C.D., Courtois, S., 2003. Cementation of a low-level radioactive waste of complex chemistry investigation of the combined action of borate, chloride, sulfate and phosphate on cement hydration using response surface methodology. Cement Concr. Res. 33, 305–316.

Cucchiella, F., D'Adamo, I., Gastaldi, M., 2017. Sustainable waste management: waste to energy plant as an alternative to landfill. Energy Convers. Manag. 131, 18–31.

Cyr, M., Idir, R., Escadeillas, G., 2012. Use of metakaolin to stabilize sewage sludge ash and municipal solid waste incineration fly ash in cement-based materials. J. Hazard Mater. 243, 193–203.

Diaz-Loya, I., Juenger, M., Seraj, S., Minkara, R., 2019. Extending supplementary cementitious material resources: reclaimed and remediated fly ash and natural pozzolans. Cement Concr. Compos. 101, 44–51.

Diliberto, C., Meux, E., Diliberto, S., Garoux, L., Marcadier, E., Rizet, L., Lecomte, A., 2020. A zero-waste process for the management of MSWI fly ashes: production of ordinary Portland cement. Environ. Technol. 41, 1199–1208.

Ferraro, A., Ducman, V., Colangelo, F., Korat, L., Spasiano, D., Farina, I., 2023. Production and characterization of lightweight aggregates from municipal solid waste incineration fly-ash through single- and double-step pelletization process. J. Clean. Prod. 383.

Ghouleh, Z., Shao, Y., 2018. Turning municipal solid waste incineration into a cleaner cement production. J. Clean. Prod. 195, 268–279.

Guo, X., Shi, H., Hu, W., Wu, K., 2014. Durability and microstructure of CSA cement-based materials from MSWI fly ash. Cement Concr. Compos. 46, 26–31.

Guo, X., Shi, H., Wu, K., Ju, Z., Dick, W.A., 2015. Performance and risk assessment of alinite cement-based materials from municipal solid waste incineration fly ash (MSWIFA). Mater. Struct. 49, 2383–2391.

Hanein, T., Galvez-Martos, J.-L., Bannerman, M.N., 2018. Carbon footprint of calcium sulfoaluminate clinker production. J. Clean. Prod. 172, 2278–2287.

He, D., Hu, H., Jiao, F., Zuo, W., Liu, C., Xie, H., Dong, L., Wang, X., 2023. Thermal separation of heavy metals from municipal solid waste incineration fly ash: a review. Chem. Eng. J. 467.

Hossain, M.U., Poon, C.S., Lo, I.M.C., Cheng, J.C.P., 2017. Comparative LCA on using waste materials in the cement industry: a Hong Kong case study. Resour. Conserv. Recycl. 120, 199–208.

Huang, T.Y., Chuieh, P.T., 2015. Life cycle assessment of reusing fly ash from municipal solid waste incineration. Procedia Eng. 118, 984–991.

Lam, C.H.K., Barford, J.P., McKay, G., 2011. Utilization of municipal solid waste incineration ash in Portland cement clinker. Clean Technol. Environ. Policy 13, 607–615.

Leckner, B., 2015. Process aspects in combustion and gasification Waste-to-Energy (WtE) units. Waste Manag. 37, 13–25.

Lederer, J., Trinkel, V., Fellner, J., 2017. Wide-scale utilization of MSWI fly ashes in cement production and its impact on average heavy metal contents in cements: the case of Austria. Waste Manag. 60, 247–258.

Li, W., Yan, D., Li, L., Wen, Z., Liu, M., Lu, S., Huang, Q., 2023. Review of thermal treatments for the degradation of dioxins in municipal solid waste incineration fly ash: proposing a suitable method for large-scale processing. Sci. Total Environ. 875, 162565.

Li, Y., Hao, L., Chen, X., 2016. Analysis of MSWI bottom ash reused as alternative material for cement production. Proc. Environ. Sci. 31, 549–553.

Liu, G., Zhan, J., Zheng, M., Li, L., Li, C., Jiang, X., Wang, M., Zhao, Y., Jin, R., 2015. Field pilot study on emissions, formations and distributions of PCDD/Fs from cement kiln co-processing fly ash from municipal solid waste incinerations. J. Hazard Mater. 299, 471–478.

Mao, Y., Wu, H., Wang, W., Jia, M., Che, X., 2020. Pretreatment of municipal solid waste incineration fly ash and preparation of solid waste source sulphoaluminate cementitious material. J. Hazard Mater. 385, 121580.

Marieta, C., Guerrero, A., Leon, I., 2021. Municipal solid waste incineration fly ash to produce eco-friendly binders for sustainable building construction. Waste Manag. 120, 114–124.

Niu, M., Li, G., Wang, Y., Li, Q., Han, L., Song, Z., 2018. Comparative study of immobilization and mechanical properties of sulfoaluminate cement and ordinary Portland cement with different heavy metals. Construct. Build. Mater. 193, 332–343.

Pan, J.R., Huang, C., Kuo, J.-J., Lin, S.-H., 2008. Recycling MSWI bottom and fly ash as raw materials for Portland cement. Waste Manag. 28, 1113–1118.

Park, K., Hyun, J., Maken, S., Jang, S., Park, J.W., 2005. Vitrification of municipal solid waste incinerator fly ash using Brown's gas. Energy Fuel. 19, 258–262.

Peysson, S., Pera, J., Chabannet, M., 2005. Immobilization of heavy metals by calcium sulfoaluminate cement. Cement Concr. Res. 35, 2261–2270.

Quina, M.J., Bordado, J.C., Quinta-Ferreira, R.M., 2008. Treatment and use of air pollution control residues from MSW incineration: an overview. Waste Manag. 28, 2097–2121.

Saikia, N., Kato, S., Kojima, T., 2007. Production of cement clinkers from municipal solid waste incineration (MSWI) fly ash. Waste Manag. 27, 1178–1189.

Sarmiento, L.M., Clavier, K.A., Paris, J.M., Ferraro, C.C., Townsend, T.G., 2019. Critical examination of recycled municipal solid waste incineration ash as a mineral source for portland cement manufacture – a case study. Resour. Conserv. Recycl. 148, 1–10.

Schneider, M., Romer, M., Tschudin, M., Bolio, H., 2011. Sustainable cement production-present and future. Cement Concr. Res. 41, 642–650.

Selih, J., Sousa, A.C.M., 2007. Life cycle assessment of construction processes. In: International Conference on Sustainable Construction, Materials and Practices, Lisbon, PORTUGAL, pp. 366–+.

Shao, N., Wei, X., Monasterio, M., Dong, Z., Zhang, Z., 2021. Performance and mechanism of mold-pressing alkali-activated material from MSWI fly ash for its heavy metals solidification. Waste Manag. 126, 747–753.

Shi, H.-s., Deng, K., Yuan, F., Wu, K., 2009. Preparation of the saving-energy sulphoaluminate cement using MSWI fly ash. J. Hazard Mater. 169, 551–555.

Shih, P.-H., Chang, J.-E., Chiang, L.-C., 2003. Replacement of raw mix in cement production by municipal solid waste incineration ash. Cement Concr. Res. 33, 1831–1836.

Shunda, L., Jiang, X., Zhao, Y., Yan, J., 2022. Disposal technology and new progress for dioxins and heavy metals in fly ash from municipal solid waste incineration: a critical review. Environ. Pollut. 311, 119878.

Singh, M., Kapur, P.C., Pradip, 2008. Preparation of calcium sulphoaluminate cement using fertiliser plant wastes. J. Hazard Mater. 157, 106–113.

Tang, J., Ylmen, R., Petranikova, M., Ekberg, C., Steenari, B.-M., 2018. Comparative study of the application of traditional and novel extractants for the separation of metals from MSWI fly ash leachates. J. Clean. Prod. 172, 143–154.

Tang, P., Florea, M.V.A., Spiesz, P., Brouwers, H.J.H., 2015. Characteristics and application potential of municipal solid waste incineration (MSWI) bottom ashes from two waste-to-energy plants. Construct. Build. Mater. 83, 77–94.

Vaitkus, A., Grazulyte, J., Vorobjovas, V., Sernas, O., Kleiziene, R., 2018. Potential of MSWI bottom ash to be used as aggregate in road buidling materials. Baltic J. Road Bridge Eng. 13, 77–86.

Viczek, S.A., Aldrian, A., Pomberger, R., Sarc, R., 2020. Determination of the material-recyclable share of SRF during co-processing in the cement industry. Resour. Conserv. Recycl. 156, 104696.

Wang, L., Huang, X., Li, X., Bi, X., Yan, D., Hu, W., Jim Lim, C., Grace, J.R., 2022. Simulation of heavy metals behaviour during Co-processing of fly ash from municipal solid waste incineration with cement raw meal in a rotary kiln. Waste Manag. 144, 246–254.

Wang, L., Jin, Y., Nie, Y., Li, R., 2010. Recycling of municipal solid waste incineration fly ash for ordinary Portland cement production: a real-scale test. Resour. Conserv. Recycl. 54, 1428–1435.

Wang, P., Hu, Y., Cheng, H., 2019. Municipal solid waste (MSW) incineration fly ash as an important source of heavy metal pollution in China. Environ. Pollut. 252, 461–475.

Wang, X., Gao, M., Wang, M., Wu, C., Wang, Q., Wang, Y., 2021. Chloride removal from municipal solid waste incineration fly ash using lactic acid fermentation broth. Waste Manag. 130, 23–29.

Wang, Y., Pan, Y., Zhang, L., Yue, Y., Zhou, J., Xu, Y., Qian, G., 2015. Can washing-pretreatment eliminate the health risk of municipal solid waste incineration fly ash reuse? Ecotoxicol. Environ. Saf. 111, 177–184.

Wong, G., Gan, M., Fan, X., Ji, Z., Chen, X., Wang, Z., 2021. Co-disposal of municipal solid waste incineration fly ash and bottom slag: a novel method of low temperature melting treatment. J. Hazard Mater. 408, 124438.

Wu, K., Shi, H., Guo, X., 2011. Utilization of municipal solid waste incineration fly ash for sulfoaluminate cement clinker production. Waste Manag. 31, 2001–2008.

Wu, K., Shi, H., Schutter, G.D., Guo, X., Ye, G., 2012. Preparation of alinite cement from municipal solid waste incineration fly ash. Cement Concr. Compos. 34, 322–327.

Wu, Q., Wu, Z., Wu, H., Wang, Q., Huang, N., Zhou, J., Li, X., Shi, L., Tian, S., Li, M., 2023. Thermal treatment of municipal solid waste incineration fly ash and nanofiltration membrane concentrate co-processing: speciation of chromium and its leachability. J. Clean. Prod. 398.

Yan, D., Peng, Z., Yu, L., Sun, Y., Yong, R., Helge Karstensen, K., 2018. Characterization of heavy metals and PCDD/Fs from water-washing pretreatment and a cement kiln co-processing municipal solid waste incinerator fly ash. Waste Manag. 76, 106–116.

Zhang, J., Mao, Y., Wang, W., Wang, X., Li, J., Jin, Y., Pang, D., 2022. A new co-processing mode of organic anaerobic fermentation liquid and municipal solid waste incineration fly ash. Waste Manag. 151, 70–80.

Zhang Li, H., 2012. Application and research progress on eco-cement. In: 2nd International Conference on Chemical, Material and Metallurgical Engineering (ICCMME 2012), Kunming, PEOPLES R CHINA, pp. 2666–2671.

Zheng, R., Wang, Y., Liu, Z., Zhou, J., Yue, Y., Qian, G., 2022. Environmental and economic performances of municipal solid waste incineration fly ash low-temperature utilization: an integrated hybrid life cycle assessment. J. Clean. Prod. 340.

Chapter 13

Recycling various slag into cement clinker

Kai Wu and Ken Yang
Key Laboratory of Advanced Civil Engineering Materials of Ministry of Education, School of Materials Science and Engineering, Tongji University, Shanghai, China

1. Introduction

Slag is a significant by-product generated in large quantities during metallurgical processes involved in the production of ferrous and nonferrous metals (Jiang et al., 2018). To promote sustainability and reduce carbon emissions in the metallurgical industry, there is a growing consensus that slag should be integrated into solid waste-processing protocols and utilized in related industries (Guo et al., 2018; Li et al., 2022; Nath et al., 2022). Fortunately, slag can be effectively incorporated into the production of cement clinker.

The decision to recycle slag in cement clinker production is primarily driven by two key factors: the similarity in chemical composition between slag and cement clinker, and the positive environmental impact achieved by partially replacing limestone with slag in the raw materials used for cement clinker production. Numerous research studies have been conducted in laboratories to explore the manufacturing of cement clinker using slag (Alp et al., 2008; Iacobescu et al., 2013, 2016). However, it is important to note that the intrinsic chemical and mineral composition of slag can vary significantly, leading to distinct properties during cement clinker production.

This chapter provides a comprehensive review of the application of slag in cement clinker manufacturing. It examines the positive and negative influences introduced by different types of slag, while also discussing the future prospects for further utilizing slag in cement clinker production. By exploring these aspects, it becomes evident that leveraging slag in cement clinker production demonstrates great potential for bolstering sustainability and reducing environmental impacts.

2. Slag used in cement clinker production

Various slags are available for their utilization in cement clinker production including ferrous slag such as blast furnace slag and steel slag, and nonferrous slag such as copper slag, nickel slag, lead slag, and magnesium slag. These types of slag have diverse compositions of oxides and mineral phases (Fig. 13.1).

2.1 Blast furnace slag

As a by-product of the iron production process, blast furnace slag is formed when iron ore is melted in a blast furnace along with limestone and coke. The reaction between limestone and impurities in the iron ore results in the formation of slag, which floats on top of the molten iron. Once the slag is removed from the furnace, it is cooled to generate blast furnace slag. The properties of blast furnace slag vary depending on the cooling conditions. Slags that are slowly air-cooled exhibit a crystalline and vesicular structure, whereas quickly air-cooled slag tends to be pelletized. Water-cooling techniques can produce expanded or foamed slag by using controlled amounts of water, or granulated slag by quenching it with high-volume, high-pressure sprays of water.

When blast furnace slag is rapidly cooled from its molten state using air or water, it undergoes vitrification, which makes it a suitable material for activation by cement clinker in cement mixtures (Piatak et al., 2015). Substituting cement

Treatment and Utilization of Combustion and Incineration Residues. https://doi.org/10.1016/B978-0-443-21536-0.00025-3

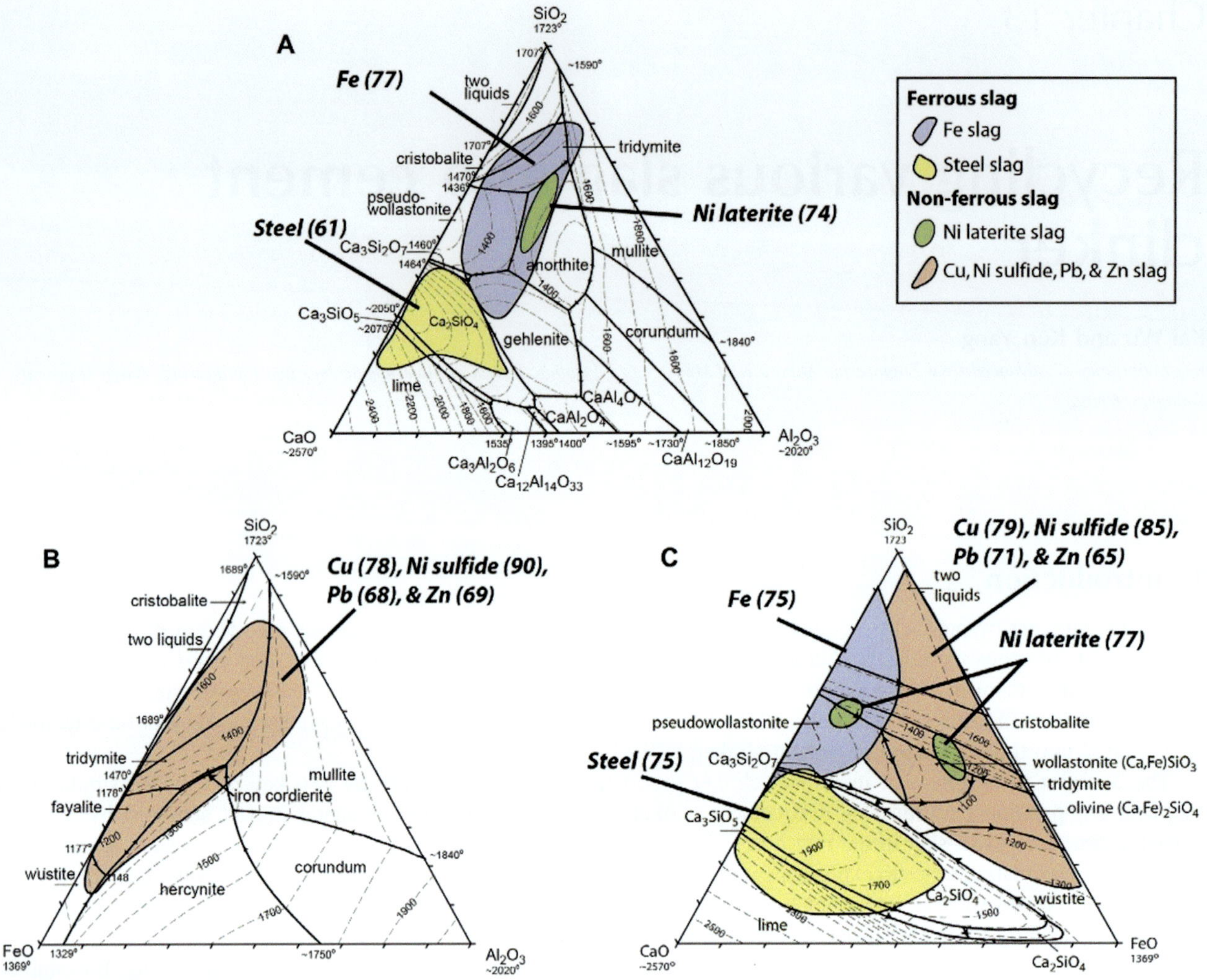

FIGURE 13.1 Bulk chemical compositions of slag in weight percent on $Al_2O_3-SiO_2-CaO$ (A), $Al_2O_3-SiO_2-FeO$ (B), and $FeO-SiO_2-CaO$ (C) ternary diagrams (Piatak et al., 2015).

clinker with blast furnace slag is regarded as a sustainable approach to recycle this solid waste. Numerous studies have been conducted on the direct replacement of cement with blast furnace slag. However, there is limited previous research on recycling blast furnace slag for cement clinker production, in comparison with its use as a cement clinker replacement.

2.2 Steel slag

Steel slag is discharged when molten steel is generated from the combination of molten iron and alloys. It is a further by-product from iron metallurgy. Steel slags are commonly categorized by the furnace used in their production such as open hearth slag, electric arc furnace (EAF) slag, and basic oxygen furnace (BOF) slag (Piatak et al., 2015), which is the same case as blast furnace slag. Steel slag mainly consists of crystalline materials due to the slow cooling process under atmospheric. In addition to the by-products in the primary steelmaking stage, slag can also be let off in secondary steel refining operations by adjusting C content and removing the remaining S, gases, or impurities. Molten Fe from the BOF and EAF process may be refined in a ladle furnace with the possible addition of alloys and fluxes to produce different grades of steel. Slag produced in this process is normally referred to ladle slag.

Most steel slags mainly consist of CaO, SiO_2, MgO, and FeO (Table 13.1). Regarding the low-phosphorus steel-making practice, the total content of these oxides in the liquid slags ranges from 88%to 92%. And the steel slag can be simply given by a $CaO-MgO-SiO_2-FeO$ quaternary system. In addition, the constitution of these oxides and other minor components are highly variable, which can be changed from batch to batch even in the same plant depending on raw

TABLE 13.1 Typical composition range of Steel slag (wt.%) (Shi, 2004).

Components	Basic oxygen furnace	Electric arc furnace (carbon steel)	Electric arc furnace (alloy/ stainless)	Ladle
SiO_2	8–20	9–20	24–32	2–35
Al_2O_3	1–6	2–9	3.0–7.5	5–35
FeO	10–35	15–30	1–6	0.1–15
CaO	30–55	35–60	39–45	30–60
MgO	5–15	5–15	8–15	1–10
MnO	2–8	3–8	0.4–2	0–5
TiO_2	0.4–2	–	–	–
S	0.05–0.15	0.08–0.2	0.1–0.3	0.1–1
P	0.2–2	0.01–0.25	0.01–0.07	0.1–0.4
Cr	0.1–0.5	0.1–1	0.1–20	0–0.5

materials, type of steel made, furnace conditions, etc. Minerals involving C_3S, β-C_2S, γ-C_2S, C_4AF, C_2F, RO phase (a type of solid solution constituted by CaO–FeO–MnO–MgO), olivine, merwinite, f-CaO, and f-MgO are also formed during steel slag generation (nomenclatures C, A, and F refer to CaO, Al_2O_3, and Fe_2O_3, respectively) (Shi, 2004).

Steel slag has been successfully utilized as a partial replacement for limestone in the kiln feed, demonstrating its potential to enhance cement clinker production (Abdul-Wahab et al., 2016; Ren et al., 2017; Tsakiridis et al., 2008). Fig. 13.2 illustrates the application of steel slag in cement production, highlighting its beneficial role in the process.

FIGURE 13.2 A schematic illustration of lab-scale recycling of steel slag into cement clinker (Zhao et al., 2023).

Studies have consistently shown that clinker produced from steel slag exhibits comparable mechanical properties, including elasticity and durability, to that of traditional clinker (Anastasiou et al., 2017; Iacobescu et al., 2013; Isteri et al., 2020).

2.3 Copper slag

Copper slag as a by-product of copper matte is mainly formed among the smelting and refining process (Wang et al., 2021a). Copper-rich matte and copper slag are formed as two separate liquid phases. The smelting charge contains sulfides and oxides of iron and copper, as well as Al_2O_3, CaO, SiO_2 and MgO, which determine the chemical compositions of the final products (Murari et al., 2015). Silica is added to form silicate anions, producing the copper slag phase, while sulfides form the matte phase (Tian et al., 2021). The structure of slag is stabilized with lime and alumina and can form a dense, hard crystalline product with a slow cooling process. Quick solidification produces the type of granulated amorphous slag (Murari et al., 2015). Generally, copper slags constitute fayalite and silicate glassy matrices as major components with others including crystalline oxides, metallic elements, sulfides, alloys, and intermetallic compounds (Ali et al., 2013).

2.4 Lead slag

Lead slag is a by-product formed during the production of lead, which can stem from either lead ore smelting or the recovery of waste lead-acid batteries (Ettler and Johan, 2014). Both primary and secondary lead slag share similar physical characteristics, such as a black color and a glassy texture. With a density ranging from 3.6 to 3.9 g/cm^3 due to its high iron oxide content, lead slag typically exhibits particle sizes ranging from 0.1 to 4 mm (Pan et al., 2019). Its specific gravity falls between 2.65 and 3.79, making it comparable to ordinary natural aggregates (Penpolcharoen, 2005).

The composition of primary lead slag varies based on factors such as ore type, fluxes used, and impurities present in the coke and iron. Primary slag typically consists of Fe_2O_3, FeO, SiO_2, CaO, Al_2O_3, MgO, PbO, ZnO, and S in varied amounts. It primarily consists of a CaO–FeO–SiO_2 glass matrix with embedded mineral phases such as wüstite, fayalite, and spinels such as franklinite and magnetite. Sulfide-rich phases such as pyrrhotite are also present. Lead exists as isolated droplets within the glass matrix (Sobanska et al., 2016; Yin et al., 2016; Zemri and Bachir Bouiadjra, 2020).

Secondary lead slag typically contains significant levels of iron compounds such as wüstite, pyrrhotite, and magnetite, along with minor amounts of fayalite. Lead is found as galena, anglesite, litharge, and metallic lead phases in secondary slag (Gomes et al., 2011; Kim et al., 2017; Lassin et al., 2007).

2.5 Magnesium slag

Magnesium slag (MS) is a type of by-product from the manufacture of metallic magnesium (Li et al., 2018). A comprehensive study involving the collection of 20 magnesium slag samples was conducted across four metallic magnesium plants located in Shanxi Province, China (Li et al., 2016a).

Chemical analysis of magnesium slag reveals that the main components include CaO, SiO_2, MgO, Al_2O_3, and Fe_2O_3, as reported by various studies (Djokic et al., 2012; Li et al., 2016a; Yang et al., 2022). Among these components, γ-C_2S has been identified as the predominant phase responsible for the decomposition of magnesium slag. Additionally, f-MgO has been observed as a secondary phase, while β-C_2S is present in trace amounts, indicating that a small quantity of β-C_2S can occur within the cooled slag particles (Yang et al., 2022).

2.6 Nickel slag

Nickel slag refers to a type of granulated slag that is formed as a result of the natural cooling or rapid quenching of molten material after the smelting of nickel metal (Wu et al., 2018). The major chemical components are SiO_2, MgO, and Fe_2O_3 (Wu et al., 2018). Within the nickel slag, various minerals can be found, including forsterite, akermanite, clinoenstatite, ferroan, and quartz (Wu et al., 2018).

In addition to the aforementioned components, nickel slag also has trace amounts of other elements, such as Cr and Ni (Wu et al., 2018). The presence of chromium (Cr) in the slag acts as a mineralizer, contributing to the crystallization of alite through solid solution. This enhances the burnability of the raw meal, which is a beneficial characteristic in certain applications (Iacobescu et al., 2016).

3. Process of cement clinker manufacture with slag

3.1 Preprocess of raw materials

The preprocessing stage is an integral part of cement production, as it prepares essential raw materials such as limestone and slag for the subsequent clinkerization process. These raw materials require milling to enhance their burnability, which refers to their ability to be efficiently and effectively calcined.

Ball milling is a widely employed method for preprocessing raw materials due to its numerous advantages (Ismail et al., 2023). Prior to kiln calcination, the blast furnace slag–limestone mixture is often milled to enhance its sintering capabilities (Ismail et al., 2023), while manual grinding with mortars and pestles is suitable for laboratory-scale studies (Prasad et al., 2018). Once the raw materials have undergone ball milling, the resulting product, known as raw meal, exhibits improved characteristics that facilitate the subsequent calcination process. By subjecting the raw materials to ball milling, the particle size distribution is finely tuned and optimized. The reduced particle size could enhance the surface area of the raw materials, providing more points of contact with the heat source during calcination. As a result, the raw materials can be more effectively and uniformly heated, resulting in a more efficient and thorough calcination process. Furthermore, ball milling promotes the homogenization of the raw materials, ensuring a consistent composition throughout the mixture. This homogeneity is crucial for achieving a uniform and predictable distribution of minerals during calcination. A well-mixed raw meal allows for more controlled and precise chemical reactions, ultimately influencing the quality and properties of the clinker formed during the calcination process.

3.2 Raw meal composition

Efforts have been made on enhancing the utilization of blast furnace slag in cement clinker production. Researchers have proposed the combination of blast furnace slag and limestone as an alternative to using limestone alone (Ismail et al., 2023; Li et al., 2016b; Qin et al., 2013; Verma et al., 2020). Blast furnace slag is valuable due to its composition, including essential components such as CaO, SiO_2, Al_2O_3, and Fe_2O_3, which play crucial roles in the calcination and sintering of clinker minerals. Incorporating blast furnace slag reduces the reliance on limestone and clay while still promoting clinker mineral formation.

Determining the appropriate content of blast furnace slag is more intricate than considering its individual components (CaO, SiO_2, and Al_2O_3) due to variations in composition. Instead, the overall content of CaO, SiO_2, and Al_2O_3 in the raw materials should be calculated considering the chemical composition and content of each material used. Additionally, common characteristic parameters used in cement clinker design, such as the lime saturation factor (LSF), lime saturation rate (KH), iron modulus, and silica modulus, can also be employed in the design of blast furnace slag-based clinker (Taylor, 1997).

3.3 Calcination

The clinker production process, which involves the use of blast furnace slag along with other essential raw materials such as limestone, typically includes calcination at temperatures ranging from 1338 to 1600°C (Ismail et al., 2023; Prasad et al., 2018; Qin et al., 2013). After a calcination period of approximately 30–120 min, clinker products are obtained. A case illustration of the heating program of the calcination of blast furnace and lime is presented in Fig. 13.3. Notably, the presence of blast furnace slag does not introduce significant new clinker minerals beyond those already present in Portland cement clinker, which mainly consists of tricalcium silicate (C_3S), dicalcium silicate (C_2S), tricalcium aluminate (C_3A), and ferrite (C_4AF). Depending on the specific composition of additional raw materials, minor amounts of calcium aluminates ($C_{12}A_7$), melilite, or spinel can also be generated (Prasad et al., 2018; Qin et al., 2013).

Calcination of clinker with steel slag may employ heating the raw meal to 1405–1450°C and holding the temperature for 30 min (Zhao et al., 2023). Including steel slag in the production process has been shown to lower the required sintering temperatures, typically reducing them to a range of 1320–1350°C (Iacobescu et al., 2016; Isteri et al., 2020). This contributes to energy efficiency and cost reduction. Extensive laboratory tests and industrial-scale trials consistently demonstrate the positive effects of steel slag on cement clinker production (Cao et al., 2019; Tsakiridis et al., 2008). By blending steel slag into the raw meal, the clinker can be fired at lower temperatures, thereby enhancing the overall quality of final product while preserving its mineralogical composition. Research indicates that steel slag can be blended up to 10% without adversely affecting the sintering or hydration properties (Iacobescu et al., 2011; Tsakiridis et al., 2008). A schematic drawing of the sintering process during calcination of steel slag–based clinker is summarized in Fig. 13.4.

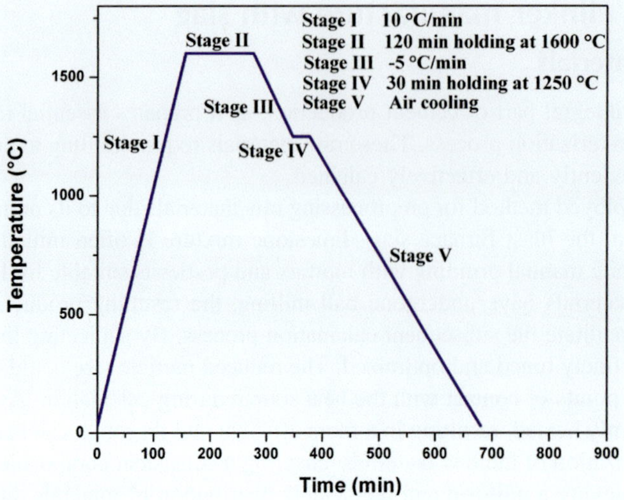

FIGURE 13.3 Heating program of clinker calcination with blast furnace slag and lime (Prasad et al., 2018).

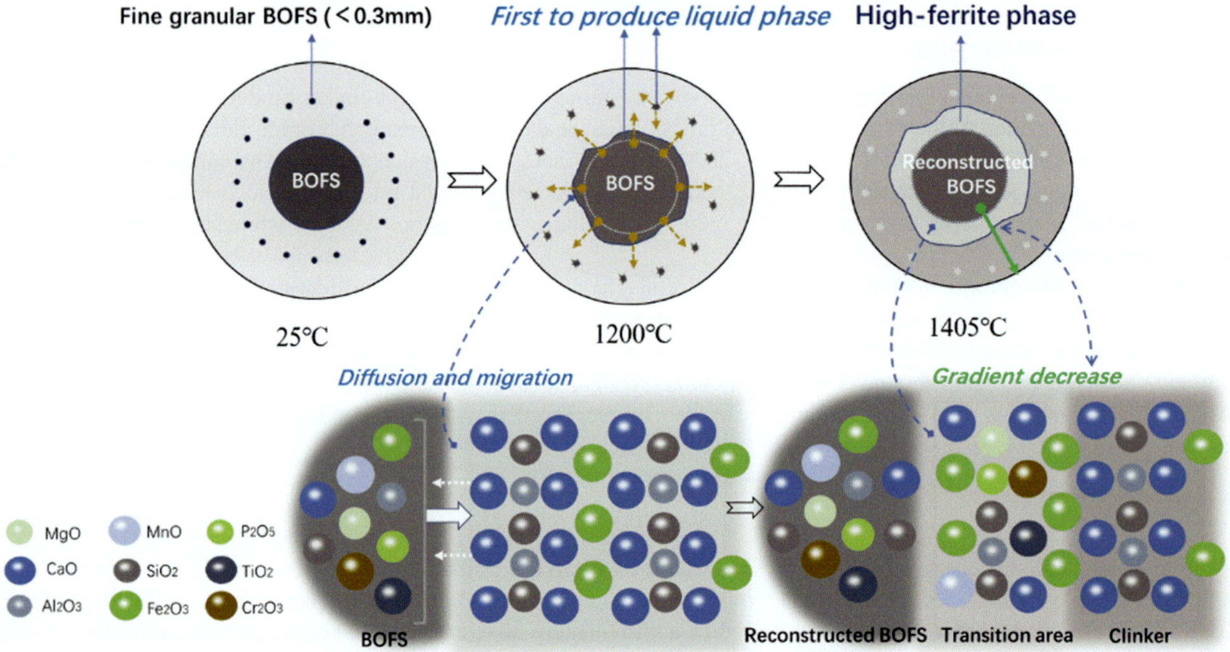

FIGURE 13.4 The ideal model diagram of the sintering mechanism of basic oxygen furnace slag (steel slag) multiphased clinker (Zhao et al., 2023).

Considering its high iron content, copper slag waste has been successfully applied as an iron adjustment material for cement clinker production. In addition, the potential utilization of flotation copper slag as an iron source for Portland cement clinker has also been confirmed (Alp et al., 2008). The main mineral of copper slag is vitreous ferrosilite ($FeSiO_3$), which is characterized by a low melting point and is able to reduce the sintering temperature of cement clinker (Wang et al., 2021a). It was reported that clinkerization of copper slag can occur at 1300−1450 with a retention time of 20 min (Ali et al., 2013). This suggests a calcination temperature lowered by 50°C with copper slag incorporation. Introducing copper slag is also beneficial for improving the grindability of Portland cement clinker (Wang et al., 2021a). It was indicated that substituting copper slag for iron powder is able to reduce or eliminate the need for mineralizer (Wang et al., 2021a).

Moreover, the lead slag can also be used as a substitute for iron ore in the production of cement clinker due to its high iron content. It is able to promote the melting and mineralization of raw materials among calcining, reducing the melting temperature, and promoting calcination. This results in a better burnability and the formation of desired compounds in the cement. A calcination temperature of raw meal incorporated with lead slag has been reported to be 1350°C (Zhanikulov et al., 2022, 2023), and the retaining time to be 45 min (Sobanska et al., 2016; Zhanikulov et al., 2023). Previous studies have shown that cement produced using lead slag performs better in terms of burnability, flexural strength, and compressive strength compared with cement produced with iron ore (Carvalho et al., 2017). Research has also explored the use of lead and zinc smelting slag as the primary ferrous raw material in the production of Portland cement, completely replacing iron ore. This substitution has improved the burnability of the cement mixture. Lead and zinc are fused into other compounds, such as alite and iron phase, thereby enhancing their activity and promoting mineralization and fusion. This improves burnability and reduces heat consumption (Pan et al., 2019).

Since the CaO content takes up more than 50% for most released magnesium slag, it can be used as an alternative to lime-based materials for cement clinker calcination (Li et al., 2018). It was estimated that part of natural limestone, sand, and gravel by 11% can be replaced (Lu et al., 2013). Due to the presence of MgO and CaF_2, magnesium slag has the potential of improving clinker burnability. Theoretical speaking, the application of magnesium slag reduced the standard coal consumption by 14% and grinding time 5%, respectively. Thus the use of magnesium slag can significantly improve the combustion properties and grindability, which leads to a remarkable energy saving (Li et al., 2018).

The MgO content in nickel slag is relatively high and primarily exists in the form of olivine minerals. Due to the low proportion of nickel slag in the raw materials, the MgO content in the resulting clinker does not exceed threshold value (Iacobescu et al., 2016). The mineralogical composition of clinker manufactured from nickel slag includes major phases such as C_3S, C_2S, C_3A, C_4AF, as well as minor constituents such as MgO (olivine), f-CaO, cordierite (α-$Mg_2Al_4Si_5O_{18}$), and diopside ($CaMg(Si_2O_6)$) (Wu et al., 2018). Research findings indicate that when a mixture of nickel slag, limestone, fly ash, and steel slag is calcined at 1350°C for 60 min, it produces clinker with f-CaO levels below 0.22%. Subsequently, incorporating gypsum leads to the production of cement with a 28-day strength measuring 52.4 MPa (Wu et al., 2018).

4. Properties of cement clinker manufactured with slag

4.1 Clinker phases composition

XRD is widely used for determining the mineralogical composition of clinker. The effect of sintering temperature on the phase compositions of clinkers made from blast furnace slag is shown in Fig. 13.5. For the clinker sintered at 1250°C, the minerals detected in the clinker are mainly composed of C_3S, β-C_2S, C_3A, C_4AF, and f-CaO with some minor components such as γ-C_2S and $C_{12}A_7$. Increasing the sintering temperature higher than 1300°C, the clinker contains less β-C_2S and less f-CaO. Different from the evolution of C_4AF, the C_3A content reduced with temperature.

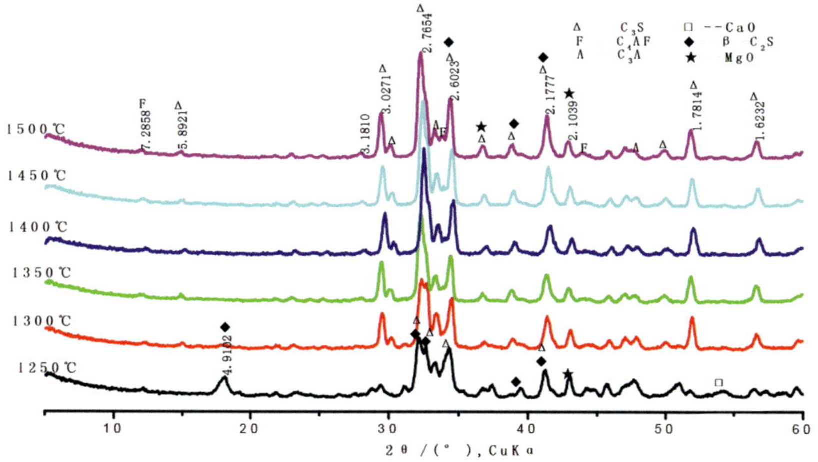

FIGURE 13.5 XRD patterns of clinkers burned at different temperatures with blast furnace in raw material.

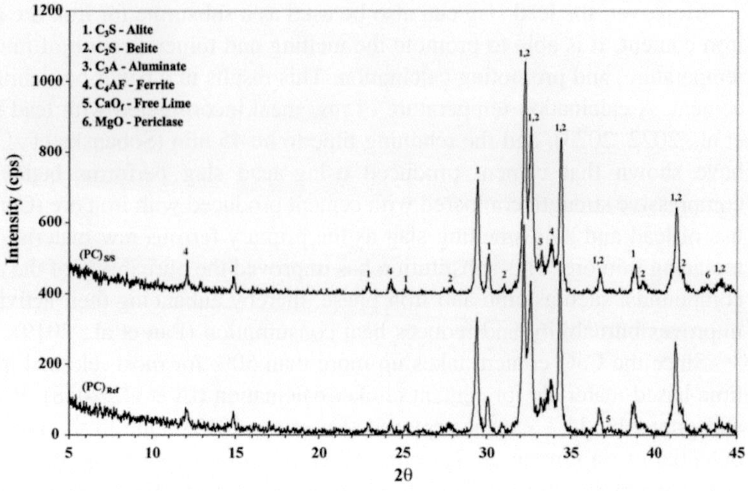

FIGURE 13.6 The XRD patterns of cement clinker manufactured with ($(PC)_{S/S}$) and without ($(PC)_{ref}$) steel slag (Tsakiridis et al., 2008).

Incorporating steel slag offers a significant advantage as it does not have a detrimental effect on the mineralogical properties of the resulting clinker (Cao et al., 2019; Tsakiridis et al., 2008). This can be observed in Fig. 13.6, where the addition of steel slag has minimal impact on the mineral composition of the clinker, ensuring the desired quality of the final product.

As is shown in Fig. 13.7, the control clinker exhibited tablet-like and tightly packed nests of C_3S crystals with interstitial matter, while the multiphased clinker with high-temperature reconstructed steel slag, a transition area mainly occupied by C_2S and an intermediate phase, along with reference clinker (Fig. 13.7B-1—D-1) (Zhao et al., 2023). During the sintering process, the emission of CO_2 resulted in increased pore formation within the clinker. The diffusion of the liquid phase formed by the initial melting of the steel slag led to the development of a transition area characterized by a high-ferrite phase at the interface. Notably, this transition area displayed an excessive amount of white mesophase, with scattered concentrated zones measuring approximately 108.6 μm in length and 37.5 μm in width (Fig. 13.7B-1). The significant content of Fe_2O_3, MnO, and MgO in the steel slag contributed to a higher liquid phase content and reduced viscosity, which in turn enhanced infiltration and expansion of the transition area, ultimately resulting in an increase in white mesophase content (Fig. 13.7C-1 and D-1).

Fe—Mn-rich inclusions were found in interstitial matter and belite (Fig. 13.7D-2). The lower CaO content of steel slag diluted surrounding areas, resulting in the prevalence of belite in the transition area. Roundish belite crystals with clear boundaries and uniform sizes were observed (Fig. 13.7B-3 and C-3). The contact between reconstructed steel slag and raw meal promoted the formation of well-crystallized elliptical belite crystals (Fig. 13.7C-2). The reconstructed steel slag exhibited diverse crystal characteristics, with modifications decreasing from the periphery to the core (Fig. 13.7B-2). Clinker mineral boundaries near the transition area contained inclusions, while interstitial matter showed uneven distribution of black and white mesophases (Fig. 13.7D-3). Alite crystals had clear boundaries and sizes between 10 and 30 μm. Periclase formation and MgO incorporation were observed in the interstitial matter (Fig. 13.7B-4—D-4). These findings shed light on clinker structure and the influence of steel slag interactions.

4.2 Mechanical strength

Mechanical strength of the clinker manufactured with slag generally fluctuates due to the variation the composition of different slag. An appropriate amount of magnesium slag is reported to shorten the setting times and improve the compressive strength of so-prepared cement, which can be attributed that more C_3S is produced in the clinker (Li et al., 2018).

However, it is important to note that the mechanical properties of the clinker tend to weaken with increasing steel slag admixture (Cao et al., 2019). Fig. 13.8 showcases the relationship between steel slag content and compressive strength, revealing that an optimal admixture of 15% represents the highest compressive strength (Gao et al., 2021). Beyond this threshold, a further increase in steel slag content may lead to a gradual decline in the clinker's mechanical performance. Another study (Fig. 13.9) also presents the reduction in compressive strength of the hydrated cements made from steel slag (Iacobescu et al., 2011).

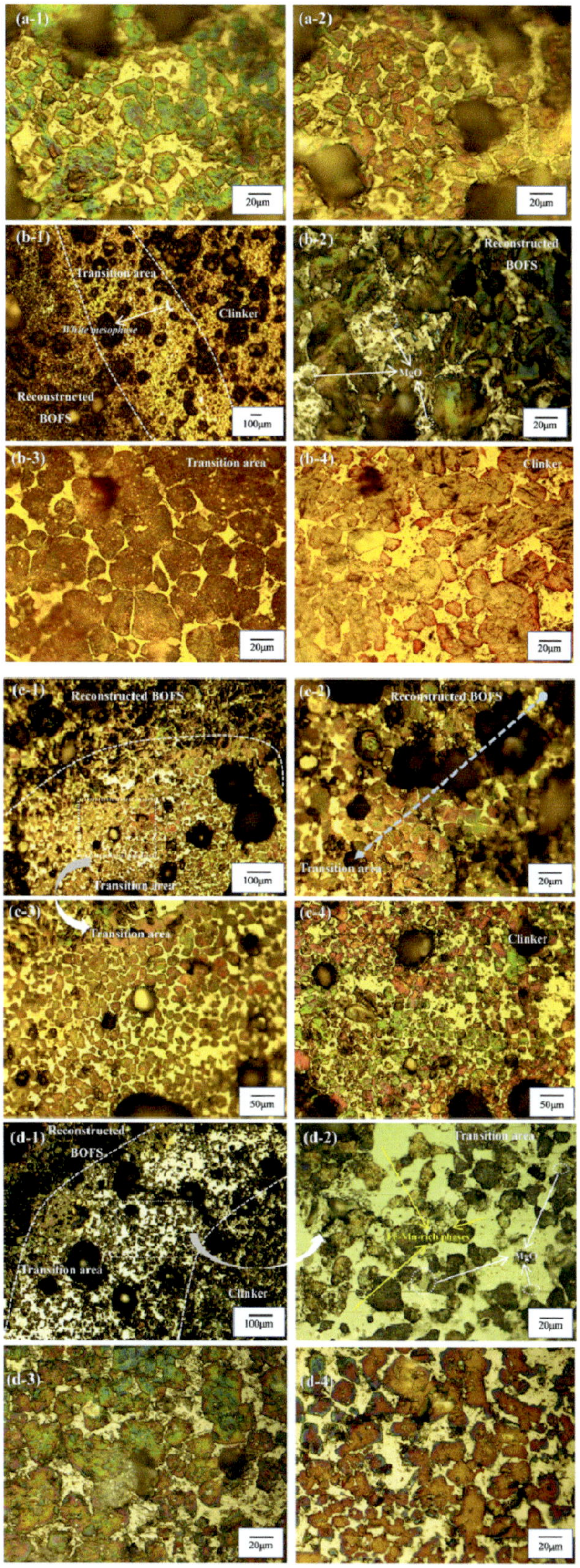

FIGURE 13.7 Microscopic morphology of clinker in relation to steel slag and sintering temperature [(A) control—1450°C; (B) 10% steel slag—1420°C; (C) 15% steel slag—1420°C; (D) 15% steel slag—1405°C] (Zhao et al., 2023).

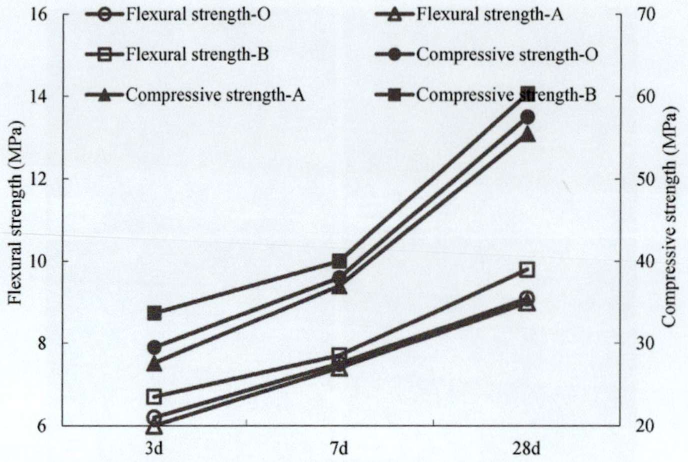

FIGURE 13.8 Compressive strength of cements made from raw materials incorporated with 0%–15% steel slag (O for 0%, A for 10%, and B for 15%) (Gao et al., 2021).

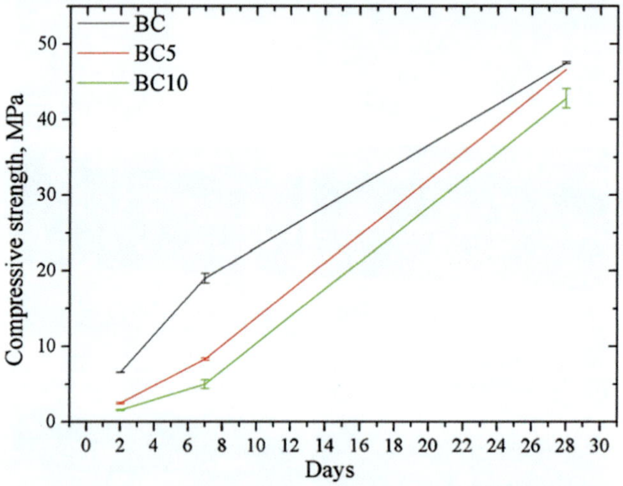

FIGURE 13.9 Compressive strength of cements made from raw materials including 0%–10% electric arc furnace steel slag (Iacobescu et al., 2011).

With respect to the mechanical properties, the 28-day compressive strength of cement made with magnesium slag can reach as high as 58.9 MPa, and a significant increase in early compressive strength is attractive. An opposite trend was observed for setting time, but the setting time was relatively long which may be related to the variation of crystalline states. The physical properties of final produced cement using MS clinker could meet the requirements of the 52.5 strength class specified in GB 175–2007. The disposal of the MS via the production of Portland cement is also environmentally friendly. The MS can be used for preventing dust pollution. Therefore, MS can be used for calcining clinker with a good quality (Cai et al., 2011).

4.3 Hydration heat

It was observed that the addition of steel slag caused a delay in the acceleratory phase and prolonged the dormant phase of hydration. This suggests that steel slag has an impact on the overall hydration process and alters its kinetics. One notable finding was that the presence of steel slag resulted in an increase in the ferrite phase content while reducing the C_3A content. This change in phase composition could have significant implications for the properties of the final cement product. The increased ferrite phase content may contribute to improved durability and strength, while the decrease in C_3A content could potentially reduce the risk of early-age cracking.

As is shown in Fig. 13.10, the faster hydration rate of C_3A in the presence of steel slag leads to a prolonged induction period (Zhao et al., 2023). This finding suggests that steel slag plays a role in promoting the early stages of cement

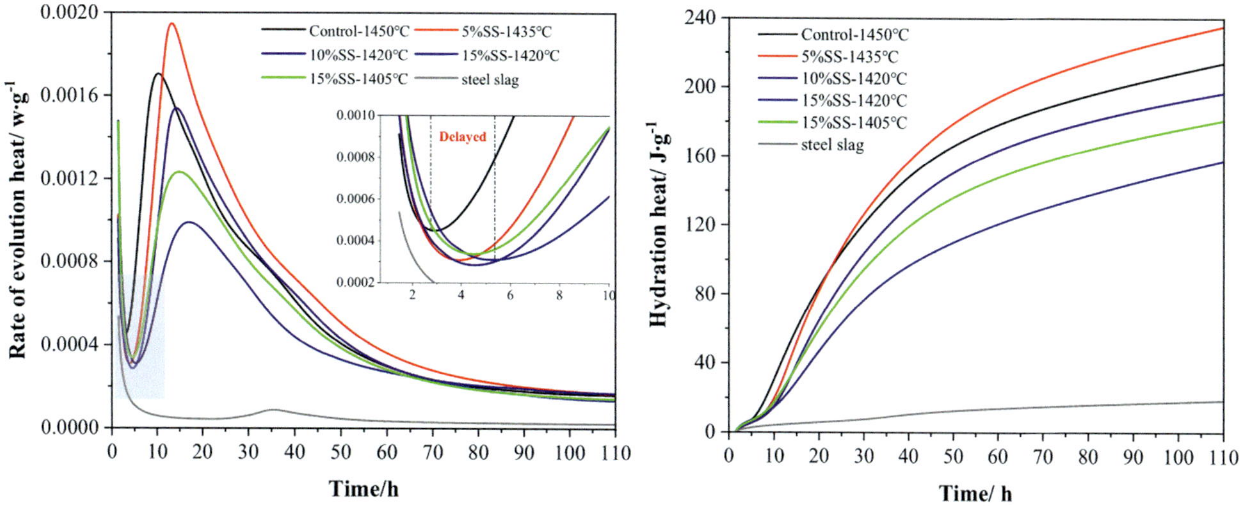

FIGURE 13.10 Heat of hydration and cumulative heat of original steel slag and multiphased linker cement over a period of 110 h (Zhao et al., 2023).

hydration. However, the presence of gypsum was found to have a significant retarding effect on the hydration of the ferrite phase. This interaction between gypsum and ferrite indicates the complex nature of the cement hydration process and the importance of considering the influence of different components.

The inhibitory effect of the ferrite phase on the hydration of C_3S was also observed, resulting in an extended induction period and decreased early hydration exothermic rate. This finding suggests that the presence of the ferrite phase, which is influenced by the addition of steel slag, can influence the reactivity of C_3S and alter the overall hydration behavior.

5. Hazard control by clinkerization of slag

It is quite easy to detect various types of heavy metals in slag. If not properly treated prior to its use in cement clinker production, these heavy metals can pose significant health risks to humans. Additionally, these heavy metals can also negatively impact the properties of the clinker, limiting its potential applications. Consequently, effectively managing and controlling the presence of hazardous heavy metals in slag becomes a crucial concern that needs to be addressed.

Heavy metals can be stabilized by the clinkerization process. It has been reported that copper slag incorporated in cement clinker production poses lower environmental threat than the substituted iron ore in the raw material (Alp et al., 2008). This can be attributed to the incorporation of heavy metals in the clinker phases such as C_3S and C_4AF, the formation of heavy metal–doped calcium oxides, and the substitution for interstitial calcium ion in crystal lattice (Wang et al., 2022; Yang et al., 2014, 2023). In the case of steel slag, as is shown in Fig. 13.11, heavy metals such as Cr and Mn are stabilized in ferrous clinker phases by calcination (Wang et al., 2021b).

However, the leaching behaviors present differences between heavy metal types. The leached quantity of Cr is rather low for the calcined clinker compared with the raw steel slag indicating a stabilized existence of Cr in the clinker phases. Nevertheless, the leached quantity increases in the case of Mn, Ni, As, etc. It is thus suggested that the clinkerization of different heavy metals should be discussed, respectively.

The presence of toxic elements such as lead in lead slag restricts its applicability in cement production due to potential adverse effects on the properties of cement. These elements can interfere with hydration reactions, especially in the early stages, thereby increasing the vulnerability of the cement to acid attack. Of particular concern is the leaching of lead into the environment from the alkaline cement matrix. To ensure environmental safety, strict regulations controlling the codisposal of solid waste in cement kilns have been implemented in China. These regulations establish maximum allowable limits for heavy metals such as lead and zinc (De Angelis and Medici, 2012). Consequently, the threshold amount of lead slag recycled in cement production is typically limited to a maximum of 3%–5% of the raw ingredients. Furthermore, careful control or pretreatment measures are essential when incorporating lead slag as inert materials in cement to ensure compliance with environmental regulations (Pan et al., 2019).

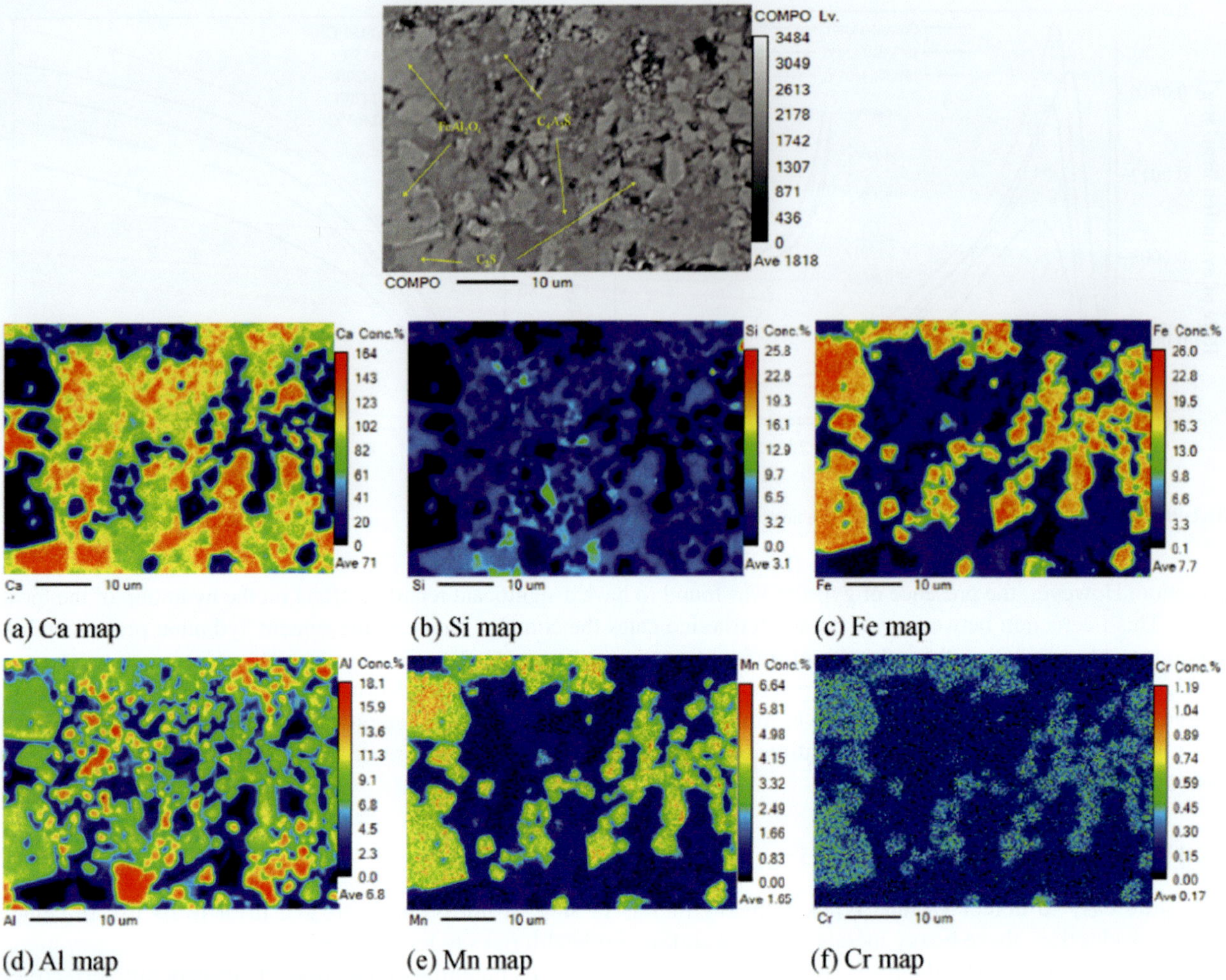

FIGURE 13.11 BSE image and X-ray maps of clinker calcined from steel slag and other raw materials: (*top*) BSE image of clinker; (*bottom*) X-ray elemental maps of Ca, Si, Fe, Al, Mn, and Cr (Wang et al., 2021b).

6. Conclusions

In conclusion, while the direct incorporation of slag into cement clinker has received more attention in the field of cement production, the recycling of various slag for clinker production should not be overlooked. The former approach capitalizes on the benefits of a secondary hydration reaction that occurs when the alkaline cement clinker encounters the slag matrices in an aqueous environment. On the other hand, the latter approach focuses on the reuse of the valuable oxides present in slag to create clinker. Despite being a less explored avenue, the utilization of slag clinkerization to recycle solid waste metallurgical slag holds great promise. By implementing this approach, not only can we effectively address the issue of slag disposal, but we can also optimize resource utilization in the cement industry. The potential of this innovative technique should be further explored and harnessed to achieve sustainable cement production.

References

Abdul-Wahab, S.A., Al-Rawas, G.A., Ali, S., Al-Dhamri, H., 2016. Impact of the addition of oil-based mud on carbon dioxide emissions in a cement plant. J. Clean. Prod. 112, 4214–4225. https://doi.org/10.1016/j.jclepro.2015.06.062.

Ali, M.M., Agarwal, S.K., Pahuja, A., 2013. Potentials of copper slag utilisation in the manufacture of ordinary Portland cement. Adv. Cem. Res. 25, 208–216. https://doi.org/10.1680/adcr.12.00004.

Alp, İ., Deveci, H., Süngün, H., 2008. Utilization of flotation wastes of copper slag as raw material in cement production. J. Hazard Mater. 159, 390–395. https://doi.org/10.1016/j.jhazmat.2008.02.056.

Anastasiou, E.K., Liapis, A., Papachristoforou, M., 2017. Life cycle assessment of concrete products for special applications containing EAF slag. Procedia Environ. Sci. 38, 469–476. https://doi.org/10.1016/j.proenv.2017.03.138.

Cai, J.W., Gao, G.L., Bai, R.Y., Lu, F., Li, L., 2011. Research on slaked magnesium slag as a raw material and blend for Portland cement. AMR (Adv. Magn. Reson.) 335–336, 1246–1249. https://doi.org/10.4028/www.scientific.net/AMR.335-336.1246.

Cao, L., Shen, W., Huang, J., Yang, Y., Zhang, D., Huang, X., Lv, Z., Ji, X., 2019. Process to utilize crushed steel slag in cement industry directly: multi-phased clinker sintering technology. J. Clean. Prod. 217, 520–529. https://doi.org/10.1016/j.jclepro.2019.01.260.

Carvalho, S.Z., Vernilli, F., Almeida, B., Demarco, M., Silva, S.N., 2017. The recycling effect of BOF slag in the Portland cement properties. Resour. Conserv. Recycl. 127, 216–220. https://doi.org/10.1016/j.resconrec.2017.08.021.

De Angelis, G., Medici, F., 2012. Reuse of slags containing lead and zinc as aggregate in a Portland cement matrix. J. Solid Waste Technol. Manag. 38, 117–123. https://doi.org/10.5276/JSWTM.2012.17.

Djokic, J., Minic, D., Kamberovic, Z., Petkovic, D., 2012. Impact analysis of airborn pollution due to magnesium slag deposit and climatic changes condition. Ecol. Chem. Eng. S 19, 439–450. https://doi.org/10.2478/v10216-011-0034-7.

Ettler, V., Johan, Z., 2014. 12 years of leaching of contaminants from Pb smelter slags: geochemical/mineralogical controls and slag recycling potential. Appl. Geochem. 40, 97–103. https://doi.org/10.1016/j.apgeochem.2013.11.001.

Gao, T., Dai, T., Shen, L., Jiang, L., 2021. Benefits of using steel slag in cement clinker production for environmental conservation and economic revenue generation. J. Clean. Prod. 282, 124538. https://doi.org/10.1016/j.jclepro.2020.124538.

Gomes, G.M.F., Mendes, T.F., Wada, K., 2011. Reduction in toxicity and generation of slag in secondary lead process. J. Clean. Prod. 19, 1096–1103. https://doi.org/10.1016/j.jclepro.2011.01.006.

Guo, J., Bao, Y., Wang, M., 2018. Steel slag in China: treatment, recycling, and management. Waste Manage. (Tucson, Ariz.) 78, 318–330. https://doi.org/10.1016/j.wasman.2018.04.045.

Iacobescu, R.I., Angelopoulos, G.N., Jones, P.T., Blanpain, B., Pontikes, Y., 2016. Ladle metallurgy stainless steel slag as a raw material in ordinary Portland cement production: a possibility for industrial symbiosis. J. Clean. Prod. 112, 872–881. https://doi.org/10.1016/j.jclepro.2015.06.006.

Iacobescu, R.I., Koumpouri, D., Pontikes, Y., Saban, R., Angelopoulos, G.N., 2011. Valorisation of electric arc furnace steel slag as raw material for low energy belite cements. J. Hazard Mater. 196, 287–294. https://doi.org/10.1016/j.jhazmat.2011.09.024.

Iacobescu, R.I., Pontikes, Y., Koumpouri, D., Angelopoulos, G.N., 2013. Synthesis, characterization and properties of calcium ferroaluminate belite cements produced with electric arc furnace steel slag as raw material. Cement Concr. Compos. 44, 1–8. https://doi.org/10.1016/j.cemconcomp.2013.08.002.

Ismail, A.A.M., Rosdin, M.R.H., Rozhan, A.N., Purwanto, H., Hamid, A.M.A., Din, M.F.M., Yasin, M.F.M., Ani, M.H., 2023. Blast furnace slag cement clinker production using limestone-hot blast furnace slag mixture. In: Maleque, MdA., Ahmad Azhar, A.Z., Sarifuddin, N., Syed Shaharuddin, S.I., Mohd Ali, A., Abdul Halim, N.F.H. (Eds.), Proceeding of 5th International Conference on Advances in Manufacturing and Materials Engineering, Lecture Notes in Mechanical Engineering. Springer Nature, Singapore, pp. 539–545. https://doi.org/10.1007/978-981-19-9509-5_71.

Isteri, V., Ohenoja, K., Hanein, T., Kinoshita, H., Tanskanen, P., Illikainen, M., Fabritius, T., 2020. Production and properties of ferrite-rich CSAB cement from metallurgical industry residues. Sci. Total Environ. 712, 136208. https://doi.org/10.1016/j.scitotenv.2019.136208.

Jiang, Y., Ling, T.-C., Shi, C., Pan, S.-Y., 2018. Characteristics of steel slags and their use in cement and concrete—a review. Resour. Conserv. Recycl. 136, 187–197. https://doi.org/10.1016/j.resconrec.2018.04.023.

Kim, E., Roosen, J., Horckmans, L., Spooren, J., Broos, K., Binnemans, K., Vrancken, K.C., Quaghebeur, M., 2017. Process development for hydrometallurgical recovery of valuable metals from sulfide-rich residue generated in a secondary lead smelter. Hydrometallurgy 169, 589–598. https://doi.org/10.1016/j.hydromet.2017.04.002.

Lassin, A., Piantone, P., Burnol, A., Bodénan, F., Chateau, L., Lerouge, C., Crouzet, C., Guyonnet, D., Bailly, L., 2007. Reactivity of waste generated during lead recycling: an integrated study. J. Hazard Mater. 139, 430–437. https://doi.org/10.1016/j.conbuildmat.2021.123371.

Li, H., Huang, Y., Yang, X., Jiang, Z., Yang, Z., 2018. Approach to the management of magnesium slag via the production of Portland cement clinker. J. Mater. Cycles Waste Manage. 20, 1701–1709. https://doi.org/10.1007/s10163-018-0735-4.

Li, L., Ling, T.-C., Pan, S.-Y., 2022. Environmental benefit assessment of steel slag utilization and carbonation: a systematic review. Sci. Total Environ. 806, 150280. https://doi.org/10.1016/j.scitotenv.2021.150280.

Li, Y., Fan, Y., Chen, Z., Cheng, F., Guo, Y., 2016a. Chemical, mineralogical, and morphological characteristics of pidgeon magnesium slag. Environ. Eng. Sci. 33, 290–297. https://doi.org/10.1089/ees.2015.0097.

Li, Y., Liu, Y., Gong, X., Nie, Z., Cui, S., Wang, Z., Chen, W., 2016b. Environmental impact analysis of blast furnace slag applied to ordinary Portland cement production. J. Clean. Prod. 120, 221–230. https://doi.org/10.1016/j.jclepro.2015.12.071.

Lu, F., Bai, R.Y., Cai, J.W., 2013. Study on clinker production using magnesium slag on a 4500tpd line. Adv. Mater. Res. 690–693, 724–727. https://doi.org/10.4028/www.scientific.net/AMR.690-693.724.

Murari, K., Siddique, R., Jain, K.K., 2015. Use of waste copper slag, a sustainable material. J. Mater. Cycles Waste Manag. 17, 13–26. https://doi.org/10.1007/s10163-014-0254-x.

Nath, S.K., Randhawa, N.S., Kumar, S., 2022. A review on characteristics of silico-manganese slag and its utilization into construction materials. Resour. Conserv. Recycl. 176, 105946. https://doi.org/10.1016/j.resconrec.2021.105946.

Pan, D., Li, L., Tian, X., Wu, Y., Cheng, N., Yu, H., 2019. A review on lead slag generation, characteristics, and utilization. Resour. Conserv. Recycl. 146, 140–155. https://doi.org/10.1016/j.resconrec.2019.03.036.

Penpolcharoen, M., 2005. Utilization of secondary lead slag as construction material. Cement Concr. Res. 35, 1050–1055. https://doi.org/10.1016/j.cemconres.2004.11.001.

Piatak, N.M., Parsons, M.B., Seal, R.R., 2015. Characteristics and environmental aspects of slag: a review. Appl. Geochem. 57, 236–266. https://doi.org/10.1016/j.apgeochem.2014.04.009.

Prasad, K., Srishilan, C., Shukla, A.K., Kaza, M., 2018. Thermodynamic assessment and experimental validation of clinker formation from blast furnace slag through lime addition. Ceram. Int. 44, 19434–19441. https://doi.org/10.1016/j.ceramint.2018.07.180.

Qin, S.W., Shen, J.J., Wang, H.F., Xiao, Z.R., 2013. Utilization of blast furnace slag, steel slag in the production of clinker. Mater. Sci. Forum 743–744, 334–338. https://doi.org/10.4028/www.scientific.net/MSF.743-744.334.

Ren, C., Wang, W., Mao, Y., Yuan, X., Song, Z., Sun, J., Zhao, X., 2017. Comparative life cycle assessment of sulfoaluminate clinker production derived from industrial solid wastes and conventional raw materials. J. Clean. Prod. 167, 1314–1324. https://doi.org/10.1016/j.jclepro.2017.05.184.

Shi, C., 2004. Steel slag—its production, processing, characteristics, and cementitious properties. J. Mater. Civil Eng. 16, 230–236. https://doi.org/10.1061/(ASCE)0899-1561(2004)16:3(230).

Sobanska, S., Deneele, D., Barbillat, J., Ledésert, B., 2016. Natural weathering of slags from primary Pb–Zn smelting as evidenced by Raman microspectroscopy. Appl. Geochem. 64, 107–117. https://doi.org/10.1016/j.apgeochem.2015.09.011.

Taylor, H.F.W., 1997. Cement Chemistry, 2nd. Thomas Telford Publishing.

Tian, H., Guo, Z., Pan, J., Zhu, D., Yang, C., Xue, Y., Li, S., Wang, D., 2021. Comprehensive review on metallurgical recycling and cleaning of copper slag. Resour. Conserv. Recycl. 168, 105366. https://doi.org/10.1016/j.resconrec.2020.105366.

Tsakiridis, P.E., Papadimitriou, G.D., Tsivilis, S., Koroneos, C., 2008. Utilization of steel slag for Portland cement clinker production. J. Hazard Mater. 152, 805–811. https://doi.org/10.1016/j.jhazmat.2007.07.093.

Verma, Y.K., Mazumdar, B., Ghosh, P., 2020. CO_2 emission reduction using blast furnace slag for the clinker manufacturing in Cement Industry. J. Indian Chem. Soc. 97.

Wang, J., Han, F., Yang, B., Xing, Z., Liu, T., 2022. A study of the solidification and stability mechanisms of heavy metals in electrolytic manganese slag-based glass-ceramics. Front. Chem. 10.

Wang, R., Shi, Q., Li, Y., Cao, Z., Si, Z., 2021a. A critical review on the use of copper slag (CS) as a substitute constituent in concrete. Construct. Build. Mater. 292, 123371. https://doi.org/10.1016/j.conbuildmat.2021.123371.

Wang, X., Wang, K., Li, J., Wang, W., Mao, Y., Wu, S., Yang, S., 2021b. Heavy metals migration during the preparation and hydration of an eco-friendly steel slag-based cementitious material. J. Clean. Prod. 329, 129715. https://doi.org/10.1016/j.jclepro.2021.129715.

Wu, Q., Wu, Y., Tong, W., Ma, H., 2018. Utilization of nickel slag as raw material in the production of Portland cement for road construction. Construct. Build. Mater. 193, 426–434. https://doi.org/10.1016/j.conbuildmat.2018.10.109.

Yang, N., Li, A., Liu, Q., Cui, Y., Wang, Z., Gao, Y., Guo, J., 2023. Incorporation and solidification mechanism of manganese doped cement clinker. Front. Chem. 11. https://doi.org/10.3389/fchem.2023.1165402.

Yang, X., Dong, F., Zhang, X., Li, C., Gao, Q., 2022. Review on comprehensive utilization of magnesium slag and development prospect of preparing backfilling materials. Minerals 12, 1415. https://doi.org/10.3390/min12111415.

Yang, Y., Xue, J., Huang, Q., 2014. Studies on the solidification mechanisms of Ni and Cd in cement clinker during cement kiln co-processing of hazardous wastes. Construct. Build. Mater. 57, 138–143. https://doi.org/10.1016/j.conbuildmat.2013.12.081.

Yin, N.-H., Sivry, Y., Guyot, F., Lens, P.N.L., van Hullebusch, E.D., 2016. Evaluation on chemical stability of lead blast furnace (LBF) and imperial smelting furnace (ISF) slags. J. Environ. Manage. 180, 310–323. https://doi.org/10.1016/j.jenvman.2016.05.052.

Zemri, C., Bachir Bouiadjra, M., 2020. Comparison between physical–mechanical properties of mortar made with Portland cement (CEMI) and slag cement (CEMIII) subjected to elevated temperature. Cement Concr. Res. 12, e00339. https://doi.org/10.1016/j.cscm.2020.e00339.

Zhanikulov, N., Sapargaliyeva, B., Agabekova, A., Alfereva, Y., Baidibekova, A., Syrlybekkyzy, S., Nurshakhanova, L., Nurbayeva, F., Sabyrbaeva, G., Zhatkanbayev, Y., Kozlov, P., Izbassar, A., Kolesnikova, O., 2023. Studies of utilization of technogenic raw materials in the synthesis of cement clinker from it and further production of Portland cement. J. Compos. Sci. 7, 226. https://doi.org/10.3390/jcs7060226.

Zhanikulov, N.N., Kolesnikov, A.S., Taimasov, B.T., Zhakipbayev, B.Y., Shal, A.L., 2022. Influence of industrial waste on the structure of environmentally friendly cement clinker. KIMS/CUMR/MShKP 323, 84–91. https://doi.org/10.31643/2022/6445.44.

Zhao, D., Shen, W., Wang, Y., Yang, Y., Zhang, W., Shi, Q., Deng, Y., Lu, J., Deng, Y., 2023. Direct use of original granular steel slag to prepare multi-phased clinker: sintering mechanism and properties. Construct. Build. Mater. 390, 131575. https://doi.org/10.1016/j.conbuildmat.2023.131575.

Chapter 14

Recycling of various combustion/incineration residues into calcium sulfoaluminate cementitious material (CSA)

Xujiang Wang[1,2,3,4]

[1]National Engineering Laboratory for Reducing Emissions from Coal Combustion, Shandong University, Jinan, China; [2]Shenzhen Research Institute of Shandong University, Shenzhen, Guangdong, China; [3]Shandong-Nottingham Joint Research Centre for Green Energy & Materials, Shandong University, Jinan, China; [4]School of Energy and Power Engineering, Shandong University, Jinan, China

1. Introduction

With the swift advancement of global economy and the rapid progression of industrialization, issues concerning the environment and climate, such as global warming caused by excessive greenhouse gas emissions, notably CO_2, pose significant threats to public health and human welfare (EPA, 2022). These issues also impede sustainable economic and societal development and have emerged as hotly debated topics within the international community (Barker et al., 2009; Lin and Zhang, 2016). The traditional building materials industry, especially the cement and concrete sector, are substantial contributors to carbon emissions, facing tremendous challenges in CO_2 reduction. Approximately 4 billion tons of cement are produced worldwide each year, with half of it produced in China (Zhu et al., 2014; Tao et al., 2022; Xu et al., 2022). Cement production is a high-energy consumption and high-emission process, contributing to 4%~6% of the total CO_2 emissions from the construction industry worldwide. In 2022, China produced 2.13 billion tons of cement (China, 2022), with clinker consumption exceeding 1.8 billion tons and CO_2 emissions around 1.6 billion tons (Amran et al., 2020). Carbon emissions of cement production derive from direct discharge during calcination, combustion of fossil fuels, mineral processing, transportation, and other indirect emissions (Plaza et al., 2020; Guo et al., 2024). To reduce carbon emission and achieve net-zero society, it is necessary not only to expedite the optimization of energy utilization and increase energy efficiency but also to promote the comprehensive green transformation of economic and social development, and improve resource utilization efficiency comprehensively. The comprehensive utilization of resources is also one of the critical ways for China to deeply implement sustainable development strategies, realize green, low-carbon, and circular development and meet the targets of carbon peaking and neutrality (Yao, 2021).

The utilization of bulk industrial wastes is the core of comprehensive resource utilization. However, the reality of China's large waste output and storage, coupled with insufficient comprehensive utilization, severely hinders the advancement of sustainable development strategies. According to statistics from the Ministry of Ecology and Environment, China produced 3.68 billion tons of general industrial waste (excluding waste rock) in 2020, with 2.04 billion tons utilized and 810 million tons stored. The comprehensive utilization rate is less than 60% (Li et al., 2019, 2023b). Combustion/incineration residues refer to solid residues produced during the combustion process (Jing et al., 2023), including steel slag, blast furnace slag, fly ash, municipal solid waste incineration ash, etc. The extensive accumulation of these industrial wastes occupies land resources and poses environmental threats and societal issues. Therefore, the comprehensive utilization of industrial waste, particularly combustion/incineration residues, is of paramount significance for sustainable development of ecological civilization and socio-economic (Sun et al., 2022). The primary volumes of combustion/incineration residues in China are shown in Fig. 14.1.

Treatment and Utilization of Combustion and Incineration Residues. https://doi.org/10.1016/B978-0-443-21536-0.00007-1

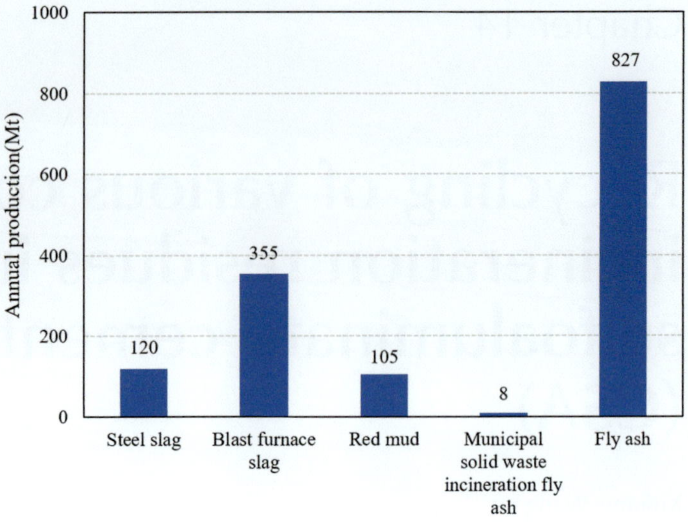

FIGURE 14.1 Production of several combustion/incineration wastes in China in 2021 (Li, 2022; Feng, 2022; Gongyan-Net, 2023; ZhongCheng-Environment, 2023).

Calcium sulfoaluminate cementitious material (CSA), as a type of low-carbon ecological cement, is characterized by high early strength and quick hardening. In recent years, it has garnered widespread attentions, finding extensive applications in emergency repairs, rapid restoration, and marine engineering. A burgeoning interest in producing low-carbon CSA from industrial wastes such as combustion/incineration residues has arisen among many researchers. Ren et al. (2017) utilized industrial solid wastes such as red mud and desulphurization gypsum to prepare CSA clinker, carrying out environmental impact assessments and comparing it with standard production methods. The research revealed that, compared with conventional production processes, the environmental burden of producing CSA clinker from comprehensive utilization of industrial solid waste could be reduced by 38.62%. Xue et al. (2016) prepared CSA clinker using steel slag and investigated the impact of various mix proportions and calcination temperatures on clinker performance. The study found that when the calcination temperature is 1350°C, and the suitable proportion of steel slag is more than 14%, with a designed $C_4A_3\overline{S}$ content of more than 20%, the clinker exhibits excellent hydration behaviors. Guo et al. (2014) prepared CSA with municipal solid waste incineration fly ash (MSWI-FA). The 28-day compressive strength reached 110 MPa and demonstrated superior corrosion resistance and durability. Using incineration/combustion slag to prepare CSA not only enables the resource recovery of waste, reducing waste landfill and harmful substance emissions, but also lessens reliance on natural resources, improves cement performance, and reduces carbon emissions (Amran et al., 2021; Gencel et al., 2021). This chapter mainly introduces the effects of various combustion/incineration wastes on CSA performance, hydration process, carbonization, etc., to comprehensively understand their potential applications in the cement industry, and manufacture green and sustainable concrete composite materials, to achieve green emission reduction and sustainable development in the construction industry.

2. Sources and physicochemical properties of combustion/incineration residues

2.1 Steel slag

Steel slag is a by-product produced in the process of steel smelting due to oxidation and slag formation of metal raw materials. It is characterized by a high content of free oxides, low cementitious properties, and high heavy metal content. China is the world's largest producer of steel, generating at least 80 million tons of steel slag per year (Kang et al., 2018), with the reuse rate about 30% (Gencel et al., 2021). Steel slag is generally hard and dense with low natural moisture content. The main chemical components include f-SiO_2, CaO, Fe_2O_3, FeO, Al_2O_3, MgO, MnO, P_2O_5, etc. (Jiang et al., 2018).

2.2 Blast furnace slag

Blast furnace slag is discharged from the blast furnace during the smelting of pig iron. When the furnace temperature reaches 1400–1600°C, a high-temperature reaction of iron ore occurs to produce pig iron and blast furnace slag (Amran et al., 2021). The slag is an angular, irregularly shaped sand-like material. The main chemical compositions include CaO, SiO_2, Al_2O_3, MgO, MnO, Fe_2O_3, and minor sulfides such as CaS and MnS (Amran et al., 2021).

2.3 Red mud

Red mud is generated during the production of Al_2O_3 using bauxite as raw material. Depending on the technology of Al_2O_3 production and the grade of the ore used, red mud can be divided into sintered red mud, Bayer red mud, and combined red mud. The most commonly used method in industrial production of Al_2O_3 is the Bayer process, but due to the lack of high-temperature calcination in this process, Bayer red mud has a higher alkalinity, more heavy metals, and poor permeability, which could potentially cause severe environmental pollution (Hertel et al., 2021). The main chemical components include CaO, SiO_2, Al_2O_3, Fe_2O_3, MnO, TiO_2, etc.

2.4 Aluminum ash

Aluminum ash is produced during the production of electrolytic aluminum, with 30–50 kg of aluminum ash produced per ton of electrolytic aluminum. The total discharge of aluminum ash in China is about 11.5 million tons per year, most of which is directly discarded or landfilled, causing significant environmental impact due to the harmful substances it contains (Wu et al., 2023). The main components of aluminum ash include SiO_2, Cl, Al_2O_3, and Na_2O, with minor amounts of F, Fe, among which Cl and F are harmful to the environment (Wu et al., 2023).

2.5 Municipal solid waste incineration fly ash

MSWI-FA is an alkaline by-product (pH = 11.1–12.3) (Wang et al., 2023a) generated from the incineration treatment of municipal solid waste. The global production of MSWI-FA is expected to reach 370~440 million tons per year by 2025. The particle size of MSWI-FA is generally 4–100 μm, which contains a lot of heavy metals, organic substances, and toxic ingredients, making its treatment very difficult. The main chemical compositions of MSWI-FA include CaO, SiO_2, Cl, Na_2O, etc.

2.6 Fly ash

Fly ash is the fine ash captured from the smoke after coal combustion and is the main solid waste discharged by coal-fired power plants, also known as coal smoke. Fly ash is one of the industrial wastes with a large discharge volume in China, with an annual output of more than 500 million tons. The particle size of fly ash usually ranges between 1 and 100 μm, and the morphology is mainly spherical. The main chemical compositions are SiO_2, Al_2O_3, Fe_2O_3, etc. (John et al., 2021; Mathapati et al., 2022).

3. The influence of combustion/incineration residues as raw materials on cement clinker and performance

As raw materials, solid waste can be broadly divided into calcium-sulfate-based and silicon-aluminum-ferrite-based. Combustion/incineration residues mainly belong to silicon–aluminum–ferrite-based raw materials, used to adjust the contents of Ca, Si, Al, Fe. Their combination with calcium–sulfate-based raw materials such as desulfurized gypsum and phosphogypsum can basically meet the composition requirements for preparing CSA. Since CSA has lower requirements for Ca than Portland cement (PC) and has no strict requirements for Fe content, it has better acceptance capacity for solid waste materials. The production of solid waste–based CSA is an effective way to utilize solid waste (Ben Haha et al., 2019).

3.1 Steel slag

Steel slag is a by-product of steel production, and the preparation of CSA with steel slag and other industrial solid wastes is an effective way to handle steel slag on a large scale (Iacobescu et al., 2013; Xue et al., 2016). Wang et al. (2021) used steel slag, desulphurization gypsum, aluminum ash, and carbide slag to prepare an environmentally friendly steel slag–based cementitious material (ESS-CM), with steel slag accounting for 38.47 wt.%. It indicated that the optimal sintering conditions were maintaining at 1220°C for 30 min. Compared with ordinary CSA, ESS-CM is characterized by a lower

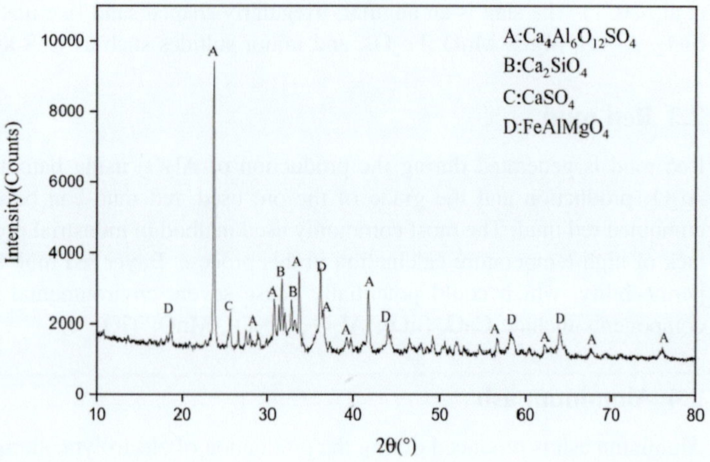

FIGURE 14.2 XRD spectrum of ESS-CM clinker (Wang et al., 2021).

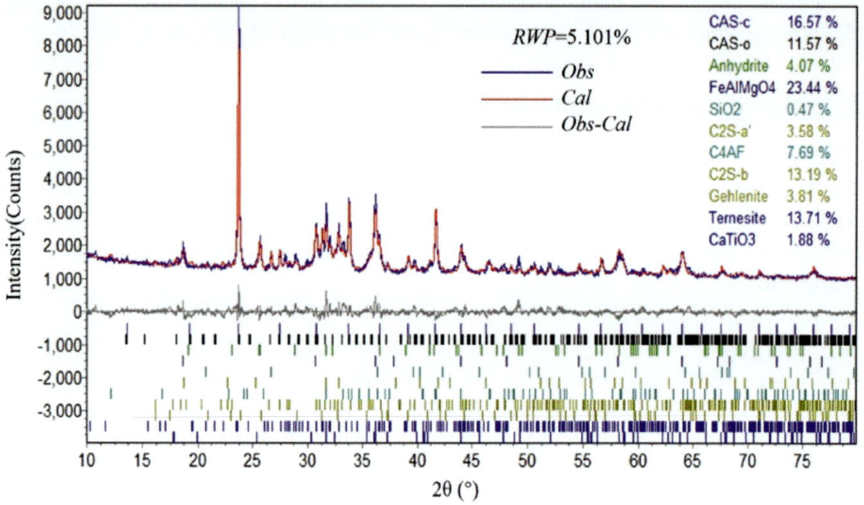

FIGURE 14.3 Quantitative Rietveld diagram of mineral phases in the clinker (Wang et al., 2021).

proportion of Al_2O_3 and higher proportions of Fe_2O_3 and SiO_2. As shown in Figs. 14.2 and 14.3, the main minerals of ESS-CM clinker are $2CaO \cdot SiO_2$ (C_2S), $3CaO \cdot 3Al_2O_3 \cdot CaSO_4$ ($C_4A_3\overline{S}$), $4CaO \cdot Al_2O_3 \cdot Fe_2O_3$ (C_4AF), $CaSO_4$, and $FeAlMgO_4$. After mixing the clinker with 5 wt.% gypsum, the 1-, 3-, and 28-day compressive strengths of ESS-CM at a water/cement ratio of 0.28 were 47.1, 53.2, and 69.0 MPa, respectively, with the early strength being comparable with the long-term strength of PC. In addition, the leaching concentrations of heavy metals in ESS-CM clinker were much lower than the standard limits, showing a good solidification capacity for heavy metals.

3.2 Blast furnace slag

Blast furnace slag can also be used as a source of Ca, Al, and Si to prepare CSA in conjunction with other solid wastes. Gao et al. (2020) used Bayer red mud, blast furnace slag, steel slag, desulphurization flue, and carbide slag as raw materials to prepare high-belite CSA, with blast furnace slag accounting for 24.14 wt.% in the raw materials. The main minerals in the clinker are 10 wt.% $C_4A_3\overline{S}$, 60 wt.% β-C_2S, 9 wt.% C_4AF, and 21 wt.% $C_{12}A_7$, respectively. As shown in Fig. 14.4, the optimal sintering temperature and holding time of the clinker are 1300°C and 60 min, respectively. With the extension of holding time, the Ca^{2+} in the mineral crystal will be replaced by impurity ions in solid waste (especially in red mud), causing disorder of the mineral crystal structure. It can be seen from Fig. 14.4, the optimal gypsum content for the preparation of cement is 20 wt.%. When the water/cement ratio is 0.5, the 1-, 3-, 7-, and 28-day compressive strengths of the cementitious material can reach 16.3, 19.0, 20.8, and 29.3 MPa, respectively, making it a viable rapid repair material.

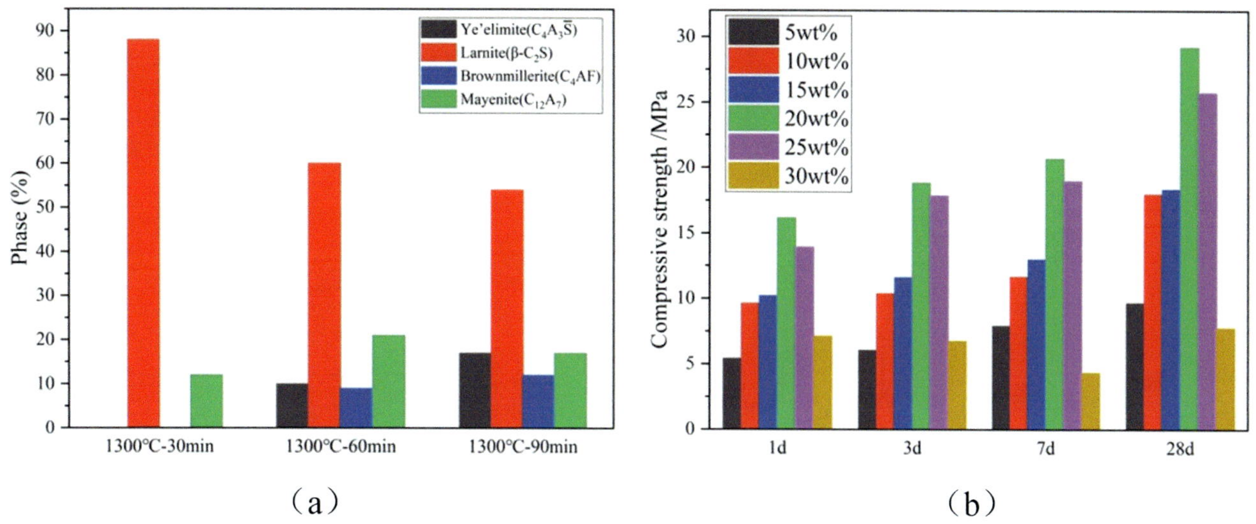

FIGURE 14.4 (A) The mineralogical compositions of high-belite CSA clinker; (B) compressive strength of high-belite CSA clinker with different gypsum contents (Gao et al., 2020).

3.3 Red mud

Red mud is the waste slag discharged during the refining of Al_2O_3 from bauxite, with a huge annual discharge. Preparing cement is an important direction for the use of red mud in the field of building materials. Due to its high content of Fe_2O_3, red mud can be used as a raw material for preparing ferric-rich CSA. Wang et al. (2013) used the dealkalized red mud to provide Si, Al, Fe, and most of the Ca, and in collaboration with limestone and bauxite to adjust the chemical composition of the raw material, sintering at 1250–1300°C for 30 min to obtain Fe-rich CSA. As shown in Fig. 14.5, the main minerals in the clinker include C_2S, $C_4A_3\overline{S}$, C_4AF, calcium aluminate, and a small amount of TiO_2. Significant incorporation of impurity elements was observed in all clinker phases. Desulphurization gypsum and red mud can account for 70–90 wt.% of the raw material. When the clinker is mixed with 5 wt.% gypsum and the water/cement ratio is 0.35, the 28-day compressive strength of the material reaches 61.3 MPa, showing excellent performance.

3.4 Aluminum ash

Aluminum ash is a waste slag discharged from the aluminum and aluminum products processing industry. It has a high content of effective aluminum components. Secondary aluminum ash has an Al_2O_3 content of 50 ~ 80 wt.%, and it has the

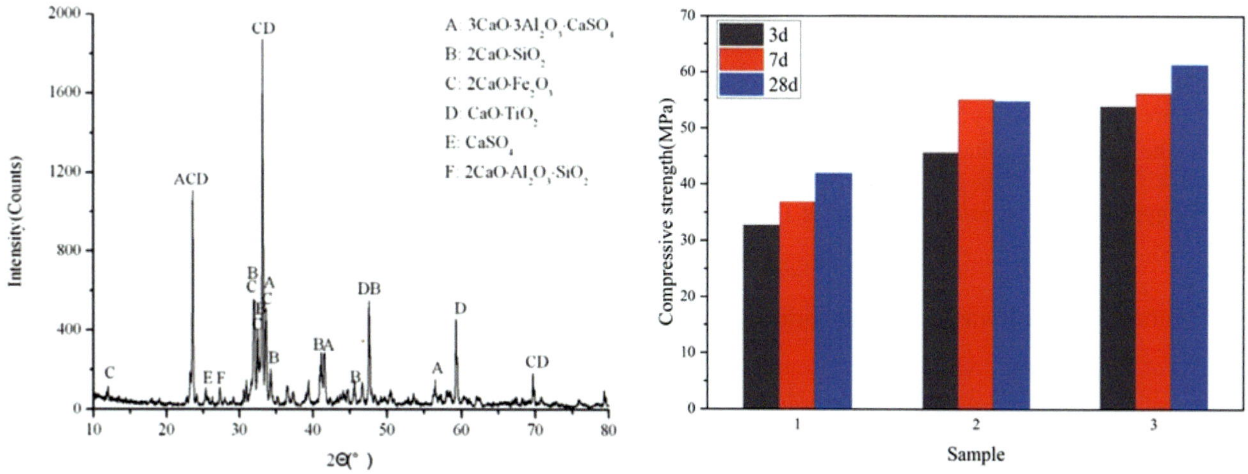

FIGURE 14.5 (A) XRD pattern of ferric-rich CSA clinker; (B) compressive strength results for clinkers (Wang et al., 2013).

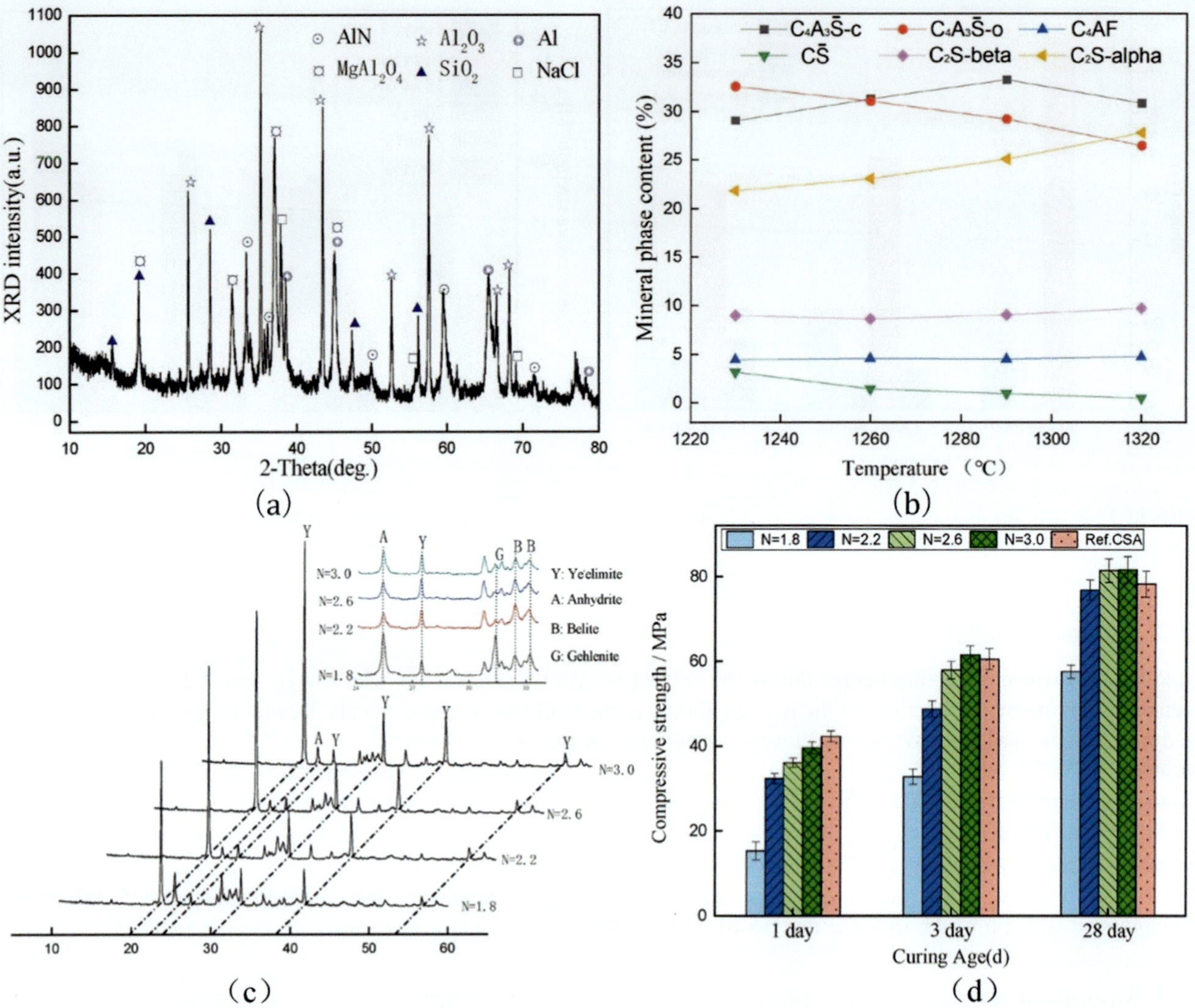

FIGURE 14.6 (A)XRD pattern of untreated secondary aluminum ash; (B) phase transition of the samples with Al/Si ratio was 3 at different calcination temperatures; (C) XRD patterns of clinker with different Al/Si ratio after calcination at 1290°C; (D) compressive strengths of samples with different Al/Si ratios (Ren et al., 2022).

potential to replace bauxite as Al source to prepare cementitious materials. Ren et al. (2022) used secondary aluminum ash with 71.64 wt.% Al_2O_3 as the Al source, together with coal gangue, desulphurization gypsum, and the mixture was maintained at 1230−1320°C for 60 min to obtain CSA clinker. As shown in Fig. 14.6A, the main mineral phases of the clinker are $C_4A_3\overline{S}$-c, $C_4A_3\overline{S}$-o, C_4AF, β-C_2S, α-C_2S, etc. With the increase of sintering temperature, the content of $C_4A_3\overline{S}$-o in the minerals gradually decreases, while the content of α-C_2S increases. It can be seen from Fig. 14.6B, the compressive strength of the material increases with the increase of Al/Si ratio. By mixing clinker calcined at 1290°C with 5 wt.% gypsum at a water/cement ratio of 0.28, the 28-day compressive strength of the cement paste reached about 80 MPa.

3.5 Municipal solid waste incineration fly ash

The use of MSWI FA to prepare green cementitious materials is in line with the concept of sustainable development (Guo et al., 2014; Clavier et al., 2020). The main chemical compositions of MSWI-FA are CaO, SiO_2, and chlorides, which can provide fundamental elements for the preparation of CSA. Mao et al. (2020) used pre-treated and dechlorinated MSWI-FA as the main raw material (up to 35 wt.%), combined with desulphurization gypsum and aluminum ash, and the mixture was sintered at 1240−1300°C for 30 min to obtain CSA clinker. Based on Fig. 14.7A, the main minerals in the clinker are

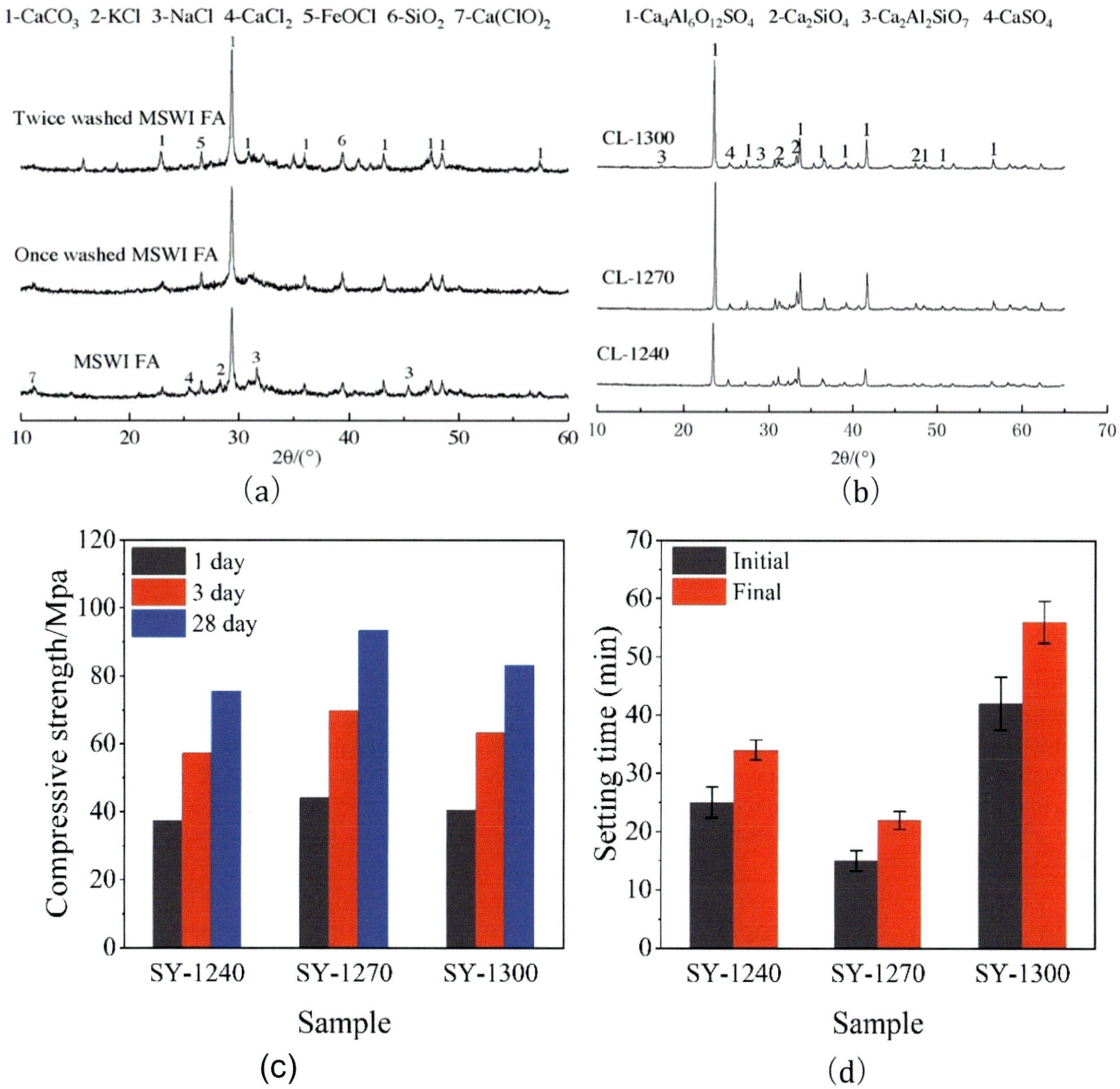

FIGURE 14.7 (A) XRD spectrum of unwashed MSWI-FA and washed MSWI-FA; (B) XRD spectrum of cementitious materials produced at different calcination temperatures; (C) compressive strength of MSWI-FA-based CSA; (D) setting time of MSWI-FA-based CSA (Mao et al., 2020).

$C_4A_3\bar{S}$, C_2S, and $2CaO \cdot Al_2O_3 \cdot SiO_2$ (C_2AS). The presence of elements such as Cr, P, and Cd in the raw materials can promote the formation of $C_4A_3\bar{S}$ at lower calcination temperatures. The cementitious material obtained by mixing the clinker with 5 wt.% gypsum has the shortest setting time when the sintering temperature is 1270°C, and its 28-day compressive strength reaches 93.4 MPa, as shown in Fig. 14.7B. In addition, the material has a good solidification effect on heavy metals.

3.6 Fly ash

Fly ash is released during the coal-burning process in power plants; its chemical compositions mainly include SiO_2 and Al_2O_3. Alumina-rich fly ash contains a high content of Al_2O_3, which can be used as a substitute for bauxite in the preparation of CSA, alleviating the pressure of bauxite resource shortage. Ma et al. (2013) cosynthesized high-belite CSA using alumina-rich fly ash, desulfurization gypsum, and limestone with calcination temperature of 1250°C and holding

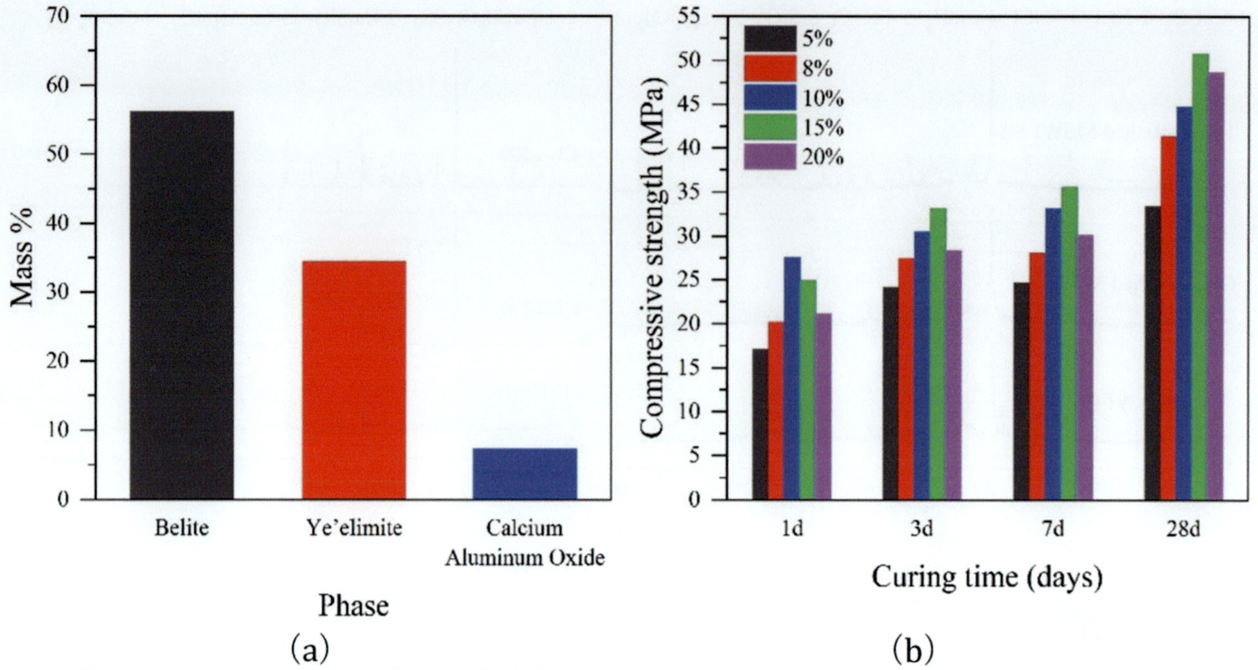

FIGURE 14.8 (A) Measured phase composition of the clinker; (B) compressive strength of fly ash–based CSA with varied gypsum.

time of 30 min. The clinker minerals contain 56.2 wt.% C_2S, 34.4 wt.% $C_4A_3\overline{S}$, and a small amount of $C_{12}A_7$. The f-CaO content in the clinker is very low, and neither increasing the sintering temperature nor extending the sintering time has a significant effect on the mineral content. Fly ash and desulfurization gypsum account for about 40 wt.% in the raw materials. According to Fig. 14.8, the optimal gypsum content is 15 wt.%. The 28-day compressive strength of the mortar reaches 50.8 MPa, with 90-day and 180-day strengths reaching 63.6 and 74.4 MPa, respectively, demonstrating a steady increase in long-term strength.

4. The impact of combustion/incineration waste used as supplementary cementitious materials on CSA

Currently, the exploration of carbon emission reduction paths in the cement industry is actively underway, aiming to achieve the sustainable development of the building materials industry. In the cement production process, methods to reduce its carbon footprint mainly include usage of green energy (Madlool et al., 2013), development of low-carbon cementitious materials (Maddalena et al., 2018), and adoption of supplementary cementitious materials (SCMs) (Juenger et al., 2011). Generally, SCMs are soluble siliceous, aluminosiliceous, or calcium aluminosiliceous powders used as partial replacements of clinker in cements or as partial replacements of PC in concrete mixtures. It can react with hydroxyl ions in concrete to produce secondary pozzolanic reactions. This practice is favorable to the industry, generally resulting in concrete with lower cost, lower environmental impact, higher long-term strength, and improved long-term durability. Thus, using SCMs to reduce the intensity of clinker in cement is considered as a very effective method in the cement industry's carbon-reduction path.

SCMs need to satisfy two characteristics: pozzolanic activity (the property of reacting with the alkaline hydration products of cement paste to produce gelling products) and gelation (reacting directly with water to produce gelling products) (Juenger and Siddique, 2015). These two characteristics depend on the contents of silicates and aluminosilicates in SCMs. Therefore, not all raw materials can be used as SCMs. It should be noted that aluminum ash, which is mainly composed of alumina, magnesium aluminum spinel, and aluminum nitride, is not suitable as SCMS because it does not have potential volcanic gray activity and gelation. In addition, MSWI-FA consists of calcium, aluminum, and silicon, making it a potential material for SCM. However, due to the high content of heavy metals and chloride, it cannot be directly used as SCMs. Preprocessing of MSWI-FA may cause the losses of volcanic ash activity (Zhang et al., 2021). Therefore, there are few literature explaining their impacts on cement. Currently, commonly used SCMs can be divided into two categories: natural materials, including volcanic ash, kaolinite, etc., and industrial by-products, including steel slag, blast furnace slag, fly ash, carbonized red mud, etc. With the scarcity of natural materials, industrial by-products have gradually become the main source of SCMs. Although there has been considerable literature exploring the compatibility of SCMs with cementitious materials, most are for PC + SCM systems, and exploration of CSA + SCM systems is relatively scarce. With the addition of solid wastes, the workability, strength, hydration heat, durability, and other macroproperties of CSA will inevitably be significantly affected. Therefore, the following part describes the impact of several types of combustion/incineration waste on the macroproperties of CSA.

4.1 Steel slag

The main mineral components of steel slag are C_3S and C_2S, which have certain hydration activity. The fineness of steel slag is an important indicator of steel slag activity. Different fineness of steel slag has different effects on cementitious materials. On the one hand, the hydration activity of steel slag powder varies with different fineness. On the other hand, the changes in the microstructures of the paste caused by the filling effect of the steel slag powder are also different, which will have different impacts on the performance of cement. When ultrafine steel slag is adopted, the setting time of the past increases, the hydration heat increases, and the compressive strength decreases. It should also be noted that ultrafine steel slag powder has a good physical filling effect, which can refine the microstructure of the past, and to a certain extent make up for the inhibitory effect of cement hydration (Liu and Li, 2014; Pang et al., 2022). It generally can be divided into ordinary steel slag powder (specific surface area 300 ~ 500) and steel slag micropowder (specific surface area above 700). When ordinary steel slag powder is coblended with CSA, as the amount of steel slag increases, the setting time of CSA gradually extends, as shown in Fig. 14.9A. This is because when the amount of steel slag is small, the "dilution effect" of steel slag delays the setting time of the slurry (Liao et al., 2020). However, the addition of steel slag micropowder shows the opposite trend, as the setting time gradually shortens with the increase of steel slag micropowder, as shown in Fig. 14.9B. This is because the steel slag has a very small particle size after ultramicropowder grinding, so the f-CaO in it can quickly react with water to generate $Ca(OH)_2$, which quickly raises the pH of the pore solution. The rise in pH will cause CSA to set quickly, resulting in "flash set" of the material (Padilla-Encinas et al., 2021).

Since the activity of C_3S and C_2S in ordinary steel slag powder is low, the early compressive strength of CSA blended with ordinary steel slag powder decreases as the dosage increases. In addition, the decrease in CSA reduces the amount of $3CaO \cdot Al_2O_3 \cdot 3CaSO_4 \cdot 32H_2O$ (AFt) generated in the paste, which is also the main reason for the decrease of strength. When the material reaches the later stage of hydration, the impact of steel slag on strength significantly increases. When the

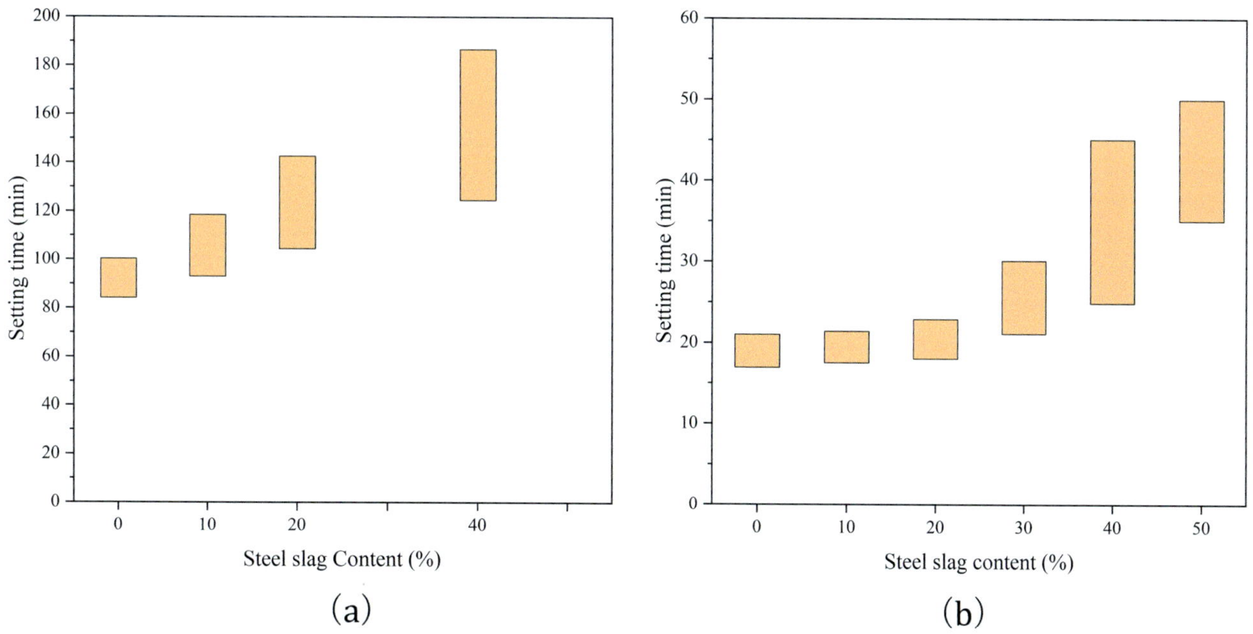

FIGURE 14.9 Setting time of CSA pastes blended with steel slag, (A) ordinary steel slag powder (Liao et al., 2020), (B) steel slag micropowder (Yu, 2022).

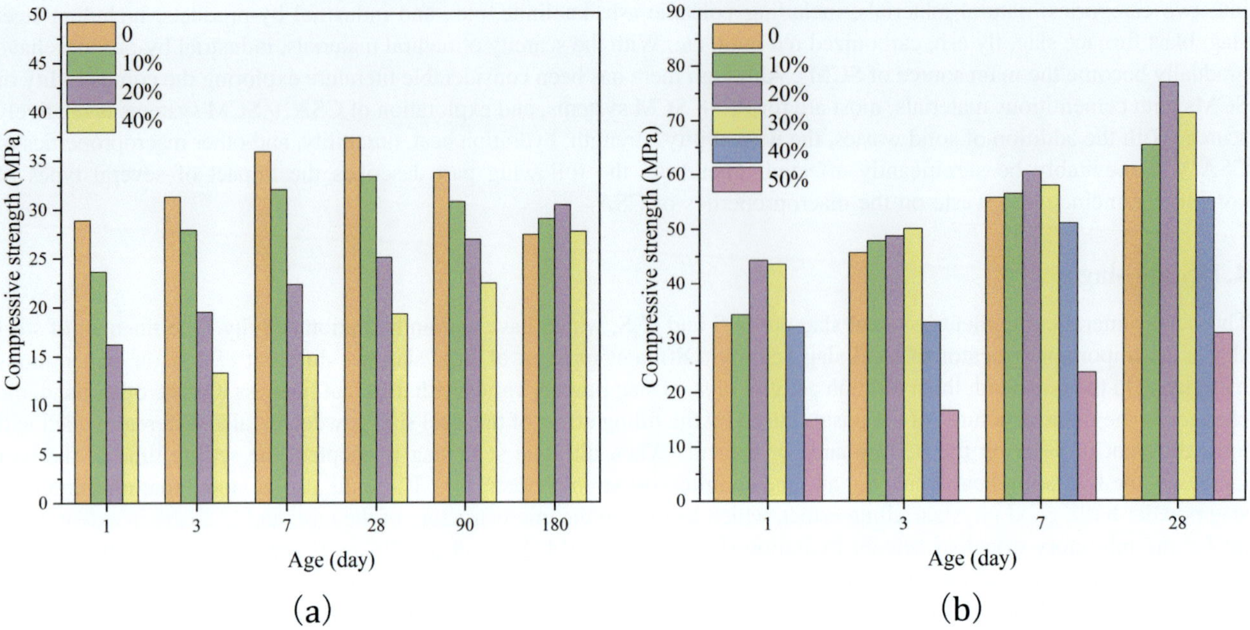

FIGURE 14.10 Compressive strength of CSA pastes blended with steel slag, (A) ordinary steel slag powder (Liao et al., 2020), (B) steel slag micropowder (Yu, 2022).

ordinary steel slag powder content does not exceed 20 wt.%, the hydration products of steel slag cannot compensate for the strength loss caused by the late decomposition and transformation of AFt, resulting in a decrease in the strength of the paste (Liao et al., 2020). When the steel slag content is 20∼40 wt.%, however, the hydration products of steel slag are enough to offset the effect of AFt decomposition, thereby enhancing the compressive strength of the cement paste. The 28-day compressive strength has increased by about 5−10 MPa, as shown in Fig. 14.10A. Unlike ordinary steel slag powder, the impact of steel slag micropowder on strength is characterized by an increase followed by a decrease as the mixing amount increases. The compressive strength peaks when the dosing is 20 wt.%., as shown in Fig. 14.10B. This is because the source of CSA's early hydration strength is mainly the skeleton structure formed by AFt lapping. Although the addition of steel slag micropowder can promote CSA hydration to form AFt, when the amount of steel slag micropowder is too large, the total amount of CSA in the material is reduced, which leads to a decrease in the final amount of AFt generated (Yu, 2022).

However, since steel slag contains a large amount of free CaO and MgO, its poor stability due to water expansion has greatly limited the application of steel slag in this area. After steel slag is added, the antichloride ion permeability of the slurry worsens, but this can be improved to some extent by increasing the specific surface area of the steel slag. In addition, the sulfate resistance of the slurry also shows a similar pattern. Some studies have shown that when steel slag is coblended with blast furnace slag, the durability of CSA can be significantly improved. The early hydration of steel slag would increase the pH and conductivity of the pore solution, thus promoting the dissolution of aluminosilicate glass in GBFS and pozzolanic ash reaction. At the same time, the consumption of $Ca(OH)_2$ by the pozzolanic ash reaction could promote the further hydration of steel slag. As a result, the synergistic hydration effect would promote hydration and generate more hydration products to fill the pore structure, and the material shows good durability (Zhao et al., 2023).

4.2 Blast furnace slag

Blast furnace slag is also widely used as an SCM. As the amount of blast furnace slag increases, the setting time of CSA gradually increases, accompanied by a decrease in fluidity, as shown in Fig. 14.11 (Gao et al., 2019). This is mainly because the water demand of blast furnace slag is larger than that of CSA, resulting in an increase in the setting time of the slurry under a certain water/cement ratio.

Unlike steel slag, changing the particle size of blast furnace slag does not significantly affect the strength of the cement paste. As shown in Fig. 14.12A, as the amount of blast furnace slag increases, the flexural strength of CSA gradually decreases, and the late flexural strength decreases with the extension of the hydration age. At the same time, the

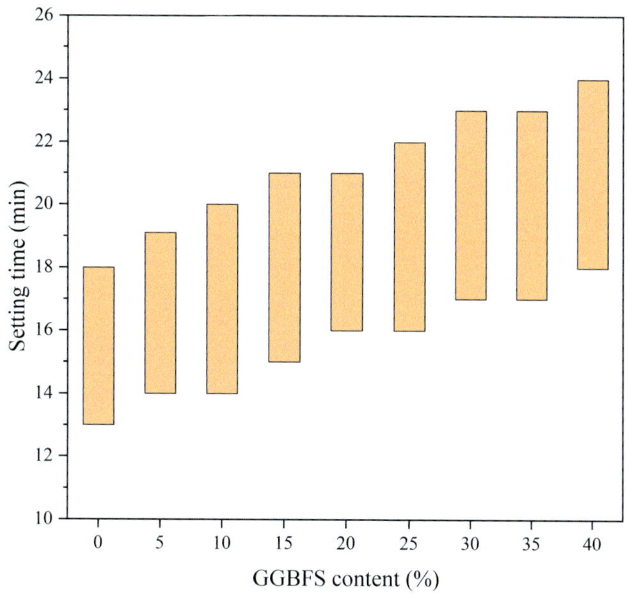

FIGURE 14.11 Setting time of CSA pastes blended with blast furnace slag (Gao et al., 2019).

compressive strength also shows a similar trend, especially when the content of blast furnace slag exceeds 10 wt.%, as shown in Fig. 14.12B. This is mainly because CSA cannot provide a large amount of $Ca(OH)_2$ needed by blast furnace slag, and its dilution effect is stronger than the pozzolanic effect, resulting in the loss of strength (Yoon et al., 2021).

Although the help to the strength of CSA is not obvious, blast furnace slag can significantly improve the erosion resistance of CSA. Fig. 14.13 shows the microscopic morphology of CSA paste containing blast furnace slag after soaking in sodium sulfate. Before sulfate erosion, as shown in Fig. 14.13A, a large number of aggregated needle-like AFt are surrounded by gel; in addition, petal-shaped AFm is also produced. After soaking for 30 days, numerous flocculants and foil-like products can be found mixed together, as shown in Fig. 14.13B; they are C—S—H and AH_3 gel. When the soaking time reaches 360 days, a plenty of well-developed columnar AFt and other hydration products are densely interwoven, and the microstructure is dense, as shown in Fig. 14.13C. The microstructure of the hardened cement pastes after soaking in Na_2SO_4 solution for 360 days provides direct evidence for the improvement of macroproperties. At the same time, the sample shows good dimensional stability after sulfate erosion, and the content of AFt and C—S—H gel increases, while the total porosity decreases (Gao et al., 2022).

4.3 Red mud

Red mud is less used as SCMs, but research shows that carbonized red mud shows potential as an CSA admixture. Sintered red mud contains a large amount of C_2S, which can react with CO_2 to form $CaCO_3$ crystals and highly polymerized SiO_2 gel. It can bind soluble Na^+through physical surface adsorption and chemical bonding (Fang et al., 2021). In addition, $CaCO_3$ can react with the aluminum phase to form hydration products, such as monocarbante, further enhancing CSA performance. As the amount of carbonized red mud increases, the fluidity of the slurry decreases, which is related to the calcite produced by carbonation. Since the particle size of calcite is smaller than that of CSA particles, calcite fills the fine pores between CSA particles, increasing the friction between solid particles, leading to a significant decrease in fluidity, as shown in Fig. 14.14. At the same time, with the increase of carbonized red mud, the setting time of the paste is drastically shortened, accompanied by an acceleration of hydration heat release, which can be explained by the fact that calcite, due to its nucleation role, absorbs the precipitate produced by CSA hydration, accelerating the hydration reaction (Liu et al., 2023).

As shown in Fig. 14.15, the addition of carbonized red mud increases the compressive strength of the paste. As the content increases, the compressive strength of the CSA slurry first increases and then decreases. This may be because when the addition amount is small, the calcite particles produced by carbonation provide nucleation sites for the hydration products, accelerating the hydration reaction rate of the cement clinker and increasing its early strength (Liu et al., 2023); too much red mud will produce a dilution effect, causing a reduction in hydration products, increasing the porosity of the hardened paste, and resulting in a decrease in compressive strength.

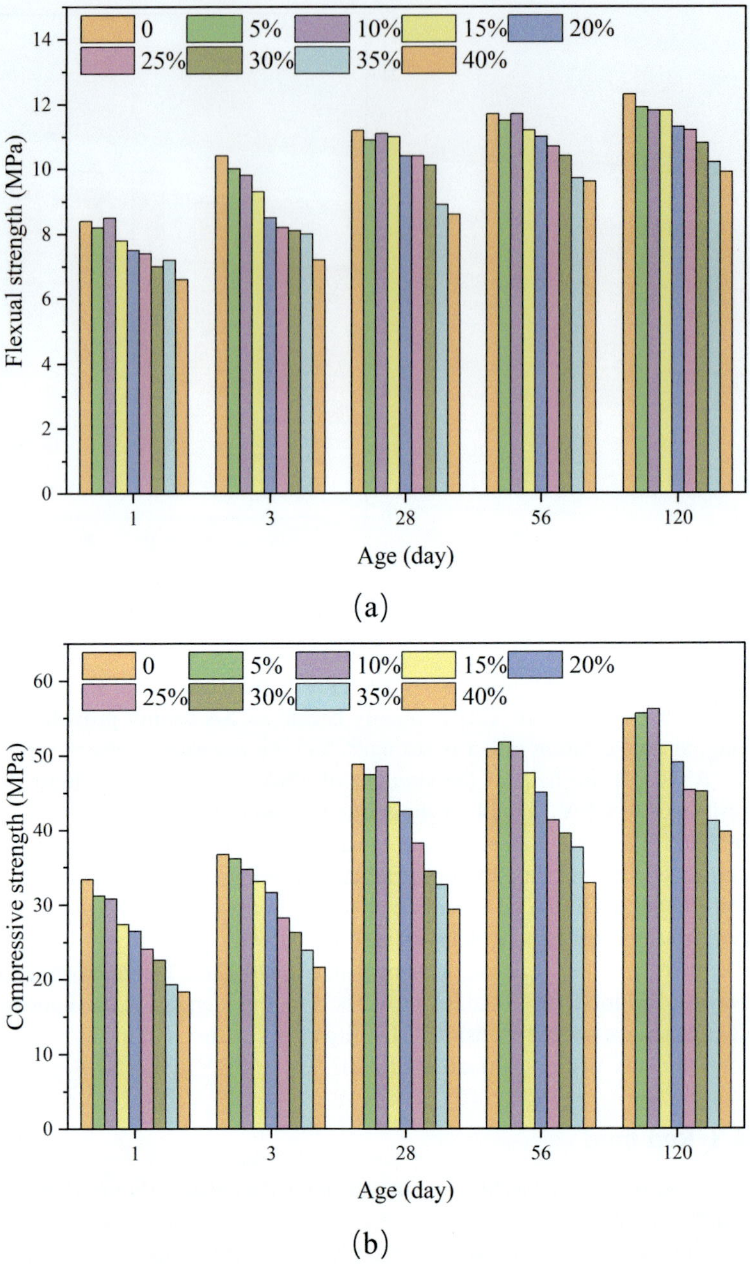

FIGURE 14.12 Effect of blast furnace slag content on the strength of mortar (Gao et al., 2019).

4.4 Fly ash

Fly ash, as a mature SCM, is widely used in PC. However, when fly ash is mixed with CSA, its reactivity is greatly reduced, with a maximum reaction degree of only 20%. This is because its hydration mainly relies on its pozzolanic activity in cementitious materials. However, due to the slow hydration of C_2S in CSA, the hydration system has a low alkalinity and cannot fully exploit the pozzolanic activity of fly ash. Studies have shown that mixing fly ash with CSA activated by C_2S can increase the activity of the system. The activated C_2S promotes the reaction of the fly ash with $Ca(OH)_2$, producing calcium aluminate and C—S—H gel, thus improving the performance of the paste. Moreover, because fly ash has a small particle size, it can provide a ball effect in the paste, thereby increasing the its fluidity. After fly ash is added, the setting time of the cement paste is shortened, as shown in Table 14.1.

As shown in Fig. 14.16, after the mixing of CSA with fly ash, the compressive strength of the cementitious materials decreases, and the larger the amount of fly ash is mixed, the more the specimen's compressive strength decreases

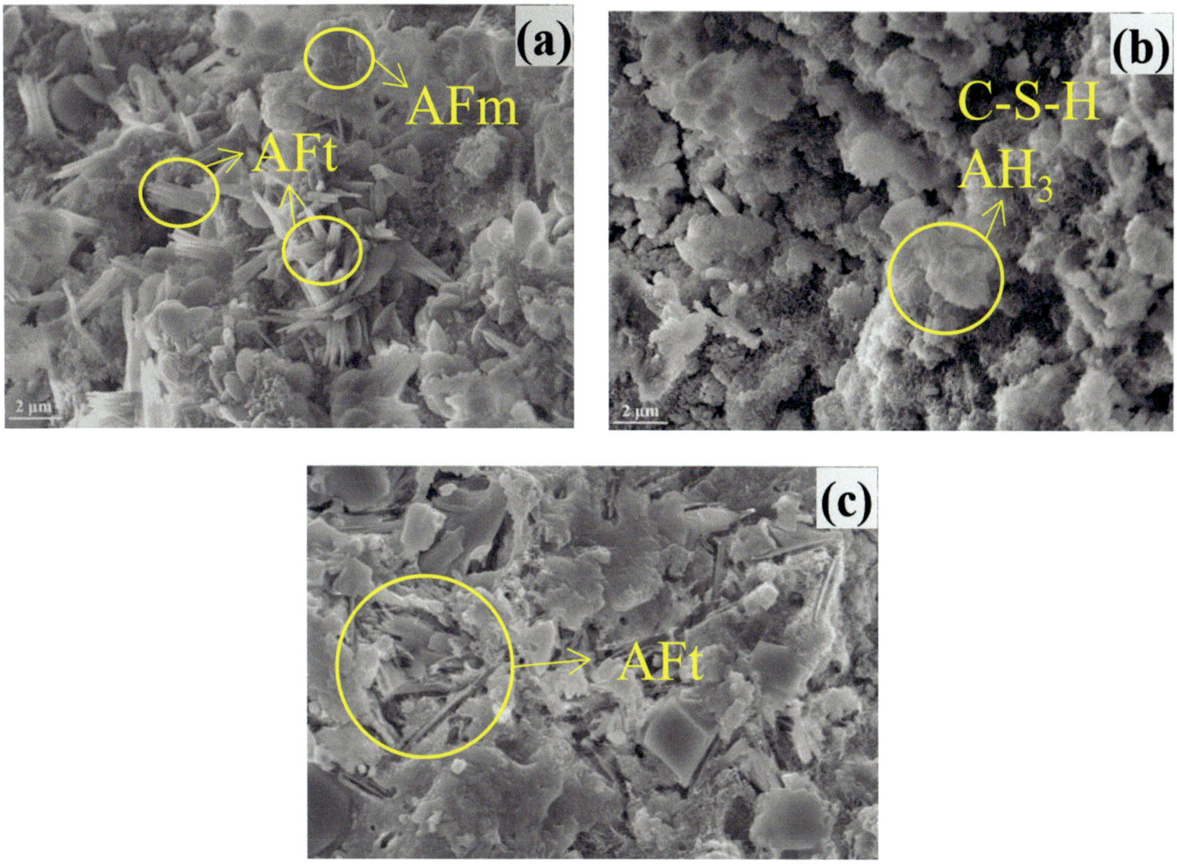

FIGURE 14.13 SEM images of the hardened cement paste with 20 wt.% blast furnace slag exposed to 5% Na_2SO_4 solution for different time: (A) 0 day, (B) 30 days, and (C) 360 days (Gao et al., 2022).

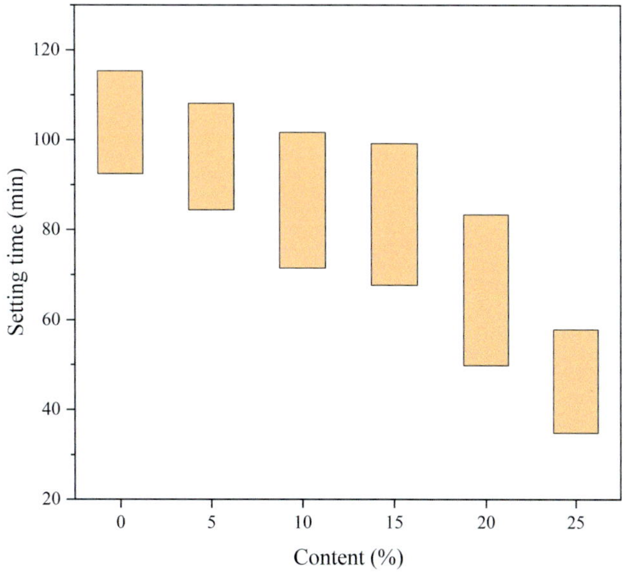

FIGURE 14.14 Setting time of CSA clinker pastes blended with carbonized red mud (Liu et al., 2023).

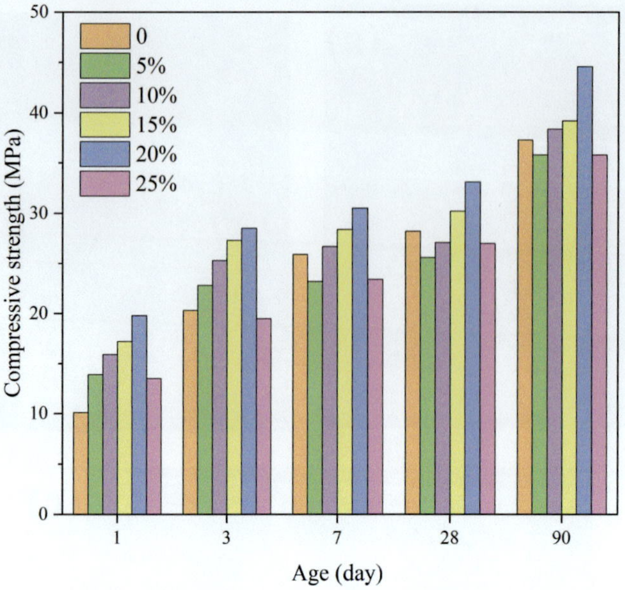

FIGURE 14.15 Compressive strength development of CSA pastes blended with carbonized red mud (Liu et al., 2023).

TABLE 14.1 Setting time of CSA blend with different dosages of fly ash (Jiang et al., 2016).

FA content	Initial setting time/min	Final setting time/min
0	279	349
20%	255	314
40%	203	255

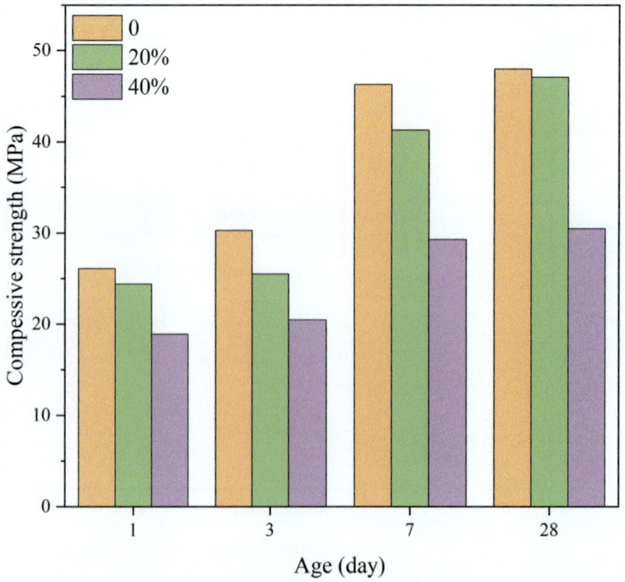

FIGURE 14.16 The compressive strength and variance of cement pastes with different contents of fly ash (Jiang et al., 2016).

(Jiang et al., 2016). However, some researches show that a small amount of fly ash (no more than 10 wt.%) will improve the strength of CSA, but the increase is limited (García-Maté et al., 2013). As the amount of fly ash increases, the chemical shrinkage of the cement paste decreases, and the early strength of the cement will also decrease significantly, which may be due to the dilution effect and low activity of fly ash, making it difficult to play an active role in the low alkalinity hydration environment of CSA.

Currently, various combustion/incineration residues, such as steel slag powder, mineral powder, etc., can show relatively good effects, but their addition is still limited in many ways. Compared with traditional SCMs, due to the low alkalinity of CSA, it limits the pozzolanic activity; therefore, there is still a need to further develop SCMs suitable for the CSA mineral system.

5. Impact of combustion/incineration residue on the hydration characteristics of cement clinker

Similar to traditional CSA clinker, the main hydration products of CSA prepared from combustion/incineration residues are also AFt, AFm, $Al(OH)_3$, $Fe(OH)_3$ gel, etc. However, due to the complex elemental composition of combustion/incineration waste, preparing cement clinker from them will inevitably introduce a large amount of multicomponent elements such as Fe, Na, K, F, and Cl into the cement system. The introduction of Fe leads to the appearance of a large number of iron phases and $C_4A_{3-x}F_x\overline{S}$ solid solutions in the clinker. These Fe-containing minerals are often considered to have higher hydration activities (Cuesta et al., 2014, 2015), and the discovery of Fe-containing hydration products such as Fe-AFt (Dilnesa et al., 2014) also shows that the introduction of Fe changes the hydration mineral system of the material. In addition, to solve the problem of harmful soluble element leaching and cement kiln skinning during the waste preparation of CSA, some research has proposed cothermal treatment methods of various wastes to reduce the retention rates of Na, K, and Cl in the raw materials and to reduce the leaching of harmful element below the detection limit, as shown in Table 14.2 (Li et al., 2023a). However, even though most impurities can be removed during the calcination process or sealed in minerals, the residual impurities in the waste will still affect the hydration characteristics of cement clinker in the form of changing the pH value of the pore solution in the slurry. Therefore, this section discusses the impact on the hydration characteristics of cement clinker when using various combustion/incineration residues as raw materials.

Red mud, aluminum ash, and MSWI-FA often contain a large amount of alkali metals Na and K, which have volatile characteristics and usually volatilize partially in the form of NaCl and KCl during high-temperature calcination. The residual Na and K are dissolved in the clinker minerals and act as activators during cement hydration, promoting the hydration of clinker minerals. Therefore, cement clinker prepared from these wastes often has higher hydration activity. Fig. 14.17 (Canbek et al., 2020) shows the heat release curve of red mud–based CSA hydration. It can be seen that except for the wetting heat release peak of cement slurry, the hydration acceleration period of red mud–based CSA is advanced, and this effect becomes more apparent as the amount of red mud increases. When the amount of red mud increases to 20 wt.%, the hydration acceleration period of the material is advanced to 0.25 h, which corresponds to the rapid reaction of $C_4A_3\overline{S}$, as shown in Eq. (14.1). The hydration acceleration period of CSA prepared from natural minerals appears after 2 h, and its peak value is far lower than that of red mud–based CSA. This is related to the higher alkali metal content in red mud. During the hydration process, the alkali metal ions Na^+ and K^+ dissolved in the clinker minerals are released into the pore solution and increase the pH value of the pore solution, thereby causing the dissolution rate of the clinker minerals to increase, accelerating the early hydration of the cement slurry. This phenomenon also caused the split of the $C_4A_3\overline{S}$ hydration heat release peak, and the peak value of the first split heat release peak gradually increased with the increase in

TABLE 14.2 Leaching concentrations of F and Na from aluminate dross, incineration fly ash, and mixed sample before and after thermal treatment (mg/L) (Li et al., 2023a).

	Aluminate dross	Treated aluminate dross	Incineration fly ash	Treated incineration fly ash	Treated mixed sample	Theoretical value of mixed sample
F	237	1250	N.D.	N.D.	N.D.	625
Na	1286	2764	948	287	270.6	1525

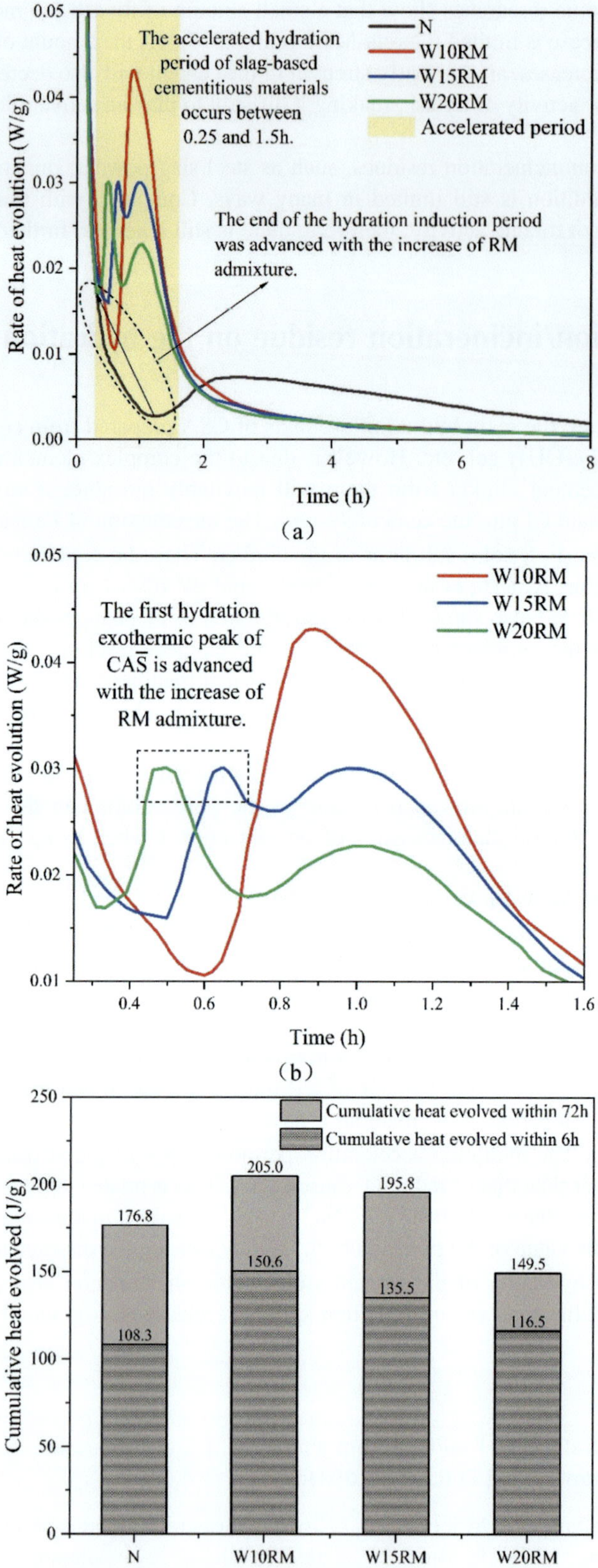

FIGURE 14.17 (A) Heat flow curve, (B) heat flow curve during hydration acceleration period, and (C) accumulated heat release of cement slurry. The markings in the figure represent: N = CSA slurry prepared from natural minerals, W10/15/20RM = CSA slurry prepared by replacing 10/15/20 wt.% bauxite with red mud (Canbek et al., 2020).

the amount of red mud. In addition, the accelerating effect of alkali metal ions Na^+ and K^+ on the early hydration of the material also makes the hydration heat release of red mud–based CSA in the first 6 h significantly higher than that of CSA prepared from natural minerals. When the red mud content is 10 wt.%, the heat release of the material in the first 6 h of hydration is increased by nearly 40%. However, as hydration continues, the hydration reaction of red mud–based CSA significantly slows down. After 72 h of hydration, the heat release of CSA prepared from natural minerals is very close to, or even exceeds, that of sample W20RM. On the one hand, due to the excessive impurities in red mud, the content of $C_4A_3\overline{S}$ in red mud–based CSA is slightly lower than that of CSA prepared from natural minerals; on the other hand, due to the rapid early hydration of red mud–based CSA, an AFt protective layer is formed on the unhydrated $C_4A_3\overline{S}$ particles, inhibiting further hydration of $C_4A_3\overline{S}$.

$$\begin{aligned} &3CaO\cdot Al_2O_3\cdot CaSO_4 + 2(CaSO_4\cdot 2H_2O) \\ &+34H_2O \rightarrow AFt + Al_2O_3\cdot 3H_2O(gel) \end{aligned} \tag{14.1}$$

The XRD spectra of different cement slurries also show similar rules, as shown in Fig. 14.18. Due to the influence of alkali metal ions dissolved in clinker minerals, the two types of cement made from red mud show the most excellent hydration performance, and $C_4A_3\overline{S}$ has basically reacted completely within 3 days of hydration. The early hydration rate of cement prepared with 45 wt.% fly ash as a raw material is similar to that of cement prepared from natural minerals. However, during the hydration period from 3 to 28 days, the diffraction peak intensity of AFt in the CSA slurry prepared

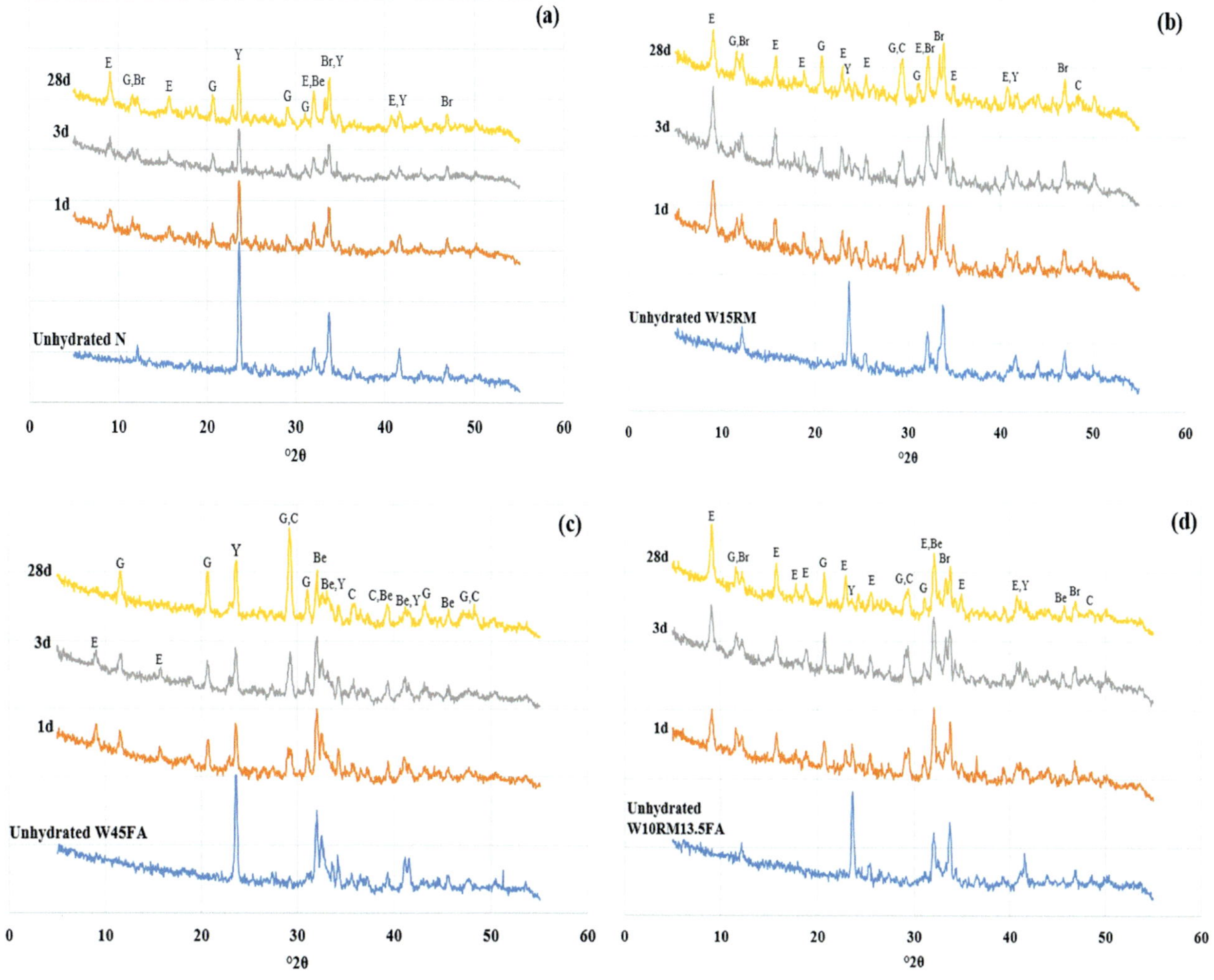

FIGURE 14.18 XRD spectra of CSA slurries prepared by hydration for 1 day, 3 days, 28 days, and (A) natural minerals, (B) 15 wt.% red mud replacing bauxite, (C) 45 wt.% fly ash replacing bauxite and gypsum, and (D) 10 and 13.5 wt.% fly ash replacing bauxite and gypsum as raw materials. The markings in the figure represent: E = AFt G = dihydrate gypsum Y = $C_4A_3\overline{S}$ Br = C_4AF Be = C_2S C = $CaCO_3$ (Canbek et al., 2020).

solely with red mud and fly ash rapidly decreases, accompanied by the appearance of $CaCO_3$ diffraction peaks. This indicates that the carbonation reaction of the CSA slurry prepared from a single type of solid waste is promoted, and the carbonation resistance of the hydration product AFt is reduced. However, the degree of carbonation of cement made by jointly using fly ash and red mud is lower than that of cement made from a single type of waste. The aforementioned phenomenon shows that CSA prepared from red mud and fly ash has a similar or even faster early hydration rate compared with CSA prepared from natural ore, and the problem of reduced carbonation resistance of hydration products can be alleviated by the synergistic use of the two wastes.

In addition to containing a large amount of alkali metals Na and K, incineration residues such as MSWI-FA, alumina ash, and fly ash usually also contain Cl^-. Although most of the Cl^- in the raw materials can be removed through pre-treatment methods such as heat treatment, the release of residual Cl^- in the cement clinker during hydration is still considered detrimental to the development of concrete strength in practical engineering. It is generally believed that Cl^- in the pore solution will undergo intense electrochemical reactions with steel bars, causing rebar corrosion (Loser et al., 2010; Ji et al., 2021), leading to concrete cracking, and seriously affecting the service life. However, some studies show that Cl^- can change the system of hydration products in CSA slurry and be fixed in some hydration products, which can avoid electrochemical reactions with steel bars and make the material show better performance. Yao (2020) believes that due to the lower pH of the pore solution of the solid waste-based CSA slurry, the C—S—H generated by C_2S hydration will react with Cl^- in a liquid phase to generate a small amount of slightly expansive $3CaO \cdot CaCl_2 \cdot 15H_2O$, which plays a role in filling the pores in the mortar specimens and improving the cement—sand interface. This makes the interior of the mortar specimens denser, thereby improving the mechanical properties of the material. In addition, when the sulfate in the pore solution during hydration is exhausted, Cl^- can combine with the aluminate phase in CSA to form Friedel's salt ($Ca_2Al(OH)_6(Cl,OH) \cdot 2H_2O$) (Shi et al., 2017).

The Fe content in steel slag is relatively high, and the Fe content of cement clinker prepared from steel slag will inevitably be higher than that of ordinary CSA, so the whereabouts of Fe in steel slag is a key factor affecting the hydration characteristics of cement clinker. Fig. 14.19 shows the heat of hydration curves of steel slag—based cement clinker, traditional CSA clinker, and PC clinker. Steel slag—based cement clinker has higher heat of hydration and faster early hydration heat, and its 72-h hydration heat reaches 396 J/g, which is much higher than traditional CSA and PC. It is generally believed that in the process of preparing cement clinker from steel slag, some of the Fe in the steel slag will replace Al^{3+} in the $C_4A_3\overline{S}$ crystal in the form of Fe^{3+}, generating $C_4A_{3-x}F_x\overline{S}$ solid solution, which exists stably as a cubic crystal at room temperature. This solid solution has a faster early hydration rate than

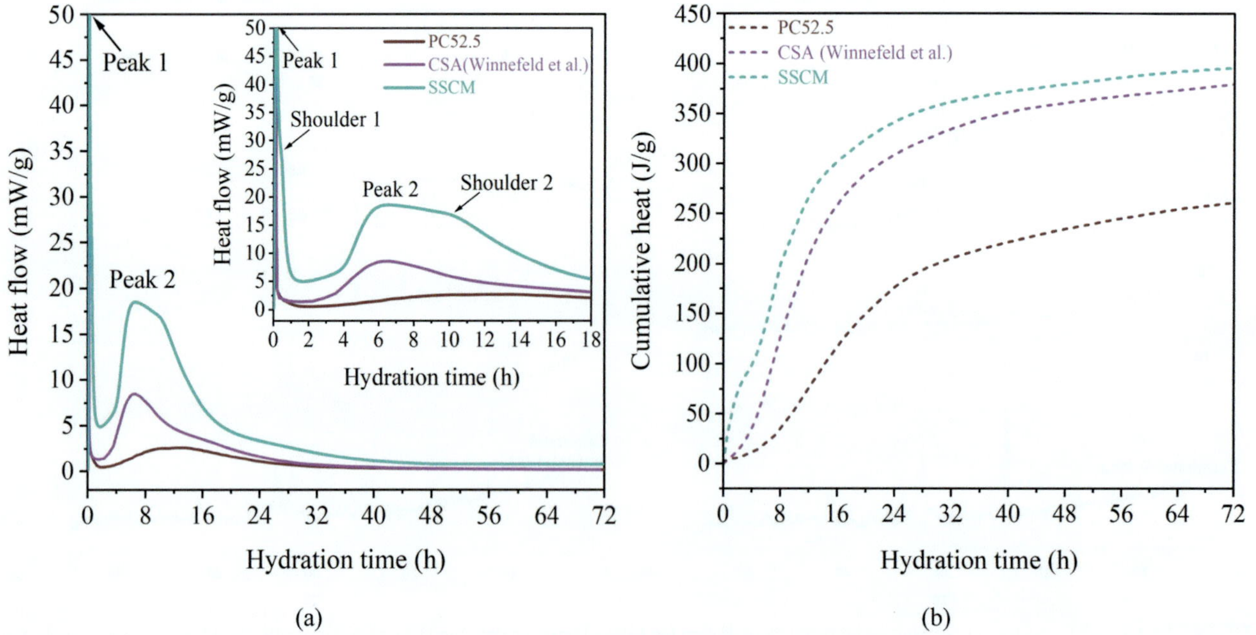

FIGURE 14.19 Hydration heat release curves of steel slag—based CSA clinker (SSCM), traditional CSA (CSA), and PC (PC52.5) (Winnefeld and Lothenbach, 2010; Wang et al., 2023b).

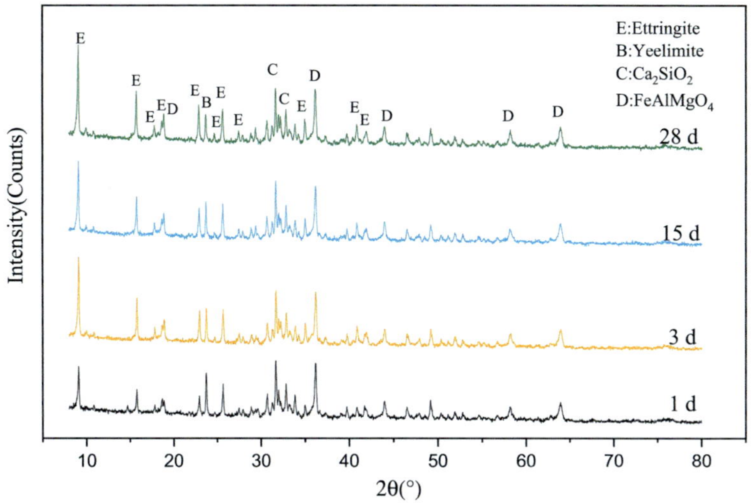

FIGURE 14.20 XRD patterns of cement pastes prepared from steel slag and aluminum ash at different hydration times (Wang et al., 2021).

orthorhombic $C_4A_3\overline{S}$. Therefore, cement clinker prepared from steel slag has a higher early hydration heat and a faster early hydration rate.

High-iron CSA clinker prepared from steel slag and alumina ash, in conjunction with desulfurization gypsum and calcium carbide slag, also exhibits good hydration characteristics. Benefit from the solid solution of Fe^{3+} in $C_4A_3\overline{S}$ crystals, as well as the promotion of hydration by alkali metals Na and K from alumina ash, the $C_4A_3\overline{S}$ in the clinker, has basically completed its reaction by 3 days of hydration. The hydration product AFt has good stability, and no carbonation phenomenon was observed at 28 days of hydration, as shown in Fig. 14.20.

The aforementioned phenomena indicate that although the hydration products of waste slag-based cement are very similar to traditional CSA, they differ due to the unique chemical composition of the waste slag and the influence of multiple components such as Na and K. Steel slag is often used to prepare high-iron cement, and its Fe^{3+} can be solidly dissolved into $C_4A_3\overline{S}$ crystals, while a large amount of alkali metals Na and K in red mud, fly ash, and alumina ash are beneficial to increase the dissolution rate of cement clinker mineral phases. Therefore, the cement prepared with these waste slags has a higher early hydration rate. Cl^- in waste slag such as MSWI-FA and aluminum ash changes the system of hydration products of the cementitious material, generating $3CaO\cdot CaCl_2\cdot 15H_2O$ and Friedel's salt. Cement clinker prepared with multiple waste slags often has excellent hydration characteristics, and AFt produced after hydration of cement clinker prepared by fly ash and red mud has higher stability.

6. Impact of heavy metals in combustion/incineration residues on CSA

At present, in the relevant applications of combustion/incineration residues in CSA cement materials, special attention should also be given to the impact of heavy metals in the residues on CSA. Since the immobilization of heavy metals will occur in both CSA clinker minerals and hydration products, research on CSA's immobilization of heavy metals is mainly focused on the two stages of calcination and hydration. Therefore, this section will discuss the introduction of residues from both the raw material end and the clinker end.

6.1 Introducing residue containing heavy metals from the raw material end

When using combustion/incineration residues as raw materials to prepare CSA, a large amount of heavy metals in the residues (such as Hg in fly ash, Cr and Mn in steel slag, Cr, As, Pb and Mn in red mud, and Zn and Cu in sludge) participate in the reaction of mineral formation at high temperatures and are well solidified in the clinker. There are two ways of immobilizing heavy metals in the clinker: one is to replace the elements such as Ca, Al, Fe, and S in the clinker minerals at high temperatures, and the other is to interstitial solid solution, such as $C_4A_3\overline{S}$, C_4AF, etc., which provide excellent space for heavy metals. Research shows that Zn and Cu can replace Ca to enter Ca-containing minerals and can also enter C_4AF by replacing Fe atoms (Tao et al., 2018; Zhu et al., 2021). Mn and Co ions can enter the iron phase (Kolovos et al., 2002). Cr can not only replace S to enter $C_4A_3\overline{S}$, but also replace Fe to enter Fe-containing phases (Sun et al., 2021). Cd and Pb are immobilized by replacing the Ca ions in the clinker and interstitial solid solution (Zhu et al., 2020). Ni will be solidified in C_4AF by replacing Fe or interstitial solid solution in C_4AF (Yang et al., 2014). The order of Ti substitution in the clinker is: Fe > Al > Ca > Si. V will enter C_2S in the clinker (Andrade et al., 2003).

At the same time, during the high-temperature calcination process, since heavy metals will be solidified in the formation stage of mineral phases, it will have a certain impact on the formation of mineral phases. Benarchid and Rogez (2005) proved through experiments that compared with pure substance preparation, there are more $C_4A_3\overline{S}$ crystals in samples doped with Cr_2O_3 and P_2O_5, indicating that Cr_2O_3 and P_2O_5 can promote the formation of $C_4A_3\overline{S}$ at 1100−1250s °C. Ma et al. (2006) added 1.0% CuO to the raw material; the results showed that CuO adding can not only reduce the mineral formation temperature of $C_4A_3\overline{S}$ in the clinker but also reduce its decomposition temperature. An appropriate amount of CuO can promote the formation of $C_4A_3\overline{S}$ and increase the content of $C_4A_3\overline{S}$ mineral. Ract et al. (2003) designed an experiment to add electroplating sludge containing Cu and Ni to cement raw material and found that adding electroplating sludge to the raw material reduced the formation temperature of C_2S.

The solidification and stabilization technology based on cement has been widely used in waste treatment. Yao et al. (2020) used red mud, desulfurization gypsum, alumina ash, and limestone tailings to prepare iron-rich CSA, which not only has high strength, corrosion resistance, and a heavy metal solidification rate of more than 90%, but the leaching concentration of heavy metals also meets national standards. Mao et al. (2020) used MSWI-FA in conjunction with other industrial wastes to prepare CSA, and the leaching concentrations of the eight heavy metals enriched in it all meet national standards. Wang et al. (2021) used steel slag, alumina ash, desulfurization gypsum, and limestone tailings to prepare steel slag−based CSA. Although the Mn and Cr contents in the steel slag are high, the leaching concentrations of heavy metals in the clinker and cementitious materials at different hydration ages are lower than the TCLP limits; after calcination, Mn and Cr are solidified in $FeAl_2O_4$ in the clinker, which can reduce the potential risk of heavy metal migration to the environment.

6.2 Introduction of residue containing heavy metals from the clinker end

Cement-based solidification/stabilization (s/s) is now considered one of the most effective techniques for dealing with industrial waste containing heavy metals. The mechanisms of heavy metal immobilization in cement-based s/s processes are (1) adsorption by hydration products; (2) chemical incorporation (surface complexation, precipitation, coprecipitation, and particle replacement); and (3) micro- or macroencapsulation (Wu et al., 2014). Because PC hydrates relatively slowly in the early stages, and the presence of heavy metals will seriously slow down the hydration process, its application in s/s is limited. CSA, which has a rapid hydration rate, is more suitable for s/s technology. For CSA, the structure of its main hydration product AFt is hexagonal prism $\{Ca_6[Al(OH)_6]_2 \cdot 24H_2O\}^{6+}$, with H_2O and SO_4^{2-} occupying the channels between the prisms (Guo et al., 2017). The ability of AFt to solidify heavy metal ions is based on this mineral being an effective ion exchanger. As shown in Fig. 14.21 (Wang and Wang, 2022), Ca^{2+}, Al^{3+}, and SO_4^{2-} can be replaced in the matrix by various cations and oxyanions with similar ion charges: trivalent ions Fe^{3+}, Cr^{3+}, Ni^{3+}, and Mn^{3+} can replace Al^{3+}, divalent ions Mg^{2+}, Zn^{2+}, Mn^{2+}, Fe^{2+}, Co^{2+}, Pb^{2+}, Cd^{2+}, and Ni^{2+} can replace Ca^{2+}, anions CO_3^{2-}, AsO_4^{3-}, CrO_4^{2-}, SeO_3^{2-}, and SeO_4^{2-} can replace SO_4^{2-} (Wu et al., 2014; Piekkari et al., 2020). Because oxyanions are usually most soluble at high pH, AFt formed at high pH is very suitable for solidifying oxyanions, but the solidification mechanism of anions is based on hydrogen bonds, while metal cations are covalently bonded to the solid crystal matrix. Therefore, the solidification effect of oxyanions in the AFt structure is weaker than that of metal cations (Piekkari et al., 2020).

Similar to the calcination stage, the presence of heavy metals also has a certain impact on the hydration process. Stephan et al. (1999) found that the addition of high concentrations of Cr would accelerate the hydration of cement and shorten the initial setting time, leading to a reduction in material strength. Zn has a strong delaying effect on hydration, leading to an extension of the initial setting time, but it increases the compressive strength of the material. However, Ni has little effect on hydration. Lu et al. (2017) found that the addition of Cd^{2+} and Pb^{2+} accelerates the hydration of C_3S, while the addition of Cr^{3+} has different effects. When the added amount reaches 1.0 wt.%, Cr^{3+} accelerates the hydration of C_3S such as Cd^{2+} and Pb^{2+}. However, when the added amount reaches 1.5 wt.%, Cr^{3+} significantly inhibits the hydration process. Yang et al. (2022) found that Cu^{2+}, Cr^{3+}, and Cd^{2+} weaken the binding energy of Ca−O and Al−O bonds, leading to a decrease in the stability of the AFt crystal structure and a decrease in compressive strength; the hydration reaction is delayed due to the presence of Cu^{2+} and Cd^{2+}, while Cr^{3+} reduces the reaction activation energy, accelerating the hydration process, which is also the reason for the shortened setting time.

Due to the different chemical properties and presence forms of heavy metals in different solid wastes, this also leads to significant differences in their leaching ability and solidification ability in materials. Some researchers have studied the

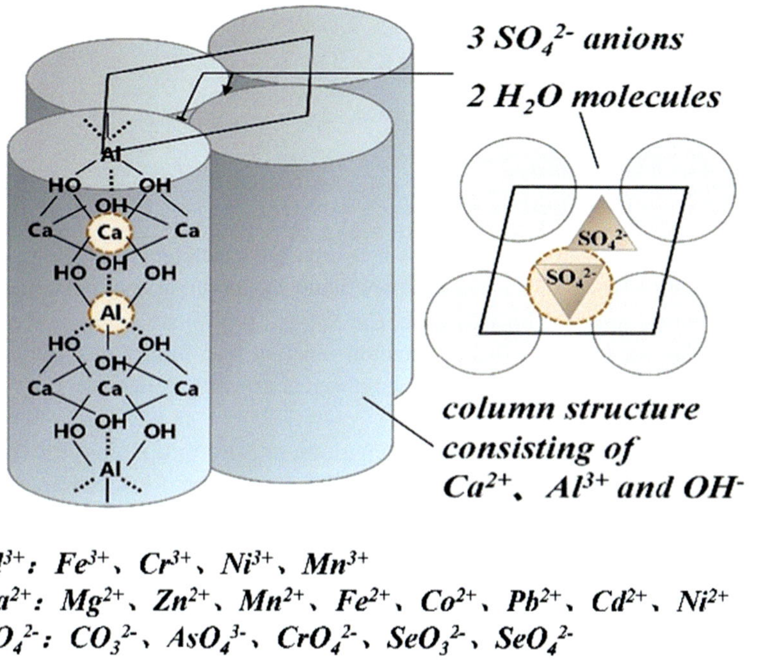

FIGURE 14.21 Solidification of heavy metal ions in AFt (Wang and Wang, 2022).

application of AFt in solidifying heavy metals by mixing $C_4A_3\bar{S}$ with gypsum doped with nitrates of Cu, Cr, Cd, Pb, Zn, and Fe. The leaching results show that the solidification process of the AFt system largely depends on the leaching medium and the properties of the doped metals. In deionized water and a HNO_3 solution with a pH of 4, the amount of heavy metals released from AFt is very small (Wu et al., 2014). The performance study of fly ash-CSA matrix and CSA mixture shows that the fly ash-CSA matrix can effectively fix high concentration heavy metals such as Pb and Zn, significantly reducing their TCLP leaching amount (Qian et al., 2008).

7. Combustion/incineration residues and the carbonation characteristics of CSA cement

Typically, the natural carbonation of concrete is considered detrimental as it reduces the material's durability properties. This risk is even more pronounced with CSA, where the natural carbonation reaction can lead to the decomposition of the material matrix. However, accelerated carbonation curing is being widely studied as a promising method to utilize and storage carbon. This type of curing transforms gaseous CO_2 into carbonate minerals for sequestration, and studies on accelerated carbonation curing for CSA are not uncommon. This part summarizes the relevant carbonation theories and research progress of combustion/incineration residues and CSA, systematically introduces the carbonation products of combustion/incineration residues and CSA and the degree of carbonation of different materials and influencing factors, and outlines the impact of carbonation on material mechanical properties.

7.1 Impact of carbonation on mineral phase

CO_2 curing of cement materials, as a carbon capture and storage process, mainly involves three steps (Ashraf and Olek, 2016): (1) The leaching of mineral ions (mainly Ca^{2+}) from solid to liquid; (2) the dissolution of CO_2 and the formation of carbonate ions; and (3) the precipitation of the carbonation reaction. The carbonation reaction of combustion/incineration residues also follows these steps, but the reactivity of the carbonation not only depends on the Ca content but is also influenced by the mineral phase that fixes Ca^{2+}. Studies have shown (Li and Wu, 2022) that fly ash, blast furnace slag, and steel slag in the waste have considerable CO_2 sequestration potential, with the involved carbonation reaction equations as follows:

$$Ca(OH)_2 + CO_2 \rightarrow CaCO_3 + H_2O \qquad 14.2$$

$$C-S-H+CO_2 \rightarrow CaCO_3+SiO_2+nH_2O \quad 14.3$$

$$C_3S+(3-x)CO_2+yH_2O \rightarrow C_xSH_y+(3-x)CaCO_3 \quad 14.4$$

$$C_2S+(2-x)CO_2+yH_2O \rightarrow C_xSH_y+(2-x)CaCO_3 \quad 14.5$$

$$3CaO\cdot Al_2O_3\cdot 3CaSO_4\cdot 32H_2O+3CO_2 \rightarrow 3CaCO_3+3(CaSO_4\cdot 2H_2O)+Al_2O_3\cdot xH_2O+(26-x)H_2O \quad 14.6$$

In these reactions, $CaCO_3$, as the main product of the carbonation reaction, forms with different crystallographic phases. If kinetic factors dominate during the carbonation reaction, $CaCO_3$ will precipitate as vaterite or aragonite, both of which will eventually transform into calcite. If thermodynamic factors dominate, $CaCO_3$ will precipitate as calcite (Šavija and Luković, 2016). In the CSA system, the $CaCO_3$ generated by the carbonation reaction will further react with the hydration product AFm to generate a small amount of monocarbonate, which contributes to strength increase (Gastaldi et al., 2018).

When fly ash and $C_4A_3\overline{S}$ cocarbonate, the resulting AFt forms spherical agglomerates, a morphology different from the needle-like AFt in traditional cement (Hertel et al., 2021). The carbonation reaction of AFt decomposes it into gypsum, $CaCO_3$, and $Al(OH)_3$. Due to the high sulfate concentration, carbonate cannot be incorporated into the structure of AFt, but it can be incorporated into the structure of gypsum, thereby accelerating the formation process of AFt.

Researchers have used thermodynamic simulations to investigate the changes in mineral phases during the carbonation process after mixing blast furnace slag, fly ash, and CSA (Wu et al., 2023). In the mixed system, the initial substance of the carbonation reaction is AFm, which transforms into monocarbonate in the mixed system. As more monocarbonate is formed, the overall volume will increase, leading to a significant decrease in the pH of the pore solution. The type and amount of SCMs used determine the amount of C-A-S-H and zeolites formed during the carbonation process. In the system of fly ash mixed with CSA cement, as the fly ash content increases, the amount of these substances also increases. The simulation results show that amorphous SiO_2 can remain stable in the mixed system at a very high degree of carbonation, thereby further reducing the pH to 8.0, significantly lower than the expected value of the CSA system.

7.2 Carbon sequestration of combustion/incineration residues

The CO_2 absorption capacity of different fly ashes ranges from 0.77% to 26.4%. Such significant differences are due to factors such as chemical composition, temperature, CO_2 concentration, and pressure. Among them, the chemical composition is the most influential factor, because the elemental content of fly ash from different sources varies greatly, with the Ca content ranging from 3.35% to 40 wt.%. Studies have shown that raising the reaction temperature from 20 to 80°C can significantly accelerate the carbonation rate of fly ash, but the final degree of carbonation increases slightly (Ukwattage et al., 2015). This is because raising temperature increases the carbonation reaction rate, which could speed up CO_2 absorption. At the same time, the rapid formation of a calcite layer moderates the accelerated leaching of Ca^{2+} caused by the temperature rise, resulting in a small increase in the final degree of carbonation. As the carbonation temperature rises to 600°C (Liu et al., 2018), both the carbonation rate and degree of fly ash are significantly improved, which might be due to the dramatic increase in the leaching rate of mineral ions (Ca^{2+}/Mg^{2+}) and the kinetic drive of the carbonation reaction at high temperatures overwhelming the obstructive effect caused by calcite layer formation. Although an increase in pressure can promote the solubility of CO_2 in the liquid and accelerate the carbonation reaction rate, the final degree of carbonation of fly ash is often slightly affected (Montes-Hernandez et al., 2009; Ukwattage et al., 2013), as the diffusion of Ca^{2+} and the kinetics of the carbonation reaction are hardly influenced by pressure changes.

The carbonation degree of steel slag and blast furnace slag is usually over 25%, and the factors affecting the carbonation degree are similar to those of fly ash. It has also been found to be closely related to particle size and microstructure. With the increase of specific surface area of combustion/incineration residues, carbonation activity increases, leading to a higher final degree of carbonation. As particle size decreases from 1000 to 80 μm, the amount of CO_2 that can be absorbed increases exponentially (Polettini et al., 2016). Then, as the particle size further decreases, the ability to absorb CO_2 decreases. This is because the accelerated carbonation reaction of the waste slag caused by reduced particle size is offset by the rapid formation of a calcite layer.

7.3 Impact of carbonation on the macroscopic properties of cementitious materials

Proper carbonation curing can improve the early strength of cement-based composites, which is mainly due to the pore-filling effect of calcite particles produced during carbonation curing and the densification of microstructures. The formation of amorphous calcium aluminate phase and $Al(OH)_3$ gel can also lead to a rapid increase in strength after carbonation. However, excessive carbonation curing often reduces the development of early compressive strength, which is due to the delay in the hydration process caused by carbonation curing and potential cracking of the microstructure caused by carbonation shrinkage.

Concrete mixed with fly ash and CSA exhibits lower porosity and higher 28-day compressive strength and demonstrates strong resistance to salt, sulfate, and acid erosion. However, some researchers have pointed out that long-term carbonation of fly ash mortar samples shows a lower strength gain (Ukwattage et al., 2015), which might be due to the fact that long-term carbonation curing reduces the reactivity of carbonation products, thus hindering the reaction between carbonation products and cement paste.

After carbonation, CSA mixed with blast furnace slag shows a noticeable increase in strength due to the formation of C-(A)-S-H gel and the precipitation of $CaCO_3$. The strength of samples with 50 wt.% blast furnace slag increases by 173.4% after carbonation (Seo et al., 2021), but as the amount of blast furnace slag increases, the strength of the material decreases significantly.

After carbonation curing, CSA mixed with steel slag shows an increase in early strength, but as the calcium silicate in CSA and steel slag gradually hydrates, the overall strength firstly decreases and then increases. This is due to the destruction of the $CaCO_3$ structure formed by short-term carbonation curing by hydration products.

So, for CSA materials mixed with combustion/incineration residues, there exist a CO_2 absorption threshold. Exceeding this threshold can be harmful to the material properties. Therefore, choosing an appropriate material ratio and curing regimen can greatly impact the performance of the material.

8. Summary and future trends

To mitigate global climate change, fostering green, low-carbon, cyclic application of combustion/incineration residues in cement and concrete domains is instrumental. CSA, characterized by extensive adaptability to raw materials and minimal calcination temperature, is pivotal in solid waste resource recovery and low-carbon transition within the cement industry. CSA's primary components, such as CaO, SiO_2, Al_2O_3, Fe_2O_3, and SO_3, are amenable to incineration residues from diverse sectors including steel, metallurgy, coal-powered electricity, and mining. This compatibility paves the way for a comprehensive, multiroute resource utilization of these residues. At the same time, CSA embodies advanced attributes, including accelerated strength development, prompt setting, increased strength, superior impermeability, and increased resistance to corrosion compared with its counterpart, PC. Nonetheless, the introduction of residues—comprising main minerals, trace elements, and heavy metals—can notably affect CSA's calcination scheme, hydration reaction, and carbonation properties. This chapter scrutinizes the prevalent state of research and prospective trends in the usage of combustion/incineration residues in CSA, culminating in the following conclusions:

(1) Utilizing combusted residues as raw materials for the preparation of CSA clinker permits a wide range of elemental composition and the effects of trace elements on mineral formation cannot be ignored. Constituents such as Fe, F, Na, and K, found in residues such as steel slag, aluminum ash, fly ash, blast furnace slag, and MSWI-FA, could curtail the formation temperature of $C_4A_3\overline{S}$, augment the dissolution rate of clinker minerals, and further promote energy-efficient production. However, elements such as F and Cl, present in the residues, can disrupt the calcination process, subsequently affecting the final clinker's performance.

(2) In the role of supplementary cementitious materials, ash-active residues, such as fly ash and blast furnace slag, find limited application in CSA due to their low alkalinity. Conversely, cementitious residues such as steel slag manifest a broader potential for application. Nevertheless, their inclusion is contingent upon factors, such as raw material source, grinding method, and cement mineral composition.

(3) Influenced by the residues' chemical composition and multiple components such as Na and K, residues-based CSA displays elevated early hydration rates. Some residues can even boost the stability of AFt. However, the presence of Cl^- in the residues alters the system of hydration products, resulting in the formation of $3CaO \cdot CaCl_2 \cdot 15H_2O$ and Friedel's salt, thus deteriorating the performance of the paste.

(4) CSA can stabilize heavy metals in the residues through two channels: clinker minerals and hydration products. The former encompasses substitution reactions and interstitial solid solutions, despite heavy metals reducing the calcination temperature of clinker minerals, which results in more intermediate products. The latter involves the absorption of hydration products, chemical insertion, and micro- or macroencapsulation. However, the addition of heavy metals impedes the hydration progression of CSA, prolongs the paste's setting time, and diminishes the performance of CSA.

(5) Residues such as fly ash, blast furnace slag, and steel slag exhibit considerable CO_2 sequestration potential. The type and content of residue are the definitive elements affecting CSA carbonation. Proper carbonation curing can expedite the densification of the material's microstructure through the filling effect of calcite particles, thereby enhancing the material's early strength. Conversely, excessive carbonation curing may result in delayed hydration and shrinkage cracking during the carbonation process, which leads to a decrease in compressive strength.

Although extensive research affirms the application potential of combusted residues in CSA, most applications remain limited to conventional fields and low-value utilization. To operationalize these residues' high-value utilization, based on the aforementioned discussions, the future research directions and requirements are recommend as follows: Firstly, considering the extensive variability and diversity of residue compositions, it is critical to devise comprehensive, effective utilization methods tailored to the properties of residues + CSA system. The impacts of impurities typically present in residues on the calcination process, hydration characteristics, and long-term performance of cement clinker demand thorough investigation. Secondly, the low compatibility of residues used as supplementary cementitious materials with CSA, which modifies CSA's water requirement and setting time, necessitates the use of admixtures, thus altering the system performance. Consequently, research focusing on the properties of materials postadmixture incorporation is needed. Thirdly, due to the risk of heavy metal leaching from residues, further exploration of the immobilization mechanism of heavy metals in residue-based CSA is required, particularly the impact of multiple heavy metals on CSA's hydration mechanism and long-term performance. Lastly, considering CSA's lower carbonation resistance and higher Cl^- permeability, additional research on the long-term performance and durability of residue-based CSA in a carbonation environment is needed.

With respect to residue utilization technology, research and development should primarily focus on high-value material utilization and the scientific prevention of secondary pollution. It is essential to augment the application of residues in high-value construction material domains, delve deeper into the mechanism of hazardous component removal, and realize harmless, resourceful, and beneficial treatment of residues.

References

Amran, M., Murali, G., Khalid, N.H.A., Fediuk, R., Ozbakkaloglu, T., Lee, Y.H., Haruna, S., Lee, Y.Y., 2021. Slag uses in making an ecofriendly and sustainable concrete: a review. Construct. Build. Mater. 272.

Amran, Y.H.M., Alyousef, R., Alabduljabbar, H., El-Zeadani, M., 2020. Clean production and properties of geopolymer concrete; a review. J. Clean. Prod. 251, 119679.

Andrade, F.R.D., Maringolo, V., Kihara, Y., 2003. Incorporation of V, Zn and Pb into the crystalline phases of Portland clinker. Cement Concr. Res. 33, 63–71.

Ashraf, W., Olek, J., 2016. Carbonation behavior of hydraulic and non-hydraulic calcium silicates: potential of utilizing low-lime calcium silicates in cement-based materials. J. Mater. Sci. 51, 6173–6191.

Barker, D.J., Turner, S.A., Napier-Moore, P.A., Clark, M., Davison, J.E., 2009. CO_2 capture in the cement industry. Energy Proc. 1, 87–94.

Ben Haha, M., Winnefeld, F., Pisch, A., 2019. Advances in understanding ye'elimite-rich cements. Cement Concr. Res. 123.

Benarchid, M.Y., Rogez, J., 2005. The effect of Cr_2O_3 and P_2O_5 additions on the phase transformations during the formation of calcium sulfoaluminate $C_4A_3\bar{S}$. Cement Concr. Res. 35, 2074–2080.

Canbek, O., Shakouri, S., Erdoğan, S.T., 2020. Laboratory production of calcium sulfoaluminate cements with high industrial waste content. Cement Concr. Compos. 106.

China, N.B.o.S.o.t.P.s.R.o., 2022. 2022 National Cement Report.

Clavier, K.A., Paris, J.M., Ferraro, C.C., Townsend, T.G., 2020. Opportunities and challenges associated with using municipal waste incineration ash as a raw ingredient in cement production – a review. Resour. Conserv. Recycl. 160.

Cuesta, A., Álvarez-Pinazo, G., Sanfélix, S.G., Peral, I., Aranda, M.A.G., De la Torre, A.G., 2014. Hydration mechanisms of two polymorphs of synthetic ye'elimite. Cement Concr. Res. 63, 127–136.

Cuesta, A., Santacruz, I., Sanfélix, S.G., Fauth, F., Aranda, M.A.G., De la Torre, A.G., 2015. Hydration of C_4AF in the presence of other phases: a synchrotron X-ray powder diffraction study. Construct. Build. Mater. 101, 818–827.

Dilnesa, B.Z., Wieland, E., Lothenbach, B., Dähn, R., Scrivener, K.L., 2014. Fe-containing phases in hydrated cements. Cement Concr. Res. 58, 45–55.

EPA, 2022. Sources of Greenhouse Gas Emissions.

Fang, X., Xuan, D., Zhan, B., Li, W., Poon, C.S., 2021. A novel upcycling technique of recycled cement paste powder by a two-step carbonation process. J. Clean. Prod. 290.

Feng, Y., 2022. Progress in high temperature and high efficiency recovery and utilization of blast furnace slag. In: Launch Meeting of the Three-Year Action Programme for Energy Efficiency Benchmarking in the Steel Industry, Zhanjiang, China, December 9, 2022.

Gao, D., Che, Q., Meng, Y., Yang, L., Xie, X., 2022. Properties evolution of calcium sulfoaluminate cement blended with ground granulated blast furnace slag suffered from sulfate attack. J. Mater. Res. Technol. 17, 1642–1651.

Gao, D., Meng, Y., Yang, L., Tang, J., Lv, M., 2019. Effect of ground granulated blast furnace slag on the properties of calcium sulfoaluminate cement. Construct. Build. Mater. 227.

Gao, Y., Li, Z., Zhang, J., Zhang, Q., Wang, Y., 2020. Synergistic use of industrial solid wastes to prepare belite-rich sulphoaluminate cement and its feasibility use in repairing materials. Construct. Build. Mater. 264.

García-Maté, M., De la Torre, A.G., León-Reina, L., Aranda, M.A.G., Santacruz, I., 2013. Hydration studies of calcium sulfoaluminate cements blended with fly ash. Cement Concr. Res. 54, 12–20.

Gastaldi, D., Bertola, F., Canonico, F., Buzzi, L., Mutke, S., Irico, S., Paul, G., Marchese, L., Boccaleri, E., 2018. A chemical/mineralogical investigation of the behavior of sulfoaluminate binders submitted to accelerated carbonation. Cement Concr. Res. 109, 30–41.

Gencel, O., Karadag, O., Oren, O.H., Bilir, T., 2021. Steel slag and its applications in cement and concrete technology: a review. Construct. Build. Mater. 283.

Gongyan-Net, 2023. Analysis of Fly Ash Market in China in 2023.

Guo, B., Sasaki, K., Hirajima, T., 2017. Selenite and selenate uptaken in ettringite: immobilization mechanisms, coordination chemistry, and insights from structure. Cement Concr. Res. 100, 166–175.

Guo, X., Shi, H., Hu, W., Wu, K., 2014. Durability and microstructure of CSA cement-based materials from MSWI fly ash. Cement Concr. Compos. 46, 26–31.

Guo, Y., Luo, L., Liu, T., Hao, L., Li, Y., Liu, P., Zhu, T., 2024. A review of low-carbon technologies and projects for the global cement industry. J. Environ. Sci. 136, 682–697.

Hertel, T., Van den Bulck, A., Onisei, S., Sivakumar, P.P., Pontikes, Y., 2021. Boosting the use of bauxite residue (red mud) in cement - production of an Fe-rich calciumsulfoaluminate-ferrite clinker and characterisation of the hydration. Cement Concr. Res. 145.

Iacobescu, R.I., Pontikes, Y., Koumpouri, D., Angelopoulos, G.N., 2013. Synthesis, characterization and properties of calcium ferroaluminate belite cements produced with electric arc furnace steel slag as raw material. Cement Concr. Compos. 44, 1–8.

Ji, Y., Pel, L., Zhang, X., Sun, Z., 2021. Cl^- and Na^+ ions binding in slag and fly ash cement paste during early hydration as studied by 1H, 35Cl and 23Na NMR. Construct. Build. Mater. 266.

Jiang, Y., Ling, T., Shi, C., Pan, S., 2018. Characteristics of steel slags and their use in cement and concrete—a review. Resour. Conserv. Recycl. 136, 187–197.

Jiang, Z., Lei, X., Liao, Y., Liao, G., 2016. Influence of fly ash on hydration process of sulphoaluminate cement. Bull. Chin. Ceram. Soc. 35, 4088–4092+4103.

Jing, X., Wu, S., Qin, J., Li, X., Liu, X., Zhang, Y., Mao, J., Nie, W., 2023. Multiscale mechanical characterizations of ultrafine tailings mixed with incineration slag. Front. Earth Sci. 11, 1123529.

John, S.K., Nadir, Y., Girija, K., 2021. Effect of source materials, additives on the mechanical properties and durability of fly ash and fly ash-slag geopolymer mortar: a review. Construct. Build. Mater. 280.

Juenger, M.C.G., Siddique, R., 2015. Recent advances in understanding the role of supplementary cementitious materials in concrete. Cement Concr. Res. 78, 71–80.

Juenger, M.C.G., Winnefeld, F., Provis, J.L., Ideker, J.H., 2011. Advances in alternative cementitious binders. Cement Concr. Res. 41, 1232–1243.

Kolovos, K., Tsivilis, S., Kakali, G., 2002. The effect of foreign ions on the reactivity of the $CaO-SiO_2-Al_2O_3-Fe_2O_3$ system Part II: cations. Cement Concr. Res. 32, 463–469.

Kang, L., Du, H.L., Zhang, H., Ma, W.L., 2018. Systematic research on the application of steel slag resources under the background of big data. Complexity 2018, 1–12.

Li, J., Jia, A., Hou, X., Wang, X., Mao, Y., Wang, W., 2023a. Thermal co-treatment of aluminum dross and municipal solid waste incineration fly ash: mineral transformation, crusting prevention, detoxification, and low-carbon cementitious material preparation. J. Environ. Manag. 329, 117090.

Li, J., Xiao, F., Zhang, L., Amirkhanian, S.N., 2019. Life cycle assessment and life cycle cost analysis of recycled solid waste materials in highway pavement: a review. J. Clean. Prod. 233, 1182–1206.

Li, L., Wu, M., 2022. An overview of utilizing CO_2 for accelerated carbonation treatment in the concrete industry. J. CO_2 Util. 60.

Li, Q., Li, J., Zhang, S., Huang, X., Wang, X., Wang, Y., Ni, W., 2023b. Research progress of low-carbon cementitious materials based on synergistic industrial wastes. Energies 16.

Li, Y., 2022. The Road to "Double Carbon" in the Steel Industry Has a Long Way to Go. Chinese People's Political Consultative Conference News.

Liao, Y., Jiang, G., Wang, K., Al Qunaynah, S., Yuan, W., 2020. Effect of steel slag on the hydration and strength development of calcium sulfoaluminate cement. Construct. Build. Mater. 265.

Lin, B., Zhang, Z., 2016. Carbon emissions in China's cement industry: a sector and policy analysis. Renew. Sustain. Energy Rev. 58, 1387–1394.

Liu, S., Li, L., 2014. Influence of fineness on the cementitious properties of steel slag. J. Therm. Anal. Calorim. 117, 629–634.

Liu, S., Pan, C., Zhang, H., Yao, S., Shen, P., Guan, X., Shi, C., Li, H., 2023. Development of novel mineral admixtures for sulphoaluminate cement clinker: the effects of wet carbonation activated red mud. J. Build. Eng. 67.

Liu, W., Su, S., Xu, K., Chen, Q., Xu, J., Sun, Z., Wang, Y., Hu, S., Wang, X., Xue, Y., Xiang, J., 2018. CO_2 sequestration by direct gas–solid carbonation of fly ash with steam addition. J. Clean. Prod. 178, 98–107.

Loser, R., Lothenbach, B., Leemann, A., Tuchschmid, M., 2010. Chloride resistance of concrete and its binding capacity − comparison between experimental results and thermodynamic modeling. Cement Concr. Compos. 32, 34−42.

Lu, L.N., Xiang, C.Y., He, Y.J., Wang, F.Z., Hu, S.G., 2017. Early hydration of C_3S in the presence of Cd^{2+}, Pb^{2+} and Cr^{3+} and the immobilization of heavy metals in pastes. Construct. Build. Mater. 152, 923−932.

Ma, B., Li, X., Mao, Y., Shen, X., 2013. Synthesis and characterization of high belite sulfoaluminate cement through rich alumina fly ash and desulfurization gypsum. Ceramics 57, 7−13.

Ma, S.H., Shen, X.D., Gong, X.P., Zhong, B.Q., 2006. Influence of CuO on the formation and coexistence of $3CaO \cdot SiO_2$ and $3CaO \cdot 3Al_2O_3 \cdot CaSO_4$ minerals. Cement Concr. Res. 36, 1784−1787.

Maddalena, R., Roberts, J.J., Hamilton, A., 2018. Can Portland cement be replaced by low-carbon alternative materials? a study on the thermal properties and carbon emissions of innovative cements. J. Clean. Prod. 186, 933−942.

Madlool, N.A., Saidur, R., Rahim, N.A., Kamalisarvestani, M., 2013. An overview of energy savings measures for cement industries. Renew. Sustain. Energy Rev. 19, 18−29.

Mao, Y., Wu, H., Wang, W., Jia, M., Che, X., 2020. Pretreatment of municipal solid waste incineration fly ash and preparation of solid waste source sulphoaluminate cementitious material. J. Hazard Mater. 385, 121580.

Mathapati, M., Amate, K., Durga Prasad, C., Jayavardhana, M.L., Hemanth Raju, T., 2022. A review on fly ash utilization. Mater. Today: Proc. 50, 1535−1540.

Montes-Hernandez, G., Pérez-López, R., Renard, F., Nieto, J.M., Charlet, L., 2009. Mineral sequestration of CO_2 by aqueous carbonation of coal combustion fly-ash. J. Hazard Mater. 161, 1347−1354.

Padilla-Encinas, P., Palomo, A., Blanco-Varela, M.T., Fernández-Carrasco, L., Fernández-Jiménez, A., 2021. Monitoring early hydration of calcium sulfoaluminate clinker. Construct. Build. Mater. 295.

Pang, L., Liao, S., Wang, D., An, M., 2022. Influence of steel slag fineness on the hydration of cement-steel slag composite pastes. J. Build. Eng. 57, 104866.

Piekkari, K., Ohenoja, K., Isteri, V., Tanskanen, P., Illikainen, M., 2020. Immobilization of heavy metals, selenate, and sulfate from a hazardous industrial side stream by using calcium sulfoaluminate-belite cement. J. Clean. Prod. 258.

Plaza, P., Martin, S., Sancristobal, E., Blázquez, M., Castro, M., Díaz, G., Labarias, M.J., García-Loro, F., Quintana, B., Franco, J.P., Perez, C., 2020. Science and technology educational quality scaling in Spain. Front. Educ. Conf. 1−8.

Polettini, A., Pomi, R., Stramazzo, A., 2016. Carbon sequestration through accelerated carbonation of BOF slag: influence of particle size characteristics. Chem. Eng. J. 298, 26−35.

Qian, G.R., Shi, J., Cao, Y.L., Xu, Y.F., Chui, P.C., 2008. Properties of MSW fly ash−calcium sulfoaluminate cement matrix and stabilization/solidification on heavy metals. J. Hazard Mater. 152, 196−203.

Ract, P.G., Espinosa, D.C.R., Tenorio, J.A.S., 2003. Determination of Cu and Ni incorporation ratios in Portland cement clinker. Waste Manage. (Tucson, Ariz.) 23, 281−285.

Ren, C., Hua, D., Bai, Y., Wu, S., Yao, Y., Wang, W., 2022. Preparation and 3D printing building application of sulfoaluminate cementitious material using industrial solid waste. J. Clean. Prod. 363.

Ren, C., Wang, W., Mao, Y., Yuan, X., Song, Z., Sun, J., Zhao, X., 2017. Comparative life cycle assessment of sulfoaluminate clinker production derived from industrial solid wastes and conventional raw materials. J. Clean. Prod. 167, 1314−1324.

Šavija, B., Luković, M., 2016. Carbonation of cement paste: understanding, challenges, and opportunities. Construct. Build. Mater. 117, 285−301.

Seo, J., Kim, S., Park, S., Yoon, H.N., Lee, H.K., 2021. Carbonation of calcium sulfoaluminate cement blended with blast furnace slag. Cement Concr. Compos. 118.

Shi, Z., Geiker, M.R., Lothenbach, B., De Weerdt, K., Garzón, S.F., Enemark-Rasmussen, K., Skibsted, J., 2017. Friedel's salt profiles from thermogravimetric analysis and thermodynamic modelling of Portland cement-based mortars exposed to sodium chloride solution. Cement Concr. Compos. 78, 73−83.

Stephan, D., Mallmann, R., Knofel, D., Hardtl, R., 1999. High intakes of Cr, Ni, and Zn in clinker Part II. Influence on the hydration properties. Cement Concr. Res. 29, 1959−1967.

Sun, C., Zhang, J., Yan, C., Yin, L., Wang, X., Liu, S., 2022. Hydration characteristics of low carbon cementitious materials with multiple solid wastes. Construct. Build. Mater. 322.

Sun, Y., Li, F., He, W., Wang, T., Gao, L., Li, Y., Gao, J., Zhao, R., Fang, M., 2021. Investigation on CO2 mineralization curing of aerated concretes. Clean Coal Technol. 27, 237−245.

Tao, M., Nie, K., Cheng, S., Zhang, X., 2022. Environmental impact analysis of ecological cementitious material production based on LCA method. J. Saf. Environ. 22, 2176−2183.

Tao, Y., Zhang, W., Li, N., Shang, D., Xia, Z., Wang, F., 2018. Fundamental principles that govern the copper doping behavior in complex clinker system. J. Am. Ceram. Soc. 101, 2527−2536.

Ukwattage, N.L., Ranjith, P.G., Wang, S.H., 2013. Investigation of the potential of coal combustion fly ash for mineral sequestration of CO_2 by accelerated carbonation. Energy 52, 230−236.

Ukwattage, N.L., Ranjith, P.G., Yellishetty, M., Bui, H.H., Xu, T., 2015. A laboratory-scale study of the aqueous mineral carbonation of coal fly ash for CO_2 sequestration. J. Clean. Prod. 103, 665−674.

Wang, D.Q., Wang, Q., 2022. Clarifying and quantifying the immobilization capacity of cement pastes on heavy metals. Cement Concr. Res. 161, 20.

Wang, S., Xue, C., Zhao, Q., Bai, Y., Guo, W., Shi, Y., Qiu, Y., Pan, H., 2023a. A novel binder prepared from municipal solid waste incineration fly ash and phosphogypsum. J. Build. Eng. 71.

Wang, W., Wang, X., Zhu, J., Wang, P., Ma, C., 2013. Experimental investigation and modeling of sulfoaluminate cement preparation using desulfurization gypsum and red mud. Ind. Eng. Chem. Res. 52, 1261–1266.

Wang, X., Sun, D., Li, J., Wang, W., Mao, Y., Song, Z., 2023b. Properties and hydration characteristics of an iron-rich sulfoaluminate cementitious material under cold temperatures. Cement Concr. Res. 168.

Wang, X., Wang, K., Li, J., Wang, W., Mao, Y., Wu, S., Yang, S., 2021. Heavy metals migration during the preparation and hydration of an eco-friendly steel slag-based cementitious material. J. Clean. Prod. 329.

Winnefeld, F., Lothenbach, B., 2010. Hydration of calcium sulfoaluminate cements — experimental findings and thermodynamic modelling. Cement Concr. Res. 40, 1239–1247.

Wu, F., He, M., Qu, G., Zhang, T., Liu, X., 2023. Synergistic densification treatment technology of phosphogypsum and aluminum ash. Process Saf. Environ. Protect. 173, 847–858.

Wu, K., Shi, H., Xu, L., Guo, X., De Schutter, G., Xu, M., 2014. Influence of heavy metals on the early hydration of calcium sulfoaluminate. J. Therm. Anal. Calorim. 115, 1153–1162.

Xu, X., Huang, B., Yuan, J., Chen, Y., 2022. Embodied environmental impacts of cement export in China. J. Fudan Univ. Nat. Sci. 61, 485–494.

Xue, P., Xu, A., He, D., Yang, Q., Liu, G., Engström, F., Björkman, B., 2016. Research on the sintering process and characteristics of belite sulphoaluminate cement produced by BOF slag. Construct. Build. Mater. 122, 567–576.

Yang, F.M., Pang, F.J., Xie, J.T., Wang, W.J., Wang, W.L., Wang, Z.M., 2022. Leaching and solidification behavior of Cu^{2+}, Cr^{3+} and Cd^{2+} in the hydration products of calcium sulfoaluminate cement. J. Build. Eng. 46, 10.

Yang, Y., Xue, J., Huang, Q., 2014. Studies on the solidification mechanisms of Ni and Cd in cement clinker during cement kiln co-processing of hazardous wastes. Construct. Build. Mater. 57, 138–143.

Yao, X., 2021. Experimental Study on Preparation of Solid Waste-Based Ferric-Rich Sulfoaluminate Cementitious Material and Non-autoclaved Lightweight Concrete. Shandong University.

Yao, Y., 2020. Preparation of Solid-Waste-Based Ferric-Rich Sulfoaluminate Marine Engineering Cementitious Material and Research on its Anti-erosion Property. Shandong University.

Yao, Y., Wang, W., Ge, Z., Ren, C., Yao, X., Wu, S., 2020. Hydration study and characteristic analysis of a sulfoaluminate high-performance cementitious material made with industrial solid wastes. Cement Concr. Compos. 112.

Yoon, H.N., Seo, J., Kim, S., Lee, H.K., Park, S., 2021. Hydration of calcium sulfoaluminate cement blended with blast-furnace slag. Construct. Build. Mater. 268.

Yu, M., 2022. Study on Composite Modification and Application of Solid Waste Based Sulfoaluminum Based Cementitious Materials. Shandong University.

Zhang, Y., Wang, L., Chen, L., Ma, B., Zhang, Y., Ni, W., Tsang, D.C.W., 2021. Treatment of municipal solid waste incineration fly ash: state-of-the-art technologies and future perspectives. J. Hazard Mater. 411.

Zhao, J., Li, Z., Wang, D., Yan, P., Luo, L., Zhang, H., Zhang, H., Gu, X., 2023. Hydration superposition effect and mechanism of steel slag powder and granulated blast furnace slag powder. Construct. Build. Mater. 366.

ZhongCheng-Environment, 2023. The Resource Utilization of Waste Incineration Ash Has Been Supported by the Policy, and the Technical Guidelines and Standards Have Been Launched Successively. CE Carbon Technology.

Zhu, B., Jiang, D., Chen, D., Li, Q., Wang, W., 2014. SFA-based resource consumption analysis on China's cement and cement-based materials industry. J. Tsinghua Univ. Sci. Technol. 54, 839–845.

Zhu, J., Chen, Y., Zhang, L., Yang, K., Guan, X., Zhao, R., 2020. Insights on substitution preference of pb ions in sulfoaluminate cement clinker phases. Materials 14.

Zhu, J., Yang, K., Chen, Y., Fan, G., Zhang, L., Guo, B., Guan, X., Zhao, R., 2021. Revealing the substitution preference of zinc in ordinary Portland cement clinker phases: a study from experiments and DFT calculations. J. Hazard Mater. 409, 124504.

Part IV

Recycling of combustion/incineration residues into SCMs and aggregates

Chapter 15

Recycling of pulverized fuel ash as supplementary cementitious materials (SCMs) and aggregates in concrete production

Muhammed Bayram[1], Ömer Faruk Kuranlı[2], Anıl Niş[3] and Togay Ozbakkaloglu[1]
[1]*Ingram School of Engineering, Texas State University, San Marcos, TX, United States;* [2]*Civil Engineering Department, Yıldız Technical University, Istanbul, Turkey;* [3]*Civil Engineering Department, Istanbul Gelisim University, Istanbul, Turkey*

1. Introduction

Coal continued to be the leading energy source for electricity generation in 2019, accounting for 37% of global electricity production, surpassing the contribution of renewable energy sources by a substantial margin of 10% points (IEA, 2021). Regrettably, coal has emerged as a key contributor to the escalation of global carbon dioxide (CO_2) emissions, contributing more than 40% of the growth seen in 2021 (IEA, 2022). The concerning trend has resulted in a historical peak of 15.3 billion tons of CO_2, reaching an unprecedented level. Additionally, the combustion of coal significantly contributes to the deterioration of air quality, heightening environmental concerns associated with its usage (IEA, 2022). Coal ash, which is the main solid waste product produced after the combustion of coal in thermal power plants, is produced in vast amounts as a result of the extensive use of coal. Estimations suggest that the global coal industry generates an annual output of around 800 million tons of coal ash (Sibanda et al., 2016; Belviso, 2018). Out of this total amount, China alone accounts for approximately 500 million tons of coal ash production, while India generates around 140 million tons. The United States and the European Union collectively contribute to the production of 115 million tons of coal ash. The aforementioned data serve as an indication of the significant quantities of coal ash that are currently being disposed of in landfills. The data also highlights the widespread mismanagement of coal ash in various regions globally, potentially leading to a range of environmental issues. Inadequate disposal of coal fly ash (CFA) can result in serious consequences, including soil degradation, potential threats to human health, and environmental pollution. The environmental consequences mentioned, combined with the increasing global emphasis on environmental conservation, have sparked broad interest in the reutilization of coal ash (Danish and Mosaberpanah, 2021; Danish et al., 2022). Due to this interest, numerous recycling systems have been widely implemented in cement and concrete industry to efficiently use coal ash in a variety of applications (Huang et al., 2017; Gupta and Chaudhary, 2020).

Pulverized fuel ash (PFA), commonly referred to as fly ash (FA), is the residual material that remains after the combustion of fuel. Fly ash is generated during the combustion process and is transported by flue gases, ultimately being collected by precipitators or other filtering mechanisms. Over time, coal-fired thermal power plants have generated substantial amounts of both coal bottom ash (CBA) and CFA. These two types of combustion by-products constitute a significant proportion, with CBA accounting for approximately 20% and CFA constituting the remaining 80% of the total by-product generated (Rafieizonooz et al., 2016). As depicted in Fig. 15.1, the process of combustion leads to the transformation of mineral particles or mineral constituents present in coal into a state of liquefaction, vaporization, condensation, or agglomeration. FA consists of spherical, noncrystalline particles, which form due to the surface tension effect from rapid cooling in the area following combustion (Xu and Shi, 2018).

Treatment and Utilization of Combustion and Incineration Residues. https://doi.org/10.1016/B978-0-443-21536-0.00004-6

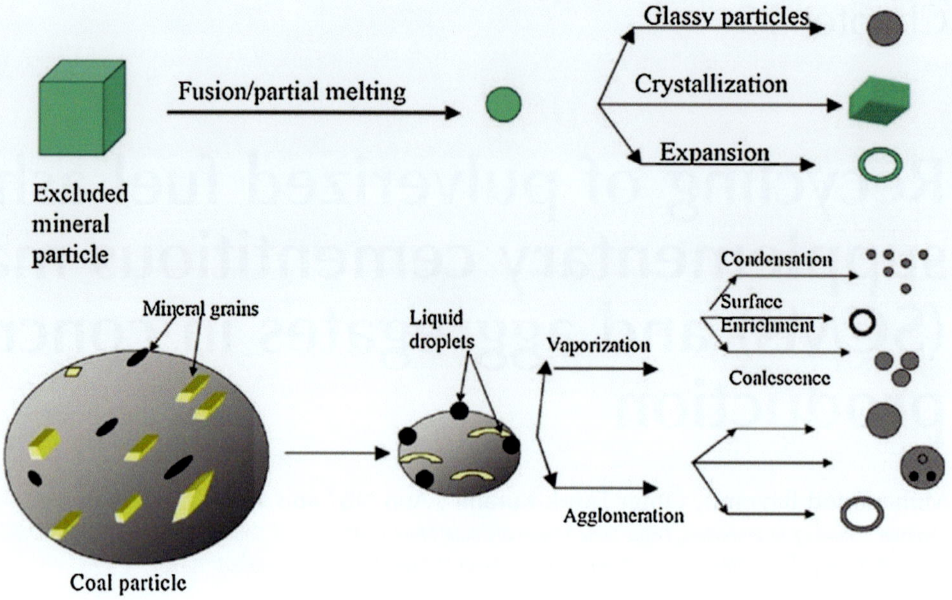

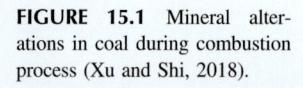

FIGURE 15.1 Mineral alterations in coal during combustion process (Xu and Shi, 2018).

The organizational structure of waste management promotes strategies that aim to prevent waste generation, minimize the amount of waste produced, explore opportunities for reusing materials, facilitate recycling processes, and harness energy from waste as preferred approaches to waste handling, with disposal being the least favorable option. Increasing apprehensions regarding the handling of PFA have sparked many studies focused on its reduction and recycling. These collective research efforts identify PFA as an industrial by-product with significant promise for repurposing, particularly as a substitute for cement or as an additive in concrete (Teixeira et al., 2016; Hemalatha and Ramaswamy, 2017).

In addition to the environmental advantages associated with waste management and CO_2 sequestration, FA offers several benefits in the realm of concrete (Ukwattage et al., 2015). Initially, it improves the concrete's workability, facilitating better mixing, placement, and finalization. Furthermore, FA aids in reducing the heat emitted during hydration, thereby decreasing the risk of thermal cracks during the initial stages of concrete setting (Dananjayan et al., 2016; Wang et al., 2020a). Moreover, the incorporation of FA in concrete is reported to enhance its mechanical properties, contributing to improved durability over time. This results in better resistance to sulfate exposure and reduced vulnerability to chloride ion penetration (Dananjayan et al., 2016; Wang et al., 2020b). FA is an essential supplemental material in the manufacturing of high-performance and environmentally friendly concrete due to these advantageous properties. Researchers have explored different approaches in utilizing FA in concrete to attain optimal advantages. To maximize the advantages of the concrete, one method calls for adding FA as a filler or as a partial substitution with cement into the cement-based materials. Another approach focuses on achieving the maximum possible incorporation of FA while ensuring that the properties of the concrete remain within acceptable standards suitable for specific applications. These varied strategies allow researchers to effectively leverage the potential of FA in concrete for diverse purposes.

To gain a comprehensive understanding of the latest technological advancements and their practical applications, it is crucial to delve into the sustainability aspect of PFA, as elucidated in the subsequent sections. This understanding will enable researchers to assess and validate the adoption of these new technologies within the industry, taking into account their potential engineering and environmental and economic impacts in relation to PFA utilization. This review aims to consolidate the findings on the effects of using PFA as SCMs and aggregates in concrete. This chapter includes characterization of PFA, exploration of concrete mix methods with PFA admixtures, systematic analysis of the performance and benefits of PFA-based SCMs and aggregates, discussion of challenges and limitations, and exploration of future prospects. By synthesizing the available research, this study offers valuable insights to researchers, unveiling new perspectives and avenues for further advancements in PFA utilization within the realm of concrete applications.

2. Properties and characteristics of PFA

PFA is the resultant substance formed through the incineration of coal, biomass, or a combination thereof, serving as a by-product of the combustion process. Numerous global studies have delved into the characterization and potential use of FA, as reflected in various academic publications (Ahmaruzzaman, 2010; Kalembkiewicz and Chmielarz, 2012; Nis and Al-Antaki, 2022; Nis and Altundal, 2023).

2.1 Composition and chemical characteristics of PFA

During coal combustion, a significant amount of solid waste, termed coal ash, is generated. The mineralogical and chemical properties of coal ash are largely influenced by the specific coal type burned, the combustion methods employed, and the cooling processes postcombustion (Yadav and Mondal, 2019; Grabias-Blicharz and Franus, 2023). The two primary forms of coal ash recognized are bottom ash and FA. Bottom ash constitutes the residue settling at the boiler's base after combustion, while finer coal ash particles rising into the exhaust stacks during combustion, subsequently captured by electrostatic precipitators, are termed FA. The crystalline phases of quartz, mullite, hematite, goethite, and calcite are among the amorphous elements that make up the inorganic fraction of coal ash (Grabias-Blicharz and Franus, 2023). These constituents contribute to the inorganic composition to varying degrees, accounting for approximately 34%−80% of the ash composition, with individual phases contributing between 17% and 63% (Ahmaruzzaman, 2010; Belviso, 2018).

FA is traditionally categorized into two types, Class F (F-FA) and Class C (C-FA), based on its chemical and mineralogical properties (Suraneni et al., 2021). F-FA is defined by ASTM C618 as having a CaO percentage less than 18%, whereas C-FA has a CaO level greater than 18% (ASTM C618, 2019). Prior to 2019, C-FA usually exhibited a combined content of SiO_2, Fe_2O_3, and Al_2O_3 between 50% and 70%. In contrast, F-FA was characterized by a combined SiO_2, Fe_2O_3, and Al_2O_3 content exceeding 70%. F-FA and C-FA must now have a minimum combined composition of SiO_2, Fe_2O_3, and Al_2O_3 that is greater than 50% according to the most recent version of ASTM C618 (Suraneni et al., 2021). F-FA has noteworthy chemical reactivity with alkaline compounds and is primarily produced after the burning of bituminous and anthracite coal. Cementitious characteristics, which have a high level of chemical reactivity and the potential for chemical activation, are a common manifestation of this reactivity. Suraneni et al. (2021) extensively documented and compared various significant global specifications for FA, such as the Canadian (CSA A3001), the European (EN 450-1), and the Australian/New Zealand (AS/NZS 3582.1) specifications. The authors provided a detailed analysis of these specifications, considering factors such as FA classes, chemical compositions, moisture content, and LOI (loss on ignition) percentages. The authors underscored the fact that the test procedures utilized to determine various parameters within each specification are not consistently uniform. These varied testing approaches hold the potential to markedly influence results, complicating direct comparisons of specification limits. Recognizing that numerous international specifications for fly ash classification rely heavily on factors like LOI and fineness, there is a suggestion to incorporate these parameters into the scope of ASTM C618. By incorporating such parameters into ASTM C618, it would align the standard with prevailing international practices and enhance the consistency and comparability of FA specifications on a global scale.

Vassilev and Vassileva (2007) proposed an innovative integrated methodology that considers the source, mineral−chemical composition, and characteristics of FA. The suggested approach begins by considering the overall bulk chemical composition as a foundational element. The primary framework of the system is established by considering the quantities of ash-forming elements, prevalent geochemical interactions, and significant positive or negative associations among these elements. By employing this classification process, the FA sample is categorized into one of four chemical types: sialic, calsialic, ferrisialic, or ferricalsialic. Additionally, the acidity level of the sample is classified into three main tendencies: high, medium, or low acid (see Fig. 15.2A). The second approach is founded on an analysis of the mineral contents, associations, correlations, properties, and behaviors that define the FA. FA is further subdivided into four phase-mineral categories according to this classification system: pozzolanic, inert, active, and mixed. In addition, three main tendencies are determined based on their pozzolanic characteristics: high, medium, and low pozzolanic (see Fig. 15.2B).

2.2 Physical properties and surface morphology of PFA

The physical properties of FA play a pivotal role in its utilization as a filler and in its pozzolanic activities. Attributes such as particle size distribution, bulk density, specific gravity, and moisture content are fundamental in evaluating FA's suitability for roles such as SCM and aggregate in concrete production (Nath and Kumar, 2019). FA generally comprises of spherical particles, which can be either hollow or filled. These particles exhibit a size range of 0.1−100 μm and possess a

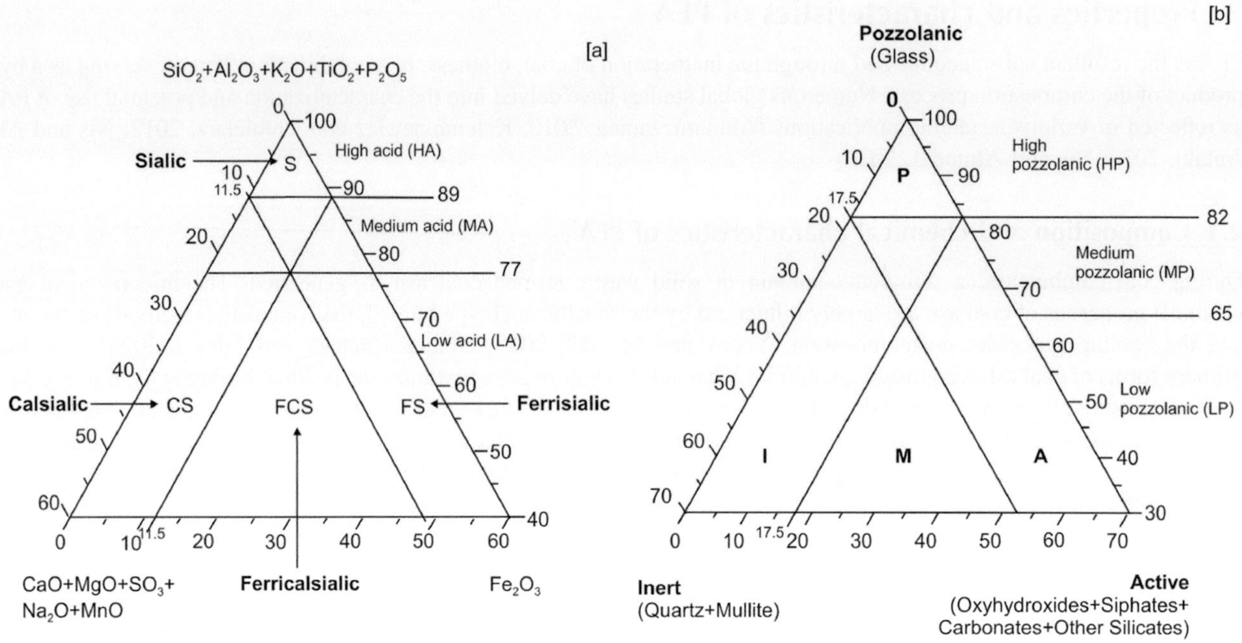

FIGURE 15.2 (A) Chemical classification system of coal fly ash, (B) phase-mineral classification system, indicating pozzolanic (P), inert (I), mixed (M), and active (A) components (Belviso, 2018).

bulk density ranging from 0.54 to 0.86 g/cm^3. The specific gravity of FA typically lies between 2.1 and 3.0, and its specific surface area usually varies from 0.3 to 1.0 m^2/g. Finer particles of FA offer a greater surface area, thereby providing increased opportunities for nucleation and facilitating improved pozzolanic reactions. The increased surface area of finer particles provides more reaction sites, facilitating a better interaction with cement compounds and elevating the pozzolanic performance of FA (Assi et al., 2018; Sevim and Demir, 2019). The ASTM C618 standards specify that for both F-FA and C-FA the content retained on a 45 μm sieve must not exceed 34%. Additionally, the FA should have a moisture content of less than 3% (ASTM C618, 2019). These requirements ensure the desired quality and suitability of the FA for various applications.

Fig. 15.3 showcases representative scanning electron microscopy (SEM) images of CFA, illustrating its surface morphology featuring cenospheres and plenospheres spanning from tens of nanometers to around 100 micrometers in size (as depicted in Fig. 15.3A–D). The spherical particles' surfaces frequently feature rough imperfections. Additionally, glassy fragments with a high degree of vesiculation, carbon blocks, and tiny, porous grains can all be seen in CFA (Fig. 15.3E–F).

3. Recycling PFA into supplementary cementitious materials

Modern concrete practice makes extensive use of SCMs, either as independent additions to the concrete-mixing process or as elements of blended cements (Lothenbach et al., 2011). Incorporating SCMs such as FA from coal combustion or blast-furnace slag, a by-product from pig iron production, provides a practical approach to reduce the consumption of ordinary Portland cement (OPC) in concrete mixtures. Using SCMs offers a promising approach to boost the sustainability and efficiency of concrete structures. The hydration of different clinker phases leads to the formation of key by-products, including calcium silicate hydrate (C–S–H), portlandite, ettringite, and AFm phases. However, when SCMs are blended with OPC, the resulting system becomes more complex. In this specific scenario, the hydration of OPC and the hydraulic reaction of the SCM happen simultaneously, which may alter how reactive one is. The interactions between these elements in the blended system can substantially affect the cementitious matrix's overall efficacy and physical characteristics.

Adding FA to OPC reduces the presence of portlandite in the hydrated mix. This reduction is particularly significant when F-FAs are incorporated, as they typically contain higher levels of alumina ranging from 15% to 35%. As a consequence, considerable volumes of Al-rich phases are formed when F-FA and OPC are combined. Thermodynamic analyses, maintaining a consistent Al/Si ratio of 0.1 in the C–S–H phase, were carried out to investigate this behavior. The results show that moderate amounts of FA lead to the destabilization of portlandite and encourage the formation of

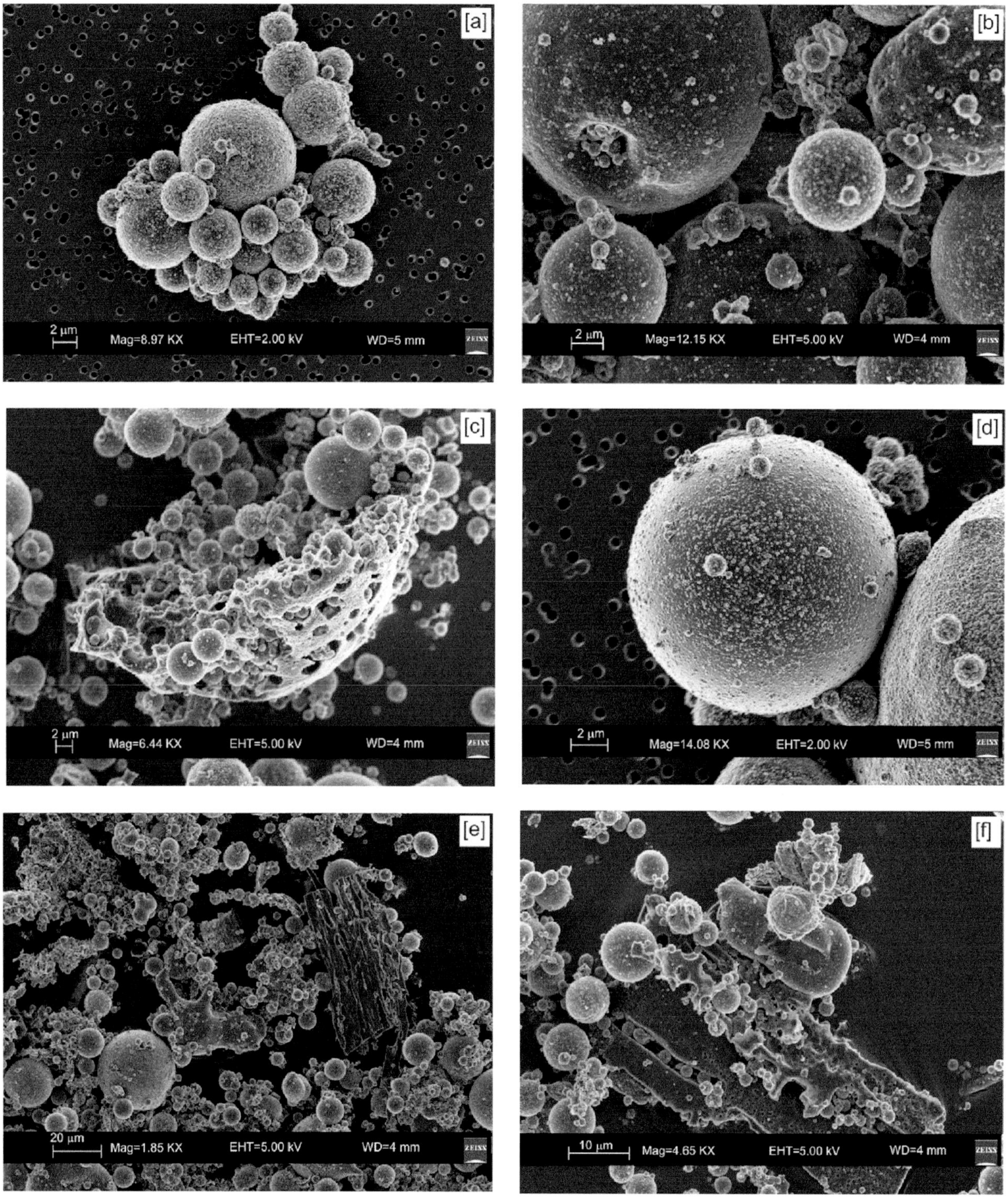

FIGURE 15.3 SEM images of coal fly ash (Belviso, 2018).

C—S—H with a reduced calcium-to-silicon ratio, as depicted in Fig. 15.4 (Lothenbach et al., 2011). These computations offer insights into the chemical reactions taking place within the system and contribute to understanding the observed alterations in hydration behavior within the mixture. The presence of significant amounts of Al_2O_3 and limited quantities of SO_3 in FAs results in a reduction in ettringite content and an increase in the content of AFm phases when OPC is blended with FA. This phenomenon is attributed to the interplay between sulfate salts derived from SO_3 in FA and their effect on

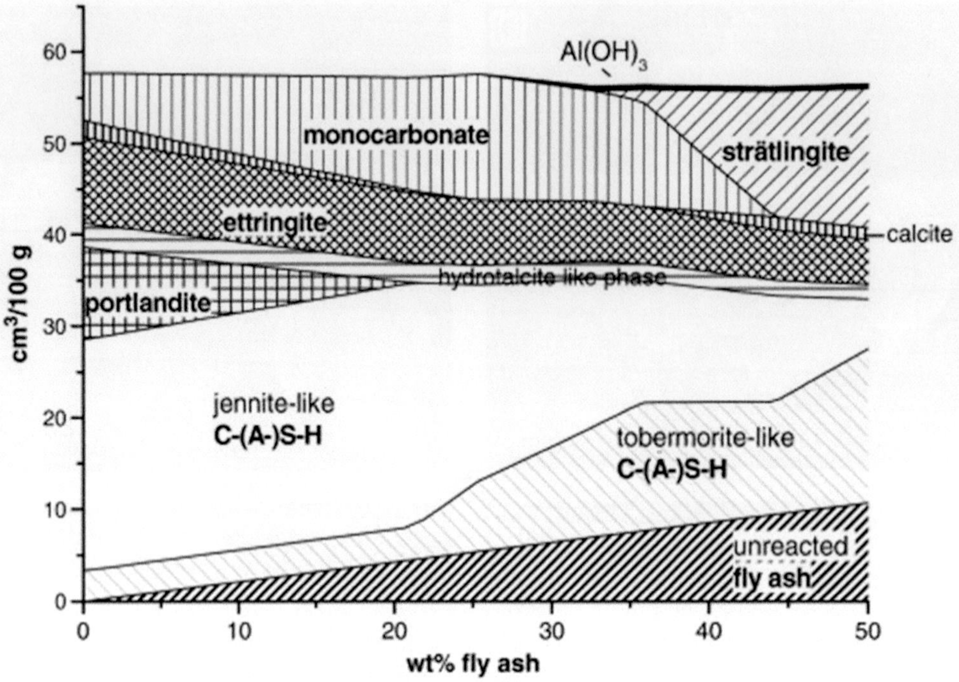

FIGURE 15.4 Morphological changes in hydrated OPC upon incorporation of fly ash, assuming full reaction of the OPC (Lothenbach et al., 2011).

ettringite formation. While the low SO_3 content in FA might be expected to inhibit the formation of ettringite, it is noteworthy that the presence of sulfate salts can have varying impacts on the reactions within the system. At the optimal SO_3 content, C_3A engages with sulfates present in the solution, leading to the formation of ettringite (Andrade Neto et al., 2021). The phase changes depicted in Fig. 15.4, as calculated, are consistent with experimental observations reported in the existing literature. In specific experimental studies referenced as Lothenbach et al. (2011) and Deschner et al. (2012), when OPC is substituted with FA at a rate of 60% or higher, portlandite is observed to completely deplete over hydration durations of 1 year or longer. These findings provide empirical evidence supporting the accuracy and relevance of the calculated phase changes, further confirming the impact of PFA on the hydration products and long-term characteristics of the cementitious system.

Existing literature consistently reports that FA reactivity is relatively slower at ambient temperatures (Ridtirud et al., 2011; Zhang et al., 2016; Özyurt et al., 2020; Niş and Altõndal, 2022). Fig. 15.5 illustrates the evolution of portlandite and

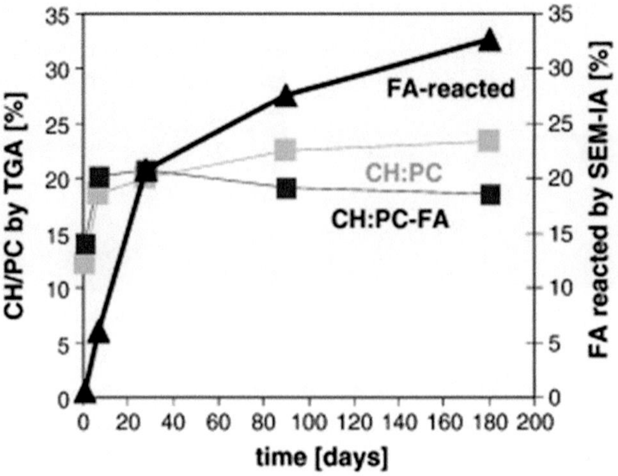

FIGURE 15.5 Comparison of portlandite concentration and extent of FA reaction in an OPC paste and a paste containing 65% OPC and 35% fly ash (Lothenbach et al., 2011).

the FA reaction degree in a paste comprising 65% OPC and 35% FA over a period of 200 days. When compared to a paste formed solely from OPC, the inclusion of FA results in a slight increase in the quantity of portlandite during the initial phases of hydration. The increased generation of portlandite is responsible for the filler effect of FA, which accelerates the hydration of the clinker. The fundamental factor behind why the hydrate assemblage in OPC-FA blends initially mimics that of pure OPC systems is the slow reaction rate of FA. This assemblage typically includes C−S−H, portlandite, ettringite, and AFm (alkali-free) phases such as monocarbonate or monosulfate. Research studies carried out by Yingliang et al. (2020) and Sun et al. (2021) have revealed that higher FA volume in the blend results in further deceleration of the reaction. However, the initial stages of the FA's pozzolanic reaction, as evidenced by the reduction in portlandite and alterations in pore solution chemistry, are undergoing a transition toward earlier hydration times under higher temperature conditions (Deschner et al., 2013). These findings highlight the dependence of FA reactivity on parameters such as its proportion in the blend and the temperature conditions during hydration. Understanding these factors makes it feasible to modify the characteristics and efficiency of FA utilization in cementitious systems.

3.1 Kinetics of hydration of PFA-based SCMs

The integration of FA as an SCM exerts a notable influence on the fundamental properties of cement-based systems. This impact is contingent upon various factors, including the quantity and quality of the FA additive. The influence of FA on concrete properties is shaped by a variety of significant elements. These encompass its chemical and mineral characteristics, level of fineness, reactivity, as well as the specific cement type utilized, the ratio of water to binder, and the conditions under which the concrete is cured (Marinković and Dragaš, 2018; Xu and Shi, 2018).

Regarding the kinetics of hydration, studies have indicated that an OPC system with an increased volume of FA exhibits a reduced rate of hydration. This retardation stems from the development of a C−S−H gel within the paste comprising FA and OPC, characterized by a reduced calcium-to-silicon ratio owing to its lower packing density and elevated diffusivity (Wang and Lee, 2010; De la Varga et al., 2018). The filler effect of the FA is unable to entirely compensate for the phenomenon of the slower rate of transformation of the low Ca/Si ratio C−S−H layer into a stable C−S−H layer (Narmluk and Nawa, 2011; Xu et al., 2017). The replacement rate of FA, the hydration level of OPC, and the curing age are all found to be positively correlated (Zeng et al., 2012; Park and Choi, 2021). The occurrence of this phenomenon can be attributed to various factors. Primarily, the utilization of high-volume fly ash (HVFA) leads to an expansion in the separation distance between OPC particles, resulting in the formation of a broader zone for hydration reactions. Consequently, this extended reaction zone enhances the physical filler effect, thereby facilitating a more comprehensive hydration of the OPC (Oey et al., 2013; Xu et al., 2017). Secondly, HVFA actively contributes to the partial deflocculation of OPC particles (De la Varga et al., 2018). Lastly, HVFA provides a substantial surface area for OPC particles and acts as novel nucleation sites for the hydration process, consequently promoting and facilitating the overall reaction (Park and Choi, 2021).

A commonly noted phenomenon is that the pozzolanic reaction of FA tends to become more noticeable beyond the initial 28-day duration (Deschner et al., 2013; Alzeebaree et al., 2021). As the hydration of OPC progresses, it generates calcium hydroxide that further reacts with FA, enhancing the overall pozzolanic reaction (Park and Choi, 2021). However, increased FA concentration limits the amount of CH that is available, thereby diminishing the intensity of the FA reactivity. Furthermore, temperature has a substantial impact on both the pozzolanic reactivity of HVFA and the hydration of OPC. Elevated temperatures tend to activate the pozzolanicity of HVFA, leading to the usage of water and the generation of C−S−H to fill the pores (Narmluk and Nawa, 2011). As a result, in the later phases of the hydration process, this could impede the cement matrix's hydration. The hydration response within HVFA blends displays diversity contingent on the type of FA applied, particularly C-FA and F-FA. C-FA, characterized by a higher CaO content and lower SiO_4 polymerization degree, exhibits cementitious characteristics, which result in faster early hydration reactions (Wattanasiriwech et al., 2021). F-FA, on the other hand, responds to hydration more slowly initially because it has more inert crystalline phases. But the greater pozzolanic activity of F-FA ultimately helps in the development of the HVFA mixture's strength (Hemalatha and Ramaswamy, 2017b; Wattanasiriwech et al., 2021).

3.2 Workability of PFA-based SCMs

The workability of cement-based systems experiences a notable impact from the proportion of FA present. As illustrated in Fig. 15.6, HVFA concrete demonstrates a more pronounced slump when contrasted with low-volume fly ash concrete (LWFA) and OPC concrete, signifying enhanced workability in the former scenario. A study by Duran-Herrera et al. (2011) found that slump loss gradually decreases as concentration of FA in concrete increases. There are several potential

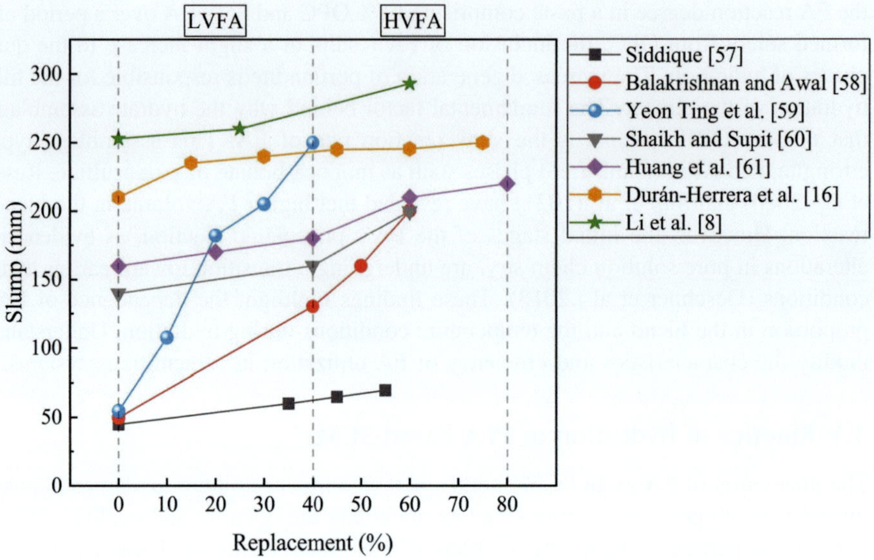

FIGURE 15.6 Slump test literature results of OPC concrete, LVFAC and HVFAC (Li et al., 2022).

causes for this phenomenon. Firstly, the low density of FA contributes to a rise in the overall volume of the mixture (Khatib, 2008). Secondly, the presence of FA hinders the flocculation of OPC particles, thereby improving the workability of the mixture (Bentz et al., 2012). Thirdly, higher FA replacement rates tend to slow down the hydration process (Bentz et al., 2012), which can help maintain the workability of the concrete for an extended period. Last but not least, FA particles' round and smooth form promotes a bearing effect and lessens friction between particles, improving the concrete mixture's overall workability (Ahari et al., 2015; Yang et al., 2018).

3.3 Strength development

This section of the chapter explores the influence of HVFA on the overall mechanical performance, considering the significant impact of the pozzolanic reaction of FA on concrete's mechanical properties. The study examines multiple variables as key factors influencing the mechanical performance of HVFA-incorporated concrete. These variables encompass mix proportion, water/binder ratio, source, type, and replacement rate. While replacing OPC with HVFA might negatively affect the initial compressive strength of concrete, it is noteworthy that HVFA concrete exhibits favorable long-term strength characteristics. This renders it well-suited for practical utilization within the realm of structural engineering. Fig. 15.7 underscores that, at a 28-day curing age, there is a noticeable reduction in the compressive strength of diverse concrete types—including OPCC, self-compacting concrete, high-performance concrete, and engineered cementitious composites—as the FA replacement rate increases, primarily due to the altered particle packing and increased porosity resulting from higher FA content. Further to that, certain types of FA might contain impurities or mineralogical components that adversely affect the overall strength gain of concrete. This variation in chemical composition can lead to differing levels of strength reduction as the FA replacement rate increases. The strength of HVFA concrete is influenced by both the physical and chemical attributes of FA, as well as the inclusion of additives. Drawing from a range of studies, it becomes evident that high-volume C-FA concrete consistently exhibits higher 28-day compressive strength when compared with its counterpart, high-volume F-FA concrete (Sumer, 2012; Uysal and Akyuncu, 2012; Huang et al., 2013). The higher levels of CaO and SO_3 in C-FA, as well as its self-cementing characteristics, are responsible for this increased strength (Ponikiewski and Gołaszewski, 2014; Zhang et al., 2019).

Hannesson et al. (2012) explored the impact of an HVFA on the development of compressive strength in self-consolidating concrete. Fig. 15.8 displays the development of concrete's compressive strength, where cement is replaced with FA at levels of 20%, 40%, 60%, 80%, and 100%, observed at varying curing intervals of 7th, 14th, 28th, 56th, 84th, and 168th days. The compressive strength of all concrete mixtures increases as they age. A careful examination of the provided data reveals that, at the 7-day interval, both the control specimen (FA-0) with a strength of 64.6 MPa and FA-20 at 64.6 MPa manifest identical compressive strengths. This observation posits that up to a 20% replacement of cement with FA did not substantially alter the early-age strength of concrete. However, with an increase in the FA content, especially from FA-40 upward, there is a marked decline in early strength, with FA-80 and FA-100 recording just 6.1 and

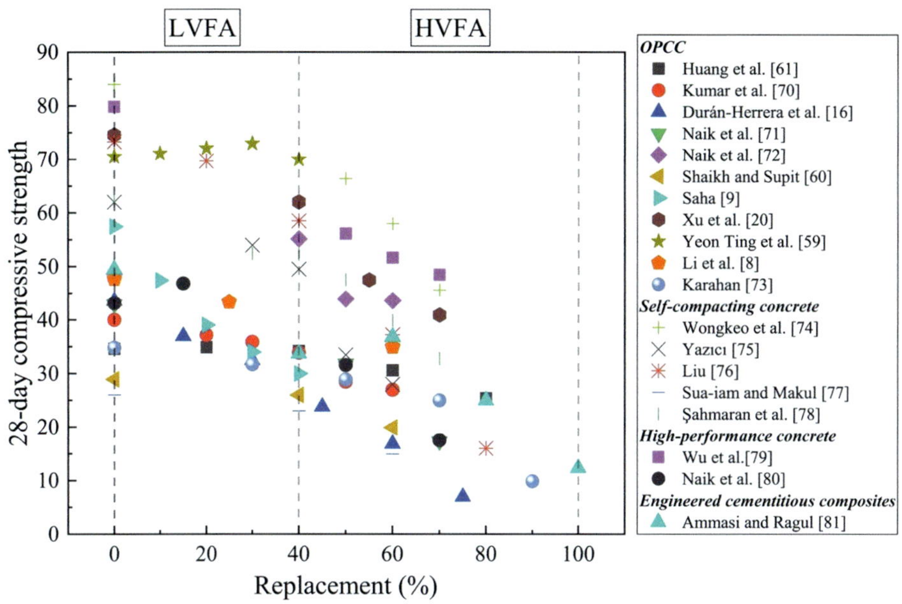

FIGURE 15.7 Changes in 28th day compressive strength with FA concentration (Li et al., 2022).

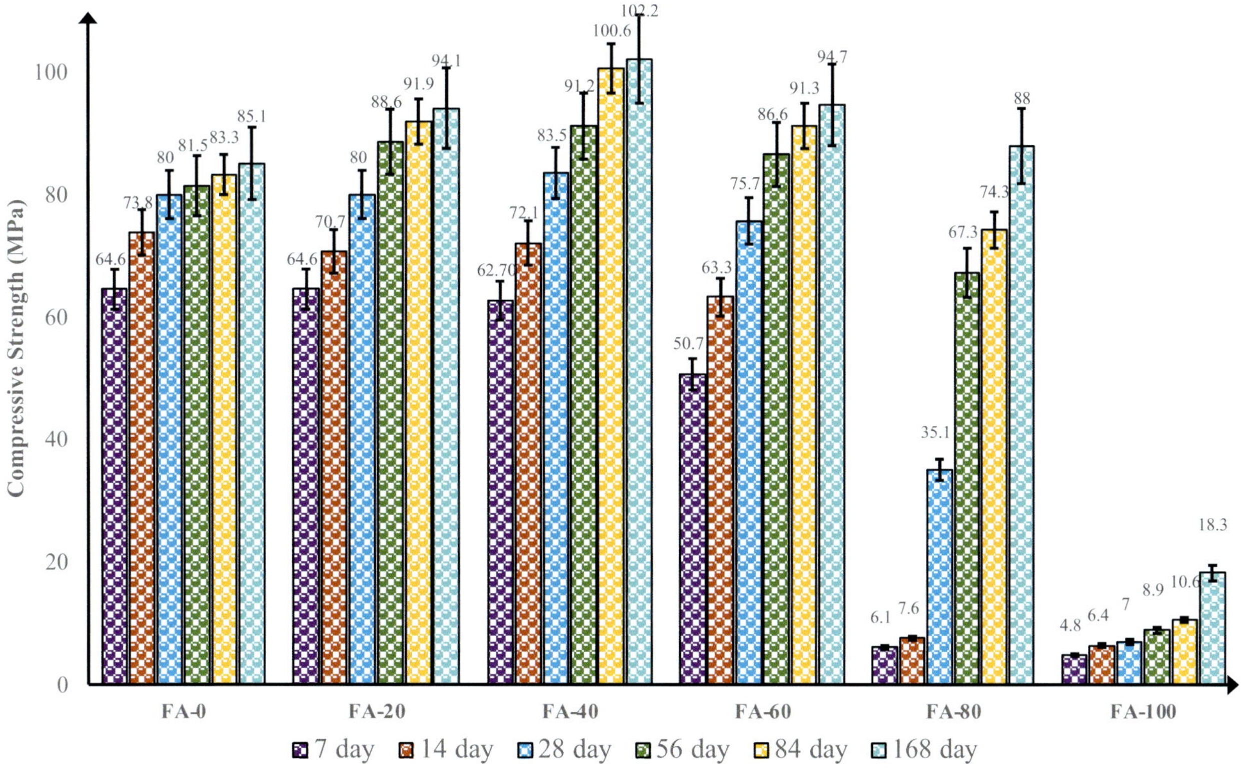

FIGURE 15.8 Compressive strength development with different FA concentrations, and curing duration. *Data adapted from Hannesson, G., Kuder, K., Shogren, R., Lehman, D., 2012. The influence of high volume of fly ash and slag on the compressive strength of self-consolidating concrete. Construct. Build. Mater. 30, 161–168.*

4.8 MPa, respectively. As the specimens age, the strength patterns change. By the 28th day, the FA-20 mixture, with a strength of 80 MPa, starts to exceed the control mixture in compressive strength. Surprisingly, even though the FA-40 mixture began weaker, by the 28th day, it achieves a strength of 83.5 MPa, surpassing both the control and FA-20 mixes. This notable increase in strength for FA-40, reaching 72.1 MPa by the 14th day, can be attributed to the slower formation of C—S—H. As more cement is replaced with FA, this chemical reaction becomes more active, possibly contributing to the growth in strength. Looking further into the 168-day mark, the compressive strength trajectory reveals that mixtures with up to 60% FA replacement, like FA-60 that reaches 94.7 MPa, not only recover their early strength shortfalls but occasionally even exceed the control's strength, which stands at 85.1 MPa. However, the FA-80 and FA-100 mixtures, although showing improvement, continue to lag behind significantly, with the 100% replacement mixture only achieving 18.3 MPa.

In summary, while the addition of higher percentages of FA may initially impact the compressive strength negatively, medium- to long-term strength can experience benefits, particularly with replacements up to 60%. Yet, there appears to be a limit beyond which the advantages diminish. These observations underline the importance of selecting the FA content to optimize the balance between early and long-term strengths. Further studies would be beneficial to explore the microstructural and chemical interactions to derive a more comprehensive understanding of the observed strength variations.

3.4 Durability properties

Considering the inherent susceptibility of concrete to durability degradation over its operational life span, it is imperative that concrete structures exhibit resistance against erosion throughout their service life. In this regard, the pozzolanic reaction of PFA plays a pivotal role in altering the microstructural characteristics of concrete, consequently yielding diverse impacts on durability-related parameters including carbonation susceptibility, abrasion resistance, water absorption, and chloride resistance. The existing literature highlights several critical factors that significantly influence the durability characteristics of PFA mixes (Mardani-Aghabaglou et al., 2013; Supit and Shaikh, 2015; Sathyan and Anand, 2019).

High compressive strength and low porosity, in general, have been found to increase resistance to carbonation in the literature (Younsi et al., 2011; Ogawa et al., 2020). Studies show that as compared with OPC concrete, HVFA concrete has a lower carbonation resistance (Van den Heede and De Belie, 2014; Singh and Singh, 2016). Herath et al. (2020) reported that a significant amount of calcium hydroxide particles is consumed by pozzolanic reaction occurring in HVFA mix, consequently reducing the CO_2 absorption capacity. Parallel discoveries have been recorded by other researchers (Ramachandran et al., 2020; Yang et al., 2021). Owing to the reduced concentration of calcium hydroxide in HVFA concrete, CO_2 can penetrate more readily into regions characterized by limited CH content. This results in an amplified carbonation of the C—S—H phase and a decrease in its calcium-to-silicon ratio. This process ultimately results in carbonation-induced shrinkage (Van den Heede and De Belie, 2014).

A consistent pattern emerges with an elevation in water absorption as the replacement rate of FA in HVFA concrete increases. This phenomenon can be clarified by the strong linear correlation between water absorption and the presence of permeable pores within the concrete matrix. The volume of permeable pores increases with FA content, which leads to an increased capacity for water absorption. In other words, as the amount of permeable pore space decreases, the water absorption of HVFA concrete diminishes as well, as reported in references Wongkeo et al. (2014) and Mardani-Aghabaglou et al. (2013). It should be noted that the HVFA cementitious systems, characterized by small pores, exhibit extensive interconnectivity. As explained in Silva and de Brito (2013), this causes substantial water absorption as a result of capillary forces within these pores.

The primary cause of drying shrinkage in concrete stems from the evaporation of water that had been absorbed by the gel-like cement paste. This water loss causes the paste's volume to decrease, which ultimately causes drying shrinkage (Hojati and Radlińska, 2017). Observations suggest that as the FA content rises, the rate of drying shrinkage in HVFA concrete appears to diminish (Li et al., 2022). In comparison with LVFA concrete and OPC concrete, HVFA concrete demonstrates higher drying shrinkage resistance when the FA replacement rate does not exceed 70%. This is attributed to the presence of a relatively modest proportion of hydrated paste within HVFA concrete. Unhydrated FA can operate as an aggregate to limit shrinkage, densify the mixture, and stop internal water from evaporating all at the same time (Hemalatha and Ramaswamy, 2017b). Research conducted by Wu et al. (2005) examined the evolution of HVFA concrete's drying shrinkage over time. The study findings indicate that HVFA concrete undergoes significant shrinkage during the initial stages, with a particularly rapid rate observed within the first 60 days. However, it is noteworthy that this shrinkage gradually decelerates over time. This observation suggests that the early stages of HVFA concrete exhibit more pronounced shrinkage, which gradually stabilizes as the concrete matures. It is important to consider this shrinkage behavior when designing and implementing HVFAC in construction projects to ensure appropriate control and management of potential shrinkage-related issues. Furthermore, Kate and Murnal (2013) exhibited that the shrinkage rate of LVFA concrete tends to achieve a more consistent pattern over time, while HVFAC typically displays a tendency of increased shrinkage beyond the initial 28 days. These findings emphasize that the shrinkage characteristics of LVFA and HVFA concrete differ, with LVFA concrete showing more consistent and predictable shrinkage behavior compared with HVFA concrete. It is essential to consider these variations in shrinkage behavior when selecting and utilizing FA concrete in construction applications.

The substitution level of FA impacts the chloride resistance of HVFAC. The total charge transferred, measured in Coulombs, tends to decrease with an increase in the FA replacement rate. HVFAC with an FA replacement rate below 70% demonstrates superior chloride resistance compared to both LWFA concrete and OPC concrete, as reported in the study referenced as Li et al. (2022). According to ASTM C1202 (ASTM, 1991), in most scenarios, HVFAC falls within the "low" or "very low" class for chloride permeation. The following list of factors can be used to explain the improvement in chloride resistance in HVFA concrete.

1. Pore structure modification: The presence of HVFA within the mixture brings about alterations in the pore structure. Consequently, the capillary network becomes more convoluted, and the connectivity between pores diminishes. These effects are a result of the pozzolanic reaction's production of C−S−H (Filho et al., 2013; Karahan, 2017).
2. Blocking of porosity: The products generated by the pozzolanic reaction act to block connected porosity. They also lower the pore diameter in the interfacial transition zone (ITZ) between the aggregate and the surrounding matrix. These effects lead to an improvement in the ITZ and the overall compactness of the concrete (Karahan, 2017; Shiyu et al., 2019)
3. Decrease in Alkali Ion Concentration: As FA replacement rates increase, the amount of alkali ions in the concrete pore solution decreases. Since the alkalinity of the concrete pore solution affects chloride ingress, this drop in alkali ion concentration helps to prevent chloride from penetrating the concrete (Şahmaran et al., 2009).
4. High Alumina Content: HVFAC typically contains a higher content of alumina due to the presence of FA. This increased alumina content leads to an elevation in the content of Tricalcium Aluminate (C3A) (Hemalatha and Ramaswamy, 2017b).

Freeze-thaw (F-T) damage is a significant factor contributing to structural deterioration and aging in infrastructure. Concrete's resistance to F-T cycles is affected in two different ways by the addition of HVFA. FA can function as a filler by altering the pore structure, thereby improving the transport attributes within the concrete matrix. Consequently, the inclusion of HVFA enhances the concrete's resistance to F-T damage (Du et al., 2019). In a study conducted by Yazõcõ (2008), it was found that following 90 cycles of F-T exposure, the remaining compressive strength of concrete containing 40%−60% HVFA surpasses that of OPC concrete. However, it should be noted that HVFA concrete exhibits certain challenges regarding F-T resistance. This is mainly because HVFA concrete hydrates slowly and has lesser strength in the early phases of the F-T cycle. As a result, HVFA concrete is more susceptible to water loss and degradation of the surface layer, which ultimately leads to a decrease in F-T resistance (Hasholt et al., 2019; Zhao et al., 2020). According to Pushpalal et al. (2022), the inclusion of C-FA with a high volume content can promote the formation of AFt compounds. These compounds possess the capacity to impede the entrained air void system within concrete. As a result, concrete with a high volume of F-FA displays superior resistance to F-T cycles when compared to concrete with a high volume of C-FA. To affirm and delve deeper into the influence of HVFA on the F-T resistance of cementitious materials, additional research is warranted.

4. Recycling PFA into aggregates

A significant milestone in effectively addressing the extensive disposal of FA has been achieved through the development of artificial aggregates utilizing FA as a raw material. These artificial aggregates offer a sustainable solution for managing FA waste by transforming it into a useful building material. In the composition of concrete, the aggregate phase typically occupies a substantial volume, ranging from 60% to 80%. As concerns regarding the depletion of natural aggregate resources continue to grow, ensuring sustainable development practices becomes paramount. Preserving finite natural aggregate resources is an imperative task. The utilization of synthetic aggregates produced from FA signifies a noteworthy advancement in addressing the issues linked to FA disposal, while also safeguarding and preserving the limited natural aggregate reserves. FA can be converted into synthetic aggregates using diverse methods, which encompass sintering, hydrothermal treatment, and cold bonding processes. The utilization rates of FA for each of these processes are approximately 90%−100%, 47%, and 60%−75% respectively, as reported by references (Van der Wegen and Bijen, 1985; Nadesan and Dinakar, 2017). These data indicate that sintering is the best method for attaining widespread FA usage. FA recycling offers flexibility in application for both structural and nonstructural uses. The integration of these synthetic aggregates brings forth numerous benefits, encompassing the reduction of dead loads, steel reinforcement requirements, transportation expenses, and the acceleration of construction processes.

The principal procedure in the production of aggregates from FA comprises three consecutive phases: raw material mixing, pelletization, and hardening. In the context of constituent materials, coal combustion residuals represent the predominant component utilized in the formation of sintered fly ash (SFA) aggregates. The production of SFA aggregates can employ coal ash derived from both bituminous and lignite-based coal (Vasugi and Ramamurthy, 2014). Furthermore, through appropriate preparatory measures, it becomes feasible to manufacture aggregates from diverse classifications of coal ash, encompassing fly ash, bottom ash, as well as C-FA or F-FA (Geetha and Ramamurthy, 2011; Vasugi and Ramamurthy, 2014). The mixing stage involves the meticulous combination of ingredients in appropriate proportions until the desired consistency is attained. Subsequently, the pelletization process entails the agglomeration of fine particles through the utilization of a suitable binding agent. The hardening of the newly formed pellets can be achieved through various methods, including sintering, autoclaving, or cold bonding. Earlier studies have provided detailed descriptions of multiple production processes for artificial fly ash aggregates, as documented by references (Kockal and Ozturan, 2010; Gomathi and Sivakumar, 2015).

The characteristics of aggregates play a pivotal role in determining the properties and behavior of concrete. To ensure the production of high-quality concrete, a thorough understanding of these variables is indispensable. Some of the key characteristics associated with FA aggregates include density, strength, and water absorption. Moreover, it is evident that several factors, including the type and quantity of the binding material employed, as well as the temperature and duration of the sintering procedure, exert significant influence on the mechanical characteristics of FA aggregates. Particle packing and the interlocking of aggregates inside the matrix are significantly influenced by the morphological structure of the aggregates (Nadesan and Dinakar, 2017). As seen in Fig. 15.9, SFA aggregates have a spherical shape, a brown color, and an interior black core. This property is a result of both the iron's oxidation state and the presence of carbon. While the macroscopic appearance of the aggregates appears smooth, microscopic analysis reveals a relatively rough surface with open pores that range in size from 10 to 200 μm (Nadesan and Dinakar, 2017). The texture of aggregate surfaces has the potential to influence the frictional properties experienced by a mixture, thereby affecting the workability of the mixture in its fresh state. Furthermore, there is a hypothesis suggesting that cement paste or the products of cement hydration could potentially permeate into larger pores or voids that exist on the surfaces of porous aggregates or those characterized by a coarse texture. This phenomenon serves as numerous "hooks," promoting an interaction that binds the paste and aggregate phases (Zhang and Gjørv, 1990).

A recent study by Dong et al. (2022) concentrated on the development and production of artificial aggregates using alkali-activated technology. In this innovative approach, the researchers achieved successful synthesis of artificial aggregates utilizing alkali-activated technology, wherein FA served as the precursor and a blended solution of Na_2SiO_3, and NaOH functioned as the activator. The synthesized samples were found to meet the prescribed engineering standards, demonstrating the effectiveness of the production process. Balapour et al. (2022) undertook a comprehensive study aimed at analyzing and characterizing the physicomechanical characteristics of lightweight aggregates (LWAs). These aggregates were produced using a thermodynamics-guided procedure that the researchers had previously developed from both low-calcium and high-calcium waste FA. The investigation concerned on evaluating the characteristics and performance of the resulting LWA derived from the two types of FA. The selection of LWAs was aimed to explore the impact of FA type and NaOH concentration on their physical appearance. Detailed images of the LWA were captured from a frontal view, and Fig. 15.10 depicts the visual characteristics of both F-FA and C-FA LWA samples. The gradual addition of NaOH caused a

FIGURE 15.9 SFA aggregates and cross section of single aggregate grain (Nadesan and Dinakar, 2017).

FIGURE 15.10 Digital images of F-FA and C-FA LWA sintered at 1160°C for 4 min; varying NaOH concentrations. *Adapted from Balapour, M., Thway, T., Moser, N., Garboczi, E.J., Grace Hsuan, Y., Farnam, Y., 2022. Engineering properties and pore structure of lightweight aggregates produced from off-spec fly ash. Construct. Build. Mater. 348, 128645.*

transition in the color of the 2% and 4% FFA samples, altering them from the initial reddish-brown of the 0% FFA sample to a darker shade of brown. Interestingly, a subtle shiny surface was observed in the FFA LWA samples with NaOH additions, suggesting that a liquid phase may have formed during the sintering process. In contrast, after the addition of NaOH, the color of the C-FA LWA was changed from yellow in the C-FA 0% sample to a lighter brown tone in the C-FA 4% sample. Notably, the sample with 0% C-FA exhibited evident surface cracking, a phenomenon ascribed to the interplay between insufficient liquid phase formation and thermal contraction during the cooling stage following sintering. On the surfaces of C-FA LWA samples made with NaOH (i.e., C-FA 2% to C-FA 4%), however, the occurrence of cracking was less noticeable.

4.1 Properties and performance of PFA-aggregate concrete

PFA aggregates offers several advantages over natural coarse aggregate (NCA), including cost reduction and reduced structural weight. Specifically, the utilization of SFA aggregates proves to be more advantageous in the manufacturing of high-strength concrete. Previous literature indicates that the durability characteristics of PFA aggregate-based concrete are comparable with those of conventional concrete (Terzić et al., 2015). Additionally, literature suggest that lightweight concrete (LWC) incorporating 100% PFA aggregate, either with or without supplementary cementitious materials, demonstrates satisfactory improvements in strength (Cui et al., 2012; Nadesan and Dinakar, 2017).

The work by Patel et al. (2019) focused on examining the durability and microstructural properties of LWC, wherein fly ash cenosphere (FAC) and SFA aggregate were utilized as substitutes for natural fine and coarse aggregates, respectively. Fig. 15.11 illustrates the changes in compressive strength observed in various concrete mixes containing varying

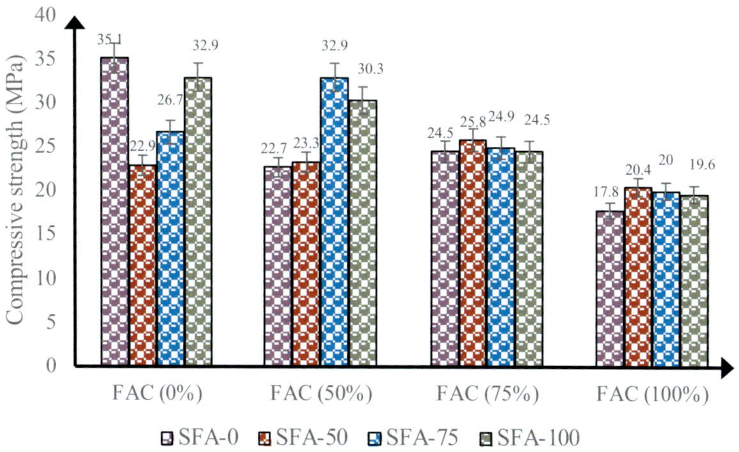

FIGURE 15.11 Twenty-eighth day compressive strength of mixes with varying SFA aggregate and FAC concentrations. *Data adapted from Patel, S.K., Majhi, R.K., Satpathy, H.P., Nayak, A.N., 2019. Durability and microstructural properties of lightweight concrete manufactured with fly ash cenosphere and sintered fly ash aggregate. Construct. Build. Mater. 226, 579–590.*

combinations of SFA aggregates and FAC after 28 days. The control mix, which does not incorporate any replacement of SFA aggregate or FAC, exhibits a compressive strength of ~35 MPa at the 28-day mark. The introduction of 50%, 75%, and 100% SFA aggregate results in reductions of around 35%, 24%, and 6% in the compressive strength of the reference specimen, respectively. This shows that the compressive strength of concrete containing any amount of SFA aggregate, whether it is 50%, 75%, or 100%, is lower than that of the control mix. This can be explained by the fact that SFA aggregate has less strength and stiffness than NCA. The loss of concrete strength does, however, gradually reduce with the addition of 50% SFA aggregate, particularly when used in combination with 75% and 100% SFA aggregate. This suggests that adding 75% and 100% SFA aggregate substantially increases the concrete's strength. The complex and multidimensional interaction between the ITZ's outstanding characteristics and the SFA aggregate's poor performance can also be attributed for the variety in strengths. The ITZ's superior characteristics are attained through a multitude of factors linked to SFA aggregate. In particular, when the amount of SFA aggregate increases, the roughened surfaces of aggregate, as described earlier, greatly improve the interfacial adhesion between SFA aggregate and the cementitious matrix. Furthermore, the pozzolanic characteristics of SFA aggregate facilitate the creation of chemical bonds between aggregate and cement paste, fostering a strong interconnection between the two components (Nadesan and Dinakar, 2018). Additionally, the incorporation of dry SFA aggregate mitigates the occurrence of the "wall effect" and enables water absorption within the concrete during later stages, unlike traditional concrete. This advantageous feature enhances the interfacial properties between SFA aggregate and the cementitious matrix. By increasing the amount of SFA aggregate, the ITZ characteristics are enhanced as outlined earlier. The strength and stiffness characteristics of SFA aggregate are significantly influenced by these advancements. As a result, the lower strength and stiffness properties of SFA aggregates were eventually offset by the stronger ITZ features (Satpathy et al., 2019). The results demonstrate that concrete mixes comprising solely of FAC as the fine aggregate exhibit a noticeable decrease in compressive strength. However, when SFA aggregate is introduced as the coarse aggregate in conjunction with FAC, there is an improvement in compressive strength. Specifically, at a 50% FAC replacement, the compressive strength increases with the incorporation of SFA aggregate up to a 75% content, after which it experiences a slight decline.

Gomathi and Sivakumar (2015) studied the mechanical performance of concrete that incorporated lightweight cold-bonded and SFA aggregates under steam and hot water curing. The results presented in Fig. 15.12 demonstrate that the strength tests exhibit a significant improvement when using a maximum substitution (62%) of SFA aggregates as opposed to cold-bonded aggregates. Specifically, concrete mixes cured with hot water and incorporating SFA aggregates demonstrated the highest compressive strength, reaching a maximum of 40 MPa. The test results indicated a reduction in both the strength and elastic modulus of the concrete as the proportion of aggregates increased. Among the various aggregates investigated, the SFA aggregate concrete exhibited the highest elastic modulus, reaching approximately 20 GPa when the aggregate was substituted to its maximum extent. It is worth mentioning that concretes formulated using cold-bonded FA aggregates consistently exhibited lower compressive strength compared with those containing high-strength SFA aggregates. The elevation in compressive strength predominantly stemmed from the superior ultimate crushing

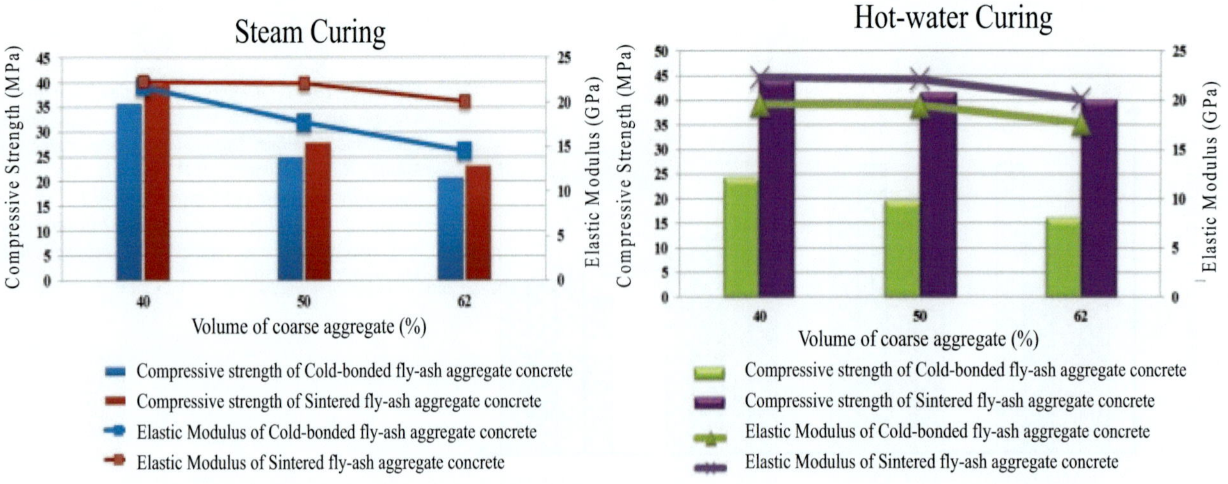

FIGURE 15.12 Compressive strength and elastic modulus of FA aggregate concrete under hot steam and water curing. *Adapted from Gomathi, P., Sivakumar, A., 2015. Accelerated curing effects on the mechanical performance of cold bonded and sintered fly ash aggregate concrete. Construct. Build. Mater. 77, 276–287.*

strength of SFA aggregates in comparison with cold-bonded aggregates. Furthermore, the concrete compositions incorporating cold-bonded FA aggregates displayed inferior performance when contrasted with concrete containing SFA aggregates, even in scenarios involving high-volume replacement of aggregates. The aforementioned test results provide compelling evidence that incorporating artificial aggregates enables a high-volume replacement (up to around 60%) of FA in concrete. This points to the potential for employment of PFA as aggregate in concrete production while still reaching ideal concrete characteristics and strength levels.

4.2 Durability properties of PFA-aggregate concrete

Along with mechanical performance, concrete's durability characteristics—particularly in the case of FA aggregate concrete—are of great significance. The permeation characteristics of concrete play a vital role in the determination of its long-term durability. The size of the pores and their connection within the concrete matrix are two elements that have an impact on these permeability capabilities. Despite lightweight aggregate concrete having a higher pore volume, recent research indicates that its durability metrics are comparable to, or in some cases even superior to, those of concrete made with normal aggregates (Dinakar, 2013; Güneyisi et al., 2013).

Kockal and Ozturan (2011) utilized a variety of aggregate types to evaluate the durability characteristics of lightweight FA aggregate concrete. According to the outcomes of the rapid chloride permeability test, the lightweight cold-bonded FA aggregate concrete exhibited the highest chloride permeability, while the lightweight FA aggregate concrete containing additional bentonite demonstrated the lowest chloride permeability. Despite the absence of significant variations in chloride permeability among the different types of concrete, the concrete containing cold-bonded aggregates exhibited relatively higher permeability. This was attributed to its increased absorption rate and greater open porosity. Notably, compared with concrete made with natural aggregates, concrete that included SFA aggregates showed much less chloride ion penetration. The efficiency of SFA lightweight aggregates in protecting reinforcement against corrosion is demonstrated by these data. The findings also underscored the discrepancy between strength and durability. Despite having the highest compressive strength, the normal weight concrete exhibited higher permeability in comparison with both the lightweight bentonite-added sintered FA aggregate concrete and the lightweight FA aggregate concrete.

Gomathi and Sivakumar (2015) conducted a comparative analysis of wetting–drying cycles and water penetration test by employing a simple sorptivity test on cold-bonded and SFA aggregate concrete. In comparison with cold-bonded FA aggregate concrete, SFA aggregate concrete showed less weight loss, according to the experimental results of the study. The reduction in weight experienced through successive wetting and drying cycles provides an indicator of the concrete's durability and ability to withstand weathering. This is evident from the fact that the dry weight of the concrete after 28 days remains notably consistent when compared with its initial weight. The test results further revealed that all concrete specimens encountered a slight weight reduction during the initial days, but this weight loss gradually escalated as the successive wetting and drying cycles advanced. The observed decline in weight can be attributed to the degradation of the microstructure, triggered by the migration of water within the pores. This phenomenon results in a significant decrease in water absorption. In summary, the results highlight the concrete's resistance to continuous moisture movement within the pores, which plays a significant role in the deterioration process. Furthermore, notable differences were observed in the water sorption properties among various concrete mixes containing FA aggregates. The concrete mixes incorporating cold-bonded FA-bentonite aggregates demonstrated the highest water sorption rate, measuring 5.67×10^{-3} cm/s$^{0.5}$, while a lower rate of 4.55×10^{-3} cm/s$^{0.5}$ was recorded for the concrete mix containing 40% cold-bonded FA-bentonite aggregate. Conversely, the concrete with SFA-bentonite aggregates showcased diminished water absorption in comparison with the cold-bonded FA aggregate concrete. This decline in water absorption can be traced back to the sintered aggregates' diminished porosity and more intricate microstructure. Furthermore, it was noted that concrete blends with a greater aggregate content displayed elevated sorptivity values, a trend ascribed to the larger exposed surface area of the aggregates.

5. Current limitations and future perspectives

Drawing from information provided by the American Coal Ash Association (ACAA), coal usage is anticipated to rise by 3.4% in the coming 2 decades, despite the retirement of numerous coal plants (Hemalatha and Ramaswamy, 2017). Considering the increasing FA production and the potential environmental repercussions, it becomes globally essential to comprehend the merits of integrating FA into concrete to enhance sustainability in the construction sector. While existing methods are adept at incorporating substantial amounts of FA into concrete to yield positive results, the full-scale utilization of high volumes of FA is hindered. This is primarily due to challenges such as increased carbonation, slow strength development, and enhanced shrinkage, among other concerns.

Furthermore, FA exhibits a diverse range of attributes including its physical, chemical, and mineralogical properties. The combustion temperature, cooling rate, and comprehensive composition significantly impact both the physical features and cementitious attributes of FA. C-FA presents greater complexities in terms of its chemical and mineral composition when compared with F-FA. Owing to its reduced reactivity, F-FA ash has frequently been employed as a concrete admixture, as it does not substantially modify the chemistry of cement hydration in the immediate term. In recent years, the utilization of C-FA has grown due to increased production. Nonetheless, its integration into concrete has encountered obstacles, notably because it can significantly influence the cement hydration process and currently lacks a well-defined reaction model within concrete. As a result, there is an evident need for additional research to advocate for the extensive use of high C-FA in concrete. Despite the challenges associated with incorporating significant quantities of FA into concrete, viable approaches—such as high-temperature curing, mechanical grinding, and chemical activation—along with complementary substances, including nano silica/nano carbonate, have been explored and have shown promising outcomes. Based on advancements in current research and the documented discoveries to date, the upcoming years could witness an increased substitution of FA, reaching proportions of 50%–60% by weight. This shift could potentially instigate modifications in prevailing standards and codes to support the extensive use of FA.

Having synthesized the insights from numerous research papers on PFA since the year 2000, the subsequent recommendations for upcoming research endeavors are put forth:

- Future studies should delve deeper into the multifaceted properties of PFA when used as SCMs and aggregates, ensuring that construction practices harness its full potential. Research could focus on establishing a universally accepted characterization method that captures the intricacies of FA's mineralogical, physical, and chemical properties.
- To further establish the environmental advantages of PFA, comprehensive studies encompassing its entire life cycle are essential. This includes its sourcing, processing, integration into construction materials, and its eventual impact postconstruction.
- Efforts should be made to identify and mitigate challenges associated with using high volumes of PFA in construction. This involves optimizing the mix design to balance strength, workability, and durability.
- While PFA's use in concrete is well acknowledged, its potential in other construction applications, such as bricks, tiles, or even as a soil stabilizer, can be further explored.
- Given the concerns about PFA's components, especially when it is sourced from different types of coal or processes, research into its potential health implications and mitigation measures is crucial. This would not only enhance safety but also build public trust in FA-based construction materials.
- A standardized testing protocol for PFA, irrespective of its source, will help in its broader acceptance. This can be backed by a dynamic classification system that's adaptive to the latest research findings.

6. Conclusions

The aim of this chapter was to offer a thorough overview of the extensive research conducted on the recycling of PFA into SCMs and aggregates. Existing experimental data and findings from previous studies pertaining to the microstructural and engineering properties of PFA concrete were gathered, compared, and analyzed. Numerous studies and research have demonstrated the advantages of utilizing FA as SCM and artificial aggregate in concrete manufacturing. Increasing its utilization appears to be a viable and promising approach for steering the industry toward a more sustainable trajectory.

The composition of PFA varies based on factors such as coal type and combustion procedures, and it consists of crystalline phases and different chemical types. Standardized methodologies for categorizing FA would enhance consistency and comparability in its utilization. Significant characteristics of FA crucial for concrete production encompass particle size distribution, bulk density, specific gravity, and moisture content. The spherical particles with a wide size range offer a larger surface area for effective pozzolanic reactions. HVFA concrete exhibits long-term strength characteristics despite potentially lower early-age compressive strength. The pozzolanic reaction becomes more pronounced over time, influenced by the replacement rate and temperature. C-FA concrete generally shows higher compressive strength due to its composition and self-cementing properties. While HVFA concrete may have lower carbonation resistance, it demonstrates improved chloride resistance and freeze-thaw resistance with higher FA replacement rates. HVFA concrete also shows reduced drying shrinkage, attributed to modifications in the pore structure. Further research is needed to explore the impact of HVFA on freeze-thaw resistance.

The recycling of FA into artificial aggregates offers a sustainable solution, with sintering being the most efficient process. FA aggregates exhibit varying characteristics influenced by binder type and sintering process. Lightweight concrete with 100% PFA aggregates demonstrates satisfactory strength improvements. SFA aggregates show lower initial compressive strength but exhibit increased strength with higher substitution levels, benefiting from improved interfacial adhesion and pozzolanic characteristics. Incorporating SFA aggregates in conjunction with fine aggregates improves compressive strength. SFA aggregates outperform cold-bonded aggregates in terms of strength and performance. PFA-aggregate concrete demonstrates comparable or superior durability properties compared with concrete with normal aggregates. SFA aggregate concrete resists chloride ion penetration and moisture movement, leading to less weight loss during wetting—drying cycles. Water sorption is influenced by aggregate type, with SFA-bentonite aggregate concrete showing lower sorption due to its refined microstructure. In summary, the utilization of FA as SCMs and artificial aggregates in concrete production offers significant environmental advantages and contributes to the industry's sustainability goals. The comprehensive research conducted in this area provides valuable insights for optimizing the use of FA and enhancing the performance and durability of PFA-aggregate concrete.

References

Ahari, R.S., Erdem, T.K., Ramyar, K., 2015. Effect of various supplementary cementitious materials on rheological properties of self-consolidating concrete. Construct. Build. Mater. 75, 89—98.

Ahmaruzzaman, M., 2010. A review on the utilization of fly ash. Prog. Energy Combust. Sci. 36, 327—363.

Alzeebaree, R., Mawlod, A.O., Mohammedameen, A., Niş, A., 2021. Using of recycled clay brick/fine soil to produce sodium hydroxide alkali activated mortars. Adv. Struct. Eng. 24 (13), 2996—3009.

Andrade Neto, J.d.S., De la Torre, A.G., Kirchheim, A.P., 2021. Effects of sulfates on the hydration of Portland cement — a review. Construct. Build. Mater. 279, 122428.

Assi, L.N., Deaver, E.E., Ziehl, P., 2018. Effect of source and particle size distribution on the mechanical and microstructural properties of fly Ash-Based geopolymer concrete. Construct. Build. Mater. 167, 372—380.

ASTM C618, A, 2019. Standard Specification for Coal Fly Ash and Raw or Calcined Natural Pozzolan for Use in Concrete. ASTM international.

ASTM, C, 1991. Standard Test Method for Electrical Indication of Concrete's Ability to Resist Chloride Ion Penetration. American society for testing and materials.

Balapour, M., Thway, T., Moser, N., Garboczi, E.J., Grace Hsuan, Y., Farnam, Y., 2022. Engineering properties and pore structure of lightweight aggregates produced from off-spec fly ash. Construct. Build. Mater. 348, 128645.

Belviso, C., 2018. State-of-the-art applications of fly ash from coal and biomass: a focus on zeolite synthesis processes and issues. Prog. Energy Combust. Sci. 65, 109—135.

Bentz, D.P., Ferraris, C.F., Galler, M.A., Hansen, A.S., Guynn, J.M., 2012. Influence of particle size distributions on yield stress and viscosity of cement—fly ash pastes. Cement Concr. Res. 42, 404—409.

Cui, H., Lo, T.Y., Memon, S.A., Xu, W., 2012. Effect of lightweight aggregates on the mechanical properties and brittleness of lightweight aggregate concrete. Construct. Build. Mater. 35, 149—158.

Dananjayan, R.R.T., Kandasamy, P., Andimuthu, R., 2016. Direct mineral carbonation of coal fly ash for CO2 sequestration. J. Clean. Prod. 112, 4173—4182.

Danish, A., Mosaberpanah, M.A., 2021. Influence of cenospheres and fly ash on the mechanical and durability properties of high-performance cement mortar under different curing regimes. Construct. Build. Mater. 279, 122458.

Danish, A., Ozbakkaloglu, T., Ali Mosaberpanah, M., Salim, M.U., Bayram, M., Yeon, J.H., Jafar, K., 2022. Sustainability benefits and commercialization challenges and strategies of geopolymer concrete: a review. J. Build. Eng. 58, 105005.

De la Varga, I., Castro, J., Bentz, D.P., Zunino, F., Weiss, J., 2018. Evaluating the hydration of high volume fly ash mixtures using chemically inert fillers. Construct. Build. Mater. 161, 221—228.

Deschner, F., Lothenbach, B., Winnefeld, F., Neubauer, J., 2013. Effect of temperature on the hydration of Portland cement blended with siliceous fly ash. Cement Concr. Res. 52, 169—181.

Deschner, F., Winnefeld, F., Lothenbach, B., Seufert, S., Schwesig, P., Dittrich, S., Goetz-Neunhoeffer, F., Neubauer, J., 2012. Hydration of Portland cement with high replacement by siliceous fly ash. Cement Concr. Res. 42, 1389—1400.

Dinakar, P., 2013. Properties of fly-ash lightweight aggregate concretes. Proc. Inst. Civil Eng. Constr. Mater. 166, 133—140.

Dong, B., Chen, C., Wei, G., Fang, G., Wu, K., Wang, Y., 2022. Fly ash-based artificial aggregates synthesized through alkali-activated cold-bonded pelletization technology. Construct. Build. Mater. 344, 128268.

Du, S., Ge, Y., Shi, X., 2019. A targeted approach of employing nano-materials in high-volume fly ash concrete. Cement Concr. Compos. 104, 103390.

Duran-Herrera, A., Juárez, C., Valdez, P., Bentz, D.P., 2011. Evaluation of sustainable high-volume fly ash concretes. Cement Concr. Compos. 33, 39—45.

Filho, J.H., Medeiros, M.d., Pereira, E., Helene, P., Isaia, G., 2013. High-volume fly ash concrete with and without hydrated lime: chloride diffusion coefficient from accelerated test. J. Mater. Civ. Eng. 25, 411—418.

Geetha, S., Ramamurthy, K., 2011. Properties of sintered low calcium bottom ash aggregate with clay binders. Construct. Build. Mater. 25, 2002—2013.

Gomathi, P., Sivakumar, A., 2015. Accelerated curing effects on the mechanical performance of cold bonded and sintered fly ash aggregate concrete. Construct. Build. Mater. 77, 276—287.

Grabias-Blicharz, E., Franus, W., 2023. A critical review on mechanochemical processing of fly ash and fly ash-derived materials. Sci. Total Environ. 860, 160529.

Güneyisi, E., Gesoğlu, M., Pürsünlü, Ö., Mermerdaş, K., 2013. Durability aspect of concretes composed of cold bonded and sintered fly ash lightweight aggregates. Compos. B Eng. 53, 258–266.

Gupta, S., Chaudhary, S., 2020. 21 - use of fly ash for the development of sustainable construction materials. In: Samui, P., Kim, D., Iyer, N.R., Chaudhary, S. (Eds.), New Materials in Civil Engineering. Butterworth-Heinemann, pp. 677–689.

Hannesson, G., Kuder, K., Shogren, R., Lehman, D., 2012. The influence of high volume of fly ash and slag on the compressive strength of self-consolidating concrete. Construct. Build. Mater. 30, 161–168.

Hasholt, M.T., Christensen, K.U., Pade, C., 2019. Frost resistance of concrete with high contents of fly ash-A study on how hollow fly ash particles distort the air void analysis. Cement Concr. Res. 119, 102–112.

Hemalatha, T., Ramaswamy, A., 2017. A review on fly ash characteristics – towards promoting high volume utilization in developing sustainable concrete. J. Clean. Prod. 147, 546–559.

Herath, C., Gunasekara, C., Law, D.W., Setunge, S., 2020. Performance of high volume fly ash concrete incorporating additives: a systematic literature review. Construct. Build. Mater. 258, 120606.

Hojati, M., Radlińska, A., 2017. Shrinkage and strength development of alkali-activated fly ash-slag binary cements. Construct. Build. Mater. 150, 808–816.

Huang, C.-H., Lin, S.-K., Chang, C.-S., Chen, H.-J., 2013. Mix proportions and mechanical properties of concrete containing very high-volume of Class F fly ash. Construct. Build. Mater. 46, 71–78.

Huang, T., Chiueh, P., Lo, S., 2017. Life-cycle environmental and cost impacts of reusing fly ash. Resour. Conserv. Recycl. 123, 255–260.

IEA, 2021. World Energy Balances: Overview. IEA, Paris, France.

IEA, 2022. Global Energy Review: CO2 Emissions in 2021. IEA, Paris.

Kalembkiewicz, J., Chmielarz, U., 2012. Ashes from co-combustion of coal and biomass: new industrial wastes. Resour. Conserv. Recycl. 69, 109–121.

Karahan, O., 2017. Transport properties of high volume fly ash or slag concrete exposed to high temperature. Construct. Build. Mater. 152, 898–906.

Kate, G.K., Murnal, P.B., 2013. Effect of addition of fly ash on shrinkage characteristics in high strength concrete. Int. J. Adv. Technol. Civil Eng. 2, 11–16.

Khatib, J.M., 2008. Performance of self-compacting concrete containing fly ash. Construct. Build. Mater. 22, 1963–1971.

Kockal, N.U., Ozturan, T., 2010. Effects of lightweight fly ash aggregate properties on the behavior of lightweight concretes. J. Hazard Mater. 179, 954–965.

Kockal, N.U., Ozturan, T., 2011. Durability of lightweight concretes with lightweight fly ash aggregates. Construct. Build. Mater. 25, 1430–1438.

Li, Y., Wu, B., Wang, R., 2022. Critical review and gap analysis on the use of high-volume fly ash as a substitute constituent in concrete. Construct. Build. Mater. 341, 127889.

Lothenbach, B., Scrivener, K., Hooton, R.D., 2011. Supplementary cementitious materials. Cement Concr. Res. 41, 1244–1256.

Mardani-Aghabaglou, A., Andiç-Çakir, Ö., Ramyar, K., 2013. Freeze–thaw resistance and transport properties of high-volume fly ash roller compacted concrete designed by maximum density method. Cement Concr. Compos. 37, 259–266.

Marinković, S., Dragaš, J., 2018. 11 - fly ash. In: Siddique, R., Cachim, P. (Eds.), Waste and Supplementary Cementitious Materials in Concrete. Woodhead Publishing, pp. 325–360.

Nadesan, M.S., Dinakar, P., 2017. Structural concrete using sintered flyash lightweight aggregate: a review. Construct. Build. Mater. 154, 928–944.

Nadesan, M.S., Dinakar, P., 2018. Influence of type of binder on high-performance sintered fly ash lightweight aggregate concrete. Construct. Build. Mater. 176, 665–675.

Narmluk, M., Nawa, T., 2011. Effect of fly ash on the kinetics of Portland cement hydration at different curing temperatures. Cement Concr. Res. 41, 579–589.

Nath, S., Kumar, S., 2019. Reaction kinetics of fly ash geopolymerization: role of particle size controlled by using ball mill. Adv. Powder Technol. 30, 1079–1088.

Niş, A., Altõndal, İ., 2022. Compressive strength performance of alkali activated concretes under different curing conditions. Period. Polytech. Civ. Eng. 65 (2), 556–565.

Niş, A., Al-Antaki, T.S.W., 2022. Pumice aggregate based lightweight concretes under sulfuric acid environment. Rev. Rom. Mater. 52 (2), 194–202.

Niş, A., Altundal, B.M., 2023. Durability performance of alkali-activated concretes exposed to sulfuric acid attack. Revista de la construcción 22 (1), 16–35.

Oey, T., Kumar, A., Bullard, J.W., Neithalath, N., Sant, G., 2013. The filler effect: the influence of filler content and surface area on cementitious reaction rates. J. Am. Ceram. Soc. 96, 1978–1990.

Ogawa, Y., Bui, P.T., Kawai, K., Sato, R., 2020. Effects of porous ceramic roof tile waste aggregate on strength development and carbonation resistance of steam-cured fly ash concrete. Construct. Build. Mater. 236, 117462.

Özyurt, N., Söylev, T.A., Özturan, T., Pehlivan, A.O., Niş, A., 2020. Corrosion and chloride diffusivity of reinforced concrete cracked under sustained flexure. Tek. Dergi 31 (6), 10315–10337.

Park, B., Choi, Y.C., 2021. Hydration and pore-structure characteristics of high-volume fly ash cement pastes. Construct. Build. Mater. 278, 122390.

Patel, S.K., Majhi, R.K., Satpathy, H.P., Nayak, A.N., 2019. Durability and microstructural properties of lightweight concrete manufactured with fly ash cenosphere and sintered fly ash aggregate. Construct. Build. Mater. 226, 579–590.

Ponikiewski, T., Gołaszewski, J., 2014. The influence of high-calcium fly ash on the properties of fresh and hardened self-compacting concrete and high performance self-compacting concrete. J. Clean. Prod. 72, 212–221.
Pushpalal, D., Danzandorj, S., Bayarjavkhlan, N., Nishiwaki, T., Yamamoto, K., 2022. Compressive strength development and durability properties of high-calcium fly ash incorporated concrete in extremely cold weather. Construct. Build. Mater. 316, 125801.
Rafieizonooz, M., Mirza, J., Salim, M.R., Hussin, M.W., Khankhaje, E., 2016. Investigation of coal bottom ash and fly ash in concrete as replacement for sand and cement. Construct. Build. Mater. 116, 15–24.
Ramachandran, D., Uthaman, S., Vishwakarma, V., 2020. Studies of carbonation process in nanoparticles modified fly ash concrete. Construct. Build. Mater. 252, 119127.
Ridtirud, C., Chindaprasirt, P., Pimraksa, K., 2011. Factors affecting the shrinkage of fly ash geopolymers. Int. J. Miner. Metall. Mater. 18, 100–104.
Şahmaran, M., Yaman, İ.Ö., Tokyay, M., 2009. Transport and mechanical properties of self consolidating concrete with high volume fly ash. Cement Concr. Compos. 31, 99–106.
Sathyan, D., Anand, K.B., 2019. Influence of superplasticizer family on the durability characteristics of fly ash incorporated cement concrete. Construct. Build. Mater. 204, 864–874.
Satpathy, H., Patel, S., Nayak, A., 2019. Development of sustainable lightweight concrete using fly ash cenosphere and sintered fly ash aggregate. Construct. Build. Mater. 202, 636–655.
Sevim, Ö., Demir, İ., 2019. Optimization of fly ash particle size distribution for cementitious systems with high compactness. Construct. Build. Mater. 195, 104–114.
Shiyu, Z., Qiang, W., Yuqi, Z., 2019. Research on the resistance to saline soil erosion of high-volume mineral admixture steam-cured concrete. Construct. Build. Mater. 202, 1–10.
Sibanda, V., Ndlovu, S., Dombo, G., Shemi, A., Rampou, M., 2016. Towards the utilization of fly ash as a feedstock for smelter grade alumina production: a review of the developments. J. Sust. Metall 2, 167–184.
Silva, P., de Brito, J., 2013. Electrical resistivity and capillarity of self-compacting concrete with incorporation of fly ash and limestone filler. Adv. Concr. Constr. 1, 65.
Singh, N., Singh, S., 2016. Carbonation resistance and microstructural analysis of low and high volume fly ash self compacting concrete containing recycled concrete aggregates. Construct. Build. Mater. 127, 828–842.
Sumer, M., 2012. Compressive strength and sulfate resistance properties of concretes containing Class F and Class C fly ashes. Construct. Build. Mater. 34, 531–536.
Sun, X., Zhao, Y., Tian, Y., Wu, P., Guo, Z., Qiu, J., Xing, J., Xiaowei, G., 2021. Modification of high-volume fly ash cement with metakaolin for its utilization in cemented paste backfill: the effects of metakaolin content and particle size. Powder Technol. 393, 539–549.
Supit, S.W.M., Shaikh, F.U.A., 2015. Durability properties of high volume fly ash concrete containing nano-silica. Mater. Struct. 48, 2431–2445.
Suraneni, P., Burris, L., Shearer, C.R., Hooton, R.D., 2021. ASTM C618 fly ash specification: comparison with other specifications, shortcomings, and solutions. ACI Mater. J. 118, 157–167.
Teixeira, E.R., Mateus, R., Camoes, A.F., Bragança, L., Branco, F.G., 2016. Comparative environmental life-cycle analysis of concretes using biomass and coal fly ashes as partial cement replacement material. J. Clean. Prod. 112, 2221–2230.
Terzić, A., Pezo, L., Mitić, V., Radojević, Z., 2015. Artificial fly ash based aggregates properties influence on lightweight concrete performances. Ceram. Int. 41, 2714–2726.
Ukwattage, N.L., Ranjith, P., Yellishetty, M., Bui, H.H., Xu, T., 2015. A laboratory-scale study of the aqueous mineral carbonation of coal fly ash for CO2 sequestration. J. Clean. Prod. 103, 665–674.
Uysal, M., Akyuncu, V., 2012. Durability performance of concrete incorporating Class F and Class C fly ashes. Construct. Build. Mater. 34, 170–178.
Van den Heede, P., De Belie, N., 2014. A service life based global warming potential for high-volume fly ash concrete exposed to carbonation. Construct. Build. Mater. 55, 183–193.
Van der Wegen, G., Bijen, J., 1985. Properties of concrete made with three types of artificial PFA coarse aggregates. Int. J. Cem. Compos. Lightweight Concr. 7, 159–167.
Vassilev, S.V., Vassileva, C.G., 2007. A new approach for the classification of coal fly ashes based on their origin, composition, properties, and behaviour. Fuel 86, 1490–1512.
Vasugi, V., Ramamurthy, K., 2014. Identification of admixture for pelletization and strength enhancement of sintered coal pond ash aggregate through statistically designed experiments. Mater. Des. 60, 563–575.
Wang, L., Guo, F., Lin, Y., Yang, H., Tang, S., 2020a. Comparison between the effects of phosphorous slag and fly ash on the CSH structure, long-term hydration heat and volume deformation of cement-based materials. Construct. Build. Mater. 250, 118807.
Wang, X., Yuan, J., Wei, P., Zhu, M., 2020b. Effects of fly ash microspheres on sulfate erosion resistance and chlorion penetration resistance in concrete. J. Therm. Anal. Calorim. 139, 3395–3403.
Wang, X.-Y., Lee, H.-S., 2010. Modeling the hydration of concrete incorporating fly ash or slag. Cement Concr. Res. 40, 984–996.
Wattanasiriwech, D., Yomthong, K., Wattanasiriwech, S., 2021. Characterisation and properties of class C-fly ash based geopolymer foams: effects of foaming agent content, aggregates, and surfactant. Construct. Build. Mater. 306, 124847.
Wongkeo, W., Thongsanitgarn, P., Ngamjarurojana, A., Chaipanich, A., 2014. Compressive strength and chloride resistance of self-compacting concrete containing high level fly ash and silica fume. Mater. Des. 64, 261–269.
Wu, J.H., Pu, X.C., Liu, F., Wang, C., 2005. High performance concrete with high volume fly ash. Key Eng. Mater. 302, 470–478.

Xu, G., Shi, X., 2018. Characteristics and applications of fly ash as a sustainable construction material: a state-of-the-art review. Resour. Conserv. Recycl. 136, 95–109.

Xu, G., Tian, Q., Miao, J., Liu, J., 2017. Early-age hydration and mechanical properties of high volume slag and fly ash concrete at different curing temperatures. Construct. Build. Mater. 149, 367–377.

Yadav, S., Mondal, S.S., 2019. A complete review based on various aspects of pulverized coal combustion. Int. J. Energy Res. 43, 3134–3165.

Yang, J., Zeng, L., He, X., Su, Y., Li, Y., Tan, H., Jiang, B., Zhu, H., Oh, S.-K., 2021. Improving durability of heat-cured high volume fly ash cement mortar by wet-grinding activation. Construct. Build. Mater. 289, 123157.

Yang, T., Zhu, H., Zhang, Z., Gao, X., Zhang, C., Wu, Q., 2018. Effect of fly ash microsphere on the rheology and microstructure of alkali-activated fly ash/slag pastes. Cement Concr. Res. 109, 198–207.

Yazõcõ, H., 2008. The effect of silica fume and high-volume class C fly ash on mechanical properties, chloride penetration and freeze–thaw resistance of self-compacting concrete. Construct. Build. Mater. 22, 456–462.

Yingliang, Z., Jingping, Q., Zhengyu, M.A., Zhenbang, G., Hui, L., 2020. Effect of superfine blast furnace slags on the binary cement containing high-volume fly ash. Powder Technol. 375, 539–548.

Younsi, A., Turcry, P., Rozière, E., Aît-Mokhtar, A., Loukili, A., 2011. Performance-based design and carbonation of concrete with high fly ash content. Cement Concr. Compos. 33, 993–1000.

Zeng, Q., Li, K., Fen-Chong, T., Dangla, P., 2012. Pore structure characterization of cement pastes blended with high-volume fly-ash. Cement Concr. Res. 42, 194–204.

Zhang, M.-H., Gjørv, O.E., 1990. Microstructure of the interfacial zone between lightweight aggregate and cement paste. Cement Concr. Res. 20, 610–618.

Zhang, N., Yu, H., Wang, N., Gong, W., Tan, Y., Wu, C., 2019. Effects of low-and high-calcium fly ash on magnesium oxysulfate cement. Construct. Build. Mater. 215, 162–170.

Zhang, Z., Li, L., Ma, X., Wang, H., 2016. Compositional, microstructural and mechanical properties of ambient condition cured alkali-activated cement. Construct. Build. Mater. 113, 237–245.

Zhao, N., Wang, S., Wang, C., Quan, X., Yan, Q., Li, B., 2020. Study on the durability of engineered cementitious composites (ECCs) containing high-volume fly ash and bentonite against the combined attack of sulfate and freezing-thawing (FT). Construct. Build. Mater. 233, 117313.

Chapter 16

Recycling of biomass combustion ash into SCMs and aggregates

Huanyu Li[1,2], Jian Yang[1], Lei Wang[3], Ning Zhang[4], Qingyuan Wang[1] and Viktor Mechtcherine[2]

[1]*School of Ocean and Civil Engineering, Shanghai Jiao Tong University, Shanghai, China;* [2]*Institute of Construction Materials, Technische Universität Dresden, Dresden, Germany;* [3]*State Key Laboratory of Clean Energy Utilization, Zhejiang University, Hangzhou, China;* [4]*Leibniz Institute of Ecological Urban and Regional Development (IOER), Dresden, Germany*

1. Introduction

With the ongoing global urbanization and industrialization, ecological issues have garnered increased attention, leading to the formulation of sustainable development goals aimed at long-term environmental sustainability (Guo et al., 2022; Xiao et al., 2022; Zhou et al., 2022). Presently, the worldwide energy sources primarily consist of approximately 80% from fossil fuels, around 2% from nuclear energy, and roughly 18% from renewable energy. Notably, solid biomass combustion accounts for over 60% of the total renewable energy supply, making it a vital contributor in the field of renewable energy (WBA, 2019). It was estimated that the annual global production of biomass combustion ashes is approximately ~170 million tons (Mt), and it was expected to be further increased up to ~1000 Mt/pro year if all biomass residues and wastes were fully utilized (Thomas et al., 2021b). In recent years, the escalating quantities of biomass incineration ash have raised significant concerns, particularly within the agricultural sector (Rithuparna et al., 2021). Typically, agricultural-based ashes are disposed of in landfills, which can pose health risks and contribute to land pollution (Thomas et al., 2021b; Adhikary et al., 2022). Hence, the need to identify an environmentally friendly approach for managing residual biomass ashes has become imperative.

The worldwide carbon dioxide (CO_2) emissions in the building sector have reached nearly 3 gigatons according to International Energy Agency (IEA, 2023), which accounted for almost 40% of global energy-related carbon emissions (GABC, 2019), resulting in ecological degradation and climate change. Concrete generally consists of cement, fine/coarse aggregates, water, and mineral admixtures (Li et al., 2023a; Zhang et al., 2023), which is the fundamental composite in civil engineering owing to its favorable properties, including cost-effectiveness, ease of preparation, and high durability (Li et al., 2019a, 2022a, 2022e). The annual consumption of concrete is estimated to be about 6.5 billion tons around the world and its annual demand is forecasted to be increased considerably up to nearly 18 billion tons by 2050, making it the second-most-consumed substance only after water (Vieira et al., 2016; Li et al., 2023b). Especially, the manufacture of ordinary Portland cement (OPC) (i.e., the most important construction material) is responsible for approximately ~36% of CO_2 emissions in the building sector (Habert et al., 2020; Adesina, 2021) and about 5%–7% of worldwide anthropogenic carbon emissions (660–820 kg CO_2/ton of cement) (Li et al., 2023c; Wang et al., 2018; Li et al., 2022c). In 2019, the global OPC production is reported to be approximately 4.1 gigatons (Association, 2020; Poudyal and Adhikari, 2021). Additionally, the production of concrete requires significant amounts of natural resources. Construction activities consume a substantial volume of mineral raw materials, including sands, stones, and gravels, accounting for nearly 40% of global consumption (Dixit et al., 2010). Annually, approximately 18 gigatons of aggregates are estimated to be used in concrete production, leading to substantial carbon emissions during the extraction and transportation processes (Miller et al., 2018). The overexploitation of natural gravels and sands also poses severe negative impacts on both human systems and the environment (Ioannidou et al., 2017; Torres et al., 2017). Consequently, the integration of sustainable materials as replacements for cementitious binders or aggregates emerges as a beneficial approach to reduce carbon footprint and conserve mineral resources.

Treatment and Utilization of Combustion and Incineration Residues. https://doi.org/10.1016/B978-0-443-21536-0.00022-8

Biomass combustion ashes are generated by burning various solid biomass wastes, such as rice husk, straw, sugarcane, palm oil fuel, corn, and wood, in open fields. The dominant chemical elements in these ashes typically include Si, Ca, K, and Al, with individual contents varying depending on the type of feedstock (Demis et al., 2014; Adhikary et al., 2022). Certain agricultural residue ashes, when incinerated under specific conditions, exhibit high pozzolanic properties due to the presence of a substantial concentration of amorphous silica. As a result, they can be employed as supplementary cementitious materials (SCMs) in cement-based composites (Thomas et al., 2021b). High-calcium (Silva et al., 2022) and high-potassium ashes (de Moraes Pinheiro et al., 2018) can also serve as precursors for the production of alkali-activated composites. Furthermore, ground ashes demonstrate potential as lightweight aggregates in mortars or concrete, reducing the density of composites and minimizing the use of sand and gravel (Memon et al., 2019; Hasnain et al., 2021).

In addition, biomass residues can be converted into biochar through the utilization of pyrolysis technology (Agbede and Oyewumi, 2022). This process involves the combustion of organic substances in an inert gas atmosphere at elevated temperatures, resulting in the production of carbon-rich materials (Apori et al., 2021; Li et al., 2023b). Biochar, characterized as a carbon-negative material, exhibits a porous structure with its pore-size distribution and density being influenced by pyrolysis temperatures and the type of feedstock (Brewer et al., 2014; Wang et al., 2023). Moreover, biochar possesses several advantageous properties, including nontoxicity, low heat conduction, high chemical stability, and low flammability. These attributes make it a promising candidate as a filler or aggregate in various applications such as structural elements, energy-saving buildings, and road construction (Gupta and Kua, 2017). Numerous studies have explored the use of biochar as a sustainable additive, replacing Portland cement (Dixit et al., 2019; Qin et al., 2021) or sand (Wang et al., 2020) in cement-based composites. Additionally, biochar has demonstrated potential in asphalt-based materials (Zhao et al., 2014) and magnesium phosphate cement (Ahmad et al., 2020). Its incorporation in these materials offers environmental benefits and widens the scope of sustainable construction practices.

Currently, the management of waste biomass has gained significant attention due to population growth and increased economic activities (Han, 2020). To address this issue, converting organic wastes into combustion ashes has emerged as an attractive alternative to landfilling, since during the open burning process, some energy can be recovered in thermal power plants, and the residual ashes can be utilized as fertilizer in farmland or as a filler in construction materials (Thomas et al., 2021b). Similarly, the conversion of solid biomass waste into biochar through pyrolysis treatment is considered a promising solution. This process allows for the regeneration of biofuels and the sequestration of significant carbon in biochar (Mattila et al., 2012; Li et al., 2023b). To achieve a low-carbon economy and zero-waste strategy, further exploration of the potential use of biomass ashes and biochar in energy-intensive construction industries is crucial. This chapter provides a comprehensive review of the application of incineration ashes and biochar derived from biomass waste in construction materials. The characteristics of various feedstocks and production processes are thoroughly summarized to fully harness their performance in practical applications. Additionally, the implications of ashes and biochar additives on the physicochemical properties of cement-based composites are overviewed. Finally, future research directions are suggested, taking into consideration the distinct properties of biomass ashes and biochar, as well as the current research gaps. By exploring these avenues, it is possible to advance the utilization of biomass residues, contribute to a sustainable construction industry, and address the challenges of waste management and carbon emissions effectively.

2. Resource of biobased combustion ash and biochar

2.1 Production of biobased combustion ash

Following the combustion of biomass in biomass power plants, two types of industrial biomass ashes can be generated: biomass fly ash (BFA) and biomass bottom ash (BBA) (Cabrera et al., 2018). BFA is typically collected outside the combustion chamber using specialized systems to prevent air pollution; therefore, it has extremely fine diameters of less than 100 μm (Lessard et al., 2017; Agrela et al., 2019). FA is primarily composed of inorganic materials, with a minor organic fraction (Agrela et al., 2019).

BBA, on the other hand, primarily consists of noncombustible heavier particles with coarser diameters exceeding 1 mm, which can be found at the base of the furnace or incinerator (Beltrán et al., 2016; Lessard et al., 2017). In the case of fluidized bed combustion (FBC), the composition of bottom ashes includes unburnt biomass, sand particles from the fluidized bed, and an inorganic fraction comprising small stones and soil (Modolo et al., 2013). BBA may also contain mineral impurities that contribute to the formation of sintered ash and slag due to their lower melting points (Agrela et al., 2019). The proportion of BBA in the overall ash composition is generally low in the FBC approach, accounting for less than 20 wt.%, whereas BBA derived from grate furnaces represents a higher fraction ranging from 60 to 90 wt.% of the total ashes produced (Obernberger and Supancic, 2009; Modolo et al., 2013).

The FBC technology is widely regarded as the most appropriate industrial way for the combustion of biomass owing to many advantages, such as flexibility of fuel, isothermal operating conditions, and low temperatures of the processing; it mainly encompasses circulating fluidized bed combustors (CFBCs) and bubbling fluidized bed combustors (BFBCs) (Modolo et al., 2013). The approaches differ in certain technical aspects, such as the bed particle size, the transfer rates of mass and heat, gas−solid hydrodynamics, and the temperature along the reactor (van Loo and Koppejan, 2007; Yin et al., 2008). These variables play an important role in the properties of the ashes produced during the incineration of biomass. A typical schematic illustration of a fluidized bed combustor in thermal plants is shown in Fig. 16.1 (Hinojosa et al., 2014). The diagram provides an overview of the key components and the flow of gases and solid particles within the combustion system.

2.1.1 Major chemical compositions of different types of biomass combustion ashes

Thermal power plants combust various types of biomass, and the selection of biomass is often closely related to the vegetation found in the local region or neighboring areas (Agrela et al., 2019). In the existing literature, researchers have investigated the utilization of ashes derived from specific types of biomass, such as rice husk (Sua-iam et al., 2016), corn cob (Shakouri et al., 2020), sugarcane bagasse (Neto et al., 2021), and palm oil fuel (Men et al., 2022), in cementitious materials. Additionally, ashes resulting from mixtures of different biomass sources, including forestry and agricultural wastes, have also been incorporated into construction materials (Carrasco-Hurtado et al., 2014; Medina et al., 2019a; Eliche-Quesada et al., 2021). To fully harness the potential of sustainable biomass combustion ashes, it is important to have a comprehensive understanding of the chemical properties of different types of ashes, which is presented in Table 16.1.

The chemical analysis of biomass ashes has provided valuable insights into their elemental composition, highlighting the predominant presence of silica in rice and wheat crops, ranging from 50% to 98%. Interestingly, the silica dioxide (SiO_2) content varies across different parts of the same plant, such as rice husk (72%−98%) and rice straw (65%−80%). Rice husk exhibits minor amounts of potassium oxide (K_2O) ranging from 0.7% to 3.3%, while rice straw contains approximately 18% of K_2O. Calcium oxide (CaO) is another notable component in rice husk (0.2%−3.3%), whereas magnesium oxide (MgO) constitutes around 8% in rice straw. Other oxides, including aluminum oxide (Al_2O_3), sodium oxide (Na_2O), iron oxide (Fe_2O_3), sulfur trioxide (SO_3), and phosphorus pentoxide (P_2O_5), are typically present in quantities below 1% in rice crops. In the case of wheat plants, CaO takes the second highest proportion (5%−11%), followed by K_2O (3%−11%), SO_3 (~6%), and Na_2O (~5%). Sugarcane ashes exhibit a dominant presence of SiO_2, accounting for approximately 54%−75% of the overall composition. Notably, sugarcane ash comprises significant amounts of CaO (2%−8%) and pozzolanic materials such as SiO_2 (54%−75%), Fe_2O_3 (3%−25%), and Al_2O_3 (3%−18%), rendering it suitable for the production of cementitious materials. Wood ashes display considerable variation in chemical composition, primarily influenced by the type of tree. These ashes are predominantly composed of calcium, silica, aluminum, and potassium. As a result, they have been extensively studied as SCMs or aggregates in the field of building

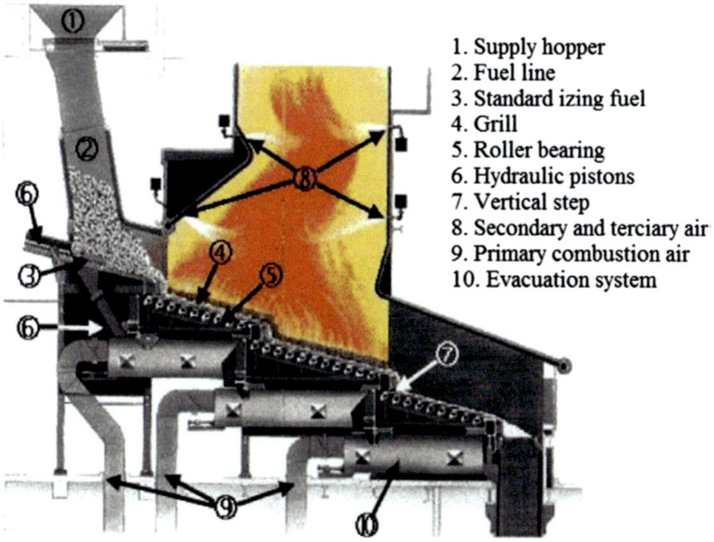

FIGURE 16.1 The representative schema of the combustion systems' details (Hinojosa et al., 2014). *Reprinted by permission of Elsevier.*

TABLE 16.1 Chemical composition of various biomass ashes in wt.%.

Biomass types	SiO_2	CaO	Al_2O_3	K_2O	Na_2O	Fe_2O_3	SO_3	MnO	MgO	TiO_2	P_2O_5	Others	LOI[a]	References
Rice husk	96.44	0.24	0.19	2.05	–	0.02	0.35	0.02	–	–	0.47	0.76	5.38	Lo et al. (2021)
	98.19	0.18	0.19	1.04	–	0.02	0.27	0.02	–	–	0.26	0.29	0.5	Lo et al. (2021)
	76.81	3.25	6.17	2.51	1.21	4.19	0.37	–	1.41	–	–	–	4.41	Abbas et al. (2017)
	93.00	1.31	0.35	1.61	0.15	0.23	0.03	–	0.41	–	–	–	1.90	Sua-iam et al. (2016)
	97.03	0.89	0.92	–	–	–	0.55	–	0.5	–	–	–	–	Liu et al. (2020)
	90.0	0.60	0.01	2.30	–	–	–	–	–	–	–	–	5.10	Almalkawi et al. (2019)
	72.25	2.32	1.12	3.34	0.82	1.28	2.11	0.24	2.09	0.034	0.75	–	12.63	Xue et al. (2014)
	83	0.30	2	0.7	–	0.096	<0.02	0.12	0.10	–	0.62	0.69	13.15	de Sensale and Viacava (2018)
	88.42	1.03	0.11	2.59	1.12	–	–	–	0.82	–	–	0.35	2.10	Tahami et al. (2018)
Rice straw	79.82	0.370	1.13	1.07	0.501	0.245	–	–	7.54	–	3.75	5.22	–	Pandey and Kumar (2019a)
	77.00	4.96	0.69	8.89	1.36	0.63	1.90	–	2.65	–	–	–	1.40	Munshi and Sharma (2019)
	65 0.92	2.4	1.78	–	–	0.2	0.69	–	3.11	–	–	–	9.71	El-Sayed and El-Samni (2006)
	76.8	2.6	–	18.2	–	–	–	–	0.10	–	–	2.30	–	Agwa et al. (2020)
Wheat straw	65.7	7.84	3.73	3.27	2.45	2.58	2.34	–	2.68	0.22	1.67	–	7.32	Qudoos et al. (2019)
	73.95	5.21	0.91	11.51	–	1.15	–	–	1.83	1.92	–	–	–	Khushnood et al. (2014)
	54.7	10.3	0.63	8.4	4.6	0.32	5.2	–	1.5	–	3.8	–	10.3	Al-Akhras (2011)
	50.7	10.6	0.48	11.4	5.41	–	6.13	–	2.20	–	4.68	–	10	Al-Akhras and Abu-Alfoul (2002)
Sugarcane	63.10	8.28	7.56	5.43	1.24	4.59	1.92	–	4.54	–	2.13	0.90	3.10	Neto et al. (2021)

	60.04	2.47	3.08	3.39	–	25.78	0.34	–	1.02	–	–	3.88	–	Anjos et al. (2020)
	71.40	6.74	3.39	8.19	–	3.51	2.25	0.18	–	–	–	–	2.28	Jittin and Bahurudeen (2022)
	53.8	3.4	18.1	4.3	–	10.8	2.7	0.2	–	1.0	1.6	–	3.9	de Siqueira and Cordeiro (2022)
	58.6	4.6	9.0	5.4	–	8.4	1.9	–	1.6	–	–	4.0	6.5	Moraes et al. (2017)
	74.63	7.04	3.54	–	–	3.67	2.35	0.19	–	–	–	–	–	Athira and Bahurudeen (2022)
Wood[a] (Eucalyptus)	38.5	16.7	14.8	5.97	1.53	5.94	2.66	0.50	3.44	0.76	1.12	–	6.39	Saeli et al. (2019)
Wood[a] (Eucalyptus, pine)	41.1	12.4	21.0	8.9	1.5	9.3	–	–	4.8	1.1	–	–	22.72	Soares and Castro-Gomes (2022)
Wood[a] (Forest residues)	52.1	15.9	13.3	4.14	–	5.30	0.45	–	3.31	–	–	0.10	10.4	Esteves et al. (2012)
Wood[a] (Eucalyptus)	1.76	43.80	5.06	4.64	0.28	2.43	–	–	8.36	0.53	3.32	–	29.72	Silva et al. (2022)
Wood[a] (Olive pruning, dry olive cake)	4.09	4.37	1.01	52.93	1.25	0.347	5.95	–	1.62	0.0479	1.06	–	20.38	Carrillo-Beltran et al. (2021)
Wood[a] (Wood chips, waste wood)	45.08	16.58	2.98	4.47	0.37	1.37	2.26	0.56	1.90	0.2	2.07	0.22	8.89	Kaminskas et al. (2015)
Wood[a] (Pine trees)	11.0	65.9	2.4	14.6	1.0	2.9	5.4	–	4.2	–	2.9	4.6	15.0 ± 0.1	Sigvardsen et al. (2021)
Wood[a] (Softwoods)	45.3	19.9	13.0	4.3	0.8	4.7	4.7	2.7	2.7	–	1.7	0.2	–	Fořt et al. (2021)
Wood[a] (Pine wood)	30.04	5.62	6.17	2.22	0.35	1.95	0.30	0.11	0.70	0.39	0.30	–	50.83	Madrid et al. (2017)
Wood[a] (Chestnut, poplar wood)	29.88	33.13	9.58	3.64	5.04	5.79	2.90	0.77	3.50	–	2.33	1.07	–	Berra et al. (2015)
Wood	5.59	42.01	0.79	7.59	5.42	1.26	–	1.76	4.08	–	1.82	–	34.10	De Souza et al. (2022)
Corn cob	20.10	2.95	0.95	38.1	0.27	0.75	0.56	0.04	2.42	0.11	5.49	4.59	23.60	Shakouri et al. (2020)
Corn straw	35.68	9.98	6.10	11.72	7.51	7.40	–	–	2.60	0.60	1.42	–	14.82	Wang et al. (2021b)
Rape straw	61.76	14.98	0.02	6.02	–	–	1.27	–	–	–	–	14.98	–	Zhang et al. (2019)
Hazelnut-shell	12.36	48.52	3.83	16.42	2.17	5.34	0.40	–	4.67	–	–	–	6.29	Yurt and Bekar (2022)

Continued

TABLE 16.1 Chemical composition of various biomass ashes in wt.%.—cont'd

Biomass types	SiO_2	CaO	Al_2O_3	K_2O	Na_2O	Fe_2O_3	SO_3	MnO	MgO	TiO_2	P_2O_5	Others	LOI[a]	References
Bamboo leaf	53.6	6.0	0.3	14.5	0.3	12.5	–	1.9	1.2	–	3.9	3.0	0.5	Vinai et al. (2022)
Palm oil fuel	48.6	16.8	1.3	16.3	0.2	1.9	3.9	–	4.0	–	–	–	10.2	Men et al. (2022)
Banana leaf	54.93	23.65	2.14	5.36	1.30	1.18	1.13	–	5.50	–	1.80	3.01	–	Tavares et al. (2022)
Olive stone	5.33	27.77	0.70	32.12	0.78	3.45	1.67	–	5.13	–	2.68	0.95	18.90	de Moraes Pinheiro et al. (2018)
Miscanthus	58.78	11.90	1.83	3.65	–	3.42	0.45	0.55	2.25	–	5.44	0.08	10.42	Lv et al. (2019)
Cattle manure	32.65	28.56	3.72	8.72	7.96	11.89	–	–	1.98	1.17	–	–	1.3	Chen et al. (2019)
	52.0	15.4	7.79	4.91	0.66	3.20	1.54	–	2.94	–	2.92	0.35	2.5	Zhou and Chen (2012)

[a]*Note: The type of wood in the brackets represents the dominant source; LOI represents the loss on ignition.*

materials (Ayobami, 2021; Martínez-García et al., 2022). Certain biomass sources, such as olive pruning (Carrillo-Beltran et al., 2021), corn cob (Shakouri et al., 2020), and olive stone (de Moraes Pinheiro et al., 2018), are characterized by their significant potassium oxide content. This particular attribute makes them highly promising as potential precursors for the production of alkali-activated binders or geopolymer materials.

It has been observed that certain plant species possess the remarkable ability to absorb soluble silica from water through their roots; this absorbed silica is then transported to various parts of the plant, where it is subsequently deposited in different locations, such as between the plasma membrane and the cellulose wall, within intercellular spaces, or even within the cell walls themselves (Prychid et al., 2003). Notably, when the plants were combusted under temperatures below 900°C, the inside amorphous silica can be transformed into highly reactive silica, while excessively high temperatures exceeding 900°C can lead to the generation of nonreactive crystalline silica (Clark et al., 2017). Both the temperature and duration of the incineration process play pivotal roles in determining the state of silica and the extent of unburned carbon present in the resulting ashes. Cao et al. (2022) conducted controlled burning experiments using highland barley straw and discovered that residual ashes still contained unburned carbonaceous compounds when incinerated at temperatures below 500°C. However, they observed that the majority of the unburned carbon could be decomposed when the incineration temperature surpassed 600°C, as illustrated in Fig. 16.2.

Memon et al. (2018) conducted a comprehensive analysis of the mineralogical phases of silica present in wheat straw ashes (WSAs), which had undergone combustion at various temperatures. They observed that at elevated temperatures exceeding 600°C, the amorphous silica in the ashes underwent crystallization, transforming into distinct crystalline forms. Conversely, when examining rice husk ashes (RHAs), it has been reported that the amorphous silica can remain in its

FIGURE 16.2 Morphologies of the highland barley straw ashes produced under different calcination temperatures (Cao et al., 2022). *Reprinted by permission of Elsevier.*

original form even at temperatures as high as 900°C, particularly when the calcination duration is short, i.e., less than 1 h (Yeoh et al., 1979; Nehdi et al., 2003). Nair et al. (2008) have incinerated rice husks at temperatures of 500−900°C for different durations (ranging from 15 min to 24 h), and their results revealed that the most reactive ashes can be obtained when burning rice husks at 500°C for 12 h.

2.2 Production of biochar

In contrast to the traditional open-burning method, residual biomass can be effectively converted into carbon-rich biochar using pyrolysis or gasification techniques in an oxygen-limited or oxygen-free environment. Among these methods, pyrolysis is the most widely employed approach, wherein biomass is subjected to thermal degradation to produce gases, liquid fuel (biooil), and solid carbon-rich char (Maschio et al., 1992; Demirbaş, 2001; Wang et al., 2021a). Conventional pyrolysis typically involves a slow heating rate (0.1−1°C/s) within a temperature range of 400−500°C, with a residence time of 5−30 min (Tripathi et al., 2016). Alternatively, fast pyrolysis is employed in the decomposition of biomass, which is often conducted at a faster heating rate of about 10°C/s and temperatures of approximately 400−800°C with less treatment time (Gopakumar, 2012; Akinyemi and Adesina, 2020). Fast pyrolysis leads to lower biochar yields and higher yields of gaseous and liquid by-products due to increased biomass ablation and fragmentation (Tripathi et al., 2016; Yuan et al., 2020). Prolonged residence time during pyrolysis promotes the development of the internal pore structure of biochar (Zhang et al., 2022). Furthermore, it was reported that a longer treatment duration improved the biochar yield at low temperatures but reduced the yield in the case of high temperatures (Tripathi et al., 2016).

Pyrolysis temperature is also a vital factor, which is highly correlated to crystalline carbon structure, pore structure, yield, and the pH value of biochar (He et al., 2021). Higher pyrolysis temperatures have been observed to enhance the surface area and promote the development of the pore structure in biochar; this is attributed to the transformation of amorphous carbon into graphite crystallites and the removal of pore-blocking substances during the pyrolysis process (Lian and Xing, 2017).

Gasification is a thermochemical process conducted at high temperatures, typically above 700°C, using various gasifying agents such as oxygen, steam, air, or gas mixtures, to produce biochar, energy, and pyrolytic gases such as CH_4, CO_2, H_2, and CO (Saxena et al., 2008; Sharma, 2019). It mainly consists of four stages: drying phase (at 100−200°C), pyrolysis phase (at 300−800°C fluidized bed, at 200−700°C for fixed bed, and at 600−1600°C for entrained flow), combustion phase (700−1500°C), and reduction phase (at 800−1000°C) (You et al., 2018). The gasification process can be tailored by adjusting various parameters such as feedstock properties, temperatures, pressures, and the types of gasifying agents used, allowing control over the production of solid biochar and biogases (Shayan et al., 2018). While high gasification temperatures generally result in reduced biochar yield, they promote the release of volatile compounds, increase gas production, enhance the degree of aromatization, and create a greater number of pores in the resulting biochar (Qi et al., 2021).

In addition to pyrolysis and gasification, there are other technologies available for the degradation of biowastes, such as hydrothermal carbonization (Cao et al., 2021) and torrefaction (Li et al., 2019b). In general, the approaches for biochar production and the treatment conditions should be carefully chosen under consideration of the specific properties and desired yield of the biochar (Li et al., 2023b).

3. Application of biomass combustion ash and biochar in construction materials

3.1 Biomass combustion ash as SCMs in cement-based composites

The large quantities of biomass combustion ashes generated and the associated environmental concerns have prompted research into utilizing these ashes as SCMs in cement-based materials. Extensive studies have been conducted by researchers investigating the potential of biomass combustion ashes as a substitute for OPC in construction applications. These ashes have been explored as precursors in alkali-activated materials, fillers in asphalt-based composites, and in magnesium-rich cement. In the forthcoming sections, we will delve into the application of some commonly used bioashes in the realm of construction materials. Specifically, we will examine the utilization of ashes derived from rice and wheat crops, sugarcane, wood, and other biomass sources.

3.1.1 Rice and wheat crop ashes

Rice and wheat are considered essential grain crops worldwide, serving as primary food resources for many populations. Following the combustion of agricultural residues from these crops, the resulting ashes predominantly consist of silica, as mentioned in Section 2.1.1. A study conducted by Pan and Sano (2005) demonstrated that the silica content in rice straw ashes (RSAs) exhibited similar results compared to WSAs. Thus, this section provides a summary of the utilization of the waste ashes from these crops.

The paddy production on the earth has reached about 752 million tons/year by 2016 (Mirmohamadsadeghi and Karimi, 2020), leading to significant amounts of plant residues in the form of rice straw after crop harvesting, as well as agro-industrial residues such as rice husk generated during the milling process (Tripathi et al., 1998). According to estimates from Lim et al. (2012) and Mirmohamadsadeghi and Karimi (2020), worldwide rice straw and rice husk production can be approximated at 752 million tons/year and 150 million tons/year, respectively. Fig. 16.3 illustrates the appearance of rice husk and its corresponding ash (Tahami et al., 2018). It was reported that in the 2021/2022 marketing year, the world consumption of wheat reached approximately 793 million metric tons (Shahbandeh, 2023). On average, for every kilogram of harvested wheat grain, there are approximately 1.3−1.4 kg of wheat straw generated; when this wheat straw undergoes the burning process, it yields approximately 9.6% of ashes by weight of the wheat straw (Pan and Sano, 2005).

The flowability/workability of fresh-state cementitious mixtures seems to be negatively affected by the addition of rice and wheat ashes. Agwa et al. (2020) incorporated RSA into concrete with OPC replacement levels of 5−20 wt.% and observed a reduction in workability as the substitution levels increased. This decrease in workability can be attributed to the irregular shape of the ashes, which results in a high specific surface area (SSA) and the ability to adsorb water during mixing, thus reducing the amount of free water in the fresh mixtures. Similar outcomes were also observed for WSA. Khushnood et al. (2014) studied the flowability of self-compacting paste with a 10% replacement of OPC with WSA. Their results demonstrated a decrease in workability due to the higher friction of WSA particles, which can be attributed to their irregular shapes and rough texture.

The strength of cement-based composites containing rice and wheat ash is closely related to the chemical composition, content, and particle size of the ashes. Sua-iam et al. (2016) incorporated 20 wt.% of RHA by weight of cement to produce self-compacting concrete, which resulted in higher compressive strengths. This improvement can be attributed to the fine ashes' ability to fill voids and the pozzolanic activity of RHA, which promotes the formation of additional calcium-silicate-hydrate (C−S−H) gel (Sua-iam et al., 2016). Agwa et al. (2020) found that compressive strengths increased at a 10 wt.% ratio of RSA, but higher RSA content led to a decline in compressive properties. In addition, the particle size of combustion ashes played a crucial role in the mechanical performance. Qudoos et al. (2019) ground the WSA using a ball mill for 30, 60, and 120 min and incorporated 10 wt.% of the treated ash into cementitious composites. Their results showed that coarse ash had a minor decrease in compressive strengths, while very fine ash particles (milling time above 60 min) contributed to the improvement of concrete strengths; see Fig. 16.4.

It was reported that the saturated water absorption of concrete can be diminished when 10 wt.% RSA was added to the cementitious mixes (Pandey and Kumar, 2019b). This is attributed to the denser packing of the mixture, which leads to a reduction in the volume of permeable voids. When RSA is added together with microsilica (MS), a greater reduction in

FIGURE 16.3 The appearance of (A) rice husk and (B) rice husk ash (Tahami et al., 2018). *Reprinted by permission of Elsevier.*

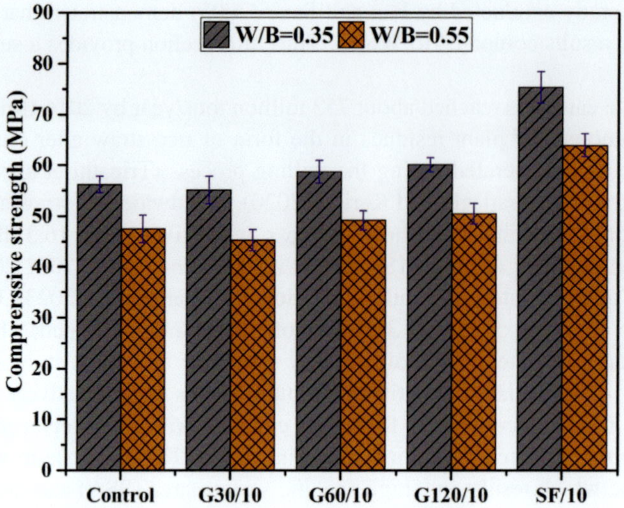

FIGURE 16.4 Compressive strengths of the concrete reinforced with ground WSA at 28 days (Qudoos et al., 2019). *Reprinted by permission of Elsevier.*

water absorption can be achieved compared to samples with only RSA or MS additives. The study also found that the addition of RSA decreases the initial/secondary water absorption rates. This is due to the reduced water-to-binder (w/b) ratio in the interfacial transition zones (ITZ) around the aggregates and a decrease in the preferred arrangement of C—S—H gels, which is a result of the pozzolanic and filler effect of RSA (Pandey and Kumar, 2019b). A similar decreasing trend in water absorption was observed in the cementitious system blended with WSA, which can be attributed to the pozzolanic reactivity of the ashes and the resulting decrease in total porosity (Khushnood et al., 2014). Thus, the silica-rich rice/wheat combustion ashes seem to endow cement-based composites with higher durability as a result of their relatively low fineness and high reactivity.

Songpiriyakij et al. (2010) have replaced 20—100 wt.% fly ash by using rice husk and bark ash (RHBA) to fabricate geopolymer materials, where sodium silicate and NaOH solutions were used as activators. An enhancement of compressive strengths was achieved for the RHBA-mixed pastes; it was explained that added bioashes enriched the silica content in the mixtures, allowing for the development of Si—O—Si bonds (Songpiriyakij et al., 2010).

In the study of Liu et al. (2020), RHA was utilized as a filler in magnesium phosphate cement (MPC) at a range of 1—7 wt.% relative to the weight of MgO. The findings revealed that the inclusion of RHA facilitated the hydration reaction and reduced the setting times of MPC. Moreover, the presence of RHA, which possesses water absorption properties, resulted in the formation of more compact hydration products with reduced porosity due to the lower water-to-binder ratio. As a result, the compressive strength of the MPC was improved by approximately 26% at 7 days (Liu et al., 2020).

In the research conducted by Xue et al. (2014), RHA particles were incorporated into asphalt-based binders at levels of 10%—20% to modify their properties. The study found that RHA increased the viscosity of the binder and had a significant impact on thermooxidative aging resistance. Furthermore, RHA-modified asphalt binders exhibited an increase in complex modulus and rutting factor, indicating improved resistance to rutting (Xue et al., 2014). Similarly, Tahami et al. (2018) demonstrated that the addition of RHA led to enhanced rutting resistance and fatigue life of hot mix asphalt. The incorporation of RHA resulted in higher stiffness modulus and improved stability of the asphalt binder. These findings suggest that the use of RHA can positively influence the performance characteristics of asphalt-based materials.

3.1.2 Sugarcane ashes

The global production of sugarcane is estimated to be nearly ~1.6 billion tons every year (Chandel et al., 2012). After sugarcane harvesting, the residual leaves are often left in the field, and bagasse is generated when sugarcane is processed into juice for producing ethanol or sugar (Singh et al., 2008; Krishnan et al., 2010). The total amount of sugarcane residues, including bagasse and leaves, reaches around 279 million metric tons per year (Chandel et al., 2012). For every ton of sugarcane, approximately 26 wt.% of bagasse and 0.62 wt.% of residual ashes are produced (Tavares and Oliveira, 2004). According to a report in 2013—14, Brazil produced approximately 4 million tons of sugarcane bagasse ash (SCBA) (Thomas et al., 2021a). The morphologies of SCBA particles are representatively shown in Fig. 16.5 (de Siqueira and Cordeiro, 2022), providing a visual representation of their characteristics. To effectively manage and utilize these bioashes, numerous scientists have explored their incorporation into cementitious mixtures as partial replacements for binders.

FIGURE 16.5 The morphological appearance of SCBA particles (de Siqueira and Cordeiro, 2022). *Reprinted by permission of Elsevier.*

Siqueira and Cordeiro (de Siqueira and Cordeiro, 2022) conducted a study on the hydration kinetics of cement paste with 15% replacement of OPC by SCBA. They observed that the presence of SCBA delayed the hydration process, resulting in an extended dormancy period and longer setting times. Additionally, the addition of SCBA reduced the flowability of the paste. This can be attributed to the high fineness and SSA of angular and elongated SCBA particles, as well as their high alkali content, which adversely affect the rheological properties (de Siqueira and Cordeiro, 2022). Jittin and Bahurudeen (Jittin and Bahurudeen (2022) reported that the yield stress and plastic viscosity of fresh cementitious materials increased with higher contents of SCBA. They explained that the substitution of OPC with SCBA, which has a lower specific gravity, led to higher powder contents in the mixture. Furthermore, the irregular shape of the ashes increased particle friction, contributing to the observed rheological changes. Anjos et al. (2020) investigated the rheology of self-leveling mortars with 15–30 wt.% OPC replacement by SCBA. They found that the incorporation of SCBA resulted in increased flow capacity, internal friction, and viscosity in the mortar.

Spósito et al. (2023) have substituted 15–50 wt.% cement clinker in mortars by SBCA and the inclusion of combustion ashes conduced to a descending tendency of compressive strengths with increasing SBCA proportions after 28-day curing. Anjos et al. (2020) have used SBCA as SCMs to replace the cement clinker in self-leveling mortars at ratios ranging from 15% to 30%. They demonstrated that the flexural and compressive performance of hardened-state mortars combined with SCBA can be diminished at various curing ages due to higher water demand (Anjos et al., 2020). When the replacement levels of cement by SCBA were less than 20 wt.%, an increase in compressive properties of concrete was recorded at 28 days and 56 days, which is due to the pozzolanic reaction of SCBA (Jittin and Bahurudeen, 2022). Yet, a higher SCBA proportion (more than 30%) triggered a decline in compressive strengths, being ascribed to the dilution effect (Jittin and Bahurudeen, 2022). Neto et al. (2021) have added 5, 10, and 15 wt.% SBCA into concrete to partially substitute the cement clinker and their outcome denoted an improvement in compressive properties with increasing replacement levels. This enhancement was explained that the fine SBCA particles could fill the pores and yielded to better packing. Additionally, the reaction of amorphous silica and alumina in SBCA with $Ca(OH)_2$ resulted in the formation of C-(A)-S-H gels, reducing porosity (Neto et al., 2021). Also, Siqueira and Cordeiro (de Siqueira and Cordeiro, 2022) have stated that the inclusion of 15 wt.% SBCA conferred higher mechanical strengths to cementitious materials. Similarly, Bayapureddy et al. (2020) have observed an augmentation in the compressive performance of concrete with below 15 wt.% of OPC replacement levels by SBCA. They suggested that when the incorporation of SCBA is over 15 wt.%, the unreacted SCBA particles in the mixture would be increased due to the dilution effect (Bayapureddy et al., 2020; Jittin and Bahurudeen, 2022), which would decrease the compressive properties. In addition, Arif et al. (2016) demonstrated that excessive SCBA could diminish the content of portlandite in cement mixes and limit the ashes' pozzolanic reaction, reducing the composite strength; see Fig. 16.6. All in all, to avoid a strength reduction of cement composites, the replacement levels of clinker by SCBA should be restricted within a certain range, which is often below 15 wt.%.

Neto et al. (2021) have studied the capillary absorption (sorptivity) of SCBA-blended concrete, and it was found that the incorporation of 5%, 10%, and 15% SCBA lowered the sorptivity by about 67%, 76%, and 81%, respectively, since the pore interconnectivity declined owing to the filler effect and pozzolanic activity of SCBA. They also recorded that SCBA caused a delay in the chlorine ions' penetration, indicating an improvement in the durability of concrete. Anjos et al. (2020)

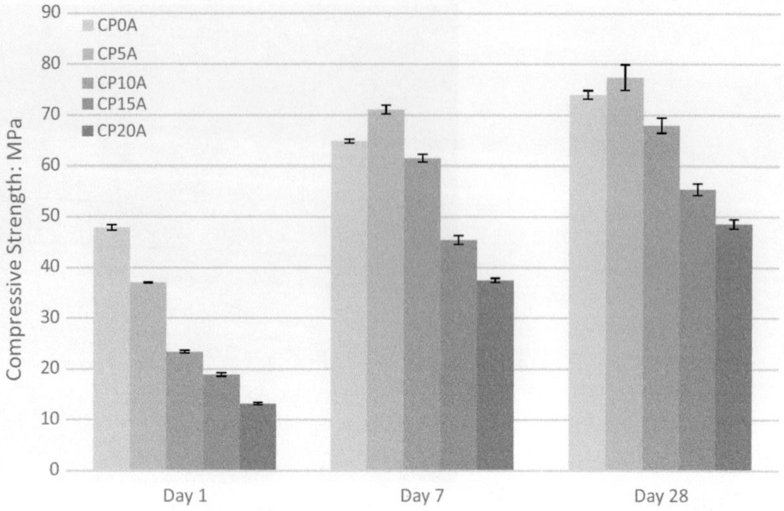

FIGURE 16.6 Compressive performance of cement pastes reinforced with SBCA at different OPC replacement ratios and curing periods (Arif et al., 2016). *Reprinted by permission of Elsevier.*

documented that when the cement replacement by SCBA was below 20 wt.%, the capillarity coefficients of mortars were increased compared with the control samples, while higher replacement levels of more than 25 wt.% gave rise to a lower capillarity coefficient, i.e., decrease in water absorption. Jittin and Bahurudeen (Jittin and Bahurudeen (2022) observed an obvious reduction in the water sorptivity with increasing content of SCBA. When the substitution ratio of OPC by SCBA reached 30% in concrete, the water sorptivity index can be diminished by up to about 14%, revealing an improvement in the resistance against unidirectional sorption (Jittin and Bahurudeen, 2022). This decline of the water sorptivity caused by SCBA addition in the cementitious system was also documented in Gopinath et al. (2018), Praveenkumar and Sankarasubramanian (2019) and Murugesan et al. (2021). Moreover, due to the pozzolanic reaction and filler effect of SCBA, the pore structures of blended concrete were refined, therefore aiding in the decline in the permeability and higher resistance against chloride ingression (Jittin and Bahurudeen, 2022).

Moraes et al. (2016) used sugarcane straw ash (SCSA) to partially replace blast furnace slag (BFS) in alkali-activated composites (AACs) at proportions of 15–50 wt.% and found that SCSA achieved a significant improvement in the compressive strength of AAC in both NaOH-activated and NaOH/sodium silicate-activated systems. XRD analysis showed that substituting BFS with SCSA increased the (Si + Al)/Na ratio and decreased the atomic ratio of Ca/Na, leading to the formation of more hydrated nepheline and hydrosodalite. The SCSA-modified AAC exhibited higher compactness and superior compressive properties compared with plain samples (Moraes et al., 2016). It was documented that among the raw materials utilized for AAC, the production process of sodium silicate has the highest carbon footprint (Mellado et al., 2014). Hence, using SCSA as a precursor replacement has promising potential for sustainable construction practices.

3.1.3 Wood ashes

Forest waste mainly consists of hardwoods and softwoods, which come from deciduous and coniferous trees, respectively (Millati et al., 2019). According to Millati et al. (2019), lumber production has yielded approximately 118 million m^3 of hardwoods and 333 million m^3 of softwoods globally. In terms of municipal solid waste, wood waste accounts for approximately 2% of the total amount generated worldwide (Kaza et al., 2018). Wood ash is typically produced during the combustion process in incinerators or furnaces used for energy production, which includes both fly ash and bottom ash (Ayobami, 2021; Wang and Haller, 2022). Werkelin et al. (2005) reported that ash content from wood tissues such as branch and stem wood ranges from 0.2% to 0.7%, while bark tissues can generate 1.9%–6.4% of combustion ashes. Ash content from bark is higher than that from stem wood, since bark contains high levels of Ca and Si, while mixed stem wood contains high concentrations of Mn, Al, and Si (Pitman, 2006). The chemical constituents of wood ash, such as quicklime, silica, alumina, and iron oxide, are essential for the properties of construction materials and depend on the tree parts and types (Ayobami, 2021). Chemical elements in the ashes of branch and root are relatively higher than those in stem wood, and bark and foliage contain 5–10 times the amount of elements found in stem wood (Pitman, 2006). Table 16.1 shows the differences in the chemical constituents of various types of wood.

The addition of wood ash could negatively impair the flowability of the fresh mixes. Cheah and Ramli (2012b) incorporated wood ash as a substitute for cement in mortar at dosages of 5%–25% and observed a decline in flowability and slump of fresh mixes with increasing replacement levels, due to the relatively high SSA of ashes. The reduction of workability of cementitious mixes by adding wood ash was also documented in Elinwa and Mahmood (2002), Udoeyo and Dashibil (2002), and Abdullahi (2006) as well. Berra et al. (2015) added woody biomass fly ash (WBFA) from chestnut virgin wood chips into concrete and observed lower workability due to the high value of SSA and the irregular shape of WBFA particles. In contrast, poplar wood-based WBFA was capable of improving flowability, which is explained by the lower ash dissolution and lower unburned carbon content of the ashes (Berra et al., 2015). In the investigation of Garcia and Sousa-Coutinho (da Luz Garcia and Sousa-Coutinho, 2013), the use of 5–10 wt.% ground bottom ash originating from wood and forest waste aided in the increase of workability of mortars. Similarly, the incorporation of wood ash can significantly improve flowability at OPC replacement levels of 20–40 wt.% (Kara et al., 2012; Raheem and Adenuga, 2013). Furthermore, Ramos et al. (2013) observed some negligible changes in flowability for the mortars with 10–20 wt.% wood ashes.

Kaminskas et al. (2015) have partially replaced the Portland cement in mortars with wood-based biomass fly ash (WBFA) at the dosages of 5, 10, and 25 wt.% and their results indicated an increase in the normal consistency and initial/final setting time as the ash content increased. The duration of the induction period was shorter after adding the wood ash, revealing that the initial cement hydration can be accelerated by the ash additives. In addition, they also showed that the mortars modified with ~10 wt.% of WBFA possessed higher or similar compressive strengths after 28-day curing in comparison with plain samples, whereas higher content (25 wt.%) of wood ash led to a decline by about 10% (Kaminskas et al., 2015). Cheah and Ramli (2012a) studied the mechanical properties of mortars with 2–10 wt.% clinker substitution by high-calcium wood ash (HCWA). They found that in the presence of silica addition, the inclusion of HCWA significantly improved the flexural strength and compressive strength of composites at the dosages of 2 and 4 wt.% because the portlandite derived from HCWA could react with the added silica and form the secondary C–S–H, ultimately densifying the particle packing and reducing the porosity (Cheah and Ramli, 2012a). However, higher replacement levels of 6–10 wt.% caused a decrease in mechanical performance as larger amounts of wood ash incurred the dilution effect and generated more pores induced by their alkali content (Cheah and Ramli, 2012a). De Souza et al. (2022) substituted OPC in cement-based composites with wood ash at contents of 20–60 wt.% and observed a significant reduction in compressive strengths. Fořt et al. (2021) observed a gradual decline in flexural and compressive properties of cement mortar with increasing WBFA proportions from 10% to 70% after 28 days, which could be attributed to the increased porosity. Interestingly, WBFA-containing mortars exhibited a higher increasing extent in compressive strengths with increasing ages compared with reference samples due to the pozzolanic reaction and the mortars' densification. Furthermore, the results of isothermal calorimetry indicated a decline in the evolution of hydration heat as the WBFA ratio increased, which may contribute to the control of thermal cracking in concrete; see Fig. 16.7 (Fořt et al., 2021).

Cheah and Ramli (2012b) investigated the durability of HCWA-modified cement mortars and found that at a curing age of 28 days, the inclusion of HCWA as a partial OPC substitution increased the water absorption and decreased the air permeability; see Fig. 16.8. They also measured the shrinkage behavior and found that lower replacement levels of 5 and 10 wt.% reduced the total shrinkage, while the 15–25 wt.% of HCWA contents resulted in an increase in the total shrinkage due to the augmentation in autogenous shrinkage (Cheah and Ramli, 2012b). Udoeyo et al. (2006) added wood waste ash as a supplement to concrete at amounts of 5–30 wt.% and reported a gradual increase in water absorption as the ash proportion was increased. Madrid et al. (2017) produced concrete masonry units using wood ash as a cement

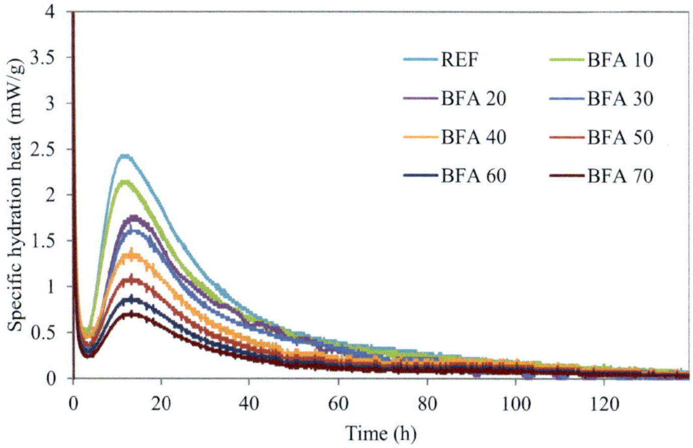

FIGURE 16.7 Hydration heat curves of cement mortars containing various contents of WBFA (Fořt et al., 2021). *Reprinted by permission of Elsevier.*

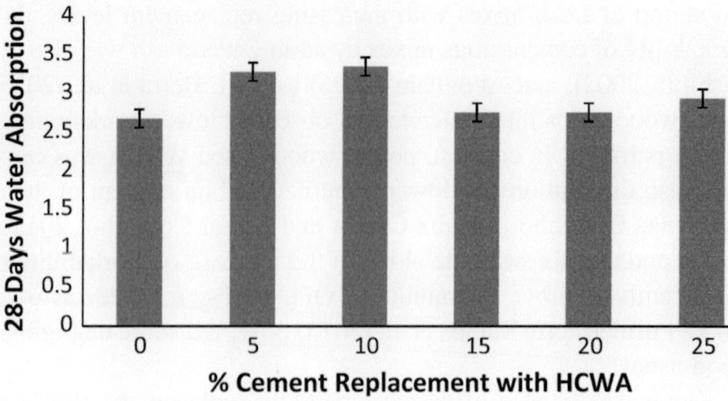

FIGURE 16.8 Water absorption of cement mortars reinforced with HCWA at various replacement levels (Cheah and Ramli, 2012b). *Reprinted by permission of Elsevier.*

replacement and found that the substitution levels of 5, 10, and 15 vol.% led to improvements in water absorption of about 16%, 29%, and 50%, respectively, due to higher porosity. Hence, the addition of wood ash could cause an increase in water absorption of cement composites, which is detrimental to their durability.

Silva et al. (2022) utilized high-calcium biomass ash derived from eucalyptus logs as solid precursors for the production of alkali-activated composites. Sodium hydroxide was used as an activator at concentrations of 5, 10, and 15 mol/L, and different dosages of eucalyptus ash (EA) were applied to partially replace silica fume (SF) at levels of 40, 50, and 60 wt.%. The alkali-activated mortars exhibited the highest compressive strengths when activated with 5M NaOH, and the strength of the composites increased with higher substitution levels of EA after 28 days. Additionally, as the content of EA increased, the mortars exhibited an increasing tendency in porosity and water absorption. Other researchers have also investigated the use of wood ash in alkali-activated materials (Zhu et al., 2022) or geopolymer composites (Novais et al., 2019; Saeli et al., 2019).

Given the significant variations in the chemical composition, particle size, and SSA of wood ash resulting from differences in wood species, combustion processes, and milling treatments, it is crucial to assess the suitability of wood ash for practical applications in construction materials through analytical investigations, as well as mechanical and durability experiments.

3.1.4 Combustion ash originating from other biomass

Olives and the produced olive oil are traditional and essential food, which are widely consumed around the world (Uylaşer and Yildiz, 2014). During the manufacture of olive oil, large amounts of solid wastes (also called "cake") can be generated, which usually contain olive stone, bark, pulp, and residual oil and possess abundant potassium and calcium (de Moraes Pinheiro et al., 2018). In Spain alone, it has been estimated that approximately 5 million metric tons of olive solid waste were generated in the 2016/2017 period (de Moraes Pinheiro et al., 2018). By combusting these dried solid wastes phase, approximately 12% of ash can be generated (Al-Akhras et al., 2009), and the biomass ash is often categorized into two types including a specific olive stone ash and a general olive waste ash (de Moraes Pinheiro et al., 2018). Rosales et al. (2022) have used 10 and 30 wt.% olive biomass bottom ash (OBBA) to substitute the cement content in concrete and observed that the unwashed OBBA prolonged the initial/final setting times of mixtures. Furthermore, the addition of this ash reduced the flexural and compressive properties, since the pozzolanic activity of OBBA is insufficient to compensate for the loss of clinker and the increase in porosity. In the study of de Moraes Pinheiro et al. (2018), high-potassium olive stone biomass ash (OBA, 32% K_2O, 28% CaO, 5% SiO_2) was employed as a precursor to partially substitute the blast furnace slag (BFS) or be added in the mixture for AAC. It was recorded that both replacement (20−35 wt.%) of BFS and addition (20−25 wt.%) of OBA presented an enhancing effect in flexural and compressive performance compared with BFS−KOH systems, being due to the lower development of zeolitic phases and the diminution of the mean pore diameter (de Moraes Pinheiro et al., 2018). Additionally, olive biomass fly ash, which primarily consists of potassium, can also serve as an alkali solution for the production of geopolymer mixtures (Carrillo-Beltran et al., 2021).

According to the Food and Agriculture Organization of the United Nations (FAO, 2021), the global production of maize is projected to be 1198 million tons in 2021, showing an explosive increase in comparison with 589 million tons in 2000 (Nations, 2016). The United States is the leading maize producer globally, with an estimated output of approximately 384 million tons, followed by China with 270 million tons and Brazil with 100 million tons (FAO, 2021). The harvesting of corn grain was reported to be accompanied by nearly 15% of corn cob wastes by weight, and this discarded corn cob residue is utilized as biofuel for energy production as a consequence of its abundant cellulose and hemicellulose (Raj et al., 2015; Gonzalez-Kunz et al., 2017). Shakouri et al. (2020) have blended ground corn cob ash (CCA; see Fig. 16.9) into

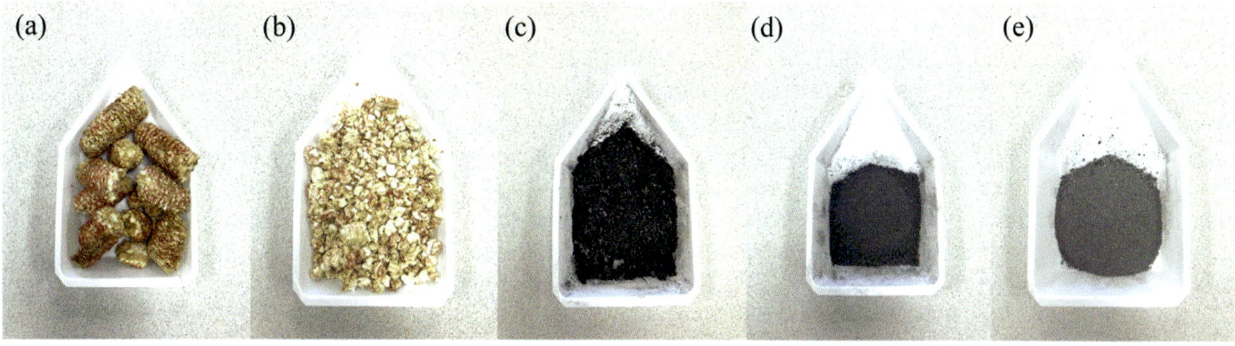

FIGURE 16.9 (A) Corn cob, (B) corn cob after grinding, (C) ground corn cob after burning, (D) CCA after heat treatment, and (E) ground CCA (Shakouri et al., 2020). *Reprinted by permission of Elsevier.*

concrete at the cement replacement ratios of 3 wt.% and 20 wt.% and revealed a decline in compressive strengths and resistance to chloride ingress due to the hindered cement hydration induced by high K_2O content of CCA. Taking into account that the CCA has abundant potassium elements in some cases (Shakouri et al., 2020), therefore, there is potential for its use in AAC.

Corn straw is a significant crop waste generated after maize harvesting, with a global output exceeding 1.0 billion tons and reaching approximately 250 million tons in China (Lu et al., 2016). Due to its natural lignocellulosic composition, corn straw is challenging to decompose, often leading to its combustion for power generation, resulting in the production of a large amount of ash (Liu et al., 2018; Lu et al., 2019). One potential approach to manage corn straw ash is its utilization in the production of cemented coal gangue backfill (Wang et al., 2021b).

In recent years, palm oil fuel ash (POFA) has aroused ascending interest in the field of civil engineering. Oil palm trees are predominantly cultivated in tropical climate regions, e.g., Southeast Asia (Hamada et al., 2020). POFA is a by-product derived from agrowaste, specifically from the combustion of oil palm residues (including palm fibers, kernel shells, and empty fruit bunches) at temperatures of approximately 800−1000°C during the energy generation process (Jaturapitakkul et al., 2007). In Thailand, over 100 million tons of POFA can be produced every year (Jaturapitakkul et al., 2007). Hamada et al. (2023) conducted research on the use of nano-POFA as a substitute for cement clinker in concrete at replacement levels of 15%−30%. They found that nano-POFA improved the workability of the concrete due to its lubricating effect among the aggregates. After a curing period of 360 days, a 15% replacement level resulted in an enhanced compressive strength of 85 MPa, attributed to the reaction between the silica in the ash and calcium hydroxide. However, a higher content of nano-POFA (30%) led to a decrease in compressive properties. Salam et al. (2018) reported that using lower replacement ratios of ground POFA (below 20 wt.%) in concrete led to beneficial strength development, while a replacement ratio of 30 wt.% had a negative impact on compressive strength. Men et al. (2022) conducted a study where they incorporated high amounts of ground POFA in concrete, replacing OPC at ratios of 60, 70, and 80 wt.%. Their results showed a significant decrease in compressive strength with high-volume POFA replacement, attributed to the lower clinker content and hindered hydration reaction. Similarly, Islam et al. (2016) observed a decline in concrete strength when using POFA to replace cement clinker at replacement levels of 10−50 wt.%, particularly at higher replacement levels above 30%, due to the reduction in free water content. Based on previous research, it is recommended to select cement substitution levels by POFA below 20% to optimize strength.

As well, some other biomass-based ashes have been investigated to fabricate the cementitious composites, such as miscanthus ash (Lv et al., 2019), switchgrass ash (Wang et al., 2014), rapeseed straw ash (Zhang et al., 2019), banana leaf ash (Tavares et al., 2022), bamboo leaf ash (Vinai et al., 2022), hazelnut shell bottom ash (Sevinç, 2022), cattle manure ash (Zhou and Chen, 2012; Chen et al., 2019), etc. Many researchers have studied the impact of the mixture containing more than one type of ash on the properties of construction materials. Carrasco-Hurtado et al. (2014) have investigated the impact of BBA on the mechanical and durability performance of calcium silicate masonry units, where the BBA is produced from a mixture including olive oil wastes, energy crops, olive, and fruit trees' pruning. Medina et al. (2019b) explored the impact of replacing 10%−20% of OPC with BBAs, which included different types of woody and nonwoody ashes obtained from electric power plants, on the properties of cement mortars. They observed an increase in water absorption, as well as a decrease in strength and drying shrinkage.

Overall, there is significant potential in utilizing biomass combustion ashes to reinforce building materials, considering the wide range of agricultural waste species and various types of residues from forestry and animal husbandry.

3.2 Biomass combustion ash as aggregates in cement-based composites

Biomass combustion ash has been broadly studied to be added as SCMs in cement-based materials, while the effective utilization of biobased ash as fine or coarse aggregates significantly lags behind; see Table 16.2. Memon et al. (2019) investigated the use of corncob ash (CCA) as a partial replacement for fine aggregates in concrete. They observed an increase in slump values as the percentage of CCA increased. This increase in fluidity can be attributed to the lubrication effect caused by the presence of more available water in the CCA mixes with a saturated surface dry condition. On the contrary, by replacing sand with RHA and bagasse ash (BA), Hasnain et al. (2021) have observed a reduction of workability in fresh concrete, since the ashes having porous and adsorptive nature brought about a higher viscosity of the mixtures.

Lessard et al. (2017) conducted a study where concrete was made by replacing 20−80 wt.% of sand with BBA. The results showed a significant reduction in compressive strength and splitting-tensile strength after a 28-day curing period. Beltrán et al. (2016) have incorporated olive BBA into mortars at the sand substitution levels of 10 and 20 vol.%, which resulted in a diminution in flexural and compressive strength as a consequence of high organic matter content and high porosity of olive BBA. On the other hand, Binici et al. (2008) observed a slight improvement in compressive strength in concrete reinforced with biomass combustion ashes. They blended corncob ash (CCA), WSA, and plane leaf ash (PLA) into mixtures to replace river sand at dosages of 2, 4, and 6 wt.% and reported that the concrete strength increased with increasing ash proportions. Soares and Castro-Gomes (2022) investigated the influence of wood bottom ash (WBA) primarily produced from eucalyptus and pine on the properties of carbonated reactive magnesia cement (CRMC). They found that by partially replacing the sand with approximately 15.5 wt.% of WBA, compressive strength could be enhanced, which could be attributed to particle adjustment induced by the filler effect of WBA. However, when the sand was completely substituted with WBA, the mortar strength declined due to the lower strengths of biobased aggregates.

The durability of CCA-modified concrete was investigated by Memon et al. (2019). They found that as the sand was replaced with CCA, the water absorption values increased. This can be attributed to the porous microstructure of saturated CCA aggregates, which resulted in a higher initial free water content and the generation of capillaries in the mixtures (Memon et al., 2019). Similarly, the results of Lessard et al. (2017) registered an augmentation in the permeable air voids and water absorption in concrete with sand replacement by BBA. Beltrán et al. (2014) also demonstrated that replacing natural sand with BBA led to an increase in water absorption, chloride penetration, and water penetration. In general, the addition of biomass combustion ash serving as aggregates seems to have detrimental effect on the durability of cementitious mixtures.

In recent studies, there has been exploration of the potential use of biomass combustion ash for preparing lightweight aggregates (Aungatichart et al., 2022; Serpell and Zwicky, 2022; Lin et al., 2023). However, further research is still needed in this area to fully understand and optimize the properties and performance of such aggregates.

3.3 Biochar as SCMs in cement-based composites

Biochar has been extensively explored as an additional filler or replacement for cement clinker in cement-based composites. Its impact on the rheology, hydration kinetics, and hardened-state properties is overviewed in this section. It was often reported that the addition of biochar could increase the plastic viscosity and yield stress of fresh mixes, triggering the decline of flowability (Sirico et al., 2020; Tan et al., 2020). Gupta and Kashani (2021) have incorporated ground biochar from waste peanut shells into mortars at cement replacement levels of 1 and 3 wt.%, and they showed that the added biochar prolonged the induction period during cement hydration and increased the cumulative heat release, which benefited the enhancement of hydration degree and the strength development at early ages. However, the addition of biochar also resulted in an evident increase in yield stress, attributed to the angular shapes of biochar particles. This increased cohesiveness of the mixtures could have a negative impact on the pumping performance. In the study by Gupta et al. (2020a), the influence of different types of biochar on the rheological properties of fresh mortar was investigated. They found that replacing 5 wt.% of ordinary Portland cement (OPC) with biochar resulted in a slight increase in plastic viscosity and dynamic yield stress. The magnitude of these increases was found to be dependent on the porosity of the biochar. Biochar with higher porosity was able to absorb more water during mixing, leading to a reduction in free water content. This, in turn, increased the interaction between flocculation and cement particles, resulting in decreased flowability of the mixtures (Gupta et al., 2020a). Interestingly, when the particle size of the biochar was extremely fine (0.10−2 μm), the biochar-modified mixtures exhibited similar or slightly higher workability compared with the control samples. This was attributed to the biochar powder filling the spaces between cement particles and providing more "excess water," which acted as a lubricating agent (Gupta and Kua, 2019).

TABLE 16.2 Selected studies of the application of biomass combustion ash as aggregates in cement composites.

References	Feedstock	Production condition	Ash type	Ash size	Ash dosage	Matrix	Curing condition	Fresh-state properties	Hardened-state properties	Durability	Characterization analysis
Al-Akhras and Abu-Alfoul (2002)	Wheat straw	Combustion in a burner at 650°C for 20 h	Ash collected from the burner	0.5–200 μm	Replacing the sand by 3.6, 7.3, 10.9 wt.%	Mortar	In molds for 24 h. In water at 23°C for 3 days before autoclaving	Initial setting time	Compressive, tensile, and flexural strength	–	SEM
Binici et al. (2008)	Corncob, wheat straw, and plane leaf	Combustion at a maximum temperature of 600°C	Ash collected from the incinerator	0.01–0.4 mm	Replacing the fine aggregate by 2, 4, 6 wt.%	Concrete	In tap water/ 5% Na_2SO_4 for 18 months, 7, 28, 90, 365 days	–	Compressive strength	Water penetration, sulfate resistance, abrasion resistance	Granulometry
Modolo et al. (2013)	Forest biomass residues (eucalyptus bark and remains from logging activities)	Burning in the bubbling bed at 800–900°C	Bottom ash	250–1000 μm	Replacing the coarse aggregates by 10 and 20 wt.%	Mortar	20°C and 65% RH for 1, 7, 28 days	Apparent density, setting time, water retentivity, flow, air content	Compressive strength and elastic modulus	Water-soluble chloride, unrestrained shrinkage	SEM-EDS, XRD, XRF
Beltrán et al. (2014)	Olive mash, olive orchard prunings, and other energy crops.	Combustion in a boiler at 403°C	Bottom ash	0–2 mm	Replacing natural sand by 3, 6 wt.%	Concrete	20°C and 100% RH for 7, 28, 90 days	Slump	Compressive and flexural strength, density	Water absorption, water penetration under pressure, chloride ion penetration, drying shrinkage	Granulometry
Beltrán et al. (2016)	Olive pruning and other plant compounds	Combustion in a chamber at 403°C	Bottom ash	0–2 mm	Replacing natural sand by 10, 20 vol.%	Mortar	20°C and 50% RH for 24 h and in water until 1, 7, 28, 90 days	–	Compressive and flexural strength, density, porosity	–	SEM, XRD
Lessard et al. (2017)	–	–	Bottom ash	5–1300 μm	Replacing the sand by 25, 40 wt.%	Dry-cast concrete	28, 91 days	Compaction index	Compressive and tensile strength, air voids, electrical resistivity	Water absorption, permeability	SEM, XRF, granulometry
Memon et al. (2019)	Corncob	Open burning (uncontrolled)	Ash collected during open burning	25–75 mm	Replacing the sand by 0, 5, 10%, 15, 20 vol.%	Concrete	In water at room temperature for 7, 28, 56, and 90 days	Slump, fresh concrete density	Compressive strength, density, ultrasonic pulse velocity	Total shrinkage, water absorption, acid attack	SEM, TGA, XRD, granulometry

Continued

TABLE 16.2 Selected studies of the application of biomass combustion ash as aggregates in cement composites.—cont'd

References	Feedstock	Production condition	Ash type	Ash size	Ash dosage	Matrix	Curing condition	Fresh-state properties	Hardened-state properties	Durability	Characterization analysis
Hasnain et al. (2021)	Rice husk and bagasse	Combustion during power production	Ash collected during power production	2–250 μm (RHA); 0.25–100 μm (BA)	Replacing the sand by 10, 20, 30 wt.% (RHA and BA)	Self-compacting concrete	Moist curing for 7, 28, 56 days	Slump flow time (T50), J-Ring test, L-Box test, V-Funnel test, air content	Compressive and split tensile strength, density	Water absorption, sulfate resistance	SEM-EDX, XRD, XRF
de Sande et al. (2022)	Sugarcane bagasse	Combustion in a traveling grate boiler at 750–800°C	Ash collected in filters	221 μm (mean, Ut-SCBA); 19.5 μm (mean, G-SCBA)	Replacing the fine aggregate by 10, 20, 30 wt.%	Mortar	In tap water for 3, 7, 28, 56, 90 days	Consistence of mortars (flow test), density of fresh mortar	Compressive and flexural strength, open porosity, apparent density	Rapid chloride migration, surface electrical resistivity, capillary water absorption	XRD, FTIR, SEM, TGA
Soares and Castro-Gomes (2022)	Forest wastes (primarily eucalyptus and pine)	Combustion in a power plant	Bottom ash	~10 mm; ~125 μm; 125 μm–4 mm	Replacing the yellow sand by 17.2, 100 wt.%	Carbonated reactive magnesia cement	In a carbonation chamber at CO_2 concentration of >99%; pressure of 0.7 bar; 50 ± 2°C; RH of >99%	–	Compressive strength, bulk density, MIP	CO_2 adsorption and carbonation degree	XRD, FTIR, SEM, TGA, pH
Serpell and Zwicky (2022)	Wood sawdust	Combustion by thermal energy supplier	Ash collected during wood combustion	0.4–500 μm (D50 = 70 μm)	The sawdust to total solids ratio (B/S) of 0.4–0.6	Lightweight aggregate	In covered plastic containers at ~20°C for 7 and 70 days	–	Particle crushing strength, particle density	Water absorption capacity, drying shrinkage	Granulometry
Aungatichart et al. (2022)	Palm oil (P-FA), wood chip (W-FA), bagasse (BA-FA), rice husk (R-FA)	–	Ash collected different locations	D50: 55.5 μm (P-FA); 58.32 μm (W-FA); 49.89 μm (Ba-FA); 41.92 μm (R-FA)	10 wt.% PC and 90 wt.% FA	Lightweight aggregate and concrete	In 100% RH at room temperature for 7 days and later in water for 21 days (aggregates); 7 and 28 days (concrete)	–	Compressive strength, bulk density (concrete)	–	SEM, XRD, XRF, granulometry
Lin et al. (2023)	Wood	Combustion in an incinerator	Ash collected from an incinerator	–	10 wt.% PC and 90 wt.% ash	Lightweight aggregate	In 90% RH at 30 ± 1°C for 3, 7, 14, 28 days	–	Single particle compressive strength, bulk density	Water absorption, toxicity characteristics leaching procedure	SEM, XRF, granulometry

Note: *BA*, bagasse ashes; *EDS*, electron dispersive spectroscopy; *EDX*, energy dispersive X-ray spectroscopy; *FTIR*, Fourier transform infrared spectroscopy; *G-SCBA*, ground bagasse ashes; *MIP*, mercury intrusion porosimetry; *SEM*, scanning electron microscope; *TGA*, thermal gravimetric analysis; *Ut-SCBA*, untreated bagasse ashes; *XRD*, X-ray diffraction; *XRF*, X-ray fluorescence spectrometry.

Numerous studies manifested that the addition of biochar at a certain content is in favor of improving the mechanical properties of cementitious composite (Gupta and Kua, 2019; Praneeth et al., 2020; Tan et al., 2020). Praneeth et al. (2020) investigated the effects of different dosages (2−8 wt.%) of corn stover biochar filler in cement-fly ash composites. They found that incorporating biochar led to an improvement in the compressive strength of M20 composites (with 20% fly ash). For composites with higher fly ash proportions (M40, M50), the strength was enhanced with an increase in biochar dosage up to 4 wt.%, but further increase in biochar content hindered the strength development. The added biochar acted as nucleation sites for the growth of calcium-silicate-hydrate (C−S−H), reduced water evaporation, and decreased the local water-to-binder ratio, resulting in higher composite strength (Praneeth et al., 2020). Similarly, Gupta et al. (2018a) added 1−2 wt.% of biochar as a filler in mortars and observed an improvement in compressive strength compared with plain samples. However, higher biochar content led to a decrease in composite strength. They also found that the addition of 1 wt.% of biochar resulted in the highest flexural performance, while excessive biochar negatively affected the flexural strength due to the introduction of air voids by porous biochar (see Fig. 16.10) (Gupta et al., 2018a).

The types of starting feedstock of biochar played a crucial role in the strength of cementitious composites (Oh et al., 2020; Sirico et al., 2020). Akhtar and Sarmah (2018) investigated the effects of biochar derived from rice husk (RH), poultry litter (PL), and pulp and paper mill (PP) on concrete by partially replacing the cement clinker at dosages of 0.1−1 vol.% of the total concrete. They found that the mechanical properties varied significantly depending on the type of biomass used. The compressive strengths of biochar-modified concrete decreased noticeably due to the high volume occupation of biochar, which hindered the formation of calcium-silicate-hydrate (C−S−H) gel. However, most of the samples exhibited an improvement in flexural behavior, particularly PL0.1, RH0.1, and RH1 specimens, as biochar

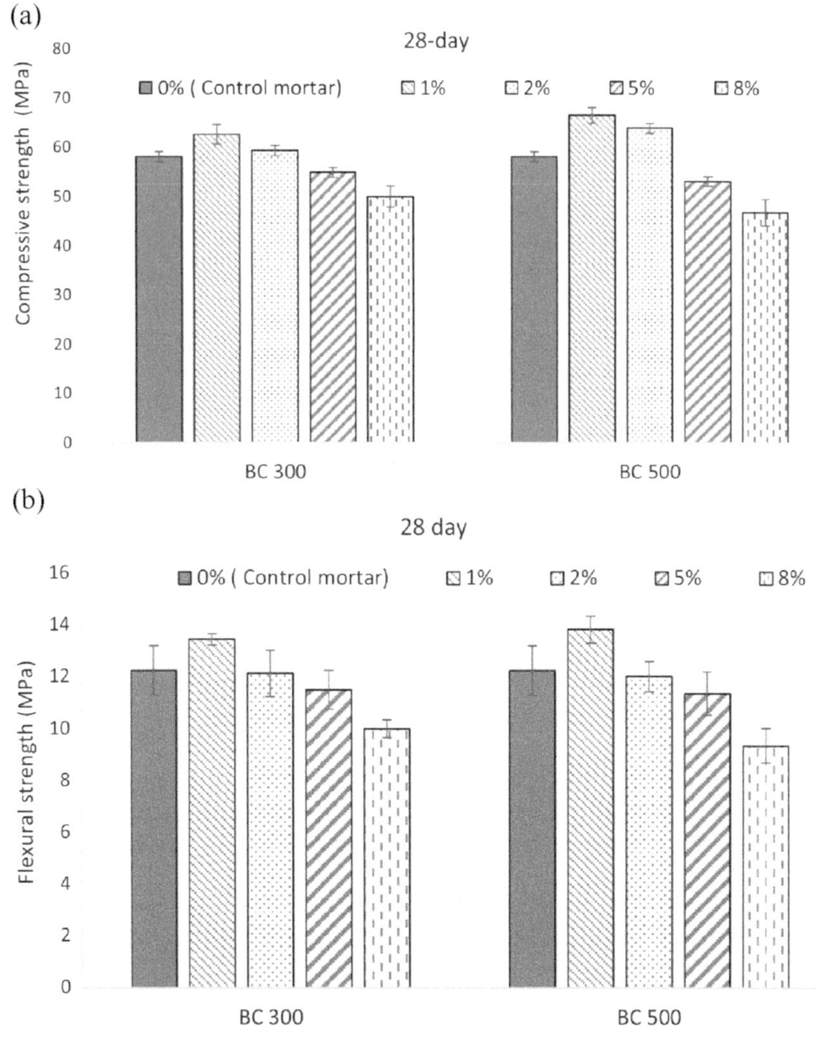

FIGURE 16.10 (A) Compressive strength and (B) flexural strength of mortars containing BC300 and BC500 after curing of 28 days (Gupta et al., 2018a). *Reprinted by permission of Elsevier.*

reduced pore size and decreased the amount of macro pores in the cross-sections. Interestingly, the sample with 0.1% RH biochar showed an increase in splitting tensile strength, indicating its potential as a candidate for concrete reinforcement (Akhtar and Sarmah, 2018). In summary, the addition of biochar within a limited range is advantageous to the strength development; nevertheless, higher contents of biochar could give rise to fewer hydration products (Akinyemi and Adesina, 2020) and the separation of cement grains (Park et al., 2021), decreasing the mechanical properties. Additionally, the presence of brittle biochar with a high proportion can facilitate the formation of microcracks (Tan et al., 2020; Wang et al., 2020).

The addition of biochar may favor the amelioration of the durability of cement composites. Gupta et al. (2018b) have disclosed that incorporating 1−2 wt.% of biochar in mortars led to a decrease in water sorptivity and water penetration depth. This improvement can be attributed to the densification of the cement mixes and the reduction of paste porosity caused by the presence of biochar fillers, resulting in reduced water absorption. Similar phenomena were observed in Gupta and Kua (2018, 2019) and Gupta et al. (2020b). However, there are contrasting results reported by Praneeth et al. (2021) and Akhtar and Sarmah (2018), who observed an increase in water absorption in cementitious composites with the addition of biochar particles. This increase in water absorption could be attributed to the presence of capillary pores introduced by the biochar particles. Additionally, it was displayed that due to water-retentive ability and internal curing action, porous biochar additives conduced to the diminution of autogenous shrinkage of cementitious materials (Mo et al., 2019; Gupta et al., 2020a).

3.4 Biochar as aggregates in cement-based composites

Indeed, the use of biochar as sustainable lightweight aggregates in mortar or concrete is a relatively less explored area compared with its use as SCMs or fillers. However, there are some studies that have investigated the incorporation of biochar as aggregates and its impact on the properties of cementitious composites. Wang et al. (2020) partially replaced river sand with biochar in mortars at varying proportions (1−5 wt.%). They observed that the addition of 1−2 wt.% of biochar resulted in an increase in compressive strength, attributed to the promotion of cement hydration. However, higher proportions of biochar negatively affected the composite strength, possibly due to the brittleness of biochar with a porous structure. Praneeth et al. (2021) replaced sand with biochar at ratios of 10−40 wt.% in mortar and reported an improvement in flexural behavior at lower biochar contents. They also found a decrease in bulk density with an increase in the amount of biochar. Chen et al. (2022) investigated the replacement of recycled fine aggregate with biochar and observed a reduction in compressive strength of cement composites as the replacement levels increased. However, they found that the incorporation of mineral SCMs (such as ground granulated BFS, SF, and metakaolin) partially compensated for the strength loss in biochar-reinforced composites. It's worth noting that the addition of 30 wt.% biochar as aggregate has the potential to achieve carbon-negative concrete, which has economic and environmental significance in the construction sector (Chen et al., 2022). Further research is needed to explore the utilization of biochar as aggregates in cementitious materials, to expand its applications and deepen our understanding of its effects on composite properties.

4. Conclusions and perspectives

Biomass combustion ash exhibits significant potential for applications in the construction industry, playing a pivotal role in advancing circular economy practices and sustainability objectives. This chapter aims to provide a comprehensive overview of the sources and production conditions associated with biomass combustion ash. The integration of bioashes as SCMs or aggregates in cement-based materials demonstrates notable enhancements in fresh properties, mechanical performance, and durability. However, it is crucial to judiciously select the dosages, surface area, types of starting biomass, processing temperatures, and particle size of ashes to optimize the microstructures of cement composites and facilitate strength development. Notably, biobased ashes derived from diverse feedstocks exhibit significant variations in chemical composition, resulting in distinct mechanisms of hydration reactions in cement-based materials. Additionally, carbon-rich biochar, a unique biomass residue obtained under controlled conditions, can serve as an SCM or aggregate in cement-based composites. The incorporation of biochar in concrete offers resource conservation benefits and aids in mitigating carbon emissions due to its carbon-negative nature.

Looking ahead, future perspectives hold immense potential for the expanded utilization of biomass combustion ash in various construction materials. Alkali-activated cement, magnesium-rich cement, and asphalt-based materials are areas where the use of biomass combustion ash can be explored. Advancements in scientific technologies, such as transmission electron microscopy, nanoindentation, and microcomputed tomography, can facilitate a deeper understanding of the interfacial reactions in bioash-modified construction materials. Furthermore, to unlock additional value-added benefits of biochar in the construction industry, its application as lightweight aggregates or nanofillers should be explored. Engineered biochar can be customized through surface functionalization, electronegativity modification, and impregnation of reactive minerals to enhance compatibility with cementitious matrices. Establishing systematic regulations for biomass ash and biochar is crucial to ensure quality control and standardized assessments, including aspects such as chemical constituents, reactivity of amorphous substances, particle sizes, and SSA. By carefully selecting raw materials, the mechanical performance and environmental advantages of building materials can be optimized. In the pursuit of engineering feasibility for bioash/biochar-reinforced composites, the addition of fiber reinforcement (Li et al., 2021a, 2022b, 2022d) or mineral SCMs (Li et al., 2020, 2021b) can be considered to enhance the matrices' properties.

Acknowledgments

The authors appreciate the support from the National Natural Science Foundation of China (Grant No. 52078293).

References

Abbas, S., Kazmi, S.M., Munir, M.J., 2017. Potential of rice husk ash for mitigating the alkali-silica reaction in mortar bars incorporating reactive aggregates. Construct. Build. Mater. 132, 61–70.

Abdullahi, M., 2006. Characteristics of wood ash/OPC concrete. Leonardo Electron. J. Pract. Technol. 8, 9–16.

Adesina, A., 2021. Performance and sustainability overview of sodium carbonate activated slag materials cured at ambient temperature. Resour. Environ. Sustain. 3, 100016.

Adhikary, S.K., Ashish, D.K., Rudžionis, Ž., 2022. A review on sustainable use of agricultural straw and husk biomass ashes: Transitioning towards low carbon economy. Sci. Total Environ. 156407.

Agbede, T.M., Oyewumi, A., 2022. Benefits of biochar, poultry manure and biochar–poultry manure for improvement of soil properties and sweet potato productivity in degraded tropical agricultural soils. Resour. Environ. Sustain. 7, 100051.

Agrela, F., Cabrera, M., Morales, M.M., Zamorano, M., Alshaaer, M., 2019. Biomass fly ash and biomass bottom ash. In: New Trends in Eco-Efficient and Recycled Concrete. Elsevier, pp. 23–58.

Agwa, I.S., Omar, O.M., Tayeh, B.A., Abdelsalam, B.A., 2020. Effects of using rice straw and cotton stalk ashes on the properties of lightweight self-compacting concrete. Construct. Build. Mater. 235, 117541.

Ahmad, M.R., Chen, B., Duan, H., 2020. Improvement effect of pyrolyzed agro-food biochar on the properties of magnesium phosphate cement. Sci. Total Environ. 718, 137422.

Akhtar, A., Sarmah, A.K., 2018. Novel biochar-concrete composites: manufacturing, characterization and evaluation of the mechanical properties. Sci. Total Environ. 616, 408–416.

Akinyemi, B.A., Adesina, A., 2020. Recent advancements in the use of biochar for cementitious applications: a review. J. Build. Eng., 101705

Al-Akhras, N.M., 2011. Durability of wheat straw ash concrete exposed to freeze–thaw damage. Proc. Inst. Civil Eng. Constr. Mater. 164, 79–86.

Al-Akhras, N.M., Abu-Alfoul, B.A., 2002. Effect of wheat straw ash on mechanical properties of autoclaved mortar. Cement Concr. Res. 32, 859–863.

Al-Akhras, N.M., Al-Akhras, K.M., Attom, M.F., 2009. Performance of olive waste ash concrete exposed to elevated temperatures. Fire Saf. J. 44, 370–375.

Almalkawi, A.T., Balchandra, A., Soroushian, P., 2019. Potential of using industrial wastes for production of geopolymer binder as green construction materials. Construct. Build. Mater. 220, 516–524.

Anjos, M.A., Araujo, T.R., Ferreira, R.L., Farias, E.C., Martinelli, A.E., 2020. Properties of self-leveling mortars incorporating a high-volume of sugar cane bagasse ash as partial Portland cement replacement. J. Build. Eng. 32, 101694.

Apori, S.O., Byalebeka, J., Murongo, M., Ssekandi, J., Noel, G.L., 2021. Effect of co-applied corncob biochar with farmyard manure and NPK fertilizer on tropical soil. Resour. Environ. Sustain. 5, 100034.

Arif, E., Clark, M.W., Lake, N., 2016. Sugar cane bagasse ash from a high efficiency co-generation boiler: applications in cement and mortar production. Construct. Build. Mater. 128, 287–297.

Association, C.T.E.C., 2020. Activity Report (Brussels).

Athira, G., Bahurudeen, A., 2022. Rheological properties of cement paste blended with sugarcane bagasse ash and rice straw ash. Construct. Build. Mater. 332, 127377.

Aungatichart, O., Nawaukkaratharnant, N., Wasanapiarnpong, T., 2022. The potential use of cold-bonded lightweight aggregate derived from various types of biomass fly ash for preparation of lightweight concrete. Mater. Lett. 327, 133019.

Ayobami, A.B., 2021. Performance of wood bottom ash in cement-based applications and comparison with other selected ashes: overview. Resour. Conserv. Recycl. 166, 105351.

Bayapureddy, Y., Muniraj, K., Mutukuru, M.R.G., 2020. Sugarcane bagasse ash as supplementary cementitious material in cement composites: strength, durability, and microstructural analysis. J. Korean Ceram. Soc. 57, 513–519.

Beltrán, M.G., Agrela, F., Barbudo, A., Ayuso, J., Ramirez, A., 2014. Mechanical and durability properties of concretes manufactured with biomass bottom ash and recycled coarse aggregates. Construct. Build. Mater. 72, 231–238.

Beltrán, M.G., Barbudo, A., Agrela, F., Jiménez, J.R., de Brito, J., 2016. Mechanical performance of bedding mortars made with olive biomass bottom ash. Construct. Build. Mater. 112, 699–707.

Berra, M., Mangialardi, T., Paolini, A.E., 2015. Reuse of woody biomass fly ash in cement-based materials. Construct. Build. Mater. 76, 286–296.

Binici, H., Yucegok, F., Aksogan, O., Kaplan, H., 2008. Effect of corncob, wheat straw, and plane leaf ashes as mineral admixtures on concrete durability. J. Mater. Civ. Eng. 20, 478–483.

Brewer, C.E., Chuang, V.J., Masiello, C.A., Gonnermann, H., Gao, X., Dugan, B., Driver, L.E., Panzacchi, P., Zygourakis, K., Davies, C.A., 2014. New approaches to measuring biochar density and porosity. Biomass Bioenergy 66, 176–185.

Cabrera, M., Rosales, J., Ayuso, J., Estaire, J., Agrela, F., 2018. Feasibility of using olive biomass bottom ash in the sub-bases of roads and rural paths. Construct. Build. Mater. 181, 266–275.

Cao, Y., He, M., Dutta, S., Luo, G., Zhang, S., Tsang, D.C., 2021. Hydrothermal carbonization and liquefaction for sustainable production of hydrochar and aromatics. Renewable Sustainable Energy Rev. 152, 111722.

Cao, F., Qiao, H., Li, Y., Shu, X., Cui, L., 2022. Effect of highland barley straw ash admixture on properties and microstructure of concrete. Construct. Build. Mater. 315, 125802.

Carrasco-Hurtado, B., Corpas-Iglesias, F., Cruz-Pérez, N., Terrados-Cepeda, J., Pérez-Villarejo, L., 2014. Addition of bottom ash from biomass in calcium silicate masonry units for use as construction material with thermal insulating properties. Construct. Build. Mater. 52, 155–165.

Carrillo-Beltran, R., Corpas-Iglesias, F.A., Terrones-Saeta, J.M., Bertoya-Sol, M., 2021. New geopolymers from industrial by-products: olive biomass fly ash and chamotte as raw materials. Construct. Build. Mater. 272, 121924.

Chandel, A.K., da Silva, S.S., Carvalho, W., Singh, O.V., 2012. Sugarcane bagasse and leaves: foreseeable biomass of biofuel and bio-products. J. Chem. Technol. Biotechnol. 87, 11–20.

Cheah, C.B., Ramli, M., 2012a. Load capacity and crack development characteristics of HCWA–DSF high strength mortar ferrocement panels in flexure. Construct. Build. Mater. 36, 348–357.

Cheah, C.B., Ramli, M., 2012b. Mechanical strength, durability and drying shrinkage of structural mortar containing HCWA as partial replacement of cement. Construct. Build. Mater. 30, 320–329.

Chen, X., Zhou, S., Zhang, H., Hui, Y., 2019. Alkali silicate reaction of cement mortar with cattle manure ash. Construct. Build. Mater. 203, 722–733.

Chen, L., Zhang, Y., Wang, L., Ruan, S., Chen, J., Li, H., Yang, J., Mechtcherine, V., Tsang, D.C., 2022. Biochar-augmented carbon-negative concrete. Chem. Eng. J. 431, 133946.

Clark, M.W., Despland, L.M., Lake, N.J., Yee, L.H., Anstoetz, M., Arif, E., Parr, J.F., Doumit, P., 2017. High-efficiency cogeneration boiler bagasse-ash geochemistry and mineralogical change effects on the potential reuse in synthetic zeolites, geopolymers, cements, mortars, and concretes. Heliyon 3, e00294.

da Luz Garcia, M., Sousa-Coutinho, J., 2013. Strength and durability of cement with forest waste bottom ash. Construct. Build. Mater. 41, 897–910.

de Moraes Pinheiro, S.M., Font, A., Soriano, L., Tashima, M.M., Monzó, J., Borrachero, M.V., Payá, J., 2018. Olive-stone biomass ash (OBA): an alternative alkaline source for the blast furnace slag activation. Construct. Build. Mater. 178, 327–338.

de Sande, V.T., Sadique, M., Bras, A., Pineda, P., 2022. Activated sugarcane bagasse ash as efficient admixture in cement-based mortars: mechanical and durability improvements. J. Build. Eng. 59, 105082.

de Sensale, G.R., Viacava, I.R., 2018. A study on blended Portland cements containing residual rice husk ash and limestone filler. Construct. Build. Mater. 166, 873–888.

de Siqueira, A.A., Cordeiro, G.C., 2022. Properties of binary and ternary mixes of cement, sugarcane bagasse ash and limestone. Construct. Build. Mater. 317, 126150.

De Souza, D., Antunes, L., Sanchez, L., 2022. The evaluation of Wood Ash as a potential preventive measure against alkali-silica reaction induced expansion and deterioration. J. Clean. Prod. 358, 131984.

Demirbaş, A., 2001. Biomass resource facilities and biomass conversion processing for fuels and chemicals. Energy Convers. Manag. 42, 1357–1378.

Demis, S., Tapali, J., Papadakis, V., 2014. An investigation of the effectiveness of the utilization of biomass ashes as pozzolanic materials. Construct. Build. Mater. 68, 291–300.

Dixit, M.K., Fernández-Solís, J.L., Lavy, S., Culp, C.H., 2010. Identification of parameters for embodied energy measurement: a literature review. Energy Build. 42, 1238–1247.

Dixit, A., Gupta, S., Dai Pang, S., Kua, H.W., 2019. Waste Valorisation using biochar for cement replacement and internal curing in ultra-high performance concrete. J. Clean. Prod. 238, 117876.

El-Sayed, M.A., El-Samni, T.M., 2006. Physical and chemical properties of rice straw ash and its effect on the cement paste produced from different cement types. J. King Saud Univ. Eng. Sci. 19, 21–29.

Eliche-Quesada, D., Calero-Rodríguez, A., Bonet-Martínez, E., Pérez-Villarejo, L., Sánchez-Soto, P., 2021. Geopolymers made from metakaolin sources, partially replaced by Spanish clays and biomass bottom ash. J. Build. Eng. 40, 102761.

Elinwa, A.U., Mahmood, Y.A., 2002. Ash from timber waste as cement replacement material. Cement Concr. Compos. 24, 219–222.

Esteves, T., Rajamma, R., Soares, D., Silva, A., Ferreira, V., Labrincha, J., 2012. Use of biomass fly ash for mitigation of alkali-silica reaction of cement mortars. Construct. Build. Mater. 26, 687–693.

FAO, F., 2021. Food Outlook—Biannual Report on Global Food Markets. Food and Agriculture Organization of the United Nations, Rome.

Fořt, J., Šál, J., Ševčík, R., Doleželová, M., Keppert, M., Jerman, M., Záleská, M., Stehel, V., Černý, R., 2021. Biomass fly ash as an alternative to coal fly ash in blended cements: functional aspects. Construct. Build. Mater. 271, 121544.

GABC, 2019. Global Alliance for Buildings and Construction (GABC) towards a Zero-Emission, Efficient, and Resilient Buildings and Construction Sector-Global Status Report 2019.

Gonzalez-Kunz, R.N., Pineda, P., Bras, A., Morillas, L., 2017. Plant biomass ashes in cement-based building materials. Feasibility as eco-efficient structural mortars and grouts. Sustain. Cities Soc. 31, 151–172.

Gopakumar, S.T., 2012. Bio-oil Production through Fast Pyrolysis and Upgrading to "Green" Transportation Fuels. A Dissertation: Doctor of Philosophy—the Graduate Faculty of Auburn University, Alabama, p. 196.

Gopinath, A., Bahurudeen, A., Appari, S., Nanthagopalan, P., 2018. A circular framework for the valorisation of sugar industry wastes: review on the industrial symbiosis between sugar, construction and energy industries. J. Clean. Prod. 203, 89–108.

Guo, H., Huang, L., Liang, D., 2022. Further promotion of sustainable development goals using science, technology, and innovation. Innovation 3.

Gupta, S., Kashani, A., 2021. Utilization of biochar from unwashed peanut shell in cementitious building materials–Effect on early age properties and environmental benefits. Fuel Process. Technol. 218, 106841.

Gupta, S., Kua, H.W., 2017. Factors determining the potential of biochar as a carbon capturing and sequestering construction material: critical review. J. Mater. Civ. Eng. 29, 04017086.

Gupta, S., Kua, H.W., 2018. Effect of water entrainment by pre-soaked biochar particles on strength and permeability of cement mortar. Construct. Build. Mater. 159, 107–125.

Gupta, S., Kua, H.W., 2019. Carbonaceous micro-filler for cement: effect of particle size and dosage of biochar on fresh and hardened properties of cement mortar. Sci. Total Environ. 662, 952–962.

Gupta, S., Kua, H.W., Dai Pang, S., 2018a. Biochar-mortar composite: manufacturing, evaluation of physical properties and economic viability. Construct. Build. Mater. 167, 874–889.

Gupta, S., Kua, H.W., Koh, H.J., 2018b. Application of biochar from food and wood waste as green admixture for cement mortar. Sci. Total Environ. 619, 419–435.

Gupta, S., Krishnan, P., Kashani, A., Kua, H.W., 2020a. Application of biochar from coconut and wood waste to reduce shrinkage and improve physical properties of silica fume-cement mortar. Construct. Build. Mater. 262, 120688.

Gupta, S., Palansooriya, K.N., Dissanayake, P.D., Ok, Y.S., Kua, H.W., 2020b. Carbonaceous inserts from lignocellulosic and non-lignocellulosic sources in cement mortar: preparation conditions and its effect on hydration kinetics and physical properties. Construct. Build. Mater. 264, 120214.

Habert, G., Miller, S.A., John, V.M., Provis, J.L., Favier, A., Horvath, A., Scrivener, K.L., 2020. Environmental impacts and decarbonization strategies in the cement and concrete industries. Nat. Rev. Earth Environ. 1, 559–573.

Hamada, H.M., Jokhio, G.A., Al-Attar, A.A., Yahaya, F.M., Muthusamy, K., Humada, A.M., Gul, Y., 2020. The use of palm oil clinker as a sustainable construction material: a review. Cement Concr. Compos. 106, 103447.

Hamada, H.M., Al-Attar, A., Shi, J., Yahaya, F., Al Jawahery, M.S., Yousif, S.T., 2023. Optimization of sustainable concrete characteristics incorporating palm oil clinker and nano-palm oil fuel ash using response surface methodology. Powder Technol. 413, 118054.

Han, T., 2020. Properties of biochar from wood and textile. IOP Publishing, 042060. IOP Conference Series: Earth and Environmental Science.

Hasnain, M.H., Javed, U., Ali, A., Zafar, M.S., 2021. Eco-friendly utilization of rice husk ash and bagasse ash blend as partial sand replacement in self-compacting concrete. Construct. Build. Mater. 273, 121753.

He, M., Xu, Z., Sun, Y., Chan, P., Lui, I., Tsang, D.C., 2021. Critical impacts of pyrolysis conditions and activation methods on application-oriented production of wood waste-derived biochar. Bioresour. Technol. 341, 125811.

Hinojosa, M., Galvín, A., Agrela, F., Perianes, M., Barbudo, A., 2014. Potential use of biomass bottom ash as alternative construction material: conflictive chemical parameters according to technical regulations. Fuel 128, 248–259.

IEA, 2023. CO_2 Emissions in 2022. IEA, Paris. https://www.iea.org/reports/co2-emissions-in-2022.

Ioannidou, D., Meylan, G., Sonnemann, G., Habert, G., 2017. Is gravel becoming scarce? Evaluating the local criticality of construction aggregates. Resour. Conserv. Recycl. 126, 25–33.

Islam, M.M.U., Mo, K.H., Alengaram, U.J., Jumaat, M.Z., 2016. Durability properties of sustainable concrete containing high volume palm oil waste materials. J. Clean. Prod. 137, 167–177.

Jaturapitakkul, C., Kiattikomol, K., Tangchirapat, W., Saeting, T., 2007. Evaluation of the sulfate resistance of concrete containing palm oil fuel ash. Construct. Build. Mater. 21, 1399–1405.

Jittin, V., Bahurudeen, A., 2022. Evaluation of rheological and durability characteristics of sugarcane bagasse ash and rice husk ash based binary and ternary cementitious system. Construct. Build. Mater. 317, 125965.

Kaminskas, R., Cesnauskas, V., Kubiliute, R., 2015. Influence of different artificial additives on Portland cement hydration and hardening. Construct. Build. Mater. 95, 537–544.

Kara, P., Korjakins, A., Stokmanis-Blaus, V., 2012. Evaluation of properties of concrete incorporating ash as mineral admixtures. Constr. Sci. 13, 17–25.

Kaza, S., Yao, L., Bhada-Tata, P., Van Woerden, F., 2018. What a Waste 2.0: A Global Snapshot of Solid Waste Management to 2050. World Bank Publications.

Khushnood, R.A., Rizwan, S.A., Memon, S.A., Tulliani, J.-M., Ferro, G.A., 2014. Experimental investigation on use of wheat straw ash and bentonite in self-compacting cementitious system. Adv. Mater. Sci. Eng. 2014.

Krishnan, C., Sousa, L.d.C., Jin, M., Chang, L., Dale, B.E., Balan, V., 2010. Alkali-based AFEX pretreatment for the conversion of sugarcane bagasse and cane leaf residues to ethanol. Biotechnol. Bioeng. 107, 441–450.

Lessard, J.-M., Omran, A., Tagnit-Hamou, A., Gagne, R., 2017. Feasibility of using biomass fly and bottom ashes in dry-cast concrete production. Construct. Build. Mater. 132, 565–577.

Li, H., Liebscher, M., Ranjbarian, M., Hempel, S., Tzounis, L., Schröfl, C., Mechtcherine, V., 2019a. Electrochemical modification of carbon fiber yarns in cementitious pore solution for an enhanced interaction towards concrete matrices. Appl. Surf. Sci. 487, 52–58.

Li, L., Yang, M., Lu, Q., Zhu, W., Ma, H., Dai, L., 2019b. Oxygen-rich biochar from torrefaction: a versatile adsorbent for water pollution control. Bioresour. Technol. 294, 122142.

Li, H., Liebscher, M., Curosu, I., Choudhury, S., Hempel, S., Davoodabadi, M., Dinh, T.T., Yang, J., Mechtcherine, V., 2020. Electrophoretic deposition of nano-silica onto carbon fiber surfaces for an improved bond strength with cementitious matrices. Cement Concr. Compos. 114, 103777.

Li, H., Liebscher, M., Michel, A., Quade, A., Foest, R., Mechtcherine, V., 2021a. Oxygen plasma modification of carbon fiber rovings for enhanced interaction toward mineral-based impregnation materials and concrete matrices. Construct. Build. Mater. 273, 121950.

Li, L., Yang, J., Li, H., Du, Y., 2021b. Insights into the microstructure evolution of slag, fly ash and condensed silica fume in blended cement paste. Construct. Build. Mater. 309, 125044.

Li, H., Liebscher, M., Ly, K.H., Ly, P.V., Köberle, T., Yang, J., Fan, Q., Yu, M., Weidinger, I.M., Mechtcherine, V., 2022a. Effect of electrophoretic deposition of micro-quartz on the microstructural and mechanical properties of carbon fibers and their bond performance toward cement. J. Mater. Sci. 57, 21885–21900.

Li, H., Liebscher, M., Micusik, M., Yang, J., Sun, B., Yin, B., Yu, M., Mechtcherine, V., 2022b. Role of pH value on electrophoretic deposition of nano-silica onto carbon fibers for a tailored bond behavior with cementitious matrices. Appl. Surf. Sci. 600, 154000.

Li, H., Liebscher, M., Yang, J., Davoodabadi, M., Li, L., Du, Y., Yang, B., Hempel, S., Mechtcherine, V., 2022c. Electrochemical oxidation of recycled carbon fibers for an improved interaction toward alkali-activated composites. J. Clean. Prod. 368, 133093.

Li, H., Liebscher, M., Zhao, D., Yin, B., Du, Y., Yang, J., Kaliske, M., Mechtcherine, V., 2022d. A review of carbon fiber surface modification methods for tailor-made bond behavior with cementitious matrices. Prog. Mater. Sci. 132, 101040.

Li, H., Zhao, D., Liebscher, M., Yin, B., Yang, J., Kaliske, M., Mechtcherine, V., 2022e. An experimental and numerical study on the age depended bond-slip behavior between nano-silica modified carbon fibers and cementitious matrices. Cement Concr. Compos. 128, 104416.

Li, H., Schamel, E., Liebscher, M., Zhang, Y., Fan, Q., Schlachter, H., Köberle, T., Mechtcherine, V., Wehnert, G., Söthje, D., 2023a. Recycled carbon fibers in cement-based composites: influence of epoxide matrix depolymerization degree on interfacial interactions. J. Clean. Prod. 411, 137235.

Li, H., Wang, L., Zhang, Y., Yang, J., Tsang, D.C., Mechtcherine, V., 2023b. Biochar for Sustainable Construction Industry. Current Developments in Biotechnology and Bioengineering. Elsevier, pp. 63–95.

Li, H., Yang, J., Wang, L., Li, L., Xia, Y., Köberle, T., Dong, W., Zhang, N., Yang, B., Mechtcherine, V., 2023c. Multiscale assessment of performance of limestone calcined clay cement (LC3) reinforced with virgin and recycled carbon fibers. Construct. Build. Mater. 406, 133228.

Lian, F., Xing, B., 2017. Black carbon (biochar) in water/soil environments: molecular structure, sorption, stability, and potential risk. Environ. Sci. Technol. 51, 13517–13532.

Lim, J.S., Manan, Z.A., Alwi, S.R.W., Hashim, H., 2012. A review on utilisation of biomass from rice industry as a source of renewable energy. Renewable Sustainable Energy Rev. 16, 3084–3094.

Lin, J., Mo, K.H., Goh, Y., Onn, C.C., 2023. Potential of municipal woody biomass waste ash in the production of cold-bonded lightweight aggregates. J. Build. Eng. 63, 105392.

Liu, H., Ou, X., Yuan, J., Yan, X., 2018. Experience of producing natural gas from corn straw in China. Resour. Conserv. Recycl. 135, 216–224.

Liu, R., Pang, B., Zhao, X., Yang, Y., 2020. Effect of rice husk ash on early hydration behavior of magnesium phosphate cement. Construct. Build. Mater. 263, 120180.

Lo, F.-C., Lee, M.-G., Lo, S.-L., 2021. Effect of coal ash and rice husk ash partial replacement in ordinary Portland cement on pervious concrete. Construct. Build. Mater. 286, 122947.

Lu, Z., Zhao, Z., Wang, M., Jia, W., 2016. Effects of corn stalk fiber content on properties of biomass brick. Construct. Build. Mater. 127, 11–17.

Lu, J., Yang, Z., Xu, W., Shi, X., Guo, R., 2019. Enrichment of thermophilic and mesophilic microbial consortia for efficient degradation of corn stalk. J. Environ. Sci. 78, 118–126.

Lv, Y., Ye, G., De Schutter, G., 2019. Utilization of miscanthus combustion ash as internal curing agent in cement-based materials: effect on autogenous shrinkage. Construct. Build. Mater. 207, 585–591.

Madrid, M., Orbe, A., Rojí, E., Cuadrado, J., 2017. The effects of by-products incorporated in low-strength concrete for concrete masonry units. Construct. Build. Mater. 153, 117–128.

Martínez-García, R., Jagadesh, P., Zaid, O., Şerbănoiu, A.A., Fraile-Fernández, F.J., de Prado-Gil, J., Qaidi, S.M., Grădinaru, C.M., 2022. The present state of the use of waste wood ash as an eco-efficient construction material: a review. Materials 15, 5349.

Maschio, G., Koufopanos, C., Lucchesi, A., 1992. Pyrolysis, a Promising Route for Biomass Utilization. Bioresource Technology, United Kingdom, p. 42.

Mattila, T., Grönroos, J., Judl, J., Korhonen, M.-R., 2012. Is biochar or straw-bale construction a better carbon storage from a life cycle perspective? Process Saf. Environ. Protect. 90, 452–458.

Medina, J.M., del Bosque, I.S., Frías, M., de Rojas, M.S., Medina, C., 2019a. Design and properties of eco-friendly binary mortars containing ash from biomass-fuelled power plants. Cement Concr. Compos. 104, 103372.

Medina, J.M., del Bosque, I.S., Frías, M., de Rojas, M.S., Medina, C., 2019b. Durability of new blended cements additioned with recycled biomass bottom ASH from electric power plants. Construct. Build. Mater. 225, 429–440.

Mellado, A., Catalán, C., Bouzón, N., Borrachero, M., Monzó, J., Payá, J., 2014. Carbon footprint of geopolymeric mortar: study of the contribution of the alkaline activating solution and assessment of an alternative route. RSC Adv. 4, 23846–23852.

Memon, S.A., Wahid, I., Khan, M.K., Tanoli, M.A., Bimaganbetova, M., 2018. Environmentally friendly utilization of wheat straw ash in cement-based composites. Sustainability 10, 1322.

Memon, S.A., Javed, U., Khushnood, R.A., 2019. Eco-friendly utilization of corncob ash as partial replacement of sand in concrete. Construct. Build. Mater. 195, 165–177.

Men, S., Tangchirapat, W., Jaturapitakkul, C., Ban, C.C., 2022. Strength, fluid transport and microstructure of high-strength concrete incorporating high-volume ground palm oil fuel ash blended with fly ash and limestone powder. J. Build. Eng. 56, 104714.

Millati, R., Cahyono, R.B., Ariyanto, T., Azzahrani, I.N., Putri, R.U., Taherzadeh, M.J., 2019. Agricultural, industrial, municipal, and forest wastes: an overview. In: Sustainable Resource Recovery and Zero Waste Approaches, pp. 1–22.

Miller, S.A., John, V.M., Pacca, S.A., Horvath, A., 2018. Carbon dioxide reduction potential in the global cement industry by 2050. Cement Concr. Res. 114, 115–124.

Mirmohamadsadeghi, S., Karimi, K., 2020. Recovery of silica from rice straw and husk. In: Current Developments in Biotechnology and Bioengineering. Elsevier, pp. 411–433.

Mo, L., Fang, J., Huang, B., Wang, A., Deng, M., 2019. Combined effects of biochar and MgO expansive additive on the autogenous shrinkage, internal relative humidity and compressive strength of cement pastes. Construct. Build. Mater. 229, 116877.

Modolo, R., Ferreira, V., Tarelho, L., Labrincha, J., Senff, L., Silva, L., 2013. Mortar formulations with bottom ash from biomass combustion. Construct. Build. Mater. 45, 275–281.

Moraes, J., Tashima, M., Akasaki, J.L., Melges, J., Monzó, J., Borrachero, M., Soriano, L., Payá, J., 2016. Increasing the sustainability of alkali-activated binders: the use of sugar cane straw ash (SCSA). Construct. Build. Mater. 124, 148–154.

Moraes, J., Tashima, M., Akasaki, J.L., Melges, J., Monzó, J., Borrachero, M., Soriano, L., Payá, J., 2017. Effect of sugar cane straw ash (SCSA) as solid precursor and the alkaline activator composition on alkali-activated binders based on blast furnace slag (BFS). Construct. Build. Mater. 144, 214–224.

Munshi, S., Sharma, R.P., 2019. Utilization of rice straw ash as a mineral admixture in construction work. Mater. Today: Proc. 11, 637–644.

Murugesan, T., Vidjeapriya, R., Bahurudeen, A., 2021. Sustainable use of sugarcane bagasse ash and marble slurry dust in crusher sand based concrete. Struct. Concr. 22, E183–E192.

Nair, D.G., Fraaij, A., Klaassen, A.A., Kentgens, A.P., 2008. A structural investigation relating to the pozzolanic activity of rice husk ashes. Cement Concr. Res. 38, 861–869.

Nations, F., 2016. Food Outlook Biannual Report on Global Food Markets (Italy).

Nehdi, M., Duquette, J., El Damatty, A., 2003. Performance of rice husk ash produced using a new technology as a mineral admixture in concrete. Cement Concr. Res. 33, 1203–1210.

Neto, J.d.S.A., de França, M.J.S., de Amorim Junior, N.S., Ribeiro, D.V., 2021. Effects of adding sugarcane bagasse ash on the properties and durability of concrete. Construct. Build. Mater. 266, 120959.

Novais, R.M., Saeli, M., Caetano, A.P., Seabra, M.P., Labrincha, J.A., Surendran, K.P., Pullar, R.C., 2019. Pyrolysed cork-geopolymer composites: a novel and sustainable EMI shielding building material. Construct. Build. Mater. 229, 116930.

Obernberger, I., Supancic, K., 2009. Possibilities of ash utilisation from biomass combustion plants. In: Proceedings of the 17th European biomass conference and exhibition. Hamburg Germany.

Oh, S.Y., Shin, M., Seo, Y.D., Cha, S.W., 2020. Evaluation of commercial biochar in South Korea for environmental application and carbon sequestration. Environ. Prog. Sustain. Energy 39, e13440.

Pan, X., Sano, Y., 2005. Fractionation of wheat straw by atmospheric acetic acid process. Bioresour. Technol. 96, 1256–1263.

Pandey, A., Kumar, B., 2019a. Effects of rice straw ash and micro silica on mechanical properties of pavement quality concrete. J. Build. Eng. 26, 100889.

Pandey, A., Kumar, B., 2019b. Evaluation of water absorption and chloride ion penetration of rice straw ash and microsilica admixed pavement quality concrete. Heliyon 5, e02256.

Park, J.H., Kim, Y.U., Jeon, J., Yun, B.Y., Kang, Y., Kim, S., 2021. Analysis of biochar-mortar composite as a humidity control material to improve the building energy and hygrothermal performance. Sci. Total Environ. 775, 145552.

Pitman, R.M., 2006. Wood ash use in forestry—a review of the environmental impacts. Forestry 79, 563–588.

Poudyal, L., Adhikari, K., 2021. Environmental sustainability in cement industry: an integrated approach for green and economical cement production. Resour. Environ. Sustain. 4, 100024.

Praneeth, S., Guo, R., Wang, T., Dubey, B.K., Sarmah, A.K., 2020. Accelerated carbonation of biochar reinforced cement-fly ash composites: enhancing and sequestering CO_2 in building materials. Construct. Build. Mater. 244, 118363.

Praneeth, S., Saavedra, L., Zeng, M., Dubey, B.K., Sarmah, A.K., 2021. Biochar admixtured lightweight, porous and tougher cement mortars: mechanical, durability and micro computed tomography analysis. Sci. Total Environ. 750, 142327.

Praveenkumar, S., Sankarasubramanian, G., 2019. Mechanical and durability properties of bagasse ash-blended high-performance concrete. SN Appl. Sci. 1, 1–7.

Prychid, C.J., Rudall, P.J., Gregory, M., 2003. Systematics and biology of silica bodies in monocotyledons. Bot. Rev. 69, 377–440.

Qi, Q., Sun, C., Cristhian, C., Zhang, T., Zhang, J., Tian, H., He, Y., Tong, Y.W., 2021. Enhancement of methanogenic performance by gasification biochar on anaerobic digestion. Bioresour. Technol. 330, 124993.

Qin, Y., Pang, X., Tan, K., Bao, T., 2021. Evaluation of pervious concrete performance with pulverized biochar as cement replacement. Cement Concr. Compos. 119, 104022.

Qudoos, A., Kim, H.G., Jeon, I.K., Ryou, J.-S., 2019. Influence of the particle size of wheat straw ash on the microstructure of the interfacial transition zone. Powder Technol. 352, 453–461.

Raheem, A.A., Adenuga, O.A., 2013. Wood ash from bread bakery as partial replacement for cement in concrete. Int. J. Sustain. Constr. Eng. Technol. 4, 75–81.

Raj, T., Kapoor, M., Gaur, R., Christopher, J., Lamba, B., Tuli, D.K., Kumar, R., 2015. Physical and chemical characterization of various Indian agriculture residues for biofuels production. Energy Fuels 29, 3111–3118.

Ramos, T., Matos, A.M., Sousa-Coutinho, J., 2013. Mortar with wood waste ash: mechanical strength carbonation resistance and ASR expansion. Construct. Build. Mater. 49, 343–351.

Rithuparna, R., Jittin, V., Bahurudeen, A., 2021. Influence of different processing methods on the recycling potential of agro-waste ashes for sustainable cement production: a review. J. Clean. Prod. 316, 128242.

Rosales, M., Rosales, J., Agrela, F., de Rojas, M.S., Cabrera, M., 2022. Design of a new eco-hybrid cement for concrete pavement, made with processed mixed recycled aggregates and olive biomass bottom ash as supplementary cement materials. Construct. Build. Mater. 358, 129417.

Saeli, M., Tobaldi, D.M., Seabra, M.P., Labrincha, J.A., 2019. Mix design and mechanical performance of geopolymeric binders and mortars using biomass fly ash and alkaline effluent from paper-pulp industry. J. Clean. Prod. 208, 1188–1197.

Salam, M.A., Safiuddin, M., Jumaat, M.Z., 2018. Durability indicators for sustainable self-consolidating high-strength concrete incorporating palm oil fuel ash. Sustainability 10, 2345.

Saxena, R., Seal, D., Kumar, S., Goyal, H., 2008. Thermo-chemical routes for hydrogen rich gas from biomass: a review. Renewable Sustainable Energy Rev. 12, 1909–1927.

Serpell, R., Zwicky, D., 2022. Low-energy lightweight aggregates by cold bonding of biomass wastes: effects of raw material proportion adjustments on product properties. Construct. Build. Mater. 346, 128392.

Sevinç, A.H., 2022. Investigating the properties of GGBFS hazelnut ash–based cement-free mortars produced at ambient temperature under different curing conditions. J. Mater. Civ. Eng. 34, 04022268.

Shahbandeh, M., 2023. Global Wheat Consumption 2017/18-2022/23. https://www.statista.com/statistics/1094056/total-global-rice-consumption/. (Accessed May 2023).

Shakouri, M., Exstrom, C.L., Ramanathan, S., Suraneni, P., 2020. Hydration, strength, and durability of cementitious materials incorporating untreated corn cob ash. Construct. Build. Mater. 243, 118171.

Sharma, T., 2019. Biochar and Other Properties Resulting from the Gasification and Combustion of Biomass with Different Components. University of Iowa.

Shayan, E., Zare, V., Mirzaee, I., 2018. Hydrogen production from biomass gasification; a theoretical comparison of using different gasification agents. Energy Convers. Manag. 159, 30–41.

Sigvardsen, N.M., Geiker, M.R., Ottosen, L.M., 2021. Phase development and mechanical response of low-level cement replacements with wood ash and washed wood ash. Construct. Build. Mater. 269, 121234.

Silva, T.H., Lara, L.F., Silva, G.J., Provis, J.L., Bezerra, A.C., 2022. Alkali-activated materials produced using high-calcium, high-carbon biomass ash. Cement Concr. Compos. 132, 104646.

Singh, P., Suman, A., Tiwari, P., Arya, N., Gaur, A., Shrivastava, A., 2008. Biological pretreatment of sugarcane trash for its conversion to fermentable sugars. World J. Microbiol. Biotechnol. 24, 667–673.

Sirico, A., Bernardi, P., Belletti, B., Malcevschi, A., Dalcanale, E., Domenichelli, I., Fornoni, P., Moretti, E., 2020. Mechanical characterization of cement-based materials containing biochar from gasification. Construct. Build. Mater. 246, 118490.

Soares, E.G., Castro-Gomes, J., 2022. The role of biomass bottom ash in Carbonated Reactive Magnesia Cement (CRMC) for CO_2 mineralisation. J. Clean. Prod. 380, 135092.

Songpiriyakij, S., Kubprasit, T., Jaturapitakkul, C., Chindaprasirt, P., 2010. Compressive strength and degree of reaction of biomass-and fly ash-based geopolymer. Construct. Build. Mater. 24, 236–240.

Spósito, C., Fazzan, J., Rossignolo, J., Bueno, C., Spósito, F., Akasaki, J., Tashima, M., 2023. Ecodesign: approaches for sugarcane bagasse ash mortars a Brazilian context. J. Clean. Prod. 385, 135667.

Sua-iam, G., Sokrai, P., Makul, N., 2016. Novel ternary blends of Type 1 Portland cement, residual rice husk ash, and limestone powder to improve the properties of self-compacting concrete. Construct. Build. Mater. 125, 1028–1034.

Tahami, S.A., Arabani, M., Mirhosseini, A.F., 2018. Usage of two biomass ashes as filler in hot mix asphalt. Construct. Build. Mater. 170, 547–556.

Tan, K., Pang, X., Qin, Y., Wang, J., 2020. Properties of cement mortar containing pulverized biochar pyrolyzed at different temperatures. Construct. Build. Mater. 263, 120616.

Tavares, M.M., Oliveira, C.H., 2004. Influence of mechanical grinding on the pozzolanic activity of residual sugarcane bagasse ash. In: PRO 40: International RILEM Conference on the Use of Recycled Materials in Buildings and Structures, vol. 2. RILEM Publications, p. 731.

Tavares, J.C., Lucena, L.F., Henriques, G.F., Ferreira, R.L., dos Anjos, M.A., 2022. Use of banana leaf ash as partial replacement of Portland cement in eco-friendly concretes. Construct. Build. Mater. 346, 128467.

Thomas, B.S., Yang, J., Bahurudeen, A., Abdalla, J., Hawileh, R.A., Hamada, H.M., Nazar, S., Jittin, V., Ashish, D.K., 2021a. Sugarcane bagasse ash as supplementary cementitious material in concrete–A review. Mater. Today Sustain. 15, 100086.

Thomas, B.S., Yang, J., Mo, K.H., Abdalla, J.A., Hawileh, R.A., Ariyachandra, E., 2021b. Biomass ashes from agricultural wastes as supplementary cementitious materials or aggregate replacement in cement/geopolymer concrete: a comprehensive review. J. Build. Eng. 40, 102332.

Torres, A., Brandt, J., Lear, K., Liu, J., 2017. A looming tragedy of the sand commons. Science 357, 970–971.

Tripathi, A.K., Iyer, P., Kandpal, T.C., Singh, K., 1998. Assessment of availability and costs of some agricultural residues used as feedstocks for biomass gasification and briquetting in India. Energy Convers. Manag. 39, 1611–1618.

Tripathi, M., Sahu, J.N., Ganesan, P., 2016. Effect of process parameters on production of biochar from biomass waste through pyrolysis: a review. Renewable Sustainable Energy Rev. 55, 467–481.

Udoeyo, F.F., Dashibil, P.U., 2002. Sawdust ash as concrete material. J. Mater. Civ. Eng. 14, 173–176.

Udoeyo, F.F., Inyang, H., Young, D.T., Oparadu, E.E., 2006. Potential of wood waste ash as an additive in concrete. J. Mater. Civ. Eng. 18, 605–611.

Uylaşer, V., Yildiz, G., 2014. The historical development and nutritional importance of olive and olive oil constituted an important part of the Mediterranean diet. Crit. Rev. Food Sci. Nutr. 54, 1092–1101.

van Loo, S., Koppejan, J., 2007. The Handbook of Combustion and Co-firing Biomass.

Vieira, D.R., Calmon, J.L., Coelho, F.Z., 2016. Life cycle assessment (LCA) applied to the manufacturing of common and ecological concrete: a review. Construct. Build. Mater. 124, 656–666.

Vinai, R., Ntimugura, F., Cutbill, W., Evans, R., Zhao, Y., 2022. Bio-derived sodium silicate for the manufacture of alkali-activated binders: use of bamboo leaf ash as silicate source. Int. J. Appl. Ceram. Technol. 19, 1235–1248.

Wang, R., Haller, P., 2022. Applications of wood ash as a construction material in civil engineering: a review. Biomass Convers. Bioref. 1–21.

Wang, Y., Shao, Y., Matovic, M.D., Whalen, J.K., 2014. Recycling of switchgrass combustion ash in cement: characteristics and pozzolanic activity with chemical accelerators. Construct. Build. Mater. 73, 472–478.

Wang, L., Yu, K., Li, J.-S., Tsang, D.C., Poon, C.S., Yoo, J.-C., Baek, K., Ding, S., Hou, D., Dai, J.-G., 2018. Low-carbon and low-alkalinity stabilization/solidification of high-Pb contaminated soil. Chem. Eng. J. 351, 418–427.

Wang, L., Chen, L., Tsang, D.C., Guo, B., Yang, J., Shen, Z., Hou, D., Ok, Y.S., Poon, C.S., 2020. Biochar as green additives in cement-based composites with carbon dioxide curing. J. Clean. Prod. 258, 120678.

Wang, F., Harindintwali, J.D., Yuan, Z., Wang, M., Wang, F., Li, S., Yin, Z., Huang, L., Fu, Y., Li, L., 2021a. Technologies and perspectives for achieving carbon neutrality. Innovation 2, 100180.

Wang, H., Qi, T., Feng, G., Wen, X., Wang, Z., Shi, X., Du, X., 2021b. Effect of partial substitution of corn straw fly ash for fly ash as supplementary cementitious material on the mechanical properties of cemented coal gangue backfill. Construct. Build. Mater. 280, 122553.

Wang, J., Fu, J., Zhao, Z., Bing, L., Xi, F., Wang, F., Dong, J., Wang, S., Lin, G., Yin, Y., 2023. Benefit analysis of multi-approach biomass energy utilization toward carbon neutrality. Innovation 4.

WBA, 2019. Global Bioenergy Statistics 2019. World Bioenergy Association, Stockholm, Sweden.

Werkelin, J., Skrifvars, B.-J., Hupa, M., 2005. Ash-forming elements in four Scandinavian wood species. Part 1: summer harvest. Biomass Bioenergy 29, 451–466.

Xiao, H., Xu, Z., Ren, J., Zhou, Y., Lin, R., Bao, S., Zhang, L., Lu, S., Lee, C.K., Liu, J., 2022. Navigating Chinese cities to achieve sustainable development goals by 2030. Innovation 3, 100288.

Xue, Y., Wu, S., Cai, J., Zhou, M., Zha, J., 2014. Effects of two biomass ashes on asphalt binder: dynamic shear rheological characteristic analysis. Construct. Build. Mater. 56, 7–15.

Yeoh, A., Bidin, R., Chong, C., Tay, C., 1979. The relationship between temperature and duration of burning of rice-husk in the development of amorphous rice-husk ash silica. In: Proceedings of UNIDO/ESCAP/RCTT, Follow-Up Meeting on Rice-Husk Ash Cement. Alor Setar, Malaysia.

Yin, C., Rosendahl, L.A., Kær, S.K., 2008. Grate-firing of biomass for heat and power production. Prog. Energy Combust. Sci. 34, 725–754.

You, S., Ok, Y.S., Tsang, D.C., Kwon, E.E., Wang, C.-H., 2018. Towards practical application of gasification: a critical review from syngas and biochar perspectives. Crit. Rev. Environ. Sci. Technol. 48, 1165–1213.

Yuan, T., He, W., Yin, G., Xu, S., 2020. Comparison of bio-chars formation derived from fast and slow pyrolysis of walnut shell. Fuel 261, 116450.

Yurt, Ü., Bekar, F., 2022. Comparative study of hazelnut-shell biomass ash and metakaolin to improve the performance of alkali-activated concrete: a sustainable greener alternative. Construct. Build. Mater. 320, 126230.

Zhang, Q., Li, Y., Xu, L., Lun, P., 2019. Bond strength and corrosion behavior of rebar embedded in straw ash concrete. Construct. Build. Mater. 205, 21–30.

Zhang, Y., He, M., Wang, L., Yan, J., Ma, B., Zhu, X., Ok, Y.S., Mechtcherine, V., Tsang, D.C., 2022. Biochar as construction materials for achieving carbon neutrality. Biochar 4, 59.

Zhang, N., Konyalõoğlu, A.K., Duan, H., Feng, H., Li, H., 2023. The impact of innovative technologies in construction activities on concrete debris recycling in China: a system dynamics-based analysis. Environ. Dev. Sustain. 1–26.

Zhao, S., Huang, B., Shu, X., Ye, P., 2014. Laboratory investigation of biochar-modified asphalt mixture. Transport. Res. Rec. 2445, 56–63.

Zhou, S., Chen, X., 2012. Pozzolanic activity of feedlot biomass (cattle manure) ash. Construct. Build. Mater. 28, 493–498.

Zhou, H., Bhattarai, R., Li, Y., Si, B., Dong, X., Wang, T., Yao, Z., 2022. Towards sustainable coal industry: turning coal bottom ash into wealth. Sci. Total Environ. 804, 149985.

Zhu, C., Pundienė, I., Pranckevičienė, J., Kligys, M., 2022. Effects of Na2CO3/Na2SiO3 ratio and curing temperature on the structure formation of alkali-activated high-carbon biomass fly ash pastes. Materials 15, 8354.

Chapter 17

Recycling of incineration sewage sludge ash as SCM and aggregate

Miao Lu[1], Zhenhao Song[1], Yan Xia[1], Guoqing Geng[2] and Lei Wang[1]
[1]State Key Laboratory of Clean Energy Utilization, Zhejiang University, Hangzhou, China; [2]Department of Civil and Environmental Engineering, National University of Singapore, Singapore, Singapore

1. Introduction

Sewage sludge is an unavoidable outcome of sewage treatment processes. Reports indicate that all the nations in the world witness a total production of roughly 70–75 million tons of dried sludge annually, underscoring the pressing concern of ongoing sewage sludge generation within the wastewater treatment sector (Chakraborty et al., 2017). Sewage sludge is rich in essential elements, such as nitrogen (N), phosphorus (P), and potassium (K), which have found application as organic fertilizers to enhance soil nutrient levels (Benassi et al., 2019). Nevertheless, growing environmental consciousness has led to the gradual imposition of restrictions and eventual bans on the agricultural utilization of sewage sludge due to safety concerns (Mejdi et al., 2020). Similarly, due to limited available land resources, the use of landfill for ultimate disposal has also been diminishing (Teoh and Li, 2020). At present, incineration has emerged as a prevalent method for solid waste disposal (Chang et al., 2022a). The thorough breakdown of organic substances during thermal treatment and the potential for energy recovery have propelled the rapid advancement of sewage sludge incineration technology, constituting more than 50% of total sewage sludge management (Chang et al., 2022c). However, sewage sludge disposal does not culminate with incineration. There is still approximately 30% by mass of residues that require further treatment after incineration, and are commonly transferred to landfill sites (Benassi et al., 2019). Recently, the construction field has presented a promising pathway for the recycling of incineration sewage sludge ash (ISSA) (Mejdi et al., 2020). Particularly, the cement and concrete industry holds significant potential for accommodating high volumes of solid waste (Gu et al., 2021). Whether integrated as supplementary cementitious materials (SCMs) or aggregates, these approaches hold the potential to serve as a waste management strategy that effectively harmonizes environmental sustainability with economic viability (Tang et al., 2020; Gu et al., 2021; Wang et al., 2023).

SCMs have achieved significant application maturity in the field of construction. Partially replacing cement with a small amount of SCMs can effectively reduce the carbon footprint of cement-based materials (Yang et al., 2015). The production of blended cements or eco-cements by mixing industrial by-products such as pulverized fly ash, ground granulated blast furnace slag (GGBS), and silica fume with cement has gained widespread industrial acceptance (Skibsted and Snellings, 2019). However, this popularity has raised concerns about potential future shortages of these pozzolanic materials (Juenger et al., 2019). Therefore, there is an urgent need to explore new types of SCM for enriching the diversity of the SCM portfolio. Recent research indicated that ISSA contains abundant aluminosilicate minerals, with a chemical composition similar to that of pulverized fly ash, suggesting the potential for ISSA to be recycled as an SCM (Chakraborty et al., 2017; Chang et al., 2022c). However, it's essential to acknowledge that the inclusion of SCMs can introduce heightened intricacy into the cementitious system. This intricacy is influenced by the inherent attributes of the SCM as well as the interaction between the SCM and the cement (Lothenbach et al., 2011). In addition to the general mechanisms of SCMs that enhance strength by generating extra hydration products and providing nucleation sites for hydration reactions, ISSA was demonstrated to further facilitate cement hydration through reducing the effective water-to-binder ratio and enabling the gradual release of adsorbed water (Chen and Poon, 2017a). Furthermore, the abundant presence of reactive Al_2O_3 in ISSA promotes the formation of ettringite (AFt), monosulfoaluminate (AFm), and calcium aluminate hydrates (C-

Treatment and Utilization of Combustion and Incineration Residues. https://doi.org/10.1016/B978-0-443-21536-0.00003-4

A-H), contributing to the early strength development of cementitious materials (Mejdi et al., 2020). Apart from these hydration products, the ISSA-OPC system also demonstrated the formation of brushite, which is believed to be beneficial for the development of matrix strength (Chang et al., 2022b). Therefore, the distinctive chemical composition of ISSA introduces complex effects on cementitious systems when used as SCMs. A comprehensive summary and investigation are required to fully understand the influence of ISSA-based SCM on cement-based materials.

On the other hand, aggregates constitute a substantial 60%−70% of concrete volume and have conventionally stemmed from the extraction of natural minerals, leading to considerable natural resource overuse in the concrete industry (Qian et al., 2022). Utilizing recycled aggregates or construction demolition slag for secondary reuse is a straightforward and feasible measure (Ren et al., 2021). However, a significant portion of the generated solid waste exists in a powdered form, which poses challenges for widespread integration into concrete. While these powders can be introduced in minor quantities as SCMs, their extensive incorporation may adversely affect the mechanical properties of cement-based materials (Gu et al., 2021; Zhou et al., 2022a). As a result, from a dual perspective of environment and demand, recycling powdered solid wastes to produce artificial aggregates emerges as another effective option for achieving sustainable construction. The common manufacturing methods for artificial aggregates primarily include sintering and cold-bonding technologies (Bekkeri et al., 2023). The appropriate content of SiO_2, Al_2O_3, and Fe_2O_3 within ISSA imparts it with promising pozzolanic potential and sintering capabilities, prompting researchers to explore its potential application in the production of artificial aggregates (Donatello and Cheeseman, 2013; Tang et al., 2020).

This chapter presents a comprehensive overview of the recent advancements in recycling ISSA as both SCMs and artificial lightweight aggregates. It critically examines the recent progress in ISSA recycling, emphasizing the inherent properties and the influence on the reactivity of ISSA. This chapter also highlights the physiochemical properties of different construction materials derived from ISSA and the factors influencing these properties.

2. Characterization of ISSA

ISSA is a residue derived from the incineration of sewage sludge, a by-product of wastewater treatment. ISSA exhibits a diverse range of physical and microstructural properties that significantly influence its potential applications (Danish and Ozbakkaloglu, 2022). In this section, the main focus is on the physical and chemical properties of ISSA, including particle size, chemical composition, and microstructure, among other significant features (Godoy et al., 2019; Gu et al., 2022; Chang et al., 2022b). It also discusses various activation methods for utilizing it as SCMs and aggregates.

2.1 Physical and microstructural characteristics

ISSA particles typically exhibit a fine particulate nature, with particle sizes spanning from nanometers to micrometers. The specific particle size distribution depends on factors such as the original composition of sewage sludge, combustion conditions, and the efficiency of the incineration process. The particle size distribution of ISSA ranges from 1 to 100 μm, and some ISSA particle sizes can even reach 194.2 μm or higher (Chen et al., 2018a; Gu et al., 2022). According to the literature, the average particle size of ISSA is around 47 μm (Danish and Ozbakkaloglu, 2022). Particle size significantly affects properties such as reactivity, porosity, and surface area. A larger surface area allows for greater contact between ISSA particles and other materials, influencing chemical reactions and adsorption capacity. The bulk density of ISSA ranges from 2.58 to 2.63 g/cm^3, and the density of these materials is comparable with natural sand and high-grade kaolin clay (de Azevedo Basto et al., 2019). The specific gravity of ISSA ranges from 1.8 to 2.9, with an average value of 2.5 (Lynn et al., 2015). The density of ISSA is influenced by both its chemical composition and physical structure. The removal of organic materials during incineration generally results in a reduction in density compared with the original sewage sludge. Bulk density, which considers the overall volume occupied by ISSA, is crucial for estimating the material's packing behavior and transportation considerations. Both density and bulk density can impact the suitability of ISSA for specific applications, such as lightweight aggregate production. The microstructure and morphology of ISSA particles reveal valuable insights into their reactivity and potential uses. As shown in Fig. 17.1, the sludge ash particles present irregular morphology with a rough surface. The loose and porous layered structures result in their significantly larger BET specific surface area than cement.

2.2 Chemical compositions

The chemical composition of ISSA is a critical factor that influences its potential applications and behavior. This section delves into the essential chemical constituents of ISSA and their significance in determining its suitability for various utilization avenues.

FIGURE 17.1 SEM image of ISSA es calcined at 600 and 800°C (Chang et al., 2022b).

ISSA comprises a variety of inorganic elements that are present in the raw sewage sludge and are retained in the ash after incineration. Common oxides found in ISSA include silica (SiO_2), alumina (Al_2O_3), iron oxide (Fe_2O_3), calcium oxide (CaO), and traces of other minerals (Chang et al., 2022b). The presence and concentrations of these oxides vary depending on the composition of the sewage sludge, the incineration process, and the incinerator's operating conditions. Fig. 17.2 displays a ternary diagram of the main oxide contents (CaO, Al_2O_3, and SiO_2) in ISSA from previous studies, compared with typical contents of other cementitious materials (Chang et al., 2020). Compared with the raw sewage sludge, the incineration process enhances the content of SiO_2, CaO, and Al_2O_3 in the sludge ash. Additionally, ISSA also contains oxides such as phosphorus oxide (P_2O_5), sulfur trioxide (SO_3), titanium dioxide (TiO_2), manganese oxide (MnO), and chloride ions (Cl^-) (Danish and Ozbakkaloglu, 2022).

Another key concern regarding the chemical composition of ISSA is the presence of heavy metals. These toxic elements, such as lead (Pb), cadmium (Cd), chromium (Cr), and others, are often found in sewage sludge due to industrial and domestic emissions (Chen et al., 2021). While incineration reduces the volume of sludge, heavy metals tend to concentrate in the resulting ash. Certain elements such as lead (Pb) and cadmium (Cd) exhibit high volatility, leading to a reduction in their concentrations in ISSA during the incineration process (Naamane et al., 2016). Moreover, chlorine (Cl) in sewage sludge readily combines with heavy metals to form chlorides, which volatilize as gases at high temperatures, the presence

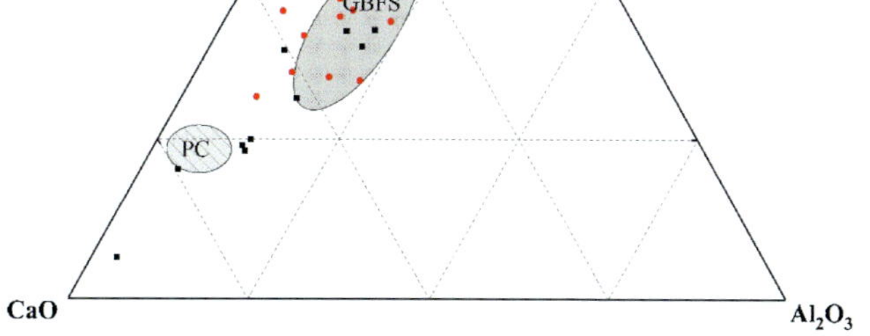

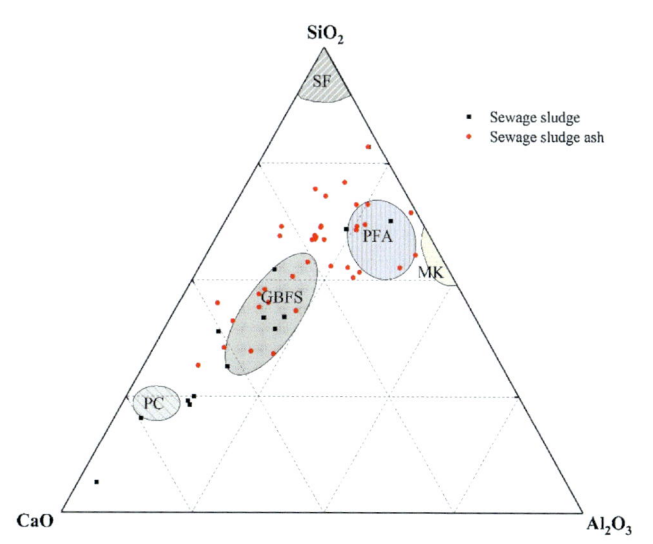

FIGURE 17.2 SiO_2, Al_2O_3, and CaO contents in sewage sludge and ISSA (Chang et al., 2020). *GBFS*, granulated blast furnace slag; *MK*, metakaolin; *PC*, Portland cement; *PFA*, pulverized fuel ash; *SF*, silica fume.

of phosphorus (P) in sludge inhibits their volatilization (Yu et al., 2017; Liu et al., 2018). It is necessary to closely monitor the levels of heavy metals in ISSA to assess the potential environmental and health risks associated with its use (Danish and Ozbakkaloglu, 2022).

The chemical composition of ISSA also dictates its mineralogical phases, which influence its reactivity and behavior. ISSA can contain crystalline phases such as quartz, mullite, hematite, and various calcium silicates. These phases contribute to the potential of ISSA to react with other materials, such as in cementitious systems or geopolymer synthesis (Danish and Ozbakkaloglu, 2022). The incineration process leads to the combustion of organic matter in sewage sludge, resulting in a reduction in carbon content in ISSA. The remaining carbon content in ISSA is typically in the form of carbonates, which can influence its alkalinity and buffering capacity. This aspect is particularly relevant when considering the potential applications of ISSA in stabilizing acidic or contaminated soils. The chemical composition of ISSA significantly affects its reactivity when combined with other materials. For instance, its content of reactive silica and alumina contributes to its pozzolanic reactivity, making it suitable for enhancing the properties of cementitious materials. However, the overall chemical composition must be considered alongside other factors, such as activation processes, to harness its full potential.

2.3 Activation

Activation processes play a pivotal role in modifying the properties and enhancing the reactivity of ISSA. These processes involve various physical and chemical treatments that transform ISSA into a material with improved characteristics for specific applications. This section explores the different activation methods and their effects on ISSA.

Thermal activation involves subjecting sewage sludge to elevated temperatures in controlled conditions (Chang et al., 2022c). This process induces phase changes and influences the microstructure of ISSA. Thermal activation can enhance the crystallinity of ISSA and promote the formation of specific mineral phases, thus altering its reactivity (Godoy et al., 2019). The pozzolanic activity of ISSA significantly depends on the sewage sludge incineration temperature. Increasing the incineration temperature alters the chemical composition of ISSA, such as increasing amorphousness, resulting in higher pozzolanic activity of ISSA. Researchers have indicated that ISSA produced at temperatures between 600 and 900°C exhibits varying levels of pozzolanic activity, with 800°C being considered the optimal incineration temperature (Tantawy et al., 2012; Chin et al., 2016). Mineralogical analysis of ISSA has also shown that an increase in the calcination temperature may lead to a decrease in the ash's reactivity (de Azevedo Basto et al., 2019). This is primarily attributed to the crystallization of amorphous silica. Researchers regulated the sewage sludge incineration process to stimulate the reactivity of ISSA by introducing cocombustion agents such as rice husks (Wang et al., 2017; Xia et al., 2023a). Cocombustion of sewage sludge and rice husks resulted in ash with higher amorphous silica content compared with sludge ash, leading to increased reactivity in the volcanic ash reactions. However, careful control of temperature and duration is essential to prevent excessive sintering or phase transformation. In addition to heat treatment, grinding can also enhance the reactivity of ISSA.

Alkaline hydroxides, sulfates, and chlorides are commonly employed as activators to enhance the reactivity of ISSA in pozzolanic reactivity (Zhou et al., 2021a,b; Xia et al., 2022, 2023b). When using CaO as an activator, silica and alumina present in ISSA react with calcium hydroxide to form calcium silicate hydrate (C−S−H) or calcium aluminosilicate hydrates (C−A−S−H) gel (Zhou et al., 2020). This gel contributes to increased mechanical strength and durability in cementitious materials, making ISSA a potential SCM. The extent of pozzolanic reactivity depends on factors such as the composition of ISSA, the type of activator, and curing conditions. Some researchers have also utilized oxalic acid (OA) to activate ISSA and applied it for the remediation of highly lead-contaminated soil. Studies indicate that OA facilitates the release of phosphate from ISSA through acid leaching and also releases lead from the soil. These two components combine and precipitate into stable lead phosphate minerals, thereby reducing the leaching of Pb (Zeng et al., 2017; Li et al., 2021a).

In conclusion, activation processes are essential tools for optimizing ISSA for various applications. Pozzolanic, alkaline, thermal, and combined activation methods offer means to tailor the properties of ISSA to meet specific requirements. Careful selection and optimization of activation methods are critical for fully utilizing the potential of ISSA, while considering economic and environmental factors. Further research can delve into fine-tuning activation protocols to maximize the utility of ISSA in sustainable materials and waste management.

3. Utilization of ISSA as SCMs

SCMs play a crucial role in enhancing the properties of hardened concrete by exhibiting hydraulic or pozzolanic activity. Conventional SCMs, such as fly ash, silica fume, and ground granulated blast furnace slag, are derived from industrial by-

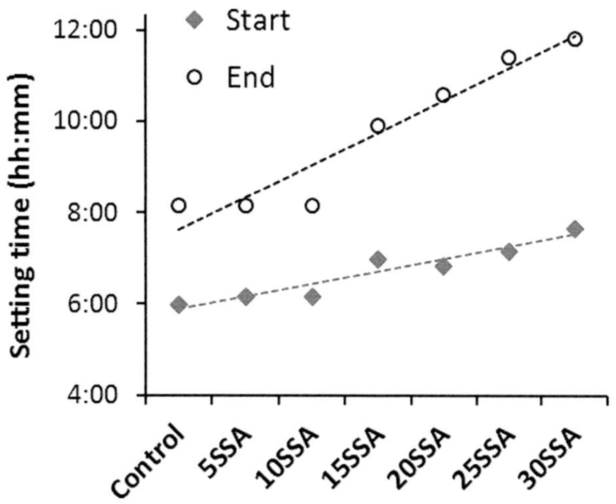

FIGURE 17.3 Influence of ISSA substitution rate on the setting time (Mejdi et al., 2020).

products (Danish and Ozbakkaloglu, 2022; Song et al., 2022). Utilizing sludge or sludge ash for producing low-carbon cementitious materials holds immense potential due to their relatively high aluminosilicate content (Zhou et al., 2022b). Aluminosilicates present in sludge can react with portlandite, a by-product of cement hydration, resulting in the formation of supplementary hydration products. However, the original sludge, characterized by high moisture and organic matter levels, adversely affects the mechanical properties, workability, and durability of cementitious systems (Wu et al., 2021). These negative impacts can be mitigated through the incineration process, which enriches the ash with aluminosilicates and activates them at high temperatures. Consequently, in this section, we will explore the utilization of ISSA in the production of SCMs, thereby offering insights into its application.

3.1 Effect of ISSA on fluidity and setting time

The incorporation of ISSA leads to an increase in the setting time of blended cements, as observed by Lin et al. (2008a). On average, there is a 35% lengthening of setting times for every 10% replacement, both for initial and final setting times. The presence of P_2O_5, SO_3, and C_2S in cement derived from sewage sludge contributes to an elevated water demand and extended setting time upon incorporating ISSA (Naamane et al., 2016). Additionally, the delays in setting times may also be attributed to the presence of trace elements (e.g., Zn, Cr) in ISSA, which can dissolve in the pore solution and subsequently impact cement hydration (Lin et al., 2012).

According to the results presented in Fig. 17.3, an increase in the content of ISSA caused a significant prolongation of both the initial and final setting times compared with a control mortar (Mejdi et al., 2020). The delay in the cement hydration and setting time can be attributed to the presence of orthophosphate ions (PO_4^{3-}), which are known to impede crystals growth (Scrivener and Nonat, 2011). Several studies have documented the retarding effect of orthophosphate on cement hydration (Cau Dit Coumes and Courtois, 2003; Bénard et al., 2005). Similar to gypsum's role in slowing down the reaction of C_3A, the adsorption of phosphate ions on reactive sites of C_3S impedes its dissolution rate (Scrivener and Nonat, 2011).

Moreover, Bénard et al. (2008) have observed a discontinuity in the retarding effect beyond a critical phosphate concentration of 0.2 mol/L. This change in behavior has been attributed to the precipitation of hydroxyapatite due to the saturation of silicate dissolution sites. However, it is worth noting that the solubility of P_2O_5 in ISSA is relatively low. Therefore, it is more likely that phosphate ions are adsorbed onto the different phases comprising the cementitious matrix rather than forming calcium phosphate compounds.

Zdeb et al. (2022) conducted a study on the utilization of ISSA in the field of cement production. They found that adding up to 30% of ISSA by weight of binder led to a consistent prolongation of both initial and final setting times. The initial setting time approximately doubled, while the final setting time increased by around 60%. These findings provide further evidence of the retarding effect of phosphate ions, which impede the hydration reaction of cement particles. When the ISSA content exceeded 11%, there was a significant rise in the concentrations of heavy metals, sulfates, and especially phosphorus. This excessive accumulation resulted in a further increase in setting times and a suppression of strength development in the cement. To mitigate this issue, it is advisable to treat the ISSA prior to usage and extract phosphorus.

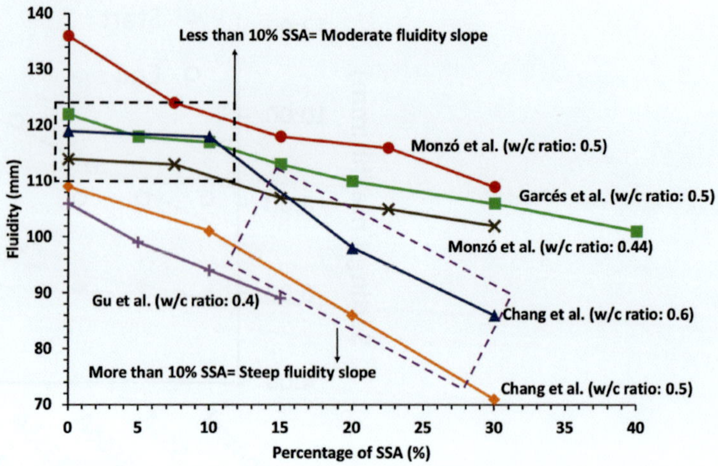

FIGURE 17.4 Fluidity of ISSA-modified cementitious composites (Danish and Ozbakkaloglu, 2022).

Such treatment can be considered a reasonable approach, as it not only improves cement performance by reducing the negative impact of phosphorus but also provides a valuable resource for agricultural applications (Lynn et al., 2015).

The incorporation of ISSA resulted in a decrease in the workability of mortars. For instance, when the water-to-binder ratio was 0.6, the inclusion of 5%, 10%, and 15% of ISSA led to a reduction in the workability of mortar from 293 to 287 mm, 266 mm, and 247 mm, respectively (Gu et al., 2023). Compared with cement particles, ISSA particles exhibit higher porosity. Consequently, during the mixing process, ISSA absorbed a greater amount of water, leading to a decrease in the workability of mortars as the ISSA content increased.

Danish and Ozbakkaloglu (2022) summarized the comparison of the effect of ISSA on the fluidity of different cementitious composites (such as cement paste and mortar) and water–cement ratio (Monzó et al., 2003; Garcés et al., 2008; Chang et al., 2010; Gu et al., 2021), as shown in Fig. 17.4. The results indicate that incorporating 5%–10% SCMs, such as ISSA, has a negligible impact on the fluidity of different cementitious composites. However, a significant reduction in fluidity is observed when higher dosages of ISSA are used. Furthermore, similar to conventional and modified cementitious composites, the fluidity of ISSA is also influenced by the water-to-binder ratio. Increasing the water content enhances the flowability of ISSA due to improved saturation of the surface. This consistency in the behavior of ISSA as a binder material highlights its potential in enhancing workability.

3.2 Effect of ISSA on hydration kinetics and phase assemblage

The addition of ISSA as a partial replacement for cement causes a delay in the setting time of the cement paste. According to Mejdi et al. (2020) in Section 3.1, this delay is approximately 2 h when using 20 wt% ISSA. The reason behind this delay is the decreased rate of hydration of C_3S, which can be attributed to the adsorption of orthophosphate ions on the dissolution sites of this phase. This adsorption retards the reaction of Alite, particularly during the initial days of hydration. Fig. 17.5 illustrates that the overall hydration of cement appears unaffected in later stages. Additionally, the use of ISSA leads to the appearance of a second peak on the heat flow curve. This heat liberation is caused by the formation of calcium aluminate hydrates, which occurs due to the continuous supply of aluminum to the system as the amorphous phase of ISSA dissolves. However, most of the phosphates present in ISSA exist in the form of whitlockite, a calcium phosphate mineral with low solubility (Jang et al., 2015). Only a small portion (approximately 6.84 wt%) is included in the amorphous phase. It is worth noting that the dissolution of amorphous P_2O_5 is very limited, thus making the formation of hydroxyapatite or other calcium phosphate compounds unlikely.

Aluminum-based sludge ash contains compounds known as aluminates, which play a crucial role in the creation of aluminate products (Chang et al., 2022d; Xia et al., 2023c). Similarly, when iron-rich sludge ash dissolves, the iron content can replace aluminum in the aluminate products, promoting the formation of AFt and AFm phases (Xia et al., 2023b). The classification of aluminate products depends on the relative amounts of aluminates and sulfates in the sludge ash. When sulfates are predominant, the combined action of aluminates and sulfates encourages the significant development of AFt.

On the other hand, when blended cements lack sulfates, the dissolution of aluminates from sludge ash triggers the conversion of AFt to AFm. The phosphates found in sludge ash contribute to the formation of whitlockite ($Ca_3(PO_4)_2$) (Chen and Poon, 2017a). These additional hydration products improve the refinement of the pore structure and promote the

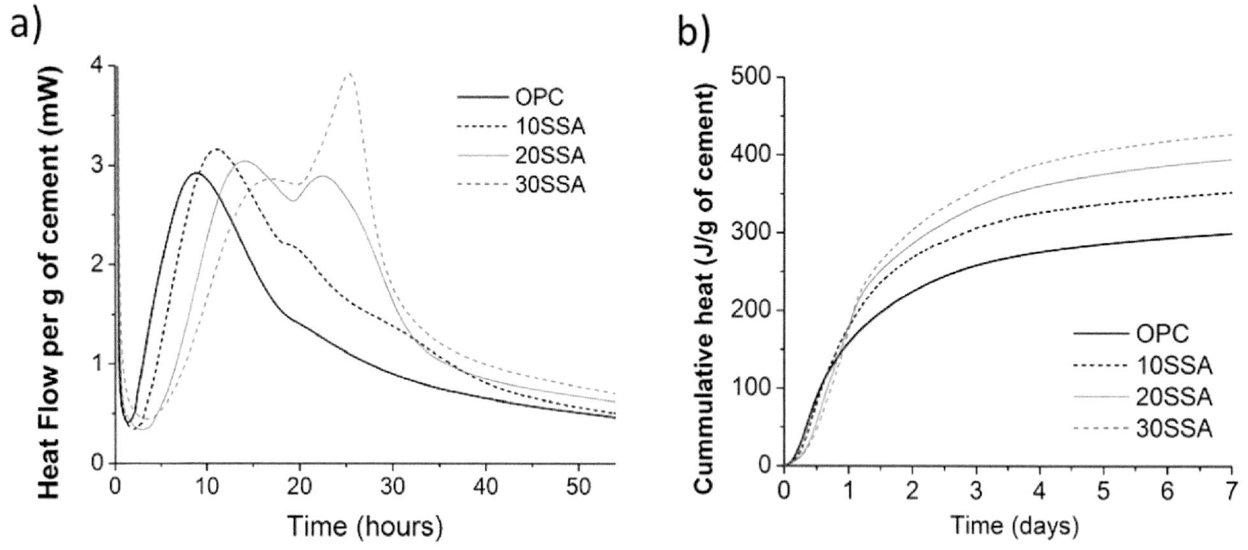

FIGURE 17.5 Influence of the ISSA substitution rate on the heat liberation: (A) heat flow and (B) cumulative heat per g of cement (Mejdi et al., 2020).

strength development of cementitious materials containing sludge ash. The porous nature of sludge ash particles increases the water requirement in blended cements, while the presence of phosphate hinders the dissolution and hydration of cement clinker, thereby prolonging the setting time of blended cements.

3.3 Effect of ISSA on mechanical properties

Numerous researchers have extensively studied the performance of cement-based composites incorporating ISSA, and it is widely acknowledged that the addition of properly calcined ISSA has minimal impact on the mechanical properties of these composites (Chen and Poon, 2017a). Oliva et al. (2019) conducted a study on blended cement with ISSA, examining the influence of ISSA treated at various temperatures (550−790°C). They discovered that the series comprising 80% ordinary Portland cement (OPC) and 20% ISSA exhibited compressive strengths comparable with or higher than those of reference pastes made entirely of OPC, while maintaining the same water-to-cement ratio. Consequently, the cementing capacity of all the ISSAs produced was found to surpass that of OPC, irrespective of the incineration treatment employed. This performance improvement could be attributed to the enhanced nucleation facilitated by the particle size and specific surface area of the ash (Oey et al., 2013). Similar outcomes were reported by Gastaldini et al. (2015), who observed that even at the lowest temperature and shortest residence time (400°C for 1 h) with only 75% cement, the 28-day compressive strength results (44 MPa) were comparable with those of the reference mix, consisting of 100% Portland cement (45 MPa). Moreover, compressive strength results for higher treatment temperatures were significantly higher than those of the control group. Naamane et al. (2016) investigated the effects of ISSA obtained under different incineration conditions on cement properties and found that cement mortars containing 15% ISSA calcined at 700 and 800°C exhibited superior compressive strength compared with the control mortar at 90 days. Notably, the strength did not diminish with the increasing curing period, regardless of the amount of cement replaced by ISSA.

Furthermore, a study conducted by Zdeb et al. (2022) revealed that the addition of 10% SCMs known as ISSA to the cement binder yields positive effects on the mechanical properties. When 10% and 20% ISSA were incorporated, the porosity of the resulting mortars decreased significantly to 16.9% and 18.2%, respectively, compared with the reference mortar's porosity of 21.1%. This indicates a more compact material structure. In another investigation by Chen et al. (2013), it was observed that substituting 10% of cement with sludge ash resulted in no more than a 25% reduction in flexural and compressive strengths when compared with samples without ash. The particle fineness of the ISSA played a significant role in the workability and mechanical properties of the cement mortars (Pan et al., 2003). The strength activity index of cement mortars blended with ISSA increased by approximately 5% for every 100 m^2/kg increase in ISSA fineness. Enhancing the fineness of ISSA particles contributed to improved compressive strength in samples with the same ISSA content, owing to the higher reactivity of smaller ISSA particles (Lin et al., 2008b; Vouk et al., 2017). It is important to note that different researchers have obtained varying results regarding strength changes. However, Chang et al. (2010) discovered that the water adsorption capability of ISSA resulted in reduced workability and compressive strength of mortars and concretes. The increased presence of mesopores, caused by the inclusion of ISSA, led to higher drying shrinkage in mortars. Consequently, the recommended dosage of ISSA in mortars or concretes is a substitution rate of 10% of the cement content.

Additionally, various research studies have explored the impact of ISSA on the fire resistance characteristics of hardened cementitious pastes. These investigations have revealed that the introduction of ISSA, up to a concentration of 20 wt%, reduces the thermal damage experienced by cement pastes when exposed to high treatment temperatures. This reduction can be attributed to the presence of cross-linked fibrous calcium silicate, which enhances the strength of the cement (Tantawy et al., 2013). Liu et al. (2019) conducted a comprehensive evaluation of cement pastes incorporating ISSA under elevated temperatures. The researchers prepared cement pastes mixed with ISSA and subjected them to thermal treatment ranging from 600 to 1000°C. The findings demonstrated that compared with plain cement paste, the ISSA-blended cement pastes exhibited enhanced resistance to high temperatures, as evidenced by their improved relative residual compressive strength. These promising results suggest a novel approach for the development of fire-resistant cementitious pastes.

The research findings indicated that the reactive abilities of ISSA were limited. This was primarily due to the porous and friable nature of ISSA, which hindered any significant strength development in the accompanying concrete. Therefore, additional processing might be necessary for effective utilization of ISSA (Halliday et al., 2012). A comprehensive summary of the impact of using incinerated sludge ash as an SCM on the compressive strength of concrete was provided by Chang et al. (2020) who summarized the effect of incinerated sludge ash used as SCM on compressive strength of concrete. Although different reports present varying relationships between compressive strength and sludge substitution, the general trend suggests a decline in compressive strength with an increasing amount of sludge added.

Furthermore, when compared with other SCMs, the pozzolanic reactivity of sludge ash is found to be inferior, imposing limitations on its potential as a cement replacement. Once the dosage of sludge ash reaches 30 wt%, there is a significant decline in strength, surpassing 35% (Donatello and Cheeseman, 2013). Consequently, to maintain satisfactory mechanical properties, the substitution rate of sludge ash is restricted to 15 wt% in accordance with relevant regulations. To enhance the substitution percentage of sludge ash as SCMs, researchers have extensively explored various modification techniques. Alkaline hydroxides, sulfates, and chloride salts are commonly utilized as early strength agents and activators in blended cements that incorporate a substantial amount of SCMs (Zhou et al., 2021a; Xia et al., 2022, 2023b).

Alkaline hydroxides play a crucial role in increasing the alkalinity of the pore solution, which facilitates the hydration of cement clinker and the dissolution of sludge ash. Sulfates, such as Na_2SO_4, react with aluminates and ferrites present in sludge ash, leading to the formation of AFt phases. This phenomenon contributes to the improvement of the pore structure and promotes the strength development of blended cements. Chloride salts, especially NaCl, are essential for promoting the formation of Friedel's salt and accelerating the early stages of hydration (Zhou et al., 2021b). To enhance the pozzolanic reactivity of sludge ash, researchers have focused on improving the sludge incineration process in conjunction with chemical activators (Chen et al., 2018b; Wang et al., 2020). Coincineration of sludge and rice husk has shown promise as a viable approach (Wang et al., 2017; Xia et al., 2023a). The high calorific value of rice husk allows for sludge incineration without additional energy consumption. Furthermore, the ash resulting from the cocombustion of sewage sludge and rice husk contains a higher content of amorphous silicon compared with sludge ash alone, thereby enhancing its pozzolanic reactivity. The maximum allowable dosage of cocombustion ash for strength requirements can reach up to 30 wt%. Incorporating multiple SCMs further enhances the utilization of sludge ash in blended cements. The synergistic effect between sludge ash and other SCMs optimizes their reactivity and promotes strength development (Duan et al., 2022). For example, the reaction between aluminates and ferrites present in sludge ash with limestone results in the formation of carboaluminate phases, namely hemicarboaluminate and monocarboaluminate (Chang et al., 2022d; Liu et al., 2022). These carboaluminate phases become integrated within the adjacent C—S—H matrix, significantly contributing to the refinement of pore structure and strength development in blended cements.

3.4 Effect of ISSA on durability

In general, the inclusion of sludge ash in cement systems tends to negatively affect their durability, leading to issues such as drying shrinkage, carbonation, and chloride resistance (Krejcirikova et al., 2019; Gu et al., 2023). Moreover, the addition of ISSA also impacts the dimensional changes in mortars and concretes (Gu et al., 2023). Tay (1987) observed a slight reduction in drying shrinkage when the content of ISSA increased in concrete. For instance, replacing 20% of cement with ISSA resulted in a decrease from 855 to 765 $\mu\varepsilon$. However, Sasaoka et al. (2006) found that concrete with 10% ISSA had approximately 1.5 times the drying shrinkage of the control concrete. Chen and Poon (2017a,b) also discovered that replacing 20% of cement with ISSA increased the drying shrinkage of mortars by about 25%, with smaller-sized ISSA particles leading to higher shrinkage. Li et al. (2017) observed an increase in drying shrinkage of mortars due to the presence of ISSA. In addition, Gu et al. (2023) found that compared with control mortars with water-to-binder ratios of 0.4, 0.5, and 0.6, mortars containing 15% ISSA exhibited an increase in drying shrinkage of 162, 261, and 313 $\mu\varepsilon$, respectively. It is worth noting that mortars with higher water-to-binder ratios displayed greater drying shrinkage, and the influence of water-to-binder ratio on drying shrinkage became more significant with higher ISSA contents.

Sulfates present in cement-based systems can lead to volume instability issues in cement-based materials, increasing their susceptibility to environmental conditions (Gu et al., 2023). The resistance to sulfate corrosion primarily relies on the cement type, specifically its calcium aluminate content, particularly the C_3A phase (Zdeb et al., 2022). Irrespective of the quantity of sludge ash added, it will adversely affect the mortar's ability to withstand sulfate ions, causing its resistance to deteriorate.

Moreover, the substitution of ordinary Portland Cement (OPC) with silica slag aggregate leads to a significant decrease in permeability. A recent study by Motisariya et al. (2023) demonstrated that after 28 days of curing, the permeability is reduced by approximately 26%−28%, and after 90 days of curing, it decreases by up to 25%. This reduction can be attributed to the presence of pores in the mortar mixture. Furthermore, the inclusion of silica fume, with its fine particles, has been shown to decrease chloride permeability. The addition of alkali in silica fume further contributes to a decrease in alkali reaction and carbonation, resulting in improved durability of cement and reduced levels of rapid chloride permeability test (RCPT). These enhanced properties enhance structural durability, as reported by Dave et al. (2016). Additionally, when the cement composite contains 10% ash content, a significant decrease in the diffusion coefficient of chloride ions is observed, as demonstrated by Zdeb et al. (2022). This reduction, amounting to approximately 62%, can be attributed to the reduced material porosity observed in the aforementioned tests. These findings establish a strong correlation between total porosity and the diffusion coefficient of Cl^- ions. However, it is worth noting that the sample size of the results obtained is limited, making it challenging to formulate a credible model that describes the correlation between these characteristics.

4. Utilization of ISSA into aggregates

Artificial lightweight aggregates are engineered construction materials produced by industrial by-products or natural materials. As a substitute for natural aggregates, they provide concrete with unique properties such as lightweightness, thermal inertness, and sound insulation (Zhou et al., 2022a; Bekkeri et al., 2023). These aggregates are primarily manufactured through sintering, vitrification, or cold-bonding techniques and have great potential for incorporating high volumes of solid wastes. It was reported that the performance of artificial aggregates is significantly influenced by the manufacturing process and source of the raw materials utilized (Ren et al., 2021). The adequate contents of SiO_2, Al_2O_3, CaO, Fe_2O_3 in ISSA imparted its potential pozzolanic or sintering properties, and the highly porous microstructure enables it to exhibit high adsorption capabilities toward moisture and some heavy metals (Chiou et al., 2006; Nie et al., 2021). These physicochemical properties of ISSA promoted researchers to investigate its potential applications in the production of artificial aggregates (Donatello and Cheeseman, 2013; Tang et al., 2020).

4.1 The production process of artificial aggregates

To address the challenges of using fine-powdered ISSA as aggregate, the granulation process should be employed first to transform such fine particles into aggregates with particle sizes ranging from 2 to 20 mm (Ren et al., 2021). A detailed review of the granulation techniques for artificial aggregates can be found in the publication by Qian et al. (2022). Fresh pellets generally exhibit low compressive strength and require additional hardening processes such as sintering, vitrification, or cold bonding to improve their mechanical performance (Ren et al., 2021). Both sintering and vitrification are thermal treatment methods, which are considered as high-energy-consuming techniques. Sintering involves heating the fresh pellets to a specific temperature below the melting point (generally 1000−1200°C) (Chiou et al., 2006; Takeuchi and Yoshimura, 2022). During the sintering process, fine particles in the matrix fuse together, forming a dense structure with improved mechanical properties (Polettini et al., 2004). Vitrification, on the other hand, requires higher heating temperatures (typically 1250−1500°C) to obtain a uniform amorphous glass phase (Zhang et al., 2013, 2015; Borowski, 2015). Cold-bonding technology achieves matrix solidification via cement hydration or geopolymerization reactions without the need for high-temperature processes (Tajra et al., 2019). As a result, it is considered a low-carbon and eco-friendly approach. Artificial aggregates produced through various hardening techniques demonstrate different properties, which will be discussed in Sections 4.3 and 4.4.

4.2 Preprocessing technology

4.2.1 Mechanical grinding pretreatment

The properties of artificial aggregates can be influenced by the particle sizes of the raw materials (Ren et al., 2021). On the one hand, the high fineness of raw materials is preferred for aggregate production process due to the higher granulation efficiency and strength of fresh pellets (Peys et al., 2022). On the other hand, several studies have demonstrated that grinding treatment improved the pozzolanic reactivity of ISSA based on the strength activity index or Frattini test results (Donatello et al., 2010; Kappel et al., 2018). Pan et al. (2003) investigated the influence of grinding treatment on the mechanical properties of ISSA mortar. The results showed that the compressive strength of ISSA mortar increased with the fineness of ISSA particles. Finer particles with large specific surface area can provide additional nucleation sites for the hydration reaction of cementitious materials, thereby promoting the formation and development of hydration products (Lothenbach et al., 2011). For the sintering process, the higher specific surface area of finer particles provides a greater driving force for achieving sintering at the point contact sites during the process (Chen et al., 2010).

4.2.2 Phosphorus recovery

Phosphates with high thermal stability tend to accumulate in the incineration residues during sewage sludge incineration, resulting in a relatively high phosphorus content in ISSA (Husek et al., 2022). Previous studies indicated that the content of P_2O_5 in ISSA is generally 10%–20%, close to the content in virgin phosphorite (25%–37%) (Ottosen et al., 2016). Given the limited availability of phosphorus resources, it is advisable to consider prerecovering phosphorus from ISSA before recycling into artificial aggregates (Ottosen et al., 2020; Wang et al., 2021). Different technologies for phosphorus recovery from ISSA are under development, including wet extraction process (e.g., acid extraction), electrodialytic separation, etc. (Husek et al., 2022)

Electrodialysis technology involves the electrolysis of water to acidify the ISSA suspension, thereby achieving the dissolution of phosphate and some heavy metals (Ottosen et al., 2016). Additionally, the application of two-compartment electrodialytic cell with a cation exchange membrane enables the simultaneous extraction and separation of phosphate and heavy metals (as shown in Fig. 17.6). After electrolysis, the separation of the phosphate-containing solution from the remaining solid can be achieved through filtration. Previous studies demonstrated that the phosphorus recovery rate of 80%–90% was obtained at a laboratory scale (Ottosen et al., 2020). The residues after electrodialysis treatment constitute approximately 90 wt% of the original ISSA (Wang et al., 2021). Moreover, the treated ISSA exhibits a substantial increase in SiO_2 and Fe_2O_3 content, while the content of Al_2O_3 experiences a slight decrease (Ottosen et al., 2020). In comparison with the original ISSA, electrolysis-treated ISSA exhibited some distinct characteristics that can be advantageous for sintered materials. When the firing temperature exceeds 1000°C, the porosity of the matrix can be effectively increased through the decomposition of Fe_2O_3. Additionally, the high content of Fe_2O_3 (hematite) imparts a unique reddish color to the product (Ottosen et al., 2020). In another study, when 20 wt% cement was substituted by electrolysis-treated ISSA, the compressive strength of the mortar is only reduced by 8% compared with the reference, but still give a strength over 50 MPa (Kappel et al., 2018).

The acid extraction process is another widely used phosphorus recovery technique, which involves using strong acids (such as H_2SO_4, HCl, and HNO_3) as extracting agents (Donatello et al., 2010; Lee and Kim, 2017). For example, H_2SO_4 is the most commonly adopted acid in industrial-scale applications due to its affordability and widespread availability

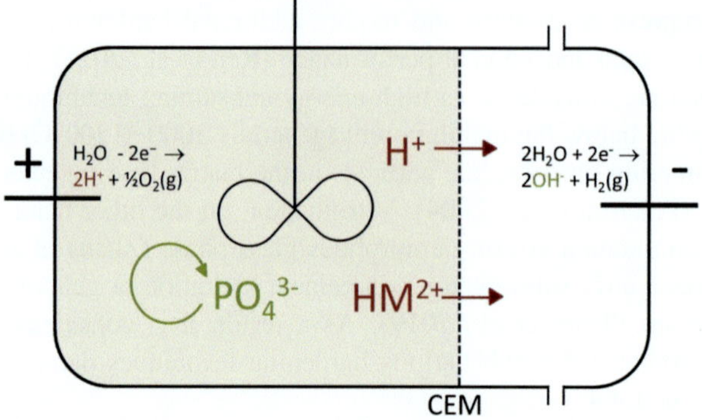

FIGURE 17.6 Two-compartment electrodialytic cell for simultaneous P recovery and heavy metal separation from ISSA (Ottosen et al., 2016).

(Donatello et al., 2010). However, it was reported that acid washing reduced the pozzolanic reactivity of ISSA (Donatello et al., 2010). Additionally, the formation of gypsum during H_2SO_4 treatment may hinder the further utilization of ISSA as construction materials, as it could increase the volume instability of the matrix (Ottosen et al., 2013).

4.3 Physical and mechanical properties of ISSA-based aggregates

The properties of artificial aggregates are assessed based on various physical parameters, including loose bulk density, saturated-surface-dry particle density, water absorption, cylinder compressive strength, single particle compressive strength, etc. Table 17.1 summarizes the mechanical properties of ISSA-derived aggregate reported in the literature. It can be observed that the loose bulk density, particle density, and water absorption of the produced aggregate are in the range of 836−1410 kg/m^3, 0.78−2.88 g/cm^3 and 0.42%−38.5%, respectively. The single particle compressive strength of spherical pellets ranges from 0.48 to 8.1 MPa. However, the compressive strength of cylindrical samples may vary significantly due to the size effect, making them not directly comparable. Certainly, the mechanical properties of the aggregates presented in Table 17.1 confirm the effectiveness of using ISSA in the production of artificial aggregates.

4.3.1 Density and its influencing factors

Almost all ISSA-derived aggregates (both sintered and cold-bonded) have achieved loose bulk density below 1200 kg/m^3, which can be classified as lightweight aggregate according to GB/T 17431 or EN 13055 (Tajra et al., 2019; Zhou et al., 2022a). The low density also offers additional advantages, such as enhanced thermal inertness and sound insulation, making them suitable for various applications including lightweight concrete, insulation products, adsorbent, etc. (Cheeseman and Virdi, 2005). In terms of cold-bonded aggregates, the bulk density increases linearly with the increasing content of binder (Tang et al., 2020). The hydration products generated from cement hydration reaction, such as hydrated calcium silicate (C−S−H) gel, ettringite, and calcium hydroxide, play an important role in filling the inner pores and resulting in the densification of the matrix (Xia et al., 2023d). However, it is important to note that high porosity is a prominent characteristic of ISSA, indicating its potential for reducing the density of cold-bonded aggregates (Donatello and Cheeseman, 2013). Besides, incorporating porous additives into the raw material system is an effective approach to reduce the density of aggregate. Tang et al. (2020) investigated the influence of wood chips on the bulk density of cold-bonded aggregates. The results revealed that the inclusion of 1%−2% wood chips led to a reduction in bulk density by 1%−7%.

In the case of sintered aggregates, the addition of pore-forming agents is a common practice to induce matrix expansion as well as reduce density. Various substances can serve as pore-forming agents, including water, carbonates, organic compounds, Fe_2O_3, etc. (Moreno-Maroto et al., 2019; Ren et al., 2021). Sodium carbonate, with a melting temperature of 850°C and a decomposition temperature of 900°C, has been identified as a fluxing and pore-forming agent for sintered aggregates (Hu et al., 2013). The incorporation of 22 wt% sodium carbonate into the ISSA-based aggregates leads to the production of lightweight aggregates with water absorption of 2.84% and compressive strength of 87.3 MPa when sintered at a temperature of 900°C (Hu et al., 2013). Recently, the research on using sewage sludge as an additive to produce lightweight aggregates has gained widespread attention due to its high organic content (Bouachera et al., 2021; Li et al., 2021b).

4.3.2 Water absorption and its influencing factors

Due to the porous nature of ISSA particles, cold-bonded aggregates prepared with ISSA typically have a higher water absorption (Zhou et al., 2022a). Studies indicate that grinding treatment can effectively reduce particle size but have a limited impact on porosity and specific surface area of ISSA particles; thus, it may not have a significant improvement on water absorption rate (Donatello et al., 2010; Ottosen et al., 2020). It was reported that accelerated carbonation can effectively reduce water absorption by forming a compact carbonated layer on the surface of cold-bonded aggregates (as shown in Fig. 17.7) (Tang et al., 2020). Different from the cold-bonding method, a dense vitrified layer can be generated on the surface of aggregates through high-temperature sintering, thereby alleviating the high water absorption caused by porosity (Chiou et al., 2006). Similarly, Li et al. (2021b) suggested that the presence of a glassy outer layer plays a crucial role in determining the water absorption of aggregates. In the absence of a glassy surface, the water absorption is influenced by the pore structure of the aggregates.

4.3.3 Mechanical properties and its influencing factors

As previously discussed, the strength development of cold-bonded aggregates is achieved by hydration or geopolymerization reactions, which are highly reliant on the chemical composition and reactivity of the raw materials (Qian

TABLE 17.1 Physical and mechanical properties of ISSA-derived artificial aggregates.

	Dosage of ISSA (wt%)	Hardening method	Morphology	Mechanical properties					Application	References
				Loose bulk density (kg/m^3)	Particle density (g/cm^3)	Water absorption (%)	Single particle compressive strength (MPa)	Cylinder compressive strength (MPa)		
CSW, ISSA	35–45	Cold-bonding	Pellet	836–1011		14–22	5.2–8.1		Aggregate	Tang et al. (2020)
OPC, GGBS, ISSA	70	Cold-bonding	Pellet	655		24.58 (24 h)		1.78–1.86	Aggregate	Zhou et al. (2022a)
Bentonite, ISSA	25	Sintering	Pellet	806			0.48		Adsorbent	Nie et al. (2021)
ISSA, Bentonite	25	Sintering	Pellet	1210	1.7	38.5 (24 h)	4.9		Adsorbent, aggregate	Wang et al. (2023)
ISSA, Na_2CO_3	78	Sintering	Cylinder	1410		4.85 (24 h)	87.3		Aggregate	Hu et al. (2013)
ISSA, SiO_2, CaO		Vitrification	Cylinder		2.88	0.42	247		Glass–ceramic	Zhang et al. (2015)

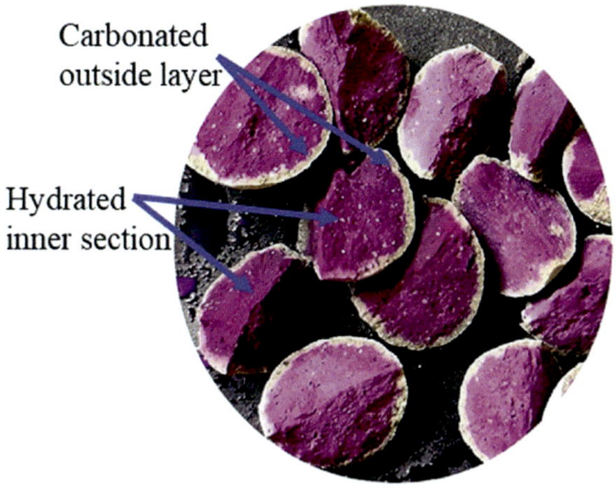

FIGURE 17.7 Carbonated outside layer characterized by phenolphthalein (Tang et al., 2020).

et al., 2022). However, excessive incorporation of binders undoubtedly increases the cost of waste disposal and contradicts the principles of sustainable development. Therefore, the current challenges for cold-bonding technology are still the low reactivity of raw materials and the compressive strength development of products. ISSA is primarily composed of SiO_2, Al_2O_3, CaO, Fe_2O_3, and P_2O_5, which exhibits a chemical composition similar to commonly used pozzolanic materials such as coal fly ash and GGBS (Chang et al., 2022b). However, the oxides with reactive potential are reported mainly exist in the form of inert phases (such as quartz or feldspar), resulting in a relatively lower pozzolanic reactivity of ISSA (Mejdi et al., 2020). To overcome this challenge, OPC and GGBS were added to enhance the reactivity of the alkali-activated system (Zhou et al., 2022a). The results showed that the combination of GGBS and cement improves the bonding performance of the aggregate, and GGBS fully participates in the geopolymerization reaction, as a result effectively enhancing the strength of the aggregates (Zhou et al., 2022a).

The mechanical properties of sintered aggregates are more dependent on the chemical composition of the raw materials compared. The appropriate ratio of SiO_2, Al_2O_3, and fluxing agents (e.g., CaO, MgO, Fe_2O_3, K_2O, and Na_2O) is necessary to form a viscous fused phase at high temperatures (Riley, 1951). Silica and alumina can be transformed into silica–aluminate polymers with high viscosity at high temperatures to form the skeleton structure of sintered aggregates, which is the main source of their strength. However, some studies have reported a decrease in the strength of sintered aggregates when the content of ISSA is increased (Nie et al., 2021). As a result, supplementing Si, Al-containing chemicals or wastes is indispensable for balancing element deficiency of ISSA. Cosintering with different additives (such as waste bentonite, pork bone, etc.) has been widely investigated to obtain the optimal SiO_2-Al_2O_3-fluxes ratio for improved mechanical properties of aggregates. Wang et al. (2023) successfully produced sintered aggregates composed of 25% ISSA, 25% peanut shell, and 50% waste bentonite with a sintering temperature of 1050°C. The resulting aggregate exhibited a good compressive strength of 4.9 MPa and bulk density of 1210 kg/m^3, demonstrating the promising potential for practical applications.

4.4 Microstructure and mineralogy characteristics of ISSA-based aggregates

4.4.1 Cold-bonded aggregates

The hydration and geopolymerization of cementitious materials contribute to the microstructure of cold-bonded aggregates (Qian et al., 2022). The hydration products, such as C–S–H gel, can fill the pores and repair microcracks, resulting in a denser microstructure of the aggregate (Zhou et al., 2022a). Chang et al. (2022b) investigated the influence of ISSA obtained at different incineration temperatures on the hydration characteristics of cementitious materials. It was observed that the high concentration of aluminum in the pore solution hindered the hydration of tricalcium silicate, leading to ineffective pore filling. However, higher incineration temperatures (800°C) increased the proportion of soluble silica and reduced the solubility of aluminum, which had a positive impact on improving the density of the matrix.

The chemical composition of ISSA gives an important influence on the mineralogical characteristics of cold-bonded aggregates (Ottosen et al., 2020). Chemical additives utilized in wastewater treatment, such as acidulating agents and precipitants containing Fe or Al compounds, introduce notable alterations in the chemical properties of the resulting ISSAs. These variations lead to the production of ash with distinct characteristics, including Al-rich, Fe-rich, and sulfate-rich compositions (Chen and Poon, 2017a; Gu et al., 2021). On the one hand, it has been shown that ISSA has no hydraulic properties, but rather involves the geopolymerization reaction in combination with $Ca(OH)_2$, and is therefore considered as a potential pozzolanic material (Chen and Poon, 2017a). Mejdi et al. (2020) investigated the hydration mechanism of blended cement incorporating ISSA. The results indicate that the presence of ISSA promotes the formation of calcium aluminate hydrates, such as AFt and AFm, due to the additional dissolved Al derived from the ash. In addition, the dissolved Al also promoted the transformation of C−S−H gel to C−A−S−H gel. Gu et al. (2021) studied the effects of sulfate-rich ISSA on the microstructure of cement-based materials, showing that the uptake of sulfates changed the morphology of hydration products in cement paste, resulting in a radial arrangement of the C−S−H gel. On the other hand, the presence of phosphates in ISSA makes it cannot be simply considered as a pozzolanic material (Mejdi et al., 2020). Brushite is generated through the reaction between the phosphates present in the ISSA and the $Ca(OH)_2$, which plays a significant role in enhancing the long term strength of the matrix (Chang et al., 2022b).

4.4.2 Sintered aggregates

The microstructure of sintered aggregates is influenced by several factors such as the chemical composition of the raw materials, the presence of pore-forming agents, and sintering parameters (Ren et al., 2021). Research has shown that three conditions are necessary to achieve the desired expansion of sintered aggregates: (1) the presence of aluminosilicate materials that can form a molten phase with appropriate viscosity at high temperatures; (2) the presence of pore-forming agents that can generate gases at high temperatures; (3) suitable temperature range for simultaneous occurrence of conditions (1) and (2) (Riley, 1951). The chemical composition of the raw materials refers to the appropriate ratio between SiO_2, Al_2O_3, and fluxing agents (e.g., CaO, MgO, Fe_2O_3, K_2O, and Na_2O). SiO_2 and Al_2O_3 can be transformed into aluminosilicate polymers with high viscosity at high temperatures to form the skeleton structure of sintered aggregates. The fluxing agents can significantly reduce the sintering temperature and accelerate the formation of the glassy phase via the liquid-state sintering mechanism (Wei et al., 2018). The chemical composition of ISSA closely resembles that of expansive clay used in commercial sintered aggregate production (Chiou et al., 2006). The composition of ISSA meets the recommended ranges proposed by Riley (1951) for expansive clay, which include SiO_2 (48%−70%), Al_2O_3 (8%−25%), and fluxing agents (4.5%−31%). Cheeseman and Virdi (2005) investigated the variation of pore structure in ISSA-based aggregates with sintering temperature. The results showed that the pore structure of the aggregates changed with increasing sintering temperature. At lower sintering temperatures, the aggregates exhibited a high number of interconnected and permeable continuous pores. As the temperature exceeded 1060°C, a well-structured and compact framework with numerous isolated, approximately spherical pores was observed, leading to a decrease in density and water absorption of the resulting product. With further temperature increase (1070°C), the isolated pores within the aggregate started to merge, resulting in irregular closed pores with larger diameters. Furthermore, the particle size of the raw materials also affects the microstructure of the aggregate. Research has indicated that when the particle size of ISSA exceeds 0.59 mm, the internal structure of the sintered aggregate becomes excessively loose. Therefore, it is recommended to use particle sizes smaller than 0.59 mm for the production of sintered aggregates using ISSA. Within this particle size range, gases can be efficiently trapped within the internal pores, leading to the formation of an optimal pore structure in the aggregates (Hu et al., 2013). While SiO_2 and Al_2O_3 can be transformed into aluminosilicate glassy phase materials under high-temperature conditions, research has shown that ISSA does not undergo significant crystallographic transformations during the sintering process (Nie et al., 2021). Cheeseman and Virdi (2005) investigated the characteristics and microstructure of ISSA-based lightweight aggregates produced at different sintering temperatures. The results showed that the main crystalline phases present in both the raw and sintered ISSA were quartz (SiO_2), whitlockite, and hematite (Fe_2O_3).

5. Perspectives

The recycling of ISSA holds the promise of offering multifaceted benefits to both the environment and the field of sustainable construction. As the world grapples with increasing waste generation and environmental concerns, repurposing ISSA as SCM and aggregate could mark a pivotal step toward achieving more sustainable practices.

From an environmental standpoint, the utilization of ISSA in construction materials presents an opportunity to reduce the demand for traditional raw materials, such as natural aggregates and cement clinker. By substituting these resources with ISSA, we not only conserve finite natural resources but also decrease the carbon footprint associated with their extraction and production. Moreover, the incorporation of ISSA in concrete formulations can contribute to the mitigation of landfill disposal, a critical issue in waste management. This recycling approach aligns with the principles of the circular economy, where waste materials are transformed into valuable resources, thus minimizing their environmental impact.

In the realm of sustainable construction, ISSA as an SCM and aggregate introduces a new dimension of innovation. The unique properties of ISSA, such as its potential to enhance compressive strength and durability in cementitious materials, offer opportunities for the development of high-performance and long-lasting construction products. The delayed release effect of moisture in sludge ash can alleviate the shrinkage of ultrahigh-performance concrete. Additionally, ISSA's fine particulate nature and controlled particle size distribution enable tailored engineering of construction materials with specific properties, addressing diverse project requirements. To ensure the performance of sludge-derived construction materials, the dosage of sludge should be limited to within 10 wt%. In future research, efforts should be intensified in the chemical modification of sludge to increase the amount of sludge added to construction materials.

However, the successful integration of ISSA into sustainable construction practices requires a comprehensive approach. It involves a thorough assessment of chemical and physical characteristics of ISSA, optimization of production processes, and adherence to quality standards. Furthermore, collaborations among researchers, policymakers, industries, and communities are essential to drive the implementation of such recycling strategies. Ensuring proper monitoring and regulation is crucial to guarantee the safety and performance of construction materials containing ISSA.

The recycling of ISSA as SCM and aggregate presents a compelling avenue for advancing environmental responsibility and sustainable construction. By harnessing its potential to reduce waste, conserve resources, and enhance construction materials, we can usher in an era where waste is no longer just discarded but transformed into valuable assets, contributing to a greener and more resilient built environment. This evolving landscape calls for ongoing research, innovation, and collaboration to realize the full potential of ISSA in shaping the future of construction practices.

6. Conclusions

The recycling of ISSA as SCM and aggregate presents a promising avenue toward sustainable waste management and resource utilization. The incorporation of ISSA as an SCM in cementitious blends not only mitigates the environmental impact of its disposal but also harnesses its pozzolanic reactivity to improve the mechanical strength, durability, and overall performance of concrete. Moreover, the versatile nature of ISSA permits its application as a partial or complete replacement of natural aggregates in concrete production, thereby reducing the demand for virgin materials and lessening the ecological strain associated with traditional construction practices.

The variability in ISSA properties, attributed to factors such as combustion conditions and original sludge composition, necessitates a comprehensive understanding of its characteristics to optimize its application. Effective quality control measures must be established to ensure consistent performance in different concrete formulations. Additionally, the interplay between ISSA's particle size distribution, chemical composition, and reactivity underscores the need for tailored mix design strategies to attain desired concrete properties. Collaboration between researchers, waste management entities, and construction industries is pivotal in developing standardized guidelines and specifications for incorporating ISSA in construction materials.

Although significant progress has been made in exploring the potential of ISSA, further research is warranted to delve deeper into the long-term behavior and durability of concrete containing ISSA. Long-term studies and field applications are essential to assess the performance and stability of ISSA-enhanced concrete under varying environmental conditions and loading scenarios. Furthermore, advancements in processing techniques, such as activation methods and particle size optimization, could unlock additional opportunities for enhancing the reactivity and application versatility of ISSA.

The integration of ISSA as SCM and aggregates is a crucial step in achieving circular economy and sustainable construction. By harnessing its latent potential and addressing challenges through continued research and collaborative efforts, the construction industry can make significant strides toward resource efficiency, reduced environmental impact, and the realization of a greener built environment.

Acknowledgments

The authors gratefully acknowledge the financial support from the Zhejiang Provincial Natural Science Foundation of China (Grant No. LZ23E060004) for this study.

References

Bekkeri, G.B., Shetty, K.K., Nayak, G., 2023. Synthesis of artificial aggregates and their impact on performance of concrete: a review. J. Mater. Cycles Waste Manag. 25, 1988–2011.

Bénard, P., Garrault, S., Nonat, A., Cau-Dit-Coumes, C., 2005. Hydration process and rheological properties of cement pastes modified by orthophosphate addition. J. Eur. Ceram. Soc. 25, 1877–1883.

Bénard, P., Garrault, S., Nonat, A., Cau-dit-Coumes, C., 2008. Influence of orthophosphate ions on the dissolution of tricalcium silicate. Cement Concr. Res. 38, 1137–1141.

Benassi, L., Zanoletti, A., Depero, L.E., Bontempi, E., 2019. Sewage sludge ash recovery as valuable raw material for chemical stabilization of leachable heavy metals. J. Environ. Manag. 245, 464–470.

Borowski, G., 2015. Using vitrification for sewage sludge combustion ash disposal. Pol. J. Environ. Stud. 24, 1889–1896.

Bouachera, R., Kasimi, R., Ibnoussina, M., Hakkou, R., Taha, Y., 2021. Reuse of sewage sludge and waste glass in the production of lightweight aggregates. Mater. Today: Proc. 37, 3866–3870.

Cau Dit Coumes, C., Courtois, S., 2003. Cementation of a low-level radioactive waste of complex chemistry: investigation of the combined action of borate, chloride, sulfate and phosphate on cement hydration using response surface methodology. Cement Concr. Res. 33, 305–316.

Chakraborty, S., Jo, B.W., Jo, J.H., Baloch, Z., 2017. Effectiveness of sewage sludge ash combined with waste pozzolanic minerals in developing sustainable construction material: an alternative approach for waste management. J. Clean. Prod. 153, 253–263.

Chang, F., Lin, J., Tsai, C., Wang, K., 2010. Study on cement mortar and concrete made with sewage sludge ash. Water Sci. Technol. 62, 1689–1693.

Chang, H., Zhao, Y., Zhao, S., Damgaard, A., Christensen, T.H., 2022a. Review of inventory data for the thermal treatment of sewage sludge. Waste Manag. 146, 106–118.

Chang, Z., Long, G., Xie, Y., Zhou, J.L., 2022b. Chemical effect of sewage sludge ash on early-age hydration of cement used as supplementary cementitious material. Construct. Build. Mater. 322, 126116.

Chang, Z., Long, G., Xie, Y., Zhou, J.L., 2022c. Pozzolanic reactivity of aluminum-rich sewage sludge ash: influence of calcination process and effect of calcination products on cement hydration. Construct. Build. Mater. 318, 126096.

Chang, Z., Long, G., Xie, Y., Zhou, J.L., 2022d. Recycling sewage sludge ash and limestone for sustainable cementitious material production. J. Build. Eng. 49, 104035.

Chang, Z., Long, G., Zhou, J.L., Ma, C., 2020. Valorization of sewage sludge in the fabrication of construction and building materials: a review. Resour. Conserv. Recycl. 154, 104606.

Cheeseman, C.R., Virdi, G.S., 2005. Properties and microstructure of lightweight aggregate produced from sintered sewage sludge ash. Resour. Conserv. Recycl. 45, 18–30.

Chen, H.J., Wang, S.Y., Tang, C.W., 2010. Reuse of incineration fly ashes and reaction ashes for manufacturing lightweight aggregate. Construct. Build. Mater. 24, 46–55.

Chen, M., Blanc, D., Gautier, M., Mehu, J., Gourdon, R., 2013. Environmental and technical assessments of the potential utilization of sewage sludge ashes (SSAs) as secondary raw materials in construction. Waste Manage. (Tucson, Ariz.) 33, 1268–1275.

Chen, R., Ma, X., Yu, Z., Chen, L., Chen, X., Qin, Z., 2021. Study on synchronous immobilization technology of heavy metals and hydrolyzed nitrogen during pyrolysis of sewage sludge. J. Environ. Chem. Eng. 9, 106079.

Chen, Y., Wang, T., Zhou, M., Hou, H., Xue, Y., Wang, H., 2018b. Rice husk and sewage sludge co-combustion ash: leaching behavior analysis and cementitious property. Construct. Build. Mater. 163, 63–72.

Chen, Z., Li, J., Zhan, B., Sharma, U., Poon, C.S., 2018a. Compressive strength and microstructural properties of dry-mixed geopolymer pastes synthesized from GGBS and sewage sludge ash. Construct. Build. Mater. 182, 597–607.

Chen, Z., Poon, C.S., 2017a. Comparative studies on the effects of sewage sludge ash and fly ash on cement hydration and properties of cement mortars. Construct. Build. Mater. 154, 791–803.

Chen, Z., Poon, C.S., 2017b. Comparing the use of sewage sludge ash and glass powder in cement mortars. Environ. Technol. 38, 1390–1398.

Chin, S.C., Doh, S.I., Andri, K., Yih, K.W., Ahmad, A.S.W., 2016. Characterization of sewage sludge ash (SSA) in cement mortar. J. Eng. Appl. Sci. 11, 2242–2247.

Chiou, I.J., Wang, K.S., Chen, C.H., Lin, Y.T., 2006. Lightweight aggregate made from sewage sludge and incinerated ash. Waste Manage. (Tucson, Ariz.) 26, 1453–1461.

Danish, A., Ozbakkaloglu, T., 2022. Greener cementitious composites incorporating sewage sludge ash as cement replacement: a review of progress, potentials, and future prospects. J. Clean. Prod. 371, 133364.

Dave, N., Misra, A.K., Srivastava, A., Kaushik, S.K., 2016. Experimental analysis of strength and durability properties of quaternary cement binder and mortar. Construct. Build. Mater. 107, 117–124.

de Azevedo Basto, P., Savastano Junior, H., de Melo Neto, A.A., 2019. Characterization and pozzolanic properties of sewage sludge ashes (SSA) by electrical conductivity. Cem. Concr. Compos. 104, 103410.

Donatello, S., Cheeseman, C.R., 2013. Recycling and recovery routes for incinerated sewage sludge ash (ISSA): a review. Waste Manage. (Tucson, Ariz.) 33, 2328–2340.

Donatello, S., Freeman-Pask, A., Tyrer, M., Cheeseman, C.R., 2010. Effect of milling and acid washing on the pozzolanic activity of incinerator sewage sludge ash. Cem. Concr. Compos. 32, 54–61.

Duan, W., Zhuge, Y., Chow, C.W.K., Keegan, A., Liu, Y., Siddique, R., 2022. Mechanical performance and phase analysis of an eco-friendly alkali-activated binder made with sludge waste and blast-furnace slag. J. Clean. Prod. 374, 134024.

Garcés, P., Pérez Carrión, M., García-Alcocel, E., Payá, J., Monzó, J., Borrachero, M.V., 2008. Mechanical and physical properties of cement blended with sewage sludge ash. Waste Manage. (Tucson, Ariz.) 28, 2495–2502.

Gastaldini, A.L.G., Hengen, M.F., Gastaldini, M.C.C., Do Amaral, F.D., Antolini, M.B., Coletto, T., 2015. The use of water treatment plant sludge ash as a mineral addition. Construct. Build. Mater. 94, 513–520.

Godoy, L.G.G.D., Rohden, A.B., Garcez, M.R., Costa, E.B.D., Da Dalt, S., Andrade, J.J.D.O., 2019. Valorization of water treatment sludge waste by application as supplementary cementitious material. Construct. Build. Mater. 223, 939–950.

Gu, C., Ji, Y., Yao, J., Yang, Y., Liu, J., Ni, T., Zhou, H., Tong, Y., Zhang, X., 2022. Feasibility of recycling sewage sludge ash in ultra-high performance concrete: volume deformation, microstructure and ecological evaluation. Construct. Build. Mater. 318, 125823.

Gu, C., Ji, Y., Zhang, Y., Yang, Y., Liu, J., Ni, T., 2021. Recycling use of sulfate-rich sewage sludge ash (SR-SSA) in cement-based materials: assessment on the basic properties, volume deformation and microstructure of SR-SSA blended cement pastes. J. Clean. Prod. 282, 124511.

Gu, C., Shuang, Y., Ji, Y., Wei, H., Yang, Y., Xu, Y., Qian, R., Cui, D., Zhou, H., 2023. Effect of environmental conditions on the volume deformation of cement mortars with sewage sludge ash. J. Build. Eng. 65, 105720.

Halliday, J.E., Jones, M.R., Dyer, T.D., Dhir, R.K., 2012. Potential use of UK sewage sludge ash in cement-based concrete. Proc. Inst. Civ. Eng.-Wast Resour. Manag. 165, 57–66.

Hu, S.-H., Hu, S.-C., Fu, Y.-P., 2013. Recycling technology-Artificial lightweight aggregates synthesized from sewage sludge and its ash at lowered comelting temperature. Environ. Prog. Sustain. Energy 32, 740–748.

Husek, M., Mosko, J., Pohorely, M., 2022. Sewage sludge treatment methods and P-recovery possibilities: current state-of-the-art. J. Environ. Manag. 315, 115090.

Jang, H.L., Lee, H.K., Jin, K., Ahn, H., Lee, H., Nam, K.T., 2015. Phase transformation from hydroxyapatite to the secondary bone mineral, whitlockite. J. Mat. Chem. B 3, 1342–1349.

Juenger, M.C.G., Snellings, R., Bernal, S.A., 2019. Supplementary cementitious materials: new sources, characterization, and performance insights. Cement Concr. Res. 122, 257–273.

Kappel, A., Viader, R.P., Kowalski, K.P., Kirkelund, G.M., Ottosen, L.M., 2018. Utilisation of electrodialytically treated sewage sludge ash in mortar. Waste Biomass Valorization 9, 2503–2515.

Krejcirikova, B., Rode, C., Peuhkuri, R., 2019. Determination of hygrothermal properties of cementitious mortar: the effect of partial replacement of cement by incinerated sewage sludge ash. J. Build. Phys. 42, 771–787.

Lee, M., Kim, D.-J., 2017. Identification of phosphorus forms in sewage sludge ash during acid pre-treatment for phosphorus recovery by chemical fractionation and spectroscopy. J. Ind. Eng. Chem. 51, 64–70.

Li, J., Guo, M., Xue, Q., Poon, C.S., 2017. Recycling of incinerated sewage sludge ash and cathode ray tube funnel glass in cement mortars. J. Clean. Prod. 152, 142–149.

Li, J., Wang, Q., Chen, Z., Xue, Q., Chen, X., Mu, Y., Poon, C.S., 2021a. Immobilization of high-Pb contaminated soil by oxalic acid activated incinerated sewage sludge ash. Environ. Pollut. 284, 117120.

Li, X., He, C., Lv, Y., Jian, S., Jiang, W., Jiang, D., Wu, K., Dan, J., 2021b. Effect of sintering temperature and dwelling time on the characteristics of lightweight aggregate produced from sewage sludge and waste glass powder. Ceram. Int. 47, 33435–33443.

Lin, D.F., Lin, K.L., Chang, W.C., Luo, H.L., Cai, M.Q., 2008a. Improvements of nano-SiO_2 on sludge/fly ash mortar. Waste Manage. (Tucson, Ariz.) 28, 1081–1087.

Lin, K.L., Chang, W.C., Lin, D.F., Luo, H.L., Tsai, M.C., 2008b. Effects of nano-SiO_2 and different ash particle sizes on sludge ash–cement mortar. J. Environ. Manag. 88, 708–714.

Lin, Y., Zhou, S., Li, F., Lin, Y., 2012. Utilization of municipal sewage sludge as additives for the production of eco-cement. J. Hazard Mater. 213–214, 457–465.

Liu, J., Zhuo, Z., Xie, W., Kuo, J., Lu, X., Buyukada, M., Evrendilek, F., 2018. Interaction effects of chlorine and phosphorus on thermochemical behaviors of heavy metals during incineration of sulfur-rich textile dyeing sludge. Chem. Eng. J. 351, 897–911.

Liu, M., Zhao, Y., Xiao, Y., Yu, Z., 2019. Performance of cement pastes containing sewage sludge ash at elevated temperatures. Construct. Build. Mater. 211, 785–795.

Liu, Y., Zhuge, Y., Duan, W., Huang, G., Yao, Y., 2022. Modification of microstructure and physical properties of cement-based mortar made with limestone and alum sludge. J. Build. Eng. 58, 105000.

Lothenbach, B., Scrivener, K., Hooton, R.D., 2011. Supplementary cementitious materials. Cement Concr. Res. 41, 1244–1256.

Lynn, C.J., Dhir, R.K., Ghataora, G.S., West, R.P., 2015. Sewage sludge ash characteristics and potential for use in concrete. Construct. Build. Mater. 98, 767–779.

Mejdi, M., Saillio, M., Chaussadent, T., Divet, L., Tagnit-Hamou, A., 2020. Hydration mechanisms of sewage sludge ashes used as cement replacement. Cement Concr. Res. 135, 106115.

Monzó, J., Payá, J., Borrachero, M.V., Girbés, I., 2003. Reuse of sewage sludge ashes (SSA) in cement mixtures: the effect of SSA on the workability of cement mortars. Waste Manage. (Tucson, Ariz.) 23, 373–381.

Moreno-Maroto, J.M., Uceda-Rodríguez, M., Cobo-Ceacero, C.J., de Hoces, M.C., MartínLara, M.Á., Cotes-Palomino, T., López García, A.B., Martínez -García, C., 2019. Recycling of 'alperujo' (olive pomace) as a key component in the sintering of lightweight aggregates. J. Clean. Prod. 239, 118041.

Motisariya, K., Agrawal, G., Baria, M., Srivastava, V., Dave, D.N., 2023. Experimental analysis of strength and durability properties of cement binders and mortars with addition of microfine sewage sludge ash (SSA) particles. Mater. Today: Proc. 85, 24–28.

Naamane, S., Rais, Z., Taleb, M., 2016. The effectiveness of the incineration of sewage sludge on the evolution of physicochemical and mechanical properties of Portland cement. Construct. Build. Mater. 112, 783–789.

Nie, J., Wang, Q., Gao, S., Poon, C.S., Zhou, Y., Li, J.S., 2021. Novel recycling of incinerated sewage sludge ash (ISSA) and waste bentonite as ceramsite for Pb-containing wastewater treatment: performance and mechanism. J. Environ. Manag. 288, 112382.

Oey, T., Kumar, A., Bullard, J.W., Neithalath, N., Sant, G., 2013. The filler effect: the influence of filler content and surface area on cementitious reaction rates. J. Am. Ceram. Soc. 96, 1978–1990.

Oliva, M., Vargas, F., Lopez, M., 2019. Designing the incineration process for improving the cementitious performance of sewage sludge ash in Portland and blended cement systems. J. Clean. Prod. 223, 1029–1041.

Ottosen, L.M., Bertelsen, I.M.G., Jensen, P.E., Kirkelund, G.M., 2020. Sewage sludge ash as resource for phosphorous and material for clay brick manufacturing. Construct. Build. Mater. 249, 118684.

Ottosen, L.M., Jensen, P.E., Kirkelund, G.M., 2016. Phosphorous recovery from sewage sludge ash suspended in water in a two-compartment electrodialytic cell. Waste Manage. (Tucson, Ariz.) 51, 142–148.

Ottosen, L.M., Kirkelund, G.M., Jensen, P.E., 2013. Extracting phosphorous from incinerated sewage sludge ash rich in iron or aluminum. Chemosphere 91, 963–969.

Pan, S., Tseng, D., Lee, C., Lee, C., 2003. Influence of the fineness of sewage sludge ash on the mortar properties. Cement Concr. Res. 33, 1749–1754.

Peys, A., Snellings, R., Peeraer, B., Vayghan, A.G., Sand, A., Horckmans, L., Quaghebeur, M., 2022. Transformation of mine tailings into cement-bound aggregates for use in concrete by granulation in a high intensity mixer. J. Clean. Prod. 366, 132989.

Polettini, A., Pomi, R., Trinci, L., Muntoni, A., Lo Mastro, S., 2004. Engineering and environmental properties of thermally treated mixtures containing MSWI fly ash and low-cost additives. Chemosphere 56, 901–910.

Qian, L., Xu, L., Alrefaei, Y., Wang, T., Ishida, T., Dai, J., 2022. Artificial alkali-activated aggregates developed from wastes and by-products: a state-of-the-art review. Resour. Conserv. Recycl. 177, 105971.

Ren, P., Ling, T.-C., Mo, K.H., 2021. Recent advances in artificial aggregate production. J. Clean. Prod. 291, 125215.

Riley, C.M., 1951. Relation of chemical properties to the bloating of clays. J. Am. Ceram. Soc. 34, 121–128.

Sasaoka, N., Yokoi, K., Yamanaka, T., 2006. Basic study of concrete made using ash derived from the incinerating sewage sludge. Int. J. Mod. Phys. B 20, 3716–3721.

Scrivener, K.L., Nonat, A., 2011. Hydration of cementitious materials, present and future. Cement Concr. Res. 41, 651–665.

Skibsted, J., Snellings, R., 2019. Reactivity of supplementary cementitious materials (SCMs) in cement blends. Cement Concr. Res. 124, 105799.

Song, Q., Guo, M., Ling, T., 2022. A review of elevated-temperature properties of alternative binders: supplementary cementitious materials and alkali-activated materials. Construct. Build. Mater. 341, 127894.

Tajra, F., Elrahman, M.A., Stephan, D., 2019. The production and properties of cold-bonded aggregate and its applications in concrete: a review. Construct. Build. Mater. 225, 29–43.

Takeuchi, N., Yoshimura, A., 2022. Fabrication of foamed porous ceramics from mixtures of pork bone and incinerated sewage sludge ash. Mater. Trans. 63, 389–393.

Tang, P., Xuan, D., Li, J., Cheng, H.W., Poon, C.S., Tsang, D.C.W., 2020. Investigation of cold bonded lightweight aggregates produced with incineration sewage sludge ash (ISSA) and cementitious waste. J. Clean. Prod. 251, 119709.

Tantawy, M.A., El-Roudi, A.M., Abdalla, E.M., Abdelzaher, M.A., 2013. Fire resistance of sewage sludge ash blended cement pastes. J. Eng. 2013, 1–7.

Tantawy, M.A., El-Roudi, A.M., Abdalla, E.M., Abdelzaher, M.A., 2012. Evaluation of the pozzolanic activity of sewage sludge ash. ISRN Chem. Eng. 2012, 1–8.

Tay, J.H., 1987. Sludge ash as filler for Portland cement concrete. J. Environ. Eng.-ASCE 113, 345–351.

Teoh, S.K., Li, L.Y., 2020. Feasibility of alternative sewage sludge treatment methods from a lifecycle assessment (LCA) perspective. J. Clean. Prod. 247, 119495.

Vouk, D., Nakic, D., Stirmer, N., Cheeseman, C., 2017. Use of sewage sludge ash in cementitious materials. Rev. Adv. Mater. Sci. 49, 158–170.

Wang, Q., Li, J., Poon, C.S., 2021. Novel recycling of phosphorus-recovered incinerated sewage sludge ash residues by co-pyrolysis with lignin for reductive/sorptive removal of hexavalent chromium from aqueous solutions. Chemosphere 285, 131434.

Wang, Q., Li, J., Poon, C.S., 2023. Production of sorptive granules from incinerated sewage sludge ash and upcycling in cement mortar. Sep. Purif. Technol. 309.

Wang, T., Xue, Y., Zhou, M., Liang, A., Liu, J., Mei, M., Lao, X., Hou, H., Li, J., 2020. Effect of addition of rice husk on the fate and speciation of heavy metals in the bottom ash during dyeing sludge incineration. J. Clean. Prod. 244, 118851.

Wang, T., Xue, Y., Zhou, M., Lv, Y., Chen, Y., Wu, S., Hou, H., 2017. Hydration kinetics, freeze-thaw resistance, leaching behavior of blended cement containing co-combustion ash of sewage sludge and rice husk. Construct. Build. Mater. 131, 361–370.

Wei, Y.L., Weng, S.D., Xie, X.Q., 2018. Reduction of sintering energy by application of calcium fluoride as flux in lightweight aggregate sintering. Construct. Build. Mater. 190, 765–772.

Wu, Z., Jiang, Y., Guo, W., Jin, J., Wu, M., Shen, D., Long, Y., 2021. The long-term performance of concrete amended with municipal sewage sludge incineration ash. Environ. Technol. Innov. 23, 101574.

Xia, Y., Liu, M., Zhao, Y., Chi, X., Guo, J., Du, D., Du, J., 2023d. Hydration of ternary blended cements with sewage sludge ash and limestone: hydration mechanism and phase assemblage. Construct. Build. Mater. 375.

Xia, Y., Liu, M., Zhao, Y., Chi, X., Lu, Z., Tang, K., Guo, J., 2023c. Utilization of sewage sludge ash in ultra-high performance concrete (UHPC): microstructure and life-cycle assessment. J. Environ. Manag. 326, 116690.

Xia, Y., Liu, M., Zhao, Y., Guo, J., Chi, X., Du, J., Du, D., Shi, D., 2023a. Hydration mechanism and environmental impacts of blended cements containing co-combustion ash of sewage sludge and rice husk: compared with blended cements containing sewage sludge ash. Sci. Total Environ. 864, 161116.

Xia, Y., Liu, M., Zhao, Y., Ma, X., 2022. Microstructure of Portland cement blended with high dosage of sewage sludge ash activated by Na2SO4. J. Clean. Prod. 351, 131568.

Xia, Y., Zhao, Y., Liu, M., Guo, J., Du, J., Du, D., 2023b. Hydration mechanism and phase assemblage of ternary blended cements based on sewage sludge ash and limestone: modified by Na2SO4. Construct. Build. Mater. 364, 129982.

Yang, K.-H., Jung, Y.-B., Cho, M.-S., Tae, S.-H., 2015. Effect of supplementary cementitious materials on reduction of CO_2 emissions from concrete. J. Clean. Prod. 103, 774–783.

Yu, S., Zhang, B., Wei, J., Zhang, T., Yu, Q., Zhang, W., 2017. Effects of chlorine on the volatilization of heavy metals during the co-combustion of sewage sludge. Waste Manage. (Tucson, Ariz.) 62, 204–210.

Zdeb, T., Tracz, T., Adamczyk, M., 2022. Physical, mechanical properties and durability of cement mortars containing fly ash from the sewage sludge incineration process. J. Clean. Prod. 345, 131055.

Zeng, G., Wan, J., Huang, D., Hu, L., Huang, C., Cheng, M., Xue, W., Gong, X., Wang, R., Jiang, D., 2017. Precipitation, adsorption and rhizosphere effect: the mechanisms for Phosphate-induced Pb immobilization in soils—a review. J. Hazard Mater. 339, 354–367.

Zhang, Z., Li, A., Yin, Y., Zhao, L., 2013. Effect of crystallization time on behaviors of glass-ceramic produced from sludge incineration ash. Procedia Environ. Sci. 18, 788–793.

Zhang, Z., Zhang, L., Yin, Y., Liang, X., Li, A., 2015. The recycling of incinerated sewage sludge ash as a raw material for $CaO\text{-}Al_2O_3\text{-}SiO_2\text{-}P_2O_5$ glass-ceramic production. Environ. Technol. 36, 1098–1103.

Zhou, X., Chen, Y., Liu, C., Wu, F., 2022a. Preparation of artificial lightweight aggregate using alkali-activated incinerator bottom ash from urban sewage sludge. Construct. Build. Mater. 341, 127844.

Zhou, Y., Cai, G., Cheeseman, C., Li, J., Poon, C.S., 2022b. Sewage sludge ash-incorporated stabilisation/solidification for recycling and remediation of marine sediments. J. Environ. Manag. 301, 113877.

Zhou, Y., Li, J., Lu, J., Cheeseman, C., Poon, C.S., 2020. Recycling incinerated sewage sludge ash (ISSA) as a cementitious binder by lime activation. J. Clean. Prod. 244, 118856.

Zhou, Y., Lu, J., Li, J., Cheeseman, C., Poon, C.S., 2021a. Influence of seawater on the mechanical and microstructural properties of lime-incineration sewage sludge ash pastes. Construct. Build. Mater. 278, 122364.

Zhou, Y., Lu, J., Li, J., Cheeseman, C., Poon, C.S., 2021b. Effect of NaCl and $MgCl_2$ on the hydration of lime-pozzolan blend by recycling sewage sludge ash. J. Clean. Prod. 313, 127759.

Chapter 18

Recycling of municipal solid waste incineration fly ash into SCMs and aggregates

Zhenhao Song[1], Yuying Zhang[2], Yan Xia[1], Chen Sun[1] and Lei Wang[1]
[1]*State Key Laboratory of Clean Energy Utilization, Zhejiang University, Hangzhou, China;* [2]*Department of Civil and Environmental Engineering, The Hong Kong University of Science and Technology, Hong Kong, China*

1. Introduction

The increasing generation of municipal solid wastes (MSW) and the widespread adoption of incineration technology worldwide have led to the production and accumulation of hazardous secondary solid by-products, including municipal solid waste incineration fly ash (MSWI FA) and bottom ash (MSWI BA). As a consequence, the proper management of these waste materials has become a pressing challenge for modern society (Yang et al., 2021). Compared with MSWI BA, the disposal of MSWI FA (categorized as hazardous wastes in most countries) poses particular disposal challenges due to its high content of potentially toxic elements (PTEs, e.g., Pb, Zn, Cu, Cr, Cd, etc.), soluble salts (chlorides), and toxic organic pollutants (polychlorinated dibenzodioxins and polychlorinated dibenzofurans, PCDD/Fs) (Phua et al., 2019). Therefore, the safe treatment and resourceful utilization of MSWI FA have attracted increasing attention (Zhang et al., 2021). In addition to those hazardous components, MSWI FA is characterized by elevated levels of CaO, contributing to its alkaline properties and making it possible to be recycled as construction materials for sustainable engineering (Liu et al., 2021).

Concrete is widely utilized as a primary structural material in various buildings and infrastructures, typically manufactured using ordinary Portland cement (OPC) as the binder, along with natural or artificial aggregates (Yang et al., 2015; Fan et al., 2023). However, the production of OPC is associated with a significant carbon footprint ranging from 0.66 to 0.82 tons per ton of OPC produced (Chen et al., 2019a), resulting in it being responsible for 6%−8% of the global anthropogenic CO_2 emissions (Skibsted and Snellings, 2019; Habert et al., 2020). To address this environmental concern, the utilization of industrial by-products or natural materials as supplementary cementitious materials (SCMs) in the form of blended cement has gained attention due to its potential to effectively reduce CO_2 emissions. Incorporating SCMs has been shown to enhance the workability and long-term durability of cement-based materials (Juenger et al., 2019). Coal fly ash, metakaolin, ground granulated blast furnace slag (GGBFS), and silica fume are the most used SCMs (Wu et al., 2021; Jin et al., 2023). In recent years, there has been increasing interest in exploring the potential of MSWI FA as an SCM in construction materials, given its inherent alkaline properties that can activate pozzolanic activities in silicon or aluminum-enriched waste materials (Wan et al., 2018). Therefore, studies on MSWI FA serving as SCMs in construction materials have attracted increasing interest recently.

Furthermore, the overexploitation of natural resources for concrete production has led to the depletion of natural ingredients. The global production of natural aggregates was reported to be around 40.0 billion tons in 2018, placing immense pressure on crude ore resources. From a dual perspective of environment and demand, finding alternative aggregates as a natural substitute is crucial (Deng et al., 2022). The utilization of MSWI FA in artificial aggregate production is an innovative approach to achieving sustainable waste management and eco-friendly construction materials production, which could prevent the irreversible depletion of natural resources (Tajra et al., 2019).

Treatment and Utilization of Combustion and Incineration Residues. https://doi.org/10.1016/B978-0-443-21536-0.00030-7

This chapter presents a comprehensive overview of the recent advancements in recycling MSWI FA as both SCMs and artificial lightweight aggregates in concrete production. It critically examines the recent progress and advancements in MSWI FA recycling, emphasizing the physiochemical properties of different construction materials derived from MSWI FA and the factors influencing these properties. This chapter also highlights the relevant environmental properties associated with the use of MSWI FA-derived construction materials, providing insights into their potential environmental impact.

2. Recycling MSWI FA into supplementary cementitious material

SCMs refer to industrial by-products or natural materials with potential pozzolanic or hydraulic reactivity, which can be used to partially substitute cement (Juenger et al., 2019; Skibsted and Snellings, 2019). The incorporation of SCMs can reduce the carbon footprint of cementitious materials indirectly and improve the workability and durability of the matrix through the "filler effect" and "pozzolanic effect" (Lothenbach et al., 2011; Wu et al., 2021). MSWI FA typically consists of fine particles with a particle size ranging from 1 to 300 μm and has a high calcium content (Li, 2021), while its chemical composition can vary considerably due to the different MSW sources, regions and incineration conditions (Li, 2021; Jin et al., 2023). Some MSWI FA has suitable contents of aluminosilicates, making it a potential material for SCMs (Zhang et al., 2021). Table 18.1 summarizes recent progress on the recycling of MSWI FA as SCMs.

2.1 Pozzolanic effect

Pozzolanic reaction refers to a chemical reaction that occurs between the reactive silica or alumina present in SCMs particles and the portlandite formed during the hydration of cement in the presence of water at ambient temperature. This reaction leads to the formation of additional calcium silicate hydrate (C—S—H) gel, which contributes to the densification and strength development of the cementitious matrix (Chen et al., 2019a). Many studies have confirmed that MSWI FA has low pozzolanic reactivity, and it contains a large amount of interfering substances, such as chlorides and PTEs, which

TABLE 18.1 Recycling MSWI FA as SCMs.

SCMs	MSWI FA content	Pretreatment	Properties	Impacts of MSWI FA	References
MSWI FA/gasification fly ash, ground waste glass	15—30 wt%	Water washing and carbonation	Specimens with 30% SCM substitution reach compressive strength >15 MPa after 90-day curing	Incorporation of wastes delayed cement hydration	Bui Viet et al. (2020)
MSWI FA, MSWI BA	6.8 wt%	—	Compressive strength of mortar reached 32.5 MPa after 28-day curing	Chloride ions reduce the immobilization efficiency of Pb/Cr	Garcia-Lodeiro et al. (2016)
MSWI FA	50 wt%	Catalytic pyrolysis, water-washing, mechanochemical pretreatment	Mortar exhibited compressive strength >40 MPa after 28-day curing	Mechanochemical-treated MSWI FA has a filler effect and pozzolanic effect	Jin et al. (2023)
MSWI FA	20—80 wt%	—	Compressive strength decreased from 55.7 to 8.74 MPa, with the MSWI FA content increased from 20% to 80%	Compressive strength decreases with an increase in MSWI FA	Wan et al. (2018)
MSWI FA, MSWI BA	10—50 wt%	Water washing	With the 10 wt% washed MSWI FA blending, the flexural and compressive strength reached 5.5 and 29 MPa after 3-day	Washed MSWI FA and MSWI BA reduced the mechanical strength	Yang et al. (2018)

result in delayed cement hydration and mechanical property degradation (Wan et al., 2018). As a result, pretreatment of MSWI FA is necessary. Water washing is a common pretreatment method to eliminate most chlorides and soluble metals, but it does not enhance the reactivity of MSWI FA and may result in concentrated heavy metal contents and wastewater generation (Yang et al., 2018). To address this issue, a three-step pretreatment process consisting of catalytic pyrolysis, water washing, and mechanochemical treatment has been proposed to simultaneously remove harmful substances and modify the physiochemical properties of MSWI FA. This pretreatment method has shown promising results, with the compressive strength of concrete blocks mixed with treated MSWI FA (50 wt% cement replacement) exceeding 40 MPa after a 28-day curing (Jin et al., 2023).

2.2 Filler effect

The filler effect refers to using fine SCMs, such as coal fly ash and ground granulated blast furnace slag (GGBFS), as fillers in cementitious systems. These SCMs, despite having lower reactivity than cement (Juenger et al., 2019; Skibsted and Snellings, 2019), can act as diluents for clinkers and provide nucleation sites for the formation of calcium silicate hydrate (C−S−H) gel through the heterogenous nucleation effect, thereby accelerating the clinker hydration reaction at the early age (Lothenbach et al., 2011). Previous experience indicated that the particle size distributions of SCMs need to be smaller than that of cement ($D_{50} \approx 16$ μm) to exhibit the nucleation effect, and it was reported that the filler effect is roughly linearly related to the fineness of SCMs (Berodier et al., 2014; Scrivener et al., 2015). However, the larger particle size distributions of MSWI FA ($D_{50} = 12.7-49.6$ μm) compared with cement indicated its limited potential as fillers (Chen et al., 2019b; Ebert et al., 2020). Consequently, milling treatment, which could reduce the particle size of MSWI FA, would be helpful before serving as filler in cement-based materials.

From the perspective of producing green and sustainable cement-based materials, increasing the content of low-carbon SCMs in the cement system has emerged as a viable option. To promote the widespread application of MSWI FA as SCMs, further research is urgently needed on modification methods, pretreatment techniques, and the underlying mechanisms governing its hydration reactions.

3. Recycling MSWI FA into artificial lightweight aggregates

Artificial aggregates are engineered construction materials that offer a viable alternative to natural aggregates in concrete. These aggregates are typically manufactured by sintering or cold-bonding technologies. Artificial aggregates have the great potential to incorporate high volumes of solid waste. The performance of artificial aggregates is assessed based on various physical parameters, including loose bulk density, saturated-surface-dry particle density, water absorption, cylinder compressive strength, etc. (Chen et al., 2010; Tajra et al., 2019; Shao et al., 2022). Table 18.2 summarizes the standard methods reported in the literature for determining these parameters. It should be noted that compressive strength can be measured as bulk (also called cylinder) compressive strength or single particle compressive strength.

3.1 Granulation method

To address the challenge of using MSWI FA as aggregate due to its fine powdery nature, the granulation process has been employed to convert such fine particles into aggregates with particle sizes ranging from 2 to 20 mm (Ren et al., 2021). The granulation methods are generally divided into agglomeration granulation, crushing, mold casting, etc. (Qian et al., 2022a).

TABLE 18.2 Standard methods for determining physical parameters of artificial aggregate.

Parameters	Standard methods
Loose bulk density	EN 1097-3, ASTM C29, GB/T 17431.2-2010, CNS 1163
Saturated-surface-dry particle density	EN 1097-6, ASTM C127, GB/T17431.2-2010
Water absorption	EN 1097-6, ASTM C127, GB/T 17431.2, CNS 488
Compressive strength	EN 13055, GB/T 17431.2-2010, CNS 14779

3.1.1 Agglomeration granulation

Agglomeration granulation, also known as disc pelletization, is a broadly used granulation method in producing sintered aggregate. This process involves wetting fine particles and allowing them to form cohesive pellets by collision and bonding while rolling in the pelletizer disc (Tajra et al., 2019), as illustrated in Fig. 18.1. Wetting of raw materials results in forming a thin liquid membrane around the surface of each particle. When the wetted particles driven by gravitational and centrifugal forces come into contact, liquid bridges form at the contact point. The surface tension of the liquid bridges causes the particles to adhere to each other, leading to the gradual development of fresh particles (Nadesan and Dinakar, 2017).

Generally, the agglomeration granulation process consists of several steps, including dry-mixing, wetting, and pelletizing, as shown in Fig. 18.1C. The granulation efficiencies and mechanical properties of fresh pellets are influenced by various factors, including water content, the particle size of precursors, and operating parameters of the pelletizer (tilt angle, rotation speed and duration) in the granulation process (Ren et al., 2021). These parameters are crucial in determining the size, strength, and uniformity of the granules produced during the agglomeration process. Optimization of these factors is essential to achieve desired granulation efficiencies and obtain high-quality artificial lightweight aggregates suitable for concrete applications.

Water content plays a significant role during the pelletization process as it acts as a bridging medium. Insufficient moisture can result in small-sized pellets and significant amounts of debris, while excessive water content can lead to large agglomerates or even slurry formation. An appropriate water content of about 20–28 wt% of the total mixture is necessary to ensure that the fresh pellets with desirable spherical shapes and diameter (Chen et al., 2010; Quina et al., 2014b; Ferraro et al., 2023). Concerning particle size, high fineness of raw materials is preferred for granulation, as they typically yield

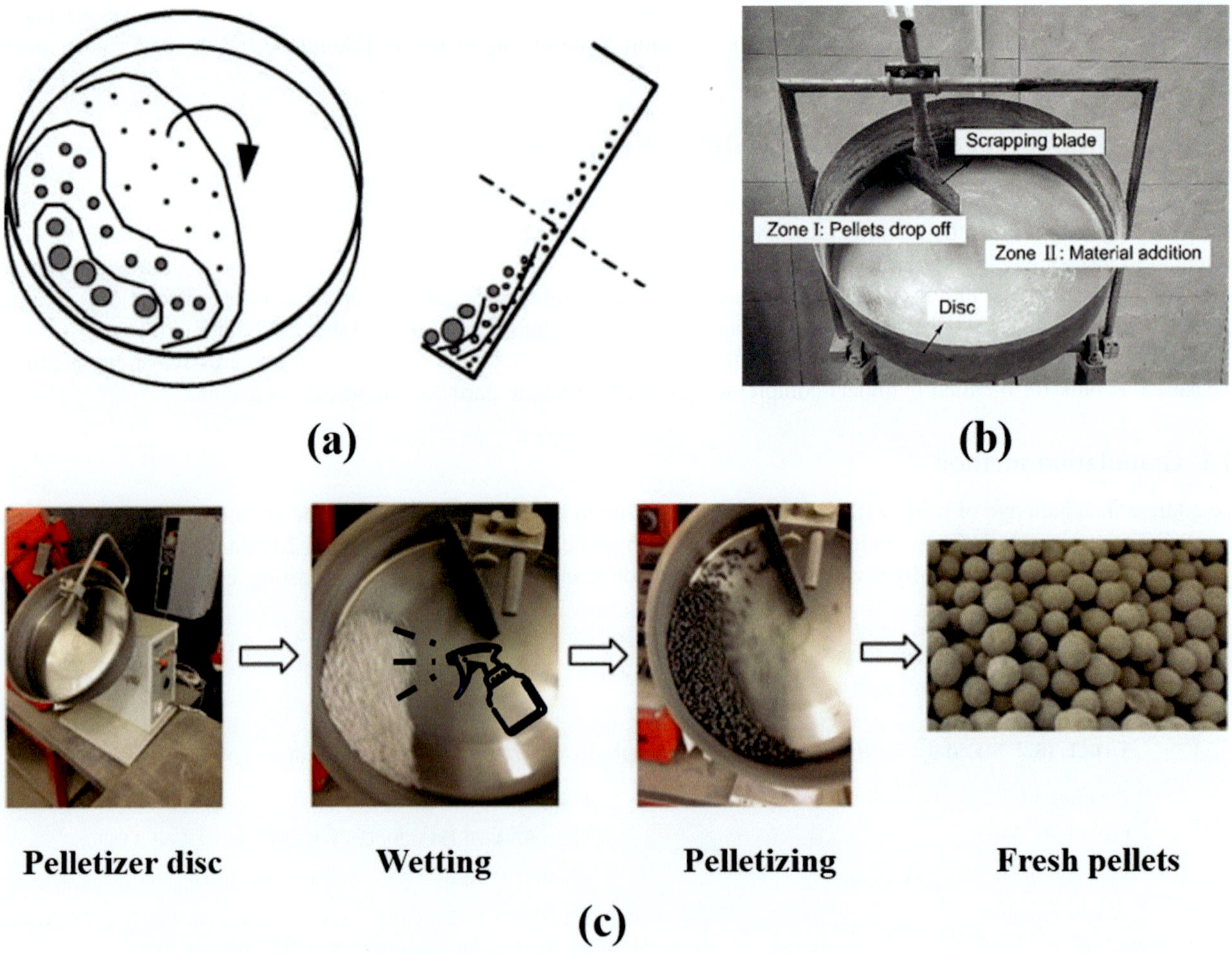

FIGURE 18.1 Agglomeration method: (A) Schematic (Shi et al., 2019), (B) the pelletizer disc (Dong et al., 2022), (C) flow diagram of fresh pellets production (Tajra et al., 2018).

stronger fresh pellets. The higher specific surface area of finer particles provides a greater driving force for achieving sintering at the point contact sites during the process (Chen et al., 2010). The particle size of MSWI FA smaller than 73 μm has been reported for producing sintered cylindrical specimens with sufficient compressive strength greater than 52 MPa (Wang et al., 2002).

The properties of the aggregate can be influenced by the particle sizes and moisture contents used in the granulation process, but the granulation efficiency also relies on the mechanical parameters of the pelletizer. The operating parameters employed in the literature, including tilt angle, rotation speed, disc diameter, and duration, are shown in Table 18.3. Previous studies have shown that small tilt angles and low rates are unfavorable for pellet formation due to reduced rolling paths and the collision chance between particles (Qian et al., 2022b). Conversely, the increase in tilt angle rotation speed, or granulation duration can contribute to the prolonging of the rolling path and higher tumbling forces, thus forming aggregates with a dense structure, low water absorption, and high strength (Vasugi and Ramamurthy, 2014; Tian et al., 2021). Besides, previous studies have explored certain relationships between the disc diameter, tilt angle, and critical speed (N_c) of the pelletizer (Shi et al., 2019). The critical speed (N_c) refers to the rate at which the raw materials roll steadily in the rotating disc without splashing and is determined by Eq. (18.1). When the diameter and tilt angle are fixed, this equation can provide a reference for rotation speed. Shi et al. (2019) concluded that the optimum inclination angle was around 40 degrees and 45 degrees, with corresponding rotation speeds of 31.0–43.3 rpm and 32.5–45.4 rpm. Regarding granulation time, a sufficient rotation duration is necessary to ensure adequate growth of pellets. Previous studies have shown that a minimum duration of 5–8 min is required for the formation of spherical pellets. Additionally, a duration of 10 min is sufficient for fresh pellets to acquire suitable initial strength (the pellets do not break when dropping from a certain height) (Harikrishnan and Ramamurthy, 2006). Azrar et al. (2016) also reported that a rotation time of 10 min is optimal after a 2-min rotation of water addition.

$$N_c = \frac{42.3\sqrt{\sin\beta}}{\sqrt{D}} \tag{18.1}$$

where N_c is the critical speed, β is the inclination angle, and D is the diameter of the disc.

Recently, an innovative core–shell structure has been proposed to improve the mechanical properties of artificial aggregates and reduce the leachabilities of PTEs (Fig. 18.2). The core structure usually consists of a mixture of hazardous waste powder and binders and is coated by a dense outer shell formed by a pure binder, a thermoplastic material, etc. (Ren et al., 2021; Ye et al., 2022). Two types of plastic powders (polypropylene and linear low-density polyethylene) were melted at temperatures ranging from 180 to 200°C to achieve sufficient fluidity and were then coated on the surface of the artificial aggregates (Ye et al., 2022). The results demonstrated that coating the aggregates with plastic materials could form a core–shell structure, resulting in improved mechanical properties and enhanced protection against water damage for the aggregates. A similar core–shell structure was reported by Li et al. (2020), where the outer shell layer was made of coal fly ash, impurities-removed soil, and sodium borate (Fig. 18.3). After sintering, the densified shell effectively encapsulated the chromium pollutant in the core and prevented its leaching. These core–shell structures offer promising strategies to enhance the performance of artificial aggregates by providing robust mechanical performances and durability while mitigating the environmental risks associated with PTEs.

TABLE 18.3 Mechanical parameters used in agglomeration granulation.

Materials	Angle (°)	Speed (rpm)	Duration (min)	Diameter (cm)	Type	References
RM, FA	40	30–55	15	–	Disc	Tian et al. (2021)
FA	45	10–30	9	50	Disc	Qian et al. (2022b)
MIBA, FA, GGBFS	45	35	–	100	Disc	Liu et al. (2022)
ISSA, CSW	45	36	10	50	Disc	Tang et al. (2020b)
QT, FA	45	30	–	–	Disc	Wang et al. (2022a)
MSWI FA, GGBFS	45	45	–	80	Disc	Ferraro et al. (2023)

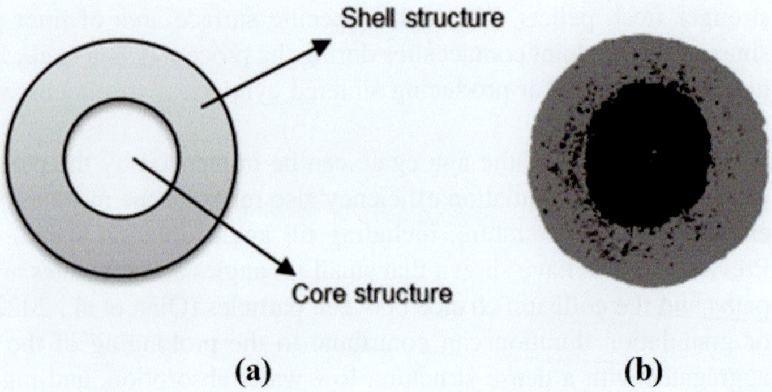

FIGURE 18.2 (A) Schematic diagram of core–shell structure, (B) micro-CT imaging of a sample aggregate (Tajra et al., 2018).

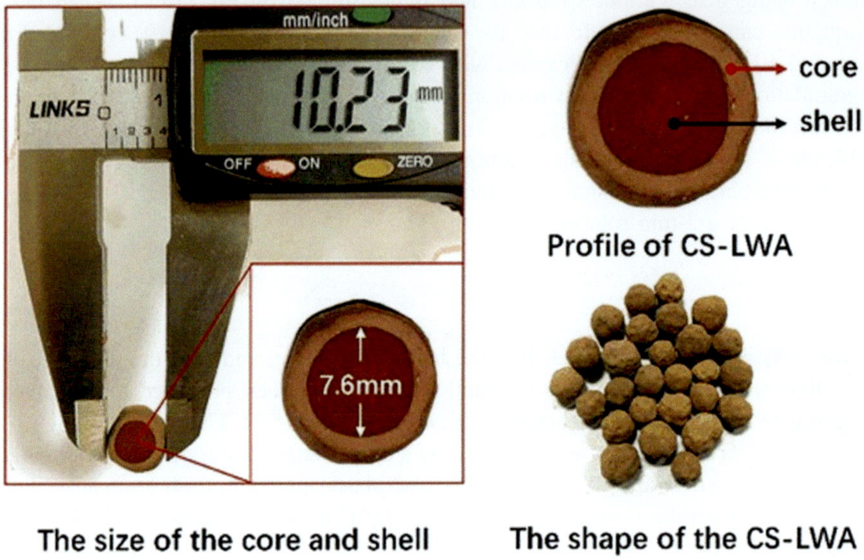

FIGURE 18.3 The appearance and core–shell structure of sintered aggregates produced by Li et al. (2020).

3.1.2 Crushing method

The crushing process, commonly employed in producing natural coarse aggregates, has recently been investigated in mainly cold-bonded artificial aggregates production (Shahane and Patel, 2021). Due to its compatibility with existing natural aggregate crushing production systems, the crushing method is particularly suitable for large-scale industrial production. In the crushing process, the powdered raw materials are generally transformed into artificial aggregates through five steps, including mixing, pouring, demolding, crushing, and curing. Crushing can be performed using a compressive loading machine (at laboratory scale) either before or after the curing process. Frankovič et al. (2017) concluded that the aggregates produced by crushing have higher density and strength compared with granulated ones. Besides, the polygonally shaped aggregates obtained from the crushing method can improve the interlocking effect with mortar in the concrete. Compared with crushing after curing, crushing before curing can be advantageous in terms of reducing production energy consumption due to the relatively low strength of pastes. Additionally, the minimum limits of early strength of above 1.0 MPa are also required to ensure the integrity of the crushed aggregates (Xu et al., 2021).

3.1.3 Mold casting method

The mold casting method enables the manufacture of aggregates with desired particle size by adjusting the mold, thereby addressing the issue of material loss during disk pelletization or crushing process (Fig. 18.4). In addition, the polycondensation reaction of slurry also ensures that the synthesized aggregates exhibit a spherical shape, making this method suitable for laboratory-scale research but may not for industrial production (Peyne et al., 2018; Qian et al., 2022a).

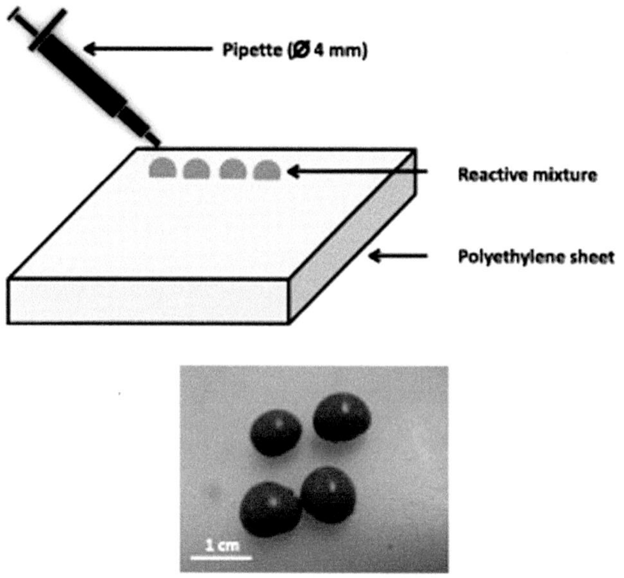

FIGURE 18.4 Schematic diagram of the mold casting method (Peyne et al., 2018).

In conclusion, there is no substantial difference between the different methods. However, it is important to note that the water demand during the aggregate production may vary slightly depending on the manufacturing technique. As mentioned earlier, the moisture content of aggregates produced through the disc pelletization process is generally controlled at 20%−28%. On the other hand, the moisture content for the crushing method typically needs to be 30%−50% to ensure sufficient hydration of the cementitious material (Frankovič et al., 2017; Xu et al., 2021). Agglomeration granulation and crushing methods are more suitable for large-scale production, while the mold casting method is more suitable for laboratory-scale research due to the lower production efficiency.

3.2 Sintering technology

Sintering technology is a thermal treatment method used to recycle MSWI FA for producing sintered aggregates. It involves heating the fresh pellets to a specific temperature below the melting point of the mixture. During sintering, fine particles in the matrix bond together, resulting in a dense structure with improved mechanical properties (Polettini et al., 2004). Melting/vitrification involves high-temperature melting and quenching to form a homogenous amorphous glass phase. Compared with sintering, melting/vitrification requires higher operating temperatures, typically 1250−1500°C (Peng et al., 2020). However, the high energy consumption associated with melting/vitrification limits its applicability in producing high value-added solid waste−derived construction materials, although it is commonly used for producing glass ceramics (Huang et al., 2022). Sintered lightweight aggregates, also known as ceramsites, have been widely studied and used as a primary construction material for lightweight concrete and envelope materials due to their low density, high strength, high porosity, high sound insulation, low thermal conductivity, and relatively low treatment temperatures (Dondi et al., 2016; Zhao et al., 2022). This section mainly discusses the latest progress on recycling MSWI FA for producing sintered aggregates. Table 18.4 summarizes the sintering parameters and mechanical properties of sintered aggregates reported in the literature.

3.2.1 Washing pretreatment before sintering

Water-washing pretreatment is an effective step in recycling MSWI FA as sintered lightweight aggregates. This pretreatment helps eliminate soluble salts (mainly chlorides), which prevents the second generation of PCDD/Fs in the sintering flue gas and avoids pipe corrosion (Huang et al., 2023). A washing time of at least 10 min is typically required to ensure the dissolution of high quantities of soluble elements, such as K^+, Na^+, Cl^-, and SO_4^{2-}. The addition of chemicals, such as phosphoric acid or phosphate, during the washing process can enhance the treatment efficiency by facilitating the immobilization of PTEs (Quina et al., 2014a; Wu et al., 2015). Adjusting the pH of the washing liquid through alkali-washing or acid-washing methods can produce different effects, such as increasing the leaching of metal ions or their precipitation (Chen et al., 2022a).

TABLE 18.4 Physical properties of sintered lightweight aggregates.

Raw materials	Dosage of MSWI FA	Sintering parameters		Mechanical properties						References
		Sintering temperature (°C)	Duration (min)	Loose bulk density (kg/m^3)	Particle density (g/cm^3)	Water absorption (%)	Bloating index (%)	Single particle compressive strength (MPa)	Cylinder compressive strength (MPa)	
MSWI FA reservoir sediments	10–30 wt%	1150–1175	10–15		0.94–1.86	0.30–10.10 (24 h)	−7.0–69.0			Chen et al. (2010)
MSWI FA, MSWI BA, reservoir sediment	5 wt%	1050	15	1600–1800		15.00 (24 h)	0–5.0	57.0		Chuang et al. (2018)
MSWI FA, bentonite, SiC	30 wt%	1120	30	211		5.00–10.00 (1 h)		0.8		Han et al. (2022)
MSWI FA, reservoir sediment	30 wt%	1120	20		1.48	8.40 (24 h)		13.4		Hwang et al. (2012)
Water-washed MSWI FA, clay	5 wt%	1100	8	178	0.50					Quina et al. (2014a)
MSWI FA, civil sludge, contaminated soil, flint clay	30 wt%	1150	20	998.7		0.97 (1 h)			37.8	Shao et al. (2022)
MSWI FA, washed MSWI FA, coal fly ash,	41 wt%	1160	12	792		8.12 (1 h)			5.9	Zhan et al. (2021)
MSWI FA, reservoir sediment	5–15 wt%	1100	15	1400–1500		1.58–6.30 (24 h)	3.1–3.4	2.7–9.9		Lu et al. (2015)
MSWI FA, sewage sludge	20 wt%	1100	8	580		16.50			5.4	Wei (2015)

3.2.2 Factors affecting the properties of sintered aggregates

3.2.2.1 Effect of raw materials

The properties of sintered aggregates are influenced by various factors, including the raw materials used in their production. The chemical composition of the raw materials plays a crucial role in determining the expansion properties of lightweight aggregates. The appropriate ratio of SiO_2, Al_2O_3, and fluxing agents (e.g., CaO, MgO, Fe_2O_3, FeO, K_2O, and Na_2O) is necessary to form a viscous fused phase at high temperatures (Riley, 1951). Silica and alumina can be transformed into silica−aluminate polymers with high viscosity at high temperatures to form the skeleton structure of sintered aggregates. However, both the melting point of alumina (2043°C) and silica (1715°C) are too high to meet the requirements of energy-saving production (Peng et al., 2020). As a result, the flux content significantly reduces the sintering temperature and accelerates the formation of the glassy phase via the liquid-state sintering mechanism (Wei et al., 2018). The second significant factor affecting the properties of sintered aggregates is the presence of pore-forming agents (Riley, 1951). These agents are responsible for generating gas during the sintering process, resulting in the formation of pores within the aggregates (Lee et al., 2019). Various substances can serve as pore-forming agents, including water, carbonates, organic compounds (such as sawdust), Fe_2O_3, sulfides, etc. (Quina et al., 2014b; Moreno-Maroto et al., 2019). The selection and incorporation of suitable pore-forming agents play a vital role in controlling the porosity and physical properties of the sintered aggregates. Recent studies have explored the use of alternative materials, such as dewatered lake sediment and dewatered algal sludge rich in organic matter, as pore-forming agents (Zhao et al., 2022). However, the dosage of sludge needs to be controlled strictly to ensure optimal performance.

Commercial sintered aggregates are mainly produced using clay, shale, or perlite (Dondi et al., 2016). With the increased adoption of thermal treatment technologies in the realm of solid waste management and recycling, various solid wastes have emerged as viable alternatives to natural materials in the preparation of sintered lightweight aggregates (Han et al., 2022). These waste materials are evaluated based on their chemical components, specifically SiO_2, Al_2O_3, and fluxing agents. A ternary diagram is commonly employed to normalize and plot the chemical components of these waste materials and compare them to the optimal expansion range proposed by Riley (1951) (see Fig. 18.5).

Fig. 18.5B reveals that the new precursors derived from various solid wastes, in particular MSWI FA, exhibit significant differences in their chemical composition compared with clays traditionally used to produce expanded clay aggregates. This variation in physiochemical properties necessitates a reevaluation of the proportion of raw materials and sintering parameters of solid waste−based lightweight aggregates compared with conventional raw materials. Consequently, Riley's theory of optimum expansion may be less reliable when designing solid waste−derived aggregates. Nonetheless, formulation design based on chemical composition remains popular (Dondi et al., 2016).

MSWI FA is rich in flux components such as CaO, Na_2O, and K_2O, making it suitable for sintering processes. An increase of MSWI FA proportion in raw materials would magnify the fluxing effect, allowing more molten components to be produced and provide a dense glaze layer for the surface of ceramsite (Polettini et al., 2004). Moreover, the sintering process has the benefits of decomposing refractory organic pollutants and immobilizing PTEs (Zhan et al., 2021), further highlighting the potential for recycling MSWI FA for producing sintered aggregates. However, the relatively low content of SiO_2 and Al_2O_3 in MSWI FA poses challenges. Excessive dosage of MSWI FA in the mixing proportion can result in inadequate production of the viscous phase at high temperatures, impeding the formation of the skeleton structure and leading to reduced strength. Additionally, interfering substances, such as chlorides and PTEs, restrict he extensive recycling of MSWI FA (Polettini et al., 2004). As a result, the mixing ratio of MSWI FA is typically kept below 25 wt% in the raw materials (Zhan et al., 2021).

To ensure that suitable expanding properties are obtained, the $(SiO_2+Al_2O_3)$/fluxing ratio should be within the range of 3.5 and 10.0 (Chen et al., 2010). Higher ratios require higher sintering temperatures, while lower ratios result in a lower melting temperature and reduced melt phase viscosity. In addition, the molar ratio of CaO/SiO_2 should also be controlled, as excess CaO can destroy the tetrahedron structure of $[SiO_4]$ in the aggregate skeleton, decreasing the compressive strength (Zhang et al., 2020). Therefore, supplementing Si, Al-containing chemicals or wastes, and gas-forming agents is indispensable for balancing element deficiency of MSWI FA. Cosintering with different additives, such as sludge, sediments, and coal fly ash, has been widely investigated to obtain the optimal SiO_2-Al_2O_3-fluxes ratio for improved mechanical properties of aggregates. Chen et al. (2010) successfully produced sintered lightweight aggregates using MSWI FA and reservoir sediments at 1150−1175°C sintering temperatures. The 24-h water absorption of the obtained lightweight aggregates ranged from 0.5%-10.1%. To alleviate the degradation of swelling properties caused by partial replacement of MSWI FA, SiC was added to the MSWI FA-bentonite system as a bloating agent; the results showed that the bulk density of lightweight aggregates decreased to below 500 kg/m^3 at 1120°C with 0.1−0.5 wt% SiC addition, which meets the requirement of superlightweight aggregates (Han et al., 2022). Hwang et al. (2012) successfully produced sintered

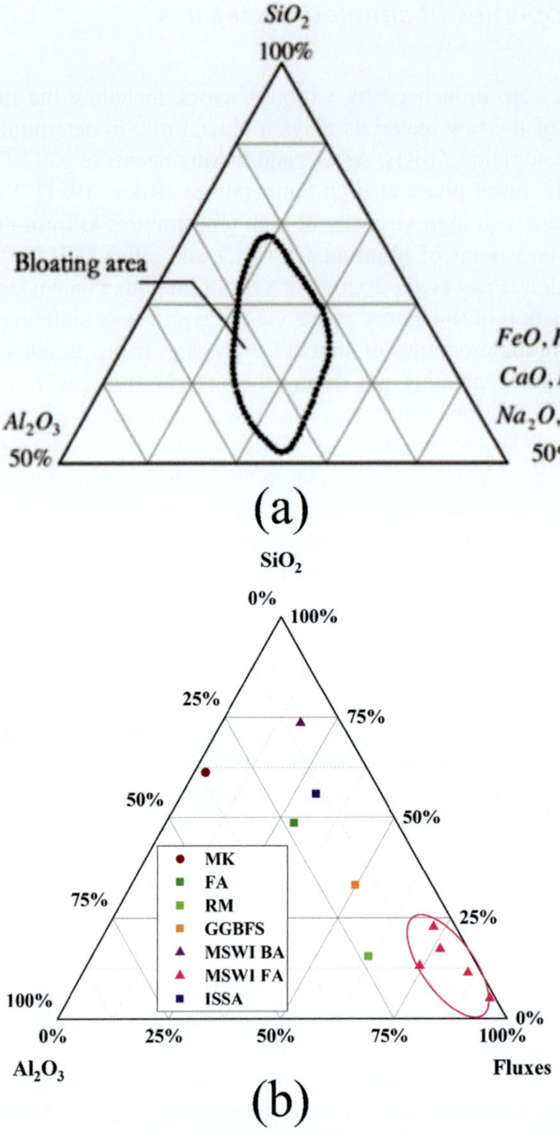

FIGURE 18.5 Ternary diagram of (A) the optimal expansion range proposed by Riley (1951), Ren et al. (2021) and (B) chemical components of solid wastes (Colangelo et al., 2015; Kiventerä et al., 2018; Chen et al., 2019a, 2019b; Shi et al., 2019; Bui Viet et al., 2020; Tang et al., 2020b; Dong et al., 2022; Liu et al., 2022). *FA*, coal fly ash; *GGBFS*, ground blast furnace slag; *ISSA*, incineration sewage sludge ash; *MK*, metakaolin; *RM*, red mud.

aggregates composed of 30 wt% MSWI FA and 70 wt% reservoir sediment with a sintering temperature of 1120°C. The resulting aggregate exhibited a cylinder compressive strength of 13.42 MPa and water absorption of 8.4%, demonstrating the promising potential for practical applications.

3.2.2.2 Effect of temperature

To ensure the quality of sintered aggregates and prevent the aggregate from cracking, exploding, and black cores, the sintering process typically involves three steps: oven-drying, presintering, and sintering (González-Corrochano et al., 2014; Ren et al., 2021). As seen in Table 18.3, presintering and sintering temperatures typically range from 400 to 500°C and 1050 to 1150°C, respectively. The optimal sintering temperature ensures that gas generation and sufficient viscosity occur simultaneously, thus preventing under- or overexpansion or melting. Chen et al. (2010) sintered lightweight aggregate using MSWI FA and reservoir sediment as raw materials and measured the initial sintering temperature and softening temperature of the aggregates in the temperature range of 1100−1200°C. They found that all the experimental groups had a narrow sintering temperature range of only 25−50°C, and softening occurred when the temperature exceeded this range.

However, the high sintering temperature poses a significant challenge due to substantial energy consumption. Although cosintering with other solid wastes can improve the mechanical properties, a high sintering temperature is still required to ensure mass transfer processes and chemical elements diffusion (Yang et al., 2021). Great efforts have been made to reduce the sintering temperature without damaging the mechanical properties of sintered aggregates. Wei et al. (2018) explored the use of CaF_2 in reducing energy consumption during sintered aggregate production. Partially replacing raw materials with 10%−20% CaF_2 can successfully reduce the sintering temperature by 200°C. A relatively low sintering temperature of 300−500°C was achieved through the application of the pressure-assisted sintering technique (Wang et al., 2022b), which enhanced the mass transfer and densification process during sintering, improving the mechanical properties and immobilizations of PTEs.

3.2.3 Environmental impacts during sintering process

During the sintering process of MSWI FA, environmental impacts need to be considered, particularly regarding the formation of dioxins and the emission of volatile metals. The chlorides in MSWI FA may cause the secondary formation of dioxins in the flue gas at temperatures of 200−400°C. Additionally, volatile metals, such as Cd, Zn, and Pb, may be emitted into the flue gas during high-temperature sintering (Cheng and Chen, 2003). To mitigate secondary atmospheric contamination, it is necessary to appraise and analyze the pollutant emissions during the sintering process.

Peng et al. (2020) investigated an industrial-scale fly ash sintering disposal facility equipped with a flue gas control system in China to study the migration, emission, and distribution characteristics of pollutants, including PCDD/Fs and PTEs, during the sintering process of MSWI FA (as shown in Fig. 18.6). The flue gas control system mainly consisted of a deacidification tower, where the flue gas can react with sprayed semidry lime slurry to remove the acid gas such as SO_3 and HCl and generate a mixture of different calcium salts such as $CaSO_4$. The addition of activated carbon can effectively adsorb toxic pollutants such as PCDD/Fs and Hg, which were captured by a bag filter and deacidification products, heavy metal chlorides, etc., recognized as secondary fly ash. The concentrations of all pollutants, including PTEs and PCDD/Fs, in the flue gas were below the national standard limits in China, indicating the effective purification by the flue gas control system. However, most PCDD/Fs were only desorbed from the MSWI FA during sintering and reenriched in the secondary fly ash, with only 8.9% thermally decomposed. Similarly, around 70% of Cl transformed into the gas phase during the thermal treatment and was captured by lime and collected in a bag filter, while around 30% of Cl was fixed in the sintering products. After high-temperature sintering, PTEs primarily migrated to the secondary fly ash and sintering products, with the heavy metal contents in the flue gas accounting for less than 0.5% of the total. Notably, a large amount of volatile PTEs, such as Hg, Cd, Cu, Pb, and Zn, were enriched in the secondary fly ash and presented high leachabilities, requiring careful attention during further disposal of the secondary fly ash.

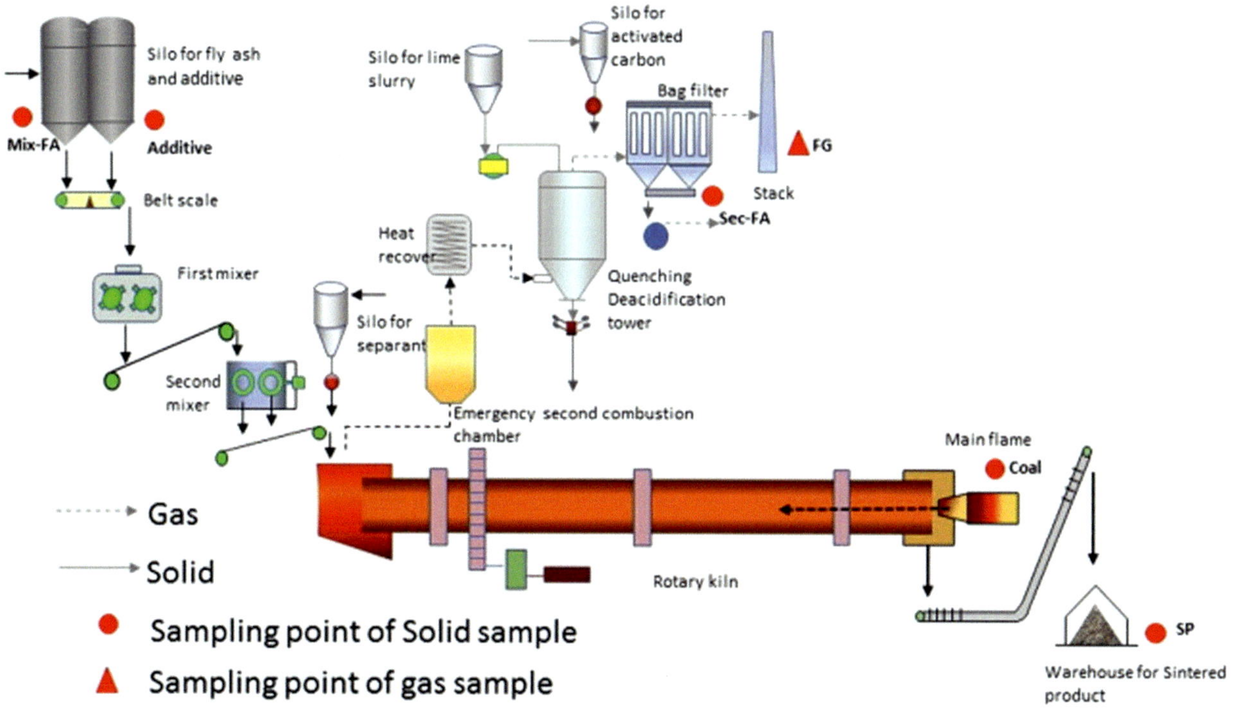

FIGURE 18.6 Schematic of the MSWI fly ash sintering facility (Peng et al., 2020).

3.3 Cold-bonding technology

Cold-bonding is an eco-friendly and low-carbon approach that has been extensively studied for artificial aggregate production. Cold-bonded aggregates achieve strength increase and immobilization of PTEs via hydration or geopolymerization reactions, which are highly reliant on the chemical composition and reactivity of the raw materials (Qian et al., 2022a). Solid wastes containing a moderate amount of reactive CaO, SiO_2, and Al_2O_3 or possessing potential cementitious properties can be applied as main binders or SCMs in cold-bonded aggregates. Recently, challenges for cold-bonding technology are still the low reactivity of raw materials and the compressive strength development of products. To address these challenges and effectively convert the combustion residues into value-added products, various methods have been raised to improve the physicochemical properties of cold-bonded aggregates. These methods can be categorized into three strategies, including pretreatment, mixing with the binder, and activating the reactivity of raw materials through different curing strategies (accelerated carbonation). Table 18.5 summarizes the physicochemical properties of the cold-bonded aggregates manufactured in previous studies.

3.3.1 Pretreatment in cold-bonding method

Pretreatment can be used in cold-bonded aggregate production to activate the pozzolanic reactivity of MSWI FA and remove toxic substances, thereby improving the physicochemical properties of MSWI FA and reducing the environmental risks associated with the final products. The pretreatment methods mainly include low-temperature thermal treatment, alkaline washing, and mechanochemical treatment, among others.

3.3.1.1 Low-temperature thermal treatment (pyrolysis)

Compared with the sintering process, cold-bonding technology does not contain thermal treatment capable of decomposing dioxins (PCDD/Fs). Therefore, dioxins can only be diluted and encapsulated within the cold-bonded matrix. Although alkali activation treatment of MSWI FA has shown some positive effects on the decomposition of PCDD/Fs, the removal efficiency remains relatively low, at only 64.8% (Liu et al., 2021). Therefore, PCDD/Fs remain a great concern for applying MSWI FA in cold-bonded aggregates. High-temperature calcination between 700 and 1000°C can decompose the toxic organic matters and enhance the reactivity of MSWI FA (Huang et al., 2020). However, the high energy consumption limits its application. Low-temperature pyrolysis provides an energy-efficient and effective pretreatment solution for PCDD/Fs removal. Song et al. (2008) showed that the total destruction efficiency of PCDD/Fs was above 95% at the treatment temperature of 450°C for 3 h under a nitrogen atmosphere.

3.3.1.2 Alkaline washing

The use of alkaline washing with various alkaline solutions, such as NaOH, Na_2CO_3, and $NaHCO_3$, is preferred over pure water washing in cold-bonding process (Chen et al., 2023). Alkaline washing not only effectively removes chlorine but also facilitates the transformation of Ca^{2+} from soluble $CaCl_2$ into insoluble $Ca(OH)_2$, thereby promoting the calcium retention. Chen et al. (2022b) achieved a chlorine removal rate of 89% and nearly 100% calcium retention under the condition of an L/S ratio of 6 mL/g and 0.20 g NaOH/g ash. Aluminum waste feed, incineration conditions, and denitrification processes in MSW incineration plants may result in the presence of trace amounts of metallic aluminum and residual ammonia in MSWI FA. These substances can react in strong alkaline environment such as alkali activation system to release a large volume of gas (H_2 and NH_3, respectively), causing the product to expand and even crack. Alkaline washing effectively consumes the aluminum and promotes ammonia release in the pretreatment stage, preventing undesirable expansion of the matrix during curing (Chen et al., 2023).

3.3.1.3 Mechanochemical treatment

Mechanochemical treatment induces solid-state chemical reactions between treated materials and additives through high-intensity mechanical forces (e.g., collision, friction, etc.) (Zhang et al., 2021). This treatment triggers the formation of crystal defects, transforming the crystal structure from crystalline to amorphous, thereby enhancing the reactivity of solid particles and reducing their particle size (Chen et al., 2019b; Jin et al., 2023). Mechanochemical treatment has been demonstrated to have positive on the degradation of PCDD/Fs and stabilization of PTEs in MSWI fly ash (Yan et al., 2007; Lu et al., 2012; Li et al., 2017; Chen et al., 2019d). For example, the water- and acid-soluble fractions with high mobility of PTEs (including Cd, Cr, Cu, Ni, Pb, and Zn) were significantly reduced via mechanochemical treatment with $Ca_3(PO_4)_2$ (Chen et al., 2019c). Research on the degradation of PCDD/Fs through mechanochemical treatment has also revealed that different additives have similar detoxification mechanisms, which attack the C–Cl bond via the electron transfer action of the additives, leading to the release of Cl (Chen et al., 2018, 2019d).

TABLE 18.5 Physical properties of cold-bonded aggregates.

					Mechanical properties			
Raw materials	Binder/Activator	Pretreatment	Content of MSWI FA	Curing method	Particle density (g/cm^3)	Water absorption (%)	Bulk compressive strength (MPa)	References
MSWI FA, GGBFS	OPC	Washing	70–80 wt%	Wet curing	1.54-–2.29	10.68–14.70 (80%)		Colangelo et al. (2015)
MSWI FA, GGBFS	OPC	Washing	80 wt%	20°C	1.63–1.75	9.99–21.63 (24 h)	10.92–10.94	Farina et al. (2022)
MSWI FA	OPC, $Ca(OH)_2$, coal fly ash	Washing	50–70 wt%	50°C	1.00–1.60	7.00–16.00 (24 h)	1.30–6.20	Ferraro et al. (2023)
MSWI FA, coal fly ash	Binder: OPC Activator: NaOH, water glass	Washing	35 wt%	150°C for 4 h.	1.13	18.50 (1 h)	5.50	Wang et al. (2023)

3.3.2 Effect of raw materials and binders

3.3.2.1 Cement-based system

In the cold-bonding method, the use of binders (such as cement and lime) in combination with mono- or multisolid wastes is a widely adopted approach to attain matrix consolidation, strength enhancement, and PTEs immobilization. Ordinary Portland cement (OPC) is commonly used in cold-bonded aggregate production due to its cost-effectiveness and favorable performance for engineering applications (Zhang et al., 2021). MSWI FA has a suitable chemical composition, which can be used as an SCM to partially substitute cement. For instance, OPC and MSWI FA were used to produce cold-bonded lightweight aggregate with water adsorption of 7%−16% and crushing strength of 1.3−6.2 MPa, and a core−shell structure allows the proportion of MSWI FA in cement-based aggregates up to 70 wt% (Colangelo et al., 2015).

While MSWI FA can be used as a kind of SCMs, it is unsuitable to be added alone or untreated to the cement-based system. Previous studies have shown that the compressive strength of cement-based artificial aggregates decreased with the increasing content of MSWI FA (Bie et al., 2016). Additionally, the separate use of OPC and MSWI FA has shown low compatibility (Chen et al., 2019a). To overcome these challenges, ground waste glass (GWG) was used as a tertiary SCM to improve the chemical composition of washed carbonated MSWI FA as well as to mitigate the negative effects of MSWI FA on cementitious mixtures (Bui Viet et al., 2020). The results showed that the synergistic combination of GWG and MSWI FA improved the performance of mortars. Fig. 18.7 shows the aggregates resulting from unwashed and washed MSWI FA. Almost all the aggregates made from unwashed MSWI FA exhibited a widely cracked surface, and washing pretreatment can effectively improve the physical properties of MSWI FA-based aggregates.

To further reduce resource consumption and the corresponding CO_2 emissions related to cement utilization, some researchers have begun to explore the use of cementitious wastes as alternative binders in artificial aggregates. Fresh concrete slurry waste (CSW) is a cementitious waste consisting of two main components, including hydration products (e.g., C−S−H gel, ettringite, calcium hydroxide, etc.) and the residue unhydrated clinkers. These compositional

FIGURE 18.7 Cold-bonded aggregates manufactured from (A) washed MSWI FA and (B) unwashed MSWI FA (Colangelo et al., 2015).

characteristics allow it to have a certain residual hydration capacity with sufficient alkalinity for simultaneous hydration and pozzolanic reactions with potential pozzolanic materials. Increased amounts of CSW resulted in an enhancement in pellet strength (Tang et al., 2020b).

3.3.2.2 Alkali activation system

Alkali activation is a low-carbon cement-free cold-bonding technology (Duxson et al., 2006; Qian et al., 2022a). Alkali-activated aggregates gain strength through the geopolymeric reaction of aluminosilicate precursors and alkali activators (Zhang et al., 2021; Ye et al., 2022).

3.3.2.2.1 Alkaline activators During the alkali activation reaction, the alkaline solution in the internal pores can dissolve the aluminosilicates in the precursors, thereby releasing [SiO_4] and [AlO_4] tetrahedral units into pore solution (Kiventerä et al., 2018). Then polymerization occurs, wherein adjacent ions are linked together by bridging oxygen to form a new amorphous three-dimensional network structure. Sodium hydroxide (NaOH) and sodium silicate (Na_2SiO_3) or their combinations are the conventional alkali activators due to their availability and cost-effectiveness (Qian et al., 2022a). Cristelo et al. (2020) investigated the activation efficiency of pastes based on MSWI FA and MSWI BA using sodium silicate or sodium hydroxide solution showing that the low-cost alkali-activated S/S method competes with up to 30% cement usage. Potassium silicate (K_2SiO_3) and potassium hydroxide (KOH) combination solution was also used as an activator for the alkali activation method in a previous study (Yliniemi et al., 2016).

3.3.2.2.2 One-part mixing versus two-part mixing The mixing procedure markedly impacts the mechanical properties of fresh and hardened alkali-activated aggregates (Alrefaei et al., 2021). Two common approaches in the production of alkali-activated aggregates are “one-part mixing” and “two-part mixing,” which differ in introducing the alkali activators into the matrix. In the one-part mixing method, the precursors and alkali activator particles are dry-mixed to form a uniform mixture, similar to Portland cement clinker in cast-in situ applications (Ma et al., 2018). Subsequently, the mixed powders are added to the rotating disc, accompanied by a continuous wetting effect with water spray. In contrast, the two-part mixing method involves dissolving the alkali activator particles into the water to form an alkaline solution, which would be sprayed onto the precursors during the disc rotation (Balapour et al., 2020).

The one-part mixing method has been reported to generate a significant amount of heat during the initial dissolution of solid activators, resulting in the faster setting and hardening of geopolymers (Ma et al., 2018). However, clustering of precursors with high pozzolanic reactivity may occur later in the granulation process due to the gradual heat release during the dissolution of solid activators, leading to accelerated stickiness development. In contrast, the two-part mixing process avoids clustering due to the advanced heat release, and the higher stickiness of alkaline solution improved the pelletization efficiency at the early stage (Qian et al., 2022b). However, high viscosity and corrosiveness are also the main disadvantages of the two-part mixing method, which increase the safety requirement for the mass production (Ma et al., 2018). The alkaline solution with 8% Na_2O concentration was also too viscous to be sprinkled by the sprayer (Qian et al., 2022b). Furthermore, research has reported the chemical failure of commercial admixtures, such as superplasticizers, in highly alkaline media, underscoring the poor compatibility of the two-part mixing method with commercial addictive mixtures (Palacios and Puertas, 2005).

3.3.2.2.3 Effect of precursors Previous studies have shown that various by-products or waste materials containing aluminosilicate minerals can be utilized as precursors for alkali-activated aggregates (Qian et al., 2022a). However, considering that the chemical composition of solid waste precursors derived from different regions and industrial fields varies considerably, the multivariate compounds of different precursors are generally adopted to obtain the desired mechanical properties (Xiao et al., 2020). The mechanical properties of the geopolymer matrix are highly related to the reaction conditions and the properties of the raw materials. The Si/Al molar ratio of the precursor mixtures should be <3 to produce a rigid three-dimensional geopolymer network, and the highest strength for the geopolymer matrix was obtained with a Si/Al ratio of 1.5−2.5 and a Na/Al ratio of 1.0−1.29 (Kiventerä et al., 2018). Besides, the activation effect of the alkali activation method is more prone to act on the amorphous phase instead of the crystal phase in the precursors. For instance, the addition of 20% GGBS had a remarkable positive effect and increased the one-day strength from less than 3−29.8 MPa, owing to the high contents of amorphous phases in GGBS (Xu et al., 2021). For MSWI FA, the replacement of 10 wt% metakaolin in alkali-activated MSWI FA could significantly improve the 28-day and 90-day compressive strengths of the matrix, which was almost 200% higher than that of a single-component matrix of 100% MSWI FA (Liu et al., 2021). This enhancement is attributed to the additional active SiO_2 and Al_2O_3 components provided by metakaolin, which contribute to the formation of a denser C-A-S-H gel structure.

3.3.3 Accelerated carbonation method

The accelerated carbonation method has attracted widespread attention recently, which refers to a method accelerating the reaction between dissolved CO_2 and basic cations (Ca^{2+} and Mg^{2+}) by exposing the fresh aggregate to a concentrated and sometimes pressurized CO_2 atmosphere (Gunning et al., 2010; Sorrentino et al., 2022). During the carbonation process, calcium- or magnesium-bearing minerals are transformed into carbonate precipitates and amorphous phase, filling the pores and densifying the microstructure (Shah et al., 2018; Deng et al., 2022), therefore enhancing the properties of the cold-bonded aggregates. Besides, accelerated carbonation can also contribute to carbon capture and storage, resulting in green and sustainable artificial aggregates (Tang et al., 2020a,b).

Many solid waste materials, especially MSWI FA, contain sufficient CaO (35 ~ 45 wt%) that can react with dissolved CO_2. When these materials are cured under a CO_2-rich environment, the hydration products and raw materials, such as calcium hydroxide, calcium silicate hydrate (C−S−H), C_3S, C_2S, and ettringite, undergo carbonation reactions to form calcium carbonate and silica gel. This results in mass increase and densification, especially on the aggregate surface. Carbonation can only take place in the presence of moisture, as CO_2 gas needs first to hydrate to form carbonic acid. The reaction process of CO_2 curing is shown in Eqs. (18.2−18.6) (Gunning et al., 2010; Ghouleh et al., 2017; Deng et al., 2022; Hanifa et al., 2023).

$$\mathbf{CO_2+H_2O \rightarrow H_2CO_3} \tag{18.2}$$

$$\mathbf{Ca(OH)_2+H_2CO_3 \rightarrow CaCO_3 + 2H_2O} \tag{18.3}$$

$$\mathbf{C\text{-}S\text{-}H + CO_2 \rightarrow CaCO_3+SiO_2 + nH_2O} \tag{18.4}$$

$$\mathbf{C_3S + (3\text{-}x)\ CO_2 + y\ H_2O \rightarrow C_xSH_y + (3\text{-}x)\ CaCO_3} \tag{18.5}$$

$$\mathbf{C_2S + (2\text{-}x)\ CO_2\ + y\ H_2O \rightarrow C_xSH_y+(2\text{-}x)\ CaCO_3} \tag{18.6}$$

Accelerated carbonation significantly reduces the water absorption of aggregates, rapid strength increase, and densification of the aggregate surface. Tang et al. (2020b) compared the accelerated CO_2 curing (0.1 bar CO_2 at 50% relative humidity) with the steam curing (60°C) and found that the accelerated CO_2 curing increased loose bulk density (LBD) by 4%−9% and decreased the water absorption by 20%−24% compared with the steam-cured ones. However, carbonation primarily occurs on the surface, forming a dense calcite layer that limits the internal deep carbonation (Tang et al., 2020a). Therefore, improving the depth of carbonation of aggregates is currently the most important issue to address.

The carbonation of aggregates is affected by various factors, including aggregate chemical composition, additives, carbon dioxide concentration, pressure, relative humidity, and temperature (Wang et al., 2019; Gholizadeh-Vayghan et al., 2020; Steiner et al., 2020). Higher CO_2 concentration and pressure enhance CO_2 uptake, while water content is also crucial (Hanifa et al., 2023). Pressurized CO_2 curing allows for deeper carbonation of the pellets. Incorporating porous materials, such as biomass, into the aggregate matrix can improve the gas permeation and promote deeper carbonation. For example, adding small amounts of miscanthus powders (MP) was found to provide a gas pathway for deeper permeation of CO_2 into the compacted matrix through the intraparticle pores, benefiting aggregates' mechanical performance (Chen et al., 2021). Wang et al. (2019) have demonstrated that high temperatures, such as 100°C, significantly increase the CO_2 absorption rate compared with normal temperatures. Moreover, the periodic addition of water enhanced CO_2 absorption at high temperatures. Regarding additives, Jiang et al. (2022) proposed a magnesium-modified carbonation method in which the aggregates were treated with a magnesium nitrate solution before carbonation, resulting in improved microstructures compared with pure carbonation.

4. Immobilization mechanisms of PTEs from MSWI FA

MSWI FA commonly contains PTEs such as Zn, Pb, Hg, Cu, Cr, Cd, and Ni (Luo et al., 2019). When incorporating MSWI FA into construction materials, it is important to evaluate the leaching behavior and immobilization mechanisms of PTEs to ensure safety and prevent environmental contamination.

4.1 PTEs in MSWI FA

Generally, the chemical species of PTEs can be divided into six states, including the water-soluble state, the exchangeable state, the carbonate-bound state, the Fe−Mn oxide-bound state, the organic matter-bound state, and the residue state (Jiao et al., 2016; Zhu et al., 2020). For instance, approximately 7.5 wt% of Pb was present in the water-soluble form in MSWI

FA, which poses a significant challenge for the heavy metal immobilization in MSWI FA-derived construction materials. Besides, metal chlorides present in MSWI FA are the main reason for the high leachabilities of Pb, Zn, Cd, and Cu, while chromates (such as K_2CrO_4) consisting of hexavalent Cr(VI) are the primary form of leachable Cr. In addition, Zn, Pb, and Cr exhibit typical amphoteric leaching characteristics with the most stabilized pH of 9.5 ~ 10.5. The leaching behavior of Cd and Cu follows the cationic leaching pattern, which decreased with an increase in pH value. However, some researchers argue that the leaching characteristics of Zn and Pb also follow a cationic leaching pattern (Zhang et al., 2016; Luo et al., 2019), possibly due to the different pH ranges selected.

4.2 Sintered aggregates

Sintered aggregates have shown effective immobilization of PTEs due to volatilization and chemical immobilization during the sintering process (Lane et al., 2020; Ardit et al., 2022). Han et al. (2022) found that sintering at 1120°C for 30 min resulted in high volatilization rates of Pb (95.7%), Cr (93.9%), Zn (78.9%), and Cu (57.0%) from MSWI FA-based aggregates, respectively. The remaining unvolatilized fraction and other PTEs were immobilized within the aggregate through the sintering process, resulting in a significant reduction in the concentration of PTEs compared with the raw MSWI FA. Except for those volatilizing during sintering, the residual amounts of PTEs in the treated material were efficiently immobilized within the aluminosilicate matrix (Polettini et al., 2004). The crystal phases in the aggregates were decomposed and reconstructed after high-temperature sintering, which encapsulated PTEs in the new phase (Shao et al., 2022). In conclusion, sintering technology can significantly reduce the leaching concentrations of PTEs in aggregates. However, flue gas treatment must be considered when producing MSWI FA-based sintered aggregates to prevent secondary pollution (Shao et al., 2022).

4.3 SCMs and cold-bonded aggregates

When recycling MSWI FA as SCMs to produce composite cementitious materials or cold-bonded aggregates, the immobilization of PTEs occurs through physical encapsulation and chemical stabilization such as pozzolanic reactions, incorporation into a solid matrix, or redox conditions (del Valle-Zermeno et al., 2013).

The binder selection in the cement-based cold-bonding method significantly affects the efficiency of heavy metal immobilization. Cement-based aggregates demonstrate better immobilization performances than the corresponding lime-based aggregates (Ren et al., 2021). Alkali-activated aggregates have a similar heavy metal immobilization mechanism to cement-based aggregates, where interaction between the metals and the hydration products, such as ettringite and various double-layered hydroxides of calcium aluminum/ferric hydrates, contributes to the significant reduction of leachabilities of PTEs, including Cd, Cr, Cu, Ni, Pb, and Zn (Li et al., 2019). The chloride contained in MSWI FA was prone to react with the hydrated calcium aluminate phase to form hydrocalumite (also known as Friedel's salt), which can release OH^- as well as immobilize chloride ions, resulting in a facilitating effect on the immobilization of PTEs (Nag and Shimaoka, 2023). However, alkali activation process was less effective in immobilizing Cr than other PTEs (Liu et al., 2021). This is because Cr has a higher oxidation state and a lower solubility limit in an alkaline environment, which makes it more challenging to be immobilized by the alkali-activated binders. To further diminish the leachabilities of PTEs, incorporating phosphates or sulfates as additives have been reported to react with metal cations to form stable precipitates (Chen et al., 2019a).

5. Conclusions and perspectives

This chapter discussed the potential recycling technologies for MSWI FA through SCMs and artificial aggregates. It has highlighted the importance of pretreatment techniques to improve the quality of MSWI FA for recycling and the challenges associated with its physicochemical properties and environmental impacts. The following conclusions can be drawn from the reviewed literature:

- MSWI FA possesses high alkaline properties, similar to cement, but untreated MSWI FA has low pozzolanic reactivity and contains interfering substances that can delay cement hydration and reduce mechanical properties.
- Washing can effectively remove chlorine and soluble PTEs, while mechanochemical treatment and low-temperature thermal treatment improve PCDD/Fs decomposition and the reactivity improvement of MSWI FA. Alkaline washing can effectively preserve calcium and consume the metallic aluminum and promote the release of ammonia and hydrogen in the pretreatment stage, preventing undesirable expansion of the matrix during curing.

- Codisposal of multiple types of solid waste can achieve element complementarity and enhance the physical and mechanical properties of value-added products derived from MSWI FA.
- Sintered aggregates derived from MSWI FA exhibit higher strength and lower water absorption, but the substantial energy consumption is the primary obstacle hindering their application.
- Cold-bonding technology has the advantages of being more energy-efficient and carbon-negative, but the mechanical properties of aggregates are highly dependent on the chemical composition and reactivity of the raw materials. To overcome this drawback, feasible methods for enhancing the mechanical properties include alkali activation and accelerated carbonation.
- Proper optimization of pelletizing processes and production environments is crucial to ensure efficient production and high-quality pellets.
- Air pollution control systems are imperative in MSWI FA sintering treatment facilities to prevent the transfer of chlorides, volatile PTEs, and PCDD/Fs to the flue gas under high temperatures. The secondary fly ash collected by the air pollution control system still owns a high leaching toxicity risk, and its further disposal should be given close attention.

Currently, there is still limited research on the application of MSWI fly ash in cold-bonded lightweight aggregates. Future research is needed to explore the application of MSWI FA in cold-bonded lightweight aggregates, assess the long-term safety of MSWI FA-derived construction materials, understand the immobilization mechanisms and environmental risks of PCDD/Fs without thermal treatment, and develop detoxification procedures with environmental and economic benefits.

Acknowledgments

The authors gratefully acknowledge the financial support from the National Natural Science Foundation of China (Grant No. 52206174 and 52236008) and the Zhejiang Provincial Natural Science Foundation of China (Grant No. LZ23E060004) for this study.

References

Alrefaei, Y., Wang, Y.S., Dai, J.G., 2021. Effect of mixing method on the performance of alkali-activated fly ash/slag pastes along with polycarboxylate admixture. Cement Concr. Compos. 117, 103917.

Ardit, M., Zanelli, C., Conte, S., Molinari, C., Cruciani, G., Dondi, M., 2022. Ceramisation of hazardous elements: benefits and pitfalls of the inertisation through silicate ceramics. J. Hazard Mater. 423, 126851.

Azrar, H., Zentar, R., Abriak, N.E., 2016. The effect of granulation time of the Pan granulation on the characteristics of the aggregates containing dunkirk sediments. Procedia Eng. 143, 10–17.

Balapour, M., Zhao, W., Garboczi, E.J., Oo, N.Y., Spatari, S., Hsuan, Y.G., Billen, P., Farnam, Y., 2020. Potential use of lightweight aggregate (LWA) produced from bottom coal ash for internal curing of concrete systems. Cement Concr. Compos. 105, 103428.

Berodier, E., Scrivener, K., Scherer, G., 2014. Understanding the filler effect on the nucleation and growth of C-S-H. J. Am. Ceram. Soc. 97, 3764–3773.

Bie, R., Chen, P., Song, X., Ji, X., 2016. Characteristics of municipal solid waste incineration fly ash with cement solidification treatment. J. Energy Inst. 89, 704–712.

Bui Viet, D., Chan, W.P., Phua, Z.H., Ebrahimi, A., Abbas, A., Lisak, G., 2020. The use of fly ashes from waste-to-energy processes as mineral CO_2 sequesters and supplementary cementitious materials. J. Hazard Mater. 398, 122906.

Chen, D., Zhang, Y., Xu, Y., Nie, Q., Yang, Z., Sheng, W., Qian, G., 2022a. Municipal solid waste incineration residues recycled for typical construction materials-a review. RSC Adv. 12, 6279–6291.

Chen, H., Zhao, R., Zuo, W., Dong, G., He, D., Zheng, T., Liu, C., Xie, H., Wang, X., 2022b. Preparation of alkali activated cementitious material by upgraded fly ash from MSW incineration. Int. J. Environ. Res. Publ. Health 19, 13666.

Chen, H.J., Wang, S.Y., Tang, C.W., 2010. Reuse of incineration fly ashes and reaction ashes for manufacturing lightweight aggregate. Construct. Build. Mater. 24, 46–55.

Chen, L., Wang, L., Cho, D.W., Tsang, D.C.W., Tong, L.Z., Zhou, Y.Y., Yang, J., Hu, Q., Poon, C.S., 2019a. Sustainable stabilization/solidification of municipal solid waste incinerator fly ash by incorporation of green materials. J. Clean. Prod. 222, 335–343.

Chen, Y.X., Liu, G., Schollbach, K., Brouwers, H.J.H., 2021. Development of cement-free bio-based cold-bonded lightweight aggregates (BCBLWAs) using steel slag and miscanthus powder via CO_2 curing. J. Clean. Prod. 322, 129105.

Chen, Z., Li, J.S., Poon, C.S., Jiang, W.H., Ma, Z.H., Chen, X., Lu, J.X., Dong, H.X., 2023. Physicochemical and pozzolanic properties of municipal solid waste incineration fly ash with different pretreatments. Waste Manag. 160, 146–155.

Chen, Z., Lu, S., Tang, M., Ding, J., Buekens, A., Yang, J., Qiu, Q., Yan, J., 2019b. Mechanical activation of fly ash from MSWI for utilization in cementitious materials. Waste Manag. 88, 182–190.

Chen, Z., Lu, S., Tang, M., Lin, X., Qiu, Q., He, H., Yan, J., 2019c. Mechanochemical stabilization of heavy metals in fly ash with additives. Sci. Total Environ. 694, 133813.

Chen, Z., Tang, M., Lu, S., Alfons, B., Ding, J., Qiu, Q., Yan, J., 2019d. Mechanochemical degradation of PCDD/Fs in fly ash within different milling systems. Chemosphere 223, 188—195.

Chen, Z., Tang, M., Lu, S., Ding, J., Qiu, Q., Wang, Y., Yan, J., 2018. Evolution of PCDD/F-signatures during mechanochemical degradation in municipal solid waste incineration filter ash. Chemosphere 208, 176—184.

Cheng, T.W., Chen, Y.S., 2003. On formation of CaO-Al(2)O(3)-SiO_2 glass-ceramics by vitrification of incinerator fly ash. Chemosphere 51, 817—824.

Chuang, K.H., Lu, C.H., Chen, J.C., Wey, M.Y., 2018. Reuse of bottom ash and fly ash from mechanical-bed and fluidized-bed municipal incinerators in manufacturing lightweight aggregates. Ceram. Int. 44, 12691—12696.

Colangelo, F., Messina, F., Cioffi, R., 2015. Recycling of MSWI fly ash by means of cementitious double step cold bonding pelletization: technological assessment for the production of lightweight artificial aggregates. J. Hazard Mater. 299, 181—191.

Cristelo, N., Segadaes, L., Coelho, J., Chaves, B., Sousa, N.R., de Lurdes Lopes, M., 2020. Recycling municipal solid waste incineration slag and fly ash as precursors in low-range alkaline cements. Waste Manag. 104, 60—73.

del Valle-Zermeno, R., Formosa, J., Chimenos, J.M., Martinez, M., Fernandez, A.I., 2013. Aggregate material formulated with MSWI bottom ash and APC fly ash for use as secondary building material. Waste Manag. 33, 621—627.

Deng, S.X., Ren, P.F., Jiang, Y., Shao, X., Ling, T.C., 2022. Use of CO_2-active BOFS binder in the production of artificial aggregates with waste concrete powder. Resour. Conserv. Recycl. 182, 106332.

Dondi, M., Cappelletti, P., D'Amore, M., de Gennaro, R., Graziano, S.F., Langella, A., Raimondo, M., Zanelli, C., 2016. Lightweight aggregates from waste materials: reappraisal of expansion behavior and prediction schemes for bloating. Construct. Build. Mater. 127, 394—409.

Dong, B., Chen, C., Wei, G., Fang, G., Wu, K., Wang, Y., 2022. Fly ash-based artificial aggregates synthesized through alkali-activated cold-bonded pelletization technology. Construct. Build. Mater. 344, 128—268.

Duxson, P., Fernández-Jiménez, A., Provis, J.L., Lukey, G.C., Palomo, A., van Deventer, J.S.J., 2006. Geopolymer technology: the current state of the art. J. Mater. Sci. 42, 2917—2933.

Ebert, B.A.R., Steenari, B.-M., Geiker, M.R., Kirkelund, G.M., 2020. Screening of untreated municipal solid waste incineration fly ash for use in cement-based materials: chemical and physical properties. SN Appl. Sci. 2, 802.

Fan, L.F., Gao, J.W., Zhang, Y.H., Zhong, W.L., 2023. Investigation of micro-structure and compression behavior of cement mortar with artificial geopolymer sand. Construct. Build. Mater. 376, 130947.

Farina, I., Moccia, I., Salzano, C., Singh, N., Sadrolodabaee, P., Colangelo, F., 2022. Compressive and thermal properties of non-structural lightweight concrete containing industrial byproduct aggregates. Materials 15, 4029.

Ferraro, A., Ducman, V., Colangelo, F., Korat, L., Spasiano, D., Farina, I., 2023. Production and characterization of lightweight aggregates from municipal solid waste incineration fly-ash through single- and double-step pelletization process. J. Clean. Prod. 383, 135275.

Frankovič, A., Bokan Bosiljkov, V., Ducman, V., 2017. Lightweight aggregates made from fly ash using the cold-bond process and their use in lightweight concrete. Mater. Tehnol. 51, 267—274.

Garcia-Lodeiro, I., Carcelen-Taboada, V., Fernandez-Jimenez, A., Palomo, A., 2016. Manufacture of hybrid cements with fly ash and bottom ash from a municipal solid waste incinerator. Construct. Build. Mater. 105, 218—226.

Gholizadeh-Vayghan, A., Bellinkx, A., Snellings, R., Vandoren, B., Quaghebeur, M., 2020. The effects of carbonation conditions on the physical and microstructural properties of recycled concrete coarse aggregates. Construct. Build. Mater. 257, 119486.

Ghouleh, Z., Guthrie, R.I.L., Shao, Y., 2017. Production of carbonate aggregates using steel slag and carbon dioxide for carbon-negative concrete. J. CO_2 Util. 18, 125—138.

González-Corrochano, B., Alonso-Azcárate, J., Rodas, M., 2014. Effect of prefiring and firing dwell times on the properties of artificial lightweight aggregates. Construct. Build. Mater. 53, 91—101.

Gunning, P.J., Hills, C.D., Carey, P.J., 2010. Accelerated carbonation treatment of industrial wastes. Waste Manag. 30, 1081—1090.

Habert, G., Miller, S.A., John, V.M., Provis, J.L., Favier, A., Horvath, A., Scrivener, K.L., 2020. Environmental impacts and decarbonization strategies in the cement and concrete industries. Nat. Rev. Earth Environ. 1, 559—573.

Han, S., Song, Y., Ju, T., Meng, Y., Meng, F., Song, M., Lin, L., Liu, M., Li, J., Jiang, J., 2022. Recycling municipal solid waste incineration fly ash in super-lightweight aggregates by sintering with clay and using SiC as bloating agent. Chemosphere 307, 135895.

Hanifa, M., Agarwal, R., Sharma, U., Thapliyal, P.C., Singh, L.P., 2023. A review on CO2 capture and sequestration in the construction industry: emerging approaches and commercialised technologies. J. CO_2 Util. 67, 102292.

Harikrishnan, K.I., Ramamurthy, K., 2006. Influence of pelletization process on the properties of fly ash aggregates. Waste Manag. 26, 846—852.

Huang, B., Gan, M., Ji, Z., Fan, X., Wang, G., Sun, Z., Zhao, Q., Wu, Y., Lu, S., 2023. Co-treating MSWI fly ash in iron ore sintering process: influence of water-washing and roll forming pretreatment on dioxins emission. Process Saf. Environ. Protect. 173, 143—153.

Huang, B., Gan, M., Ji, Z., Fan, X., Zhang, D., Chen, X., Sun, Z., Huang, X., Fan, Y., 2022. Recent progress on the thermal treatment and resource utilization technologies of municipal waste incineration fly ash: a review. Process Saf. Environ. Protect. 159, 547—565.

Huang, G., Yang, K., Chen, L., Lu, Z., Sun, Y., Zhang, X., Feng, Y., Ji, Y., Xu, Z., 2020. Use of pretreatment to prevent expansion and foaming in high-performance MSWI bottom ash alkali-activated mortars. Construct. Build. Mater. 245, 118471.

Hwang, C.L., Bui, L.A.T., Lin, K.L., Lo, C.T., 2012. Manufacture and performance of lightweight aggregate from municipal solid waste incinerator fly ash and reservoir sediment for self-consolidating lightweight concrete. Cement Concr. Compos. 34, 1159—1166.

Jiang, Y., Li, L., Lu, J.-x., Shen, P., Ling, T.-C., Poon, C.S., 2022. Enhancing the microstructure and surface texture of recycled concrete fine aggregate via magnesium-modified carbonation. Cement Concr. Res. 162, 106967.

Jiao, F., Zhang, L., Dong, Z., Namioka, T., Yamada, N., Ninomiya, Y., 2016. Study on the species of heavy metals in MSW incineration fly ash and their leaching behavior. Fuel Process. Technol. 152, 108–115.

Jin, L., Chen, M., Wang, Y., Peng, Y., Yao, Q., Ding, J., Ma, B., Lu, S., 2023. Utilization of mechanochemically pretreated municipal solid waste incineration fly ash for supplementary cementitious material. J. Environ. Chem. Eng. 11, 109112.

Juenger, M.C.G., Snellings, R., Bernal, S.A., 2019. Supplementary cementitious materials: new sources, characterization, and performance insights. Cement Concr. Res. 122, 257–273.

Kiventerä, J., Lancellotti, I., Catauro, M., Poggetto, F.D., Leonelli, C., Illikainen, M., 2018. Alkali activation as new option for gold mine tailings inertization. J. Clean. Prod. 187, 76–84.

Lane, D.J., Jokiniemi, J., Heimonen, M., Peraniemi, S., Kinnunen, N.M., Koponen, H., Lahde, A., Karhunen, T., Nivajarvi, T., Shurpali, N., Sippula, O., 2020. Thermal treatment of municipal solid waste incineration fly ash: impact of gas atmosphere on the volatility of major, minor, and trace elements. Waste Manag. 114, 1–16.

Lee, K.H., Lee, J.H., Wie, Y.M., Lee, K.G., 2019. Bloating mechanism of lightweight aggregates due to ramping rate. Adv. Mater. Sci. Eng. 2019, 1–12.

Li, B., Deng, Z., Wang, W., Fang, H., Zhou, H., Deng, F., Huang, L., Li, H., 2017. Degradation characteristics of dioxin in the fly ash by washing and ball-milling treatment. J. Hazard Mater. 339, 191–199.

Li, H., Yang, Y., Zheng, W., Chen, L., Bai, Y., 2020. Immobilization of high concentration hexavalent chromium via core-shell structured lightweight aggregate: a promising soil remediation strategy. Chem. Eng. J. 401, 126044.

Li, J., 2021. Municipal solid waste incineration ash-incorporated concrete: one step towards environmental justice. Buildings 11, 495.

Li, Y., Min, X., Ke, Y., Liu, D., Tang, C., 2019. Preparation of red mud-based geopolymer materials from MSWI fly ash and red mud by mechanical activation. Waste Manag. 83, 202–208.

Liu, J., Hu, L., Tang, L., Ren, J., 2021. Utilisation of municipal solid waste incinerator (MSWI) fly ash with metakaolin for preparation of alkali-activated cementitious material. J. Hazard Mater. 402, 123451.

Liu, J., Li, Z., Zhang, W., Jin, H., Xing, F., Chen, C., Tang, L., Wang, Y., 2022. Valorization of municipal solid waste incineration bottom ash (MSWIBA) into cold-bonded aggregates (CBAs): feasibility and influence of curing methods. Sci. Total Environ. 843, 157004.

Lothenbach, B., Scrivener, K., Hooton, R.D., 2011. Supplementary cementitious materials. Cement Concr. Res. 41, 1244–1256.

Lu, C.H., Chen, J.C., Chuang, K.H., Wey, M.Y., 2015. The different properties of lightweight aggregates with the fly ashes of fluidized-bed and mechanical incinerators. Construct. Build. Mater. 101, 380–388.

Lu, S., Huang, J., Peng, Z., Li, X., Yan, J., 2012. Ball milling 2,4,6-trichlorophenol with calcium oxide: dechlorination experiment and mechanism considerations. Chem. Eng. J. 195–196, 62–68.

Luo, H., Cheng, Y., He, D., Yang, E.H., 2019. Review of leaching behavior of municipal solid waste incineration (MSWI) ash. Sci. Total Environ. 668, 90–103.

Ma, C., Long, G., Shi, Y., Xie, Y., 2018. Preparation of cleaner one-part geopolymer by investigating different types of commercial sodium metasilicate in China. J. Clean. Prod. 201, 636–647.

Moreno-Maroto, J.M., Uceda-Rodríguez, M., Cobo-Ceacero, C.J., de Hoces, M.C., MartínLara, M.Á., Cotes-Palomino, T., López García, A.B., Martínez -García, C., 2019. Recycling of 'alperujo' (olive pomace) as a key component in the sintering of lightweight aggregates. J. Clean. Prod. 239, 118041.

Nadesan, M.S., Dinakar, P., 2017. Structural concrete using sintered flyash lightweight aggregate: a review. Construct. Build. Mater. 154, 928–944.

Nag, M., Shimaoka, T., 2023. A novel and sustainable technique to immobilize lead and zinc in MSW incineration fly ash by using pozzolanic bottom ash. J. Environ. Manag. 329, 117036.

Palacios, M., Puertas, F., 2005. Effect of superplasticizer and shrinkage-reducing admixtures on alkali-activated slag pastes and mortars. Cement Concr. Res. 35, 1358–1367.

Peng, Z., Weber, R., Ren, Y., Wang, J., Sun, Y., Wang, L., 2020. Characterization of PCDD/Fs and heavy metal distribution from municipal solid waste incinerator fly ash sintering process. Waste Manag. 103, 260–267.

Peyne, J., Gautron, J., Doudeau, J., Rossignol, S., 2018. Development of low temperature lightweight geopolymer aggregate, from industrial Waste, in comparison with high temperature processed aggregates. J. Clean. Prod. 189, 47–58.

Phua, Z., Giannis, A., Dong, Z.L., Lisak, G., Ng, W.J., 2019. Characteristics of incineration ash for sustainable treatment and reutilization. Environ. Sci. Pollut. Res. Int. 26, 16974–16997.

Polettini, A., Pomi, R., Trinci, L., Muntoni, A., Lo Mastro, S., 2004. Engineering and environmental properties of thermally treated mixtures containing MSWI fly ash and low-cost additives. Chemosphere 56, 901–910.

Qian, L., Xu, L., Alrefaei, Y., Wang, T., Ishida, T., Dai, J., 2022a. Artificial alkali-activated aggregates developed from wastes and by-products: a state-of-the-art review. Resour. Conserv. Recycl. 177, 105971.

Qian, L.P., Xu, L.Y., Huang, B.T., Dai, J.G., 2022b. Pelletization and properties of artificial lightweight geopolymer aggregates (GPA): one-part vs. two-part geopolymer techniques. J. Clean. Prod. 374, 133933.

Quina, M.J., Almeida, M.A., Santos, R., Bordado, J.M., Quinta-Ferreira, R.M., 2014a. Compatibility analysis of municipal solid waste incineration residues and clay for producing lightweight aggregates. Appl. Clay Sci. 102, 71–80.

Quina, M.J., Bordado, J.M., Quinta-Ferreira, R.M., 2014b. Recycling of air pollution control residues from municipal solid waste incineration into lightweight aggregates. Waste Manag. 34, 430–438.

Ren, P., Ling, T.-C., Mo, K.H., 2021. Recent advances in artificial aggregate production. J. Clean. Prod. 291, 125215.

Riley, C.M., 1951. Relation of chemical properties to the bloating of clays. J. Am. Ceram. Soc. 34, 121–128.

Scrivener, K.L., Lothenbach, B., De Belie, N., Gruyaert, E., Skibsted, J., Snellings, R., Vollpracht, A., 2015. TC 238-SCM: hydration and microstructure of concrete with SCMs. Mater. Struct. 48, 835–862.

Shah, V., Scrivener, K., Bhattacharjee, B., Bishnoi, S., 2018. Changes in microstructure characteristics of cement paste on carbonation. Cement Concr. Res. 109, 184–197.

Shahane, H.A., Patel, S., 2021. Influence of curing method on characteristics of environment-friendly angular shaped cold bonded fly ash aggregates. J. Build. Eng. 35, 101997.

Shao, Y., Shao, Y., Zhang, W., Zhu, Y., Dou, T., Chu, L., Liu, Z., 2022. Preparation of municipal solid waste incineration fly ash-based ceramsite and its mechanisms of heavy metal immobilization. Waste Manag. 143, 54–60.

Shi, M., Ling, T.-C., Gan, B., Guo, M.-Z., 2019. Turning concrete waste powder into carbonated artificial aggregates. Construct. Build. Mater. 199, 178–184.

Skibsted, J., Snellings, R., 2019. Reactivity of supplementary cementitious materials (SCMs) in cement blends. Cement Concr. Res. 124, 105799.

Song, G.J., Kim, S.H., Seo, Y.C., Kim, S.C., 2008. Dechlorination and destruction of PCDDs/PCDFs in fly ashes from municipal solid waste incinerators by low temperature thermal treatment. Chemosphere 71, 248–257.

Sorrentino, G.P., Zanoletti, A., Ducoli, S., Zacco, A., Iora, P., Invernizzi, C.M., Di Marcoberardino, G., Depero, L.E., Bontempi, E., 2022. Accelerated and natural carbonation of a municipal solid waste incineration (MSWI) fly ash mixture: basic strategies for higher carbon dioxide sequestration and reliable mass quantification. Environ. Res. 217, 114805.

Steiner, S., Lothenbach, B., Proske, T., Borgschulte, A., Winnefeld, F., 2020. Effect of relative humidity on the carbonation rate of portlandite, calcium silicate hydrates and ettringite. Cement Concr. Res. 135, 106116.

Tajra, F., Elrahman, M.A., Chung, S.-Y., Stephan, D., 2018. Performance assessment of core-shell structured lightweight aggregate produced by cold bonding pelletization process. Construct. Build. Mater. 179, 220–231.

Tajra, F., Elrahman, M.A., Stephan, D., 2019. The production and properties of cold-bonded aggregate and its applications in concrete: a review. Construct. Build. Mater. 225, 29–43.

Tang, P., Xuan, D., Cheng, H.W., Poon, C.S., Tsang, D.C.W., 2020a. Use of CO_2 curing to enhance the properties of cold bonded lightweight aggregates (CBLAs) produced with concrete slurry waste (CSW) and fine incineration bottom ash (IBA). J. Hazard Mater. 381, 120951.

Tang, P., Xuan, D., Li, J., Cheng, H.W., Poon, C.S., Tsang, D.C.W., 2020b. Investigation of cold bonded lightweight aggregates produced with incineration sewage sludge ash (ISSA) and cementitious waste. J. Clean. Prod. 251, 119709.

Tian, K., Wang, Y., Hong, S., Zhang, J., Hou, D., Dong, B., Xing, F., 2021. Alkali-activated artificial aggregates fabricated by red mud and fly ash: performance and microstructure. Construct. Build. Mater. 281, 122552.

Vasugi, V., Ramamurthy, K., 2014. Identification of design parameters influencing manufacture and properties of cold-bonded pond ash aggregate. Mater. Des. 54, 264–278.

Wan, S., Zhou, X., Zhou, M., Han, Y., Chen, Y., Geng, J., Wang, T., Xu, S., Qiu, Z., Hou, H., 2018. Hydration characteristics and modeling of ternary system of municipal solid wastes incineration fly ash-blast furnace slag-cement. Construct. Build. Mater. 180, 154–166.

Wang, D., Noguchi, T., Nozaki, T., 2019. Increasing efficiency of carbon dioxide sequestration through high temperature carbonation of cement-based materials. J. Clean. Prod. 238, 117980.

Wang, K.-S., Sun, C.-J., Yeh, C.-C., 2002. The thermotreatment of MSW incineration fly ash for use as an aggregate: a study of the characteristics of size-fractioning. Resour. Conserv. Recycl. 35, 177–190.

Wang, S., Yu, L., Qiao, Z., Deng, H., Xu, L., Wu, K., Yang, Z., Tang, L., 2023. The toxic leaching behavior of MSWI fly ash made green and non-sintered lightweight aggregates. Construct. Build. Mater. 373, 130809.

Wang, S., Yu, L., Yang, F., Zhang, W., Xu, L., Wu, K., Tang, L., Yang, Z., 2022a. Resourceful utilization of quarry tailings in the preparation of non-sintered high-strength lightweight aggregates. Construct. Build. Mater. 334, 127444.

Wang, X., Zhu, K., Zhang, L., Li, A., Chen, C., Huang, J., Zhang, Y., 2022b. Mechanical property and heavy metal leaching behavior enhancement of municipal solid waste incineration fly ash during the pressure-assisted sintering treatment. J. Environ. Manag. 301, 113856.

Wei, N., 2015. Leachability of heavy metals from lightweight aggregates made with sewage sludge and municipal solid waste incineration fly ash. Int. J. Environ. Res. Publ. Health 12, 4992–5005.

Wei, Y.L., Weng, S.D., Xie, X.Q., 2018. Reduction of sintering energy by application of calcium fluoride as flux in lightweight aggregate sintering. Construct. Build. Mater. 190, 765–772.

Wu, M., Sui, S., Zhang, Y., Jia, Y., She, W., Liu, Z., Yang, Y., 2021. Analyzing the filler and activity effect of fly ash and slag on the early hydration of blended cement based on calorimetric test. Construct. Build. Mater. 276, 122201.

Wu, S., Xu, Y., Sun, J., Cao, Z., Zhou, J., Pan, Y., Qian, G., 2015. Inhibiting evaporation of heavy metal by controlling its chemical speciation in MSWI fly ash. Fuel 158, 764–769.

Xiao, R., Jiang, X., Zhang, M., Polaczyk, P., Huang, B., 2020. Analytical investigation of phase assemblages of alkali-activated materials in CaO-SiO_2-Al_2O_3 systems: the management of reaction products and designing of precursors. Mater. Des. 194, 108975.

Xu, L.Y., Qian, L.P., Huang, B.T., Dai, J.G., 2021. Development of artificial one-part geopolymer lightweight aggregates by crushing technique. J. Clean. Prod. 315, 128200.

Yan, J.H., Peng, Z., Lu, S.Y., Li, X.D., Ni, M.J., Cen, K.F., Dai, H.F., 2007. Degradation of PCDD/Fs by mechanochemical treatment of fly ash from medical waste incineration. J. Hazard Mater. 147, 652–657.

Yang, G., Ren, Q., Xu, J., Lyu, Q., 2021. Co-melting properties and mineral transformation behavior of mixtures by MSWI fly ash and coal ash. J. Energy Inst. 96, 148–157.

Yang, K.-H., Jung, Y.-B., Cho, M.-S., Tae, S.-H., 2015. Effect of supplementary cementitious materials on reduction of CO2 emissions from concrete. J. Clean. Prod. 103, 774–783.

Yang, Z., Ji, R., Liu, L., Wang, X., Zhang, Z., 2018. Recycling of municipal solid waste incineration by-product for cement composites preparation. Construct. Build. Mater. 162, 794–801.

Ye, W., Li, S., Qiu, T., Cong, X., Tan, Y., 2022. Effects of plastic coating on the physical and mechanical properties of the artificial aggregate made by fly ash. J. Clean. Prod. 360, 132187.

Yliniemi, J., Nugteren, H., Illikainen, M., Tiainen, M., Weststrate, R., Niinimäki, J., 2016. Lightweight aggregates produced by granulation of peat-wood fly ash with alkali activator. Int. J. Miner. Process. 149, 42–49.

Zhan, X., Wang, L., Wang, L., Gong, J., Wang, X., Song, X., Xu, T., 2021. Co-sintering MSWI fly ash with electrolytic manganese residue and coal fly ash for lightweight ceramisite. Chemosphere 263, 127914.

Zhang, Y., Cetin, B., Likos, W.J., Edil, T.B., 2016. Impacts of pH on leaching potential of elements from MSW incineration fly ash. Fuel 184, 815–825.

Zhang, Y., Wang, L., Chen, L., Ma, B., Zhang, Y., Ni, W., Tsang, D.C.W., 2021. Treatment of municipal solid waste incineration fly ash: state-of-the-art technologies and future perspectives. J. Hazard Mater. 411, 125132.

Zhang, Z., Wang, J., Liu, L., Ma, J., Shen, B., 2020. Preparation of additive-free glass-ceramics from MSW incineration bottom ash and coal fly ash. Construct. Build. Mater. 254, 119345.

Zhao, L., Hu, M., Muslim, H., Hou, T., Bian, B., Yang, Z., Yang, W., Zhang, L., 2022. Co-utilization of lake sediment and blue-green algae for porous lightweight aggregate (ceramsite) production. Chemosphere 287, 132145.

Zhu, J., Hao, Q., Chen, J., Hu, M., Tu, T., Jiang, C., 2020. Distribution characteristics and comparison of chemical stabilization ways of heavy metals from MSW incineration fly ashes. Waste Manag. 113, 488–496.

Chapter 19

Pretreatments of municipal solid waste incineration bottom ash for the engineering utilizations as aggregates and cementitious materials

Lufan Li[1], Tung-Chai Ling[2], Pengfei Ren[2] and Pei Tang[3]

[1]*Department of Civil Engineering, Hangzhou City University, Hangzhou, China;* [2]*College of Civil Engineering, Hunan University, Changsha, China;* [3]*State Key Laboratory of Silicate Materials for Architectures, Wuhan University of Technology, Wuhan, China*

1. Introduction

The growth of population and urbanization results in huge increase of food consumption, economic, and industrial development, followed by massive amount of municipal solid waste (MSW), the total amount of which is expected to reach 3.4 billion by 2050 (Kaza et al., 2018). Incineration technique is the most suitable and effective approach for recycling MSW, especially in regions or countries with scarce land resources. The mass and volume of MSW after incineration can be reduced by ~70% and ~90%, respectively. In China, the percentage of MSW incineration rose dramatically in recent years, from only 3% in 2004 to 62% in 2020 (Chen et al., 2023). Meanwhile, the energy produced during incineration is treated as strategic asset, because it can be recovered to generate electricity, which is counted in the form of renewable energy sources in most waste-to-energy plants (Xin-gang et al., 2016; Zakir Hossain et al., 2014).

Solid residues from the incineration process include incinerated fly ashes (IFA), air pollution control residues (APCr), and incinerated bottom ashes (IBA). Fresh IBA can be cooled either with water quenching or drying in air. Water quenching helps the cooling of IBA and may cause the agglomeration of particles and formation of glassy components, which is believed to influence the metal recovery in a negative way (Astrup et al., 2016). IBA particles are generally in light to dark gray color with irregular granular shape, with a gravel-like particle size distributions. The particle sizes typically range between 0.1 and 100 mm, with little portion of fine powders <63 μm or oversize particles >40 mm (Tang et al., 2015).

As presented in Fig. 19.1, fresh IBA usually consists of incombustible components such as ceramic, glass, and mineral particles (Clavier et al., 2020), as well as nonferrous metals (mainly aluminum (Al) and stainless steel), ferrous metals (iron and steel, at a percentage between 5% and 13%.), and heavy nonferrous metals (mainly copper (Cu) and zinc (Zn), at a percentage between 2% and 5%) (Berkhout et al., 2011). Detailed oxide compositions are listed in Table 19.1.

As the amount of MSW to be incinerated is increasing and IBA accounts approximately 80% of the ash resides, the pressure falls on the future treatment and reutilization of IBA. Currently, IBA is normally disposed at landfilling area, but this is not sustainable considering the potential pollution caused by the leaching of heavy metals and soluble salts (Verbinnen et al., 2017). Regarding the recycling of IBA, both the reutilization of minerals and recovery of metals are essential in achieving green environment and circular economy. The minerals contained can be reused as unbound aggregates or embedded in concrete or asphalt in various engineering applications, while the metals can be recycled as raw materials in metal industry (Chen et al., 2023; Saikia et al., 2015; Tang et al., 2020b).

The treatment of IFA is trickier as it contains highly contaminated components (e.g., dioxins, furans, chloride, acids, etc.). The amount of heavy metal and chloride content contained in IBA are much lower than that in IFA, since most hazardous elements/components evaporate during combustion process (Lam et al., 2010). The compositions of IBA vary

Treatment and Utilization of Combustion and Incineration Residues. https://doi.org/10.1016/B978-0-443-21536-0.00006-X

FIGURE 19.1 Basic compositions of IBA.

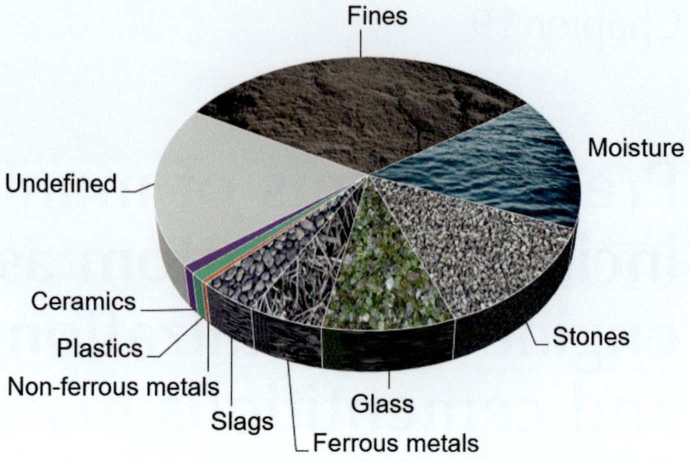

TABLE 19.1 Oxide composition of IBA (wt.%).

Oxides (%)	SiO_2 (%)	Al_2O_3 (%)	CaO (%)	Fe_2O_3 (%)	MgO (%)	K_2O (%)	Na_2O (%)	SO_3 (%)	P_2O_5 (%)	TiO_2 (%)
wt.	5–50	1–10	13–50	1–9	1–3	1–5	3–17	0–13	0–7	0–2

Data collected from Lam, C.H.K., Ip, A.W.M., Barford, J.P., McKay, G., 2010. Use of incineration MSW ash: a review. Sustainability 2, 1943–1968. https://doi.org/10.3390/su2071943.

over time and from country to region, due to the differences in waste disposal policies, resident lifestyle, and recycling procedures.

This chapter starts with a brief introduction on the production of fresh IBA and emphasizes on the state-of-the-art progress of IBA, mainly from the following two points: (1) current pretreatment methods for the safe reutilization of IBA and (2) the engineering properties and performances of IBA as replacement of traditional construction materials.

2. Pretreatment techniques

Using IBA as construction products has been greatly promoted worldwide, especially in European countries (Keulen et al., 2016; Tang et al., 2015). Although IBA is classified as nonhazardous, regulations or standards are published to strictly inhibit its direct applications without proper pretreatment (Verbinnen et al., 2017).

2.1 Mechanical grinding and metal separation

Normally, fresh IBA will go through a series of operations, including the reduction of particle size, the extraction of metals, and natural weathering, cleaning, and screening, etc. The metal recovery process usually requires crushing, grinding, or further pulverization in advance. Grinding can be processed either in the wet or dry conditions (Bertolini et al., 2004; Keppert et al., 2015), aiming to obtain a homogenous distribution of IBA. The low-speed dry milling process helps crack the brittle components into small fractions and press the ductile metals into plate shape particles, making the metals easier to be sieved out (Tang et al., 2016). The removal rate of metallic Al reaches 80 wt.% under proper grinding parameters and milling equipment. Magnetic separator is used to extract the ferrous components, while other conductive metals can be recovered using eddy current separator (Šyc et al., 2020). Advanced sensor detective technology can even identify all kinds of metals in particles larger than 4 mm, including stainless steel, metal, glass, etc.

2.2 Thermal treatment

Thermal treatment refers to the heating of MSWI residuals at a relative high temperature. Heating between 500 and 900°C aims at burning out the organic components and oxidizing metallic Al and Zn. High temperature heating between 100 and

1500°C not only provides the previous low temperature heating effects, but also help the immobilization of heavy metals and formation of amorphous phases (Tang et al., 2016). Particularly, the leaching of chloride, Cu, and lead (Pb) can be reduced by at least one order of magnitude (Hyks et al., 2011), but the leaching of some heavy metals were found to increase by more than two orders of magnitude. The disadvantages of thermal treatment are obvious. The release of gas pollutants (dioxins and PCDD/Fs) could be the major concern (Schabbach et al., 2012). The glassy phases produced during sintering may induce alkaline silica reaction when applied as aggregate in concrete. Most importantly, the heating process is energy and cost consuming. Hence, the heating process is more recommended to be implemented in the rotary kiln at the downstream of the combustion chamber or integrated in the combustion chamber unit.

2.3 Chemical treatment

The chemical treatment is always performed after plant-scale treatment of IBA to further achieve less heterogenous compositions and prevent heavy metal leaching, for the purpose of safely using in construction materials. Water washing is the simplest and most effective step to remove most soluble contaminants. Results show that the contaminants are usually concentrated in small particles with sizes less than 4 mm but can be effectively removed by water washing. Chemical treatment refers to the soaking of IBA in either alkaline or acid solutions, but they offer different effects (Chen et al., 2019; Huang et al., 2020; Mathews et al., 2020).

Alkaline solution treatment aims at reducing the metallic Al and Zn (Xuan and Poon, 2018), normally with NaOH solution. Wet grinding with water shares similar principles by creating alkaline solution through the dissolution of alkalis from IBA (Bertolini et al., 2004). It was also reported that Na_2CO_3 can reduce not only metallic Al and Zn but also other heavy metals such as barium (Ba), selenium (Se), chromium (Cr), and Cu (Saikia et al., 2015); however, this process is time-consuming and may take days to complete (Astrup, 2007). The critical factors determining the removal rate are the particle size of IBA, concentration of alkaline solution, liquid-to-solid ratio, etc. (Liu et al., 2018). The pozzolanic reactivity was improved after alkaline solution treatment, but additional water washing is required to remove the remaining alkaline solutions before it can be used as supplementary cementitious material (SCM) (Liu et al., 2018).

IBA can be prewashed with water and then followed by acid to remove the chloride and sulfate contents. Sulfate can be stabilized using $NaHCO_3$ or CO_2 during washing, by the precipitation of Ca as carbonate instead of sulfate forms (Astrup et al., 2016). The chloride content was found to reduce from 2.78 wt.% to 0.09 wt.% after 0.1M acetic acid washing and heating at 1100°C (Lo et al., 2020). Unfortunately, few studies focused on the removal rate of chloride and sulfate content. The application of HCl and NaCl at certain concentrations also presents excellent results for Zn, Cd, Cu, and Pb removal (Tang and Steenari, 2016).

2.4 Natural weathering and carbonation

Fresh IBA can be stabilized under atmospheric conditions by simply exposed to wind and rain in the open air, which is called natural weathering or aging. The chemical and mineralogy compositions of IBA change with time, temperature, and moisture condition due to the synergistic reactions of hydration, carbonation, hydrolysis, leaching, etc. Reaction products generally include carbonate minerals, hydrous sulfate minerals, and amorphous gel phases. A duration of 1−3 months is always adopted from both the material and economic perspectives.

The most important advantage of natural weathering is the stabilization of heavy metals, as presented in Table 19.2. The natural weathering helps transform the IBA from high alkaline to almost neutral, which inhibits the leaching of most heavy metals, such as barium (Ba) and molybdenum (Mo). Fresh IBA possesses a pH around 12. The pH is found to decrease to around 10 after 1 month of natural weathering, and further decreases and reaches the lowest value between 8 and 8.5 after one and half years (Meima and Comans, 1999). The leaching of Pb was reported to be pH independent, as it was more likely to be controlled by the sorption process and formation of stable minerals.

TABLE 19.2 Leaching of heavy metals after carbonation (Brück et al., 2018; Lin et al., 2015; Van Gerven et al., 2005).

Metals	Cu	Cr	Ni	Pb	Zn	Ba	Co	Mo	Sb	V
Leaching results	↓	↓	→	↓	↑↓?	↓	↑	↓	↓	↓

As stated before, contaminants are usually concentrated in small particles and can also be reduced to an acceptable range (As 0.1 mg/L; Cd 0.1 mg/L; Cu 2 mg/L; Cr 0.1 mg/L; Pb 0.5 mg/L; and Zn 2 mg/L) after 50 days' natural weathering, which suggests that even the small particles can also be reused as value-added materials (Chimenos et al., 2003). However, the leaching of Co (cobalt) was found to slightly increase, and questionable results were obtained for the leaching of Zn. It should be noticed that carbonation does not help control the leaching of chloride and may cause negative impact on the leaching of sulfate with prolonging carbonation duration.

Natural weathering is considered as the most cost-effective and straightforward method, but it is also time-consuming and requires land spaces for stockpiling. Moreover, the carbonation often happens in the surface layer of IBA heap, while the core of IBA heap remains uncarbonated. Based on the principle of natural weathering process, acceleration carbonation was developed to enhance the weathering effect (Chang et al., 2015; Tang et al., 2020b). With the help of accelerated carbonation, IBA can be reutilized as potential binder material in the production of artificial aggregates and SCMs, which will be introduced in detail in the following sections.

3. Engineering utilization of IBA

3.1 Application as aggregate

The most direct application of IBA involves the aggregate in both unbound (e.g., railway ballast, lower layer of road pavement) and bound (i.e., mixed with asphalt or cementitious materials) forms. In addition to directly using IBA as a substitute for natural aggregates, corecycling IBA with other industrial by-products or waste materials through pelletization technology has also been introduced, especially for fine IBA fraction.

3.1.1 Unbound aggregate

IBA can be applied in various cases mainly involving the replacement of natural aggregates in both unbound and bound forms (i.e., to be added in cement or asphalt mixtures). Road pavement is a typical case including the base and subbase layers of unbound granular compacted materials and pavement layers of cement/asphalt-based bound materials, and 20%−30% of IBA were applied in road construction (Huang et al., 2023). The particle size distribution of IBA meets the standard to be used in unstructural applications such as embankment, fill, and subbase materials, while other engineering properties (e.g., permeability, shear strength, and elastic modulus, bearing capacity) of IBA were proved to be comparable as natural sand (Lynn et al., 2017). In France, IBA has been used as unbound granular subbase with a California Bearing Ratio (CBR) value > 120% and has been proven to be in a good condition for 20 years (Cho et al., 2020). Moreover, IBA had an average Los Angeles abrasion value of around 45, which has the potential to be applied as lightweight aggregates.

3.1.2 Bound aggregate

The application of IBA in asphalt mixtures can be traced long back when Musselman et al. (1994) used 50% NA with 50% IBA to blend with 7% of bitumen. Recent researchers can even increase the replacement level up to 100% in hot-mix asphalt with 4% bitumen content. However, the high porosity of IBA may still exhibit negative impacts on the strength performance, and poorer durability performance was always noticed (Ding et al., 2022).

Comparative life cycle assessment (LCA) studies were conducted on IBA as landfill or as unbound aggregate in subbase layer for secondary road by previous researchers. Silva et al. (2019) reviewed the published LCA studies and concluded that from the perspectives of global warming potential, acidification potential, eutrophication potential, the application of IBA as aggregate is more environmentally friendly compared to be disposed as landfill, despite the fact that the long-term leaching of heavy metal in groundwater systems would negate the positive effect by using IBA as unbound aggregate.

With the properties of granular shape, high stiffness, and a wide particle size distribution, recycling IBA as fine or coarse aggregates in mortar, concrete, and road construction fields is another common practice (Xuan et al., 2018a). Using fine IBA as an alternative to sand has been investigated in the literature, where Woo et al. (2005) demonstrated that the IBA after proper pretreatment can be used as fine aggregate in mortar up to 20% of substitution without compromising 28 days compressive strength. The properties with IBA as fine aggregate also vary with different mixing methods, for example, the dry-mixed mortar shows an increase in mechanical strength and mitigation of expansion after using 100% IBA as fine sand (Xuan et al., 2018b).

In addition, owing to the porous structure of IBA, it may provide an internal curing effect, thus improving the mechanical and durability performance of concrete. Minane et al. (2017) produced concrete using fine IBA particles to replace sand with 100% and reported that concrete containing IBA sand showed comparable performance to that of control group, attributed to the internal curing effect caused by the porous structure of IBA. Shen et al. (2020) reused fine IBA particles to produce ultrahigh-performance concrete (UHPC), and it was found that the incorporation of IBA particles up to 25% could achieve higher compressive strength than that of control group. However, over this replacement limit, IBA had an adverse effect.

As for the coarser IBA particles, it is usually used to replace coarse aggregates. Müller and Rübner (2006) produced concrete with IBA as an aggregate component and showed poor durability due to the reaction of metallic aluminum (Al) and glass cullet with cement paste to cause cracks and spalling on the concrete surface. To avoid this problem, permeable concrete with relatively high pore structure was investigated. Wu et al. (2016) produced pervious concrete using IBA in place of sandstone and found that with an appropriate particle size and content, the compressive strength of previous concrete exceeded that of traditional red brick by 14 MPa. Shen et al. (2021) produced high-strength pervious concrete with IBA and reported that 25% IBA incorporation could achieve compressive strength of 50 MPa, and excessive replacement (25%−100%) reduced compressive strength.

The unbound applications have been well developed in many countries, but the application as bound aggregate is still facing many technical problems. Main problems for both unbound and bound aggregate are the leaching of alkaline metals and chloride, as well as the dissolution of salts. In the form of bound aggregate, the leaching of heavy metals highly depends on the amount of binder materials. Most studies indicated that the concentrations of heavy metals in asphalt remain a low level and usually within the safety limits according to the Toxicity Characteristic Leaching Procedure (TCLP) test results (Lo et al., 2020; Zhu et al., 2021). The leaching levels of Cl^- and SO_4^{2-} also vary in a wide range of 720−4597 mg/kg and 1470−5100 mg/kg, respectively (Xuan et al., 2018a).

Another major concern is the soundness issue of structure containing IBA. Such extensive cracking and spalling are associated with (1) the slow hydration of calcium and magnesium oxides and formation of ettringite (Alam et al., 2017), which results in the increase of volume (Saikia et al., 2015), (2) expansion induced by the ASR effect due to the existence of glass cullet, which varies between 7% and 60% by mass, (3) the corrosion of steel due to the high concentration of chloride and sulfate content (Van Der Wegen et al., 2013), and most importantly (4) the oxidation of aluminum particles emits hydrogen gas, which influences the volume stability in a very short time (as stated in Eq. 19.1). The amount of metallic aluminum in IBA ranges between 0.4% and 2.3% by mass (Xuan et al., 2018a). Hence, removing or inactivating the reaction of metallic aluminum (by using eddy current separation technique (Alderete et al., 2021), wet grinding (Bertolini et al., 2004) or alkaline solution immersion (Saikia et al., 2015)) is the potential solution to mitigate the swelling effect.

$$Al + 3H_2O + OH^- \rightarrow Al(OH)_4^- + 3/2H_2\uparrow \tag{19.1}$$

To summarize, the overall mechanical performance of concrete incorporation of IBA as aggregate is comparable with the replacement controlled at relative low level. However, even if the mechanical strengths can be adjusted at comparable values, negative effects were commonly found with porosity, water absorption (mainly caused by the higher water absorption of IBA), drying shrinkage, sulfate attack, and freeze thaw resistance (Huynh and Ngo, 2022; Jurič et al., 2006; Van Der Wegen et al., 2013). Fortunately, in most cases, the concrete made with IBA can still fulfill the basic requirement for construction applications, and it is recommended to use IBA aggregate in less demanding or unistructural applications, such as lightweight concrete blocks or other low-strength construction materials.

3.1.3 Artificial aggregate by pelletization

Cold-bonded pelletizing technique to produce artificial aggregate is a practical solution to reduce the environmental impact caused by the landfilling of IBA and conserve natural resources by producing alternative aggregates (Li et al., 2023; Ren et al., 2020; Tang et al., 2020b). Cioffi et al. (2011) produced IBA-based artificial aggregates by adding hydraulic binders such as cement, lime, and coal fly ash, in which 60%−90% IBA powder was incorporated. The results showed that artificial aggregates with bulk density between 1170 and 1330 kg/m^3, water absorption in the range of 11.13%−13.46%, and crushing strength between 1.9 and 4.5 MPa were obtained. The mechanical strength of artificial aggregates is strongly dependent on the type of binder used, and the concrete-adopted artificial aggregates can be used for the manufacture of standard concrete blocks.

Tang et al. (2017) applied cold-bonded pelletizing technique to produce artificial lightweight aggregates using IBA particles (0−2 mm) and other solid waste powders, including coal fly ash, paper sludge ash, washing aggregate sludge, and ordinary Portland cement. The results showed that compared with aggregates fully pelletized by cementitious powders, IBA particles can replace part of the cementitious powders and serve as skeleton to support larger external loading to achieve comparable mechanical performance. Besides, it was found that the heavy metals in IBA can be efficiently solidified in the artificial aggregates. The increasing of binder content significantly reduced the leaching of Sb, Cu, Mo, chloride, and sulfate (Tang and Brouwers, 2017).

Recently, Tang et al. (2019) combined concrete slurry waste (CSW) and IBA particles (0−2 mm) into a cold-bonded lightweight artificial aggregate. It was found that the produced aggregate showed excellent mechanical performance even without additional binders ascribed to the residual hydration behavior of CSW. The incorporation of small amount of cement and slag can largely increase the crushing strength of aggregates, as well as to immobilize the heavy metals, especially for slag. In addition, the effect of curing method on IBA-CSW-based artificial aggregates was also investigated. It was found that steam curing at 60°C had the highest pellet strength, and the aggregates cured by 100% CO_2 (either placed in flow-through chamber with 2.5L/min flow rate or in hermetic chamber with 0.1 bar pressure) for 24 h showed the lower water absorption (Tang et al., 2020b). The CO_2 uptake of 3.5%−4.1% also indicated that CO_2 curing can serve as a sustainable CO_2 sequestration process to produce artificial aggregates. Also, Yaphary et al. (2022) produced artificial aggregates by combining CSW and IBA particles, and the results showed that after 8h of CO_2 curing at 0.1 bar, the aggregate had individual strength of 2.01−2.31 MPa, and the produced artificial aggregate can be used for the production of semidry lightweight concrete.

Unlike cold bonding and accelerated carbonation granulation technologies, pelletization by sintering can be realized within 1 h. Chuang et al. (2018) produced sintered artificial aggregate by combining IFA, IBA, and reservoir sediment (RS). The results showed that the performance of artificial aggregates was largely influenced by sintering temperature and mixing ratio of raw materials. With the temperature increased from 1100 to 1150°C, individual pellet strength increased form 10 MPa up to 50 MPa with 65% IBA, 15% IFA, and 20% RS incorporation. Giro-Paloma et al. (2019) produced artificial aggregate using IBA larger than 8 mm by a rapid sintering process, in which rice hush (RH) was used as bloating agent. It was reported that the aggregate strength sharply decreased with 5% RH addition, and as RH continuously increased to over 10%, the aggregate strength remained stable of ~3 MPa. A lightweight ceramsite was developed by sintering technology using red mud (RM) and IBA as raw materials (Sun et al., 2021). The results showed that at the temperature of 1070°C and the ratio of 1:1 of IBA and RM, the produced ceramsites reached the best performance, with a loose bulk density of 1046.73 kg/m^3 and crushing strength of 27.11 MPa.

3.2 Cement clinker cocombustion

The application of IBA as a kiln feed material has been largely investigated, as it processes similar elements such as Al, Si, Ca, and Fe, which are the key components in the production of cement clinker (Clavier et al., 2020; Sarmiento et al., 2019). Extensive literature had covered the raw material compositions, mechanical and durability properties, and environmental impacts, particularly from the life cycle perspective. Due to the retardation effect and less C_3S contained, the early age strength (before 7 days) of cement incorporating with IBA all performs relatively lower compressive strength, but the differences of compressive strength after 7 days can be compensated. It was also confirmed that the main mineralogical assemblage of the hydrated cement remains the same (Kleib et al., 2021).

The aluminum content in IBA-incorporated cements is generally higher than ordinary Portland cement, especially with untreated IBA, where a 8% replacement level of cement with IBA presents over 15% aluminate content; thus, additional gypsum is required to adjust the setting time (Clavier et al., 2021). The maximum proportion of IBA depends on the Fe_2O_3 content; otherwise, it may lead to different clinker compositions comparing to theoretical compositions of target Portland cement clinker (Kleib et al., 2021). With proper mix design, the mineralogical compositions can remain the same as Portland clinker, but some parameters differ slightly. Similar grindability was observed in lab-scale studies, but considering the heterogenous structure of IBA, results obtained at a smaller scale may not reflect the real situation and pilot scale analysis is suggested for future research (Clavier et al., 2021).

The mechanical properties of IBA-incorporated cement yielded diverse performances (Pan et al., 2008), but most studies proved that the replacement of raw materials for clinker with IBA has a negative impact on the formation of clinker phase and the corresponding reactivity, especially at a higher replacement level (Pan et al., 2008). Clavier et al. (2021) found that each replacement percentage of washed and unwashed IBA incorporation resulted in an approximate 45% and 35%, of tricalcium silicate content, respectively. Krammart and Tangtermsirikul (2004) found lower compressive strength and longer setting time after replacing 5% and 10% clinker with IBA, and followed by 1450°C combustion, but 25−49 kg of CO_2 emissions can be reduced per ton of clinker produced. In addition to global warming potential, lower environmental impacts were found in other categories such as abiotic depletion, acidification potential, eutrophication potential, etc. (Margallo. et al., 2013).

Since the metallic and nonmetallic elements are intractable issues restricting the application of IBA in clinker production, pretreatment of IBA is necessary to remove the detrimental components such as chloride, alkali, sulfate, and heavy metals (Silva et al., 2019). Cement manufactures usually set a chloride limits of 0.02% or 100 ppm, and pretreatment of IBA (e.g., water/acid washing) is required to reduce the chloride and alkali content to a maximum extent (Huang et al., 2023; Pan et al., 2008). The contained alkali (e.g., sodium and potassium) may also influence the cement hydration and the subsequent strength development. Special attention should be paid to the heavy metal leaching (e.g., As, Ba, and Cu) of IBA-incorporated clinker. Although most of these heavy metals can also be noticed in ordinary Portland cement, they exceed the threshold after certain amount of IBA added (Clavier et al., 2019).

3.3 Low carbon cementitious materials

Blended cement with Portland cement and supplementary cementitious materials is a common approach to achieve low carbon and cost-efficient cementitious material. Many previous researchers attempt to apply IBA as SCM by taking advantage of its pozzolanic reactivity (Fan et al., 2022). Those IBAs with higher SiO_2 but lower CaO content are more encouraged as SCMs to be mixed in cement pastes (Li et al., 2012; Tang et al., 2020a). Results showed that the pozzolanic reactivity of IBA slightly decreased after weathering, but was still comparable with that of Class F coal fly ash or natural pozzolans (Joseph, 2021), and higher than quartz at the same particle sizes (Caprai, 2019). The properties of mortar produced using fine IBA particles and pulverized IBA powder to replace sand and cement was investigated by Cheng (2012). Mortars containing IBA as cement or sand replacement present poorer flowability and may induce bleeding. Large amount of replacement as IBA will greatly increase shrinkage as well. Generally, it was found that incorporation IBA caused a reduction in both physical and mechanical properties when used as sand and cement substitution. However, after melting treatment and quenching treatment, IBA powders can partially replacement cement since it regains higher pozzolanic reactivity. Cement paste with water cooling pulverized IBA as replacement leads to better strength performance, lower total, and capillary porosity compared with raw IBA or air-cooled IBA.

Tang et al. (2020a) manually separated broken glass, ceramic, mineral, and various slags from IBS and investigated the reactivity of each component. It was concluded that the two most reactive fractions are broken glass, ceramic, and finer IBA. However, the incorporation of broken glass, ceramic or ground finer IBA all had negative impact on the early age compressive strength. After 90 days' curing, the compressive strength or mortar samples with 20% finer IBA or glass show comparable strength as that of GGBS. For the further valorization of IBA, it is suggested to increase the milling duration to refine the IBA particles. With proper separation techniques, the glass and ceramic components can be adopted as active SCMs, while other minerals and slags are encouraged to be used purely as inert filler.

Li et al. (2012) found that the cementitious properties of IBA are certainly lower than Portland cement and thus will cause the retardation of cement hydration, but a denser microstructure was formed with more C−S−H gel being produced. Similar result was obtained by Alderete et al. (2021), where delayed pozzolanic reaction of IBA contributes more to the later age strength. Furthermore, the incorporation of IBA showed better resistance to carbonation and chloride ingression compared with OPC reference group, owing to the formation of C-A-S-H and thus enlarged chloride binding (Thomas et al., 2012). The detailed changes of material properties have been summarized in Table 19.3. Overall, the upper limit of IBA incorporation is suggested below 30%.

TABLE 19.3 Changes of material properties after incorporating IBA as SCMs (Alderete et al., 2021).

Properties	Variation trend	Notes
Fresh properties	↓	The loss can be compensated by adding plasticizers
Compressive strength	↓	Lower early age strength, but comparable later age strength
Flexural strength	↑	Higher strength at all ages
Tensile splitting strength	↑?	Higher strength at 28 days but fluctuations observed at 91 days
Open porosity	↓	Lower porosity due to progression of hydration
Chloride diffusion	↓	In agreement with closed porosity and higher tortuosity of pore structure
Carbonation	↓	In agreement with closed porosity and higher tortuosity of pore structure

3.4 Alkaline-activated material

Alkaline-activated material (AAM) is recognized as a low carbon cementitious material compared with ordinary Portland cement (Robayo-Salazar et al., 2018). Since IBA contains certain amount of Al, IBA can be used as the precursor of alkali-activated slag at a weight ratio of 0%–12% (Zhang et al., 2023). Applying untreated IBA solely as precursor often leads to relatively lower performance; hence, pretreatment is unavoidable. The compressive strength of IBA based AAM ranges between 0.3 and 160 MPa, depending on several factors, e.g., content of precursor, type of alkaline activator, liquid to solid ratio, etc. IBA is often mixed with other common precursors (e.g., GGBS or FA), and the compressive strength development highly depends on the proportion of common precursors added. The high temperature resistance was found to be significantly enhanced, owing to the improved pore connectivity and thereby reduced water vapor pressure and shrinkage.

Similar volume expansion phenomenon was noticed in IBA-based AAM owing to the existence of Al. The density of alkaline-activated IBA is commonly lower (612–1036 kg/m^3) because of the air voids generated during the reaction between the precursor and the alkaline solution (Kurda et al., 2020). Interestingly, Xuan et al. (2019) used IBA and waste glass to produce aerated alkali-activated concrete, and IBA not only can play the role as main precursor but can also act as a forming agent owing to the contained metallic Al (as stated in Eq.19.1).

4. Conclusions and future perspectives

The pretreatment approaches and the current engineering utilizations of IBA for construction materials application were already well explored. Based on previous literature review, the following conclusion can be drawn:

1. Natural weathering is the very initial treatment of IBA, as it helps alter the pH and immobilize some heavy metals. Mechanical separation of metal components should also be performed prior all recycling cases as most metals are valuable and the separation process could help improve the subsequent properties of engineering materials (i.e., the removal of Al would benefit the volume stability).
2. Among all washing methods, water washing is highly recommended considering the easy removal of most heavy metals, chlorides, sulfates, as well as the very fine factions. Washing with alkaline is considered as an assisted option to remove the residual Al in IBA. It should still be noted that pretreatment is necessary but not effective enough to remove all the hazard components IBA contained.
3. IBA can be reutilized as aggregates or ground into powdery form as supplementary cementitious material in cement concrete, or as precursor in alkaline-activated material. Taking advantage of the alkaline minerals contained in IBA, the pelletization together with carbonation process can be an effective method to produce environmentally friendly aggregates. However, they share common problems such as the leaching of heavy metals and the volume stability issue.
4. Application of IBA as unbound/bound aggregate is still the most common utilization approach. Comparable mechanical properties can be achieved with lower amount of IBA used, but negative effects can still be noticed on the physical and long-term durability properties.

Even though the utilization of IBA in constriction field had been largely investigated, the industrial applications are still very limited, attributing to the low efficiency and high cost of pretreatment, as well as the quality variation between each type of products. The chemical compositions and reactivities of IBA considerably vary with source and feeding materials, which cause barriers to the standardization of product quality and the future commercialization, especially concerning the leaching behaviors of potential toxic elements. To conclude, the pretreatment of IBA, manufacturing process of IBA-incorporated material, and the corresponding materials properties still require holistic investigations.

Acknowledgments

Financial supports from the Research Fund for International Senior Scientists (RFIS-III) from NSFC, entitled "*Wastes recycling and CO_2 sequestration for sustainable construction materials*" (52250710158) and "Science and Technology Innovation Plan" of Hunan Province (2022WZ1010), are greatly appreciated.

References

Alam, Q., Florea, M.V.A., Schollbach, K., Brouwers, H.J.H., 2017. A two-stage treatment for Municipal Solid Waste Incineration (MSWI) bottom ash to remove agglomerated fine particles and leachable contaminants. Waste Manag. 67, 181–192. https://doi.org/10.1016/j.wasman.2017.05.029.

Alderete, N.M., Joseph, A.M., Van Den Heede, P., Matthys, S., De Belie, N., 2021. Effective and sustainable use of municipal solid waste incineration bottom ash in concrete regarding strength and durability. Resour. Conserv. Recycl. 167, 105356. https://doi.org/10.1016/j.resconrec.2020.105356.

Astrup, T., 2007. Pretreatment and utilization of waste incineration bottom ashes: Danish experiences. Waste Manag. 27, 1452–1457. https://doi.org/10.1016/j.wasman.2007.03.017.

Astrup, T., Muntoni, A., Polettini, A., Pomi, R., Van Gerven, T., Van Zomeren, A., 2016. Treatment and reuse of incineration bottom ash. In: Environmental Materials and Waste. Elsevier, pp. 607–645. https://doi.org/10.1016/B978-0-12-803837-6.00024-X.

Berkhout, S.P.M., Oudenhoven, B.P.M., Rem, P.C., 2011. Optimizing non-ferrous metal value from MSWI bottom ashes. J. Environ. Prot. 02, 564–570. https://doi.org/10.4236/jep.2011.25065.

Bertolini, L., Carsana, M., Cassago, D., Quadrio Curzio, A., Collepardi, M., 2004. MSWI ashes as mineral additions in concrete. Cement Concr. Res. 34, 1899–1906. https://doi.org/10.1016/j.cemconres.2004.02.001.

Brück, F., Schnabel, K., Mansfeldt, T., Weigand, H., 2018. Accelerated carbonation of waste incinerator bottom ash in a rotating drum batch reactor. J. Environ. Chem. Eng. 6, 5259–5268. https://doi.org/10.1016/j.jece.2018.08.024.

Caprai, V., 2019. Applications in Cement-based Materials.

Chang, E.-E., Pan, S.-Y., Yang, L., Chen, Y.-H., Kim, H., Chiang, P.-C., 2015. Accelerated carbonation using municipal solid waste incinerator bottom ash and cold-rolling wastewater: performance evaluation and reaction kinetics. Waste Manag. 43, 283–292. https://doi.org/10.1016/j.wasman.2015.05.001.

Chen, B., Perumal, P., Illikainen, M., Ye, G., 2023. A review on the utilization of municipal solid waste incineration (MSWI) bottom ash as a mineral resource for construction materials. J. Build. Eng. 71, 106386. https://doi.org/10.1016/j.jobe.2023.106386.

Chen, B., Sun, Y., Jacquemin, L., Zhang, S., Blom, K., Luković, M., Ye, G., 2019. Pre-treatments of Mswi Bottom Ash for the Application as Supplementary Cememtitious Matreial in Blended Cement Paste.

Cheng, A., 2012. Effect of incinerator bottom ash properties on mechanical and pore size of blended cement mortars. Mater. Des. 1980-2015 36, 859–864. https://doi.org/10.1016/j.matdes.2011.05.003.

Chimenos, J.M., Fernández, A.I., Miralles, L., Segarra, M., Espiell, F., 2003. Short-term natural weathering of MSWI bottom ash as a function of particle size. Waste Manag. 23, 887–895. https://doi.org/10.1016/S0956-053X(03)00074-6.

Cho, B.H., Nam, B.H., An, J., Youn, H., 2020. Municipal solid waste incineration (MSWI) ashes as construction materials—a review. Materials 13, 3143. https://doi.org/10.3390/ma13143143.

Chuang, K.-H., Lu, C.-H., Chen, J.-C., Wey, M.-Y., 2018. Reuse of bottom ash and fly ash from mechanical-bed and fluidized-bed municipal incinerators in manufacturing lightweight aggregates. Ceram. Int. 44, 12691–12696. https://doi.org/10.1016/j.ceramint.2018.04.070.

Cioffi, R., Colangelo, F., Montagnaro, F., Santoro, L., 2011. Manufacture of artificial aggregate using MSWI bottom ash. Waste Manag. 31, 281–288. https://doi.org/10.1016/j.wasman.2010.05.020.

Clavier, K.A., Paris, J.M., Ferraro, C.C., Bueno, E.T., Tibbetts, C.M., Townsend, T.G., 2021. Washed waste incineration bottom ash as a raw ingredient in cement production: implications for lab-scale clinker behavior. Resour. Conserv. Recycl. 169, 105513. https://doi.org/10.1016/j.resconrec.2021.105513.

Clavier, K.A., Paris, J.M., Ferraro, C.C., Townsend, T.G., 2020. Opportunities and challenges associated with using municipal waste incineration ash as a raw ingredient in cement production – a review. Resour. Conserv. Recycl. 160, 104888. https://doi.org/10.1016/j.resconrec.2020.104888.

Clavier, K.A., Watts, B., Liu, Y., Ferraro, C.C., Townsend, T.G., 2019. Risk and performance assessment of cement made using municipal solid waste incinerator bottom ash as a cement kiln feed. Resour. Conserv. Recycl. 146, 270–279. https://doi.org/10.1016/j.resconrec.2019.03.047.

Ding, Y., Xi, Y., Gao, H., Wang, J., Wei, W., Zhang, R., 2022. Porosity of municipal solid waste incinerator bottom ash effects on asphalt mixture performance. J. Clean. Prod. 369, 133344. https://doi.org/10.1016/j.jclepro.2022.133344.

Fan, X., Li, Z., Zhang, W., Jin, H., Chen, C., Liu, J., Xing, F., Tang, L., 2022. Effects of different supplementary cementitious materials on the performance and environment of eco-friendly mortar prepared from waste incineration bottom ash. Construct. Build. Mater. 356, 129277. https://doi.org/10.1016/j.conbuildmat.2022.129277.

Giro-Paloma, J., Mañosa, J., Maldonado-Alameda, A., Quina, M.J., Chimenos, J.M., 2019. Rapid sintering of weathered municipal solid waste incinerator bottom ash and rice husk for lightweight aggregate manufacturing and product properties. J. Clean. Prod. 232, 713–721. https://doi.org/10.1016/j.jclepro.2019.06.010.

Huang, G., Yang, K., Chen, L., Lu, Z., Sun, Y., Zhang, X., Feng, Y., Ji, Y., Xu, Z., 2020. Use of pretreatment to prevent expansion and foaming in high-performance MSWI bottom ash alkali-activated mortars. Construct. Build. Mater. 245, 118471. https://doi.org/10.1016/j.conbuildmat.2020.118471.

Huang, Y., Wang, L., Wu, T., Liu, W., Tang, Q., 2023. Mechanical properties and heavy metal leaching behaviors of municipal solid waste incineration bottom ash as road embankment fillings. J. Clean. Prod. 394, 136355. https://doi.org/10.1016/j.jclepro.2023.136355.

Huynh, T.-P., Ngo, S.-H., 2022. Waste incineration bottom ash as a fine aggregate in mortar: an assessment of engineering properties, durability, and microstructure. J. Build. Eng. 52, 104446. https://doi.org/10.1016/j.jobe.2022.104446.

Hyks, J., Nesterov, I., Mogensen, E., Jensen, P.A., Astrup, T., 2011. Leaching from waste incineration bottom ashes treated in a rotary kiln. Waste Manag. Res. J. Sustain. Circ. Econ. 29, 995–1007. https://doi.org/10.1177/0734242X11417490.

Joseph, A.M., 2021. Processed Bottom Ash Based Sustainable Binders for Concrete. PhD Thesis).

Jurič, B., Hanžič, L., Ilić, R., Samec, N., 2006. Utilization of municipal solid waste bottom ash and recycled aggregate in concrete. Waste Manag. 26, 1436–1442. https://doi.org/10.1016/j.wasman.2005.10.016.

Kaza, S., Yao, L., Bhada-Tata, P., Van Woerden, F., 2018. What a Waste 2.0: A Global Snapshot of Solid Waste Management to 2050, Urban Development. World Bank Publications.

Keppert, M., Siddique, J.A., Pavlík, Z., Černý, R., 2015. Wet-treated MSWI fly ash used as supplementary cementitious material. Adv. Mater. Sci. Eng. 2015, 1–8. https://doi.org/10.1155/2015/842807.

Keulen, A., Zomeren, A. van, Harpe, P., Aarnink, W., Simons, H.A.E., Brouwers, H.J.H., 2016. High performance of treated and washed MSWI bottom ash granulates as natural aggregate replacement within earth-moist concrete. Waste Manag. 49, 83–95. https://doi.org/10.1016/j.wasman.2016.01.010.

Kleib, J., Aouad, G., Abriak, N.-E., Benzerzour, M., 2021. Production of portland cement clinker from French municipal solid waste incineration bottom ash. Case Stud. Constr. Mater. 15, e00629. https://doi.org/10.1016/j.cscm.2021.e00629.

Krammart, P., Tangtermsirikul, S., 2004. Properties of cement made by partially replacing cement raw materials with municipal solid waste ashes and calcium carbide waste. Construct. Build. Mater. 18, 579–583. https://doi.org/10.1016/j.conbuildmat.2004.04.014.

Kurda, R., Silva, R.V., De Brito, J., 2020. Incorporation of alkali-activated municipal solid waste incinerator bottom ash in mortar and concrete: a critical review. Materials 13, 3428. https://doi.org/10.3390/ma13153428.

Lam, C.H.K., Ip, A.W.M., Barford, J.P., McKay, G., 2010. Use of incineration MSW ash: a review. Sustainability 2, 1943–1968. https://doi.org/10.3390/su2071943.

Li, X.-G., Lv, Y., Ma, B.-G., Chen, Q.-B., Yin, X.-B., Jian, S.-W., 2012. Utilization of municipal solid waste incineration bottom ash in blended cement. J. Clean. Prod. 32, 96–100. https://doi.org/10.1016/j.jclepro.2012.03.038.

Li, Z., Zhang, W., Jin, H., Fan, X., Liu, J., Xing, F., Tang, L., 2023. Research on the durability and Sustainability of an artificial lightweight aggregate concrete made from municipal solid waste incinerator bottom ash (MSWIBA). Construct. Build. Mater. 365, 129993. https://doi.org/10.1016/j.conbuildmat.2022.129993.

Lin, W.Y., Heng, K.S., Sun, X., Wang, J.-Y., 2015. Influence of moisture content and temperature on degree of carbonation and the effect on Cu and Cr leaching from incineration bottom ash. Waste Manag. 43, 264–272. https://doi.org/10.1016/j.wasman.2015.05.029.

Liu, Y., Sidhu, K.S., Chen, Z., Yang, E.-H., 2018. Alkali-treated incineration bottom ash as supplementary cementitious materials. Construct. Build. Mater. 179, 371–378. https://doi.org/10.1016/j.conbuildmat.2018.05.231.

Lo, F.-C., Lo, S.-L., Lee, M.-G., 2020. Effect of partially replacing ordinary Portland cement with municipal solid waste incinerator ashes and rice husk ashes on pervious concrete quality. Environ. Sci. Pollut. Res. 27, 23742–23760. https://doi.org/10.1007/s11356-020-08796-z.

Lynn, C.J., Ghataora, G.S., Dhir Obe, R.K., 2017. Municipal incinerated bottom ash (MIBA) characteristics and potential for use in road pavements. Int. J. Pavement Res. Technol. 10, 185–201. https://doi.org/10.1016/j.ijprt.2016.12.003.

Margallo, M., Aldaco, R., Irabien, A., 2013. Life cycle assessment of bottom ash management from a municipal solid waste incinerator (mswi). Chem. Eng. Trans. 35, 871–876. https://doi.org/10.3303/CET1335145.

Mathews, G., Moazeni, F., Smolinski, R., 2020. Treatment of reclaimed municipal solid waste incinerator sands using alkaline treatments with mechanical agitation. J. Mater. Cycles Waste Manag. 22, 1630–1638. https://doi.org/10.1007/s10163-020-01053-y.

Meima, J.A., Comans, R.N.J., 1999. The leaching of trace elements from municipal solid waste incinerator bottom ash at different stages of weathering. Appl. Geochem. 14, 159–171. https://doi.org/10.1016/S0883-2927(98)00047-X.

Minane, J.R., Becquart, F., Abriak, N.E., Deboffe, C., 2017. Upgraded mineral sand fraction from MSWI bottom ash: an alternative solution for the substitution of natural aggregates in concrete applications. Procedia Eng. 180, 1213–1220. https://doi.org/10.1016/j.proeng.2017.04.282.

Müller, U., Rübner, K., 2006. The microstructure of concrete made with municipal waste incinerator bottom ash as an aggregate component. Cement Concr. Res. 36 (8), 1434–1443.

Musselman, C.N., Killeen, M.P., Eighmy, T.T., Gress, D.L., Presher, J.R., Sills, M.H., 1994. The New Hampshire Bottom Ash Paving Demonstration US Route 3 (Laconia, New Hampshire).

Pan, J.R., Huang, C., Kuo, J.-J., Lin, S.-H., 2008. Recycling MSWI bottom and fly ash as raw materials for Portland cement. Waste Manag. 28, 1113–1118. https://doi.org/10.1016/j.wasman.2007.04.009.

Ren, P., Ling, T.-C., Mo, K.H., 2020. Recent advances in artificial aggregate production. J. Clean. Prod. 19.

Robayo-Salazar, R., Mejía-Arcila, J., Mejía De Gutiérrez, R., Martínez, E., 2018. Life cycle assessment (LCA) of an alkali-activated binary concrete based on natural volcanic pozzolan: a comparative analysis to OPC concrete. Construct. Build. Mater. 176, 103–111. https://doi.org/10.1016/j.conbuildmat.2018.05.017.

Saikia, N., Mertens, G., Van Balen, K., Elsen, J., Van Gerven, T., Vandecasteele, C., 2015. Pre-treatment of municipal solid waste incineration (MSWI) bottom ash for utilisation in cement mortar. Construct. Build. Mater. 96, 76–85. https://doi.org/10.1016/j.conbuildmat.2015.07.185.

Sarmiento, L.M., Clavier, K.A., Paris, J.M., Ferraro, C.C., Townsend, T.G., 2019. Critical examination of recycled municipal solid waste incineration ash as a mineral source for portland cement manufacture – a case study. Resour. Conserv. Recycl. 148, 1–10. https://doi.org/10.1016/j.resconrec.2019.05.002.

Schabbach, L.M., Bolelli, G., Andreola, F., Lancellotti, I., Barbieri, L., 2012. Valorization of MSWI bottom ash through ceramic glazing process: a new technology. J. Clean. Prod. 23, 147–157. https://doi.org/10.1016/j.jclepro.2011.10.029.

Shen, P., Zheng, H., Lu, J., Poon, C.S., 2021. Utilization of municipal solid waste incineration bottom ash (IBA) aggregates in high-strength pervious concrete. Resour. Conserv. Recycl. 174, 105736. https://doi.org/10.1016/j.resconrec.2021.105736.

Shen, P., Zheng, H., Xuan, D., Lu, J.-X., Poon, C.S., 2020. Feasible use of municipal solid waste incineration bottom ash in ultra-high performance concrete. Cem. Concr. Compos. 114, 103814. https://doi.org/10.1016/j.cemconcomp.2020.103814.

Silva, R.V., De Brito, J., Lynn, C.J., Dhir, R.K., 2019. Environmental impacts of the use of bottom ashes from municipal solid waste incineration: a review. Resour. Conserv. Recycl. 140, 23–35. https://doi.org/10.1016/j.resconrec.2018.09.011.

Sun, Y., Li, J., Chen, Z., Xue, Q., Sun, Q., Zhou, Y., Chen, X., Liu, L., Poon, C.S., 2021. Production of lightweight aggregate ceramsite from red mud and municipal solid waste incineration bottom ash: mechanism and optimization. Construct. Build. Mater. 287, 122993. https://doi.org/10.1016/j.conbuildmat.2021.122993.

Šyc, M., Simon, F.G., Hykš, J., Braga, R., Biganzoli, L., Costa, G., Funari, V., Grosso, M., 2020. Metal recovery from incineration bottom ash: state-of-the-art and recent developments. J. Hazard Mater. 393, 122433. https://doi.org/10.1016/j.jhazmat.2020.122433.

Tang, J., Steenari, B.-M., 2016. Leaching optimization of municipal solid waste incineration ash for resource recovery: a case study of Cu, Zn, Pb and Cd. Waste Manag. 48, 315–322. https://doi.org/10.1016/j.wasman.2015.10.003.

Tang, P., Brouwers, H.J.H., 2017. Integral recycling of municipal solid waste incineration (MSWI) bottom ash fines (0–2mm) and industrial powder wastes by cold-bonding pelletization. Waste Manag. 62, 125–138. https://doi.org/10.1016/j.wasman.2017.02.028.

Tang, P., Chen, W., Xuan, D., Zuo, Y., Poon, C.S., 2020a. Investigation of cementitious properties of different constituents in municipal solid waste incineration bottom ash as supplementary cementitious materials. J. Clean. Prod. 258, 120675. https://doi.org/10.1016/j.jclepro.2020.120675.

Tang, P., Florea, M.V.A., Brouwers, H.J.H., 2017. Employing cold bonded pelletization to produce lightweight aggregates from incineration fine bottom ash. J. Clean. Prod. 165, 1371–1384. https://doi.org/10.1016/j.jclepro.2017.07.234.

Tang, P., Florea, M.V.A., Spiesz, P., Brouwers, H.J.H., 2016. Application of thermally activated municipal solid waste incineration (MSWI) bottom ash fines as binder substitute. Cem. Concr. Compos. 70, 194–205. https://doi.org/10.1016/j.cemconcomp.2016.03.015.

Tang, P., Florea, M.V.A., Spiesz, P., Brouwers, H.J.H., 2015. Characteristics and application potential of municipal solid waste incineration (MSWI) bottom ashes from two waste-to-energy plants. Construct. Build. Mater. 83, 77–94. https://doi.org/10.1016/j.conbuildmat.2015.02.033.

Tang, P., Xuan, D., Cheng, H.W., Poon, C.S., Tsang, D.C.W., 2020b. Use of CO2 curing to enhance the properties of cold bonded lightweight aggregates (CBLAs) produced with concrete slurry waste (CSW) and fine incineration bottom ash (IBA). J. Hazard Mater. 381, 120951. https://doi.org/10.1016/j.jhazmat.2019.120951.

Tang, P., Xuan, D., Poon, C.S., Tsang, D.C.W., 2019. Valorization of concrete slurry waste (CSW) and fine incineration bottom ash (IBA) into cold bonded lightweight aggregates (CBLAs): feasibility and influence of binder types. J. Hazard Mater. 368, 689–697. https://doi.org/10.1016/j.jhazmat.2019.01.112.

Thomas, M.D.A., Hooton, R.D., Scott, A., Zibara, H., 2012. The effect of supplementary cementitious materials on chloride binding in hardened cement paste. Cement Concr. Res. 42, 1–7. https://doi.org/10.1016/j.cemconres.2011.01.001.

Van Der Wegen, G., Hofstra, U., Speerstra, J., 2013. Upgraded MSWI bottom ash as aggregate in concrete. Waste Biomass Valoriz. 4, 737–743. https://doi.org/10.1007/s12649-013-9255-6.

Van Gerven, T., Van Keer, E., Arickx, S., Jaspers, M., Wauters, G., Vandecasteele, C., 2005. Carbonation of MSWI-bottom ash to decrease heavy metal leaching, in view of recycling. Waste Manag. 25, 291–300. https://doi.org/10.1016/j.wasman.2004.07.008.

Verbinnen, B., Billen, P., Van Caneghem, J., Vandecasteele, C., 2017. Recycling of MSWI bottom ash: a review of chemical barriers, engineering applications and treatment technologies. Waste Biomass Valorization 8, 1453–1466. https://doi.org/10.1007/s12649-016-9704-0.

Woo, L.Y., Wansom, S., Ozyurt, N., Mu, B., Shah, S.P., Mason, T.O., 2005. Characterizing fiber dispersion in cement composites using AC-Impedance Spectroscopy. Cem. Concr. Compos. 27, 627–636. https://doi.org/10.1016/j.cemconcomp.2004.06.003.

Wu, M.-H., Lin, C.-L., Huang, W.-C., Chen, J.-W., 2016. Characteristics of pervious concrete using incineration bottom ash in place of sandstone graded material. Construct. Build. Mater. 111, 618–624. https://doi.org/10.1016/j.conbuildmat.2016.02.146.

Xin-gang, Z., Gui-wu, J., Ang, L., Yun, L., 2016. Technology, cost, a performance of waste-to-energy incineration industry in China. Renew. Sustain. Energy Rev. 55, 115–130. https://doi.org/10.1016/j.rser.2015.10.137.

Xuan, D., Poon, C.S., 2018. Removal of metallic Al and Al/Zn alloys in MSWI bottom ash by alkaline treatment. J. Hazard Mater. 344, 73–80. https://doi.org/10.1016/j.jhazmat.2017.10.002.

Xuan, D., Tang, P., Poon, C.S., 2019. MSWIBA-based cellular alkali-activated concrete incorporating waste glass powder. Cem. Concr. Compos. 95, 128–136. https://doi.org/10.1016/j.cemconcomp.2018.10.018.

Xuan, D., Tang, P., Poon, C.S., 2018a. Limitations and quality upgrading techniques for utilization of MSW incineration bottom ash in engineering applications – a review. Construct. Build. Mater. 190, 1091–1102. https://doi.org/10.1016/j.conbuildmat.2018.09.174.

Xuan, D., Tang, P., Poon, C.S., 2018b. Effect of casting methods and SCMs on properties of mortars prepared with fine MSW incineration bottom ash. Construct. Build. Mater. 167, 890–898. https://doi.org/10.1016/j.conbuildmat.2018.02.077.

Yaphary, Y.L., Lu, J.-X., Chengbin, X., Shen, P., Ali, H.A., Xuan, D., Poon, C.S., 2022. Characteristics and production of semi-dry lightweight concrete with cold bonded aggregates made from recycling concrete slurry waste (CSW) and municipal solid waste incineration bottom ash (MSWIBA). J. Build. Eng. 45, 103434. https://doi.org/10.1016/j.jobe.2021.103434.

Zakir Hossain, H.M., Hasna Hossain, Q., Uddin Monir, MdM., Ahmed, MdT., 2014. Municipal solid waste (MSW) as a source of renewable energy in Bangladesh: revisited. Renew. Sustain. Energy Rev. 39, 35–41. https://doi.org/10.1016/j.rser.2014.07.007.

Zhang, B., Ma, Y., Yang, Y., Zheng, D., Wang, Y., Ji, T., 2023. Improving the high temperature resistance of alkali-activated slag paste using municipal solid waste incineration bottom ash. J. Build. Eng. 72, 106664. https://doi.org/10.1016/j.jobe.2023.106664.

Zhu, J., Wei, Z., Luo, Z., Yu, L., Yin, K., 2021. Phase changes during various treatment processes for incineration bottom ash from municipal solid wastes: a review in the application-environment nexus. Environ. Pollut. 287, 117618. https://doi.org/10.1016/j.envpol.2021.117618.

Chapter 20

Recycling of various types of slags as SCMs and aggregates

Ömer Faruk Kuranlı[1], Muhammed Bayram[2], Anıl Niş[3], Mucteba Uysal[1] and Togay Ozbakkaloglu[2]
[1]*Civil Engineering Department, Yıldız Technical University, Istanbul, Turkey;* [2]*Ingram School of Engineering, Texas State University, San Marcos, TX, United States;* [3]*Istanbul Gelisim University, Civil Engineering Department, Istanbul, Turkey*

1. Introduction

Concrete, due to its versatility as a construction material, ability to create durable structures that can withstand long periods, and wide range of applications, ranks among the most consumed materials worldwide. The increasing production of concrete and its environmental impacts have led to a growing importance of sustainable and more environmentally friendly solutions specifically within the realm of concrete materials. The rising concerns over carbon footprint, aggregate consumption, depletion of natural resources such as water, and the responsibility of Portland cement for approximately 7%–8% of CO_2 emissions have prompted a global quest for transformation in the industry. Despite these challenges, the construction industry is continuously researching and developing more sustainable practices and methods in concrete to minimize environmental impacts. Researchers aim to minimize resource consumption, extend the service life of structures, and reduce environmental effects (Nilimaa, 2023).

In the literature, the term "green concrete" has been developed to address the concerns mentioned before, which incorporates recycled materials and waste products. Concrete formulations that utilize cementitious and by-product materials such as fly ash and slag, as well as recycled aggregates, exemplify green concrete applications that reduce environmental impacts. Additionally, the use of local materials in concrete formulations can significantly reduce emissions associated with transportation. Local materials offer notable opportunities for enhancing concrete properties such as durability and strength. Examples of local materials include recycled aggregates and natural fibers (Saad Agwa et al., 2022; Kuranlõ et al., 2022).

It is widely recognized that a significant amount of diverse waste materials is generated from various sectors globally, including rural, urban, agricultural, and industrial sectors. Improper storage and disposal of waste materials can cause serious environmental harm. In the construction sector, it has long been known that many industrial wastes, such as slag, fly ash, silica fume, and red mud, can be successfully utilized in building materials such as concrete or mortar. The addition of these waste materials, either fully or partially, to Portland cement provides satisfactory solutions to the environmental issues (Nilimaa, 2023).

Slag material is typically a by-product obtained through the melting process of metals. In the broader perspective of the construction industry, slag can be categorized into two main groups: those derived from the production of iron and steel, specifically ferrous ores, and those obtained through the recovery of certain valuable nonferrous ores. Slag can be generated through the process of reclaiming raw materials, such as lead scrap and alkaline battery recycling, as well as during the vitrification of both municipal and nuclear waste. When examining the research conducted on slag, it can generally be categorized into two main areas: environmental effects and reuse focus. In terms of reuse, the research primarily focuses on using slag as a construction material, environmental improvement, and metal recovery from slag, as observed in the literature. Research in the area of reuse primarily revolves around iron slag, with a particular emphasis on its characterization and testing from an engineering perspective, especially within the field of civil engineering. Environmental studies mainly concentrate on the potential environmental impacts and the mineralogical properties of slag materials (Piatak et al., 2015).

Treatment and Utilization of Combustion and Incineration Residues. https://doi.org/10.1016/B978-0-443-21536-0.00034-4

While extracting metals from ore, a significant amount of slag is generated, commonly found in modern smelting facilities. In the past, there was limited scientific understanding of the environmental impact of slag material, and the lack of regulatory measures resulted in minimal public concern regarding slag waste and its reuse. It is evident that over time, awareness has increased and efforts have been directed toward the importance of recycling and reusing slag materials, primarily due to the limited scientific knowledge and inadequate legal regulations regarding the environmental effects of slag. This shift in focus can be attributed to growing global concerns regarding the environmental impact of slag materials and recognizing their potential as a valuable resource. As a result, slag has been used in construction materials, driven by the need to address environmental concerns and explore ways to characterize it as a sustainable material (Piatak and Seal, 2010).

Records of the use of slag in the construction sector date back to the 1930s, with notable examples including the Empire State Building, and its utilization has been increasing in the industry ever since. Slag-blended cement has been employed in various concrete projects, particularly in the United States and Europe, over the past few decades. It has been demonstrated that incorporating slag into concrete or mortar as a reinforcement can reduce the pore volume of the material, leading to the production of more durable materials that can be effectively utilized in construction. The utilization of slag material, along with supplementary cementitious materials (SCMs), has enabled the development of environmentally friendly and economically advantageous materials with higher mechanical properties, improved durability, and reduced permeability. This approach offers the advantage of obtaining a cost-effective material that is environmentally friendly, characterized by enhanced mechanical properties, improved durability, and reduced permeability (Aprianti, 2017).

One of the most commonly used slag types as a SCM in the construction sector is ground granulated blast furnace slag (GGBFS), obtained from pig iron production. It is utilized in clinker blends, often up to 95%, due to its hydraulic properties, and is known to significantly enhance concrete durability. As long as logistical and economic feasibility allows, GGBFS is readily used as an SCM by concrete producers. While GGBFS is practically used as an SCM, this is not the case for other metallurgical slags due to reasons such as low reactivity and the presence of incompatible phases. Nevertheless, research on the utilization of both ferrous and nonferrous metallurgical slags as SCMs is increasing, and slags from processes such as basic oxygen furnace (BOF) and electric arc furnace (EAF) are of particular interest in this regard (Juenger et al., 2019).

Furthermore, in addition to being used as an SCM, slag material can also be incorporated and used as an aggregate in concrete. For instance, GGBFS has been widely utilized for some time now, serving both as an SCM and an aggregate. When air-cooled blast furnace slag (BFS), which is a by-product of iron production, is slowly cooled, it solidifies into a gray, crystalline material with a stone-like characteristic and can be crushed to obtain coarse aggregates suitable for use in concrete (Wang, 2016a). Likewise, the dense and hard nature of Fe and steel slags, even if they contain voids, makes them suitable for use as construction aggregates. In a study conducted by Meslehuddin et al., it has been claimed that the use of steel slag as an aggregate results in a more durable product compared with some other aggregates, such as limestone aggregate (Piatak et al., 2015).

Understanding the concept of sustainability related to slag and having comprehensive knowledge about the latest technological advancements and their practical applications in the construction industry are crucial aspects of optimizing the utilization of existing material resources in construction. This knowledge and understanding will enable researchers to promote, validate, and evaluate the adoption of slag usage in economically, environmentally, and promising engineering practices. This chapter provides insights into the usage and characterization of various types of slag, systematic analysis of the performance and benefits of slag-based SCMs and aggregates, their limitations, and future expectations. It aims to present new perspectives and approaches for advancing technological aspects in slag utilization, offering pathways for further advancements in the field.

1.1 Bibliometric analysis

In the scope of this text, a search for "slag" in the Scopus database resulted in a list of 17,950 documents where the term appeared in the title, abstract, or keywords. The examined documents were filtered to include research articles, review articles, and book chapters. The documents were limited to the past 5 years, from 2018 to 2023, and a total of 32,027 keywords were listed. To visualize the connections and relationships among the keywords, the obtained file was analyzed using the VOSviewer software. The maximum number of keywords allowed in the analysis was set to 100. VOSviewer is a scientific academic mapping program that enables the visualization of bibliometric networks. Fig. 20.1 represents the network of keywords and their connections derived from the slag term search in the Scopus database. The thickness of the line connecting two keywords indicates the strength of the overall connection between them. The distance between keywords represents the strength of the relationship and how frequently they appear together in the literature.

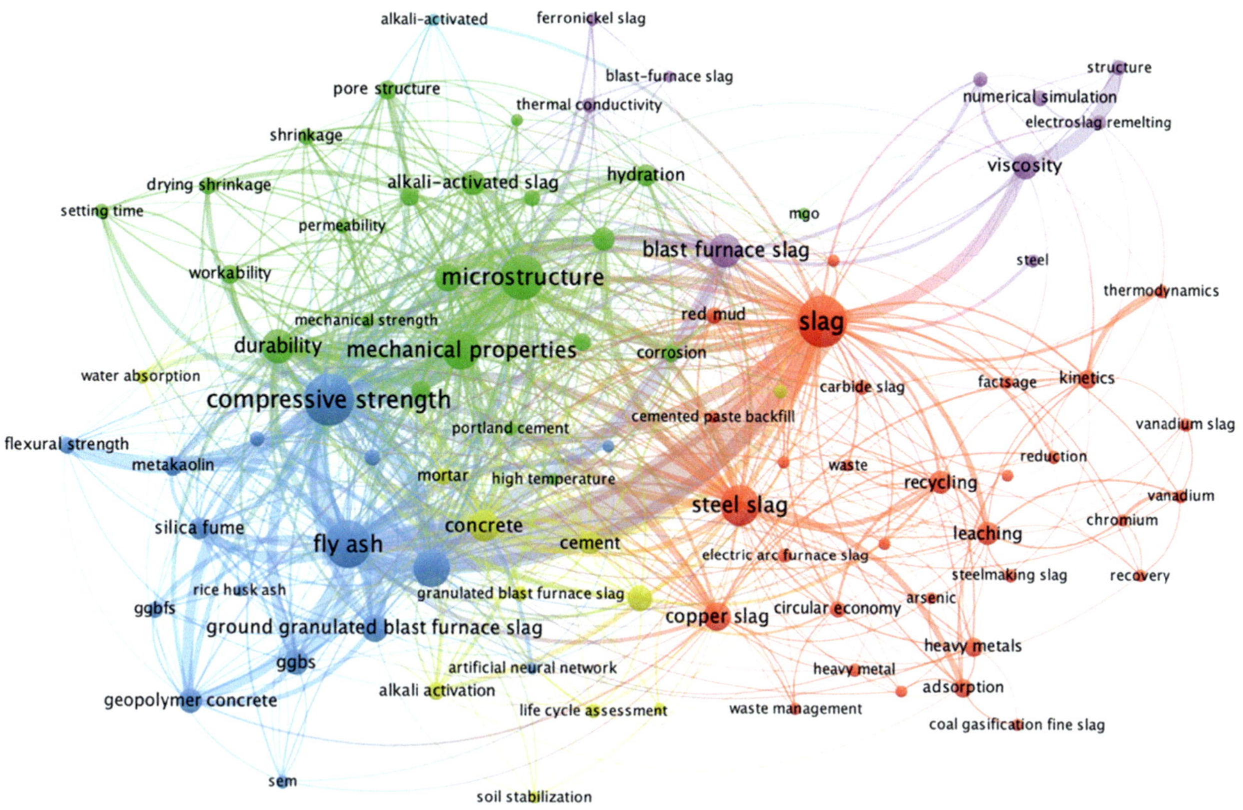

FIGURE 20.1 Network of keywords and their connections derived from the slag term search in the Scopus database.

According to the obtained keyword network, the existing literature focuses primarily on slag materials such as GGBFS, steel slag, and copper slag, highlighting their properties and applications. The documents are also closely related to emerging sustainable material types such as geopolymers, as well as other types of SCMs, particularly fly ash, metakaolin, and red mud. Additionally, the analysis reveals the strong association of slag materials with concepts such as circular economy, waste, and waste management. The literature explores various aspects, including life cycle analysis, microstructure, and various mechanical and durability properties of slag materials.

On the other hand, Table 20.1 illustrates the frequency of keyword usage in the documents listed under the focus of slag. The "Occurrences" attribute indicates how many times a keyword appears in the documents, representing the number of occurrences within the examined database. The "Total link strength" criterion captures the frequency of association (i.e., connectedness) between the specified keyword and other keywords in the database. From this perspective, it becomes apparent that slag materials are frequently evaluated in conjunction with another SCM, fly ash. They are also extensively researched within the context of new generation materials, such as geopolymers or alkali-activated materials. The literature explores various material properties, including durability and strength, and also delves into the relationship between slag materials and artificial intelligence applications, such as machine learning.

According to the dataset obtained from Scopus, the number of documents related to the "slag" material (appearing in abstracts, titles, or keywords) has significantly increased each year within the specified time range. Fig. 20.2 illustrates the document count for leading academic sources and journals worldwide, showing a consistent upward trend since 2018. This observation aligns with the earlier statement in this chapter, highlighting the growing attention of researchers from various disciplines (see Fig. 20.3 and Fig. 20.4) to slag materials due to their environmental impact and potential as a valuable resource. The data obtained from the keywords further supports this trend, indicating the association of sustainable new technologies with the topic. Furthermore, the analysis of important journals in the field demonstrates a continuous increase, affirming the rising interest in slag materials year by year.

TABLE 20.1 Frequency of keyword usage in the documents listed under the "slag" search focus.

No.	Keyword	Occurrences	Total link strength	No	Keyword	Occurrences	Total link strength	No.	Keyword	Occurrences	Total link strength
1	Fly ash	988	1966	**16**	Geopolymer concrete	264	425	**31**	Supplementary cementitious materials	192	226
2	Compressive strength	1132	1835	**17**	Cement	233	410	**32**	Pore structure	167	216
3	Slag	1128	1487	**18**	Sustainability	278	382	**33**	Alkali activation	143	212
4	Microstructure	868	1271	**19**	Hydration	213	354	**34**	Recycling	242	211
5	Geopolymer	619	1012	**20**	Copper slag	348	344	**35**	Leaching	223	208
6	Durability	512	907	**21**	Carbonation	224	343	**36**	ggbfs	126	207
7	Mechanical properties	636	870	**22**	Workability	152	327	**37**	Setting time	100	197
8	Concrete	416	708	**23**	Alkali-activated slag	263	295	**38**	Red mud	127	190
9	Steel slag	721	593	**24**	Flexural strength	124	265	**39**	Rheology	123	185
10	Strength	314	575	**25**	Viscosity	309	247	**40**	Self-compacting concrete	118	185
11	Blast furnace slag	514	565	**26**	Mortar	126	244	**41**	Water absorption	91	182
12	Ground granulated blast furnace slag	317	487	**27**	Alkali-activated materials	172	241	**42**	Ground granulated blast-furnace slag	107	159
13	Silica fume	209	476	**28**	Drying shrinkage	143	233	**43**	Permeability	91	157
14	ggbfs	280	459	**29**	Shrinkage	124	232	**44**	Granulated blast furnace slag	102	152
15	Metakaolin	187	447	**30**	Porosity	132	230	**45**	Mechanical strength	92	152

46	Corrosion	139	147	**61**	Autogenous shrinkage	81	107	**76**	Soil stabilization	68	86
47	Rice husk ash	67	144	**62**	xrd	68	107	**77**	Carbide slag	100	85
48	Thermal conductivity	107	143	**63**	Limestone	61	103	**78**	Crystallization	110	85
49	Sem	81	142	**64**	Unconfined compressive strength	89	103	**79**	Temperature	76	84
50	Geopolymers	101	137	**65**	Ferronickel slag	85	100	**80**	Waste management	71	84
51	Kinetics	152	136	**66**	Heavy metals	151	97	**81**	Waste	73	80
52	Portland cement	83	131	**67**	High temperature	66	97	**82**	Stabilization	69	77
53	Alkali-activated	75	124	**68**	Phosphogypsum	70	96	**83**	Industrial waste	66	77
54	Adsorption	155	120	**69**	Structure	109	96	**84**	Heavy metal	78	62
55	Circular economy	114	118	**70**	Density	54	93	**85**	Solid waste	76	59
56	Life cycle assessment	105	112	**71**	Machine learning	69	93	**86**	Reduction	83	57
57	Hydration products	73	111	**72**	Cemented paste backfill	66	91	**87**	Recovery	73	55
58	Lime	62	111	**73**	Blast-furnace slag	65	89	**88**	Electroslag remelting	95	49
59	Alkali-activated concrete	63	108	**74**	Artificial neural network	70	86	**89**	Steelmaking slag	90	41
60	Electric arc furnace slag	96	108	**75**	Optimization	77	86	**90**	Vanadium slag	80	38

FIGURE 20.2 The number of documents on the topic of "slag" in globally leading academic sources and journals.

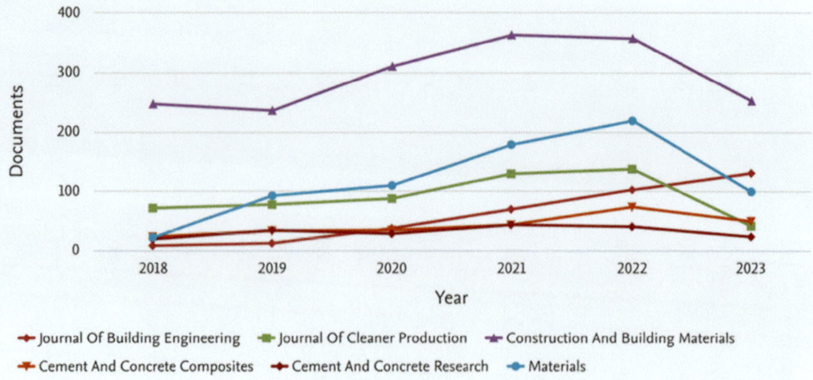

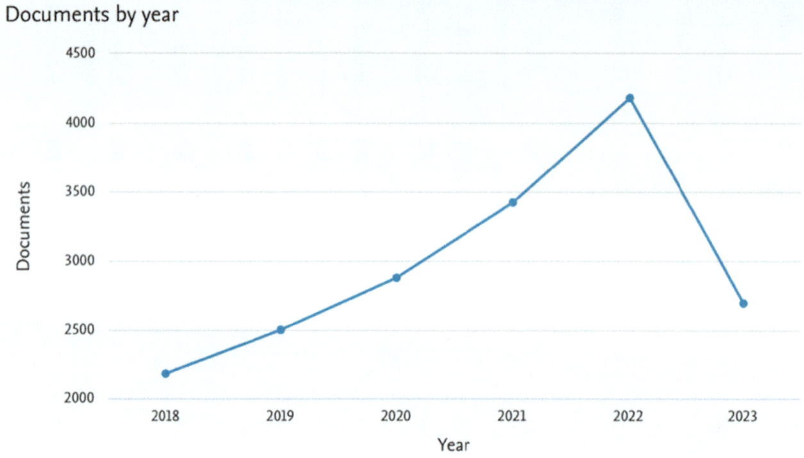

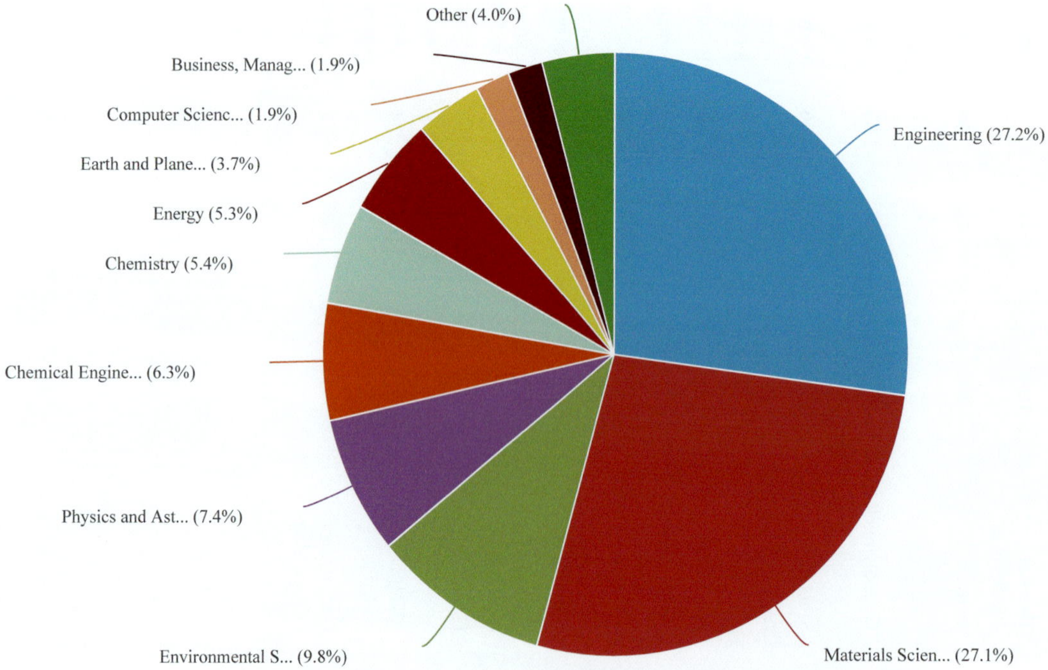

FIGURE 20.3 Pie chart distribution of documents related to the subject area of "slag."

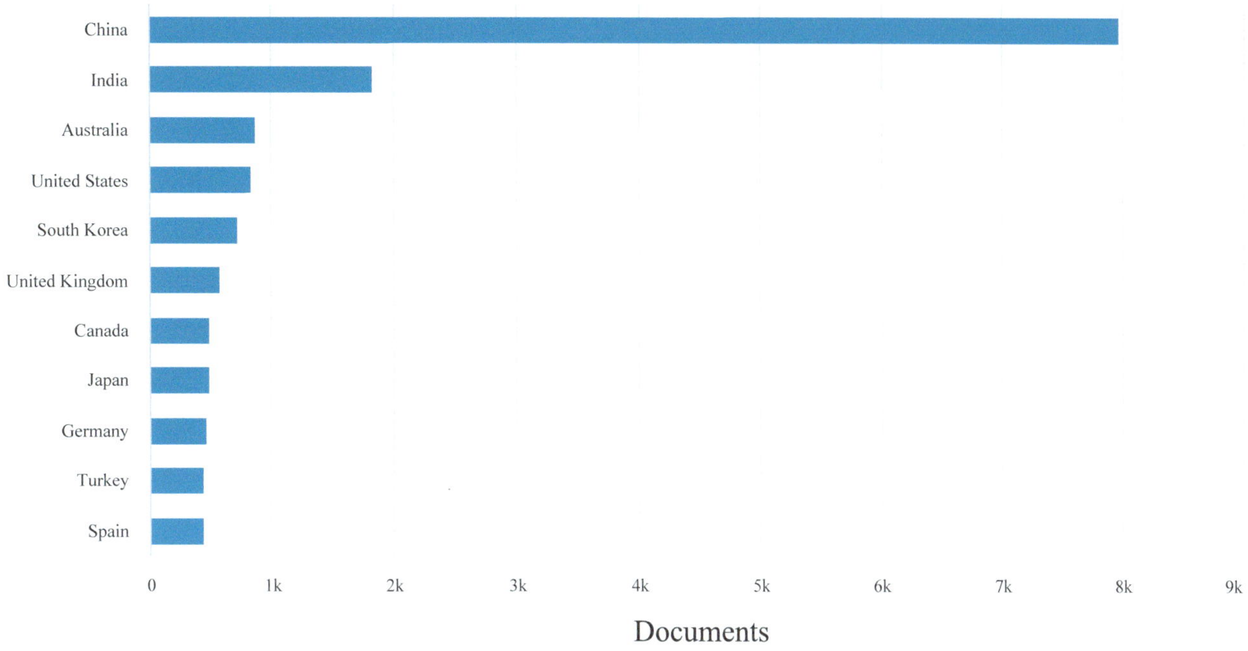

FIGURE 20.4 Distribution of documents related to the subject of "slag" by countries.

1.2 Slag types

In general, slag is known as a valuable by-product that occurs in powdered form during metal refining processes. It can be used in various applications, particularly in construction, where it is commonly utilized as an additive in mortar and concrete (Sajedi et al., 2012). Slag can be divided into two main types: ferrous slag, derived from the primary production of iron-bearing ores, and nonferrous slag, obtained from non—iron-bearing ores or the recovery of nonferrous metals. As previously mentioned, slag can also be obtained in a nonferrous form during processes such as vitrification of nuclear or domestic waste or recycling of certain raw materials such as scrap and alkaline substances. However, due to limited research available on these types of slags in the literature, this chapter does not focus on this particular aspect of slag materials (Piatak et al., 2015).

1.2.1 Ferrous slags

Steel, which is a product of carbon and iron alloy, can be defined and expressed within the ferrous metals category. Steel is a widely encountered and well-known iron-based metal, extensively used in various industries, particularly in the construction sector. The production of steel involves an integrated process that includes BOFs, EAFs, and blast furnaces (BFs) (Wang 2016b).

Ferrous slag, which refers to the slag produced from the steel or iron production sector, can generally be categorized into two types: ironmaking slag (BFS) and steelmaking slag (BOF, EAF, and ladle slag) (Yildirim and Prezzi, 2011). After the EAF or BOF process, the molten steel obtained, as shown in Fig. 20.5, can be transferred to a ladle furnace for additional refining or can be sent directly to a continuous casting facility for further casting. On the other hand, BFS is obtained during the melting of iron scrap, coke, and ore as part of the iron production process. BFS is formed after the melting process and is discharged. Steel slag, on the other hand, is discharged as molten slag during steel production and then remelted and discharged again with the addition of steel scraps, fluxes, and ferroalloys. The molten slag is processed through various cooling processes after discharge, resulting in various slag products (Wang 2016b). Furthermore, it is known that each cycle can produce up to 300 tons of steel, and the duration of this cycle is approximately 3 h. EAF, despite being more expensive than the BOF process, was preferred for the production of high-quality steel materials. However, it has been noted that with the increase in the size of EAFs and the ability to produce various steel grades, the method has gained a competitive position in the US industry, with its share in steelmaking reaching around 70.6% (Yildirim and Prezzi, 2011; American Iron and Steel Institution, 2018).

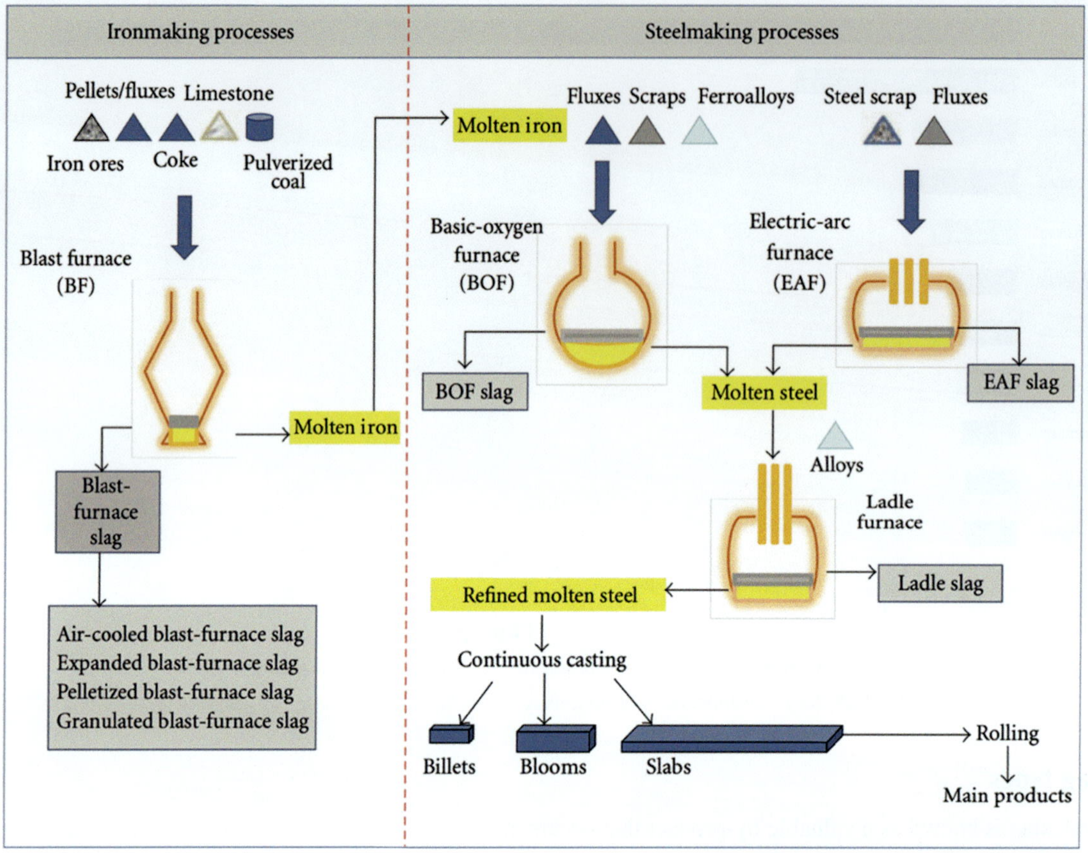

FIGURE 20.5 Illustration of the processes involved in the production of iron and steel (Yildirim and Prezzi, 2011).

According to the United States Geological Survey, ferrous slag production can typically be estimated based on slag-to-metal ratios. For a typical iron ore with a slag-to-metal ratio of 60%−66%, BFs producing pig iron yield 0.25−0.30 metric tons of slag per metric ton of pig iron. It has been reported that lower-grade ores, which are below the average, produce more slag, resulting in 1−1.2 tons of slag per metric ton of crude iron. According to the United States Geological Survey, global BFS production is estimated to be around 312 million metric tons, while the average steel furnace slag production is estimated to be 225 million metric tons (U.S. Geological Survey, 2023).

The cooling method employed for ferrous slags actually influences their commercial usability. BFS can be cooled into four main product types: expanded BFS, pelletized BFS, granulated BFS, and air-cooled BFS. Air-cooled BFS is slowly cooled and is suitable for use as a construction aggregate. GBFS, when finely ground, produces GGBFS, which possesses cementitious properties and can be incorporated into concrete mixtures. Pelletized (expanded) slag, similar to GGBFS in cementitious characteristics, is cooled by a water jet and is commonly used as lightweight aggregate. On the other hand, steel furnace slag (SFS), a by-product of the steelmaking process, is cooled in a manner similar to air-cooled slag and can be used for similar purposes. SFS can also be utilized in environmental applications such as water filtration (U.S. Geological Survey, 2023).

1.2.2 Nonferrous slag

Nonferrous slags are produced from primary or secondary raw materials during the recovery and processing of nonferrous metals. Nonferrous slags are molten by-products and are used to separate nonferrous metals from other components. Nonferrous slags are generally obtained from the processing of nickel, phosphorus, copper, and zinc. For example, in copper production, the ore is crushed and separated through flotation to concentrate the copper content. The resulting concentrates are then melted with a focus on removing impurities to obtain the primary form of copper material. To achieve a higher level of purity, copper is remelted at high temperatures. In the mentioned process, during the flotation stage, along with various sediment materials, copper slag is obtained from the smelting stage (Kovacs et al., 2017). The schematic information regarding the general production logic of nonferrous slags is depicted in Fig. 20.6.

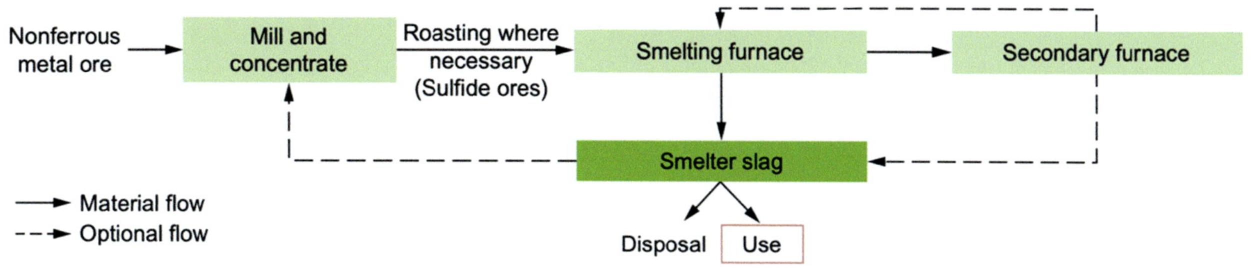

FIGURE 20.6 The schematic information regarding the general production logic of nonferrous slags (Lewis, 1982; Kovacs et al., 2017).

Gorai et al. (2003) claimed that approximately 2.2 tons of slag is produced per ton of copper produced. Nonferrous slags cannot be produced in large volumes such as ferrous slags. It is believed that the construction of engineering sector is the main end-use application for nonferrous slags in terms of large-volume consumption. Industry reports have indicated that the global volume of nonferrous slag usage in the construction engineering sector is approximately 35% in previous years. Due to the decrease in aggregate demand in various regions of the construction sector and environmental concerns, nonferrous slags such as copper, nickel, chromium, and lead are seen as significant sources of supply for aggregates and reinforcement materials (Smithers, 2023).

2. Properties and characteristics of various slags

2.1 Composition and chemical characteristics of slags

Table 20.2 illustrates the typical chemical compositions of ferrous slag types. It can be observed that ferrous slags are predominantly dominated by Ca and Si, with varying amounts of Mg, Al, and Fe. It has been noted in the literature that certain slag components such as SiO_2 and MgO are partially introduced from the ash of the carbon source or the refractory material of the BF (Yang et al., 2022).

The chemical compositions of ferrous slags vary depending on the quality of the iron ore, operational characteristics, and parameters of the carbon material. For example, some iron ores obtained from different locations may contain low amounts of Al_2O_3, such as in Brazil, while others may have high amounts of Al_2O_3, such as in Malaysia. Additionally, the scarcity of high-quality iron ores and the utilization of secondary sources of iron waste contribute to the complexity of BFS compositions. In slag compositions, typical components such as SiO_2, MnO, and Al_2O_3 are predominantly determined using X-ray fluorescence (XRF) and inductively coupled plasma (ICP) methods. The chemical titration process is known to be used for detecting the reduction of iron oxides. Additionally, X-ray diffraction (XRD) methods are preferred to better determine the cementitious properties of the material (Yang et al., 2022).

Table 20.3 presents the typical chemical compositions of common nonferrous slags. Nonferrous slags, despite containing trace amounts, do contain heavy metal oxides in comparison to the flagship ferrous slags such as BF and steel slags. As a result, some nonferrous slags are classified as hazardous waste. The disposal processes of these hazardous waste slags from storage facilities automatically increase the costs. The utilization of nonferrous slags and the obligations and challenges imposed by their hazardous nature create research areas for researchers (Wang, 2016c).

Air-cooled copper slag, which is a type of nonferrous slag, has a glassy appearance and a matte black color. On the other hand, granulated copper slag is more porous compared with air-cooled slag, resulting in a lower specific gravity. The specific gravity of copper slag can vary from 2.8 to 3.8, with the iron content being the determining factor in this range.

TABLE 20.2 Typical chemical compositions of ferrous slag types (Yang et al., 2022).

Slag type	CaO	SiO_2	Al_2O_3	MgO	MnO	FeO	P_2O_5
BFS	30–45	28–40	8–24	1–18	<2	<1	–
BOFS	35–60	7–18	0.5–5	0.4–14	2–6	10–38	1–4
EAFS	22–60	5–34	3–18	3–13	–	<40	2–5
LFS	30–60	2–35	5–35	1–10	0–5	0.1–1.5	≤0.9

TABLE 20.3 Typical chemical compositions of common nonferrous slags (Wang, 2016c).

	Fe (total)	SiO_2	Al_2O_3	CaO	MgO	MnO	S	Cu	Ni	PbO	ZnO
Copper slag	29–51	24–38.9	2.94–15.6	2.0–5.9	0.1–3.5	0.4–1.7	0.28–0.98	0.2–2.1	–		–
Nickel slag	20.7–53.06	29–50.5	<0.10	1.8–3.96	1.56–26.7	0.35–0.36	0.36	–	0.16–0.17		–
Lead slag	20–28.7	21.39–35	3.56–10	16–23.11	5.44–10	1.44–5	0.37–2	–	–	2–6	–
Zinc slag	25–52	19–40	2–10	15–23	0.5–5	–	1.1	–	–		0–5

Nickel slag, on the other hand, is a highly siliceous material, and this value can reach up to 55%. Compared with copper slag, nickel slag has lower alumina, iron, and lime content. As it contains a certain amount of alumina and lime in its chemical composition, nickel slag cannot form the primary silicate forms responsible for the hydration of BFS. Therefore, the literature indicates that nickel slag has low hydraulic activity. Nickel slag, which has an amorphous massive glassy texture, has a black color with a brownish tint and can have a specific gravity of up to 3.5. The BF method is known to be the most popular furnace method for smelting lead. Based on known XRD patterns, lead slag consists of kirschsteinite ($CaFe(SiO_4)$), franklinite ($(ZnFeMn)(FeMn)_2O_4$), metallic lead wustite (FeO), and spinel ($MgAl_2O_4$). However, secondary lead slag is classified as hazardous waste. Lead slag has a glassy appearance and can range in color from black to red. Its particles are cubic and have a sharp form. Granulated lead slag has a wide specific gravity range of 2.65–3.79 and tends to be porous (Wang, 2016c).

On the other hand, SEM images of some ferrous slags are shown in Fig. 20.7A and C as BOF slag, and in Fig. 20.7B as EAF slag. As can be observed from Fig. 20.7, it can be stated that both BOF and EAF slags are composed of irregularly shaped fine particles and are coarse. Researchers have suggested that this condition may arise due to irregular thermodynamic conditions during slag formation. Some researchers have noted that BOF slags have a rough surface texture due to the low brittleness of the crystalline material, as well as a slightly rounded irregular morphology. Furthermore, it has been stated that this morphological characteristic of BOF slag could enhance the interaction between cement paste and concrete aggregates, thereby potentially improving certain mechanical properties (Carvalho et al., 2017). SEM images of EAF slags in the literature have shown a tendency for EAF slags to have a plate-like and crystalline structure.

SEM images of copper slag are presented in Fig. 20.8A and B, while zinc slag is shown in Fig. 20.8C among the nonferrous slag types. Generally, copper slag exhibits an angular and sharp-edged appearance. Additionally, it can be stated that the surface of copper slag is glassy and dry (Wu et al., 2010). It has been stated that copper slags predominantly consist of particles smaller than 100 μm and exhibit a very smooth surface morphology (Chun et al., 2016). On the other hand, zinc slag is primarily observed in the form of block-like and small particles. It is noted that the particle diameter of zinc slag varies between 1 and 100 μm, and its structure is loose (Duan et al., 2021; Kanneboina et al., 2023).

FIGURE 20.7 SEM images of some ferrous slags (A), (C) BOF slag, and (B) EAF slag (Li et al., 2023; Carvalho et al., 2017).

FIGURE 20.8 SEM images of (A, B) copper slag and (C) zinc slag (Wu et al., 2010; Chun et al., 2016; Duan et al., 2021).

3. Recycling various slags into supplementary cementitious materials

The demand for concrete material, which is an indispensable material of modern society, is increasing day by day. Consequently, as mentioned in previous sections, various resource consumption and environmental issues have been brought to the forefront, and the drawbacks of the concrete industry have become a subject of discussion. To minimize the negative effects of the industry and provide sustainable solutions, the use of cementitious materials and their recycling will greatly contribute to environmental sustainability. As mentioned, slag material stands out as a prominent SCM due to its nature as a by-product and its abundant availability worldwide. Its ability to be recycled and used with satisfactory engineering properties further enhances its status as an SCM material.

The provided text summarizes the existing studies on the use of various ferrous and nonferrous slag types as SCMs, and these studies are presented for ferrous slag types in Table 20.4 and for nonferrous slag types in Table 20.5. These studies primarily focus on research conducted within the past 3 years. In the case of ferrous slag, ground granulated blast furnace slag (GGBFS) and steel slag are the most commonly used types. Researchers have explored substituting these mentioned ferrous slag types for ordinary Portland cement (OPC), as well as incorporating other materials such as fly ash, aluminum powder, rice husk ash, waste glass powder, silica fume, and limestone powder filler. It is observed that the majority of the studies presented in the tables investigate the mechanical properties such as compressive strength, tensile strength, flexural strength, as well as the characterization, microstructure, and durability properties such as high-temperature resistance, acid-attack resistance, freeze-thaw resistance, and others.

Regarding nonferrous slag types used as SCMs, fly ash remains a popular by-product choice, and in addition to this, other materials such as iron ore tailings, silica fume, and quartz powder have been considered. These studies have employed both ferrous and nonferrous slag types in various applications such as cement paste, mortars, high-strength concrete, ultrahigh-strength concrete, self-compacting concrete, 3D printed concrete, and similar applications, leading to a wide range of outcomes and results.

TABLE 20.4 Various studies on the utilization of different types of ferrous slag as supplementary cementitious materials (SCMs) and their respective parameters have been conducted.

Authors	Slag binder type	Other materials	Product	Aggregate	Curing (days)	Tests conducted	Year
Rajesh Ram et al. (2023)	GGBFS	FA, aluminum powder, OPC	Autoclaved aerated concrete	Unspecified	28	Compressive strength, water absorbtion, acid resistance	2023
Alomayri et al. (2023)	GGBFS	OPC, polypropylene fibers	Seawater concrete	Natural fine and coarse	7, 28, 91, 182	Compressive strength, split tensile, flexural strength, water absorbtion, drying shrinkage, rapid chloride ion permeability, wetting cycle	2023
Kumar and Deep (2023)	GGBFS	OPC, FA	Concrete	Natural fine and coarse	7, 12, 28	clump, compressive strength, flexural strength, acid attack	2023
Shahas et al. (2023)	GGBFS	OPC, rice husk ash	Mortar	Natural fine	7, 28	Setting time, flow test, compressive strength	2023
Ramakrishnan et al. (2017)	GGBFS	OPC, waste glass powder	Concrete	Natural fine and coarse	3, 7, 28	Compressive strength, split tensile, sorptivity, water absorption, flexural strength, bond strength	2017
Raj et al. (2023)	GGBFS	OPC	Concrete	Natural fine and coarse	7, 14, 28	Compressive strength, split tensile, flexural strength	2023
Gholampour et al. (2021)	GGBFS	OPC, FA	Concrete	Recycled fine, foundry sand, natural fine and coarse	7, 28	Compressive strength, flexural strength, split tensile, water absorbtion	2021
Kanavaris et al. (2023)	GGBFS	OPC	Concrete, mortar	Natural fine and coarse	3, 7, 28	Compressive strength, elevated temperature	2023
Li G. et al. (2020)	GGBFS	OPC, crystalline admixture	Mortar	Natural fine	28, 56	Compressive strength, water absorption, self-healing capability	2020
Ramakrishnan et al. (2023)	GGBFS	OPC, granite powder	Concrete	Natural fine	7, 14, 28	Compressive strength, tensile strength	2023
Gokulakannan et al. (2023)	GGBFS	OPC, metakaolin	Concrete	Natural fine and coarse, fine copper slag	28	Workability/slump, water absorption, porosity, acid attack, flexural behavior	2023
He. et al. (2023)	GGBFS	OPC, FA	Paste	Natural fine	–	Carbonation degree, compressive strength	2023
Mohan and Tabish Hayat (2021)	GGBFS	OPC, silica fume	Concrete	Natural fine and coarse	3, 7, 28	Compressive strength, flexural strength, split tensile	2023

Continued

TABLE 20.4 Various studies on the utilization of different types of ferrous slag as supplementary cementitious materials (SCMs) and their respective parameters have been conducted.—cont'd

Authors	Slag binder type	Other materials	Product	Aggregate	Curing (days)	Tests conducted	Year
Huang et al. (2023)	GGBFS	OPC	Concrete	Steel slag coarse aggregate, natural fine and coarse	28	Heating test, compressive test, mass loss, surface change	2023
Nicula et al. (2020)	GGBFS	OPC	Road concrete	Natural fine and coarse	7, 50, 90	Dynamic elasticity, freeze-thaw resistance, water absorption	2020
Dhivya et al. (2023)	Steel slag (IFS)	OPC	Concrete	Natural fine and coarse	28	Compressive strength, flexural strength, chloride penetrability, drying shrinkage, resistance to carbonation	2023
Wang and Suraneni (2019)	Steel slag (BFS, BOFS, and LFS	OPC	Paste	–	1.7, 28, 91	Pozzolanic test, isothermal calorimetry, thermogravimetric analysis, compressive strength	2019
Liu Q. et al. (2016)	Steel slag (BOF)	OPC	Paste		7, 28	Compressive strength, materials characterization tests	2016
Carvalho et al. (2023)	Steel slag (BOF)	OPC	Mortar	Natural fine	3, 7, 28, 90	Water absorbtion, porosity, void index, pore distribution, compressive strength, hydration process	2023
Gu et al. (2023)	Steel slag and lithium slag	OPC	Mortar	Natural fine	7, 28, 90	Flowability, phase analysis, surface fractal characteristics, mercury intrusion porosimetry, compressive strength	2023
Liu P. et al. (2023)	Granular steel slag	OPC	Paste, mortar	Natural fine	3, 7, 28, 90	Compressive strength, microstructures, and hydration reactivities	2023
Srivastava et al. (2023)	Steel furnace slag, basic oxygen furnace slag and desulfurized slag	OPC	Paste	–	3, 7, 28	Compressive strength, microstructural analysis, thermal decomposition	2023
Fan et al. (2023)	Steel slag	OPC, silica fume, limestone powder	Ultrahigh strength concrete	River sand	3, 28	Packing density, flowability, compressive strength, flexural strength	2023

TABLE 20.4 Various studies on the utilization of different types of ferrous slag as supplementary cementitious materials (SCMs) and their respective parameters have been conducted.—cont'd

Authors	Slag binder type	Other materials	Product	Aggregate	Curing (days)	Tests conducted	Year
Weng et al. (2021)	Steel slag	OPC	Paste	–	1	Hydration expansion, SEM	2020
Yu et al. (2023)	Steel slag	OPC, silica fume, and FA	3D printed concrete	River sand	7, 28	Flexural and compressive strength, mechanical anisotropy analysis, cost analysis, rheology properties	2023
Roslan et al. (2020)	Steel slag (EAF)	OPC, steel sludge	Concrete	River sand, crushed granite coarse aggregate	7, 28, 60, 90	Workability, compressive strength, initial surface absorption test (ISAT), water absorption, leaching test, morphology	2020
Baalamurugan et al. (2022)	Induction furnace steel slag	OPC	Concrete	Natural fine and coarse	28	Surface morphology, structural analysis, pH change, compressive strength	2022
Matos et al. (2020)	Air-cooled blast furnace slag, GBFS	OPC, limestone filler	Paste, self-compacting concrete	Natural fine and coarse	3, 28, 91	Rheological tests, slump flow, V-funnel, isothermal calorimetry, compressive strength	2020
Fang et al. (2022)	Ladle furnace slag	OPC	Paste	–	1, 3, 28	Isothermal calorimetric reaction measurement, setting time, hydration property tests, compressive strength, microstructure	2022
Sultana and Islam (2023)	Ladle furnace slag	OPC	Concrete	Natural fine and coarse	3, 7, 28, 56, 90, 180	Workability, compressive strength, tensile strength, pulse velocity, gas permeability, sorptivity	2023

While some nonferrous slags are used for construction purposes, the majority of ferrous slags can be utilized in the construction industry. When slag material, especially in granulated form, has a glassy nature and combines with free lime during hydration, it imparts cementitious properties that significantly enhance the material's strength. In the literature, it is noted that a significant portion of Fe slag is granulated, and its most common application is in cement production. It has been stated that the average prices of granulated slags are considerably higher than those of other types of ferrous slag (Van Oss, 2013; Ali et al., 2013).

As mentioned, nonferrous slags can also be reused in the construction sector. It is known that zinc slag can be used for ceramic tile production and can also be preferred as asphalt aggregate. Similarly, copper and Ni slags, which are other types of nonferrous slag, have been shown in the literature to be suitable for roof covering, ballast, or asphalt aggregate applications. On the other hand, the positive effect of adding a certain amount of Cu to cement material on its burnability has been addressed by Ali et al. (2013). The researchers conducted a study on the effect of adding a typical copper slag by-product at a ratio of 1.5%–2.5% to cement raw mixtures prepared from two different limestone samples with free silica

TABLE 20.5 Various studies on the utilization of different types of nonferrous slag as supplementary cementitious materials (SCMs) and their respective parameters have been conducted.

Authors	Slag binder type	Other materials	Product	Aggregate	Curing (days)	Tests conducted	Year
Revathi et al. (2023)	Copper slag	OPC, coconut fiber	Concrete	Natural fine and coarse	7, 28	Compressive strength, split tensile, flexural strength, impact test	2023
Edwin et al. (2022)	Copper slag	OPC, silica fume, quartz powder	Ultrahigh-performance concrete	Basalat aggregate, quickly cooled granulated copper slag	7, 28, 56	Workability, compressive strength, flexural strength, isothermal calorimetry, Tg analysis, mercury intrusion porosimetry, leaching test	2022
Oyejobi et al. (2023)	Copper slag	OPC, FA	Mortar	River sand	1, 3, 7, 14, 28	Flow characteristics, setting time, compressive strength	2023
Tiwary and Bhatia (2022)	Copper slag	OPC, FA	Concrete	Natural fine and coarse	7, 14, 28, 56, 91	Compressive strength, split tensile, flexural strength, rebound hammer, UPV	2022
He et al. (2021)	Copper slag, blast furnace slag	OPC, FA	Paste	–	3, 28	Leaching tests, pH-dependent leaching behavior, reactivity tests, sequential extraction tests, exchangeable fraction, carbonate fraction, compressive strength	2022
Wang D. et al. (2020)	Copper slag	OPC	Paste	–	7, 28	Chemical properties, physical properties, soundness, fluidity and reactivity index, radioactivity, heavy metals properties, pH-dependent leaching test, compressive strength	2020
Han F. et al. (2023)	Nickel slag, blast furnace slag	OPC, FA	Paste	–	–	Rheological tests, hydration heat	2023
Yang H. et al. (2021)	Nickel slag, GGBFS	OPC, pulverized FA	Paste	–	–	Elevated temperature, residual compressive strength, hydration products (XRD, TG, and DTG)	2021
Wu Q. et al. (2018)	Nickel slag, steel slag	OPC, FA, clay, fluoride	Paste	–	3, 7, 28	Characterization of clinker (XRD, XRF), compressive strength, bending strength	2018
Zhang et al. (2023)	Phosphate slags, lithium slags	OPC, iron ore tailings	Concrete	Iron tailings sand	7, 14, 28	Compressive strength, mercury intrusion porosimetry tests, backscattering electron tests	2023

contents of 5.52 and 10.0. The results showed an increase in lime assimilation rate and a decrease in clinkering temperature by 50°C. Gorai et al. (2003) stated that copper slag, as a type of nonferrous slag, can be used not only for metallurgical recycling or recovery but also in the production of materials such as aggregates, glass, tiles, and cement.

GGBFS is typically one of the most commonly used materials among SCMs. It can largely replace clinker and significantly improve the durability of concrete to a great extent. Compared with other slags, BOF carbon steel slags are known to contain a significantly lower amount of heavy metals. However, they still contain calcium silicates that do not possess free lime and hydraulic properties. In the literature, it has been reported that carbon steel slags cause a delay in early hydration of cement and reduce early-age strength development. Studies have shown that carbonation curing or curing at high temperatures can activate steel slag and improve material performance. Nonferrous slags derived from the production of lead, copper, or other metals are generally rich in iron. Copper slag, as a nonferrous slag material, has limited pozzolanic activity and is often utilized as a filling material in concrete. In general, fine grinding, high-temperature processing, and the addition of calcium enhance the potential use of slag as SCM. However, it is known that the presence of heavy metals such as Pb or Zn in nonferrous slags delays hydration (Juenger et al., 2019).

The hydraulic property of steel slag is weak due to its low hydraulic component content. Therefore, various techniques have been developed in the literature to enhance the cementitious activity and activate the slag. For example, Carvalho et al. (2017) produced and characterized basic oxygen furnace and similar materials as powders with different particle size ranges. They expanded the concept of high binding efficiency, good mechanical performance, and low-cost supplementary material using this mechanical activation method. Qian et al. (2002) subjected nonhydraulic kirschsteinite-based steel slag, which does not behave like a typical hydraulic material, to autoclave processes and evaluated the hydrothermal products and binding properties. They demonstrated that thermal activation methods can modify the binding behavior of kirschsteinite-based steel slag. Qi et al. (2016), on the other hand, improved the binding capacity and hydration behavior of steel slag using a chemical activation method by utilizing the compound effect of $CaCO_3$ and $CaSO_4 \cdot 2H_2O$. Furthermore, adding mineral admixtures to the slag material has increased the content of cementitious materials and improved the hydraulic activity of cementitious materials. For example, Liu and Wang (2017) blended steel slag and silica fume and milled them to obtain a composite mineral admixture material. The researchers stated that the composite mineral admixture improved durability properties such as chloride resistance and carbonation resistance, as well as certain strength properties.

3.1 Mechanical performance of cement-based materials produced with various slags

There are numerous current studies that investigate the mechanical properties of various types of slag when used as SCMs:

Alomayri et al. (2023) examined the drying shrinkage, tidal erosion, and some mechanical properties of high-volume ground granulated blast furnace slag (GGBFS) and polypropylene fiber–incorporated seawater concrete. The researchers replaced 50% of OPC with GGBFS. As shown in Fig. 20.9, the early-age strengths of high-strength concrete produced with freshwater decreased with the replacement of GGBFS. However, the strength development increased in the later stages, yet it was still observed to be 5% lower than the control series at 182 days. Researchers attributed this result to the pozzolanic reaction of GGBFS, indicating that the slow consumption of $Ca(OH)_2$ led to a denser microstructure in later ages. In the study, a similar trend in flexural and split tensile strengths was observed for mixtures produced with both freshwater and seawater, attributed to the low nature of pozzolanic reactions.

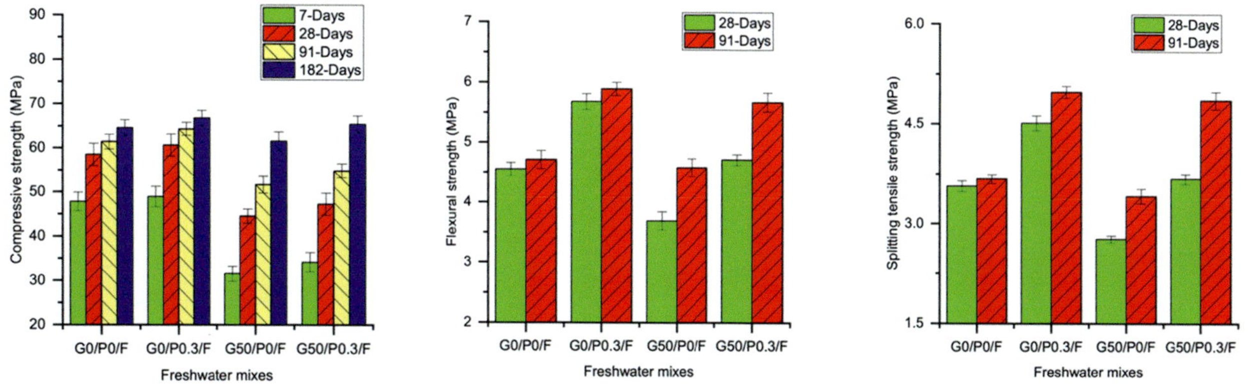

FIGURE 20.9 Early-age strengths of high-strength concrete produced with freshwater (notation in the study, G50 represents 50% GGBFS replacement, P0.3 represents 0.3% polypropylene fiber reinforcement, and F represents freshwater) (Alomayri et al., 2023).

Dhivya et al. (2023) investigated the mechanical properties and durability characteristics of concrete when steel slag is used as an SCM. The researchers focused on two different conditions: the water/binder ratio and the compressive strength at 28 days, as variable parameters. The results indicated that an increase in the percentage of steel slag as a mineral additive in concrete led to a decrease in compressive strength, especially in early strength. Based on durability results and other parameters, the researchers recommended an optimum steel slag substitution rate of 30%. The split tensile strength also showed a similar trend to the compressive strength.

Zhang et al. (2023) undertook a comprehensive investigation with the primary objective of enhancing the mechanical characteristics of cement–steel slag mortar. This was accomplished through the incorporation of specific alkali activators, namely NaOH, Na_2CO_3/NaOH, and water glass. Notably, all activators were administered proportionally based on the mass of the steel slag. The proportions employed were as follows: NaOH at levels of 4% and 6%, Na_2CO_3 at 2% within particular series, and water glass at 6% within designated series. The researchers examined some mechanical properties such as flexural and compressive strength to explore how the alkali activators affected the hydration of the cement–steel slag system. The study demonstrated that alkali activators can be effectively used in the cement–steel slag system. An increase in the alkali activator ratio resulted in strength gain in pastes and mortars, and the 30-day compressive strengths showed an average strength level of 35 MPa, which was 6.30%, 7.47%, and 9.03% higher than the strengths of the control samples.

On the other hand, unlike traditional concrete production methods, Yu et al. (2023) have investigated the mechanical and rheological properties of 3D printed eco-friendly concrete containing steel slag. The researchers also used silica fume and fly ash as additional substitute materials for OPC, apart from steel slag as a binder type. In the study, the compressive and flexural strengths of environmentally friendly 3D printed concrete containing steel copper slag were presented in Fig. 20.10A–D, considering different percentages of steel slag substitution. Furthermore, to shed light on the mechanical properties concerning material anisotropy, a mechanical analysis was provided in Fig. 20.10E. The content of steel slag and the curing time were found to have no significant impact on the material's mechanical anisotropy, as the highest mechanical strengths were achieved in the Y direction, while the lowest strengths were observed in the X direction. The results of the study indicated that the substitution of steel slag and its increasing content gradually reduced the mechanical properties of 3D printed concrete. The addition of steel slag hindered the hydration reaction at an early stage, but it led to a decrease in mechanical anisotropy. When the slag content was kept below 20%, the mechanical properties were well preserved, leading to reduced environmental impacts and enhanced cost-effectiveness in the final product.

Another focus of interest in the current literature is ladle furnace slag (LFS), which is a ferrous slag type and a secondary by-product of the steel industry. Although LFS is not as common and abundant as other types of ferrous slag, it still holds a notable position in research and studies. Sultana and Islam (2023) have conducted a study on the potential use of

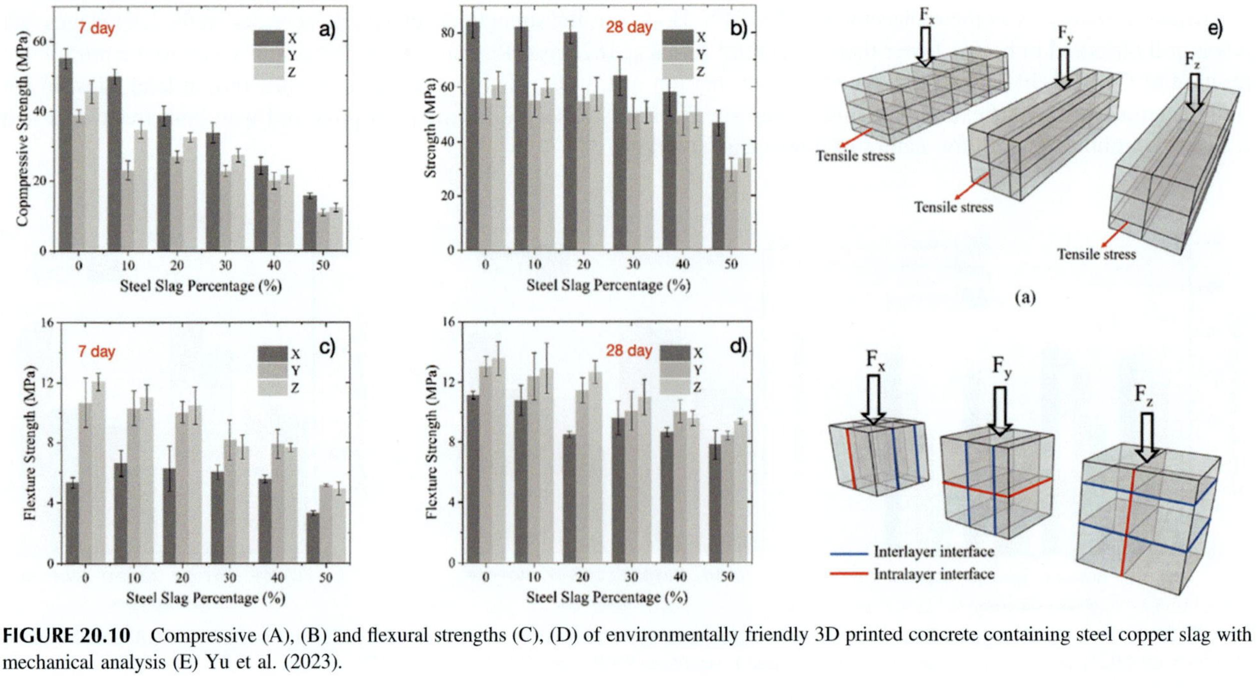

FIGURE 20.10 Compressive (A), (B) and flexural strengths (C), (D) of environmentally friendly 3D printed concrete containing steel copper slag with mechanical analysis (E) Yu et al. (2023).

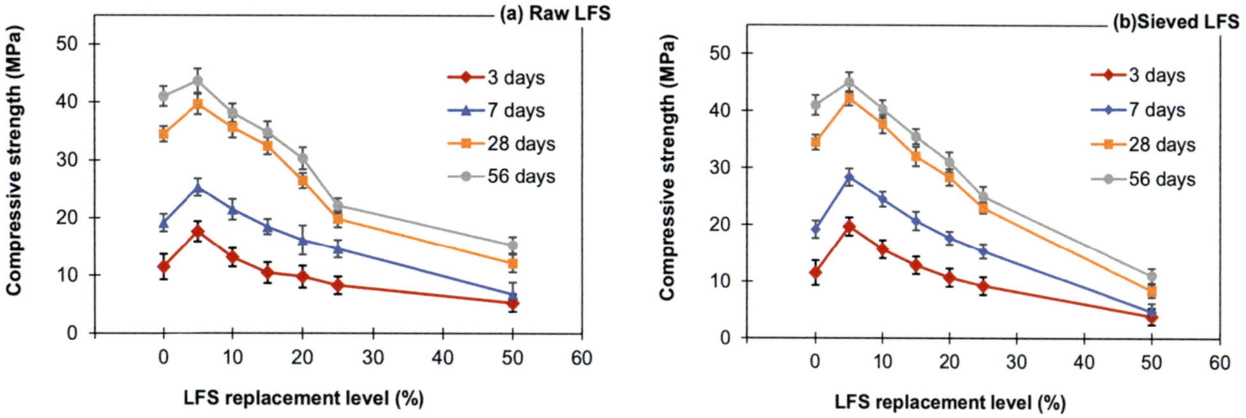

FIGURE 20.11 The variation in compressive strength according to the replacement rate of LFS with OPC Sultana and Islam (2023).

LFS as an SCM in concrete materials. In the study, LFS was substituted with OPC at various weight percentages of 5%, 10%, 15%, 20%, 25%, and 50%, both in raw and sieved forms, and various mechanical and nondestructive properties were investigated. Along with mechanical properties, durability characteristics such as nitrogen gas permeability and sorptivity were also examined. The results of compressive strength obtained from the research are shown in Fig. 20.11. In both types of LFS, an increase was observed at the 5% and 10% substitution range, followed by a decreasing trend. The literature suggests that the optimal substitution rate for LFS slag falls within the range of 5%–10%. Furthermore, the literature indicates that aluminate products can be formed by combining dispersed Al^{3+} with Ca^{2+} and SO_4^{2-} through silicate materials' hydration. It is also stated that adding an appropriate amount of LFS during this process can enrich the structural components of the material.

Oyejobi et al. (2023) have investigated the performance of a modified cement mortar by substituting copper slag, a nonferrous slag type, with fly ash in ordinary Portland cement. As a result of the research, it was reported that cement with 10% fly ash, cement with copper slag, and cement with a ternary blend achieved maximum compressive strengths of 47.5 MPa, 41.8 MPa, and 41.9 MPa, respectively. In the Cu–copper slag combination, the presence of heavy metals may have caused delayed hydration, leading to potential adverse effects on mechanical properties. Another reason indicated is that the surface area of copper slag is relatively smaller compared with fly ash. In Fig. 20.12, the compressive strength of

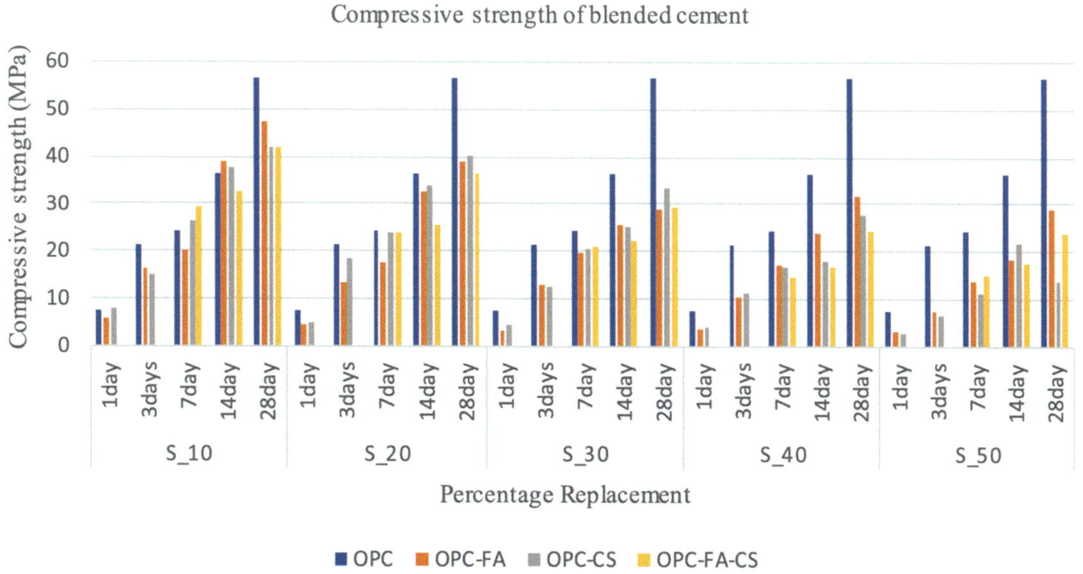

FIGURE 20.12 The strength under compression of mortar made from binary blended cement, and also the compressive strength results for mortar mixes containing three different components. (In terms of notation in the study, the numerical values in the mixture codes (e.g., S_10 or S_50) represent the substitution rate of the relevant SCMs (copper slag and fly ash) with respect to OPC) (Oyejobi et al., 2023).

mortar is shown for binary blended cement, as well as the compressive strength results for mortar mixes containing three different components.

He et al. (2021) have conducted research focusing on heavy metal leaching behavior and the applicability of copper slag as an SCM. Through compressive strength tests conducted on cement pastes after 3 and 28 days of curing, it was revealed that the substitution of copper slag led to a reduction in strength due to its low reactivity. However, with longer curing periods, the effect of copper slag on strength properties was observed to decrease. Chemical stabilization involves the transformation of heavy metals into less soluble or less toxic forms. Heavy metals can be incorporated into cement hydration gel structures, such as C—S—H and Aft. In the literature, there are reports on the immobilization mechanism of Zn, which involves the precipitation of metal silicates and the inclusion of Zn in C—S—H. The researchers in this study stated that the investigated heavy metals exhibited satisfactory immobilization. However, they recommended addressing properties such as sulfate attack under varying geochemical conditions in future studies involving the OPC—copper slag combination. The compressive strength results obtained in the study have been presented in Fig. 20.13 (in the notation of the study, FA represents fly ash and BFS represents blast furnace Slag) based on the substitution rate of copper slag.

Yang et al. (2021) conducted a study aimed at evaluating the enhancement of fire resistance performance through the addition of ferronickel slag in a high-temperature environment. Following a 28-day curing period at room temperature, the cement pastes underwent exposure to elevated temperatures (200, 400, 600, and 800°C). Additionally, the cement pastes were substituted not only with nickel slag but also with other SCMs such as pulverized fly ash and GGBFS. According to the compressive strength results, the OPC control sample exhibited the highest strength, while pastes produced with nickel slag showed the lowest strength results. This is attributed to the low CaO/SiO_2 nature of nickel slag powder, which hinders its direct participation in hydration reactions and results in lower strength performance compared with other SCM-substituted series. Nickel slag contains fayalite and forsterite, which are low-reactivity compounds, thus contributing to strength development through ion exchange in the C—S—H gel rather than direct involvement in hydration reactions. The obtained strength results from the study are shown in Fig. 20.14.

3.2 Durability properties

One of the main important aspects of concrete and SCMs is the durability feature, which arises from the transport of harmful substances or moisture penetration. When it comes to durability, the porous structure of the material allows the transport of water and other chemicals, leading to various deteriorations. These attacks or deteriorations are based on physical and chemical reactions. Concrete durability is assessed in terms of service life when exposed to the mentioned reactive environmental conditions (Al-Jabari, 2022). In this context, it is crucial for structures to resist throughout their service life in terms of economic viability, sustainability, recycling, and recovery. The pozzolanic reaction of slags, as mentioned earlier, plays a significant role in altering microstructural properties, which, in turn, has a substantial impact on

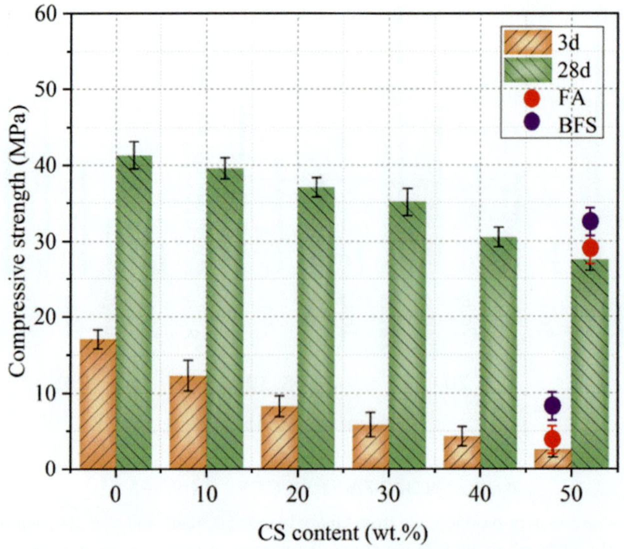

FIGURE 20.13 Compressive strength results of copper slag—OPC mortars according to the percentage of copper slag (He et al., 2021).

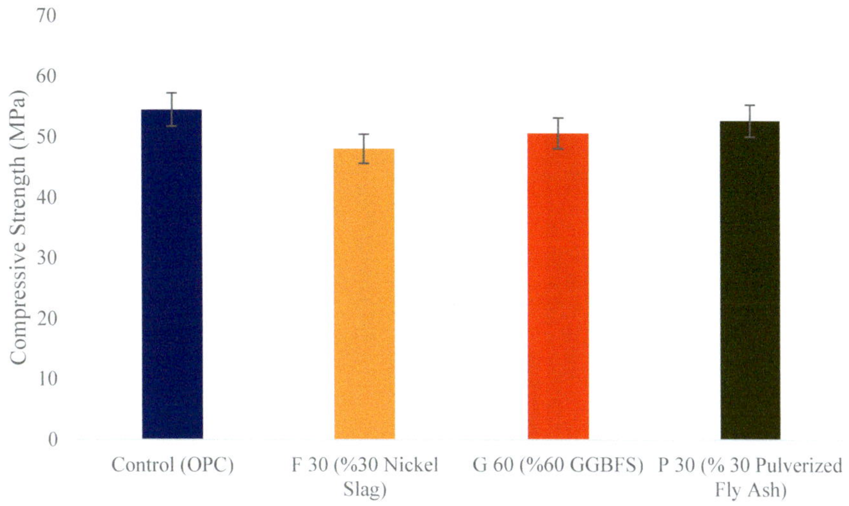

FIGURE 20.14 Compressive strength of paste specimens cured for 28 days (Yang et al., 2021).

various durability parameters such as freeze-thaw resistance, water absorption, chloride resistance, and high-temperature performance. These important effects have been effectively addressed in the literature.

Gokulakannan et al. (2023) conducted experimental studies on the durability and deformation behavior of concrete with OPC replaced by GGBFS and metakaolin. Additionally, copper slag was substituted for fine aggregate (sand). Copper slag was replaced with sand at percentages ranging from 20% to 60%, while GGBFS and metakaolin were substituted in OPC at levels between 5% and 10%. The study focused on acid attack and sorptivity resistance. Two different types of acid were evaluated in the acid attack tests: hydrochloric acid and sulfuric acid. According to the results obtained from the study, the concrete specimen containing 60% copper slag, 10% GGBFS, and 10% metakaolin showed a gradual weight loss after exposure to hydrochloric acid, which was 27% higher than the control sample. Furthermore, as the copper slag content increased, the specimen's weight loss after hydrochloric acid exposure also increased. This was attributed to the larger porosity of the specimens resulting from the addition of copper slag, which led to a higher weight loss of the concrete specimens. On the other hand, it was observed that the strength and mass loss caused by sulfuric acid were greater than those caused by hydrochloric acid. The strength losses in sulfuric acid attack followed a similar trend to the other acid type.

Sultana and Islam (2023) have researched the potential of LFS as an SCM and conducted sorptivity and nitrogen gas permeability tests as part of the durability evaluation. In the study, LFS was substituted for OPC at weight percentages of 5%, 10%, 15%, 20%, 25%, and 50%. The results showed that the substitution of LFS in OPC had variable effects on nitrogen gas permeability compared with normal strength concrete. However, in general, the positive effect of LFS on nitrogen gas permeability was observed, which is believed to be due to the nucleation effect resulting from the OPC−LFS combination.

On the other hand, Alomayri et al. (2023) investigated the durability properties of GGBFS-substituted concrete, including tidal erosion, rapid chloride ion permeability (RCIP), and wetting cycle. According to ASTM standards, RCIP values below 1000 coulombs are classified as "very low" permeability category (impermeable concrete), while mixtures with values between 1000 and 2000 coulombs fall under the "low" permeability category. Concrete mixtures with chloride ions, such as those in seawater, are not suitable for reinforced concrete structures. Therefore, RCIP values can provide insights into the overall durability of concrete mixtures when considering seawater exposure. As depicted in Fig. 20.15, the RCIP value varied depending on the percentage of GGBFS and the use of polypropylene fibers in the study. The presence of polypropylene fibers notably reduced the RCIP values. Similarly, the inclusion of GGBFS in the mixtures also significantly decreased the RCIP value. The substitution of GGBFS reduced the RCIP value from 1390 coulombs to 512 coulombs after 28 days, as reported.

Yang et al. (2021) examined the performance of cement pastes containing nickel slag under the influence of high temperatures. As expected, the compressive strength decreased with increasing temperature, irrespective of the mixture design. After 400°C, it was observed that the residual strengths of pastes containing nickel slag exhibited relatively small losses. This trend was attributed to the reduction in porosity due to the substitution of nickel slag with OPC. Furthermore, it was stated that the modification of the pore structure by nickel slag also contributed to changes in thermal conductivity. In the study, it was noted that the self-healing ability of iron-containing slag had a more significant impact on the increase in

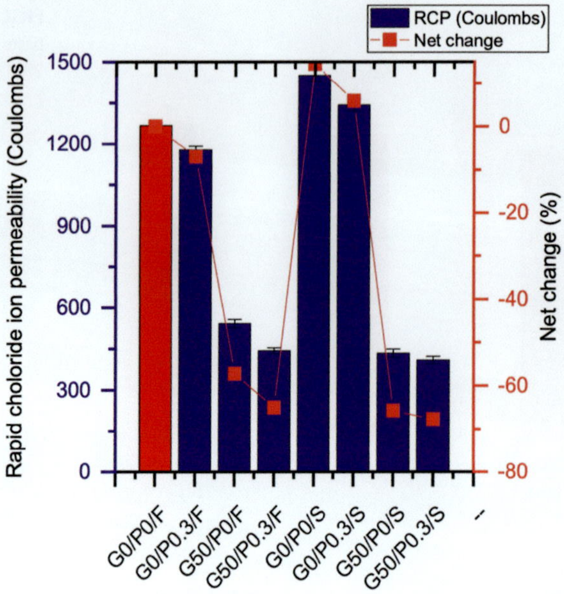

FIGURE 20.15 RCIP capacity of investigated mixes (In terms of notation in the study, G50 represents 50% GGBFS replacement, P0.3 represents 0.3% polypropylene fiber reinforcement, and F represents freshwater and S for seawater.) (Alomayri et al. (2023).

durability before/after exposure to high temperatures. This increase in fire resistance has also been identified in the current results, where the paste substituted with nickel slag exhibited the highest residual durability rate even after exposure to 800°C. In Fig. 20.16, the residual compressive strengths of cement pastes containing the respective SCMs after exposure to high temperatures are presented.

Various studies on the durability properties of different ferrous and nonferrous slag types, including acid attack, sorptivity, high temperature, freeze-thaw, and others, are summarized in Tables 20.4 and 20.5, which represents the most up-to-date research in this chapter.

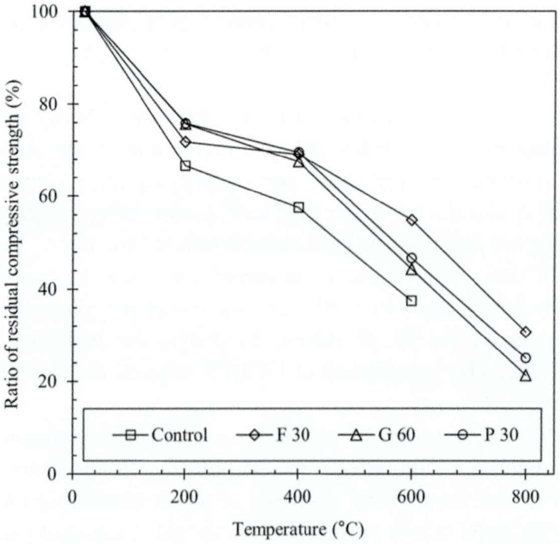

FIGURE 20.16 Residual compressive strengths of cement pastes containing the respective SCMs after exposure to high temperatures (In terms of notation in the study, F30 represents 30% nickel slag substitution in OPC, G60 represents 60% GGBFS substitution in OPC, and P30 represents 30% fly ash substitution in OPC.) (Yang et al., 2021).

4. Recycling various slags into aggregates

The mixture of concrete consists of approximately 65%−85% aggregate, which plays a critical role in the strength, durability, economy, and sustainability of concrete. Large-scale extraction of natural fine and coarse aggregates has led to significant environmental issues, prompting the implementation of stringent environmental protection policies (Lai et al., 2022). The existing disparities between demand and supply in the market have driven up prices, creating an urgent need for alternative materials to replace natural fine and coarse aggregates in concrete.

Steel slag is a major by-product of the steel industry, with approximately 10%−20% steel slag generated for every ton of steel produced. In China alone, over 400 million tons of slag are produced annually, and the recycling rate is reported to be around 30% (Lai et al., 2022). Another type of slag, nickel slag, is obtained in the production of ferronickel, with 1 ton of nickel slag generated for every 14 tons of ferronickel. It has even been reported that nickel slag is the fourth-largest industrial slag in China. Although nickel slag is not fully recycled, it can potentially be used as an alternative aggregate in cement and concrete production due to its low absorption, high hardness, and specific structure (Chen et al., 2023).

Copper slag, on the other hand, can yield amorphous, glassy, and smooth granules resembling natural fine aggregate when rapidly cooled. China is the largest producer of copper slag, producing 20.30 million tons annually. However, copper slag contains toxic metals such as arsenic and cadmium, posing potential risks not only to water and air pollution but also to human health. Considering its production figures and environmental impacts, it is essential to explore alternative environmentally friendly methods, such as using copper slag as aggregate or SCM, to address its disposal and potential hazards (Singh et al., 2022).

Tables 20.6 and 20.7 present a summary of recent publications from the last 3 years, focusing on the recycling of various ferrous and nonferrous slag types as aggregates.

TABLE 20.6 **Recent publications from the past 3 years, focusing on the recycling of various ferrous slag types as aggregates.**

Authors	Slag aggregate type	Normal aggregate	Other materials	Product	Curing (days)	Tests conducted	Year
Lai et al. (2022)	Coarse steel slag (BOF) aggregate	Natural fine and coarse	OPC, FA, silica fume	Concrete	7, 28, 91, 182	Compressive strength, elevated temperature, weight loss, failure behavior	2023
Costa et al. (2022)	Steel slag (BOF) fine and coarse aggregate	Natural fine and coarse	OPC	Concrete	28, 42	Specific gravity, water absorption, UPV, compressive strength, tensile strength, chloride ions penetration test, DTG, microstructure	2022
Santillán et al. (2022)	Steel slag (EAF) fine aggregate	Natural fine and coarse	OPC, steel fiber	Concrete	Unspecified	Compressive strength, the indirect traction test (Barcelona test), electrical conductivity, electrical resistivities	2022
Murmu et al. (2023)	Fine and coarse steel slag aggregate	Natural fine and coarse	OPC	Concrete	7, 28, 56	Compressive strength, flexural strength, splitting tensile strength, ultrasonic pulse velocity, rebound hammer test	2023
Huang et al. (2023)	Coarse steel slag aggregate	Natural fine and coarse	OPC, GGBFS	Concrete	28	Flowability, compressive strength, elevated temperature, mass loss, appearance and color changes, residual compressive strength, XRD	2023
Lai et al. (2023)	Coarse steel slag (BOF) aggregate	Natural fine and coarse	OPC, FA, silica fume	Concrete	28	Wet packing density, slump, elevated temperature, weight loss, postfire residual compressive strength	2023

TABLE 20.7 Recent publications from the past 3 years, focusing on the recycling of various nonferrous slag types as aggregates.

Authors	Slag aggregate type	Normal aggregate	Other materials	Product	Curing (days)	Tests conducted	Year
Jenifer et al. (2023)	Fine copper slag aggregate	Natural fine and coarse	OPC, basalt fiber	Reinforced concrete	7, 28	Compressive strength, split tensile, flexural strength, microstructure	2023
Afshoon et al. (2023)	Coarse copper slag	Natural fine and coarse	OPC, limestone powder, steel fiber	Self-compacting concrete	3, 7, 28	Slump flow, V-funnel, L-box, GTM screen stability, compressive strength, microstructural analyses, flexural strength, load–displacement test, flexural toughness test	2023
Edwin et al. (2022)	Quickly cooled granulated copper slag	Coarse basalt	OPC, silica fume, quartz powder	Ultrahigh-performance concrete	7, 28, 56	Workability, compressive strength, flexural strength, isothermal calorimetry, TG analysis, mercury intrusion porosimetry, leaching test	2022
Reddy et al. (2022)	Fine copper slag aggregate	Natural fine and coarse	OPC, egg shell powder	Concrete	7, 14, 28	Compressive strength, split tensile, acid attack	2022
Velumani et al. (2023)	Fine copper slag aggregate	Natural fine	OPC	Concrete	7, 14, 21, 28	Physical properties, slump, compressive strength, tensile strength, flexural strength	2023
Mirnezami et al. (2023)	Fine and coarse copper slag aggregate, fine and coarse steel slag	Natural fine and coarse	OPC	Concrete	28, 90	Slump, thermal conductivity, ultrasonic pulse velocity, abrasion resistance, compressive strengths, splitting tensile strength, flexural strength	2023
Panda and Sarkar (2022)	Fine copper slag aggregate	Natural fine and coarse	OPC	Concrete	91	Abrasion resistance, Taguchi tests, compressive strength	2023
Chaitanya and Sivakumar (2020)	Fine copper slag aggregate	Natural fine and coarse	OPC, FA	Self-compacting concrete	28	Slump flow, V-funnel, L-box, compressive strength, flexural strength	2021
Gong et al. (2021)	Fine copper slag aggregate	Natural fine and coarse	OPC, FA, steel bar	Concrete	28	Elevated temperature, accelerated corrosion, Bond slip, chloride ion distribution, compressive strength, bond strength, peak slip displacement, bond stiffness	2021
Nainwal et al. (2021)	Fine copper slag aggregate	Natural fine and coarse	OPC, metakaolin	Concrete	7, 28, 56	Water absorption, Initial surface absorption, compressive strength, split tensile strength	2021
Gupta and Siddique (2020)	Fine copper slag aggregate	Natural fine and coarse	OPC	Self-compacting concrete	7, 28, 90, 365	Slump flow, V-funnel, L-box, U-box, compressive strength, sorptivity, microstructure, rapid chloride permeability	2020

Siddique et al. (2020)	Fine copper slag aggregate	Natural fine and coarse	OPC, steel fiber	Concrete	7, 28, 56	Workability, compressive strength, splitting tensile strength, water penetration, water sorptivity, bond strength, microstructure	2020
Chen et al. (2023)	Fine nickel slag aggregate	Natural fine	OPC	Mortar	3, 14, 28	Methylene blue, workability, porosity, absorption tests, resistivity tests, compressive strength, flexural strength, mercury intrusion porosimetry, microstructure	2023
Saha et al. (2021)	Fine nickel slag aggregate	Natural fine and coarse, recycled coarse aggregate	OPC	Concrete	28	Compressive strength, ultrasonic pulse velocity, microstructure	2021
Gao et al. (2023)	Fine nickel slag aggregate	–	OPC, FA, blast furnace slag	Mortar (cement and geopolymer)	7, 120	Accelerated mortar bar test, alkali–silica reaction expansion, lab-simulated hydrothermal dissolution test, microstructure	2020

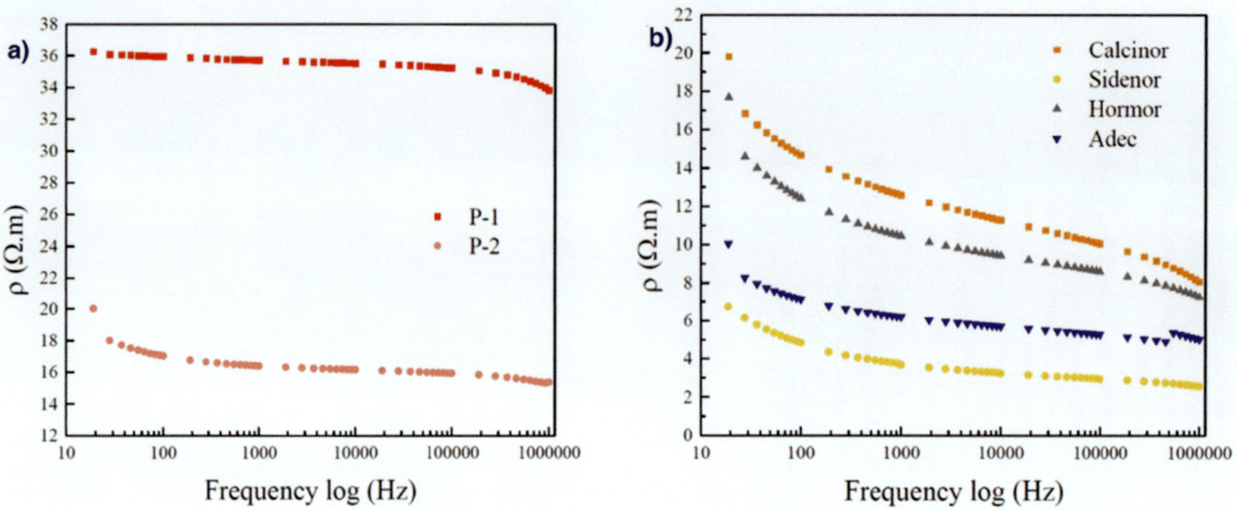

FIGURE 20.17 Electrical behavior of the samples (In the study, the notation represents P-1 as the fiberless series, P-2 as the series with fibers, and the names Sidenor, Hormor, Calcinor, and Adec represent the companies from which the steel slag was obtained.) (Santillán et al., 2022).

4.1 Mechanical and durability properties in the recycling of various slag types as aggregates

In general, it is known that the use of steel slag as fine aggregate, depending on the particle size distribution, leads to higher compressive strength in terms of mechanical properties. Due to its significant environmental threats when released into nature, steel slag can now be utilized for concrete as both fine and coarse aggregate (Gencel et al., 2021).

Santillán et al. (2022) conducted a study to examine the effects of using steel slag as a fine aggregate in concrete to respond to the growing awareness of sustainability. In addition to mechanical properties, electrical conductivity and electrical resistance were also investigated in the study. It was reported that the use of steel slag aggregates as a fine aggregate caused a loss in strength compared with the reference concrete mix series, attributed to the low workability resulting in high entrapped air content and porosity. On the other hand, the electrical properties of the steel slag–containing series were found to be influenced by frequencies in AC measurements, as shown in Fig. 20.17. In Fig. 20.17A, the reference series exhibited no change in impedance at any frequency, behaving like an insulator. However, the inclusion of steel slag altered the electrical behavior of the samples, reducing impedance with increasing current frequency. In general, electrical conductivity capacity increased by approximately 70%, and the mechanical performance improved by about 14% in fiber-reinforced series. The researchers emphasized that sustainability focus does not only pertain to the use of waste materials but also include the potential for enhancing electrical conductivity capacity, which could have various applications, ranging from providing ice or snow-free pedestrian crossings, seating areas, or roads, to contributing to the generation of greener energies that assist in operating traffic lights, or even ambitiously charging a small vehicle.

Huang et al. (2023) conducted a study to improve the fire performance by adding GGBFS when using steel slag as coarse aggregate. Table 20.8 presents the compressive strength results of the specimens after exposure to high temperatures. The mixtures in the table are divided into three groups. The first group represents the traditional concrete mixture, the second group includes mixtures with steel slag partially replacing coarse aggregate up to a maximum of 50%, and the third group consists of mixtures with steel slag fully replacing coarse aggregate. In the second group, the GGBFS volumetric replacement ratios are listed as 0% (M2), 15% (M3), 25% (M4), and 35% (M5). In the third group, the GGBFS volumetric replacement ratios are 0% (M6), 15% (M7), 25% (M8), and 35% (M9). After exposure to 600 and 800°C, it was observed that the relative strength rapidly decreased due to the decomposition of hydration products. On the other hand, as the steel slag coarse aggregate replacement ratio increased, the relative strength of concrete first decreased and then increased, as natural aggregates have higher thermal conductivity compared with steel slag aggregates. This led to greater thermal stress and more serious damage in the cubes of M2.

Reddy et al. (2022) investigated the durability and strength properties of concrete using copper slag as fine aggregate and eggshell as a partial replacement. In addition to mechanical properties such as compressive strength and split tensile strength, durability properties such as acid attack were also examined. The strength showed a significant increase at the optimum 10% eggshell powder replacement and 15% copper slag fine aggregate replacement, but beyond these ratios, the strength deteriorated. On the other hand, in the series without acid attack, the average compressive strength was around

TABLE 20.8 Compressive strength results of the specimens after exposure to high temperatures (Huang et al., 2023).

Group	Specimens	25°C (MPa)	400°C (MPa)	Relative Strength (%)	600°C (MPa)	Relative Strength (%)	800°C (MPa)	Relative Strength	1000°C (MPa)	Relative Strength (%)
1	M1	47.1	51.6	110.0	29.2	62.0	10.5	22.0	5.5	12.0
2	M2	48.9	47.4	97.0	24.5	50.0	7.5	15.0	4.0	8.0
	M3	48.1	48.4	101.0	31.7	66.0	12.7	26.0	5.3	11.0
	M4	46.2	48.2	104.0	34.2	74.0	12.9	28.0	5.4	12.0
	M5	45.0	49.3	110.0	36.7	82.0	13.2	29.0	5.5	12.0
3	M6	59.9	52.2	87.0	39.2	65.0	12.1	20.0	4.7	8.0
	M7	58.2	51.1	88.0	38.1	66.0	15.3	26.0	5.3	9.0
	M8	55.1	50.6	92.0	36.8	67.0	15.1	27.0	6.2	11.0
	M9	49.2	48.5	99.0	37.2	75.0	15.9	32.0	6.0	12.0

35.95 MPa, while in the series subjected to acid attack, the average strength was found to be 31.55 MPa. The strength results of the study are shown in Fig. 20.18.

Numerous studies (Murmu et al., 2023; Costa et al., 2022; Lai et al., 2022; Lai et al., 2023; Afshoon et al., 2023; Chaitanya and Sivakumar, 2020; Saha et al., 2021; Siddique et al., 2020). have been carried out in the literature to explore the use of various types of slag as recycled aggregates, with a particular emphasis on a diverse set of mechanical and durability properties. These properties encompass bending behavior, failure behavior, bond strength, leaching test, abrasion resistance, bond stiffness, accelerated corrosion, bond slip, and chloride ion distribution, among others.

5. Current limitations and future perspectives

Steel slag, ferronickel slag, copper slag, and GGBFS hold promise as partial substitutes for cement and fine aggregates in concrete, simultaneously allowing for the separation of heavy metals within the composite derived from slag types, while enhancing mechanical strength. Ferronickel slag, copper slag, and copper residues, when mixed with cement via pozzolanic reactions, can exhibit a high content of SiO_2, which forms secondary calcium silicate hydrates. Steel slag, being a dominant chemical component with CaO, bears resemblance to Portland cement. Nevertheless, despite these attributes, the need for grinding processes cannot be disregarded to activate highly crystalline materials and transform them into the amorphous phase.

On the other hand, from an environmental impact perspective, literature has indicated the safety of BF, BOF, and EAF slags based on leaching tests (Kurniati et al., 2023). The detected quantities of leaking toxic elements align with certain standards, thus being deemed safe. However, due to exceeding safe limits in leaching test results for copper wastes, these are stored in large outdoor ponds. Understanding the degradation of these wastes becomes essential to identify former storage sites suitable for material extraction. Future research efforts should focus on their characterization and utilization to reduce their represented footprint and economic burden. Additionally, the effects of weather conditions and the degradation of pozzolanic activity due to prolonged storage of slags and tailings could become focal points of forthcoming investigations. Moreover, a comparison of greenhouse gas emissions and comprehensive analyses such as life cycle assessment are imperative for a holistic evaluation.

Furthermore, despite the diverse range of comprehensive applications for steel slag, its utilization in industries such as construction materials, agriculture, ceramics, and wastewater augmentation remains limited. Enhancing the effectiveness of steel slag in the construction sector while maintaining cost efficiency remains a significant question. Simultaneously, based on their known advantages, exploring how slag and steel slag composite can synergistically leverage each other's respective benefits becomes crucial. Defining the optimal particle size of steel slag, comprehensive methods to enhance its

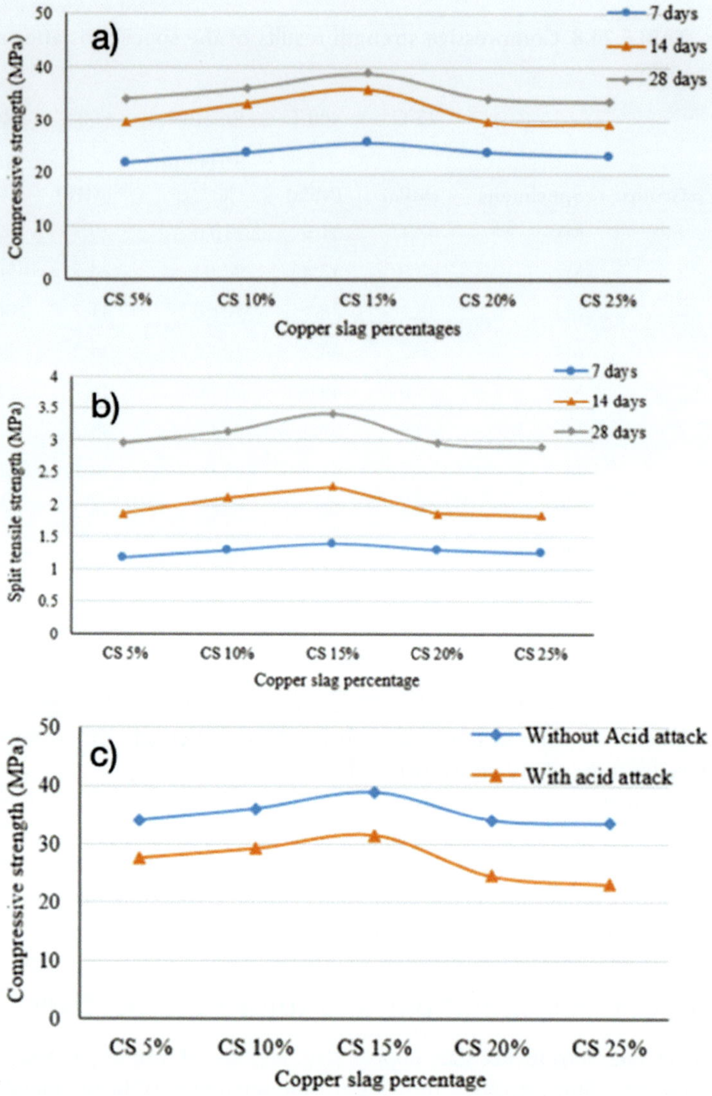

FIGURE 20.18 Strength results with respect to the durability effect and mechanical tests after copper slag aggregate replacement (Reddy et al., 2022).

activity, optimal blending ratios with other industrial by-products, and proposing a high-value steel slag concrete technology are all pivotal aspects that require attention in future studies.

Incorporating slags into concrete additives holds promise as a solution for industrial waste management. However, as previously mentioned, in the endeavor to develop environmentally friendly and eco-sustainable materials, it is imperative not to overlook the investigation of the leachability of metals present in the slags and their tendencies toward radioactivity. This subject necessitates future research efforts to ensure that the initiative of developing such materials does not halt prematurely. Natural resources are rapidly depleting day by day. Therefore, as always emphasized in this study, directing attention toward waste materials, slags, and similar industrial products is one of the most crucial endeavors. Particularly in developing countries, efforts to recycle industrial by-products and to environmentally sensitively treat waste materials require attention. Collaborating with companies and researchers engaged in this field in developed countries and receiving governmental support through legislations for the implementation domains are of paramount importance.

6. Conclusions

The purpose of this chapter was to provide a comprehensive overview of the extensive research conducted on the reclamation of various slag types as SCMs and aggregates. Beforehand, a bibliometric analysis was undertaken on the term "slag," and thereafter, a comprehensive compilation, comparison, and analysis of existing experimental data and previous research findings pertaining to the microstructural, mechanical, and durability properties, as well as the environmental advantages of pastes, mortars, and concretes incorporating various slag types, were carried out. Numerous studies and research endeavors have convincingly showcased the benefits of incorporating slag as SCMs and aggregates in composite material fabrication. Augmenting the utilization of slag emerges as a practicable and promising approach to steer the industry toward a more sustainable trajectory. By synthesizing the findings from a multitude of published articles on slag research, the following conclusions can be drawn:

The bibliometric analysis conducted over the past 5 years reveals a robust interrelation between the term "slag" and key keywords such as cement, fly ash, circular economy, waste management, life cycle analysis, recycling, and geopolymers/alkali. Notably, the extensive association of cement with diverse slag types (e.g., copper slag, steel slag, GGBFS) in the context of emerging material technologies such as geopolymers, along with the intertwining concepts of circular economy and recycling, underscores the pivotal role of slag materials in terms of recovery and utilization. Across the globe, numerous slag varieties, including copper slag, nickel slag, and others, continue to be amassed in substantial quantities as extensive stockpiles. Particularly, certain types such as nickel slag encounter difficulties in achieving sufficient recycling rates. While investigations have showcased the potential of utilizing nonmainstream slag types as alternative aggregates in cement and concrete manufacturing, a more comprehensive research effort is warranted to establish a consensus on the optimal utilization ranges and to delve into specific applications for select slag materials. In the recent research spanning the past 3–5 years, researchers have predominantly favored GGBFS and steel slag as preferred substitutes for OPC within the category of ferrous slag. These materials were often utilized either in combination or individually with silica fume and fly ash, and occasionally with metakaolin, forming binary or ternary mixtures. As for nonferrous slag types, the research trend has focused on copper slag and nickel slag. Studies in this domain have primarily centered on mortar and paste compositions rather than concrete materials. Regarding aggregate utilization, considerable emphasis has been placed on steel slag within ferrous slag types, whereas copper slag has been significantly studied among nonferrous slag types, closely followed by nickel slag. These slag varieties have predominantly been employed as replacements for fine aggregates, while noteworthy and promising outcomes have also been achieved in the context of their application as substitutes for coarse aggregates. The diverse products derived from various slag types can be effectively employed as construction materials in a plethora of building applications, encompassing mortar, concrete, and beyond. The literature substantiates this claim, as the products obtained from these slag variants have demonstrated commendable outcomes in numerous demanding durability tests, showcasing their resilience against acid aggression, high-temperature exposure, abrasion, and corrosion challenges.

In essence, the incorporation of slag as SCMs and aggregates in concrete and mortar fabrication bestows noteworthy environmental benefits, bolstering the industry's strides toward sustainability. The extensive research endeavors in this domain yield invaluable knowledge, facilitating the optimization of diverse slag types' utilization and the augmentation of slag–aggregate concrete and mortar's performance and durability, thereby elevating the industry's commitment to sustainable practices.

References

Afshoon, I., Miri, M., Mousavi, S.R., 2023. Evaluating the flexural behavior of green copper slag-contained steel fiber reinforced SCC beams with/without initial notches. Construct. Build. Mater. 395 (January), 132316. https://doi.org/10.1016/j.conbuildmat.2023.132316.

Al-Jabari, M., 2022. Concrete durability problems: physicochemical and transport mechanisms. In: Integral Waterproofing of Concrete Structures. Elsevier, pp. 69–107. https://doi.org/10.1016/B978-0-12-824354-1.00003-9.

Ali, M.M., Agarwal, S.K., Pahuja, A., 2013. Potentials of copper slag utilisation in the manufacture of ordinary Portland cement. Adv. Cement Res. 25 (4), 208–216. https://doi.org/10.1680/adcr.12.00004.

Alomayri, T., Amir, M.T., Ali, B., Raza, S.S., Hamad, M., 2023. Mechanical, tidal erosion, and drying shrinkage behaviour of high performance seawater concrete incorporating the high volume of GGBS and polypropylene fibre. J. Build. Eng. 76. https://doi.org/10.1016/j.jobe.2023.107377.

American Iron and Steel Institution, 2018. Facts About American Steel Sustainability. https://www.steel.org/wp-content/uploads/2021/11/AISI_Fact-Sheet_SteelSustainability-11-3-21.pdf.

Aprianti, S.,E., 2017. A huge number of artificial waste material can be supplementary cementitious material (SCM) for concrete production – a review part II. J. Clean. Prod. 142, 4178–4194. https://doi.org/10.1016/j.jclepro.2015.12.115.

Baalamurugan, J., Ganesh Kumar, V., Naveen Prasad, B.S.N., Padmapriya, R., Karthick, V., Govindaraju, K., 2022. Recycling of induction furnace steel slag in concrete for marine environmental applications towards ocean acidification studies. Int. J. Environ. Sci. Technol. 19 (6), 5039–5048. https://doi.org/10.1007/s13762-021-03362-7.

Carvalho, S.Z., Vernilli, F., Almeida, B., Demarco, M., Silva, S.N., 2017. The recycling effect of BOF slag in the portland cement properties. Resour. Conserv. Recycl. 127, 216–220. https://doi.org/10.1016/j.resconrec.2017.08.021.

Carvalho, V.R., Costa, L.C.B., Elói, F.P. da F., Bezerra, A.C. da S., Carvalho, J. M. F. de, Peixoto, R.A.F., 2023. Performance of low-energy steel slag powders as supplementary cementitious materials. Construct. Build. Mater. 392 (March), 1–12. https://doi.org/10.1016/j.conbuildmat.2023.131888.

Chaitanya, B.K., Sivakumar, I., 2020. Influence of waste copper slag on flexural strength properties of self compacting concrete. Mater. Today: Proc. 42, 671–676. https://doi.org/10.1016/j.matpr.2020.11.059.

Chen, S.-C., Wang, M.-T., Gu, L.-S., Lin, W.-T., Liang, J.-F., Korniejenko, K., 2023. Effects of incorporating large quantities of nickel slag with various particle sizes on the strength and pore structure of cement-based materials. Construct. Build. Mater. 393, 132034. https://doi.org/10.1016/j.conbuildmat.2023.132034.

Chun, T., Ning, C., Long, H., Li, J., Yang, J., 2016. Mineralogical characterization of copper slag from tongling nonferrous metals group China. J. Occup. Med. 68 (9), 2332–2340. https://doi.org/10.1007/s11837-015-1752-6.

Costa, L.C.B., Nogueira, M.A., Andrade, H.D., Carvalho, J. M. F. de, Elói, F.P. da F., Brigolini, G.J., Peixoto, R.A.F., 2022. Mechanical and durability performance of concretes produced with steel slag aggregate and mineral admixtures. Construct. Build. Mater. 318 (December 2021). https://doi.org/10.1016/j.conbuildmat.2021.126152.

Dhivya, K., Anusha, G., Vinoth, S., Shanjai, P.R., Sreedharan, A., Rahuma, V.M., 2023. Steel slag's effect on concrete mechanical properties and durability. Mater. Today: Proc. xxxx. https://doi.org/10.1016/j.matpr.2023.05.330.

Duan, X., Li, X., Li, Y., Qi, X., Li, G., Lu, Z., Yang, N., 2021. Separation and stabilization of arsenic in copper smelting wastewater by zinc slag. J. Clean. Prod. 312, 127797. https://doi.org/10.1016/j.jclepro.2021.127797.

Edwin, R.S., Gruyaert, E., De Belie, N., 2022. Valorization of secondary copper slag as aggregate and cement replacement in ultra-high performance concrete. J. Build. Eng. 54 (May), 104567. https://doi.org/10.1016/j.jobe.2022.104567.

Fan, D., Zhang, C., Lu, J.X., Liu, K., Yin, T., Dong, E., Yu, R., 2023. Recycling of steel slag powder in green ultra-high strength concrete (UHSC) mortar at various curing conditions. J. Build. Eng. 70 (March), 106361. https://doi.org/10.1016/j.jobe.2023.106361.

Fang, K., Zhao, J., Wang, D., Wang, H., Dong, Z., 2022. Use of ladle furnace slag as supplementary cementitious material before and after modification by rapid air cooling: a comparative study of influence on the properties of blended cement paste. Construct. Build. Mater. 314 (PA), 125434. https://doi.org/10.1016/j.conbuildmat.2021.125434.

Gao, X., Yang, T., Wang, A., Zhu, Y., Zhu, H., Zhuang, P., Wu, Q., 2023. Alkali-silica reaction of high-magnesium nickel slag fine aggregate in alkali-activated ground granulated blast-furnace slag mortar. Construct. Build. Mater. 406, 133374. https://doi.org/10.1016/j.conbuildmat.2023.133374.

Gencel, O., Karadag, O., Oren, O.H., Bilir, T., 2021. Steel slag and its applications in cement and concrete technology: a review. Construct. Build. Mater. 283, 122783. https://doi.org/10.1016/j.conbuildmat.2021.122783.

Gholampour, A., Zheng, J., Ozbakkaloglu, T., 2021. Development of waste-based concretes containing foundry sand, recycled fine aggregate, ground granulated blast furnace slag and fly ash. Construct. Build. Mater. 267, 121004. https://doi.org/10.1016/j.conbuildmat.2020.121004.

Gokulakannan, S., Harikaran, M., Mounesh, K., Sankaranarayanan, S., Sathishkumar, R., Sivakumar, R., 2023. Experimental studies on durability and deflection behavior of concrete in metakaolin cement concrete. Mater. Today: Proc. https://doi.org/10.1016/j.matpr.2023.05.236.

Gong, W., Chen, Q., Miao, J., 2021. Bond behaviors between copper slag concrete and corroded steel bar after exposure to high temperature. J. Build. Eng. 44 (June), 103312. https://doi.org/10.1016/j.jobe.2021.103312.

Gorai, B., Jana, R.K., Premchand, 2003. Characteristics and utilisation of copper slag—a review. Resour. Conserv. Recycl. 39 (4), 299–313. https://doi.org/10.1016/S0921-3449(02)00171-4.

Gu, X., Wang, H., Zhu, Z., Liu, J., Xu, X., Wang, Q., 2023. Synergistic effect and mechanism of lithium slag on mechanical properties and microstructure of steel slag-cement system. Construct. Build. Mater. 396 (March), 131768. https://doi.org/10.1016/j.conbuildmat.2023.131768.

Gupta, N., Siddique, R., 2020. Durability characteristics of self-compacting concrete made with copper slag. Construct. Build. Mater. 247, 118580. https://doi.org/10.1016/j.conbuildmat.2020.118580.

Han, F., Li, Y., Jiao, D., 2023. Understanding the rheology and hydration behavior of cement paste with nickel slag. J. Build. Eng. 73 (March), 106724. https://doi.org/10.1016/j.jobe.2023.106724.

He, P., Drissi, S., Hu, X., Liu, J., Shi, C., 2023. Investigation on the influential mechanism of FA and GGBS on the properties of CO2-cured cement paste. Cement Concr. Compos. 142 (February), 105186. https://doi.org/10.1016/j.cemconcomp.2023.105186.

He, R., Zhang, S., Zhang, X., Zhang, Z., Zhao, Y., Ding, H., 2021. Copper slag: the leaching behavior of heavy metals and its applicability as a supplementary cementitious material. J. Environ. Chem. Eng. 9 (2), 105132. https://doi.org/10.1016/j.jece.2021.105132.

Huang, Z.C., Ho, J.C.M., Cui, J., Ren, F.M., Cheng, X., Lai, M.H., 2023. Improving the post-fire behaviour of steel slag coarse aggregate concrete by adding GGBFS. J. Build. Eng. 76 (May), 1–17. https://doi.org/10.1016/j.jobe.2023.107283.

Jenifer, J.V., Brindha, D., Annie Sweetlin Jebarani, J.P., Venkadapriya, S., Pandieswari, M., 2023. Mechanical and microstructure properties of copper slag based basalt fiber reinforced concrete. Mater. Today: Proc. xxxx. https://doi.org/10.1016/j.matpr.2023.03.505.

Juenger, M.C.G., Snellings, R., Bernal, S.A., 2019. Supplementary cementitious materials: new sources, characterization, and performance insights. Cement Concr. Res. 122, 257–273. https://doi.org/10.1016/j.cemconres.2019.05.008.

Kanavaris, F., Soutsos, M., Chen, J.F., 2023. Enabling sustainable rapid construction with high volume GGBS concrete through elevated temperature curing and maturity testing. J. Build. Eng. 63 (PA), 105434. https://doi.org/10.1016/j.jobe.2022.105434.

Kanneboina, Y.Y.,T.,J.S., Kabeer, K.I.S.A., Bisht, K., 2023. Valorization of lead and zinc slags for the production of construction materials - a review for future research direction. Construct. Build. Mater. 367, 130314. https://doi.org/10.1016/j.conbuildmat.2023.130314.

Kovacs, T., Bator, G., Schroeyers, W., Labrincha, J., Puertas, F., Hegedus, M., Nicolaides, D., Sanjuán, M.A., Krivenko, P., Grubeša, I.N., Sas, Z., Michalik, B., Anagnostakis, M., Barisic, I., Nuccetelli, C., Trevisi, R., Croymans, T., Schreurs, S., Todorović, N., Doherty, R., 2017. From raw materials to NORM by-products. In: Naturally Occurring Radioactive Materials in Construction. Elsevier, pp. 135–182. https://doi.org/10.1016/B978-0-08-102009-8.00006-2.

Kumar, A., Deep, K., 2023. Experimental investigation of concrete with cementitious waste material such as GGBS and fly ash over conventional concrete. Mater. Today: Proc. 74, 953–961. https://doi.org/10.1016/j.matpr.2022.11.297.

Kuranlõ, Ö.F., Uysal, M., Abbas, M.T., Cosgun, T., Niş, A., Aygörmez, Y., Canpolat, O., Al-mashhadani, M.M., 2022. Evaluation of slag/fly ash based geopolymer concrete with steel, polypropylene and polyamide fibers. Construct. Build. Mater. 325, 126747. https://doi.org/10.1016/j.conbuildmat.2022.126747.

Kurniati, E.O., Pederson, F., Kim, H., 2023. Resources, conservation & recycling application of steel slags, ferronickel slags , and copper mining waste as construction materials : a review. Resour. Conserv. Recycl. 198, 107175. https://doi.org/10.1016/j.resconrec.2023.107175.

Lai, M.H., Chen, Z.H., Wang, Y.H., Ho, J.C.M., 2022. Effect of fillers on the mechanical properties and durability of steel slag concrete. Construct. Build. Mater. 335 (February). https://doi.org/10.1016/j.conbuildmat.2022.127495.

Lai, M.H., Chen, Z.H., Cui, J., Zhong, J.P., Wu, Z.R., Ho, J.C.M., 2023. Enhancing the post-fire behavior of steel slag normal-strength concrete by adding SCM. Construct. Build. Mater. 398 (April). https://doi.org/10.1016/j.conbuildmat.2023.132336.

Lewis, D.W., 1982. Properties and Uses of Iron and Steel Slags, pp. 1–11.

Li, G., Liu, S., Niu, M., Liu, Q., Yang, X., Deng, M., 2020. Effect of granulated blast furnace slag on the self-healing capability of mortar incorporating crystalline admixture. Construct. Build. Mater. 239, 117818. https://doi.org/10.1016/j.conbuildmat.2019.117818.

Li, Y., Mehdizadeh, H., Mo, K.H., Ling, T.-C., 2023. Co-utilization of aqueous carbonated basic oxygen furnace slag (BOFS) and carbonated filtrate in cement pastes considering reaction duration effect. Cement Concr. Compos. 138, 104988. https://doi.org/10.1016/j.cemconcomp.2023.104988.

Liu, J., Wang, D., 2017. Influence of steel slag-silica fume composite mineral admixture on the properties of concrete. Powder Technol. 320, 230–238. https://doi.org/10.1016/j.powtec.2017.07.052.

Liu, P., Mo, L., Zhang, Z., 2023. Effects of carbonation degree on the hydration reactivity of steel slag in cement-based materials. Construct. Build. Mater. 370 (January), 130653. https://doi.org/10.1016/j.conbuildmat.2023.130653.

Liu, Q., Liu, J., Qi, L., 2016. Effects of temperature and carbonation curing on the mechanical properties of steel slag-cement binding materials. Construct. Build. Mater. 124, 999–1006. https://doi.org/10.1016/j.conbuildmat.2016.08.131.

Matos, P.R., Oliveira, J.C.P., Medina, T.M., Magalhães, D.C., Gleize, P.J.P., Schankoski, R.A., Pilar, R., 2020. Use of air-cooled blast furnace slag as supplementary cementitious material for self-compacting concrete production. Construct. Build. Mater. 262, 120102. https://doi.org/10.1016/j.conbuildmat.2020.120102.

Mirnezami, S.M., Hassani, A., Bayat, A., 2023. Evaluation of the effect of metallurgical aggregates (steel and copper slag) on the thermal conductivity and mechanical properties of concrete in jointed plain concrete pavements (JPCP). Construct. Build. Mater. 367 (June 2022), 129532. https://doi.org/10.1016/j.conbuildmat.2022.129532.

Mohan, A., Tabish Hayat, M., 2021. Characterization of mechanical properties by preferential supplant of cement with GGBS and silica fume in concrete. Mater. Today: Proc. 43, 1179–1189. https://doi.org/10.1016/j.matpr.2020.08.733.

Murmu, M., Ranjan Mohanta, N., Bapure, N., 2023. Study on the fresh and hardened properties of concrete with steel slag as partial replacement for natural aggregates. Mater. Today: Proc. https://doi.org/10.1016/j.matpr.2023.03.151.

Nainwal, A., Emani, P.K., Shah, M.C., Negi, A., Kumar, V., Negi, P., 2021. The influence of Metakaolin on the copper slag substituted concrete with the fine aggregate of Beas river. Mater. Today: Proc. 46, 10425–10432. https://doi.org/10.1016/j.matpr.2020.12.981.

Nicula, L.M., Corbu, O., Iliescu, M., 2020. Influence of blast furnace slag on the durability characteristic of road concrete such as freeze-thaw resistance. Procedia Manuf. 46 (2019), 194–201. https://doi.org/10.1016/j.promfg.2020.03.029.

Nilimaa, J., 2023. Smart materials and technologies for sustainable concrete construction. Developments in the Built Environment 15, 100177. https://doi.org/10.1016/j.dibe.2023.100177.

Oyejobi, D.O., Adewuyi, A.P., Yusuf, S.O., Oyebanji, Y.O., Suleiman, I., Hassan, I.A., 2023. Performance of blended cement mortar modified with fly ash and copper slag. Mater. Today: Proc. 86, 104–110. https://doi.org/10.1016/j.matpr.2023.03.294.

Panda, S., Sarkar, P., 2022. Abrasion resistance of copper slag aggregate concrete designed by Taguchi method. Mater. Today: Proc. 65, 434–441. https://doi.org/10.1016/j.matpr.2022.02.545.

Piatak, N.M., Parsons, M.B., Seal, R.R., 2015. Characteristics and environmental aspects of slag: a review. Appl. Geochem. 57, 236–266. https://doi.org/10.1016/j.apgeochem.2014.04.009.

Piatak, N.M., Seal, R.R., 2010. Mineralogy and the release of trace elements from slag from the Hegeler Zinc smelter, Illinois (USA). Appl. Geochem. 25 (2), 302–320. https://doi.org/10.1016/j.apgeochem.2009.12.001.

Qi, L., Liu, J., Liu, Q., 2016. Compound effect of $CaCO_3$ and $CaSO_4$.$_2H_2O$ on the strength of steel slag - cement binding materials. Mater. Res. 19 (2), 269–275. https://doi.org/10.1590/1980-5373-MR-2015-0387.

Qian, G., Sun, D.D., Tay, J.H., Lai, Z., Xu, G., 2002. Autoclave properties of kirschsteinite-based steel slag. Cement Concr. Res. 32 (9), 1377–1382. https:/a/doi.org/10.1016/S0008-8846(02)00790-1.

Raj, G., Rai, D., Singh, R.S., Sofi, A., 2023. Study on properties of concrete using slag as partial replacement of cement. Mater. Today: Proc. 1–5. https://doi.org/10.1016/j.matpr.2023.04.040.

Rajesh Ram, A., Sai Kumar, C., Shirisha, A., Anosh Kumar, R., Mounika Naidu, G., 2023. Experimental investigation on compressive strength and acid resistance of autoclaved aerated concrete using GGBS. Mater. Today: Proc. https://doi.org/10.1016/j.matpr.2023.05.593 xxxx.

Ramakrishnan, U., Mounika, G., Likhitha, M., Harikrishna, M., Sachin, K., 2023. Experimental investigation of strength of concrete with GGBS, granite powder and treated wastewater. Mater. Today: Proc. 3–7. https://doi.org/10.1016/j.matpr.2023.03.610.

Ramakrishnan, K., Pugazhmani, G., Sripragadeesh, R., Muthu, D., Venkatasubramanian, C., 2017. Experimental study on the mechanical and durability properties of concrete with waste glass powder and ground granulated blast furnace slag as supplementary cementitious materials. Construct. Build. Mater. 156, 739–749. https://doi.org/10.1016/j.conbuildmat.2017.08.183.

Reddy, Y.R., Prasad, J.S.R., Kumar Balguri, P., 2022. Durability and strength properties of concrete by using egg shell powder and copper slag. Mater. Today: Proc. 62, 2996–3000. https://doi.org/10.1016/j.matpr.2022.02.625.

Revathi, S., Dinesh, M., Suba Sri Varsan, S.G., 2023. Mechanical properties of concrete incorporating coconut fibers and copper slag. Mater. Today: Proc. https://doi.org/10.1016/j.matpr.2023.02.276.

Roslan, N.H., Ismail, M., Khalid, N.H.A., Muhammad, B., 2020. Properties of concrete containing electric arc furnace steel slag and steel sludge. J. Build. Eng. 28 (October 2019), 101060. https://doi.org/10.1016/j.jobe.2019.101060.

Saad Agwa, I., Zeyad, A.M., Tayeh, B.A., Adesina, A., de Azevedo, A.R.G., Amin, M., Hadzima-Nyarko, M., 2022. A comprehensive review on the use of sugarcane bagasse ash as a supplementary cementitious material to produce eco-friendly concretes. Mater. Today: Proc. 65, 688–696. https://doi.org/10.1016/j.matpr.2022.03.264.

Saha, A.K., Majhi, S., Sarker, P.K., Mukherjee, A., Siddika, A., Aslani, F., Zhuge, Y., 2021. Non-destructive prediction of strength of concrete made by lightweight recycled aggregates and nickel slag. J. Build. Eng. 33, 101614. https://doi.org/10.1016/j.jobe.2020.101614.

Sajedi, F., Razak, H.A., Mahmud, H.B., Shafigh, P., 2012. Relationships between compressive strength of cement–slag mortars under air and water curing regimes. Construct. Build. Mater. 31, 188–196. https://doi.org/10.1016/j.conbuildmat.2011.12.056.

Santillán, N., Speranza, S., Torrents, J.M., Segura, I., 2022. Evaluation of conductive concrete made with steel slag aggregates. Construct. Build. Mater. 360 (June). https://doi.org/10.1016/j.conbuildmat.2022.129515.

Shahas, S., Girija, K., Nazeer, M., 2023. Evaluation of pozzolanic activity of ternary blended supplementary cementitious material with rice husk ash and GGBS. Mater. Today: Proc. 2–7. https://doi.org/10.1016/j.matpr.2023.01.073.

Siddique, R., Singh, M., Jain, M., 2020. Recycling copper slag in steel fibre concrete for sustainable construction. J. Clean. Prod. 271, 122559. https://doi.org/10.1016/j.jclepro.2020.122559.

Singh, N., Gupta, A., Haque, M.M., 2022. A review on the influence of copper slag as a natural fine aggregate replacement on the mechanical properties of concrete. Mater. Today Proc. 62, 3624–3637. https://doi.org/10.1016/j.matpr.2022.04.414.

Smithers, 2023. Global non-ferrous slag volumes to reach 133.7 million tonnes by 2029e. https://www.smithers.com/resources/2019/may/global-non-ferrous-slag-market-to-reach-133-7-m.

Srivastava, S., Cerutti, M., Nguyen, H., Carvelli, V., Kinnunen, P., 2023. Carbonated steel slags as supplementary cementitious materials : reaction kinetics and phase evolution. Cement Concr. Compos. 142 (May), 105213. https://doi.org/10.1016/j.cemconcomp.2023.105213.

Sultana, I., Islam, G.M.S., 2023. Potential of ladle furnace slag as supplementary cementitious material in concrete. Case Stud. Constr. Mater. 18 (April), e02141. https://doi.org/10.1016/j.cscm.2023.e02141.

Tiwary, A.K., Bhatia, S., 2022. A study incorporating the influence of copper slag and fly ash substitutions in concrete. Mater. Today Proc. 48, 1476–1483. https://doi.org/10.1016/j.matpr.2021.09.293.

U.S. Geological Survey, 2023. Minerals Yearbook. https://doi.org/10.3133/mybvI.

Van Oss, H.G., 2013. Slag, Iron and Steel: U.S. Geological Survey, 2011 Minerals Yearbook.

Velumani, M., Gowtham, S., Dhananjayan, M.P., Tamil Eniyan, G., 2023. Strength assessment of concrete with copper slag as fine aggregates. Mater. Today: Proc. https://doi.org/10.1016/j.matpr.2023.05.700.

Wang, D., Wang, Q., Huang, Z., 2020. Reuse of copper slag as a supplementary cementitious material: reactivity and safety. Resour. Conserv. Recycl. 162 (April), 105037. https://doi.org/10.1016/j.resconrec.2020.105037.

Wang, G.C., 2016a. Slag use as an aggregate in concrete and cement-based materials. In: The Utilization of Slag in Civil Infrastructure Construction. Elsevier, pp. 239–274. https://doi.org/10.1016/B978-0-08-100381-7.00011-2.

Wang, G.C., 2016b. Ferrous metal production and ferrous slags. In: The Utilization of Slag in Civil Infrastructure Construction. Elsevier, pp. 9–33. https://doi.org/10.1016/B978-0-08-100381-7.00002-1.

Wang, G.C., 2016c. Nonferrous metal extraction and nonferrous slags. In: The Utilization of Slag in Civil Infrastructure Construction. Elsevier, pp. 35–61. https://doi.org/10.1016/B978-0-08-100381-7.00003-3.

Wang, Y., Suraneni, P., 2019. Experimental methods to determine the feasibility of steel slags as supplementary cementitious materials. Construct. Build. Mater. 204, 458–467. https://doi.org/10.1016/j.conbuildmat.2019.01.196.

Weng, Y., Liu, Y., Liu, J., 2021. Study on mathematical model of hydration expansion of steel slag-cement composite cementitious material. Environ. Technol. 42, 2776–2783. https://doi.org/10.1080/09593330.2020.1713906.

Wu, Q., Wu, Y., Tong, W., Ma, H., 2018. Utilization of nickel slag as raw material in the production of Portland cement for road construction. Construct. Build. Mater. 193, 426–434. https://doi.org/10.1016/j.conbuildmat.2018.10.109.

Wu, W., Zhang, W., Ma, G., 2010. Mechanical properties of copper slag reinforced concrete under dynamic compression. Construct. Build. Mater. 24 (6), 910–917. https://doi.org/10.1016/j.conbuildmat.2009.12.001.

Yang, H.J., Lee, C.H., Shim, S.H., Kim, J.H.J., Lee, H.J., Park, J.W., 2021. Performance evaluation of cement paste incorporating ferro-nickel slag powder under elevated temperatures. Case Stud. Constr. Mater. 15 (September), e00727. https://doi.org/10.1016/j.cscm.2021.e00727.

Yang, J., Firsbach, F., Sohn, I., 2022. Pyrometallurgical processing of ferrous slag "co-product" zero waste full utilization: a critical review. Resour. Conserv. Recycl. 178, 106021. https://doi.org/10.1016/j.resconrec.2021.106021.

Yildirim, I.Z., Prezzi, M., 2011. Chemical, mineralogical, and morphological properties of steel slag. Adv. Civ. Eng. 2011, 1–13. https://doi.org/10.1155/2011/463638.

Yu, Q., Zhu, B., Li, X., Meng, L., Cai, J., Zhang, Y., Pan, J., 2023. Investigation of the rheological and mechanical properties of 3D printed eco-friendly concrete with steel slag. J. Build. Eng. 72 (February), 106621. https://doi.org/10.1016/j.jobe.2023.106621.

Zhang, Y., Zhang, L., Wang, Q., 2023. Iron ore tailings, phosphate slags, and lithium slags as ternary supplementary cementitious materials for concrete: study on compression strength and microstructure. Mater. Today Commun. 106644. https://doi.org/10.1016/j.mtcomm.2023.106644.

Part V

Recycling of combustion/incineration residues into functional materials

Chapter 21

Recycling of combustion/incineration residues (fly ash) into zeolites and ceramics

Qili Qiu[1] and Yunan Zhou[2]
[1]*School of Environmental Engineering, Nanjing Institute of Technology, Nanjing, China;* [2]*Aviation Key Laboratory of Science and Technology on Aero Electromechanical System Integration, Nanjing Engineering Institute of Aircraft System, Nanjing, China*

1. Introduction

Fly ash, including coal fly ash (CFA) and municipal solid waste incineration (MSWI) fly ash, is a kind of an incineration residue with huge output (Zhuang et al., 2016; Sun et al., 2022). As the scale of incineration power generation in China increases, the emission of fly ash emissions continues to increase significantly. According to the National Bureau of Statistics of China, the total installed capacity is more than 2.2 billion kilowatts by 2020, while the thermal power installed capacity accounts for 55%. Therefore, huge impacts and challenges have been brought to the environment from the disposal of the large amount of fly ash emissions (Ren et al., 2020; Boycheva et al., 2021).

The utilization of fly ash is very extensive, and the traditional resource utilization path mainly includes the fields of building materials, agriculture, environmental protection, chemical industry, etc. The main component of fly ash belongs to the $CaO-SiO_2-Al_2O_3-Fe_2O_3$ system, which is very similar to the composition of ordinary Portland cement (OPC), making it very suitable as a raw material for cement production. In addition, fly ash can also directly replace some cement to prepare concrete. Hamernik and Frantz (1991) found that replacing some cement with fly ash as an auxiliary cementitious material can improve the compressive strength of concrete: when the replacement amount was 15 wt.% of the cement dosage, the compressive strength significantly increased. The direct preparation of concrete using fly ash can not only achieve solidification of heavy metals but also reduce the amount of cement used, leading to the decrease of the production cost. Furthermore, mixing fly ash with cement can produce a certain degree of hardness to meet the requirements of roadbed materials, so fly ash can be used as roadbed and embankment (Tang et al., 2018). Moreover, the use of fly ash has the advantages of reducing the expansion of materials and improving the California bearing ratio (CBR) and elastic modulus (Vizcarra et al., 2014). The application in agriculture mainly utilizes the loose and porous nature of fly ash, which also contains multiple trace elements, to achieve soil improvement and provide nutrients for plant growth (Sirikunpitak et al., 2022). In the field of chemical engineering, the main method is to extract useful substances from fly ash, such as extracting a large amount of Al_2O_3, heavy metals from fly ash (Hettenkofer et al., 2020). The adsorption characteristics of fly ash can also be utilized for the removal of pollutants (Sun et al., 2019).

At present, the main use of fly ash is in the form of low-end building materials, such as cement/concrete. With the increasingly stringent environmental requirements and the decreasing demand for low-end building materials, it is a major trend to recycle fly ash for the synthesis of high-value products such as zeolites and ceramics. Compared with traditional building materials, fly ash—based zeolites/ceramics are more convenient and cost-effective to transport. Therefore, this utilization method can facilitate the transportation and sales of fly ash products.

Treatment and Utilization of Combustion and Incineration Residues. https://doi.org/10.1016/B978-0-443-21536-0.00036-8

2. Characteristics of fly ash

2.1 Coal fly ash

CFA, as an industrial solid waste from coal combustion, contains a variety of heavy metal elements (Bukhari et al., 2015). According to the Statistical Review of World Energy, it is estimated that the global CFA production in recent years has been approximately 1 billion tons. With the increase of thermal power installed capacity, fly ash has become one of the main solid wastes in China. According to China Statistical Yearbook data, the annual output of CFA has exceeded 650 million tons, and it shows an increasing trend year by year. Therefore, it is of great significance to explore the effective way of CFA disposal to realize high value utilization. CFA is a kind of powdery solid waste, and the color is generally white, gray, or grayish black, whose main components are Al_2O_3, SiO_2, Fe_2O_3, CaO, and other oxides. The composition of CFA depends on many aspects, such as the mineral composition of the coal, the degree of pulverized coal, the type of furnace, and the oxidation conditions (Bukhari et al., 2015). Typically, CFA has a 15%–40% content of Al_2O_3, and the content of SiO_2 is between 40% and 60% (Quan et al., 2021). The micromorphology is composed of spherical and amorphous particles with rough and uneven surfaces. Meanwhile, various heavy metal elements, mainly including As, Cr, Pb, Cd, Ni, etc., can be detected in CFA. Comparatively, the content of these heavy metals is low and easy to solidify. Due to the influence of coal production location, quality, and combustion conditions, the uniformity and composition of CFA vary greatly.

CFA is generally classified as a nonhazardous waste, and it is commonly used in the production of zeolites and ceramics due to its high content of silica and alumina, which are key ingredients in these materials. The synthesized zeolites can find applications in areas such as adsorption, gas separation, catalysis, and ion exchange. Additionally, CFA can be utilized as a supplementary cementitious material in concrete production, thereby contributing to the construction industry as well. Various applications of CFA are presented in Fig. 21.1.

2.2 MSWI fly ash

MSWI fly ash refers to the ash collected by the flue gas purification system and settled at the bottom of the flue and chimney, which enriches more heavy metals and organic pollutants, compared with CFA. The main components of MSWI fly ash are CaO, Al_2O_3, and SiO_2, while Hg, Pb, Zn, Cd, Cr, Cu, and Ni are several common heavy metals. The regional

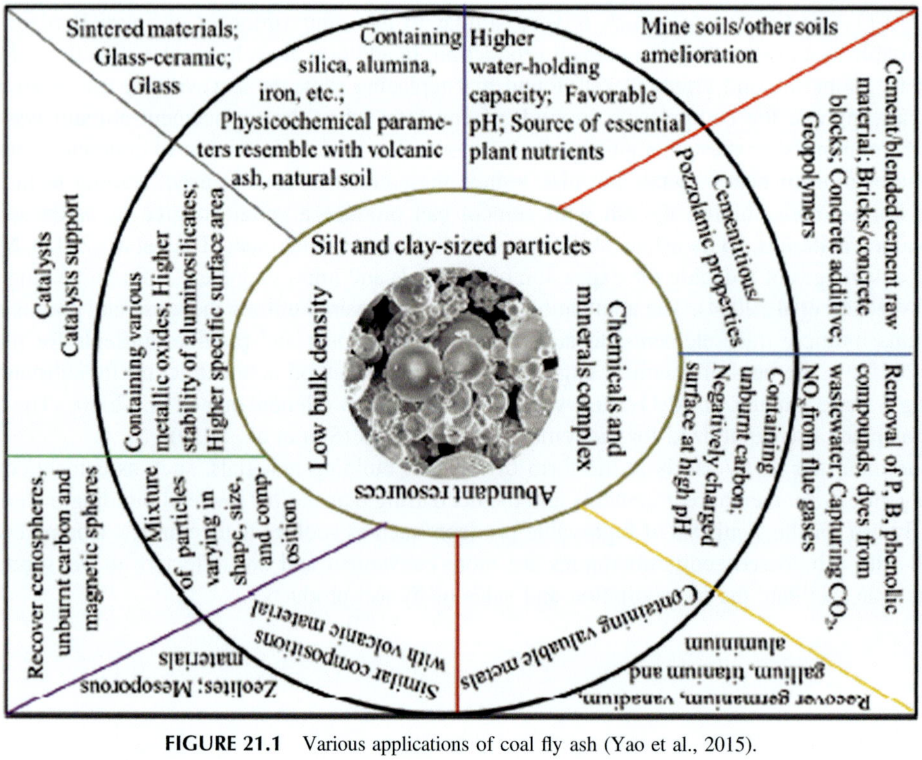

FIGURE 21.1 Various applications of coal fly ash (Yao et al., 2015).

differences and treatment process differences of the waste result in a large fluctuation range of its composition. Although heavy metals only exist in trace amounts (accounting for 0.5% of the weight of fly ash), fly ash has a certain leaching toxicity. If directly landfilled, it will cause serious pollution to the surrounding water bodies, soil, etc. Furthermore, MSWI fly ash also contains soluble organic matter, which accounts for 1%−4% by weight. Due to the potential presence of these contaminants, it is often classified as a hazardous waste. Consequently, the application of MSWI fly ash for zeolite and ceramic synthesis is generally more limited compared with CFA.

Generally, the room temperature treatment has major drawbacks in the disposal of MSWI fly ash, such as poor stability of heavy metals, susceptibility to secondary leaching, and inability to degrade dioxins. Compared with CFA, whose content of high silicon and aluminum is up to 40% (Song et al., 2014), the content of silicon and aluminum in MSWI fly ash is often less than 20% (Long et al., 2020). Overall, utilizing MSWI fly ash for zeolite and ceramic production typically involves specialized treatment processes to address the hazardous characteristics. These processes may focus on capturing and controlling the potential release of heavy metals and other contaminants. Once the hazardous components are effectively managed, the fly ash can be processed to synthesize zeolites or incorporated into ceramics, taking advantage of its inherent properties such as silica and alumina content.

Therefore, while both CFA and MSWI fly ash can be used for zeolite and ceramic synthesis, the application of MSWI fly ash requires additional considerations and treatment steps to ensure the safe utilization of the hazardous waste.

3. Zeolites

Zeolite is a kind of aluminosilicate minerals with special three-dimensional frameworks, which contains various uniform pores and channels (Joseph et al., 2020). Zeolites are generally composed of $[SiO_4]^{4-}$ tetrahedra and $[AlO_4]^{5-}$ tetrahedra, forming a three-dimensional skeleton structure in different arrangements by sharing an oxygen atom (Liu et al., 2015). Secondary circular structures of different sizes can be formed by $[SiO_4]^{4-}$ tetrahedra and $[AlO_4]^{5-}$ tetrahedra through various connection methods. Then secondary circular structures can be combined into different cage structures, whose arrangement can affect the structure and performance of zeolites (Li et al., 2021). The unique pore structure of zeolite leads to its large specific surface area, good ion exchange performance, and excellent adsorption performance. Additionally, due to its strong acidic centers on the surface and strong Coulomb field/polarity within the pores, zeolite exhibits excellent catalytic properties (Schumann et al., 2012). The crystal structure of zeolite can be divided into three components: (1) Aluminosilicate framework, (2) framework containing channels and cavities, and (3) water molecules of potential phase. Due to its potential molecular sieve, high specific surface area, and good thermal/chemical stability, zeolite is widely used as adsorbent, ion exchanger, and catalyst (Jiang et al., 2011; Li et al., 2021). The basic formula of structure cell of zeolite is $M_{x/n}[(AlO_2)_x(SiO_2)_y]\cdot \omega H_2O$, where M and n represent the metal cations and their valence states outside the framework, and ω represents the number of water molecules in the crystal cell. Additionally, x and y represent the number of aluminum and silicon tetrahedra in the cell.

Zeolite minerals can be classified into different groups based on their structure and properties (Yan et al., 2022). Here are some common classifications:

(1) Framework type:
Aluminosilicate zeolites: These zeolites have a framework composed of aluminum, silicon, and oxygen atoms. Examples include zeolite A, zeolite X, and zeolite Y.
Silicoaluminophosphate (SAPO) zeolites: These zeolites contain aluminum, silicon, phosphorus, and oxygen in their framework. Examples include SAPO-34 and SAPO-18.

(2) Pore size and shape:
Microporous zeolites: These zeolites have small micropores, typically less than 2 nm in diameter. Examples are zeolite A, zeolite X, and zeolite Y.
Mesoporous zeolites: These zeolites have larger mesopores, typically ranging from 2 to 50 nm in diameter. Examples include MCM-41 and MCM-48.
Hierarchical zeolites: These zeolites feature both micropores and mesopores, offering improved mass transfer and accessibility. They are often prepared by postsynthetic treatments of conventional zeolites.

(3) Ion-exchange capacity:
Cation-exchange zeolites: These zeolites can exchange cations in their framework, such as sodium (Na^+), potassium (K^+), and calcium (Ca^{2+}). Examples include zeolite A, zeolite X, and zeolite Y.
Anion-exchange zeolites: These zeolites can exchange anions in their framework, such as chloride (Cl^-) and nitrate (NO^{3-}). Examples include zeolite ZM-5 and zeolite ETA-15.

(4) Application-specific:
Adsorbent zeolites: These zeolites have a high affinity for adsorbing gases, liquids, and molecules of certain sizes. Examples include zeolite 4A and zeolite 13X, which are used in gas separation and dehydration.
Catalytic zeolites: These zeolites exhibit catalytic activity due to their shape-selective pores and acidic properties. Examples include zeolite beta and zeolite ZSM-5, widely used in petrochemical and refining processes.

In general, zeolites can be divided into two categories based on their sources: naturally formed zeolites and artificially synthesized zeolites. Due to its low purity and poor quality, the application scope of natural zeolites is limited. There are far more types of artificial zeolites than natural zeolites. Moreover, artificial zeolite has the advantages of high purity, large specific surface area, and adjustable pore size and acidity, making its application range more extensive.

The synthesis of zeolites is usually from mineral resources and pure chemicals, which can be replaced by silica-alumina-rich materials such as CFA and MSWI fly ash (Abdullahi et al., 2017). The main components in fly ash are similar to those of zeolite and can be used for the synthesis of zeolite. Therefore, when introducing fly ash into zeolite synthesis, not only can the cost of zeolite synthesis be reduced, but also the high-value recovery of fly ash can be achieved (Gross et al., 2007; Liu and Lu, 2020). The synthesis of coal fly ash zeolites (CFA zeolites) has a history of over 30 years. New ideas are continuously explored by researchers over the world, through changing methods and parameters, and various zeolites have been successfully synthesized.

3.1 Coal fly ash

Most of the metals in CFA can be solidified and stabilized during the synthesis process of zeolite. The main toxic elements are not detected in the leachate, indicating that synthesis process greatly reduces the migration rate of these elements inherited from fly ash (Feng et al., 2018). Therefore, the production of zeolite using CFA is considered safe, and the synthesis of zeolite from CFA is a way to broaden the resource utilization, which has both economic and environmental benefits. This approach has also become a key research direction for the high value-added utilization of fly ash in recent years.

The synthesis methods of fly ash zeolite include conventional hydrothermal synthesis (Kobayashi et al., 2020), alkali melting hydrothermal method (Joseph et al., 2020), microwave-assisted hydrothermal method (Qiu et al., 2018), supercritical hydrothermal method (Ma et al., 2019), etc. Zeolite synthesized from fly ash is considered a promising low-cost water treatment material. Compared with commercial zeolites, CFA zeolites have similar removal efficiency for acidic mining wastewater (Cardoso et al., 2015). Even though the cation-exchange capacity (CEC) of CFA zeolites is only half that of commercial zeolites, similar treatment effects can be achieved under the same conditions. Moreover, both the type of fly ash and modification technology can affect the removal efficiency of heavy metals, dyes, and other ions in wastewater or gas.

3.1.1 Conventional hydrothermal synthesis

The application of conventional hydrothermal methods in CFA mainly includes zeolite synthesis and heavy metal stabilization. The latter achieves the stabilization of heavy metals through the synthesis of zeolite-like substances during hydrothermal treatment, which have the functions of ion adsorption, exchange, precipitation, and physical encapsulation on heavy metals. Therefore, not only heavy metals in CFA can be effectively prevented from infiltrating into the residual solution during the hydrothermal process, but the solidified fly ash has acid resistance and can also be applied as an acidic neutralizer. Many kinds of zeolites have been hydrothermally synthesized, such as cristobalite, analcime, zeolite P, scolecite, chabazite, etc. (Feng et al., 2018; Liu et al., 2023). The zeolitization is mainly divided into four steps: the dissolution of Al and Si in the fly ash; the aggregation of polymers (dimers, trimers, etc.) formed by Al and Si on the surface of fly ash; zeolite nucleation; the growth of zeolite crystals. For hydrothermal synthesis of zeolites, there are two main methods: one-step synthesis and two-step synthesis.

One-step hydrothermal method is to synthesize zeolites using fly ash and alkaline solution as raw materials under hydrothermal conditions directly. Under the action of alkaline solution, the dissolution of Si and Al in fly ash is accelerated to form silicon aluminum gel, thus accelerating the formation and growth of zeolite crystals. The Si/Al molar ratio and alkaline condition have a significant impact on the types, crystal structure, and surface structure of zeolite synthesized (Ameh et al., 2017). Therefore, the difference in silicon and aluminum content in fly ash will affect the type of zeolite products, and different target zeolite product targets can be achieved by adding silicon or aluminum sources. The most commonly used alkaline solutions include sodium hydroxide, potassium hydroxide, sodium carbonate, etc. For example, Na ions can promote the formation of Zeolite P, and also it helps to obtain the zeolite with higher yield and relative crystallinity (Murayama et al., 2002). The one-step hydrothermal method is simple to operate. However, the drawbacks, including long reaction time, high energy consumption, low product purity, many by-products, etc., cannot be ignored.

Two-step synthesis of pure zeolites from CFA is an extension of the one-step hydrothermal synthesis, which was first proposed by Hollman et al. (1999). The procedure of two-step synthesis is to filter the filtrate after one-step hydrothermal treatment, add silicon aluminum sources, adjust a certain proportion of Si/Al, and synthesize zeolite with relatively high crystallinity and purity. This method can maximize the utilization of silicon and aluminum components in fly ash from the waste liquid in one-step hydrothermal process. Besides, a single phase with higher larger surface area and crystallinity can be obtained by increasing the induction time and optimizing the process route (Iqbal et al., 2019). In summary, the operation of this method is relatively more complex and time-consuming, but the obtained zeolite has higher purity and better crystallization.

3.1.2 Alkaline fusion-hydrothermal method

The alkaline fusion-hydrothermal method is a combination of melting and hydrothermal methods. Firstly, CFA and alkaline substances (NaOH, Na_2CO_3, KOH, etc.) are premixed in a certain proportion; secondly, the mixture is calcined at high temperature (450−850°C) in a muffle furnace for 1−2 h, thus effectively destroying the dense glass bead structure on the surface of fly ash; after alkali melting, the product processed through a hydrothermal reaction to obtain zeolites. Under the combined action of melting and alkaline activator, the inert components, such as quartz stone and mullite, in fly ash can be completely destroyed to release the alkali-soluble nepheline phase of amorphous Si and Al elements (Panitchakarn et al., 2014). Thus, more aluminum silicate can be dissolved into the solution, ultimately improving the conversion rate of CFA and BET surface area of the crystallized product (Sakthivel et al., 2013). In addition, thus, the reactivity can be well improved in terms of quality of zeolite, such as CEC and crystallinity (up to over 95%), through the subsequent hydrothermal reaction (Molina and Poole, 2004; Huber et al., 2018). It has been reported that fusion-hydrothermal method favored the production of zeolite A and zeolite X (Shigemoto et al., 1993). It follows that alkaline fusion-hydrothermal method has higher zeolite conversion rate, as well as zeolite purity and quality compared with traditional hydrothermal methods. In the process, organic pollutants in fly ash can also be removed at high temperatures, but some heavy metal pollutants may migrate to the gas phase. Moreover, this method requires more economic investment, which is also an important limiting factor for promotion.

In addition to this sequential fusion and hydrothermal method, direct fusion synthesis has also been attempted. This method involves directly mixing fly ash with a mixture of alkali or salt for fusion disposal to obtain zeolites, such as cancrinite and sodalite (Park et al., 2000). The salt, NaCl, not only has a melting aid effect in this method but also reduces the combustion temperature of carbon. Obviously, this method can avoid the generation of secondary wastewater during the hydrothermal process, but it has been reported that the overall conversion rate of its zeolite is not high (Choi et al., 2001). Moreover, the fusion process requires high temperatures and a large amount of salt or alkali, and the product has a low purity. Therefore, this technology has not been widely used at present.

3.1.3 Microwave-assisted hydrothermal method

Microwave hydrothermal-assisted synthesis is a method of heating the reaction system by microwave radiation. Microwaves are electromagnetic waves with wavelengths ranging from 1 mm to 1 m, and frequencies ranging from 300 MHz to 300 GHz (Chandrasekaran et al., 2013). There is a significant difference in the principle of solution heating between microwave heating and conventional heating. Ions in the solution will move and collide in the microwave field, leading to the conversion of kinetic energy into thermal energy. Therefore, as the concentration of ions in the solution increases, collisions will become more severe, and water, as a polar ion, can achieve rapid heating in the microwave field. Thus, microwave heating has the characteristic of simultaneity, avoiding the generation of internal and external temperature gradients. Therefore, under the same conditions, compared with traditional hydrothermal methods, microwave heating has obvious advantages of fast reaction speed, short reaction time, uniform heating, low energy consumption, and high economy (Makul et al., 2014). There is data indicating that microwave hydrothermal treatment can heat the temperature to 373 K after 3 min, while traditional hydrothermal treatment requires more than 15 min (Fukui et al., 2006). The earliest application of this method was the synthesis of NaP1 zeolite by Querol et al. (1997). Besides, various zeolites have been synthesized in succession, such as kalsilite, phillipsite, analcime, tobermorite, and hydroxysodalite (Inada et al., 2005; Fukui et al., 2006; Ansari et al., 2014). In addition to these advantages, microwave hydrothermal technology has unique advantages in the field of nanozeolite synthesis, such as nano-NaX zeolite (Ansari et al., 2014). In terms of synthesis mechanism, microwave hydrothermal method is almost identical to traditional hydrothermal method, with the only difference being the heating source. Therefore, under these two heating forms, the types of zeolites synthesized are almost identical and the structural properties of zeolites synthesized are almost the same. The only difference is that KM zeolite is not synthesized under microwave irradiation (even if the reaction temperature and reagent concentration are changed) (Querol et al., 1997). Under microwave radiation, the initial efficiency of zeolite synthesis is higher than the overall efficiency. The main reason is that the dissolution of SiO_2 and Al_2O_3 is accelerated at the initial stage of radiation (0−20 min), which greatly speeds up the synthesis of zeolite. In the middle stage (20−40 min), microwave seriously impedes the synthesis of zeolite, for the reason that microwave slows down the nucleation of silica alumina gel. In the later stage, microwave hydrothermal has little impact on the formation of zeolite.

In summary, conventional heating methods, including hydrothermal and fusion methods, consume a large amount of energy and require longer reaction time, while microwave-assisted hydrothermal process has the advantages of higher energy efficiency (Hermassi et al., 2020). The microwave-assisted method accelerates the entire zeolite synthesis process, through effectively promoting the dissolution of Si and Al phases in raw materials to promote the crystallization induction period. Accordingly, the microwave-assisted hydrothermal synthesis can achieve a sudden decrease in reaction time from several days to only a few hours or even tens of minutes. However, continuity problem is limited by microwave radiation; there are certain difficulties in achieving large-scale production of microwave hydrothermal synthesis at present. Meanwhile, this technology also has some problems similar to conventional hydrothermal methods, such as low product purity and low conversion rate.

3.1.4 Ultrasonic-assisted method

Ultrasound is a type of sound wave with a frequency higher than 20,000 Hz, characterized by high frequency and short wavelength. In addition, it has characteristics such as good directionality, high energy, and strong penetration and can cause cavitation during the propagation process. The cavitation effect refers to the dynamic process of growth and collapse of microgas nucleus cavitation bubbles that exist in liquids, which vibrate under the action of sound waves and occur when the sound pressure reaches a certain value. Therefore, the cavitation effect of ultrasound is often accompanied by mechanical and thermal effects, which can instantly generate extremely high temperatures (>5000 K) and pressures (>20 MPa), making it possible to trigger some special reactions under the action of ultrasound radiation (Pal et al., 2013). In the field of zeolite synthesis, ultrasound can accelerate chemical reactions, effectively reduce the temperature required for the reaction, and promote the reaction between liquid and solid. Ultrasonic chemistry is considered as a promising technology that can be applied to synthetic reactions.

Ultrasonic treatment of fly ash into zeolite mainly involves zeolite A, zeolite P, and other zeolites (Belviso et al., 2013). According to a research report in 2016 (Bukhari et al., 2016), ultrasound-assisted hydrothermal treatment can produce Na-A zeolite with higher crystallinity than conventional hydrothermal treatment, laying the foundation for subsequent research on zeolite synthesis using ultrasound. Furthermore, sonochemistry is also an important application field of ultrasound. Research (Wang and Zhu, 2005) has found that the specific surface area of fly ash treated with ultrasound can be significantly increased (from 5.6 to 30.4 m^2/g), and it exhibits significant adsorption for methyl blue. In addition, ultrasonic waves can promote the formation of free radicals that can rapidly crystallize the zeolite phase (Pal et al., 2013). In some cases, microwave and ultrasound can also be combined to prepare high-quality zeolites.

Ultrasound has the characteristics of violent liquid movement and molecular vibrations and plays an important role in the synthesis of fly ash zeolite. Many studies have shown that ultrasound has the following effects on the synthesis of fly ash zeolites:

(1) Accelerate reaction speed: Ultrasound can promote the mixing and diffusion of reactants, accelerate the reaction speed, and shorten the synthesis time of fly ash zeolite.
(2) Improve product performance: Due to the effect of ultrasound, the product can have a more uniform pore size and higher specific surface area, improving its adsorption and catalytic performance.
(3) Enhance crystallinity: Ultrasound can promote the formation of zeolite crystals through intense cavitation and vibration, resulting in better crystallinity and lattice integrity of the product.
(4) Energy saving: Ultrasound can replace traditional mechanical stirring, reduce energy consumption during the reaction process, and improve synthesis efficiency.

In summary, ultrasound has broad application prospects and research value in the synthesis of fly ash zeolite, and also provides new ideas and methods for the comprehensive utilization of fly ash.

3.1.5 Seed-induced method

High-quality zeolite materials can be quickly and efficiently prepared by the seed-induced method, and meanwhile the morphology and crystal structure of zeolites are also controlled. Therefore, this method has been widely applied in fields such as high-performance adsorbents and ion exchangers. After fully mixing specific zeolite crystal seeds with CFA and alkaline solution, they are placed in a tank reactor to induce crystal growth, thereby shortening the nucleation time and reaction cycle (Kacirek and Lechert, 1976). Ultimately, zeolites of the same type as the crystal seeds will be produced. Moreover, crystal seeds can inhibit the formation of other heterocrystals, which helps to select and synthesize specific types of zeolites.

It is found that under different reaction conditions such as alkali concentration, silicon aluminum ratio, and solid–liquid ratio, there is no significant effect on the type of synthesized product. This is due to the addition of trace amounts of molecular sieve seeds, which have a guiding effect on the synthesis of molecular sieves and reduce the occurrence of impurities. Accordingly, this zeolite synthesis method has the characteristics of short period, high crystallinity, and complete crystal phase, possessing certain development prospects in large-scale applications.

3.1.6 Other methods

In addition to the relatively extensively studied methods for synthesizing zeolites mentioned before, there is stepwise heating method, dialytic hydrothermal method, etc.

The stepwise heating method is a new synthesis method derived from the hydrothermal method, which is similar in steps to the hydrothermal method. However, the crystallization process is divided into two stages. Firstly, the reaction is accelerated at low temperature for a period of time to activate fly ash in an alkali solution, and then the crystallization reaction is continued at high temperature to improve the conversion rate of zeolite products and shorten the crystallization time. This method has the problem of low conversion rate.

The principle of dialysis hydrothermal method is to fully mix alkali solution and fly ash into a test tube made of a semipermeable membrane, and then place the test tube in a NaOH solution. The Si^{4+} and Al^{3+} dissolved in the test tube enter NaOH through the semipermeable membrane. In the solution, the NaOH solution is analyzed for components, and the silicon aluminum ratio is adjusted. After hydrothermal reaction, zeolite molecular sieve is obtained. In this way, the substances involved in the reaction are controllable, so the purity of the zeolite product is high. However, a large amount of dialyzed waste liquid is generated, which poses a new problem of secondary pollution.

3.2 MSWI fly ash

Under the high-temperature conditions of hydrothermal process, the degradation of dioxins in MSWI fly ash can be achieved through efficient dechlorination and carbon ring decomposition reaction (Alterary and Marei, 2021). At the same time, zeolite-like substances can be formed with the existence of Al_2O_3 and SiO_2 in fly ash, thereby achieving the stability of heavy metals. The process of hydrothermal synthesis of zeolite from MSWI fly ash is similar to that of CFA, and the zeolite-like minerals (silica aluminate) can also be detected in the product. The treatment effect of hydrothermal reaction on MSWI fly ash is shown in Fig. 21.2. And the products from hydrothermal reaction have high stability and adsorption,

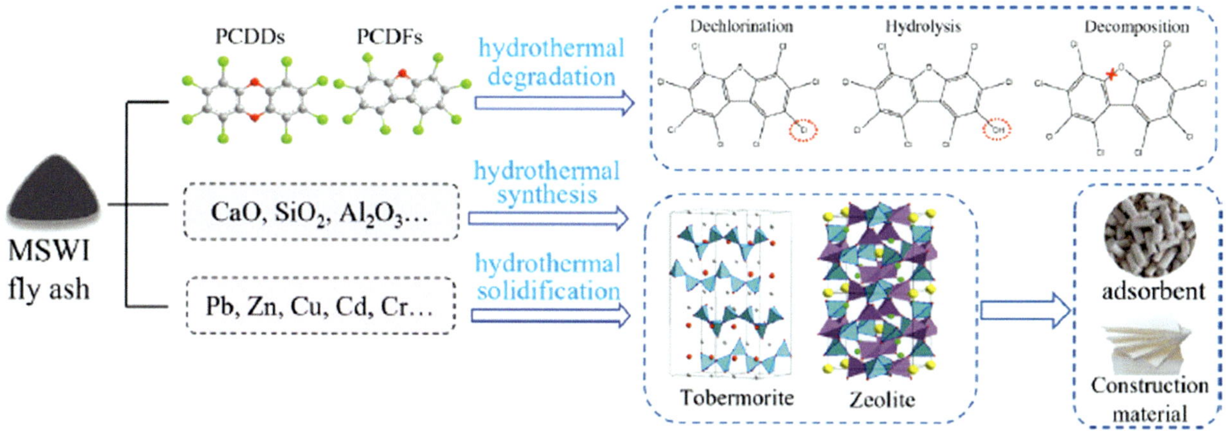

FIGURE 21.2 Treatment effect of hydrothermal reaction on MSWI fly ash (Fan et al., 2023).

leading to good prospects for resource utilization. But the silicon and aluminum content of fluidized bed waste incineration fly ash is often less than 20% (Long et al., 2020), while CFA has a high silicon and aluminum content of up to 40% (Wang et al., 2009; Song et al., 2014). Therefore, it is difficult to prepare zeolite-like substances from MSWI fly ash, and the synthesized zeolites often have low quality.

The concentration of NaOH solution, solid−liquid ratio, reaction time, and reaction temperature all affect the synthesis amount and type of zeolite. With the change of reaction conditions, sodium chabazite, orthorhombic scolecite, cancrinite, and X zeolite can be synthesized. Research has found that when the hydrothermal time is extended to 48 h and at 180°C, a large amount of aluminum-containing tobermorite and katonite can be obtained (Bayuseno et al., 2009). Thus, it is necessary to enhance the synthesis efficiency.

The commonly used measures to enhance zeolitization reaction are as follows:

(1) Add silicon aluminum additives
The effect of MSWI fly ash to synthesize silicate aluminate minerals depends on the composition of fly ash, especially the content of silicon and aluminum. The content of silicon and aluminum in MSWI fly ash is relatively low, and sufficient silicon aluminate minerals cannot be obtained during the hydrothermal treatment. Commonly used silicon−aluminum additives include quartz, CFA, bentonite, kaolin, silica fume, metakaolin, etc., which can be used to adjust the element composition of the reaction system. Through hydrothermal treatment, several zeolitic products can be formed, such as boehmite, katonite, tobermorite, analcime, boehmite, etc. And by adjusting the proportion of additives, it can further realize the synthesis of specific types of silicon aluminate minerals. Research (Shi et al., 2017; Tian et al., 2020) has found that by adding quartz or $Ca(OH)_2$ to adjust the Ca/Si molar ratio, more tobermorite can be synthesized.

(2) Adjust reaction time and temperature
The influence of reaction temperature and reaction time on zeolite synthesis is mainly reflected in two aspects: one is the dissolution of silicon aluminum materials, and the other is the type and characteristics of zeolite. In general, prolonging the reaction time and increasing the reaction temperature is conducive to the dissolution of silicon and aluminum components in raw materials and will promote the formation of silicon aluminate minerals. In low-temperature hydrothermal conditions (about below 70°C), it is difficult to form zeolite-like substances. When the temperature exceeds 90°C, the hydrothermal system gradually begins to form zeolite. Generally, it is necessary to adjust the temperature to 150°C or above to obtain relatively stable zeolite (Bayuseno et al., 2009). In terms of reaction time, it usually takes several hours or even 24−48 h.

(3) Alkaline activator and concentration
Alkaline hydrothermal environment is essential to synthesize silicate aluminate minerals from waste incineration fly ash. The selection and concentration of alkaline activator will affect the type and output of the ultimately formed zeolites. From existing research, NaOH is a very suitable alkaline activator. Moreover, from the perspective of hydrothermal stabilization of heavy metals in fly ash, its effectiveness is significantly superior to Na_2CO_3, KOH, K_2CO_3, etc. (Qiu et al., 2016). The influence of NaOH concentration is also very significant. When the concentration is too low, the synthesis amount of zeolite is low, while when the NaOH concentration increases to the appropriate concentration range, the synthesis amount and the specific surface area of the zeolite increase, resulting in an increase in CEC. However, when the concentration increases too high (over 4 M), the specific surface area and CEC actually show a downward trend. This phenomenon may be caused by changes in the type of zeolite synthesized as the concentration increases. In addition, it is worth noting that the effect of the liquid−solid ratio is often consistent with the effect of the concentration of alkaline activator.

3.3 Applications of zeolites synthesized

Zeolites synthesized from CFA and MSWI fly ash can find diverse applications due to their unique properties. Zeolites derived from CFA may be preferred for applications requiring high thermal stability, while those derived from MSWI fly ash may offer greater potential for waste remediation due to their ability to immobilize contaminants. It's important to note that the specific properties of zeolites depend on the specific fly ash characteristics and the synthesis conditions employed. Further research and characterization are often needed to fully understand and optimize the properties of zeolites obtained from different fly ash sources.

Here are some applications of zeolites:

(1) Adsorbents: Zeolites can be used as adsorbents for removing pollutants and heavy metals from water and wastewater. The porous structure of zeolites enables them to selectively trap and remove contaminants, such as heavy metals (Qiu et al., 2018), agricultural wastewater, organic pollutants (Zhao et al., 2020a), etc.

(2) Catalysis: Zeolites can act as catalysts in various chemical processes (Zhang et al., 2017). The unique structure of zeolites provides a high surface area and acidity, making them effective catalysts for reactions such as cracking, isomerization, and oxidation.
(3) Gas separation: Zeolites have the ability to selectively adsorb gases (Lee et al., 2017). Zeolites can be used in gas separation processes to separate specific gases from mixtures such as removing carbon dioxide and sulfur dioxide.

4. Ceramics

4.1 Coal fly ash

The research on the preparation of ceramics from CFA is mainly focused on the field of structural ceramics, which can be divided into two categories: ordinary structural ceramics and advanced structural ceramics. Structural ceramics are advanced ceramics with excellent mechanical, thermal, and chemical properties such as high temperature resistance, erosion resistance, corrosion resistance, high hardness, high strength, and low creep rate, commonly used in various structural components. Therefore, under extremely harsh environmental or engineering application conditions, the high stability and excellent mechanical properties exhibited have attracted great attention in the material industry, and their range of use is also expanding day by day.

Compared with the conventional building materials applications of CFA, such as cement and concrete, structural ceramics are materials with high density and added value in the materials industry. Among ordinary structural ceramics, fly ash is the most widely used in building ceramics. Besides, fly ash has also been applied and studied in the produce of some advanced structural ceramics, such as mullite ceramics (Dong et al., 2008), wollastonite ceramics (Lu et al., 2014), cordierite ceramics, microcrystalline glass (Fan et al., 2019), etc.

4.1.1 Building ceramics

Building ceramics, also known as ceramic bricks, are used to protect and decorate buildings, walls, and floors in construction projects. The production of architectural ceramics is the largest structural ceramics industry in China, and the demand for ceramic tiles in the market is rapidly increasing, occupying an important position in the national economy. They are mainly prepared from three traditional silicate raw materials: clay, feldspar, and quartz (Carty and Senapati, 1998). The high-temperature stability characteristics of quartz enable it to serve as an inert filler to reduce ceramic deformation and shrinkage, while feldspar serves as a flux to promote matrix densification. Clay, as an adhesive in the green body, provides the required plasticity for molding, and another more important role is to form mullite crystals during the sintering process to enhance the strength of ceramics (Reinosa et al., 2015; Turkmen et al., 2015).

The research on the preparation of ceramics using CFA is mainly focused on the field of building ceramics, and the main focus is on the substitution mechanism of CFA for ceramic raw materials. Due to the inertness of most CFA particles during the sintering process, they are usually used to replace quartz-based raw materials. In a traditional triaxial porcelain composition consisting of kaolinitic clay, quartz, and feldspar, quartz is completely replaced by CFA (Mukhopadhyay et al., 2010). By measuring the relevant properties fired at 1100−1300°C, it was found that using a sample containing 30 wt.%, CFA can achieve the best performance. Moreover, at 1300°C, the bending strength of the ceramic body reached 72.3 MPa, while the apparent porosity was almost zero. Compared with traditional triaxial bodies, the bending strength has increased by about 20%. In ceramic products, the content of mullite increases with the increase of CFA ratio, and the close interlocking of fine mullite needles is the reason for the higher bending strength of CFA added to the ceramic body. When quartz was gradually replaced by 5, 10, and 15 wt.% CFA, the results showed that samples containing CFA achieved higher density (up to 2.46 g/cm^3) throughout the entire firing temperature range (1150−1300°C) and matured earlier than traditional ceramic components (Dana et al., 2004). The increase in the replacement amount of CFA leads to not only an increase in the linear shrinkage rate and bulk density of ceramics but also a significant decrease in mechanical properties. Compared with 61.1 MPa obtained in the traditional porcelain sample, the maximum flexural strength (70.5 MPa) was also obtained at 1300°C in the 15 wt.% fly ash containing sample. In addition, in some cases, CFA is used to replace feldspar in ceramic formulations. However, the allowed amount of addition is often limited, usually not exceeding 10 wt.%. The appropriate addition of CFA can aid in melting and improve ceramic properties, but excessive amounts often lead to macroscopic defects and deterioration of mechanical properties of the product (Olgun et al., 2005; Kockal, 2012). At present, the process of replacing clay with CFA is not yet mature, as it is not possible to achieve large-scale addition and application in ceramic production processes. Luo et al. (2018) successfully developed a new type of preparation technology for all CFA-based building ceramics with a flexural strength of 50.10 MPa, a bulk density of 2.50 g/cm^3, and a water absorption rate of 0%, which was far superior to the requirements of the Chinese national standard GB/T 4100-2015 at the optimal sintering temperature of 1100°C.

By using CFA together with various solid wastes, such as shale, sludge, iron tailings, coal gangue, etc., it can also compensate for the defects in CFA properties and obtain high-performance and low-cost ceramic products (Luo et al., 2020). For some special CFA with high aluminum content, high-strength ceramic bricks can be synthesized (Wang et al., 2017).

4.1.2 Mullite ceramics

The main raw materials for synthesizing mullite include industrial raw materials and natural mineral raw materials. The main components in high alumina CFA are $A1_2O_3$ and SiO_2. High alumina CFA is a new type of fly ash, which is generally believed that CFA with an alumina content greater than 37 wt.% is called high alumina fly ash. The main crystalline mineral phases are mullite, corundum, etc., accounting for about 45 wt.%. The amorphous phase mainly consists of vitreous aluminum silicate and amorphous silica, accounting for about 50 wt.% of the total amount of fly ash (He et al., 2014). In some typical CFA, the content of $A1_2O_3$ exceeds 50 wt.%, leading to significant advantages in the production of mullite ceramics. Mullite ceramics refer to ceramics with mullite as the main crystalline phase. Due to the high temperature and low pressure conditions required for the formation of mullite in nature, the natural production of mullite is scarce, mainly through artificial synthesis. Owing to the presence of a certain amount of mullite crystal nuclei or microcrystals in CFA, the conditions required for the formation of crystal nuclei in the early stage of firing can be eliminated. Compared with natural ores, synthesizing mullite from CFA has advantages such as low energy consumption, short time, and secondary utilization of waste.

Mullite is the only stable binary compound in the $Al_2O_3{-}SiO_2$ system, and its composition can vary from $3Al_2O_3 \cdot 2SiO_2$ to $2Al_2O_3 \cdot SiO_2$. The crystal configuration belongs to the rhombic crystal system, similar to sillimanite. It can be seen that the alumina content in mullite is often higher than that of high alumina CFA. Therefore, in the process of synthesizing mullite ceramics, additional alumina or other substances with high alumina content need to be added. Moreover, in the preparation of mullite ceramics, compared with building ceramics, the addition of CFA can be significantly increased to over 70 wt.%. Dong et al. (2008) prepared low-cost mullite ceramics using natural bauxite and industrial waste fly ash as raw materials. After sintering at 1600°C for 4 h, the average fracture strength of the sample was 186.19 MPa. The significant improvement in fracture strength is attributed not only to the enhancement of densification but also to the network structure of a large number of the compact interlocked rod-like mullite crystals embedded in the glass phase. It was also observed that the increase in sintering temperature led to the precipitation of more coarse grains of mullite. The digital photos of mullite ceramics produced in lab are shown in Fig. 21.3.

Research (Dong et al., 2011) has found that adding MgO to the CFA/bauxite mixture can promote densification by reducing volume expansion, especially at higher temperatures. Moreover, the results of shrinkage, bulk density, and pore structure effectively confirm the sintering aid function of magnesium oxide, which allows the formation of f a magnesia-containing silica-rich ternary liquid phase. The effect of V_2O_5 doping on the properties of mullite ceramics was introduced (Li et al., 2009). The apparent porosity and water absorption decreased with the increase of V_2O_5 content at 1500°C.

The development of mullite ceramics will add enormous economic value to the large amount of industrial waste residue and CFA that exist around the world. In addition, the cheap natural raw mineral bauxite has been proven to be an effective substitute for industrial alumina in the production of mullite (Dong et al., 2008).

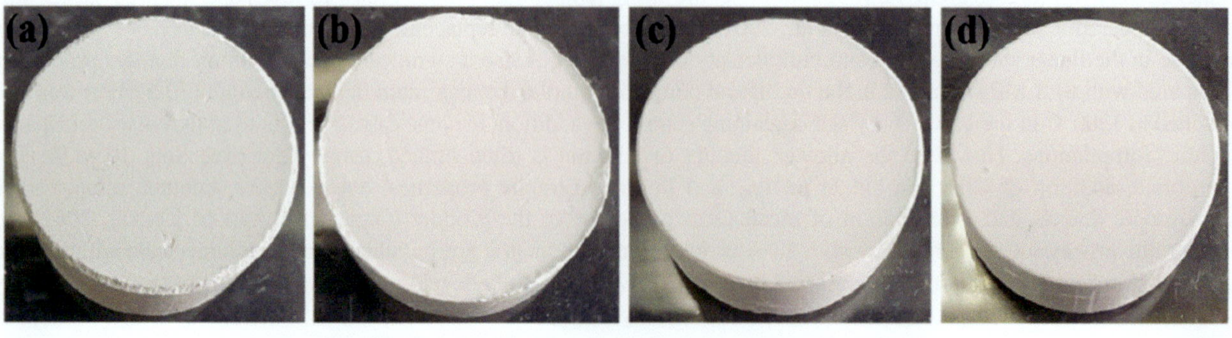

FIGURE 21.3 Digital photos of mullite ceramics (Huo et al., 2023).

4.1.3 Wollastonite ceramics

Wollastonite ceramics refer to the ceramics whose main crystal phase is calcium silicate ($CaO \cdot SiO_2$). In the process of preparing wollastonite ceramics from fly ash, the common and simple application is to mix with waste glass, glass slag, etc. for sintering. The sintering temperature is generally lower than the preparation temperature of mullite ceramics, ranging from approximately 1000 to 1100°C. The study of disposal temperature between 850 and 1050°C found that acicular crystal phase crystallization can be detected at the heat treatment temperature of 1000 and 1050°C, and wollastonite ($CaSiO_3$) was found in the main crystal phase (Yoon et al., 2013). Moreover, the mechanical properties of glass ceramics obtained at this heat treatment temperature, including density, compressive strength, bending strength, and chemical durability, are superior to those obtained at other heat treatment temperatures. When the preparation temperature is raised to 1100°C, it is found that as the temperature increases, the mechanical properties of the product can be further improved (Lu et al., 2014). Compared with the raw materials in coarse powder, fine powder sintered microcrystalline glass shows the characteristics of fast sintering speed, high stacking density, and low sintering activation energy. According to linear shrinkage analysis, it is found that the crystallization process begins at 800°C.

In addition to direct utilization of CFA, wollastonite glass ceramics can also be prepared by using the residual tailings after extracting alumina from high alumina CFA, supplemented by basic glass raw materials (Chen et al., 2016). The main components of the tailings are CaO and SiO_2, and their molar ratio is close to 1:1, which is close to the theoretical composition of wollastonite. Thus, it is an ideal raw material for preparing wollastonite glass ceramics.

Some experts believe that the optimization of compositional uniformity, particle size grading, and heating rate would pave the way to the practical utilization of waste materials in the construction materials (Lu et al., 2014).

4.1.4 Cordierite ceramics

The synthesis of cordierite has a history of nearly 100 years. In 1995, Sampathkumar et al. first synthesized α-cordierite by CFA, talc, and Al_2O_3 at 1370°C (Sampathkumar et al., 1995). Cordierite-based ceramics have unique properties, such as low thermal expansion coefficient, low dielectric constant, high thermal shock resistance, high chemical stability, and strong mechanical strength (Harrati et al., 2022). Therefore, cordierite is widely used in the ceramic industry, including thermal shock resistant materials, catalyst carriers, foam ceramics, printed circuit boards, refractories, etc. The chemical composition of cordierite is $2MgO \cdot 2Al_2O_3 \cdot 5SiO_2$, so it can be prepared according to the stoichiometric ratio, that is, MgO 13.8 wt.%, Al_2O_3 34.9 wt.%, and SiO_2 51.3 wt.% (Ma et al., 2014). Cordierite is synthesized from CFA, mainly by solid-reactant sintering method. Generally, at about 1100°C, cordierite begins to form, and at about 1250°C, the content of cordierite increases sharply, while the synthesis temperature is generally above 1250°C. Effects of sintering temperature on the properties of cordierite ceramics are shown in Fig. 21.4.

The cordierite can be obtained from CFA directly, but the purity of the product is relatively low. Ma et al. (2014) found that 75 wt.% high alumina CFA, 17.4 wt.% industrial sintered magnesia, and 7.6 wt.% silica powder are the optimal ratios. The volume density and compressive strength of this formula are relatively high, reaching 1.84 g/cm^3 and 72.8 MPa, respectively. In addition, samples synthesized at a sintering temperature of 1280°C exhibit significant content of α-cordierite and β-cordierite, according to XRD results. Cheng et al. (2023) used CFA, clay, bauxite, and talc as main raw materials to prepare cordierite-based low thermal expansion ceramics by sintering reaction. The addition of talc presents a significant effect on the microstructure and properties of cordierite ceramics. The excessive talc content promoted the formation of forsterite, resulting in the increase of coefficient of thermal expansion of the sample, which reached $2.82 \times 10^{-6}/°C$ with 33.14 wt.% talc sintered at 1300°C. The sintering time also shows a certain influence on the microstructure and properties of cordierite ceramics. Proper prolongation of sintering time is conducive to promoting the densification of cordierite ceramics, reducing the thermal expansion coefficient of materials, and improving the mechanical properties of ceramics.

In the process of synthesizing cordierite, different additives will result in different products. Senthil Kumar et al. (2019) studied the sintering synthesis of cordierite with CFA and MgO as the main raw materials, supplemented by dopants such as ZrO_2, CeO_2, and TiO_2 with different compositions (5–20 wt.%). It is found that the addition of dopant in cordierite matrix can improve the hardness, fracture toughness and flexural strength. The Vickers hardness of cordierite-ZrO_2 ceramics is 7.04 GPa, and the fracture toughness is 3.47 MPa m½. Therefore, using CFA as the raw material, supplemented by dopants, is suitable for use as a catalytic substrate material with improved mechanical properties.

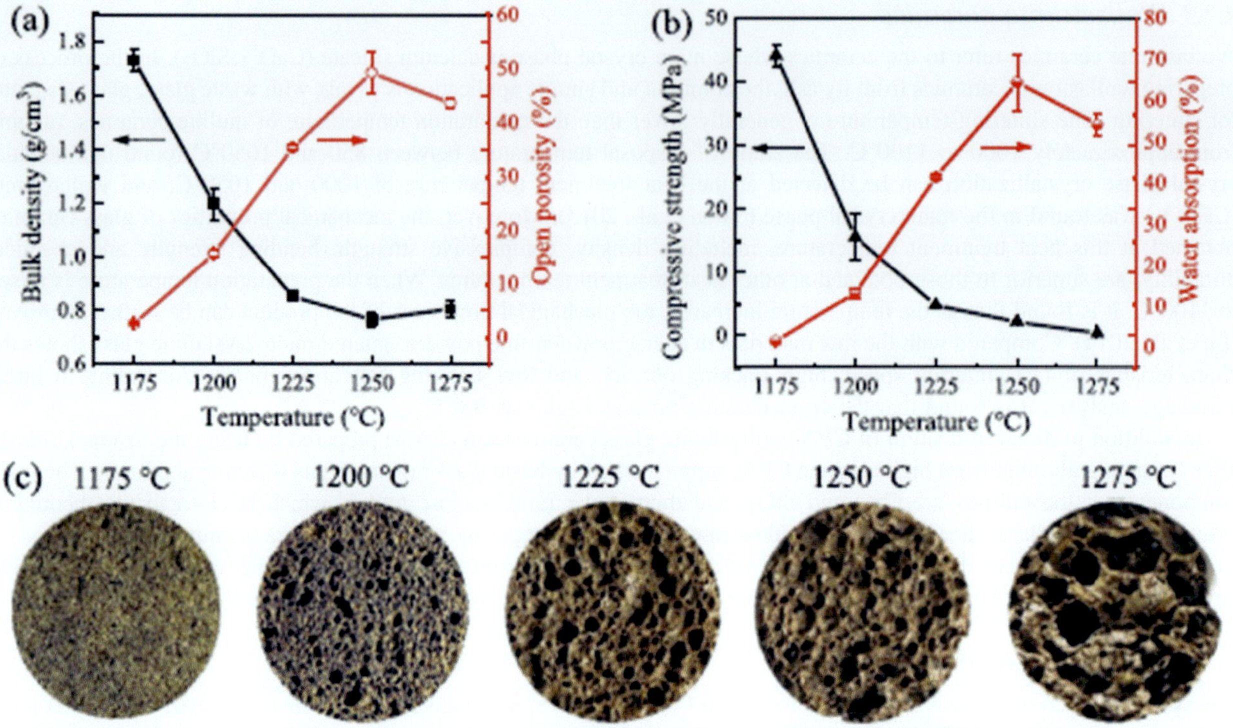

FIGURE 21.4 Effects of sintering temperature on the ceramic properties (Hui et al., 2021).

4.2 MSWI fly ash

MSWI fly ash can also be used to produce building ceramics (Deng et al., 2018). In addition, there is also a considerable amount of research on microcrystalline glass. Microcrystalline glass is a type of polycrystalline solid material containing a large amount of microcrystalline and glass phases, obtained by controlling the crystallization of a specific composition of basic glass during the heating process. Microcrystalline glass combines the characteristics of glass and ceramics in terms of structure and performance, with excellent mechanical properties, chemical corrosion resistance, thermal stability, etc. (Deng et al., 2021). The chemical components of fly ash are basically consistent with the raw materials required for microcrystalline production. Thus, fly ash can be used as a raw material, and as long as a small amount of auxiliary raw materials are introduced and adjusted, high value-added microcrystalline glass can be prepared. In the process of preparing microcrystalline glass from fly ash, while the dioxins in fly ash are completely decomposed, the vast majority of heavy metals are also solidified in the glass slag. Then the fused glass slag can be heat-treated to obtain microcrystalline glass, which meets the requirements of zero emission for waste treatment and can improve additional economic benefits and reduce the treatment cost of fly ash. The preparation processes of microcrystalline glass mainly include sintering method and melting method.

Fan et al. (2019) synthesized microcrystalline glass for heavy metal solidification and waste recovery using MSWI fly ash as raw material. The produced microcrystalline glass exhibits good performance (density 3.42 g/cm^3, Vickers hardness 6.91 GPa). In addition, the leaching concentration of heavy metal elements meets the allowable value of TCLP. The group also prepared microcrystalline glass using stainless steel slag, MSWI fly ash, and CFA as raw materials from industrial solid waste (Fan et al., 2018). The crystallization behavior and mechanical properties of microcrystalline glass with different $(CaO + MgO)/(SiO_2 + Al_2O_3)$ content ratios (alkalinity) were studied through adjusting the dosage of MSWI fly ash. It is found that the activation energy increases about 25 kJ/mol with the decrease of alkalinity. This formula is cost-effective and has advantages over other production methods for using industrial steel slag, MSWI fly ash, and CFA as raw materials. MSWI fly ash–based microcrystalline glass systems can also be prepared by a one-step method (Zhao et al., 2020b), By controlling the amount of sludge added, the basic glass undergoes nucleation and crystallization at the same temperature. Therefore, the preparation process of microcrystalline glass is simplified and energy consumption is reduced. When the content of MSWI fly ash is 50 wt.%, the crystallization activation energy is relatively low (Zhao et al., 2021). In addition, the microcrystalline glass composed of spherical diopside, cuspidine, and glass phase is obtained, whose hardness and flexural strength are 7.97 GPa and 114.86 MPa, respectively. Furthermore, the migration track and occurrence state of chlorine were revealed. It is believed that part of chlorine is volatilized during the heating process of fly ash preparation glass, and residual chlorine is dispersed in the glass ceramics, mainly in the form of solid solution in the amorphous phase and calcium chlorosilicate phase. Due to the excellent acid and alkali resistance of microcrystalline glass, chlorine exhibits good stability in microcrystalline glass. Therefore, not only does it increase the added value of MSWI fly ash, but it also solves the problems caused by high chlorine and heavy metals in MSWI fly ash. In addition, this study on the effect of alkalinity on the synthesis of microcrystalline glass also yields similar conclusions. It is believed that the increase in alkalinity promotes the transformation of bridging oxygen bonds into nonbridging oxygen bonds. Microcrystalline glass can also be produced by the mixture of MSWI fly ash, Al-rich CFA, and Si-rich waste glass, and the scheme is given in Fig. 21.5.

In addition to conventional melting techniques, there are also thermal plasma melting methods. The temperature of thermal plasma is very high, about 10^3-10^5 k, so it can effectively melt, decompose, and phase-change substances. The main methods of generating thermal plasma include direct current discharge ionization, three-phase AC discharge ionization, and high-frequency discharge ionization. Thermal plasma treatment is regard as an optional and high effective technology to dispose MSWI fly ash, and the produced slag is stable and safe enough to the surroundings (Wang et al., 2010). After treatment, the slag melted by porous fly ash is amorphous, glassy, completely dense, and uniform. In slag, the decomposition rate in PCDD/Fs is 99.32% (Wang et al., 2009). After further preparation of microcrystalline glass, the main crystal phases obtained are anorthite and diopside. The mechanical properties and acid–base corrosion resistance are superior to the basic glass. The mechanical properties are comparable with the advanced natural stones such as granite and marble. Therefore, the products can be used as advanced building materials to realize the resource reuse of incineration fly ash.

4.3 Applications of ceramics synthesized

Fly ash can be used as a raw material for synthesizing ceramics through processes such as sintering, which involves heating the ash to high temperatures to achieve densification and bonding. The properties of ceramics, such as strength, porosity, and thermal and chemical resistance, can be tailored by controlling the sintering conditions and composition of the fly ash. Ceramics, such as building ceramics, mullite ceramics, wollastonite ceramics, cordierite ceramics, and microcrystalline glass, have been extensively studied and utilized in various applications.

(1) Building ceramics can be used in the manufacturing of bricks, blocks, tiles, and decorative elements for construction purposes (Luo et al., 2017). They offer advantages such as improved strength, durability, fire resistance, and thermal insulation properties.

(2) Mullite ceramics are widely used in the production of refractory materials, particularly in high-temperature applications. Their high melting point, low thermal expansion, and excellent thermal shock resistance make them suitable for electroceramics, lining furnaces, kilns, and other equipment used in industries such as steel, glass, and cement manufacturing.

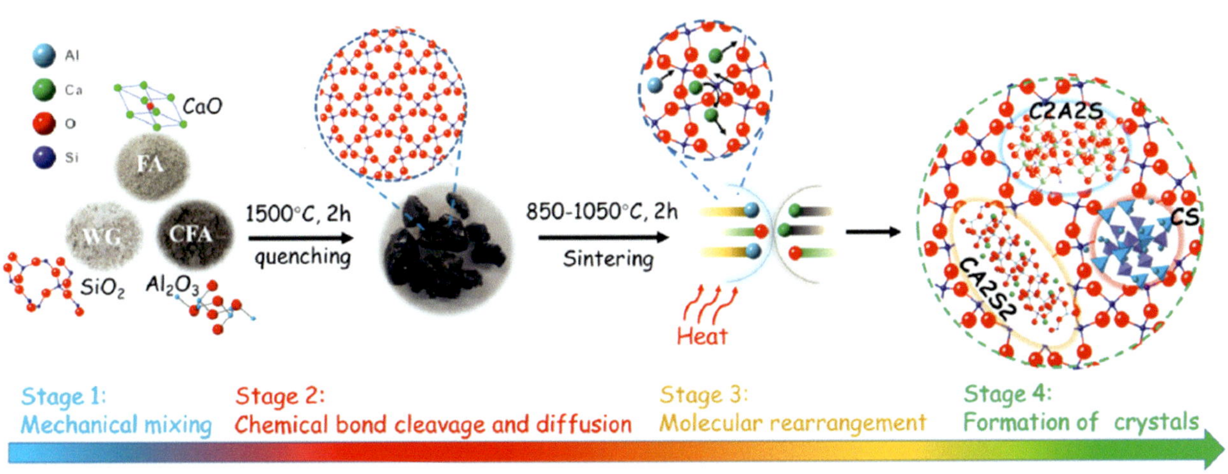

FIGURE 21.5 Scheme of the synthesis of fly ash–based glass-ceramics (Zhang et al., 2022).

(3) Wollastonite ceramics is a calcium silicate mineral that can be processed into a ceramic material with unique properties. Wollastonite can be used in building materials, automobile industry, and adsorption materials. In addition to being used as building materials, wollastonite ceramics can also use its porous properties as a filter and adsorbent to remove impurities and pollutants in liquids and gases (Obradovic et al., 2017).

(4) Cordierite ceramics are commonly used in the manufacturing of catalytic converters due to their excellent thermal shock resistance, low thermal expansion coefficient, and high mechanical strength (Senthil Kumar et al., 2019). Besides, cordierite ceramics are also used in kiln furniture, cookware and bakeware, insulation, and foundry applications.

(5) Microcrystalline glass is a material that combines the properties of both glass and ceramics. It possesses fine-grained microstructures with crystalline phases embedded in a glassy matrix. Microcrystalline glass is an ideal high-end environmentally friendly decorative material in the modern construction industry (Luan et al., 2017), which can be widely used in buildings such as squares, hotels, financial buildings, shopping malls, entertainment facilities, and family residences. It is a good substitute product for glass, ceramics, and high-end natural stone materials.

5. Limitations and future trends

Recycling of fly ash into zeolites and ceramics is a kind of effective technical means for reducing capacity and volume, and for increasing the value of products. The major control factors, such as solid—liquid ratio, silicon aluminum ratio, alkali concentration, reaction time, and temperature in the synthesis process of fly ash—based zeolites, have important effects on the purity and type of synthesized zeolite. Therefore, how to regulate the reaction conditions in actual production will directly affect the quality of synthetic zeolite, and the universality and complexity of the source of fly ash bring great challenges to this, especially MSWI fly ash. At the same time, it is also necessary to focus on the long-term leaching toxicity of zeolite products, including heavy metals and dioxins, as well as whether the adsorption process of fly ash—based zeolites will generate secondary pollution. In addition, plenty of studies currently focus mainly on the synthesis and adsorption effects of various zeolites, but how to remove pollutants from the adsorbed zeolites should also be a key research direction. The oriented synthesis of high-purity zeolites with application orientation is the main development direction of artificial zeolite synthesis in the future.

Compared with zeolites, the preparation conditions of ceramic products are more demanding, and the temperature is generally higher than 1000°C, or even as high as 1600°C. Therefore, the product is less polluting and has more reliable stability. For CFA, especially the high alumina CFA, it exhibits superior characteristics in the preparation of ceramics, such as mullite ceramics, wollastonite ceramics, etc. However, with the addition of fly ash, the temperature required for ceramic synthesis also increases, resulting in greater consumption of cosolvent, and also affecting some properties of ceramics, such as flexural strength, product purity, etc. Moreover, in the process of ceramic production, the additive amount of fly ash is relatively small. Accordingly, further studies should address the possible alternative approaches to increase the proportion of fly ash in ceramic preparation, further simplifying the preparation process.

6. Summary

This chapter summarized the technology progress of applying fly ash with large output to the production of zeolites and ceramics, which are currently regarded as the two major paths of obtaining high-value products. The technologies of zeolite synthesis mainly include traditional hydrothermal method, alkaline fusion-hydrothermal method, microwave-assisted hydrothermal method, and ultrasonic-assisted method. Compared with traditional hydrothermal method, microwave-assisted method and ultrasonic-assisted method are more economical, but it is difficult to achieve large-scale disposal. Many kinds of zeolites have been synthesized, such as cristobalite, analcime, zeolite P, scolecite, chabazite, etc. (Feng et al., 2018; Liu et al., 2023). The hydrothermal zeolitization is mainly divided into four steps (Bukhari et al., 2015): the dissolution of Al and Si in the fly ash; the aggregation of polymers (dimers, trimers, etc.) formed by Al and Si on the surface of fly ash; zeolite nucleation; and the growth of zeolite crystals. The commonly used measures to enhance zeolitization reaction include adding silicon aluminum additives, adjusting reaction time and temperature, and selecting efficient alkaline activator and suitable concentration. However, the silicon and aluminum content of fluidized bed waste incineration fly ash is often less than 20 wt.% (Long et al., 2020), while CFA has a high silicon and aluminum content of up to 40 wt.% (Wang et al., 2009; Song et al., 2014). Therefore, during the preparation of zeolites from MSWI fly ash, it is often necessary to add additional silicon and aluminum sources.

In the aspect of ceramic preparation, fly ash is mainly applied to the synthesis of building ceramics, mullite ceramics, wollastonite ceramics cordierite ceramics, microcrystalline glass, etc. In ceramic preparation, it is necessary to strictly control the content of fly ash. The increase in the replacement amount of fly ash leads to not only an increase in the linear shrinkage rate and bulk density of ceramics but also a significant decrease in mechanical properties. For some special CFA with high aluminum content, high-strength ceramic bricks can be synthesized (Wang et al., 2017). MSWI fly ash is mainly used in the synthesis of microcrystalline glass. The crystallization behavior and mechanical properties of microcrystalline glass with different $(CaO + MgO)/(SiO_2 + Al_2O_3)$ content ratios (alkalinity) can be adjusted through the dosage of MSWI fly ash. Within a certain range, it is found that the activation energy increases with the decrease of alkalinity.

Acknowledgments

This research was funded by Jiangsu Provincial Natural Science Foundation of China, grant number BK20201032, and Jiangsu Province Industry University Research Cooperation Project, grant number BY20221211.

References

Abdullahi, T., Harun, Z., Othman, M.H.D., 2017. A review on sustainable synthesis of zeolite from kaolinite resources via hydrothermal process. Adv. Powder Technol. 28, 1827–1840.

Alterary, S.S., Marei, N.H., 2021. Fly ash properties, characterization, and applications: a review. J. King Saud Univ. Sci. 33, 101536.

Ameh, A.E., Fatoba, O.O., Musyoka, N.M., Petrik, L.F., 2017. Influence of aluminium source on the crystal structure and framework coordination of Al and Si in fly ash-based zeolite NaA. Powder Technol. 306, 17–25.

Ansari, M., Aroujalian, A., Raisi, A., Dabir, B., Fathizadeh, M., 2014. Preparation and characterization of nano-NaX zeolite by microwave assisted hydrothermal method. Adv. Powder Technol. 25, 722–727.

Bayuseno, A.P., Schmahl, W.W., Müllejans, T., 2009. Hydrothermal processing of MSWI fly ash-towards new stable minerals and fixation of heavy metals. J. Hazard Mater. 167, 250–259.

Belviso, C., Cavalcante, F., Fiore, S., 2013. Ultrasonic waves induce rapid zeolite synthesis in a seawater solution. Ultrason. Sonochem. 20, 32–36.

Boycheva, S., Zgureva, D., Lazarova, H., Popova, M., 2021. Comparative studies of carbon capture onto coal fly ash zeolites Na-X and Na–Ca-X. Chemosphere 271, 129505.

Bukhari, S.S., Behin, J., Kazemian, H., Rohani, S., 2015. Conversion of coal fly ash to zeolite utilizing microwave and ultrasound energies: a review. Fuel 140, 250–266.

Bukhari, S.S., Rohani, S., Kazemian, H., 2016. Effect of ultrasound energy on the zeolitization of chemical extracts from fused coal fly ash. Ultrason. Sonochem. 28, 47–53.

Cardoso, A.M., Paprocki, A., Ferret, L.S., Azevedo, C.M.N., Pires, M., 2015. Synthesis of zeolite Na-P1 under mild conditions using Brazilian coal fly ash and its application in wastewater treatment. Fuel 139, 59–67.

Carty, W.M., Senapati, U., 1998. Porcelain—raw materials, processing, phase evolution, and mechanical behavior. J. Am. Ceram. Soc. 81, 3–20.

Chandrasekaran, S., Ramanathan, S., Basak, T., 2013. Microwave food processing—a review. Food Res. Int. 52, 243–261.

Chen, J., Ma, Z., Jiang, H., Zhang, P., 2016. Preparation of wollastonite glass-ceramics by using silicate-calciumslag generated in process of extracting alumina from high-alumina fly ash. Bull. Chin. Ceram. Soc. 35, 2898–2903.

Cheng, L., Xiao, Z., Xiao, X., Li, X., Dong, H., Kong, L., 2023. Cordierite-based ceramics with low thermal expansion coefficient from clay and coal fly ash. J.Ceram. 44, 272–278.

Choi, C.L., Park, M., Lee, D.H., Kim, J., Park, B., Choi, J., 2001. Salt-Thermal zeolitization of fly ash. Environ. Sci. Technol. 35, 2812–2816.

Dana, K., Das, S., Das, S.K., 2004. Effect of substitution of fly ash for quartz in triaxial kaolin–quartz–feldspar system. J. Eur. Ceram. Soc. 24, 3169–3175.

Deng, L., Lu, W., Zhang, Z., Fu, Z., Li, H., Chen, H., Du, Y., Ma, Y., Wang, W., 2021. Crystallization behavior and structure of CaO–MgO–Al2O3–SiO2 glass ceramics prepared from Cr-bearing slag. Mater. Chem. Phys. 261, 124249.

Deng, Y., Gong, B., Chao, Y., Dong, T., Yang, W., Hong, M., Shi, X., Wang, G., Jin, Y., Chen, Z.G., 2018. Sustainable utilization of municipal solid waste incineration fly ash for ceramic bricks with eco-friendly biosafety. Mater. Today Sustain. 1–2, 32–38.

Dong, Y., Feng, X., Feng, X., Ding, Y., Liu, X., Meng, G., 2008. Preparation of low-cost mullite ceramics from natural bauxite and industrial waste fly ash. J. Alloys Compd. 460, 599–606.

Dong, Y., Hampshire, S., Zhou, J., Ji, Z., Wang, J., Meng, G., 2011. Sintering and characterization of flyash-based mullite with MgO addition. J. Eur. Ceram. Soc. 31, 687–695.

Fan, W., Liu, B., Luo, X., Yang, J., Guo, B., Zhang, S., 2019. Production of glass–ceramics using municipal solid waste incineration fly ash. Rare Met. 38, 245–251.

Fan, W., Yang, Q., Guo, B., Liu, B., Zhang, S., 2018. Crystallization mechanism of glass-ceramics prepared from stainless steel slag. Rare Met. 37, 413–420.

Fan, X., Yuan, R., Gan, M., Ji, Z., Sun, Z., 2023. Subcritical hydrothermal treatment of municipal solid waste incineration fly ash: a review. Sci. Total Environ. 865, 160745.

Feng, W., Wan, Z., Daniels, J., Li, Z., Xiao, G., Yu, J., Xu, D., Guo, H., Zhang, D., May, E.F., Li, G.K., 2018. Synthesis of high quality zeolites from coal fly ash: mobility of hazardous elements and environmental applications. J. Clean. Prod. 202, 390–400.

Fukui, K., Arai, K., Kanayama, K., Yoshida, H., 2006. Phillipsite synthesis from fly ash prepared by hydrothermal treatment with microwave heating. Adv. Powder Technol. 17, 369–382.

Gross, M., Soulard, M., Caullet, P., Patarin, J., Saude, I., 2007. Synthesis of faujasite from coal fly ashes under smooth temperature and pressure conditions: a cost saving process. Micropor. Mesopor. Mat. 104, 67–76.

Hamernik, J.D., Frantz, G.C., 1991. Physical and chemical properties of municipal solid waste fly ash. ACI Meter. J. 88, 294–301.

Harrati, A., Arkame, Y., Manni, A., El Haddar, A., Achiou, B., El Bouari, A., El Amrani El Hassani, I., Sdiri, A., Sadik, C., 2022. Cordierite-based refractory ceramics from natural halloysite and peridotite: insights on technological properties. J. Indian Chem. Soc. 99, 100496.

He, S., Li, H., Li, S., Li, Y., Xie, Q., 2014. Kinetics of desilication process of fly ash with high aluminum from pulverized coal fired boiler in alkali solution. Chin. J. Nonferrous Metals 24, 1888–1894.

Hermassi, M., Valderrama, C., Font, O., Moreno, N., Querol, X., Batis, N.H., Cortina, J.L., 2020. Phosphate recovery from aqueous solution by K-zeolite synthesized from fly ash for subsequent valorisation as slow release fertilizer. Sci. Total Environ. 731, 139002.

Hettenkofer, C., Fromm, S., Schuster, M., 2020. Municipal solid waste as secondary resource: selectively separating Cu(II) from highly saline fly ash extracts by polymer-assisted ultrafiltration. Processes 8, 1662–1680.

Hollman, G.G., Steenbruggen, G., Janssen-Jurkovičová, M., 1999. A two-step process for the synthesis of zeolites from coal fly ash. Fuel 78, 1225–1230.

Huber, F., Herzel, H., Adam, C., Mallow, O., Blasenbauer, D., Fellner, J., 2018. Combined disc pelletisation and thermal treatment of MSWI fly ash. Waste Manage. (Tucson, Ariz.) 73, 381–391.

Hui, T., Sun, H.J., Peng, T.J., 2021. Preparation and characterization of cordierite-based ceramic foams with permeable property from asbestos tailings and coal fly ash. J. Alloys Compd. 885, 160967.

Huo, X., Xia, B., Hu, T., Zhang, M., Guo, M., 2023. Effect of MoO3 addition on fly ash based porous and high-strength mullite ceramics: in situ whisker growth and self-enhancement mechanism. Ceram. Int. 49, 21069–21077.

Inada, M., Tsujimoto, H., Eguchi, Y., Enomoto, N., Hojo, J., 2005. Microwave-assisted zeolite synthesis from coal fly ash in hydrothermal process. Fuel 84, 1482–1486.

Iqbal, A., Sattar, H., Haider, R., Munir, S., 2019. Synthesis and characterization of pure phase zeolite 4A from coal fly ash. J. Clean. Prod. 219, 258–267.

Jiang, N., Yang, G., Zhang, X., Wang, L., Shi, C., Tsubaki, N., 2011. A novel silicalite-1 zeolite shell encapsulated iron-based catalyst for controlling synthesis of light alkenes from syngas. Catal. Commun. 12, 951–954.

Joseph, I.V., Tosheva, L., Doyle, A.M., 2020. Simultaneous removal of Cd(II), Co(II), Cu(II), Pb(II), and Zn(II) ions from aqueous solutions via adsorption on FAU-type zeolites prepared from coal fly ash. J. Environ. Chem. Eng. 8, 103895.

Kacirek, H., Lechert, H., 1976. Rates of crystallization and a model for the growth of sodium-Y zeolites. J. Phys. Chem. 80, 1291–1296.

Kobayashi, Y., Ogata, F., Takehiro, N., Kawasaki, N., 2020. Synthesis of novel zeolites produced from fly ash by hydrothermal treatment in alkaline solution and its evaluation as an adsorbent for heavy metal removal. J. Environ. Chem. Eng. 8, 103687.

Kockal, N., 2012. Utilisation of different types of coal fly ash in the production of ceramic tiles. Mater. Des. 51.

Lee, Y.R., Soe, J.T., Zhang, S., Ahn, J.W., Park, M.B., Ahn, W.S., 2017. Synthesis of nanoporous materials via recycling coal fly ash and other solid wastes: a mini review. Chem. Eng. J. 317, 821–843.

Li, J., Ma, H., Huang, W., 2009. Effect of V2O5 on the properties of mullite ceramics synthesized from high-aluminum fly ash and bauxite. J. Hazard Mater. 166, 1535–1539.

Li, Y., Yang, L., Li, X., Miki, T., Nagasaka, T., 2021. A composite adsorbent of ZnS nanoclusters grown in zeolite NaA synthesized from fly ash with a high mercury ion removal efficiency in solution. J. Hazard Mater. 411, 125044.

Liu, Y., Lu, H., 2020. Synthesis of ZSM-5 zeolite from fly ash and its adsorption of phenol, quinoline and indole in aqueous solution. Mater. Res. Express 7.

Liu, Y., Zhou, T., Chen, X., Li, H., Xu, X., Dou, J., Yu, J., 2023. Synthesis of a coal fly Ash-Based NaP zeolite using the microwave-ultrasonic assisted method: preparation, growth mechanism, and kinetics. ChemistrySelect 8, e202204353.

Liu, Z., Zhang, Z., Zhai, J., Liu, P., Yang, C., 2015. Preparation and catalytic performance of high dispersion of Y zeolite treated with alkali solution. Appl. Petrochem. Res. 5, 263–267.

Long, L., Jiang, X., Lv, G., Chen, Q., Liu, X., Chi, Y., Yan, J., Zhao, X., Kong, L., 2020. Characteristics of fly ash from waste-to-energy plants adopting grate-type or circulating fluidized bed incinerators: a comparative study. Energy Sources, Part A Recovery, Util. Environ. Eff. 1–17.

Lu, J., Lu, Z., Peng, C., Li, X., Jiang, H., 2014. Influence of particle size on sinterability, crystallisation kinetics and flexural strength of wollastonite glass-ceramics from waste glass and fly ash. Mater. Chem. Phys. 148, 449–456.

Luan, J.D., Chai, M.Y., Li, R.D., Yao, P.F., Wang, L., Li, S.B., 2017. Crystalline phase evolution behavior and physicochemical properties of glass-ceramics from municipal solid waste incineration fly ash. J. Mater. Cycles Waste 19, 1204–1210.

Luo, L., Li, K., Fu, W., Liu, C., Yang, S., 2020. Preparation, characteristics and mechanisms of the composite sintered bricks produced from shale, sewage sludge, coal gangue powder and iron ore tailings. Constr. Build. Mater. 232, 117250.

Luo, Y., Ma, S.H., Liu, C.L., Zhao, Z.Q., Zheng, S.L., Wang, X.H., 2017. Effect of particle size and alkali activation on coal fly ash and their role in sintered ceramic tiles. J. Eur. Ceram. Soc. 37, 1847–1856.

Luo, Y., Zheng, S., Ma, S., Liu, C., Wang, X., 2018. Preparation of sintered foamed ceramics derived entirely from coal fly ash. Constr. Build. Mater. 163, 529–538.

Ma, L., Han, L., Chen, S., Hu, J., Chang, L., Bao, W., Wang, J., 2019. Rapid synthesis of magnetic zeolite materials from fly ash and iron-containing wastes using supercritical water for elemental mercury removal from flue gas. Fuel Process. Technol. 189, 39–48.

Ma, L., Xu, D., Yang, P., 2014. Effects of additives on structure and properties of synthesized cordierite. China's Refract. 23, 11–14.

Makul, N., Rattanadecho, P., Agrawal, D.K., 2014. Applications of microwave energy in cement and concrete – a review. Renew. Sust. Energ. Rev. 715–733.

Molina, A., Poole, C., 2004. A comparative study using two methods to produce zeolites from fly ash. Miner. Eng. 17, 167–173.

Mukhopadhyay, T.K., Ghosh, S., Ghosh, J., Ghatak, S., Maiti, H.S., 2010. Effect of fly ash on the physico-chemical and mechanical properties of a porcelain composition. Ceram. Int. 36, 1055–1062.

Murayama, N., Yamamoto, H., Shibata, J., 2002. Mechanism of zeolite synthesis from coal fly ash by alkali hydrothermal reaction. Int. J. Miner. Process. 64, 1–17.

Obradovic, N., Filipovic, S., Markovic, S., Mitric, M., Rusmirovic, J., Marinkovic, A., Antic, V., Pavlovic, V., 2017. Influence of different pore-forming agents on wollastonite microstructures and adsorption capacities. Ceram. Int. 43, 7461–7468.

Olgun, A., Erdogan, Y., Ayhan, Y., Zeybek, B., 2005. Development of ceramic tiles from coal fly ash and tincal ore waste. Ceram. Int. 31, 153–158.

Pal, P., Das, J.K., Das, N., Bandyopadhyay, S., 2013. Synthesis of NaP zeolite at room temperature and short crystallization time by sonochemical method. Ultrason. Sonochem. 20, 314–321.

Panitchakarn, P., Laosiripojana, N., Viriya-Umpikul, N., Pavasant, P., 2014. Synthesis of high-purity Na-A and Na-X zeolite from coal fly ash. J. Air Waste Manage. 64, 586–596.

Park, M., Choi, C.L., Lim, W.T., Kim, M.C., Choi, J., Heo, N.H., 2000. Molten-salt method for the synthesis of zeolitic materials: II. Characterization of zeolitic materials. Micropor. Mesopor. Mat. 37, 91–98.

Qiu, Q., Jiang, X., Lu, S., Ni, M., 2016. Effects of microwave-assisted hydrothermal treatment on the major heavy metals of municipal solid waste incineration fly ash in a circulating fluidized bed. Energ. Fuel. 30, 5945–5952.

Qiu, Q., Jiang, X., Lv, G., Chen, Z., Lu, S., Ni, M., Yan, J., Deng, X., 2018. Adsorption of heavy metal ions using zeolite materials of municipal solid waste incineration fly ash modified by microwave-assisted hydrothermal treatment. Powder Technol. 335, 156–163.

Quan, L., Zhi-Min, B., Dong, W., Wang, Y., 2021. Chemical composition and physicochemical properties of fly ash and its application. China Non-Metallic Mater. Ind. 1–9.

Querol, X., Alastuey, A., Soler, A.L., Plana, F., 1997. A fast method for recycling fly ash: microwave-assisted zeolite synthesis. Environ. Sci. Technol. 31, 2527–2533.

Reinosa, J.J., Del Campo, A., Fernández, J.F., 2015. Indirect measurement of stress distribution in quartz particles embedded in a glass matrix by using confocal Raman microscopy. Ceram. Int. 41, 13598–13606.

Ren, X., Qu, R., Liu, S., Zhao, H., Wu, W., Song, H., Zheng, C., Wu, X., Gao, X., 2020. Synthesis of zeolites from coal fly ash for removal of harmful gaseous pollutants: a review. Aerosol Air Qual. Res. 20, 1127–1144.

Sakthivel, T., Reid, D.L., Goldstein, I., Hench, L., Seal, S., 2013. Hydrophobic high surface area zeolites derived from fly ash for oil spill remediation. Environ. Sci. Technol. 47, 5843–5850.

Sampathkumar, N.N., Umarji, A.M., Chandrasekhar, B.K., 1995. Synthesis of α-cordierite (indialite) from flyash. Mater. Res. Bull. 30, 1107–1114.

Schumann, K., Unger, B., Brandt, A., Scheffler, F., 2012. Investigation on the pore structure of binderless zeolite 13× shapes. Micropor. Mesopor. Mat. 154, 119–123.

Senthil Kumar, M., Vanmathi, M., Senguttuvan, G., Mangalaraja, R.V., Sakthivel, G., 2019. Fly ash constituent-silica and alumina role in the synthesis and characterization of cordierite based ceramics. Silicon 11, 2599–2611.

Shi, D., Hu, C., Zhang, J., Li, P., Zhang, C., Wang, X., Ma, H., 2017. Silicon-aluminum additives assisted hydrothermal process for stabilization of heavy metals in fly ash from MSW incineration. Fuel Process. Technol. 165, 44–53.

Shigemoto, N., Hayashi, H., Miyaura, K., 1993. Selective formation of Na-X zeolite from coal fly ash by fusion with sodium hydroxide prior to hydrothermal reaction. J. Mater. Sci. 28, 4781–4786.

Sirikunpitak, S., Choomkong, A., Chowdhury, S., Hua, C.C., Jamal, M.S., Kaewthongrach, R., Techato, K., Sreesawet, S., 2022. Biomass fly ash from multiple sources and the composition of intertidal mangrove soil: a systematic review. Fresen. Environ. Bull. 31, 10550–10555.

Song, H., Cheng, H., Zhang, Z., Cheng, F., 2014. Adsorption properties of zeolites synthesized from coal fly ash for Cu (II). J. Environ. Biol. 35, 983–988.

Sun, J., Wang, L., Yu, J., Guo, B., Chen, L., Zhang, Y., Wang, D., Shen, Z., Tsang, D.C.W., 2022. Cytotoxicity of stabilized/solidified municipal solid waste incineration fly ash. J. Hazard Mater. 424, 127369.

Sun, X.L., Guo, Y., Yan, Y.B., Li, J.S., Shen, J.Y., Han, W.Q., Sun, X.Y., Wang, L.J., 2019. Co-processing of MSWI fly ash and copper smelting wastewater and the leaching behavior of the co-processing products in landfill leachate. Waste Manage. (Tucson, Ariz.) 95, 628–635.

Tang, Q., Gu, F., Chen, H., Lu, C., Zhang, Y., 2018. Mechanical evaluation of bottom ash from municipal solid waste incineration used in roadbase. Adv. Civ. Eng. 2018.

Tian, X., Rao, F., Morales-Estrella, R., Song, S., 2020. Effects of aluminum dosage on gel formation and heavy metal immobilization in Alkali-Activated municipal solid waste incineration fly ash. Energ. Fuel. 34, 4727–4733.

Turkmen, O., Kucuk, A., Akpinar, S., 2015. Effect of wollastonite addition on sintering of hard porcelain. Ceram. Int. 41, 5505–5512.

Vizcarra, G.O.C.O., Dal Toé Casagrande, M., Motta, A.L.M.G., 2014. Applicability of municipal solid waste incineration ash on base layers of pavements. J. Mater. Civ. Eng. 26, 1–7.

Wang, C., Li, J., Sun, X., Wang, L., Sun, X., 2009. Evaluation of zeolites synthesized from fly ash as potential adsorbents for wastewater containing heavy metals. J. Environ. Sci. China 21, 127–136.

Wang, H., Zhu, M., Sun, Y., Ji, R., Liu, L., Wang, X., 2017. Synthesis of a ceramic tile base based on high-alumina fly ash. Constr. Build. Mater. 155, 930–938.

Wang, Q., Yan, J., Chi, Y., Li, X., Lu, S., 2010. Application of thermal plasma to vitrify fly ash from municipal solid waste incinerators. Chemosphere 78, 626–630.

Wang, Q., Yan, J., Tu, X., Chi, Y., Li, X., Lu, S., Cen, K., 2009. Thermal treatment of municipal solid waste incinerator fly ash using DC double arc argon plasma. Fuel 88, 955–958.

Wang, S., Zhu, Z.H., 2005. Sonochemical treatment of fly ash for dye removal from wastewater. J. Hazard Mater. 126, 91–95.

Yan, P., Xiao, Z., Xiao, G., Pan, Q., Hui, H., Wu, Y., Ma, Y., Xu, Y., 2022. Undetection of australasian microtektites in the Chinese loess plateau. Palaeogeogr. Palaeoclimatol. Palaeoecol. 585, 110721.

Yao, Z.T., Ji, X.S., Sarker, P.K., Tang, J.H., Ge, L.Q., Xia, M.S., Xi, Y.Q., 2015. A comprehensive review on the applications of coal fly ash. Earth Sci. Rev. 141, 105–121.

Yoon, S., Lee, J., Lee, J., Yun, Y., Yoon, W., 2013. Characterization of wollastonite glass-ceramics made from waste glass and coal fly ash. J. Mater. Sci. Technol. 29, 149–153.

Zhang, Y.J., He, P.Y., Zhang, Y.X., Chen, H., 2017. A novel electroconductive graphene/fly ash-based geopolymer composite and its photocatalytic performance. Chem. Eng. J. 334, S1034785050.

Zhang, Z., Li, Z., Yang, Y., Shen, B., Ma, J., Liu, L., 2022. Preparation and characterization of fully waste-based glass-ceramics from incineration fly ash, waste glass and coal fly ash. Ceram. Int. 48, 21638–21647.

Zhao, S., Liu, B., Ding, Y., Zhang, J., Wen, Q., Ekberg, C., Zhang, S., 2020a. Study on glass-ceramics made from MSWI fly ash, pickling sludge and waste glass by one-step process. J. Clean. Prod. 271, 122674.

Zhao, S., Zhang, X., Liu, B., Zhang, J., Shen, H., Zhang, S., 2021. Preparation of glass–ceramics from high-chlorine MSWI fly ash by one-step process. Rare Met. 40, 3316–3328.

Zhao, Y.P., Guo, D.X., Li, S.F., Cao, J.P., Wei, X.Y., 2020b. Removal of methylene blue by NaX zeolites synthesized from coal gasification fly ash using an alkali fusion-hydrothermal method. Desalin. Water Treat. 355–363.

Zhuang, X.Y., Chen, L., Komarneni, S., Zhou, C.H., Tong, D.S., Yang, H.M., Yu, W.H., Wang, H., 2016. Fly ash-based geopolymer: clean production, properties and applications. J. Clean. Prod. 125, 253–267.

Chapter 22

Recycling incinerated water treatment residue into alkaline-activated materials

Weiwei Duan, Yan Zhuge and Yue Liu

Sustainable Infrastructure and Resource Management (SIRM), UniSA STEM, University of South Australia, Adelaide, SA, Australia

1. Introduction

The water industry plays a vital role in modern society and the overall ecosystem by ensuring the supply of safe and clean water to homes, businesses, and public establishments. Simultaneously, it is responsible for wastewater treatment, which involves treating sewage to remove harmful contaminants before returning it to the environment. Two types of water treatment residue (WTR) are generated in these operations: drinking water treatment sludge and sewage sludge (Lynn et al., 2015; Liu et al., 2022b).

Drinking water treatment sludge is produced when coagulants interact with suspended solids, dissolved colloids, organic matter, and microorganisms present in raw water. In mainland Australia alone, there are approximately 400 drinking water treatment plants, with each site generating up to 2000 ton (dry weight) of sludge annually (Duan et al., 2022b). Exact figures for global sludge production are challenging to determine due to a lack of comprehensive documentation. However, sludge produced from the drinking water treatment process is typically estimated to constitute 1% −3% by volume of the raw water used (Dassanayake et al., 2015). Sewage sludge, on the other hand, is the end product of sewage treatment processes. The annual production of sewage sludge is substantial, estimated at 3 million wet tons in Australia and a combined 240 million wet tons in Europe, the United States, and China (Pritchard et al., 2010). Projections indicate wastewater volume will increase by 24% by 2030 and 51% by 2050, leading to continued rapid growth in sewage sludge production (Di Giacomo and Romano, 2022).

Most drinking water treatment sludge is currently disposed of in landfills, and some are disposed of together with sewage sludge (Maiden et al., 2015). Disposal in landfills raises significant concerns, including land wastage and secondary pollution. In addition, conventional landfill systems contribute an estimated 29.4 ton of CO_2 equivalent emissions for each ton of sludge disposed of (Zhang et al., 2013), a higher emission factor than cement production (Liu et al., 2023). Unlike drinking water treatment sludge, sewage sludge contains various toxic constituents, including heavy metals, antibiotics, and pathogens (Ihsanullah et al., 2022). The presence of these substances can have a devastating impact on the environment and human health. Consequently, the management of sewage sludge poses greater challenges and pressures compared with that of drinking water treatment sludge. Depending on the chemical composition of sewage sludge, low-hazardous and phosphorus-rich sewage sludge can be disposed of in landfills or used in agriculture, while high-risk sewage sludge is typically incinerated (Wang et al., 2017). Fig. 22.1 illustrates the sludge disposal methods in European countries in 2020.

Although incineration of WTR eliminates organic matter and pathogens, incinerated WTR can still contain significant amounts of heavy metals, such as lead, antimony, and cobalt. Improper disposal of incinerated WTR can result in financial and environmental issues. Therefore, the increasing amount of incinerated WTR necessitates more efficient and environmentally friendly engineering solutions for its management. Incorporating incinerated WTR in the production of alkaline-activated materials (AAMs) offers a sustainable and safe solution for WTR disposal.

AAMs have garnered significant attention as alternative binders with the potential to provide environmental benefits in the construction industry. The primary binding phases of AAMs are formed through the reaction of aluminosilicate sources, such as calcined clays, coal fly ash, and metallurgical slags, with an alkaline solution. This solution accelerates the reaction process and induces the formation of robust, insoluble binding phases (Provis and Bernal, 2014). WTR has been

Treatment and Utilization of Combustion and Incineration Residues. https://doi.org/10.1016/B978-0-443-21536-0.00015-0

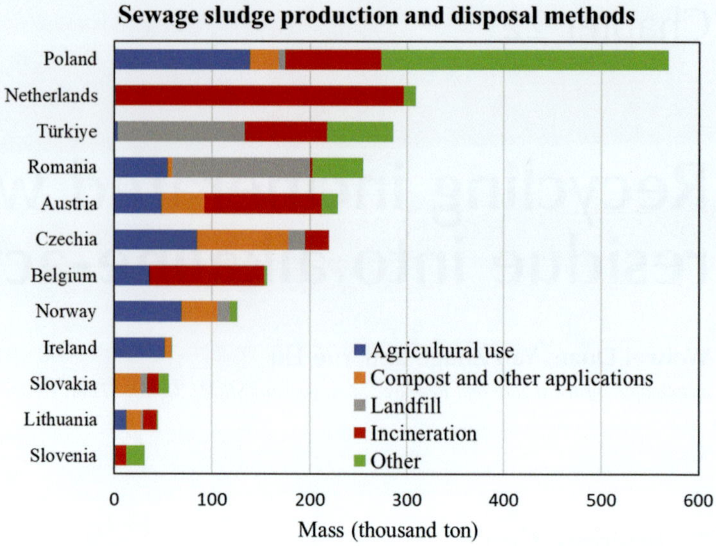

FIGURE 22.1 Sludge disposal solutions in European countries in 2020. *Modified from Ducoli, S., Zacco, A., Bontempi, E., 2021. Incineration of sewage sludge and recovery of residue ash as building material: a valuable option as a consequence of the COVID-19 pandemic. J. Environ. Manag. 282, 111966.*

proven feasible as a solid precursor for AAMs (Sun et al., 2018; Duan et al., 2022a), presenting a potential solution for safely managing incinerated WTR. This chapter provides a comprehensive review of recent research advancements in reusing WTR for the production of AAMs. It begins with a summary of WTR characterization, followed by a review and discussion of the reaction mechanism, mechanical performance, and hazardous metal immobilization ability of WTR-derived AAMs. Finally, the phase analysis and microstructure of AAM binders are elucidated.

2. Physical and chemical properties of WTR

Raw WTR from sewage sludge comprises various components, including organic matter, nitrogen- and phosphorus-containing compounds, microbiological pollutants, and metal elements such as Al, Fe, Ca, and Mg, along with other heavy metals (Rulkens, 2008; Hu et al., 2022). During the incineration process, a significant portion of the volatile matter, which constitutes approximately 20wt% to 50wt% of the sewage sludge, can be effectively removed (Huang and Yuan, 2016). Additionally, the incineration process involves the addition of sand to assist with the combustion, resulting in an increase in silica (Zhang et al., 2016). The chemical composition and trace element analysis of incinerated WTR sourced from sewage sludge is outlined in Table 22.1 (Cyr et al., 2007). It is essential to note the considerable variability in the chemical composition of raw sludge, owing mainly to the diversity of wastewater sources. For example, the content of silica has been found to vary widely, ranging from 14% to 65% in different studies. As such, the summary of incinerated WTR composition is presented here based on estimated mean values. Predominantly, incinerated WTR is composed of silicon, calcium, phosphorus, and aluminum. The quantities of CaO and P_2O_5 are higher than other mineral admixtures such as fly ash, silica fume, and metakaolin (Djobo and Stephan, 2022; Ke et al., 2022; Liu et al., 2022a). These two elements are crucial for forming whitlockite, a type of calcium phosphate that is relatively insoluble in atmospheric conditions. The quantities of Al_2O_3 and SiO_2 constitute less than 50% of WTR. These oxides, which contribute significantly to the reactivity of materials, are found in lesser quantities than other pozzolans. The mineralogy investigation of incinerated WTR from sewage sludge indicates that in addition to whitlockite, most sludge contains minerals such as quartz, calcium sulfate, and feldspars. Additional minerals such as calcite, moganite, hematite, and magnetite have also been observed. The amorphous phase in sludge varies between 50% and 74%. Fig. 22.2A presents the morphology of incinerated WTR particles, where both the irregular shape and porous structure are clearly identifiable (Zhou et al., 2020).

The components of WTR from drinking water treatment sludge include a range of solid substances, the composition of which largely depends on the quality of the raw water and the dosage of the aluminum-based coagulant used (Duan et al., 2020; Zhuge et al., 2022). For example, sludge derived from groundwater generally exhibits consistent properties due to the stable quality of the water, while sludge from surface water exhibits more noticeable variations. Generally, the primary constituents are SiO_2, Al_2O_3, and organic matter. They are followed by lesser quantities of Fe_2O_3 and CaO. Trace amounts of other oxides, such as K_2O, Na_2O, MgO, P_2O_5, and SO_3, can also be detected in the composition (Ahmad et al., 2016). The organic matter can be significant reduced to less than 10wt% during the incineration process (Liu et al., 2020b).

TABLE 22.1 Chemical composition and trace elements of WTR from sewage sludge.

Chemical composition (wt%)				Trace elements (mg/kg)			
Oxide	Mean	Min	Max	Elements	Mean	Min	Max
SiO_2	36.1	14.4	65	As	87	0.4	726
Al_2O_3	14.2	4.4	34.2	Ba	4142	90	14,600
Fe_2O_3	9.2	2.1	30	Cd	20	4	94
CaO	14.8	1.1	40.1	Co	39	19	78
P_2O_5	11.6	0.3	26.7	Cr	452	16	2100
SO_3	2.8	0.01	12.4	Cu	1962	200	5420
Na_2O	0.9	0.01	6.8	Ni	671	79	2000
K_2O	1.3	0.1	3.1	Pb	600	93	2055
TiO_2	1.1	0.3	1.9	Sb	35	35	35
MgO	2.4	0.02	23.4	Sn	400	183	617
MnO	0.3	0.03	0.9	Sr	539	539	539
LOI	6.1	0.2	41.8	V	35	14	66
				Zn	3512	1084	10,000

Modified from Cyr, M., Coutand, M., Clastres, P., 2007. Technological and environmental behavior of sewage sludge ash (SSA) in cement-based materials. Cement Concr. Res. 37, 1278–1289.

FIGURE 22.2 Scanning electron microscope images of incinerated WTR from (A) sewage treatment and (B) drinking water treatment. *Modified from Zhou, Y.-F., Li, J.-S., Lu, J.-X., Cheeseman, C., Poon, C.S., 2020. Recycling incinerated sewage sludge ash (ISSA) as a cementitious binder by lime activation. J. Clean. Prod. 244. Liu, Y., Zhuge, Y., Chow, C.W.K., Keegan, A., Pham, P.N., Li, D., Oh, J.-A., Siddique, R., 2021. The potential use of drinking water sludge ash as supplementary cementitious material in the manufacture of concrete blocks. Resour. Conserv. Recycl. 168.*

Contrary to sewage sludge, the quantities of Al_2O_3 and SiO_2 are typically higher than 50%. Given these proportions, it is reasonable to anticipate that WTR from drinking water treatment might exhibit superior performance as a precursor for AAMs compared with that from sewage sludge. For the mineralogy, the main crystalline peaks of sludge detected are quartz and kaolinite, and some peaks of calcite, hematite, and feldspars are also observed (Owaid et al., 2014; Hagemann et al., 2019; Liu et al., 2021). The morphology is similar to sewage sludge, both exhibiting porous structure and irregular shape (see Fig. 22.2B) (Liu et al., 2021).

3. Reaction mechanism of alkaline-activated materials

AAMs are produced by blending solid precursors with an alkaline solution. Depending on the reactive calcium (Ca) content in the precursors, AAMs can be categorized into high-Ca systems and low-Ca systems, also known as geopolymers. The presence of calcium has a substantial influence on the reaction mechanism, resulting in significant differences in strength development, mechanical performance, and durability properties of the resulting AAM products.

3.1 High-Ca AAM

Ground granulated blast-furnace slag (GGBS) is commonly used as the primary precursor in the production of high-Ca AAMs. The activation process of GGBS involves the use of sodium hydroxide (NaOH) and sodium silicate (Na_2SiO_3), and it can be divided into three stages, shown in Fig. 22.3.

In the first stage, the precursor undergoes dissolution as a result of an alkaline (OH^-) attack. During this stage, the dissolution rate of calcium and aluminum species is higher compared with silicon (Sun et al., 2022a). Consequently, the solution experiences a rapid concentration increase of Ca and Al ions. However, the high concentration of these cations hinders the dispersion of Si species, leading to their accumulation on the surface of the precursor particles. This accumulation of Si species on the surface acts as a barrier, slowing down the precursor dissolution (Sun and Vollpracht, 2018). The second stage is gel formation. With the increase of ion concentration approaching supersaturation, nuclei of C−S−H will initialize. The initially formed C-S-H gels consume the ions in the solution, reducing the accumulated Si layer, and more ions could be released from the precursor surface into the solution. Extra C-S-H gels will precipitate at this stage with Na ions as a catalyzer to exchange Ca from the gel products (Pacheco-Torgal et al., 2008; Sun et al., 2022a). With continuing of the reaction, Al tetrahedrons gradually replace Si tetrahedrons in bridging positions, transferring C-S-H to C-A-S-H. The bridging effect increases the chain length. Na ions are absorbed into the chain to balance the charge. The microstructure of the slag-based AAM in the study by Duan et al. (2022a) exhibited the two-layer structure of the reaction products in Fig. 22.4. The darker reaction products surrounding the unreacted slag indicated the initially formed C-S-H gels, while the outer gels indicated C-A-S-H gels. In the final stage, with the growth of reaction products surrounding the unreacted slag particles, alkaline becomes difficult to transport through the C-(A)-S-H shell, and the reaction is gradually stopped.

The reaction products in high-Ca AAM exhibit a tobermorite-like C-(A)-S-H structure, which closely resembles the hydration products found in ordinary Portland cement (Provis et al., 2015). One significant benefit of high-Ca AAM is that these reaction products can be formed at room temperature, allowing for an ambient curing condition. However, it is

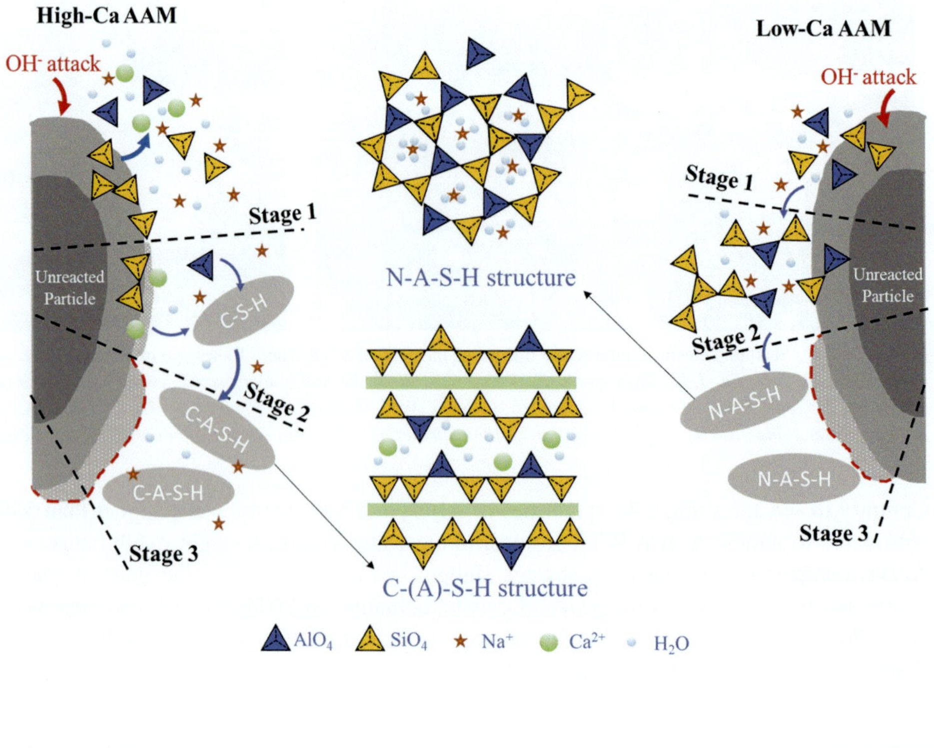

FIGURE 22.3 Reaction mechanism of AAM. *Modified from Duan, W., Zhuge, Y., Chow, C.W.K., Keegan, A., Liu, Y., Siddique, R., 2022a. Mechanical performance and phase analysis of an eco-friendly alkali-activated binder made with sludge waste and blast-furnace slag. J. Clean. Prod. 374, 134024. Sun, B., Ye, G., de Schutter, G., 2022a. A review: reaction mechanism and strength of slag and fly ash-based alkali-activated materials. Construct. Build. Mater. 326, 126843.*

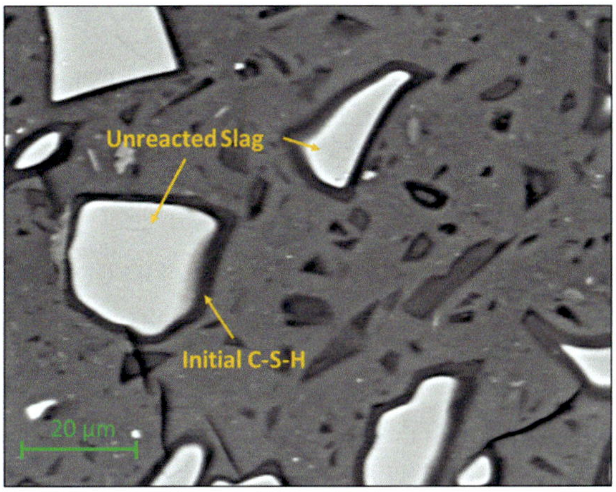

FIGURE 22.4 Backscatter electron images of slag-based AAM mortar.

important to note that the C−(A)−S−H phases in high-Ca AAM are relatively susceptible to acid attack, which results in lower corrosion resistance compared to the low-Ca system, where the primary phase is N−A−S−H (Provis et al., 2015; Albitar et al., 2017; Koenig et al., 2017).

3.2 Low-Ca AAM (geopolymer)

In the case of low-Ca system AAMs, aluminosilicate materials such as class F fly ash and metakaolin are commonly used as precursors. As illustrated in Fig. 22.3, the initial stage is similar to that of high-Ca AAM, involving the dissolution of the solid precursor. Alkali OH^- anions attack the Si−O and Al−O bonds in the precursors, resulting in an increased concentration of Al and Si tetrahedrons ($[SiO_4]^{4-}$ and $[AlO_4]^{5-}$) in the solution. Several studies (Duxson et al., 2005; Fernández-Jiménez et al., 2005; Hajimohammadi et al., 2010) have indicated that Al species tend to exhibit a higher dissolution rate compared to Si species. Additionally, Al monomers have a tendency to attach to the surface of silica particles, which slows down the dissolution rate of Si species. In the subsequent stage, as the concentration of Si and Al species reaches supersaturation, nucleation of short-chain products occurs near the precursor surface. These short-chain products accumulate and connect with other Si or Al tetrahedrons in a process known as polycondensation. As this process continues, the short chains grow and form a three-dimensional structure known as N-A-S-H gels. In the third stage, the precipitation of N-A-S-H shells on the precursor surface, coupled with the depletion of alkaline contents, leads to a decrease in the mass transportation of reactive species. Ultimately, this halts the geopolymerization process (Davidovits, 1991; Krivenko and Kovalchuk, 2007; Shi et al., 2011).

Unlike the layered structure observed in high-Ca AAM, the N-A-S-H gels in low-Ca AAM exhibit a zeolite-like structure, as shown in Fig. 22.3. The formation of this phase typically requires an elevated temperature when there is insufficient reactive calcium present. Consequently, low-Ca AAMs are commonly cured in a chamber maintained above 40°C, and additional precautions must be taken to maintain relative humidity and prevent water evaporation from the samples (Zhang et al., 2020; Ruengsillapanun et al., 2021).

3.3 Hybrid phase AAM

In addition to utilizing a single precursor, the use of binary blended precursors containing both high-Ca and low-Ca sources has been explored for the production of hybrid phase AAM. Several studies have demonstrated that blending GGBS and fly ash (FA) can enhance the mechanical performance of AAM products, both under ambient and elevated temperature curing conditions (Zhang et al., 2020; Abhishek et al., 2021; Athira et al., 2021). This improvement is attributed to the release of free calcium ions from GGBS, which accelerates the dissolution of FA and enhances the overall reaction extent (Puligilla and Mondal, 2013). Furthermore, research has also identified the coexistence of N-A-S-H and C-A-S-H gels when employing a blend of low-Ca FA and GGBFS (Yip et al., 2008; Garcia-Lodeiro et al., 2011; Walkley et al., 2016). The N-A-S-H and C-A-S-H gels formed together in hybrid phase AAM and are different to those in the AAM with a single precursor in terms of Ca/Si and Al/Si (Garcia-Lodeiro et al., 2011). The interaction of these two phases contributes to enhancing the mechanical performance and durability properties of AAMs.

In comparison to FA and GGBS, the dissolution and reaction rate of Si and Al species in WTR is generally slower (Abdalqader et al., 2016; Chakraborty et al., 2017; Chang et al., 2020). Consequently, AAMs manufactured solely with incinerated WTR may not exhibit satisfactory mechanical performance. To address this limitation, it is common to mix WTR with other highly reactive materials. When blended with aluminosilicate precursors, WTR-incorporated AAMs are categorized as low-Ca systems. On the other hand, when blended with Ca-rich precursors, it contributes to the formation of hybrid phase AAMs.

4. Pretreatment techniques of WTR for use as precursors

Depending on the variation of raw WTR properties and incineration technologies, pretreatment of the as-received WTR may be required to improve the reactivity before using it as a precursor. To achieve considerable reactivity, the particle size of the precursor should usually be under 75 μm. The studies by Cyr et al. (2007), Donatello (2009), and Donatello and Cheeseman (2013) took different incinerated WTR samples from the United Kingdom and reported that the average particle size ranged between 8 and 263 μm. Therefore, for the WTR particles larger than 75 μm, further milling is necessary before it can be successfully utilized in making AAMs. Donatello et al. (2010) used dry ring mill technology to evaluate the average particle diameter on the reactivity of the incinerated sewage sludge. The study found that milling for 1 min could significantly reduce the average particle diameter from 141 to 8.6 μm, and a more extended milling period had a limited effect on further reducing the particle size. A comparison of the particle size of WTR ash and GGBS is shown in Fig. 22.5. The reduced particle size significantly improved the reactivity of the incinerated sewage sludge based on the Frattini test and strength activity index test described in the study (Donatello et al., 2010).

5. Activators

To enhance the activation performance of the material, binary alkaline activators such as sodium hydroxide (NaOH) and sodium silicate (Na_2SiO_3) are commonly used. A high concentration of NaOH improves the dissolution rate of structural elements such as calcium (Ca) and silicon (Si), while Na_2SiO_3 solution provides sufficient dissolved silica initially (Provis, 2018; Wen et al., 2019). Therefore, the activator modulus, represented by the molar ratio of SiO_2/Na_2O, is an important parameter in the production of WTR-based AAMs. In a study conducted by Chen et al. (2018), a mixture of NaOH and Na_2SiO_3 solution with a silica modulus ranging from 0.6 to 1.3 was used to activate the precursors based on GGBS and WTR. The results showed that samples produced with a 4% Na_2O alkali content and a silica modulus of 0.95 exhibited the highest compressive strength. Similarly, Duan et al. (2022a) found that an activator modulus of 0.9 resulted in the best mechanical performance for samples containing calcined drinking water treatment sludge and blast furnace slag. Another study (Zhao et al., 2023), which utilized incinerated sewage sludge ash as the precursor, evaluated the silica modulus ranging from 0.95 to 1.4. The results indicated that the optimal modulus value for achieving the highest 28-day compressive strength was between 0.95 and 1.2. Mineralogical analysis suggested that activators with such modulus

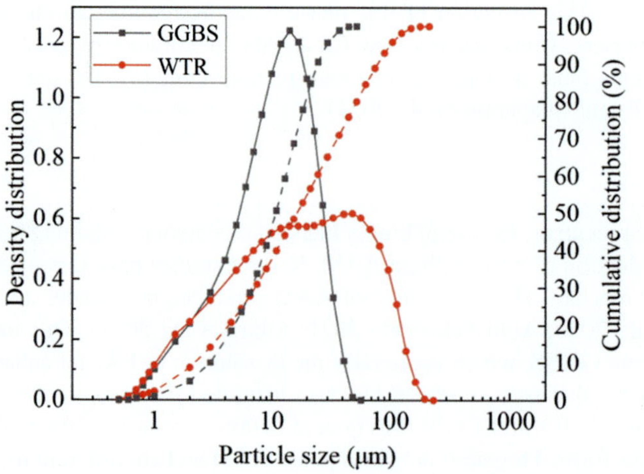

FIGURE 22.5 Particle size distribution of GGBS and incinerated WTR. *Adapted from the study Zhao, Q., Ma, C., Huang, B., Lu, X., 2023. Development of alkali activated cementitious material from sewage sludge ash: two-part and one-part geopolymer. J. Clean. Prod. 384, 135547.*

values optimized the reaction process by facilitating a fast dissolution rate without excessive accumulation of silica species on the surface of unreacted particles. Moreover, the dissolved silica from Na_2SiO_3 was beneficial for the polycondensation process. However, higher modulus values, while maintaining a higher concentration of dissolved Si in the pore solution, also indicated a lower alkaline concentration, which may result in a slower release rate of Al and Ca species. Additionally, the high cost of Na_2SiO_3 solution can reduce the economic feasibility of utilizing incinerated WTR-derived AAMs (Chang et al., 2020; Zhao et al., 2023).

6. Mechanical performance

Table 22.2 provides a summary of the mechanical performance of AAM products incorporating incinerated WTR as a precursor. To ensure satisfactory mechanical performance, incinerated WTR is commonly mixed with other highly reactive precursors, such as FA and GGBS. This blending approach helps overcome the slow dissolution and reaction rate of Si and Al in incinerated WTR when manufacturing AAMs.

6.1 WTR-incorporated low-Ca AAM

In the case of low-Ca AAM, some studies have reported adverse effects on the mechanical performance of samples when WTR is used as a precursor. However, these negative impacts can be mitigated by using a high silica modulus activator. A study by Istuque et al. (2019) examined the mechanical performance of the mortar samples made with WTR and metakaolin. The addition of 10% WTR by mass in the precursors resulted in a reduction in compressive strength from 31 to 26 MPa compared with the neat metakaolin-based AAM. However, by optimizing the silica modulus to 1.6, this strength reduction issue was effectively addressed, increasing the compressive strength to 48 MPa, similar to that of samples using neat metakaolin. A similar result was observed in another study by Suksiripattanapong et al. (2015), where a high volume of incinerated WTR, generated from a drinking water treatment plant, was used to replace 70% of FA by mass. The results showed that the highest compressive strength achieved was approximately 20 MPa when the silica modulus was set to 1.6.

On the other hand, other studies concluded that using less than 20% of incinerated WTR in precursors has a neglectable impact or slightly improves the mechanical performance of AAM. In the study by Istuque et al. (2016), the mortar using 10% WTR as a replacement for metakaolin exhibited a considerable compressive strength of 28 MPa compared with the neat metakaolin-based sample. However, increasing the WTR content to 20% slightly reduced the strength to 22 MPa. In a study by Kozai et al. (2021), the compressive strength of the AAM paste with WTR content ranging from 20% to 50% blended with metakaolin was investigated. The results showed that 20% WTR could slightly improve the strength to 70 MPa, but a significant reduction in strength occurred when the WTR content exceeded this value. The variation in the mechanical performance of AAM with WTR content may be attributed to the varying mineralogical characteristics of the WTR. The composition of WTR compounds can vary due to fluctuations in the quantity of wastewater produced, which in turn affects the properties of the resulting AAM (Kozai et al., 2021).

6.2 WTR-incorporated hybrid phase AAM

One limitation of blending WTR with low-Ca precursors is the requirement for curing the samples at elevated temperatures, typically above 40°C, to achieve a high degree of geopolymerization. This elevated temperature curing reduces the economic feasibility of using WTR in applications. However, when WTR is blended with high-Ca precursors, elevated temperature curing is not necessary. Furthermore, the release of calcium in the pore solution is beneficial for increasing the dissolution rate of Si species from WTR, thereby enhancing its reactivity. Consequently, AAMs that blend WTR with high-Ca sourced precursors exhibit relatively better mechanical performance compared with those that blend WTR with aluminosilicate precursors. For example, a study by Chen et al. (2018) mixed 50% WTR with GGBS using a dry-mixing method to manufacture AAM. The compressive strength achieved was 33 MPa when an activator with a silica modulus of 0.95 was used. To further enhance the reaction degree and mechanical performance in AAMs incorporating high volumes of WTR, Chakraborty et al. (2017) used GGBS and quicklime as additional sources of Ca mixed with 50%–100% of WTR sourced from sewage sludge. The results showed that the neat WTR-based AAM mortar had a compressive strength of only 5 MPa, but replacing 20% WTR with quicklime significantly increased the strength to 20 MPa. Moreover, when a ternary blend of precursors containing 10% GGBS, 20% quicklime, and 70% WTR was used, the compressive strength further improved to 30 MPa. In another study by Duan et al. (2022a), mortar samples activated with 40% WTR from drinking water treatment achieved a compressive strength of 56 MPa when the silica modulus was set to be 0.9. Increasing the silica modulus to 1.2 further increased the strength to approximately 70 MPa. The lower mechanical performance of the

TABLE 22.2 Summary of the mechanical performance of incinerated WTR-incorporated AAMs.

Product	Precursors	Percentage of WTR (%)	Activators	Mechanical performance	Ref
Paste	WTR (S) GGBS	50	SS + SH M = 0.6, 0.95, 1.3	• The highest compressive strength is 33 MPa using the dry-mix method. • Increasing silica modulus up to 0.95 is beneficial for improving mechanical performance.	Chen et al. (2018)
Paste	WTR (S) FA	70	SS, SS + SH M = 3.4, 2.7, 2.1, 1.6, 1.0	• The highest compressive strength is 20 MPa under 75°C curing for 120 h. • Using the mix of SS and SH as an activator is more beneficial for strength development than using SS alone. • The optimal silica modulus for mechanical performance is 1.6.	Suksiripattanapong et al. (2015)
Paste	Alkali-treated WTR (S) GGBS	20 30 40 50	SS + SH M = 1.2, 1.3, 1.4, 1.5	• The wet alkalinizing method as a pretreatment for WTR could improve mechanical performance. • Samples with 20% pretreated WTR with a silica modulus of 1.3 exhibits the compressive strength of 90 MPa. • A high proportion of WTR significantly reduced the strength.	Sun et al. (2018)
Paste	WTR (S) Metakaolin	20 30 50	SS + SH M = 1.6 and 2.0	• 20% WTR improved the compressive strength to 71 MPa. • Increasing WTR content by more than 20% resulted in a significant loss of strength. • Mechanical performance varied depending on the sources of WTR.	Kozai et al. (2021)
Mortar	WTR (S) Metakaolin	10	SS + SH M = 0.8 and 1.6	• Increasing silica modulus from 0.8 to 1.6 improved compressive strength from 25 to 48 MPa. • 10% WTR slightly reduced strength compared with using metakaolin alone. • WTR slows down the rate of strength development.	Istuque et al. (2019)
Mortar	WTR (S) Metakaolin	10 20	SS + SH M = 2.0	• The sample with 10% WTR exhibited a compressive strength of 28 MPa, which is comparable with that of the neat-metakaolin-based samples. • 20% WTR reduced strength slightly to 22 MPa.	Istuque et al. (2016)
Mortar	WTR (S) GGBS Quicklime	60 70 80 90 100	SH (solid) mixed with precursors Water was added after the mixing	• The compressive strength was less than 5 MPa when using WTR alone. • Adding 20% quicklime could increase compressive strength to 20 MPa. • The sample with ternary blended precursors exhibited the highest compressive strength of 31 MPa.	Chakraborty et al. (2017)
Mortar	WTR (D) GGBS	20 40 60 80	SS + SH M = 0.6, 0.9, 1.2	• Increasing WTR up to 40% could improve compressive strength to 56 MPa using an activator with a silica modulus of 0.9. • Further increasing the silica modulus to 1.2 improved the maximum strength to 67 MPa.	Duan et al. (2022a)

D, sourced from drinking water treatment sludge; *S*, sourced from sewage sludge; *SH*, sodium hydroxide; *SS*, sodium silicate.

AAM made of the sewage sludge compared with those made with drinking water treatment sludge is possibly due to the lower amount of reactive Si species present in incinerated sewage sludge. To improve the reactivity of sewage sludge-based WTR, Sun et al. (2018) introduced a wet alkalinizing method as a pretreatment for sewage sludge. The result showed that the compressive strength of the sample with 20% pretreated sewage sludge was approximately 90 MPa. However, the additional cost for using alkaline in the pretreatment process may reduce the economic feasibility of AAMs.

7. Immobilization of hazardous metals

WTR can contain significant amounts of hazardous metals, including heavy metals and radioactive metals, depending on the source of the wastewater. Improper disposal of WTR without suitable treatment can lead to soil and water pollution, posing a risk to drinking water supplies and aquatic ecosystems. Furthermore, if not effectively managed, these contaminants can leach into groundwater sources, intensifying the environmental impact.

To address this issue, several studies have highlighted the benefits of incorporating WTR in the production of AAMs as a means to immobilize heavy metals present in WTR. Sun et al. (2022b) conducted a study to investigate the hazardous metal leaching behavior of AAM containing up to 7% incinerated WTR using the toxicity characteristic leaching procedure (TCLP) method. The AAM was found to leach several hazardous metals, including Ag, Pb, Sb, Co, Cr, and Zn. The concentration of these leached metal elements ranged from 0.02 to 3.28 mg/L, all of which were below the threshold values set by the US Environmental Protection Agency. In a subsequent study (Sun et al., 2023) by the same authors, water washing was employed to remove soluble salts from the WTR. The results indicated that the washing process had negligible effects on the leaching behavior of WTR, suggesting that the heavy metals were bonded into the sintered phases. The leaching concentration of hazardous metals from WTR itself was much higher compared with that from the WTR-based AAM, suggesting that the manufacturing process of AAM is effective in immobilizing the heavy metals present in WTR. A similar outcome was observed in a study by Sun et al. (2018), where the use of WTR and GGBS in making AAM resulted in an immobilization efficiency of heavy metals exceeding 80%, as depicted in Fig. 22.6. The immobilization of heavy metals can be attributed to the aluminosilicate phases present in GGBS, which generate silicate metallic compounds that exhibit strong bonding ability to the heavy metals. However, it was noted that the leaching of heavy metals significantly increased when the WTR-based AAM was exposed to acidic solutions. This finding raises concerns regarding the safety of WTR-based AAM products in low-pH environments, such as sewage pipes, where microbiological acidification commonly occurs (Zhuge et al., 2021).

In addition to reducing the leaching of heavy metals, the use of WTR in AAM manufacturing has been found to minimize the risks associated with radioactive-contaminated WTR. In a study by Kozai et al. (2021), 30% of radiocaesium (^{137}Cs)-contaminated WTR was blended with metakaolin as a precursor. The results showed that only 1% of ^{137}Cs was leached out from the WTR-incorporated AAM, while 9% was leached out from the as-received WTR when soaked at 60°C for 28 days. Further investigation revealed that the disconnected pore structure of the AAM is beneficial in preventing the transportation of Cs from the matrix or pore solution to the external environment. Additionally, the AAM matrix was found to have a strong adsorption capacity for Cs, and the leaching rate was less dependent on the environmental pH compared with using WTR as a supplementary cementitious material in ordinary Portland cement (OPC) (Li et al., 2013). However,

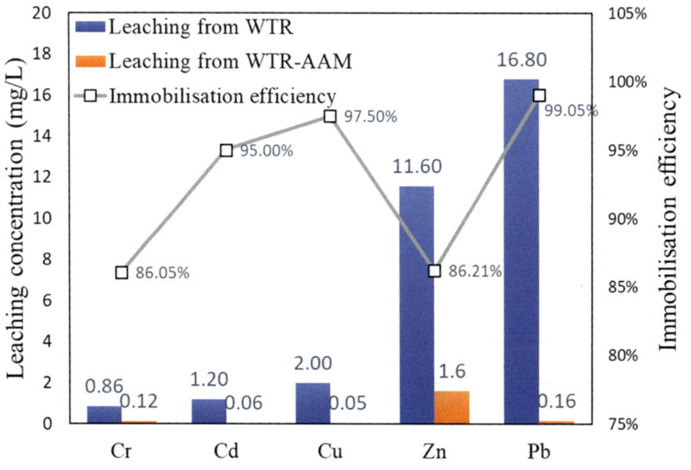

FIGURE 22.6 Leaching concentration and immobilization efficiency of using WTR in AAM. *Modified from Sun, S., Lin, J., Zhang, P., Fang, L., Ma, R., Quan, Z., Song, X., 2018. Geopolymer synthetized from sludge residue pretreated by the wet alkalinizing method: compressive strength and immobilization efficiency of heavy metal. Construct. Build. Mater. 170, 619–626.*

some studies indicate that the leaching rate of radioactive metals may increase with the concentration of external salt solution (El-Naggar and Amin, 2018; He et al., 2020; Kozai et al., 2021), highlighting a potential risk associated with using AAM for immobilizing radioactive metals.

8. Microstructural characteristics

To assess the impact of incinerated WTR on the properties of AAMs and to examine the reaction mechanisms involved, researchers typically employ mineralogical, elemental, and microstructural analyses. Techniques such as X-ray diffraction (XRD), Fourier transform infrared spectroscopy (FTIR), scanning electron microscope (SEM), and energy-dispersive X-ray spectroscopy (EDS) are commonly used to investigate the AAM matrix and the resulting reaction products in these studies.

8.1 X-ray diffraction analysis of WTR-incorporated AAM products

For XRD analysis of the reaction products, broad humps ranging from 20 degrees to 40 degrees 2θ are usually detected for the patterns obtained using Cu Kα radiation, as shown in Fig. 22.7. These humps were widely reported as a fingerprint of geopolymerization due to the existence of amorphous N-A-S-H gels (Tashima et al., 2013; Fernández-Jiménez et al., 2017; Yaseri et al., 2017). Some studies (Chen et al., 2018; Kozai et al., 2021) reported no new crystalline phases can be detected in the reaction products. The remaining peaks, such as quartz, hematite, and goethite, were attributed to either unreacted WTR or other precursors. In addition, intensity reduction of these peaks was observed, indicating the crystalline phases were consumed in the geopolymerization reactions and the crystalline minerals in the WTR were integrated into the AAM products. On the other hand, some other studies (Suksiripattanapong et al., 2015; Istuque et al., 2019; Duan et al., 2022a) found newly formed zeolite group phases in the reaction products, such as Na P-type and FAU-type zeolite. The new phases can be the reaction intermediate due to incomplete geopolymerization, possibly attributed to insufficient alkaline or low reaction temperature.

8.2 Fourier transform infrared spectroscopy analysis of WTR-incorporated AAM products

FTIR is a technique used to evaluate the molecular chains of the incinerated WTR-based AAM products by analyzing characteristic chemical bonds in products, such as T-OH and Si-O-T (T stands for Al or Si). T-OH bond can usually be detected in the WTR precursor at the wavenumber of 3544−3475 cm^{-1}, indicating the uncondensed Al or Si species (Lin et al., 2017). High absorbance bands ranging between 1300−900 cm^{-1} can usually be found in both the WTR precursor and its AAM products, which is caused by asymmetric stretching vibration of Si-O-T bonds (Rovnaník, 2010; Ozer and

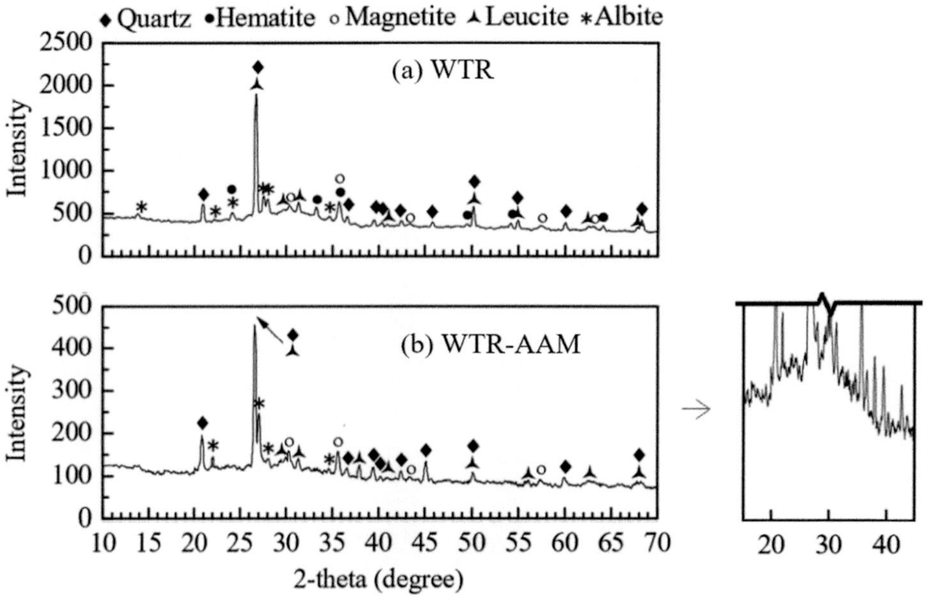

FIGURE 22.7 XRD patterns of (A) incinerated WTR and (B) WTR-incorporated AAM. *Modified from Chen, Z., Li, J.-S., Zhan, B.-J., Sharma, U., Poon, C.S., 2018. Compressive strength and microstructural properties of dry-mixed geopolymer pastes synthesized from GGBS and sewage sludge ash. Construct. Build. Mater. 182, 597−607.*

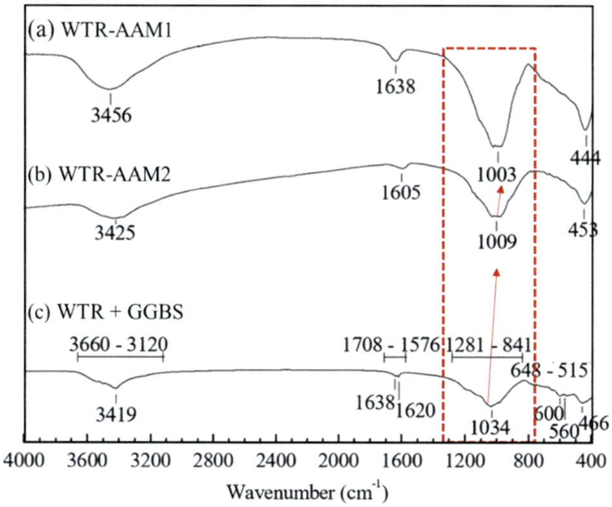

FIGURE 22.8 FTIR patterns of (A) WTR-AAM sample 1, (B) WTR-AAM sample 2, and (C) mix of WTR and GGBS. *Modified from Chen, Z., Li, J.-S., Zhan, B.-J., Sharma, U., Poon, C.S., 2018. Compressive strength and microstructural properties of dry-mixed geopolymer pastes synthesized from GGBS and sewage sludge ash. Construct. Build. Mater. 182, 597–607.*

Soyer-Uzun, 2015). A high degree of geopolymerization reaction will reduce the absorbance of the T-OH, increase the absorbance of Si-O-T, and shift its band to a lower wavenumber. This is because the growth of N-A-S-H gels consumes the uncondensed Al and Si species in precursors (Liu et al., 2020a; Mañosa et al., 2021). The N-A-S-H gels exhibited a lower wavenumber around 1000 cm^{-1} (Wan et al., 2017), thus causing the Si-O-T band to shift to a lower wavenumber. For example, in the study by Chen et al. (2018), the FTIR spectrum of the sample in Fig. 22.8A with better mechanical performance than the sample in Fig. 22.8B exhibited a higher absorbance and a lower wavenumber of the Si-O-T band.

8.3 Scanning electron microscope and elemental analysis of WTR-incorporated AAM products

SEM coupled with backscattered electron (BSE) observation is commonly conducted to evaluate the morphology of the AAM microstructure. The AAM matrix with a high degree of reaction usually exhibits a compact and homogenous structure. In the study by Duan et al. (2022a), the BSE image of the samples with 40% WTR exhibits a dense matrix compared with that of the samples, which had lower compressive strength, as shown in Fig. 22.9A and B. Due to the irregular shape of the incinerated WTR particles and similar atomic density to the AAM matrix, identification of the unreacted WTR contents may be difficult via SEM and BSE observation. Therefore, EDS analysis is used to further determine the products. For example, the unreacted WTR particles in Fig. 22.9A were confirmed using the EDS mapping in Fig. 22.9C according to the high-Al feature of the WTR. The quantitative EDS analysis showed the Al/Si ratio of the WTR-based AAM products varying between 0.2 and 0.86, and the Ca/Si ratio ranging from 0.25 to 0.8.

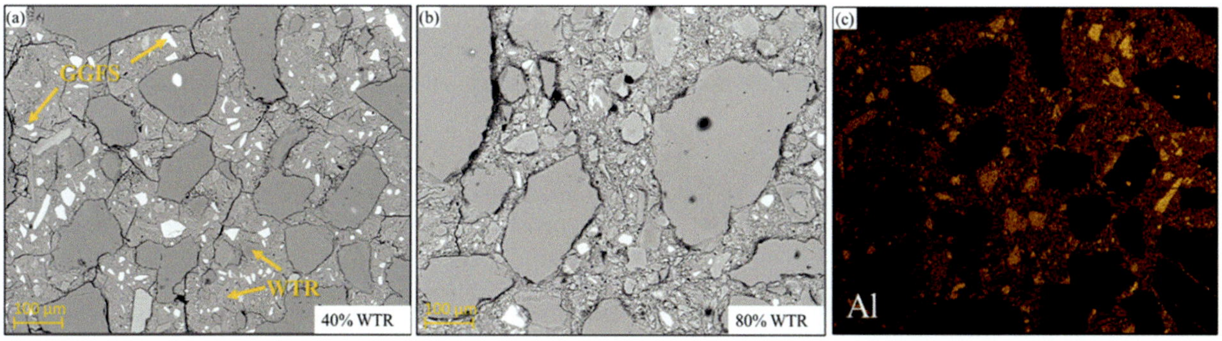

FIGURE 22.9 SEM-BSE images of (A) the sample with 40% WTR, (B) the sample with 80% WTR, and (C) the EDS mapping of Al in the sample with 40% WTR. *Modified from Duan, W., Zhuge, Y., Chow, C.W.K., Keegan, A., Liu, Y., Siddique, R., 2022a. Mechanical performance and phase analysis of an eco-friendly alkali-activated binder made with sludge waste and blast-furnace slag. J. Clean. Prod. 374, 134024.*

9. Conclusions

The use of incinerated WTR as a precursor for producing AAMs offers a cost-effective and sustainable solution for waste disposal. AAMs, when used as a substitute construction material for cement, have the potential to significantly reduce the consumption of traditional cement and contribute to a decrease in greenhouse gas emissions. This reuse of WTR presents an opportunity for both the construction and water industries to mitigate the environmental impact associated with their production processes.

Incinerated WTR can be sourced from drinking water treatment sludge and sewage sludge. The morphology of incinerated WTR particles exhibits a porous structure, resulting from the pyrolysis of organic matter at high temperature. Generally, WTR contains silicon, calcium, aluminum, and phosphor elements. The chemical composition of WTR can vary significantly due to variations in water or wastewater quality. Because of the presentence of aluminosilicate compounds, WTR has the potential to be used in the manufacturing of AAMs through alkaline activation.

The mechanical performance of samples using pure WTR as a precursor may not be comparable with samples activated with typical precursors due to the lower reactivity of the silica contents in WTR. Therefore, WTR is commonly mixed with other highly reactive industrial wastes, such as coal FA and GGBS, at a mass ratio of under 50%. In high-Ca AAMs, a higher proportion of WTR can be used to achieve considerable compressive strength compared with samples made with OPC. However, in low-Ca AAMs, using more than 10% WTR may result in a strength reduction. Increasing the silicate modulus of the activator could slightly mitigate this reduction.

Incinerated WTR may contain significant amounts of hazardous metals, including Pb, Sb, and Co, which pose an environmental risk if improperly disposed of. Incorporating WTR into AAM production can effectively immobilize these hazardous metals due to the disconnected pore structure and the strong bonding ability of the AAM products. However, an increase in salt concentration and a decrease in pH in the external environment may increase the leaching rate of these hazardous metals.

Mineralogical analysis reveals the formation of N-A-S-H gels in the WTR-incorporated AAMs. Microstructural observations indicate that an appropriate volume of WTR in precursors can form a dense AAM matrix, while an excess of WTR may lead to a porous structure. Quantitative elemental analysis shows that the products have similar element ratios to hybrid C-A-S-H and N-A-S-H gels.

Using incinerated WTR as a precursor for manufacturing AAM has garnered significant research interests in recent years. Researchers have gained insights into the physical, chemical, and microstructural characteristics of WTR. Many studies have explored the mixing of WTR with other industrial waste to produce AAMs as a sustainable substitute for cement binders. Extensive evaluations have been conducted on the mechanical performance and the ability of WTR-incorporated AAMs to immobilize hazardous metals. However, the chemical composition of WTR can vary significantly based on the source and treatment processes of the water. This variability can affect the performance and properties of the resulting AAM, making it challenging to achieve consistent and predictable outcomes. Further research is needed to understand the impact of different WTR compositions on the properties of AAM and develop strategies to mitigate the effects of this variability. Additionally, while extensive evaluations have been conducted on the mechanical performance of WTR-incorporated AAMs, more research is needed to assess their long-term durability and resistance to environmental factors such as freeze-thaw cycles, chemical attack, and other degradation mechanisms. Understanding the aging behavior and potential deterioration of AAMs over extended periods is crucial for their practical application and acceptance in real-world construction projects.

References

Abdalqader, A.F., Jin, F., Al-Tabbaa, A., 2016. Development of greener alkali-activated cement: utilisation of sodium carbonate for activating slag and fly ash mixtures. J. Clean. Prod. 113, 66–75.

Abhishek, H.S., Prashant, S., Kamath, M.V., Kumar, M., 2021. Fresh mechanical and durability properties of alkali-activated fly ash-slag concrete: a review. Innov. Infrastruct. Solut. 7, 116.

Ahmad, T., Ahmad, K., Alam, M., 2016. Sustainable management of water treatment sludge through 3'R' concept. J. Clean. Prod. 124, 1–13.

Albitar, M., Mohamed Ali, M.S., Visintin, P., Drechsler, M., 2017. Durability evaluation of geopolymer and conventional concretes. Construct. Build. Mater. 136, 374–385.

Athira, V.S., Bahurudeen, A., Saljas, M., Jayachandran, K., 2021. Influence of different curing methods on mechanical and durability properties of alkali activated binders. Construct. Build. Mater. 299, 123963.

Chakraborty, S., Jo, B.W., Jo, J.H., Baloch, Z., 2017. Effectiveness of sewage sludge ash combined with waste pozzolanic minerals in developing sustainable construction material: an alternative approach for waste management. J. Clean. Prod. 153, 253–263.

Chang, Z., Long, G., Zhou, J.L., Ma, C., 2020. Valorization of sewage sludge in the fabrication of construction and building materials: a review. Resour. Conserv. Recycl. 154, 104606.

Chen, Z., Li, J.-S., Zhan, B.-J., Sharma, U., Poon, C.S., 2018. Compressive strength and microstructural properties of dry-mixed geopolymer pastes synthesized from GGBS and sewage sludge ash. Construct. Build. Mater. 182, 597–607.

Cyr, M., Coutand, M., Clastres, P., 2007. Technological and environmental behavior of sewage sludge ash (SSA) in cement-based materials. Cement Concr. Res. 37, 1278–1289.

Dassanayake, K.B., Jayasinghe, G.Y., Surapaneni, A., Hetherington, C., 2015. A review on alum sludge reuse with special reference to agricultural applications and future challenges. Waste Manag. 38, 321–335.

Davidovits, J., 1991. Geopolymers: Inorganic polymeric new materials. J. Therm. Anal. Calorim. 37, 1633–1656.

Di Giacomo, G., Romano, P., 2022. Evolution and prospects in managing sewage sludge resulting from municipal wastewater purification. Energies 15, 5633.

Djobo, J.N.Y., Stephan, D., 2022. The reaction of calcium during the formation of metakaolin phosphate geopolymer binder. Cement Concr. Res. 158.

Donatello, S., 2009. Characteristics of Incinerated Sewage Sludge Ashes: Potential for Phosphate Extraction and Re-use as a Pozzolanic Material in Construction Products. Department of Civil and Environmental Engineering, Imperial College London.

Donatello, S., Cheeseman, C.R., 2013. Recycling and recovery routes for incinerated sewage sludge ash (ISSA): a review. Waste Manag. 33, 2328–2340.

Donatello, S., Freeman-Pask, A., Tyrer, M., Cheeseman, C.R., 2010. Effect of milling and acid washing on the pozzolanic activity of incinerator sewage sludge ash. Cement Concr. Compos. 32, 54–61.

Duan, W., Zhuge, Y., Chow, C.W.K., Keegan, A., Liu, Y., Siddique, R., 2022a. Mechanical performance and phase analysis of an eco-friendly alkali-activated binder made with sludge waste and blast-furnace slag. J. Clean. Prod. 374, 134024.

Duan, W., Zhuge, Y., Pham, P.N., Liu, Y., Kitipornchai, S., 2022b. A ternary blended binder incorporating alum sludge to efficiently resist alkali-silica reaction of recycled glass aggregates. J. Clean. Prod. 349, 131415.

Duan, W., Zhuge, Y., Pham, P.N., Chow, C.W.K., Keegan, A., Liu, Y., 2020. Utilization of drinking water treatment sludge as cement replacement to mitigate alkali–silica reaction in cement composites. J. Compos. Sci. 4, 171.

Duxson, P., Lukey, G.C., Separovic, F., van Deventer, J.S.J., 2005. Effect of alkali cations on aluminum incorporation in geopolymeric gels. Ind. Eng. Chem. Res. 44, 832–839.

El-Naggar, M.R., Amin, M., 2018. Impact of alkali cations on properties of metakaolin and metakaolin/slag geopolymers: microstructures in relation to sorption of 134Cs radionuclide. J. Hazard Mater. 344, 913–924.

Fernández-Jiménez, A., Cristelo, N., Miranda, T., Palomo, Á., 2017. Sustainable alkali activated materials: precursor and activator derived from industrial wastes. J. Clean. Prod. 162, 1200–1209.

Fernández-Jiménez, A., Palomo, A., Criado, M., 2005. Microstructure development of alkali-activated fly ash cement: a descriptive model. Cement Concr. Res. 35, 1204–1209.

Garcia-Lodeiro, I., Palomo, A., Fernández-Jiménez, A., Macphee, D.E., 2011. Compatibility studies between N-A-S-H and C-A-S-H gels. Study in the ternary diagram $Na_2O-CaO-Al_2O_3-SiO_2-H_2O$. Cement Concr. Res. 41, 923–931.

Hagemann, S.E., Gastaldini, A.L.G., Cocco, M., Jahn, S.L., Terra, L.M., 2019. Synergic effects of the substitution of Portland cement for water treatment plant sludge ash and ground limestone: technical and economic evaluation. J. Clean. Prod. 214, 916–926.

Hajimohammadi, A., Provis, J.L., van Deventer, J.S.J., 2010. Effect of alumina release rate on the mechanism of geopolymer gel formation. Chem. Mater. 22, 5199–5208.

He, P., Cui, J., Wang, M., Fu, S., Yang, H., Sun, C., Duan, X., Yang, Z., Jia, D., Zhou, Y., 2020. Interplay between storage temperature, medium and leaching kinetics of hazardous wastes in Metakaolin-based geopolymer. J. Hazard Mater. 384, 121377.

Hu, M., Hu, H., Ye, Z., Tan, S., Yin, K., Chen, Z., Guo, D., Rong, H., Wang, J., Pan, Z., Hu, Z.-T., 2022. A review on turning sewage sludge to value-added energy and materials via thermochemical conversion towards carbon neutrality. J. Clean. Prod. 379, 134657.

Huang, H.-j., Yuan, X.-z., 2016. The migration and transformation behaviors of heavy metals during the hydrothermal treatment of sewage sludge. Bioresour. Technol. 200, 991–998.

Ihsanullah, I., Khan, M.T., Zubair, M., Bilal, M., Sajid, M., 2022. Removal of pharmaceuticals from water using sewage sludge-derived biochar: a review. Chemosphere 289, 133196.

Istuque, D.B., Reig, L., Moraes, J.C.B., Akasaki, J.L., Borrachero, M.V., Soriano, L., Payá, J., Malmonge, J.A., Tashima, M.M., 2016. Behaviour of metakaolin-based geopolymers incorporating sewage sludge ash (SSA). Mater. Lett. 180, 192–195.

Istuque, D.B., Soriano, L., Akasaki, J.L., Melges, J.L.P., Borrachero, M.V., Monzó, J., Payá, J., Tashima, M.M., 2019. Effect of sewage sludge ash on mechanical and microstructural properties of geopolymers based on metakaolin. Construct. Build. Mater. 203, 95–103.

Ke, Y., Liang, S., Hou, H., Hu, Y., Li, X., Chen, Y., Li, X., Cao, L., Yuan, S., Xiao, K., Hu, J., Yang, J., 2022. A zero-waste strategy to synthesize geopolymer from iron-recovered Bayer red mud combined with fly ash: roles of Fe, Al and Si. Construct. Build. Mater. 322.

Koenig, A., Herrmann, A., Overmann, S., Dehn, F., 2017. Resistance of alkali-activated binders to organic acid attack: Assessment of evaluation criteria and damage mechanisms. Construct. Build. Mater. 151, 405–413.

Kozai, N., Sato, J., Osugi, T., Shimoyama, I., Sekine, Y., Sakamoto, F., Ohnuki, T., 2021. Sewage sludge ash contaminated with radiocesium: solidification with alkaline-reacted metakaolinite (geopolymer) and Portland cement. J. Hazard Mater. 416, 125965.

Krivenko, P.V., Kovalchuk, G.Y., 2007. Directed synthesis of alkaline aluminosilicate minerals in a geocement matrix. J. Mater. Sci. 42, 2944–2952.

Li, Q., Sun, Z., Tao, D., Xu, Y., Li, P., Cui, H., Zhai, J., 2013. Immobilization of simulated radionuclide 133Cs+ by fly ash-based geopolymer. J. Hazard Mater. 262, 325–331.

Lin, Y., Liao, Y., Yu, Z., Fang, S., Ma, X., 2017. A study on co-pyrolysis of bagasse and sewage sludge using TG-FTIR and Py-GC/MS. Energy Convers. Manag. 151, 190–198.

Liu, X., Jiang, J., Zhang, H., Li, M., Wu, Y., Guo, L., Wang, W., Duan, P., Zhang, W., Zhang, Z., 2020a. Thermal stability and microstructure of metakaolin-based geopolymer blended with rice husk ash. Appl. Clay Sci. 196, 105769.

Liu, Y., Zhuge, Y., Chen, X., Duan, W., Fan, R., Outhred, L., Wang, L., 2023. Micro-chemomechanical properties of red mud binder and its effect on concrete. Compos. B Eng. 258.

Liu, Y., Zhuge, Y., Chow, C.W.K., Keegan, A., Li, D., Pham, P.N., Huang, J., Siddique, R., 2020b. Properties and microstructure of concrete blocks incorporating drinking water treatment sludge exposed to early-age carbonation curing. J. Clean. Prod. 261, 121257.

Liu, Y., Zhuge, Y., Chow, C.W.K., Keegan, A., Li, D., Pham, P.N., Yao, Y., Kitipornchai, S., Siddique, R., 2022a. Effect of alum sludge ash on the high-temperature resistance of mortar. Resour. Conserv. Recycl. 176.

Liu, Y., Zhuge, Y., Chow, C.W.K., Keegan, A., Pham, P.N., Li, D., Oh, J.-A., Siddique, R., 2021. The potential use of drinking water sludge ash as supplementary cementitious material in the manufacture of concrete blocks. Resour. Conserv. Recycl. 168.

Liu, Y., Zhuge, Y., Duan, W., Huang, G., Yao, Y., 2022b. Modification of microstructure and physical properties of cement-based mortar made with limestone and alum sludge. J. Build. Eng. 58.

Lynn, C.J., Dhir, R.K., Ghataora, G.S., West, R.P., 2015. Sewage sludge ash characteristics and potential for use in concrete. Construct. Build. Mater. 98, 767–779.

Maiden, P., Hearn, M.T.,W., Boysen, R.I., Chier, P., Warnecke, 2015. Alum Sludge Re-use, Investigation (10OS-42) Prepared by GHD and Centre for Green Chemistry (Monash University) for The Smart Water Fund. Victoria. ACTEW Water & Seawater, Melbourne, Australia.

Mañosa, J., Cerezo-Piñas, M., Maldonado-Alameda, A., Formosa, J., Giro-Paloma, J., Rosell, J.R., Chimenos, J.M., 2021. Water treatment sludge as precursor in non-dehydroxylated kaolin-based alkali-activated cements. Appl. Clay Sci. 204, 106032.

Owaid, H.M., Hamid, R., Taha, M.R., 2014. Influence of thermally activated alum sludge ash on the engineering properties of multiple-blended binders concretes. Construct. Build. Mater. 61, 216–229.

Ozer, I., Soyer-Uzun, S., 2015. Relations between the structural characteristics and compressive strength in metakaolin based geopolymers with different molar Si/Al ratios. Ceram. Int. 41, 10192–10198.

Pacheco-Torgal, F., Castro-Gomes, J., Jalali, S., 2008. Alkali-activated binders: a review: Part 1. Historical background, terminology, reaction mechanisms and hydration products. Construct. Build. Mater. 22, 1305–1314.

Pritchard, D., Penney, N., McLaughlin, M., Rigby, H., Schwarz, K., 2010. Land application of sewage sludge (biosolids) in Australia: risks to the environment and food crops. Water Sci. Technol. 62, 48–57.

Provis, J.L., 2018. Alkali-activated materials. Cement Concr. Res. 114, 40–48.

Provis, J.L., Bernal, S.A., 2014. Geopolymers and related alkali-activated materials. Annu. Rev. Mater. Res. 44, 299–327.

Provis, J.L., Palomo, A., Shi, C., 2015. Advances in understanding alkali-activated materials. Cement Concr. Res. 78, 110–125.

Puligilla, S., Mondal, P., 2013. Role of slag in microstructural development and hardening of fly ash-slag geopolymer. Cement Concr. Res. 43, 70–80.

Rovnaník, P., 2010. Effect of curing temperature on the development of hard structure of metakaolin-based geopolymer. Construct. Build. Mater. 24, 1176–1183.

Ruengsillapanun, K., Udtaranakron, T., Pulngern, T., Tangchirapat, W., Jaturapitakkul, C., 2021. Mechanical properties, shrinkage, and heat evolution of alkali activated fly ash concrete. Construct. Build. Mater. 299, 123954.

Rulkens, W., 2008. Sewage sludge as a biomass resource for the production of energy: overview and assessment of the various options. Energy & Fuels 22, 9–15.

Shi, C., Jiménez, A.F., Palomo, A., 2011. New cements for the 21st century: the pursuit of an alternative to Portland cement. Cement Concr. Res. 41, 750–763.

Suksiripattanapong, C., Horpibulsuk, S., Chanprasert, P., Sukmak, P., Arulrajah, A., 2015. Compressive strength development in fly ash geopolymer masonry units manufactured from water treatment sludge. Construct. Build. Mater. 82, 20–30.

Sun, B., Ye, G., de Schutter, G., 2022a. A review: reaction mechanism and strength of slag and fly ash-based alkali-activated materials. Construct. Build. Mater. 326, 126843.

Sun, K., Ali, H.A., Ji, W., Ban, J., Poon, C.S., 2023. Utilization of contaminated air pollution control residues generated from sewage sludge incinerator for the preparation of alkali-activated materials. Resour. Conserv. Recycl. 188, 106665.

Sun, K., Ali, H.A., Xuan, D., Ban, J., Poon, C.S., 2022b. Utilization of APC residues from sewage sludge incineration process as activator of alkali-activated slag/glass powder material. Cement Concr. Compos. 133, 104680.

Sun, S., Lin, J., Zhang, P., Fang, L., Ma, R., Quan, Z., Song, X., 2018. Geopolymer synthetized from sludge residue pretreated by the wet alkalinizing method: compressive strength and immobilization efficiency of heavy metal. Construct. Build. Mater. 170, 619–626.

Sun, Z., Vollpracht, A., 2018. Isothermal calorimetry and in-situ XRD study of the NaOH activated fly ash, metakaolin and slag. Cement Concr. Res. 103, 110–122.

Tashima, M.M., Akasaki, J.L., Melges, J.L.P., Soriano, L., Monzó, J., Payá, J., Borrachero, M.V., 2013. Alkali activated materials based on fluid catalytic cracking catalyst residue (FCC): influence of SiO_2/Na_2O and H_2O/FCC ratio on mechanical strength and microstructure. Fuel 108, 833–839.

Walkley, B., San Nicolas, R., Sani, M.-A., Rees, G.J., Hanna, J.V., van Deventer, J.S.J., Provis, J.L., 2016. Phase evolution of C-(N)-A-S-H/N-A-S-H gel blends investigated via alkali-activation of synthetic calcium aluminosilicate precursors. Cement Concr. Res. 89, 120–135.

Wan, Q., Rao, F., Song, S., García, R.E., Estrella, R.M., Patiño, C.L., Zhang, Y., 2017. Geopolymerization reaction, microstructure and simulation of metakaolin-based geopolymers at extended Si/Al ratios. Cement Concr. Compos. 79, 45–52.

Wang, Q., Wei, W., Gong, Y., Yu, Q., Li, Q., Sun, J., Yuan, Z., 2017. Technologies for reducing sludge production in wastewater treatment plants: state of the art. Sci. Total Environ. 587–588, 510–521.

Wen, N., Zhao, Y., Yu, Z., Liu, M., 2019. A sludge and modified rice husk ash-based geopolymer: synthesis and characterization analysis. J. Clean. Prod. 226, 805–814.

Yaseri, S., Hajiaghaei, G., Mohammadi, F., Mahdikhani, M., Farokhzad, R., 2017. The role of synthesis parameters on the workability, setting and strength properties of binary binder based geopolymer paste. Construct. Build. Mater. 157, 534–545.

Yip, C.K., Lukey, G.C., Provis, J.L., van Deventer, J.S.J., 2008. Effect of calcium silicate sources on geopolymerisation. Cement Concr. Res. 38, 554–564.

Zhang, J., Zhang, J., Tian, Y., Li, N., Kong, L., Sun, L., Yu, M., Zuo, W., 2016. Changes of physicochemical properties of sewage sludge during ozonation treatment: correlation to sludge dewaterability. Chem. Eng. J. 301, 238–248.

Zhang, P., Gao, Z., Wang, J., Guo, J., Hu, S., Ling, Y., 2020. Properties of fresh and hardened fly ash/slag based geopolymer concrete: a review. J. Clean. Prod. 270, 122389.

Zhang, X., Yan, S., Tyagi, R.D., Surampalli, R.Y., 2013. Energy balance and greenhouse gas emissions of biodiesel production from oil derived from wastewater and wastewater sludge. Renew. Energy 55, 392–403.

Zhao, Q., Ma, C., Huang, B., Lu, X., 2023. Development of alkali activated cementitious material from sewage sludge ash: two-part and one-part geopolymer. J. Clean. Prod. 384, 135547.

Zhou, Y.-F., Li, J.-S., Lu, J.-X., Cheeseman, C., Poon, C.S., 2020. Recycling incinerated sewage sludge ash (ISSA) as a cementitious binder by lime activation. J. Clean. Prod. 244.

Zhuge, Y., Fan, W., Duan, W., Liu, Y., 2021. The durability and rehabilitation technologies of concrete sewerage pipes: a state-of-the-art review. J. Asian Concr. Fed. 7, 1–16.

Zhuge, Y., Liu, Y., Pham, P.N., 2022. Sustainable utilization of drinking water sludge. Low Carbon Stabilization and Solidification of Hazardous Wastes. Elsevier, pp. 303–320.

Chapter 23

Recycling of incineration bottom ash into soil stabilization

Xinlei Sun and Yaolin Yi
School of Civil and Environmental Engineering, Nanyang Technological University, Singapore, Singapore

1. Introduction

The growing global economy and rapid urbanization have led to a substantial increase in municipal solid waste (MSW) generation (Kaza et al., 2018). Incineration has emerged as a critical strategy for MSW management in land-scarce countries and regions. It plays a vital role in substantially reducing the mass and volume of waste by 70%–90% (Lam et al., 2010; Dou et al., 2017). However, this process produces a significant by-product known as incineration bottom ash (IBA) (Wiles, 1996). IBA exhibits characteristics of an aggregate-like material with a diverse particle size distribution ranging from fine particles (<2 mm) to coarse particles (>2 mm) (Zekkos et al., 2013; Loginova et al., 2019). To date, the utilization of coarse or total IBA as construction materials in civil engineering has seen significant interest and exploration. Notably, researchers have investigated its application as aggregates for subbase/base in road construction (Forteza et al., 2004; Xie et al., 2016; Lynn et al., 2017; Schafer et al., 2019a). Additionally, its potential as lightweight aggregates for concrete production has been explored (Zhang and Zhao, 2014; Schafer et al., 2019b). Furthermore, IBA has been considered as a viable option for filling materials in land reclamation projects (Chan et al., 2018; Yin et al., 2018a, 2019). However, due to the leaching of heavy metals, landfilling is still the main disposal method for IBA in many countries (Luo et al., 2019). For instance, all IBA is landfilled in Singapore and Switzerland, 90% of IBA is landfilled in the United States, and the percentages of landfilled IBA are 80% in Italy and 48% in Norway (Bai and Sutanto, 2002; Kahle et al., 2015; He et al., 2017). In particular, the fine fraction is usually separated and directed to landfills (Bosmans et al., 2013; Bourtsalas et al., 2015; Simon and Holm 2016). This is because fine IBA possesses a higher concentration of heavy metals and lower strength compared with coarse fraction, making it less suitable for aforementioned applications (Tang et al., 2017; Zhu et al., 2019). Consequently, finding alternative ways to recycle and utilize fine IBA is significant to address the challenges associated with its disposal and environmental impact. One potential application is to recycle fine IBA into soil stabilization practices.

Soil stabilization encompasses various methods to improve the engineering properties of soil, enhancing its strength, durability, and overall performance (Sherwood, 1993; Afrin, 2017). Mechanical and chemical techniques are commonly employed methods for soil stabilization. The mechanical technique of soil stabilization is based on decreasing the void rate through compaction or physical alteration of the gradation involving the adjustment of the particle size composition of soil (Makusa, 2012). Chemical stabilization improves soil properties with chemical additives (Sherwood, 1993). By introducing agents such as lime, cement, and industrial byproducts/wastes, the characteristics of soil are enhanced, resulting in increased load-bearing capacity, reduced compressibility, improved shear strength, and overall stability. Fine IBA may have potentials to be used in soil stabilization because it has aggregate-like properties similar to sand, and contains lime and cementitious constituents (Sun and Yi, 2020, 2022b). It may be used as a substitute for sand and as a source of lime. By incorporating fine IBA into soil stabilization techniques, it is possible to create a closed-loop economy, conserving natural aggregates, and reducing the environmental burden of waste disposal. This approach offers promising benefits, such as enhancing soil mechanical properties, reducing heavy metal leaching, and providing a cost-effective alternative for construction materials in various projects, including road pavements, embankments, and building foundations. Despite the challenges in fine IBA recycling, the concept holds immense potential in transforming waste into a valuable resource and

Treatment and Utilization of Combustion and Incineration Residues. https://doi.org/10.1016/B978-0-443-21536-0.00033-2

advancing sustainability in the construction industry. Therefore, this chapter focuses on recycling fine IBA into soil stabilization.

2. Characterization of IBA

2.1 Particle size distribution

The particle size distribution curves of IBA from existing literature (Berg and Neal, 1998; Izquierdo et al., 2001; Bendz et al., 2007; Hjelmar et al., 2007; Travar et al., 2009; Zekkos et al., 2013; Tang et al., 2015; Weng et al., 2015) are plotted in Fig. 23.1. IBA is an aggregate-like material with a wide range of particle sizes from fine (<2 mm) to coarse (>2 mm), encompassing the sizes of silt, sand and gravel. It is noted that the fine fraction accounts for 30%−60% of the mass of IBA. The effective size (D_{10}) of IBA is in the range of 0.085−0.35 mm. The coefficient of uniformity (C_u) and coefficient of curvature (C_c) are 17.65−25.71 and 0.31−1.03, respectively. As a reference, soils with $C_c = 1-3$ and $C_u > 4$ for gravels and $C_u > 6$ for sands are well-graded (Holtz et al., 1981), otherwise are gap-graded. Hence, IBA can be considered well-graded, making it a viable substitute for natural aggregates.

2.2 Chemical composition

The oxide composition of IBA with different particle sizes from available literature, measured by X-ray fluorescence (XRF), is plotted in Table 23.1. It can be pointed out that the main oxides in the IBA are CaO, SiO_2, Al_2O_3, and Fe_2O_3. The contents of these main oxides are comparable with certain recognized pozzolanic latent hydraulic cementitious materials such as ground granular blast furnace slag (GGBS) (Sun and Yi, 2022a). The pozzolanic potential of IBA was also reported by some researchers (Lin et al., 2008; Lynn et al., 2017). The XRF results show that IBA with the coarse particle size tends to have more SiO_2, whereas finer IBA has more CaO. Moreover, the content of SO_3 in IBA is also high and increases as the particle size decreases.

2.3 Total and leaching concentrations of heavy metals

The source of heavy metals in IBA can be traced back to MSW. Some components of MSW contain heavy metals, such as batteries, pigments, and preservatives (Chandler et al., 1997). After incineration, these heavy metals are concentrated into IBA. Among these heavy metals, lead (Pb), zinc (Zn), chromium (Cr), cadmium (Cd), nickel (Ni), and copper (Cu) are

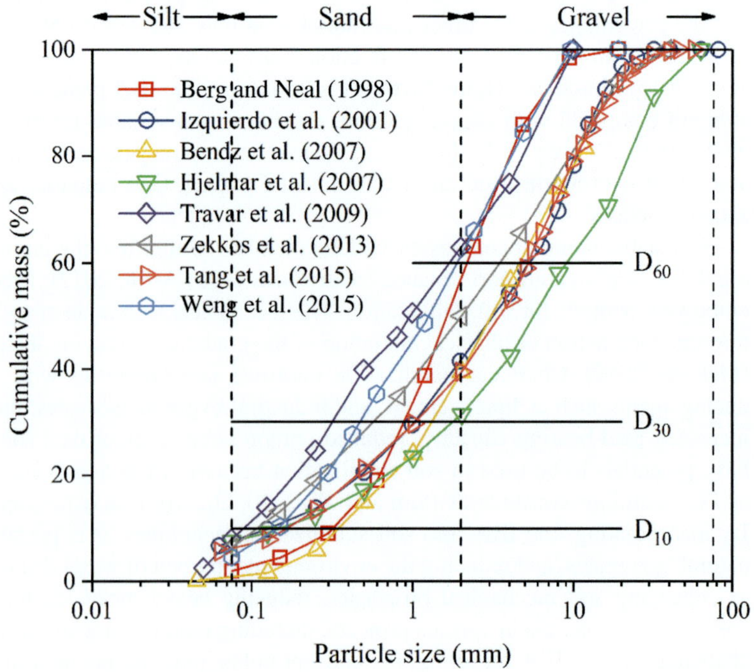

FIGURE 23.1 Particle size distribution curves of IBA.

TABLE 23.1 Chemical composition of IBA with different particle sizes (% of mass).

References	Particle sizes (mm)	SiO_2	CaO	Al_2O_3	Fe_2O_3	SO_3	P_2O_5	Na_2O	MgO	ZnO	CuO	PbO
Kirby and Rimstidt (1993)	<0.841	33.4	9.1	13.4	8.7	3.5	1.1	4.2	1.5	1.3	0.22	0.41
Speiser et al. (2000)	<0.2	28.3	37.4	8.2	7.5	–	–	1.1	3.6	–	–	–
	0.2–1	44.1	22.1	9.8	12.4	–	–	1.6	2.8	–	–	–
	1–4	42.4	18.8	10.0	16.2	–	–	2.5	2.7	–	–	–
	4–20	46.1	13.0	13.3	16.1	–	–	4.1	2.1	–	–	–
	>20	40	15.0	12.7	19.3	–	–	2.3	2.3	–	–	–
Ecke and Åberg (2006)	<4	37	15	13	15	–	1	2.8	2.5	–	–	–
Liu et al. (2008)	<1	27.6	30.6	8.5	3.6	–	3.1	2.3	4.1	–	–	–
	1–4.5	33.8	27.6	10.6	5.6	–	2.9	3.5	3.8	–	–	–
	4.5–30	35.2	25.6	12.3	8.6	–	2.6	3.6	3.1	–	–	–
Qiao et al. (2008)	14–40	51.8	14.2	11.9	6.0	–	1.1	1.3	3.3	0.2	–	–
Etoh et al. (2009)	<13	31.1	30.0	12.7	5.7	–	2.6	3.8	3.1	–	–	–
	<2	24.8	35.5	12.2	4.3	–	2.3	4	3.1	–	–	–
Schabbach et al. (2011)	0–2	30.3	23.1	13.0	10.0	–	2.0	1.9	2.8	0.7	0.7	0.4
	2–8	47.4	18.8	11.1	4.4	–	1.3	4.5	3	0.3	0.3	0.5
Cheng (2012)	<4.75	50.3	15.3	16.4	7.7	1.8	–	1.3	–	–	–	–
Chiang et al. (2012)	<2	28.6	40.9	8.3	8.9	–	1.1	2	3	0.7	0.4	–
Vichaphund et al. (2012)	<1[a]	27.7	41.5	4.6	6.2	4.4	4.2	3.3	1.6	0.5	–	0.1
	<1[a]	29.9	37.5	7.4	4.6	1.7	5.2	4.4	1.2	0.3	–	0.2
Bourtsalas et al. (2015)	<4	30.4	26.8	11.5	13.1	3.1	2.2	3.3	2.6	1	0.4	–
Tang et al. (2015)	<2[a]	54.2	13.5	7.9	13.8	1.3	0.8	2.8	1.8	0.6	0.4	0.3
	<2[a]	51.4	14.0	7.9	15.1	1.2	1.1	3.1	1.8	0.6	0.4	0.3
Zhang et al. (2015)	<0.105	22.0	42.9	6.0	5.2	5.3	3.7	1.8	4	–	–	–
Chen and Yang (2017)	0.075	6.9	46.2	4.7	2.8	13.6	2.1	1.6	1.9	0.5	0.1	0.1
	0.075–0.3	17	35.3	6.7	5	8.5	4.1	2.8	1.9	0.5	0.2	0.1
	0.3–1.18	34.8	24.4	6.7	9.1	5	4.3	3.1	1.5	0.5	0.3	0.1
	1.18–6.3	51.5	15.9	7	13.4	1.8	3.3	5.4	1.6	0.4	0.2	0.2
	>6.3	57.6	10.3	5.9	6.4	0.8	0.7	8.6	1.5	0.2	0.1	0.1

Continued

TABLE 23.1 Chemical composition of IBA with different particle sizes (% of mass).—cont'd

References	Particle sizes (mm)	SiO_2	CaO	Al_2O_3	Fe_2O_3	SO_3	P_2O_5	Na_2O	MgO	ZnO	CuO	PbO
Luo et al. (2017)	<0.3	8.2	54.1	6.8	4.6	7.9	4.8	2.6	2.4	0.8	0.3	0.1
Alam et al. (2019)	<0.125	19.6	25	12.6	6.3	3.7	1.6	–	–	–	–	–
Sun and Yi (2022a)	<1.18	26.7	36.3	6.9	10.2	7.2	–	–	1.5	0.7	0.4	0.1
Sun and Yi (2022b)	<1.18	25.7	34.6	5.7	10.2	12.9	–	1.3	–	0.7	0.4	0.1
Huang et al. (2023)	<20	56.7	9.7	13.8	6.1	0.4	–	1.9	2.9	–	–	–

[a]*Stand for two different types of IBA.*

most easily found in IBA, and Zn and Cu usually exist in the largest amount (Yin et al., 2018b). The concentrations of these heavy metals versus the mean particle sizes of IBA are illustrated in Fig. 23.2. It can be found that finer size particles of IBA tend to have higher concentrations of these heavy metals (Speiser et al., 2000; Chimenos et al., 2003; Baciocchi et al., 2010; Lin et al., 2015; Loginova et al., 2019).

The leachability of heavy metals is the key environmental concern before IBA reuse. Some studies reported that the leaching concentrations of Pb, Zn, Cr, Ni, and Cu increased as the particle size of IBA reduced (Loginova et al., 2019; Luo et al., 2019). As a matter of fact, the leaching of heavy metals is often influenced by many factors, such as pH, liquid to solid (L/S) ratio, organic matters, and so on (Luo et al., 2019). Among them, pH is a significant factor for controlling the leaching concentrations of heavy metals. The fresh IBA usually possesses pH in the alkaline range of 11−12, and finer IBA tends to have even higher pH values (Chimenos et al., 2003). This can be attributed to the presence of quicklime and the formation of hydrated lime, with finer IBA tending to have a higher lime content (You et al., 2006).

To illustrate the effect of pH on the leaching concentrations of heavy metals, the leaching concentrations versus pH for Pb, Zn, Cd, Cr, Cu, and Ni from available literature (Johnson et al., 1996; Polettini and Pomi, 2004; Baciocchi et al., 2010; Yin et al., 2017; Sun and Yi, 2021) have been plotted in Fig. 23.3, along with the solubility of corresponding metal hydroxides (Boardman et al., 2004). Some heavy metals may be in a large amount in IBA, but the leaching concentration is not high in some pH condition and vice versa. As shown in Figs. 23.3(a)−3(d), the concentrations of Pb, Zn, Cd, and Cr versus pH profiles follow the trends of the solubility curves of their hydroxides. Other than pH, the existence of carbonates and organic matter may also influence the leaching of heavy metals.

2.4 Mineralogical property

Some common minerals in IBA with different particle sizes, detected by X-ray diffraction (XRD), from available literature are listed in Table 23.2. Anhydrite, calcite, ettringite, and portlandite are more likely to be found in fine IBA. Calcite ($CaCO_3$) is usually found as a relic in the incinerator. During the incineration process, calcite is decomposed to form quicklime (CaO) due to high temperatures (>700°C). Portlandite ($Ca(OH)_2$) is formed due to the hydration of quicklime during the quenching process. Anhydrite is another phase widespread in IBA, and it can come from the oxidation of the SO_2 and S (Clozel-Leloup et al., 1999), and the decomposition of gypsum (Bayuseno and Schmahl, 2010). The alkaline environment created by lime and the presence of SO_4^{2-} ions are likely to foster the formation of ettringite, as long as there is Al available in the environment (Inkaew et al., 2016). Water-treated IBA experienced a notable increase in calcite contents. Simultaneously, some of the ettringite underwent transformations through dissolution and precipitation, resulting in gypsum formation. The remaining portion of ettringite reacted with chloride, leading to the creation of hydrocalumite ($Ca_2Al(OH)_6[Cl_{1-x}(OH)_x]\cdot 3(H_2O)$) (Bayuseno and Schmahl, 2010). As the particle sizes of IBA decreased, minerals such as ettringite, gypsum, lime, and portlandite tended to be found.

3. Soil stabilization methods

Soil stabilization plays a crucial role in enhancing the engineering properties of soil, including its strength, compressibility, and durability. The most used methods for soil stabilization involve both mechanical and chemical techniques. Understanding soil stabilization is essential when considering the recycling of IBA for soil stabilization purposes.

3.1 Mechanical stabilization

The mechanical technique of soil stabilization is based on decreasing the void rate through compaction or physical alteration of the gradation involving the adjustment of the particle size composition of soil (Makusa, 2012; Afrin, 2017; Archibong et al., 2020). Cohesionless soils such as sand, typically uniformly graded, are mixed with cohesive soil to facilitate easy compaction using conventional road rollers. Soil compaction involves increasing the soil density by reducing void spaces through the application of mechanical energy. This results in improved load-bearing capacity and reduced susceptibility to settlement, making it suitable for constructing foundations, roads, and embankments (Holtz et al., 1981; Evans et al., 2021). Compaction of shallow depths utilizes rollers, tampers, and other mechanical means. However, for deeper layers of soil, dynamic compaction is employed, which includes vibrocompaction, heavy weight compaction, and blast densification. Through a combination of soil blending and compaction, mechanical soil stabilization plays a pivotal role in optimizing the performance of the ground for construction projects.

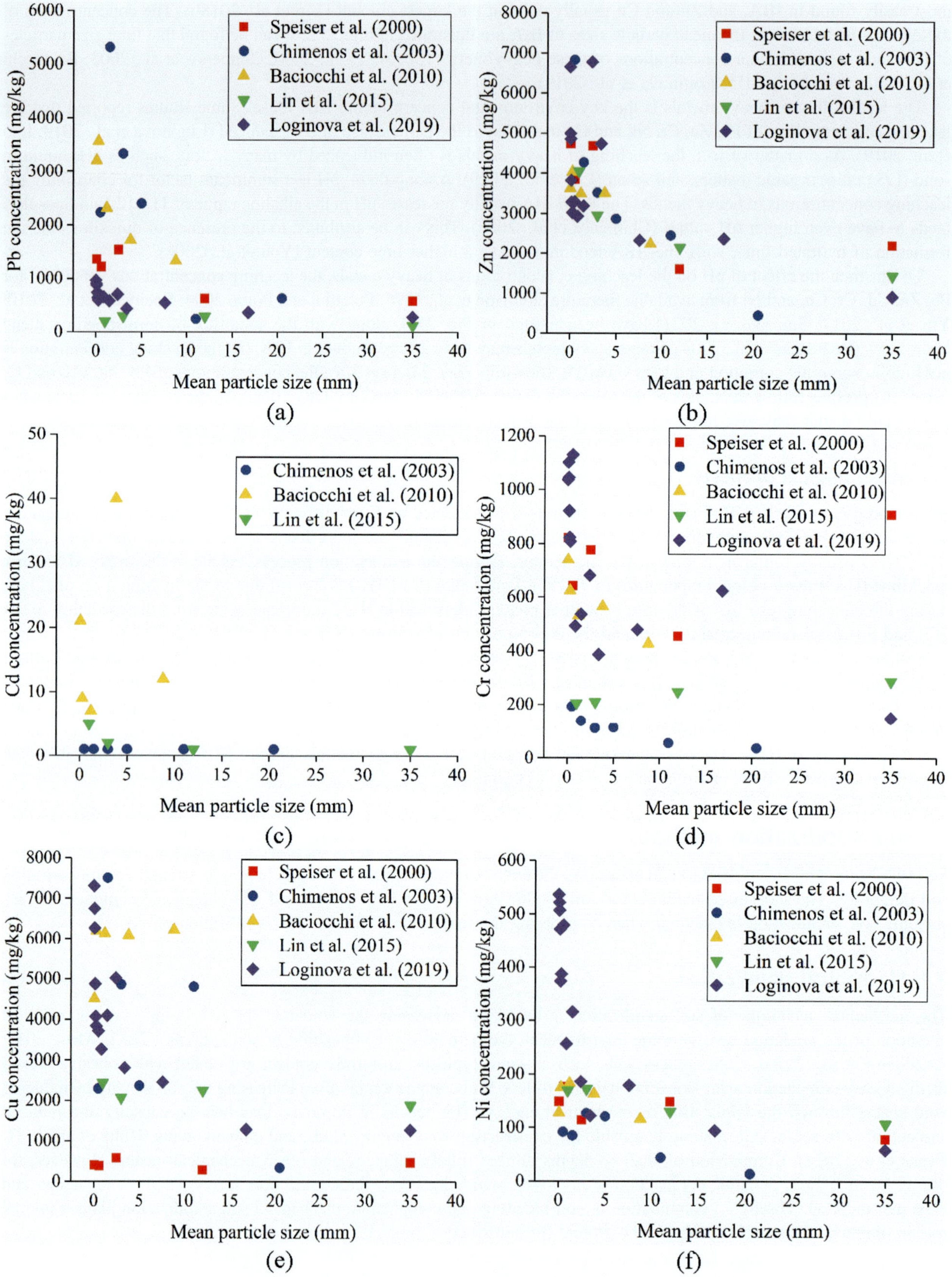

FIGURE 23.2 Concentrations of heavy metals in IBA with different particle sizes (A) Pb, (B) Zn, (C) Cr, (D) Cd, (E) Cu, and (F) Ni.

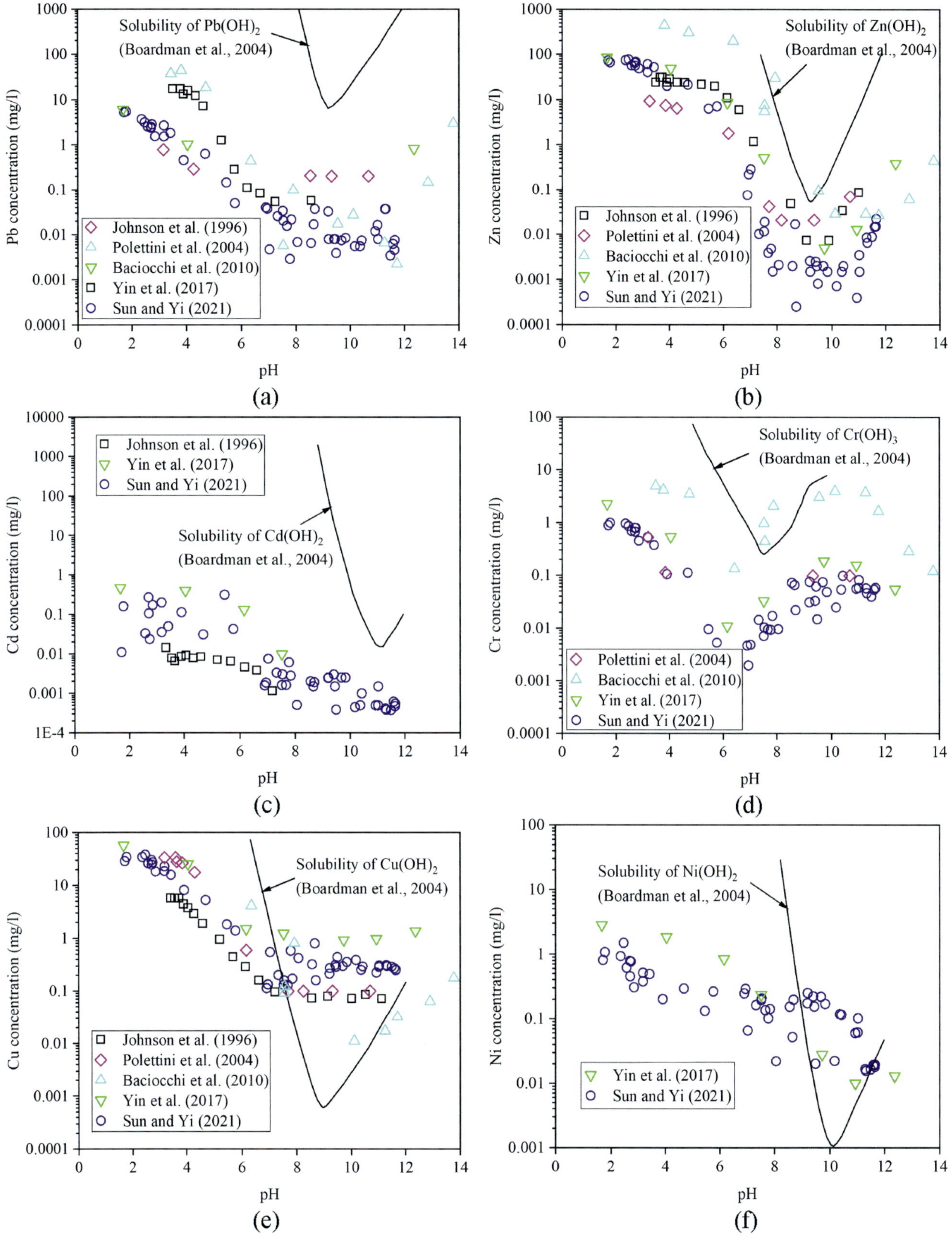

FIGURE 23.3 Concentrations of heavy metals and solubility of their hydroxides versus pH: (A) Pb, (B) Zn, (C) Cr, (D) Cd, (E) Cu, and (F) Ni.

TABLE 23.2 Common minerals found in XRD patterns for different sizes of IBA.

References	Particle sizes (mm)	Anhydrate	Calcite	Ettringite	Hematite	Gehlenite	Gypsum	Quicklime	Portlandite	Quartz
Kirby and Rimstidt (1993)	<0.841	–	√	–	√	–	√	–	–	√
Chimenos et al. (2003)	<1[a]	√	√	–	–	–	–	–	√	√
	<1[b]	–	–	√	–	√	–	–	–	–
Piantone et al. (2004)	<0.05	√	√	√	√	–	√	–	–	√
Vegas et al. (2008)	4–8	–	√	√	–	√	√	–	–	√
Etoh et al. (2009)	<13	–	√	√	√	√	√	√	√	√
Baciocchi et al. (2010)	<0.150	√	√	–	√	√	–	–	√	–
	0.150–0.425	√	√	–	√	√	–	–	√	–
	0.425–2	√	√	–	√	√	–	–	–	–
	2–5.6	√	√	–	√	√	–	–	–	–
	5.6–12	√	√	–	√	√	–	–	–	–
Lin et al. (2015)	<2	√	√	–	–	–	–	–	√	√
	2–4	√	√	–	–	–	–	–	√	√
	4–20	√	√	–	–	–	–	–	–	√
	20–50		√	–	–	–	–	–	–	√
Wang et al. (2016)	<0.212	–	√	–	–	√	–	–	–	√
	0.212–0.6	–	√	–	–	√	–	–	–	√
	0.6–1.0	–	√	–	–	√	–	–	–	√
	1.0–1.4	–	√	–	–	√	–	–	–	√
	1.4–1.7	–	√	–	–	√	–	–	–	√
Chen and Yang (2017)	0.075	√	√	√	–	–	–	–	√	√
	0.075–0.3	√	√	–	–	–	–	–	–	√
	0.3–1.18	–	√	–	–	–	–	–	–	√
	1.18–6.3	–	√	–	–	–	–	–	–	√
	>6.3	–	√	–	–	–	–	–	–	√
Alam et al. (2017)	<0.125	–	–	–	–	–	√	–	–	–
	0.125–1	–	–	–	–	–	–	–	–	–
	1–4	–	–	–	–	–	–	–	–	–

Loginova et al. (2019)	<0.18	√	√	–	–	–	–	–	–	√
	0.18–0.5	–	√	–	–	–	–	–	–	√
	0.5–1	–		–	–	–	–	–	–	√
	1–4	–	√	–	–	–	–	–	–	√
	4–22	–	–	–	–	–	–	–	–	–
	>22	–	–	–	–	–	–	–	–	√
Alam et al. (2019)	<0.125	–	√	√	–	–	√	–	–	√
Sun and Yi (2022a)	<1.18	√	√	√	–	√	√	√	√	√

$CaSO_4$, anhydrate; $CaCO_3$, calcite; $Ca_6Al_2(SO4)_3(OH)_{12}\cdot 26H_2O$, ettringite; $CaO\cdot Al_2O_3\cdot 2SiO_2$, feldspar; $Ca_2Al_2SiO_7$, gehlenite; $CaSO_4\cdot 2H_2O$, gypsum; NaCl, halite; Fe_2O_3, hematite; CaO, quicklime; $Ca(OH)_2$, portlandite; SiO_2, quartz.
[a] *Freshly quenched.*
[b] *165 d weathered.*

3.2 Chemical stabilization

Chemical stabilization improves soil properties for construction projects with chemical additives (Sherwood, 1993). By introducing agents such as lime, cement, industrial by-products/wastes, the soil's characteristics are enhanced, resulting in increased load-bearing capacity, reduced compressibility, improved shear strength, and overall stability. This technique is particularly versatile, as it can address various soil types and conditions, making it effective for stabilizing expansive clays, soft soils, and contaminated soil (Seco et al., 2011; Yi et al., 2014; Li et al., 2019). Chemical soil stabilization also offers the advantage of being time-efficient compared with mechanical methods, ensuring faster project completion and cost-effectiveness.

3.2.1 Lime

Lime stabilization is a widely adopted soil improvement technique that involves adding lime (quicklime or hydrated lime) to the soil to enhance its engineering properties. The process relies on two essential mechanisms: cation exchange and the pozzolanic reaction (Little and Nair, 2009). Through cation exchange, lime replaces exchangeable cations such as sodium with calcium ions, promoting better soil particle bonding and reducing the swelling potential of expansive clay soils (Nalbantoglu, 2006; Thyagaraj et al., 2012). Additionally, the pozzolanic reaction occurs when the lime reacts with the clay particles, causing the formation of calcium—silicate—hydrate (CSH) compounds and calcium—aluminate—hydrate (CAH) compounds. These newly formed compounds bind the soil particles together, leading to increased cohesion and improved strength of the soil (Boardman et al., 2001). Lime also acts as a drying agent, reducing the moisture content in the soil and decreasing its plasticity, enhancing soil workability and compaction, and thus simplifying construction operations (Sakr et al., 2009). Furthermore, lime stabilization enhances the durability and resistance of soil against weathering and erosion (Bell, 1996; Sakr et al., 2009).

3.2.2 Ordinary Portland cement

Cement stabilization is a widely employed soil improvement technique in the construction industry that enhances the engineering properties of soils (Fang et al., 1994; Horpibulsuk et al., 2003). This process involves incorporating cement, typically ordinary Portland cement (OPC), into the soil to create a cementitious matrix, significantly enhancing the strength and stability of soil. During cement hydration, a chemical reaction occurs when water mixes with cement, leading to the formation of CSH and CAH gels (Chrysochoou et al., 2010). These gels bind the soil particles together, resulting in a more robust and cohesive soil structure. Moreover, portlandite ($Ca(OH)_2$) is generated during hydration and participate in the pozzolanic reaction with clay particles, which further strengthens the soil—cement matrix and enhances the overall stability of the treated soil. Cement stabilization also reduces the permeability of soil, making it more resistant to the penetration of water and improving its overall durability.

Currently, OPC serves as the conventional binder for chemical stabilization, effectively enhancing the engineering properties of soft clay and contaminated soil (Chew et al., 2004; Lee et al., 2005; Zhang and Tao, 2008). However, the production of OPC comes at a significant environmental cost, as it emits a substantial amount of CO_2 (approximately 0.92 t/t OPC) and consumes a high level of energy (around 3300 MJ/t OPC) (Schneider et al., 2011). These environmental impacts mainly arise from the decomposition of limestone and the combustion of fossil fuels during production (Benhelal et al., 2021). Thus, there is an urgent need to explore and adopt alternative low-carbon binders.

3.2.3 Industrial byproducts/wastes

The use of industrial by-products and wastes in soil stabilization is an environmentally friendly approach gaining popularity in geotechnical engineering. By-products such as fly ash, GGBS, or silica fume, are incorporated into the soil to improve its engineering properties for construction projects (Sabat and Pati, 2014). When mixed with the soil, they react with its particles, creating stronger bonds and enhancing load-bearing capacity, shear strength, and overall stability. This sustainable practice not only reduces waste disposal but also offers a cost-effective and eco-friendly alternative to traditional stabilizing agents such as lime and cement. By utilizing industrial by-products, engineers can contribute to sustainable construction practices and promote a more efficient and environmentally conscious approach to infrastructure development. This chapter focuses on the introduction of GGBS.

GGBS is a by-product of the iron and steel industry, generated during the extraction of iron from iron ore in blast furnaces. The production process involves the formation of molten slag, a mixture of oxides and silicates, which is rapidly cooled and granulated to create fine granules. These granules are then further processed through grinding to produce a fine powder (GGBS) (Nidzam and Kinuthia, 2010). The production process of GGBS emits low CO_2 footprint (only 0.07 t/t

GGBS) and consumes low energy (1300 MJ/t GGBS) (Higgins, 2007; Cabeza et al., 2013). This eco-friendly characteristic makes GGBS an attractive alternative to traditional binders like OPC. However, the hydration process of GGBS is slow when used alone, necessitating the combination with alkaline activators. They help in breaking Si—O and Al—O bonds in GGBS and facilitate its hydration, which creates CSH and CAH binding the stabilized soil (Shi et al., 2006). Furthermore, the fine particles of GGBS fill void spaces, reducing permeability and improving resistance to water infiltration and erosion.

GGBS has demonstrated its efficacy in stabilizing soft clay and contaminated soils, as evident from various studies (Yi et al., 2015a,b,c; Salimi and Ghorbani, 2020; Li et al., 2022; Zhang et al., 2022). The findings from these studies reveal that GGBS-treated soil exhibits superior 28-day strength, surpassing that achieved with OPC-treated soil by up to two times. Moreover, GGBS-treated soil shows better resilience to the adverse effects of heavy metal contamination, making it a more sustainable and robust solution for soil stabilization. Some studies have employed alkaline chemicals such as, sodium hydroxide (NaOH), lime (CaO and $Ca(OH)_2$), and magnesia (MgO) to activate the hydration of GGBS; however, it has led to increased treatment costs. Recent advancements have shown promising results by utilizing certain industrial waste materials, such as ladle slag and carbide slag, as alkaline activators for GGBS (Xu and Yi, 2019; Li and Yi, 2020). These waste materials possess alkaline compositions, making them suitable alternatives to activate GGBS hydration effectively. Embracing the usage of alkaline industrial wastes as activators not only enhances the sustainability of the GGBS stabilization process but also promotes waste recycling and reduces the environmental burden.

4. Recycling IBA into soil stabilization

4.1 State-of-the-art of utilization of IBA and soil

For the past two decades, the utilization of IBA and waste soil as composite materials with or without binders has been studied on a laboratory scale. The parameters of related research have been summarized in Table 23.3, including soil type, water content, particle sizes and content of IBA, binder type and content, as well as the applications.

Marine clay (MC) and dredged sediments were the most common soils mixed with IBA. These soils usually have high natural water content (>50%), leading to their poor engineering performance such as low strength and high compressibility (Bo et al., 2015; Chu and Guo, 2016; Lam et al., 2018). The addition of IBA may improve the engineering performance of soft clay for different applications. The compressibility and permeability of the mixture of IBA and Ariake clay were found significantly reduced as IBA content increased up to 70% (Omine et al., 2002). Specifically, the coefficient of permeability (k) of Ariake clay was in the order of 10^{-8} cm/s; with the IBA content increasing to 70%, k increased to the order of 10^{-4} cm/s. Meanwhile, the compression index decreased from 0.8 to 0.22 as IBA content increased to 70%, indicating the compressibility decreased with the increase in IBA.

The addition of the IBA into soil changed the compaction characteristics of the soil. The optimal water content of IBA varies from 9% to 28.8% and maximum dry density varies from 1530 kg/m^3 to 2120 kg/m^3 (Sun, 2022). As the content of fine fraction increased, the optimal water content increased while the maximum dry density decreased. The optimal water content of the mixture of IBA and low plastic clay with IBA content of 10%—40% was around 13%—16%, and the maximum dry density was around 1600—1900 kg/m^3 (Vizcarra et al., 2014; Vaitkus et al., 2019).

The strength of compacted soil samples was also affected by the IBA addition. Vaitkus et al. (2019) found that the addition 15% IBA with particle size less than 2 mm significantly increased the CBR of low plastic clay by around 1.5 times. Melese (2022) reported that CBR increased with the addition of IBA, up to approximately 25%, after which it started to decrease. The compressive strength of the soil—IBA increased from ~400 to ~1000 kPa after curing from 7 to 28 days (Vaitkus et al., 2019). Similar trend was also observed in Wang et al. (2015a,b). The mixture of 20% cement, 60% IBA, and 20% dredged sediments can reach a compressive strength of ~8000 kPa at 28 days. Li et al. (2020) carried out triaxial shear tests under unconsolidated and undrained (UU) condition, and they found that soil—IBA had a higher deviatoric stress and a stiffer stress-strain curves than pure clay. Under a confining pressure of 600 kPa, the peak strength of soil—IBA with 75% IBA content exhibited a significant increase from approximately 900 kPa to around 1500 kPa when compared with the strength of pure clay.

The addition of IBA into soil may also affect the workability of IBA—soil mixture such as flowability. Yan et al. (2014) used cement, IBA, and dredged sediments to produce controlled low-strength material (CLSM). They designed CLSMs with hardened densities between 1525 and 1657 kg/m^3 and flowabilities of 210—250 mm. It is found that CLSMs manufactured with dredged sediment demanded a greater quantity of water to maintain their desired flowability, compared with those containing MSW bottom ash. This disparity can be attributed to the low water-absorbing capacity and nonplasticity

TABLE 23.3 Summary of research related to the utilization of IBA and soil with or without binders.

References	Soil type	Water content[a] (%)	IBA particle size (mm)	IBA content[b] (%)	Binder	Binder content[c] (%)	Application
Omine et al. (2002)	Ariake clay	160	0–26.5	60–100	–	–	Land reclamation
Naganathan et al. (2010)	Kaolin clay	–	0–10	80–100	Ordinary Portland cement (OPC)	9–25	Controlled low-strength material (CLSM)
Toraldo et al. (2013)	Gravel, sand	–	0–30	10–30	OPC, asphalt	3–5	Road construction
Vizcarra et al. (2014)	Clay	–	0–10	20–40	–	–	Road construction
Yan et al. (2014)	Dredged sediment	–	0–5	50–100	OPC	10–30	CLSM
Wang et al. (2015a)	Dredged sediment	68.9	–	50–75	OPC	20	Filling materials
Wang et al. (2015b)	Dredged sediment	50.8	2.36–5	21–75	OPC, fly ash, CaO	30	Filling materials
Guo et al. (2015)	Marine clay	110	–	30	Polymer-based cementitious additive	3	Land reclamation
Quek et al. (2016)	Marine clay	158	–	10–30	Polymer-based cementitious additive	4.6–6	Land reclamation
Vaitkus et al. (2019)	Clay	–	0–22	5–15	–	–	Road construction
Li et al. (2020)	Clay, fiber	23.6	0–20	50–100	–	–	Road construction
Melese (2022)	Expansive soil	–	–	10–30	–	–	Road construction
Sun and Yi (2022b)	Marine clay	50	0–1.18	10–100	–	–	Road construction
Sun and Yi (2022c)	Marine clay	50	0–1.18	30–70	OPC, ground granular blast furnace slag	2.5–7.5	Construction materials

[a] *Water content is defined as the mass of water over the dry mass of soil.*
[b] *IBA content is defined as the mass of IBA over the dry mass of IBA and soil.*
[c] *Binder content is defined as the mass of binder over the dry mass of IBA and soil.*

exhibited by IBA. The nonplastic property of IBA also reduced the liquid limit and plastic limit of the clay (Melese, 2022; Sun and Yi, 2022b).

Although the addition of IBA improved the mechanical properties of soft clay, the strength of the IBA and soil mixture was still low in the short term. Moreover, the leaching of heavy metals from IBA was also an environmental concern. Guo et al. (2015) and Quek et al. (2016) investigated the leaching properties of the mixture of IBA and MC. They conducted seawater percolation tests and found that heavy metals were not exchanged and mobile when IBA proportion was less than 30%. The mixture of IBA and MC could be considered as a nonhazardous material based on the toxicity characteristic leaching procedure (TCLP) tests. Some of the studies in Table 23.3 used binders to treat the mixtures of IBA and soil to further increase strength and lower the leaching of heavy metals. Among different binders, OPC is the main binder used to stabilize/solidify IBA and soil. With the use of binders, the leaching of heavy metals from IBA and contaminated sediments is also reduced. This is because the hydration products had a strong affinity and high capacity for metal sequestration (Wang et al., 2015a). The binder content used for soil stabilization depends on the applications. For reusing as controlled low-strength material and flowable filling material, the binder content can increase up to 30%. For road construction and land reclamation, binder content is typically less than 7.5%.

In summary, most studies used total IBA or coarse IBA (>2 mm) to mix with soil and concluded that IBA increased the mechanical properties of soft clay such as compressibility, permeability, and strength. Although leaching of heavy metals is a major concern of reusing IBA, it is not widely studied when coarse IBA was used with soil. This may be due to three reasons. Firstly, coarse IBA has a lower leaching potential than fine IBA. Especially, when used coarse IBA separately, the leaching of heavy metals may not be a serious problem. Secondly, soil possesses adsorption properties to bind various chemicals such as heavy metals. Clay minerals, such as kaolinite, illite, and montmorillonite, are one of the factors responsible for adsorption of heavy metals (Li et al., 2020). Heavy metal adsorbed by clay minerals results from pH-dependent inner-sphere surface coordination with edge hydroxyl groups and outer-sphere ion exchange with permanent negative surface sites (Wahba and Zaghloul, 2007). Therefore, it is valuable to reuse coarse or total IBA in soil stabilization. Thirdly, the permeability of soil—IBA is much lower than that of IBA, which slows down the percolation of leachates from IBA.

4.2 Potential of using fine IBA in soil stabilization

In most of the aforementioned studies related to IBA and soil, the total or coarse IBA was used as a substitution for natural coarse aggregates. The mechanical properties of the mixture of IBA and soil were improved with addition of coarse or total IBA. However, the fine fraction that takes up to 30%—60% of the total IBA was seldom used separately. Although fine IBA (<2 mm) has less favorable mechanical performance than coarse IBA, it also has potential to be used in soil stabilization, such as substitution for sand and source of lime.

Fine IBA may have the potential to substitute sand in the sand blending method. Compared with sand, the high porosity of fine IBA endows low density and high water absorption (Yao et al., 2023). Moreover, fine IBA has a greater frictional resistance because of the high angularity and rough surface texture (Gupta et al., 2021). Additionally, the permeability of fine IBA is comparable with the sand. Adding fine IBA to soft clay can increase the silt and sand fraction content in the mixture. The nonplasticity of fine IBA may reduce the plasticity of soft clay and increase the workability of the mixture. This substitution can alleviate the burden of sand depletion.

In addition to replacing sand, fine IBA may be also used in lime stabilization of soil. Fine IBA has a high pH and contains a high content of CaO, which suggests that it can be a source of lime (quicklime or hydrated lime) (Sun and Yi, 2020). Hence, IBA can replace lime to treat soil with high water content such as soft clay. The short-term workability of the soft soil may be improved after the reduction of water content. The compaction can then be conducted to the mixture of fine IBA and soil by using heavy machinery. In the long term, the hydration of IBA and the pozzolanic reaction between IBA and soil can further improve the strength of the mixture of fine IBA and soil. Moreover, since fine IBA contains alkaline minerals such as quicklime and portlandite, it may also be able to activate GGBS hydration. The hydration of GGBS can generate the cementitious products and further enhance the strength properties of the soil-IBA mixture. In conclusion, the use of fine IBA in soil stabilization provides several potential benefits, including the reduction of water content, improved compactability, and the potential for activating GGBS hydration.

However, the leaching of heavy metals of fine IBA was much higher than that of coarse fraction, which is a major environmental concern of reusing fine IBA. It needs to be pointed out that reusing fine IBA with soil already has a significant effect on reducing the leaching of heavy metals (Sun and Yi, 2022b). Nevertheless, to further reduce the heavy metals of fine IBA and enhance its feasibility in soil stabilization applications, some pretreatment for total IBA can also be

TABLE 23.4 Summary of common treatments for IBA.

Methods	Specific treatment	Advantages	Disadvantages	References
Washing	Water washing	Effective removal of soluble salts.	Insignificant removal of heavy metals. Wastewater to be treated.	Stegemann et al. (1997); Kim et al. (2003); Yang et al. (2012); Cossu and Lai (2012); Sun and Yi (2020)
	Acid washing	Effective removal of soluble salts and heavy metals.	Wastewater and acid residue to be treated.	Ito et al. (2006); Sun and Yi (2021)
Thermal	Vitrification	Effective removal of organic matters. Effective immobilization of heavy metals.	High energy and cost. Release of gas pollutants. Small-scale operation.	Ferraris et al. (2009); Schabbach et al. (2011); Andreola et al. (2019)
Stabilization/ solidification (S/S)	S/S by OPC	Effective immobilization of heavy metals. Strength improvement.	Energy intense and CO_2 emission. Less effective on removal of soluble salts and heavy metals.	Lynn et al. (2017); Sun and Yi (2022a)
	Carbonation	Effective immobilization of heavy metals. HN practical and cost-effective.	Less effective on removal of soluble salts and heavy metals.	Meima et al. (2002); Kim et al. (2003); Polettini and Pomi (2004); Van Gerven et al. (2005); Arickx et al. (2006, 2010); Rendek et al. (2006); Baciocchi et al. (2010); Santos et al. (2013); Lin et al. (2015); Um and Ahn (2017); Brück et al. (2018)

applied to fine IBA. Table 23.4 is the summary of common pretreatments for IBA, including washing, thermal treatment, and stabilization/solidification (S/S).

It is suggested that the pretreatment should be selected based on the application of fine IBA. For the usage of substituting sand, washing is an economic and easy handling way to deal with a large quantity of fine IBA. It can reduce not only the leaching but also the total amount of heavy metals in IBA. Nevertheless, there is still a large amount of wastewater to be treated. Thermal treatment also can be used to treat fine IBA. It can effectively remove the organic matters and immobilize the heavy metals in fine IBA. Moreover, calcium carbonates abundant in fine IBA can decompose to calcium oxide at high temperatures (>700°C), which increases the content of lime in treated IBA. Finally, stabilization/solidification (S/S) is also an effective way to treat fine IBA. S/S with chemical binders such as OPC can immobilize the heavy metals in fine IBA and increase the strength of soil−IBA mixture. Carbonation is also an effective way to deal with large quantities of fine IBA. It can not only turn fine IBA into an inert aggregate but also encourage the fixation of carbon dioxide, which is more environmentally friendly.

To date, the research on the recycling of fine IBA in soil stabilization is limited. Hence, the contents of Sun and Yi (2022b,c) are presented as a case study in this chapter to introduce the properties of excavated MC amended by fine IBA with and without binders.

4.3 Case study: excavated soft marine clay stabilized using fine IBA, with and without other binders (Sun and Yi, 2022b,c)

4.3.1 Materials and tests

MC, excavated from a construction site of the mass rapid transit (MRT) station in Singapore, had natural water content (w), liquid limit (w_L), and plastic limit (w_P) of 50%, 70%, and 29%, respectively. The IBA used in this study was collected from a local incineration plant and was fine, with a particle size of less than 1.18 mm. GGBS and OPC were obtained from a local company in Singapore. X-ray fluorescence (XRF) analysis was conducted to determine the chemical compositions of MC, IBA, GGBS, and OPC. Table 23.5 presents the XRF results, indicating that MC primarily consisted of SiO_2 and Al_2O_3, while IBA, GGBS, and OPC were mainly composed of CaO, SiO_2, and Al_2O_3.

TABLE 23.5 XRF results of MC, IBA, GGBS, and OPC (% of mass).

Oxide	CaO	SiO_2	SO_3	Fe_2O_3	Al_2O_3	TiO_2	MgO	Na_2O	K_2O	Others
MC	1.8	64.6	0.8	5.8	20.4	1.2	1.5	0.8	2.2	0.8
IBA	34.6	25.7	12.9	10.2	5.7	1.7	1.7	1.3	0.8	3.1
GGBS	37.0	32.7	4.7	–	15.3	1.1	8.1	0.4	0.5	0.2
OPC	61.3	21.3	3.1	2.3	6.8	1.0	3.0	0.4	0.4	0.4

TABLE 23.6 Mix design and test program for the first part of the case study.

	Mix proportion	
Mix type	IBA/MC	Tests
IBA-MC	0/10, 1/9, 2/8, 3/7, 4/6, 5/5, 6/4, and 10/0	Atterberg's limits, visual examination

The case study consisted of two main parts, presenting the properties of a mixture of MC and fine IBA, both with and without the incorporation of other binders. In the first part (Sections 4.2.2), the objective was to enhance the compactability of IBA-MC by utilizing IBA to reduce the high water content of MC. The mixing proportions of IBA-MC and the testing methods employed are detailed in Table 23.6. To begin, different IBA/MC ratios of 0/10, 1/9, 2/8, 3/7, 4/6, 5/5, 6/4, and 10/0 (based on dry mass) were employed, and the MC was thoroughly mixed with the fine IBA for a duration of 20 min. Afterward, a portion of the mixture was enclosed in plastic film and allowed to mature for one day for conducting Atterberg's limits tests (ASTM, 2010). Meanwhile, another portion of the mixture was compacted using a cylindrical mold measuring 25 mm in diameter and 50 mm in height. A mini compactor, as reported by Li et al. (2019), was used for this purpose, with a 2.5-kg hammer dropped from a height of 20 cm. Each specimen was prepared with three layers and compacted with one blow per layer, providing a compaction energy of 598.93 kJ/m^3, which closely resembled the ASTM standard proctor compaction test (ASTM, 2012).

The second part of the case study (Sections 4.2.3–4.2.5) involved stabilizing IBA-MC with different ratios of IBA/MC, by using GGBS and OPC. The aim was to examine the impact of GGBS and OPC on the strength and leaching of IBA-MC. Table 23.7 outlines the mixing proportions of IBA, MC, GGBS, or OPC. The study used three IBA/MC ratios, namely 3/7, 5/5, and 7/3 (based on dry mass), and incorporated four binder contents, expressed as the dry mass of binder relative to the dry mass of IBA-MC, set at 0%, 2.5%, 5%, and 7.5% for each IBA/MC ratio. The treated IBA-MC samples with GGBS and OPC were named GGBS-IBA-MC and OPC-IBA-MC, respectively. The specimen preparation procedures were similar to those described in the first part. The compacted specimens were then cured for 28 days in sealed bags. Unconfined compressive strength (UCS) tests (ASTM, 2017a), void ratio measurements, and batch leaching tests (EN 12457-1) (BSI, 2002) were conducted for both IBA-MC and stabilized IBA-MC to evaluate their properties.

TABLE 23.7 Mix design and test program for the second part of the case study.

	Mix proportion		
Mix type	IBA/MC	GGBS or OPC/(IBA + MC)	Tests
IBA-MC	3/7, 5/5, 7/3	0%	UCS, water content, void ratio, leaching of heavy metals
GGBS-IBA-MC		2.5%, 5%, 7.5%	
OPC-IBA-MC			

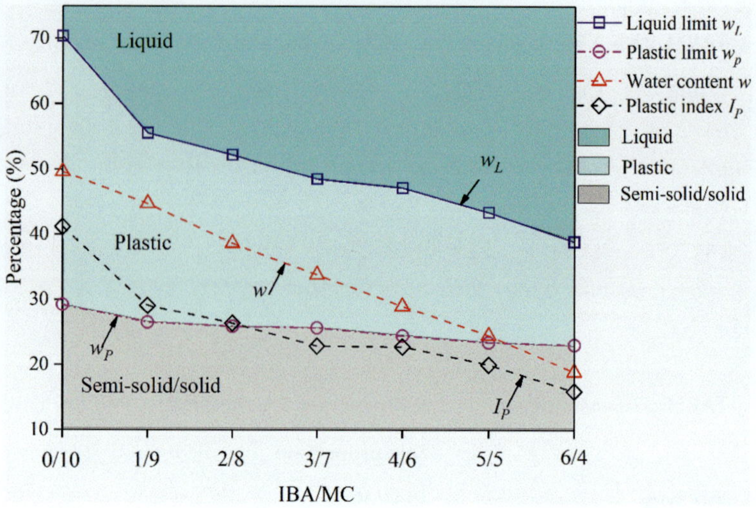

FIGURE 23.4 Consistency limits and water content of IBA-MC.

4.3.2 Atterberg limits, classification, and compactability of IBA-MC

The relationships between the liquid limit (w_L) and plastic limit (w_P) of IBA-MC with different IBA/MC ratios are illustrated in Fig. 23.4. It was not possible to measure the w_L and w_P for pure IBA (without MC) because IBA showed no plasticity. This finding aligns with previous studies conducted by Weng et al. (2010), Lin et al. (2012), and Alhassan and Tanko (2012). In the case of IBA-MC mixtures, the values of w_L and w_P demonstrated a decreasing trend as the proportion of IBA increased. However, this relationship was not linear. Precisely, with an increase in the IBA/MC ratio from 0/10 to 6/4, the w_L decreased from 70.4% to 38.9%, while the w_P decreased from 29.2% to 23.0%. Moreover, there was a notable drop in the plasticity index (I_P) from 41.2% to 15.9%. These variations indicate a shift in the plasticity of the mixtures from highly plastic to moderately plastic as the proportion of IBA increased.

The results of the Atterberg limits for IBA-MC mixtures are plotted on the Casagrande plasticity chart to classify them according to the Unified Soil Classification System (USCS) (ASTM, 2017b) as shown in Fig. 23.5. In the USCS, MC is categorized as high plastic clay (CH), while IBA falls under well-graded sand (SW) or poorly graded sand (SP). According to the USCS classification, IBA-MC mixtures with IBA/MC ratios of 1/9 and 2/8 are classified as high plasticity clay (CH). As the proportion of IBA increases, the data points on the chart shift from the area of high plasticity clay (CH) toward low plasticity clay (CL), gradually approaching the A-line.

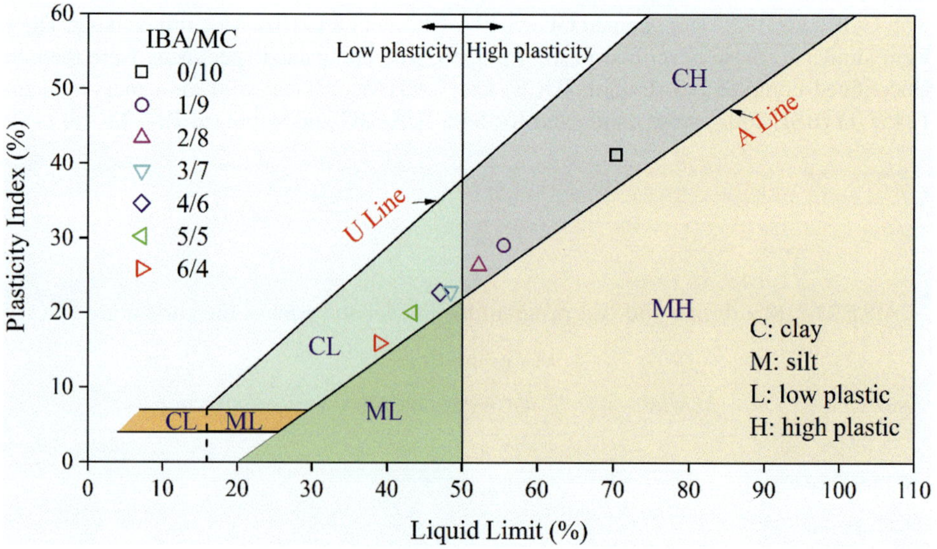

FIGURE 23.5 Plasticity index versus liquid limit for IBA-MC.

FIGURE 23.6 Images of IBA-MC with different IBA/MC ratios.

The visual observation of the variation in soil state for IBA-MC is demonstrated in Fig. 23.6. With an increase in the IBA/MC ratio, the paste-like mixture first lost its liquidity, then crumbled into pieces, and finally transformed into a granular state. To quantitatively describe this change in soil state, the liquidity index (I_L) was calculated and presented in Fig. 23.6. It is evident that as the IBA/MC ratio increased from 0/10 to 6/4, the liquidity index (I_L) dropped from 0.56 to -0.26. The negative value of the liquidity index indicates that the IBA-MC mixtures were in the semisolid or solid state. This means that the mixture lost its ability to flow and became more solid in nature as the proportion of IBA increased.

The changes in soil plasticity and state significantly influenced the compactability of IBA-MC. Mixtures with IBA/MC ratios of 0/10 and 1/9 were too soft to be effectively compacted, and the dry IBA alone could not bind together without the presence of MC. However, in Fig. 23.7, it is evident that as the IBA/MC ratios increased from 2/8 to 6/4, IBA-MC became much more compactable, exhibiting fewer defects on the specimen surface. Remarkably, the increase in IBA/MC ratios from 2/8 to 6/4 led to the successful compaction of IBA-MC into intact cylindrical specimens. Through compaction, the compressibility of IBA-MC can be effectively reduced, and its strength can be increased, making it more suitable for various engineering applications.

4.3.3 Water content and void ratio of IBA-MC and stabilized IBA-MC

The water contents of IBA-MC and stabilized IBA-MC, after 0 and 28 days of curing, are illustrated in Fig. 23.8. The variation between the 0-day and 28-day water content is attributed to the hydration of IBA, GGBS, and OPC, along with minimal evaporation. As the GGBS or OPC content increases, the difference between the 28-day and 0-day water content becomes larger, as more binders in the mixture consume additional water.

It is essential to note that the extent of this decrease varies between GGBS and OPC, owing to their distinct hydration behavior. For the IBA/MC ratio of 3/7, OPC-IBA-MC consumes more water compared with GGBS-IBA-MC, and this difference becomes more pronounced as the GGBS or OPC content increases. However, as the IBA/MC ratio is increased from 3/7 to 5/5, the difference between GGBS-IBA-MC and OPC-IBA-MC becomes smaller. Surprisingly, when the IBA/MC ratio reaches 7/3, the water content of GGBS-IBA-MC becomes smaller than that of OPC-IBA-MC. This result can be attributed to the limited hydration of GGBS in the presence of insufficient alkaline minerals as GGBS activators for IBA/

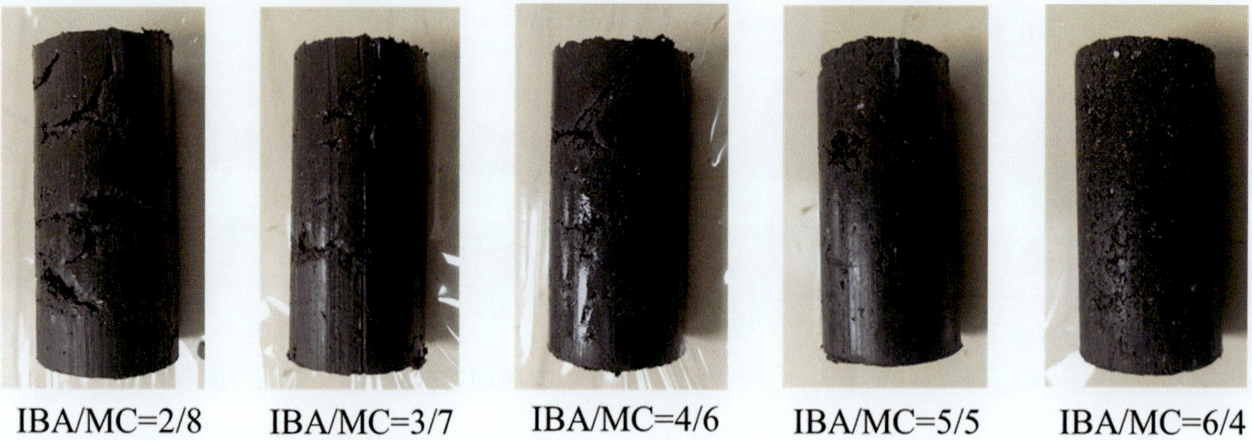

FIGURE 23.7 Compacted IBA-MC specimens with different IBA/MC ratios.

FIGURE 23.8 Water content of IBA-MC, GGBS-IBA-MC, and OPC-IBA-MC.

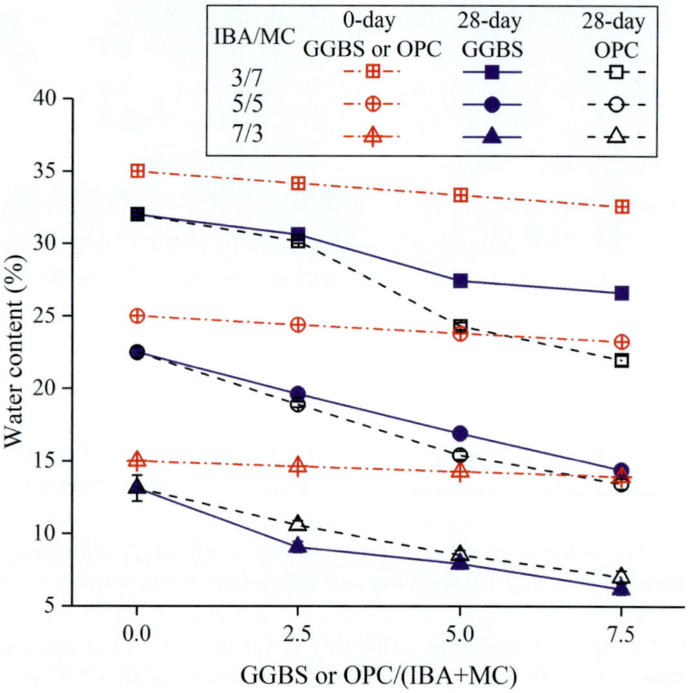

MC ratios of 5/5 and 3/7. However, when the IBA/MC ratio is increased to 7/3, the higher alkaline activator content from IBA efficiently activates the hydration of GGBS, leading to more water consumption.

The void ratio values of IBA-MC and stabilized IBA-MC before and after 28 days of curing can be observed in Fig. 23.9. Despite minimal variation in water content, the void ratio fluctuates due to the differing specific gravity of the samples. The void ratio of MC with natural water content is approximately 1.50, while for IBA-MC, the values ranged from 0.76 to 0.96. This indicates that the addition of IBA made IBA-MC more compactable and resulted in a lower void ratio.

After 28 days of curing, the void ratios of GGBS-IBA-MC and OPC-IBA-MC decreased compared with those of the 0-day samples after compaction. These values were lower than those of IBA-MC. This reduction can be attributed to the hydration process of GGBS and OPC during the curing period. As these materials underwent hydration, they consumed water and produced hydration products, filling the voids in the mixtures. The void ratio range for GGBS-IBA-MC (28 days) was 0.60−0.92, while for OPC-IBA-MC (28 days), it was 0.62−0.90, which decreased as GGBS or OPC content

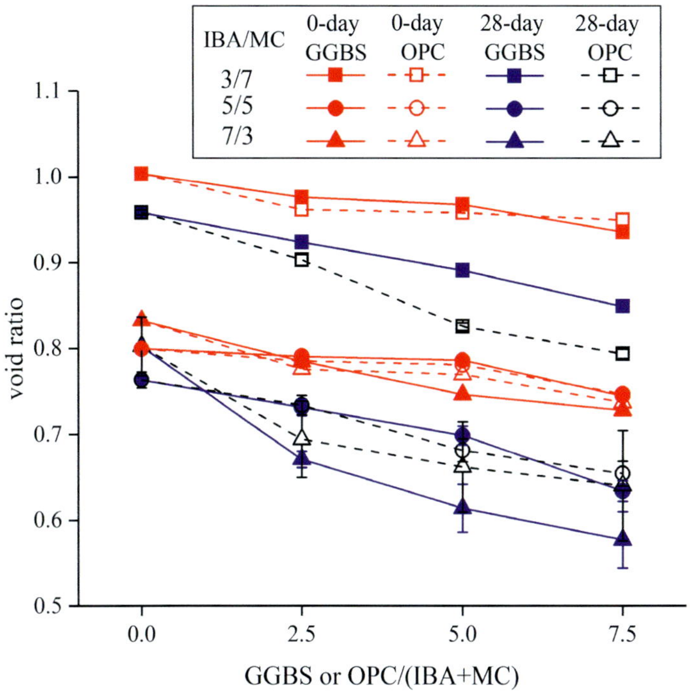

FIGURE 23.9 Void ratio of IBA-MC, GGBS-IBA-MC, and OPC-IBA-MC.

increased. At the IBA/MC ratio of 3/7, the void ratio values of GGBS-IBA-MC were slightly greater than those of OPC-IBA-MC. Conversely, with an IBA/MC ratio of 5/5, as the binder content increased from 0.0% to 5.0%, the void ratio values of GGBS-IBA-MC remained higher than those of OPC-IBA-MC. However, when the binder content was increased to 7.5%, the void ratio of GGBS-IBA-MC became lower compared with that of OPC-IBA-MC.

4.3.4 UCS of IBA-MC and stabilized IBA-MC

Fig. 23.10 displays the UCS values of IBA-MC and stabilized IBA-MC. The trend for IBA-MC reveals that the UCS value increases as the IBA/MC ratio goes from 3/7 to 5/5, but subsequently decreases as the ratio increases further to 7/3. The reason behind this behavior is closely related to the void ratio of IBA-MC. Notably, IBA-MC with an IBA/MC ratio of 5/5 exhibits the smallest void ratio among the three ratios (Fig. 23.9), which consequently results in the highest UCS.

Both GGBS-IBA-MC and OPC-IBA-MC exhibit significantly higher UCS values compared with IBA-MC. For instance, GGBS-IBA-MC with an IBA/MC ratio of 7/3 and GGBS content of 7.5% demonstrates an impressive UCS of 2805 kPa, which is more than 10 times greater than the UCS of IBA-MC (247 kPa). The substantial enhancement in strength can be attributed to the hydration of GGBS or OPC, leading to the formation of additional hydration products such as CSH (Chew et al., 2004; Yi et al., 2016). These hydration products effectively bond with IBA-MC, thereby increasing the overall strength of the mixture.

The UCS of GGBS-IBA-MC exhibited an increase as both the IBA/MC ratio and GGBS content increased, as shown in Fig. 23.10. For GGBS-IBA-MC with IBA/MC ratios of 3/7, 5/5, and 7/3, the UCS ranges were 252–617 kPa, 905–2492 kPa, and 1505–2805 kPa, respectively, when the GGBS content increased from 2.5% to 7.5%. Notably, GGBS-IBA-MC with an IBA/MC ratio of 7/3 displayed the highest UCS value. This outcome can be attributed to the increasing amount of alkaline minerals in GGBS-IBA-MC, which effectively activates the hydration of GGBS (Song and Jennings, 1999; Li et al., 2022; Sun et al., 2024). Consequently, higher strength development occurs in GGBS-IBA-MC due to this enhanced hydration process.

The UCS of OPC-IBA-MC increased as the OPC content increased, but the trend with IBA/MC ratio was different from that of GGBS-IBA-MC. For IBA/MC ratios of 3/7, 5/5, and 7/3, the UCS ranges were 661–1505 kPa, 995–2165 kPa, and 818–1030 kPa, respectively, as the OPC content increased from 2.5% to 7.5%. The UCS of OPC-IBA-MC increased as the IBA/MC ratio increased from 3/7 to 5/5. However, in contrast to GGBS-IBA-MC, as the IBA/MC ratio increased from 5/5 to 7/3, the UCS of OPC-IBA-MC decreased. This discrepancy can be explained by the fact that unlike GGBS-IBA-MC, the increasing amount of IBA in OPC-IBA-MC cannot facilitate the hydration of OPC.

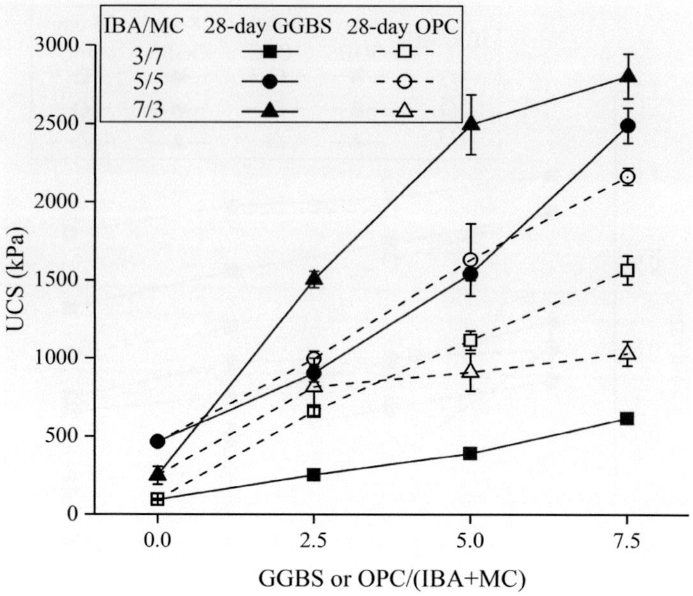

FIGURE 23.10 UCS of IBA-MC, GGBS-IBA-MC, and OPC-IBA-MC.

Instead, it increases the total content of heavy metals, which significantly and detrimentally affects the hydration of OPC (Trussell and Spence, 1994; Chen et al., 2007, 2009).

4.3.5 pH and leaching of IBA-MC and stabilized IBA-MC

Fig. 23.11 presents the pH of the leachate from IBA-MC and stabilized IBA-MC. The pH of IBA-MC increased with an increase in the IBA/MC ratio, primarily due to the rising amount of alkaline minerals, such as quicklime, portlandite, and magnesium (Alam et al., 2019; Sun and Yi, 2020), resulting from the higher IBA content. Consequently, the leachate pH was positively affected by the higher alkalinity in the mixture. For OPC-IBA-MC, the pH exhibited an increase with both the IBA/MC ratio and OPC content. This can be attributed to the hydration of OPC, which generated extra portlandite ($Ca(OH)_2$) (Taylor, 1997), consequently elevating the pH of OPC-IBA-MC. Notably, the pH of OPC-IBA-MC (ranging from 11.43 to 12.25) was higher than that of GGBS-IBA-MC (ranging from 11.27 to 11.56).

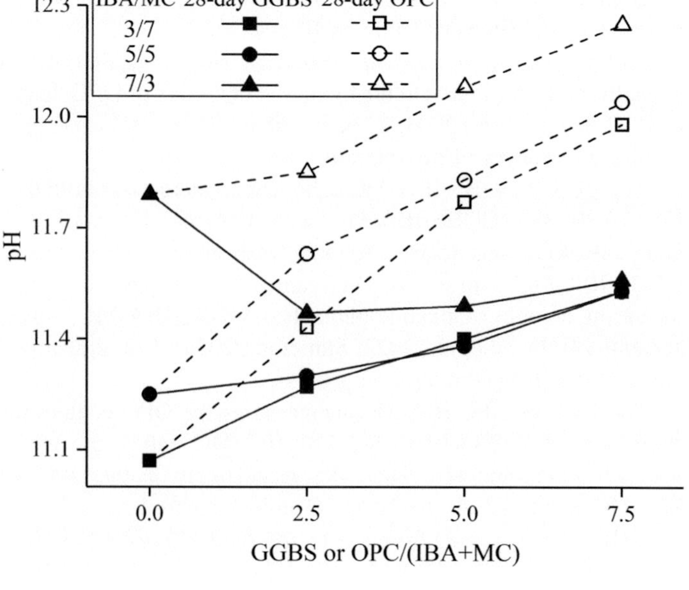

FIGURE 23.11 Leachate pH of IBA-MC, GGBS-IBA-MC, and OPC-IBA-MC.

The pH of GGBS-IBA-MC increased with higher IBA/MC ratios for both 3/7 and 5/5, surpassing the pH of IBA-MC. However, with an IBA/MC ratio of 7/3, the pH of GGBS-IBA-MC was lower compared to IBA-MC. This can be attributed to the consumption of alkaline minerals, like quicklime and portlandite, in IBA to activate the hydration of GGBS at the higher IBA/MC ratio of 7/3. Moreover, as the binder content increased, the pH difference between GGBS-IBA-MC samples with various IBA/MC ratios reduced. At a binder content of 7.5%, the final pH values of GGBS-IBA-MC were essentially the same.

Fig. 23.12 displays the leaching concentrations of heavy metals from IBA-MC and stabilized IBA-MC. The DNK-Category 3 represents the limits for leaching concentrations of heavy metals for Category 3 construction materials according to the Danish regulation (Statutory Order No. 1662/2010) (Hjelmar et al., 2016). The comparison with the DNK-Category 3 limits enables us to evaluate the effectiveness of the treatment methods in reducing the leaching of heavy metals from the IBA-MC mixtures.

In Fig. 23.12, the leaching profile of each heavy metal exhibits distinct characteristics, which can be attributed to two main reasons. Firstly, different heavy metals have varying optimal pH levels for achieving the lowest leaching concentration. For instance, the optimal pH values for Pb, Zn, Cr, and Cd are 9.2, 9.2, 7.5, and 11.2, respectively (Boardman et al., 2004). Secondly, the leaching of heavy metals, such as Cu and Ni, may also be influenced by complexation with dissolved organic carbon (DOC) (Hyks et al., 2009; Di Gianfilippo et al., 2018).

It is worth noting that the leaching concentrations of Cd, Cr, and Ni in fine IBA exceeded the limits of DNK-Category 3. However, after the addition of MC, the leaching concentrations of heavy metals from IBA-MC decreased and were brought below the regulated limits of DNK-Category 3, except for Ni. This improvement can be attributed to the negative-charged surface of MC, which enables it to adsorb heavy metals (Lim et al., 1997; Chalermyanont et al., 2009; George et al., 2014).

The leaching concentrations of heavy metals in IBA-MC were also significantly influenced by the addition of GGBS and OPC. In the case of OPC-IBA-MC, the leaching of Pb and Zn increased with an increase in OPC content, while the leaching of Cd and Ni decreased. On the other hand, for GGBS-IBA-MC, the leaching of heavy metals decreased as GGBS content increased and was lower than those observed in IBA-MC and OPC-IBA-MC. Notably, when the GGBS content was increased from 5% to 7.5%, the leaching concentrations of all studied heavy metals in GGBS-IBA-MC, with IBA/MC ratios of 5/5 and 7/3, satisfied the requirements for DNK-Category 3.

The difference in leaching concentrations of heavy metals between OPC-IBA-MC and GGBS-IBA-MC can be attributed to two main reasons. Firstly, the pH of GGBS-IBA-MC was lower than that of OPC, which was closer to the optimal pH range (9−11) for some of the studied heavy metals, such as 9.2 for Pb and Zn (Boardman et al., 2004). This lower pH in GGBS-IBA-MC resulted in a higher tendency for heavy metals to precipitate as metal hydroxides, reducing their leaching concentrations. Secondly, the main hydration product of both GGBS and OPC was calcium silicate hydrate (CSH). It has been reported that the hydration of GGBS consumed $Ca(OH)_2$ and tended to generate a lower Ca/Si ratio CSH (Wang and Scrivener, 1995), which could adsorb more cations through addition or substitution reactions (Shi and Spence, 2004). This leads to a higher capacity for heavy metal retention in GGBS-IBA-MC, further reducing their leaching concentrations compared to OPC-IBA-MC.

5. Conclusions and perspectives

This chapter illustrates an overview of recycling fine IBA into soil stabilization. IBA is a granular material containing both coarse (>2 mm) and fine fractions (<2 mm). The fine fraction that takes up to 30%−60% of the total IBA is usually landfilled. However, based on the analysis of the characterization of fine IBA, it is possible to utilize fine IBA in soil stabilization. For the usage of sand substitution, some pretreatments such as washing and carbonation are suggested for fine IBA. When used in chemical soil stabilization, fine IBA can be regarded as a source of lime. This means it can be used in lime stabilization or used as alkaline activator for GGBS.

The state-of-the-art of IBA for soil stabilization has also been systematically discussed. Most studies used total IBA or coarse IBA to mix with soil and concluded that IBA increased the mechanical properties of soft clay such as compressibility, permeability, and strength. However, the research on the recycling of fine IBA in soil stabilization is limited. Hence, a case study was used to elaborate on the properties of fine IBA and MC with and without binders. The combination of fine IBA and MC effectively addressed their individual limitations. MC demonstrated the ability to adsorb and retain heavy metals, leading to a reduction in the leaching of heavy metals from IBA. Additionally, the incorporation of dry fine IBA helped lower the water content of MC, thereby enhancing its compactability. Both GGBS-IBA-MC and OPC-IBA-MC exhibited significantly higher strength performance than IBA-MC. Specifically, for higher IBA/MC ratios, GGBS-IBA-MC displayed superior strength performance compared with OPC-IBA-MC. This outcome can be attributed to

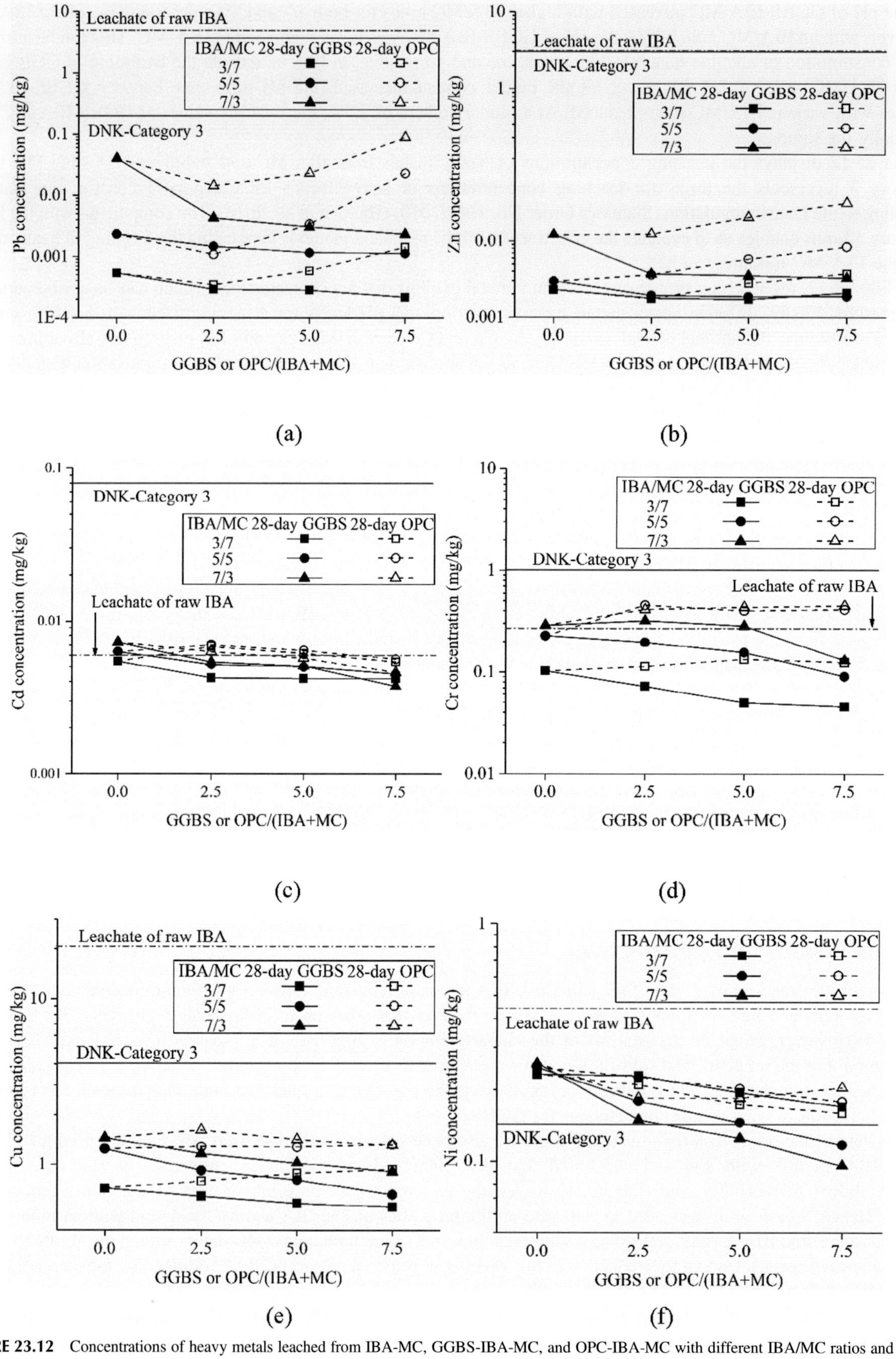

FIGURE 23.12 Concentrations of heavy metals leached from IBA-MC, GGBS-IBA-MC, and OPC-IBA-MC with different IBA/MC ratios and binder contents: (A) Pb, (B) Zn, (C) Cd, (D) Cr, (E) Cu, (F) Ni.

the adverse impact of heavy metals present in fine IBA on the hydration process of OPC. Conversely, IBA acted as an alkaline activator, facilitating the hydration of GGBS. Furthermore, the leaching concentrations of heavy metals in GGBS-IBA-MC were lower in comparison with those found in IBA-MC and OPC-IBA-MC. This indicates a more effective immobilization of heavy metals in the GGBS-IBA-MC mixture, highlighting its potential as a more environmentally friendly option for managing heavy metal leaching.

To date, research on stabilized mixtures of IBA and soils has been limited to air-cured conditions. In practical applications, these stabilized mixtures may come into contact with groundwater or even seawater. Surprisingly, the effect of water environments on the stability of the composite material of IBA and soil remains unexplored. To ensure the durability of such mixtures in real-world scenarios, it is crucial to investigate the effects of water environments on their stability. Therefore, further research in this area is essential to assess and optimize the performance of IBA and soil mixtures in different water environments.

Acknowledgments

This research is supported by the Ministry of Education, Singapore, under its Academic Research Fund Tier 1 (RG139/20).

References

Afrin, H., 2017. A review on different types soil stabilization techniques. Int. J. Transport. Eng. Technol. 3 (2), 19–24.

Alam, Q., Florea, M.V.A., Schollbach, K., Brouwers, H.J.H., 2017. A two-stage treatment for Municipal Solid Waste Incineration (MSWI) bottom ash to remove agglomerated fine particles and leachable contaminants. Waste Manag. 67, 181–192.

Alam, Q., Schollbach, K., Rijnders, M., van Hoek, C., van der Laan, S., Brouwers, H.J.H., 2019. The immobilization of potentially toxic elements due to incineration and weathering of bottom ash fines. J. Hazard Mater. 379, 120798.

Alhassan, H.M., Tanko, A.M., 2012. Characterization of solid waste incinerator bottom ash and the potential for its use. Int. J. Eng. Res. Afr. 2 (4), 516–522.

Andreola, F., Barbieri, L., Soares, B.Q., Karamanov, A., Schabbach, L.M., Bernardin, A.M., Pich, C.T., 2019. Toxicological analysis of ceramic building materials—tiles and glasses—obtained from post-treated bottom ashes. Waste Manag. 98, 50–57.

Archibong, G.A., Sunday, E.U., Akudike, J.C., Okeke, O.C., Amadi, C., 2020. A review of the principles and methods of soil stabilization. Int. J. Adv. Acad. Res. Sci. 6 (3), 2488–9849.

Arickx, S., Van Gerven, T., Vandecasteele, C., 2006. Accelerated carbonation for treatment of MSWI bottom ash. J. Hazard Mater. 137 (1), 235–243.

Arickx, S., De Borger, V., Van Gerven, T., Vandecasteele, C., 2010. Effect of carbonation on the leaching of organic carbon and of copper from MSWI bottom ash. Waste Manag. 30 (7), 1296–1302.

ASTM, D 4318, 2010. Standard Test Methods for Liquid Limit, Plastic Limit, and Plasticity Index of Soils. ASTM International, West Conshohocken, PA.

ASTM, D 698, 2012. Standard Test Methods for Laboratory Compaction Characteristics of Soil Using Standard Effort (12,400 Ft-Lbf/ft3(600 kN-M/m3)). ASTM International, West Conshohocken, PA.

ASTM, D 1633, 2017a. Standard Method for Compressive Strength of Molded Soil-Cement Cylinders. ASTM International, West Conshohocken, PA.

ASTM, D 2487, 2017b. Standard Practice for Classification of Soils for Engineering Purposes (Unified Soil Classification System). ASTM International, West Conshohocken, PA.

Baciocchi, R., Costa, G., Lategano, E., Marini, C., Polettini, A., Pomi, R., Postorino, P., Rocca, S., 2010. Accelerated carbonation of different size fractions of bottom ash from RDF incineration. Waste Manag. 30, 1310–1317.

Bai, R., Sutanto, M., 2002. The practice and challenges of solid waste management in Singapore. Waste Manag. 22 (5), 557–567.

Bayuseno, A.P., Schmahl, W.W., 2010. Understanding the chemical and mineralogical properties of the inorganic portion of MSWI bottom ash. Waste Manag. 30, 1509–1520.

Bell, F.G., 1996. Lime stabilization of clay minerals and soils. Eng. Geol. 42 (4), 223–237.

Bendz, D., Tüchsen, P.L., Christensen, T.H., 2007. The dissolution kinetics of major elements in municipal solid waste incineration bottom ash particles. J. Contam. Hydrol. 94, 178–194.

Benhelal, E., Shamsaei, E., Rashid, M.I., 2021. Challenges against CO_2 abatement strategies in cement industry: a review. J. Environ. Sci. 104, 84–101.

Berg, E.R., Neal, J.A., 1998. Municipal solid waste bottom ash as Portland cement concrete ingredient. J. Mater. Civ. Eng. 10, 168–173.

Bo, M.W., Arulrajah, A., Sukmak, P., Horpibulsuk, S., 2015. Mineralogy and geotechnical properties of Singapore marine clay at Changi. Soils Found. 55, 600–613.

Boardman, D.I., Glendinning, S., Rogers, C.D.F., 2001. Development of stabilisation and solidification in lime—clay mixes. Geotechnique 51 (6), 533–543.

Boardman, D.I., Glendinning, S., Rogers, C.D.F., 2004. The influences of iron (III) and lead (II) contaminants on lime-stabilised clay. Geotechnique 54, 467–486.

Bosmans, A., Vanderreydt, I., Geysen, D., Helsen, L., 2013. The crucial role of Waste-to-Energy technologies in enhanced landfill mining: a technology review. J. Clean. Prod. 55, 10–23.

Bourtsalas, A., Vandeperre, L.J., Grimes, S.M., Themelis, N., Cheeseman, C.R., 2015. Production of pyroxene ceramics from the fine fraction of incinerator bottom ash. Waste Manag. 45, 217–225.

Brück, F., Schnabel, K., Mansfeldt, T., Weigand, H., 2018. Accelerated carbonation of waste incinerator bottom ash in a rotating drum batch reactor. J. Environ. Chem. Eng. 6 (4), 5259–5268.

BSI (British Standard Institute), 2002. Characterisation of Waste—Leaching—Compliance Test for Leaching of Granular Waste Materials M and Sludges—Part 1: One Stage Batch Test at a Liquid to Solid Ratio of 2 L/kg for Materials with High Solid Content and with Particle Size below 4 Mm (Without or with Size Reduction). EN 12457-1. BSI, London.

Cabeza, L.F., Barreneche, C., Miró, L., Morera, J.M., Bartolí, E., Fernández, A.I., 2013. Low carbon and low embodied energy materials in buildings: a review. Renew. Sustain. Energy Rev. 23, 536–542.

Chalermyanont, T., Arrykul, S., Charoenthaisong, N., 2009. Potential use of lateritic and marine clay soils as landfill liners to retain heavy metals. Waste Manag. 29, 117–127.

Chan, W.P., Ren, F., Dou, X., Yin, K., Chang, V.W.C., 2018. A large-scale field trial experiment to derive effective release of heavy metals from incineration bottom ashes during construction in land reclamation. Sci. Total Environ. 637–638, 182–190.

Chandler, A.J., Eighmy, T.T., Hjelmar, O., Kosson, D.S., Sawell, S.E., Vehlow, J., Van der Sloot, H.A., Hartlén, J., 1997. Municipal Solid Waste Incinerator Residues. Elsevier.

Chen, Q.Y., Hills, C.D., Tyrer, M., Slipper, I., Shen, H.G., Brough, A., 2007. Characterisation of products of tricalcium silicate hydration in the presence of heavy metals. J. Hazard Mater. 147 (3), 817–825.

Chen, Q.Y., Tyrer, M., Hills, C.D., Yang, X.M., Carey, P., 2009. Immobilisation of heavy metal in cement-based solidification/stabilisation: a review. Waste Manag. 29, 390–403.

Chen, Z., Yang, E.H., 2017. Early age hydration of blended cement with different size fractions of municipal solid waste incineration bottom ash. Construct. Build. Mater. 156, 880–890.

Cheng, A., 2012. Effect of incinerator bottom ash properties on mechanical and pore size of blended cement mortars. Mater. Des. 36, 859–864.

Chew, S.H., Kamruzzaman, A.H.M., Lee, F.H., 2004. Physicochemical and engineering behavior of cement treated clays. J. Geotech. Geoenviron. Eng. 130, 696–706.

Chiang, Y.W., Ghyselbrecht, K., Santos, R.M., Meesschaert, B., Martens, J.A., 2012. Synthesis of zeolitic-type adsorbent material from municipal solid waste incinerator bottom ash and its application in heavy metal adsorption. Catal. Today 190 (1), 23–30.

Chimenos, J.M., Fernandez, A.I., Miralles, L., Segarra, M., Espiell, F., 2003. Short-term natural weathering of MSWI bottom ash as a function of particle size. Waste Manag. 23, 887–895.

Chrysochoou, M., Grubb, D.G., Drengler, K.L., Malasavage, N.E., 2010. Stabilized dredged material. III: mineralogical perspective. J. Geotech. Geoenviron. Eng. 136 (8), 1037–1050.

Chu, J., Guo, W., 2016. Land reclamation using clay slurry or in deep water: challenges and solutions. Jpn. Geotech. Soc. Spec. Publ. 2 (51), 1790–1793.

Clozel-Leloup, B., Bodénan, F., Piantone, P., 1999. Bottom ash from municipal solid waste incineration. Mineralogy and distribution of metals. In: Mehu, J., Keck, G., Navarro, A. (Eds.), STAB and ENV 99. Proceedings of the International Conference on Waste Stabilization and Environment. Société Alpine de Publications, Lyone Villeurbanne, France, pp. 46–51.

Cossu, R., Lai, T., 2012. Washing of waste prior to landfilling. Waste Manag. 2 (5), 869–878.

Di Gianfilippo, M., Hyks, J., Verginelli, I., Costa, G., Hjelmar, O., Lombardi, F., 2018. Leaching behaviour of incineration bottom ash in a reuse scenario: 12 years-field data vs. lab test results. Waste Manag. 73, 367–380.

Dou, X., Ren, F., Nguyen, M.Q., Ahamed, A., Yin, K., Chan, W.P., Chang, V.W.C., 2017. Review of MSWI bottom ash utilization from perspectives of collective characterization, treatment and existing application. Renew. Sustain. Energy Rev. 79, 24–38.

Ecke, H., Åberg, A., 2006. Quantification of the effects of environmental leaching factors on emissions from bottom ash in road construction. Sci. Total Environ. 362 (1–3), 42–49.

Etoh, J., Kawagoe, T., Shimaoka, T., Watanabe, K., 2009. Hydrothermal treatment of MSWI bottom ash forming acid-resistant material. Waste Manag. 29, 1048–1057.

Evans, J., Ruffing, D., Elton, D., 2021. Fundamentals of Ground Improvement Engineering. CRC Press.

Fang, Y.S., Liao, J.J., Sze, S.C., 1994. An empirical strength criterion for jet grouted soilcrete. Eng. Geol. 37 (3–4), 285–293.

Ferraris, M., Salvo, M., Ventrella, A., Buzzi, L., Veglia, M., 2009. Use of vitrified MSWI bottom ashes for concrete production. Waste Manag. 29 (3), 1041–1047.

Forteza, R., Far, M., Seguõ, C., Cerdá, V., 2004. Characterization of bottom ash in municipal solid waste incinerators for its use in road base. Waste Manag. 24, 899–909.

George, S., Paul, J., Jacob, J., 2014. Heavy metal retention of Cochin marine clay. Int. J. Eng. Sci. 9, 54–59.

Guo, L., Xu, W.Y., Quek, A., Wu, D.Q., 2015. Leaching assessment of matrix land reclamation material. Environ. Geotech. 2, 349–358.

Gupta, G., Datta, M., Ramana, G.V., Alappat, B.J., 2021. MSW incineration bottom ash (MIBA) as a substitute to conventional materials in geotechnical applications: a characterization study from India and comparison with literature. Construct. Build. Mater. 308, 124925.

He, P.J., Pu, H.X., Shao, L.M., Zhang, H., 2017. Impact of co-landfill proportion of bottom ash and municipal solid waste composition on the leachate characteristics during the acidogenesis phase. Waste Manag. 69, 232–241.

Higgins, D.D., 2007. GGBS and sustainability. Construct. Mater. 160, 99–101.

Hjelmar, O., Holm, J., Crillesen, K., 2007. Utilisation of MSWI bottom ash as sub-base in road construction: first results from a large-scale test site. J. Hazard Mater. 139 (3), 471–480.

Hjelmar, O., Hansen, J.B., Wahlström, M., Wik, O., 2016. End-of-Waste Criteria for Construction and Demolition Waste. Nordic Council of Ministers.

Holtz, R.D., Kovacs, W.D., Sheahan, T.C., 1981. An Introduction to Geotechnical Engineering. Prentice-Hall, Englewood Cliffs, NJ.

Horpibulsuk, S., Miura, N., Nagaraj, T.S., 2003. Assessment of strength development in cement-admixed high water content clays with Abrams' law as a basis. Geotechnique 53 (4), 439–444.

Huang, Y., Wang, L., Wu, T., Liu, W., Tang, Q., 2023. Mechanical properties and heavy metal leaching behaviors of municipal solid waste incineration bottom ash as road embankment fillings. J. Clean. Prod. 394, 136355.

Hyks, J., Astrup, T., Christensen, T.H., 2009. Leaching from MSWI bottom ash: evaluation of non-equilibrium in column percolation experiments. Waste Manag. 29, 522–529.

Inkaew, K., Saffarzadeh, A., Shimaoka, T., 2016. Modeling the formation of the quench product in municipal solid waste incineration (MSWI) bottom ash. Waste Manag. 52, 159–168.

Ito, R., Fujita, T., Sadaki, J., Matsumoto, Y., Ahn, J.W., 2006. Removal of chloride in bottom ash from the industrial and municipal solid waste incinerators. Int. J. Soc. Mater. Eng. Resour. 13 (2), 70–74.

Izquierdo, M., Vazquez, E., Querol, X., Barra, M., Lopez, A., Plana, F., 2001. Use of bottom ash from municipal solid waste incineration as a road material. In: International Ash Utilization Symposium. Kentucky, USA: Center for Applied Energy Research, University of Kentucky.

Johnson, C.A., Kersten, M., Ziegler, F., Moor, H.C., 1996. Leaching behaviour and solubility-Controlling solid phases of heavy metals in municipal solid waste incinerator ash. Waste Manag. 16 (1–3), 129–134.

Kahle, K., Kamuk, B., Kallesøe, J., Fleck, E., Lamers, F., Jacobsson, L., Sahlén, J., 2015. Bottom Ash from WTE Plants: Metal Recovery and Utilization. Ramböll, Copenhagen, Denmark.

Kaza, S., Yao, L., Bhada-Tata, P., Van Woerden, F., 2018. What a Waste 2.0: A Global Snapshot of Solid Waste Management to 2050. World Bank Publications.

Kirby, C.S., Rimstidt, J.D., 1993. Mineralogy and surface properties of municipal solid waste ash. Environ. Sci. Technol. 27 (4), 652–660.

Kim, S.Y., Matsuto, T., Tanaka, N., 2003. Evaluation of pre-treatment methods for landfill disposal of residues from municipal solid waste incineration. Waste Manag. Res. 21 (5), 416–423.

Lam, C.H., Ip, A.W., Barford, J.P., McKay, G., 2010. Use of incineration MSW ash: a review. Sustainability 2, 1943–1968.

Lam, K.P., Kou, H.L., Xie, B., Chu, J., He, J., 2018. Use of a waste-based binder for high water content soil treatment. J. Mater. Civ. Eng. 30 (8), 06018009.

Lee, F.H., Lee, Y., Chew, S.H., Yong, K.Y., 2005. Strength and modulus of marine clay-cement mixes. J. Geotech. Geoenviron. Eng. 131 (2), 178–186.

Li, L., Zang, T., Xiao, H., Feng, W., Liu, Y., 2020. Experimental study of polypropylene fibre-reinforced clay soil mixed with municipal solid waste incineration bottom ash. Eur. J. Environ. Civil Eng. 1–17.

Li, W., Ni, P., Yi, Y., 2019. Comparison of reactive magnesia, quick lime, and ordinary Portland cement for stabilization/solidification of heavy metal-contaminated soils. Sci. Total Environ. 671, 741–753.

Li, W., Yi, Y., 2020. Use of carbide slag from acetylene industry for activation of ground granulated blast-furnace slag. Construct. Build. Mater. 238, 117713.

Li, W., Yi, Y., Puppala, A.J., 2022. Comparing carbide sludge-ground granulated blastfurnace slag and ordinary Portland cement: different findings from binder paste and stabilized clay slurry. Construct. Build. Mater. 321, 126382.

Lim, T.T., Tay, J.H., Teh, C.I., 1997. Sorption and speciation of heavy metals from incinerator fly ash in a marine clay. J. Environ. Eng. 123, 1107–1115.

Lin, C.L., Weng, M.C., Chang, C.H., 2012. Effect of incinerator bottom-ash composition on the mechanical behavior of backfill material. J. Environ. Manag. 113, 377–382.

Lin, K.L., Chang, W.C., Lin, D.F., 2008. Pozzolanic characteristics of pulverized incinerator bottom ash slag. Construct. Build. Mater. 22, 324–329.

Lin, W.Y., Heng, K.S., Sun, X., Wang, J.Y., 2015. Accelerated carbonation of different size fractions of MSW IBA and the effect on leaching. Waste Manag. 41, 75–84.

Little, D.N., Nair, S., 2009. Recommended Practice for Stabilization of Subgrade Soils and Base Materials. National Cooperative Highway Research Program. Transportation Research Board of the National Academies.

Liu, Y., Li, Y., Li, X., Jiang, Y., 2008. Leaching behavior of heavy metals and PAHs from MSWI bottom ash in a long-term static immersing experiment. Waste Manag. 28 (7), 1126–1136.

Loginova, E., Volkov, D.S., Van de Wouw, P.M.F., Florea, M.V.A., Brouwers, H.J.H., 2019. Detailed characterization of particle size fractions of municipal solid waste incineration bottom ash. J. Clean. Prod. 207, 866–874.

Luo, H., Cheng, Y., He, D., Yang, E.H., 2019. Review of leaching behavior of municipal solid waste incineration (MSWI) ash. Sci. Total Environ. 668, 90–103.

Luo, H., Wu, Y., Zhao, A., Kumar, A., Liu, Y., Cao, B., Yang, E.H., 2017. Hydrothermally synthesized porous materials from municipal solid waste incineration bottom ash and their interfacial interactions with chloroaromatic compounds. J. Clean. Prod. 162, 411–419.

Lynn, C.J., Ghataora, G.S., Dhir, R.K., 2017. Municipal incinerated bottom ash (MIBA) characteristics and potential for use in road pavements. Int. J. Pavement Res. Technol. 10, 185–201.

Makusa, G.P., 2012. Soil Stabilization Methods and Materials. Lulea University of Technology.

Meima, J.A., van der Weijden, R.D., Eighmy, T.T., Comans, R.N., 2002. Carbonation processes in municipal solid waste incinerator bottom ash and their effect on the leaching of copper and molybdenum. Appl. Geochem. 17 (12), 1503–1513.

Melese, D.T., 2022. Utilization of waste incineration bottom ash to enhance engineering properties of expansive subgrade soils. Adv. Civ. Eng. 2022.

Naganathan, S., Razak, H.A., Hamid, S.N.A., 2010. Effect of kaolin addition on the performance of controlled low-strength material using industrial waste incineration bottom ash. Waste Manag. Res. 28 (9), 848–860.

Nalbantoglu, Z., 2006. Lime Stabilization of Expansive Clay. CRC Press.

Nidzam, R.M., Kinuthia, J.M., 2010. Sustainable soil stabilisation with blastfurnace slag—a review. Construct. Mater. 163, 157–165.

Omine, K., Ochiai, H., Yasufuku, N., 2002. Utilization of MSW Incineration Bottom Ash and Clay as a Composite Geomaterial.

Piantone, P., Bodénan, F., Chatelet-Snidaro, L., 2004. Mineralogical study of secondary mineral phases from weathered MSWI bottom ash: implications for the modelling and trapping of heavy metals. Appl. Geochem. 19, 1891–1904.

Polettini, A., Pomi, R., 2004. The leaching behavior of incinerator bottom ash as affected by accelerated ageing. J. Hazard Mater. 113 (1–3), 209–215.

Qiao, X.C., Ng, B.R., Tyrer, M., Poon, C.S., Cheeseman, C.R., 2008. Production of lightweight concrete using incinerator bottom ash. Construct. Build. Mater. 22, 473–480.

Quek, A., Wu, D., Xu, W., Guo, L., 2016. Feasibility of Singapore IBA waste for land reclamation. Environ. Geotech. 4, 56–64.

Rendek, E., Ducom, G., Germain, P., 2006. Carbon dioxide sequestration in municipal solid waste incinerator (MSWI) bottom ash. J. Hazard Mater. 128 (1), 73–79.

Santos, R.M., Mertens, G., Salman, M., Cizer, Ö., Van Gerven, T., 2013. Comparative study of ageing, heat treatment and accelerated carbonation for stabilization of municipal solid waste incineration bottom ash in view of reducing regulated heavy metal/metalloid leaching. J. Environ. Manag. 128, 807–821.

Sabat, A.K., Pati, S., 2014. A review of literature on stabilization of expansive soil using solid wastes. Electron. J. Geotech. Eng. 19 (6), 251–256.

Sakr, M.A., Shahin, M.A., Metwally, Y.M., 2009. Utilization of lime for stabilizing soft clay soil of high organic content. Geotech. Geol. Eng. 27, 105–113.

Salimi, M., Ghorbani, A., 2020. Mechanical and compressibility characteristics of a soft clay stabilized by slag-based mixtures and geopolymers. Appl. Clay Sci. 184, 105390.

Schafer, M.L., Clavier, K.A., Townsend, T.G., Ferraro, C.C., Paris, J.M., Watts, B.E., 2019a. Use of coal fly ash or glass pozzolan addition as a mitigation tool for alkali-silica reactivity in cement mortars amended with recycled municipal solid waste incinerator bottom ash. Waste Biomass Valoriz. 9, 2733–2744.

Schafer, M.L., Clavier, K.A., Townsend, T.G., Kari, R., Worobel, R.F., 2019b. Assessment of the total content and leaching behavior of blends of incinerator bottom ash and natural aggregates in view of their utilization as road base construction material. Waste Manag. 98, 92–101.

Schabbach, L.M., Andreola, F., Lancellotti, I., Barbieri, L., 2011. Minimization of Pb content in a ceramic glaze by reformulation the composition with secondary raw materials. Ceram. Int. 37, 1367–1375.

Schneider, M., Romer, M., Tschudin, M., Bolio, H., 2011. Sustainable cement production—present and future. Cement Concr. Res. 41 (7), 642–650.

Seco, A., Ramírez, F., Miqueleiz, L., García, B., 2011. Stabilization of expansive soils for use in construction. Appl. Clay Sci. 51 (3), 348–352.

Sherwood, P., 1993. Soil Stabilization with Cement and Lime. Her Majesty Stationary Office.

Shi, C., Krivenko, P.V., Roy, D., 2006. Alkali-activated Cements and Concretes. Taylor and Francis, London.

Shi, C., Spence, R., 2004. Designing of cement-based formula for solidification/stabilization of hazardous, radioactive, and mixed wastes. Crit. Rev. Environ. Sci. Technol. 34 (4), 391–417.

Simon, F.G., Holm, O., 2016. Exergetic considerations on the recovery of metals from waste. Int. J. Exergy 19 (3), 352–363.

Song, S., Jennings, H.M., 1999. Pore solution chemistry of alkali-activated ground granulated blast-furnace slag. Cement Concr. Res. 29 (2), 159–170.

Speiser, C., Baumann, T., Niessner, R., 2000. Morphological and chemical characterization of calcium-hydrate phases formed in alteration processes of deposited municipal solid waste incinerator bottom ash. Environ. Sci. Technol. 34, 5030–5037.

Stegemann, J.A., Shi, C., Caldwell, R.J., 1997. Response of various solidification systems to acid addition. Stud. Environ. Sci. 71, 803–814.

Sun, X., 2022. Treating Fine Incineration Bottom Ash (IBA) of Municipal Solid Waste Using Marine Clay (MC) and Ground Granulated Blastfurnace Slag (GGBS) for Land Reclamation. Nanyang Technological University.

Sun, X., Ting, M.Z.Y., Yi, Y., 2024. Feasibility of using three solid wastes/byproducts to produce pumpable materials for land reclamation. Can. Geotec. J. https://doi.org/10.1139/cgj-2023-0580.

Sun, X., Yi, Y., 2020. pH evolution during water washing of incineration bottom ash and its effect on removal of heavy metals. Waste Manag. 104, 213–219.

Sun, X., Yi, Y., 2021. Acid washing of incineration bottom ash of municipal solid waste: effects of pH on removal and leaching of heavy metals. Waste Manag. 120, 183–192.

Sun, X., Yi, Y., 2022a. Stabilization and solidification of fine incineration bottom ash of municipal solid waste using ground granulated blast-furnace slag. J. Mater. Civ. Eng. 34 (6), 04022076.

Sun, X., Yi, Y., 2022b. Amending excavated soft marine clay with fine incineration bottom ash as a fill material for construction of transportation infrastructure. Transport. Geotech. 35, 100796.

Sun, X., Yi, Y., 2022c. Utilization of incineration bottom ash, waste marine clay, and ground granulated blast-furnace slag as a construction material. Resour. Conserv. Recycl. 182, 106292.

Tang, P., Florea, M.V.A., Spiesz, P., Brouwers, H.J.H., 2015. Characteristics and application potential of municipal solid waste incineration (MSWI) bottom ashes from two waste-to-energy plants. Construct. Build. Mater. 83, 77–94.

Tang, P., Florea, M.V.A., Brouwers, H.J.H., 2017. Employing cold bonded pelletization to produce lightweight aggregates from incineration fine bottom ash. J. Clean. Prod. 165, 1371–1384.

Taylor, H.F., 1997. Cement Chemistry, vol. 2. Thomas Telford, London.

Thyagaraj, T., Rao, S.M., Sai Suresh, P., Salini, U., 2012. Laboratory studies on stabilization of an expansive soil by lime precipitation technique. J. Mater. Civ. Eng. 24 (8), 1067–1075.

Toraldo, E., Saponaro, S., Careghini, A., Mariani, E., 2013. Use of stabilized bottom ash for bound layers of road pavements. J. Environ. Manag. 121, 117–123.

Travar, I., Lidelow, S., Andreas, L., Tham, G., Lagerkvist, A., 2009. Assessing the environmental impact of ashes used in a landfill cover construction. Waste Manag. 29, 1336–1346.

Trussell, S., Spence, R.D., 1994. A review of solidification/stabilization interferences. Waste Manag. 14 (6), 507–519.

Um, N., Ahn, J.W., 2017. Effects of two different accelerated carbonation processes on MSWI bottom ash. Process Saf. Environ. Protect. 111, 560–568.

Vaitkus, A., Gražulytė, J., Šernas, O., Vorobjovas, V., Kleizienė, R., 2019. An algorithm for the use of MSWI bottom ash as a building material in road pavement structural layers. Construct. Build. Mater. 212, 456–466.

Van Gerven, T., Van Keer, E., Arickx, S., Jaspers, M., Wauters, G., Vandecasteele, C., 2005. Carbonation of MSWI-bottom ash to decrease heavy metal leaching, in view of recycling. Waste Manag. 25 (3), 291–300.

Vegas, I., Ibañez, J.A., San José, J.T., Ursela, A., 2008. Construction demolition wastes, Waelz slag and MSWI bottom ash: a comparative technical analysis as material for road construction. Waste Manag. 28, 565–574.

Vichaphund, S., Jiemsirilers, S., Thavarniti, P., 2012. Sintering of municipal solid waste incineration bottom ash. Int. J. Eng. Sci. 8, 51–59.

Vizcarra, G.O.C., Casagrande, M.D.T., da Motta, L.M.G., 2014. Applicability of municipal solid waste incineration ash on base layers of pavements. J. Mater. Civ. Eng. 26 (6), 06014005.

Wahba, M.M., Zaghloul, A.M., 2007. Adsorption characteristics of some heavy metals by some soil minerals. J. Appl. Sci. Res. 3 (6), 421–426.

Wang, S.D., Scrivener, K.L., 1995. Hydration products of alkali activated slag cement. Cement Concr. Res. 25 (3), 561–571.

Wang, L., Tsang, D.C., Poon, C.S., 2015a. Green remediation and recycling of contaminated sediment by waste-incorporated stabilization/solidification. Chemosphere 122, 257–264.

Wang, L., Kwok, J.S., Tsang, D.C., Poon, C.S., 2015b. Mixture design and treatment methods for recycling contaminated sediment. J. Hazard Mater. 283, 623–632.

Wang, Y., Huang, L., Lau, R., 2016. Conversion of municipal solid waste incineration bottom ash to sorbent material for pollutants removal from water. J. Taiwan Inst. Chem. Eng. 60, 275–286.

Weng, M.C., Lin, C.L., Ho, C.I., 2010. Mechanical properties of incineration bottom ash: the influence of composite species. Waste Manag. 30, 1303–1309.

Weng, M.C., Wu, M.H., Lin, C.L., Syue, D.K., Hung, C., 2015. Long-term mechanical stability of cemented incineration bottom ash. Construct. Build. Mater. 93, 551–557.

Wiles, C.C., 1996. Municipal solid waste combustion ash: state-of-the-knowledge. J. Hazard Mater. 47, 325–344.

Xie, R., Xu, Y., Huang, M., Zhu, H., Chu, F., 2016. Assessment of municipal solid waste incineration bottom ash as a potential road material. Road Mater. Pavement Des. 18, 992–998.

Xu, B., Yi, Y., 2019. Soft clay stabilization using ladle slag-ground granulated blastfurnace slag blend. Appl. Clay Sci. 178, 105136.

Yan, D.Y., Tang, I.Y., Lo, I.M., 2014. Development of controlled low-strength material derived from beneficial reuse of bottom ash and sediment for green construction. Construct. Build. Mater. 64, 201–207.

Yang, R., Liao, W.P., Wu, P.H., 2012. Basic characteristics of leachate produced by various washing processes for MSWI ashes in Taiwan. J. Environ. Manag. 104, 67–76.

Yao, J., Song, H., Li, Y., Cui, Y., Chai, M., Ling, W., 2023. Mechanism of macro-and microscopic performance of cement mortars influenced by municipal solid waste incineration bottom ash as sand substitution. Construct. Build. Mater. 397, 132317.

Yi, Y., Liska, M., Al-Tabbaa, A., 2014. Properties of two model soils stabilized with different blends and contents of GGBS, MgO, lime, and PC. J. Mater. Civ. Eng. 26 (2), 267–274.

Yi, Y., Gu, L., Liu, S., Puppala, A.J., 2015a. Carbide slag–activated ground granulated blastfurnace slag for soft clay stabilization. Can. Geotech. J. 52, 656–663.

Yi, Y., Gu, L., Liu, S., 2015b. Microstructural and mechanical properties of marine soft clay stabilized by lime-activated ground granulated blastfurnace slag. Appl. Clay Sci. 103, 71–76.

Yi, Y., Li, C., Liu, S., 2015c. Alkali-activated ground-granulated blast furnace slag for stabilization of marine soft clay. J. Mater. Civ. Eng. 27, 04014146.

Yi, Y., Liska, M., Jin, F., Al-Tabbaa, A., 2016. Mechanism of reactive magnesia–ground granulated blastfurnace slag (GGBS) soil stabilization. Can. Geotech. J. 53 (5), 773–782.

Yin, K., Chan, W.P., Dou, X., Ren, F., Chang, V.W.C., 2017. Measurements, factor analysis and modeling of element leaching from incineration bottom ashes for quantitative component effects. J. Clean. Prod. 165, 477–490.

Yin, K., Chan, W.P., Dou, X., Ren, F., Chang, V.W.C., 2018a. Cr, Cu, Hg and Ni release from incineration bottom ash during utilization in land reclamation–based on lab-scale batch and column leaching experiments and a modeling study. Chemosphere 197, 741–748.

Yin, K., Dou, X., Ren, F., Chan, W.P., Chang, V.W.C., 2018b. Statistical comparison of leaching behavior of incineration bottom ash using seawater and deionized water: significant findings based on several leaching methods. J. Hazard Mater. 344, 635–648.

Yin, K., Chan, W.P., Dou, X., Lisak, G., Chang, V.W.C., 2019. Vertical distribution of heavy metals in seawater column during IBA construction in land reclamation–Re-exploration of a large-scale field trial experiment. Sci. Total Environ. 654, 356–364.

You, G.S., Ahn, J.W., Han, G.C., Cho, H.C., 2006. Neutralizing capacity of bottom ash from municipal solid waste incineration of different particle size. Korean J. Chem. Eng. 23 (2), 237–240.

Zekkos, D., Kabalan, M., Syal, S.M., Hambright, M., Sahadewa, A., 2013. Geotechnical characterization of a municipal solid waste incineration ash from a Michigan monofill. Waste Manag. 33, 1442–1450.

Zhang, Z., Tao, M., 2008. Durability of cement stabilized low plasticity soils. J. Geotech. Geoenviron. Eng. 134 (2), 203–213.

Zhang, T., Zhao, Z., 2014. Optimal use of MSWI bottom ash in concrete. Int. J. Concrete Struct. Mater. 8, 173–182.

Zhang, Y., Ong, Y.J., Yi, Y., 2022. Comparison between CaO-and MgO-activated ground granulated blast-furnace slag (GGBS) for stabilization/solidification of Zn-contaminated clay slurry. Chemosphere 286, 131860.

Zhang, Z., Zhang, L., Li, A., 2015. Development of a sintering process for recycling oil shale fly ash and municipal solid waste incineration bottom ash into glass ceramic composite. Waste Manag. 38, 185–193.

Zhu, W., Teoh, P.J., Liu, Y., Chen, Z., Yang, E.H., 2019. Strategic utilization of municipal solid waste incineration bottom ash for the synthesis of lightweight aerated alkali-activated materials. J. Clean. Prod. 235, 603–612.

Part VI

Resource recovery from combustion/incineration residues

Chapter 24

Resource recovery from pulverized fly ash and bottom ash

Fangqin Cheng[1,2], Zhibin Ma[1,2], Jian-ming Gao[1], Huiping Song[1], Yuan Fan[1], Hongyu Gao[1], Jinglei Cui[1], Quan An[1], Jianbo Li[3] and Dongke Zhang[1,2]

[1]*State Key Laboratory of Technologies for Efficient Utilization of Coal Waste Resources, Shanxi University, Taiyuan, China;* [2]*Centre for Energy (M473), The University of Western Australia, Crawley, WA, Australia;* [3]*Key Laboratory of Low-grade Energy Utilization Technologies and Systems, Ministry of Education of PRC, Chongqing University, Chongqing, China*

1. Introduction

Coal is one of the major fossil energy sources in the world. The total consumption of world shows an increasing trend from 2001 to 2017 (Fig. 24.1). Most of these coals Are used to generate electricity by coal combustion. In addition to combustible organic matter, coal also contains 5%–30% inorganic matter, which is transferred to fine fly ash and coarse bottom ash. Therefore, a large amount of coal fly ash (CFA) is generated during the combustion of pulverized coal in coal-fired power stations every year. The growing demand for energy at the world level has greatly increased the use of coal as a fuel for thermoelectric power plants, thus resulting in an increased accumulation of fly ash in the environment. India and China are two countries that have a large production of CFA. China generated approximately 620 million tons of CFA in 2015, and annual generation is still increasing (Sun et al., 2017). Approximately 166 million tons of CFA is generated annually in India (Nadesan and Dinakar, 2017). In addition, Australia is also heavily reliant upon the use of coal for electric power generation with approximately 84% of electricity being generated in this manner, which results in the annual production of approximately 12 million tons of fly ash. CFA has become one of Australia's biggest waste problems. However, technical limitations mean that large amounts of CFA cannot be fully utilized at present. Current CFA utilization ratio is ~70% in China, ~44% in the Australia, ~39% in the United States, and ~47% in Europe. The global average is estimated to be close to 25% (Blissett and Rowson, 2012). The accumulated of the total CFA has been so far more than 2500 million tons in China (Duan et al., 2018). A substantial amount of CFA is still disposed of in landfills or in ash dam. The irregular accumulation and inappropriate disposal of CFA endanger both human health and the environment. It may cause air, water, and soil pollution; disrupt ecological cycles; and pose environmental hazards if it cannot be properly disposed. The growing accumulation of CFA has led to many studies aiming at developing possible industrial uses.

CFA is generally considered to contain three major constituents: the major one is noncrystalline aluminosilicate glass, with moderate content of crystalline minerals, and a small proportion of unburned carbon particles (Hemalatha and Ramaswamy, 2017). The CFA contains abundant SiO_2 and Al_2O_3, moderate contents of Fe_2O_3 and CaO, and minor content of other oxides. The CFA as the main raw material can be used to prepare construction materials, ceramics, refractory materials, cement, concrete, and zeolite based on its chemical composition. In recent years, large quantities of coal that contains abundant Al, Li, Ga, and REY (REE + Y) have been found in many mining districts located in northern Shanxi Province and in the middle-western region of Inner Mongolia, China (Dai et al., 2012). These metals can be further enriched during the combustion process, and most of them are transferred to coal ash. Many previous studies (Tripathy et al., 2015; Ding et al., 2016; Xu et al., 2016) corroborated that high-alumina fly ash can be utilized as a substitute for bauxite for Al production. Recycling these valuable metals from CFA can be one of the efficient ways to utilization of it. Besides, the CFA can also be used as a fertilizer/soil amendment in agriculture. It can buffer the pH of soil to improve the quality of soil (Elseewi et al., 1978).

Treatment and Utilization of Combustion and Incineration Residues. https://doi.org/10.1016/B978-0-443-21536-0.00014-9

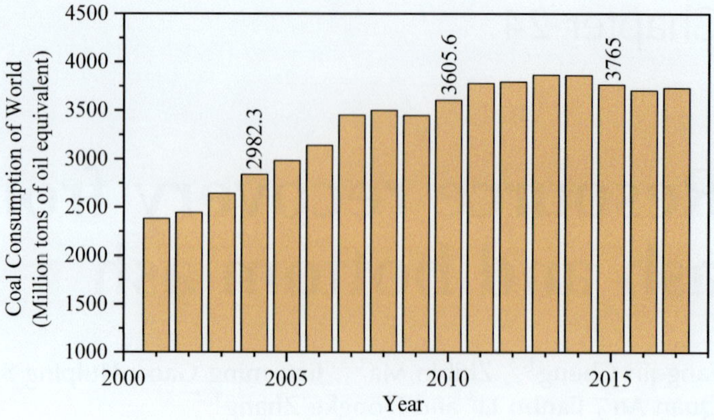

FIGURE 24.1 Variation of total coal consumption of world from 2001 to 2017 (BP world energy statistics yearbook, 2011–18).

Although several reviews of the utilization of CFA have been conducted (Ahmaruzzaman, 2010; Blissett and Rowson, 2012; Yao et al., 2014, 2015), many new and valuable knowledge and technologies on the utilization of CFA appear in recent years. This chapter systematically summarizes the application progress of CFA in construction materials, chemical industry, metallurgical industry, and agriculture. This chapter will add to the further understanding of current utilizations of CFA and identifying the promising applications.

2. Formation and properties of CFA

A basic understanding on the formation and properties of CFA is always essential before any of its utilization. This section describes how CFA is generated during combustion, how its chemical, mineralogical, and morphological properties differ from the mineral matter and inorganic constituents in coal, and how its properties direct its utilization.

2.1 Mineralogy and geochemistry of coal

Coal is a complex compound that consists of not only organic matters (i.e., C, H, O, N, and S), but also ash-forming mineral matter and inorganic constituents including Si, Al, Fe, Ca, Na, Mg, Ti and heavy metals Cr, Se, Pb, etc. (Mclennan et al., 2000). These mineral matter and inorganic constituents, depending on their intimacy with coal matrix, could be further divided into excluded minerals, included minerals, and organically bound inorganic constituents. Representative minerals identified in coal often include but not limit to silicates (quartz, kaolinite, illite, etc.), oxides and hydroxides (hematite, corundum, etc.), sulfides (pyrite, marcasite, etc.), sulfates (gypsum, melanterite, etc.), and carbonates (calcite, siderite), etc. (Mukherjee and Srivastava, 2006; Vuthaluru and French, 2008). Typically, organically bound inorganic constituents include Na, K, Ca, and S. The heavy metals in coal are also in these modes of occurrence; however, their contents are relatively low. Nonetheless, the mode of occurrence and concentration of these elements are dependent on the rank, origin, and even particle size of one given coal.

2.2 Formation of CFA during coal combustion

The minerals and inorganic constituents in coal undergo a series of transformation during combustion and eventually form the ash residue. The physical and mineralogical transformation pathways of the mineral matter and inorganic constituents are illustrated in Fig. 24.2. The excluded minerals would undergo fragmentation, generating ash particles with sizes larger than 1 μm. The included minerals or organically bound inorganic constituents would experience coalescence or fragmentation, forming fine ash in the same size as excluded minerals. For minerals or organically bound inorganic constituents with low melting points and high mobility, vaporization of these elements would occur, generating ash aerosols in the size less than 1 μm. Apart from morphological changes, these mineral matter and inorganic constituents also undergo chemical decomposition (Fig. 24.2B), transformation, and interactions, leading to the mineral phases that differ from their occurrence in coal.

It is noted that the transformation pathways of the mineral matter and inorganic constituents are dependent on the temperature and atmosphere they subject to. In pulverized coal-fired boilers at temperatures above 1200°C, minerals with low melting points would melt, generating cenosphere in shape and glass phases, whereas in circulating fluidized bed

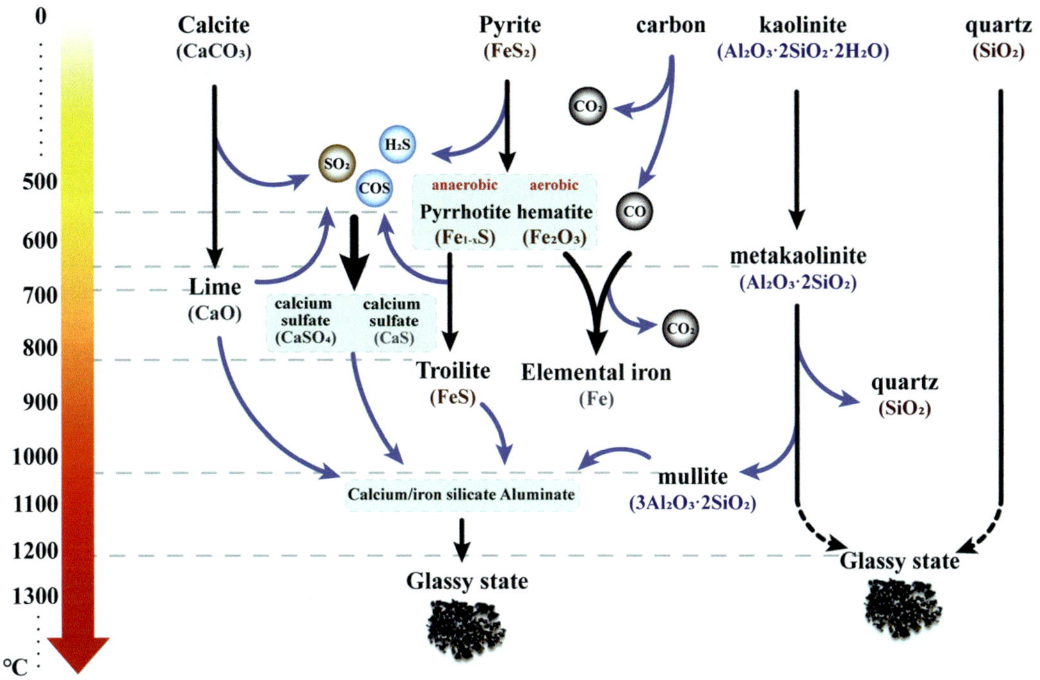

FIGURE 24.2 Mineralogical transformation of the minerals and inorganic constituents during combustion.

boilers, most ash particles would retain their irregular shape, with less glass phase being formed. In addition, the mineral transformation would be altered when the surrounding atmosphere (reducing, inert, and oxidizing) varied. A typical example is iron oxide, which exists as Fe (III) at oxidizing atmosphere but became Fe (II) at reducing temperature. In practical combustion facilities, partitioning of these ash-forming elements occurs during combustion. The bottom ash is often enriched in aluminosilicate with large particle sizes, whereas the fine fly ash (CFA) is found rich in aluminosilicate, soluble salts as well as heavy metals.

2.3 Chemical and mineralogical properties of CFA

CFA is heterogenous in nature and comprises both organic constituents, i.e., unburned carbon, and inorganic crystalline and noncrystalline (glass phase or amorphous) constituents. CFA contains a certain amount of unburnt carbon due to incomplete combustion of coal. Besides, the major inorganic crystalline and noncrystalline (glass phase or amorphous) constituents in CFA, while presented as their oxide forms, include both acidic oxides, i.e., SiO_2, Al_2O_3, and TiO_2 and basic oxides, i.e., Fe_2O_3, CaO, MgO, Na_2O, and K_2O. Based on the contents of SiO_2, Al_2O_3, and Fe_2O_3, the American Society for Testing and Materials (ASTMs) groups CFA as Class C when these contents in CFA are higher than 50%, and Class F when above 70%. In specific cases, the content of valuable Al in CFA can as high as 40%, making it suitable for metal recovery. Nonetheless, CFA also contains a certain amount of heavy metals such as Hg, As, Pb, Zn, Cu, Cr, Cd, etc. (Fu et al., 2019), as most of which were prone to be concentrated in fine ash particles during combustion (Tang et al., 2013). The chemical composition of one given CFA depends on coal geochemistry and the transformation pathways during combustion as aforementioned.

Apart from oxides, the inorganic constituents in CFA are also in the forms of other minerals and even glass phases. Representative crystalline minerals that are commonly identified in CFA include mullite ($3Al_2O_3 \cdot 2SiO_2$), quartz (SiO_2), corundum (Al_2O_3) magnesioferrite ($MgFe_2O_4$), anorthite ($CaAl_2Si_2O_8$), and gehlenite ($Ca_2Al_2SiO_7$) (Yu et al., 2012; Yõlmaz, 2015). These minerals account for 25–70 wt% of the overall ash, and the remaining 30%–75% are glass phases that are generated during combustion.

2.4 Physical and morphological characteristics of CFA

CFAs are generally fine powdery particles with 0.5–100 μm in size, 2.1–3.0 in specific gravity, and 170–1000 m^2/kg in specific surface area. Its apparent color, depending on its the chemical composition (and carbon), can be tan, gray, and black (Ahmaruzzaman, 2010). With respect to its morphology, most ash particles from p.f. combustion are spherical in shape with a solid or hollow structure. Those particles with hollow lightweight microspheres are cenospheres with light weight, excellent mechanical strength, good thermal resistance, and electrical properties (Ngu et al., 2007).

2.5 Handing, storage, and utilization of CFA

CFA is mostly disposed in China either by land-filling or dumping in a lagoon or a pond in the vicinity of a power plant. However, this is by no means a sustainable strategy. More plants would suffer from the transportation, storage, and disposal expenditure once the adjacent disposal sites are saturated (Rashidi and Yusup, 2016). Moreover, such a disposal method would bring in serious human health issues as well as economic and environmental challenges. The fine CFA particulates $PM_{2.5}$ with large surface area and strong surface activity would carry toxic compounds and bring harm to human health once being exhaled into human lungs (Oliveira et al., 2017). Moreover, the soluble salts and heavy metals in CFA could be leached out by water, which would contaminate the surrounding land and water resources.

Alternative and cleaner practices are therefore required to better utilize CFA. These include utilizing CFA in construction materials, function materials, agriculture field, valuable metal recovery, and high-value consumer products, which would be reviewed in Sections 3–7. An ash evaluation system is developed in Section 8 for instructing CFA utilization with the ideology of cleaner production.

3. Utilization of CFA in function materials

Using CFA in functional materials fabrication has been investigated for its potential to reduce the consumption of materials that have limited reserves or that are costly to manufacture. Fabrication of functional materials including mullite ceramics, glass-ceramics, zeolite, and coatings using CFA is reviewed in this section.

3.1 Mullite refractory

Mullite refractories have been widely used in the fields of glass, metallurgy, chemical engineering due to the high thermal stability, low thermal conductivity, and good chemical resistance (Luo et al., 2018), which are mainly prepared from the calcination of bauxite. In recent years, preparation of mullite refractory using CFA as raw materials have attracted much attention in China with the decreasing bauxite reserves and increasing demand for mullite refractory (Dong et al., 2010). The content of SiO_2 and Al_2O_3 in high-alumina CFA can reach more than 80%, and the main crystalline is mullite, making it suitable to produce mullite refractory materials from CFA. However, the SiO_2/Al_2O_3 ratio (2:1–1:1) in CFA is much higher than that of mullite (1:2–1:3), and certain amounts of CaO, Fe_2O_3, Na_2O, MgO, and K_2O are contained in CFA, which are detrimental to the preparation of mullite using CFA and also decrease the strength of refractory materials (Sukkae et al., 2018). Therefore, it is necessary to adjust the chemical composition of CFA to meet the requirements of SiO_2/Al_2O_3 ratio and remove impurities to produce mullite refractory materials with excellent performance.

So far, predesilication and addition of exotic aluminum source are the effective pathways to increase the SiO_2/Al_2O_3 ratio for producing mullite materials from CFA. The CFA contains approximately 30%–40% of amorphous glassy phase, which adversely affects the mechanical properties and alkali corrosion resistance of the ceramic products (Zhang et al., 2017). The alkali and acid leaching process have been employed to remove the glassy phase and adjust the chemical compositions of high-alumina fly ash (HAFA), and some glassy phase could be removed from HAFA through the formation of soluble sodium silicate during alkali leaching process and dissolution of liberative aluminum in HCl solution (Lin et al., 2015). The Al_2O_3 content in the treated powder can reach up to 65.35 wt%, and most of the impurities were removed after the alkali and acid treatment. The beneficiated sintering sample exhibited a mullite content of 88.33%, an apparent porosity of 1.20%, a bulk density of 2.78 g/cm^3, and a compressive strength of 169 MPa. To deeply remove most of the glassy phase, the mechanical–chemical synergistic activation method was developed to diminish the particle size and destroy the structure of encapsulation (Zhang et al., 2018a). The impurities and active Al were efficiently leached to increase the reactivity of amorphous silicate, and then amorphous silicate were absolutely leached by alkali leaching process. The fine grains grow to rod-like mullite, which exhibit excellent properties (bulk density >2.85 g/cm^3, apparent porosity $<0.5\%$).

Besides predesilication, addition of exotic aluminum source to adjust the chemical compositions of CFA for preparation of mullite is also an effective pathway. So far, boehmite sol (Li et al., 2018), $Al(OH)_3$, and Al_2O_3 (Zhu and Yan, 2017) have been used as the exotic aluminum sources to react with the amorphous glassy phase and produce mullite materials from CFA. The mullite powder consisting of 96.4% mullite and 3.6% quartz could be achieved after being treated in alkaline solution and acidic solution, which calcined at 1200 °C using the mixtures of milled CFA and 12% boehmite sol as raw materials. Furthermore, porous mullite ceramics can be prepared using CFA slurry samples with certain content of Al_2O_3 by a dipping-polymer-replica approach. It is found that the mullite-based porous ceramic with an Al/Si mole ratio of 2.40 had a much higher compressive strength and exhibited greater shrinkage than the other samples. Li et al. (2012) fabricated self-reinforced porous mullite ceramics by starch consolidation method with CFA, different aluminum sources ($Al(OH)_3$, Al_2O_3), and the additive AlF_3 as raw materials. AlF_3 as the additive is helpful to form mullite whiskers at a low temperature. These needle-like mullite whiskers constructed an interlocking structure, which enhanced the bending strength of the porous mullite ceramics effectively, and the porous mullite ceramics with bending strength about 100 MPa and apparent porosity about 55% were obtained at 1550°C.

In addition, Han et al. (2018) recovered mullite from CFA using a flotation and metallurgy process as illustrated in Fig. 24.3. Through a series of condition optimization, approximately 93.2% of the unburned carbon was firstly removed from the CFA assisted with effective humic acid surfactant in the froth flotation step. Then approximately 97.4% of glass phase in carbon-free powder was dissolved in the mixed acid solutions (HF: H_2SO_4 = 3:1) in the metallurgy step. The mineral matter in the obtained residue is mainly mullite. The performance of the refractory obtained from the water-cleaned mullite-enriched product meets the production requirements of refractory materials. Luo et al. (2017) prepared fully ash-based ceramic tiles using the mixtures of untreated CFA and alkali (NaOH solution) activation pretreatment of CFA. NaOH solution was used to adjust the chemical composition of CFA and enhance the plasticity of material surface. The obtained tiles that were sintered at 1100°C exhibited excellent performance with the highest rupture modulus of 50.1 MPa, largest bulk density of 2.5 g/cm^3, and lowest water absorption of 0%.

Comprehensive utilization of CFA for preparation of mullite refractory is not helpful with conservation of the environment, but benefit for the production cost reduction of mullite refractory. The CFA with high Al_2O_3 content, low alkali and alkaline earth metals content, and low ratio of SiO_2 to Al_2O_3 is more suitable for the preparation of mullite refractory. However, the components of CFA are complex and unfixed, of which the physical and chemical properties depend on the type of raw coal, as well as combustion conditions, which greatly influence the performance of CFA-based mullite refractory. In addition, many chemical reagents such as HCl, NaOH, etc. are generally used in preparing mullite from CFA, which greatly increases the synthesis cost of mullite. Replacing these reagents by other industrial solid wastes should be

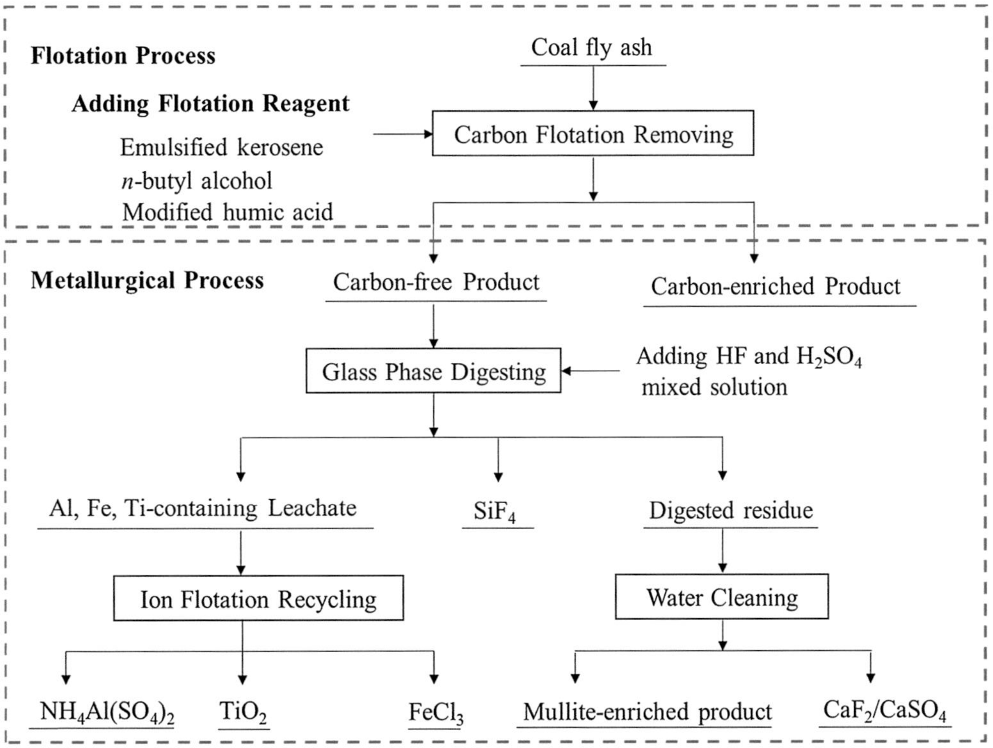

FIGURE 24.3 Schematic of a process flow diagram for CFA separation and mullite recovering.

focused on in the future. Economic evaluation of this application would give a greater indication of the area in the research that should also be focused on.

3.2 Glass-ceramics

Glass-ceramics is a new kind of thermal insulation materials, which consists of glass phase and crystalline phase. It has been widely used in the interior and exterior walls of buildings, floors and corridors, and other advanced decorative surfaces, due to its characteristics of high strength, abrasion resistance, corrosion resistance, weathering resistance, non-water absorption, and no radioactive pollution. The typical chemical composition of glass-ceramics contains SiO_2 (55%−65%), Al_2O_3 (7%−10%), CaO(13%−17%), Na_2O/K_2O, and a small amount of Fe_2O_3, B_2O_3, ZnO, or BaO (Ibáñez et al., 2013). The performance of glass-ceramics greatly depends on the chemical composition of raw materials, single CFA is not suitable to be used to prepare glass ceramics, and a few of materials including quartz (SiO_2) and dolomite ($CaMg(CO_3)_2$) are required (Rawlings et al., 2006). Moreover, soda (Na_2CO_3) and borax ($Na_2B_4O_7 \cdot 10H_2O$) are typically added in the raw materials during preparation of glass-ceramics due to the high melting temperature of CFA (Zeng et al., 2022). In 1983, DeGulro and Risbud prepared glass-ceramics for the first time using CFA from Illinois power plant by conventional melt quenching method. The glass-ceramics contains 23% of ferroaugite [$(Ca, Fe^{2+})(Al, Fe^{3+})_2SiO_6$] and potassium melilite [$KCaAlSi_2O_7$] (Deguire and Risbud, 1984). Recent research showed that CFA is mainly used to prepare the glass-ceramics of $CaO-Al_2O_3-SiO_2$ and $MgO-Al_2O_3-SiO_2$. The $CaO-Al_2O_3-SiO_2$ type mainly contains wollastonite (β-$CaSiO_3$) and diopside ($CaMg(SiO_3)_2$ or $CaFe(SiO_3)_2$), and the $MgO-Al_2O_3-SiO_2$ type contains cordierite ($Mg_2Al_4Si_5O_{18}$) (Deng et al., 2006). These crystalline minerals in glass-ceramics are formed by controlling the parameters during the synthesis process (Yao et al., 2015). Two methods are applied for the preparation process: body crystallization method and particle sintering crystallization method (An et al., 2022). The body crystallization method contains the mixing of raw materials, melting, and annealing to obtain the glass-ceramics with fine grains and uniform structure (Rawlings et al., 2006). Particle sintering crystallization method includes the melting of raw materials into glass liquid, water quenching, crystallizing according to a certain particle size distribution, trimming, smoothing, and polishing (Hill and Wood, 1995; Deng et al., 2006). The advantages and disadvantages for these two methods are shown in Table 24.1 (Hai et al., 2014).

The particle size of raw materials will affect the performance of glass-ceramics. The suitable particle size of raw material is less than 178 μm. Particularly, for the particle sintering crystallization method, the maximum particle size after melting is 5 mm. Heat treatment temperature also has an important effect on the properties of products. It can directly affect the type, size, and quantity of the crystalline phase of glass-ceramics. Thermal treatment system of glass-ceramics can be divided into stepped temperature system and isothermal temperature system. The optimum melting temperature, annealing temperature, and crystallization temperature of raw materials for preparing the glass-ceramics are 1300−1500°C, ~550°C, and 850−1100°C, respectively (Hai et al., 2014).

In China, the research and development of glass-ceramics began in the middle of 1970s, and its industrial production dates from the beginning of 1990s (Feng, 2004). To date, many scholars have successfully mastered the key technology of producing glass-ceramics using solid wastes such as CFA, coal gangue, various tailings, smelting slag, river sediment, and waste glass. The amount of CFA added in the raw materials varies from 24% to 60%, and the total amount of various industrial solid wastes is over 90% (Yang et al., 2004; Rawlings et al., 2006).

With the rapid development of China's economy, the demand for wear resistance and corrosion resistance building materials will be increasing. CFA-based glass-ceramics will have stronger market-competitive advantage and broad development prospects. However, the following aspects should be studied in the future: (1) enriching the tone and color series of fly ash glass-ceramics; (2) reducing or eliminating the air bubbles of fly ash glass-ceramics produced by sintering

TABLE 24.1 Comparison of the two methods to produce glass ceramics.

Method	Advantage	Disadvantage
Body crystallization method	1. Homogeneous and compact 2. Transparent materials 3. Convenient to mechanization	1. Limit of thickness 2. Only monochrome patterns 3. High cost of raw materials
Particle sintering crystallization method	1. Low cost of raw materials 2. Unlimit of thickness 3. Natural pattern can be formed	1. Easy to produce bubbles 2. Surface roughness 3. Cutting cracking

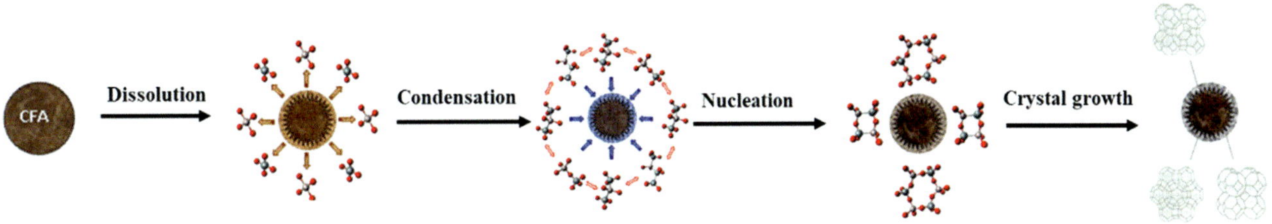

FIGURE 24.4 Schematic of the CFA zeolitization mechanism.

method; (3) eliminating the influence of composition changes of fly ash on the performance of fly ash glass-ceramics; (4) formulating the national standard of fly ash glass-ceramics products in order to facilitate the production, inspection, and application of enterprises.

3.3 Utilization of CFA in zeolite production

Zeolites comprise an important group of crystalline aluminosilicate minerals; they possess an infinitely extending three-dimensional anion network made up of $(SiO_4)^{4-}$ and $(AlO_4)^{5-}$ tetrahedra that link at the corners via their shared oxygen atoms (Belviso, 2018). The compositional similarity of CFA to some volcanic material, precursor of natural zeolites, is the main driving force to synthesis zeolites from CFA (Höller and Wirsching, 1985). Although this potential application may consume only a small proportion of the CFA production, the final products obtained may reach a relatively higher added value than when applied in current applications. Ojha et al. reported that the cost of synthesized 13X zeolite from CFA was less than one-fifth of that in the market taking into account the costs of all steps in the preparing process (Ojha et al., 2004). Moreover, Panitchakarn et al. reported that the cost of preparing zeolite from CFA in a large-scale was estimated to 19.74 US$ per kg of zeolite, whereas the price of 4A is at US$83.87 per kg in the market (Panitchakarn et al., 2014).

In the synthesis process, the aluminosilicates of CFA were converted to zeolite crystal by alkali hydrothermal reactions. The soluble silicate and aluminate form CFA cluster into geopolymers, nucleate, and grow into crystals, which was shown in Fig. 24.4 (Bukhari et al., 2015). In the process, the chemical component of CFA has significant influence on the synthesis of zeolite. The properties of CFA greatly depend on the feed coal and combustion conditions. CFA contains a series of phases including quartz, mullite, tricalcium aluminate, calcium silicate, and hematite. Among them, quartz and mullite are the major crystalline compounds in the CFA. Besides the minerals, amorphous solid such as glass consist the majority of CFA. Catalfamo et al. found that the mullite phase inside CFA is hard to dissolve at the suitable conditions for zeolite synthesis, and it hindered the conversion to zeolite at the beginning stage of the hydrothermal process (Catalfamo et al., 1993). It is reported that mullite showed the lowest reactivity, glass, and quartz were reactive in the synthesis of zeolite from CFA. The components of CFA also may affect the morphology of the prepared zeolite (Querol et al., 1995). In contrast, the amorphous aluminosilicate would dissolve in the alkaline condition and provides the silica and aluminum sources for the formation of zeolite. The concentration of Si and Al in the reaction system is varied by means of the properties of CFA and reaction conditions. Meanwhile, the type of zeolite can be tuned by adjusting the Si/Al ratio with adding constituents. In addition to SiO_2 and Al_2O_3, CFA also contains a certain amount of Fe_2O_3 and CaO, depending on the coal used in the thermal power plant. In the hydrothermal process, Fe^{3+} and Ca^{2+} can enter into the lattice of zeolite along with the Al^{3+} and $Si^{4+}$and compete with Al^{3+} in the formation of zeolite framework; these two ions would reduce the crystallinity of synthesized zeolites (Kuwahara et al., 2008). To obtain a zeolite with high purity and enhance the crystallinity of product, the acid pretreatment is performed to remove most of Fe_2O_3, CaO, and other impurities in CFA (Panitchakarn et al., 2014). In the acid pretreatment process, HCl, H_2SO_4, and HNO_3 can be selected as acid sources, which would remove the calcite phase directly (Li et al., 2006).

The direct hydrothermal process is the earliest method for the conversion of CFA to zeolite. Specifically, CFA is uniformly dispersed in the appropriate concentration of alkali solution, aged at a certain temperature for a period of time, then crystallized in the appropriate temperature range, finally filtered, washed, dried, and finally obtained zeolites (Steenbruggen and Hollman, 1998). The synthesis parameters including alkali concentration, liquid–solid ratio, reaction time, and reaction temperature determined the resulted type of zeolite. It has been reported that a series of zeolite, such as A, X, P, analcime, chabazite, or hydrosodalite, was synthesized by direct hydrothermal method. Murayama et al. (2002) proposed that the hydrothermal synthesis of P-type and rhombohedral zeolites from CFA requires three steps: (1) the dissolution of Si^{4+} and Al^{3+} in CFA; (2) the formation of silicon–aluminum gel; (3) crystallization to form zeolite crystals.

Considering the shortcomings of one-step hydrothermal synthesis method, the pretreatment of CFA using alkaline fusion method is employed to increase the efficiency of synthesis and the yield of zeolite. The alkali (typically NaOH) and CFA are mixed and fused at high temperature (typically >500°C), and the mineral structures of mullite and quartz are destructed in this process. Therefore, more content of Si and Al in CFA can be introduced into the subsequent hydrothermal process, which not only enhanced the conversion of CFA but also increased the surface area of resulted zeolites (Rayalu et al., 2000). The conversion of CFA to sodium silicate and sodium aluminate by an alkali-fusion treatment depends on the amount of NaOH, reaction temperature, and reaction time. Shigemoto et al. (1993) were one of the early teams that synthesized zeolite using alkali fusion pretreatment. In the synthesis process, quartz dissolves and takes part in the reaction, while mullite remains unchanged. Na-X zeolite with 62% crystallinity was synthesized when the base fusion temperature reached 550°C and the mass ratio of NaOH to CFA is 1.2. Berkgaut and Singer (1996) added a small amount of water to CFA and NaOH mixture, which results in the full decomposition of mullite.

Microwave heating is more efficient than bulk heating, so microwave heating instead of conventional heating is considered to improve the synthesis efficiency. CFA can be aged by microwave heating, crystallized for a period, then filtered, washed, and dried to obtain zeolite. Inada et al. (2005) prepared NaPl zeolite using CFA as raw material by microwave-assisted method. It is found that the dissolution of aluminosilicate is promoted in the early stage of microwave heating, but the formation of zeolite is obviously hindered in the intermediate stage due to formation of crystal nucleus. Therefore, they combined the microwave heating in the early stage and traditional heating in the later stage, and pointed out that the microwave frequency from high to low is advantageous to the synthesis zeolite. Additionally, the ultrasonic wave is applied in the synthesis of zeolite from CFA. The ultrasonic wave can spread around in liquid phase, and the tiny vesicular nucleus in the liquid is activated to form zeolite. The combination of ultrasonic and microwave synthesis would have better synthesis effect. Musyoka et al. (2012) found that in situ ultrasonic method can be used to observe the conversion process of CFA to zeolite, verifying that preparing zeolite by CFA is from liquid and solid phases.

After a long period of exploration, hydrothermal method has been developed rapidly for synthesis zeolite from CFA. According to the specific properties of CFA generated at different places, the synthesis method can be reasonably selected to obtain zeolite with higher added value, in which the optimizing of technology parameters is still of great significance. The reported zeolite produced from CFA focused on the types with low Si/Al ratio; reliable technology to synthesis SiO_2-rich zeolite such as ZSM and mordenite is required. A few of drawbacks still remain in the large-scale production of zeolite by CFA, including high consumption of water and alkali, low purity, and high cost. Therefore, developing more efficient and greener technology for the zeolite production is necessary.

3.4 Fly ash—based coating materials

Traditional coating fillers include porcelain powder, light calcium carbonate, talc, quartz powder, etc. With the development of the coatings industry, the demand for fillers is gradually increasing. Suitable and cheap substitutes to reduce the cost of coating fillers are needed at home and abroad. CFA has good dispersity, applicable density and fluidity of globular particles, and CFA is active and has a certain gel ability. It is potential be used as a cementing material for coatings, or as a substitute for pigments and fillers.

Based on mechanisms of CFA, it can be divided into multiple functional coatings. Based on the low thermal conductivity of cenosphere particles, zeolite and geopolymer made from fly ash, CFA can be used to prepare heat-insulating coatings. CFA has a relatively dense structure due to the rapid cooling in a high-temperature molten state. Therefore, it can play a role in flame retardant and fireproofing in coatings. According to the pozzolanic activity of fly ash, it can be used as a filler for anticorrosive coating, waterproof coating, sealing coating, building exterior wall coating, casting coating, etc. The aluminosilicates in the CFA react with the polar group to form an inorganic gel based on the principle of geopolymer, which has good adhesion, strength and stability, and good resistance to acid—alkali, aging, and freeze-thaw. Due to the good shape effect and microaggregate effect, the addition of CFA not only enhances the fluidity of the coating system and reduces the water consumption, but also improves the microstructure and the performance of the coating. As a cheap and good quality filler, CFA has certain industrial utilization value.

Many international researchers have done a lot of work on CFA fillers in coatings. CFA has been studied more in anticorrosive coatings and thermal insulation coatings. The corrosion-resistant coating for steel substrate was prepared with CFA without carbon content and dust impurities, and the dosage of CFA is about 10% (Ruhi et al., 2015). The metal—ceramic composite corrosion-resistant coatings are developed using fly ash premixed with aluminum powder, and the CFA dosage is 5%—35% (Sahu et al., 2010). Thermal insulation coatings based on CFA of Class F on the mild steel plates were stable up to a temperature of 500°C (Deshmukh et al., 2017).

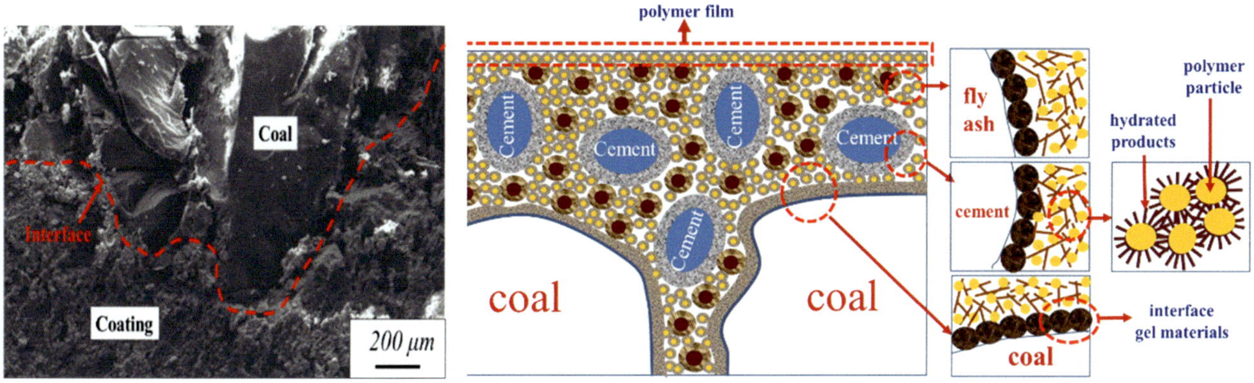

FIGURE 24.5 SEM image of the microscopic penetration and combination of sealing coating on coal (left), the composite mechanism model for sealing coating after hardening (right) (Song et al., 2018).

In the 1990s, the Shanghai Coating Research Institute paid attention to the application of CFA in coating fillers earlier (Wang et al., 1994). Subsequently, Hua Jianshe of Xi'an University of Architecture and Technology started the research on CFA casting coatings since 2002 (Hua, 2002). Since 2009, CFA has been gradually applied in waterproof coatings (Cheng et al., 2012a), fireproof coatings (Cheng et al., 2012b), and gas sealing coatings (Xue et al., 2017) in the group of Professor Cheng Fangqin in Shanxi University.

However, the amount of CFA added is still very low due to technical limits. The high volume of CFA may affect the crystallization and slow down the pozzolanic reaction. One of the most effective approaches is to grind CFA to ultrafine CFA (UCFA) by the steam mill or other supermicromill, which is beneficial in reducing the harm of residual carbon and minimizing the variability of constituents in typical CFA (Obla et al., 2003).

The dosage of UCFA in gas sealing coating is up to 60%, because the UCFA could form good gradation with other powder, could have great compatibility with polymer emulsion, and could increase the strength characteristics, the suspension property, and dispersibility (Song et al., 2016).

The interfacial compatibility between the CFA-based sealing coating and coal is very good, and the bonding between them is tight shown in Fig. 24.5 (left). The curing mechanism and synthetic model of CFA-based gas sealing coating had been summarized expressed in Fig. 24.5 (right), providing a new theoretical basis for the development of gas-sealing coatings (Song et al., 2018).

Based on the research, the pilot test of this gas sealing coating had been done on the underground tunnel wall (more than 1500 m) in the returning roadway of 3108# fully machined mining face of Lu'an Group in Shanxi Province, China (Song et al., 2019). The results showed that gas sealing coating based on UCFA with dense structure could seal the small cracks in the surface of the coal wall, which effectively delayed gas gush.

In a word, UCFA is a kind of good fillers for coating. Low whiteness of gray CFA results in an undesirable appearance to the final coating product (Yang et al., 2006). It is more suitable for coal mine wall, tunnel wall, warehouse, bridges, and somewhere else with low color requirement. Additionally, large dosage of UCFA as filler can reduce the cost of the coatings. Therefore, there are great development potential and broad application prospect of UCFA in these kinds of coatings.

4. Applications of CFA in agricultural field

Soils are dynamic and diverse natural systems and lie at the interface among the earth, air, water, and life, which are also critical ecosystem service providers for the sustenance of humanity. Soil degradation is a decline in soil physical, chemical, and biological quality. It can be the loss of organic matter, decline in soil fertility, structural condition, and accumulation of salinity and toxic chemicals. According to UNCCD (Dessertification, 2012), 1.9×10^9 ha of lands have been degraded. The total area of degraded land has increased from 15% in 1991 to 25% in 2011 (Gnacadja, 2013). Degraded soils reduce the ecosystem function, and even worse, some soils and their runoff water are harmful to our environment and human health. At present, repairing the degraded soil is among the great efforts of soil scientists around the world.

Due to the similarities between soils and CFA (Page et al., 1979; Jala and Goyal, 2006), CFA could be used as a fertilizer/soil amendment in agriculture to improve the growth of plants and reduce the pollution (Singh et al., 2010; Parab et al., 2012). Some criteria should be met before CFA was applied in the agriculture field. Firstly, heavy metal content in

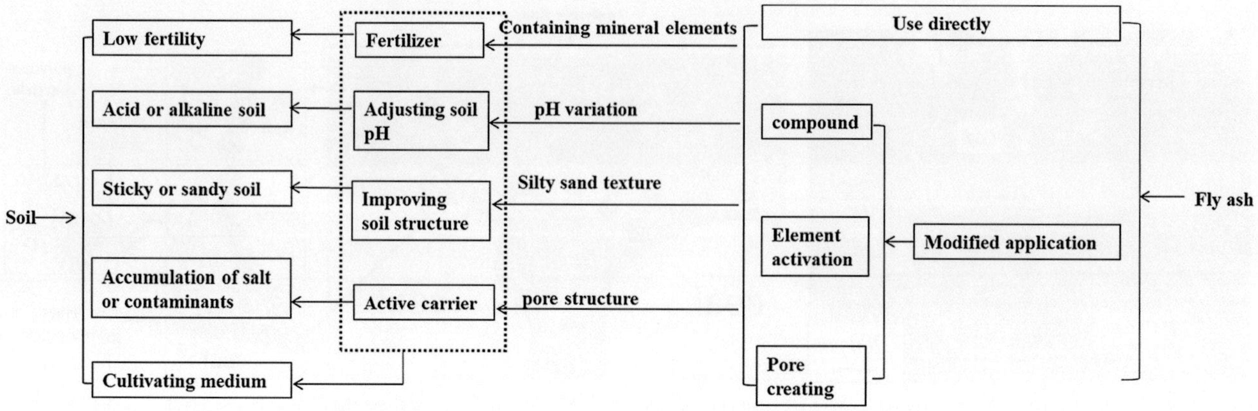

FIGURE 24.6 Schematic diagram of agricultural utilization of CFA.

CFA should be below the standard limits, and the contents of water and carbon should be lower than 10% and 15%, respectively. Particle size of CFA should be uniform and does not contain sundries. CFA containing high content of nutrients is more desirable in agriculture.

According to the physical and chemical properties of CFA, it can be used in the following ways (Fig. 24.6).

4.1 Use directly

4.1.1 Effect of CFA on the soil physical properties

The reason that CFA can be used to alter the soil texture is that its lower bulk density, higher water-holding capacity, and lower hydraulic conductivity than soil alone (Rechcigl, 1996; Jayasinghe and Tokashiki, 2012). CFA can change the bulk density of soil due to its similar particle size range as silt. Prabakar et al. (2004) reported that application of CFA at different ratio in clay soil could significantly reduce the bulk density and improve the soil structure (Prabakar et al., 2004). Soil texture altered by application of CFA would to some extent affect the plant growth, nutrient retention, and biological activity of the soil directly or indirectly. Pathan et al. used CFA in degraded soil and found that soil water content was increased progressively, and extractable P content in the soil was increased and NO_3^-, NH_4^+, and P leaching were reduced in the sandy soil (Pathan et al., 2003). Yan and Dong (2001) found that surface encrustation was reduced, and soil aeration was enhanced in acid clay soils by a high dose of CFA; meanwhile the germinations of plants were improved.

4.1.2 Effect of CFA on the soil pH

The CFA can either be acidic or alkaline, relating to the source of the coal used for combustion, especially the S content of the parent coal, and the operating condition of the plant (Ram and Masto, 2010). When CFA was applied to soil, the soil pH change is related to the pH difference between the CFA and soil, the buffering capacity of soil, and the neutralizing capacity of CFA depending on the amount of CaO, MgO, and Al_2SiO_5 in it (Basu et al., 2009). Fan et al. (2016) found that CFA with low B and salt content can be used as liming agents in acid soils at dose of no more than the 40 mg/ha. Acidic CFA has also been used for reclamation of sodic soils, resulting in a slight decrease in soil pH and a significant decrease in soil conductivity. Fan et al. used CFA and vinegar residue, to reduce the pH and improve the soil structure and fertility in saline–alkaline soil. The changes of soil pH would be beneficial for the growth of plants by improving nutrient availability and reducing metal toxicity, etc.

4.1.3 Effect of CFA on the soil heavy metals

CFA contains some toxic metals such as As, Cd, Co, Cr, Cu, Hg, Mo, Ni, Pb, V, Zn, etc. The application of CFA could pose a contamination risk to soil, plants, surface, and groundwater due to elevated concentrations of potentially toxic heavy metals (Wong and Wong, 1986; Ram and Masto, 2010). Thus, the potential risk of CFA will be assessed before it is applied in the soils.

The pH value has been reported as the most influential parameter for leaching of many heavy metal elements (Zandi and Russell, 2007). Hence, CFA could be used to immobilize the heavy metals in soils due to its high pH. This was demonstrated by Jurate et al. (2007), who found that the exchangeable metal form was reduced due to the formation of new copper- and lead-containing minerals by CFA. Besides raising soil pH, CFA could enhance soil metal sorption through supplying metal sorbents (e.g., SiO_2, Al_2O_3) and by increasing soil surface area. This interpretation is consistent with the observations that the hydrous silicon oxide and aluminum oxide in CFA form complexes with metal ions by chemical bonding (Pathan et al., 2003).

4.1.4 Effect of CFA on the soil fertility

CFA contains large amounts of elements involving K, Na, Zn, Ca, Mg, and Fe, which has a positive effect on the yield of agricultural crops. Seshadri et al. found increase in P adsorption and enhanced the P retention capacity in acidic and neutral soils by application of alkaline CFA (Seshadri et al., 2013). In addition to increasing the soil fertility, CFA could also reduce the fertilizer loss in soil water (Stout et al., 1999). According to Stout et al. (2000), the soluble inorganic P could be converted to more tightly bound Fe and Al, less amounts of Ca-bound soil inorganic P so that it could be precipitated, and will not release to water.

4.1.5 Effect of CFA on the soil microbes

Soil biota are fundamental for soil function and the growth of plants. CFA exerts influence on soil microbes by affecting the physicochemical properties of soils (Ram et al., 2011). Schutter and Fuhrmann (2001) found an improvement in microbial community numbers, including an increase in bacterial and actinomycetes counts, but no effect on fungi in clayey and sandy soils amended with cocomposted FA. According to Kumpiene et al. (2009), CFA increased microbial biomass and respiration, meanwhile increased key soil enzyme activities in a Cu−Pb-contaminated soil. But some research reported that application of CFA can inhibit microbial respiration, enzyme activity, and restricted occurrence of organic matter in amended soil attributing to high pH and high concentration of soluble salts (Klubek et al., 1992; Sims et al., 1995).

4.1.6 Effect of CFA on the growth of plants

The agronomic benefits of CFA applications are primarily associated with improved physicochemical and biological characteristics of the soil. Luo et al. (2021) reported that application of moderate amounts of CFA could contribute to better growth, dry matter production, and increased photosynthetic pigments in *Lactuca sativa*. Besides, CFA could effectively control various pests infesting several vegetables and finally enhanced the fruit yield (Eswaran and Manivannan, 2007).

4.2 Modified application

4.2.1 Blending CFA with other inorganic and organic materials

Coapplication of CFA with organic/inorganic materials has many advantages than single CFA utilization (Fig. 24.7). When using CFA in agriculture as a soil ameliorant, it is better to seek the locally available fitting blend materials for exploiting the benefits from their synergistic interaction. Fan et al. (2016) compounded CFA with vinegar residue, the main waste in Shanxi Province, and had been proved to be an effective way to improve saline−alkaline soil and plant growth. Spark and Swift (2008) found that CFA has the potential to be used as a base material of soil amendment due to its changeable pH and fertility.

4.2.2 Activation of mineral elements in CFA

Although CFA contains many kinds of mineral elements, most of them are inactive and could not be absorbed by plants directly. Many modification methods have been reported to activate mineral elements in CFA. A merlinoite was synthesized from CFA employing the KOH direct conversion method and was proved to be an efficient slow release K-fertilizer for plant growth in nutrient-limited soils and in opencast coal mine areas around the power plants (Li et al., 2014). A potassium silicate fertilizer was prepared with CFA and KOH, and the growth performance of wheat was better than that of the control (Goto et al., 2000).

4.2.3 Remediation of heavy metal-polluted soil used CFA-based zeolite

CFA has the similar elements composition as zeolite. CFA was prepared to zeolite with high performance, which can adsorb and fix heavy metals. Querol et al. (2006) investigated the use of zeolitic material synthesized from CFA for the immobilization of pollutants in contaminated soils and found that the zeolitic material considerably decreases the leaching of Cd, Co, Cu, Ni, and Zn in contaminated soils. The increase of soil pH caused by the zeolite application

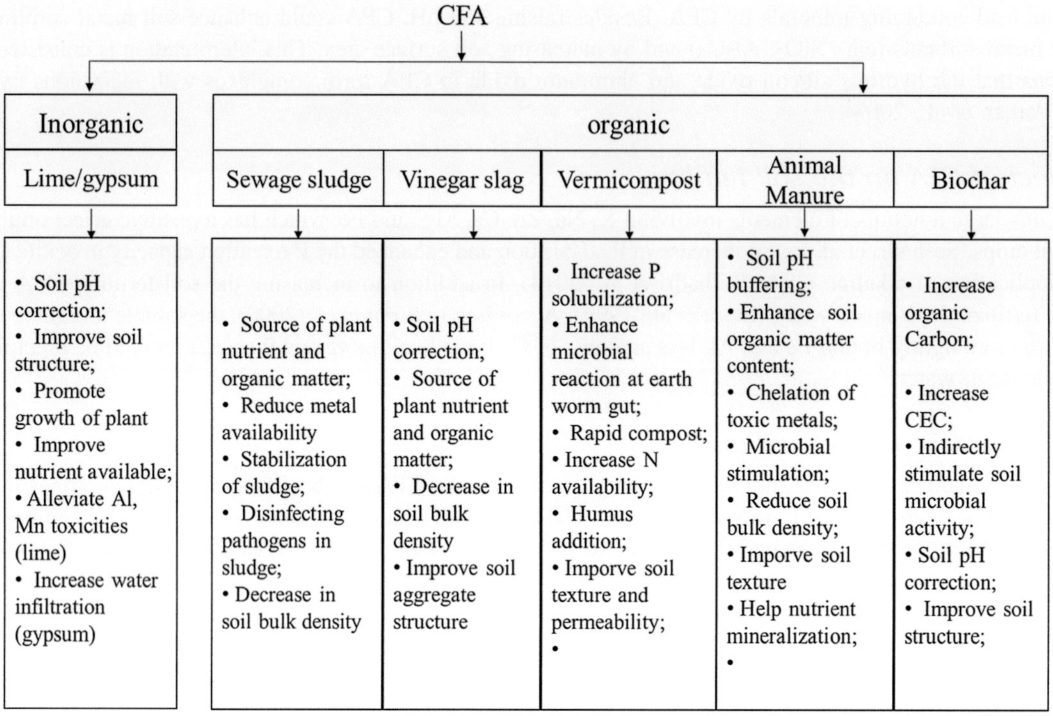

FIGURE 24.7 Beneficial role of other inorganic and organic amendments used along with CFA modified (Ram and Masto, 2014).

seems to be the factor responsible for metal immobilization, favoring metal adsorption onto illite surfaces and precipitation of metal hydroxides. Zeolite X was synthesized at low temperature (around 30°C) directly in Zn- or Pb-contaminated soil treated with CFA and the presence of the generated mineral reduces the availability of the toxic elements (Belviso et al., 2012).

4.2.4 Preparation of soilless culture substrates

CFA can also be used to prepare soilless culture substrates to meet the requirements of plant growth under specific conditions. Menzies and Aitken (1996) investigated the potential for incorporation of CFA in soilless potting substrates and found that CFA can be successfully incorporated in potting substrates at low rates ($\leq$20% of the mix volume) without loss of plant yield or quality. In addition, the preparation of porous ceramsite from CFA is another important way to utilize CFA. Shi et al. (2018) studied the possibility of cultivating ornamental plants with ceramsite and found that pod anthracene in Europe can grow normally on ceramsite matrix. Zhang and Liu et al. developed a new plant nutrition matrix, which was prepared by CFA and waste fungus sticks, straw, livestock and poultry manure, sawdust and other agricultural wastes as the main raw materials, which solves the problem of waste storage and reduced the pressure of environmental protection and was applied on a large scale in China (Zhang et al., 2018b).

Overall, positive effect is shown as the CFA/amendment applications at lower doses or combined with other organic amendments. However, excess amounts of CFA would have negative impact on the growth of plants, due to its acidic/alkaline and high pH value (Yang et al., 2011).

There is a great potential for the consumption of CFA in the field of soil remediation. However, most of the application of CFA in agriculture remains on the laboratory scale and is rarely used in large-scale applications.

For the use of CFA in agriculture, the following issues should be paid more attention to:

(1) Emphasis should be laid on the foundational research for the long-term effects of CFA with various properties on different kinds of soil and plant.
(2) The classification system of fly ash should be established according to different origin and nature, so CFA can be classified and applied according to different properties.
(3) Related technical standards and regulations should be established aiming to use the CFA moderately without harming the environment.
(4) Related policy should be released to encourage the application of CFA in agriculture.

5. Metals recovery

In the northwest of China, especially in Inner Mongolia and Shanxi province, the coal is rich in aluminum (10%–15%), and the aluminum is further enriched in CFA with the alumina content of 35%–50%, close to that of midgrade bauxite, making it as a potential raw resource for aluminum recovery. In addition, the CFA in some areas in China associated with some rare metals, such as germanium, lithium, gallium, and rare earth elements (REEs), offering an opportunity for valuable metals recovery and value-added utilization of CFA. In recent years, more and more efforts have been focused on extraction of valuable metals especially alumina from CFA in China.

5.1 Aluminum

In the world, most of the alumina are produced from bauxite by the Bayer process. However, with the restriction on mining and dependent on import of bauxite resources as well as the increasing demand for aluminum, strategic replacement of bauxite is of great significance to the sustainable development of aluminum resources (Yao et al., 2014). So far, a number of processes, including sintering process, acid leaching process, and chlorination process for alumina recovery from CFA, have been reported and proposed.

Unlike bauxite, alumina in CFA exists in the form of chemically stable mullite, leading to low recovery efficiency of aluminum and high recovery cost. To overcome the problems, sinter processes including limestone sintering process (Fig. 24.8) and lime-soda sintering process were proposed to recovery aluminum from CFA. The procedures of limestone sintering process are as follows: sintering of CFA and limestone, alkali leaching, desilication of the crude $NaAlO_2$ solution, and carbonization (Lin et al., 2012). In the sintering step, CFA and limestone mixture is sintered at 1300–1400°C to

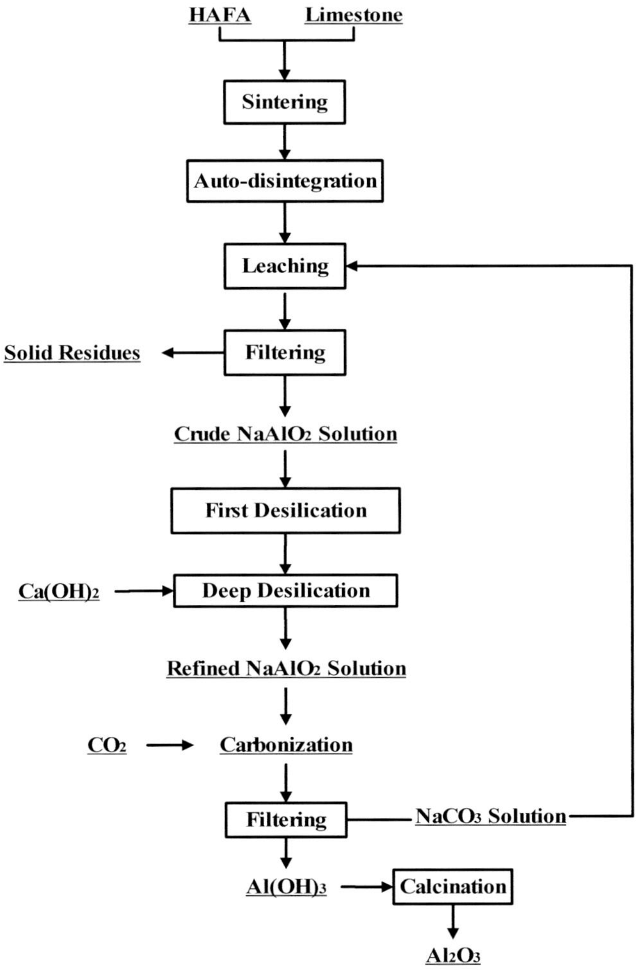

FIGURE 24.8 The schematic diagram of the limestone sintering process.

transform mullite and quartz into $12CaO \cdot 7Al_2O_3$ and $2CaO \cdot SiO_2$, respectively, and $12CaO \cdot 7Al_2O_3$ could easily dissolve in the recyclable Na_2CO_3 solution to form soluble $NaAlO_2$ while $2CaO \cdot SiO_2$ is difficult to dissolve. About 80% alumina recovery could be obtained at the optimal conditions using limestone sintering process. However, the process has some disadvantages, such as high energy consumption and a large discharge amount of silicon−calcium slag. As a result, the lime−soda sintering process was developed (Ding et al., 2017), and the mixture of quicklime and soda with CFA is sintered at relatively low temperatures of 1100−1400°C to produce soluble sodium aluminate ($NaAlO_2$) and insoluble calcium silicate ($2CaO \cdot SiO_2$). Even so, 2−3 tons of silicon−calcium residues could be discharged when 1 ton of alumina was generated. Therefore, a novel process named predesilication and lime−soda sintering process was reported and developed by the Datang International Group, and various researches on the predesilication have been conducted as discussed in Section 4.1 (Sun and Chen, 2013). A pilot test with an annual output of 3000 tons of alumina was built and passed a technical appraisal in 2008. A project with an annual production capacity of 200,000 t was then completed in 2010 and continuously maintained stable operation since 2012.

Direct acid leaching at atmospheric pressure cannot obtain high leaching efficiency of alumina from CFA yet. Hence, the pressure hydrochloric acid leaching process (as shown in Fig. 24.9) was proposed in 2004 by Shenhua Group Co., Ltd. together with Jilin University, which involves dissolving aluminum in hydrochloric acid, filtering and purifying the crude $AlCl_3$ solution, evaporation, crystallization, and calcining into alumina. A pilot-scale plant generating 4000 t of alumina with a recovery efficiency of 85% was constructed in August 2011, commissioned since 2012. However, the high corrosion resistance for reaction medium limited the industrial application besides low extraction efficiency and difficulty in removal of impurities. To reduce the corrosion resistance, sulfuric acid pressure leaching process was reported by Xu et al. (2016) to extract metallurgical level alumina from CFA including acid leaching, removing iron, crystallizing, and roasting processes. The leaching efficiency of Al_2O_3 reached 93.1%, and the purity of Al_2O_3 product reached above 98% under the optimum conditions. Bai et al. (2011) described a concentrated sulfuric acid roasting process. Under the optimum conditions of sulfuric acid concentration at 1.2 times the HAFA and a calcination temperature of 300°C for 2 h, the alumina extraction efficiency reached up to 85%. To date, this process remains at laboratory- or pilot-plant-scale level.

Besides alkali sintering process and acid leaching process, chlorination process for alumina extraction from CFA has been developed and reported by Wang et al. (2019). With increasing time, temperature, and carbon, the chlorination rate of alumina and silica increased significantly. Under the optimal conditions, the highest chlorination rates of alumina and silica were 85.8% and 74.0%, respectively.

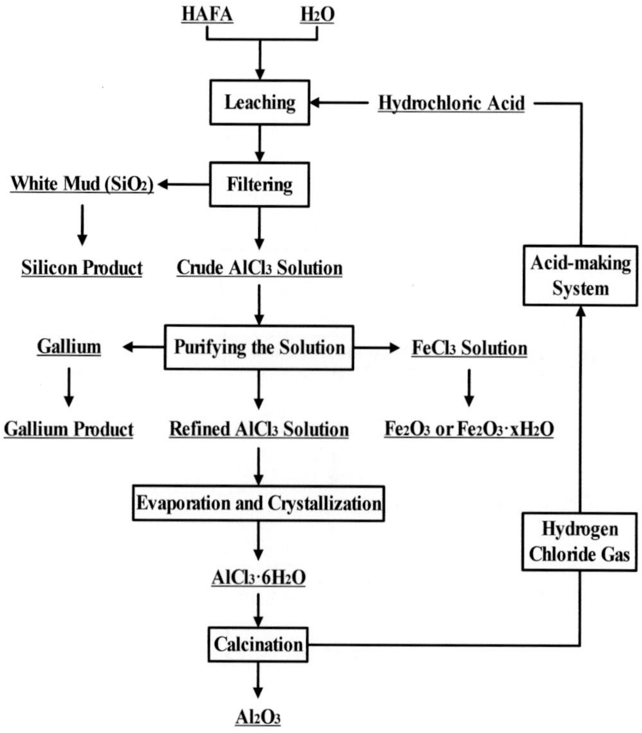

FIGURE 24.9 The schematic diagram of one-step direct hydrochloric acid leach process.

Most of the processes and technologies for aluminum extraction from CFA are in the laboratory experiment stage, and there are drawbacks for most of the aforementioned methods. For sintering processes, the energy consumption is relatively high, and the process is complex, restricting their large-scale application. In addition, a generous amount of calcium silicate residues is generated, but not utilized in an effective way except by manufacturing cement. Other problems, however, such as filtration and the washing of desilicated liquor, are also confronted in the industrial application of these processes. The recovery cost of each ton of alumina from CFA is greatly higher than that of alumina produced from bauxite using Bayer process. The acid leaching process can dissolve Al and other metals, and relatively small amount of silicon-rich residue could be generated, resulting in comprehensive utilization of Al and Si. However, aluminum recovery from leach liquor is quite complex, due to the pretreatment for purifying the aluminum, leaching to low recovery and purity. Furthermore, the acid leaching processes require acid-resistant and pressure-resistant equipment. The drawbacks also limited their industrial application.

5.2 Rare-earth metals

Besides aluminum, some associated metals such as germanium, gallium, lithium, REEs are concentrated in certain coals, and they are enriched in fly ash during combustion and gasification processes, where the contents are even 5−10 times greater than in raw coal (Bai et al., 2011). When the contents of associated metals reach industrial mining grade, such as germanium ($\geq$80 μg/g), gallium ($\geq$60 μg/g), lithium ($\geq$200 μg/g), and REEs ($\geq$300 μg/g), it is economic and feasible to recover the associated metals from fly ash.

Germanium (Ge) is a valuable element used in the manufacture of infrared optics, light-emitting diodes, photovoltaic cells, and catalyst for polyethylene terephthalate. Several studies have focused on Ge recovery from CFA using acid leaching followed by a process separating Ge from the other elements. The separation methods mainly included precipitation with tannin (Arroyo et al., 2009), distillation of $GeCl_4$ (Moskalyk, 2004), flotation (Hernández-Expósito et al., 2006), adsorption onto activated carbon (Marco-Lozar et al., 2007), solvent extraction (Kamran Haghighi et al., 2018), and adsorption onto chelating exchange resins (Torralvo and Fernández-Pereira, 2011). Arroyo Torralvo et al. studied the Ge recovery from CFA leach liquor using ion-exchange processes. The procedure was based on Ge complexation with catechol in an aqueous solution followed by the retention of the Ge-catechol complex onto a conventional strongly basic anionic resin. The maximum amount of retained Ge was calculated to be 215.5 mg/g resin. The elution of the germanium complex was achieved with HCl in 50% ethanol solution. They also studied the Ge recovery from CFA leach liquor by solvent extraction processes. The recovering 1.3 g/h of GeO_2 could be obtained from a 5 kg/h pilot plant. The semiindustrial implantation was estimated a capital cost of €1.512 million without any solvent recovery unit included.

Gallium (Ga) is an important metal and widely used in the applications of optoelectronics, telecommunications, photovoltaic, aerospace, alloys, and computers. So far, some research group have focused on Ga recovery from CFA (Lu et al., 2017). Fang and Gesser (1996) recovered Ga from CFA by using the processes including acid leaching, impurity removal, foam extraction of Ga, and further purification. For the two-stage leaching process, 2 M hydrochloric acid was used sufficiently to recover 95% of the Ga from CFA. The processes required the impurities removal, such as SiO_3 ions, Fe^{3+}, and Ca^{2+} in the leach liquor. Gutiérrez et al. (1997) also studied the Ga recovery from CFA by acid leaching and solvent extraction with commercial extractants of Amberlite LA-2 and LIX 54 dissolved in kerosene. The leach liquor was first contacted with Amberlite LA-2 to extract Ga and Fe, and then the iron was precipitated with sodium hydroxide, while Ga remained in the leach liquor. Finally, LIX 54 was used for Ga-selective extraction from the iron-removal solution, and 83% of the Ga present in the leach liquor was obtained in the resulting stripped solution. Besides acid leaching process, several NaOH-based extraction tests were performed on the IGCC fly ash to recover Ga (Font et al., 2007). Under the optimal conditions, Ga extraction yield of 60−86% could be obtained, and Ga concentration in the leach liquor could increase from 25−38 mg/L to 188−215 mg/L with the recirculation of leachates. Recently, stepwise separation using P507 and Cyanes272 was also reported to recover Ga from sulfuric acid leach liquor of coal fly ash (Zhao et al., 2020). At equilibrium pH $\leq$ 0.8, the impurities Ti (IV) and Fe (III) in the liquor could be preferably removed by a two-stage cross-current extraction with P507, and then Ga (III) can be selectively separated against Al (III) using Cyanex 272 at equilibrium pH of 2.4−2.6.

Lithium (Li) is electrochemically active, having the highest redox potential value and the highest specific heat capacity of any solid element make it the hottest commodity for modern life and a key element for modern electric vehicle revolution. It is reported that Li has been anomalously enriched in some coal deposits in China, and the CFA is further enrichment of Li, which is a promising source of Li (Qin et al., 2015; Hu et al., 2018). A novel technique was developed by Li et al. (2017) for the recovery of Li from CFA using a combination of predesilication and an intensified acid leaching process. The leaching efficiencies of Li and Al were 82.23% and 76.72%, respectively, under the optimal acid leaching conditions with 6 mol/L HCl, solid to liquid ratio of 1:20, 120°C, and 4 h.

REEs are indispensable strategic resources, which consist 17 elements that are chemically like each other, including 15 lanthanides as well as yttrium and scandium. The growing demand for REEs has stimulated the development of original processes to recover REEs from secondary resources. In some areas of the world, REEs are enriched in CFA, which have reached the industrial mining grade (Franus et al., 2015). Direct acid leaching of PC fly ash and CFB fly ash exhibited relatively low leaching efficiency (about 20%) and medium leaching efficiency (60%−70%) (Cao et al., 2018), and a process named roasting activation−acid leaching was developed to recover REEs from PC fly ash with leaching efficiency of REEs reaching about 70% (Taggart et al., 2018; Tang et al., 2019). After leaching, various methods including nano-filtration (Kose Mutlu et al., 2018), TEHDGA (N,N,N′,N′-tetrakis-2-ethylhexyldiglycolamide) impregnated XAD-7 resin adsorption (Mondal et al., 2019), and solvent extraction (Hiskey and Copp, 2018) were employed to separate REEs from leach liquors of CFA, and the concentrated solution containing rare-earth ions was obtained for REE production.

In summary, recovery valuable metals from high alumina fly ash have the vital significance for sustainable development of mineral resources and the supply of strategic metals. Several recovery processes have been developed; however, much effort should be taken to reduce energy consumption and figure out how to turn the research into industrial applications. Simultaneous recovery of associated metals (Ge, Ga, Li, etc.) during alumina recovery processes should be considered to increase economic benefits. Besides, CFA with higher grade and less impurities should be collected for valuable metals recovery, leading to very good economy benefits.

6. Consumer goods and artwork

Consumer goods and artwork based on CFA as a kind of green environmentally construction, furniture, decoration materials, and rockery stone have attracted considerable attention due to its low cost, no secondary pollution, recycling property, corrosion resistance, friction resistance, waterproof, and fire prevention. The composite for consumer goods could be made by the process of compression molding or extrusion molding. Particularly, the activated CFA, polyvinyl chloride resin (PVC), and other assistants including plasticizers, stabilizers, and lubricants were mixed according to a certain proportion to knead. After that, the composite materials could be created by the following process: sanding, four-side planer, two-end miller, sanding, repairing, sanding primers, sanding UV paint, topcoat, inspection, and packing.

It was reported that the properties of CFA-PVC composites plates greatly depend on the structure, particle size, and surface chemistry properties of CFA, which affected the dispersity and compatibility. The processing rheological properties of composite plates were improved obviously due to the uniform dispersity and lower agglomeration degree between glass beads existed in CFA, including the hollow microspheres and glass microspheres. Furthermore, studies have shown that the composite plates filled with CFA have the higher tensile strength than the ones with larger particle size. It was also found that the composite plates produced with 74−147 μm CFA have the desirable hardness, impact strength, and wear rate. The aforementioned enhanced properties can be attributed to the better filler−polymer interfacial adhesion among smaller particles, which possessed higher surface area and excellent dispersion. The matching degree of the free volume between CFA and PVC increased, and the impurity effect of CFA decreased with decreasing the particle size of CFA. The interaction between CFA and PVC macromolecules was enhanced due to the higher surface energy of smaller particle. The addition amounts were also affected by the particle size of CFA. The hardness and wear resistance of the composites decreased when the amount of CFA added was higher than 50%.

The properties of composite plates and addition amounts of CFA were also related to the surface chemistry of CFA. The CFA is hydrophobic because of hydroxyl on the particle surface and has poor compatibility with PVC and other organic substances. Thus, the dispersion of CFA is less homogenous in hydrophobic PVC mixtures due to the greater CFA−CFA interactions. The dispersion of CFA in the mixture is improved with coupling agents, which react with both CFA and PVC mixtures. For practical applications, coupling agents are always used with CFA-reinforced composites. Studies demonstrated that the properties of composite plates filled with CFA including rheological properties, mechanical properties, and corrosion resistance were improved by modifying with titanite coupling agent or silane coupling agent.

On March 26, 2013, the composite plates owned by ShuozhouRunzhen new technology development Co., Ltd. won the domestic and foreign businessmen at 15th China International Floor, Material and Paving Technology Exhibition in Shanghai. The appearance of composite plates not only created great economic benefits and ornamental value, but also solved the environmental pollution brought by CFA. Furthermore, it solved the shortcomings of over standard formaldehyde and high cost of traditional plates. However, the further application of CFA used as filler has been limited by the poor compatibility of CFA and the polymer matrix with different interface properties. It is necessary to activate CFA before preparing composites from CFA. Thus, it is the direction of future efforts to improve the addition amount and activity of

CFA. Besides, another limiting factor that the composite plates based on CFA have not been completely recognized by the market is the safety consideration. It is suggested to issue a series of standards for the composite plates and further improve the production–detection–market structure system of the composite plates.

The high simulation coal ash rockery stone was first proposed by Institute of Kaifeng Jinlu Industrial Waste Residue in 2003. It was developed on the basis of CFA-cement and CFA-concrete. Thus, the rockery stone has the advantages of low production cost, sturdy, durable, various specifications, easy to scale production, and large utilization of CFA. The main raw materials of rockery stone are CFA, cement, and gravel. The rockery stone is prepared by the design and sculpture after the process of batching, mixing, molding, consolidation, and drying of CFA-based concrete and cement products.

The rockery stone can be divided into heavyweight and lightweight products according to the density. The density of heavy and light products is 1600–2500 kg/m^3 and 350–600 kg/m^3, respectively. Heavy products are mainly used to prepare outdoors with high strength, weather resistance, and long lasting. The lightweight products are mainly used to prepare indoor scenery, especially high-rise buildings.

The literature shows that the chemical composition of CFA used in the production of rockery stone was $SiO_2+Al_2O_3>70\%$, $SO_3<3\%$, and loss on ignition $<15\%$. The addition amounts of CFA are up to 40–55% in the production process of rockery stone. The consumption of CFA is about 50,000 tons when the annual output of rockery stone is 100,000 tons. Moreover, it has the characters of high value-added products. For example, the profit is dozens of times of the cost of a 5t CFA-based rockery stone (less than $ 57) compared with the natural rockery stone of same weight (price $ 1416). Thus, the appearance of CFA-based rockery stone can be used for decorative materials, garden construction, and bonsai design in the industry, which has broad application prospects in the increasingly developing society.

7. Ash evaluation system and future R&D needs

At present, the rapid development of fly ash resource utilization has been applied in multiple fields; however, there are still some problems. Firstly, the geographical distribution of fly ash is uneven, with its origin mostly located in the northwest of China and its consumption mostly located in the southeastern coastal areas of China, resulting in high transportation costs and causing transportation pollution. Secondly, the utilization of fly ash resources is not rich enough, and the level of utilization technology needs to be improved. The degree of high value-added utilization is low, and the utilization methods are mostly in low value-added fields. The high value-added utilization process is cumbersome, and the cost is high, making it difficult to achieve large-scale commercial applications. Thirdly, the utilization cost of low-quality fly ash is high. Before resource utilization, decarbonization, grinding, and other processes are usually required to increase the utilization cost.

In the future, it is necessary to accurately grasp the development trend and promote the resource utilization of fly ash. Firstly, based on the geographical distribution of fly ash, it is necessary to coordinate the development relationship between different regions, tilt technology and capital toward the ash source area, reduce transportation costs, and drive the economic development of the ash source area. In addition, it is necessary to explore joint utilization methods of fly ash and other local waste, leverage the synergistic effects of different types of waste, and achieve comprehensive utilization of multiple types of waste. The physical and chemical characteristics of fly ash continue to explore the generation mechanism of fly ash, improve the refined classification of fly ash, explore the separation methods of each component of fly ash, and provide theoretical support for the application of fly ash in different fields. The third is to explore more diverse utilization modes based on the utilization of fly ash, improve the overall utilization level, simplify the utilization process, reduce utilization costs, and focus on the application in high value-added fields, so that the high-value utilization of fly ash does not stay in the laboratory stage, and ultimately achieve large-scale commercial application. Overall, the resource utilization of fly ash is developing toward multichannel, high utilization rate, high value, low cost, low pollution, commercialization, and sustainability, as illustrated in Fig. 24.10.

8. Concluding remarks

Up to date, there are many reported methods to utilize the coal fly ash in various literature. Preparing the construction materials is the most effective way to consume large amounts of coal fly ash, and it is currently the only commercial utilization. However, the difficulty in transporting construction materials greatly limits the application of coal fly ash in this aspect. The use of coal fly ash to prepare the alkali-activated materials such as geopolymer, zeolite, etc. is an emerging technology, which has broad prospects. To reach the full potential of the material, further research needs to be continued on the decrease in the consumption of alkali and improving the durability. Although there are some reports on the soil remediation using coal fly ash, the effect of heavy metals in the coal fly ash on the soil needs further evaluation. Presently, most of the current studies on the utilization of CFA are in the laboratory stage, and more results should be verified on the

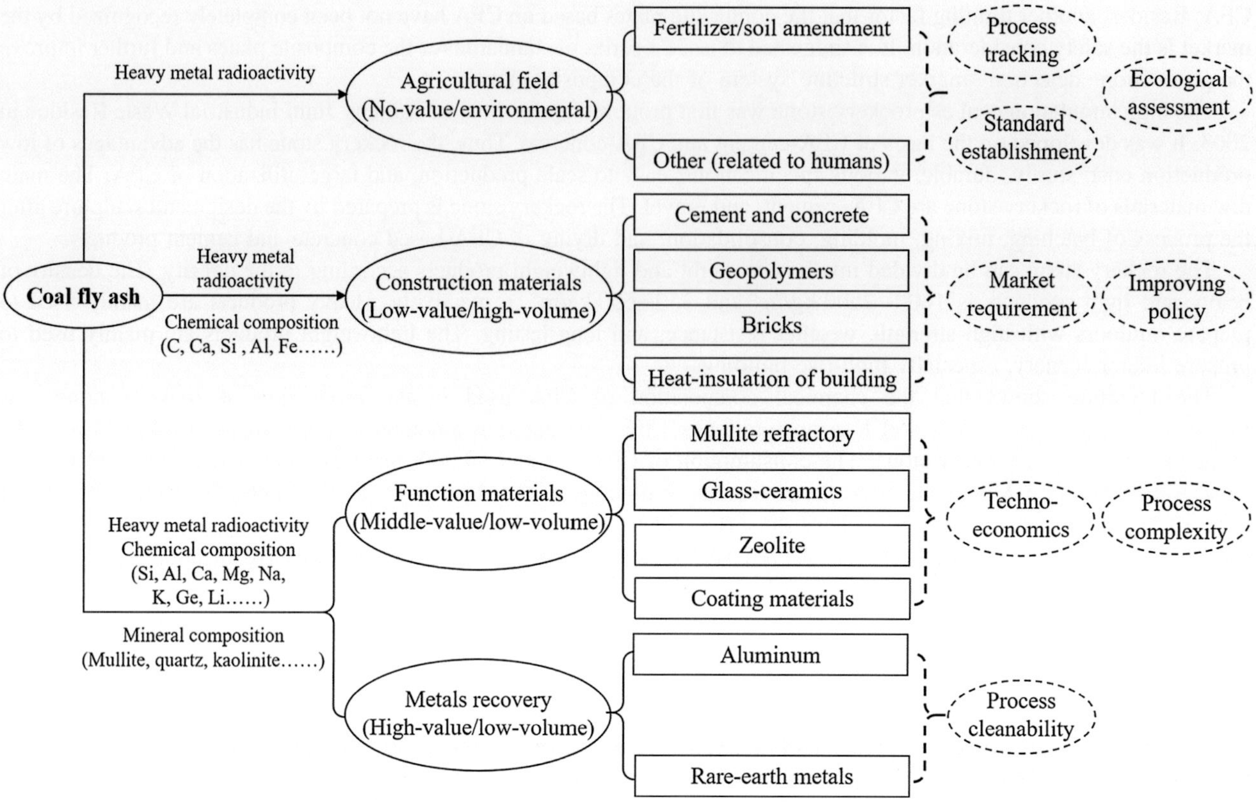

FIGURE 24.10 Ash evaluation system and future R&D needs.

pilot plant. Research on the environmental pollution and economic evaluation during the utilization of CFA is still lacking. A cradle-to-grave assessment of CFA-based utilization technologies should also be studied to ensure effective reuse, reduce, and disposal of CFA.

Acknowledgment

This work was supported by National Key R& D Program of China (2017YFB0603100) and the Australia Research Council under the ARC Linkage Projects Scheme (Project Number: LP100200135 and LP160100035) in partnership with "whoever."

References

Ahmaruzzaman, M., 2010. A review on the utilization of fly ash. Prog. Energy Combust. Sci. 36, 327−363.

An, Z.Q., Wang, Y.C., Zhang, S., Li, Q.H., Guo, W.M., 2022. Review of research progress on preparing glass-ceramics from metallurgical solid wastes. China Ceram. 58, 9−20.

Arroyo, F., Font, O., Fernandez-Pereira, C., Querol, X., Juan, R., Ruiz, C., Coca, P., 2009. Germanium recovery from gasification fly ash: evaluation of end-products obtained by precipitation methods. J. Hazard Mater. 167, 582−588.

Bai, G.H., Qiao, Y.H., Shen, B., Chen, S.L., 2011. Thermal decomposition of coal fly ash by concentrated sulfuric acid and alumina extraction process based on it. Fuel Process. Technol. 92, 1213−1219.

Basu, M., Pande, M., Bhadoria, P.B.S., Mahapatra, S.C., 2009. Potential fly-ash utilization in agriculture: a global review. Prog. Nat. Sci. 19, 1173−1186.

Belviso, C., 2018. State-of-the-art applications of fly ash from coal and biomass: a focus on zeolite synthesis processes and issues. Prog. Energy Combust. Sci. 65, 109−135.

Belviso, C., Cavalcante, F., Ragone, P., Fiore, S., 2012. Immobilization of Zn and Pb in polluted soil by in situ crystallization zeolites from fly ash. Water Air Soil Pollut. 223, 5357−5364.

Berkgaut, V., Singer, A., 1996. High capacity cation exchanger by hydrothermal zeolitization of coal fly ash. Appl. Clay Sci. 10, 369−378.

Blissett, R.S., Rowson, N.A., 2012. A review of the multi-component utilisation of coal fly ash. Fuel 97, 1−23.

Bukhari, S.S., Behin, J., Kazemian, H., Rohani, S., 2015. Conversion of coal fly ash to zeolite utilizing microwave and ultrasound energies: a review. Fuel 140, 250−266.

Cao, S.S., Zhou, C.C., Pan, J.H., Liu, C., Tang, M.C., Ji, W.S., Hu, T.T., Zhang, N.N., 2018. Study on influence factors of leaching of rare earth elements from coal fly ash. Energy Fuel. 32, 8000−8005.
Catalfamo, P., Corigliano, F., Primerano, P., Di Pasquale, S., 1993. Study of the pre-crystallization stage of hydrothermally treated amorphous aluminosilicates through the composition of the aqueous phase. J. Chem. Soc., Faraday Trans. 89, 171−175.
Cheng, F.Q., Shang, J.G., Yang, F.L., Xue, F.B., 2012a. Coal Mine Waterproof Coatings for Coal Mines. Invention Patent.
Cheng, F.Q., Shang, J.G., Zhang, W.P., Yang, F.L., Xue, F.B., 2012b. Coal Mine Fireproof Foatings for Coal Mines. Invention Patent, 200910227964.200910227969.
Dai, S.F., Jiang, Y.F., Ward, C.R., Gu, L.D., Seredin, V.V., Liu, H.D., Zhou, D., Wang, X., Sun, Y.Z., Zou, J.H., Ren, D.Y., 2012. Mineralogical and geochemical compositions of the coal in the Guanbanwusu Mine, Inner Mongolia, China: further evidence for the existence of an Al (Ga and REE) ore deposit in the Jungar Coalfield. Int. J. Coal Geol. 98, 10−40.
Deguire, E.J., Risbud, S.H., 1984. Crystallization and properties of glasses prepared from Illinois coal fly ash. J. Mater. Sci. 19, 1760−1766.
Deng, Y.S., Yang, M., Xing, X.L., 2006. Developing double-layer glass-ceramics with coal-solid-waste as major. Raw Material. Coal Eng. 84−87.
Deshmukh, K., Parsai, R., Anshul, A., Singh, A., Bharadwaj, P., Gupta, R., Mishra, D., Sitaram Amritphale, S., 2017. Studies on fly ash based geopolymeric material for coating on mild steel by paint brush technique. Int. J. Adhesion Adhes. 75, 139−144.
Dessertification, 2012. Message of UNCCD Executive Secretary on World Day to Combat Desertification, 17 June 2012.
Ding, J., Ma, S.H., Shen, S., Xie, Z.L., Zheng, S.L., Zhang, Y., 2017. Research and industrialization progress of recovering alumina from fly ash: a concise review. Waste Manage. 60, 375−387.
Ding, J., Ma, S.H., Zheng, S.L., Zhang, Y., Xie, Z.L., Shen, S., Liu, Z.K., 2016. Study of extracting alumina from high-alumina PC fly ash by a hydrochemical process. Hydrometallurgy 161, 58−64.
Dong, Y.C., Hampshire, S., Zhou, J.E., Lin, B., Ji, Z.L., Zhang, X.Z., Meng, G.Y., 2010. Recycling of fly ash for preparing porous mullite membrane supports with titania addition. J. Hazard Mater. 180, 173−180.
Duan, S.Y., Liao, H.Q., Ma, Z.B., Cheng, F.Q., Fang, L., Gao, H.Y., Yang, H.Q., 2018. The relevance of ultrafine fly ash properties and mechanical properties in its fly ash-cement gelation blocks via static pressure forming. Construct. Build. Mater. 186, 1064−1071.
Elseewi, A.A., Bingham, F.T., Page, A.L., 1978. Availability of sulfur in fly ash to plants. J. Environ. Qual. 7, 69−73.
Eswaran, A., Manivannan, K., 2007. Effect of foliar application of lignite fly ash on the management of papaya leaf curl disease. Acta Hortic. 740, 271−276.
Fan, Y., Ge, T., Zheng, Y.L., Li, H., Cheng, F.Q., 2016. Use of mixed solid waste as a soil amendment for saline-sodic soil remediation and oat seedling growth improvement. Environ. Sci. Pollut. Res. 23, 21407−21415.
Fang, Z., Gesser, H.D., 1996. Recovery of gallium from coal fly ash. Hydrometallurgy 41, 187−200.
Feng, X.P., 2004. Application of fly ash in the glass manufacture industry. Coal Ash 24−26.
Font, O., Querol, X., Juan, R., Casado, R., Ruiz, C.R., Lopez-Soler, A., Coca, P., Garcia Pena, F., 2007. Recovery of gallium and vanadium from gasification fly ash. J. Hazard Mater. 139, 413−423.
Franus, W., Wiatros-Motyka, M.M., Wdowin, M., 2015. Coal fly ash as a resource for rare earth elements. Environ. Sci. Pollut. Res. 22, 9464−9474.
Fu, B., Liu, G.J., Mian, M.M., Sun, M., Wu, D., 2019. Characteristics and speciation of heavy metals in fly ash and FGD gypsum from Chinese coal-fired power plants. Fuel 251, 593−602.
Gnacadja, L., 2013. Desertification: To Care or Not to Care Realizations of the UNCCD since its Establishment in 1992.
Goto, S., Aoki, M., Long, C.D., Takada, C., Hayashi, H., Chino, M., 2000. Potassium silicate fertilizer using Chinese fly ash and a fertilizer response test. J. Soil Sci. Plant Nutr. 71, 378−384.
Gutiérrez, B., Pazos, C., Coca, J., 1997. Recovery of gallium form coal fly ash by a dual reactive extraction process. Waste Manag. Res. 15, 371−382.
Hai, Y.S., An, Z.H., Sheng, F.D., Min, S.Y., Li, L.X., Lun, L.K., Yun, Z., Qing, L.M., 2014. Review on the preparation of glass-ceramics from fly ash. Bull. Chin. Ceram. Soc. 33, 2902−2907.
Han, G.H., Yang, S.Z., Peng, W.J., Huang, Y.F., Wu, H.Y., Chai, W.C., Liu, J.T., 2018. Enhanced recycling and utilization of mullite from coal fly ash with a flotation and metallurgy process. J. Clean. Prod. 178, 804−813.
Hemalatha, T., Ramaswamy, A., 2017. A review on fly ash characteristics−towards promoting high volume utilization in developing sustainable concrete. J. Clean. Prod. 147, 546−559.
Hernández-Expósito, A., Chimenos, J.M., Fernández, A.I., Font, O., Querol, X., Coca, P., García Peña, F., 2006. Ion flotation of germanium from fly ash aqueous leachates. Chem. Eng. J. 118, 69−75.
Hill, R., Wood, D., 1995. Apatite-mullite glass-ceramics. J. Mater. Sci. Mater. Med. 6, 311−318.
Hiskey, J.B., Copp, R.G., 2018. Solvent extraction of yttrium and rare earth elements from copper pregnant leach solutions using Primene JM-T. Miner. Eng. 125, 265−270.
Höller, H., Wirsching, U., 1985. Zeolite formation from fly ash. Fortschr. Mineral. 63, 21−43.
Hu, P.P., Hou, X.J., Zhang, J.B., Li, S.P., Wu, H., Damø, A.J., Li, H.Q., Wu, Q.S., Xi, X.G., 2018. Distribution and occurrence of lithium in high-alumina-coal fly ash. Int. J. Coal Geol. 189, 27−34.
Hua, J.S., 2002. Study on the coal ash content mold coating. Foundry Technol. 219−220.
Ibáñez, J., Font, O., Moreno, N., Elvira, J.J., Alvarez, S., Querol, X., 2013. Quantitative Rietveld analysis of the crystalline and amorphous phases in coal fly ashes. Fuel 105, 314−317.
Inada, M., Tsujimoto, H., Eguchi, Y., Enomoto, N., Hojo, J., 2005. Microwave-assisted zeolite synthesis from coal fly ash in hydrothermal process. Fuel 84, 1482−1486.

Jala, S., Goyal, D., 2006. Fly ash as a soil ameliorant for improving crop production—a review. Bioresour. Technol. 97, 1136–1147.

Jayasinghe, G.Y., Tokashiki, Y., 2012. Influence of coal fly ash pellet aggregates on the growth and nutrient composition of brassica campestris and physicochemical properties of grey soils in Okinawa, Japan. J. Plant Nutr. Soil Sci. 35, 453–470.

Jurate, K., Solvita, O., Anders, L., Christian, M., 2007. Stabilization of Pb- and Cu- contaminated soil using coal fly ash and peat. Environ. Pollut. 145, 365–373.

Kamran Haghighi, H., Irannajad, M., Fortuny, A., Sastre, A.M., 2018. Recovery of germanium from leach solutions of fly ash using solvent extraction with various extractants. Hydrometallurgy 175, 164–169.

Klubek, B., Carison, C.L., Oliver, J., Adriano, D.C., 1992. Characterization of microbial abundance and activity from three coal ash basins. Soil Biol. Biochem. 24, 1119–1125.

Kose Mutlu, B., Cantoni, B., Turolla, A., Antonelli, M., Hsu-Kim, H., Wiesner, M.R., 2018. Application of nanofiltration for Rare Earth Elements recovery from coal fly ash leachate: performance and cost evaluation. Chem. Eng. J. 349, 309–317.

Kumpiene, J., Guerri, G., Landi, L., Pietramellara, G., Nannipieri, P., Renella, G., 2009. Microbial biomass, respiration and enzyme activities after in situ aided phytostabilization of a Pb- and Cu- contaminated soil. Ecotoxicol. Environ. Saf. 72, 115–119.

Kuwahara, Y., Ohmichi, T., Mori, K., Katayama, I., Yamashita, H., 2008. Synthesis of zeolite from steel slag and its application as a support of nano-sized TiO2 photocatalyst. J. Mater. Sci. 43, 2407–2410.

Li, J., Zhuang, X., Font, O., Moreno, N., Vallejo, V.R., Querol, X., Tobias, A., 2014. Synthesis of merlinoite from Chinese coal fly ashes and its potential utilization as slow release K-fertilizer. J. Hazard Mater. 265, 242–252.

Li, S.H., Du, H.Y., Guo, A.R., Xu, H., Yang, D., 2012. Preparation of self-reinforcement of porous mullite ceramics through in situ synthesis of mullite whisker in flyash body. Ceram. Int. 38, 1027–1032.

Li, S.Y., Qin, S.J., Kang, L.W., Liu, J.J., Wang, J., Li, Y.H., 2017. An efficient approach for lithium and aluminum recovery from coal fly ash by pre-desilication and intensified acid leaching processes. Met. Mater. Int. 7, 272.

Li, Y.D., Lu, J.S., Zeng, Y.P., Liu, Z.Y., Wang, C.L., 2018. Preparation and characterization of mullite powders from coal fly ash by the mullitization and hydrothermal processes. Mater. Chem. Phys. 213, 518–524.

Li, Y.Z., Liu, C.J., Luan, Z.K., Peng, X.J., Zhu, C.L., Chen, Z.Y., Zhang, Z.G., Fan, J.H., Jia, Z.P., 2006. Phosphate removal from aqueous solutions using raw and activated red mud and fly ash. J. Hazard Mater. 137, 374–383.

Lin, B., Li, S.P., Hou, X.J., Li, H.Q., 2015. Preparation of high performance mullite ceramics from high-aluminum fly ash by an effective method. J. Alloys Compd. 623, 359–361.

Lin, H.Y., Wan, L., Yang, Y.F., 2012. Aluminium hydroxide ultrafine powder extracted from fly ash. Adv. Mater. Res. 1790, 1548–1553.

Lu, F.H., Xiao, T.F., Lin, J., Ning, Z.P., Long, Q., Xiao, L.H., Huang, F., Wang, W.K., Xiao, Q.X., Lan, X.L., Chen, H.Y., 2017. Resources and extraction of gallium: a review. Hydrometallurgy 174, 105–115.

Luo, Y., Wu, Y., Ma, S., Zheng, S., Zhang, Y., Chu, P.K., 2021. Utilization of coal fly ash in China: a mini-review on challenges and future directions. Environ. Sci. Pollut. Res. 28, 18727–18740.

Luo, Y., Zheng, S.L., Ma, S.H., Liu, C.L., Wang, X.H., 2017. Ceramic tiles derived from coal fly ash: preparation and mechanical characterization. Ceram. Int. 43, 11953–11966.

Luo, Y., Zheng, S.L., Ma, S.H., Liu, C.L., Wang, X.H., 2018. Preparation of sintered foamed ceramics derived entirely from coal fly ash. Construct. Build. Mater. 163, 529–538.

Marco-Lozar, J.P., Cazorla-Amorós, D., Linares-Solano, A., 2007. A new strategy for germanium adsorption on activated carbon by complex formation. Carbon 45, 2519–2528.

Mclennan, A.R., Bryant, G.W., Stanmore, B.R., Wall, T., 2000. Ash Formation mechanisms during pf combustion in reducing conditions. Energy Fuel. 14, 150–159.

Menzies, N.W., Aitken, R.L., 1996. Evaluation of fly ash as a component of potting substrates. Sci. Hortic. 67, 87–99.

Mondal, S., Ghar, A., Satpati, A.K., Sinharoy, P., Singh, D.K., Sharma, J.N., Sreenivas, T., Kain, V., 2019. Recovery of rare earth elements from coal fly ash using TEHDGA impregnated resin. Hydrometallurgy 185, 93–101.

Moskalyk, R.R., 2004. Review of germanium processing worldwide. Miner. Eng. 17, 393–402.

Mukherjee, S., Srivastava, S., 2006. Minerals transformations in northeastern region coals of India on heat treatment. Energy Fuel. 20.

Murayama, N., Yamamoto, H., Shibata, J., 2002. Mechanism of zeolite synthesis from coal fly ash by alkali hydrothermal reaction. Int. J. Miner. Process. 64, 1–17.

Musyoka, N.M., Petrik, L.F., Hums, E., Baser, H., Schwieger, W., 2012. In situ ultrasonic monitoring of zeolite A crystallization from coal fly ash. Catal. Today 190, 38–46.

Nadesan, M.S., Dinakar, P., 2017. Structural concrete using sintered flyash lightweight aggregate: a review. Construct. Build. Mater. 154, 928–944.

Ngu, L.N., Wu, H.W., Zhang, D.K., 2007. Characterization of ash cenospheres in fly ash from Australian power stations. Energy Fuel. 21, 3437–3445.

Obla, K.H., Hill, R.L., Thomas, M.D.A., Shashiprakash, S.G., Perebatova, O.O., 2003. Properties of concrete containing ultra-fine fly ash. ACI Mater. J. 100, 426–433.

Ojha, K., Pradhan, N.C., Samanta, A.N., 2004. Zeolite from fly ash: synthesis and characterization. Bull. Mater. Sci. 27, 555–564.

Oliveira, M.L.S., Navarro, O.G., Crissien, T.J., Tutikian, B.F., da Boit, K., Teixeira, E.C., Cabello, J.J., Agudelo-Castañeda, D.M., Silva, L.F.O., 2017. Coal emissions adverse human health effects associated with ultrafine/nano-particles role and resultant engineering controls. Environ. Res. 158, 450–455.

Page, A.L., Elseewi, A.A., Straughan, I.R., 1979. Physical and chemical properties of fly ash from coal-fired power plants with reference to environmental impacts. In: Residue Reviews: Residues of Pesticides and Other Contaminants in the Total Environment. Springer, pp. 83–120.

Panitchakarn, P., Laosiripojana, N., Viriya-Umpikul, N., Pavasant, P., 2014. Synthesis of high-purity Na-A and Na-X zeolite from coal fly ash. J. Air Waste Manag. Assoc. 64, 586–596.

Parab, N., Mishra, S., Bhonde, S.R., 2012. Prospects of bulk utilization of fly ash in agriculture for integrated nutrient management. Bull. Natl. Inst. Ecol. 23, 31–46.

Pathan, S.M., Aylmore, L.A.G., Colmer, T.D., 2003. Soil properties and turf growth on a sandy soil amended with fly ash. Plant Soil 256, 103–114.

Prabakar, J., Dendorkar, N., Morchhale, R.K., 2004. Influence of fly ash on strength behavior of typical soils. Construct. Build. Mater. 18, 263–267.

Qin, S.J., Zhao, C.L., Li, Y.H., Zhang, Y., 2015. Review of coal as a promising source of lithium. Int. J. Oil Gas Coal Technol. 9, 215–229.

Querol, X., Alastuey, A., Fernández-Turiel, J., López-Soler, A., 1995. Synthesis of zeolites by alkaline activation of ferro-aluminous fly ash. Fuel 74, 1226–1231.

Querol, X., Alastuey, A., Moreno, N., Alvarez-Ayuso, E., García-Sánchez, A., Cama, J., Ayora, C., Simón, M., 2006. Immobilization of heavy metals in polluted soils by the addition of zeolitic material synthesized from coal fly ash. Chemosphere 62, 171–180.

Ram, L.C., Masto, R.E., 2010. An appraisal of the potential use of fly ash for reclaiming coal mine spoil. J. Environ. Manag. 91, 603–617.

Ram, L.C., Masto, R.E., 2014. Fly ash for soil amelioration: a review on the influence of ash blending with inorganic and organic amendments. Earth Sci. Rev. 128, 52–74.

Ram, L.C., Masto, R.E., Singh, S., Tripathi, R.C., Jha, S.K., Srivastava, N.K., Sinha, A.K., Selvi, V.A., Sinha, A., 2011. An appraisal of coal fly ash soil amendment technology (FASAT) of Central Institute of Mining and Fuel Research (CIMFR). Int. J. Environ. Chem. Ecol. Geoecol. Geophys. Eng. 5, 255–266.

Rashidi, N.A., Yusup, S., 2016. Overview on the potential of coal-based bottom ash as low-cost adsorbents. ACS Sustainable Chem. Eng. 4, 1870–1884.

Rawlings, R.D., Wu, J.P., Boccaccini, A.R., 2006. Glass-ceramics: their production from wastes—a review. J. Mater. Sci. 41, 733–761.

Rayalu, S., Meshram, S.U., Hasan, M.Z., 2000. Highly crystalline faujasitic zeolites from flyash. J. Hazard Mater. 77, 123–131.

Rechcigl, J.E., 1996. Soil amendments and environmental quality. J. Environ. Qual. 25, 926–927.

Ruhi, G., Bhandari, H., Dhawan, S.K., 2015. Corrosion resistant polypyrrole/flyash composite coatings designed for mild steel substrate. J. Polym. Sci. 5, 18–27.

Sahu, S.P., Satapathy, A., Patnaik, A., Sreekumar, K.P., Ananthapadmanabhan, P.V., 2010. Development, characterization and erosion wear response of plasma sprayed fly ash–aluminum coatings. Mater. Des. 31, 1165–1173.

Schutter, M.E., Fuhrmann, J.J., 2001. Soil microbial community responses to fly ash amendment as revealed by analyses of whole soils and bacterial isolates. Soil Biol. Biochem. 33, 1947–1958.

Seshadri, B., Bolan, N.S., Kunhikrishnan, A., 2013. Effect of clean coal combustion products in reducing soluble phosphorus in soil I. Adsorption study. Water Air Soil Pollut. 224, 1–11.

Shi, L., Lv, N., Nie, J., Tan, D., Zheng, J., Luo, G., Li, Y., Xiong, Y., Xie, Y., Wang, J.W., 2018. Effects of culture media on yield, growth, photosynthetic capacity and antioxidant enzyme of different tomato (Solanum lycopersicum L.) varieties. Appl. Ecol. Environ. Res. 16, 5909–5919.

Shigemoto, N., Hayashi, H., Miyaura, K., 1993. Selective formation of Na-X zeolite from coal fly ash by fusion with sodium hydroxide prior to hydrothermal reaction. J. Mater. Sci. 28, 4781–4786.

Sims, J.T., Vasilas, B.L., Ghodrati, M., 1995. Evaluation of fly ash as a soil amendment for the Atlantic Coastal Plain: II. Soil chemical properties and crop growth. Water Air Soil Pollut. 81, 363–372.

Singh, R.P., Gupta, A.K., Ibrahim, M.H., Mittal, A.K., 2010. Coal fly ash utilization in agriculture: its potential benefits and risks. Rev. Environ. Sci. Biotechnol. 9, 345–358.

Song, H.P., Cao, Z.Y., Xie, W.S., Cheng, F.Q., Gasem, K.A.M., Fan, M.H., 2019. Improvement of dispersion stability of filler based on fly ash by adding sodium hexametaphosphate in gas-sealing coating. J. Clean. Prod. 235, 259–271.

Song, H.P., Liu, J.Q., Xue, F.B., Cheng, F.Q., 2016. The application of ultra-fine fly ash in the seal coating for the wall of underground coal mine. Adv. Powder Technol. 27, 1645–1650.

Song, H.P., Xie, W.S., Liu, J.Q., Cheng, F.Q., Gasem, K.A.M., Fan, M.H., 2018. Effect of surfactants on the properties of a gas-sealing coating modified with fly ash and cement. J. Mater. Sci. 53, 15142–15156.

Spark, K.M., Swift, R.S., 2008. Use of alkaline flyash-based products to amend acid soils: plant growth response and nutrient uptake. Soil Res. 46, 578–584.

Steenbruggen, G., Hollman, G.G., 1998. The synthesis of zeolites from fly ash and the properties of the zeolite products. J. Geochem. Explor. 62, 305–309.

Stout, W.L., Sharpley, A.N., Gburek, W.J., Pionke, H.B., 1999. Reducing phosphorus export from croplands with FBC fly ash and FGD gypsum. Fuel 78, 175–178.

Stout, W.L., Sharpley, A.N., Landa, J., 2000. Effectiveness of coal combustion by-products in controlling phosphorus export from soils. J. Environ. Qual. 29, 1239–1244.

Sukkae, R., Suebthawilkul, S., Cherdhirunkorn, B., 2018. Utilization of coal fly ash as a raw material for refractory production. J. Met. Mater. Miner. 28.

Sun, J., Chen, P., 2013. Resourcing utilization of high alumina fly ash. Adv. Mater. Res. 652–654, 2570–2575.

Sun, L.Y., Luo, K., Fan, J.R., Lu, H.L., 2017. Experimental study of extracting alumina from coal fly ash using fluidized beds at high temperature. Fuel 199, 22–27.

Taggart, R.K., Hower, J.C., Hsu-Kim, H., 2018. Effects of roasting additives and leaching parameters on the extraction of rare earth elements from coal fly ash. Int. J. Coal Geol. 196, 106–114.

Tang, M.C., Zhou, C.C., Pan, J.H., Zhang, N.N., Liu, C., Cao, S.S., Hu, T.T., Ji, W.S., 2019. Study on extraction of rare earth elements from coal fly ash through alkali fusion—acid leaching. Miner. Eng. 136, 36–42.

Tang, Q., Liu, G.J., Zhou, C.C., Sun, R.Y., 2013. Distribution of trace elements in feed coal and combustion residues from two coal-fired power plants at Huainan, Anhui, China. Fuel 107, 315–322.

Torralvo, F.A., Fernández-Pereira, C., 2011. Recovery of germanium from real fly ash leachates by ion-exchange extraction. Miner. Eng. 24, 35–41.

Tripathy, A.K., Sarangi, C.K., Tripathy, B.C., Sanjay, K., Bhattacharya, I.N., Mahapatra, B.K., Behera, P.K., Satpathy, B.K., 2015. Aluminium recovery from NALCO fly ash by acid digestion in the presence of fluoride ion. Int. J. Miner. Process. 138, 44–48.

Vuthaluru, H.B., French, D., 2008. Ash chemistry and mineralogy of an Indonesian coal during combustion. Fuel Process. Technol. 89, 595–607.

Wang, L., Zhang, T.A., Lv, G.Z., Dou, Z.H., Zhang, W.G., Niu, L.P., 2019. Carbochlorination kinetics of high-alumina fly ash. J. Occup. Med. 71, 492–498.

Wang, X.T., Xu, L.L., Zhong, B.Y., Yang, N.R., Guo, Y., 1994. Development of porous adsorbent for fly ash matrix. Fly Ash Compr. Util. 33–36.

Wong, M.H., Wong, J.W.C., 1986. Effects of fly ash on soil microbial activity. Environ. Pollut. Ecol. Biol. 40, 127–144.

Xu, D.H., Li, H.Q., Bao, W.J., Wang, C.Y., 2016. A new process of extracting alumina from high-alumina coal fly ash in NH4HSO4+H2SO4 mixed solution. Hydrometallurgy 165, 336–344.

Xue, F.B., Yang, F.L., Song, H.P., Cheng, F.Q., 2017. Gas Sealing Coating for Mine Roadway Wall. Invention Patent, 201510078403.201510078402.

Yan, C.X., Dong, J.M., 2001. Fly ash utilization in agriculture. Fly Ash Compr. Util. 41–44.

Yang, J.K., Zhang, D.D., Xiao, B., Wang, X.P., 2004. Study on glass-ceramics mostly made from red mud and fly ash. Glass Enamel. Ophthalmic Opt. 9–11.

Yang, Y., Zhi Hui, C., Shu Xia, Y., 2011. Influence of fly ash and soilless substrates on chemical properties of their compound medium and tall fescue growth. In: 2011 International Conference on Electric Technology and Civil Engineering (ICETCE), pp. 5118–5122.

Yang, Y.F., Gai, G.S., Cai, Z.F., Chen, Q.R., 2006. Surface modification of purified fly ash and application in polymer. J. Hazard Mater. 133, 276–282.

Yao, Z.T., Ji, X.S., Sarker, P.K., Tang, J.C., Ge, L.Q., Xia, M.S., Xi, Y.Q., 2015. A comprehensive review on the applications of coal fly ash. Earth Sci. Rev. 141, 105–121.

Yao, Z.T., Xia, M.S., Sarker, P.K., Chen, T., 2014. A review of the alumina recovery from coal fly ash, with a focus in China. Fuel 120, 74–85.

Yõlmaz, H., 2015. Characterization and comparison of leaching behaviors of fly ash samples from three different power plants in Turkey. Fuel Process. Technol. 137, 240–249.

Yu, J.L., Li, X.C., Fleming, D., Meng, Z.Q., Wang, D.M., Tahmasebi, A., 2012. Analysis on characteristics of fly ash from coal fired power stations. Energy Proc. 17, 3–9.

Zandi, M., Russell, N.V., 2007. Design of a leaching test framework for coal fly ash accounting for environmental conditions. Environ. Monit. Assess. 131, 509–526.

Zeng, L., Sun, H.J., Peng, T.J., 2022. Effect of borax on sintering kinetics, microstructure and mechanical properties of porous glass-ceramics from coal fly ash by direct overfiring. Front. Chem. 10, 839680.

Zhang, J.B., Li, H.Q., Li, S.P., Hu, P.P., Wu, W.F., Wu, Q.S., Xi, X.G., 2018a. Mechanism of mechanical—chemical synergistic activation for preparation of mullite ceramics from high-alumina coal fly ash. Ceram. Int. 44, 3884–3892.

Zhang, J.B., Li, S.P., Li, H.Q., Wu, Q.S., Xi, X.G., Li, Z.B., 2017. Preparation of Al—Si composite from high-alumina coal fly ash by mechanical—chemical synergistic activation. Ceram. Int. 43, 6532–6541.

Zhang, Q., Zhu, Y.H., Zhao, Y.Z., Qu, M.S., Yan, F., Wang, S.L., Liu, K.F., 2018b. Using the fly ashas plant substrate to grow cucumber and baby Chinese cabbage. J. Beijing Univ. Agric. 33, 38–42.

Zhao, Z.S., Cui, L., Guo, Y.X., Li, H.Q., Cheng, F.Q., 2020. Recovery of gallium from sulfuric acid leach liquor of coal fly ash by stepwise separation using P507 and Cyanex 272. Chem. Eng. J. 381, 122699.

Zhu, J.B., Yan, H., 2017. Microstructure and properties of mullite-based porous ceramics produced from coal fly ash with added Al2O3. Int. J. Miner. Metall. Mater. 24, 309–315.

Chapter 25

Resource recovery of phosphorus from incinerated sewage sludge ash

Yanjun Hu, Qianqian Guo, Lingqin Zhao, Yanan Wang and Fan Yu
Institute of Energy & Power Engineering, Zhejiang University of Technology, Hangzhou, China

1. Introduction

Phosphorus (P) is an important element in human cellular activity and is a resource with limited storage capacity. The P cycle in nature is presented in Fig. 25.1 (Elser, 2012). Nowadays, it was reported that the natural P resource would be depleted within only 50–60 years (Tao and Xia, 2007). China has listed the P as one of the 24 national strategic minerals in 2016. Assessing usage and recycling measures is helpful for extending the P resources reserves to the 23rd century. Therefore, researches are gradually focusing on recycling P sources from solid waste, such as sewage sludge and its resulted by-products (Tansel et al., 2018).

Municipal sewage sludge is a common P-containing solid wastes; the P in it was concentrated from waste water by municipal sewage treatment process (Kruger and Adam, 2015). Therefore, municipal sewage sludge was considered as the final destination of the majority of the P in human daily life and animal farming, and has been regarded as a reliable recovery for P source. Besides, the moisture content of municipal sewage sludge is approximately 80%, and rich in nitrogen and pathogens, heavy metals, etc. (Appels et al., 2008).

In recent years, incineration is increasingly chosen to be one of the common disposal methods for municipal sewage sludge all over the world. The high temperature of incineration treatment is beneficial for the organic pollutants decomposition and pathogens reduction in municipal sewage sludge (Bloem et al., 2017). It leads to a reduced volume of obtained sewage sludge incineration ash (SSIA), which is nonhazardous and very favorable for the following transport operations as well as resource recovery. During the incineration process, P was remained to SSIA, accounting for higher than 90% of the P in municipal sewage sludge.

On the basis of the latest reports, P recovery should focus on the SSIA. Sewage sludge incineration is most often conducted at about 800–1000°C. At this range, P releases as volatile oxides and then condenses at 400–600°C to form P_4O_{10} thus being a composition of the ash. More than 5–10 times of P can be recovered from the SSIA than from the municipal sewage sludge. Therefore, more study that addresses the issues of P recovery from SSIA is currently published.

With the widespread use of incineration technology for municipal sewage sludge treatment and the urgency of P resource recovery, the challenge of obtaining P resources has also become an urgent need to be addressed. In this chapter, the sewage sludge mentioned represents the municipal sewage sludge. The characteristics of P in SSIA was discussed. The methods of P recovery from SSIA, including the thermochemical extraction, wet extraction, electrodialytic extraction, and bioleaching extraction, were systematically summarized and compared. The high-value utilization of P recovered from SSIA was addressed. The case analysis, policy, and legal level of P extraction from SSIA was reviewed. Finally, some perspectives for future studies on the P recovery from incineration residues are proposed.

2. Distribution characteristics of P species in SSIA

Incineration is an effective method for achieving P enrichment in ash. P enrichment in SSIA can reach 60%–90% (Bogdan et al., 2022). The mass fraction of P in SSIA is usually about 4.9%–11.9%, with an average value of about 8.9% (converted to P_2O_5 of about 20.4%), which is equivalent to medium- to high-grade natural phosphate ore (Hao et al., 2021).

Treatment and Utilization of Combustion and Incineration Residues. https://doi.org/10.1016/B978-0-443-21536-0.00026-5

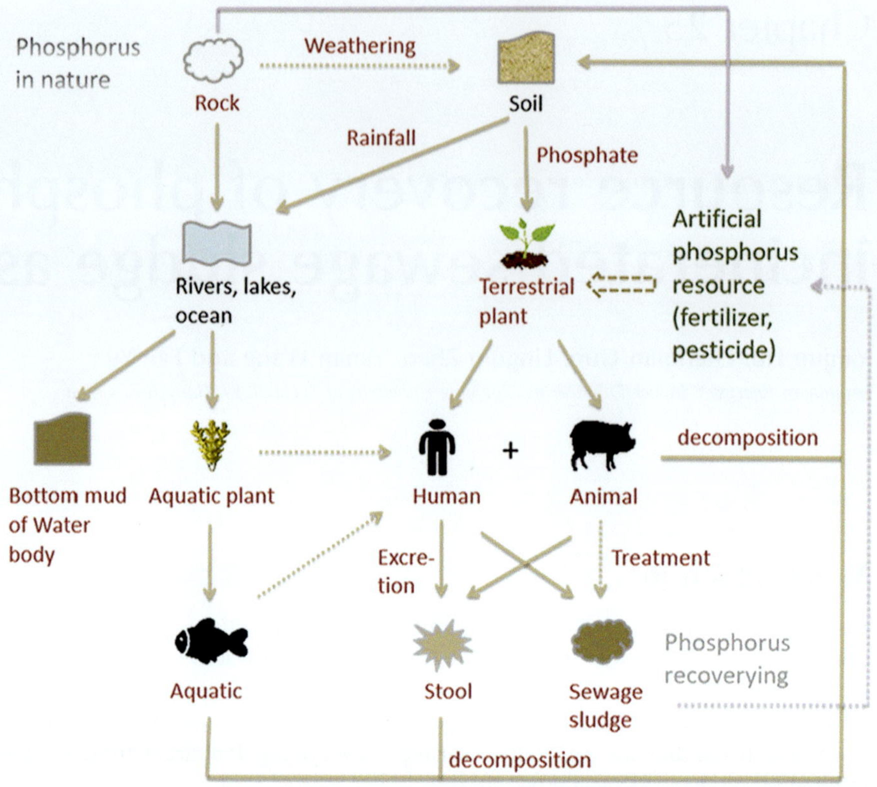

FIGURE 25.1 Phosphorus cycle in nature.

2.1 Chemical forms of P in SSIA

The chemical form of P is one of the most critical properties for P resources recovery, which directly affects the bioavailability of recovered products. Based on the method of standards measurements and testing (SMT) (Pardo et al., 2003), the chemical forms of P species in sewage sludge can be classified as total phosphorus (TP), organic phosphorus (OP), and inorganic phosphorus (IP) (Zakaria et al., 2022). IP could be divided into apatite (AP) as well as nonapatite (NAIP). The TP is the sum of all P in the sewage sludge according to the SMT method: TP = OP + IP. The OP is that of P in the form of organic matter. It can be released into the water by mineralization of organic matter and by biological metabolism. The inorganic phosphorus is that of P in the form of inorganic compounds. IP is equal to NAIP plus AP according to the SMT method. The NAIP is that of mainly P in the form of Fe/Al/Mn oxides and their hydroxides. The bioavailability of NAIP is lower than AP (Pang, 2004). The AP is that of mainly phosphate in the form of Ca/Mg combined P, with strong bioavailability, and is the best industrial raw material for phosphate fertilizer production (Donnert and Salecker, 1999).

It was found that P in sludge was mainly dominated by IP, which accounted for 75% of TP, while NAIP accounted for 80% of IP (Xie et al., 2011; Marta et al., 2012). P enrichment is achieved by the incineration treatment of sewage sludge, and P morphology is transformed. Characterization studies of P in SSIA showed that most of the P was bound in the Ca, Fe, Al, and Si structures (Cieslik and Konieczka, 2017), and IP was the main form. Calcium phosphate ($Ca_9Al(PO_4)_7$), iron phosphate ($FePO_4$), and aluminum phosphate ($AlPO_4$) are the main components in the SSIA (Ohbuchi et al., 2008). The TP content in SSIA ranged from 36 mg/g to 131 mg/g (Hao et al., 2020). Meng et al. (2018) showed that the low-temperature incineration of sewage sludge increased the TP content of bottom ash by 45.6%, and some NAIP converted into AP leading to a 46.3% increase in AP content.

2.2 Migration and transformation of P species during the sewage sludge incineration process

The chemical forms of P species in SSIA usually depend on the migration and transformation of P species during the incineration process. However, the P species in SSIA is influenced by the additives, incineration temperature, and incineration types (e.g., single combustion/cocombustion). Of all the influencing factors, the most significant driving force of P conversion and enrichment is temperature (Gu et al., 2021). Fig. 25.2 shows concentrations of various P species in SSIA. The increase in incineration temperature promoted P enrichment, and most of the P exists in the form of IP. This is because during the incineration process, OP decomposes and is converted into IP (Li et al., 2014). The transformations of P are mainly due to changes in complexation state and mineralogy. At lower incineration temperature, P in sewage sludge is mainly converted into NAIP. Higher incineration temperature promotes the conversion of NAIP into AP (Ca/Mg−P), since the thermal stability of NAIP (Fe/Al/Mn−P) is weaker than AP (Ca/Mg−P) and is easily decomposed at higher temperature. Meanwhile, the Ca/Mg tends to replace the sites of Fe/Al/Mn to form AP. The previous study found that the ratio of NAIP/IP of 71.9% at 600°C decreased to 53.7% at 900°C (Liang et al., 2019). The proportion of AP would increase by 46.3%, and the amount of bioavailable P would increase by 2.9 times (Lim and Kim, 2017). After the treatment, the P species in SSIA mainly exist in the form of AP (Ca−P and Mg−P), which behave a higher bioavailability than NAIP (Fe/Al/Mn−P) (Fang et al., 2023).

The additives also affect the migration and transformation of P species. For example, CaO addition can improve the release of P and the conversion of NAIP into AP (Li et al., 2015). Some studies have shown that higher temperatures would promote NAIP volatilization, while CaO addition could suppress NAIP volatilization (Li et al., 2020a,b). The AP/TP ratio increased to 92.23% at 950°C by adding CaO (Saleh et al., 2018). CaO will react with Al/Fe−P to form Ca−P. Although CaO addition promoted the transformation of NAIP into AP, some studies have shown that it is not possible to completely convert NAIP to AP even with the addition of excessive CaO.

Cl-based additives such as PVC and MCl_x ($M = K^+, Ca^{2+}$; Na^+, Mg^{2+}, Al^{3+}, Fe^{3+}, H^+, $NH^+{}_4$) can also affect the species of P. $AlPO_4$ can react with $MgCl_2$ to generate $Mg_3(PO_4)_2$ with an addition ratio of 3% and a temperature of 900°C; the enrichment rate of P reached 98.5% (Yang et al., 2019). Xia et al. found that when $CaCl_2$ was used as an additive, the enrichment rate of P was 97.8%, and $AlPO_4$ would react with $CaCl_2$ to generate Ca_2PO_4Cl. When the addition ratio of $CaCl_2$ was 40%, 82% of P volatilized into the gas phase at 1100°C. It was correlated with that $CaCl_2$ behaved molten salt ability and changed the crystal structure to form $Ca_5(PO_4)_3Cl$ and $Ca_5(PO_4)_3Cl$, and these chemical forms could be reduced by the carbon in the sludge to form gas-phase P_2 (Xia et al., 2020).

Kaolin ($Al_2O_3 \cdot 2SiO_2 \cdot 2H_2O$) could also effectively enhance transformation of P species (Herzel et al., 2016). When temperature was lower than 850°C, kaolin addition could increase the TP and NAIP content in SSIA. However, when temperature increased to 950°C, TP and AP content decreased (Li et al., 2015). The effect of kaolinite minerals on P enrichment and P morphology changes is limited and less effective than that of CaO.

The alkali metals such as K in biomass significantly affected the transformation of P species during incineration process. Researches showed that adding biomass into sewage sludge incineration can reduce the enrichment ratio of P in the ash. The enrichment ration of P was only 78% under the coincineration of sewage sludge and chip (Fournie et al., 2022). Moreover, the reactive site provided by Cl, Ca, and Mg matters in biomass will promote the transformation of NAIP into AP. Moreover, the developed pore structure of fly ash could provide a larger reaction contact area for the generation of $Ca_{18}Mg_2H_2(PO_4)_{14}$, $Ca_2P_2O_7$, $Ca_5(PO_4)_3Cl$, $Mg_3(PO_4)_6$, and $Ca_4Mg_5(PO_4)_6$ (Nilsson et al., 2022). This promotion effect

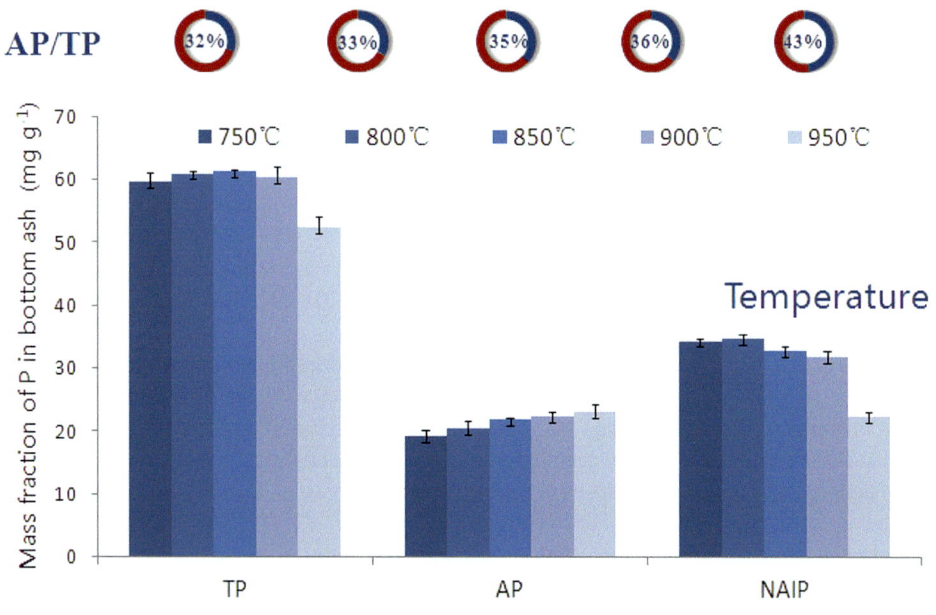

FIGURE 25.2 Concentrations of various P species (mg/g) in SSIA (Li et al., 2015).

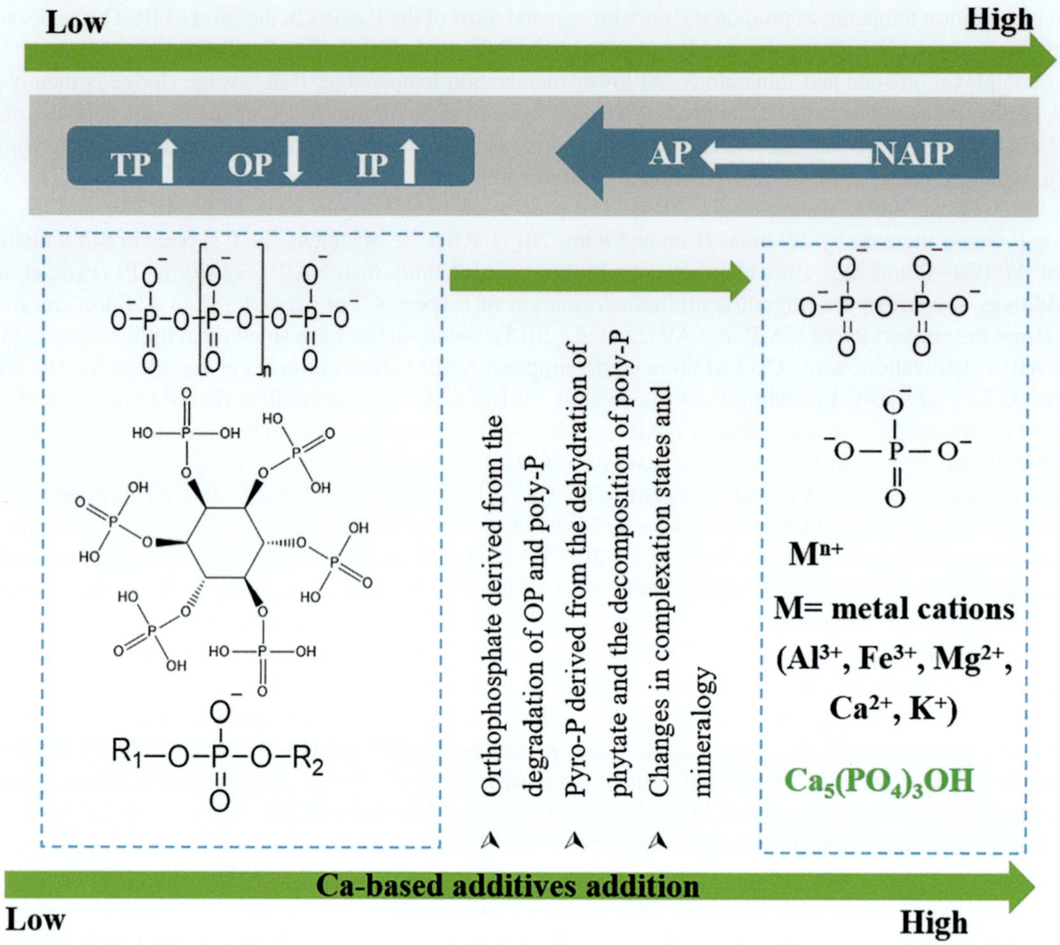

FIGURE 25.3 The transformation mechanism of chemical form of P species during sewage sludge incineration process (Fang et al., 2023).

gradually strengthens as the addition ration of biomass increases. Adding rice husks, corn cob meal, or soybean straws could enhance the bioavailability of P in ash (Meng et al., 2021). Zhao et al. (2018) found that Ca and Mg in cotton straw could react with P in sewage sludge leading to a higher AP content. It was linked to that OP and pyrophosphates were gradually transformed into orthophosphates. They can further react with Ca, Fe, and Al metal cations to form phosphates during coincineration process (Fig. 25.3).

3. Thermochemical extraction of P from SSIA

Thermochemical extraction is a primary alternative for P recovery from SSIA, and the main processes of which are the gasification or liquefaction of heavy metals and their compounds at high temperatures of 900–2000°C, and the heavy metals separation from the P via a gas-phase separation (or density separation) device. Therefore, the theoretical thermochemical extraction can effectively remove toxic heavy metals such as Cd, Cu, Pb, Zn, Mo, Sn, As, etc. and retain available fractions such as Mg, Al, and K, leading to a pure P-enriched product (Fraissler et al., 2009). In addition, the higher temperature increases the phytoavailability of phosphate by breaking down the original mineral phase in the ash and forming a new phosphate mineral phase (Ca–P). According to Fig. 25.4, the initial stage of the EU research project SUSAN's sustainable and safe reuse of municipal sewage sludge for nutrient recovery is the monoincinerated of sewage sludge at 800–900°C. The second stage of P recovery from sewage sludge utilized thermochemical extraction. In Leoben, Austria, the procedure had already undergone successful industrial testing (Mattenberger et al., 2010). However, the products recovered by this process have poorly fertilizer in all soil types.

If the Fe content in SSIA is low and satisfies the Fe/P < 0.2 condition, SSIA is potentially to produce white phosphorus (P_4). This can be achieved by avoiding to use Fe salts in the sewage sludge treatment process of wastewater treatment

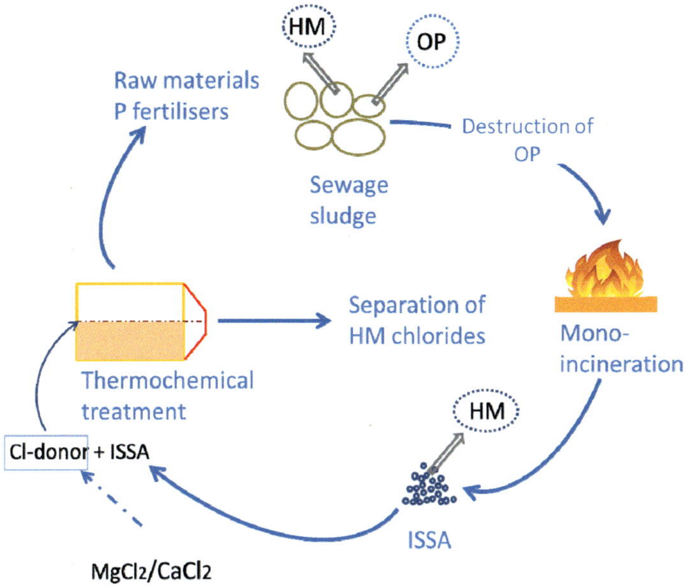

FIGURE 25.4 Principle of P recovery from SSIA by thermochemical extraction. *Adapted from Adam, C., Peplinski, B., Michaelis, M., Kley, G., Simon, F.G., 2009. Thermochemical treatment of sewage sludge ashes for phosphorus recovery. Waste Management 29, 1122–1128.*

plants to somewhat control the iron content to prepare high-purity white phosphorus (P_4) products. However, the presence of the heavy metals such as Cu and Zn in the ash is still seen to be of concern. By adding chlorides such as KCl or $MgCl_2$ to react with the heavy metals in SSIA, it results in high volatility and removal of Cd, Cu, Pb, Zn, Mo, Sn, As, etc. at about 900–1000°C (Mattenberger et al., 2008). However, this process can also lose up to 30% of P in SSIA and is ineffective in removing the nonvolatile Cr and Ni, which remain in the ash. To reduce the loss of P from SSIA, the authors improved the process by replacing the original ash with SSIA pellets (Vogel et al., 2010, 2013; Stemann et al., 2015) (Fig. 25.5).

The advantage of the thermochemical extraction is that the toxic heavy metals in the ash will volatilize at high temperatures (>1000°C), and the residual heavy metals content will lower than the limits imposed by agricultural regulations (Havukainen et al., 2016; Adam et al., 2009). The products generated by the thermochemical extraction retain almost P of the original SSIA and can be directly utilized as P fertilizer on the soil. However, the high energy consumption, operation complexity, and high-cost constraints of this method hinder its rapid development. The energy cost of the thermochemistry technique and incineration may be extremely high for the sewage sludge. Currently, the parameters influencing the process effectiveness are not fully researched, and there is still a need for further exploration at the laboratory stage.

4. Wet extraction of P from SSIA

Different extraction reagents were used for wet extraction processes to dissolve P in the SSIA. Wet extraction of P from SSIA is to transfer P from the solid phase into the liquid phase by directly inputting acid or alkali solution into the ash, thus changing the acid–base conditions of the ash to increase its P solubility. Further the P in the liquid phase is separated from heavy metals to obtain high additional-value P products. Wet chemical method can be divided into alkali wet chemical method, acid wet chemical method, and so on (Fang et al., 2021a,b; Liu et al., 2021).

4.1 Acid extraction of P from SSIA

The most widely used reagent to extract P from SSIA are acids (Petzet et al., 2011). P is released when metal-P bonds by acid, especially in strongly acidic conditions (pH range of 1–2) (Biswas et al., 2009). Both organic and inorganic acids have been employed.

These inorganic acids have demonstrated a high capacity for extraction P by leaching P and dissolving alkali-metal oxides from most phases, while organic acids enhance the release of metals/metalloids and metal-bound P via producing chelating effects (Fang et al., 2018a). Among these acids, H_2SO_4 is cost-effective and is predominantly employed at a commercial scale, while H_3PO_4 is expensive (Cieslik and Konieczka, 2017). Based on the fact that P in SSIA mainly exists in the form of Ca–P, Al–P, and Fe–P, the potential reactions (simplified) during acid extraction are listed in Eqs. (25.1)–(25.4) (Petzet et al., 2012).

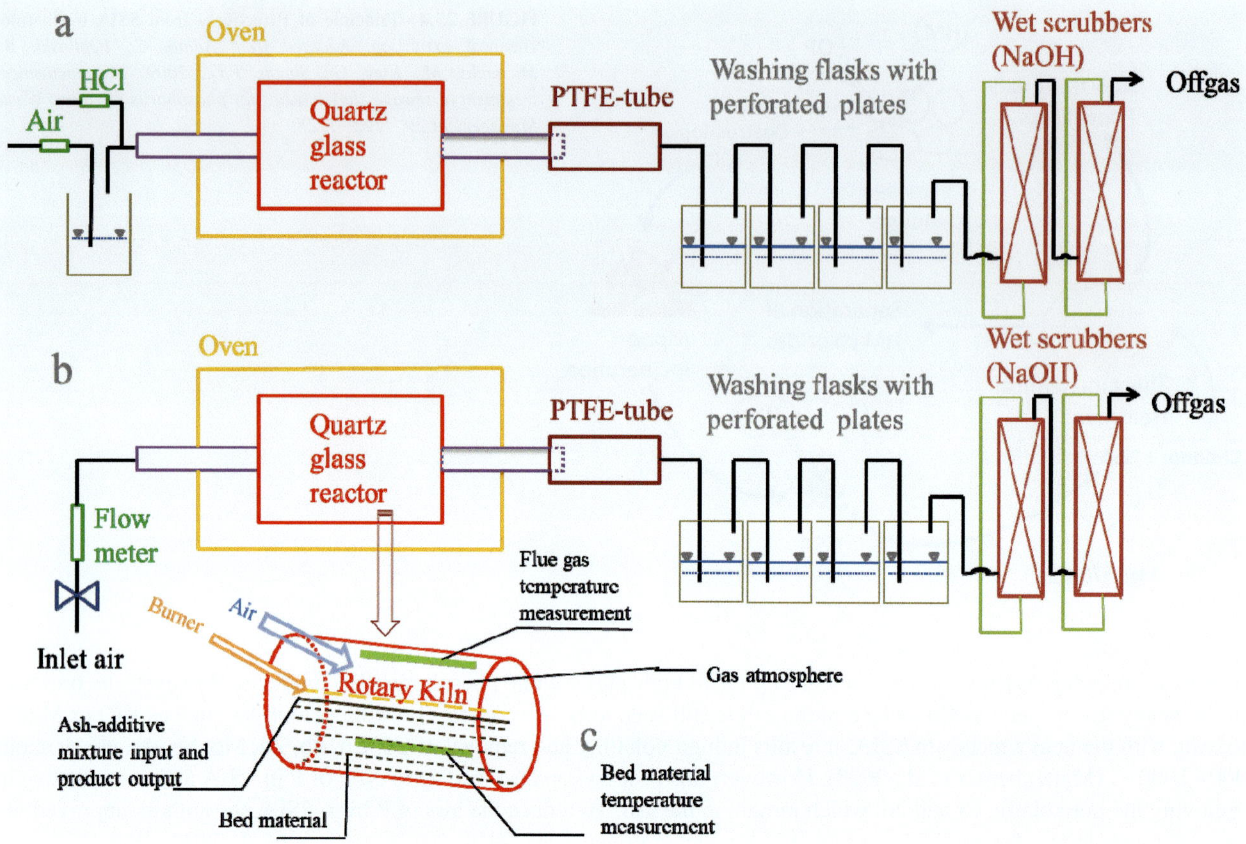

FIGURE 25.5 (A) and (B) are the experimental device diagrams that have appeared in the literature (Vogel et al., 2010; Adam et al., 2009).

$$AlPO_4 + 3H^+ \rightleftharpoons Al^{3+} + H_3PO_4, K_{sp} = 9.8 \times 10^{-21} \tag{25.1}$$

$$FePO_4 + 3H^+ \rightleftharpoons Fe^{3+} + H_3PO_4, K_{sp} = 1.3 \times 10^{-22} \tag{25.2}$$

$$Ca_3(PO_4)_2 + 6H^+ \rightleftharpoons 3Ca^{2+} + 2H_3PO_4, K_{sp} = 2.1 \times 10^{-33} \tag{25.3}$$

$$Fe_3(PO_4)_2 + 6H^+ \rightleftharpoons 3Fe^{2+} + 2H_3PO_4, K_{sp} = 1.6 \times 10^{-36} \tag{25.4}$$

Based on the stoichiometric calculations of these chemical reactions, P can theoretically be entirely extracted from SSIA with the H^+/P molar ratio of 3.0 (Liang et al., 2019; Lim and Kim, 2017). However, experiments revealed that more H^+ often needs to be applied to lead to higher P extraction efficiency. Process parameters significantly affect P extraction efficiency, including the liquid–solid ratios (L/S ratios), and final pH of the extraction solutions (Fang et al., 2018b; Li et al., 2018). Considering the energy and costs, a lower L/S ratio (2.3 mL/g) and short extraction time (10 min) with an intensive mixing process (1200 rpm) are preferred on an industrial scale to achieve a concentrated P solution with over 90% extraction efficiency (Gorazda et al., 2016; Luyckx et al., 2020).

Heavy metal impurities must be separated throughout the P recovery process because the acid leaching process converts P and heavy metals from SSIA into the leaching solution. Diverse wet chemical methods have been utilized to recover P from SSIA in Europe, where the fraction of sewage sludge incineration is largest, resulting in various forms of P products at varied scales (Fig. 25.6). In the process of phosphorus extraction from SSIA using inorganic and organic acid extractants, heavy metals are inevitably extracted such as Cu, Pb, Cr, Cd, As, Zn, etc. Studies have revealed that inorganic acids leach more Cd and organic acids leach more heavy metals such as Pb, As, Cu, and Zn from SSIA (Fang et al., 2018b). Heavy metals such as Pb, As, Cu, Zn, Cr, Ni, and Cd were discovered to leach quickly from HCl (Katsuura et al., 1996). It was discovered that H_3PO_4 and citric acid were found to leach Ni more readily (Kasina, 2022; Luyckx et al., 2020). Furthermore, despite extracting the majority of the P, HNO_3, and oxalic acid removed more than 39% of the Pb (Fang et al., 2020).

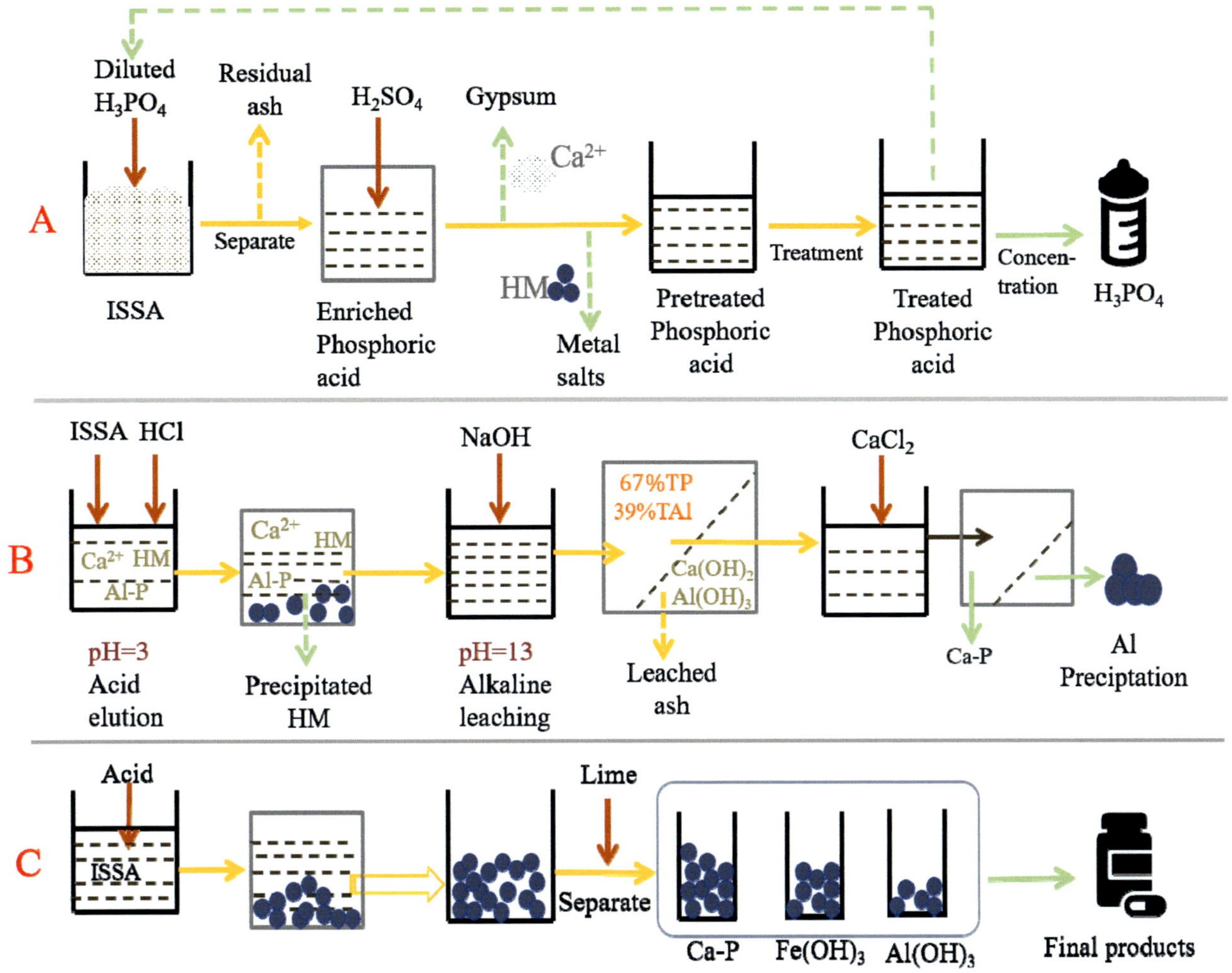

FIGURE 25.6 Recover P from SSIA with different forms of P products at different pilot plant scales (Remondis, 2018; Petzet et al., 2011; Mining, 2019).

4.2 Alkaline extraction of P from SSIA

According to the characteristics that NAIP such as Al—P dissolves under alkaline conditions and heavy metal elements and their compounds are almost insoluble under alkaline conditions (Ph > 12), alkaline solutions such as NaOH and KOH can be added to extract NAIP, while heavy metals are retained in SSIA (Eqs. 25.5–25.7) (Cao et al., 2019; Falayi, 2019). As a result, this process is highly selective.

$$AlPO_4 + 4OH^- \rightleftharpoons [Al(OH)_4]^- + PO_4^{3-} \tag{25.5}$$

$$Fe_3(PO_4)_2 + 6OH^- \rightleftharpoons 3Fe(OH)_2\downarrow + 2PO_4^{3-} \tag{25.6}$$

$$FePO_4 + 3OH^- \rightleftharpoons Fe(OH)_3\downarrow + PO_4^{3-} \tag{25.7}$$

Alkalic extraction's P extraction efficiency increases with increased L/S ratios, extractant concentrations, and contact time. It was demonstrated that 10% P was extracted from hydrochar at low pH level (pH = 10.5), whereas there was no significant increase in extraction efficiency at pH levels above 12 (L/S = 100 mL/g, 16 h) (Li et al., 2020a,b). Sufficient contact time achieved maximum extraction efficiency. Luyckx et al. observed that the amount of P leached from ash increased from 9% to 42% when contact time increased from 10 to 120 min with 0.5 N NaOH at L/S = 50 mL/g (Luyckx et al., 2020). As presented in Fig. 25.7, the extraction efficiency of phosphorus in SSIA was also related to the P/Ca molar ratio, which significantly decreased when P/Ca molar ratio decreased in the ash. Additionally, after the ash's CaO content reached more than 20%, the alkaline extraction of phosphorus was rendered impossible (Petzet et al., 2012). Therefore, alkaline leaching would not be appropriate for P extraction from Ca-rich ash.

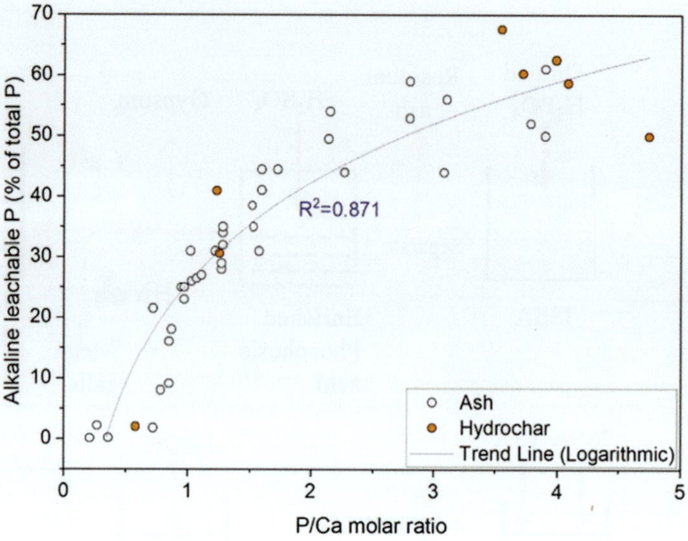

FIGURE 25.7 Molar ratio of alkaline-extractable P versus the corresponding P/Ca in SSIA (Liu et al., 2021).

4.3 Other extraction methods

Other reagents have also been employed for P extraction from SSIA. Fang et al. (2018a,b) used ethylene diamine tetra acetic acid (EDTA) and ethylene diamine tetramethylene phosphonate (EDTMP); the overall P extraction efficiency was only about 50%. However, the pretreatment of SSIA with EDTA under acidic conditions eliminated sizable amounts of metal(loid)s without leaching P, suggesting that EDTA might be utilized before the P extraction procedure.

To improve the quality of the recovered P products, metals/metalloids dissolved with P should also be separated, which needs additional energy or chemical input. Due to the restricted supply of NAIP, direct alkalic extraction does not apply to all ash and hydrochar formed from sludge. Direct alkalic extraction can prevent the interfering elements dissolution. To optimize extracted P and reduce leached metals/metalloids, various research studies suggested integrated sequential extraction method (Fig. 25.8) (Lim and Kim, 2017; Petzet et al., 2012; Zin and Kim, 2019; Fang et al., 2018a,b; Wang et al., 2018). Although sequential extraction techniques appear to be most compatible with heavy metal and hazardous element removal for SSIA, they do have some disadvantages. Strong acids and alkaline extractants are advised against using since they are seen to be very forceful and do not produce clearly defined fractions. They also cause heavy metals to dissolve together. The procedures still need to be optimized, which includes the pH levels, duration time, liquid-to-solid ratio, and chemical reagents and their concentrations.

5. Electrodialytic extraction of P from SSIA

The electrodialysis separation technology separates anionic phosphate ions and cationic heavy metals on the basis of electromigration, electrophoresis, and electroosmosis mechanisms. The electrodialysis process entails at least four phases, including the use of electrodes, ion exchange membranes, ion exchange membranes, and electrolytes. Ion transport is strengthened by the direct current supply, and the rate of ion transportation can be altered by varying the current density. Direct current is carried through electrodes into the system. The ions are separated using ion exchange membranes. The current between the anode and cathode compartments is carried by electrolytes (Huang et al., 2007). Transport of cations through the cation exchange membrane to the anode is triggered by the application of an electric current. As a result, the anode and cathode solutions get concentrated, while the solution between the membranes loses its ions (Altin et al., 2017). The electrolysis of water is the primary reaction involved in the electrodialysis procedure used to recover P (Guedes et al., 2014). The reactions for the anode and cathode are as follows:

$$\text{Cathode}: 4H_2O + 4e^- \rightarrow 2H_2\uparrow + 4OH^- \tag{25.8}$$

$$\text{Anode}: 2H_2O - 4e^- \rightarrow 4H^+ + O_2\uparrow \tag{25.9}$$

As shown in Eqs. (25.8) and (25.9), the anode reactions produce acid, and cathode reactions produce a base.

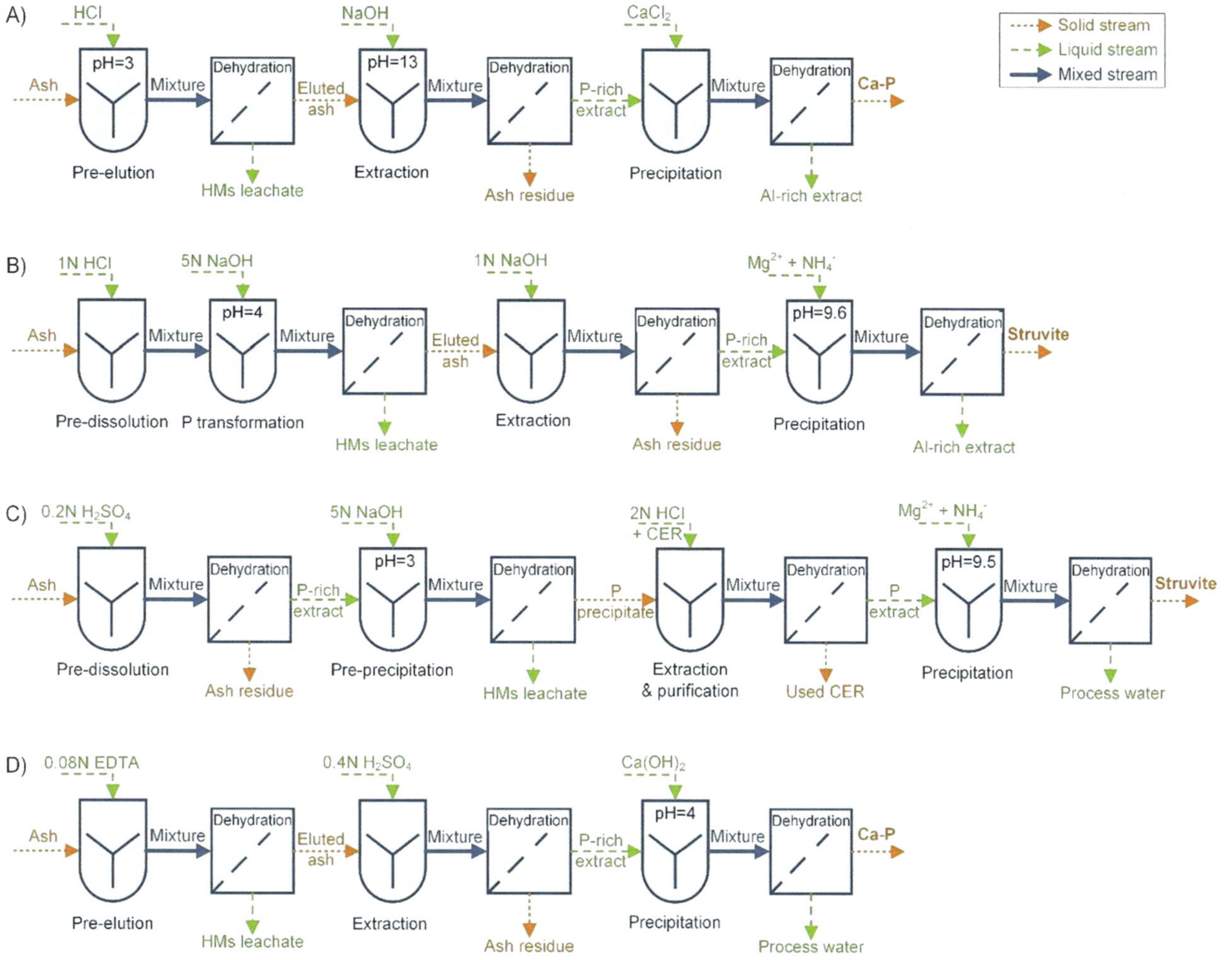

FIGURE 25.8 Schematic diagrams of sequential P extraction from SSIA and hydrochar. (A) SESAL-Phos; (B) acid–base leaching; (C) three-step extraction; (D) two-step extraction. *CER,* cation exchange resin; *EDTA,* ethylenediaminetetraacetic acid; *HMs,* heavy metals.

The setting principle of the three-chamber (3C) electrodialysis device and two-chamber electrodialysis is illustrated in Fig. 25.9 (Ebbers et al., 2015; Nystroem et al., 2005). The regulation of pH is important because it has a significant impact on the chemical form of P in the ash suspension. With sulfuric acid, the initial pH of the ash suspension is about 3.2–3.6. For the recovered phosphates, four forms H_3PO_4, $H_2PO_4^-$, HPO_4^{2-}, and PO_4^{3-} need to be considered. And the acidity constant states that when the pH value is less than 2, the P element is primarily present as phosphoric acid. It exists as monovalent, divalent, and trivalent orthophosphate when the pH value is greater than 2. A variety of measurements revealed a significant correlation between the conductivity and pH of the polar water (Sturm et al., 2010).

The main cause of the lower P recovery rate is the migration of P species in the electrodialysis suspension to the cathode, which leads to a small number of waste P resources and is a problem for which there is no viable solution. Most significantly, electrodialysis recovery of P is excessively expensive as compared with the conventional wet chemical approach due to its expensive operating steps and relatively high energy requirements.

6. Bioleaching extraction of P from SSIA

Bioleaching (Fig. 25.10), which is the conversion of inorganic or secreted organic acids produced by the metabolic activity of microorganisms from solid compounds to soluble and extractable forms under certain process conditions, is a promising option for the extraction of phosphorus from SSIA (Mehta et al., 2015; Priha et al., 2014). Microorganisms that can be used for bioleaching in SSIA include bacteria (*Thiobacillus ferrooxidans*, *Thiobacillus oxidans*, *Bacillus lipolyticus*, etc.) and fungi (*Aspergillus niger*, Penicillium gray mold, *Penicillium flavum*, etc.) (Mahmoud et al., 2017; Wu and Ting, 2006; Xiao et al., 2011). Fungal bioleaching of phosphorus-enriched SSIA is preferable to acidic *Thiobacillus*- and

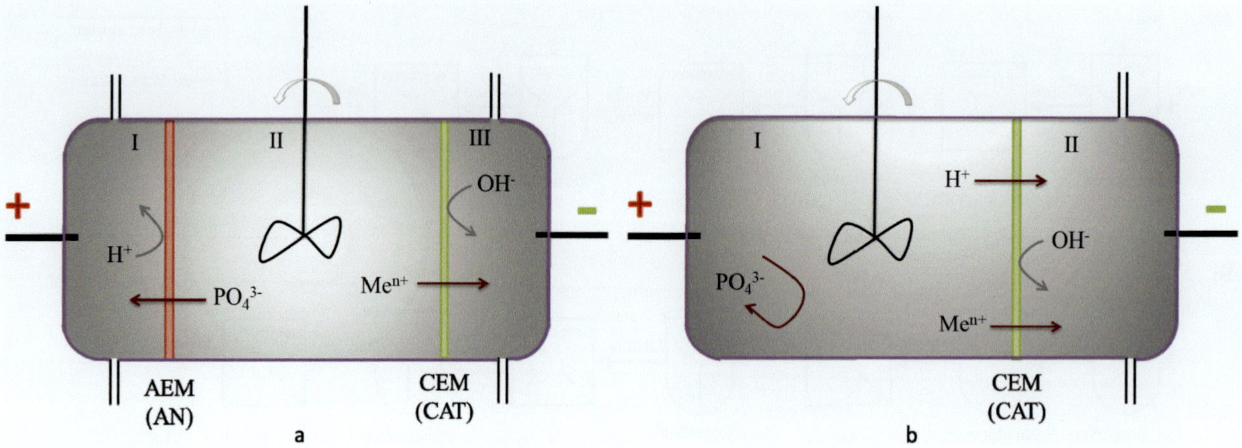

FIGURE 25.9 (A) Commonly used three-chamber (3C) electrodialysis device, (B) two-chamber electrodialysis setting.

Pseudomonas-type bacteria from the perspective of microbial adaptability. This is because it can grow at higher pH levels, has a faster rate of growth, and requires less energy (just 30% of the oxygen that bacteria do) (Cairns et al., 2018; More et al., 2010; Zhang et al., 2019). In addition, *T. ferrooxidans* is able to oxidize ferrous iron or oxidize sulfide to monomeric sulfur for proliferative metabolism; *Thiobacillus thiosulfuricus* is able to use reduced sulfur and monomeric sulfur as a substrate to grow and produce sulfuric acid, and the two microorganisms can work synergistically to produce sulfuric acid to leach out phosphorus and heavy metals. The two microorganisms can work synergistically to produce sulfuric acid, which can leach out phosphorus and heavy metals (More et al., 2010; Tao et al., 2021). Low-molecular-weight organic acids such as citric, oxalic, and gluconic acids produced by fungi could dissolve P from SSIA and alleviate the inhibition of dissolved heavy metals to microorganisms via complexing them (Gadd 1999; Santhiya and Ting, 2005).

However, the results of conventional bioleaching using *Acidithiobacillus thiooxidans* are far from adequate, with a recovery efficiency of up to 50%. Difficulties may also arise from the inclusion of costly chemicals (such as sucrose and beet molasses) required for fungus development and organic acids excretion (Chroumpi et al., 2020; Kim et al., 2006), making fungal extraction of SSIA unsustainable in large-scale application.

7. High-value utilization of recovered P

7.1 Liquid P fertilizer

According to Regulation (EC) NO. 2003/2003 of the European Commission (Directive, 2003), there are two types of liquid fertilizers, including solution fertilizers and suspension fertilizers. While suspension fertilizers have particles suspended in the fertilizer, solution fertilizers do not contain any solid particles. Specifically used in developed regions, liquid fertilizers account for 30% of the fertilizer market (Huang and Tang, 2016).

Liquid phosphorus fertilizers are highly soluble, stable, and diffusible, which allow the formation of a homogenous distribution of phosphorus in calcareous soils. But this is not significant in alkaline noncalcareous soils (Gorazda et al., 2016). The chemical/thermal P extraction procedure, the purification technique, and the composition of SSIA will all affect the composition of liquid fertilizers. The most preferred product, aside from P fertilizers, for the recovery of P from SSIA can be made from the liquid P-extract if the purity is high enough. A cation exchange column was created by Donatello et al. (2010) to purify the P-extracts and recover technical-grade H_3PO_4, but it is unclear whether the method is cost-effective (Donatello et al., 2010). A purifying procedure such as nanofiltration (Paltrinieri et al., 2019) is needed to produce H_3PO_4 from ash leftover from the burning of sewage sludge, according to other studies, and additional study is needed to create straightforward, inexpensive procedures.

7.2 Solid P fertilizers

After wet extraction of P from SSIA, P is usually recovered in the form of solid fertilizers by subsequent chemical precipitation. The solid mineral P fertilizers include $Ca_3(PO_4)_2$, monoammonium phosphate (MAP), and struvite. The manufacturing of solid P fertilizers using various P precipitation procedures is widely employed because they are highly recyclable, economical, and convenient to transport and store (Gorazda et al., 2013).

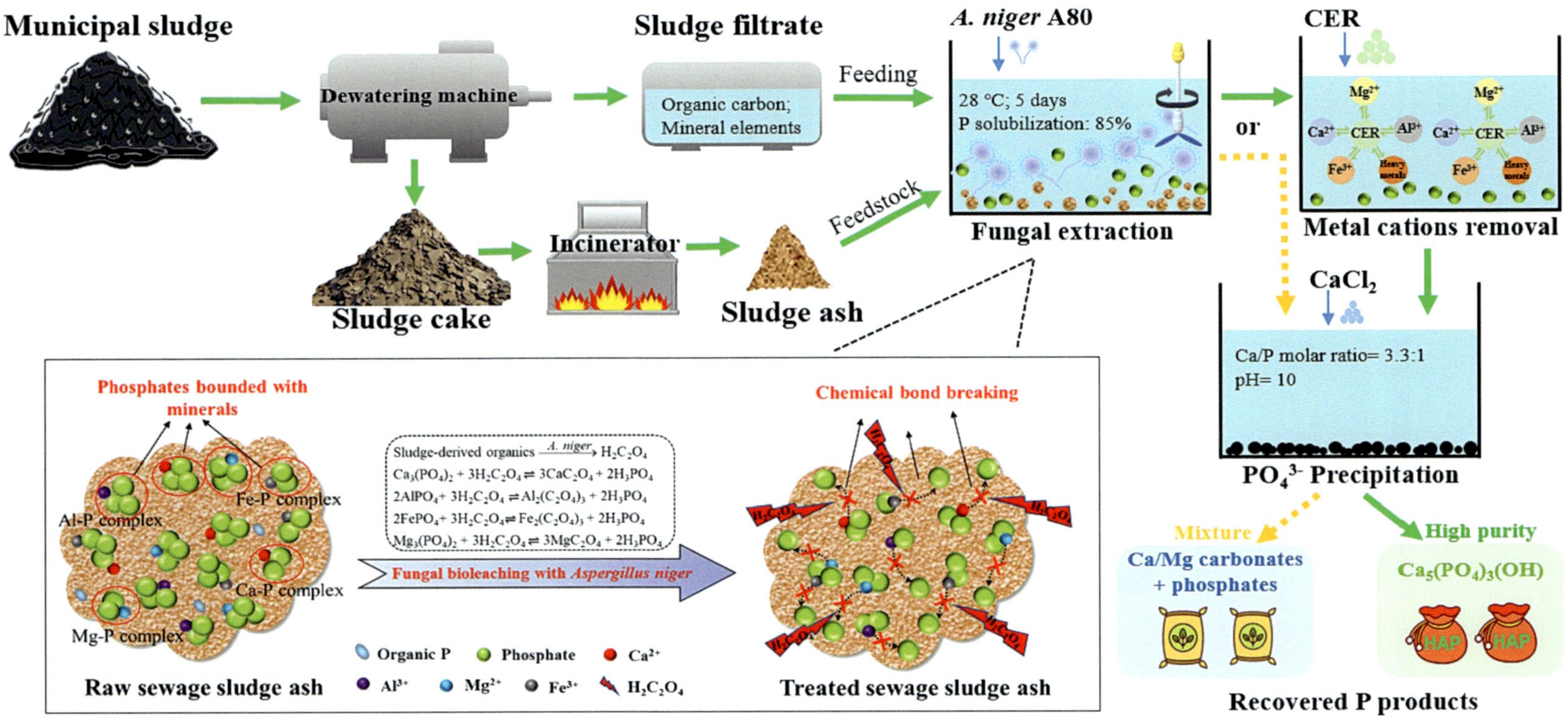

FIGURE 25.10 A possible technological strategy for P recovery from SSIA by integrated fungal extraction, CER purification, and HAP precipitation (Su et al., 2023).

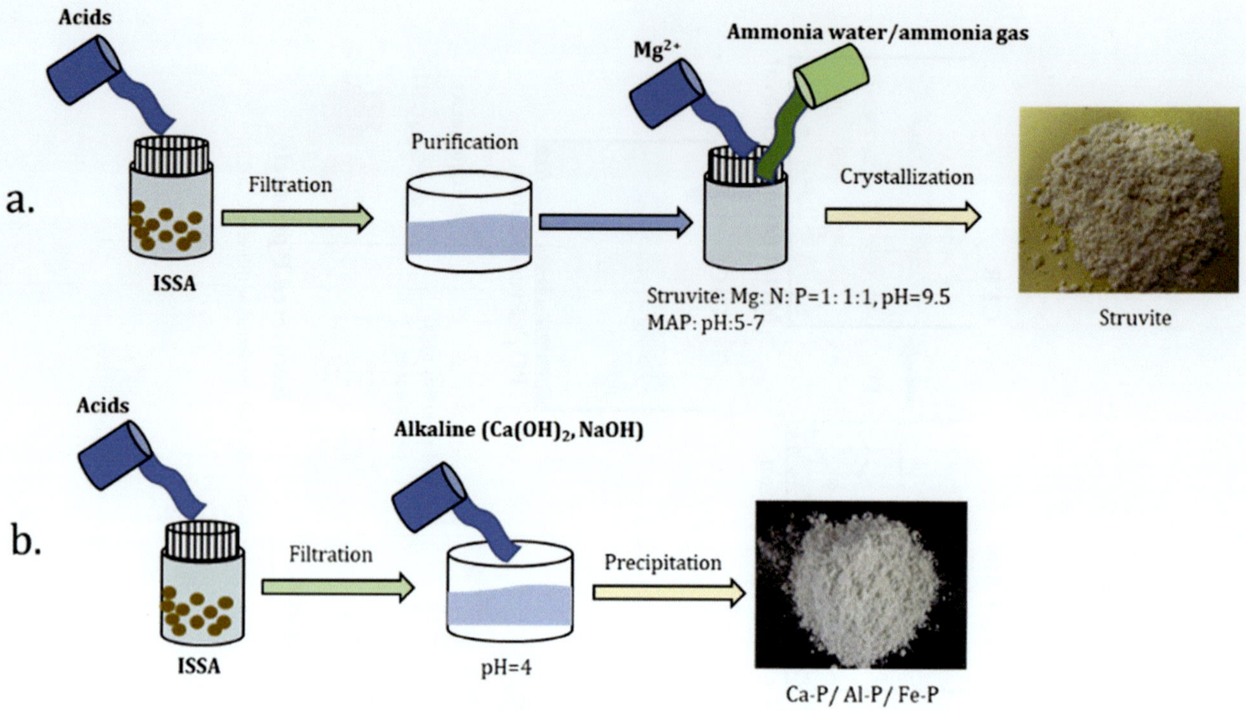

FIGURE 25.11 Production processes of two major kinds of solid P fertilizers.

The production processes used for the production of MAP and struvite are shown in Fig. 25.11A, and those for Ca—P and Al—P are shown in Fig. 25.11B. Struvite is produced by reducing the pH to <2, which results in the almost complete dissolution of phosphorus and heavy metals in the ash, and then the impurities are purified using phosphorus purification procedures such as cation exchange absorption (Meng et al., 2019; Wang et al., 2018), sulfur ion precipitation, and liquid phase extraction (Kataki et al., 2016; Zin and Kim, 2019). Subsequently, struvite crystals formed at pH 9.5 with a molar ratio of Mg^{2+}, NH_4^+, and PO_4^{3-} of 1:1:1 (Wang et al., 2018). However, the disadvantage is high purity requirement of the P-extract and high ammonium requirement (Cieslik and Konieczka, 2017; Xue and Huang, 2007). MAP is produced similarly to struvite by adding ammonia solution or ammonia gas into a high-purity P extract at a pH between 5 and 7 (Gorazda et al., 2017). For Al—P, Fe—P, and Ca—P, by maintaining the pH of the P-extract between 3 and 4, it is possible to selectively precipitate $Fe(II)_3(PO_4)_2(H_2O)_8$ and $Al_3(PO_4)_2(OH)_3(H_2O)_5$, while Ca—P ($Ca_3(PO_4)_2$) will precipitate above pH 4.

Since P-extracts required the complex purification processes and the precipitation/crystallization processes for the production of P fertilizer, Fang et al. investigated the simplification of this complicated processes by selectively adsorbing P from acid extracts of SSIA using plant-available media such as biochar (Fang et al., 2018a,b). This study combined complex purification and precipitation/crystallization steps for adsorption. The phosphorus-enriched biochar after adsorption could be used directly as a P fertilizer.

7.3 Vivianite

The conditions of low ORP, moderate pH (6—9), and abundant Fe and P elements required for the vivianite formation can be realized in the sludge anaerobic digestion system. Under the condition of sufficient iron content, 80%—90% of the phosphorus in the digested sludge will be in the form of the vivianite. So the sludge vivianite recycling has been widely concerned and studied (Li et al., 2021). Vivianite not only has a high phosphorus content (28.3%, P_2O_5), also is the key raw material for the lithium batteries production, and is a more convenient and economical phosphorus recovery product (Wu et al., 2019). Lab-scale studies have demonstrated that feasibility of recovering P products crystallized from vivianite from SSIA in complex environments, with increased precipitation efficiency (nearly 100%) and high purity (e.g., 82% vivianite and 10% silica) of vivianite crystallization pH 5—8 and Fe:P ratio of 1—2. Temperature (25—55°C) was shown to be insignificant (Liu et al., 2018). There is a lack of studies on the recovery of P as vivianite from SSIA. This may be due to several challenges, including the requirement to add expensive iron (II) salts, the absence of reducing and anoxic conditions for recovery, and the presence of impurities in the prepared vivianite crystal product that need to be further isolated and recovered. Vivianite is the only metastable at ambient conditions, while impurities may be fully oxidized within 2 days (Wu et al., 2019). Future researches should concentrate on improving operating parameters, suitable purifying techniques, potential risks of pollutants in recovered products, and comprehensive economic and life cycle assessments, to further improve this innovative and profitable precipitation technology.

8. Case study of P extraction methods

8.1 ICL Fertilizers, the Netherlands

ICL Fertilizers, a phosphate fertilizer producer, develops technologies for the use of secondary phosphate raw materials such as SSIA, phosphate to be recovered, and meat and bone meal ash. ICL Fertilizers signed a contract with the Dutch government in 2011 to replace phosphate ore with SSIA recovery technology by 2025 (European Sustainable Phosphorus Platform, 2011). Fig. 25.12 shows ICL schematic process for phosphate recovery from SSIA. There is no shortage of phosphorus resources and even excess phosphate production in the countries where ICL's European plants are located (Germany, the Netherlands). Therefore, the company mainly exports most of its recycled phosphate products to phosphorus-deficient countries, aiming to address the domestic phosphorus surplus problem (Langeveld, 2019) and achieve the goal of "sustainable development and environmental protection in parallel." The plant starts with single or multiple acid phosphate etching to release phosphorus and then continues with the addition of potassium chloride, potassium sulfate, or other trace metal elements (Mg, Zn, Mn, Cu, Mo, etc.) to produce different forms of compound fertilizers. Finally, ammonium phosphate is added to produce NPK fertilizer. In addition, ICL is developing downstream operations in Europe and the United States, such as a pilot plant in Terneuzen, the Netherlands, to produce white phosphorus (P_4) through the RecoPhos and Tenova processes. (P_4) in Terneuzen, the Netherlands, and a food-grade phosphoric acid production project that is undergoing feasibility studies (Langeveld, 2018).

8.2 Meta Water Group, Japan

Unlike the phosphorus resource situation in Europe, Japan does not have sufficient phosphorus ore and relies on imports for all phosphorus ore. However, Japanese regulations are extremely strict on the environmentally sound treatment of sludge, and the law prohibits the direct agricultural use of surplus sludge from sewage treatment (Nättorp et al., 2019). Therefore, to reduce the cost of sludge treatment and to solve the shortage of phosphorus resources, Japan promotes the P recovery from anaerobically digested sludge, dewatering fluid, or sludge incineration ash from sewage treatment plants (Hao et al., 2018). The Meta Water Group operates two sludge incineration ash phosphate recovery plants in Japan, one in Gifu (established in 2010) and the other in Tottori (established in 2013). These plants produced 300 and 150 ton of phosphate fertilizer annually, respectively (Buttmann, 2017). The phosphorus extraction process (refer to Fig. 25.13) involves several steps, firstly, washing the ash phosphorus with NaOH; secondly, recovering phosphorus through calcium hydroxyapatite (HAP) precipitation by adding Ca^{2+}; finally, dewatering, drying, and granulating the recovered product to produce phosphorus fertilizer for local farmers. The remaining ash residue undergoes a cleaning process using weak acid to eliminate heavy metals. Subsequently, it is converted into odorless brown particles that comply with soil pollution prevention standards. These particles have multiple applications, including serving as a substitute for stone powder in road base material or asphalt filler, as well as functioning as a soil conditioner (Nakagawa and Ohta, 2019).

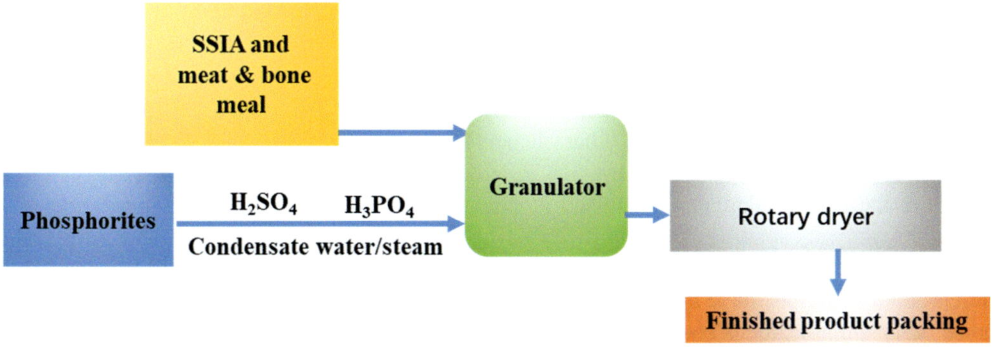

FIGURE 25.12 ICL schematic process for phosphate recovery from SSIA.

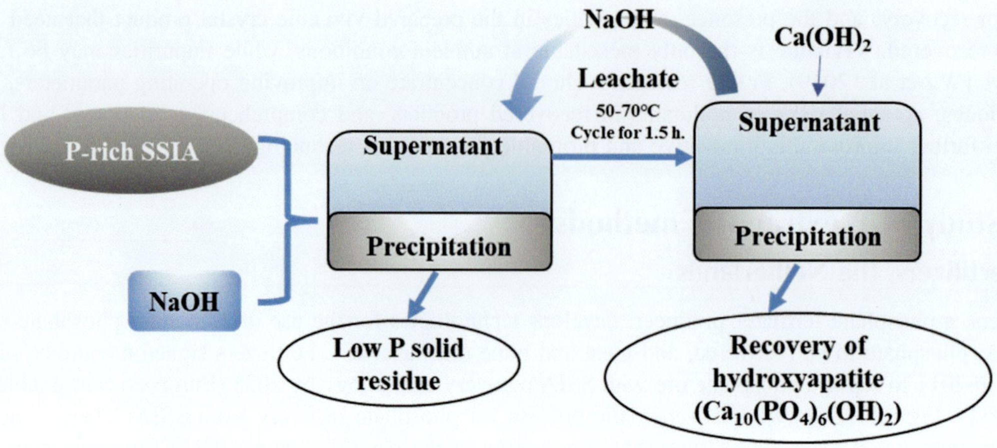

FIGURE 25.13 Alkaline-leaching schematic process of phosphate recovery from SSIA.

8.3 Ash2Phos (EasyMining), Sweden

Ash2Phos (EasyMining) utilizes SSIA obtained from wastewater projects that employ biological and/or chemical phosphorus removal as the primary input material. The process flow is first that SSIA dissolved in hydrochloric acid (ambient temperature, no pressure). The acid-insoluble ash consists of inorganic silicates, which are separated, washed, and can be used, for example, in the cement or concrete industry. Phosphorus, aluminum, and iron compounds are separated from acid leach solution by specific dissolution and precipitation reactions, a process characterized by internal cycling of chemicals. The remaining acid solution is neutralized and treated to remove heavy metals. The product is calcium phosphate, which can be future processed calcium superphosphate, dicalcium phosphate (DCP), and MAP. Currently, operational plants include Fei Uppsala in Sweden, handing 50 kg/day of SSIA, and Helsingborg, processing 600 kg/day of SSIA. Full-scale plants under construction are the Helsingborg in Sweden (license pending) with an expected SSIA capacity of 30,000 tons/year; Kemira with an expected SSIA capacity of 60,000 tons/year; and Bitterfeld-Wolfen located near Berlin.

8.4 TetraPhos (Remondis), Germany

TetraPhos (Remondis) uses SSIA from wastewater projects as input material for biological and/or chemical P removal. The SSIA is first submerged in phosphoric acid, which dissolves P and calcium but not the majority of iron or heavy metals. Sulfide is next added to precipitate heavy metals and to maximize the remained solid fraction in the leached ash. The solids are then partially separated from the phosphoric acid–rich liquid by the addition of sulfuric acid, and the calcium precipitates as gypsum, which is then separated out by vacuum belt filters and water washing. Part of the phosphoric acid generated is added back to the leaching procedure. By running it through an ion exchanger and, optionally, a nanofiltration membrane, the extra acid created is cleaned up. The purified phosphoric acid obtained is preferably concentrated by secondary heating, for example, from a sludge incinerator. Finally, the ion exchange resin is regenerated using hydrochloric acid, and the resulting metal salt solution is potentially recycled to a wastewater treatment plant for phosphorus removal. Products produced include phosphoric acid, gypsum, iron and aluminum salts, and mineral ash. Pilot plants currently in operation include full-scale plants in Elflingsen, Germany (with an SSIA processing capacity of 50 kg/h), and Hamburg, Germany (currently under construction), with a planned processing capacity of 20,000 tons/year.

8.5 PHOS$_4$Green (Glatt), Germany

PHOS$_4$Green (Glatt) utilizes SSIA as its raw material. The ash is reacted with phosphoric acid to increase its content of biologically effective phosphorus. Additionally, other elements such as N, K, Mg, S, and trace elements can be incorporated into the suspension. The resulted material is then pelletized to produce fertilizer granules. Heavy metals, Fe, Al, Si, and other minerals in the sewage sludge, are retained in the end product. The end product can be either a phosphorus (P) fertilizer or a nitrogen–phosphorus–potassium (NPK) fertilizer. Currently, laboratory and pilot scale of plants at Glatt's Technology Center in Weimar are conducting tests with input ash capacities of up to 30 kg/h. These pilot runs, involving different input materials, have been carried out continuously for several consecutive days. Furthermore, a full-scale plant named Seraplant in Haldensleben, Germany, with an annual SSIA processing of 30,000 tons, is currently in the commissioning phase.

9. International policy related to the application of P recovery technologies

The shortage of P resources makes it necessary to support the recycling of P products at policy and legal level. As early as 2004 to 2011, the Federal Ministry of Education and Research, Federal Ministry for the Environment, and Nature Conservation and Nuclear Safety carried out large-scale implementation of P recovery technologies. The material flows considered were municipal sewage sludge, municipal wastewater, residual manure, etc. Germany decided in 2013 to stop using sewage sludge as fertilizer and instead required the recovery of P and other nutrients from it, which highlighted sewage, sewage sludge, and SSIA as the focus of P recovery technology development. In 2016, the EU introduced a new draft fertilizer regulation that officially includes phosphate products such as guano stone recovered from sewage treatment plants for fertilizer production and sets such standard specifications for phosphate fertilizers. The EU guidelines shifted their focus from considering P as a low priority to prioritizing its prevention and control as a significant pollutant, while also emphasizing the recycling and utilization of P as a nutrient (Hu et al., 2020). On June 5, 2019, the EU revised the EU Fertilizer Regulation (2019/1009) to classify three P recovery products, namely STRUBIAS (guano, biochar, and incineration ash combined), as secondary P for fertilizer production. Fertilizers that meet the necessary utilization and safety criteria are eligible for unrestricted marketing throughout Europe (European Commission, 2021).

Besides the aforementioned summary of the EU's initiatives to dismantle trade barriers in the P recycling market, member states have cooperated in formulating suitable laws and regulations. Switzerland and the Netherlands were the pioneering nations in implementing the "Swiss Phosphorus Closed Cycle Construction" and the "2050 Dutch Cycle Plan," respectively. Their objective is to accomplish full nutrient recovery and establish a closed-loop system within their respective territories. Furthermore, the United Kingdom, France, and Germany put forward proposals to foster the sustainable growth of the P recycling industry, aiming to establish a regulatory framework and network system for effective P recycling. Additionally, other Nordic countries have implemented policies to broaden the scope and depth of P recycling initiatives (European Sustainable Phosphorus Platform, 2018). European countries have established the European Sustainable P Platform (ESPP) to facilitate the exchange of academic findings and management insights regarding P recycling technologies across Europe. Additionally, the platform serves as a communication forum for managers, consumers, and markets to foster the multifaceted development of the P recycling market (Hu et al., 2020).

The Swedish government awarded 51 million kronor (equivalent to approximately RMB 38.5 million) on March 24, 2021, to endorse the efforts of EasyMinings, a Swedish P recycling company, in conserving natural resources through initiatives like P recovery from sewage sludge and the reduction of P mining activities in Europe (EasyMining, 2021). In 2018, Denmark established a goal of achieving 80% P recovery from sludge, thereby advancing the progress of P recovery technologies and applications. This objective is pursued through the implementation of a tax on discharged effluent P (€22 per kgP) and a tax on sludge landfill (€63 per ton of sludge) (Gaard, 2019).

For the recycling of P from SSIA, the following are the main international policies and regulations: European Union (EU) Fertiliser Directive: The directive calls for increased harvesting of nutrients such as P and K, which are high in SSIA, by 2030, and encourages the use of SSIA as an alternative fertilizer or for reuse in agriculture and building materials. Switzerland first enacted legislation on sludge treatment in 1986, known as the Sewage Sludge Ordinance. This ordinance was subsequently amended in 1998 and 2008 to form the current system of sludge management regulations. The law stipulates that sludge must be reused or disposed of and prohibits direct disposal. Phosphorus in sludge incineration ash must be recycled. Japan encourages the use of SSIA as a raw material for concrete and is actively researching P recovery technology. Germany's circular economy law: SSIA belongs to industrial waste and must be treated and reused in accordance with the requirements of the Waste Act. The United States water quality law: The United States does not have a policy of explicitly requiring the recovery of P from SSIA, but encourages the reuse of pollutants to reduce dependence on new resources. Regulations in Taiwan, China: Taiwan requires wastewater treatment plants to recover more than 70% of the P in SSIA. The Environmental Protection Bureau of Taiwan also provides subsidies for recovery and treatment equipment to promote the recycling of P resources. According to China's "Urban Sewage Treatment Plant Sludge Disposal and Utilization Management Measures," the main contents of the recycling of phosphorus resources in sludge are as follows: Encourage sewage treatment plants to adopt incineration and other methods to recycle phosphorus, nitrogen, organic matter, etc. in sludge resource. For resources such as phosphorus, calcium, silicon, and aluminum enriched in sludge ash, sewage treatment plants should take effective measures to recycle them.

In general, the international community generally requires the recycling of P resources in SSIA and has introduced a variety of policies, regulations, and economic means to promote the recycling of P. This is of great significance to the protection of P resources and the implementation of circular economy.

10. Summary

P in sewage sludge was mainly dominated by IP, while NAIP accounted for 80% of IP. After incineration treatment, the P species in SSIA mainly also existed in the form of IP, while the higher bioavailability of AP was the main component in IP. Additives in incineration process could affect the migration and transformation of P. The addition of CaO can promote P release and the conversion of NAIP into AP, and the biomass with higher content of alkali metals such as K could also enhance the bioavailability of P in ash.

For P recovery from SSIA, the wet extraction method is favored because of its simplicity as well as low energy consumption, while the purity of the recovered P products is low and further purification is needed. Theoretically, thermochemical extraction and electrodialytic extraction can separate heavy metals from SSIA to obtain nontoxic and higher content of P products. However, the disadvantages of these two methods are complex operating conditions and high energy consumption. These two methods applied to the P recovery are neglected in recent research. If the recovered P is desired to be directly used in agricultural, thermochemical extraction would be an option worth considering.

With the continuous improvement of environmental protection requirements in various countries, the regulations and standards for waste management are also continuously strengthened. This will promote the development of P resource recovery from SSIA and will also put forward higher requirements and challenges for recovery technology and management. It is necessary to continuously explore and try new technologies and methods to achieve maximum utilization and recycling of P resources, so as to achieve environmental protection and sustainable development.

11. Future trends

Some researchers noted that additives utilization could transform P species into a new bioavailable mineral phase while removing heavy metals. The main incineration additives currently being studied are CaO, chlorides, and some biomass such as wheat straw and cotton stalk. Compared with monoincineration, the P characteristics, and recovery efficiency, dissolution of heavy metals and the suitable extraction methods are new hot spots in recent years. However, coincineration will make the existence of P in the SSIA more complex, thus leading to P extraction difficulty. This requires a more in-depth study of coincineration and the development of pretreatment processes and optimizes the extraction technology to achieve efficient and economical P recovery.

After incineration, the pathogens and organic matter are removed from the sewage sludge, while the fate of the remaining heavy metals varies with different recovery methods. Although different methods to recover P have been explored, specific experimental parameters are still not sufficiently mapped out. Addressing these factors could provide a better idea to explore the environmentally sound disposal of sludge and P recovery. And the future technology of P recovery from SSIA will pay more attention to the refinement and high efficiency of the extraction of P.

Acknowledgments

The authors gratefully acknowledge the financial support from the National Natural Science Foundation of China (Grant No. 52236008, 52170141, and 52206178) for this study.

References

Adam, C., Peplinski, B., Michaelis, M., Kley, G., Simon, F.G., 2009. Thermochemical treatment of sewage sludge ashes for phosphorus recovery. Waste Manag. 29, 1122–1128.

Altin, S., Oztekin, E., Altin, A., 2017. Comparison of electrodialysis and reverse electro-dialysis process in the removal of Cu (II) from dilute solutions. Kor. J. Chem. Eng. 34, 2218–2224.

Appels, L., Baeyens, J., Degr` eve, J., Dewil, R., 2008. Principles and potential of the anaerobic digestion of waste-activated sludge. Prog. Energy Combust. Sci. 34 (6).

Biswas, B.K., Inoue, K., Harada, H., Ohto, K., Kawakita, H., 2009. Leaching of phosphorus from incinerated sewage sludge ash by means of acid extraction followed by adsorption on orange waste gel. J. Environ. Sci. 21, 1753–1760.

Bloem, E., Albihn, A., Elving, J., Hermann, L., Lehmann, L., Sarvi, M., Schaaf, T., Schick, J., Turtola, E., Ylivainio, K., 2017. Contamination of organic nutrient sources with potentially toxic elements, antibiotics and pathogen microorganisms in relation to P fertilizer potential and treatment options for the production of sustainable fertilizers: a review. Sci. Total Environ. 607–608, 225–242.

Bogdan, A., Robles-aguilar, A.A., Liang, Q., Pap, S., Michels, E., Meers, E., 2022. Substrate-driven phosphorus bioavailability dynamics of novel inorganic and organic fertilizing products recovered from municipal wastewater—tests with ryegrass. Agronomy 12 (2), 292.

Buttmann, M., 2017. Industrial scale plant for sewage sludge treatment by hydro-thermal carbonization in Jining/China and phosphate recovery by TerraNova® Ultra HTC process. In: European Biosolids and Organic Resources Conference.

Cairns, T.C., Nai, C., Meyer, V., 2018. How a fungus shapes biotechnology: 100 years of Aspergillus Niger research. Fungal Biol. Biotechnol. 5 (1), 1−14.

Cao, J.S., Wu, Y., Zhao, J.N., Jin, S., Aleem, M., Zhang, Q., Fang, F., Xue, Z.X., Luo, J.Y., 2019. Phosphorus recovery as vivianite from waste activated sludge via optimizing iron source and ph value during anaerobic fermentation. Bioresour. Technol. 293.

Chroumpi, T., Makelä, M.R., de Vries, R.P., 2020. Engineering of primary carbon metabolism in filamentous fungi. Biotechnol. Adv. 43, 107551.

Cieslik, B., Konieczka, P., 2017. A review of phosphorus recovery methods at various steps of wastewater treatment and sewage sludge management. The concept of "no solid waste generation" and analytical methods. J. Clean. Prod. 142, 1728−1740.

Directive, E.U., 2003. Regulation (EC) No 2003/2003 of the European Parliament and of the Council of 13 October 2003 Relating to Fertilizers. European Parliament and European Council.

Donatello, S., Tong, D., Cheeseman, C.R., 2010. Production of technical grade phosphoric acid from incinerator sewage sludge ash (ISSA). Waste Manage. 30, 1634−1642.

Donnert, D., Salecker, M., 1999. Elimination of phosphorus from wastewater by crystallisation. Environ. Technol. 20, 735−742.

EasyMining, 2021. Our Phosphorus Recovery Solution Receives Multi-Million grant[EB/OL]. https://www.easymining.se/newsroom/articles-news/51-msek-grant/.

Ebbers, B., Ottosen, L.M., Jensen, P.E., 2015. Comparison of two different electrodialytic cells for separation of phosphorus and heavy metals from sewage sludge ash. Chemosphere 125, 122−129.

Elser, J.J., 2012. Phosphorus: a limiting nutrient for humanity? Curr. Opin. Biotechnol. 23 (6), 833−838.

European Sustainable Phosphorus Platform, 2011. Dutch Phosphate Value Chain Agreement. https://www.phosphorusplatform.eu/images/download/Dutch_phosphate_value_chain_agreement_-_Oct_4th_2011.pdf.

European Sustainable Phosphorus Platform, 2018. Summary of the 3rd European Sustainable Phosphorus Conference (ESPC3)[EB/OL]. https://phosphorusplatform.eu/images/scope/scopenewsletter127.

European Commission, 2021. Published Initiatives-Fertilizing Products-Technical Update [EB/OL]. https://ec.europa.eu/info/law/better-regulation/have-your-say/initiatives/12135-Fertilising-products-technical-update_en.

Falayi, T., 2019. Alkaline recovery of phosphorous from sewage sludge and stabilisation of sewage sludge residue. Waste Manage. 84, 166−172.

Fang, L., Li, J.S., Donatello, S., Cheeseman, C.R., Wang, Q.M., Poon, C.S., Tsang, D.C.W., 2018a. Recovery of phosphorus from incinerated sewage sludge ash by combined two-step extraction and selective precipitation. Chem. Eng. J. 348, 74−83.

Fang, L., Wang, Q.M., Li, J.S., Poon, C.S., Cheeseman, C.R., Donatello, S., Tsang, D.C.W., 2020. Feasibility of wet-extraction of phosphorus from incinerated sewage sludge ash (ISSA) for phosphate fertilizer production: a critical review. Crit. Rev. Environ. Sci. Technol. 51, 939−971.

Fang, L.E., Li, J.S., Guo, M.Z., Cheeseman, C.R., Tsang, D.C.W., Donatello, S., Poon, C.S., 2018b. Phosphorus recovery and leaching of trace elements from incinerated sewage sludge ash (ISSA). Chemosphere 193, 278−287.

Fang, Z.Q., Zhuang, X.Z., Zhang, X.H., Li, Y.L., Li, R.D., Ma, L.L., 2023. Influence of paraments on the transformation behaviors and directional adjustment strategies of phosphorus forms during different thermochemical treatments of sludge. Fuel 333 (2), 126544.

Fournie, T., Rashwan, T.L., Switzer, C., Gerhard, J.I., 2022. Phosphorus recovery and reuse potential from smouldered sewage sludge ash. Waste Manag. 137, 241−252.

Fraissler, G., Joller, M., Mattenberger, H., Brunner, T., Obernberger, I., 2009. Thermodynamic equilibrium calculations concerning the removal of heavy metals from sewage sludge ash by chlorination. Chem. Eng. Process. Process Intensif. 48, 152−164.

Gaard, J.J., 2019. Danish National Taxes on Phosphorous Discharges and Onsludge Ash landfill[EB/OL]. https://phosphorusplatform.eu/images/download/P-removal-workshop-2019/Gaard_Denmark_Ministry_9_11_19_Liege.pdf.

Gadd, G.M., 1999. Fungal production of citric and oxalic acid: importance in metal speciation, physiology and biogeochemical processes. Adv. Microb. Physiol. 41, 47−92.

Gorazda, K., Tarko, B., Wzorek, Z., Kominko, H., Nowak, A.K., Kulczycka, J., Henclik, A., Smol, M., 2017. Fertilisers production from ashes after sewage sludge combustion−a strategy towards sustainable development. Environ. Res. 154, 171−180.

Gorazda, K., Tarko, B., Wzorek, Z., Nowak, A.K., Kulczycka, J., Henclik, A., 2016. Characteristic of wet method of phosphorus recovery from polish sewage sludge ash with nitric acid. Open Chem. 14, 37−45.

Gorazda, K., Wzorek, Z., Tarko, B., Nowak, A.K., Kulczycka, J., Henclik, A., 2013. Phosphorus cycle-possibilities for its rebuilding. Acta Biochim. Pol. 60, 725−730.

Gu, S., Fu, B., Ahn, J.W., Fang, B., 2021. Mechanism for phosphorus removal from wastewater with fly ash of municipal solid waste incineration. Seoul Korea J. Clean. Prod. 280, 124430.

Guedes, P., Couto, N., Ottosen, L.M., Ribeiro, A.B., 2014. Phosphorus recovery from sewage sludge ash through an electrodialytic process. Waste Manag. 34 (5), 886−892.

Hao, X.D., Yu, W.B., Shi, C., Cheng, Z.H., 2021. Analysis of phosphorus recovery potential from sludge incineration ash and its market prospects. China Water Wastewater, 37-4.

Hao, X.D., Yu, J.L., Liu, R.B., 2020. Advances of phosphorus recovery from the incineration ashes of excess sludge and its associated technologies. Acta Sci. Circumstantiae 40 (4), 1149−1159.

Hao, X.D., Zhou, J., Wang, C.C., 2018. New product of phosphorus recovery—vivianite. Acta Sci. Circumstantiae 38 (11), 4223−4234.

Havukainen, J., Nguyen, M.T., Hermann, L., Horttanainen, M., Mikkila, M., Deviatkin, I., Linnanen, L., 2016. Potential of phosphorus recovery from sewage sludge and manure ash by thermochemical treatment. Waste Manag. 49, 221−229.

Herzel, H., Kruger, O., Hermann, L., Adam, C., 2016. Sewage sludge ash−A promising secondary phosphorus source for fertilizer production. Sci. Total Environ. 542, 1136−1143.

Hu, W.J., Zhao, Y.C., Zhen, G.Y., 2020. Analysis and inspiration of sewage sludge treatment and disposal policy and phosphorus recovery technology in Germany. Water Wastewater Eng. 46 (6), 15−20.

Huang, R.X., Tang, Y.Z., 2016. Evolution of phosphorus complexation and mineralogy during (hydro) thermal treatments of activated and anaerobically digested sludge: insights from sequential extraction and p k-edge XANES. Water Res. 100, 439−447.

Huang, C., Xu, T., Zhang, Y., Xue, Y., Chen, G., 2007. Application of electrodialysis to the production of organic acids: state-of-the-art and recent developments. J. Membr. Sci. 288, 1−12.

Kasina, M., 2022. The assessment of phosphorus recovery potential in sewage sludge incineration ashes—a case study. Environ. Sci. Pollut. Res. 30, 13067−13078.

Kataki, S., West, H., Clarke, M., Baruah, D.C., 2016. Phosphorus recovery as struvite from farm, municipal and industrial waste: feedstock suitability, methods and pre-treatments. Waste Manage. (Tucson, Ariz.) 49, 437−454.

Katsuura, H., Inoue, T., Hiraoka, M., Sakai, S., 1996. Full-scale plant study on fly ash treatment by the acid extraction process. Waste Manage. 16, 491−499.

Kim, J.-W., Barrington, S., Sheppard, J., Lee, B., 2006. Nutrient optimization for the production of citric acid by Aspergillus Niger NRRL 567 grown on peat moss enriched with glucose. Process Biochem. 41 (6), 1253−1260.

Kruger, O., Adam, C., 2015. Recovery potential of German sewage sludge ash. Waste Manag. 45, 400−406.

Langeveld, K., 2018. The Recophos/Inducarb process (the Netherlands). SCHAUM C. Phosphorus: Polluter and Resource of the Future Removal and Recovery from Wastewater. IWA Publishing, London, pp. 443−446.

Langeveld, K., 2019. Phosphorus recovery into fertilizers and industrial products by ICL in Europe. In: Ohtake, H., Tsuneda, S. (Eds.), PhosphorusRecovery and Recycling. Springer Singapore, Singapore, pp. 235−255.

Li, C.Y., Sheng, Y.Q., Xu, H.D., 2021. Phosphorus recovery from sludge by pH enhanced anaerobic fermentation and vivianite crystallization. J. Environ. Chem. Eng. 9.

Li, J.S., Chen, Z., Wang, Q.M., Fang, L., Xue, Q., Cheeseman, C.R., Donatello, S., Liu, L., Poon, C.S., 2018. Change in re-use value of incinerated sewage sludge ash due to chemical extraction of phosphorus. Waste Manage. 74, 404−412.

Li, R., Yin, J., Wang, W., Li, Y., Zhang, Z., 2014. Transformation of phosphorus during drying and roasting of sewage sludge. Waste Manag. 34 (7), 1211−1216.

Li, R., Zhang, Z., Li, Y., Teng, W., Wang, W., Yang, T., 2015. Transformation of apatite phosphorus and non-apatite inorganic phosphorus during incineration of sewage sludge. Chemosphere 141, 57−61.

Li, S.S., Zeng, W., Jia, Z.Y., Wu, G.D., Xu, H.H., Peng, Y.Z., 2020a. Phosphorus species transformation and recovery without apatite in fecl3- assisted sewage sludge hydrothermal treatment. Chem. Eng. J. 399.

Li, Y., Fang, Z., Teng, W., Shen, S., Li, R., 2020b. Comprehensive evaluation of the control efficiency of heavy-metal emissions during two-step thermal treatment of sewage sludge. ACS Omega 5 (38), 24467−24476.

Liang, S., Chen, H., Zeng, X., Li, Z., Yu, W., Xiao, K., 2019. A comparison between sulfuric acid and oxalic acid leaching with subsequent purification and precipitation for phosphorus recovery from sewage sludge incineration ash. Water Res. 159, 242−251.

Lim, B.H., Kim, D.J., 2017. Selective acidic elution of ca from sewage sludge ash for phosphorus recovery under ph control. J. Ind. Eng. Chem. 46, 62−67.

Liu, H., Hu, G.J., Basar, I.A., Li, J.B., Lyczko, N., Nzihou, A., Eskicioglu, C., 2021. Phosphorus recovery from municipal sludge-derived ash and hydrochar through wet-chemical technology: a review towards sustainable waste management. Chem. Eng. J. 417.

Liu, J.Q., Cheng, X., Qi, X., Li, N., Tian, J.B., Qiu, B., Xu, K.N., Qu, D., 2018. Recovery of phosphate from aqueous solutions via vivianite crystallization: thermodynamics and influence of ph. Chem. Eng. J. 349, 37−46.

Luyckx, L., Geerts, S., Van Caneghem, J., 2020. Closing the phosphorus cycle: multi-criteria techno-economic optimization of phosphorus extraction from wastewater treatment sludge ash. Sci. Total Environ. 713.

Mahmoud, A., Cézac, P., Hoadley, A.F., Contamine, F., d'Hugues, P., 2017. A review of sulfide minerals microbially assisted leaching in stirred tank reactors. Int. Biodeterior. Biodegrad. 119, 118−146.

Marta, G.A., Azucena, M., Cartagena, M.C., 2012. Fractionation of phosphorus biowastes: characterisation and environmental risk. Waste Manag. 32 (6), 1061−1068.

Mattenberger, H., Fraissler, G., Brunner, T., Herk, P., Hermann, L., Obernberger, I., 2008. Sewage sludge ash to phosphorus fertiliser: variables influencing heavy metal removal during thermochemical treatment. Waste Manag. 28, 2709−2722.

Mattenberger, H., Fraissler, G., Joller, M., Brunner, T., Obernberger, I., Herk, P., Hermann, L., 2010. Sewage sludge ash to phosphorus fertiliser (II): influences of ash and granulate type on heavy metal removal. Waste Manag. 30, 1622−1633.

Mehta, C.M., Khunjar, W.O., Nguyen, V., Tait, S., Batstone, D.J., 2015. Technologies to recover nutrients from waste streams: a critical review. Crit. Rev. Environ. Sci. Technol. 45 (4), 385−427.

Meng, X., Huang, Q., Gao, H., Tay, K., Yan, J., 2018. Improved utilization of phosphorous from sewage sludge (as Fertilizer) after treatment by Low-Temperature combustion. Waste Manag. 80, 349−358.

Meng, X.D., Liu, X.J., Huang, Q.X., Gao, H.P., Tay, K., Yan, J.H., 2019. Recovery of phosphate as struvite from low-temperature combustion sewage sludge ash (LTCA) by cation exchange. Waste Manage. 90, 84–93.

Meng, X.D., Wang, Q., Wan, B., Xu, J., Huang, Q.X., Yan, J.H., 2021. Transformation of phosphorus during low-temperature co-combustion of sewage sludge with biowastes. ACS Sustain. Chem. Eng. 9 (10), 3661–3669.

Mining, E., 2019. Ash2Phos from Sewage Sludge Ash to Clean Phosphorus Products.

More, T., Yan, S., Tyagi, R., Surampalli, R., 2010. Potential use of filamentous fungi for wastewater sludge treatment. Bioresour. Technol. 101 (20), 7691–7700.

Nakagawa, H., Ohta, J., 2019. Phosphorus recovery from sewage sludge ash: a case study in Gifu, Japan. In: Ohtake, H., Tsuneda, S. (Eds.), Phosphorus Recovery and Recycling. Springer Singapore, Singapore, pp. 149–155.

Nättorp, A., Kabbe, C., Matsubae, K., 2019. Development of phosphorus recycling in Europe and Japan. In: Ohtake, H., Tsunedas, S. (Eds.), Phosphorus Recovery and Recycling. Springer Singapore, Singapore, pp. 3–27.

Nilsson, C., Sjöberg, V., Grandin, A., Karlsson, S., Allard, B., Von, K.T., 2022. Phosphorus speciation in sewage sludge from three municipal wastewater treatment plants in Sweden and their ashes after incineration. Waste Manag. Res. 40 (8), 1267–1276.

Nystroem, G.M., Ottosen, L.M., Villumsen, A., 2005. Acidification of harbor sediment and removal of heavy metals induced by water splitting in electrodialytic remediation. Separ. Sci. Technol. 40 (11), 2245–2264.

Ohbuchi, A., Sakamoto, J., Kitano, M., Nakamura, T., 2008. X-ray fluorescence analysis of sludge ash from sewage disposal plant. X Ray Spectrom. 37, 544–550.

Paltrinieri, L., Remmen, K., Muller, B., Chu, L.Y., Koser, J., Wintgens, T., Wessling, M., de Smet, L.C.P.M., Sudholter, E.J.R., 2019. Improved phosphoric acid recovery from sewage sludge ash using layer-by-layer modified membranes. J. Membr. Sci. 587.

Pang, Y., 2004. Study on the Phosphorus Morphology of Great Lakes Sediments and its Phosphorus Sorption Characteristics. Chinese Research Academy of Environmental Sciences.

Pardo, P., López-Sánchez, J.F., Rauret, G., 2003. Relationships between phosphorus fractionation and major components in sediments using the SMT harmonized extraction procedure. Anal. Bioanal. Chem. 376 (2), 248–254.

Petzet, S., Peplinski, B., Bodkhe, S.Y., Cornel, P., 2011. Recovery of phosphorus and aluminium from sewage sludge ash by a new wet chemical elution process (SESAL-Phos-recovery process). Water Sci. Technol. 64, 693–699.

Petzet, S., Peplinski, B., Cornel, P., 2012. On wet chemical phosphorus recovery from sewage sludge ash by acidic or alkaline leaching and an optimized combination of both. Water Res. 46, 3769–3780.

Priha, O., Sarlin, T., Blomberg, P., Wendling, L., Makinen, ¨ J., Arnold, M., Kinnunen, P., 2014. Bioleaching phosphorus from fluorapatites with acidophilic bacteria. Hydrometallurgy 150, 269–275.

Remondis, A., 2018. Acid Extraction of Phosphorus from Sewage Sludge Incineration Ash. REMONDIS TetraPhos®.

Saleh, B., Li, R.D., Li, Y., Gao, H.X., Sema, T., Teng, W.C., Kumar, S., Liang, Z., 2018. New advancement perspectives of chloride additives on enhanced heavy metals removal and phosphorus fixation during thermal processing of sewage sludge. J. Clean. Prod. 188, 185–194.

Santhiya, D., Ting, Y.-P., 2005. Bioleaching of spent refinery processing catalyst using Aspergillus Niger with high-yield oxalic acid. J. Biotechnol. 116 (2), 171–184.

Stemann, J., Peplinski, B., Adam, C., 2015. Thermochemical treatment of sewage sludge ash with sodium salt additives for phosphorus fertilizer production–analysis of underlying chemical reactions. Waste Manag. 45, 385–390.

Sturm, G., Weigand, H., Marb, C., Weiß, W., Huwe, B., 2010. Electrokinetic phosphorus recovery from packed beds of sewage sludge ash: yield and energy demand. J. Appl. Electrochem. 40 (6), 1069–1078.

Su, L., Hu, L.Y., Sui, Q.H., Ding, C.C., Fang, D., Zhou, L.X., 2023. Improvement of fungal extraction of phosphorus from sewage sludge ash by aspergillus Niger using sludge filtrate as nutrient substrate. Waste Manage. 157, 25–35.

Tansel, B., Lunn, G., Monje, O., 2018. Struvite formation and decomposition characteristics for ammonia and phosphorus recovery: a review of magnesium-ammonia-phosphate interactions. Chemosphere 194, 504–514.

Tao, X., Xia, H., 2007. Releasing characteristics of phosphorus and other substances during thermal treatment of excess sludge. J. Environ. Sci. 19, 1153–1158.

Tao, N., Wu, X., Zhang, F., Pi, Z.L., Wen, J.Q., Fang, D., Zhou, L.X., 2021. Enhancement of sewage sludge dewaterability by fungal conditioning with Penicillium simplicissimum NJ12: from bench-to pilot-scale consecutive multi-batch investigations. Environ. Sci. Pollut. Res. 28 (44), 62255–62265.

Vogel, C., Adam, C., Peplinski, B., Wellendorf, S., 2010. Chemical reactions during the preparation of P and NPK fertilizers from thermochemically treated sewage sludge ashes. Soil Sci. Plant Nutr. 56, 627–635.

Vogel, C., Exner, R.M., Adam, C., 2013. Heavy metal removal from sewage sludge ash by thermochemical treatment with polyvinylchloride. Environ. Sci. Technol. 47, 563–567.

Wang, Q.M., Li, J.S., Tang, P., Fang, L., Poon, C.S., 2018. Sustainable reclamation of phosphorus from incinerated sewage sludge ash as value-added struvite by chemical extraction, purification and crystallization. J. Clean. Prod. 181, 717–725.

Wu, Y., Luo, J.Y., Zhang, Q., Aleem, M., Fang, F., Xue, Z.X., Cao, J.S., 2019. Potentials and challenges of phosphorus recovery as vivianite from wastewater: a review. Chemosphere 226, 246–258.

Wu, H.-Y., Ting, Y.-P., 2006. Metal extraction from municipal solid waste (MSW) incinerator fly ash—chemical leaching and fungal bioleaching. Enzym. Microb. Technol. 38 (6), 839–847.

Xia, Y., Tang, Y., Shih, K., Li, B., 2020. Enhanced phosphorus availability and heavy metal removal by chlorination during sewage sludge pyrolysis. J. Hazard Mater. 382, 121110.

Xiao, C.Q., Chi, R.A., Li, W.S., Zheng, Y., 2011. Biosolubilization of phosphorus from rock phosphate by moderately thermophilic and mesophilic bacteria. Miner. Eng. 24 (8), 956–958.

Xie, C.S., Zhao, J., Tang, J., 2011. The phosphorus fractions and alkaline phosphatase activities in sludge. Bioresour. Technol. 102 (3), 2455–2461.

Xue, T., Huang, X., 2007. Releasing characteristics of phosphorus and other substances during thermal treatment of excess sludge. J. Environ. Sci. 19, 1153–1158.

Yang, F., Chen, J., Yang, M., Wang, X., Sun, Y., Xu, Y., 2019. Phosphorus recovery from sewage sludge via incineration with chlorine-based additives. Waste Manag. 95, 644–651.

Zakaria, K.A., Yatim, N.I., Ali, N., Rastegari, H., 2022. Recycling phosphorus and calcium from aquaculture waste as a precursor for hydroxyapatite (HAp) production: a review. Environ. Sci. Pollut. Res. Int. 29 (31), 46471–46486.

Zhang, L., Song, X.W., Shao, X.Q., Wu, Y.L., Zhang, X.Y., Wang, S.M., Pan, J.J., Hu, S.J., Li, Z., 2019. Lead immobilization assisted by fungal decomposition of organophosphate under various pH values. Sci. Rep. 9 (1), 1–9.

Zhao, Y.Z., Ren, Q.Q., Na, Y.J., 2018. Promotion of cotton stalk on bioavailability of phosphorus in municipal sewage sludge incineration ash. Fuel 214, 351–355.

Zin, M.M.T., Kim, D.J., 2019. Struvite production from food processing wastewater and incinerated sewage sludge ash as an alternative n and p source: optimization of multiple resources recovery by response surface methodology. Process Saf. Environ. 126, 242–249.

Further reading

Hao, X.D., Zhan, S., Li, J., 2022. Global applied cases and technical summary of phosphate recovery from wastewater. Chin. J. Environ. Eng. 16 (11), 3507–3516.

Teng, W.C., 2016. Mechanism of Phosphorus Enrichment during Fluidised Bed Incineration of Sludge. Shenyang Aerospace University.

Chapter 26

Resource recovery from municipal solid waste incineration fly ash

Jinpeng Wu[1], Yuying Zhang[2], Xinni Xiong[3] and Lei Wang[1]
[1]*State Key Laboratory of Clean Energy Utilization, Zhejiang University, Hangzhou, China;* [2]*Department of Civil and Environmental Engineering, The Hong Kong University of Science and Technology, Hong Kong, China;* [3]*School of Environmental Science and Engineering, Guangzhou University, Guangzhou, China*

1. Introduction

With the rapid development of industries and modern technologies, increasing amounts of solid waste are being generated. Excessive waste accumulation may leach into the natural environment, severely contaminating water, air, and farmland soil. The incineration of municipal solid waste is becoming a promising and effective method of disposing of this waste, while fly ash production must be addressed as well (Huang et al., 2022). After treatment by municipal solid waste incineration (MSWI), the volume of waste can be reduced by about 85%–90%, the mass reduced by 60%–90%, and the organic content reduced by nearly 100% (Leckner, 2015).

The composition of the waste, the design of the incineration furnace, and the air pollution control devices significantly affect the physical and chemical properties of municipal solid waste incineration fly ash (MSWI FA) (Zhang et al., 2021; Phua et al., 2019). For example, unsorted kitchen waste, wood, and polyvinyl chloride (PVC) can contribute to higher concentrations of Na, K, and Cl in MSWI (Sun et al., 2016). Depending on the composition of the incineration feedstock, MSWI FA may contain hazardous compounds, including high concentrations of metals (Taušováet al., 2019) and persistent organic pollutants (Van Caneghem et al., 2019). Based on previous studies, the main elements in FA include Ca, Si, Al, Fe, Mg, Na, and K, among which Ca, Si, and Al are the dominant ones (Table 26.1). The main phases in MSWI FA include rock salt (NaCl), calcium hydroxide ($Ca(OH)_2$), chlorinated calcium hydroxide (CaClOH), calcium carbonate ($CaCO_3$), quartz (SiO_2), and potassium salt (KCl) (Chen et al., 2019; Liu et al., 2018). Furthermore, Zn, Pb, Cd, Cr, Cu, Hg, Ni, As, and Sb are the most commonly detected potentially toxic elements (PTEs) in MSWI FA, among which Zn and Pb are generally found with high concentrations (Table 26.2) (Zhang et al., 2021). The unburned organic carbon proportion of MSWI FA is usually high, with a combustion loss range of 11%–35%. Although the composition of MSWI FA varies under different conditions, high alkalinity with pH values ranging from 10.5 to 13.5 can be generally observed (Cristelo et al., 2020; Quina et al., 2018).

Resource recovery from MSWI FA is a promising route to simultaneously achieve the goal of pollutant control and sustainable development. Currently, the main methods of heavy metals (HMs) recovery from FA include hydrometallurgy (Tang et al., 2019), electrodialysis (Zha et al., 2023), and thermal separation technologies (He et al., 2023). Meanwhile, hydrometallurgy can be further divided into chemical extraction and bioleaching. During the hydrometallurgy process, inorganic solvents, such as hydrochloric acid and sulfuric acid, are the commonly used leaching agents. Additionally, to a less extent, microorganisms have been used in the extraction of HMs. Currently, there are some processes for recovering resources from FA, such as the FLUWA process, which uses acidic and neutral solutions to extract some HMs (Quina et al., 2018). HALOSEP process has been developed to remove/recover chloride from APCr/FA generated by MSWI. Despite the potential for extractable resources with a total content comparable with low-grade active mines (Funari et al., 2015), there is still a lack of mature technology for their recovery. Recently, bioleaching has emerged as a promising hydrometallurgical method that utilizes microorganisms to solubilize metals, which is being explored as a potential route for the valorization of waste streams. Therefore, this chapter mainly elaborates on the extraction methods of salts, HMs, precious metals, and rare earth metals in MSWI FA.

Treatment and Utilization of Combustion and Incineration Residues. https://doi.org/10.1016/B978-0-443-21536-0.00017-4

TABLE 26.1 Major elements composition in MSWI FA.

Major components (%)		References
Ca	2.71–40.43	Huang et al. (2022); Qiu et al. (2019); Zhang et al. (2022a–c); Ge et al. (2022); Tian et al. (2021); Chuai et al. (2022); Lv et al. (2022)
Si	0.69–24.43	
Al	0.20–18.32	
Fe	0.32–5.07	
Mg	0.30–3.48	
Na	0.59–10.41	
K	0.83–6.72	
S	0.56–5.81	
Cl	5.48–48.08	
LOI	1.51–19.05	

LOI, loss on ignition, mass loss from 440 to 1200°C.

TABLE 26.2 Trace elements composition in MSWI FA.

Trace elements (mg/kg)		References
Zn	1,831–18,400	Nag and Shimaoka (2023); Zheng et al. (2022); Long et al. (2023a,b); Gu et al. (2021); Ali et al. (2022a,b); Zhang et al. (2022a,c); Zhao et al. (2023)
Cu	265.83–1451	
Pb	187–3900	
Cr	54.9–863.7	
Cd	43.8–306	
Ni	21.06–118.22	
As	11.15–119.3	
Hg	1.07–13.4	
Mn	197–1092.72	
Ba	927–2020	
Sb	1.53–1235	
Se	2.9–6.99	

2. Extraction of salts (chlorine salt, sulfate)

Of the elements in FA, Ca is found in the highest content, followed by soluble Cl, Na, and K (Atanes et al., 2019). HMs, which mainly include Zn and Pb, constitute minor components (Song et al., 2023). Chlorine is detrimental to the thermal recovery and following utilization of FA as building material, which should be removed.

2.1 Water washing

Washing is a simple separation technology that uses water as a leaching agent to remove soluble salts. Studies have shown that washing by water could reduce over 90% of chloride and over 25% of sulfate, with the removal rates of Na_2O and K_2O reaching 77.69%–79.93% (Ma et al., 2023). However, about 4% of the Cl content that remains in the residues after washing exists in the forms of insoluble chloride salts, such as Friedel's salt (Xue and Liu, 2021). HMs are also leached out in very small amounts (0.04 wt%) (Tang et al., 2014). The filtrate obtained from water washing can be separated using technologies such as evaporation, membrane distillation, and electrodialysis in the field of seawater desalination and brine

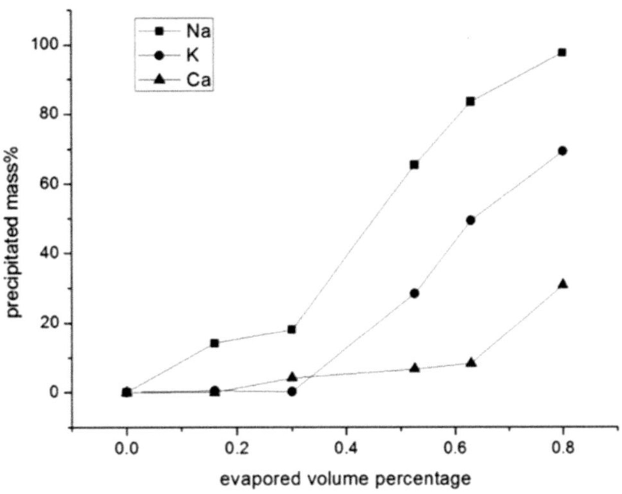

FIGURE 26.1 Variation of the mass percentage of Na^+, K^+, and Ca^{2+} in the precipitate obtained after evaporation with the water evaporation percentage (Tang et al., 2014).

separation. The soluble salts found in waste streams consist mainly of NaCl, KCl, and $CaCl_2$, with the potential for use as industrial salts. NaCl can be employed as an industrial raw material, KCl as a fertilizer, and $CaCl_2$ as a road deicing agent or industrial refrigerant, among other applications.

Currently, there are several methods for extracting and separating alkaline metal chlorides in water washing solution, for example, three-stage evaporation precipitation process and one-step ethanol precipitation method (Tang et al., 2014). According to the different precipitation ratios of NaCl, KCl, and $CaCl_2$ at different water evaporation percentages, it can be separated into three stages (Fig. 26.1). In the first stage, at a water evaporation percentage (WEP) of 30%, about 20% pure NaCl can be obtained. In the second stage, at a WEP of 60%, an additional 60% of Na^+, along with 50% of K^+ and a small amount of Ca^{2+}, can be precipitated. In the third stage, at a WEP of 80%, 20% of Na^+, K^+, and Ca^{2+} can be precipitated and returned to the washing solution, leaving only K^+ and Ca^{2+} in the remaining solution. Then the NaCl and KCl crystals precipitated in the second stage are dissolved in water, which can be added with ethanol for stepwise precipitation. This process can recover 80% NaCl, 50% KCl, and 80% $CaCl_2$, where the purity of the crude product (i.e., NaCl, KCl, $CaCl_2$) can reach about 90%.

Electrodialysis concentration is another method for separating the soluble salts in FA. Wang et al. (2022) showed that after two-stage washing of FA at a liquid-to-solid ratio of 4 L kg^{-1}, the concentration of soluble chloride in FA can be reduced to meet the limit specified in the Technical Specification for Pollution Control of MSWI FA (HJ 1134-2020) of China. The first-stage washing solution of FA was added to the dilute solution chamber and concentrate chamber of the electrodialysis unit. The concentrated solution obtained by evaporation and crystallization can yield industrial-grade potassium chloride and deicing agent with a purity of 92.2%. Moreover, 68% of the potassium leached out from the FA can be recovered as industrial-grade potassium chloride.

Amine-based capture agents can precipitate potassium chloride. Xie et al. (2022) demonstrated that ball milling and water washing can achieve higher extraction rates of chloride salts than stirring and water washing. By taking advantage of the similar solubility of sodium chloride and potassium chloride that significantly differs from that of calcium chloride, a mixed salt form of sodium chloride and potassium chloride can be selectively precipitated by evaporation. Then, an amine-based capture agent with a similar radius to potassium ions can be added to improve the hydrophobicity of the potassium chloride surface through ion exchange with amine ions and potassium ions. Finally, flotation technology can be used to separate potassium chloride from sodium chloride. In practice, dodecylamine was used, reaching the recovery rates of Na and K over 90%. The obtained potassium chloride product with a purity of 98.3% and sodium chloride with a purity of 96.6% meet the agricultural fertilizer-grade standard and the industrial product requirement, respectively.

2.2 Other washing methods

Since simple water washing cannot remove insoluble Cl, some studies have reported alternative washing methods for FA. Ultrasound-assisted water washing is considered as an effective way to promote the absorption of the solidification of Cr and CO_2 by Ca in MSWI FA. The propagation of ultrasound waves in liquids generates cavitation effects, which involve the formation, growth, and rupture of microbubbles. This leads to the creation of local high temperature, high pressure, and high-speed turbulence, thereby enhancing mass transfer to the solid surface and increasing the reaction rate.

Research using organic acids and microwave treatment on FA showed that double washing combined with microwave and citric acid (CA) treatment could remove 42.9% and 47.2% of insoluble chlorides, respectively (Wang et al., 2023a–c). This method can also promote the leaching of HMs and make the remaining FA more stable, which is beneficial for the combination of HM recovery and industrial salt recovery processes.

By subjecting FA to electron-enhanced oxalic acid washing, more escape channels for internal Cl and HMs can be created through the changing direction of electron impacts on the FA surface. A recent study found the removal efficiency of insoluble Cl reached 95.32% within 4-h reaction time, at the condition of 40 Hz frequency of electrode exchange, with a density of 50 mA/cm^2 and 0.5 mol/L $H_2C_2O_4$, respectively (Zhao et al., 2023).

Suthatta Dontriros et al. (2020) used deionized water, 0.01 and 0.1 M HNO_3, and 0.1 and 0.25 M Na_2CO_3 to treat MSWI FA and compared the efficiency of chloride and sulfate removal. The study showed that each treatment solution could remove about 250,000 mg/kg chloride, and Na_2CO_3 showed the strongest ability to remove sulfate (15,821 mg/kg), since Na_2CO_3 could combine with $CaSO_4$ to form more soluble Na_2SO_4. It has been proven that carbonation can effectively reduce the leaching of HMs (Atanes et al., 2019). It has also been studied the use of Na_2CO_3 solution as a stabilizer to treat MSWI FA, and the mass balance showed that most HMs were retained in the processed FA, with a fixation rate of more than 87%. After stirring with Na_2CO_3 solution for 5 min at 30°C, the insoluble and nonleachable salt in FA was significantly promoted to form metal carbonates, thereby improving the stability of HMs in the study range. This method achieves the separation of chloride and fixation of HMs in one step, which is very promising at the industrial level and has significant advantages compared with the carbonation process using CO_2 (Atanes et al., 2019).

Xie et al. (2020) studied a molten salt thermal treatment method to remove salts from FA. The method involves using a mixture of molten carbonate and chloride to remove salts from MSW incineration FA at temperatures ranging from 773 to 1073 K. Specifically, the mixed carbonate/chloride was heated for 6 h at 773 K to form a single eutectic for preparing the molten salt, and then the molten salt was subject to thermal treatment after fully mixed with the ash. Results show that the mixed product was mainly composed of the upper hydrocarbons, the middle salts, and the bottom residues. Analysis showed that molten carbonate dissolved more SO_4^{2-} and chloride salts, and molten chloride showed better selectivity in alkaline metal chloride extraction. This is because during the molten chloride extraction experiment, HMs in the FA undergo the following reactions Eqs. (26.1)–(26.2), strengthening the interaction among Ca^{2+}, Cl^-, and Si/Al/Fe compounds, and improving the stability of Ca and Cl.

$$3Ca^{2+} + Al_2O_3 + 2SiO_2 + CO_3^{2-} + 4Cl^- \rightarrow Ca_3Al_2(SiO_4)_2Cl_4 + CO_2 \tag{26.1}$$

$$3Ca^{2+} + Fe_2O_3 + 3SiO_2 + 3CO_3^{2-} \rightarrow Ca_3Fe_2(SiO_4)_3 + 3CO_2 \tag{26.2}$$

The remaining ash residue mainly consists of Si/Al and undissolved Ca. The middle layer of salt produced can be recovered or recycled.

3. HMs extraction

MSWI FA contains a significant amount of HMs and is defined as hazardous waste, which requires extraction or solidification of HMs to avoid environmental pollution. Different types of FA have similar HM species, with Zn and Pb showing the highest concentrations, followed by Cu, Cr, Cd, and Ni. These HMs mainly exist in the form of oxides, chlorides, hydroxides, and carbonates (Wang et al., 2021). Currently, landfill is the widely used method to dispose of FA after HM solidification/stabilization, which, however, cannot guarantee the solidification effect of HMs in the long term. The recovery of HMs is a necessary measure to promote both ecological sustainability and economic sustainability. The current methods for HMs recovery mainly include thermal separation, hydrometallurgical methods, and electrochemical processes.

3.1 Electrochemical processes

Electrochemical process is a promising technology for HMs recovery, with electrolysis as the main reaction. This process involves the application of an electric field to induce the oxidation of substances at the anode and reduction at the cathode, causing the movement of ion-loaded charges. Besides, this process was time-consuming.

Currently, electrodialysis is widely studied due to its advantages in the remediation of HM-contaminated soils, sludge, tailings, and other materials, which can effectively remove relatively high valence state HM ions (such as Cd and Cu) and pollutants from MSWI FA. It has been widely used in small-scale treatment of HMs in MSWI FA (Zha et al., 2023). For example, in laboratory-scale electrodialysis remediation experiments, electrodialysis was used to remove cadmium from four different biomass combustion FAs, and the Cd concentration eventually dropped to below 2.0 mg/kg within 2 weeks (Kirkelund et al., 2019). When the electrodialysis cell is powered, metal ions released during electrolysis carry the current, and the electrolysis and ion migration processes lead to changes in pH and redox potential in the solution, thereby improving further reactions.

Electrodialysis remediation can usually be enhanced through two major means, namely, adjusting process parameters (such as voltage gradient, treatment time, electrode material, and layout) and coupling with other technologies (such as chemical leaching and bioremediation). Huang et al. (2018a) studied the electrochemical treatment of FA using three-dimensional electrodes instead of two-dimensional electrodes. Three-dimensional electrodes refer to electrode structures with complex surface structures or volume shapes, such as particle electrodes in the sample chamber, as shown in Fig. 26.2. Results showed that the presence of particle electrodes led to smaller pH change in the sample chamber than in the two-dimensional electrode sample chamber. The oxidation and reduction occurring on the particle electrode promoted the dissolution and migration of HMs in the solid matrix pores and enhanced hydrolysis as well as ion permeability, which was inferred to be the primary reason for improving the extraction of HMs during electrodialysis. The specific reaction formula for the process involves the anode surface (E), the target contaminants (R), and the metal elements (M). It is worth noting that hydrolysis (Eqs. 26.3 and 26.4) and anode oxidation (Eqs. 26.5–26.7) may occur on the surface of both the main electrode and particle electrode, as demonstrated in similar content.

$$2H_2O - 4e^- \rightarrow 4H^+ + O_2\uparrow \tag{26.3}$$

$$4H_2O + 4e^- \rightarrow 4OH^- + 2H_2\uparrow \tag{26.4}$$

$$E + H_2O \rightarrow E(\cdot OH) + H^+ + e^- \tag{26.5}$$

$$E(\cdot OH) \rightarrow EO + H^+ + e^- \tag{26.6}$$

$$EO + R \rightarrow RO + E \tag{26.7}$$

$$M^{n+} + ne^- \rightarrow M \tag{26.8}$$

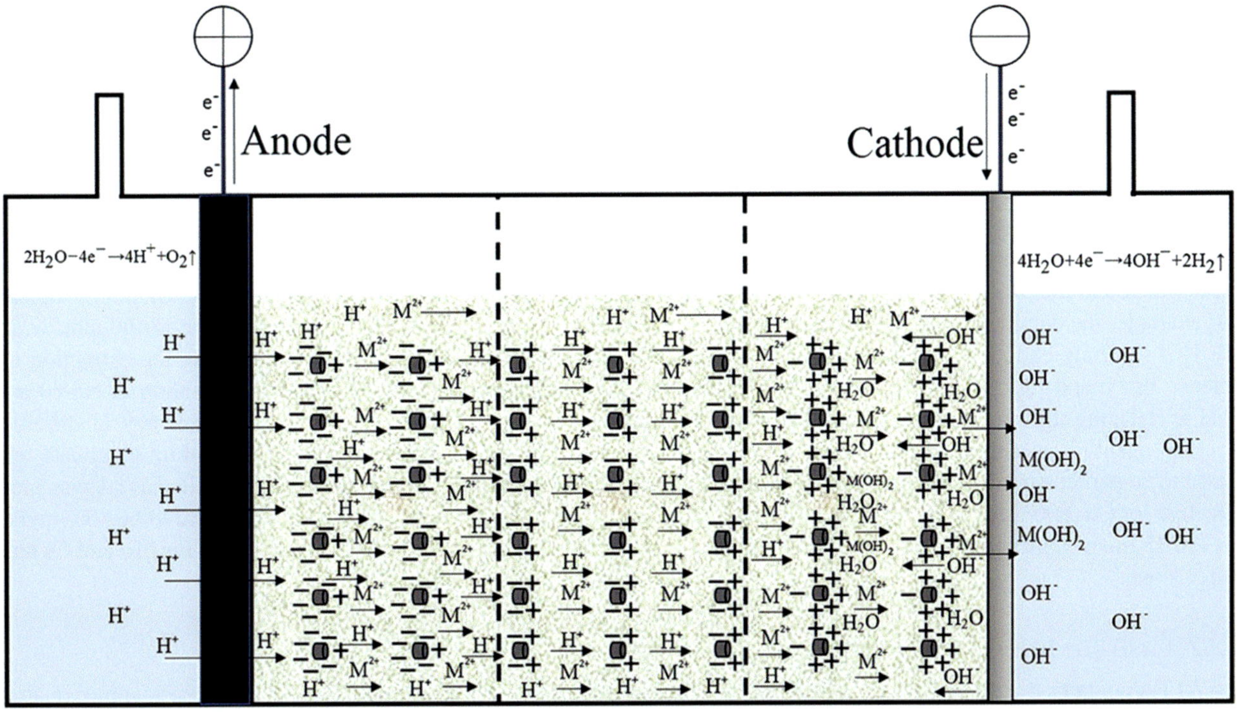

FIGURE 26.2 Schematic diagram of the rectangular three-dimensional electrode reactor (TDER) (Huang et al., 2018a).

The pH significantly affects the migration of heavy metals and determines the types of charged ions present. Most metals form soluble cations (such as Zn^{2+}) under acidic pH values, while acidic and neutral metals (such as Pb, Zn) also have soluble hydroxides (such as $Zn(OH)_3^-$, $Zn(OH)_4^{2-}$) under alkaline pH values. Some metals, especially those with oxygen anions, such as Cr, exist in different mobile phases, depending on the pH value as well as redox potential (Cornelis et al., 2008).

Cylindrical electrolysis cells have been found to enhance the electrodialysis remediation process by increasing the migration rates of Cu, Cd, Pb, and Zn, compared with traditional rectangular instruments. For instance, Huang et al. (2018c) reported an increase in migration rates of 43. 5%, 41.3%, 17.3%, and 46.0%, respectively, when using cylindrical electrolysis cells instead of rectangular ones.

In addition, the electrochemical process can also be significantly enhanced by coupling with other technologies. Acidification and ultrasonic pretreatment can enhance the solubility and mobility of Zn (69.8%), Pb (64.2%), Cu (67.7%), and Cd (59.9%) in electrokinetic remediation (Huang et al., 2018b). Integration of bioleaching and electrodialysis techniques resulted in the removal of heavy metals (Pb, Mn, Cu, and Cd) from fly ash (Zha et al., 2023). The leaching efficiency of residual ash for heavy metal migration decreased compared with other heavy metal removal techniques, resulting in relative energy savings of 6.72, 15.61, 8.99, and 17.46 kWh/kg. However, the high technical requirements and power demands of electrodialysis technology make the electrochemical process expensive, limiting its application in HM recovery from FA.

3.2 Hydrometallurgical process

The hydrometallurgical process involves mixing MSWI FA with other solutions to leach out heavy metal ions through various chemical reactions. This process yields a metal-rich solution, which can be further processed through precipitation, crystallization, or other methods, including electrochemical techniques, to recover the metals.

3.2.1 Chemical leaching

Chemical leaching includes acid/alkaline leaching and chelating agent leaching. Acid extraction is mainly influenced by the pH and the type of leaching reagent used. Inorganic acids (such as HCl, HNO_3, and H_2SO_4) and organic acids (such as CA and malic acid) are commonly used chemical reagents. Due to the low pH value, virtually every heavy metal can be leached with the aid of strong inorganic acids. For example, treatment with HCl for 7 days can effectively leach Cr (100%), Cd (100%), Cu (98%), Pb (90%), and Zn (80%) (Kuboňová et al., 2013). Organic acids, on the other hand, have limited leaching effects, such as Cu and Pb, but could be improved by lowering the pH or increasing the liquid–solid ratio. CA is an effective organic acid extractant, with optimized conditions achieving effective extraction of Pb (96.9%), Fe (67%), Zn (100%), and Cu (100%) (Huang et al., 2011). Alkaline leaching is typically effective for amphoteric HMs. For example, 3M sodium hydroxide (NaOH) has been shown to extract Pb (84%), Zn (75%), Cr (71%), Cu (41%), as well as Ni and Cd (approximately 20%) (Kuboňová et al., 2013).

Leaching with a chelating agent involves the formation of soluble complexes with HMs through coordination, extracting metals from FA into the solution. Chelating agents have high selectivity for specific HMs, and their efficiency is independent of pH. Ethylenediaminetetraacetic acid disodium salt (EDTA) and sodium gluconate (Na-Gl) are particularly effective for Zn and Pb (Ferreira et al., 2002).

It was found that the combined use of reagents can achieve higher extraction efficiency than single chemical reagent. For example, the combination of NaOH and HCl can extract 30% of Zn and 90% of Pb when leaching for 60 min, while 5% HCl can only extract less than 63% and less than 10%. When these two processes are combined, the extraction efficiency of Pb and Zn can reach 98% and 68.6%, respectively (Nagib and Inoue, 2000). The combination of Na-Gl and EDTA chelating agents was also proven effective, as illustrated in Fig. 26.3, by comparing the efficiency of different leaching agents in washing out HMs from WFA (Loginova et al., 2019). As shown in Fig. 26.4, sodium gluconate was found to be particularly effective in washing out Cd, Cu, and Zn (10–200 times higher than water), while EDTA was most effective for Cd (800 times higher than water). The combined treatment of washing out metals was found to be 40% higher for Pb, 15 times higher for Zn, 1800 times higher for Cd, 115 times higher for Cu, and 2 times higher for Mo and Cr than water alone.

3.2.2 Case study—FLUWA and FLUREC

MSWI FA contains a great amount of valuable metals. For example, the total Zn content was found to reach approximately 60,000 mg/kg (Fellner et al., 2015). To recover Zn, Pb, Cd, and Cu from MSWI FA, nowadays several emerging processes

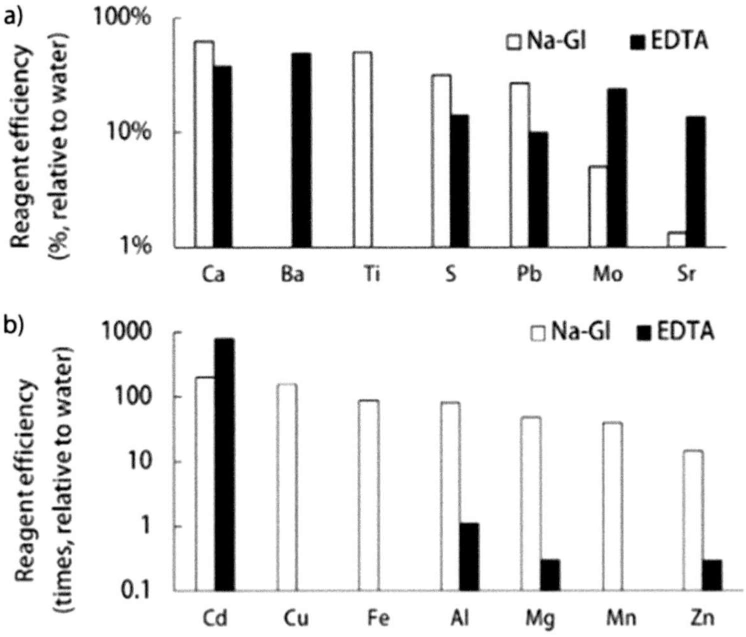

FIGURE 26.3 Reagent efficiency relative to water: (A) For the element, it is comparable to water. (B) For the element, it is much higher than water (Loginova et al., 2019).

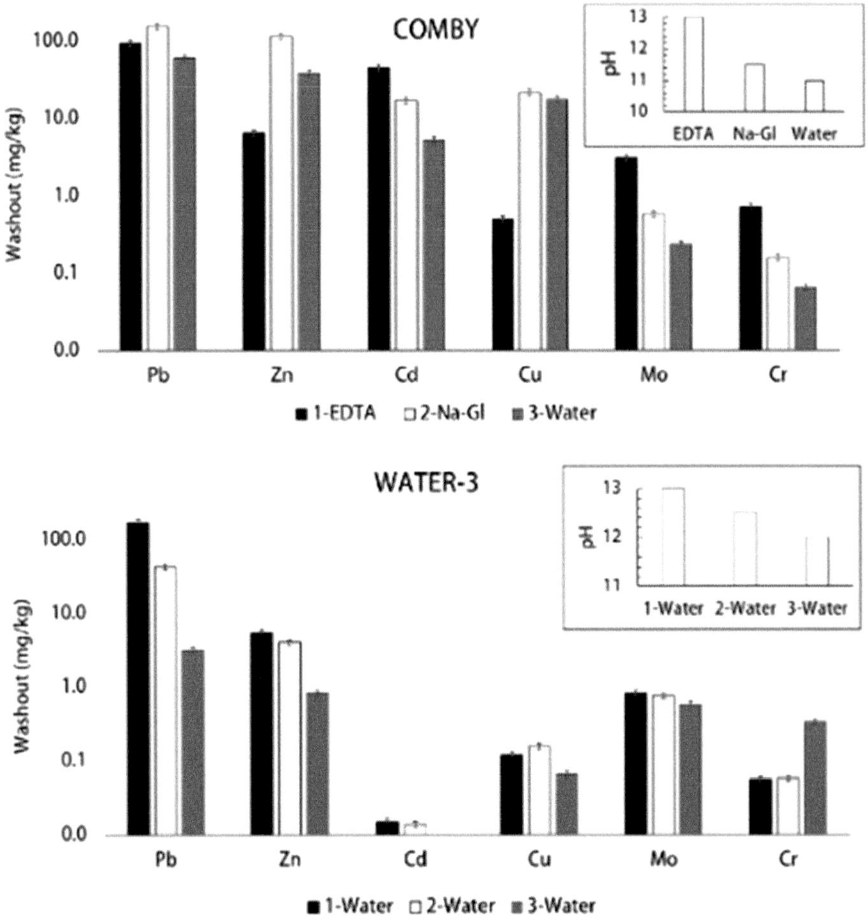

FIGURE 26.4 Elution of metals during WATER-3 and COMBY treatments for each step; insets: pH of the washing waters during each step (Loginova et al., 2019).

have been investigated, such as the Fluidized Bed Waste-to-Energy and Materials Recovery (FLUWA) process and its extended process Fluidized Bed Reactor for Enhanced Contaminant Removal (FLUREC) (Bühler and Schlumberger, 2010).

The FLUWA process is ineffective for leaching FA, as it requires the use of alkaline washing solutions. The characteristics of FA, the alkalinity of washing water, L/S ratio, redox potential, temperature, and leaching time have little effect on the extractability of metals such as Zn, Pb, Cu, and Cd. FLUWA process cannot extract any significant amount of these metals. Adding hydrogen peroxide to the FLUWA process does not retain any redox-sensitive metals, where Fe^{2+} fails to convert to Fe^{3+} and precipitate in the form of iron hydroxide without accumulation in the remaining filter cake. After sufficient extraction time, the suspension cannot be separated into metal-poor filter cake and metal-enriched filtrate by vacuum belt filtration.

The FLUWA process, as illustrated in Fig. 26.5, provides the basis for extension methods such as the FLUREC process. This process can achieve the recovery of high-purity Zn (Zn > 99.995%) from HM-enriched filtrates, where the acidic washing water from the waste incineration plant would be mixed with MSWI FA and used as an extractant. After sufficient contact time, solid—liquid separation could be performed, and the remaining solid materials would be treated in a sanitary landfill. Zn powder would be added as a reducing agent to exchange Cd, Cu, and Pb in the form of metals. Subsequently, Zn could be recovered from the solution by reactive extraction and electrolytic deposition. The more valuable metals than zinc are separated as metal cement. This cement, which contains 50%—70% Pb, can be directly sent to a lead smelter to recover metals during lead production.

Solvent extraction steps selectively extract Zn in the water-insoluble organic phase. An organic phase can extract 99.5% of Zn at a pH of 2.7—3. Other metals and impurities in the organic phase were removed during the washing step. By regenerating the organic phase with sulfuric acid, a by-product of the electrolytic extraction process was obtained. On the aluminum cathode, special high-grade Zn is recovered using direct current applied to the obtained high-purity Zinc sulfate

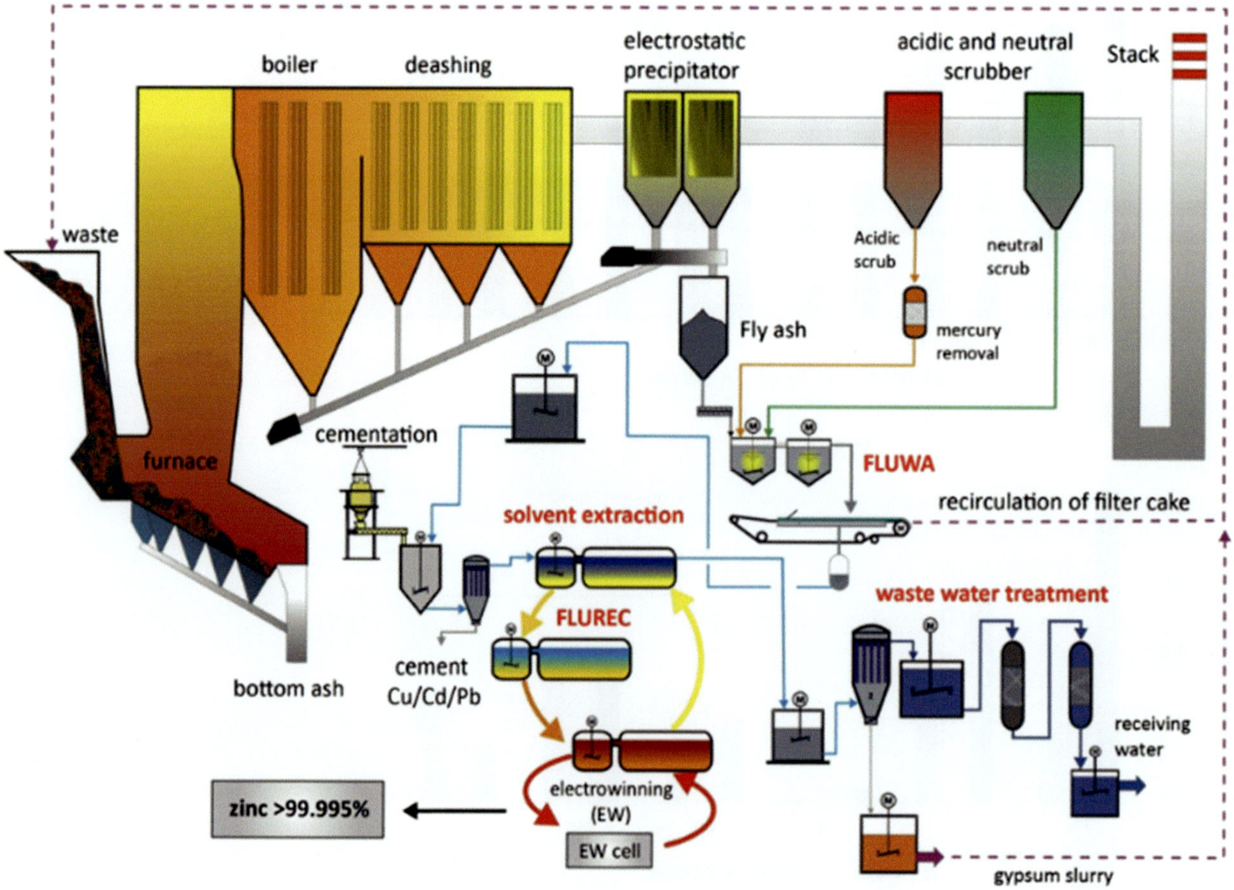

FIGURE 26.5 Process diagram of waste incineration, FA leaching (FLUWA) and recovery of recyclable materials (FLUREC) (Quina et al., 2018).

solution. If metals cannot be recovered directly as in the FLUWA process, the metal-containing filtrate should be sent to the wastewater treatment plant where the metal hydroxide sludge is deposited.

3.2.3 Bioleaching

Bioleaching originated from biohydrometallurgy technology and is commonly used in metals recovery from low-grade ores or insoluble metal sulfide minerals (Xu et al., 2014). Bioleaching utilizes the natural ability of microorganisms to convert insoluble HM components in solids into soluble and extractable forms (Xu et al., 2014). In the bioleaching process, it is generally believed that metals are released through acid formation, direct enzymatic reduction, or indirect interactions with metabolites produced extracellularly (Wang et al., 2021). For example, acidophilic bacteria reduce sulfides (S_8, H_2S, or polysulfides) to sulfuric acid and oxidize divalent iron (Fe^{2+}) to trivalent iron (Fe^{3+}) (Srichandan et al., 2019). Fe^{3+} and H^+ (from H_2SO_4) dissolve the metals in the secondary waste through contact and noncontact mechanisms (such as the thiosulfate pathway and polysulfide pathway) (Srichandan et al., 2019). In small-scale municipal solid waste incineration residue experiments, the metal bioleaching rates for Mg and Zn can be over 90%, while those for Al and Mn can exceed 85%, and those for Cr, Ga, Ce, Nd, Pb, and Co can be between 65% and 50% (Funari et al., 2017). Other studies have shown that using a mixed bacterial population of iron sulfide oxidizing bacteria and sulfur-oxidizing bacteria can remove 90% of Zn, Cu, and Ni (Li et al., 2023). Bioleaching has the advantages of low energy input, low capital cost, and overall improvement of workplace health. The use of fungi, mixed acidophilic bacteria (Lee and Pandey, 2012), and mixed alkaliphilic bacteria (Ramanathan and Ting, 2016) for bioleaching of MSWI FA have good economic benefits, but the bioleaching process has not yet been fully applied.

The efficiency of bioleaching is affected by several factors, including the source of carbon, microbial species, pH, temperature and duration, substrate concentration, oxygen, and leaching method (Wang et al., 2021). Heterotrophic fungi, autotrophic bacteria, and heterotrophic bacteria are the primary microorganisms used in bioleaching. Heterotrophic fungi can survive in both acidic and alkaline environments, making them appropriate for the leaching of alkaline waste. Autotrophic bacteria have a relatively low operating cost (Xu et al., 2014). The most commonly utilized bioleaching agents include *Aspergillus niger*, iron-oxidizing bacteria (IOB), and sulfur-oxidizing bacteria (SOB).

Currently, bioleaching includes single-step and two-step methods. In single-step bioleaching, the waste residue to be treated is inoculated with microorganisms, and the growth of microorganisms and the leaching of metals occur simultaneously. Two-step bioleaching involves filtering the suspension after microorganisms have grown to their maximum and metabolites have reached their maximum, and using only the supernatant for leaching. The advantage of two-step bioleaching is that the generation of the leaching agent is independent, eliminating the connection between the biological and chemical processes so that each process can be optimized independently to maximize productivity (Mishra and Rhee, 2010). Studies have found that *Aspergillus niger* is more effective in extracting HMs Cu, Zn, and Pb in the two-step method at low FA concentrations, with efficiencies of 40%, 55%, and 57%, respectively, compared with 18.9%, 19%, and 44% in the one-step method (Wang et al., 2009).

For SOB and IOB, H_2SO_4 is produced by oxidizing metal sulfide phases and elemental sulfur as a substrate to extract HMs (Sand et al., 2001). SOB has a high resistance to ash content and can grow in 8% (w/v) FA (Ishigaki et al., 2005), but cannot produce trivalent iron for oxidization and reduction of metal compounds, thus limiting its ability to leach metals. SOB can only leach 3.7% Cu, 60.8% Zn, and 15% As at 2% (w/v) FA (Ishigaki et al., 2005). Comparatively, IOB has strong oxidation−reduction reaction ability and high metal leaching ability, but low resistance to ash content (Ishigaki et al., 2005; Mäkinen et al., 2019). Through IOB, 92% of Zn, 39% of Cu, and 30% of As can be reached at 2% (w/v) FA (Ishigaki et al., 2005). When SOB and IOB are cocultured, a high metal leaching rate (Cu $>$ 70%, Zn $>$ 90%, and Cd 100%) can be achieved (Funari et al., 2019; Ishigaki et al., 2005). Since the alkaline environment of FA is not conducive to the growth of acidophilic bacteria, the reaction time is relatively long, which usually requires 1−3 months to reach a low pH value (Brandl and Faramarzi, 2006). To bioleach, alkaline microorganisms are being used in place of acidophilic bacteria due to their weakness. The results showed that the alkaline bacteria sp.TRTYP6 could extract 52% of Cu (Ramanathan and Ting, 2016).

Bioleaching has the advantages of mild reaction conditions, low energy input, and low operating costs, while the main drawbacks are the requirement on long leaching time and the cultivation of strains. This technology is still in the development stage for recovering HMs from MSWI FA and has not been widely applied yet (Quina et al., 2018).

3.3 Heat treatment

The main purpose of the heat treatment process is to separate hazardous trace metals from FA through evaporation and concentrate HMs for easy recovery. During the process, FA is heated to a high temperature, and the volatile metal vapors can be collected in a gas processing system for generating secondary fly ash (SFA), which can be recycled as metallurgical material. Studies have shown that increasing the process temperature, heating time, and airflow rate may improve the efficiency of HM volatilization (Wang et al., 2021). In addition, chlorinating agents and atmosphere also affect the volatilization of HMs.

HMs can be divided into volatile, semivolatile, and low-volatile categories. More volatile metals such as Cd and Pb are often evaporated and captured in the form of filter ash during the combustion process in MSW incinerators, while low-volatile HMs are not easily collected as secondary FA. The composition of the gas phase in the furnace and ash-forming substances in the fuel can affect volatility, and the formation of metal chlorides can significantly increase the volatilization of many metals. For semivolatile metals such as Zn, the composition of FA from different MSW incinerators may affect the volatilization process. Under reductive conditions, Zn can be volatilized in the form of atomic Zn. If the chlorine content is high, it will volatilize in the form of $ZnCl_2$. Comparatively, under oxidative conditions, ZnO will be formed, which is difficult to volatilize. Therefore, the addition of chloride-containing additives or heating treatment under reductive conditions (e.g., adding reducible carbon) can enhance the volatilization of HMs in the ash.

The boiling points of various chlorides (e.g., 732°C of $ZnCl_2$, 993°C of $CuCl_2$, 950°C of $PbCl_2$, 960°C of $CdCl_2$, 801°C of NaCl, and 770°C of KCl) indicate their feasibility of significant volatilization below 1000°C (Song et al., 2023). Common chlorinating agents used in industrial processes include chlorine gas (Cl_2), hydrogen chloride (HCl) smoke, and inorganic chlorides such as sodium chloride (NaCl), calcium chloride ($CaCl_2$), and magnesium chloride ($MgCl_2$) (Wang et al., 2021). Cl_2 and HCl can directly and efficiently combine with HM elements to generate metal chlorides, but the equipment required for their application is more demanding (Kurashima et al., 2019). Moreover, among the three inorganic chlorides, $CaCl_2$ and $MgCl_2$ can first react with O_2 and/or water to generate Cl_2 and/or HCl (indirect chlorination), which is more effective for HM chlorination than NaCl (direct chlorination).

The study by Lane et al. (2020) investigated the production of Zn-rich metal concentrate from MSWI FA through heat treatment and selective volatilization and condensation of Zn and evaluated the effect of reductive atmosphere on the volatility of Zn, Pb, and Cu, as illustrated in Fig. 26.6. Research found that the concentrate collected by condensing elements volatilized from ash at 900°C was mainly composed of particles (based on mass), contained the concentrations of Zn (55.9 ± 12.5 wt%), Pb (4.7 ± 1.1 wt%) and Sn (0.68 ± 0.13 wt%). These metals become highly enriched in the metal

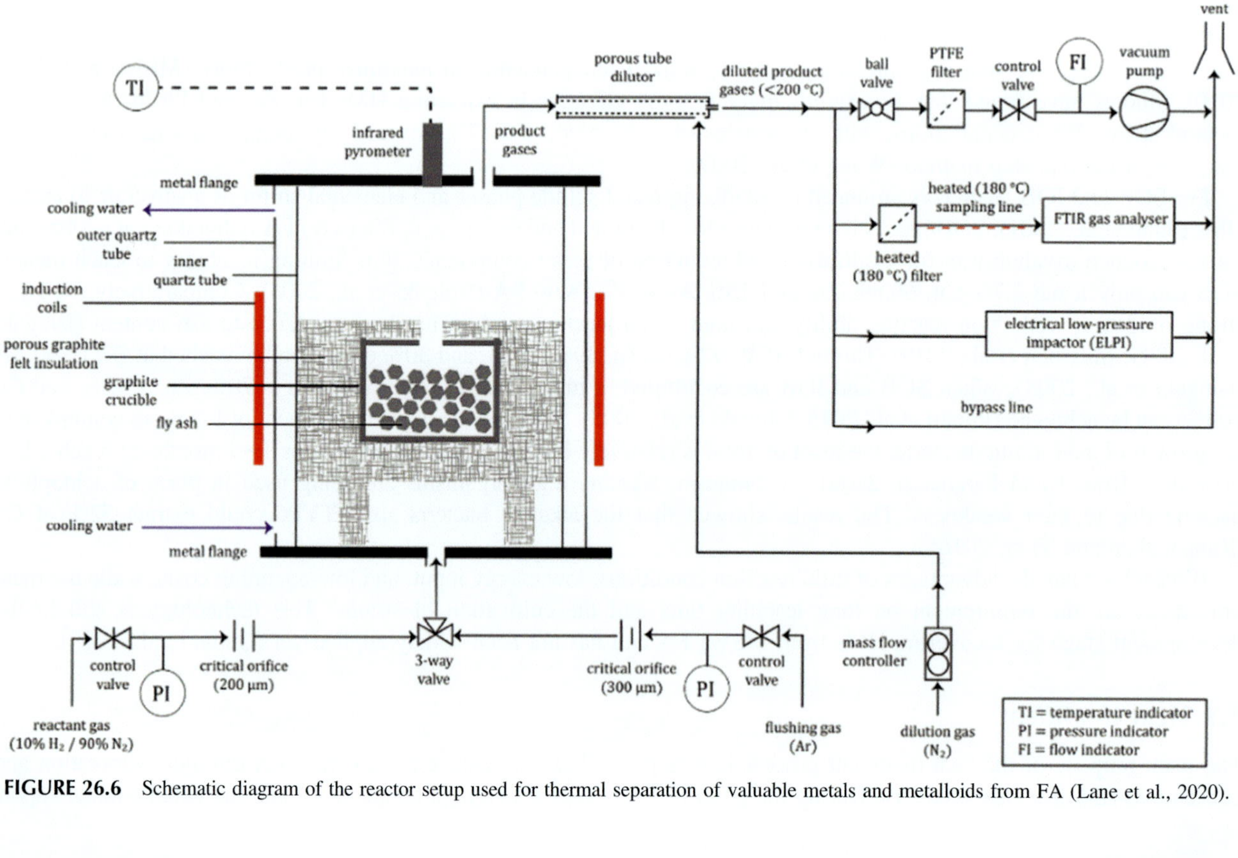

FIGURE 26.6 Schematic diagram of the reactor setup used for thermal separation of valuable metals and metalloids from FA (Lane et al., 2020).

concentrate during heat treatment, which is promising for developing methods of separating valuable and metalloid metals from FA.

Studies have shown that HMs in FA are transferred to slag as nonvolatile metals such as Fe, Al, Ca, Cr, Ni, Mn, and Mo, while volatile metals such as Cd, Pb, As, and Hg are transferred to SFA based on their volatility during the smelting process (Chen et al., 2017; Okada and Tomikawa, 2016). The volatilization behavior of Zn and Cu is affected by the SiO_2 content and the atmosphere during the smelting process. Therefore, there are still a large number of nonvolatile metals that are difficult to be extracted through thermal treatment. To address this, some studies have proposed joint heating and reduction of FA and red mud (RM). Due to the main components of RM being Fe_2O_3 (10%−40%), SiO_2 (20%−40%), and Al_2O_3 (5%−20%) (Mukiza et al., 2019), the reduction is conducive to the formation of iron-loving metals Cu, Cr, Ni, Co, etc. (Yu et al., 2019; Geng et al., 2020a,b) into alloys, which is beneficial for subsequent separation. Results found that while the ratio of FA and RM is 3:7, the recovery rates of Fe, Cu, Ni, Cr, in the iron alloy were 76.06%, 74.53%, 83.40%, and 58.44%, respectively. The recovery rates of Cd, Pb, and Zn in SFA could reach 98.95%, 98.21%, and 86.78%, respectively (Geng et al., 2020a,b).

The main disadvantage of thermal processes is high energy consumption, which causes high environmental impacts in life cycle assessment (LCA) of different treatment schemes (Fruergaard et al., 2010). In addition, some impurity elements, especially K, Na, and Cd, are also enriched in the concentrate during heat treatment. Separation of these elements remains a significant challenge in FA heat treatment.

3.4 Supercritical fluid extraction

Supercritical carbon dioxide (SC−CO_2) is defined as carbon dioxide that is above its critical temperature of 31°C and critical pressure of 7.39 MPa. Under these conditions, the density and solvation ability of CO_2 are similar to many liquid solvents (Sinclair et al., 2019).

Conventional leaching procedures that remove HMs from soil, sludge, or compost under atmospheric pressure require a large amount of solvents, which must be pretreated with neutralizing agents before disposal. This turns out to be both an economic and environmental problem. In view of this, SFE by CO_2 has become a promising choice for traditional solvent extraction to purify environmental samples. SFE is relatively faster and has a more controllable selectivity than conventional solvent extraction. When applied in conjunction with appropriate chelating agents, SC-CO_2 is a strong solvent with high diffusivity and low viscosity. Additionally, it can be easily recovered and reused by avoiding additional treatment of solvent waste, which is a major advantage. Since supercritical CO_2 could only dissolve electrically neutral substances and coordination must be satisfied, the metal−ligand complex formed by bonding with negatively charged ligands or anions and neutral ligands must be nonpolar.

Supercritical CO_2 extraction can flexibly adjust the solubility of different substances by adjusting the operating pressure and temperature, and achieve selective extraction compared with traditional extraction processes (Jitpinit et al., 2022), avoiding the mixing of a large amount of impurities in the extracted products, and reducing the use of organic solvents for subsequent extraction. In 2009, a research demonstrated the supercritical extraction of uranium from incineration FA (Koegler, 2010). Christof Kersch et al. (2004) mixtures used tributyl phosphate (TBP), di(2-ethylhexyl) phosphoric acid (D2EHPA), and their mixtures to investigate the extraction efficiency of Zn, Pb, Cu, V, Mn, Cd, Sb, Ni, Mo, Cr, and Co in supercritical CO_2. It was found that when using D2EHPA or TBP-D2EHPA mixtures, the extraction efficiency for V, Sb, Ni, Mo, Cr, and Co exceeded 90%, while the EE of Pb, Mn, and Cu was only 40%−60%. Compared with using each extractant alone, the equimolar mixture of TBP-D2EHPA enhanced extraction (Mn, Cd) and exhibited a synergistic effect (Pb, Cu), while a higher proportion of TBP-D2EHPA showed lower extraction efficiency. Yao et al. (2018) demonstrated that TBP, as a chelating agent, can achieve recovery rates of 86%, 90%, 86%, and 88% for Ce, Nd, La, and Pr.

4. Rare earth metal extraction and rare metal extraction

Rare earth elements (REEs) are a group of 17 elements including lanthanides, yttrium, and scandium. According to their different physical and chemical properties and slight differences in ionic radii, they are roughly divided into light REEs (LREEs) and heavy REEs (HREEs). Elements from lanthanum (La) to europium (Eu) are considered as LREEs, while other elements from gadolinium (Gd) to lutetium (Lu) are considered as HREEs. Compared with LREEs, HREEs are more prone to desorption by acidic solutions, showing stronger organic affinity as well as higher stability in their complex forms (Xiong et al., 2019). Studies have shown that the total content of REEs in solid residues generated from MSWI is 100−102 mg/kg, which is relatively low in absolute concentration, but the total amount is comparable with that of low-grade active mines (Funari et al., 2015). The sources of REEs in MSWI ash include electronic devices, lighting fixtures, rare earth magnets, cosmetics, food packaging, and others. Elemental characterization analysis of different incinerator ashes has also confirmed the presence of a significant proportion of rare metals in MSWI ash, such as Y, La, Ce, Pr, Nd, Dy, Yb, Ho, Er, Tm, Lu, Ag, Bi, Ga, Ge, Pd, In, Sb, Sn, Te, and Tl (Hasegawa et al., 2014). Material flow studies of MSWI FA have shown that the flow rates of Mg (79 t/a), Sb (2.4 t/a), Cr (1 t/a), Ce (0.05 t/a), and Co (0.04 t/a) are relatively high (Funari et al., 2023). Currently, there is little research on the recovery of REEs from MSWI FA.

It is common for REEs to be present as insoluble oxalates and chlorides when rare earth minerals or secondary resources are extracted or recovered. To obtain a single rare earth element, it should firstly be separated from the non−rare earth element precipitates before subsequent processing (Zhou et al., 2020). In previous studies, Funari et al. (2017) removed most of the chloride and mineral salts (such as Na, Ca, and K) by water washing to reduce the obstruction of target metals recovery from MSWI FA. Then, they used a mixed culture of sulfur and iron-oxidizing bacteria for bioleaching and studied the bioleaching behavior of various metals in MSWI FA. The results showed that the leaching efficiency of Ce could reach nearly 50%, and that of Nd could reach 87%. Compared with the method of chemical leaching via H_2SO_4 after water washing, bioleaching reduced the consumption of H_2SO_4 by half, which largely saved the reagent cost. This may be the result of the acid produced by microorganisms and the increase in metal solubility.

Although the total amount of REEs in MSWI FA is considerable, the limited content can yield lower economic value than that of e-waste, tailings, and other materials. Despite time consumption, bioleaching technology is generally applied due to its low cost. Bioadsorption technology is favorable at selectively adsorbing individual REEs and reducing the difficulty of later separation, which may be the future direction of REE recovery from FA.

Bioleaching is a promising approach for extracting REEs from waste using microorganisms and their metabolic byproducts. This method involves several pathways such as iron/sulfur oxidation, organic acid production, and hydrogen cyanide formation (Valix, 2017), which can solubilize the REEs. Iron- and sulfur-oxidizing bacteria, such as thermophiles or hyperthermophiles, generate Fe^{3+} and H_2SO_4 by oxidizing Fe^{2+} and S, respectively. The REEs are then oxidized by Fe^{3+} and H_2SO_4 and released into the culture medium (Dev et al., 2020). Organic acids produced by heterotrophic microorganisms using external carbon sources during growth can acidify and form soluble chelates with the metal fraction of solid REE waste through chelation (Barnett et al., 2018). Cyanogenic bacteria produce hydrogen cyanide, which can form water-soluble cyanide complexes with metals and REEs present in the waste (Arab et al., 2020).

Soluble REEs in bioleaching solutions can be separated and recovered using various biological or chemical methods, as illustrated in Fig. 26.7. Common chemical methods for separating REEs and metals from solutions include precipitation and solvent extraction (Li et al., 2013). Another method, diffusion dialysis, involves driving diffusion through a membrane by the concentration difference between two compartments (Hammache et al., 2021). Bioadsorption is a sustainable approach that uses different types of biomaterials and bioligands to bind and concentrate REEs or metal ions.

Biological adsorption is a low-cost and highly efficient method for separating and recovering REEs from solutions, which produces minimal waste and allows for the regeneration of the adsorbent (Torres, 2020). The process involves interactions between REEs and organic biomass such as bacteria, fungi, plants, and algae, as shown in Fig. 26.8A. The

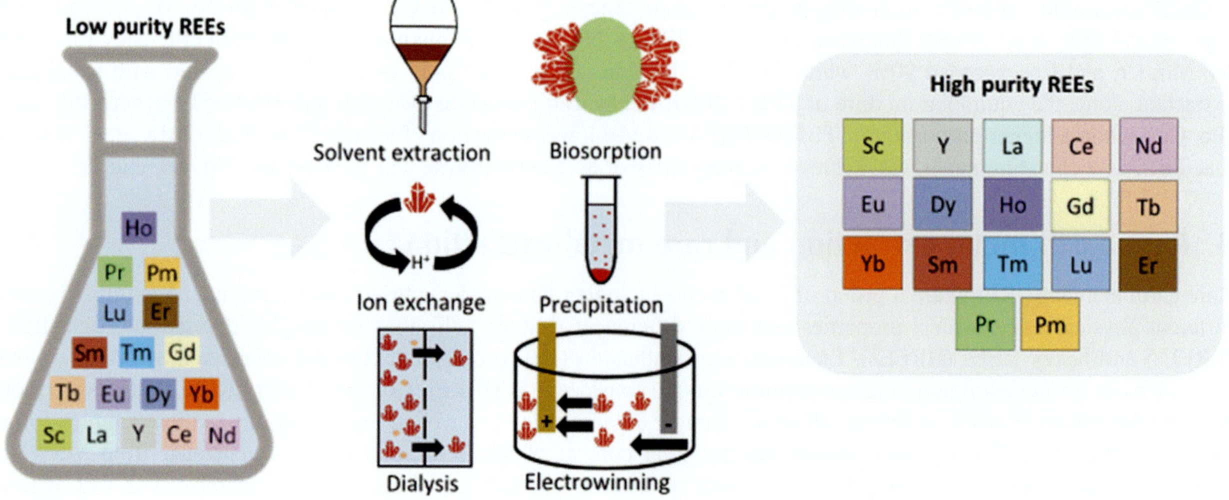

FIGURE 26.7 Bioleaching REEs separation methods (Brown et al., 2023).

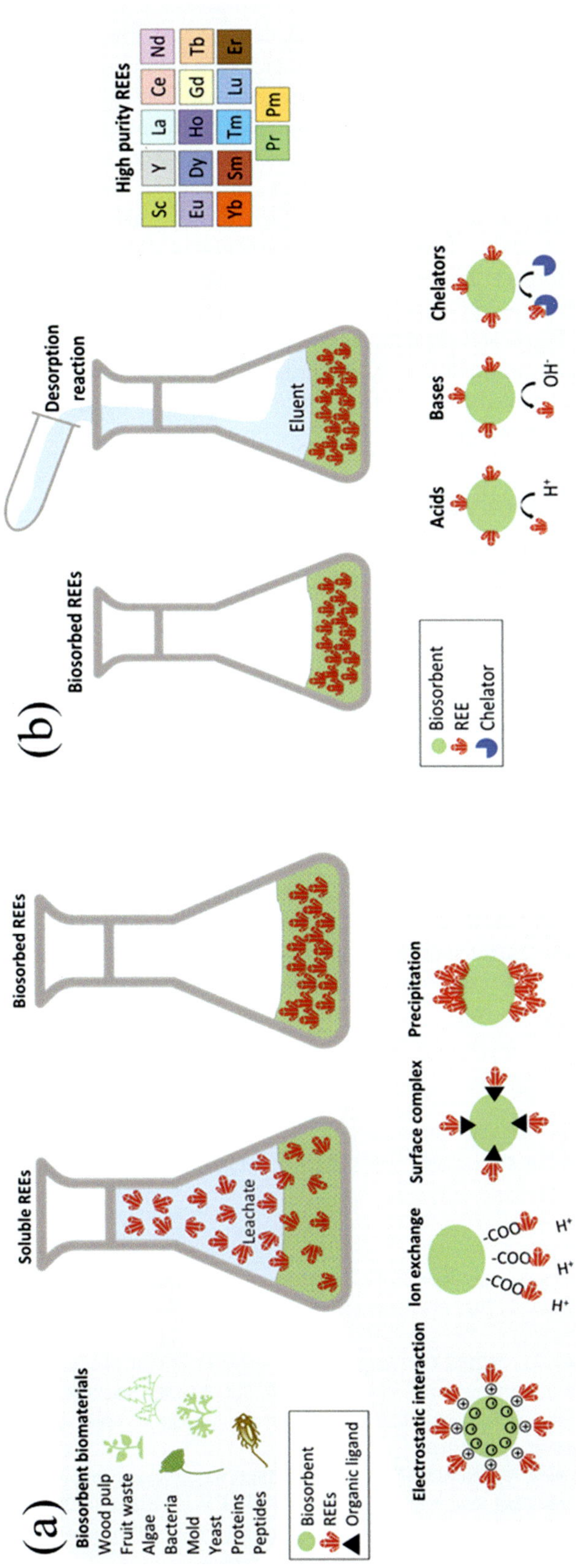

FIGURE 26.8 Schematic of reaction process: (A) biosorption process schematic; (B) desorption process schematic (Brown et al., 2023).

recovered REEs can be separated from the filtrate using a combination of electrostatic interactions, ion exchange, surface complexation, and precipitation (Ali, 2020). Electrostatic interactions occur when opposite charges on the surface of the adsorbent weakly attract metal ions in the solution (Michalak et al., 2013). The ion exchange method involves displacing ions on the surface of the biomaterial with metal ions present in the solution (Gadd, 2009). Surface complexes can be formed when metal ions interact with organic molecules or ligands on the surface of the adsorbent (Fomina and Gadd, 2014). During the precipitation process, insoluble metal substances are formed on the adsorbent due to the metabolism of live biomass or unrelated to metabolic activity in dead biomass (Javanbakht et al., 2014). The extracted REEs should be removed from the adsorbent through the desorption process, as illustrated in Fig. 26.8B. Desorption involves mechanisms such as proton exchange, ion exchange, and chelation, which is typically carried out using acidic or alkaline solutions, or chelating agents (Wang and Chen, 2009). It is ideal for the desorption eluent to completely desorb the solute without damaging the structure of the adsorbent. The binding strength and desorption efficiency of REEs depend on the physical properties of the adsorbent and the biosorption mechanisms involved (Chatterjee and Abraham, 2019). The most common desorption eluents include acids, bases, and chelating agents, such as sodium hydroxide, sodium carbonate, EDTA, EDDS, hydrochloric acid, HNO_3, sulfuric acid, and CA (Kołodyńska et al., 2017; Kanamarlapudi et al., 2018). Both of these technologies have been studied for extracting REEs from e-waste and ores, which may provide scientific insights for future research on REE extraction from MSWI FA.

5. Conclusions and perspectives

There is still significant room for further improvement of the available technologies for resource recovery from MSWI FA. Extraction technologies using supercritical CO_2 have been developed, which can easily extract lanthanide metal nitrates, oxides, hydroxides, and nitrate solutions. However, further research is needed to investigate the effects of temperature and pressure on metal extraction equilibrium and relative impurity selectivity. Electrodialysis and thermal treatment technologies with extensive energy consumption may increase the environmental burden. In the leaching of rare earth metals, FA contains a high total amount of REEs with low concentrations, and there is currently limited research on its recovery technology. Future approaches that balance recovery capacity and cost may involve biological leaching and biological adsorption technologies. However, current biological technologies have the disadvantages of long cycles and strict requirements for the cultivation of specific strains, and further research is needed to fully explore their industrial applications. Chemical leaching is currently the most mature and widely used technology, but it also faces the challenge of waste liquid treatment.

In addition to the demand for further technological improvements, the low value of MSWI FA itself limits the driving force for developing technologies on resource recovery, and small- and medium-sized enterprises face significant barriers to participation. Therefore, the government should provide funding support and strengthen the daily supervision of MSWI FA management. This will promote the resource utilization of FA and enable the development of corresponding technical product standards, which can achieve a balance on economic costs, environmental protection, and social benefits.

Acknowledgments

The authors gratefully acknowledge the financial support from the National Natural Science Foundation of China (Grant No. 52206174; 52236008) and the Zhejiang Provincial Natural Science Foundation of China (Grant No. LZ23E060004) for this study.

References

Ali, R.A., 2020. Removal of heavy metals from aqueous media by biosorption. Arab. J. Basic Appl. Sci. 27, 183−193.

Ali, H.A., Xuan, D., Zhang, B., Xiao, C., Zhao, B., 2022a. Cementitious characteristics and environmental behaviour of vitrified MSW incineration fly ash slag. Cleaner Mater. 4, 100092.

Ali, M.U., Liu, Y., Yousaf, B., Wong, M.H., Li, P., Liu, G., Wang, R., Wei, Y., Lu, M., 2022b. Morphochemical investigation on the enrichment and transformation of hazardous elements in ash from waste incineration plants. Sci. Total Environ. 828, 154490.

Arab, B., Hassanpour, F., Arshadi, M., Yaghmaei, S., Hamedi, J., 2020. Optimized bioleaching of copper by indigenous cyanogenic bacteria isolated from the landfill of e-waste. J. Environ. Manag. 261, 110124.

Atanes, E., Cuesta-García, B., Nieto-Márquez, A., Fernández-Martínez, F., 2019. A mixed separation-immobilization method for soluble salts removal and stabilization of heavy metals in municipal solid waste incineration fly ash. J. Environ. Manag. 240, 359−367.

Barnett, M., Palumbo-Roe, B., Gregory, S., 2018. Comparison of heterotrophic bioleaching and ammonium sulfate ion exchange leaching of rare earth elements from a Madagascan ion-adsorption clay. Minerals 8, 236.

Brandl, H., Faramarzi, M., 2006. Microbe–metal-interactions for the biotechnological treatment of metal-containing solid waste. China Particuol. 4, 93–97.

Brown, R.M., Mirkouei, A., Reed, D., Thompson, V., 2023. Current nature-based biological practices for rare earth elements extraction and recovery: bioleaching and biosorption. Renew. Sustain. Energy Rev. 173, 113099.

Bühler, A., Schlumberger, S., 2010. Schwermetalle aus der Flugasche zurückgewinnen «Saure Flugaschewäsche – FLUWA-Verfahren» ein zukunftsweisendes Verfahren in der Abfallverbrennung. KVA-Rückstände in der Schweiz - Der Rohstoff mit Mehrwert. Federal Office for the Environment (FOEN), pp. 185–192.

Chatterjee, A., Abraham, J., 2019. Desorption of heavy metals from metal loaded sorbents and e-wastes: a review. Biotechnol. Lett. 41, 319–333.

Chen, W., Kirkelund, G.M., Jensen, P.E., Ottosen, L.M., 2017. Comparison of different MSWI fly ash treatment processes on the thermal behavior of As, Cr, Pb and Zn in the ash. Waste Manage. 68, 240–251.

Chen, L., Wang, L., Cho, D.W., Tsang, D.C.W., Tong, L., Zhou, Y., Yang, J., Hu, Q., Poon, C.S., 2019. Sustainable stabilization/solidification of municipal solid waste incinerator fly ash by incorporation of green materials. J. Clean. Prod. 222, 335–343.

Chuai, X., Yang, Q., Zhang, T., Zhao, Y., Wang, J., Zhao, G., Cui, X., Zhang, Y., Zhang, T., Xiong, Z., Zhang, J., 2022. Speciation and leaching characteristics of heavy metals from municipal solid waste incineration fly ash. Fuel 328, 125338.

Cornelis, G., Johnson, C.A., Gerven, T.V., Vandecasteele, C., 2008. Leaching mechanisms of oxyanionic metalloid and metal species in alkaline solid wastes: a review. Appl. Geochem. 23, 955–976.

Cristelo, N., Segadães, L., Coelho, J., Chaves, B., Sousa, N.R., de Lurdes Lopes, M., 2020. Recycling municipal solid waste incineration slag and fly ash as precursors in low-range alkaline cements. Waste Manag. 104, 60–73.

Dev, S., Sachan, A., Dehghani, F., Ghosh, T., Briggs, B.R., Aggarwal, S., 2020. Mechanisms of biological recovery of rare-earth elements from industrial and electronic wastes: a review. Chem. Eng. J. 397, 124596.

Dontriros, S., Likitlersuang, S., Janjaroen, D., 2020. Mechanisms of chloride and sulfate removal from municipal-solid-waste-incineration fly ash (MSWI FA): effect of acid-base solutions. Waste Manage. 101, 44–53.

Fellner, J., Lederer, J., Purgar, A., Winterstetter, A., Rechberger, H., Winter, F., Laner, D., 2015. Evaluation of resource recovery from waste incineration residues—the case of zinc. Waste Manag. 37, 95–103.

Ferreira, C., Ribeiro, A.B., Ottosen, L.M., 2002. Study of different assisting agents for the removal of heavy metals from MSW fly ashes. Waste Manag. Environ. 171–179.

Fomina, M., Gadd, G.M., 2014. Biosorption:current perspectives on concept, definition and application. Bioresour. Technol. 160, 3–14.

Fruergaard, T., Hyks, J., Astrup, T., 2010. Life-cycle assessment of selected management options for air pollution control residues from waste incineration. Sci. Total Environ. 408, 4672–4680.

Funari, V., Braga, R., Bokhari, S.N., Dinelli, E., Meisel, T., 2015. Solid residues from Italian municipal solid waste incinerators: a source for "critical" raw materials. Waste Manag. 45, 206–216.

Funari, V., Makinen, J., Salminen, J., Braga, R., Dinelli, E., Revitzer, H., 2017. Metal removal from municipal solid waste incineration fly ash: a comparison between chemical leaching and bioleaching. Waste Manag. 60, 397–406.

Funari, V., Gomes, H.I., Cappelletti, M., Fedi, S., Dinelli, E., Rogerson, M., Mayes, W.M., Rovere, M., 2019. Optimization routes for the bioleaching of MSWI fly and bottom ashes using microorganisms collected from a natural system. Waste Biomass Valor 10, 3833–3842.

Funari, V., Toller, S., Vitale, L., Santos, R.M., Gomes, H.I., 2023. Urban mining of municipal solid waste incineration (MSWI) residues with emphasis on bioleaching technologies: a critical review. Environ. Sci. Pollut. Res. 30, 59128–59150.

Gadd, G.M., 2009. Biosorption: critical review of scientific rationale, environmental importance and significance for pollution treatment. J. Chem. Technol. Biotechnol. 84, 13–28.

Ge, X., Hu, X., Shi, C., 2022. Impact of micro characteristics on the formation of high-strength Class F fly ash-based geopolymers cured at ambient conditions. Construct. Build. Mater. 352, 129074.

Geng, C., Chen, C., Shi, X., Wu, S., Jia, Y., Du, B., Liu, J., 2020a. Recovery of metals from municipal solid waste incineration fly ash and red mud via a co-reduction process. Resour. Conserv. Recycl. 154, 104600.

Geng, C., Liu, J., Wu, S., Yu, F., Du, B., Yu, S., 2020b. Novel method for comprehensive utilization of MSWI fly ash through co-reduction with red mud to prepare crude alloy and cleaned slag. J. Hazard Mater. 384, 1–9.

Gu, Q., Wang, T., Wu, W., Wang, D., Jin, B., 2021. Influence of pretreatments on accelerated dry carbonation of MSWI fly ash under medium temperatures. Chem. Eng. J. 414, 128756.

Hammache, Z., Bensaadi, S., Berbar, Y., Audebrand, N., Szymczyk, A., Amara, M., 2021. Recovery of rare earth elements from electronic waste by diffusion dialysis. Sep. Purif. Technol. 254, 117641.

Hasegawa, H., Rahman, I.M.M., Egawa, Y., Sawai, H., Begum, Z.A., Maki, T., Mizutani, S., 2014. Recovery of the rare metals from various waste ashes with the aid of temperature and ultrasound irradiation using chelants. Water Air Soil Pollut. 225, 2112.

He, D., Hu, H., Jiao, F., Zuo, W., Liu, C., Xie, H., Dong, L., Wang, X., 2023. Thermal separation of heavy metals from municipal solid waste incineration fly ash: a review. Chem. Eng. J. 467, 143344.

Huang, K., Inoue, K., Harada, H., Kawakita, H., Ohto, K., 2011. Leaching of heavy metals by citric acid from fly ash generated in municipal waste incineration plants. J. Mater. Cycles Waste Manag. 13, 118–126.

Huang, T., Liu, L., Zhou, L., Yang, K., 2018a. Operating optimization for the heavy metal removal from the municipal solid waste incineration fly ashes in the three-dimensional electrokinetics. Chemosphere 204, 294–302.

Huang, T., Zhou, L., Liu, L., Xia, M., 2018b. Ultrasound-enhanced electrokinetic remediation for removal of Zn, Pb, Cu and Cd in municipal solid waste incineration fly ashes. Waste Manage. 75, 226–235.

Huang, T., Zhou, L., Tao, J., Liu, L., 2018c. Cylindrical electrolyser enhanced electrokinetic remediation of municipal solid waste incineration fly ashes. IOP Conf. Ser. Mater. Sci. Eng. 301, 12097.

Huang, B., Gan, M., Ji, Z., Fan, X., Zhang, D., Chen, X., Sun, Z., Huang, X., Fan, Y., 2022. Recent progress on the thermal treatment and resource utilization technologies of municipal waste incineration fly ash: a review. Process Saf. Environ. Protect. 159, 547–565.

Ishigaki, T., Nakanishi, A., Tateda, M., Ike, M., Fujita, M., 2005. Bioleaching of metal from municipal waste incineration fly ash using a mixed culture of sulfur-oxidizing and iron-oxidizing bacteria. Chemosphere 60, 1087–1094.

Javanbakht, V., Alavi, S.A., Zilouei, H., 2014. Mechanisms of heavy metal removal using microorganisms as biosorbent. Water Sci. Technol. 69, 1775–1787.

Jitpinit, S., Siraworakun, C., Sookklay, Y., Nuithitikul, K., 2022. Enhancement of omega3 content in sacha inchi seed oil extracted with supercritical carbon dioxide in semicontinuous process. Heliyon 8, e08780.

Kanamarlapudi, S.L.R.K., Chintalpudi, V.K., Muddada, S., 2018. Application of biosorption for removal of heavy metals from wastewater. Biosorption 18, 70–116.

Kersch, C., Woerlee, G.F., Witkamp, G.J., 2004. Supercritical fluid extraction of heavy metals from fly ash. Ind. Eng. Chem. Res. 43, 190–196.

Kirkelund, G.M., Jensen, P.E., Ottosen, L.M., Pedersen, K.B., 2019. Comparison of two- and three-compartment cells for electrodialytic removal of heavy metals from contaminated material suspensions. J. Hazard Mater. 367, 68–76.

Koegler, S., 2010. Development of a unique process for recovery of uranium from incinerator ash. ACS Symp. Ser. 1046, 65–78.

Kołodyńska, D., Krukowska, J., Thomas, P., 2017. Comparison of sorption and desorption studies of heavy metal ions from biochar and commercial active carbon. Chem. Eng. J. 307, 353–363.

Kuboňová, L., Langová, Š., Nowak, B., Winter, F., 2013. Thermal and hydrometallurgical recovery methods of heavy metals from municipal solid waste fly ash. Waste Manage. 33, 2322–2327.

Kurashima, K., Matsuda, K., Kumagai, S., Kameda, T., Saito, Y., Yoshioka, T., 2019. A combined kinetic and thermodynamic approach for interpreting the complex interactions during chloride volatilization of heavy metals in municipal solid waste fly ash. Waste Manage. 87, 204–217.

Lane, D.J., Hartikainen, A., Sippula, O., Lähde, A., Mesceriakovas, A., Peräniemi, S., Jokiniemi, J., 2020. Thermal separation of zinc and other valuable elements from municipal solid waste incineration fly ash. J. Clean. Prod. 253, 120014.

Leckner, B., 2015. Process aspects in combustion and gasification waste-to-energy (WtE) units. Waste Manag. 37, 13–25.

Lee, J.C., Pandey, B.D., 2012. Bio-processing of solid wastes and secondary resources for metal extraction—a review. Waste Manag. 32, 3–18.

Li, L., Dunn, J.B., Zhang, X.X., Gaines, L., Chen, R.J., Wu, F., Amine, K., 2013. Recovery of metals from spent lithium-ion batteries with organic acids as leaching reagents and environmental assessment. J. Power Sources 233, 180–189.

Li, W., Wang, W., Wu, D., Yang, S., Fang, H., Sun, S., 2023. Mechanochemical treatment with red mud added for heavy metals solidification in municipal solid waste incineration fly ash. J. Clean. Prod. 398, 136642.

Liu, Z., Zhang, T., Zhang, J., Xiang, H., Yang, X., Hu, W., Liang, F., Mi, B., 2018. Ash fusion characteristics of bamboo, wood and coal. Energy 161, 517–522.

Loginova, E., Proskurnin, M., Brouwers, H.J.H., 2019. Municipal solid waste incineration (MSWI) fly ash composition analysis: a case study of combined chelatant-based washing treatment efficiency. J. Environ. Manag. 235, 480–488.

Long, L., Zhao, Y., Lv, G., Duan, Y., Liu, X., Jiang, X., 2023a. Improving stabilization/solidification of MSWI fly ash with coal gangue based geo-polymer via increasing active calcium content. Sci. Total Environ. 854, 158594.

Long, Y., Pu, K., Yang, Y., Huang, H., Fang, H., Shen, D., Geng, H., Ruan, J., Gu, F., 2023b. Preparation of high-strength ceramsite from municipal solid waste incineration fly ash and clay based on CaO-SiO_2-Al_2O_3 system. Construct. Build. Mater. 368, 130492.

Lv, Y., Yang, L., Wang, J., Zhan, B., Xi, Z., Qin, Y., Liao, D., 2022. Performance of ultra-high-performance concrete incorporating municipal solid waste incineration fly ash. Case Stud. Constr. Mater. 17, e01155.

Ma, X., He, T., Da, Y., Xu, Y., Luo, R., Yang, R., 2023. Improve toxicity leaching, physicochemical properties of incineration fly ash and performance as admixture by water washing. Construct. Build. Mater. 386, 131568.

Mäkinen, J., Salo, M., Soini, J., Kinnunen, P., 2019. Laboratory scale investigations on heap (bio)leaching of municipal solid waste incineration bottom ash. Minerals 9, 290.

Michalak, I., Chojnacka, K., Witek-Krowiak, A., 2013. State of the art for the biosorption process—a review. Appl. Biochem. Biotechnol. 170, 1389–1416.

Mishra, D., Rhee, Y., 2010. Current research trends of microbiological leaching for metal recovery from industrial wastes. Curr. Res. Technol. Educ. Topics Appl. Microbiol. Microb. Biotechnol. 2.

Mukiza, E., Zhang, L., Liu, X., Zhang, N., 2019. Utilization of red mud in road base and subgrade materials: a review. Resour. Conserv. Recycl. 141, 187–199.

Nag, M., Shimaoka, T., 2023. A novel and sustainable technique to immobilize lead and zinc in MSW incineration fly ash by using pozzolanic bottom ash. J. Environ. Manag. 329, 117036.

Nagib, S., Inoue, K., 2000. Recovery of lead and zinc from fly ash generated from municipal incineration plants by means of acid and/or alkaline leaching. Hydrometallurgy 56, 269–292.

Okada, T., Tomikawa, H., 2016. Efficiencies of metal separation and recovery in ash-melting of municipal solid waste under non-oxidative atmospheres with different reducing abilities. J. Environ. Manage. 166, 147–155. https://doi.org/10.1016/j.jenvman.2015.10.010.

Phua, Z., Giannis, A., Dong, Z.L., Lisak, G., Ng, W.J., 2019. Characteristics of incineration ash for sustainable treatment and reutilization. Environ. Sci. Pollut. Res. 26, 16974–16997.

Qiu, Q., Jiang, X., Lü, G., Chen, Z., Lu, S., Ni, M., Yan, J., Deng, X., 2019. Degradation of PCDD/Fs in MSWI fly ash using a microwave-assisted hydrothermal process. Chin. J. Chem. Eng. 27, 1708–1715.

Quina, M.J., Bontempi, E., Bogush, A., Schlumberger, S., Weibel, G., Braga, R., Funari, V., Hyks, J., Rasmussen, E., Lederer, J., 2018. Technologies for the management of MSW incineration ashes from gas cleaning: new perspectives on recovery of secondary raw materials and circular economy. Sci. Total Environ. 635, 526–542.

Ramanathan, T., Ting, Y., 2016. Alkaline bioleaching of municipal solid waste incineration fly ash by autochthonous extremophiles. Chemosphere 160, 54–61.

Sand, W., Gehrke, T., Jozsa, P., Schippers, A., 2001. (Bio)chemistry of bacterial leaching—direct vs. indirect bioleaching. Hydrometallurgy 59, 159–175.

Sinclair, L.K., Tester, J.W., Thompson, J.F.H., Fox, R.V., 2019. Supercritical extraction of lanthanide tributyl phosphate complexes: current status and future directions. Ind. Eng. Chem. Res. 58, 9199–9211.

Song, Z., Zhang, X., Tan, Y., Zeng, Q., Hua, Y., Wu, X., Li, M., Liu, X., Luo, M., 2023. An all-in-one strategy for municipal solid waste incineration fly ash full resource utilization by heat treatment with added kaolin. J. Environ. Manag. 329, 117074.

Srichandan, H., Mohapatra, R.K., Parhi, P.K., Mishra, S., 2019. Bioleaching approach for extraction of metal values from secondary solid wastes: a critical review. Hydrometallurgy 189, 105122.

Sun, X., Li, J., Zhao, X., Zhu, B., Zhang, G., 2016. A review on the management of municipal solid waste fly ash in American. Proc. Environ. Sci. 31, 535–540.

Tang, H., Erzat, A., Liu, Y., 2014. Recovery of soluble chloride salts from the wastewater generated during the washing process of municipal solid wastes incineration fly ash. Environ. Technol. 35, 2863–2869.

Tang, J., Su, M., Wu, Q., Wei, L., Wang, N., Xiao, E., Zhang, H., Wei, Y., Liu, Y., Ekberg, C., Steenari, B.-M., Xiao, T., 2019. Highly efficient recovery and clean-up of four heavy metals from MSWI fly ash by integrating leaching, selective extraction and adsorption. J. Clean. Prod. 234, 139–149.

Taušová, M., Mihaliková, E., Čulková, K., Stehlíková, B., Tauš, P., Kudelas, D., Štrba, Ľ., 2019. Recycling of communal waste: current state and future potential for sustainable development in the EU. Sustainability 11, 2904.

Tian, X., Rao, F., León-Patiño, C.A., Song, S., 2021. Co-disposal of MSWI fly ash and spent caustic through alkaline-activation consolidation. Cem. Concr. Compos. 116, 103888.

Torres, E., 2020. Biosorption: a review of the latest advances. Processes 8, 1584.

Valix, M., 2017. Bioleaching of electronic waste. Processes 8, 1584.

Van Caneghem, J., Van Acker, K., De Greef, J., Wauters, G., Vandecasteele, C., 2019. Waste-to-energy is compatible and complementary with recycling in the circular economy. Clean Technol. Environ. Policy 21, 925–939.

Wang, J., Chen, C., 2009. Biosorbents for heavy metals removal and their future. Biotechnol. Adv. 27, 195–226.

Wang, Q., Yang, J., Wang, Q., Wu, T., 2009. Effects of water-washing pretreatment on bioleaching of heavy metals from municipal solid waste incinerator fly ash. J. Hazard Mater. 162, 812–818.

Wang, H., Zhu, F., Liu, X., Han, M., Zhang, R., 2021. A mini-review of heavy metal recycling technologies for municipal solid waste incineration fly ash. Waste Manag. Res. 39, 1135–1148.

Wang, R., He, P., Lu, F., Shao, L., Zhang, H., 2022. Concentration of fly ash eluate by electrodialysis and recovery of industrial salt. Chinese J. Environ. Eng. 2365–2373.

Wang, H., Zhao, B., Zhu, F., Chen, Q., Zhou, T., Wang, Y., 2023a. Study on the reduction of chlorine and heavy metals in municipal solid waste incineration fly ash by organic acid and microwave treatment and the variation of environmental risk of heavy metals. Sci. Total Environ. 870, 161929.

Wang, S., Xue, C., Zhao, Q., Bai, Y., Guo, W., Shi, Y., Qiu, Y., Pan, H., 2023b. A novel binder prepared from municipal solid waste incineration fly ash and phosphogypsum. J. Build. 71, 106486.

Wang, Y., Ma, J., Qing, L., Liu, L., Shen, B., Li, S., Zhang, Z., 2023c. Accelerated carbonation pretreatment of municipal solid waste incineration fly ash and its conversion to geopolymer with coal fly ash. Construct. Build. Mater. 383, 131363.

Xie, K., Hu, H., Cao, J., Yang, F., Liu, H., Li, A., Yao, H., 2020. A novel method for salts removal from municipal solid waste incineration fly ash through the molten salt thermal treatment. Chemosphere 241, 125107.

Xie, Q., Wang, D., Liang, G., Tao, H., Liu, S., 2022. Separation of inorganic chlorine salts from municipal solid waste incineration fly ash. Min. Metall. Eng. 42 (1), 81–84.

Xiong, X., Liu, X., Iris, K.M., Wang, L., Zhou, J., Sun, X., Rinklebe, J., Shaheen, S.M., Ok, Y.S., Lin, Z., Tsang, D.C.W., 2019. Potentially toxic elements in solid waste streams: fate and management approaches. Environ. Pollut. 253, 680–707.

Xu, T., Ramanathan, T., Ting, Y., 2014. Bioleaching of incineration fly ash by Aspergillus niger–precipitation of metallic salt crystals and morphological alteration of the fungus. Biotechnol. Rep. 3, 8–14.

Xue, Y., Liu, X., 2021. Detoxification, solidification and recycling of municipal solid waste incineration fly ash: a review. Chem. Eng. J. 420, 130349.

Yao, Y., Farac, N.F., Azimi, G., 2018. Supercritical fluid extraction of rare earth elements from nickel metal hydride battery. ACS Sustain. Chem. Eng. 6, 1417–1426.

Yu, W., Sun, Y., Lei, M., Chen, S., Qiu, T., Tang, Q., 2019. Preparation of micro-electrolysis material from flotation waste of copper slag and its application for degradation of organic contaminants in water. J. Hazard Mater. 361, 221–227.

Zha, F., Wang, S., Liu, Z., Dai, J., Yue, S., Qi, W., Xue, X., Wang, X., Zhang, S., 2023. Removal of heavy metals from fly ash using electrodialysis driven by a bioelectrochemical system: a case study of Pb, Mn, Cu and Cd. Environ. Technol. 1−12.

Zhang, Y., Wang, L., Chen, L., Ma, B., Zhang, Y., Ni, W., Tsang, D.C.W., 2021. Treatment of municipal solid waste incineration fly ash: state-of-the-art technologies and future perspectives. J. Hazard Mater. 411, 125132.

Zhang, Z., Li, Z., Yang, Y., Shen, B., Ma, J., Liu, L., 2022c. Preparation and characterization of fully waste-based glass-ceramics from incineration fly ash, waste glass and coal fly ash. Ceram. Int. 48, 21638−21647.

Zhang, J., Mao, Y., Wang, W., Wang, X., Li, J., Jin, Y., Pang, D., 2022a. A new co-processing mode of organic anaerobic fermentation liquid and municipal solid waste incineration fly ash. Waste Manage. 151, 70−80.

Zhang, Z., Wang, Y., Zhang, Y., Shen, B., Ma, J., Liu, L., 2022b. Stabilization of heavy metals in municipal solid waste incineration fly ash via hydrothermal treatment with coal fly ash. Waste Manage. 144, 285−293.

Zhao, H., Yang, F., Wang, Z., Li, Y., Guo, J., Li, S., Shu, J., Chen, M., 2023. Chlorine and heavy metals removal from municipal solid waste incineration fly ash by electric field enhanced oxalic acid washing. J. Environ. Manag. 340, 117939.

Zheng, R., Wang, Y., Liu, Z., Zhou, J., Yue, Y., Qian, G., 2022. Environmental and economic performances of municipal solid waste incineration fly ash low-temperature utilization: an integrated hybrid life cycle assessment. J. Clean. Prod. 340, 130680.

Zhou, Y., Xue, X., Yang, H., Song, S., Huang, X., 2020. Novel harmless utilization of Bayan Obo tailings: separation and recovery of iron and rare earth. Ind. Eng. Chem. Res. 59, 13682−13695.

Chapter 27

Resource recovery from municipal solid waste incineration bottom ash

Valerio Funari[1,2], Junaid Ghani[1,3] and Luciana Mantovani[4]

[1]*Institute of Marine Sciences, National Research Council of Italy, Venice, Italy;* [2]*Department of Ecosustainable Marine Biotechnologies, Stazione Zoologica Anton Dohrn, Naples, Italy;* [3]*Department of Biological Geological and Environmental Science, University of Bologna, Bologna, Italy;* [4]*Department of Chemistry, Life Sciences and Environmental Sustainability, University of Parma, Parma, Italy*

1. Introduction

Municipal solid waste incineration (MSWI) allows the recovery of various valuable resources and is considered as a fundamental part of circular economy (Van Caneghem et al., 2019). Typically, the MSWI technologies are designed because of regional policies and economic evaluation, local socioeconomic conditions, expected waste feed, and composition (Moya et al., 2017). Worldwide, air pollution control systems are used to control the emissions generated relying on adsorption and injection of particulates (Chang et al., 2009; Jones and Harrison, 2016; Leckner and Lind, 2020). Eventually, the acid gases, such as, SO_2 and HCl, and particulates are removed using Ca-rich, carbonatic additives, hydrated or dehydrated, frequently nebulized, contributing to pollutant emission abatement at the smokestack. Regarding the incineration of solid materials, "grate furnace technology" is the most used in medium to large capacity facilities (3–6 t waste/h), with waste residence time of 2–3 h at temperatures between 750 and 1100°C. "Rotating furnace technology" consists of a tilted furnace, generally cylindrical, reaching up to 1200°C temperature, where waste is introduced longitudinally by aeration, allowing the optimization of the mix waste/air and low residence time to complete ignition. These furnaces are typically better suited for medium-capacity plants in terms of waste volume (about 3 t waste/h). "Fluidized bed technology" entails three types of fluidized bed: dense, rotating, and circulating. Firstly developed for coal ash burning and better suited for specialized incineration, it implies closed semivertical system proceeding stepwise in small fraction of waste treated, with a bed of sand at a temperature between 750 and 800°C. Fluidized bed furnaces would allow better recoveries of resource since the recoverable fractions occur in larger particles and metals are less oxidized compared with grate furnace technologies, but they are very few in operation for municipal waste treatment because they produce around five times more fly ashes per unit of waste incinerated than grate furnaces (Blasenbauer et al., 2023).

Many parameters, especially temperature, have important effects in the production of bottom ash (BA) and the residual presence of organic matter. The type and composition of the residual, solid ashes vary among different incinerators, depending upon countries, seasons, the incinerator technology employed, and waste collection efficiency (Chuang et al., 2018; Wang et al., 2019). Moreover, divergences and nonhomogeneity arise in regulatory and legislative framework of countries where waste incineration technology is used: the experimental protocols for evaluating the environmental quality for BA reuse in the construction industry differ locally. During inorganic chemistry analysis required by different state-of-the-art characterization methods to quantify metal solubility and environmental impacts, a selection of analytes is measured in solid and liquid phases of MSWI BA (Table 27.1).

In MSWI plants, BA is the most significant solid residue, accounting for 85%–95% of the total ashes generated. Among developed countries, the United States incinerates around 258 million t/a waste, while China 249 million t/a, followed by Japan 43.98 million t/a, Canada 35.62 million t/a, and Italy 30 million t/a. In China, there are about 500 incineration plants operational and some are under construction. Around 100 million t/a municipal waste surplus are expected to be incinerated in coming years in mainland China (Youcai, 2017). For the United States, only 12.7% of the 243 million tons of municipal solid waste underwent to incineration in 2017 (USEPA, 2019). In 2015, above 50% of the

Treatment and Utilization of Combustion and Incineration Residues. https://doi.org/10.1016/B978-0-443-21536-0.00011-3

TABLE 27.1 A summary of analytes sought (marked with "X") in MSWI BA characterization routine (standard leaching tests coupled with analysis of the eluates or analysis on bulk samples) by different countries along with the minimum recommended by European Union (EN/EU) and Environmental Protection Agency of the United States of America (US EPA).

	Belgium	Denmark	EN/EU	France	Germany	Italy	The Netherlands	US EPA
Al	X							
As	X	X	X	X		X	X	X
Ba		X		X		X		X
Br							X	
Cd	X	X	X	X	X	X	X	X
Cl	X	X		X	X	X	X	
Co	X					X	X	
Cr	X	X	X	X	X	X	X	X
Cu	X	X		X	X	X	X	
F	X			X		X	X	
Hg	X	X	X	X	X	X	X	X
In							X	
K	X							
Mn		X				X		
Mo	X			X		X	X	
Na		X						
Ni	X	X		X	X	X	X	
Pb	X	X	X	X	X	X	X	X
pH					X	X		
S	X	X		X	X	X	X	
Sb	X			X		X	X	
Se				X		X	X	X
Sn						X	X	
Soluble			X	X				
Ti	X							
TREE							X	
Zn	X	X	X	X	X	X	X	

TREE is the sum of rare earth elements (REEs), while soluble refers to the mobilized fraction of metals after water washing or other, generally mild, chemical treatments of leaching.
Readapted after Minane, J.R., Vinai, R., 2023. Bottom ash: production, characterisation, and potential for recycling. In: Tribaudino, M., Vollprecht, D., Pavese, A. (Eds.), Minerals and Waste. Earth and Environmental Sciences Library. Springer.

total urban waste was incinerated in several countries including Japan, Denmark, Norway, Sweden, and the Netherlands, while about 25% was incinerated in Germany, Switzerland, the United Kingdom, France, and South Korea (OECD, 2015). The mass of MSWI BA generated is typically from 40% to 60% of the total input mass that is, in turn, the reason of a substantial interest in incineration technologies for waste reduction. The landfilling or treatment for recovery of valuable resources and reuse as secondary raw materials (e.g., as construction material) are BA management options that vary from country to country. BA is a heterogenous solid material, containing hard pebbles, ceramics, metals, slag, cullet glass, and unburnt materials, and various mineralogical phases (Mantovani et al., 2021). Glass and materials with predominant mineralogical phases aluminosilicates make up around 90% of BA (Huber et al., 2019), which is well suited for reuse in some countries as a construction material with minimal treatment (Blasenbauer et al., 2020). See Part I of this book for more details on generation and characteristics of BA. Resource recovery options are summarized in Table 27.2 based on the main processing method.

Most operations (dry or wet processes) aim at recovering secondary raw materials for direct reuse, sometimes energy (e.g., waste pyrolysis), and ferrous and nonferrous fractions (e.g., Back and Sakanakura, 2022). Many commercially available treatment trains enable the recovery of 6.3 wt% ferrous metals and 1.7 wt% nonferrous metals on average (CEWEP, 2017), even though significant losses of resources are estimated at large (Enzner et al., 2017; Abis et al., 2020). Treatments off site MSWI plants can favor mineral beneficiation and subsequent metal recovery, for example, by separating inhomogenous fractions stepwise or through simply aging, likely to reduce the BA polluting potential. The material is exposed to air and rain up to 12 months in a controlled environment so that hydration and carbonation reactions (Nørgaard et al., 2019) stabilize BA into a more stable material to ambient leaching at new usage and can generate reduce carbon dioxide emission of the MSWI system (Aricks et al., 2006). Finely tuned metal recovery usually relies on advanced screening and separation methods (Bunge et al., 2015; Šyc et al., 2018). The typical recovery is a lighter, Al-rich fraction than the ferrous metal one. Agglomerates of metals including metallic (alloyed) particles and valuable minerals occur in heavier fraction of the BA (Back and Sakanakura, 2021). However, the full potential of resource recovery from this ferrous fractions is not achieved widely, with potential resource losses estimated at 40–50 wt% (Enzner et al., 2017). An urgent breakthrough is required to counteract waste production and anthropogenic load on Earth.

In the following paragraphs, bibliometric analysis is used deliberately to evaluate the impacts of the scientific literature on resource recovery from MSWI BA and the flow of substance of proposed actions, respectively. Bibliometric clustering map and bibliometric citation map were prepared.

2. A sight over 30-year scientific literature

A search in Scopus database was carried out to retrieve bibliographic publication records related to resources recovery of BA and then elaborated with VOSviewer software (Van Eck and Waltman, 2014). The Scopus search was limited to scientific research and review articles using suitable query[1] for bibliometric analysis. According to Fig. 27.1A, the trend of scientific publications from 1990 to 2022 (total 2072) on BA treatment for resources recovery is substantially increasing. In detail, the number of publications was low (210 articles; 17.1% contribution) prior to 2005 and then increased (1013 articles; 82.8% contribution) with peaks in 2007, 2013, and 2019, suggesting research and technology in the general trend has emerged and improved within the latest socioeconomic crisis (2006, 2011, 2017, 2021). The country-wise, cumulative number of publications from 2010 to 2023 is reported in Fig. 27.1B with larger font size in the function of occurrences (previously normalized to the number of publications for a single country by VOSviewer). China has major contribution, followed by the United States, Germany, Italy, the United Kingdom, Denmark, India, and Japan. In the past 13 years, BA treatment and utilization for resources recovery (especially metals and minerals) has drawn increasing attention worldwide. Among the top 10 journals, the majority of articles (121 publications) were published in the scientific journal "*Waste Management*," followed by "*Journal of Hazardous Materials*" (47 articles), "*Waste Management and Research*" (34 articles), "*Journal of Cleaner Production*" (31 articles), and "*Chemosphere*" (28 articles). Regarding Journals' main field, "Environmental Science" has the major contribution (779 articles), followed by "Engineering" (270 articles), "Energy" (262 articles), "Chemical Engineering" (208 articles), "Chemistry" (188 articles), and "Materials Science" (135 articles), "Earth and Planetary Sciences" (69 articles), "Agricultural and Biological Sciences" (61 articles), and "Biochemistry, Genetics and Molecular Biology" (34 articles). The other fields of study count less than 65 articles altogether.

A pilot, four-cluster map from bibliometric analysis (Waltman and Van Eck, 2013; Van Eck and Waltman, 2014) targeting resources recovery from BA (Fig. 27.2) visualizes the principal components in the scientific panorama. One cluster focalizes on MSWI systems and BA characterization analysis (RED), a branch of "soft" resources recovery (GREEN), one of metal resource recovery (BLUE), and the yellow cluster visualizes scientific contribution to energy- and

1. The following keywords were used in the combined fields of search as defined by the authors: TITLE-ABS-KEY ("Municipal Solid Waste" OR "Municipal Solid Waste Incineration Plants" OR "Waste To Energy" OR "Bottom Ash" AND "Bottom Ash Characterization" OR "Heavy Metals" OR "Major Elements" OR "Trace Elements" OR "Rare Earth Elements" OR "Critical Elements" OR "Metal Recovery" OR "Metals" OR "elements" OR "resource" OR "mineral resource" AND "Urban Waste Treatment" OR "Urban Waste Management" OR "Resource Recovery" OR "Secondary Raw Materials" OR "Materials Recovery" OR "Critical Materials" OR "Critical Raw Materials" OR "Minerals" OR "Raw materials" OR " hydroxide" OR "Carbonates" OR " Sulfates" OR "Phosphates" OR "Oxides" OR "Gypsum" OR "Chloride" OR "Zeolite" OR "Silicates" OR "Organic Matter" OR "Oil" OR "Fuel" OR "Glass" OR "Ceramics" OR "Plastic" OR "Brass" AND NOT "Coal" OR "Coal Ash" OR "Fly Ash" OR "Wastewater" OR "Soil" OR "Mud" OR "Water") AND PUBYEAR: 2010–2023, using the VOSviewer software (version 1.6.18).

TABLE 27.2 A simplified list of suitable techniques to treat MSWI BA, available to application not only for waste minimization but also for secondary resource recovery.

Gravity	Screening/separation	Other physics-based
Sieving Size reduction and separation Separation based on elastic properties (e.g., ballistic separators) Separation based on surface properties (e.g., flotation, agglomeration)	Magnetic separation (e.g., magnetic drums) Electrical separation (e.g., electrostatic separator, Eddy current separator) Wet separation with wet air, water, and fixed density substances (e.g., centrifuges, cyclones) Thermic treatments	Reflecting power detection by color Reflecting power detection by shape X-ray transmission X-ray fluorescence Near-infrared sensors Detection of isotopes

Chemistry-based	Biology-based	Hybrid
Sequential extraction, chemical leaching Sequential precipitation, coprecipitation Cementation, chemical stabilization	Bioleaching, bioaccumulation Phyco/phytoremediation Biostabilization, bio-cementation Bioremediation	Waste maturation (e.g., accelerated carbonation) Advanced oxidation processes Advanced reduction processes Recovery processes coupled with energy production

pollution-related matter (YELLOW). Dioxin, silicate, chlorides, scrubbers, concrete, plastics, ceramics, soft minerals, lanthanide, Mg, Ba, rare earth elements (REEs), and thermoplastic materials such as polyvinyl chloride, polypropene, polyethylene, and polypropylenes were highly cited (>40). In contrast, the association of terms such as copper (Cu), zinc (Zn), iron (Fe), nickel (Ni), "chlorine compounds," "sulfur compounds," "aluminum compounds," "iron oxides," and "aluminum oxides" occurred and impacted mainly the earlier literature, demonstrating a constant interest with time. Notably, As, Sb, V of BLUE cluster appeared in earlier literature but are currently at low citations (<40), while REE/lanthanide fell in the GREEN cluster with high recent citations (>40). The high occurrence of these keywords indicates the search for critical raw materials for industry, including organic compounds, is increasing with time by recycling urban waste streams such as MSWI BA.

3. Developments in characterization and assessment for resource recovery

Since early studies (Chandler et al., 1997; Eusden et al., 1999), it was clear that the absence of ancillary mineralogical phases attached on glass, ceramics, and slag, the low corrosions, and limited intermineral reaction edges typical of fast cooling, can be an advantage for resource recovery from BA. BA production in MSWI plants needs to adhere minimum requirements set by Directive (2010)/75/EU before direct reuse. For example, it is mandatory that final BAs have LOI equal to or lower than 5% (see Table 27.1), and the evaluation of the leachate is of paramount importance. The European recommended procedures to evaluate ambient leaching are EN 12457/1 2002, EN 12457/2 2002, EN 12457/3 2002, and EN 12457/4 2002, with various liquid-to-solid (L/S) ratios and experimental configurations (batch and column leaching tests), following corresponding criteria and concentration thresholds adopted by each EU Member State (Table 27.1). Insights from chemical and mineralogical data on MSWI residues informed recovery of secondary raw material and marketable ore metals from BA produced by different incineration technologies. The mineralogical analysis is as useful as chemical data for evaluating any strategic metal enrichments because it can tell how it occurred and which mineralogical phase is more easily recoverable. A potential treatment toward Cu and Zn recovery can be affordable and align the residues to the regulated limit on leachates, which are particularly stringent for Cu and Zn. Cu and Zn seem to be the most suitable metals to be mined in particular because good separability can be achieved using commercially available treatments (e.g., separation of ferrous metal fraction). For the study of the partition of these elements in minerals, the difficulty lies in the precise identification of the main mineralogical phases in which metals are present. Significant concentrations of Zn were found in willemite that is a Zn orthosilicate, with stoichiometric formula Zn_2SiO_4, with significant Mg content exchanging for Zn in solid solution and a Na-rich hardystonite, a Zn-bearing melilite, with nominal composition $Ca_2ZnSi_2O_7$. Zn was also found as gahnite spinel ($ZnAl_2O_4$) and dissolved in Na- and Mg-rich silicate glass in grains with

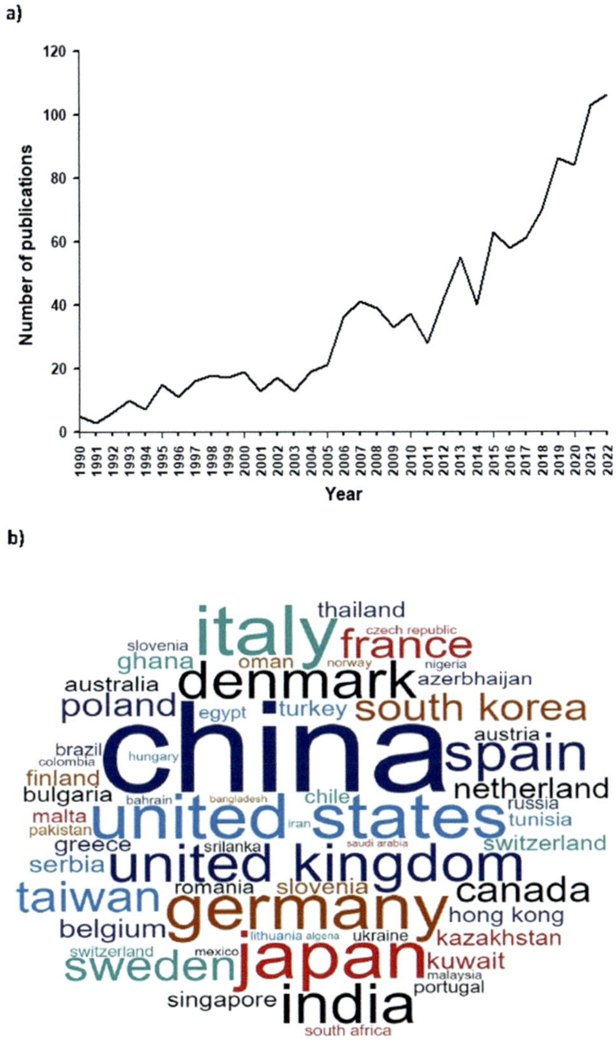

FIGURE 27.1 Preliminary bibliometric analysis using data from VOSviewer and https://www.wordclouds.com/web-app: (A) the number of publications of MSWI BA treatment for resources recovery from 1990 to 2022; (B) word cloud chart shows scientific contributions of different countries from 2010 to 2023 with randomized colors and font sizes proportional to the number of research articles and reviews provided by each country.

Zn content up to 2 wt% (De Matteis et al., 2023). Copper can be present as fuses or melted drops, often in the form of micrometric spheres easily detectable by compositional contrast because the chemical composition of these beads is close to 100 wt% Cu. Therefore, these analyzed spheres are a compelling host of metallic Cu, probably due to the poor miscibility under the system conditions (i.e., the pressures and temperatures reached in the combustor). While Cu may be present in significant quantities in some samples due to the "nugget" effect, Zn seems to be present as a secondary element diffused in various mineralogical phases. It is also possible to identify sites where Cu- and Zn-rich phases are present and form their own phase. Beside Cu and Zn, other elements showing significant concentrations (>10 g/kg) in the bulk residue are Si, Fe, Ca, Al, Mg, Na, and K, followed by Cl, Pb, Ti, Mn, Ba, and Cr. It has been suggested that among metals of strategic interest and potentially mineable from MSWI residues, Ag, Sb, Ni, V, Sb, Ce, La, Nb, and Ni are enriched in the fine fractions, while other metals such as Cr, Gd, W, Y, and Sc partition in the coarse fractions (Pan et al., 2013). Fig. 27.3 shows the mineralogical phases observed in some recent studies. Although a potential variability in elemental composition is largely dependent on the particle size, the mineralogical phases' content seems to not change significantly. A notable trend observed by many authors is the presence of carbonates in smaller particle sizes and amorphous phases in larger ones.

The average mineralogy composition of BA from MSWI systems in northern Italy is reported in Table 27.3 as a reference. Other commonly observed minerals are typical cement minerals, such as ettringite, hydrocalumite, and calcium silicates. These minerals can be reused as cementitious materials, but only after careful evaluation regarding their release of

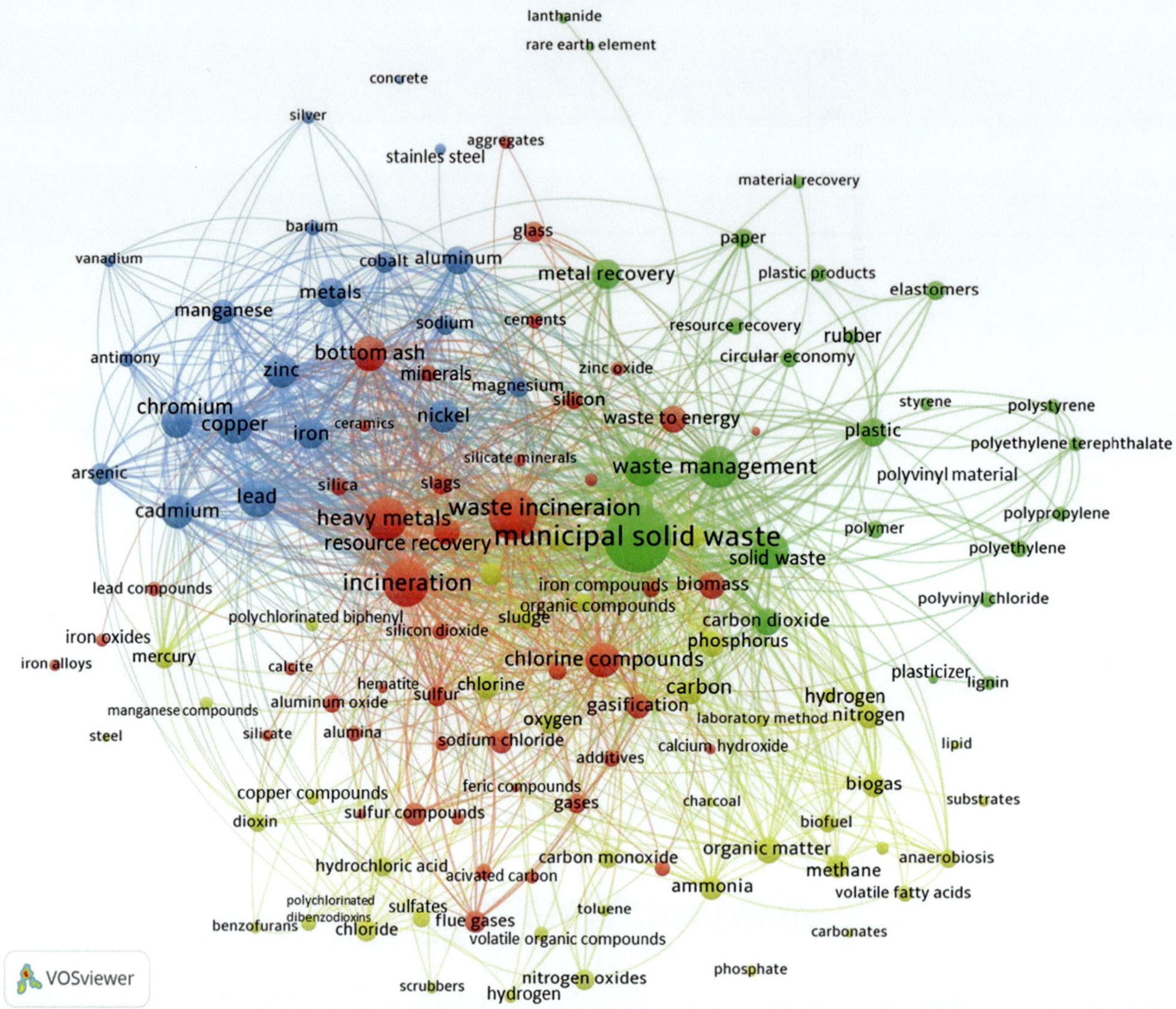

FIGURE 27.2 Bibliometric clustering map analysis of resources recovery in BA. The nodes (term-associated circles) thicknesses relate to the number of occurrences in scientific publications. The occurrence of terms with the same domain of knowledge extracted with the same color. For the clustering, terms that appeared at least eight times from the retrieved publications were extracted. A different text size is used deliberately, avoiding text overlaps.

elements into the environment. MSWI BA is primarily composed of various minerals (50%−70%), ceramics and glass (10%−30%), and unburned materials (1%−5%) (Hyks and Astrup, 2009). The solid-phase refractories of BA include several amorphous phases including ceramics, glass pieces, and silicate glass (Alorro et al., 2009; Inkaew et al., 2016; Yang et al., 2016; Mantovani et al., 2023). The presence of heavy metals and Pb and Cl compounds can be seen as a resource in terms of environmental impact savings. Chlorine-based compounds are frequently found in BA and in the municipal solid waste input. In the cycle of primary raw materials and output flows of secondary raw materials, it is worth mentioning the mass balance of chloride compounds is critical because they could hasten the rate at which equipment corrodes and the extent at which final BA fits concrete industry applications (Ma et al., 2014). Not only different mineralogical phases can be exploited in first-rate applications, but also amorphous or glassy phases find relevant afteruses (e.g., bicarbonates as buffering agents in different applications, silicates in glass/ceramic tiles, and jewelery, chlorides, and polymers in the chemical synthesis). Incineration BA in any suitable forms is virtually tested for enhanced resource recovery.

Looking at the bibliometric analysis on resource recovery from BA (Fig. 27.4), RED cluster refers to the core terminology of "solid waste," "incineration," "bottom ash," and "waste to energy," including recoverable resources, such as "heavy metals," "glass," "slags," "ceramics," "silicates," "cements," "calcite," "chlorine," "chlorine compounds," "lead compounds," "sodium chloride," "iron oxides," "iron alloys," "alumina," "additives," "activated carbon," and

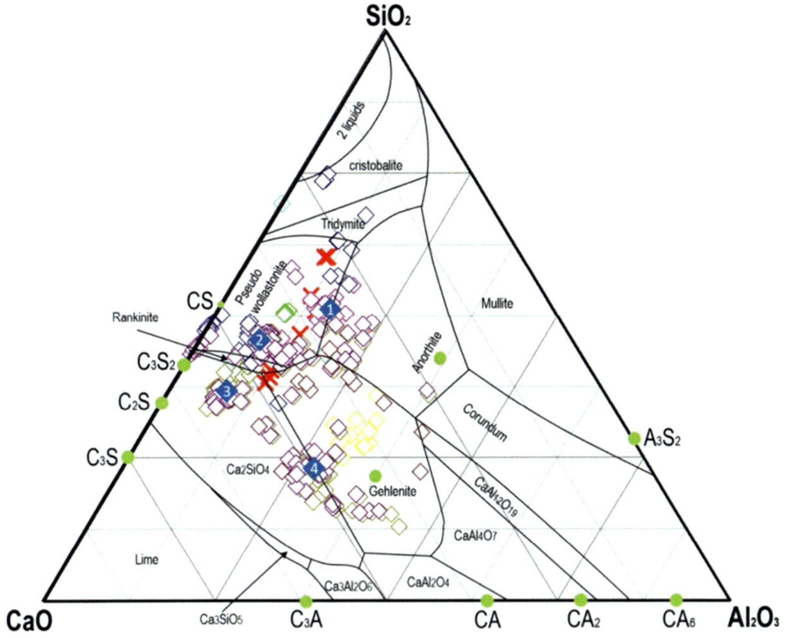

FIGURE 27.3 Phase diagram of the $CaO-SiO_2-Al_2O_3$ system applied to MSWI BA reported in Mantovani et al. (2021), originally representing the different regions where a given oxide component crystallizes from the corresponding melt (Rankin, 1915). Cement notation and stoichiometric formula of some compounds present in BA are given as green points. Colored squares and crosses represent the average composition of point analysis in different BA samples.

TABLE 27.3 Quantitative analysis (expressed in wt%) made by rietveld refinement for crystalline and amorphous phases of different MSWI BA (BA1-5).

Mineral name	Chemical formula	BA1	BA2	BA3	BA4	BA5	Total average
Calcite	$CaCO_3$	8.06	5.42	3.38	16.97	17.03	10.17
Quartz	SiO_2	4.48	4.61	7.76	12.44	7.71	7.40
Plagioclase	$(Na,Ca)\ (Si,Al)_4O_8$	2.74	2.52	5.87	8.67	7.46	5.45
Ettringite	$Ca_6Al_2(SO_4)_3(OH)_{12} \cdot 26H_2O$	3.06	5.68	3.82	7.15	13.83	6.71
Hydrocalumite	$Ca_4Al_2(OH)_{12}(Cl,CO_3,OH)_2 \cdot 4H_2O$	3.16	5.68	0.93	4.62	5.25	3.93
Strätlingite	$Ca_2Al_2SiO_7 \cdot 8H_2O$	1.30					1.30
Pyroxenes	$ABSi_2O_6$			1.59			1.59
Vaterite	$CaCO_3$	2.51	0.83		1.45	1.70	1.62
Gehlenite	$Ca_2Al(AlSiO_7)$	2.02	1.93	1.87			1.94
Akermanite	$Ca_2Mg(Si_2O_7)$				3.90	5.28	4.59
Cristobalite	SiO_2			0.51			0.51
Larnite	Ca_2SiO_4				5.52	8.31	6.92
Portlandite	$Ca(OH)_2$				4.40	5.33	4.87
Hematite	Fe_2O_3	0.81	0.88	0.41	1.09	1.25	0.89
Magnetite	$Fe^{2+}Fe^{3+}{}_2O_4$	0.13	0.28	0.31	0.61	0.67	0.40
Crystalline		27.25	27.82	25.79	66.79	73.52	44.43
Amorphous		71.75	72.18	74.21	33.21	26.48	55.57

"aggregates." In the RED cluster, moreover, this bibliographic search on BA resource recovery identifies relevant resources from BA such as iron, alumina, glass tiles, calcite, salts, and aggregates, stressing the primary applications in cements, ceramics, and as additives in binders and replacements of different raw materials (e.g., NaCl recovery for road deicing).

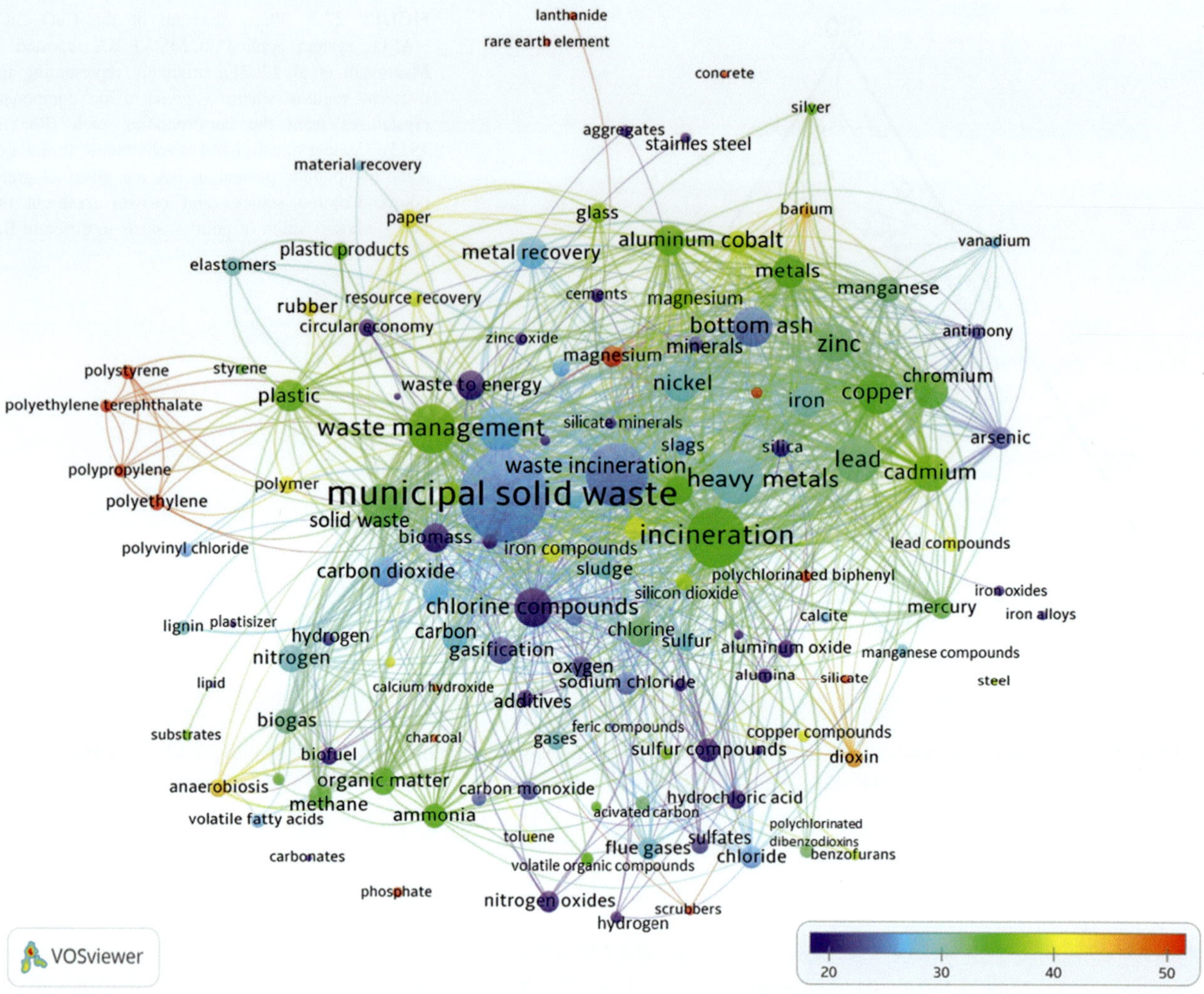

FIGURE 27.4 Average citation map based on resources recovery in MSWI BA (in published literature from 2010 to 2023). The scale shows the earliest (*blue*) and more recent (*red*) years of publication. The circles with small to big label sizes represent the terms appeared in the number of publications.

The GREEN cluster relates soft resource recovery, including thermoplastic materials such as "polyvinyl chloride," "polypropene," "polystyrene," "styrenes," other synthetic substances ("polymers," "plasticizer"), and "lignin." Besides different kinds of polymers and raw materials such as paper, cardboard and paperboard, cloth strips, plastics and synthetic fibers, bone chips, noncombustible organic material, and organic waste, the GREN cluster shows "lanthanide," "REE," and minerals such as quartz, K-feldspar, plagioclase, and biotite (Bayuseno and Schmahl, 2010; Wei et al., 2011; Inkaew et al., 2016; Youcai, 2017). The GREEN cluster suggests attention in the recovery of metal and mineral commodities parallels that in new polymeric resources. The BLUE cluster comprises terms strictly linked to the "metal resource recovery" than minerals and raw materials recovery, including Zn, Cu, Ba, Mg, Fe, and Al, and many "potentially toxic elements" such as Cd, Ni, Cr, Pb, As, Hg, Sb, and Co, implying that resource recovery can carry remediation purposes (Fig. 27.2). The recoverability of metals from different BA fractions in different countries is primarily reported for Zhejiang province, East China (Yao et al., 2013, 2017), Southwestern Germany (Abramov et al., 2018), Northern Italy (Funari et al., 2016), Iran (Nikravan et al., 2020), and Switzerland (Mehr et al., 2021). YELLOW cluster is energy-related and pollution-related keyword, showing the terms "sulfur dioxide (SO_2)," "nitrogen oxides (NOx)," "carbon monoxide (CO)," "flue gases," "volatile fatty acids," "toluene," "dioxins," "hydrogen," "ammonia," "charcoal," "substrates," and "phosphates." These terms likely relate to energy pollution issues and health risks (Kim et al., 2005; Oh et al., 2006; Walton, 2010). The high occurrence of the different keywords in the four clusters indicates the search for critical raw materials including organic compounds for industry is increasing with time by recycling urban waste streams such as MSWI BA. In Fig. 27.4, it is possible to notice that the trends of publication indicate the areas of fast development or increased research interest. The first can be the case of thermoplastic polymers and persistent pollutants (PCBs), the latter can link to the relatively recent interest in new concrete formulations and critical raw materials including REE and phosphates for the development of new and green technologies (EC, 2017).

4. Utilized treatments, transfer-ready applications, and limitations

Profoundly different technologies exist, e.g., based on the suitability at treating BA produced with or without quenching steps, which are technically demanding and risky for wholesale scale (Šyc et al., 2020). The full-scale pilot of the ABB InRec process worked between 1995 and 1996 at the GEVAG MSWI plant (Trimmis, Graubünden Canton, Switzerland), but the dry discharge technology raised limited interest among the relevant stakeholders despite the high levels of purity of recovered fractions. Dry technologies tailored for wet discharged MSWI BA are most common since most MSWI discharge systems are based on quenching stages. However, dry discharge is experiencing a renaissance tied to its ability of avoiding or minimizing detrimental reaction by-products, such as $Ca(OH)_2$, $CaCO_3$, Friedel's salt, and hydrocalumite (Inkaew et al., 2016).

As a rule, primary options for resource recovery from MSWI BA are physical–mechanical treatments irrespective of the BA relying on dry or wet discharge. This is because the distribution of elements into different grain size or density fractions varies, so material upgrading can be envisaged favoring the marketability of added-value streams of material or their acceptance at smelters at lower costs. The recovery of ferrous metal (FeF) and nonferrous metal (n-FeF) fractions is widespread and usually accomplished using Eddy current separators and other processing stages such as those based on density separation technology. Earlier data show that easily recyclable FeF and n-FeF averaged 63 and 17 kg per ton of treated BA, respectively, while the latest installations can produce almost doubled recovery of n-FeF (Simon and Holm, 2017). Commercial processing technologies can be divided into dry and wet technology. Wet technologies imply liquid media in traditional separators (e.g., wet Eddy current), but it is important to highlight that massively ground on solvent extraction methods and processes from soil cleaning and remediation. Dry technologies can render more efficient physical–mechanical processing stages such as Eddy current separators compared with wet technologies. Other advantages are savings in water consumption and, to some extent, reduced transport costs (due to reduced weight and volume). In contrast, the main drawback of dry technology is excessive dust formation, which can require costly, advanced devices to detect nanoparticles for instance. In advanced management and treatment plants, up to 6–9 fractions sorted by grain size can undergo metal recovery. However, sieving may be expensive if the water content is lower than 10% because appropriate dust control during the handling of the material must be assured. Flotation and density separation showed some limitations for BA treatment. For example, density separation does not apply for Al recovery because its density resembles that of bulk MSWI BA (2700 kg/m^3). Density separation can effectively recover different components such as copper, gold, and brass showing a significant density contrast compared with MSWI BA matrix. Multistep magnetic separation is typically employed for removing the magnetic fraction (iron oxides and magnetically susceptive agglomerates) also from presieved fractions in advanced treatment plants before the Eddy current separation stage (Smith et al., 2019). Eddy currents can divide FeF and n-FeF efficiently. Such devices require a proper calibration, based on the size fractions to be treated. Eddy current separators tailored for fine particles have been marketed in recent years, while other prototypes based on Eddy currents never reached the full scale (e.g., wet, backward operating, and Magnus Eddy current separators). In advanced dry recovery processes (e.g., supersortfine pss, DHZ, Switzerland), ballistic separation is a cutting-edge technology, developed in a cooperation between Inashco company and TU Delft (patent WO 2009/123452 A1), which mechanically separates the fine particles (<2 mm) associated with the highest moisture content (i.e., those causing the grain to stick together). A ballistic separator can couple with conventional dry separation processes, improving performances in nonferrous metals recovery. The sensor-based separation for glass recovery went industrial scale with the company Binder + Co AG, Austria, and its application requires a pretreatment (similar to ballistic separation) named cullet sublimation to remove adherent dust particles and free paper labels, which typically restrict the sensitivity of optical detectors. These treatment trains can separate the metallic fractions effectively, while optical separators sort out glassy objects with recovery rate up to 75% (Makari, 2014). An induction sorting system is operational in Denmark (Afatek) since 2015, for separation of stainless steel particles (Kallesøe and Dyhr-Jensen, 2018). Today, the Danish facility includes processing stages such as aging, sieving (into seven particle size fractions: 0.5–1, 1–2, 2–4, 4–9, 9–18, and 18–50 mm), sorting for removal of magnetic metals and n-FeF, and sensor-based separators. The outputs from the nonferrous sorting plant include stainless steel (9–50 mm), aluminum (0.5–50 mm), and a nonferrous heavy-density fraction (0.5–50 mm).

The following MSWI and treatment systems are schematized in Fig. 27.5A. In Switzerland, a centralized treatment plant was specifically devised to recover metals from BA. Here, after a first stage of magnetic separation to remove the

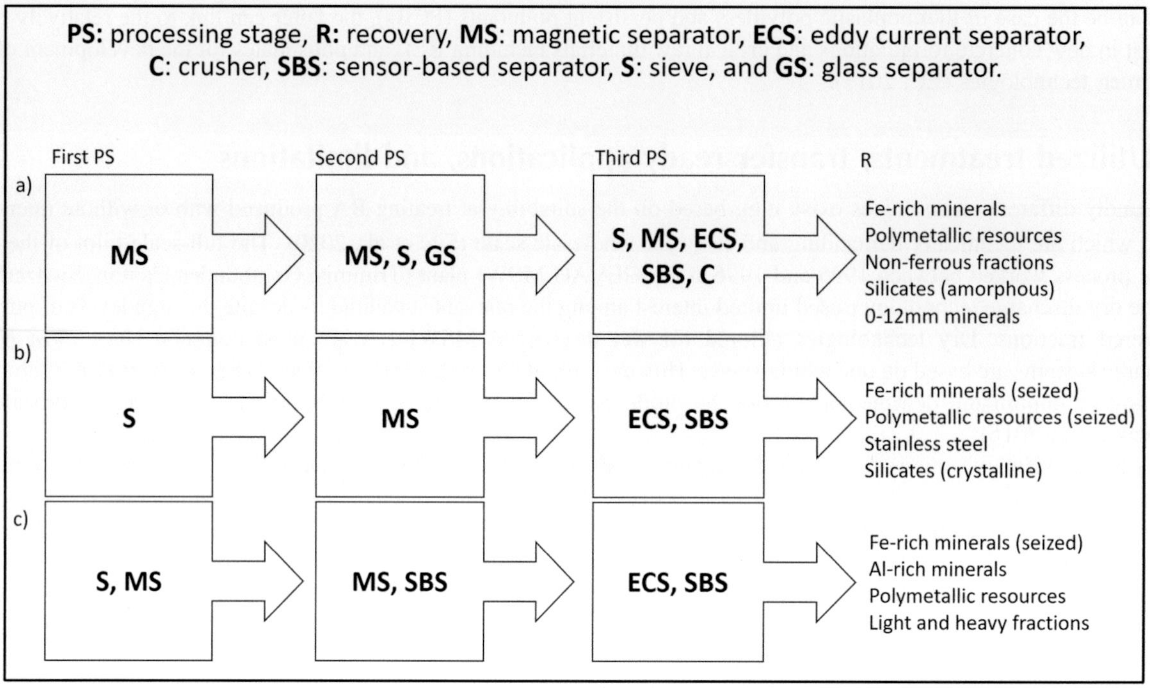

FIGURE 27.5 Scheme of multistep technology for resource recovery commercially available and implemented in (A) the Hinwii MSWI plant; (B) the Brantner or Heros MSWI plant; (C) the Alkmaar MSWI plant, providing different stream fractions at the end of the processing chain. *Readapted after Šyc, M., Simon, F.G., Hykš, J., Braga, R., Biganzoli, L., Costa, G., Funari, V., Grosso, M., 2020. Metal recovery from incineration bottom ash: state-of-the-art and recent developments. J. Hazard Mater. 393, 122433.*

metal scraps (>80 mm), the BA is sorted in four streams of predetermined grain size. The fraction >80 mm undergoes a handpicking station and then grounding to join the treatment trains of finer fractions. Each separated stream of a grain size lower than 80 mm undergoes complicated lines of physical separation (magnetic, Eddy current, and density separators) with eventual stepwise crushing of residual materials after every upgrading stages. This is one example of dry treatment of dry discharged MSWI BA that yields around 10% FeF, 4.5% n-FeF, and 1.1% glass, generating a significant revenue with a total consumption of about 16 kWh per ton of treated waste (Böni and Morf, 2018). The recovery rate increases with the number of recovery devices, for example, sometimes more than 10 Eddy current separators are deployed, influencing the overall costs. According to the technical report dated 2018 of European Integrated Pollution Prevention and Control Bureau (IPPC), the average electricity consumption for current treatment plants is 3 kWh per ton of treated waste, sometimes reaching up to 15 kWh when treatments guarantee other revenues. Full-scale dry treatment plants (seven plants in Europe) include a dry ram discharger and a wind sifter for dust removal, a vibrating conveyor provided with inlet air that can promote afterburning of organic components reducing the organic carbon levels (<0.3%). However, the numerous stages of crushing eventually lead to abundant dust formation and unfavorable grain size distribution curve for residue's reuse in the production of building materials. There is narrow line between revenue and costs in innovative installations. For instance, extensive crushing of BA can expose well mineral surfaces, and subsequently, different finely grained fractions may become more suited to metal recovery or produce streams of different BA granulometries that display adequate pozzolanic properties in cementitious mixtures. On the other hand, wet technologies for BA treatment are widely deployed, but they imply a massive use of water as a primary limiting factor. The first pilot for metal recovery came in 2005 in Amsterdam city, the Netherlands, equipped with several wet separators, for the recovery of inert granulates for building materials and marketable metal fractions of different levels of purity (Muchova et al., 2009). Although a recovery efficiency is up to 83% FeF and 73% n-FeF, the plant never went to full scale mainly due to the high water demand and costs for wastewater treatment. The Brantner & Co. plant built in 2013 in Austria (Fig. 27.5B) counts on two-step magnetic separation, separating iron scraps, and fine (>50 mm) and large (<50 mm) fractions, and a wet jig that further separates material streams by density difference (Stockinger, 2018). A floating fraction of plastics and other carbon-based materials, the heavy (density <4000 kg/m^3) n-FeF containing stainless steel, copper, brass, and precious metals, and the light n-FeF mainly composed of Al-bearing materials, and the finest (less than 2 mm) fraction are recovered. All these fractions after dewatering (by cyclones) go to smelters at virtually no costs or specialized, external facilities that pay streams of materials, with substantial savings for the MSWI plant's company. For example, the slag treatment plant of the SEPRO Company adopts other density and magnetic separators for the efficient recovery of the heavy n-FeF, including precious metals. The treatment of BA by Indaver process, developed in the Netherlands, relies on screening, washing, and separation of granular and metallic streams, optimizing removals of organic matter and soluble salts through a washing process (Vandecasteele et al., 2007). One wet technology, installed in 2016 in Alkmaar, the Netherlands, and firstly developed by the Boskalis Company in response to the Netherlands' Green Deal, is tailored to improve the leaching behavior of BA residues for safe and economically sound reuse. This technology (Fig. 27.5C) separates different fractions using dry sieving or a wet drum sieve and then washes out soluble salts and metals from each fraction. A bar sizer separates fine (>40 mm) and large (<40 mm) particles followed by magnetic separation for removing iron scrap and stainless steel as a first marketable stream. The fraction of fine particles undergoes a wet drum sieve and a vibrating screen to divide the following fractions: 20–40, 8–20, 4–8, and >4 mm. Particles larger than 20 mm are crushed and put back into the cycle. Then, wet magnetic and Eddy current separators divide FeF from n-FeF. The heavy n-FeF that includes precious metals is treated by density separation for metal recovery, reaching up to 50 mg/kg Au and 900 mg/kg Ag from particles smaller than 4 mm. Only the finest fraction from the different streams is residual and undergoes a washing treatment. The water consumption is below 0.5 m^3 per t of BA treated. This technology is found to increase the efficiency of n-FeF from 2.6% to 3.5% compared with dry technology, and an additional 0.3% of the heavy n-FeF is obtained as value-added, resulting in a total resource recovery of 11.85% of the BA input (Born, 2018). However, according to the mass balance of the Alkmaar MSWI plant, the main drawback is the production of large amounts of sludge with a high concentration of potentially toxic elements that require advanced treatments. Another example reported in Šyc et al. (2020) is the Heros plant also in the Netherlands. This advanced management and treatment plant includes aging and the separation of Fe scraps and other nine classes of grain size after different stages of grounding and sieving, with fractions sold as sustainable secondary aggregates (Granova). An Eddy current separator for each grain sizes allows efficient recovery of n-FeF, while the stainless steel is separated by handpicking and a final water washing removes chlorides, sulfates, and other metals to comply with the Netherlands' Green Deal requirements (e.g., AEB, 2015). Similar technology is available in Germany and Italy, where aging is eventually avoided.

From an economic point of view, the high investment required for advanced treatment plants steams from the particularly demanding crushing stages, the presence of multistep magnetic separation, and sensor-based sorting systems, which aim to provide enhanced recovery, especially from n-FeF. Metal recovery can take place on-site, preferably at large MSWI plants where the flow of residues can justify the investment leveraging on transportation costs, while centralized or mobile treatment plants serving several MSWI plants usually demonstrate lower efficiency than on-site plants (Kallesøe, 2014; Kallesøe and Dyhr-Jensen, 2018; Šyc et al., 2020). Although aging stages are likely detrimental for metal recovery due to the formation of mineral coatings, latest pilots aim at improving economic and environmental impacts of BA management using accelerated carbonation exploiting the exhaust lines of the MSWI plant itself. For mineral beneficiation and metal recovery, classical, hydrometallurgical methods coparticipated by biological machinery are being evaluated such as bioleaching at critical stages of BA processing (Gomes et al., 2020). At present, most MSWI plants lack an efficient recovery of the heavy n-FeF, which can increase the revenues due to its precious metals content, and several fine-grained (from <0.3 mm to lower granulometries) streams are sold at a likely depreciated value. Also, in dry treatment plants with demonstrated recoveries of high-quality metals (e.g., Holm and Simon, 2017), a residual fraction of particles <2 mm is not treated.

5. The case of Italian flows and future perspectives

New options to improve management of MSWI residues are wanted, especially those capable of metal recovery and quality enhancement of the posttreatment residue. Urban mining attempts from of MSWI residues demonstrated potential for application in integrated waste management to simultaneously increase revenues and reduce environmental impacts (Fellner et al., 2015). In 2020, the 37 Italian MSWI plants produced about $1.7 \cdot 10^5$ t of hazardous waste (EER 190111*, 190113*, and 190115*, respectively) and $1.1 \cdot 10^6$ t nonhazardous waste categories (EER 190112, 190114, and 190116, respectively). The year before, almost all nonhazardous BAs were recycled or sent to recovery treatment, but around 10% landfilled. Third parties, often from other countries, who developed suitable technologies, collected and conducted recovery of about 90% hazardous BA produced in Italy.

We take into account the Italian use case collecting chemical/mineralogical data and general information from five metropolitan incinerators serving relatively large cities of Parma, Ferrara, Piacenza, Torino, and Forlì Cesena and their surroundings in northern Italy, to provide an estimate of available resources from national MSWI BA in terms of

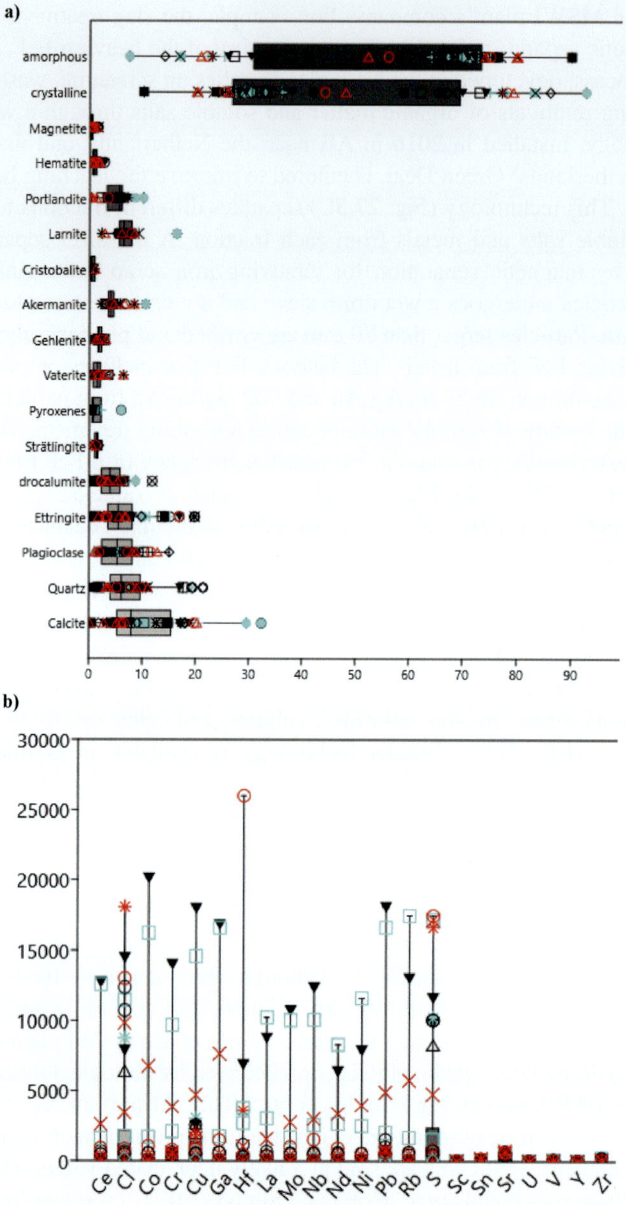

FIGURE 27.6 Box and jitter plot of (A) mineralogical composition and (B) chemical concentration of selected minor and trace elements of five N-Italian MSWI plants. Samples' relative concentration of the jitter used the following symbols as a function of grain size: dot (*light blue*) < 0.063 mm; plus (*light blue*) from 0.063 to 0.2 mm; square from 0.2 to 0.3 mm; diamond from 0.3 to 0.5 mm; fill. inverted triangle from 0.5 to 1 mm; star from 1 to 2 mm; cross from 2 to 4 mm; fill. square from 4 to 8 mm; triangle (*red*) from 8 to 16 mm; open dot (*red*) > 16 mm. In (b), open square (*light blue*) < 0.063 mm; star (*light blue*) from 0.063 to 0.2 mm; star (*red*) from 8 to 16 mm; open dot (*black*) from 2 to 4 mm; and cross (*red*) from 4 to 8 mm were used to improve readability.

secondary, nonenergy raw materials (i.e., minerals and metals). Fig. 27.6 shows the relative abundance of the mineralogical phases identified (Fig. 27.6A) and some elements of the Periodic Table (Fig. 27.6B) as the five-incinerator average of representative BA samples. Regarding the mineralogical composition, calcite, quartz, plagioclase, ettringite, larnite, portlandite, and akermanite are more abundant than other minerals, and the majority of these phases are primary constituents of glass, ceramics, and concrete. It is interesting to note that grain size plays a role in the relative enrichment of expected resources. At looking the distribution of the minerals as a function of the grain size, minerals such as portlandite, larnite, pyroxene, and calcite are more likely to occur in finer fractions than coarser ones, and uneven distribution for other mineralogical phases is apparent (Fig. 27.6A). About the chemical composition of BA samples from the five MSWI plants, Si, Al, Ca, Fe, and Mg are most abundant chemical elements, as well ascertained in previous studies always in consistency with the mineralogy.

Chemical anomalies arise in minor and trace elements composition more than that of major elements. While Sc, Sn, Sr, U, V, Y, and Zr are less variable with no outliers, likely meaning that they are consistent to geogenic samples, other elements display significant variation (Fig. 27.6B), suggesting enrichment eventually due to nugget effects. For examples, Co and Hf spikes might represent a good resource in some circumstances, and on the box plots, Cl, S, Cu, Cr, Pb, Sr, Ba, and Zn are enriched compared with the other minor and trace elements. The grain size has a slight control over the elemental abundance. Volatile elements such as Cl, S, and Hf preferably partition into the averaged ($n = 138$), coarse sized BA, as well as Zr. Ba, Co, Ga, Sn, Pb, and the lanthanides, Ce, La, and Nd, seem to occur at significant levels in the finer grain sizes.

The amount of recoverable resource from anthropogenic flows such as MSWI BA strongly depends on the output rates of the industrial waste flow (e.g., Funari et al., 2015). For instance, Torino and Forlì Cesena MSWI plants likely host the highest quantities of available calcite, quartz, ettringite, and plagioclase for recovery, in the light of their higher BA output compared with other plants. Therefore, the annual flow of substance (expressed in mass/time) for each incinerator can be calculated by multiplying the measured concentration of each element (mass/mass) and the total output (mass/time) of BA produced (Table 27.4). According to the substance flow analysis, S, Cl, Zn, Cu, Pb, Ba, Cr, Sr, Zr, and Ni concentrations (in decreasing order) are in the range of magnitude between 10^3 and 10^4 kg per year. Potential recovery of Co, Ce, V, and Sn can be marginally economic for the largest incinerators of Torino and Forlì Cesena due to their comparably high treatment capacity (and subsequent outputs). Data from subsamples such as grain size fractions or other types of separates are relevant to identify existing or potential fraction streams that can make the recovery significantly easier.

Substance flow analysis specifies that S, Cl, Zn, Cu, and Ba flows likely enrich in the finer fractions of BA with a potential recovery gain between 30% and 90% compared with the bulk sample, while Si, Al, Fe, Y, and Zr enrich in the coarse BA with no more than 40% concentration enhancement.

Certainly, the increased awareness of MSWI BA environmental relevance and exploitation potential coupled with availability of national use cases can ameliorate resource management downstream production chains in the next future. According to the Italian use case and its estimated flow of strategic substance, it can be said that the presence of metals and mineral resources depends on several variables, which can be observed in fact and used with profit. From the quantitative standpoint, more resource otherwise lost can be recovered with technically stringent and affordable arguments, especially if the options of landfilling and recycling into relatively unvalued and scarcely environment-attentive applications are dismissed. In the context of innovation, in particular, the contribution of multidisciplinary studies to MSWI BA has a high potential to provide new processes and strategies, e.g., advanced mineral beneficiation and new hybrid technologies for metal recovery, to tighten the gap between consumption and supply of earth's resources. A demonstration is the accelerated mineral carbonation as a straight countermeasure to carbon dioxide emission or biorecovery coupled with secondary production of fuel. Summarizing, it seems from the bibliographic analysis in this chapter that new data coverage, although not available for the range of MSWI BA produced by different incineration technologies, is adding new perspectives in secondary resource recovery for the global, circular economy. Although the ineluctable flaws of the query construction, the bibliometric analysis is another strategic tool that provided a full picture of the past 30 years of scientific experience in the study of MSWI BA. Experience says that more control over the input waste composition is essential to provide the resulting residues with enhanced recoverability, unless new detection and separation technologies are made available. As such, bibliometric analysis and the analysis of citation trends allow forecasting new direction of research and possibilities, with some example deriving from terms at the edge of the cloud in Fig. 27.4. Combining company factsheets, geochemical and mineralogical quantitation, and bibliography trend assessment is the flywheel for a critical evaluation of MSWI BA impacts onto the environment and society.

TABLE 27.4 Substance flow analysis (SFA) of selected major, minor, and trace elements of the Periodic Table in kg/year from five incinerators in northern Italy.

Parma							
Year built-in	Waste type treated	Feed rate (tons/year)	Waste type input (%)	Total input (tons/year)	BA generated (tons/year)	Owner last report	Sampling
2014	Undifferentiated municipal solid waste	126,317	79	159,832	32,904	2019	2019
	Special wastes	33,515	21				

	MEAN (kg/year)	RSD (%)	MIN (kg/year)	MAX (kg/year)	Parma SFA		
Si	1.12E+06	24%	7.80E+05	1.52E+06			
Ti	2.78E+04	13%	2.11E+04	3.20E+04			
Al	2.96E+05	9%	2.72E+05	3.42E+05			
Fe	1.05E+05	19%	7.28E+04	1.33E+05			
Mg	1.22E+05	12%	1.03E+05	1.46E+05			
Ca	8.63E+05	12%	7.11E+05	9.91E+05			
Hf	17	39%	10	23			
Nb	36	7%	33	40			
Nd	47	49%	7	73			
Mo	45	15%	33	57			
Th	26	29%	18	44			
Y	48	18%	37	63			
Ga	44	4%	40	47			
Sc	39	36%	18	59			
La	61	50%	18	115			
Rb	96	11%	80	111			
Co	135	32%	82	203			
Ce	137	16%	95	167			
V	255	11%	200	295			
As	165	27%	87	242			
Sn	141	52%	57	299			
Ni	494	15%	391	601			
Zr	610	18%	470	844			
Sr	1690	17%	1242	2240			
Cr	2089	17%	1429	2895			
Ba	4753	23%	2985	6182			
Pb	2621	41%	1027	4455			
Cu	5149	18%	3631	6715			
Zn	15,277	49%	4541	28,758			

TABLE 27.4 **Substance flow analysis (SFA) of selected major, minor, and trace elements of the Periodic Table in kg/year from five incinerators in northern Italy.—cont'd**

	MEAN (kg/year)	RSD (%)	MIN (kg/year)	MAX (kg/year)	Parma SFA		
Cl	28,809	25%	17,867	38,169			
S	40,979	30%	25,599	58,372			
				Piacenza			
Year built-in	**Waste type treated**	**Feed rate (tons/year)**	**Waste type input (%)**	**Total input (tons/year)**	**BA generated (tons/year)**	**Owner last report**	**Sampling**
2002	Undifferentiated municipal solid waste	110,041	96	114,231	21,135	2018	2019
	Special wastes	4190	4				
	MEAN (kg/year)	**RSD (%)**	**MIN (kg/year)**	**MAX (kg/year)**	**Piacenza SFA**		
Si	9.21E+05	27%	6.68E+05	1.38E+06			
Ti	3.46E+04	12%	2.52E+04	3.84E+04			
Al	4.36E+05	22%	2.80E+05	5.46E+05			
Fe	1.75E+05	17%	1.42E+05	2.20E+05			
Mg	1.17E+05	13%	1.00E+05	1.44E+05			
Ca	7.78E+05	20%	5.12E+05	9.43E+05			
Hf	11	65%	6	15			
Nb	23	8%	20	27			
Nd	24	50%	6	40			
Mo	40	17%	26	47			
Th	25	17%	21	33			
Y	26	38%	15	38			
Ga	32	7%	27	34			
Sc	38	39%	15	60			
La	55	24%	36	78			
Rb	66	10%	54	73			
Co	65	24%	48	90			
Ce	117	21%	73	157			
V	184	10%	146	207			
As	168	28%	80	228			
Sn	236	38%	88	375			
Ni	292	21%	224	408			
Zr	359	32%	245	531			
Sr	902	12%	666	961			
Cr	1011	10%	888	1141			

Continued

TABLE 27.4 Substance flow analysis (SFA) of selected major, minor, and trace elements of the Periodic Table in kg/year from five incinerators in northern Italy.—cont'd

	MEAN (kg/year)	RSD (%)	MIN (kg/year)	MAX (kg/year)	Piacenza SFA		
Ba	3648	24%	2395	5310			
Pb	2793	35%	1057	4248			
Cu	5568	8%	4755	6045			
Zn	8557	40%	6045	16,147			
Cl	31,539	39%	9447	44,510			
S	20,377	35%	7757	28,363			

Torino

Year built-in	Waste type treated	Feed rate (tons/year)	Waste type input (%)	Total input (tons/year)	BA generated (tons/year)	Owner last report	Sampling
2014	Undifferentiated municipal solid waste	457,603	81	562,269	118,969	2019	2019
	Special wastes	104,666	19				

	MEAN (kg/year)	RSD (%)	MIN (kg/year)	MAX (kg/year)	Torino SFA		
Si	1.12E+06	25%	6.78E+05	1.49E+06			
Ti	2.99E+04	26%	1.38E+04	3.75E+04			
Al	3.08E+05	18%	2.18E+05	3.91E+05			
Fe	2.44E+05	35%	1.55E+05	4.27E+05			
Mg	1.38E+05	35%	6.29E+04	2.40E+05			
Ca	6.39E+05	30%	2.32E+05	8.74E+05			
Hf	53	21%	43	71			
Nb	121	12%	96	136			
Nd	122	43%	59	214			
Mo	91	19%	60	117			
Th	101	62%	53	226			
Y	205	25%	131	297			
Ga	160	6%	141	170			
Sc	200	44%	93	339			
La	287	41%	71	483			
Rb	378	17%	268	452			
Co	517	43%	310	1031			
Ce	629	30%	444	1109			
V	970	13%	702	1118			
As	466	63%	155	1190			
Sn	800	64%	124	1901			
Ni	1475	47%	666	3212			

TABLE 27.4 **Substance flow analysis (SFA) of selected major, minor, and trace elements of the Periodic Table in kg/year from five incinerators in northern Italy.—cont'd**

	MEAN (kg/year)	RSD (%)	MIN (kg/year)	MAX (kg/year)	Torino SFA		
Zr	1875	25%	1563	3079			
Sr	4122	34%	2093	6736			
Cr	6769	15%	5829	9042			
Ba	7700	27%	3784	9531			
Pb	7376	78%	1309	21,176			
Cu	12,028	59%	2498	21,890			
Zn	34,049	63%	9161	71,143			
Cl	109,761	48%	34,263	194,633			
S	150,567	49%	25,935	242,935			

Ferrara

Year built-in	Waste type treated	Feed rate (tons/year)	Waste type input (%)	Total input (tons/year)	BA generated (tons/year)	Owner last report	Sampling
2008	Undifferentiated municipal solid waste	69,598	53	131,895	27,021	2020	2020
	Special wastes	62,296	47				

	MEAN (kg/year)	RSD (%)	MIN (kg/year)	MAX (kg/year)	Ferrara SFA		
Si	1.11E+06	28%	5.93E+05	1.60E+06			
Ti	2.56E+04	15%	1.92E+04	3.18E+04			
Al	2.53E+05	11%	2.08E+05	3.15E+05			
Fe	1.36E+05	25%	7.72E+04	1.76E+05			
Mg	1.32E+05	16%	1.05E+05	1.60E+05			
Ca	1.04E+06	21%	7.00E+05	1.46E+06			
Hf	12	25%	8	15			
Nb	26	10%	23	31			
Nd	20	37%	11	32			
Mo	20	19%	13	28			
Th	14	34%	8	24			
Y	38	32%	19	54			
Ga	35	7%	31	38			
Sc	42	34%	15	65			
La	56	49%	14	96			
Rb	83	21%	59	124			
Co	56	21%	39	74			
Ce	125	30%	39	171			
V	169	8%	154	195			

Continued

TABLE 27.4 Substance flow analysis (SFA) of selected major, minor, and trace elements of the Periodic Table in kg/year from five incinerators in northern Italy.—cont'd

	MEAN (kg/year)	RSD (%)	MIN (kg/year)	MAX (kg/year)	Ferrara SFA		
As	114	35%	46	159			
Sn	207	54%	63	439			
Ni	217	21%	138	303			
Zr	441	28%	300	710			
Sr	1304	15%	952	1629			
Cr	1211	18%	811	1432			
Ba	1683	25%	1314	2742			
Pb	2018	51%	594	3999			
Cu	4177	28%	1891	5647			
Zn	9604	64%	2270	20,693			
Cl	35,868	34%	16,618	54,393			
S	28,580	42%	11,430	44,909			

Forlì				Cesena			
Year built-in	Waste type treated	Feed rate (tons/year)	Waste type input (%)	Total input (tons/year)	BA generated (tons/year)	Owner last report	Sampling
2008	Undifferentiated municipal solid waste	119,215	100	119,215	53,308	2020	2020

	MEAN (kg/year)	RSD (%)	MIN (kg/year)	MAX (kg/year)	Forlì Cesena SFA		
Si	1.07E+06	37%	5.53E+05	1.86E+06			
Ti	2.27E+04	19%	1.30E+04	2.66E+04			
Al	2.53E+05	10%	1.91E+05	2.80E+05			
Fe	1.75E+05	30%	9.46E+04	2.32E+05			
Mg	1.18E+05	17%	9.39E+04	1.53E+05			
Ca	9.76E+05	24%	5.12E+05	1.31E+06			
Hf	29	49%	16	49			
Nb	51	8%	45	56			
Nd	46	51%	16	80			
Mo	44	44%	31	95			
Th	128	212%	17	742			
Y	67	32%	37	96			
Ga	85	61%	59	223			
Sc	70	44%	32	130			
La	130	35%	64	197			
Rb	169	50%	110	393			
Co	192	24%	108	248			

TABLE 27.4 Substance flow analysis (SFA) of selected major, minor, and trace elements of the Periodic Table in kg/year from five incinerators in northern Italy.—cont'd

	MEAN (kg/year)	RSD (%)	MIN (kg/year)	MAX (kg/year)	Forlì Cesena SFA
Ce	312	69%	137	847	
V	312	12%	256	368	
As	898	221%	155	6184	
Sn	588	49%	202	1135	
Ni	445	17%	373	597	
Zr	836	26%	566	1153	
Sr	2764	21%	2030	3857	
Cr	4389	127%	2239	19,191	
Ba	2728	21%	1591	3544	
Pb	19,197	233%	1866	138,601	
Cu	12,101	24%	8263	16,206	
Zn	18,510	45%	5544	30,812	
Cl	55,630	35%	34,170	96,381	
S	62,412	39%	20,257	93,236	

Acknowledgments

The authors acknowledge the kind support of Iren S.p.A. and HERAmbiente S.p.A., who provided access and assistance in sampling. V.F. acknowledges the administrative support of the national research council of Italy and the financial support from PNRR MUR project ECS_00000033_ECOSISTER.

References

Abis, M., Bruno, M., Kuchta, K., Simon, F.G., Grönholm, R., Hoppe, M., Fiore, S., 2020. Assessment of the synergy between recycling and thermal treatments in municipal solid waste management in Europe. Energies 13, 6412.

Abramov, S., He, J., Wimmer, D., Lemloh, M.L., Muehe, E.M., Gann, B., Roehm, E., Kirchhof, R., Babechuk, M.G., Schoenberg, R., 2018. Heavy metal mobility and valuable contents of processed municipal solid waste incineration residues from Southwestern Germany. Waste Manag. 79, 735–743.

AEB, 2015. Green Deal Bottom Ash Programme a Success. https://www.aebamsterdam.com/about/news/2015/green-deal-bottom-ash-programme-asuccess.

Alorro, R.D., Hiroyoshi, N., Ito, M., Tsunekawa, M.J.H., 2009. Recovery of heavy metals from MSW molten fly ash by CIP method. Hydrometallurgy 97, 8–14.

Aricks, S., Van Gerven, T., Vandecasteele, C., 2006. Accelerated carbonation for treatment of MSWI bottom ash. J. Hazard. Mater. 137 (1), 235–243.

Back, S., Sakanakura, H., 2021. Distribution of recoverable metal resources and harmful elements depending on particle size and density in municipal solid waste incineration bottom ash from dry discharge system. Waste Manag. 126, 652–663.

Back, S., Sakanakura, H., 2022. Comparison of the efficiency of metal recovery from wet-and dry-discharged municipal solid waste incineration bottom ash by air table sorting and milling. Waste Manag. 154, 113–125.

Bayuseno, A.P., Schmahl, W.W., 2010. Understanding the chemical and mineralogical properties of the inorganic portion of MSWI bottom ash. Waste Manag. 30 (8–9), 1509–1520.

Blasenbauer, D., Huber, F., Lederer, J., Quina, M.J., Blanc-Biscarat, D., Bogush, A., Bontempi, E., Blondeau, J., Chimenos, J.M., Dahlbo, H., 2020. Legal situation and current practice of waste incineration bottom ash utilisation in Europe. Waste Manag. 102, 868–883.

Blasenbauer, D., Huber, F., Mühl, J., Fellner, J., Lederer, J., 2023. Comparing the quantity and quality of glass, metals, and minerals present in waste incineration bottom ashes from a fluidized bed and a grate incinerator. Waste Manag. 161, 142–155.

Böni, D., Morf, L.S., 2018. Thermo-recycling: efficient recovery of valuable materials from dry bottom ash. In: Holm, O., Thomé-Kozmiensky, E. (Eds.), Removal, Treatment and Utilisation of Waste Incineration Bottom Ash. Thomé-Kozmiensky Verlag GmbH, pp. 25–37.

Born, J.P., 2018. Mining incinerator bottom ash for heavy non-ferrous metals and precious metal. In: Removal, treatment and utilisation of waste incineration bottom ash. TK Verlag, Neuruppin, pp. 11–24.

Bunge, R.J.R., Holm, O., Thome-Kozmiensky, E. (Eds.), 2015. Recovery of Metals from Waste Incinerator Bottom Ash, pp. 63–143. Treatment.

CEWEP, 2017. Bottom Ash Fact Sheet 19–20. https://www.cewep.eu/wp-content/uploads/2017/09/FINAL-Bottom-Ash-factsheet.pdf.

Chandler, A.J., Eighmy, T.T., Hjelmar, O., Kosson, D., Sawell, S., Vehlow, J., Van der Sloot, H., Hartlén, J., 1997. Municipal Solid Waste Incinerator Residues. Elsevier.

Chang, Y.M., Hung, C.Y., Chen, J.H., Chang, C.T., Chen, C.H., 2009. Minimum feeding rate of activated carbon to control dioxin emissions from a large-scale municipal solid waste incinerator. J. Hazard Mater. 161, 1436–1443.

Chuang, K.H., Lu, C.H., Chen, J.C., Wey, M.Y., 2018. Reuse of bottom ash and fly ash from mechanical-bed and fluidized-bed municipal incinerators in manufacturing lightweight aggregates. Ceram. Int. 44, 12691–12696.

De Matteis, C., Mantovani, L., Tribaudino, M., Bernasconi, A., Destafanis, E., Caviglia, C., Toller, S., Dinelli, E., Funari, V., 2023. Sequential extraction procedure of municipal solid waste incineration (MSWI) bottom ashes targeting grain size and the amorphous fraction. FENVS 11, 1–10. https://doi.org/10.3389/fenvs.2023.1254205.

EC, 2017. List of critical raw materials for the EU. European Commission, Brussels, Belgium.

Enzner, V., Holm, O., Abis, M., Kuchta, K., 2017. The characterisation of the fine fraction of MSWI bottom ashes for the pollution and resource potential. In: Sixteenth International Waste Management and Landfill Symposium. Cisa Publisher, pp. 1–9.

Eusden, J.D., Eighmy, T.T., Hockert, K., Holland, E., Marsella, K., 1999. Petrogenesis of municipal solid waste combustion bottom ash. Appl. Geochem. 14, 1073–1091.

Fellner, J., Lederer, J., Purgar, A., Winterstetter, A., Rechberger, H., Winter, F., Laner, D., 2015. Evaluation of resource recovery from waste incineration residues – the case of zinc. Waste Manag. 37, 95–103.

Funari, V., Bokhari, S.N.H., Vigliotti, L., Meisel, T., Braga, R., 2016. The rare earth elements in municipal solid waste incinerators ash and promising tools for their prospecting. J. Hazard Mater. 301, 471–479.

Funari, V., Braga, R., Bokhari, S.N.H., Dinelli, E., Meisel, T., 2015. Solid residues from Italian municipal solid waste incinerators: a source for "critical" raw materials. Waste Manag. 45, 206–216.

Gomes, H.I., Funari, V., Ferrari, R., 2020. Bioleaching for resource recovery from low-grade wastes like fly and bottom ashes from municipal incinerators: a SWOT analysis. Sci. Total Environ. 715, 136945.

Holm, O., Simon, F.G., 2017. Innovative treatment trains of bottom ash (BA) from municipal solid waste incineration (MSWI) in Germany. Waste Manag. 59, 229–236.

Huber, F., Blasenbauer, D., Aschenbrenner, P., Fellner, J., 2019. Chemical composition and leachability of differently sized material fractions of municipal solid waste incineration bottom ash. Waste Manag. 95, 593–603.

Hyks, J., Astrup, T.J.C., 2009. Influence of operational conditions, waste input and ageing on contaminant leaching from waste incinerator bottom ash: a full-scale study. Chemosphere 76, 1178–1184.

Inkaew, K., Saffarzadeh, A., Shimaoka, T., 2016. Modeling the formation of the quench product in municipal solid waste incineration (MSWI) bottom ash. Waste Manag. 52, 159–168.

Jones, A.M., Harrison, R.M., 2016. Emission of ultrafine particles from the incineration of municipal solid waste: a review. Atmos. Environ. 140, 519–527.

Kallesøe, J., 2014. Recovery of resources in bottom ash – second stage. In: VDI International Congress.

Kallesøe, J., Dyhr-Jensen, S., 2018. Recovery of resources in bottom ash - semi dry concept. In: Removal, Treatment and Utilisation of Waste Incineration Bottom Ash. Thomé-Kozmiensky Verlag GmbH, pp. 39–46.

Kim, B.-H., Lee, S.-J., Mun, S.-J., Chang, Y.-S.J.C., 2005. A case study of dioxin monitoring in and around an industrial waste incinerator in Korea. Chemosphere 58 (11), 1589–1599.

Leckner, B., Lind, F., 2020. Combustion of municipal solid waste in fluidized bed or on grate – a comparison. Waste Manag. 109, 94–108.

Ma, W., Chen, G., Rotter, S., Zhang, N., Du, G.J.E.P., 2014. Chloride deposit formation in a 24 MW waste to energy plant. Energy Procedia 61, 2359–2362.

Makari, C., 2014. Optical sorting for the recovery of glass from WIP slags. In: Thomé- Kozmiensky, K.J., Thiel, S. (Eds.), Waste Manage, vol. 4, pp. 345–354.

Mantovani, L., De Matteis, C., Tribaudino, M., Boschetti, T., Funari, V., Dinelli, E., Toller, S., Pelagatti, P., 2023. Grain size and mineralogical constraints on leaching in the bottom ashes from municipal solid waste incineration: a comparison of five plants in northern Italy. Front. Environ. Sci. 11.

Mantovani, L., Tribaudino, M., De Matteis, C., Funari, V., 2021. Particle size and potential toxic element speciation in municipal solid waste incineration (MSWI) bottom ash. Sustainability 13, 1911.

Mehr, J., Haupt, M., Skutan, S., Morf, L., Adrianto, L.R., Weibel, G., Hellweg, S., 2021. The environmental performance of enhanced metal recovery from dry municipal solid waste incineration bottom ash. Waste Manag. 119, 330–341.

Moya, D., Aldás, C., Jaramillo, D., Játiva, E., Kaparaju, P., 2017. Waste-To-Energy Technologies: an opportunity of energy recovery from Municipal Solid Waste, using Quito-Ecuador as case study. Energy Proc. 134, 327–336.

Muchova, L., Bakker, E., Rem, P., 2009. Precious metals in municipal solid waste incineration bottom ash. Water Air Soil Pollut. Focus 9, 107–116.

Nikravan, M., Ramezanianpour, A.A., Maknoon, R., 2020. Study on physiochemical properties and leaching behavior of residual ash fractions from a municipal solid waste incinerator (MSWI) plant. J. Environ. Manag. 260, 110042.

Nørgaard, K.P., Hyks, J., Mulvad, J.K., Frederiksen, J.O., Hjelmar, O., 2019. Optimizing large-scale ageing of municipal solid waste incinerator bottom ash prior to the advanced metal recovery: Phase I: Monitoring of temperature, moisture content, and CO_2 level. Waste Manag. 85, 95–105.

OECD, 2015. Organization for Economic Cooperation and Development, Environment at a Glance 2015: O.E.C.D. Indicators. O.E.C.D. Publishing, Paris.

Oh, J.-E., Choi, S.-D., Lee, S.-J., Chang, Y.-S.J.C., 2006. Influence of a municipal solid waste incinerator on ambient air and soil PCDD/Fs levels. Chemosphere 64 (4), 579–587.

Pan, Y., Wu, Z., Zhou, J., Ruan, X., Liu, J., Qian, G., 2013. Chemical characteristics and risk assessment of typical municipal solid waste incineration (MSWI) fly ash in China. J. Hazard. Mater. 261, 269–276.

Rankin, G.A., 1915. The ternary system $CaO-Al_2O_3-SiO_2$, with optical study by F. E. Wright. Am. J. Sci. 39, 1–79.

Simon, F.G., Holm, O., 2017. Metal recovery from wet discharged bottom ash, COST action MINEA workshop recovery technologies for waste incineration residues I, 30-31.5.2017, Bologna, Italy.

Smith, Y.R., Nagel, J.R., Rajamani, R.K., 2019. Eddy current separation for recovery of non-ferrous metallic particles: a comprehensive review. Miner. Eng. 133, 149–159.

Stockinger, G., 2018. Direct wet treatment of fresh, wet removed IBA from waste in- cinerator. In: Holm, O., Thomé-Kozmiensky, E. (Eds.), Removal, Treatment and Utilisation of Waste Incineration Bottom Ash. TK Verlag, Neuruppin, pp. 47–52.

Šyc, M., Simon, F.G., Biganzoli, L., Grosso, M., Hyks, J., 2018. Resource recovery from incineration bottom ash: basics, concepts principles, pp. 1–10.

Šyc, M., Simon, F.G., Hykš, J., Braga, R., Biganzoli, L., Costa, G., Funari, V., Grosso, M., 2020. Metal recovery from incineration bottom ash: state-of-the-art and recent developments. J. Hazard Mater. 393, 122433.

US-EPA, 2019. United states Environmental Protection Agency. In: Advancing Sustainable Materials 2017 Fact Sheet. November 2019.

Van Caneghem, J., Van Acker, K., De Greef, J., Wauters, G., Vandecasteele, C., 2019. Waste-to-energy is compatible and complementary with recycling in the circular economy. Clean Techn. Environ. Policy 21, 925–939.

Van Eck, N.J., Waltman, L., 2014. Visualizing Bibliometric Networks, pp. 285–320.

Vandecasteele, C., Wauters, G., Arickx, S., Jaspers, M., Van Gerven, T., 2007. Integrated municipal solid waste treatment using a grate furnace incinerator: the Indaver case. Waste Manag. 27, 1366–1375.

Waltman, L., Van Eck, N., 2013. A smart local moving algorithm for large-scale modularity-based community detection. Eur. Phys. J. B 86, 1–14.

Walton, H., 2010. The Impact on Health of Emissions to Air from Municipal Waste Incinerators: Advice from the Health Protection Agency.

Wang, P., Hu, Y., Cheng, H., 2019. Municipal solid waste (MSW) incineration fly ash as an important source of heavy metal pollution in China. Environ. Pollut. 252, 461–475.

Wei, Y., Shimaoka, T., Saffarzadeh, A., Takahashi, F., 2011. Mineralogical characterisation of municipal solid waste incineration bottom ash with an emphasis on heavy metal-bearing phases. J. Hazard Mater. 187, 534–543.

Yang, S., Saffarzadeh, A., Shimaoka, T., Kawano, T., Kakuta, Y., 2016. The impact of thermal treatment and cooling methods on municipal solid waste incineration bottom ash with an emphasis on Cl. Environ. Technol. 37, 2564–2571.

Yao, J., Kong, Q., Zhu, H., Long, Y., Shen, D., 2013. Content and fractionation of Cu, Zn and Cd in size fractionated municipal solid waste incineration bottom ash. Ecotoxicol. Environ. Saf. 94, 131–137.

Yao, J., Qiu, Z., Kong, Q., Chen, L., Zhu, H., Long, Y., Shen, D.J.E.E., 2017. Migration of Cu, Zn and Cr through municipal solid waste incinerator bottom ash layer in the simulated landfill. Ecol. Eng. 102, 577–582.

Youcai, Z., 2017. Municipal Solid Waste Incineration Process and Generation of Bottom Ash and Fly Ash, pp. 1–59.

Part VII

Future prospects

Chapter 28

Environmental impacts of combustion/incineration residue-derived products

Zhiliang Chen
Department of Environmental Science, College of Environment and Ecology, Chongqing University, Chongqing, China

1. Introduction

Combustion and incineration processes are widely utilized for waste management and energy generation, resulting in the production of significant quantities of residues as by-products (e.g., municipal solid waste incineration fly ash [MSWI FA], MSWI bottom ash [MSWI BA], coal fly ash [CFA], etc.). Taking China as an example, the annual production of MSWI FA, MSWI BA, and CFA has reached approximately 10 million, 40 million, and 790 million ton/y by 2021. These residues have potentials of resource utilization for the preparation of construction and building materials, glass—ceramics, and soil ameliorator, but they are usually enriched in constituents of potential concern (COPCs; e.g., heavy metals and organic contaminants). Understanding the environmental impacts associated with these residues and their derived products is crucial for sustainable waste management and resource utilization.

MSWI is a widespread technology for reducing the volume of waste (70—90 vol%) and recovering energy (Dou et al., 2017), and it is a preferred waste treatment method in many Asian and European Countries (e.g., China, Japan, Denmark, Switzerland, Sweden, etc.) (Kumar and Samadder, 2017). The MSWI FA and BA are two primary by-products from waste incineration, accounting for 20—30 wt% of municipal solid waste (MSW) (Chandler et al., 1997; Mühl et al., 2023). Technically, the MSWI FA is defined as "the particulate matter carried over from the combustion chamber and removed from the flue gas stream prior to addition of any type of sorbent material" (Chandler et al., 1997). However, the MSWI FA in literature is mostly referring to the air pollution control (APC) residues, which is the mixture of particulate matter from the combustion chamber, the solid materials captured from the acid gas neutralization units, and contaminants sorbents (e.g., activated carbon) (Quina et al., 2008).

1.1 Municipal solid waste incineration fly ash

The MSWI FA primarily contains constituents including Ca, Cl, Na, K, Si, Al, Fe, Mg, etc. and has pozzolanic properties, so MSWI FA can react with water to generate cementitious compounds. Thus, MSWI FA has the potential of being utilized to produce construction materials (e.g., cement, concrete, blocks, bricks, and aggregates), where it can reduce the carbon footprint of construction materials. However, MSWI FA is also enriched in leachable heavy metals (e.g., Pb, Zn, Cr, Cd, Cu, Ni, As, Se, etc.) and polychlorinated dibenzo-p-dioxins and dibenzofurans (PCDD/Fs), so it is usually classified as a hazardous waste in most countries (Quina et al., 2008). As such, the resource utilization of MSWI FA is associated with challenges and environmental concerns.

1.2 Municipal solid waste incineration bottom ash

MSWI also generates BA, which is the coarse residue remaining at the bottom of the incinerator (Dou et al., 2017). It mainly consists of glass, ceramics, metals, mineral agglomerates, and noncombustible inorganic substances (del Valle-Zermeño et al., 2017; Huber et al., 2020; Šyc et al., 2018). The geotechnical properties of MSWI BA (e.g., size, density, and mechanical strength) are similar to sand and gravel, making MSWI BA suitable for various civil engineering applications (Dou et al., 2017). The MSWI BA has been successfully utilized as the substitution of light aggregates in road

Treatment and Utilization of Combustion and Incineration Residues. https://doi.org/10.1016/B978-0-443-21536-0.00002-2

construction, embankments, and landfill engineering in many European countries including the Netherlands, Demark, Germany, and France (Bureau, 2005; Crillesen et al., 2006; Dou et al., 2017; Lam et al., 2010). However, MSWI BA poses unique challenges due to its heterogeneity and varying compositions, requiring proper handling and management prior to utilization (e.g., crushing, multistep sieving, and metal separation) (Šyc et al., 2020). The content of inorganic and organic in MSWI BA is much lower compared with MSWI FA, but the potential leaching of heavy metals is still a limiting factor for the utilization of MSWI BA-derived products (Blasenbauer et al., 2020).

1.3 Coal fly ash

Coal-fired power plants are a significant source of electricity generation worldwide, producing large quantities of CFA as a by-product of the combustion process (i.e., fine and powdery residue collected from the flue gases) (Blissett and Rowson, 2012). The CFA primarily comprises small particles composed of silica (SiO_2), alumina (Al_2O_3), iron oxide (Fe_2O_3), and varying amounts of calcium oxide (CaO), magnesium oxide (MgO), and other trace elements (e.g., Hg, As, Se, B, Cr, Mo, V, Pb, Cd, etc.) (Feng et al., 2018; Wang et al., 2022b, 2023). The specific composition of CFA depends on factors such as the coal type, combustion conditions, and air pollution control measures employed at the power plant (Wang et al., 2022b). The CFA has been widely used as a cementitious additive in concrete and as a component in construction materials. Its pozzolanic properties contribute to enhanced workability, strength, and reduced environmental impact of concrete (Blissett and Rowson, 2012). Additionally, CFA can be utilized in the manufacturing of ceramics, geopolymers, soil amendment, and other innovative products (Yao et al., 2015). However, a significant challenge associated with the utilization of CFA is the potential leaching of trace elements and heavy metals present in the ash, which poses a risk of groundwater contamination (Izquierdo and Querol, 2012; Wang et al., 2022a).

Closing the material loops from municipal solid waste and coal combustion is crucial to save natural resources and achieve circular economy (Abis et al., 2020; Lederer et al., 2022). The MSWI FA, MSWI BA, and CFA as typical combustion/incineration residues have good potentials of resource utilization, but the leaching of COPCs is a widely recognized challenge to the use of residue-derived products. Herein, the basic properties of these residues are presented, and the products derived from these residues are briefly reviewed. The methods for assessing the leaching and release of COPCs from residue-derived products are introduced in detail as COPCs leaching is the primary environmental impact. Moreover, greenhouse gas emission reduction, regulatory guidelines, and future perspectives regarding the utilization of combustion/incineration residues are briefly overviewed and discussed.

2. Basic properties of combustion/incineration residues

2.1 Municipal solid waste incineration fly ash and bottom ash

The total quantity of MSWI FA and BA accounts for 20–30 wt% of the incinerated MSW, and the mass distribution between FA and BA is largely dependent on the type of furnace (i.e., grate furnace vs. circulated fluidized bed furnace) (Chen et al., 2019a; Mühl et al., 2023). For the incinerators equipped with grate furnace, the generated FA is only 2–5 wt% of MSW, while the BA is 20–25 wt% of MSW (Obe et al., 2017). For the incinerators equipped with circulated fluidized bed furnace, the quantity of FA and BA is comparable (i.e., 10–15 wt% of MSW for each) (Blasenbauer et al., 2023; Fan et al., 2022). Accordingly, the chemical composition and minerology of FA and BA from incinerators with different furnaces are also largely differentiated.

The composition of MSWI FA and BA is mainly determined by the types of incineration furnace, air pollution control units, and feeding waste composition (Blasenbauer et al., 2023; Saqib and Bäckström, 2015). Tables 28.1 and 28.2 summarize the typical elemental composition of MSWI FA collected from grate incinerators (GIs) and circulated fluidized bed incinerators (CFBIs) (Chen, 2019; Chen et al., 2019a). Generally, both types of FA have high content of Ca, O, Cl, Na, K, and C. Other elements in the FA from GIs are in low content, while the content of Si, Al, Fe, and Mg in the FA from CFBIs is also considerable, which is mainly attributed to the slag particles carried by the flue gas with fast flow rate and then captured into the FA. The entrainment of the slag particles in the FA can dilute the Cl content, resulting in much lower content of Cl in the FA from CFBIs compared with the FA from GIs. For the trace elements, Zn, Pb, and Cu are the primary heavy metals in both types of FA. Specifically, the FA from CFBIs contains much more Ba, Cr, and Ni compared with FA from GIs. The crystalline minerals in MSWI FA from GIs and CFBIs have some similarities as halite, sylvite, and calcite are identified in both FAs (Fig. 28.1). Additionally, the FA from GIs has more Ca-bearing minerals (i.e., portlandite, calcium oxide, and calcium hydroxychloride), while the FA from CFBIs has high content of quartz.

TABLE 28.1 Elemental composition of MSWI FA from GIs and CFBIs (wt %) (Chen, 2019).

Major elements	MSWI FA from GIs	MSWI FA from CFBIs
Ca	**25.9–29.9**	**15.8–17**
O	**18.7–19.2**	**33.8–37**
Si	**1.05–1.26**	**9.95–12.7**
Al	**0.25–0.55**	**5.21–10.4**
Cl	**19.4–21.6**	**5.41–8.85**
C	14–17.2	5.1–8
Fe	0.21–0.55	2.15–4.01
S	1.26–1.6	1.09–1.49
Mg	0.42–0.83	1.59–5.06
Na	**7.06–7.63**	**1.25–5.31**
K	**3.19–3.54**	**0.95–2.52**
P	0.09–0.12	0.81–1.94
Ti	0.07–0.11	0.51–0.65
Br	0.16–0.8	0.01–0.04

The generation of PCDD/Fs, a group of chlorinated persistent organic pollutants, is one of the primary concerns associated with MSW incineration because of the high content of Cl in MSW. Usually, 3T method (i.e., a Temperature > 850°C, a residence Time > 2s, and sufficient Turbulence) is used to suppress the formation of PCDD/Fs during the combustion of MSW (Lin et al., 2023). However, the PCDD/Fs can still be easily formed in the postcombustion area. The primary formation pathways of PCDD/Fs are (1) de novo synthesis from carbon matrices, polycyclic aromatic hydrocarbons (PAHs) and Cl source; and (2) formation from PCDD/F-precursors (e.g., polychlorinated biphenyl, chlorophenol, chlorobenzene, etc.) (Chen et al., 2018). Usually, the de novo pathway is recognized to contribute more to PCDD/Fs formation than the precursor pathway in the MSWI system, but cases with the precursor pathway as the major one were also reported. One simple method of identifying the major formation pathway is the mass ratio between PCDFs and PCDDs (PCDF/PCDD). Specifically, a PCDF/PCDD greater than 1 usually indicates the de novo as the primary pathway because de novo synthesis reaction mainly contributes to the formation of PCDFs (Lin et al., 2023).

TABLE 28.2 Total content of trace heavy metals in MSWI FA from GIs and CFBIs (mg/kg) (Chen, 2019).

Trace heavy metals	MSWI FA from GIs	MSWI FA from CFBIs
Zn	**2950–4870**	**2793–7270**
Ba	7–11	658–2330
Cu	**474–561**	**1069–3050**
Pb	**1070–1190**	**382–1130**
Cr	44–79	253–727
Ni	22–31	92–177
Se	1–22	1–83
As	1–43	7–87
Cd	60–65	21–30

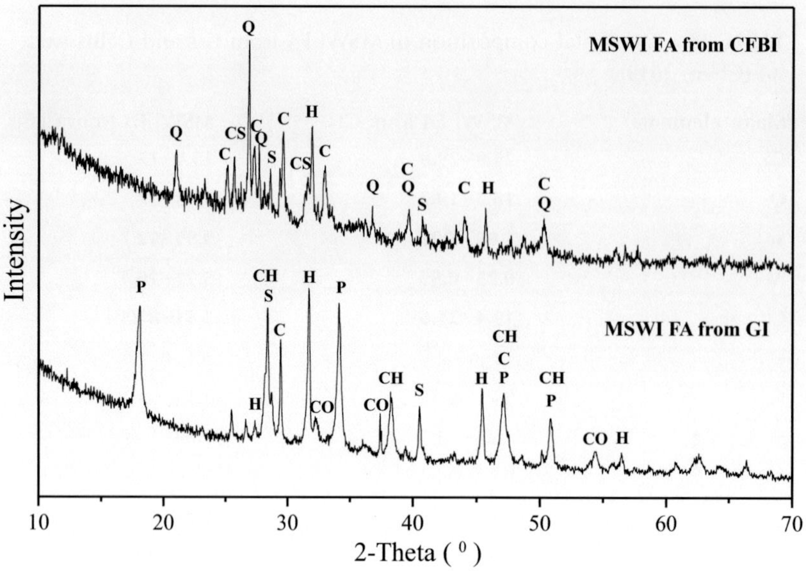

FIGURE 28.1 XRD patterns of MSWI FA from CFBI and MSWI FA from GI (Chen et al., 2019a). *C*, calcite ($CaCO_3$); *CH*, calcium hydroxychloride (CaClOH); *CO*, calcium oxide (CaO); *CS*, calcium sulfate ($CaSO_4$); *H*, halite (NaCl); *P*, portlandite ($Ca(OH)_2$); *Q*, quartz (SiO_2); *S*, sylvite (KCl).

The optimum formation temperature range of PCDD/Fs is 200−500°C, at which the heterogenous reactions at the surface of transition metals in MSWI FA (e.g., Cu, Fe, Cr, etc.) were facilitated to generate PCDD/Fs via both de novo and precursor pathways (Altarawneh et al., 2009). Besides, homogenous reactions in gas phase at temperatures greater than 600°C can also form PCDD/Fs via precursor pathway. To meet the PCDD/Fs emission standard in flue gas (i.e., 0.1 ng TEQ/Nm^3), the activated carbon is usually injected to adsorb the generated PCDD/Fs in gas phase in the postcombustion zone followed by the collection of the injected activated carbon at the baghouse filter, leading to the enrichment of PCDD/Fs in MSWI FA (Lin et al., 2020). Therefore, over 90 wt% of PCDD/Fs are usually enriched in MSWI FA, making PCDD/Fs a primary concern for MSWI FA treatment and utilization.

The content of PCDD/Fs in MSWI FA from CFBIs has been reported to be greater than that in MSWI FA from GIs (Table 28.3), because the content of catalytic metals (e.g., Cu, Fe, and Cr) in CFBIs FA is much higher than that in GIs FA

TABLE 28.3 Total and TEQ content of PCDD/Fs in MSWI FA from GIs and CFBIs (ng/g or ng I-TEQ/g) (Chen, 2019).

Congener	MSWI FA from GIs	MSWI FA from CFBIs
TCDD	0.23−6.47	9.02−59.2
PeCDD	0.32−3.91	11.3−53.2
HxCDD	0.58−4.68	18.4−147
HpCDD	0.46−2.15	8.12−57.7
OCDD	0.35−0.93	7.06−61.2
PCDD	1.94−18.1	53.9−354
TCDF	1.24−4.14	35.3−95.1
PeCDF	0.96−2.61	24.8−56.3
HxCDF	0.9−2.2	14.6−50.4
HpCDF	0.34−0.8	9.44−30.2
OCDF	0.09−0.18	1.95−9.4
PCDF	3.55−9.94	86.2−229
PCDF/PCDD	0.55−1.85	0.5−1.61
PCDD/Fs	5.48−28.1	140−531
TEQ (ng I-TEQ/g)	0.09−0.25	2.82−6.75

(Chen, 2019). The catalytic metals are crucial for PCDD/Fs generation because the formation rate of PCDD/Fs in the absence of catalytic metals has been observed to be very slow.

The MSWI BAs are highly heterogenous with mineral materials, glass, and metals (e.g., magnetic ferrous metals, weakly magnetic ferrous metals, aluminum, copper, etc.) as the major components, and the total content of mineral materials and glass is usually over 80 wt% of BA (Blasenbauer et al., 2023). The content of metals in BA largely depends on the composition waste input, but it has been observed that the CFBIs with low combustion temperature of 600−900°C (GIs with combustion temperature of 800−1100°C) are favorable to reduce oxidation processes and preserve aluminum in the metallic form (Hu et al., 2011; Leckner and Lind, 2020). Lower combustion temperature in CFBIs also prevents the melting of glass to for agglomerates with other minerals and metals. As such, the glass content in CFBIs BA is usually higher compared with GIs BA.

For mineral materials as the primary component in MSWI BA, the specific elemental compositions and mineralogical characteristics greatly vary depending on the type of furnace. The mineral materials in GIs BA have much higher content of Al, Si, and Fe than the mineral materials in CFBIs BA (Table 28.4), which is mainly because of the higher bed temperature and less pretreatment of feeding waste (e.g., removing the ferrous metals from waste prior to incineration) in GIs (Blasenbauer et al., 2023). Mineral materials in GIs BA relatively have a higher content of heavy metals compared with CFBIs BA, such as Pb, Sb, and Cd, mainly due to the shorter residence times of waste in grate furnace.

2.2 Coal fly ash

The CFA usually accounts for 5−20 wt% of feeding coal (Yao et al., 2015), and it is primarily classified Class F and Class C CFA (Committee, 2019). Class F CFA, derived from burning anthracite or bituminous coals, is typically high in Si, Al, and Fe content (Table 28.5). On the other hand, Class C CFA, generated from burning subbituminous or lignite coals, contains higher levels of Ca, in addition to Si, Al, and Fe (Blissett and Rowson, 2012). Besides glass as the primary component in CFA, common minerals found in CFA include quartz, mullite, hematite, magnetite, anhydrite, lime, portlandite, and various calcium and aluminum silicates (Vassilev and Vassileva, 2007; Wang, 2008; Wang et al., 2022b). As reported by Wang et al. (2022b), minerals identified in Class F CFA are quartz, mullite, magnetite, hematite, lime,

TABLE 28.4 Elemental composition of MSWI BA from GIs and CFBs (Blasenbauer et al., 2023).

Elements	MSWI BA from GIs	MSWI BA from CFBs
Al[a]	2.23−3.57	0.75−1.47
Ca[a]	7.45−23.7	4.48−14.3
Cl[a]	0.45−0.7	0.09−0.25
Fe[a]	2.5−5.52	1.36−6.1
K[a]	0.62−1.41	0.94−1.54
Na[a]	1.0−10.2	0.26−2.37
S[a]	0.2−0.7	0.12−0.38
Si[a]	4.46−15.1	5.36−13.2
As[b]	16−32	3.5−46
Cd[b]	7.0−43	4.9−12
Co[b]	10−113	5.1−33
Cr[b]	455−1180	385−883
Hg[b]	0.24−0.58	0.56−0.87
Ni[b]	58−217	21−336
Pb[b]	195−1160	114−1310
Sb[b]	12.0−24	0.4−0.42

[a]*wt%.*
[b]*mg/kg.*

TABLE 28.5 Elemental composition of Class F and Class C CFA (wt%) (Blissett and Rowson, 2012).

	Class F	Class C	
	CFA from bituminous	CFA from subbituminous	CFA from lignite
SiO_2	20–60	40–60	15–45
Al_2O_3	5–35	20–30	10–25
Fe_2O_3	10–40	4–10	4–15
CaO	1–12	5–30	15–40
MgO	0–5	1–6	3–10
Na_2O	0–4	0–2	0–6
K_2O	0–3	0–4	0–4
SO_3	0–4	0–2	0–10
Loss of ignition	0–15	0–3	0–5

portlandite, and anhydrite; while the minerals observed in Class C CFA are quartz, calcium aluminum oxide, lime, anhydrite, periclase, and hematite (Fig. 28.2). The specific mineral phases present in CFA contribute to its physical and chemical properties, such as pozzolanic reactivity and the potential for self-cementing.

The specific heavy metal content in CFA can vary depending on the specific coal source, combustion process, and the efficiency of the emission control technologies employed in the power plant. Based on the reported data (Kosson et al., 2009), the content of most heavy metals in Class F and Class C CFA is relatively comparable, except for a higher content of Hg and Ba in Class C CFA and a slightly higher content of As and Cr in Class F CFA.

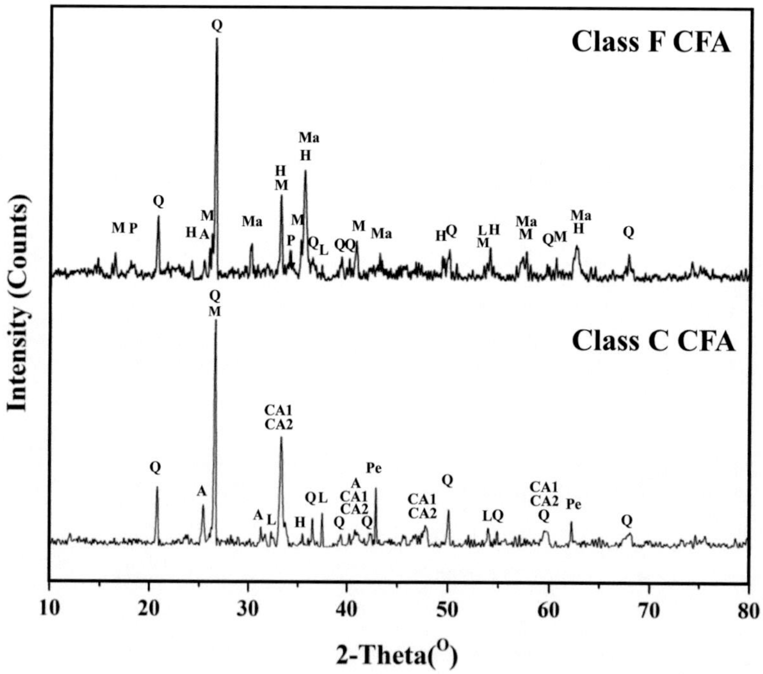

FIGURE 28.2 Mineralogical composition of Class F and Class C CFA (Wang et al., 2022b). *A*, anhydrite ($CaSO_4$); *CA*1, calcium aluminum oxide ($Ca_3Al_2O_6$); *CA*2, calcium aluminum oxide ($Ca_9Al_6O_{18}$); *H*, hematite (Fe_2O_3); *L*, lime (CaO); *M*, mullite ($Al_6Si_2O_{13}$); *Ma*, magnetite (Fe_3O_4); *P*, portlandite ($Ca(OH)_2$); *Pe*, periclase (MgO); *Q*, quartz (SiO_2).

3. Combustion/incineration residue-derived products: Types and applications

3.1 Cement-based products

Combustion/incineration residue-derived products including MSWI FA, MSWI BA, and CFA have the potential of being used as supplementary cementitious materials in cement-based products because of their pozzolanic and/or cementitious properties. These materials not only offer opportunities for waste management but also provide benefits such as improved material properties and reduced environmental impact. The overall idea is to use these residues to partially replace cement in concrete, resulting in different products including concrete, geopolymer concrete, and cast concrete products (e.g., bricks, blocks, and paving stones) (Chen et al., 2019a; Joseph and Mathew, 2012; Kuo et al., 2013; Naik et al., 2003; Siddique, 2010; Temuujin et al., 2010). Among three residues, CFA has been widely accepted for use in concrete as cement replacement due to the formation of calcium silicate hydrate from the reaction of silica in CFA and calcium hydroxide in concrete matrix (González et al., 2009). The inclusion of CFA in concrete also has benefits of long-term durability improvement, lowering the cracking risk in early stage by reducing hydration heat, and reducing the bleeding of freshly mixed concrete (Nath and Sarker, 2011; Sarker and McKenzie, 2009).

The MSWI FA and BA contain a certain amount of cement minerals, making their utilization in cement-based products possible, but pretreatment is usually necessary to amend the properties of MSWI FA and BA for more benign application (Lam et al., 2010; Xuan et al., 2018b). The major issues associated with the utilization of MSWI FA and BA are as follows: (1) the expansive cracking of cement-based products caused by the generation of hydrogen gas (i.e., H_2 generated from the reaction of the residual metallic aluminum or Zn/Al alloy in FA and/or BA under the alkaline conditions) (Bertolini et al., 2004; Xuan et al., 2018a), the formation of ettringite with high molar volume (i.e., sulfate attack due to the introduced SO_4^{2-} from FA and/or BA) (Berg and Neal, 1998; Collepardi, 2003; Müller and Rübner, 2006), and the formation of alkali-silica reaction gel with high molar volume (Karthik et al., 2016); and (2) the potential steel corrosion induced by the Cl^- from FA and/or BA (Mehta and Monteiro, 2006; Sun et al., 2022). Pretreatments of MSWI FA and BA to address these issues include the following: (1) wet ball milling to exhaust the reactions leading to the generation of H_2 before FA and/or BA are used in cement-based products (Bertolini et al., 2004; Chen et al., 2019a); (2) water washing or washing with Na_2CO_3 or $NaHCO_3$ solution to remove SO_4^{2-} and Cl^- (Xuan et al., 2018b); and (3) sorting and removal of metals and glasses from the MSWI BA.

3.2 Ceramic products

The MSWI FA, MSWI BA, and CFA can be used to produce ceramic products (i.e., glass, ceramic, and glass−ceramic) due to the high content of SiO_2, Al_2O_3, CaO, and Fe_2O_3 (Deng et al., 2018; Zhang et al., 2020b, 2022). At high temperatures (typically above 1200°C), SiO_2 and Al_2O_3, as the primary source, combine with the fluxing agents (e.g., CaO and MgO) to form a homogenous melt. Once the melt is formed, it is cooled gradually to promote the nucleation and growth of crystals initiated with the help of nucleating agents (e.g., Fe_2O_3) within the glass matrix, resulting in controlled crystal growth and phase formation (Wang et al., 2014). The overall idea of producing ceramic products using combustion/incineration residues is to transform the residue components into a glassy matrix through melting and subsequent controlled crystallization, resulting in a material with tailored properties suitable for various industrial applications (e.g., pottery, porcelain, ceramic bricks, ceramic tiles, ceramic stoneware, road base materials, embankments, and ceramic aggregates) (Cheeseman et al., 2003; Deng et al., 2018; Erol et al., 2008; Sokolar and Vodova, 2011).

3.3 Geotechnical applications

The geotechnical applications of MSWI FA, MSWI BA, and CFA are usually road pavement and embankment construction (Balaguera et al., 2018; Blissett and Rowson, 2012; González et al., 2009; Oehmig et al., 2015; Poulikakos et al., 2017). The specific applications are as follows: (1) as stabilizing agents to provide cohesion and reduce the plasticity of the soil or aggregate mixture, thus improving the strength and stability of road pavements and embankments (Nalbantoğlu, 2004; Zha et al., 2008); (2) to modify the subgrade soil by filling voids in the soil matrix and reducing permeability, enhancing the stability and durability of the road pavement or embankment (Ferreira et al., 2003); (3) as fill materials (placed and compacted in layers) to create a stable foundation for the embankment, which is helpful for preventing excessive water infiltration, protecting structural integrity, and reducing the weight load of the embankment (Yan et al., 2014; Zhang et al., 2016); and (4) as base or subbase materials (mixing with other aggregate materials and then compacted to form a stable layer beneath the asphalt or concrete surface) to improve load-bearing capacity, reduce moisture susceptibility, and enhance the overall performance of the pavement structure (Lynn et al., 2016). When using combustion/

incineration residues for road pavement and embankment applications, it is important to consider factors such as the specific characteristics of the residues, their compatibility with the existing materials, and the engineering requirements of the project.

3.4 Agricultural applications

Combustion/incineration residues including MSWI FA, MSWI BA, and CFA can also be used to ameliorate soils, improving the fertility, structure, and overall quality of soils (Ferreira et al., 2003; Lam et al., 2010; Yao et al., 2015). These residues contain nutrients (e.g., P, S, K, Ca, Mg, Cu, Mn, and Zn), and mixing residues with the soil helps improve soil fertility and support plant growth. Combustion/incineration residues are usually alkaline due to the presence of CaO and other alkaline compounds, so they can be used to adjust the acidic soils to the pH conditions favorable for plant growth and nutrient availability (i.e., neutral or slightly alkaline). Specifically, the fine particles of CFA can help to increase drainage and reduce compaction of heavy clay soils, enhancing soil aeration and root penetration. The lightweight and porous nature of CFA also improve water-holding capacity in sandy soils, preventing excessive leaching of nutrients (Basu et al., 2009; Shaheen et al., 2014).

4. Environmental impacts of combustion/incineration residue-derived products

4.1 Leaching of inorganic contaminants

4.1.1 Brief overview of leaching assessment methods

The release of heavy metals is a significant environmental concern associated with combustion/incineration residue-derived products due to their enrichment in the residues (Sun et al., 2022). Therefore, conducting the leaching assessment of residue-derived products is crucial and necessary to evaluate the potential environmental impacts of residue-derived products and, in turn, aid in optimizing the utilization methods of residues.

Table 28.6 is a brief overview of standard leaching test methods for solid materials from the United States, Europe, and China. The methods can be generally separated into two groups: (1) methods with specific extract solutions (usually with only a single leaching point) so that the test results can be used to evaluate if the solid materials can meet the standard of being disposed of in specific facilities or recycling; and (2) methods to simulate different leaching scenarios and conditions (usually with multiple leaching test points) to assess the equilibrium leaching concentrations or dynamic release rates of constituents from solid materials.

For the past decades, the Group I methods (e.g., US EPA Method 1311—TCLP and HJ/T 300-2007) have been overly used for assessing the release of COPCs, but it should be aware that these methods were developed only for specific scenarios (Kosson et al., 2002). Also, the leaching test results of Group I methods hardly provide insights into the mechanisms controlling the leaching of COPCs. To more accurately assess the release of COPCs from solid materials under a wide range of scenarios, mechanisms controlling the liquid−solid partitioning (LSP) of COPCs and the mass transport rate of COPCs within solid media are the keys (Kosson et al., 2009).

The LSP of constituents is generally controlled by the precipitation/dissolution of minerals and the adsorption to minerals and/or organic matter (Chen et al., 2021b). Therefore, many factors including pH, pe, content of iron oxides and organic matter, and elemental and mineralogical compositions can significantly affect the LSP of constituents from solid materials (Wang et al., 2022a, 2023). The pH- and liquid-to-solid ratio (L/S)-dependent leaching tests (e.g., EPA Method 1313/1316 and EN 14429:2015) in conjunction with geochemical speciation modeling are helpful to look into the mechanisms behind the LSP of constituents.

The mass transport rate of constituents is mainly determined by the pore structure (e.g., porosity, pore connectivity, pore distribution, etc.) and the water saturation of media (Chen et al., 2023c,d). The diffusion of constituents is usually described by Fick's law of diffusion, and the effective diffusion coefficient is further determined by the porosity, pore connectivity (or tortuosity), and water saturation. Monolithic diffusion and percolation column leaching test methods (e.g., US EPA Method 1314/1315, EN 14405:2017, CEN/TS 16637-2:2014, and GB/T 7023-2011) together with reactive mass transport modeling are usually used to identify the release rates of constituents in different media.

Even though the standard leaching tests in conjunction with modeling can provide insights into the mechanisms controlling the LSP of constituents and mass transport rates of constituents within media, many other factors are still needed to be considered for scenario-based assessment on the release of COPCs. For example, the redox states (or the pe) of solid materials under lab leaching conditions are usually very different from that under field leaching conditions, which is likely to dramatically affect the LSP of redox-sensitive COPCs (e.g., As, Se, Cr, etc.) (Wang et al., 2022a, 2023).

TABLE 28.6 Summary of standard leaching methods from the United States, Europe, and China.

	Method	Name	Type of leaching test	Objectives
US	EPA method 1311	Toxicity characteristic leaching procedure	Equilibrium extraction test (single point)	Designed to determine the mobility of both organic and inorganic analytes present in liquid, solid, and multiphasic wastes
	EPA method 1312	Synthetic precipitation leaching procedure	Equilibrium extraction test (single point)	Designed to determine the mobility of both organic and inorganic analytes present in liquid, solid, and multiphasic wastes
	EPA method 1313	Liquid–solid partitioning as a function of extract pH using a parallel batch extraction procedure	pH-dependent leaching test	To identify the equilibrium leaching concentration of constituents from granular solid as a function of pH
	EPA method 1314	Liquid–solid partitioning as a function of liquid–solid ratio for constituents in solid materials using an upflow percolation column procedure	Percolation column leaching test	To identify the release rate of constituents from granular solid under percolation flow conditions
	EPA method 1315	Mass transfer rates of constituents in monolithic or compacted granular materials using a semidynamic tank leaching procedure	Monolithic diffusion leaching test	To identify the mass transfer rate of constituents from monolithic or compacted granular solid to the contacting solution
	EPA method 1316	Liquid–solid partitioning as a function of liquid-to-solid ratio in solid materials using a parallel batch procedure	L/S-dependent leaching test	To identify the equilibrium leaching concentration of constituents from granular solid as a function of L/S
Europe	EN 14429:2015	Characterization of waste—leaching behavior test—influence of pH on leaching with initial acid/base addition	pH-dependent leaching test	To identify the equilibrium leaching concentration of constituents from granular solid as a function of pH
	EN 14405:2017	Characterization of waste—leaching behavior test—upflow percolation test (under specified conditions)	Percolation column leaching test	To identify the release rate of constituents from granular solid under percolation flow conditions
	CEN/TS 16637-2:2014	Construction products—assessment of release of dangerous substances—part 2: horizontal dynamic surface leaching test	Monolithic diffusion leaching test	To identify the mass transfer rate of constituents from monolithic or compacted granular solid to the contacting solution
	CEN/TS 16660:2015	Characterization of waste—leaching behavior test—determination of the reducing character and the reducing capacity	Reducing capacity determination test	To determine the reducing character and the reducing capacity of construction products, waste materials and the eluate resulting from exposure of these solids to a leachant
China	HJ/T 299-2007	Solid waste extraction procedure for leaching toxicity—sulfuric acid and nitric acid method	Equilibrium extraction test (single point)	To simulate the leaching of contaminants from solid wastes under the influence of acid rain
	HJ/T 300-2007	Solid waste extraction procedure for leaching toxicity—acetic acid buffer solution method	Equilibrium extraction test (single point)	To simulate the leaching of industrial wastes under the disposal scenario of sanitary landfill
	HJ 557-2010	Solid waste extraction procedure for leaching toxicity—horizontal vibration method	Equilibrium extraction test (single point)	To simulate the leaching of contaminants from solid wastes under the influence of surface water or underground water
	GB/T 7023-2011	Standard test method for leachability of low and intermediate level solidified radioactive waste forms	Monolithic diffusion leaching test	To identify the mass transfer rate of constituents from monolithic or compacted granular solid to the contacting solution

Besides, the pH and L/S conditions of solid materials under field scenarios should also be characterized carefully because they can greatly affect the LSP of COPCs. For the long-term leaching assessment of monolithic materials, environmental aging processes (e.g., drying, carbonation, and oxidation) should be taken into account because these aging processes can alter the pH and pe conditions and pore structure of the monolithic materials, leading to significant changes in the LSP and mass transport rates of constituents (Chen et al., 2022,2023d).

4.1.2 Release potential of heavy metals

The primary heavy metals leached from combustion/incineration residue-derived products mainly depend on the source of residues. The Pb, Zn, Cr, Cd, Cu, and Ni are the major heavy metals released from MSWI FA. For MSWI BA, the leaching of Cr, Cu, and Ni is the primary concern, and the leaching of As, Se, Cr, Mo, V, and B from CFA is usually focused for environmental risk assessment (Chen et al., 2019b; Wang et al., 2022b). Effective stabilization/solidification (usually with over 95% reduction in the equilibrium leaching concentration) of cationic heavy metals (e.g., Pb, Zn, Cd, Cu, and Ni) in cement-based products has been widely reported because their speciation can be transformed into more stable and less soluble metal-bearing minerals (i.e., chemical stabilization) (Shi and Kan, 2009; Sorlini et al., 2017; Tian et al., 2020). Aside from the chemical stabilization, the incorporation of heavy metals in cementitious materials provides a low permeable barrier to suppress the transport of heavy metals (i.e., physical solidification) (Shi and Kan, 2009; Sorlini et al., 2017). The ordinary cementitious materials are relatively ineffective for the chemical stabilization of anionic heavy metals (e.g., Cr, As, and Se), which, however, can be improved by modifying the formula of cementitious materials to generate minerals that can effectively incorporate anionic heavy metals (e.g., ettringite).

The leaching concentration of heavy metals from residue-derived ceramic products is mostly reported to be very low (mostly lower than 0.01 mg/L for Cd, Cr, Cu, Ni, Pb, and Zn with the immobilization efficiency over 99% for all heavy metals), because the glass matrix formed under high temperatures can effectively incorporate heavy metals for solidification (Deng et al., 2018; Zhang et al., 2019, 2020b, 2022). Given the extremely low permeability of the ceramic products, the release rates of heavy metals are significantly decreased. It is worth noting that the immobilization of heavy metals in ceramic products is primarily achieved by physical solidification, and so the leaching of both cationic and anionic heavy metals can be efficiently suppressed, but the leaching of heavy metals can be accelerated if the ceramic products lose its integrity (e.g., being crushed into powder).

For other residue-derived products including soil ameliorators and fill materials for road pavement and embankment, the leaching of heavy metals is hardly suppressed, so the potential impact of heavy metals from these products should be carefully assessed and evaluated prior to the application.

4.2 Polychlorinated dibenzo-p-dioxins and dibenzofurans

The PCDD/F is a potential concern specifically for the utilization of MSWI FA (Peng et al., 2023). Usually, the PCDD/Fs in FA need to be degraded prior to the utilization, and the degradation of PCDD/Fs usually requires thermal treatment of FA. As such, the MSWI FA-derived products without thermal treatment including cement-based products, filling materials for road pavement and embankment, and soil ameliorators contain almost all PCDD/Fs from the MSWI FA. However, the release potential of PCDD/Fs from FA products produced via a nonthermal process is assumed to be very limited given the low solubility of PCDD/F-congeners. On the opposite, the FA-derived ceramic products contain low content of PCDD/Fs (usually less than 10 ng I-TEQ/kg with destruction efficiency over 95%) because the thermal treatment of FA can effectively destroy PCDD/Fs (Chen et al., 2020, 2021a). In summary, FA-derived products from nonthermal processes have a potential concern of high PCDD/Fs content, but the release of PCDD/Fs is assumed to be very limited; FA-derived products from thermal processes hardly have any potential concern regarding PCDD/Fs.

4.3 Greenhouse gas emission reduction

The MSWI FA, MSWI BA, and Class C CFA as alkaline wastes are suitable for capturing and mineralizing CO_2 from the industrial flue gases (i.e., direct CO_2 emission reduction). The mineralization reactions are thermodynamically favored with fast reaction kinetics at ambient conditions. Moreover, the carbonated residues can potentially be used as the raw materials for substitution in clinker or as aggregates in construction and building materials, which can indirectly decrease the production of Portland cement and reduce the CO_2 emission from cement industry (i.e., indirect CO_2 emission reduction). The emission of a substantial amount of CO_2 (ca. 4.02 Gt per year) can be decreased by CO_2 mineralization and utilization using alkaline solid wastes, which approximately accounts for 12.5% global anthropogenic CO_2 emission (Pan et al., 2020). The utilization of combustion/incineration residues coupled with CO_2 mineralization is an important component for the CO_2 mineralization and utilization by alkaline solid wastes.

5. Regulatory framework and guidelines

The CFA has been routinely reused in construction and building materials in many countries, but the reuse of MSWI FA and BA faces many challenges. Table 28.7 is a brief summary on the US EPA best available techniques and best environmental practices (BAT and BEP) guidelines for the treatment and reuse of MSWI FA and BA. First of all, the MSWI FA is not allowed to mix with BA in many countries (except for the United States) because it will contaminate the BA. The MSWI BA is mainly disposed of in landfills but can be utilized in construction and road-building material after an appropriate pretreatment. The MSWI FA is predominantly disposed of in dedicated landfills after a pretreatment (e.g., stabilization/solidification).

The content of Hg is of special concern in terms of the reuse of MSWI FA and BA. The US EPA suggests that any residues containing or contaminated with mercury should not be recycled and should never be used as soil amendment in agricultural or similar applications if mercury concentration exceeds levels of concern. Also, the US EPA indicates that the use of waste incineration residues for construction purposes is very problematic and cannot be considered as BEP. If the reuse of FA and BA is anticipated, a careful evaluation on the potential environmental impacts of these materials should be undertaken.

Recently, China has released a guideline that contains the specific standards and guidance for the utilization of MSWI FA (i.e., HJ 1134-2020, *Technical Specification for Pollution Control of Fly-ash from Municipal Solid Waste Incineration*) (Ministry of Ecology and Environment of the People's Republic of China, 2020). The MSWI FA has to be detoxified and modified prior to the reuse. Specifically, the PCDD/Fs in the MSWI FA need to be decomposed to reduce the content lower than 50 ng-TEQ/kg, and thermal treatment methods are recommended (e.g., low temperature thermal decomposition, sintering, melting, vitrification, etc.) (Zhang et al., 2021). The leaching concentration of heavy metals from the treated FA (referring to leaching method HJ 557-2010) should meet the limits required by standard GB 8978 (Table 28.8). In addition, the content of Cl in the treated FA needs to be lower than 2 wt% (1 wt% preferred), and the water washing and high temperature thermal treatment are recommended for Cl removal (Chen et al., 2016; Lin et al., 2021; Zhang et al., 2020a).

6. Future outlook and challenges

The sustainable utilization of combustion/incineration residues is beneficial in terms of waste reduction, energy and resource conservation, and decrease in greenhouse gas emission, but it is also associated with potential concerns including heavy metals leaching, PCDD/Fs contamination, chloride introduction, and generation of secondary pollutants. Among these potential concerns, the release of heavy metals from the residue-derived products is the primary one. The group I

TABLE 28.7 Summary on the US EPA BAT and BEP for MSWI residues.

	Treatment techniques for solid residues from incineration
MSWI BA treatment techniques	• Because of the differences in pollutant concentration, **the mixing of MSWI BA with FA** will contaminate the former and **is forbidden** in many countries. • **MSWI BA** is disposed of in landfills in many countries but **may be reused in construction and road-building material following pretreatment.**
Treatment of MSWI FA	• **MSWI FA is disposed of in dedicated landfills** in many countries. • **Pretreatment is likely to be required** for this to constitute BAT, depending on national landfill acceptance criteria.
Residue reuse	• **MSWI BA and FA** should **never be used as soil amendment in agricultural** or similar applications **if mercury concentration exceeds levels of concern.** • **Use of waste incineration residues for construction purposes** is also very problematic and **cannot be considered as the best environmental practice.** • **Careful evaluation** of these materials **should be undertaken if any reuse is anticipated.**
Stabilization and solidification	• **Solidification or stabilization with Portland cement** (or other pozzolanic materials), alone or with additives or a number of thermally based treatments • **Followed by appropriate disposal** in conformance with national landfill acceptance criteria (based on anticipated releases from the treated residuals).
Final disposal of residues	• Any residues containing or **contaminated with mercury should not be recycled.** • When disposed **in a landfill, evaluation of the release potential** and the appropriateness of the landfill for this type of material **should be considered.**

TABLE 28.8 Limit concerning the concentration of heavy metals (mg/L) required by standard GB 8978.

Hg	Cd	Total Cr	Cr(VI)	As	Pb	Ni	Cu	Zn	Mn
0.05	0.1	1.5	0.5	0.5	1	1	2	5	5

leaching methods are widely used and are usually associated with the corresponding standards to assess the potential environmental risk of heavy metals in the residue-derived products. However, the group I methods cannot provide insights into the release rates of heavy metals into the environment because the leaching conditions of specific scenarios (i.e., pH, pe, L/S, effective diffusion coefficient, etc.) are not carefully taken into account.

Compared with the group I methods, the group II leaching methods are not so widely used and more complicated because these methods are usually time-consuming, labor-intensive, and costly. However, geochemical speciation modeling and reactive mass transport modeling based on the leaching test results of group II methods can provide insights into the mechanisms controlling the LSP and transport of heavy metals. Moreover, the reactive mass transport model developed based on the test results in conjunction with the field leaching conditions characterization can be used to accurately assess the release rates of different heavy metals under various disposal and utilization scenarios (Chen et al., 2022; Kosson et al., 2002), providing useful feedback to evaluate and further optimize the disposal and utilization methods. Therefore, the use of group II leaching methods is strongly encouraged for assessing the potential environmental impact of combustion/incineration residue-derived products.

Pretreatment of MSWI FA is necessary prior to utilization because of its high content of multiple leachable heavy metals, PCDD/Fs, and Cl. Thermal treatment methods including sintering, melting, and vitrification are usually preferred for the modification and detoxification because they can effectively immobilize heavy metals, decompose PCDD/Fs, and remove Cl. However, the issues associated with thermal treatment are the high energy cost and the generation of secondary pollutants including secondary fly ash, acid gases in the off-gas, and the reformation of PCDD/Fs. These issues should be paid special attention to in the future.

The utilization of MSWI FA and CFA with high Ca content in construction and building materials, coupled with CO_2 curing, potentially has the benefits of decreasing carbon footprint (i.e., directly mineralizing CO_2 and indirectly decreasing CO_2 generation from cement production due to the substitution of FA and CFA for cement), improving the mechanical strength of derived products, and further immobilizing heavy metals contributed by the formation of carbonate-bearing minerals (Chen et al., 2023a,b; Sun et al., 2023). Nevertheless, more mechanistic studies and the pilot- and full-scale studies are urgently needed to prove the good potential of this method because it is still a relatively new territory to explore.

7. Conclusions

The combustion/incineration residues including MSWI FA, MSWI BA, and CFA have the good potential to be used to produce cement-based and ceramic products or to be used in geotechnical and agricultural applications, which is beneficial for waste reduction, energy and resource conservation, and reduction of greenhouse gas emissions. However, the potential release of COPCs from the residue-derived products is of the primary concern in terms of the environmental impacts. The group I leaching methods can be used to quickly evaluate the release of heavy metals from the residue-derived products, but the results hardly provide insights into the mechanisms controlling heavy metal release and the leaching conditions of specific scenarios are not carefully considered. The group II leaching methods are more comprehensive with many factors (e.g., pH, pe, L/S, and pore structure) taken into account. The test results of group II methods in conjunction with models development can provide insights into the mechanisms controlling the LSP and transport of heavy metals, and the developed models can be used for scenario-specific assessment in terms of heavy metals release into the environment. The CFA has been routinely used in construction and building materials, but the utilization of MSWI FA and BA usually requires pretreatment to modify their properties and effectively decrease their toxicities. Among three residues, MSWI FA is the most challenging one for utilization because of the enrichment of multiple inorganic and organic contaminants and chlorine. China has released a specific guideline for the utilization of MSWI FA, and thermal treatment methods are usually recommended. More studies focusing on addressing the issues associated with MSWI FA thermal treatment (e.g., secondary pollutants and waste control) are needed in the future.

Declaration of competing interest

The authors declare that they have no known competing financial interests or personal relationships that could have appeared to influence the work reported in this chapter.

References

Abis, M., Bruno, M., Kuchta, K., Simon, F.-G., Grönholm, R., Hoppe, M., Fiore, S., 2020. Assessment of the synergy between recycling and thermal treatments in municipal solid waste management in Europe. Energies 13, 6412.

Altarawneh, M., Dlugogorski, B.Z., Kennedy, E.M., Mackie, J.C., 2009. Mechanisms for formation, chlorination, dechlorination and destruction of polychlorinated dibenzo-p-dioxins and dibenzofurans (PCDD/Fs). Prog. Energy Combust. Sci. 35, 245–274.

Balaguera, A., Carvajal, G.I., Albertí, J., Fullana-i-Palmer, P., 2018. Life cycle assessment of road construction alternative materials: a literature review. Resour. Conserv. Recycl. 132, 37–48.

Basu, M., Pande, M., Bhadoria, P., Mahapatra, S., 2009. Potential fly-ash utilization in agriculture: a global review. Prog. Nat. Sci. 19, 1173–1186.

Berg, E.R., Neal, J.A., 1998. Municipal solid waste bottom ash as Portland cement concrete ingredient. J. Mater. Civ. Eng. 10, 168–173.

Bertolini, L., Carsana, M., Cassago, D., Curzio, A.Q., Collepardi, M., 2004. MSWI ashes as mineral additions in concrete. Cement Concr. Res. 34, 1899–1906.

Blasenbauer, D., Huber, F., Lederer, J., Quina, M.J., Blanc-Biscarat, D., Bogush, A., Bontempi, E., Blondeau, J., Chimenos, J.M., Dahlbo, H., 2020. Legal situation and current practice of waste incineration bottom ash utilisation in Europe. Waste Manag. 102, 868–883.

Blasenbauer, D., Huber, F., Mühl, J., Fellner, J., Lederer, J., 2023. Comparing the quantity and quality of glass, metals, and minerals present in waste incineration bottom ashes from a fluidized bed and a grate incinerator. Waste Manag. 161, 142–155.

Blissett, R., Rowson, N., 2012. A review of the multi-component utilisation of coal fly ash. Fuel 97, 1–23.

Bureau, E., 2005. Reference Document on the Best Available Techniques for Waste Incineration. European Union, Brussels.

Chandler, A.J., Eighmy, T.T., Hjelmar, O., Kosson, D., Sawell, S., Vehlow, J., Van der Sloot, H., Hartlén, J., 1997. Municipal Solid Waste Incinerator Residues. Elsevier.

Cheeseman, C., Da Rocha, S.M., Sollars, C., Bethanis, S., Boccaccini, A., 2003. Ceramic processing of incinerator bottom ash. Waste Manag. 23, 907–916.

Chen, Z., 2019. Mechanism Study of Mechanochemistry on PCDD/Fs Degradation and on Heavy Metals Stabilization in MSWI Fly Ash. Zhejiang University, Hangzhou. http://cdmd.cnki.com.cn/article/cdmd-10335-1019028701.htm.

Chen, Z., Chang, W., Jiang, X., Lu, S., Buekens, A., Yan, J., 2016. Leaching behavior of circulating fluidised bed MSWI air pollution control residue in washing process. Energies 9, 743.

Chen, Z., Lin, X., Lu, S., Li, X., Qiu, Q., Wu, A., Ding, J., Yan, J., 2018. Formation pathways of PCDD/Fs during the Co-combustion of municipal solid waste and coal. Chemosphere 208, 862–870.

Chen, Z., Lu, S., Tang, M., Ding, J., Buekens, A., Yang, J., Qiu, Q., Yan, J., 2019a. Mechanical activation of fly ash from MSWI for utilization in cementitious materials. Waste Manag. 88, 182–190.

Chen, Z., Lu, S., Tang, M., Lin, X., Qiu, Q., He, H., Yan, J., 2019b. Mechanochemical stabilization of heavy metals in fly ash with additives. Sci. Total Environ. 694, 133813.

Chen, Z., Zhang, S., Lin, X., Li, X., 2020. Decomposition and reformation pathways of PCDD/Fs during thermal treatment of municipal solid waste incineration fly ash. J. Hazard Mater. 394, 122526.

Chen, Z., Lin, X., Zhang, S., Xiangbo, Z., Li, X., Lu, S., Yan, J., 2021a. Thermal cotreatment of municipal solid waste incineration fly ash with sewage sludge for PCDD/Fs decomposition and reformation suppression. J. Hazard Mater. 416, 126216.

Chen, Z., Zhang, P., Brown, K.G., Branch, J.L., van der Sloot, H.A., Meeussen, J.C., Delapp, R.C., Um, W., Kosson, D.S., 2021b. Development of a geochemical speciation model for use in evaluating leaching from a cementitious radioactive waste form. Environ. Sci. Technol. 55, 8642–8653.

Chen, Z., Zhang, P., Brown, K.G., van der Sloot, H., Meeussen, J., Garrabrants, A., DeLapp, R., Kosson, D., 2022. Modeling of oxidized cementitious waste forms: oxidation depths estimation and Tc leaching assessment under field scenarios. Waste Manag. Symp., 22040

Chen, J., Shen, Y., Chen, Z., Fu, C., Li, M., Mao, T., Xu, R., Lin, X., Li, X., Yan, J., 2023a. Accelerated carbonation of ball-milling modified MSWI fly ash: migration and stabilization of heavy metals. J. Environ. Chem. Eng. 11, 109396.

Chen, J., Zhu, W., Shen, Y., Fu, C., Li, M., Lin, X., Li, X., Yan, J., 2023b. Resource utilization of ultrasonic carbonated MSWI fly ash as cement aggregates: compressive strength, heavy metal immobilization, and environmental-economic analysis. Chem. Eng. J. 472, 144860.

Chen, Z., Zhang, P., Brown, K.G., van der Sloot, H.A., Meeussen, J.C., Garrabrants, A.C., Delapp, R.C., Um, W., Kosson, D.S., 2023c. Evaluating the impact of drying on leaching from a solidified/stabilized waste using a monolithic diffusion model. Waste Manag. 165, 27–39.

Chen, Z., Zhang, P., Brown, K.G., van der Sloot, H.A., Meeussen, J.C., Garrabrants, A.C., Wang, X., Delapp, R.C., Kosson, D.S., 2023d. Impact of oxidation and carbonation on the release rates of iodine, selenium, technetium, and nitrogen from a cementitious waste form. J. Hazard Mater. 449, 131004.

Collepardi, M., 2003. A state-of-the-art review on delayed ettringite attack on concrete. Cement Concr. Compos. 25, 401–407.

Committee, A., 2019. ASTM C618-19 Standard Specification for Coal Fly Ash and Raw or Calcined Natural Pozzolan for Use in Concrete. Barr Harbor Drive. Annual Book of ASTM Standards, West Conshohocken.

Crillesen, K., Skaarup, J., Bojsen, K., 2006. Management of Bottom Ash from WTE Plants. International Solid Waste Association (ISWA).

del Valle-Zermeño, R., Gómez-Manrique, J., Giro-Paloma, J., Formosa, J., Chimenos, J., 2017. Material characterization of the MSWI bottom ash as a function of particle size. Effects of glass recycling over time. Sci. Total Environ. 581, 897–905.

Deng, Y., Gong, B., Chao, Y., Dong, T., Yang, W., Hong, M., Shi, X., Wang, G., Jin, Y., Chen, Z.-G., 2018. Sustainable utilization of municipal solid waste incineration fly ash for ceramic bricks with eco-friendly biosafety. Mater. Today Sust. 1, 32–38.

Dou, X., Ren, F., Nguyen, M.Q., Ahamed, A., Yin, K., Chan, W.P., Chang, V.W.-C., 2017. Review of MSWI bottom ash utilization from perspectives of collective characterization, treatment and existing application. Renew. Sustain. Energy Rev. 79, 24–38.

Erol, M., Küçükbayrak, S., Ersoy-Mericboyu, A., 2008. Comparison of the properties of glass, glass–ceramic and ceramic materials produced from coal fly ash. J. Hazard Mater. 153, 418–425.

Fan, C., Wang, B., Ai, H., Liu, Z., 2022. A comparative study on characteristics and leaching toxicity of fluidized bed and grate furnace MSWI fly ash. J. Environ. Manag. 305, 114345.

Feng, W., Wan, Z., Daniels, J., Li, Z., Xiao, G., Yu, J., Xu, D., Guo, H., Zhang, D., May, E.F., 2018. Synthesis of high quality zeolites from coal fly ash: mobility of hazardous elements and environmental applications. J. Clean. Prod. 202, 390–400.

Ferreira, C., Ribeiro, A., Ottosen, L., 2003. Possible applications for municipal solid waste fly ash. J. Hazard Mater. 96, 201–216.

González, A., Navia, R., Moreno, N., 2009. Fly ashes from coal and petroleum coke combustion: current and innovative potential applications. Waste Manag. Res. 27, 976–987.

Hu, Y., Bakker, M., De Heij, P., 2011. Recovery and distribution of incinerated aluminum packaging waste. Waste Manag. 31, 2422–2430.

Huber, F., Blasenbauer, D., Aschenbrenner, P., Fellner, J., 2020. Complete determination of the material composition of municipal solid waste incineration bottom ash. Waste Manag. 102, 677–685.

Izquierdo, M., Querol, X., 2012. Leaching behaviour of elements from coal combustion fly ash: an overview. Int. J. Coal Geol. 94, 54–66.

Joseph, B., Mathew, G., 2012. Influence of aggregate content on the behavior of fly ash based geopolymer concrete. Sci. Iran. 19, 1188–1194.

Karthik, M.M., Mander, J.B., Hurlebaus, S., 2016. ASR/DEF related expansion in structural concrete: model development and validation. Construct. Build. Mater. 128, 238–247.

Kosson, D.S., van der Sloot, H.A., Sanchez, F., Garrabrants, A.C., 2002. An integrated framework for evaluating leaching in waste management and utilization of secondary materials. Environ. Eng. Sci. 19, 159–204.

Kosson, D., Sanchez, F., Kariher, P., Turner, L., DeLapp, R., Seignette, P., 2009. Characterization of Coal Combustion Residues from Electric Utilities—Leaching and Characterization Data. EPA-600/R-09/151.

Kumar, A., Samadder, S.R., 2017. A review on technological options of waste to energy for effective management of municipal solid waste. Waste Manag. 69, 407–422.

Kuo, W.-T., Liu, C.-C., Su, D.-S., 2013. Use of washed municipal solid waste incinerator bottom ash in pervious concrete. Cement Concr. Compos. 37, 328–335.

Lam, C.H., Ip, A.W., Barford, J.P., McKay, G., 2010. Use of incineration MSW ash: a review. Sustainability 2, 1943–1968.

Leckner, B., Lind, F., 2020. Combustion of municipal solid waste in fluidized bed or on grate—A comparison. Waste Manag. 109, 94–108.

Lederer, J., Bartl, A., Blasenbauer, D., Breslmayer, G., Gritsch, L., Hofer, S., Lipp, A.-M., Mühl, J., 2022. A review of recent trends to increase the share of post-consumer packaging waste to recycling in europe. Detritus 3, 3–17.

Lin, X., Ma, Y., Chen, Z., Li, X., Lu, S., Yan, J., 2020. Effect of different air pollution control devices on the gas/solid-phase distribution of PCDD/F in a full-scale municipal solid waste incinerator. Environ. Pollut. 265, 114888.

Lin, X., Mao, T., Chen, Z., Chen, J., Zhang, S., Li, X., Yan, J., 2021. Thermal cotreatment of municipal solid waste incineration fly ash with sewage sludge: phases transformation, kinetics and fusion characteristics, and heavy metals solidification. J. Clean. Prod. 317, 128429.

Lin, X., Wang, X., Ying, Y., Wu, A., Chen, Z., Wang, L., Yu, H., Zhang, H., Ruan, A., Li, X., 2023. Formation pathways, gas-solid partitioning, and reaction kinetics of PCDD/Fs associated with baghouse filters operated at high temperatures: a case study. Sci. Total Environ. 857, 159551.

Lynn, C.J., Ghataora, G.S., Dhir, R.K., 2016. Environmental impacts of MIBA in geotechnics and road applications. Environ. Geotech. 5, 31–55.

Mehta, P., Monteiro, P., 2006. Concrete. Microstructure, Properties and Materials. McGraw-Hill, New York, USA.

Ministry of Ecology and Environment of the People's Republic of China, 2020. Technical specification for pollution control of fly-ash from municipal solid waste incineration. https://www.mee.gov.cn/ywgz/fgbz/bz/bzwb/dqhjbh/xgbz/202009/W020200902556206483862.pdf.

Mühl, J., Skutan, S., Stockinger, G., Blasenbauer, D., Lederer, J., 2023. Glass recovery and production of manufactured aggregate from MSWI bottom ashes from fluidized bed and grate incineration by means of enhanced treatment. Waste Manag. 168, 321–333.

Müller, U., Rübner, K., 2006. The microstructure of concrete made with municipal waste incinerator bottom ash as an aggregate component. Cement Concr. Res. 36, 1434–1443.

Naik, T.R., Kraus, R.N., Chun, Y.-m., Ramme, B.W., Singh, S.S., 2003. Properties of field manufactured cast-concrete products utilizing recycled materials. J. Mater. Civ. Eng. 15, 400–407.

Nalbantoğlu, Z., 2004. Effectiveness of class C fly ash as an expansive soil stabilizer. Construct. Build. Mater. 18, 377–381.

Nath, P., Sarker, P., 2011. Effect of fly ash on the durability properties of high strength concrete. Procedia Eng. 14, 1149–1156.

Obe, R.K.D., De Brito, J., Lynn, C.J., Silva, R.V., 2017. Sustainable Construction Materials: Municipal Incinerated Bottom Ash. Woodhead Publishing.

Oehmig, W.N., Roessler, J.G., Blaisi, N.I., Townsend, T.G., 2015. Contemporary practices and findings essential to the development of effective MSWI ash reuse policy in the United States. Environ. Sci. Pol. 51, 304–312.

Pan, S.-Y., Chen, Y.-H., Fan, L.-S., Kim, H., Gao, X., Ling, T.-C., Chiang, P.-C., Pei, S.-L., Gu, G., 2020. CO2 mineralization and utilization by alkaline solid wastes for potential carbon reduction. Nat. Sustain. 3, 399–405.

Peng, Y., Ma, Y., Lin, X., Long, J., Li, X., 2023. Emission control and phase migration of PCDD/Fs in a rotary kiln incinerator: hazardous vs medical waste incineration. Waste Dispos. Sust. Energy 1–12.

Poulikakos, L., Papadaskalopoulou, C., Hofko, B., Gschösser, F., Falchetto, A.C., Bueno, M., Arraigada, M., Sousa, J., Ruiz, R., Petit, C., 2017. Harvesting the unexplored potential of European waste materials for road construction. Resour. Conserv. Recycl. 116, 32–44.

Quina, M.J., Bordado, J.C., Quinta-Ferreira, R.M., 2008. Treatment and use of air pollution control residues from MSW incineration: an overview. Waste Manag. 28, 2097–2121.

Saqib, N., Bäckström, M., 2015. Distribution and leaching characteristics of trace elements in ashes as a function of different waste fuels and incineration technologies. J. Environ. Sci. 36, 9–21.

Sarker, P., McKenzie, L., 2009. Strength and hydration heat of concrete using fly ash as a partial replacement of cement. In: Proceedings of the 24th Biennial Conference of the Concrete Institute Australia. Concrete Institute of Australia.

Shaheen, S.M., Hooda, P.S., Tsadilas, C.D., 2014. Opportunities and challenges in the use of coal fly ash for soil improvements—a review. J. Environ. Manag. 145, 249–267.

Shi, H.-S., Kan, L.-L., 2009. Leaching behavior of heavy metals from municipal solid wastes incineration (MSWI) fly ash used in concrete. J. Hazard Mater. 164, 750–754.

Siddique, R., 2010. Use of municipal solid waste ash in concrete. Resour. Conserv. Recycl. 55, 83–91.

Sokolar, R., Vodova, L., 2011. The effect of fluidized fly ash on the properties of dry pressed ceramic tiles based on fly ash—clay body. Ceram. Int. 37, 2879–2885.

Sorlini, S., Collivignarelli, M.C., Abbà, A., 2017. Leaching behaviour of municipal solid waste incineration bottom ash: from granular material to monolithic concrete. Waste Manag. Res. 35, 978–990.

Sun, C., Wang, L., Lin, X., Lu, S., Huang, Q., Yan, J., 2022. Low-carbon stabilization/solidification of municipal solid waste incineration fly ash. Waste Dispos. Sust. Energy 4, 69–74.

Sun, C., Ge, W., Zhang, Y., Wang, L., Xia, Y., Lin, X., Huang, Q., Lu, S., Tsang, D.C., Yan, J., 2023. Designing low-carbon cement-free binders for stabilization/solidification of MSWI fly ash. J. Environ. Manag. 339, 117938.

Šyc, M., Krausová, A., Kameníková, P., Šomplák, R., Pavlas, M., Zach, B., Pohořelý, M., Svoboda, K., Punčochář, M., 2018. Material analysis of Bottom ash from waste-to-energy plants. Waste Manag. 73, 360–366.

Šyc, M., Simon, F.G., Hykš, J., Braga, R., Biganzoli, L., Costa, G., Funari, V., Grosso, M., 2020. Metal recovery from incineration bottom ash: state-of-the-art and recent developments. J. Hazard Mater. 393, 122433.

Temuujin, J., van Riessen, A., MacKenzie, K., 2010. Preparation and characterisation of fly ash based geopolymer mortars. Construct. Build. Mater. 24, 1906–1910.

Tian, Y., Bourtsalas, A.T., Kawashima, S., Ma, S., Themelis, N.J., 2020. Performance of structural concrete using Waste-to-Energy (WTE) combined ash. Waste Manag. 118, 180–189.

Vassilev, S.V., Vassileva, C.G., 2007. A new approach for the classification of coal fly ashes based on their origin, composition, properties, and behaviour. Fuel 86, 1490–1512.

Wang, S., 2008. Application of solid ash based catalysts in heterogeneous catalysis. Environ. Sci. Technol. 42, 7055–7063.

Wang, S., Zhang, C., Chen, J., 2014. Utilization of coal fly ash for the production of glass-ceramics with unique performances: a brief review. J. Mater. Sci. Technol. 30, 1208–1212.

Wang, X., Garrabrants, A.C., Chen, Z., van der Sloot, H.A., Brown, K.G., Qiu, Q., Delapp, R.C., Hensel, B., Kosson, D.S., 2022a. The influence of redox conditions on aqueous-solid partitioning of arsenic and selenium in a closed coal ash impoundment. J. Hazard Mater. 428, 128255.

Wang, X., van der Sloot, H.A., Brown, K.G., Garrabrants, A.C., Chen, Z., Hensel, B., Kosson, D.S., 2022b. Application and uncertainty of a geochemical speciation model for predicting oxyanion leaching from coal fly ash under different controlling mechanisms. J. Hazard Mater. 438, 129518.

Wang, X., Garrabrants, A.C., van der Sloot, H.A., Chen, Z., Brown, K.G., Hensel, B., Kosson, D.S., 2023. Leaching and geochemical evaluation of oxyanion partitioning within an active coal ash management unit. Chem. Eng. J. 454, 140406.

Xuan, D., Tang, P., Poon, C.S., 2018a. Effect of casting methods and SCMs on properties of mortars prepared with fine MSW incineration bottom ash. Construct. Build. Mater. 167, 890–898.

Xuan, D., Tang, P., Poon, C.S., 2018b. Limitations and quality upgrading techniques for utilization of MSW incineration bottom ash in engineering applications—A review. Construct. Build. Mater. 190, 1091–1102.

Yan, D.Y., Tang, I.Y., Lo, I.M., 2014. Development of controlled low-strength material derived from beneficial reuse of bottom ash and sediment for green construction. Construct. Build. Mater. 64, 201–207.

Yao, Z.T., Ji, X., Sarker, P., Tang, J., Ge, L., Xia, M., Xi, Y., 2015. A comprehensive review on the applications of coal fly ash. Earth Sci. Rev. 141, 105–121.

Zha, F., Liu, S., Du, Y., Cui, K., 2008. Behavior of expansive soils stabilized with fly ash. Nat. Hazards 47, 509–523.

Zhang, Y., Soleimanbeigi, A., Likos, W.J., Edil, T.B., 2016. Geotechnical and leaching properties of municipal solid waste incineration fly ash for use as embankment fill material. Transport. Res. Rec. 2579, 70–78.

Zhang, Z., Wang, J., Liu, L., Shen, B., 2019. Preparation and characterization of glass-ceramics via co-sintering of coal fly ash and oil shale ash-derived amorphous slag. Ceram. Int. 45, 20058–20065.

Zhang, S., Chen, Z., Lin, X., Wang, F., Yan, J., 2020a. Kinetics and fusion characteristics of municipal solid waste incineration fly ash during thermal treatment. Fuel 279, 118410.

Zhang, Z., Wang, J., Liu, L., Ma, J., Shen, B., 2020b. Preparation of additive-free glass-ceramics from MSW incineration bottom ash and coal fly ash. Construct. Build. Mater. 254, 119345.

Zhang, Y., Wang, L., Chen, L., Ma, B., Zhang, Y., Ni, W., Tsang, D.C., 2021. Treatment of municipal solid waste incineration fly ash: state-of-the-art technologies and future perspectives. J. Hazard Mater. 411, 125132.

Zhang, Z., Li, Z., Yang, Y., Shen, B., Ma, J., Liu, L., 2022. Preparation and characterization of fully waste-based glass-ceramics from incineration fly ash, waste glass and coal fly ash. Ceram. Int. 48, 21638–21647.

Chapter 29

Life cycle and economic assessment on different utilization and treatment strategies of combustion and incineration residues

Claudia Labianca[1], Ilenia Farina[2], Francesco Colangelo[2], Narinder Singh[2], Francesco Todaro[3], Sabino De Gisi[3], Michele Notarnicola[3] and Daniel C.W. Tsang[4,5]

[1]*Department of Civil and Environmental Engineering, The Hong Kong Polytechnic University, Hong Kong, China;* [2]*Department of Engineering, University Parthenope of Naples, Centro Direzionale, Naples, Italy;* [3]*Department of Civil, Environmental, Land, Building Engineering and Chemistry (DICATECh), Polytechnic University of Bari, Bari, Italy;* [4]*State Key Laboratory of Clean Energy Utilization, Zhejiang University, Hangzhou, China;* [5]*Department of Civil and Environmental Engineering, The Hong Kong University of Science and Technology, Hong Kong, China*

List of abbreviations

A-ACI Aquatic acidification
A-ECO Aquatic ecotoxicity
A-EUT Aquatic eutrophication
ADP Abiotic depletion potential
AP Acidification potential
BAT Best available techniques
BIOA Biomass ash
CBA Coal bottom ash
CCA Cocombustion ash
CCR Coal combustion residues
CExD Cumulative exergy demand
CFA Coal fly ash
CT Copper tailings
D-CBP Double-step cold bonding pelletization
D-LWAs Double-step cold bonding pelletization
EP Eutrophication potential
EQ Ecosystem quality
FE Freshwater ecotoxicity
FGD Flue gas desulphurization
FRS Fossil fuel resource scarcity
FU Functional unit
FWE Freshwater eutrophication
GGBS Ground granulated blast furnace slag
GHGs Greenhouse gases
GWP Global warming potential
HH Human health
HTc Human carcinogenic toxicity
HTnc Human noncarcinogenic toxicity

Treatment and Utilization of Combustion and Incineration Residues. https://doi.org/10.1016/B978-0-443-21536-0.00005-8

HTP Human toxicity potential
ILCD International Reference Life Cycle Data System
IR Ionizing radiation
LCA Life cycle assessment
LCC Life cycle costing
LCI Life cycle inventory
LCIA Life cycle impact assessment
LO Land occupation
LWAs Lightweight aggregates
MDP Metal depletion potential
ME Mineral extraction
MRS Mineral resource scarcity
MS Marble sludge
MSWI BA Municipal solid waste incineration bottom ashes
MSWI FA Municipal solid waste incineration fly ashes
NRE Nonrenewable energy
ODP Ozone layer depletion potential
OPC Ordinary Portland cement
PMF Particulate matter formation
POCP Photochemical ozone creation potential
RE Resources
RI Respiratory inorganics
RO Respiratory organics
S-CBP Single-step cold bonding pelletization
S/S Stabilization/solidification
SCMs Supplementary cementitious materials
SSA Sewage sludge ash
T-ACI Terrestrial acidification
T-ECO Terrestrial ecotoxicity
WC Water consumption

1. Introduction

Climate change and sustainability have received much attention in the past decades. Residues have been and are still being disposed of in large quantities in landfill, but there are increasing efforts to recover and recycle the residues without harming the environment. These efforts are aligned with the EU's Circular Economy action plan (Commission of European Communities, 2015), which aims to reduce natural resource pressure and encourage sustainable consumption, waste prevention, and recycling. In China, the "Circular economy promotion law of the People's Republic of China" passed in 2008. This action determined a series of institutional arrangements making "reduce, reuse, and recycle" as the basic principles for the national economy (The Central People's Government of the People's Republic of China, 2008).

Nevertheless, there are many residues that can contain potentially hazardous substances, so they may need to be treated before they can be used for various applications, and/or special conditions must be set up prior their use to prevent harmful effects on the environment or human health. The Waste Framework Directive 2008/98/EC (European Union, 2008) provides a hierarchy for waste treatment: prevention > preparing for reuse > recycling > energy recovery > disposal (Fig. 29.1). In the Directive, it is stated that "*the first objective of any waste policy should be to minimize the negative effects of the generation and management of waste on human health and the environment.*"

The recycling of various combustion and incineration ashes, including municipal solid waste incineration fly and bottom ashes (MSWI FA, MSWI BA), coal fly and bottom ashes (CFA, CBA), ground granulated blast furnace slags (GGBS), biomass ashes (BIOA), cocombustion ashes (CCA), sewage sludge ashes (SSA), and more, was thoroughly examined from an environmental and economic perspective. The potential of waste materials as resources recovery is also scrutinized. To ensure the chemical reactivity of qualified waste materials, they should contain an appropriate amount of amorphous SiO_2, Al2O_3, and Fe_2O_3 along with small particle sizes so that sufficient reactive surfaces can be generated during pozzolanic reactions (Yin et al., 2018). CFA typically accounts for 70–85 wt% up to 90–95 wt% of the total ash generated (Mikulčić et al., 2016). Although CFA leachates contain traceable amounts of As, Cr, Cu, Ni, and Zn, they typically do not exhibit significant ecotoxicity; therefore, they are not regulated as hazardous wastes in the United States according to the Resource Conservation and Recovery Act (RCRA), 42 USC 6901-6991, and the Environmental Protection

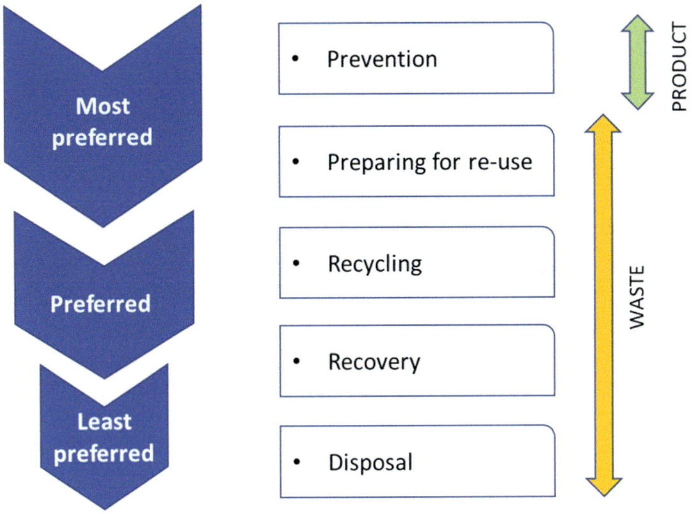

FIGURE 29.1 The waste hierarchy as described in the Waste framework directive. *Adapted from European Union, 2008. Directive 2008/98/EC of the European parliament and of the council on waste and repealing certain directives. Off. J. Eur. Union 2008;L312, 3–20.*

Agency (EPA) (Vassilev and Vassileva, 2007). About the 20% is used in concrete production, while other applications include road base construction, use as a filler in polymers, soil amendment, and zeolite synthesis. However, a large portion of CFA is still unused and disposed of as waste (Mikulčić et al., 2016).

It has been estimated that MSWI FA accounts for between 3% and 15% of input waste depending on different MSWI furnaces (e.g., fluidized bed furnace and moving grate furnace) (Dou et al., 2017; Zhang et al., 2021). Unlike CFA, MSWI FA usually contains significant amounts of toxic organic compounds such as dioxins and furans, as well as a high content of chlorides, sulfates, and heavy metals, which are carcinogenic to humans, thereby making it a hazardous waste (Huang et al., 2018). In addition, MSWI FA is extremely irregular in shape, with significant amounts of irregular particles and rods, together with filamentous crystals.

For what concerns bottom ashes, in many countries, CBA is considered as a nonhazardous waste and is used broadly in construction applications despite an official legal certainty. For waste incineration, about 30% by mass of the residues produced from the combustion process is MSWI BA. Stone, bricks, ceramics, glass, and unburned organic materials (wood, plastic, and fibers) are the main components of MSWI BA. Currently, the recycling of MSWI BA into engineering applications is common, especially in EU countries, such as the United Kingdom (45% for blended cement), France (50% for aggregates in concrete), and Portugal (50% for cement clinker) (Yin et al., 2018). The up-to-date utilization rate of steel slags varies among countries, reaching high percentages close to 100% for countries such as the United States, Germany, France, and Japan, but also low values like only 22% in China. However, many other slags from different thermal processes carry promising recycling in the cement industry, engineering applications, fertilizer production, and landfill daily cover. BIOA is collected after biomass carbonization, such as rice husks, sugarcane bagasse, corncobs, palm fruit bunch, wood chips, and any agriwaste. It can partially replace cement or substitute fine aggregates in concrete production. In some cases, they may contain contaminating elements such as Cd and sporadically As, Cu, Pb, Sn, and Zn may exceed limit values given by national regulations (Maschowski et al., 2020) for wood ash permitted for application in cement products.

Also, incineration is one of the most common disposal methods for sewage sludge since it achieves large volume reduction, odor minimization, and thermal destruction of toxic organic compounds (Fytili and Zabaniotou, 2008). Thanks to its pozzolanic properties and high Si-content, SSA can be used in concrete applications, leading to an increase of compressive strength as demonstrated in literature (Fytili and Zabaniotou, 2008). Hossain et al. (2020) used SSA as a low-carbon and sustainable strategy as an alternative to supplementary cementitious materials (SCMs) in the stabilization/solidification (S/S) treatment.

In the review by Yin et al. (2018), CaO was revealed to be the limiting component for the recycling of coal slags and GGBS in cement, with the mixing percentage ranges equal to 8.8–13.0% and 18.1–23.0%, respectively. Salt contents in MSWI BA and BIOA are the limiting component, allowing a quantity generally not higher than 7.0% and 5.5%, respectively. The chloride content in MSWI FA limits its use for construction applications. In this regard, it is common to pretreat the ashes with a water washing process, promoting the formation of hydrate phases, by converting heavy metals into less reactive forms (Zhang et al., 2022). However, although the quantity of chlorides and sulfates in MSWI FA could be curtailed by a washing pretreatment, this would also cause a loss of its pozzolanic activity. Therefore, alternative methods were proposed, such as the mechanochemical treatment as dry or wet milling process (Yuan et al., 2021) and roasting with earth elements extraction (Taggart et al., 2018). Wu et al. (2022) activated coal fly ash by NaOH roasting, effectively destroying the crystalline and amorphous phase in the ashes and with a recovery ratio of rare earth elements above 90%. According to current cement production practices, ashes are generally introduced in proportions ranging from 5% to 10%, up to 30% in some cases (Yin et al., 2018).

One of the main concerns is the environmental and economic sustainability of the ash recycling processes. In fact, the need of pretreatments with subsequent production of waste and wastewater may arise questions about the overall sustainability of the treatment. In this regard, life cycle assessment (LCA) represents a worldwide accepted and reliable tool that is able to globally quantify the total impacts of the treatment and recycling (ISO, 2006a, 2006b). An environmental LCA consists of compiling and analyzing inputs, outputs, and possible environmental impacts of a product system (Panesar et al., 2019). However, even though LCA studies on incorporating ashes into cement production or replacement provided promising results, they rarely addressed the long-term fate and transport of pollutants together with durability concerns (Huntzinger and Eatmon, 2009; Teixeira et al., 2016). In addition, there is some uncertainty about how to allocate impacts when dealing with by-products and waste materials. Several researchers have conducted LCA analyses of concrete containing PFA as a cement replacement, but disagreements persist over whether upstream production impacts should be included (Seto et al., 2017). A major benefit of recycling combustion and incineration residues comes from the reduction of greenhouse gas emissions and landfilling as well as the greater credit related to the avoided impacts (Yin et al., 2018).

In accordance with the SETAC guideline, the Environmental-Life Cycle Costing (E-LCC) is performed in compliance with the standards ISO 14040 and 14044 (ISO, 2006a, 2006b). The E-LCC usually uses the same LCA functional unit, system boundaries, and inventory to provide consistency to the analysis. By considering externalities, also the economic impact of emissions can be evaluated (Hunkeler et al., 2008). According to Liu et al. (2021), "true" costs of technologies can be revealed by estimating externalities such as environmental and social costs.

In the following sections, the environmental and economic benefits of ash recycling and resource recovering will be investigated, with a focus also on the main limitations of the treatment and application.

2. Life cycle assessment of treatment and recycle of combustion and incineration residues

2.1 Scope of the studies

The production of green aggregates from recycled ashes together with partial replacement of cement are among the main objectives of the literature studies (Dandautiya and Singh, 2019; O'Hare et al., 2020; Tosti et al., 2020). Furthermore, the recovery of materials such as aluminum, ferrous, stainless steel, and copper scrap was also considered by some authors (Allegrini et al., 2015; Arena et al., 2015).

In the study by Zanoletti and Ciacci (2022), the reduction in the environmental impacts related to the production of the flame retardant with stabilized MSWI FA as an alternative to traditional flame retardant was considered as the main scope of the study. In general, all the studies investigated the advantages of recycling and recovering combustion and incineration residues in comparison with conventional treatments and disposal scenarios (Table 29.1).

2.2 Functional units and system boundaries

The functional unit (FU) of the scenarios considered in the literature studies is represented by 1 tonne (dry basis) of residues or ashes in most of the cases (Arena et al., 2015; Boesch et al., 2014; da Costa et al., 2020; Huber and Fellner, 2018; Margallo et al., 2013; Tosti et al., 2020). However, 1 m^3 of concrete containing recycled ashes is also often chosen as FU (Dandautiya and Singh, 2019; Labianca et al., 2022; Seto et al., 2017; Teixeira et al., 2016). Lastly, 1 kg or 1 tonne of final products such as recycled composites and aggregates was considered by (O'Hare et al., 2020; Zanoletti and Ciacci, 2022).

For what concerns the system boundaries, if the main objective of the study was the replacement of a conventional material such as cement, a cradle-to-grave approach was often adopted to evaluate different mix designs starting from the production and transport of the raw materials (Dandautiya and Singh, 2019; O'Hare et al., 2020; Seto et al., 2017; Teixeira et al., 2016). Alternatively, a gate-to-grave approach was chosen when the main focus was the treatment or recycle of residues, and it included the incineration process, the transport, and processes involved in the treatment and end-of-life of residues (da Costa et al., 2020; Margallo et al., 2013). It is common to adopt the "zero-burden" assumption for all the combustion and incineration residues, i.e., the environmental impacts related to life cycle stages prior to the

TABLE 29.1 LCA studies in literature of treatment and recycle of combustion and incineration residues.

References	Type of ashes	Scope of the study	FU and system boundaries	LCI database, methodology and impact categories	Main results
Allegrini et al. (2015)	MSWI BA	1) Resource recovery instead of treatment and disposal. 2) Alternative configurations and utilization options for the treated MSWI BA.	1 tonne of MSWI BA.	Ecoinvent v.2.2. (Hauschild et al., 2013) for methodology selection. GWP, AP, MRS, HTc, HTnc and FE.	Landfilling, road subbase, and aggregate in concrete were compared as disposal methods. During operation, metal recycling and leaching from aggregates were associated with toxic impacts, whereas large nontoxic savings were found.
Arena et al. (2015)	1) APC system; 2) bottom ashes, slags and metals; 3) APC residues	1) APC system treated through hydrated lime absorption and activated carbon adsorption. 2) Bottom ashes treated with 60% for metal and inert material recovery, 40% to landfill. Full recovery of slags and metals. 3) APC residues for underground deposit.	1 tonne of residual waste.	Direct data and Ecoinvent 3.01. Impact 2002+. HH, EQ, RE. HTc, HTnc, RI, RO, IR, ODP, A-ACI, A-ECO, T-ACI, T-ECO, A-EUT, LO, GWP, NRE, and ME.	40% of bottom ash is sent to landfill with arsenic and antimony emissions into the water as leachate. The recovered inert materials were assumed to replace gravel for road backfilling, whereas metals substituted steel and aluminum.
Boesch et al. (2014)	MSWI residues	1) Fly ash treatment and disposal; 2) landfilling with 100% cement binder; 3) landfilling with 50% alternative binder; 4) underground deposit; 5) conventional acidic fly ash scrubbing (FLUWA); 6) acidic fly ash scrubbing with integrated Zn recycling (FLUREC).	1 tonne of MSW. Gate-to-grave.	Ecoinvent 2.2. ReCiPe and IPCC 2007. GWP, CExD and aggregated recipe categories.	Zinc recovery had the largest savings. The main differences in terms of impacts between FLUWA and FLUREC treatments are the different impacts and recovery rates of the processes.
da Costa et al. (2020)	Woody biomass ash from power plant	To assess the environmental impacts of woody biomass ash 1) landfilling or 2) landfarming for soil amelioration, as 2a) liming and 2b) fertilization.	1 tonne (dry basis) of woody biomass ash. Gate-to-grave.	Ecoinvent. ILCD. USEtox method. GWP, PMF, POCP, MRS, FRS, FEW, AP, HTc, HTnc, and FE.	Woody biomass ash landfarming obtained satisfactory performance compared with landfilling. By substituting hydrated lime for woody biomass ash for liming, more environmental savings were achieved.

Continued

TABLE 29.1 LCA studies in literature of treatment and recycle of combustion and incineration residues.—cont'd

References	Type of ashes	Scope of the study	FU and system boundaries	LCI database, methodology and impact categories	Main results
Dandautiya and Singh (2019)	CFA and copper tailings (CT)	Combined utilization of CFA and CT in concrete as a partial replacement of cement.	1 m^3 of concrete. Cradle-to-grave.	Ecoinvent 3.0. GWP, HTP, ODP, LO, WC, FRS, PMF, and MDP.	Environmental impacts were reduced up to: 38% in GWP, 32.6% in HTP, 33.6% in ODP, 31.9% in LO, 34.3% in WC, 34.8% in FRS, 35.4% in PMF, and 25.2% in MDP.
Huber and Fellner (2018)	MSWI FA	Disposal of MSWI FA at 1) underground deposits, 2) at aboveground landfills after cement stabilization, 3) application of the FLUREC process, 4) thermal treatment in a dedicated furnace, 5) thermal cotreatment with combustible hazardous waste.	Treatment and disposal of 1 Mg of MSWI FA.	Ecoinvent 3.2. All ReCiPe midpoint impact categories.	FLUREC process showed the lowest impacts in more than 90% of the results. When long-term emissions were neglected, the second lowest impact was related to thermal cotreatment. The most impactful treatments were stabilization with cement and thermal treatment in a dedicated furnace.
Labianca et al. (2022)	MSWI FA	To evaluate different mix designs with alkali-activated binders for the stabilization/solidification (S/S) of MSWI FA.	1 m^3 of the final product. Cradle-to-gate.	Ecoinvent 3.2. All ReCiPe midpoint impact categories.	Alkali-activated S/S allowed the reduction of CO_{2eq} between 55% and 71%. Alternative alkali production routes can reduce impacts in HTP and FRS by 23% and 28%.
Margallo et al. (2013)	W-t-E plant FA	To determine the environmental performance of FA stabilization and carbonation with different CO_2 sources, and different pressures (1–5 bar) and percentages of CO_2 excess (10%, 55%, and 100%) in the flue gas stream.	The amount of FA produced in 2014 in Cantabria, equal to 4655 t. Cradle-to-grave.	Literature data and Think step or Ecoinvent database. IPCC and CML methods and environmental sustainability assessment (ESA). GWP, AP, HTP, and EP.	The best carbonation process performance was found under flue gas pressures of 3–5 bar, since the total energy consumption decreases as pressure increases.
O'Hare et al., (2020) (in book rilem)	Power plant coal combustion residues (CCR)	To convert CBA to spherical porous reactive aggregates (SPoRA) using a sintering process and NaOH as fluxing agent of the rotary furnace to produce lightweight aggregates (LWAs).	1 tonne of lightweight aggregates. Cradle-to-gate.	CML 2015 method. GWP, ODP, EP, and AP.	CCR is a green raw material for SPoRA production.

TABLE 29.1 LCA studies in literature of treatment and recycle of combustion and incineration residues.—cont'd

References	Type of ashes	Scope of the study	FU and system boundaries	LCI database, methodology and impact categories	Main results
Petrillo et al. (2022)	MSWI FA, GGBS, and MS	To produce light-weight artificial aggregates as sustainable alternative to natural aggregates.	1 kg of double-step cold bonding pelletization (D-CBP). Cradle-to-gate.	Ecoinvent, literature data. ReCiPe midpoint categories.	As regards CO_2 emissions and water consumption, D-LWA containing the average value of 10% of OPC and GGBS represented the most advantageous mix design, with 23 kg CO_2-eq. and 0.16 m^3, respectively.
Seto et al. (2017)	CFA	Comparison of 1) no allocation, 2) mass allocation, 3) economic allocation, and 4) disposal avoidance of 10%, 25%, and 50% CFA as cement replacement.	1 m^3 of concrete. Cradle-to-grave.	Literature data. International References Life Cycle Data System (ILCD) method. AP, GWP, RE, and WC.	By increasing CFA content in concrete, environmental impacts can be reduced, but the extent of the reductions is highly dependent on the allocation.
Teixeira et al. (2016)	BIOA and CFA	Incorporating different percentages of two types of FA as cement replacement.	1 m^3 of concrete. Cradle-to-grave.	Ecoinvent 2.2. MARS—SC method. GWP, ODP, A-ACI, T-ACI, EP, POCP, and ADP.	The best results were found when 60% of cement was replaced by biomass FA. It was demonstrated that compressive strength reached a satisfactory performance even with a low OPC content.
Tosti et al. (2020)	BIOA	To reuse biomass fly ash as SCMs in cement mortars as alternative to the landfill scenario.	1 ton of biomass fly ash.	Ecoinvent, Highway_EASETECH Database, Literature data. All ReCiPe midpoint categories.	The use of BIOA in cement applications appeared beneficial in most of the impact categories, without adding any further risk.
Zanoletti and Ciacci (2022)	MSWI residues	The FA was stabilized involving the mixing of MSWI BA, flue gas desulfurization (FGD), residues, and CFA. Stabilized FA, calcite, and commercial flame retardants were compared as additives in an epoxy resin or polypropylene (PP) matrix.	1 kg of composites five and nine (30% epoxy resin, 70% stabilized fly ash (FA) or calcite, respectively). 1 kg of brominated and P-based flame retardant.	Ecoinvent, Literature data. All ReCiPe midpoint and endpoint categories.	It was demonstrated the stabilized FA self-extinguish properties to be similar to calcite. When stabilized FA was used as a replacement for calcite, the saving was 24% in epoxy resin matrixes, and 50% in PP matrixes.

Continued

TABLE 29.1 LCA studies in literature of treatment and recycle of combustion and incineration residues.—cont'd

References	Type of ashes	Scope of the study	FU and system boundaries	LCI database, methodology and impact categories	Main results
Zhang et al., 2022	MSWI FA	To comprehensively evaluate the economic and environmental impacts of MSWI FA as a resource for cement clinker, concrete block, and industrial salt.	1 ton of MSWI FA.	Ecoinvent. CML-IA categories. GWP, HTP, A-ECO, and WC.	The avoided impacts in GWP, primary energy demand, and HTP accounted for 15–53%, 26–65%, and 93–96% of the total impacts, respectively. Except for HTP, adopting salt recovery leads to 3.19–12.64 times greater environmental credit than coprocessing with cement kilns.

generation of waste are disregarded. Instead, material and energy inputs for transporting and treating the residues are usually gathered from direct laboratory measurements (Allegrini et al., 2015). In the LCA study by da Costa et al. (2020), the environmental burdens avoided thanks to the valorization of woody biomass ash were modeled through the system expansion by substitution (EC/JRC/IES, 2010). According to the application, a "credit" was given to the residues. (1) In the case of liming, woody biomass ash was replacing three conventional liming products (limestone, quicklime, and hydrated lime) (Goulding, 2016); (2) in the case of soil fertilization, woody biomass ash was replacing five conventional fertilizers (potassium chloride, potassium nitrate, potassium sulfate, single superphosphate, and triple superphosphate) (FAO, 2017). In the study by Tait and Cheung (2016), GGBS and CFA were considered by-products, and impacts were partially allocated to them. In fact, they are no longer considered as waste, since the European Union established the requirements to qualify for the by-product status with the Directive 2008/98/EC (European Union, 2008). At this regard, mass or economic allocation is commonly adopted to associate a partial burden to those materials (Hossain et al., 2021).

2.3 Life cycle inventories and software

Generally, life cycle inventories (LCIs) are based on both primary and secondary data. For key constituents, primary data are recommended, which are more accurate and are usually available directly from the manufacturers. Alternatively, LCA databases, such as Ecoinvent v. 3.9 (Fitzgerald and Sonderegger, 2017), EASETECH database (DTU, 2017), and literature data, were also used to collect inventory data on materials and processes (e.g., transportation, energy) (Table 29.1). The reviewed LCA studies were built mostly in SimaPro and GaBi software (PRé Consultants, 2019; Sphera Solutions, 2011). In addition to time and geographic coverage, scale, representativeness, consistency, reproducibility, and uncertainty are the main requirements for data (Duval-Dachary et al., 2023). A sensitivity analysis is highly recommended since each LCA phase is affected by data availability and quality.

2.4 Life cycle impact assessment

In life cycle impact assessment (LCIA), potential environmental impacts are evaluated as a result of the specific resource components from the LCI. It is a multidimensional process that categorizes these components as impacts on the environment. The most adopted LCIA methods are (1) ReCiPe and CML-IA, as multiimpact categories method, and (2) IPCC, as single impact category (Table 29.1). In particular, two main approaches are usually adopted within the ReCiPe methodology: midpoint and endpoint. The former correlates the impact to a specific change in the environment (e.g., GWP), and the latter correlates the same modification to final damage in a cause-effect relation (e.g., RE) (Olagunju and Olanrewaju, 2021).

2.5 Interpretation of the LCA results

The LCA studies indicate that ash recycling reduces raw material consumption by replacing Portland cement with ashes; it is also expected to reduce emissions to water (particularly total dissolved solids) and metal emissions to land (Babbitt and Lindner, 2008). Nevertheless, the major advantage stems from the credit associated with the avoided impacts (Labianca et al., 2022), such as reduced greenhouse gases (GHGs) and landfilling. As extensively demonstrated in literature, clinker production in Portland cement is the primary source of CO_2 emissions, making OPC responsible for 74−81% of the total carbon emissions of concrete (Jhatial et al., 2021). Therefore, the use of green binders or recycled aggregates has been investigated in the past decades. For instance, the use of alkali-activated binders instead of OPC could achieve a reduction of emissions up to 55−75% (Kajaste and Hurme, 2016; Labianca et al., 2022).

Among the different types of ashes, CFA involves a lower material emission factor (4 kg CO_2/t) compared with GGBS (52 kg CO_2/t), thanks to a reduced processing. However, in the mix design, the use of CFA is more limited compared with GGBS (7% for CFA vs. 20% for GGBS), and the overall impacts could reach higher values as well as durability studies should also be considered (García-Segura et al., 2014). A study by Hossain et al. (2017) demonstrated that the use of waste materials to replace virgin materials (clinker) can save up to 12% of the total GHGs in Hong Kong. Boesch et al. (2014) proposed an MSW management model based on the combustion with a grate incinerator, several flue gas treatment technologies, steam and electricity production from waste heat recovery, and metal recovery from slag and fly ash. They estimated that the energy recovery could result in a net CO_2-eq savings ranging from 67 to 752 kg, while recovering metals from slag and fly ash could result in a net saving of 35 kg.

Combustion and incineration residues have great potential in concrete applications. Labianca et al. (2022) demonstrated that alkali-activated binders could represent a sustainable alternative to OPC for the S/S of MSWI FA. In this study, GGBS was used as a precursor allowing up to 70% reduction of GWP. Furthermore, all the alkali-activated blocks were all suitable for landfill and reuse as fill material. Also, Tosti et al. (2020) demonstrated that the replacement of biomass ash as a secondary cementitious material in cement products performed environmentally better, when compared with the reference landfill scenario. However, only the utilization of paper sludge combustion ashes performed better than the reference case in the ecotoxicity (ET) category.

3. Technical and economic considerations

In waste handling systems, effective management of combustion and incineration residues is essential for the preservation of the environment and resource efficiency. This section explores the technical and economic factors that must be considered when evaluating various utilization and treatment strategies. By comprehending both the technical and economic viability of diverse approaches, decision-makers can make informed decisions that encourage sustainable waste management methods and minimize the negative environmental effects.

3.1 Technical aspects

Characterization and separation of waste materials is one of the first stages in the management of residues from combustion and incineration (Van der Sloot et al., 2001). The chemical composition, physical properties, and hazardous components of various residues may vary. Effective segregation permits targeted treatment and utilization strategies. Advanced analytical techniques, such as X-ray fluorescence, scanning electron microscopy, and leaching testing, can contribute to precise characterization. Furthermore, a variety of treatment technologies are available for dealing with combustion and incineration residues (Bosmans et al., 2013). Stabilization/solidification, vitrification, and thermal treatment are common techniques (Zhang et al., 2016). The choice of remediation technology is contingent on the type of residue, its hazard potential, and the intended end use. Each technique's efficacy must be thoroughly evaluated to guarantee conformance with environmental regulations and the production of safe, stable, and reusable materials. The management of combustion and incineration residues should prioritize the integration of material recovery and recycling processes (Bueno et al., 2015). Utilizing specialized recovery techniques, valuable resources such as metals and aggregates can often be extracted from these residues. By implementing recycling practices, industries can reduce their demand for virgin resources and promote a circular economy (Kurniawan et al., 2022). The viability of various application methods for combustion and incineration residues is contingent on the quality and usefulness of the outcomes. Construction materials (e.g., cement, concrete), road aggregates, landfill capping, and soil amendment are a few applications for residues. Performance, durability, and environmental impact of the products derived from these residues should be the focus of the technical evaluation.

3.2 Economic aspects

A comprehensive economic analysis is required to comprehend the financial ramifications of each utilization and treatment strategy. The cost analysis must incorporate the entire life cycle of the residues, from collection and treatment to final use (McDougall et al., 2008). To determine the overall economic viability of the chosen approach, factors such as initial investment, operating costs, transportation, and prospective revenue from recovered materials or products must be taken into account. Moreover, the market demand and value for the end products derived from combustion and incineration residues play a significant role in the economic evaluation (Shah et al., 2022). Understanding the prospective demand for construction materials, aggregate products, and other applications is crucial for determining the commercial viability and profitability of the generated materials. Compliance with local, regional, and international regulations is an essential component of the economic analysis (Dewi et al., 2019). Failure to comply with environmental standards may result in fines or additional waste management expenses. Consequently, it is essential to assess the costs associated with regulatory compliance and include them in the economic model. Long-term economic impacts are also an important factor to be considered. When evaluating various strategies, one must also consider the long-term economic implications. Evaluating the potential for long-term revenue generation, cost savings, and resource conservation aids in comprehending the long-term economic benefits of employing particular utilization and treatment strategies (Richardson et al., 2019). A comprehensive evaluation of both technical and economic factors is necessary for making informed decisions regarding the utilization and treatment of combustion and incineration residues, as stated in the conclusion. Waste management stakeholders can design sustainable strategies that mitigate environmental impact, conserve resources, and contribute to a circular economy by balancing technical feasibility and economic viability. By carefully considering these factors, society can progress toward a more sustainable and effective management of combustion and incineration residues. In one particular case of research by Petrillo et al. (2022), they prepared LWA aggregates from industrial waste by double-step cold bonding pelletization (D-CBP). Fly ashes from a municipal waste incineration plant, GGBS, and marble sludge (MS) were recycled in the mixture. Through a multicriteria decision analysis (MCDA), they compared and evaluated environmental impacts (through LCA), economic aspects (through LCC), and technical—functional aspects. The optimal mix design was identified as fly ash at 80%, blast furnace slag at 5%, and cement at 15% though various proportions were tried in the research. However, there are multiple expenses involved with the generation of recycled materials. LCA and LCC were the most significant aspects of this investigation. In the LCC section, they carried out economic analyses in accordance with the E-LCC, which corresponds to the SETAC recommendations (Hunkeler et al., 2008; Swarr et al., 2011). The guidelines contain all costs associated with a product's life cycle that must be recognized and measured. Critical aspects include which costs to incorporate and how to arrange the data, but this depends upon the study objective, subject matter application, and chosen perspective. In general, the following expenditure elements are present in this study.

- Direct production costs: raw materials, direct labor, etc.
- Indirect costs: overheads, indirect labor, waste treatment or pollutant abatement costs, etc.
- The externalities that are presumed to be internalized: costs related to CO_2 emissions (e.g., for taxes on carbon emissions or the purchase of negotiable permits).

3.3 Production costs

For the objectives of the analysis, the variable costs, which are the part of costs that varies with production variations, are of particular significance.

As a result, fixed expenses are excluded from the E-LCC analysis because they are constant irrespective of productivity or activity levels. Typically, the total expenses or costs consist of the aggregate of the initial costs, production costs, operating costs, and end-of-life (EoL) expenditures. Initial expenses consist of activities such as research and development, strategy, and design, as well as investment expenses. Consequently, the initial expenditures are predetermined. Because of this, they were omitted from the E-LCC calculation. As D-LWAs are prototype laboratories, the operating and EoL phases are not anticipated. In fact, the prototypes are meant for laboratory storage, for use in future testing. Also included in the production costs, there are fixed and variable expenses (Fig. 29.2). The cost of initial capital and labor comprises fixed production expenses. The cost of capital comprises equipment purchases. Labor costs consist of salaries paid to workers directly involved in the production of D-LWAs. These costs are excluded from the E-LCC evaluation because they do not vary with the aggregates production. Other costs not included are the fixed costs applied to energy and water expenditures, such as energy transport expenditure, system charges, and total taxes and VAT. Consequently, genuine water and electricity usage expenses are included. Resources are regarded as variable costs because they fluctuate based on production volume and are therefore included in the cost breakdown. Table 29.2 provides a summary of the model incorporated and

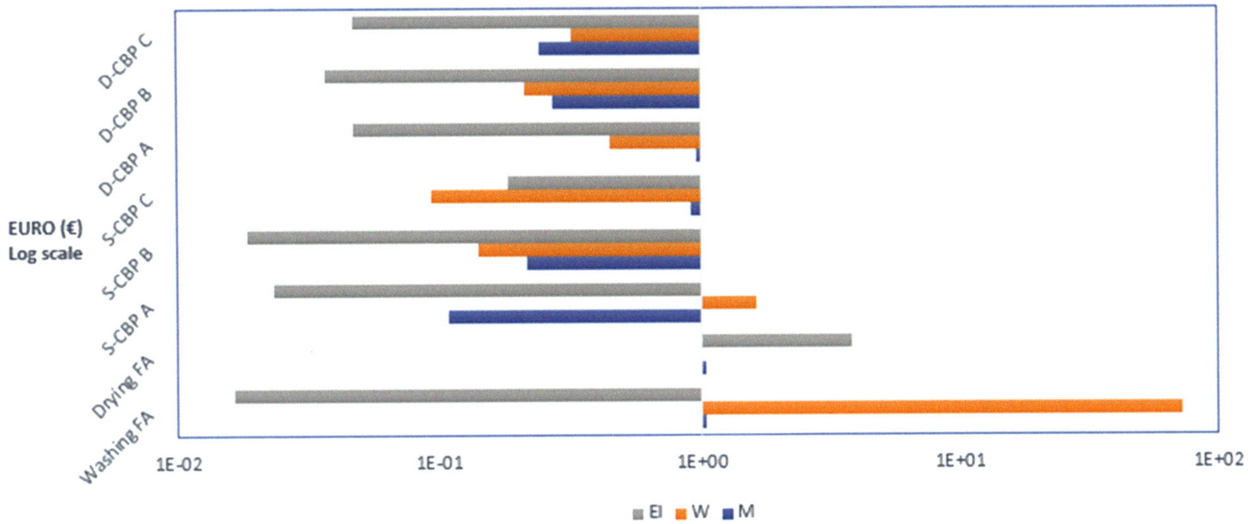

FIGURE 29.2 Production costs of each process, relating to materials (M), water resource (W), and electricity (El). *Adapted from Petrillo, A., Colangelo, F., Farina, I., Travaglioni, M., Salzano, C., Cioffi, R., 2022. Multi-criteria analysis for life cycle assessment and life cycle costing of lightweight artificial aggregates from industrial waste by double-step cold bonding palletization. J. Clean. Prod. 351, 131395. https://doi.org/10.1016/J.JCLEPRO.2022.131395.*

TABLE 29.2 Cost of items included and excluded from the analysis.

Included costs	Excluded costs
Purchase materials	Fixed production costs
Variable production costs	Initial costs
Actual water consumption	Salaries
Actual electricity consumption	Purchase of equipment
	• Electricity transport costs • System charges • Total taxes and VAT • Operation costs • EoL costs

omitted costs. For the model variable expenses, it is adequate to contemplate the purchase prices. In this cost model, the term "material" refers to the precursor (OPC, GGBS, MSWI FA, and MS); "resources" refers to the quantity of water used in the rinsing and pelletizing operations; and "energy cost" refers to the amount of electricity used in the manufacturing of aggregates. The by-products (GGBS, MSWI FA, and MS) are acquired from firms with whom they had supply agreements, whereas the OPC was acquired from unaffiliated companies. Costs for water and electricity are based on the rates established by each Italian national network with which the agency has signed supply contracts. In Table 29.3 are listed all material, resource, and electricity acquisition costs. By analyzing the costs associated with each individual process (washing, drying, single-CBP, and double-CBP), the expenses related to every component utilized for the manufacturing of lightweight artificial aggregates were determined. Materials (OPC, GGBS, MSWI FA, MS), electricity, and water make up these resources. The B blend is the finest overall, while the A blend is lowest in quality. This is due to the mix composition, specifically the OPC and GGBS amounts. In the single pelletization (S-CBP), OPC and GGBS are present at 10% concentration. The dehydration process consumes the most electricity (0.06 €), followed by the D-CBPs processes (0.04–0.05 €). In contrast, dehydration is characterized solely by the electrical component required to activate the furnace at 45°C for 24 h (Section 2.1.1), whereas double-step pelletization necessitates a higher energy consumption for the granulator. Lastly,

TABLE 29.3 Purchase costs of materials, electricity, and water resource (Petrillo et al., 2022).

Material (€/kg)	Costs per unit	Source
MSWI-FA	0.13	Southern Italy
OPC	0.56	Southern Italy
GGBS	0.23	Southern Italy
Electricity	0.0687	Enel S.p.A., Italy
MS	0.12	Southern Italy
Water	3.61	A2A S.p.A., Italy

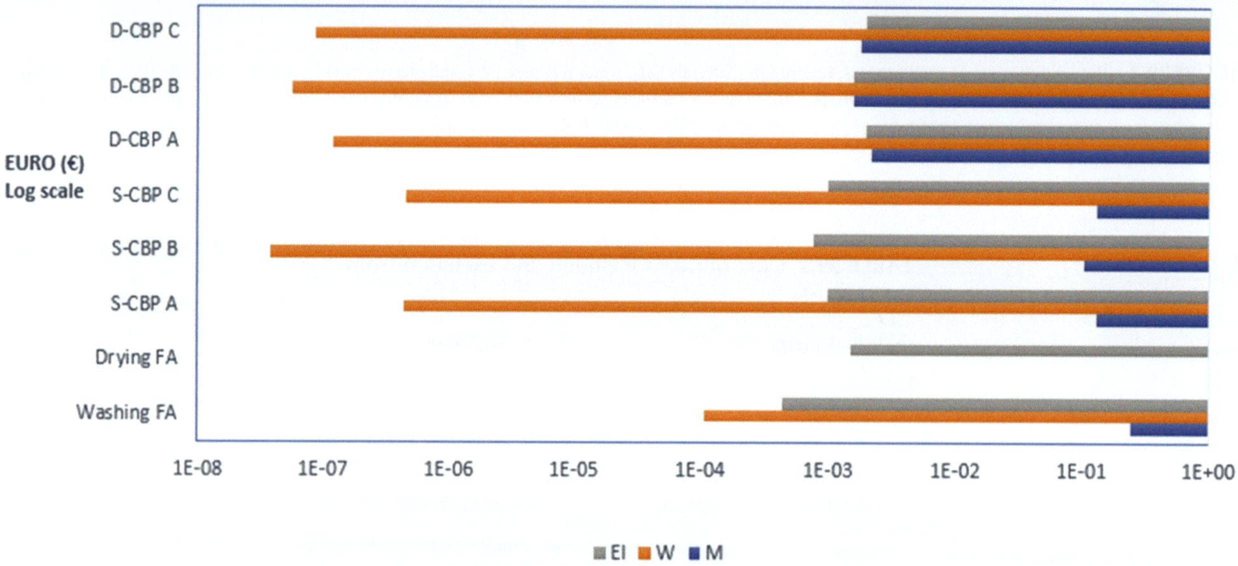

FIGURE 29.3 Externalities of each process, relating to materials (M), water resource (W), and electricity (El). *Adapted from Petrillo, A., Colangelo, F., Farina, I., Travaglioni, M., Salzano, C., & Cioffi, R., 2022. Multi-criteria analysis for Life Cycle Assessment and Life Cycle Costing of lightweight artificial aggregates from industrial waste by double-step cold bonding palletization. J. Clean. Prod., 351, 131395. https://doi.org/10.1016/J.JCLEPRO.2022.131395.*

washing MSWI FA consumes the most water (Fig. 29.3), which is the reason why washing has the highest costs (approximately 73 €) of all procedures. In both single-step and double-step cold bonding pelletization, the D-LWA B composition results in a decreasing use of water. This is due to the fact that the formation of aggregates in mixture B requires a lesser quantity of water than in other mixtures. Consider that the OPC:GGBS ratio for D-LWA B pelletization is 10%:10%.

3.4 Externalities

For the estimation of impacts, the proposed procedure uses the IPCC 2013 data as a reference. The IPCC-developed data presented here pertain to the effects of climate change, specifically the possibility of global warming over 100 years (GPW100). With respect to the GWP100, the total is indicated in EURO (€). The findings suggest that, in 100 years, investments should be made to reduce this influence on the environment to repair the harm triggered by emissions. Fig. 29.3 depicts the expenditures associated with repairing the harm to the environment caused by CO_2 emissions. Fig. 29.3 demonstrates that, for both materials and water, the costs associated with externalities are higher for washing. While electricity has significant externalities connected with the D-CBPs process, the dehydrating process has the highest values.

TABLE 29.4 Total costs of LWAs and individual processes, expressed in EURO (Petrillo et al., 2022).

Mixtures	Washing MSWIFA	SCBP	Drying MSWIFA	Total Costs	DCBP
D-LWA C	26568	2035	1395	30498	0500
D-LWA B	20752	0372	1089	22582	0369
D-LWA A	26571	1895	1395	30522	0661

3.5 Total cost

Internal expenses, referred to as manufacturing expenses, are added to outside influences, in particular emissions of carbon dioxide, to determine the total cost. Table 29.4 presents the total expenses for the three mixtures, including the cost of each process as a whole and the total cost. Based on the overall cost results (Table 29.4), it is evident that D-LWA B is the optimal aggregate, while D-LWA A and D-LWA C are equally crucial. In actuality, there is a 26% savings in costs between D-LWA B and D-LWA A (or D-LWA C). It is because of a reduced pelletization procedures for mixture B being significantly less dense than mixtures A and C. In reality, in both single and double pelletization, D-WA B requires a lower water input than the other aggregate formulations, resulting in less consumption cost and a reduced externality expenditure.

4. Future trends

The literature suggests the potential integration of numerous types of combustion and incineration residues into engineering applications together with their potential as resources recovery. Hence, this chapter aimed to explore the main insights of an LCA study when combustion and incineration residues are recycled or reused. As a result, there have been a number of "green" concrete concepts developed over the years, together with metal recovery designs, soil applications, and waste-to-energy strategies. A wide variety of techniques must be employed in this kind of sustainability assessments, including LCA, LCC, cost–benefit/effectiveness analysis, health risk assessments, multicriteria decision analysis, and social evaluations or also streamlined versions such as environmental ecological and footprint analyses (Hu et al., 2022).

Deviations from applying the directive hierarchy (Fig. 29.1) may also happen if motivated on the basis of life cycle thinking. Considerations such as no market for recycled products, high cost, high energy consumption, insufficient quality of recycled product may limit the utilization of the treated residues. Therefore, a comprehensive analysis of environmental and economic sustainability, together with compliance with technical specifications, and social acceptance are recommended to increase the utilization and to identify the best treatment of combustion and incineration residues.

References

Allegrini, E., Vadenbo, C., Boldrin, A., Astrup, T.F., 2015. Life cycle assessment of resource recovery from municipal solid waste incineration bottom ash. J. Environ. Manag. 151, 132–143. https://doi.org/10.1016/j.jenvman.2014.11.032.

Arena, U., Ardolino, F., Di Gregorio, F., 2015. A life cycle assessment of environmental performances of two combustion- and gasification-based waste-to-energy technologies. Waste Manag. 41, 60–74. https://doi.org/10.1016/j.wasman.2015.03.041.

Babbitt, C.W., Lindner, A.S., 2008. A life cycle comparison of disposal and beneficial use of coal combustion products in Florida - Part 1: methodology and inventory of materials, energy, and emissions. Int. J. Life Cycle Assess. 13 (3), 202–211. https://doi.org/10.1065/LCA2007.07.353/METRICS.

Boesch, M.E., Vadenbo, C., Saner, D., Huter, C., Hellweg, S., 2014. An LCA model for waste incineration enhanced with new technologies for metal recovery and application to the case of Switzerland. Waste Manag. 34 (2), 378–389. https://doi.org/10.1016/j.wasman.2013.10.019.

Bosmans, A., Vanderreydt, I., Geysen, D., Helsen, L., 2013. The crucial role of Waste-to-Energy technologies in enhanced landfill mining: a technology review. J. Clean. Prod. 55, 10–23.

Bueno, G., Latasa, I., Lozano, P.J., 2015. Comparative LCA of two approaches with different emphasis on energy or material recovery for a municipal solid waste management system in Gipuzkoa. Renew. Sustain. Energy Rev. 51, 449–459.

Commission of European Communities, 2015. Closing the Loop - An EU Action Plan for the Circular Economy; Communication No. 614; (COM (2015), 614). Commission of European Communities, Brussels, Belgium. https://eur-lex.europa.eu/resource.html?uri=cellar:8a8ef5e8-99a0-11e5-b3b7-01aa75ed71a1.0012.02/DOC_1&format=PDF. (Accessed 13 April 2023).

da Costa, T.P., Quinteiro, P., Tarelho, L.A.C., Arroja, L., Dias, A.C., 2020. Life cycle assessment of woody biomass ash for soil amelioration. Waste Manag. 101, 126–140. https://doi.org/10.1016/j.wasman.2019.10.006.

Dandautiya, R., Singh, A.P., 2019. Utilization potential of fly ash and copper tailings in concrete as partial replacement of cement along with life cycle assessment. Waste Manag. 99, 90–101. https://doi.org/10.1016/j.wasman.2019.08.036.

Dewi, N., Azam, S., Yusoff, S., 2019. Factors influencing the information quality of local government financial statement and financial accountability. Manag. Sci. Lett. 9 (9), 1373–1384.

Dou, X., Ren, F., Nguyen, M.Q., Ahamed, A., Yin, K., Chan, W.P., Chang, V.W.C., 2017. Review of MSWI bottom ash utilization from perspectives of collective characterization, treatment and existing application. Renew. Sustain. Energy Rev. 79, 24–38. https://doi.org/10.1016/J.RSER.2017.05.044.

DTU (Technical University of Denmark), 2017. EaseTech. Available at: http://www.easetech.dk/. (Accessed 14 April 2023).

Duval-Dachary, S., Beauchet, S., Lorne, D., Salou, T., Helias, A., Pastor, A., 2023. Life cycle assessment of bioenergy with carbon capture and storage systems: critical review of life cycle inventories. Renew. Sustain. Energy Rev. 183, 113415.

EC/JRC/IES, 2010. International reference life cycle data system (ILCD) handbook. In: General Guide for Life Cycle Assessment. Joint Research Centre - Institute for Environment and Sustainability. Luxembourg, Luxembourg, p. 417. https://doi.org/10.2788/38479.

European Union, 2008. Directive 2008/98/EC of the European parliament and of the council on waste and repealing certain directives. Off. J. Eur. Union L312, 3–20, 2008.

FAO, 2017. Resource Statistics - Fertilizers. URL. https://knoema.com/FAORSF2015/fao-resource-statistics-fertilizers?country=1001330-portugal.

Fitzgerald, D., Sonderegger, T., 2017. Documentation of Changes Implemented in the Ecoinvent Database V3. 9.1.

Fytili, D., Zabaniotou, A., 2008. Utilization of sewage sludge in EU application of old and new methods—a review. Renew. Sustain. Energy Rev. 12 (1), 116–140. https://doi.org/10.1016/J.RSER.2006.05.014.

García-Segura, T., Yepes, V., Alcalá, J., 2014. Life cycle greenhouse gas emissions of blended cement concrete including carbonation and durability. Int. J. Life Cycle Assess. 19 (1), 3–12. https://doi.org/10.1007/S11367-013-0614-0/TABLES/8.

Goulding, K.W.T., 2016. Soil acidification and the importance of liming agricultural soils with particular reference to the United Kingdom. Soil Use Manag. 32 (3), 390–399. https://doi.org/10.1111/sum.12270.

Hauschild, M.Z., Goedkoop, M., Guinée, J., Heijungs, R., Huijbregts, M., Jolliet, O., Margni, M., De Schryver, A., Humbert, S., Laurent, A., Sala, S., Pant, R., 2013. Identifying best existing practice for characterization modeling in life cycle impact assessment. Int. J. Life Cycle Assess. 18 (3), 683–697. https://doi.org/10.1007/S11367-012-0489-5/TABLES/4.

Hossain, M.U., Poon, C.S., Lo, I.M.C., Cheng, J.C.P., 2017. Comparative LCA on using waste materials in the cement industry: a Hong Kong case study. Resour. Conserv. Recycl. 120, 199–208. https://doi.org/10.1016/J.RESCONREC.2016.12.012.

Hossain, M.U., Wang, L., Chen, L., Tsang, D.C.W., Ng, S.T., Poon, C.S., Mechtcherine, V., 2020. Evaluating the environmental impacts of stabilization and solidification technologies for managing hazardous wastes through life cycle assessment: a case study of Hong Kong. Environ. Int. 145, 106139. https://doi.org/10.1016/J.ENVINT.2020.106139.

Hossain, M.U., Cai, R., Ng, S.T., Xuan, D., Ye, H., 2021. Sustainable natural pozzolana concrete – a comparative study on its environmental performance against concretes with other industrial by-products. Construct. Build. Mater. 270. https://doi.org/10.1016/j.conbuildmat.2020.121429.

Huang, T., Zhou, L., Liu, L., Xia, M., 2018. Ultrasound-enhanced electrokinetic remediation for removal of Zn, Pb, Cu and Cd in municipal solid waste incineration fly ashes. Waste Manag. 75, 226–235. https://doi.org/10.1016/J.WASMAN.2018.01.029.

Huber, F., Fellner, J., 2018. Integration of life cycle assessment with monetary valuation for resource classification: the case of municipal solid waste incineration fly ash. Resour. Conserv. Recycl. 139, 17–26. https://doi.org/10.1016/j.resconrec.2018.08.003.

Hu, G., Liu, H., Chen, C., He, P., Li, J., Hou, H., 2022. Selection of green remediation alternatives for chemical industrial sites: an integrated life cycle assessment and fuzzy synthetic evaluation approach. Sci. Total Environ. 845. https://doi.org/10.1016/j.scitotenv.2022.157211.

Hunkeler, D., Lichtenvort, K., Rebitzer, G., 2008. Environmental Life Cycle Costing. Crc press.

Huntzinger, D.N., Eatmon, T.D., 2009. A life-cycle assessment of Portland cement manufacturing: comparing the traditional process with alternative technologies. J. Clean. Prod. 17 (7), 668–675. https://doi.org/10.1016/J.JCLEPRO.2008.04.007.

ISO, 2006a. ISO 14040: Environmental Management: Life Cycle Assessment, Principles and Guidelines. International Organization for Standardization, Geneva.

ISO, 2006b. ISO 14044: International Standard. Environmental Management–Life Cycle Assessment- Requirements and Guidelines. International Organisation for Standardization, Geneva, Switzerland.

Jhatial, A.A., Goh, W.I., Mastoi, A.K., Rahman, A.F., Kamaruddin, S., 2021. Thermo-mechanical properties and sustainability analysis of newly developed eco-friendly structural foamed concrete by reusing palm oil fuel ash and eggshell powder as supplementary cementitious materials. Environ. Sci. Pollut. Control Ser. 28 (29), 38947–38968. https://doi.org/10.1007/S11356-021-13435-2/FIGURES/15.

Kajaste, R., Hurme, M., 2016. Cement industry greenhouse gas emissions – management options and abatement cost. J. Clean. Prod. 112, 4041–4052. https://doi.org/10.1016/J.JCLEPRO.2015.07.055.

Kurniawan, T.A., Othman, M.H.D., Hwang, G.H., Gikas, P., 2022. Unlocking digital technologies for waste recycling in industry 4.0 era: a transformation towards a digitalization-based circular economy in Indonesia. J. Clean. Prod. 357, 131911.

Labianca, C., Ferrara, C., Zhang, Y., Zhu, X., De Feo, G., Hsu, S.-C., You, S., Huang, L., Tsang, D.C.W., 2022. Alkali-activated binders – a sustainable alternative to OPC for stabilization and solidification of fly ash from municipal solid waste incineration. J. Clean. Prod. 380, 134963. https://doi.org/10.1016/j.jclepro.2022.134963.

Liu, F., Liu, H., Yang, N., Wang, L., 2021. Comparative study of municipal solid waste incinerator fly ash reutilization in China: environmental and economic performances. Resour. Conserv. Recycl. 169, 105541.

Margallo, M., Aldaco, R., Irabien, Á., 2013. Life cycle assessment of bottom ash management from a municipal solid waste incinerator (MSWI). Chemical Engineering Transactions 35, 871–876. https://doi.org/10.3303/CET1335145.

Maschowski, C., Kruspan, P., Arif, A.T., Garra, P., Trouvé, G., Gieré, R., 2020. Use of biomass ash from different sources and processes in cement. Journal of Sustainable Cement-Based Materials 9 (6), 350. https://doi.org/10.1080/21650373.2020.1764877.

McDougall, F.R., White, P.R., Franke, M., Hindle, P., 2008. Integrated Solid Waste Management: A Life Cycle Inventory. John Wiley & Sons.

Mikulčić, H., Klemeš, J.J., Vujanović, M., Urbaniec, K., Duić, N., 2016. Reducing greenhouse gasses emissions by fostering the deployment of alternative raw materials and energy sources in the cleaner cement manufacturing process. J. Clean. Prod. 136, 119–132. https://doi.org/10.1016/J.JCLEPRO.2016.04.145.

O'Hare, K.J., Pizzulli, G., Torelli, M., Balapour, M., Farnam, Y., Grace Hsuan, Y., Billen, P., Spatari, S., 2020. Life cycle assessment of lightweight aggregates from coal ashes: a cradle-to-gate analysis. RILEM Bookseries 26, 47–51. https://doi.org/10.1007/978-3-030-43332-1_10/TABLES/3.

Olagunju, B.D., Olanrewaju, O.A., 2021. Life cycle assessment of Ordinary Portland Cement (OPC) using both problem oriented (midpoint) approach and damage oriented approach (endpoint). In: Product Life Cycle-Opportunities for Digital and Sustainable Transformation. Intech Open, London.

Petrillo, A., Colangelo, F., Farina, I., Travaglioni, M., Salzano, C., Cioffi, R., 2022. Multi-criteria analysis for life cycle assessment and life cycle costing of lightweight artificial aggregates from industrial waste by double-step cold bonding palletization. J. Clean. Prod. 351, 131395. https://doi.org/10.1016/J.JCLEPRO.2022.131395.

Panesar, D.K., Kanraj, D., Abualrous, Y., 2019. Effect of transportation of fly ash: life cycle assessment and life cycle cost analysis of concrete. Cement Concr. Compos. 99, 214–224.

PRé Consultants, 2019. SimaPro 9.0049 Version. PRé Consultants, The Netherlands.

Richardson, L., Mammel, M., Milloy, M.J., Hayashi, K., 2019. Employment cessation, long term labour market engagement and HIV infection risk among people who inject drugs in an urban Canadian setting. AIDS Behav. 23 (12), 3267–3276.

Seto, K.E., Churchill, C.J., Panesar, D.K., 2017. Influence of fly ash allocation approaches on the life cycle assessment of cement-based materials. J. Clean. Prod. 157, 65–75. https://doi.org/10.1016/j.jclepro.2017.04.093.

Shah, A.V., Singh, A., Mohanty, S.S., Srivastava, V.K., Varjani, S., 2022. Organic solid waste: biorefinery approach as a sustainable strategy in circular bioeconomy. Bioresour. Technol. 349, 126835.

Sphera Solutions, 2011. GaBi Software. Available online: http://www.gabi-software.com. (Accessed 14 April 2023).

Swarr, T.E., Hunkeler, D., Klöpffer, W., Pesonen, H.L., Ciroth, A., Brent, A.C., Pagan, R., 2011. Environmental life-cycle costing: a code of practice. Int. J. Life Cycle Assess. 16 (5), 389–391. https://doi.org/10.1007/S11367-011-0287-5/TABLES/1.

Taggart, R.K., Hower, J.C., Hsu-Kim, H., 2018. Effects of roasting additives and leaching parameters on the extraction of rare earth elements from coal fly ash. Int. J. Coal Geol. 196, 106–114.

Tait, M.W., Cheung, W.M., 2016. A comparative cradle-to-gate life cycle assessment of three concrete mix designs. Int. J. Life Cycle Assess. 21 (6), 847–860. https://doi.org/10.1007/s11367-016-1045-5.

Teixeira, E.R., Mateus, R., Camões, A.F., Bragança, L., Branco, F.G., 2016. Comparative environmental life-cycle analysis of concretes using biomass and coal fly ashes as partial cement replacement material. J. Clean. Prod. 112, 2221–2230. https://doi.org/10.1016/J.JCLEPRO.2015.09.124.

Tosti, L., van Zomeren, A., Pels, J.R., Damgaard, A., Comans, R.N.J., 2020. Life cycle assessment of the reuse of fly ash from biomass combustion as secondary cementitious material in cement products. J. Clean. Prod. 245. https://doi.org/10.1016/j.jclepro.2019.118937.

Van der Sloot, H.A., Kosson, D.S., Hjelmar, O., 2001. Characteristics, treatment and utilization of residues from municipal waste incineration. Waste Manag. 21 (8), 753–765.

Vassilev, S.V., Vassileva, C.G., 2007. A new approach for the classification of coal fly ashes based on their origin, composition, properties, and behaviour. Fuel 86 (10–11), 1490–1512. https://doi.org/10.1016/J.FUEL.2006.11.020.

Wu, G., Wang, T., Chen, G., Shen, Z., Pan, W.P., 2022. Coal fly ash activated by NaOH roasting: rare earth elements recovery and harmful trace elements migration. Fuel 324, 124515.

Yin, K., Ahamed, A., Lisak, G., 2018. Environmental perspectives of recycling various combustion ashes in cement production – a review. Waste Manag. 78, 401–416. https://doi.org/10.1016/j.wasman.2018.06.012. Elsevier Ltd.

Yuan, Q., Zhang, Y., Wang, T., Wang, J., Romero, C.E., 2021. Mechanochemical stabilization of heavy metals in fly ash from coal-fired power plants via dry milling and wet milling. Waste Manag. 135, 428–436.

Zanoletti, A., Ciacci, L., 2022. The reuse of municipal solid waste fly ash as flame retardant filler: a preliminary study. Sustainability 14 (4). https://doi.org/10.3390/su14042038.

Zhang, Z., Li, A., Wang, X., Zhang, L., 2016. Stabilization/solidification of municipal solid waste incineration fly ash via co-sintering with waste-derived vitrified amorphous slag. Waste Manag. 56, 238–245.

Zhang, Y., Wang, L., Chen, L., Ma, B., Zhang, Y., Ni, W., Tsang, D.C.W., 2021. Treatment of municipal solid waste incineration fly ash: state-of-the-art technologies and future perspectives. J. Hazard Mater. 411 (November 2020), 125132. https://doi.org/10.1016/j.jhazmat.2021.125132.

Zhang, Z., Wang, Y., Zhang, Y., Shen, B., Ma, J., Liu, L., 2022. Stabilization of heavy metals in municipal solid waste incineration fly ash via hydrothermal treatment with coal fly ash. Waste Manag. 144, 285–293.

Chapter 30

Current bottlenecks and future directions on academic studies and industrial applications

Bojun Zhao[1], Caicai Xu[2], Hanyang Sun[3], Bin Du[1], Lei Wang[3], Bin Yang[2] and Chen Sun[2]
[1]*Key Laboratory of Special Equipment Safety Testing Technology of Zhejiang Province, Zhejiang Academy of Special Equipment Science, Hangzhou, China;* [2]*Institute of Zhejiang University-Quzhou, Quzhou, China;* [3]*State Key Laboratory of Clean Energy Utilization, Zhejiang University, Hangzhou, China*

1. Introduction

The global generation of various combustion and incineration residues is experiencing a consistent and continuous increase in the past decades, propelled by factors such as population growth, increasing demand for energy, urbanization, and improvements in lifestyle. The treatment of different kinds of combustion and incineration residues, such as fly ash (FA, collected from air pollution control device and bottom ash (BA, collected from the bottom of the furnace), has significant impacts on both the environment and the economy (Bilibio et al., 2021). Some original FAs, especially waste incineration FA without treatment, are often classified as hazardous waste in numerous countries due to their high contents of potentially toxic substances, including heavy metals (Pb, Hg, Cr, Cu, Zn, etc.) and persistent organic pollutants (POPs) (Silva et al., 2017; Yang et al., 2018; Yin et al., 2018). When the ash comes into contact with water in the environment, particularly acid rain or corrosive liquids, heavy metals and organic pollutants present in the ash residue can be released and emitted into the ecological system, leading to the contamination of groundwater, rivers, lakes, seas, and soil, finally threatening human's health (Gao et al., 2017).

Presently, there exist two main categories of purification and detoxification treatment processes for BA/FA: thermal treatment methods (such as sintering, melting, hydrothermal treatment, etc.) and nonthermal treatment methods (including cement-based immobilization, chemical agent–based immobilization, biochemical treatment, electrochemical and mechanochemical approaches, etc.), while each of these methods has its own merits and defects (Shunda et al., 2022). Nevertheless, most BA/FA after purification and detoxification still enter the landfill process, resulting in the squandering of land resources and a potential risk of leachable remaining heavy metals after a period of long time (Huang et al., 2022; Ma et al., 2021).

In recent years, the governments and researchers from various countries started to focus on the recycling resource utilization technologies of combustion/incineration residues (Kanhar et al., 2020). For example, apart from its hazardous components, FA is characterized by high levels of CaO, and thus the alkaline properties enable the potential recycling of FA as construction materials in sustainable engineering practices (Liu et al., 2021). The use of combustion/incineration residues as supplementary cementitious materials (SCMs) in blended cement industry has gained attention due to its potential to efficiently reduce CO_2 emissions (Juenger et al., 2019). Moreover, some critical valuable metal-enriched ashes could also serve as an ideal candidate for resource recovery. In addition, the BA/FA can also be used as raw materials for the preparation of functional materials, such as foaming agents, soil amendments, and zeolites (Zhao et al., 2021).

As global environmental concerns are increasing, there has been a notable enhancement in laws and regulations pertaining to environmental protection and resource recycling. To summarize, implementing green and sustainable treatment and recovery techniques for combustion/incineration residues can yield numerous advantages, including diminished ecological and human health hazards, decreased toxic air emissions, lower energy expenses, and increased

Treatment and Utilization of Combustion and Incineration Residues. https://doi.org/10.1016/B978-0-443-21536-0.00032-0

resource utilization efficiency. The objective of this chapter is to present a comprehensive overview of the current challenges regarding various utilization and recovery processes of combustion/incineration residues, and to provide future directions from the perspectives of academic research and industrial applications, with the goal of promoting future studies in this area.

2. Novel technologies for detoxification of combustion/incineration residues

Combustion or incineration residues often contain high concentrations of toxic substances, including heavy metals and hazardous organic compounds. Without treatment, these residues pose significant risks to human health and the environment. Therefore, it is of utmost importance to urgently address this pressing issue and implement effective treatment methods to mitigate the potential harm caused by these toxic residues (Luo et al., 2019).

2.1 Cement-based immobilization of combustion/incineration residues

Until now, cement solidification followed by landfill is currently the most convenient method of hazardous ashes disposal. It involves mixing the FA with cement or other gel materials, adding water, and stirring until uniform. After demolding and curing, the resulting cement blocks can be directly buried, efficiently reducing their harmful effects (Hu et al., 2015). Cementation can improve the physical and chemical stability of residues and make it a controllable material. Due to its convenience in operation, cement-based solidification has been widely utilized worldwide. Cement solidification offers significant advantages as it can immobilize heavy metals within hydration products such as calcium silicate hydrate gel (C—S—H) and ettringite (AFt), thereby inhibiting their leaching (Lam et al., 2011; Yin et al., 2018). Various cementitious materials, such as ordinary Portland cement (OPC), calcium aluminate cement (CAC), and alkali-activated cement (AAC), have been studied and identified for their ability to effectively immobilize combustion/incineration residues (Fan et al., 2018; Liu et al., 2013). OPC is the most widely utilized cement materials immobilizing FA. Extensive studies have demonstrated its effectiveness in immobilizing heavy metals, as well as reducing the leaching of both metals and organic compounds (Bogush et al., 2020). However, it is important to note that the long-term stability of OPC-based immobilization is still a significant concern, owing to the possibility of carbonation and the subsequent formation of active secondary minerals. It has been reported that the potential toxic elements in OPC-stabilized FA were no longer effectively immobilized after a natural aging period of 6 years (Du et al., 2018).

Due to its rapid setting and hardening properties, excellent resistance to acid, sulfates, and fire, CAC emerges as a promising alternative to OPC for solidification/stabilization of combustion/incineration residues, which can form more stable mineral phases that are less prone to leaching (Contessi et al., 2020). The addition of municipal solid waste incineration fly ash (MSWI FA) into CAC resulted in improved workability and higher mechanical strength, which is primarily attributed to the formation of monocarbon aluminate (López-Zaldívar et al., 2017). Moreover, AAC, a kind of low-carbon cementitious material typically produced from industrial by-products such as FA, slag, and metakaolin, is another highly promising material for immobilizing of combustion/incineration residues. In comparison with OPC, AAC demonstrates enhanced durability, particularly in aggressive environments. The estimated CO_2 savings associated with AAC vary from 9% to 97% when compared with OPC, depending on the specific design of the geopolymer binder mixture used (Provis and Bernal, 2014; Shi et al., 2019). However, although AAC can efficiently immobilize FA in laboratory studies, further investigation is necessary to optimize the performance of geopolymer cement and assess its long-term solidification/stabilization ability (Luna Galiano et al., 2011).

Overall, the immobilization of combustion/incineration residues using conventional and alternative cements shows the advantages of low cost, simple operation, and high efficiency. But its development is still faced with many problems, such as the significantly increasing volume of cement-based products, the uncertain long-term durability, and the risk of leaching heavy metals such as Cd and Zn in the long term (Lin et al., 2022). Hence, future research should incorporate the leaching behavior and environmental risks of high toxic elements in residue-based cementitious materials under complex conditions (atmospheric carbon dioxide, acid rain, various salts, etc.) and provide a better understanding of the actual landfill process and the interaction between with the cementitious materials and working environment.

2.2 Chemical agent—based immobilization of combustion/incineration residues

To mitigate the drawbacks associated with the increased mass and volume during the cement-based solidification/stabilization process, alternative chemical agent immobilization methods have been proposed for treating heavy metals in combustion/incineration residues (Du et al., 2019). Currently, the chemical agents can be divided into inorganic agents (such as phosphates, silicates, and sulfides) and organic stabilization agents (such as tetrathiobicarbamic acid, dithiocarbamates, sodium dimethyldithiocarbamate, thiourea, thiols, etc.) (Lin et al., 2022). There has been a growing interest in organic additives for their ability to effectively treat heavy metals in combustion/incineration residues, mainly attributed to their low cost and high tolerance for various environmental conditions (Luo et al., 2019). Organic reagents demonstrate stronger resistance to acid leaching when forming chelates with heavy metal ions.

However, the primary challenge faced by chemical stabilization methods is their limited effectiveness in addressing dioxin pollutants. Additionally, the chemical agents only exhibit good stability toward several specific heavy metals (Pan et al., 2022). Currently, there is a lack of a universally adaptable technology using chemical agents that can effectively address all heavy metals. Furthermore, the low environmental stability of inorganic agents and limited reusability of products pose significant challenges to their widespread adoption and industrial application (Lin et al., 2022). As chemical reagents exhibit selectivity toward specific heavy metals, it is possible to employ multiple chemical reagents or utilize a combination of chemical reagents with cement to achieve synergistic effects, reducing costs while enhancing the stability of heavy metal elements in future research.

2.3 Thermal treatment of combustion/incineration residues

Thermal treatment methods have emerged as promising technologies for the effective degradation of dioxins and immobilization of heavy metals, which have demonstrated impressive degradation rates for dioxins, reaching up to 99%, and exceptional removal rates for volatile heavy metals (Lindberg et al., 2015). At temperatures exceeding 650°C, the toxic equivalent of polychlorinated dibenzo-p-dioxins (PCDDs) and polychlorinated dibenzofurans (PCDFs) tend to decrease significantly or even undergo complete degradation (Kanhar et al., 2020). High-temperature thermal treatment includes two approaches: sintering and melting. In the sintering process, combustion/incineration residues are heated to temperatures between 700 and 1200°C (below the melting point), to promote the bonding and recrystallization of fine particles, resulting in the formation of compact polycrystalline materials. In contrast, melting treatment is typically conducted at temperatures higher than 1200°C. The operating temperature of the melting process exceeds the melting point of the incineration residues, leading to the disruption of crystalline structures and the formation of a uniform liquid phase without the addition of any substances (Lindberg et al., 2015).

During sintering and melting processes, the PCDD/F is effectively decomposed and the heavy metals are solidified simultaneously, while the products after thermal treatment could serve as raw materials of construction materials (Peng et al., 2020). However, the primary challenges associated with direct sintering and melting of combustion/incineration residues are the substantial energy consumption and the potential for secondary pollution due to the volatilization of heavy metal elements (Huang et al., 2022). Due to the extensive infrastructure requirements and high energy demands of traditional thermal treatments, they are currently not the predominant methods for the detoxification of combustion/incineration residues at present. In addition, given that heavy metals (such as Zn, Cu, Cd, Ni, Pb, Cr) in FA tend to vaporize at different extents under high-temperature conditions, finding ways to minimize heavy metal volatilization and effectively treat secondary FA remains a challenging issue that requires further research (Ma et al., 2021). The research regarding the migration behavior and microcharacteristics of heavy metals will be greatly helpful in addressing this issue. Furthermore, it is significant to highlight the importance of rapidly cooling MSWI FA following thermal treatment to prevent the resynthesis of PCDD/F. Achieving effective oxidation of PCDD/F and its precursors needs an aerobic environment, yet controlling the reformation of PCDD/F under such conditions remains challenging (Xue and Liu, 2021). Apart from solely utilizing high-temperature treatment for combustion/incineration residues, the investigation of cotreatment of combustion/incineration residues with other solid wastes under high temperature is an emerging and valuable research area to explore.

2.4 Hydrothermal treatment of combustion/incineration residues

Hydrothermal treatment is a significant thermochemical conversion technique that involves subjecting reactants to high temperature (200–374°C) and high pressure (4–22 MPa) within a closed reaction vessel (Goto, 2009). This approach utilized the favorable properties of water under subcritical condition, including its low dielectric constant, strong solubility, and the ability to degrade organic molecules (Xue and Liu, 2021). Hydrothermal treatment of combustion/incineration residues can be primarily classified into three categories: conventional hydrothermal method, additive-assisted hydrothermal method, and microwave-assisted hydrothermal method. Under alkaline hydrothermal conditions, silicon and aluminum elements can form aluminosilicate minerals. Through physical/chemical adsorption, ion exchange, and physical encapsulation, heavy metals can be immobilized within the aluminosilicate structure. Moreover, dioxins in the residues can be degraded in high-temperature and high-pressure aqueous solutions, facilitating the simultaneous treatment of heavy metals and dioxins (Jin et al., 2012). Recent research indicated that temperature exhibited the most significant influence on dioxins. By changing the rate of molecular diffusion, the degradation time of dioxins can be effectively shorten and the stability of heavy metals can be obviously enhanced (Huang et al., 2022).

In general, the treatment of combustion/incineration residues using hydrothermal methods is still undergoing exploration and development, including the research on finding the optimal temperature for dioxin degradation, improving the removal rate of toxic compounds with new additives, etc. Moreover, the investigation of the characteristics of hydrothermal solvents and mineralizers, the hydrothermal equipment, as well as the heat and mass transfer mechanisms in hydrothermal systems, will be significant in enhancing the effectiveness of hydrothermal treatment of combustion/incineration residues and promoting the industrial application of hydrothermal techniques. Simultaneously, exploring hydrothermal methods with other approaches will emerge as a prominent trend, leading to broader utilization of hydrothermal techniques for various types of residues.

2.5 Mechanochemical treatment of combustion/incineration residues

Mechanochemical (MC) treatment involves the utilization of intense mechanical force, such as collision, compression, shear, and friction, to induce chemical reactions between solid substances. Simultaneously, this process modifies the crystal structure and surface morphology of the materials (Zhang et al., 2021). MC treatment is considered a sustainable and environmentally friendly approach for waste disposal, characterized by its mild reaction conditions, simple operation, high efficiency, and absence of secondary pollution. In recent studies, MC treatment has been explored as an effective method for treating FA, offering the opportunity to simultaneously address heavy metal contamination while enhancing the quality of FA. Moreover, MC treatment enhances the reactivity of FA, leading to increased homogeneity and a reduction in particle size. In summary, MC treatment can be considered as an effective pretreatment approach for enhancing the activity of FA and mitigating the presence of potentially toxic elements (PTEs) (Jin et al., 2023; Xiang et al., 2022; Yuan et al., 2021). In addition, the presence of chlorides and sulfates in FA can be reduced through washing pretreatment, but this may lead to a decline in the pozzolanic activity of the material. MC treatment can also modify the properties of FA and mitigate the decrease in pozzolanic reactivity (Zacco et al., 2014). Nevertheless, high levels of FA can hinder the reaction process of SCMs in present research, resulting in inadequate formation of the reaction products, low mechanical strength, and increased leaching of contaminants. To facilitate the widespread use of FA-based SCMs, there is an urgent need for research on the specifications of FA-based SCMs, standardization, advanced characterization techniques, and long-term leaching behavior. These investigations are crucial to ensure the reliable and optimal performance of these materials (Zhang et al., 2021).

3. Recycling of combustion/incineration residues into construction materials

3.1 Recycling of combustion/incineration residues into cement clinker

Combustion/incineration residues contain similar mineral compositions with OPC, indicating a feasible recycling route for various ashes to the construction materials. Cement clinker production predominantly depends on the presence of critical elements (calcium, silicon, iron, and aluminum) in raw materials, which are usually obtained from natural sources such as limestone, sand, and bauxite (Viczek et al., 2020). Subsequently, the raw materials undergo the firing process in a rotary kiln, operating at temperatures between 1400 and 1500°C, resulting in the formation of cement clinker. Thus, the cement manufacturing industry is a high energy—consuming sector and a major contributor to carbon dioxide emissions (Clavier et al., 2020). Combustion/incineration residues are waste materials that have elemental compositions suitable for cement production, such as CaO, Al_2O_3, and silicate, making them potentially viable alternatives to traditional raw materials in cement manufacturing process (Tang et al., 2018). By partially substituting limestone or clay as a raw material in cement production, the utilization of combustion/incineration residues has the ability to contribute to environmental preservation. The high temperature (higher than 1450°C) and extended duration in the cement manufacturing process effectively remove toxic and hazardous pollutants such as dioxins in BA/FA. Additionally, this process aids in the stabilization of heavy metal content within the ash by forming stable compounds in the resulting clinker phase (Ghouleh and Shao, 2018).

However, the direct addition of combustion/incineration residues to raw materials is not feasible due to the complexity of their compositions and the presence of corrosive components. For example, the use of MSWI FA in cement production should be approached with caution due to its fine particle size and high levels of chloride, heavy metals, and soluble salts. To ensure that the cement maintains considerable mechanical strength, it is advisable to limit the maximum allowable addition of MSWI FA to 30% or below. This precautionary measure takes into account the potential impact of MSWI FA on the overall performance and durability of the cement (Cristelo et al., 2020). In addition, the presence of a substantial quantity of chloride (such as NaCl and KCl) in FA can result in increased porosity and decreased strength of concrete, as well as causing some degree of corrosion to the rotary kiln. As a precautionary measure, the amount of untreated FA typically added to cement is generally restricted to less than 5%. Thus, to ensure the quality of FA-modified cement clinker and mitigate the impact of hazardous substances, some pretreatments are necessary. Currently, washing pretreatment and ecological cement technology are commonly recommended in cement-based codisposal. Through washing pretreatment, a significant portion of the chloride content can be eliminated, enabling the addition of pretreated FA at levels ranging from 30% to 50% (Assi et al., 2020; Clavier et al., 2020).

Moreover, it is crucial to analyze and evaluate the properties, mineral composition, and strength of clinker products when combustion/incineration residues are included in the raw materials. This assessment allows for a comprehensive understanding of the impact of residues addition on the final cement product and provides valuable insights for optimizing its incorporation in the cement manufacturing process. Nevertheless, there are certain limitations in the stability testing and characterization of the product under restricted conditions. Additionally, the biocompatibility assessment of the product has not been adequately addressed. Further research is required to understand the impacts of chloride ions and heavy metals on the entire life cycle of the cement products (Huang et al., 2022).

Furthermore, based on previous studies, it has been evidenced that the corrosive characteristics of the kiln can be affected by the existence of chloride ions in the residues. When conducting waste treatment in cement rotary kilns, it is crucial to guarantee that the waste materials do not result in any damage to the equipment and do not cause any disturbances to the regular operation of the system. As most researches are performed on a laboratory scale, there is an urgent demand for industrial-scale experimentation in cement manufacturing using combustion/incineration residues. For example, investigation is warranted to explore the cost-effective pretreatment options at an industrial scale, along with conducting a comprehensive life cycle assessment.

Actually, incorporating combustion/incineration residues as kiln feed requires careful consideration of practical limitations. These limitations encompass challenges associated with the conveyance of the material, presence of large metallic particles, moisture levels, the inherent heterogeneity of the ash, etc. (Clavier et al., 2020) In addition, to enhance the applicability of combustion/incineration residues for reuse purposes, it is necessary to conduct further research on front-end processing techniques, particularly the utilization of combustion additives, aiming to effectively regulate the concentrations of chloride and trace elements in the ash. The formation of volatile condensation cycles can pose challenges such as the formation of crust, blockages, and clinker agglomeration within the kiln, which also need to be addressed at an industrial scale.

3.2 Recycling of combustion/incineration residues into supplementary cementitious materials and aggregates

Due to their hydraulic or pozzolanic activity, SCMs can be used for partial substitution of cement. Especially, the application of natural materials or industrial by-products as SCMs in the form of blended cement has attracted interest due to its potential to significantly reduce CO_2 emissions, and the addition of SCMs has been demonstrated to improve the workability and long-term durability of cementitious materials (Juenger et al., 2019; Lothenbach et al., 2011). Until now, coal FA, metakaolin, ground granulated blast furnace slag, and silica fume have been widely used as SCMs (Jin et al., 2023). Recently, using MSWI FA as SCMs in building materials has attracted increasing attention. The fine particle size (below 100 μm) and composition rich in calcium, aluminum, and silicon suggest that FA holds promise as a low-carbon cementitious material. In addition, utilizing FA as an SCM in building materials not only addresses the issue of excessive cement consumption but also presents additional economic advantages (Liu et al., 2021; Phua et al., 2019).

Despite its potential benefits, the presence of heavy metals remains a risk factor that suppresses the utilization of FA as an SCM. These heavy metals can greatly inhibit cement hydration and pose a potential environmental risk if they leach into the surroundings (Wang et al., 2021). The utilization of alkali-activated concrete is an effective approach that involves the reaction of alkaline activators such as NaOH, KOH, Na_2SiO_3, and K_2SiO_3 with FA, resulting in the formation of a three-dimensional network of geopolymers. This process helps to immobilize heavy metals within the formed matrix (Tian et al., 2020). However, alkali-activated concrete has a drawback wherein the ratio of silicon and aluminum content needs to meet a specific value. The silicon and aluminum content as well as the morphology of FA can vary depending on the source, particularly if it is derived from waste incineration. Therefore, it is necessary to adjust and optimize the ratio accordingly before utilizing FA for alkali-activated concrete (Jin et al., 2023).

On the other hand, concrete is commonly employed as the primary construction material in diverse buildings and infrastructure projects, which is typically produced using OPC as the binding agent, combined with natural sand and stone as aggregates (Yang et al., 2015). The excessive exploitation of natural resources for concrete manufacturing has resulted in the exhaustion of natural ingredients, and the production of OPC is linked to a substantial carbon footprint, which ranges from 0.66 to 0.82 tons per ton of OPC produced (Chen et al., 2019). Therefore, researchers have shown significant interest in investigating the viability of recycling solid waste materials, including metallurgical slags, sediments, coal FA, MSWI residues, and different types of ashes, for the production of artificial lightweight aggregates (Li et al., 2020; Tang et al., 2020). One commonly used production method for FA-based lightweight aggregates is thermal treatment. Sintering is employed to recycle FA into lightweight aggregates as the primary thermal process with the objectives of immobilizing PTEs, reducing the volume of FA, and producing suitable products for practical applications (Zhang et al., 2021). The lightweight aggregates effectively stabilize the PTEs from FA, and the solidification efficiency can be improved by increasing the sintering temperature and prolonging the sintering time (Wei, 2015).

However, the application of sintering method typically requires intensive consumption of energy, which is not suitable from an economic and environmental perspective. An alternative and effective method known as cold-bonding pelletization has been proposed for recycling powdered solid waste into artificial lightweight aggregates. This method offers the advantage of avoiding the high-energy consumption associated with sintering methods (Tang et al., 2020). The application of the pelletizing technique for solid waste recycling significantly reduces the need for landfill space and produces valuable alternative aggregates that can be used in concrete production. A two-step cold-bonding pelletization process has proven to be effective in recycling washed FA and enhancing its mechanical and leaching properties, where the second step involves using a pure binder to encapsulate lightweight aggregates and form an outer shell. Especially, these lightweight aggregates could sequester CO_2 with high environmental advantages and low carbon footprint (Zhang et al., 2021). It should be noted that the chemical composition and reactivity of the raw materials significantly influence the mechanical properties of aggregates, and it is essential to optimize the pelletizing processes and production environments to ensure the efficient production of high-quality pellets. For future research, it is of significance to investigate the pretreatment of combustion/incineration residues, incorporation contents, curing conditions (such as curing time and combined curing conditions), and pelletization parameters to develop reliable lightweight aggregate products derived from the residues. Feasible methods, such as alkali activation and accelerated carbonation, can be employed to enhance the mechanical properties of these artificial aggregates.

4. Functional materials synthesis and resource recovery

4.1 Recycling of combustion/incineration residues into functional materials

As discussed in previous sections, the synthesis of zeolites can be achieved through various methods, including the traditional hydrothermal method, alkaline fusion-hydrothermal method, microwave-assisted hydrothermal method, and ultrasonic-assisted method. While the microwave-assisted and ultrasonic-assisted methods offer economic advantages compared with the traditional hydrothermal method, they face challenges in scaling up for large-scale disposal purposes. To enhance the zeolitization reaction, several measures are commonly employed, including incorporating silicon aluminum additives, adjusting reaction time and temperature, selecting an efficient alkaline activator, and determining suitable concentrations. However, it is worth noting that the silicon and aluminum content in fluidized bed waste incineration FA is typically below 20 wt.%, whereas coal FA contains a higher silicon and aluminum content of up to 40 wt.%. Therefore, when preparing zeolites from MSWI FA, the addition of supplementary silicon and aluminum sources is often necessary.

Moreover, FA finds application in ceramic preparation, including the synthesis of building ceramics, mullite ceramics, wollastonite ceramics, cordierite ceramics, and microcrystalline glass. However, it is crucial to carefully control the FA content during ceramic synthesis. Excessive replacement of FA can lead to increased linear shrinkage rate and bulk density, as well as a significant decrease in mechanical properties. Nevertheless, certain special coal FA with high aluminum content can be used to synthesize high-strength ceramic bricks. Regarding MSWI FA, it is primarily utilized in the synthesis of microcrystalline glass. The alkalinity, determined by the $(CaO + MgO)/(SiO_2+Al_2O_3)$ content ratio, can be adjusted by varying the dosage of MSWI FA. Within a specific range, it has been observed that the activation energy increases as alkalinity decreases, influencing the crystallization behavior and mechanical properties of microcrystalline glass.

4.2 Resource recovery from combustion/incineration residues

In recent years, there have been advancements in technologies aimed at recovering resources from combustion/incineration residues. The development of resource recovery technologies for combustion/incineration residues aims to mitigate environmental concerns while also offering secondary resource opportunities. For example, the MSWI FA contains significant quantities of soluble salts and valuable metals, which can be extracted and utilized as valuable secondary resources. MSWI BAs have a substantial output quantity. Apart from secondary raw materials, there is potential for the recovery of base metals and technology-related elements from national BA flows. While metals such as Fe, Al, Cu, Zn, Sn, and precious metals hold promise for economic recovery, their extraction from the complex matrix poses considerable challenges (Zhang et al., 2021). There is still opportunity for enhancing the existing resource recovery technologies for combustion/incineration residues. Innovative extraction methods utilizing supercritical CO_2 have been developed to effectively extract lanthanide metal nitrates, oxides, hydroxides, and nitrate solutions. However, additional research is required to explore the impact of temperature and pressure on the equilibrium of metal extraction and the selective removal of impurities.

Electrodialysis and thermal treatment technologies, despite their potential for resource recovery, may impose a significant environmental burden due to their high energy consumption. When it comes to leaching rare earth metals, combustion/incineration residues present a substantial overall quantity of rare earth elements with low concentrations. However, research on the recovery technology for rare earth elements from combustion/incineration residues is currently limited (Gomes et al., 2020). At present, a washing pretreatment approach for the residues followed by chemical extraction proves to be a time-efficient method for element recovery. Additionally, the electrochemical process shows potential in improving the recovery rates of valuable elements from combustion/incineration residues. Moving forward, it is crucial to conduct an economic feasibility evaluation to compare the costs and benefits associated with various resource recovery processes. This evaluation will play a vital role in promoting widespread application of resource recovery from combustion/incineration residues (Zhang et al., 2021).

5. Conclusion

Treatment of large amount of combustion/incineration residues has become a pressing global environmental issue. Various organic pollutants and leachable potential toxic elements in the solid residues are major barriers for sustainable utilization. Simple and effective methods are urgent for fast purification of hazardous ashes, including thermal and nonthermal routes, including pyrolysis, sintering, vitrification, hydrothermal treatment, mechanochemical treatment, and stabilization/solidification. Most technologies work well in toxic elements immobilization, while only thermal routes are effective for organic pollutants degradation.

Moreover, due to the "circular economy" concepts and land shortage, it is necessary to provide effective recycling solutions for combustion/incineration residues instead of just landfill. Value added utilization including recycling to construction materials, functional materials synthesis and resource recovery. Considering that different disposal technologies have their own strengths and weaknesses, the selection of an appropriate technology should be evaluated on a case-by-case basis. Recycling to construction materials seemed to be a universal destination for all kinds of combustion/incineration residues but with low additional value. Some resource recovery route targeting valuable metals and salts could be adopted before recycling to cement industry. Meanwhile, functional zeolite and ceramics would be more value-added products. However, the properties of ash-derived functional materials should be qualified through various advanced characterization technologies.

Combined technologies are usually recommended to achieve safe disposal and value-added utilization, simultaneously. However, life cycle assessment and cost-benefit analysis would be necessary to evaluate the carbon footprint and economics of novel treatment/utilization routes. Moreover, big data and machine learning could be powerful tools to provide suitable solutions for certain combustion/incineration residues based on simple characterization results. Thus, a critical framework for safe and valuable combustion/incineration disposal could be established to guide both academicians and engineers.

Acknowledgements

The authors gratefully acknowledge the financial support from the National Natural Science Foundation of China (Grant No. 52306280, 52206174 and 52236008) for this study.

References

Assi, A., Bilo, F., Zanoletti, A., Ponti, J., Valsesia, A., La Spina, R., Zacco, A., Bontempi, E., 2020. Zero-waste approach in municipal solid waste incineration: reuse of bottom ash to stabilize fly ash. J. Clean. Prod. 245, 118779.

Bilibio, C., Retz, S., Schellert, C., Hensel, O., 2021. Drainage properties of technosols made of municipal solid waste incineration bottom ash and coal combustion residues on potash-tailings piles: a lysimeter study. J. Clean. Prod. 279, 123442.

Bogush, A.A., Stegemann, J.A., Zhou, Q., Wang, Z., Zhang, B., Zhang, T., Zhang, W., Wei, J., 2020. Co-processing of raw and washed air pollution control residues from energy-from-waste facilities in the cement kiln. J. Clean. Prod. 254, 119924.

Chen, L., Wang, L., Cho, D.-W., Tsang, D.C.W., Tong, L., Zhou, Y., Yang, J., Hu, Q., Poon, C.S., 2019. Sustainable stabilization/solidification of municipal solid waste incinerator fly ash by incorporation of green materials. J. Clean. Prod. 222, 335–343.

Clavier, K.A., Paris, J.M., Ferraro, C.C., Townsend, T.G., 2020. Opportunities and challenges associated with using municipal waste incineration ash as a raw ingredient in cement production – a review. Resour. Conserv. Recycl. 160, 104888.

Contessi, S., Calgaro, L., Dalconi, M.C., Bonetto, A., Bellotto, M.P., Ferrari, G., Marcomini, A., Artioli, G., 2020. Stabilization of lead contaminated soil with traditional and alternative binders. J. Hazard Mater. 382, 120990.

Cristelo, N., Segadães, L., Coelho, J., Chaves, B., Sousa, N.R., de Lurdes Lopes, M., 2020. Recycling municipal solid waste incineration slag and fly ash as precursors in low-range alkaline cements. Waste Manage. 104, 60–73.

Du, B., Li, J., Fang, W., Liu, J., 2019. Comparison of long-term stability under natural ageing between cement solidified and chelator-stabilised MSWI fly ash. Environ. Pollut. 250, 68–78.

Du, B., Li, J., Fang, W., Liu, Y., Yu, S., Li, Y., Liu, J., 2018. Characterization of naturally aged cement-solidified MSWI fly ash. Waste Manage. 80, 101–111.

Fan, C., Wang, B., Zhang, T., 2018. Review on cement stabilization/solidification of municipal solid waste incineration fly ash. Adv. Mater. Sci. Eng. 1–7.

Gao, X., Yuan, B., Yu, Q.L., Brouwers, H.J.H., 2017. Characterization and application of municipal solid waste incineration (MSWI) bottom ash and waste granite powder in alkali activated slag. J. Clean. Prod. 164, 410–419.

Ghouleh, Z., Shao, Y., 2018. Turning municipal solid waste incineration into a cleaner cement production. J. Clean. Prod. 195, 268–279.

Gomes, H.I., Funari, V., Ferrari, R., 2020. Bioleaching for resource recovery from low-grade wastes like fly and bottom ashes from municipal incinerators: a SWOT analysis. Sci. Total Environ. 715, 136945.

Goto, M., 2009. Chemical recycling of plastics using sub- and supercritical fluids. J. Supercrit. Fluids 47 (3), 500–507.

Hu, Y., Zhang, P., Li, J., Chen, D., 2015. Stabilization and separation of heavy metals in incineration fly ash during the hydrothermal treatment process. J. Hazard Mater. 299, 149–157.

Huang, B., Gan, M., Ji, Z., Fan, X., Zhang, D., Chen, X., Sun, Z., Huang, X., Fan, Y., 2022. Recent progress on the thermal treatment and resource utilization technologies of municipal waste incineration fly ash: a review. Process Safe. Environ. 159, 547–565.

Jin, L., Chen, M., Wang, Y., Peng, Y., Yao, Q., Ding, J., Ma, B., Lu, S., 2023. Utilization of mechanochemically pretreated municipal solid waste incineration fly ash for supplementary cementitious material. J. Environ. Chem. Eng. 11 (1), 109112.

Jin, Y.-q., Ma, X.-j., Jiang, X.-g., Liu, H.-m., Li, X.-d., Yan, J.-h., 2012. Hydrothermal degradation of polychlorinated dibenzo-p-dioxins and polychlorinated dibenzofurans in fly ash from municipal solid waste incineration under non-oxidative and oxidative conditions. Energy Fuel. 27 (1), 414–420.

Juenger, M.C.G., Snellings, R., Bernal, S.A., 2019. Supplementary cementitious materials: new sources, characterization, and performance insights. Cement Concr. Res. 122, 257–273.

Kanhar, A.H., Chen, S., Wang, F., 2020. Incineration fly ash and its treatment to possible utilization: a review. Energies 13 (24), 6681.

Lam, C.H.K., Barford, J.P., McKay, G., 2011. Utilization of municipal solid waste incineration ash in Portland cement clinker. Clean Technol. Envir. 13 (4), 607–615.

Li, J., Zhang, S., Wang, Q., Ni, W., Li, K., Fu, P., Hu, W., Li, Z., 2020. Feasibility of using fly ash–slag-based binder for mine backfilling and its associated leaching risks. J. Hazard. Mater. 400, 123191.

Lin, X., Chen, J., Xu, S., Mao, T., Liu, W., Wu, J., Li, X., Yan, J., 2022. Solidification of heavy metals and PCDD/Fs from municipal solid waste incineration fly ash by the polymerization of calcium carbonate oligomers. Chemosphere 288, 132420.

Lindberg, D., Molin, C., Hupa, M., 2015. Thermal treatment of solid residues from WtE units: a review. Waste Manage. 37, 82–94.

Liu, J., Hu, L., Tang, L., Ren, J., 2021. Utilisation of municipal solid waste incinerator (MSWI) fly ash with metakaolin for preparation of alkali-activated cementitious material. J. Hazard Mater. 402, 123451.

Liu, J., Nie, X., Zeng, X., Su, Z., 2013. Long-term leaching behavior of phenol in cement/activated-carbon solidified/stabilized hazardous waste. J. Environ. Manag. 115, 265–269.

López-Zaldívar, O., Lozano-Díez, R.V., Verdú-Vázquez, A., Llauradó-Pérez, N., 2017. Effects of the addition of inertized MSW fly ash on calcium aluminate cement mortars. Construct. Build. Mater. 157, 1106–1116.

Lothenbach, B., Scrivener, K., Hooton, R.D., 2011. Supplementary cementitious materials. Cement Concr. Res. 41 (12), 1244–1256.

Luna Galiano, Y., Fernández Pereira, C., Vale, J., 2011. Stabilization/solidification of a municipal solid waste incineration residue using fly ash-based geopolymers. J. Hazard Mater. 185 (1), 373–381.

Luo, H., Cheng, Y., He, D., Yang, E.-H., 2019. Review of leaching behavior of municipal solid waste incineration (MSWI) ash. Sci. Total Environ. 668, 90–103.

Ma, W., Shi, W., Shi, Y., Chen, D., Liu, B., Chu, C., Li, D., Li, Y., Chen, G., 2021. Plasma vitrification and heavy metals solidification of MSW and sewage sludge incineration fly ash. J. Hazard Mater. 408, 124809.

Pan, S., Yao, Q., Cai, W., Peng, Y., Luo, Y., Wang, Z., Jiang, C., Li, X., Lu, S., 2022. Characterization of dioxins and heavy metals in chelated fly ash. Energies 15 (13).

Peng, Z., Weber, R., Ren, Y., Wang, J., Sun, Y., Wang, L., 2020. Characterization of PCDD/Fs and heavy metal distribution from municipal solid waste incinerator fly ash sintering process. Waste Manage. 103, 260–267.

Phua, Z., Giannis, A., Dong, Z.-L., Lisak, G., Ng, W.J., 2019. Characteristics of incineration ash for sustainable treatment and reutilization. Environ. Sci. Pollut. Res. 26 (17), 16974–16997.

Provis, J.L., Bernal, S.A., 2014. Geopolymers and related alkali-activated materials. Annu. Rev. Mater. Res. 44 (1), 299–327.

Shi, C., Qu, B., Provis, J.L., 2019. Recent progress in low-carbon binders. Cement Concr. Res. 122, 227–250.

Shunda, L., Jiang, X., Zhao, Y., Yan, J., 2022. Disposal technology and new progress for dioxins and heavy metals in fly ash from municipal solid waste incineration: a critical review. Environ. Pollut. 311, 119878.

Silva, R.V., de Brito, J., Lynn, C.J., Dhir, R.K., 2017. Use of municipal solid waste incineration bottom ashes in alkali-activated materials, ceramics and granular applications: a review. Waste Manage. 68, 207–220.

Tang, J., Ylmén, R., Petranikova, M., Ekberg, C., Steenari, B.-M., 2018. Comparative study of the application of traditional and novel extractants for the separation of metals from MSWI fly ash leachates. J. Clean. Prod. 172, 143–154.

Tang, P., Xuan, D., Li, J., Cheng, H.W., Poon, C.S., Tsang, D.C.W., 2020. Investigation of cold bonded lightweight aggregates produced with incineration sewage sludge ash (ISSA) and cementitious waste. J. Clean. Prod. 251, 119709.

Tian, X., Rao, F., Morales-Estrella, R., Song, S., 2020. Effects of aluminum dosage on gel formation and heavy metal immobilization in alkali-activated municipal solid waste incineration fly ash. Energy Fuel. 34 (4), 4727–4733.

Viczek, S.A., Aldrian, A., Pomberger, R., Sarc, R., 2020. Determination of the material-recyclable share of SRF during co-processing in the cement industry. Resour. Conserv. Recycl. 156, 104696.

Wang, L., Chen, L., Poon, C., Wang, C.-H., Ok, Y.S., Mechtcherine, V., Tsang, D.C., 2021. Roles of biochar and CO2 curing in sustainable magnesia cement-based composites. ACS Sust. Chem. Eng. 9 (25), 8603–8610.

Wei, N., 2015. Leachability of heavy metals from lightweight aggregates made with sewage sludge and municipal solid waste incineration fly ash. Int. J. Environ. Res. Publ. Health 12 (5), 4992–5005.

Xiang, J., Qiu, J., Li, Z., Chen, J., Song, Y., 2022. Eco-friendly treatment for MSWI bottom ash applied to supplementary cementing: mechanical properties and heavy metal leaching concentration evaluation. Construct. Build. Mater. 327, 127012.

Xue, Y., Liu, X., 2021. Detoxification, solidification and recycling of municipal solid waste incineration fly ash: a review. Chem. Eng. J. 420, 130349.

Yang, K.-H., Jung, Y.-B., Cho, M.-S., Tae, S.-H., 2015. Effect of supplementary cementitious materials on reduction of CO2 emissions from concrete. J. Clean. Prod. 103, 774–783.

Yang, Z., Ji, R., Liu, L., Wang, X., Zhang, Z., 2018. Recycling of municipal solid waste incineration by-product for cement composites preparation. Construct. Build. Mater. 162, 794–801.

Yin, K., Ahamed, A., Lisak, G., 2018. Environmental perspectives of recycling various combustion ashes in cement production – a review. Waste Manage. 78, 401–416.

Yuan, Q., Zhang, Y., Wang, T., Wang, J., Romero, C.E., 2021. Mechanochemical stabilization of heavy metals in fly ash from coal-fired power plants via dry milling and wet milling. Waste Manage. 135, 428–436.

Zacco, A., Borgese, L., Gianoncelli, A., Struis, R.P.W.J., Depero, L.E., Bontempi, E., 2014. Review of fly ash inertisation treatments and recycling. Environ. Chem. Lett. 12 (1), 153–175.

Zhang, Y., Wang, L., Chen, L., Ma, B., Zhang, Y., Ni, W., Tsang, D.C.W., 2021. Treatment of municipal solid waste incineration fly ash: state-of-the-art technologies and future perspectives. J. Hazard Mater. 411, 125132.

Zhao, S.-Z., Zhang, X.-Y., Liu, B., Zhang, J.-J., Shen, H.-L., Zhang, S.-G., 2021. Preparation of glass–ceramics from high-chlorine MSWI fly ash by one-step process. Rare Met. 40 (11), 3316–3328.

Index

'*Note:* Page numbers followed by "t" indicate tables.'

J

K

L

M

T

Printed in the United States
by Baker & Taylor Publisher Services